ENCYCLOPEDIA OF

CELL TECHNOLOGY

VOLUME 1

 # WILEY BIOTECHNOLOGY ENCYCLOPEDIAS

Encyclopedia of Bioprocess Technology: Fermentation, Biocatalysis, and Bioseparation
Edited by Michael C. Flickinger and Stephen W. Drew

Encyclopedia of Molecular Biology
Edited by Thomas E. Creighton

Encyclopedia of Cell Technology
Edited by Raymond E. Spier

Encyclopedia of Ethical, Legal, and Policy Issues in Biotechnology
Edited by Thomas J. Murray and Maxwell J. Mehlman

ENCYCLOPEDIA OF CELL TECHNOLOGY
EDITORIAL BOARD

ENCYCLOPEDIA OF

CELL TECHNOLOGY

VOLUME 1

Raymond E. Spier
University of Surrey
Guildford, Surrey
United Kingdom

A Wiley-Interscience Publication

John Wiley & Sons, Inc.

New York / Chichester / Weinheim / Brisbane / Singapore / Toronto

For ordering and customer service, call 1-800-CALL-WILEY.

Library of Congress Cataloging in Publication Data

Encyclopedia of cell technology / [edited by] Raymond E. Spier.
      p.   cm.
   ISBN 0-471-16123-3 (cloth : set : alk. paper) — ISBN (invalid)
0-471-16643-X (v. 1 : alk. paper). — ISBN 0-471-16623-5 (v. 2 :
alk. paper).
   1. Animal cell biotechnology Encyclopedias.   2. Plant cell
biotechnology Encyclopedias.   3. Cell culture Encyclopedias.
4. Cytology Encyclopedias.   I. Spier, R. (Raymond)
  TP248.27.A53E53 2000
  660.6 — dc21   99-25295

Printed in the United States of America.

10 9 8 7 6 5 4 3 2 1

PREFACE

The Wiley Biotechnology Encyclopedias, composed of the *Encyclopedia of Molecular Biology*; the *Encyclopedia of Bioprocess Technology: Fermentation, Biocatalysis, and Bioseparation*; the *Encyclopedia of Cell Technology*; and the *Encyclopedia of Ethical, Legal, and Policy Issues in Biotechnology* cover very broadly four major contemporary themes in biotechnology. The series comes at a fascinating time in that, as we move into the twenty-first century, the discipline of biotechnology is undergoing striking paradigm changes.

Biotechnology is now beginning to be viewed as an informational science. In a simplistic sense there are three types of biological information. First, there is the digital or linear information of our chromosomes and genes with the four-letter alphabet composed of G, C, A, and T (the bases guanine, cytosine, adenine, and thymine). Variation in the order of these letters in the digital strings of our chromosomes or our expressed genes (or mRNAs) generates information of several distinct types: genes, regulatory machinery, and information that enables chromosomes to carry out their tasks as informational organelles (e.g., centromeric and telomeric sequences).

Second, there is the three-dimensional information of proteins, the molecular machines of life. Proteins are strings of amino acids employing a 20-letter alphabet. Proteins pose four technical challenges: (*1*) Proteins are synthesized as linear strings and fold into precise three-dimensional structures as dictated by the order of amino acid residues in the string. Can we formulate the rules for protein folding to predict three-dimensional structure from primary amino acid sequence? The identification and comparative analysis of all human and model organism (bacteria, yeast, nematode, fly, mouse, etc.) genes and proteins will eventually lead to a lexicon of motifs that are the building block components of genes and proteins. These motifs will greatly constrain the shape space that computational algorithms must search to successfully correlate primary amino acid sequence with the correct three-dimensional shapes. The protein-folding problem will probably be solved within the next 10–15 years. (*2*) Can we predict protein function from knowledge of the three-dimensional structure? Once again the lexicon of motifs with their functional as well as structural correlations will play a critical role in solving this problem. (*3*) How do the myriad of chemical modifications of proteins (e.g., phosphorylation, acetylation, etc.) alter their structures and modify their functions? The mass spectrometer will play a key role in identifying secondary modifications. (*4*) How do proteins interact with one another and/or with other macromolecules to form complex molecular machines (e.g., the ribosomal subunits)? If these functional complexes can be isolated, the mass spectrometer, coupled with a knowledge of all protein sequences that can be derived from the complete genomic sequence of the organism, will serve as a powerful tool for identifying all the components of complex molecular machines.

The third type of biological information arises from complex biological systems and networks. Systems information is four dimensional because it varies with time. For example, the human brain has 1,012 neurons making approximately 1,015 connections. From this network arise systems properties such as memory, consciousness, and the ability to learn. The important point is that systems properties cannot be understood from studying the network elements (e.g., neurons) one at a time; rather the collective behavior of the elements needs to be studied. To study most biological systems, three issues need to be stressed. First, most biological systems, three issues need to be stressed. First, most biological systems are too complex to study directly, therefore they must be divided into tractable subsystems whose properties in part reflect those of the system. These subsystems must be sufficiently small to analyze all their elements and connections. Second, high-throughput analytic or global tools are required for studying many systems elements at one time (see later). Finally the systems information needs to be modeled mathematically before systems properties can be predicted and ultimately understood. This will require recruiting computer scientists and applied mathematicians into biology—just as the attempts to decipher the information of complete genomes and the protein folding and structure/function problems have required the recruitment of computational scientists.

I would be remiss not to point out that there are many other molecules that generate biological information: amino acids, carbohydrates, lipids, and so forth. These too must be studied in the context of their specific structures and specific functions.

The deciphering and manipulation of these various types of biological information represent an enormous technical challenge for biotechnology. Yet major new and powerful tools for doing so are emerging.

One class of tools for deciphering biological information is termed high-throughput analytic or global tools. These tools can be used to study many genes or chromosome features (genomics), many proteins (proteomics), or many cells rapidly: large-scale DNA sequencing, genomewide genetic mapping, cDNA or oligonucleotide arrays, two-dimensional gel electrophoresis and other global protein separation technologies, mass spectrometric analysis of proteins and protein fragments, multiparameter, high-throughput cell and chromosome sorting, and high-throughput phenotypic assays.

A second approach to the deciphering and manipulation of biological information centers around combinatorial strategies. The basic idea is to synthesize an informational string (DNA fragments, RNA fragments, protein fragments, antibody combining sites, etc.) using all combinations of the basic letters of the corresponding alphabet, thus creating many different shapes that can be used to activate, inhibit, or complement the biological functions of designated three-dimensional shapes (e.g., a molecule in a signal transduction pathway). The power of combinational chemistry is just beginning to be appreciated.

A critical approach to deciphering biological information will ultimately be the ability to visualize the functioning of genes, proteins, cells, and other informational elements within living organisms (in vivo informational imaging).

Finally, there are the computational tools required to collect, store, analyze, model, and ultimately distribute the various types of biological information. The creation presents a challenge comparable to that of developing new instrumentation and new chemistries. Once again this means recruiting computer scientists and applied mathematicians to biology. The biggest challenge in this regard is the language barriers that separate different scientific disciplines. Teaching biology as an informational science has been a very effective means for breeching these barriers.

The challenge is, of course, to decipher various types of biological information and then be able to use this information to manipulate genes, proteins, cells, and informational pathways in living organisms to eliminate or prevent disease, produce higher-yield crops, or increase the productivity of animals for meat and other foods.

Biotechnology and its application raise a host of social, ethical, and legal questions, for example, genetic privacy, germline genetic engineering, cloning of animals, genes that influence behavior, cost of therapeutic drugs generated by biotechnology, animal rights, and the nature and control of intellectual property.

Clearly, the challenge is to educate society so that each citizen can thoughtfully and rationally deal with these issues, for ultimately society dictates the resources and regulations that circumscribe the development and practice of biotechnology. Ultimately, I feel enormous responsibility rests with scientists to inform and educate society about the challenges as well as the opportunities arising from biotechnology. These are critical issues for biotechnology that are developed in detail in the *Encyclopedia of Ethical, Legal, and Policy Issues in Biotechnology*.

The view that biotechnology is an informational science pervades virtually every aspect of this science, including discovery, reduction of practice, and societal concerns. These Encyclopedias of Biotechnology reinforce the emerging informational paradigm change that is powerfully positioning science as we move into the twenty-first century to more effectively decipher and manipulate for humankind's benefit the biological information of relevant living organisms.

Leroy Hood
University of Washington

CONTRIBUTORS

P. Marlene Absher, *Department of Medicine, University of Vermont, College of Medicine, Colchester, Vermont,* Enrichment and Isolation Techniques for Animal Cell Types

Nicholas R. Abu-Absi, *Department of Chemical Engineering and Materials Science, Biological Process Technology Institute, The University of Minnesota, St. Paul, Minnesota,* Cell Cycle Events and Cell Cycle-Dependent Processes

S. Robert Adamson, *Mammalian and Microbial Cell Sciences and Pilot Lab, Genetics Institute, Andover, Massachusetts,* Cell Stability, Animal

Motomu Akita, *Department of Biotechnological Science, Kinki University, Wakayama, Japan,* Bioreactor Culture of Plant Organs

Mohamed Al-Rubeai, *Animal Cell Technology Group, Centre for Bioprocess Engineering, School of Chemical Engineering, University of Birmingham, Edgbaston, Birmingham, U.K.,* Cell Cycle Synchronization

Eunice J. Allan, *Department of Agriculture, University of Aberdeen, Aberdeen, UK,* Culture Establishment, Plant Cell Culture

Arie Altman, *The Hebrew University of Jerusalem, Rehovot, Israel,* Micropropagation of Plants, Principles and Practices

Claude Artois, *SmithKline Beecham Biologicals, Rixensart, Belgium,* cGMP Compliance for Production Rooms in Biotechnology: A Case Study

Hiroshi Ashihara, *Department of Biology, Ochanomizu University, Tokyo 112-8610, Japan,* Physiology of Plant Cells in Culture

John G. Aunins, *Merck Research Laboratories, West Point, Pennsylvania,* Viral Vaccine Production in Cell Culture

Hans Becker, *FR 12.3 Fachrichtung Pharmakognosie und Analytische, Phytochemie Universität des Saarlandes, Postfach 15 11 50, D 66041 Saarbrücken,* Bryophyte In Vitro Cultures, Secondary Products

Erica E. Benson, *School of Molecular & Life Sciences, University of Abertay Dundee, Bell Street, Dundee DD1 1HG, Scotland, UK,* Cryopreservation of Plant Cells, Tissues and Organs

Michael J. Betenbaugh, *Department of Chemical Engineering, The Johns Hopkins University, Baltimore, MD,* Protein Processing, Processing in the Endoplasmic Reticulum and Golgi Network

John R. Birch, *Lonza Biologics plc, Slough SL1 4DY, Berkshire, United Kingdom,* Cell Products—Antibodies

N.W. Blackhall, *School of Biological Sciences, University of Nottingham, University Park, Nottingham NG7 2RD, UK,* Flow Cytometry of Plant Cells

Hendrik F.J. Bonarius, *Department of Cell Biology, Novo Nordisk Ltd., Gentofte, Denmark,* Flux Analysis of Mammalian Cell Culture: Methods and Applications

J.M. Bonga, *Natural Resources Canada, Canadian Forest Service–Atlantic Forestry Centre, Fredericton, Canada,* Culture of Conifers

Michael C. Borys, *Abbott Laboratories, Chemical and Agricultural Products Division, 1401 Sheridan Road, North Chicago, IL,* Animal Cell Culture, Physiochemical Effects of pH

Jan Brazolot, *Plant Agriculture Department, University of Guelph, Guelph, Ontario, Canada,* Toxin Resistant Plants from Plant Cell Culture and Transformation

Richard Brettell, *CSIRO Plant Industry, Darwin, Australia,* Monocot Cell Culture

Helen M. Buettner, *Department of Chemical and Biochemical Engineering, Rutgers University, Piscataway, NJ,* Cell Structure and Motion, Cytoskeleton and Cell Movement

Aquanette M. Burt, *Department of Chemical and Biochemical Engineering, Rutgers University, Piscataway, NJ,* Cell Structure and Motion, Cytoskeleton and Cell Movement

Thomas F. Busby, *Jerome H. Holland Laboratory, American Red Cross, Rockville, Maryland,* Viral Inactivation, Emerging Technologies for Human Blood Products

M. Butler, *Department of Microbiology, University of Manitoba, Winnipeg, Manitoba, Canada R3T 2N2,* Equipment and Laboratory Design for Cell Culture; Measurement of Cell Viability

Heino Büntemeyer, *Institute of Cell Culture Technology, University of Bielefeld, Bielefeld, Germany,* Off-Line Analysis in Animal Cell Culture, Methods

Marco A. Cacciuttolo, *MedImmune, Inc., Gaithersburg, Maryland,* Animal Cell Culture, Physiochemical Effects of Dissolved Oxygen and Redox Potential

R.J. Carmody, *Tumour Biology Laboratory, Department of Biochemistry, University College, Lee Maltings, Prospect Row, Cork, Ireland,* Programmed Cell Death (Apoptosis) in Culture, Impact and Modulation

Graham Carpenter, *Departments of Biochemistry and Medicine, Vanderbilt University School of Medicine, Nashville, TN,* Cell-Surface Receptors: Structure, Activation, and Signaling

Alan C. Cassells, *Department of Plant Science, National University of Ireland, Cork, Ireland,* Contamination Detection and Elimination in Plant Cell Culture

Jeff Chalmers, *Department of Chemical Engineering, The Ohio State University, Columbus, Ohio,* Animal Cell Culture, Effects of Agitation and Aeration on Cell Adaptation

Timothy S. Charlebois, *Mammalian and Microbial Cell Sciences and Pilot Lab, Genetics Institute, Andover, Massachusetts,* Cell Stability, Animal

David B. Clark, *Plasma Services, American Red Cross, Shawnee, Colorado,* Product Development, Quality and Regulatory Issues

Tim Clayton, *Glaxo Wellcome, Langley Park, Beckenham, Kent, United Kingdom,* Cell Products—Immunoregulators; Cell Products—Viral Gene Therapy Vectors

Suzanne D. Conzen, *Section of Hematology/Oncology, University of Chicago, Chicago, IL,* Cellular Transformation, Characteristics

T.G. Cotter, *Tumour Biology Laboratory, Department of Biochemistry, University College, Lee Maltings, Prospect Row, Cork, Ireland,* Programmed Cell Death (Apoptosis) in Culture, Impact and Modulation

Ruth W. Craig, *Dartmouth Medical School, Hanover, New Hampshire,* Characterization and Determination of Cell Death by Apoptosis

Wayne R. Curtis, *The Pennsylvania State University, University Park, PA,* Hairy Roots, Bioreactor Growth

M.R. Davey, *School of Biological Sciences, University of Nottingham, University Park, Nottingham NG7 2RD, UK,* Flow Cytometry of Plant Cells; Plant Protoplasts; Protoplast Fusion for the Generation of Unique Plants

John M. Davis, *Bio-Products Laboratory, Dagger Lane, Elstree, Herts., United Kingdom,* Aseptic Techniques in Cell Culture

Jean-Luc De Bouver, *SmithKline Beecham Biologicals, Rixensart, Belgium,* ICH GCP Guidelines: Preparation, Conduct and Reporting of Clinical Trials

Cornelis D. De Gooijer, *Department of Food Science, Food and Bioengineering Group, Wageningen Agricultural University, Wageningen, The Netherlands,* Flux Analysis of Mammalian Cell Culture: Methods and Applications

Geert-Jan De Klerk, *Centre for Plant Tissue Culture Research, P.O. Box 85, 2160 AB Lisse, The Netherlands,* Adventitious Organogenesis

Pierre C.A. Debergh, *University Gent, Coupure links 653, 9000 Gent, Belgium,* Micropropagation, Hyperhydricity

Scott L. Diamond, *Institute for Medicine and Engineering, Department of Chemical Engineering, Philadelphia, PA,* Receptors and Cell Signaling, Intracellular Receptors—Steroid Hormones and NO

Jean Didelez, *SmithKline Beecham Biologicals, Rixensart, Belgium,* cGMP Compliance for Production Rooms in Biotechnology: A Case Study

Pauline M. Doran, *Department of Biotechnology, University of New South Wales, Sydney, Australia,* Bioreactors, Stirred Tank

A. Doyle, *Wellcome Trust, London, England,* Cell and Cell Line Characterization; Cell Banks: A Service to Animal Cell Technology

Denis Drapeau, *Mammalian and Microbial Cell Sciences and Pilot Lab, Genetics Institute, Andover, Massachusetts,* Cell Stability, Animal

Hans G. Drexler, *DSMZ-German Collection of Microorganisms, and Cell Cultures, Department of Human and Animal Cell Cultures, Braunschweig, Germany,* Contamination of Cell Cultures, Mycoplasma

Dominique J. Dumet, *School of Molecular & Life Sciences, University of Abertay Dundee, Bell Street, Dundee, DD1 1HG, Scotland, UK,* Cryopreservation of Plant Cells, Tissues and Organs

Alan Eastman, *Dartmouth Medical School, Hanover, New Hampshire,* Characterization and Determination of Cell Death by Apoptosis

Patrick Florent, *SmithKline Beecham Biologicals, Rixensart, Belgium,* cGMP Compliance for Production Rooms in Biotechnology: A Case Study

M.R. Fowler, *The Norman Borlaug Institute for Plant Science, Research De Montfort University, Scraptoft Campus, Leicester LE7 9SU. U.K.,* Cell Cycle in Suspension Cultured Plant Cells; Plant Cell Culture, Laboratory Techniques

Hiroshi Fukui, *Department of Bioresource Science, Faculty of Agriculture, Kagawa University, Kagawa 761-0795, Japan,* Plant Cell Culture Bag

Guy Godeau, *SmithKline Beecham Biologicals, Rixensart, Belgium,* cGMP Compliance for Production Rooms in Biotechnology: A Case Study

Sandra L. Gould, *Merck Research Laboratories, Merck & Co., Inc., Rahway, NJ,* Animal Cell Culture Media

David R. Gray, *Chiron Corporation, Emeryville, California,* Bioreactor Operations—Preparation, Sterilization, Charging, Culture Initiation and Harvesting

Bryan Griffiths, *SC& P, P.O. Box 1723, Porton, Salisbury SP4 0PL, UK,* Animal Cell Products, Overview

Colin Harbour, *Department of Infectious Diseases, The University of Sydney, NSW, Australia,* Contamination Detection in Animal Cell Culture

Keith Harding, *Crop Genetics, Scottish Crop Research Institute, Invergowrie, Dundee DD2 5DA, Scotland, UK,* Cryopreservation of Plant Cells, Tissues and Organs

Scott Harrison, *Mammalian and Microbial Cell Sciences and Pilot Lab, Genetics Institute, Andover, Massachusetts,* Cell Stability, Animal

Yi He, *Department of Horticulture, Plant Science Building, University of Georgia, Athens, Georgia,* Anatomy of Plant Cells

G. Hodges, *53 Manor Court Rd., London, United Kingdom,* Animal Cell Types, Kidney Cells; Animal Cell Types, Ovarian Cells

Paul Holmes, *Animal Cell Technology Group, Centre for Bioprocess Engineering, School of Chemical Engineering, University of Birmingham, Edgbaston, Birmingham, U.K.,* Cell Cycle Synchronization

Erwin Huebner, *Department of Zoology, University of Manitoba, Winnipeg, Manitoba, Canada,* Characterization of Cells, Microscopic

Lena Häggström, *Department of Biotechnology, Royal Institute of Technology, SE-100 44 Stockholm, Sweden,* Cell Metabolism, Animal

András Kapus, *Toronto Hospital, Department of Surgery, Transplantation Research, 101 College Street, Toronto, Ontario, CANADA M5G 1L7,* Membrane Structure and the Transport of Small Molecules and Ions

K. Klimaszewska, *Natural Resources Canada, Canadian Forest Service–Laurentian Forestry Centre, Quebec, Canada,* Culture of Conifers

Katja Klockewitz, *Institut für Technische Chemie, Callinstr. 3, 30167 Hannover, Germany,* Immuno Flow Injection Analysis in Bioprocess Control

Celia D. Knight, *University of Leeds, Leeds LS2 9JT, UK,* Moss, Molecular Tools for Phenotypic Analysis

Dhinakar S. Kompala, *Department of Chemical Engineering, University of Colorado, Boulder, Colorado,* Cell Growth and Protein Expression Kinetics

Toyoki Kozai, *Laboratory of Environmental Control Engineering, Faculty of Horticulture, Chiba University, Matsudo, Chiba 271-8510, Japan,* Acclimatization

Gerlinde Kretzmer, *Institut für Technische, Chemie der Universität Hannover, Germany,* On-Line Analysis in Animal Cell Culture

Mark Leonard, *Mammalian and Microbial Cell Sciences and Pilot Lab, Genetics Institute, Andover, Massachusetts,* Cell Stability, Animal

Beth Loberant, *StePac L.A., Ltd., Tefen Industrial Park, Tefen, Israel,* Micropropagation of Plants, Principles and Practices

K.C. Lowe, *School of Biological Sciences, University of Nottingham, University Park, Nottingham NG7 2RD, UK,* Flow Cytometry of Plant Cells; Plant Protoplasts; Protoplast Fusion for the Generation of Unique Plants

Melissa J. Mahoney, *School of Chemical Engineering, Cornell University, Ithaca, New York,* Cell Structure and Motion, Extracellular Matrix and Cell Adhesion

Bernard Massie, *Institut de Recherche en Biotechnologie, Conseil National de Recherches du Canada, 6100 Avenue Royalmount, Montréal, Québec, Canada H4P 2R2,* Transcription, Translation and the Control of Gene Expression

John A. McBain, *Dartmouth Medical School, Hanover, New Hampshire,* Characterization and Determination of Cell Death by Apoptosis

Christi L. McDowell, *Department of Chemical Engineering, Northwestern University, Evanston, IL,* Animal Cell Culture, Physiochemical Effects of pH

S.L. McKenna, *Tumour Biology Laboratory, Department of Biochemistry, University College, Lee Maltings, Prospect Row, Cork, Ireland,* Programmed Cell Death (Apoptosis) in Culture, Impact and Modulation

Carol McLean, *Protein Fractionation Centre, Scottish National Blood Transfusion Service, Edinburg, Scotland,* Contamination Detection in Animal Cell Culture

Otto-Wilhelm Merten, *Généthon II, 1, rue de l'Internationale, BP 60, F-91002 Evry Cedex, France,* Cell Detachment

Shirley I. Miekka, *Jerome H. Holland Laboratory, American Red Cross, Rockville, Maryland,* Viral Inactivation, Emerging Technologies for Human Blood Products

Tetsuro Mimura, *Biological Laboratory, Hitotsubashi University, Naka 2-1, Kunitachi, Tokyo 186-8601, Japan,* Physiology of Plant Cells in Culture

Alaka Mullick, *Institut de Recherche en Biotechnologie, Conseil National de Recherches du Canada, 6100 Avenue Royalmount, Montréal, Québec, Canada H4P 2R2,* Transcription, Translation and the Control of Gene Expression

Lars Keld Nielsen, *Department of Chemical Engineering, The University of Queensland, Brisbane QLD 4072, Australia,* Virus Production from Cell Culture, Kinetics

David J. Odde, *Department of Chemical Engineering, Michigan Technological University, Houghton, MI,* Cell Structure and Motion, Cytoskeleton and Cell Movement

Cynthia Oliver, *MedImmune, Inc., Gaithersburg, Maryland,* Animal Cell Culture, Physiochemical Effects of Dissolved Oxygen and Redox Potential

Karin Øyaas, *The Norwegian Pulp and Paper Research Institute, N-7034 Trondheim, Norway,* Animal Cell Culture, Physiochemical Effects of Osmolality and Temperature

Laura A. Palomares, *Departamento de Bioingeniería, Instituto de Biotecnología, Universidad Nacional Autónoma de México, Cuernavaca, México,* Bioreactor Scale-Down; Bioreactor Scale-Up

Eleftherios T. Papoutsakis, *Department of Chemical Engineering, Northwestern University, Evanston, IL,* Animal Cell Culture, Physiochemical Effects of pH

Nancy L. Parenteau, *Organogenesis, Inc., 150 Dan Road, Canton, MA,* Cell Differentiation, Animal

Wayne A. Parrott, *Department of Crop & Soil Sciences, The University of Georgia, Athens, GA,* Embryogenesis in Angiosperms, Somatic

K. Peter Pauls, *Department of Plant Agriculture, Ontario Agricultural College, University of Guelp, Guelp Ontario N1G 2W1, Canada,* Disease Resistance in Transgenic Plants; Toxin Resistant Plants from Plant Cell Culture and Transformation

Thomas Pickardt, *Institute of Applied Genetics, FU Berlin, Albrecht-Thaer-Weg 6, 14195 Berlin, Germany,* Transformation of Plants

Birgitt Pläsier, *School of Chemical Engineering, University of Birmingham, Edgbaston, Birmingham B15 2TT, UK,* Off-Line Immunoassays in Bioprocess Control

J.B. Power, *School of Biological Sciences, University of Nottingham, University Park, Nottingham NG7 2RD, UK,* Flow Cytometry of Plant Cells; Plant Protoplasts; Protoplast Fusion for the Generation of Unique Plants

Niesko Pras, *Department of Pharmaceutical Biology, University Centre for Pharmacy, Groningen Institute for Drug Studies, University of Groningen, A. Deusinglaan 1, NL-9713AV Groningen, The Netherlands,* Plant Cell Cultures, Secondary Product Accumulation

Margaret M. Ramsay, *Royal Botanic Gardens, Kew, Richmond, Surrey, United Kingdom,* Angiosperms; Dicotyledons

Octavio T. Ramírez, *Departamento de Bioingeniería, Instituto de Biotecnología, Universidad Nacional Autónoma de México, Cuernavaca, México,* Bioreactor Scale-Down; Bioreactor Scale-Up

Govind Rao, *Medical Biotechnology Center and, Department of Chemical and, Biochemical Engineering, Baltimore, Maryland,* Animal Cell Culture, Physiochemical Effects of Dissolved Oxygen and Redox Potential

Jason E. Reynolds, *Dartmouth Medical School, Hanover, New Hampshire,* Characterization and Determination of Cell Death by Apoptosis

P.L. Roberts, *Bio Products Laboratory, Elstree, Herts, United Kingdom,* Sterilization and Decontamination

David K. Robinson, *Merck Research Laboratories, Merck & Co., Inc., Rahway, NJ,* Animal Cell Culture Media

Mary Robison, *Department of Plant Agriculture, Ontario Agricultural College, University of Guelp, Guelp Ontario N1G 2W1, Canada,* Disease Resistance in Transgenic Plants

Gregory L. Rorrer, *Department of Chemical Engineering, Oregon State University, Corvallis, OR,* Seaweeds: Cell and Tissue Suspension Cultures

Laura A. Rudolph-Owen, *Department of Biochemistry, Vanderbilt University School of Medicine, Nashville, TN,* Cell-Surface Receptors: Structure, Activation, and Signaling

W. Mark Saltzman, *School of Chemical Engineering, Cornell University, Ithaca, New York,* Cell Structure and Motion, Extracellular Matrix and Cell Adhesion

J.E. Schlatmann, *Kluyver Laboratory for Biotechnology, Delft University for Technology, Julianalaan 67, 2628 BC, Delft, The Netherlands,* Bioreactors, Recirculation Bioreactor in Plant Cell Culture

Alan Scragg, *Department of Environmental Health and Science, University of the West of England, Frenchay, Bristol BS 16 1QY, United Kingdom,* Plant Cell Culture, Effect on Growth and Secondary Product Accumulation of Organic Compounds and Surfactants; Plant Cell Cultures, Selection and Screening for Secondary Product Accumulation; Somaclonal Variation

Alan H. Scragg, *Department of Environmental Health and Science, University of the West of England, Frenchay, Bristol BS 16 1QY, U.K.,* Plant Cell Cultures, Physiochemical Effects on Growth

Kevin L. Shade, *Bio-Products Laboratory, Dagger Lane, Elstree, Herts., United Kingdom,* Aseptic Techniques in Cell Culture

John L. Sherwood, *Department of Plant Pathology, University of Georgia, Athens, GA,* Virus Removal from Plants

Johannes Siemens, *Institute of Applied Genetics, FU Berlin, Albrecht-Thaer-Weg 6, 14195 Berlin, Germany,* Transformation of Plants

Martin S. Sinacore, *Mammalian and Microbial Cell Sciences and Pilot Lab, Genetics Institute, Andover, Massachusetts,* Cell Stability, Animal

D.D. Songstad, *Monsanto GG4H, 700 Chesterfield Parkway North, St. Louis, MO,* Herbicide-Resistant Plants, Production of

Alexander Sorkin, *Department of Pharmacology, University of Colorado Health Sciences Center, Denver, CO,* Protein Processing, Endocytosis and Intracellular Sorting of Growth Factors

R.E. Spier, *University of Surrey, Guildford, Surrey GU2 5XH, England,* Ethical Issues in Animal and Plant Cell Technology; History of Animal Cell Technology

Friedrich Srienc, *Department of Chemical Engineering and Materials Science, Biological Process Technology Institute, The University of Minnesota, St. Paul, Minnesota,* Cell Cycle Events and Cell Cycle-Dependent Processes

Alison Stacey, *2 High Street, Barley Hertfordshire, SG8 8HZ, UK,* Animal Cell Types, Hybridoma Cells

G. Stacey, *National Institute of Biological Standards and Control, South Mimms, England,* Animal Cell Types, Cell Lines Used in Manufacture of Biological Products; Animal Cell Types, Liver Cells; Cell and Cell Line Characterization; Cell Banks: A Service to Animal Cell Technology

Richard Stebbings, *NIBSC, Blanche Lane, South Mimms EN6 3QG, UK,* Animal Cell Types, T Lymphocytes

Wei Wen Su, *Department of Biological Engineering, University of Missouri-Columbia, Columbia, Missouri,* Bioreactors, Perfusion

Shinsaku Takayama, *Department of Biological Science and Technology, Tokai University, Shizuoka, Japan,* Bioreactors, Airlift

Michio Tanaka, *Department of Horticulture, Faculty of Agriculture, Kagawa University, Kagawa 761-0795, Japan,* Plant Cell Culture Bag

H.J.G. ten Hoopen, *Kluyver Laboratory for Biotechnology, Delft University for Technology, Julianalaan 67, 2628 BC, Delft, The Netherlands,* Bioreactors, Continuous Culture of Plant Cells; Bioreactors, Recirculation Bioreactor in Plant Cell Culture

Leigh E. Towill, *USDA-ARS National Seed Storage Laboratory, 1111 S. Mason St., Fort Collins, CO,* Germplasm Preservation of In Vitro Plant Cultures

Johannes Tramper, *Department of Food Science, Food and Bioengineering Group, Wageningen Agricultural University, Wageningen, The Netherlands,* Flux Analysis of Mammalian Cell Culture: Methods and Applications

Yung-Shyeng Tsao, *Schering-Plough Research Institute, Union, NJ,* Animal Cell Culture Media

Richard M. Twyman, *Molecular Biotechnology Unit, John Inness Centre, Norwich NR4 7UH, U.K.,* Genetic Engineering: Animal Cell Technology

Cord C. Uphoff, *DSMZ-German Collection of Microorganisms, and Cell Cultures, Department of Human and Animal Cell Cultures, Braunschweig, Germany,* Contamination of Cell Cultures, Mycoplasma

Richard E. Veilleux, *Department of Horticulture, Virginia Polytechnic Institute and State University, Blacksburg, Virginia,* Haploid Plant Production: Pollen/Anther/Ovule Culture

P. von Aderkas, *Centre for Forest Biology, University of Victoria, Victoria, B.C., Canada,* Culture of Conifers

K.B. Walker, *National Institute of Biological Standards and Control, South Mimms, England,* Animal Cell Types, B Cell Lineage; Animal Cell Types, Monocyte, Macrophage Lineage

Pamela J. Weathers, *Biology and Biotechnology Department,* Bioreactors, Mist

Hazel Y. Wetzstein, *Department of Horticulture, Plant Science Building, University of Georgia, Athens, Georgia,* Anatomy of Plant Cells

Bruce Whitelaw, *Division of Molecular Biology, Roslin Institute, Roslin, Midlothian EH259PS, U.K.,* Genetic Engineering: Animal Cell Technology

Erik M. Whiteley, *Department of Chemical Engineering, The Johns Hopkins University, Baltimore, MD,* Protein Processing, Processing in the Endoplasmic Reticulum and Golgi Network

Jack M. Widholm, *Department of Crop Sciences, University of Illinois at Urbana-Champaign, ERML, 1201 W. Gregory Drive, Urbana, IL,* Plant Cell Cultures, Photosynthetic

John A. Wilkins, *Rheumatic Diseases Research Laboratory, Departments of Medicine, Immunology and Medical Microbiology, University of Manitoba, Winnipeg MB R3A 1M4,* Cell Fusion

Barbara E. Wyslouzil, *Chemical Engineering Department, Worcester Polytechnic Institute, Worcester, MA,* Bioreactors, Mist

Ping Zhou, *Dartmouth Medical School, Hanover, New Hampshire,* Characterization and Determination of Cell Death by Apoptosis

Sayed M.A. Zobayed, *Laboratory of Environmental Control Engineering, Faculty of Horticulture, Chiba University, Matsudo, Chiba 271-8510, Japan,* Acclimatization

COMMONLY USED ACRONYMS AND ABBREVIATIONS

ABA	abscisic acid
AC	alternating current
ACC	alpha-aminocyclopropane-1-carboxylic acid
AFLP	amplified fragment length polymorphism
AFM	atomic force microscope/microscopy
ATL	atrial natriuretic peptide receptor
BAEC	bovine aortic endothelial cell
BLA	biological license application
BOD	biological oxygen demand
BRG 1	human homologue of drosophila brahma gene (brm)
BSE	Bovine Spongiform Encephalopathy
CAD	caspase activated DNAse
CAMs	Cell-Adhesion Molecules
CBER	Center of Biologics Evaluation and Review
CB MNC	cord blood mononuclear cells
CDC	Center for Disease Control and Prevention
CDK	cyclin dependent kinase
CE	centrifugal elutriation
CEIA	capillary electrophoretic immunoassay
CFD	computational fluid dynamics
cGMP	current Good Manufacturing Practices
CHO	chinese hamster ovary cells
CIP	cleaning-in-place procedures
CJD	Creutzfeld-Jakob Disease
CMC	Chemistry, Manufacturing, and Controls studies
CMS	cytoplasmic male sterility
COP	critical oxygen pressure
CTL	cytotoxic T lymphocytes
DEPC	diethyl pyrocarbonate
DF	dark field microscopes
DHFR	dihydrofolate reductase
DIBA	dot immunobinding assays
DIC	differential interference microscopes
DMSO	dimethyl sulfoxide
DNA-PK	DNA-dependent protein kinase
DOP	dioctylphthalate
DW	dry weight
EIA	competitive binding enzyme immunoassay
ELAM	endothelial leukocyte adhesion molecule
ELISA	enzyme-linked immunosorbent assay
EMIT	enzyme-monitored immunotest
EPO	Erythropoitin (hormone)
ERE	estrogen response element
FDA	Food and Drug Administration
FDA	fluroescein diacetate
FIA	flow-injection analysis
FRALE	frangible anchor linker effector (compounds)
FRIM	fluorescence ratio imaging microscopy
FWB	fresh weight basis
GISH	genomic in situ hybridization
GMP	good manufacturing practice
GRE	glucocorticoid response element
GS	glutamine synthetase
GUS	b-glucuronidase gene
HBV	hepatitis B virus
HEPA	High Efficiency Particulate Air (filtration)
HF	hollow fiber (reactors)
HIV	human immunodeficiency virus
HRE	hormone response element
IAA	indole-3-acetic acid
IBA	indole-3-butyric acid
ICAM-1	intracellular adhesion molecule-1
ICH	International Conference on Harmonisation
IGF	insulin growth factor
IPGRI	International Plant Genetic Resources Institute
IRMA	immunoradiometric asay
LAFs	Laminar flow cabinets/hoods
LCR	ligase chain reaction
LMTD	log mean temperature difference
LNAME	N-nitro-L-arginine methyl ester
MACS	magnetic cell sorter
MAP	mitogen activated protein
MB	methylene blue
MCB	Master Cell Bank
MCP-1	monocyte chemoattractant protein-1
MEM	minimal essential media
MOI	multiplicity of infection
MSCs	microbiological safety cabinets
MTB	multipurpose tower bioreactor
MTT	3-(4,5-dimethylthiazol-2-yl)-2,5-diphenyl tetrazolium bromide
NAA	1-naphthylacetic acid
NASBA	nucleic acid sequence based amplification
NAT	nucleic acid testing
NEAAs	non-essential amino acids
Nes	naphthalene endoperoxides
NHEs	Na+/H+ antiports or exchangers
NIH	National Institutes of Health
NK	natural killer cells
NMR	nuclear magnetic resonance
NOA	2-naphthylocyacetic acid
NOS	nitric oxide synthase
OTR	oxygen transfer rate
OUR	oxygen uptake rate
PAI-1	plasminogen activator inhibitor, type 1
PAL	phenylalanine ammonia lyase
PARP	poly(ADP-ribose) polymerase
PB MNC	peripheral blood mononuclear cells
PCMBS	p-chloromercuribenzenesulfonic acid
PCNA	proliferating cell nuclear antigen
PCD	physiological cell death
PCV	packed cell volume
PDGF	platelet derived growth factor
PEG	polyethylene glycol
PFCs	perfluorocarbons
PVA	polyvinyl alcohol
rAHF	recombinant human anti-hemophiliac factor
RAPD	randomly amplified polymorphic DNA
RCM	reflection contrast microscopy

RCV	replication competent virus	TSE	transmissible spongiform encephalopathies
RFLP	restriction fragment length polymorphism	ULPA	Ultra Low Penetration Air (filters)
RH	relative humidity	VCAM-1	vascular cell adhesion molecule-1
RIA	radioimmunoassay	VFF	vortex flow filtration
RLF	replication licensing factor	WHO	world health organization
TCA	tricarboxylic acid	WCB	working cell bank
TIL	tumor infiltrating lymphocytes		

CONVERSION FACTORS, ABBREVIATIONS, AND UNIT SYMBOLS

SI UNITS (Adopted 1960)

The International System of Units (abbreviated SI), is being implemented throughout the world. This measurement system is a modernized version of the MKSA (meter, kilogram, second, ampere) system, and its details are published and controlled by an international treaty organization (The International Bureau of Weights and Measures).

SI units are divided into three classes:

BASE UNITS		SUPPLEMENTARY UNITS	
length	meter[†] (m)	plane angle	radian (rad)
mass	solid angle	steradian (sr)	kilogram (kg)
time	second (s)		
electric current	ampere (A)		
thermodynamic temperature[‡]	kelvin (K)		
amount of substance	mole (mol)		
luminous intensity	candela (cd)		

Quantity	Unit	Symbol	Acceptable equivalent
volume	cubic meter	m^3	
	cubic diameter	dm^3	L (liter) (5)
	cubic centimeter	cm^3	mL
wave number	1 per meter	m^{-1}	
	1 per centimeter	cm^{-1}	

In addition, there are 16 prefixes used to indicate order of magnitude, as follows:

Multiplication factor	Prefix	Symbol
10^{18}	exa	E
10^{15}	peta	P
10^{12}	tera	T
10^9	giga	G
10^6	mega	M
10^3	kilo	k
10^2	hecto	h^a
10	deka	da^a
10^{-1}	deci	d^a
10^{-2}	centi	c^a
10^{-3}	milli	m
10^{-6}	micro	€
10^{-9}	nano	n
10^{-12}	pico	p
10^{-15}	femto	f
10^{-18}	atto	a

[a] Although hecto, deka, deci, and centi are SI prefixes, their use should be avoided except for SI unit-multiples for area and volume and nontechnical use of centimeter, as for body and clothing measurement.

For a complete description of SI and its use the reader is referred to ASTM E380.

A representative list of conversion factors from non-SI to SI units is presented herewith. Factors are given to four significant figures. Exact relationships are followed by a dagger. A more complete list is given in the latest editions of ASTM E380 and ANSI Z210.1.

[†] The spellings "metre" and "litre" are preferred by ASTM; however, "-er" is used in the *Encyclopedia*.

[‡] Wide use is made of Celsius temperature (*t*) defined by

$$t = T - T_0$$

where T is the thermodynamic temperature, expressed in kelvin, and $T_0 = 273.15$ K by definition. A temperature interval may be expressed in degrees Celsius as well as in kelvin.

CONVERSION FACTORS TO SI UNITS

To convert from	To	Multiply by
acre	square meter (m^2)	4.047×10^3
angstrom	meter (m)	$1.0 \times 10^{-10\dagger}$
are	square meter (m^2)	$1.0 \times 10^{2\dagger}$
astronomical unit	meter (m)	1.496×10^{11}
atmosphere, standard	pascal (Pa)	1.013×10^5
bar	pascal (Pa)	$1.0 \times 10^{5\dagger}$
barn	square meter (m^2)	$1.0 \times 10^{-28\dagger}$
barrel (42 U.S. liquid gallons)	cubic meter (m^3)	0.1590
Bohr magneton (ε_B)	J/T	9.274×10^{-24}
Btu (International Table)	joule (J)	1.055×10^3
Btu (mean)	joule (J)	1.056×10^3
Btu (thermochemical)	joule (J)	1.054×10^3
bushel	cubic meter (m^3)	3.524×10^{-2}
calorie (International Table)	joule (J)	4.187
calorie (mean)	joule (J)	4.190
calorie (thermochemical)	joule (J)	$4.184^\dagger$
centipoise	pascal second (Pa · s)	$1.0 \times 10^{-3\dagger}$
centistokes	square millimeter per second (mm^2/s)	$1.0^\dagger$
cfm (cubic foot per minute)	cubic meter per second (m^3/s)	4.72×10^{-4}
cubic inch	cubic meter (m^3)	1.639×10^{-5}
cubic foot	cubic meter (m^3)	2.832×10^{-2}
cubic yard	cubic meter (m^3)	0.7646
curie	becquerel (Bq)	$3.70 \times 10^{10\dagger}$
debye	coulomb meter (C m)	3.336×10^{-30}
degree (angle)	radian (rad)	1.745×10^{-2}
denier (international)	kilogram per meter (kg/m)	1.111×10^{-7}
	tex‡	0.1111
dram (apothecaries')	kilogram (kg)	3.888×10^{-3}
dram (avoirdupois)	kilogram (kg)	1.772×10^{-3}
dram (U.S. fluid)	cubic meter (m^3)	3.697×10^{-6}
dyne	newton (N)	$1.0 \times 10^{-5\dagger}$
dyne/cm	newton per meter (N/m)	$1.0 \times 10^{-3\dagger}$
electronvolt	joule (J)	1.602×10^{-19}
erg	joule (J)	$1.0 \times 10^{-7\dagger}$
fathom	meter (m)	1.829
fluid ounce (U.S.)	cubic meter (m^3)	2.957×10^{-5}
foot	meter (m)	$0.3048^\dagger$
footcandle	lux (lx)	10.76
furlong	meter (m)	2.012×10^{-2}
gal	meter per second squared (m/s^2)	$1.0 \times 10^{-2\dagger}$
gallon (U.S. dry)	cubic meter (m^3)	4.405×10^{-3}
gallon (U.S. liquid)	cubic meter (m^3)	3.785×10^{-3}
gallon per minute (gpm)	cubic meter per second (m^3/s)	6.309×10^{-5}
	cubic meter per hour (m^3/h)	0.2271
gauss	tesla (T)	1.0×10^{-4}
gilbert	ampere (A)	0.7958
gill (U.S.)	cubic meter (m^3)	1.183×10^{-4}
grade	radian	1.571×10^{-2}
grain	kilogram (kg)	6.480×10^{-5}
gram force per denier	newton per tex (N/tex)	8.826×10^{-2}
hectare	square meter (m^2)	$1.0 \times 10^{4\dagger}$
horsepower (550 ft · lbf/s)	watt (W)	7.457×10^2
horsepower (boiler)	watt (W)	9.810×10^3
horsepower (electric)	watt (W)	$7.46 \times 10^{2\dagger}$
hundredweight (long)	kilogram (kg)	50.80
hundredweight (short)	kilogram (kg)	45.36
inch	meter (m)	$2.54 \times 10^{-2\dagger}$
inch of mercury (32 °F)	pascal (Pa)	3.386×10^3
inch of water (39.2 °F)	pascal (Pa)	2.491×10^2
kilogram-force	newton (N)	9.807
kilowatt hour	megajoule (MJ)	$3.6^\dagger$

CONVERSION FACTORS TO SI UNITS

To convert from	To	Multiply by
kip	newton (N)	4.448×10^3
knot (international)	meter per second (m/S)	0.5144
lambert	candela per square meter (cd/m³)	3.183×10^3
league (British nautical)	meter (m)	5.559×10^3
league (statute)	meter (m)	4.828×10^3
light year	meter (m)	9.461×10^{15}
liter (for fluids only)	cubic meter (m³)	$1.0 \times 10^{-3\dagger}$
maxwell	weber (Wb)	$1.0 \times 10^{-8\dagger}$
micron	meter (m)	$1.0 \times 10^{-6\dagger}$
mil	meter (m)	$2.54 \times 10^{-5\dagger}$
mile (statute)	meter (m)	1.609×10^3
mile (U.S. nautical)	meter (m)	$1.852 \times 10^{3\dagger}$
mile per hour	meter per second (m/s)	0.4470
millibar	pascal (Pa)	1.0×10^2
millimeter of mercury (0 °C)	pascal (Pa)	$1.333 \times 10^{2\dagger}$
minute (angular)	radian	2.909×10^{-4}
myriagram	kilogram (kg)	10
myriameter	kilometer (km)	10
oersted	ampere per meter (A/m)	79.58
ounce (avoirdupois)	kilogram (kg)	2.835×10^{-2}
ounce (troy)	kilogram (kg)	3.110×10^{-2}
ounce (U.S. fluid)	cubic meter (m³)	2.957×10^{-5}
ounce-force	newton (N)	0.2780
peck (U.S.)	cubic meter (m³)	8.810×10^{-3}
pennyweight	kilogram (kg)	1.555×10^{-3}
pint (U.S. dry)	cubic meter (m³)	5.506×10^{-4}
pint (U.S. liquid)	cubic meter (m³)	4.732×10^{-4}
poise (absolute viscosity)	pascal second (Pa · s)	$0.10^\dagger$
pound (avoirdupois)	kilogram (kg)	0.4536
pound (troy)	kilogram (kg)	0.3732
poundal	newton (N)	0.1383
pound-force	newton (N)	4.448
pound force per square inch (psi)	pascal (Pa)	6.895×10^3
quart (U.S. dry)	cubic meter (m³)	1.101×10^{-3}
quart (U.S. liquid)	cubic meter (m³)	9.464×10^{-4}
quintal	kilogram (kg)	$1.0 \times 10^{2\dagger}$
rad	gray (Gy)	$1.0 \times 10^{-2\dagger}$
rod	meter (m)	5.029
roentgen	coulomb per kilogram (C/kg)	2.58×10^{-4}
second (angle)	radian (rad)	$4.848 \times 10^{-6\dagger}$
section	square meter (m²)	2.590×10^6
slug	kilogram (kg)	14.59
spherical candle power	lumen (lm)	12.57
square inch	square meter (m²)	6.452×10^{-4}
square foot	square meter (m²)	9.290×10^{-2}
square mile	square meter (m²)	2.590×10^6
square yard	square meter (m²)	0.8361
stere	cubic meter (m³)	$1.0^\dagger$
stokes (kinematic viscosity)	square meter per second (m²/s)	$1.0 \times 10^{-4\dagger}$
tex	kilogram per meter (kg/m)	$1.0 \times 10^{-6\dagger}$
ton (long, 2240 pounds)	kilogram (kg)	1.016×10^3
ton (metric) (tonne)	kilogram (kg)	$1.0 \times 10^{3\dagger}$
ton (short, 2000 pounds)	kilogram (kg)	9.072×10^2
torr	pascal (Pa)	1.333×10^2
unit pole	weber (Wb)	1.257×10^{-7}
yard	meter (m)	$0.9144^\dagger$

† Exact.

NOMENCLATURE FOR BIOREACTOR OPERATIONS: PREPARATION, STERILIZATION, CHARGING, CULTURE INITIATION AND HARVESTING

τ	space time, calculated for a stirred tank
ρ	density
α	thermal conductivity
β	angle of slope
δ	thickness of layer, length of dead leg
μ	viscosity
a	width of rectangular channel
b	height of rectangular channel
B	constant
c	molar concentration
c_H	mixing time constant
Cp	specific heat capacity at constant pressure
D	diameter
d_I	diameter of nozzle on tank
f_a	fraction of original air left in system
g	gravitational acceleration
h	heat transfer coefficient
H	specific latent heat
k	thermal conductivity
K	constant
L	length
M	molecular weight
N	parameter
N_s	molal flux of water
$\wp$	binary diffusivity
P	pressure

P_{Tot}	total system pressure
Q/A	heat flux
r	spatial radial coordinate
R	ideal gas constant
R_s	thermal heat transfer resistance
T	temperature
t_H	mixing time in a stirred tank reactor for 95% homogeneity
V	volume
$\overline{V}$	partial molal volume
w	mass flow rate
x	spatial coordinate in one dimensonal moel
X	mole fraction
z	axial spatial coordinate

Subscripts

a	air
c	condensate
s	steam
T	thermal
conv	convective
INT	interface
w	wall
INS	insulated
bare	not insulated

INTRODUCTION

An encyclopedia serves many functions. The word itself exposes the multiplicity of its purposes. While it seeks to be all encompassing (as it encycles) it also takes us on a journey whereby we both discover new ideas and extend our learning experiences (rather in the manner of a child being exposed to the "circle of knowledge" in Greek times). The editors and authors of this work have followed such a tradition and have sought to provide readers with a state-of-the-art compilation of information, ideas, procedures, and guidelines so that they may enhance their abilities and understandings of cell technology. This in turn should lead to both new processes and products as well as increases in the productivity and efficiency of existing processes dependent on the cultivation of animal and plant cells.

In both the history of the origin of the idea of the cellularity of all living beings (with the exception of viruses, plasmids, and nucleic acid molecules) and the history of how we might view the way cells emerged from a proto-Earth some 4 billion years ago, the pervasive synergism between the concepts generated in the animal cell world and those rising from the plant cell world has led us to our present world view. This reciprocating reinforcement of views, visions, and experimental observations has been one of the crucial features of the way knowledge and capability have advanced as rapidly as is related in these pages. To maintain this rapid rate of progress, the *Encyclopedia of Cell Technology* has been built about the concept of the facilitation and encouragement of the transference of ideas and practical processes between animal and plant cell technologies. The editors hold that, in spite of some overlap of these areas, the differences between them are such as to stimulate and promote the use of assays or techniques that have worked in one area, say, animal cell technology, in the corresponding area of plant cell technology, and vice versa.

We have progressed considerably in the past 100 years. From the tentative experiments in the last decade of the nineteenth century to the large-scale commercially successful technologies of the last decade of the twentieth century, we can discern a dramatic transformation. Not only have we been able to all but eliminate the exogenous contamination of cultures but the equipment which we now deploy is robust, reliable, and can be used to achieve a predefined outcome within relatively close tolerance limits and with a high degree of consistency. And the scope of those capabilities has widened. Whereas initial experimentation with cells in culture was clearly focused on the solution of intellectual problems of a preponderantly analytic nature, concerning anatomy and physiology (plus or minus biochemistry), the thrust of modern endeavors has been more synthetic and has resulted in a welter of new product areas and opportunities.

Plant cell culturists struggled with in vitro axenic growth of cells for many years before the advent of antibiotics. During this time they were able to define simple nutrient media and to examine the physiology of cells under controlled conditions. While the early uses of plant cells in culture echoed cloning procedures which could be applied to whole plants, the development of techniques for the establishment of uncontaminated callus cultures, which could then be used to either form plantlets or a bulk culture of monodisperse suspension cells or clumps, became a useful technology in the 1960s. The extension of the techniques of in vitro orchid cultivation to that of tree plantlet propagation in the 1980s has become a platform from which an effective reforestation program can be mounted. The successful and commercial production of a secondary metabolite (shikonin) from large-scale suspension cultures was achieved in 1983. Currently, the use of plant cell cultures for the production of anticancer drugs based on taxol is receiving much attention as well as the production of a diverse array of plant cell enzymes, perfumes, and additional pharmaceuticals. From a virtually exclusive concentration on the use of animal cells in culture for the production of virus vaccines in the 1950s to 1970s, the introduction of two new technologies set in train an expansion of effort leading to a corresponding burgeoning of the commercially manufactured product profile. Following Kohler and Millstein's demonstration of the production of monoclonal antibodies from hybridoma cells in 1975 and the way in which animal cells in culture might be genetically engineered in the late 1970s, a second wave of products reached the marketplace. More recently, the use of animal cells in culture as replacements for animals in toxicity testing has received much attention. And the original idea of Carrel in 1913 of using human organs grown in culture to replace pathological tissues is moving from the concept stage to realization. In parallel with these recent developments, the use of animal cells to produce adeno- and lentiviruses, which are principal candidates for vectors of genetic therapeutics of whole animals and humans, is beginning to show signs of becoming a major animal cell technology application. The pluripotency (if not totipotency) of human embryo stem cells may be further explored to provide replacement cells for defunct tissues if, and when, the ethical issues, which the use of such cells engenders, may be resolved. It is the clear intention of the editors and authors of the *Encyclopedia of Cell Technology* to provide readers with a powerful new tool that will enable them to more skillfully and rapidly achieve their goals. As a comprehensive resource of information and process techniques, most practitioners active in the field will find in these pages something which is both fresh and helpful to their personal endeavors. Students and researchers entering this area for the first time will be able to obtain an essential overview of what is available and where further information can be found. We have done everything we can to make the material of this encyclopedia both accessible and of benefit to its users. The outcome we seek is the continued and more extensive development of these areas. From such a venture we are confident that we can continue to bring to the plants, animals, and humans of this planet much for their progress and advantage.

R.E. Spier

A

ACCLIMATIZATION

Toyoki Kozai
Sayed M.A. Zobayed
Chiba University
Matsudo, Chiba
Japan

OUTLINE

INTRODUCTION

Efficient commercial micropropagation for producing a large number of quality plants using limited resources of time, labor, and money largely depends upon high multiplication rates and successful acclimatization.

This article describes: the general characteristics of the in vitro environment, responses of plants to the in vitro environment, general responses of plants to the ex vitro environment, reason for the difficulty of the ex vitro acclimatization, environmental control for the ex vitro acclimatization, and the in vitro acclimatization. Aspects of physical environments and their effects on the plant growth and development in vitro are discussed. Recent research on the in vitro acclimatization is introduced in relation to the photoautotrophic micropropagation.

DEFINITION AND OBJECTIVE OF ACCLIMATIZATION

Definition

The term *acclimatization* is defined as the climatic or environmental adaptation of an organism, especially a plant, that has been moved to a new environment (1). In this article, the term acclimatization is specifically used to mean the environmental adaptation of a tissue-cultured or a micropropagated plant to a greenhouse or a field environment (2).

Acclimatization takes place under the active guidance of human beings. The term *acclimation* has a similar meaning, but it is a process of nature. In this article, the term acclimatization is used in preference to terms such as acclimation, hardening, weaning, habituation, conditioned, etc. (3).

Acclimatization is required because there is, in general, a significant difference between the tissue culture or micropropagation environment and the greenhouse or field environment. The former is called the in vitro environment and the latter the ex vitro environment. Acclimatization is mostly conducted in a greenhouse and sometimes in a field under shade, which is called ex vitro acclimatization. On the other hand, acclimatization conducted in a tissue culture vessel or a micropropagation box is called in vitro acclimatization (4–6) for ex vitro environment. Hereafter, the term acclimatization will be used to mean in vitro and ex vitro acclimatization. If in vitro acclimatization is successfully conducted, the ex vitro acclimatization can be simplified or even eliminated.

During the ex vitro acclimatization, the ex vitro environment is changed gradually with time, starting with the near in vitro environment and finishing with the near greenhouse or field environment. Besides the physical environment control, the acclimatization can be enhanced by application of some chemicals for accelerating rooting (7,8), reducing transpiration by regulation of stomatal functioning (6,9), and by application of symbiotic microorganisms (10). In general, the period needed for the ex vitro acclimatization ranges between several days and a few weeks.

In the following sections, it is assumed, unless otherwise stated, that cultures that are subject to acclimatization, such as shoots and regenerated plants, possess chlorophyll in their leaves and have photosynthetic ability to a certain degree. Also, for convenience, all types of chlorophyllous cultures including leafy explants and shoots in the culture vessel are called plants in vitro hereafter, unless otherwise stated. Microtubers and bulblets without chlorophyllous shoots, which are often produced in a bioreactor and are usually buried under the ground in the greenhouse for further growth, are not discussed extensively in this article.

Objective of Acclimatization and Steps for Achieving the Objective

Micropropagation is a technology for producing a large number of genetically identical, pathogen-free transplants by means of tissue or organ culture. The last stage of the micropropagation process, following the multiplication and/or rooting stage, is called the ex vitro acclimatization stage.

The widespread use of micropropagated plants is currently restricted, partly due to high percentages of death and damaged plants in the micropropagation process, especially at the ex vitro acclimatization stage (11). Thus the main objective of acclimatization is to provide an optimum in vitro and/or ex vitro environment for minimizing the percentages of death and damaged plants in the micropropagation process, for enhancing the plant establishment, and for promoting the plant growth at and after the acclimatization stage (6). In order to achieve this objective, the following subjects need to be discussed:

1. The general characteristics of the in vitro environment in conventional micropropagation
2. The responses of plants in vitro to the in vitro environment in conventional micropropagation
3. Improvements of the in vitro environment and modifications of the conventional micropropagation system so that plants in vitro are grown vigorously and easily acclimatized in vitro and/or ex vitro with minimum percentages of death and damaged plants and with a high growth rate at reasonable costs
4. The general characteristics of the ex vitro environment at the acclimatization stage in conventional micropropagation
5. The responses of plants at the ex vitro acclimatization stage in conventional micropropagation
6. Improvements of the ex vitro environment and modifications of the conventional acclimatization system so that plants ex vitro are grown vigorously at a high growth rate and
7. Possibilities for developing a new micropropagation system that can minimize the percentages of death and damaged plants and can enhance plant growth at minimal cost

GENERAL CHARACTERISTICS OF THE IN VITRO ENVIRONMENT

Environmental Factors Affecting the Growth and Development of Plants In Vitro

In conventional micropropagation, the in vitro environment has unique general characteristics compared with the ex vitro environment. Figure 1 is a simplified relational diagram or a schematic eco-physiological model showing the environmental factors affecting the growth and development of plants in vitro (12). It also shows the functional relationships among state variables (contents, concentration, etc.) and rate variables (flows or fluxes) of mass (or material) and energy within the tissue culture vessel, and between inside and outside the culture vessel.

State variables quantify conserved properties of the culture vessel system containing plants and culture medium. On the other hand, rate variables quantify the time rate of change of the state variables, expressing flows of material or energy per unit time.

The functional relationships among the state and rate variables in Figure 1 are similar to those for a greenhouse ecosystem or any other ecosystem in a semiclosed system. However, the numerical values of state and rate variables for the in vitro environment are significantly different from those for the ex vitro or the greenhouse environment, as will be discussed in the following.

Aerial Environment

General characteristics of the in vitro aerial environment in conventional micropropagation with respect to state variables are:

1. High relative humidity, usually 95–100% (13,14)
2. Relatively constant air temperature throughout the day, typically $25 \pm 3\,°C$ (14)
3. Low CO_2 concentration during photoperiod (15–17)
4. High CO_2 concentration during the dark period (15,18,19) and
5. High ethylene (C_2H_4) concentration (18,19)

These characteristics are largely due to the low number of air exchanges per hour of the culture vessel and the relatively small air volume of the culture vessel (14). Number of air exchanges per hour of the culture vessel is defined as the hourly ventilation rate of the culture vessel divided by the air volume of the culture vessel (14).

General characteristics of the in vitro aerial environment in conventional micropropagation with respect to rate variables are:

1. Low air movement rate (low CO_2 and water vapor diffusion coefficients), that is, stagnant air (20) due to low photosynthetic photon flux, low net thermal radiation flux, and a relatively uniform distribution of air temperatures inside and around the culture vessel
2. Low transpiration rate of plants and low evaporation rate of the culture medium due to high relative humidity and low air movement rate (or small difference in water potential between air and culture medium) (21,22)
3. Low gross photosynthetic rate of plants due to low CO_2 concentration, low air movement rate during photoperiod, and low ability of photosynthesis (23–25)

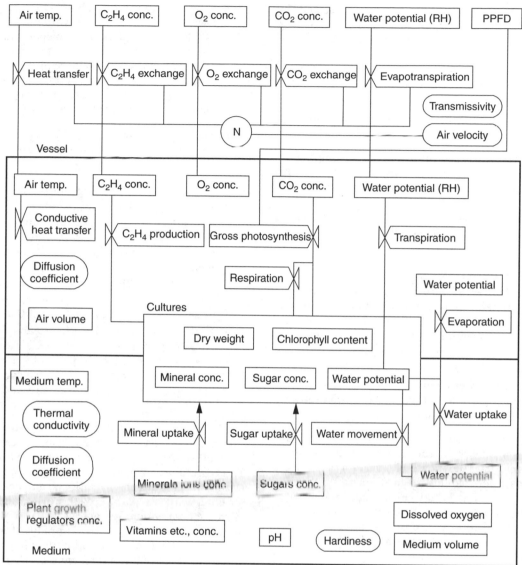

Figure 1. A simplified relational diagram showing the environmental factors affecting the growth and development of plants in vitro, and the flows of mass and energy in and around the culture vessel (14).

PPFD: Photosynthetic photon flux density
 N: Number of air exchanges per hour of the vessel
 RH: Relative humidity. Water potential of the air is determined as a function of RH and air temperature.
      Water potential of culture medium is a sum of osmotic, matric, and pressure potentials.

4. High dark respiration rate of plants due to the high sugar concentrations in plants and culture medium (14) and

5. Low or negative daily net photosynthetic rate (negative daily CO_2 balance) of plants due to low gross photosynthetic rate and high dark respiration rate (23,26–28)

As shown previously, flows of material and energy in the culture vessel are significantly restricted largely by the relatively small differences in spatial gradients of physical potentials (spatial differences in water vapor pressure, temperature, CO_2 concentration, etc.) within the culture vessel and between inside and outside the culture vessel.

Root Zone Environment

General characteristics of the in vitro root zone environment in conventional micropropagation with respect to state variables are as follows:

1. Presence of sugar (sugar is absent in normal soil)

2. Relatively high mineral ion concentrations, especially NH_4^+ concentration, (significantly higher than those of nutrient solution for hydroponics and much higher than those of normal fertile soil for horticulture and agriculture) (29–33)

3. Low dissolved oxygen concentration, especially in gelled agents such as agar (14,34,35)

4. Absence of microorganisms in most cases, and, in other cases, high density of microorganisms causing microbial contamination with the presence of sugar in the culture medium

5. Presence of phenolic compounds or other toxic substances in many cases (35)

6. Presence of exogenous amino acids, vitamins, etc.

7. Presence of exogenous plant growth regulators in many cases

As shown previously, the culture medium composition is significantly different from the composition of nutrient solution for hydroponics such as Hoagland solution and normal soil (33).

The general characteristics of the in vitro root zone environment in conventional micropropagation with respect to rate variables are:

1. Low water uptake rate of plants due to the low transpiration rate (22)

2. Low ion uptake rate of plants due to the low water uptake rate of plants and/or low ion diffusion rates in the culture medium (30,32,33)

3. Low sugar uptake rate of plants due to its low diffusion rate in the culture medium (33)

4. Low transport rates of culture medium components other than sugar due to their low diffusion rates in the culture medium (14,35)

The low transpiration rate of plants and/or slow diffusion of components in the gelled culture medium significantly restricts the flows of substances such as water and ions from the culture medium to the plants. A bioreactor containing liquid culture medium with a liquid mixing system enhances the flows of sugar and ions from the culture medium to the plants, although plants produced in a bioreactor show some disadvantageous responses (5,36).

RESPONSES OF PLANTS TO THE IN VITRO ENVIRONMENT

Typical responses of plants at the tissue level to the in vitro environment in conventional micropropagation are:

1. Little epicuticular and cuticular wax formation (37–39) largely due to high relative humidity and low photosynthetic photon flux

2. Stomatal malfunction (25,39,40,41) (stomata remain open even at a low relative humidity (42,43) and in the dark (44), resulting in wilting of plants due to the low stomatal and cuticular resistance to transpiration)

3. Low chlorophyll contents in the leaves (37,45)

4. Low percent dry matter or hyperhydrated shoots (46–48)

5. Restricted leaf area expansion (48)

6. Low stomatal density on leaves (6,49)

7. Poorly structured spongy and palisade tissues and inferior vascular connections of shoots with roots and

8. Low photosynthetic ability (43) associated with low activities of photosynthetic enzymes such as Rubisco (ribulose-1,5-bisphosphate carboxylase/oxygenase) and PEPCase (phosphoenolpyruvate carboxylase) and abnormal chlorophyll fluorescence responses (28,50). Interaction between sucrose uptake and photosynthesis of micropropagated rose plants was studied extensively by DeRiek (51).

Typical responses of plants at whole plant level to the in vitro environment in conventional micropropagation (36) are:

1. Low growth and development rates of plants

2. Succulent or hyperhydrated shoots with physiological and/or morphological disorders (curled leaves, ion deficiency in leaves etc.), which tend to die or be damaged when transplanted to the ex vitro environment

3. Incomplete rooting and few secondary roots (most roots developed in vitro tend to die within one week or so at the acclimatization stage) (34,35) and

4. High variations in size, shape, and developmental stage of plants due to the spatial variation of the in vitro environment in the culture vessel and the variations in size, shape, and developmental stage of explants, and partly due to the presence of exogenous plant growth regulators

Plants cultured under the typical in vitro environment, showing the responses shown previously, tend to be weakened or eventually die during the ex vitro acclimatization. Thus careful ex vitro acclimatization is essential for minimizing the percentages of death and damage of plants, and for enhancing the plant growth during and after the ex vitro acclimatization. In general, the responses given are more clearly observed in plants produced in a bioreactor containing liquid culture medium than in plants produced in a culture vessel containing agar or other gelled culture medium (5). On the other hand, liquid culture medium combined with a supportive system such as a membrane raft and a vent system with use of gas-permeable film reduces hyperhydration of plants in vitro significantly (5).

EX VITRO ACCLIMATIZATION AND GENERAL RESPONSES OF PLANTS IN CONVENTIONAL MICROPROPAGATION

A typical procedure at an early stage of the ex vitro acclimatization in conventional micropropagation is to provide a relatively dense shade (Fig. 2). In addition to the shade, mists or fogs are often provided during the daytime (7,52). The dense shading with or without frequent misting or fogging is necessary during the daytime to keep the relative humidity high and the temperature moderate under varying natural solar light conditions. This procedure, however, brings about a vicious circle of environmental control at the ex vitro acclimatization stage, as will be shown.

Plants cultured in vitro are sensitive to water stress (53). Thus, in some cases, a portion of leaves has to be removed before the ex vitro acclimatization to reduce excess transpiration of plants. On the other hand, some

Figure 2. Typical ex vitro acclimatization procedures. (**a**) Shading using plastic tunnels in the greenhouse. (**b**) A misting system inside the plastic tunnels for shading. (**c**) Plants being acclimatized ex vitro in the plastic tunnels. Many plants are dead during the acclimatization ex vitro. (**d**) Retransplanting of successfully ex vitro acclimatized plants into the plastic trays for shipping. This procedure is labor intensive.

portion of roots is often removed for easy transplanting at the ex vitro acclimatization stage. In fact, roots produced in vitro are often not functional and die after they are transferred to the ex vitro environment (4,5). Thus, in a sense, there is no essential need to keep such abnormal roots when transplanting.

Furthermore, in many cases, plants cultured in vitro are considered not to have developed full photosynthetic potential at an early stage of the ex vitro acclimatization. Therefore, especially under shade, photosynthesis of plants with reduced leaf area and roots is suppressed. Suppression of photosynthesis retards the emergence of new functional leaves and roots of plants. Then, water and nutrient uptake are suppressed due to the poor root development and reduced leaf area, resulting in low transpiration rates. Thus the plants remain sensitive to water stress and light intensity. Under high light intensity, the photosynthetic organs in leaves are sometimes damaged due to the photoinhibition (54).

Some chemicals are used to reduce the water stress of plants at the ex vitro acclimatization stage. Addition of abscisic acid (ABA) as an antitranspirant into the culture medium decreased stomatal conductance without noticeable negative effects on plant photosynthetic growth (8,55). Growth retardants, inhibitors of gibberellin biosynthesis, were reported to reduce shoot elongation and improve environmental stress resistance (5,56). Paclobutrazol was found to increase the resistance to low relative humidity for micropropagated chrysanthemum, rose, and grapevine (57–59).

Typically, during the first week of ex vitro acclimatization, a large part of the soluble carbohydrates in plants is utilized to develop the root system (51). Once the root system is established, the upper part of a plant starts growing exponentially. On the other hand, if the plant fails to develop a root system within about one week, the plant does not grow anymore and eventually dies.

In order to minimize the delay of growth and death of plants ex vitro and overcome this vicious circle of environmental control at the ex vitro acclimatization stage, it is necessary to consider the fundamental reasons why plants in vitro are difficult to acclimatize ex vitro.

REASONS FOR THE DIFFICULTY OF EX VITRO ACCLIMATIZATION IN CONVENTIONAL MICROPROPAGATION

The reason why plants cultured in vitro are difficult to acclimatize ex vitro will be discussed here from the aspects of trophic phases, photosynthesis, and transpiration.

Trophic Phases — Heterotrophy, Photomixotrophy, and Photoautotrophy

In conventional micropropagation, plants in vitro uptake carbon-containing compounds for synthesizing their carbohydrates from two sources: one is sugar (mainly sucrose, glucose, and fructose) in the culture medium, and the other is carbon dioxide (CO_2) in the air. Growth which depends upon sugar in the culture medium as the sole carbon source is called *heterotrophic growth*, and growth dependent upon CO_2 in the air as the sole carbon source for photosynthesis is called *photoautotrophic growth*. Photoautotrophic plants require inorganic energy sources only: primarily, light energy (photosynthetic photons), CO_2, water, and minerals. Seedlings of almost all higher plants grow photoautotrophically in the field. Growth which depends upon sugar and CO_2 is called *photomixotrophic growth* (60–62), regardless of the ratio of carbon uptake from sugar to the total carbon uptake.

In conventional micropropagation, chlorophyllous plants are cultured on sugar-containing culture medium. Thus they grow photomixotrophically. In this case, at the beginning of the ex vitro acclimatization stage, plants encounter a drastic change in a trophic or nutritional phase — from a photomixotrophic to a photoautotrophic phase, because no sugar is present in the culture medium (substrate or soil) at the ex vitro acclimatization stage. During the ex vitro acclimatization stage, plants are forced to develop photoautotrophy.

Photosynthetic Ability, CO_2 Concentration in Culture Vessel, and Net Photosynthetic Rate of Plants in Vitro

Heterotrophic and photomixotrophic plants in vitro tend not to develop their leaves and roots fully, especially under high relative humidity, low photosynthetic photon flux, and low CO_2 concentration during the photoperiod (5).

It had been believed that plants in vitro did not have sufficient photosynthetic ability to develop photoautotrophy. However, recent research revealed that plants in vitro have sufficient photosynthetic ability to develop photoautotrophy (15,23,26–29,62–68). However, they have not developed their full photosynthetic potential; that is, the photosynthetic ability measured as the maximum net photosynthetic rate at saturated CO_2 concentration and photosynthetic photon flux is significantly lower in young leaves of plants cultured in vitro than in those of plants grown in the field.

Since the photomixotrophic plants in vitro start absorbing CO_2 mainly through stomata at the onset of photoperiod, the CO_2 concentration in the airtight culture vessel decreases sharply within one to two hours nearly to a CO_2 compensation point (ca. 80 μmol mol^{-1}) (15,26). The CO_2 compensation point is a CO_2 concentration at which the net photosynthetic rate of plants is balanced to be zero (gross photosynthesis rate = respiration rate), even at optimum photosynthetic photon fluxes and temperatures.

Namely, the low net photosynthetic rate of plants in vitro is largely attributed to the low CO_2 concentration in the culture vessel during the photoperiod, and only partly to the low photosynthetic ability of plants (69–71). In other words, plants in vitro can develop photoautotrophy

and may develop their full photosynthetic potential, provided that the in vitro environment is controlled properly for promoting photosynthesis. It is noted that the photosynthetic ability is often significantly lower in photomixotrophic plants in vitro than in photoautotrophic plants in vitro (24,72,73).

Reason for High Relative Humidity in Vitro

In conventional micropropagation, relative humidity in the culture vessel is always higher than about 95% (47) because the culture vessel, containing liquid water, is sealed and the temperature is approximately constant with time. However, the purpose of sealing the culture vessel at the multiplication and rooting stages is not to keep the relative humidity high. The culture vessel is sealed to prevent microbes from entering it. Once microbes enter the culture vessel, they grow rapidly and cause microbial contamination from the sugar in the culture medium, resulting in the loss of plants. Thus, the high relative humidity is an adverse side effect resulting from the sealing of the culture vessel for the prevention of microbial contamination.

Therefore, if the relative humidity can be reduced without any microbial contamination at a reasonable cost, nonhyperhydrated plants with normal stomata and cuticular wax layers (74) can be obtained. Then, percentages of death and damaged plants due to excess transpiration at the ex vitro acclimatization can be significantly reduced.

Conventional and Future Approaches for Acclimatization

Considering the preceding discussion, we can consider four approaches to solve the acclimatization problem.

1. Developing a sophisticated environment control system for the ex vitro acclimatization of plants cultured in vitro photomixotrophically in a conventional way

2. Growing plants in vitro having a large amount of carbohydrate reserve (mostly soluble starch and sugar)

   The plants cultured in vitro with a higher sucrose concentration in the culture medium show a significantly higher content of carbohydrates in plants, mostly in leaves (51,54). In this case, the leaves act as storage organs. Then the plants start developing new and functional roots and leaves ex vitro, using the carbohydrate reserve, within about one week from the start of the ex vitro acclimatization stage (75). The plants with newly developed normal roots and leaves can grow quickly. In this case, a large portion of the carbohydrate reserve in old shoots is translocated to form new shoots and roots ex vitro. Thus death of old roots and leaves produced in vitro causes less damage for the subsequent plant growth at the ex vitro acclimatization stage. Of course, plants in vitro with high carbohydrate contents can be better acclimatized ex vitro if the sophisticated environment control system for the ex vitro acclimatization mentioned above is used.

3. Developing a system to grow plants in vitro photoautotrophically under low relative humidity, high CO_2 concentration, and high photosynthetic photon flux conditions at the multiplication and/or rooting stages.

   These plants do not encounter a change in a trophic phase, and have less environmental and nutritional stresses at the ex vitro acclimatization stage.

4. Acclimatizing the plants ex vitro under artificial light in a closed room where the ex vitro environment can be controlled precisely as desired (76,77).

   This method can be applied for plants grown in vitro photomixotrophically and also for those grown in vitro photoautotrophically.

The first two approaches are suitable for the improvements or modifications of the conventional ex vitro acclimatization methods. The third one is for the in vitro acclimatization in combination with the photoautotrophic micropropagation, which seems to be a future acclimatization method. The fourth one is applied both for the conventional and the future ex vitro acclimatization methods.

ENVIRONMENTAL CONTROL UNIT FOR THE EX VITRO ACCLIMATIZATION

A sophisticated microcomputer-controlled acclimatization unit has been developed and applied to give a more appropriate environment than the conventional one (78,79). With this acclimatization unit, water vapor saturation deficit (which is more directly related to transpiration rate than relative humidity) is accurately controlled at a desired level using an ultrasonic humidifier under the changing solar radiation. Then, plants do not face excess transpiration and are not wilted during the ex vitro acclimatization.

At the same time, plants receive a high enough solar radiation to achieve a high net photosynthetic rate. The ex vitro environment is controlled on the basis of acclimatization curves. Figure 3 shows a schematic diagram of the acclimatization curve for air temperature control. The cooling system is turned on when the air temperature in the unit is higher than the upper limit, and is turned off when it reaches the set point. The heating system is operated in a similar way. The set point of the average air temperature on the first day of acclimatization is almost constant, simulating the conditions of the in vitro environment. The set point for the final day is set to be similar to the expected air temperature fluctuation of the greenhouse or field environment where the plants will be transplanted to the soil.

Using the acclimatization curve, the diurnal amplitude of the air temperature is magnified gradually day by day. The diurnal amplitude of the acclimatization curves for solar radiation, and water vapor saturation deficit are also changed in a similar way. The daily average and the magnitude of diurnal amplitude for each environmental factor are modified, depending upon the crop species, the season, etc. CO_2 can be enriched with this unit. Using this acclimatization unit, hyperhydrated strawberry plants that were cultured submerged in a liquid culture medium were successfully acclimatized ex vitro (78).

Apart from the acclimatization unit, in general, CO_2 enrichment at the ex vitro acclimatization stage gives positive effects on rooting and growth of grapevine (80),

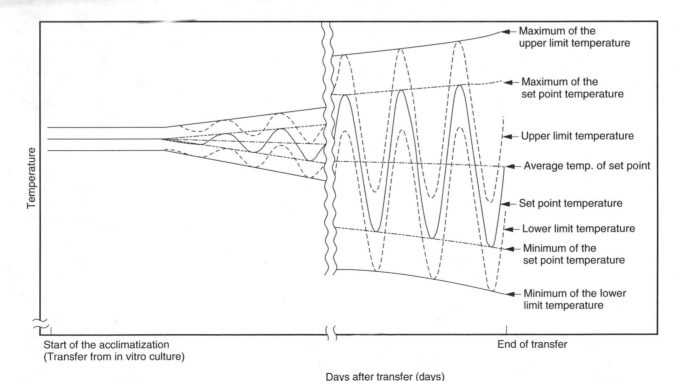

Figure 3. Schematic diagram of the acclimatization curve for air temperature control (78).

strawberry (78,81), raspberry (81) and asparagus (81) plants. CO_2 enrichment can be a common commercial practice in the future because the cost of CO_2 enrichment is minimal. In 1998, the cost of pure liquid CO_2 in a high-pressure container was about $0.12 per kg in Japan and $0.30 per kg in China. Using one kilogram of pure CO_2, more than 10,000 plants can be acclimatized ex vitro under CO_2-enriched conditions. A CO_2 controller for the ex vitro acclimatization costs about $1,000 in 1998.

IN VITRO ACCLIMATIZATION

Plants can also be acclimatized in vitro, provided that the relative humidity, CO_2 concentration, and photosynthetic photon flux (PPF) in the culture vessel are properly controlled. Furthermore, if the plants are cultured in vitro on the appropriate supporting material with high air porosity and containing no sugar under low relative humidity, high CO_2 concentration, and high PPF conditions, they are acclimatized in vitro, and then the ex vitro acclimatization can be simplified or even eliminated. The methods of the in vitro acclimatization are discussed in more detail in the following.

Relative Humidity

Morphological and physiological characteristics of plants in vitro can be improved significantly by reducing the relative humidity (RH) in the culture vessel to about 90% under a photosynthetic photon flux of 150 μmol m^{-2}s^{-1} and higher (47). The transpiration rate and ion uptake rates of plants in vitro increases with reducing relative humidity (48).

This reduction in relative humidity can be achieved by use of the gas-permeable film such as a microporous polypropylene film. For example, a few holes with a diameter of 10 mm are made on each of the cap of a Magenta-type culture vessel. Then, a 14 mm disk of the film is placed and glued on each of the holes (14,30,60,63). Several kinds of culture vessels with the gas-permeable film on the culture vessel caps are commercially available now. Use of the gas-permeable film generally increases CO_2 concentration in the culture vessel during the photoperiod, decreases CO_2 concentration during the dark period, and decreases ethylene concentration throughout the culture period, compared with those without the gas-permeable film.

At the same time, the loss of water from the culture vessel by evaporation from the culture medium and transpiration from plants in vitro is enhanced by the gas permeable film. This loss of water causes desiccation of the gelled culture medium and retards the plant growth in vitro in certain cases. The loss of water from the culture vessel can be reduced significantly by increasing the relative humidity in the culture room to 70–80%.

The relative humidity in the culture room during the photoperiod is typically around 40%. This is because the water vapor of the room air is condensed at the cooling coils of the cooling system, and thus the room air is dehumidified when the cooling system is operated. The cooling system needs to be operated during the photoperiod

to remove heat from the fluorescent lamps even in the winter. Theoretically, the loss of water from the culture vessel can be reduced by about 60–70% in case the relative humidity in the culture room is raised from 40% to 80% at temperatures of about 25 °C.

Another method to prevent the desiccation of the culture medium due to excess water loss from the culture vessel is to increase the volume of the culture medium about twice the volume of the standard formulation. Then, the change in water content (percent of water volume per culture medium volume) is approximately halved compared with the change in water content for standard culture medium volume when an equal volume of water is lost from the culture medium.

CO_2 Concentration

The use of the gas-permeable film for the culture vessels in a culture room is also effective for raising CO_2 concentration in the culture vessel. However, in this case, the CO_2 concentration in the culture vessel during the photoperiod is lower than the atmospheric CO_2 concentration (350 μmol mol^{-1}), although it is higher than that in the vessel without a gas-permeable film.

CO_2 concentration in the culture vessel during the photoperiod can be raised higher than the atmospheric CO_2 concentration by keeping the CO_2 concentration in the culture room significantly higher than the atmospheric CO_2 concentration. CO_2 concentration in the culture room is generally kept at 1000 μmol mol^{-1} or higher to keep the CO_2 concentration in the culture vessel higher than the atmospheric concentration.

CO_2 enrichment by use of the gas-permeable film or a large culture box promotes the growth, rooting, and/or development of turfgrass (82), potato (83), and many other plant species (27–29,61,62,65–68,72–74,83,84). CO_2 concentration in the culture vessel with use of the gas-permeable film changes with time, even when CO_2 concentration in the culture room is constant over time, depending upon the plant growth in vitro, etc. On the other hand, by using a large culture box with a forced ventilation system (62,66,67,85), CO_2 concentration in the box can be controlled accurately at a desired level.

Photosynthetic Photon Flux

In conjunction with relative humidity, PPF is an important environmental factor in the acclimatization stage. To acclimatize plantlets, in the conventional method, efforts have been given so far to control the environment in the ex vitro acclimatization mainly by maintaining low light intensity and high relative humidity at the early phase and gradually decreasing the relative humidity and increasing light intensity towards the level in the outer environment. However, these practices under low light usually suppress photosynthesis, and hence the development of autotrophy and rooting of the plantlets (3). As a result the survival percentage of the micropropagated plantlets became significantly reduced in many plant species.

By controlling PPF and RH, rapid acclimatization (24 h) was achieved in *Eucalyptus* plantlets (67). Efforts were also made to control the culture vessel environment in the multiplication and/or rooting stage. As a result, the drastic change in the environment after transplantation of the plantlets outside has been reduced. This is termed in vitro acclimatization, and the photoautotrophic micropropagation system (under high PPF; 150 μmol m^{-2} s^{-1} and enriched CO_2) is able to facilitate this condition specially under low relative humidity. It was also revealed that under high PPF, net photosynthetic rates of the in vitro grown plantlets increased, which was reflected in the growth (root and shoot). Therefore those plantlets survived and grew well in the ex vitro condition.

Supporting Material

Use of fibrous material such as cellulose plugs (67), rockwool (61), and artificial soil with high air porosity such as vermiculite (66,67) as supporting materials generally gives better root development and then enhanced plant growth in vitro. Plants in vitro have more secondary roots with normal vascular systems when cultured with fibrous materials than when cultured with gelled material such as agar (66,67).

When plants are cultured in vitro photoautotrophically using a plug tray with cells containing the artificial substrate, each plant can be pulled up together with the substrate and removed from the cell with minimum damage of roots. Then the plant with substrate can be transplanted into the soil with minimum transplanting damage and can be acclimatized and established ex vitro easily.

Use of Mycorrhizal Fungi and Bacteria

Micropropagation is widely believed to be a procedure that produces pathogen free plantlets. However, the persistence of latent microbial contaminants in micropropagated plantlets has been suggested in many reports to increase the vigor of the micropropagules (shoot and root growth) and also the hardiness of micropropagated plantlets without showing any symptom of disease (9,10). For example, inoculation of Verticillium strain (Pseudomonas) into potato nodal cultures resulted in improved root formation and made it possible to eliminate the ex vitro acclimatization step, thus allowing the direct transfer of the plantlets from the tissue culture vessel to the field. The inoculated bacteria carried through at least sixteen generations of mass propagation, without reinoculation (10). Culturing the potato plantlets with Pseudomonas is also reported to increase root and shoot growth and the lignin deposition; it also reduced the percent of water loss from the detached leaves after exposing to the ex vitro condition (20% of total water content was lost compared to 55% in control) (86). When kiwifruit's shoot tips were cultured with two types of bacteria (Aerococcus sp. and Bacillus fastidiosus), over 97% of the plantlets were successfully acclimatized in the field (87). Eucalyptus clones were also inoculated with Agrobacterium rhizogenes, and extensive root formation was observed, plantlets were well acclimatized, developing a firm root plug in the field (88). Inoculation of fungus (Sphaeropsis tumefaciens) is also known to improve the growth of in vitro grown citrus plant and orchids (Cattleyea and Phalaenopsis) (75). Co-inoculation of bacterium (P. fluorescens) together with an extomycorrhizal fungus (Laccaria bicolor) enhanced the production of high-quality Douglas fir planting stocks (89). Therefore, it appears that in vitro or ex vitro co-inoculation with mycorrhizal fungi and bacteria can improve acclimatization and growth of micropropagules.

The reason for the improved growth and the acclimatization of the plantlets inoculated with microbes is not yet fully clear. It has been speculated that some microbes may stimulate the growth by altering the phytohormone levels in the host plants, secreting growth-promoting substances and/or increasing photosynthetic and nutrient absorption efficiencies (10). Also increased phosphate uptake and antagonism through production of antibiotics and siderophores have been shown to be involved (86). Whatever may be the cause, this technique will be a useful acclimatization technique in the near future.

ADVANTAGES OF PHOTOAUTOTROPHIC MICROPROPAGATION

In relation to the in vitro acclimatization, the photoautotrophic micropropagation under a low relative humidity, a high CO_2 concentration, and a high photosynthetic photon flux has many advantages over conventional, photomixotrophic micropropagation (60–62). They are:

1. Growth and development of shoots/plantlets in vitro are promoted
2. Physiological/morphological disorders such as hyperhydration are reduced, and plant quality is improved
3. Relatively uniform growth in size and shape and uniform development are expected
4. Leafy single or multinode cuttings can be used as explants
5. Procedures for rooting and ex vitro acclimatization are simplified or eliminated
6. Application of growth regulators and other organic substances such as amino acids and vitamins can be minimized
7. Losses of plants due to microbial contamination in vitro and during the ex vitro acclimatization stage are reduced
8. A large culture vessel with or without a forced ventilation system can be used with minimum risk of microbial contamination
9. Then environmental control of the culture vessel becomes easier
10. The control of growth and development by means of environmental control becomes easier
11. Asepsis in the culture vessel is not required if no pathogen is guaranteed
12. Automation, robotization, and computerization become easier
13. Propagation and production of plants can be done rationally based upon standard plant eco-physiology.

ACCLIMATIZATION IN FUTURE MICROPROPAGATION SYSTEMS

In vitro acclimatization will become more and more important and popular in future micropropagation systems. It would be ideal if plants cultured in vitro had functional stomata and roots, had fully developed photosynthetic organs, and had been acclimatized in vitro. Then the plants can be transplanted to the ex vitro environment without any special ex vitro acclimatization. The in vitro acclimatization in photoautotrophic micropropagation will become increasingly important in the future micropropagation systems. Finding optimum combinations of bioreactor systems with in vitro acclimatization will be an important and practical research subject for future micropropagation systems.

The use of artificial substrate (culture medium) with high air porosity in photoautotrophic micropropagation is an interesting application. A multicell tray (ca. 30 cm wide × 60 cm long) with several hundred cells can be used to hold the artificial substrate in each cell of the multicell tray. Then plants can be removed with the artificial substrate from the cell and can be transplanted into soil together with the substrate so that mechanical damage of roots at transplanting can be minimized. An automatic transplanting machine handling the multicell trays can also be used.

Different prototypes of photoautotrophic micropropagation systems for enhancing growth and development and for in vitro acclimatization have been developed. One is a large culture vessel with a forced ventilation system and a multicell tray containing the artificial substrate (62,83,85). With this system, the forced ventilation rate of the culture vessel can be controlled. Furthermore, asepsis can be maintained easily by the use of a gas-permeable filter at the air inlet and outlet of the culture vessel. Another is a kind of microhydroponic system or aseptic vegetative production system using cuttings in an aseptic room (76).

In the future, we may use a sterile culture room in which people cannot enter. Its appearance looks like an automated warehouse with a full automatic handling and transportation system. Sterile multishelves with fluorescent lamps are installed in the culture room. Multicell trays are placed on each of the shelves. Its inside view looks like a clean vegetative (cutting) propagation system under artificial lighting.

In fact, the major differences between conventional vegetative (cutting) propagation and micropropagation are: (1) The former is not considered to be pathogen free, but the latter is considered to be pathogen free; (2) explants are larger in the former than in the latter; and (3) natural solar light is used in the former, but artificial lamps are used in the latter (76).

However, these differences are not essential. For examples, pathogen-free plants can be propagated in an aseptic propagation room with the use of aseptic plug trays, instead of using aseptic culture vessels with caps in a septic culture room. Smaller chlorophyllous explants can be used in conventional vegetative propagation if the environment is properly controlled. Artificial light can be used for conventional vegetative propagation commercially if it is competitive in cost with natural light. It is shown that the cost of electricity for lighting and air conditioning is approximately $0.01–0.015 per plant in Japan if the lighting and air conditioning systems are well designed (76).

In future micropropagation systems for the growth of explants to plantlets, in vitro rooting and in vitro acclimatization will be conducted concurrently using a large culture box or in a culture room equipped with a control facility for modulating the physical parameters of the environment, resulting in minimum ex vitro acclimatization.

BIBLIOGRAPHY

1. C.M. Conover and R.T. Poole, *Hortic. Rev.* **6**, 120–154 (1984).
2. D.J. Donnelly and W.E. Vidaver, *Glossary of Plant Tissue Culture*, Belhaven Press, London, 1988, p. 9.
3. T. Kozai, in Y.P.S. Bajaj, ed., *Biotechnology in agriculture and forestry*, Vol. 17, Springer-Verlag, New York, 1991, pp. 127–141.
4. K. Wardel, E.B. Dobbs, and K.C. Short, *J. Am. Soc. Hortic. Sci.* **108**, 386–389 (1983).
5. M. Ziv, in L.A. Withers and P.G. Alderson, eds., *Plant Tissue Culture and its Agricultural Application*, Butterworth, London, 1986, pp. 187–196.
6. M. Ziv, in J. Aitken-Christie, T. Kozai, and M.A.L. Smith, eds., *Automation and Environmental Control in Plant Tissue Culture*, Kluwer Academic Publishers, Dordrecht, The Netherlands, 1995, pp. 493–516.
7. J.E. Preece and E.G. Sutter, in P.C. Debergh and R.H. Zimmerman, eds., *Micropropagation: Technology and Application*, Kluwer Academic Publishers, Dordrecht, The Netherlands, 1991, pp. 71–93.
8. P.E. Read and C.D. Fellman, *Acta Hortic.* **166**, 15–20 (1985).
9. J. Pospisilova et al., *J. Exp. Bot.* **49**, 863–869 (1998).
10. E.B. Herman, *Recent Adv. Plant Tissue Culture* **3**, 109–126 (1995); **5**, 74–80 (1997).
11. J.A. Marin, R. Gella, and M. Herrero, *Ann. Bot. (London)* **62**, 663–670 (1988).
12. T. Kozai, K. Fujiwara, and Y. Kitaya, *Acta Hortic.* **393**, 63–73 (1995).
13. K.E. Brainerd and L.H. Fuchigami, *J. Am. Soc. Hortic. Sci.* **106**, 515–512 (1981).
14. K. Fujiwara and T. Kozai, in J. Aitken-Christie, T. Kozai, and M.A.L. Smith, eds., *Automation and Environmental Control in Plant Tissue Culture*, Kluwer Academic Publishers, Dordrecht, The Netherlands, 1995, pp. 319–369.
15. K. Fujiwara, T. Kozai, and I. Watanabe, *J. Agric. Meteorol. (Tokyo)* **43**, 21–30 (1987) (in Japanese, with English abstract and captions).
16. J. Catsky, J. Pospisilova, J. Solarova, and H. Synkova, *Biol. Plant.* **37**, 35–48 (1995).
17. J. Pospisilova et al., *Photosynthetica* **22**, 205–213 (1998).
18. M.P. De Proft, M.L. Maene, and P. Debergh, *Physiol. Plant.* **65**, 375–379 (1985).
19. M.B. Jackson et al., *Ann. Bot. (London)* **67**, 229–237 (1991).
20. K. Ohyama and T. Kozai, *Environ. Control Biol.* **35**, 197–202 (1997) (in Japanese, with English abstract and captions).

21. H. Sallanon and A. Goudret, *C.R. Seances Acad. Sci.*, **310**, 607–613 (1990).
22. K. Tanaka, K. Fujiwara, and T. Kozai, *Acta Hortic.* **319**, 59–64 (1992).
23. T. Kozai, Oki, and K. Fujiwara, *Plant Cell, Tissue Organ Cult.* **22**, 205–211 (1990).
24. M. Nakayama, T. Kozai, and K. Watanabe, *Plant Tissue Cult. Lett.* **8**, 105–109 (1991) (in Japanese, with English abstract and captions).
25. J. Pospisilova, *Biol. Plant.* **38**, 605–609 (1996)
26. J. Pospisilova, J. Catsky, J. Solarova, and I. Ticha, *Biol. Plant.* **29**, 415–421 (1987).
27. Y.C. Desjardins, F. Laforge, C. Lussier, and A. Gosselin, *Acta Hortic.* **230**, 45–53 (1988).
28. Y.C. Desjardins, C. Hdider, and J. DeRiek, in J. Aitken-Christie, T. Kozai, and M.A.L. Smith, eds., *Automation and Environmental Control in Plant Tissue Culture*, Kluwer Academic Publishers, Dordrecht, The Netherlands, 1995, pp. 441–471.
29. T. Kozai, N. Ohde, and C. Kubota, *Plant Cell, Tissue Organ Cult.* **24**, 181–86 (1991).
30. T. Kozai, Iwabuchi, K. Watanabe, and I. Watanabe, *Plant Cell, Tissue Organ Cult.* **25**, 107–115 (1991).
31. R.R. Williams, *Acta Hortic.* **289**, 165–166 (1991).
32. R.R. Williams, in K. Kurata and T. Kozai, eds., *Transplant Production Systems*, Kluwer Academic Publishers, Dordrecht, The Netherlands, 1992, pp. 213–229.
33. R.R. Williams, in J. Aitken-Christie, T. Kozai, and M.A.L. Smith, eds., *Automation and Environmental Control in Plant Tissue Culture*, Kluwer Academic Publishers, Dordrecht, The Netherlands, 1995, pp. 405–439.
34. M.A.L. Smith and M.T. McClelland, *In Vitro Cell Dev. Biol. — Plant* **27**, 52–56 (1991).
35. M.A.L. Smith and L.A. Spomer, in J. Aitken-Christie, T. Kozai, and M.A.L. Smith, eds., *Automation and Environmental Control in Plant Tissue Culture*, Kluwer Academic Publishers, Dordrecht, The Netherlands, 1995, pp. 371–404.
36. M. Ziv, in P.C. Debergh and R.H. Zimmerman, eds., *Micropropagation: Technology and Application*, Kluwer Academic Publishers, Dordrecht, The Netherlands, 1991, pp. 45–69.
37. B.W.W. Grout and M.J. Aston, *Hortic. Res.* **17**, 1–7 (1977).
38. E.G. Sutter and R.W. Langhans, *Can. J. Bot.* **60**, 2896–2902 (1982).
39. K.E. Brainerd and L.H. Fuchigami, *J. Exp. Bot.* **33**, 388–392 (1982).
40. M. Ziv, G. Meir, and A.H. Halevy, *Plant Cell, Tissue Organ Cult.* **2**, 55–65 (1983).
41. M. Cappellades, R. Fontarnau, C. Carulla, and P. Debergh, *J. Am. Soc. Hortic. Sci.* **115**, 141–145 (1990).
42. D.J. Donnelly and W.E. Vidaver, *J. Am. Soc. Hortic. Sci.* **109**, 172–176 (1984).
43. D.J. Donnelly and W.E. Vidaver, *J. Am. Soc. Hortic. Sci.* **109**, 177–181 (1984).
44. M. Ziv, *In Vitro Cell Dev. Biol. — Plant* **27**, 64–69 (1991)
45. N. Lee, H.Y. Wetzstein, and H.E. Sommer, *Plant Physiol.* **78**, 637–641 (1985).
46. M. Paques and P.H. Boxus, *Acta Hortic.* **230**, 155–166 (1987).
47. T. Kozai, K. Tanaka, B.R. Jeong, and K. Fujiwara, *J. Jpn. Soc. Hortic. Sci.* **62**, 413–417 (1993).
48. T. Kozai, Fujiwara, M. Hayashi, and J. Aitken-Christie, in K. Kurata and T. Kozai, eds., *Transplant Production Systems*, Kluwer Academic Publishers, Dordrecht, The Netherlands, 1992, pp. 247–282.
49. M.A.L. Smith and B.H. McCown, *Plant Sci. Lett.* **28**, 149–151 (1983).
50. C. Hdider and Y. Desjardins, *Plant Cell, Tissue Organ Cult.* **36**, 27–33 (1994).
51. J. DeRiek, Ph.D. Thesis, Gent University, Belgium (1995).
52. E.G. Sutter and M. Hutzell, *Sci. Hortic.* **23**, 303–312 (1984).
53. L.H. Fuchigami, T.Y. Cheng, and A. Soeldner, *J. Am. Soc. Hortic. Sci.* **106**, 519–522 (1981).
54. J. DeRiek and J. Van Huylenbroeck, in P.J. Lumsden, J.R. Nicholas, and W.J. Davies, eds., *Physiology, Growth and Development of Plants in Culture*, Kluwer Academic Publishers, Dordrecht, The Netherlands, 1994, pp. 309–313.
55. J. Pospisilova et al., *J. Exp. Bot.* **49**, 863–869 (1998).
56. M. Ziv, *Acta Hortic.* **319**, 119–124 (1992).
57. A.V. Roberts et al., *Acta Hortic.* **319**, 153–158 (1992).
58. E.F. Smith, A.V. Roberts and J. Mottley, *Plant Cell, Tissue Organ Cult.* **21**, 129–132 (1990).
59. E.F. Smith, A.V. Roberts, and J. Mottley, *Plant Cell, Tissue Organ Cult.* **21**, 133–140 (1990).
60. T. Kozai, in Y.P.S. Bajaj, ed., *Biotechnology in Agriculture and Forestry*, Vol. 17, Springer-Verlag, New York, 1991, pp. 313–343.
61. T. Kozai, in P.C. Debergh and R.H. Zimmerman, eds., *Micropropagation: Technology and Application*, Kluwer Academic Publishers, Dordrecht, The Netherlands, 1991, pp. 447–469.
62. K. Fujiwara, T. Kozai, and I. Watanabe, *Acta Hortic.* **230**, 153–158 (1988).
63. T. Kozai, C. Kubota, and I. Watanabe. *Acta Hortic.* **230**, 159–166 (1988).
64. L. Cournac, I. Cirier, and P. Chagvardieff, *Acta Hortic.* **319**, 53–58 (1992).
65. C. Kirdmanee, Y. Kitaya, and T. Kozai, *Acta Hortic.* **393**, 111–118 (1995).
66. C. Kirdmanee, Y. Kitaya, and T. Kozai, *In Vitro Cell Dev. Biol. — Plant* **31**, 144–149 (1995).
67. C. Kirdmanee, Y. Kitaya, and T. Kozai, *Environ. Control Biol.* **33**, 123–132 (1995).
68. S.M.A. Zobayed, C. Kubota, and T. Kozai, *In Vitro Cell Dev. Biol. — Plant* **34**(4), In press (1999).
69. B.W.W. Grout, *Plant Sci. Lett.* **5**, 401–405 (1975).
70. B.W.W. Grout and M.J. Aston, *Ann. Bot. (London)* **42**, 993–995 (1978).
71. D.J. Donnelly, W.E. Vidaver, and K. Colbow, *Plant Cell, Tissue Organ Cult.* **3**, 313–317 (1984).
72. G. Niu and T. Kozai, *Trans. ASAE* **401**, 255–260 (1997).
73. G. Niu and T. Kozai, *Acta Hortic.* **456**, 37–43 (1998).
74. B.R. Jeong, K. Fujiwara, and T. Kozai, *Hortic. Rev.* **17**, 125–172 (1995).
75. R.T. McMillan et al., *Proc. Fla. State Hortic. Soc.* **101**, 336 (1988).
76. T. Kozai et al., *Proc. Int. Workshop Sweetpotato Prod. Sys. Toward 21st Century*, 1998, pp. 201–214.
77. T. Kozai, *Proc. 3rd Asian Crop Sci. Conf.*, Taiwan, 1998, pp. 296–308.
78. M. Hayashi and T. Kozai, *Proc. Symp. Florizel Plant Micropropagation Hortic. Ind.*, 1987, pp. 123–134.
79. M. Hayashi, M. Nakayama, and T. Kozai, *Acta Hortic.* **230**, 189–194 (1988).

80. A.N. Lakso, B.I. Reisch, J. Montensen, and M.H. Roberts, *J. Am. Soc. Hortic. Sci.* **111**, 634–638 (1987).

81. Y.C. Desjardins, F. Laforge, and A. Gosselin, *Proc. Symp. Florizel Plant Micropropagation Hortic. Ind.*, 1987, pp. 176–184.

82. Y. Seko and T. Kozai, *Acta Hortic.* **440**, 600–605 (1998).

83. T.D. Roche, R.D. Long, and M.J. Hennerty, *Acta Hortic.* **440**, 515–520 (1988).

84. M. Tanaka, S. Nagae, S. Fukai, and M. Goi, *Acta Hortic.* **314**, 139–140 (1992).

85. C. Kubota and T. Kozai, *HortScience* **27**, 1312–1314 (1992).

86. M.I. Frommel, J. Nowak, and G. Lazarovits, *Plant Physiol.* **96**, 929–936 (1991).

87. P.L. Monette, *Plant Cell, Tissue Organ Cult.* **6**, 73–82 (1986).

88. S. MacRae and J.V. Staden, *Tree Physiol.* **12**, 411 (1993).

89. B. Herman, *Agricell Rep.* **31**(6), 42–47 (1998).

See also CRYOPRESERVATION OF PLANT CELLS, TISSUES AND ORGANS; EMBRYOGENESIS IN ANGIOSPERMS, SOMATIC; MICROPROPAGATION OF PLANTS, PRINCIPLES AND PRACTICES; MICROPROPAGATION, HYPERHYDRICITY; PLANT CELL CULTURE BAG.

ADVENTITIOUS ORGANOGENESIS

GEERT-JAN DE KLERK
Centre for Plant Tissue Culture Research
The Netherlands

OUTLINE

INTRODUCTION

In plants, differentiated, somatic cells that are at an advanced stage of the ontogenetic cycle may reinitiate the developmental program and give rise to adventitious shoots, roots, or embryos. This phenomenon is called *regeneration*, a term originating from animal developmental biology, where it refers to the formation of new organs from somatic cells. The formation of adventitious shoots (caulogenesis) and roots (rhizogenesis) are specified by the term *adventitious organogenesis*, whereas *somatic embryogenesis* refers to the formation of adventitious (somatic) embryos. Regeneration occurs frequently during the natural life of plants but may be achieved at very high frequencies in tissue culture. For basic science, regeneration is highly interesting because it facilitates experimentation on the mechanisms acting during developmental processes. In agriculture, the capability of plants to form new organs from somatic cells is of utmost importance for propagators and breeders. Root regeneration from cuttings has been used for more than 2000 years in vegetative propagation. Regeneration of adventitious shoots forms part of many micropropagation protocols, and the formation of somatic embryos from cell suspensions will, if broadly applicable, revolutionize plant propagation. Biotechnological breeding methods also include adventitious organ formation: Genetic engineering and haploid plant production, for example, involve adventitious regeneration of complete plants from somatic cells.

This article deals with adventitious organogenesis. It discusses the topic from the perspective of developmental biology, stressing that adventitious organogenesis is composed of successive, distinct phases. Unfortunately, most research has been carried out from a practical point of view and does not deal with underlying mechanisms. Basic biochemical and molecular studies suffer from the complication that during the first steps of the process, only very few cells in an explant are actually involved.

FORMATION OF ADVENTITIOUS MERISTEMS AND ORGANS

Preexisting and Adventitious Meristematic Tissues

An organism begins its existence as a single, morphologically simple cell, the zygote. During ontogenetic development, a complete organism with distinct organs, tissues, and cell types is formed from this cell. In mammals, there are more than 200 clearly recognizable distinct cell types, for example, muscle, nerve, and blood cells. In plants, this number is smaller, possibly some 50. The process during which cells become different from one another is called *differentiation*.

During the initial steps in the life cycle of higher plants, an embryo is formed from the zygote. This involves division and differentiation of cells and the organization of cells into tissues and systems of tissues. The embryo consists of a shoot (composed of an apical meristem, a hypocotyl, and cotyledons) and a root (composed of an apical meristem and the root body) and has, as compared to adult plants, still a simple structure. Thus the embryo contains two main types of meristematic cells: the shoot and the root meristem. In many plants, both meristems remain present during all of the plant's life unless they are damaged. Meristems usually consist of relatively small cells with small vacuoles that are scattered throughout the protoplasm. In flowering plants they are rarely more than 0.25 mm in diameter. In addition to the two apical meristems, in adult plants other types of meristematic tissues occur that have been formed directly from the apical meristem: axillary meristems in the axils of petioles, vascular cambium, cork cambium, and intercalary meristems. The latter occur, for example, in the base of internodes and leaf sheaths of many monocotyledons and ascertain elongation. Most of the cell divisions in plants occur in meristems and in meristematic tissues. In adult animals, localized regions with relatively undifferentiated meristematic cells are not present. When plants reach the reproductive stage, the microsporogenic and megasporogenic cells are formed from the shoot meristems. So in plants these meristems function as the germ lines. In animals, the germ line is a group of cells that are usually specified and set aside very early in an animal's development. In contrast to the apical meristems in plants, the cells of the germ line in animals are inactive in the somatic body of animals.

During their natural development, plants form new meristems and meristematic tissues from somatic cells. Secondary cambium can be formed to achieve secondary thickening of stems. Lateral root meristems are not produced from the root apical meristem but from cells in the pericycle of the root. (As noted before, axillary buds, the analoges of lateral roots in shoots, arise directly from the shoot apical meristem.) Adventitious roots may regenerate from stems, usually from cells in between the vascular tissues. This happens frequently in many species during natural development and during exposure to certain environmental conditions, for example, during partial submersion. In monocotyledons, the initial root derived from the root meristem in the embryo usually dies early, and new root meristems are formed from the basal part stem. The occurrence of leaves, shoots, or inflorescences on leaves is known as *epiphylly*. Shoots growing from leaves remain attached to the parent leaf and grow out as epiphyllous branches. Alternatively, they may be released and function as propagules. Well-documented examples are *Bryophyllum* and *Kalanchoë*.

Under natural or seminatural conditions, plants are capable of the regeneration of lost parts. When a stem is partly damaged by an incision, the vascular tissues are repaired, and when a small portion of a root or shoot meristem has been removed, the meristem is repaired. Němec, for example, reported in 1905 that when a piece of 1 mm is removed from the root tips of broad beans at first callus is formed on the cut surface and subsequently a new tip that later produces a root cap. If most of the meristem has been removed, the meristem is not repaired. In shoots usually an axillary meristem (occasionally an adventitiously formed meristem) grows out to form a new main shoot, and in roots a lateral root is formed from the pericycle close to the removed apical meristem. When roots are removed from shoots, the stem forms new roots relatively easily. The formation of adventitious roots on cuttings is known from ancient times and is used for vegetative propagation of elite plants that have either been selected from natural populations or obtained in breeding programs. It has been shown that a signal from the shoot, *viz.*, an auxin produced in the apex and transported downward in the stem, induces the response (Fig. 1). A root from which the shoot has been removed may form a new, adventitious shoot, usually from pericycle cells. In this case, the inducing signal is likely cytokinin

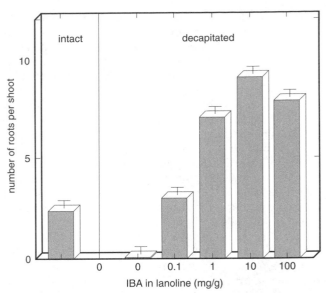

Figure 1. Root regeneration from intact and decapitated micro-cuttings cultured on auxin-free medium. In decapitated shoots, the apex was either replaced by lanolin without any addition or by lanolin with $0.1-100$ mg g^{-1} IBA. The data show that the apex can be replaced by auxin-releasing lanolin and thereby present evidence that auxin produced by the apex is a root-inducing signal in microcuttings.

produced in the root tips. Plant leaves are also able to form a complete plant. Hagemann reported in 1931 that out of 1,196 species, 308 formed both buds and roots from isolated leaves (1). Because in this case the definition of regeneration as repair does not seem to be appropriate, some authors reject the term regeneration but use instead, for example, vegetative propagation.

The mechanisms that control regeneration under natural conditions in whole plants are often studied in tissue culture experiments in which presumed regulating factors are added to isolated plant tissues. A relatively simple regenerative event is the formation of new vascular tissues (2). In horticulture, grafting of a scion on a rootstock, which involves the regeneration of connecting vascular tissues between scion and root stock, is a method of propagating plants already known to the ancient Greeks. It has been found in tissue-culture experiments that the plant hormones (growth regulators) auxin and cytokinin play a major role. For example, after grafting a *Syringa* bud into callus in vitro, vascular tissue is formed from callus cells underlying the site of implantation. The same effect is achieved by grafting agar wedges containing auxin and sucrose (3). It should be noted that in shoots the apex produces auxin and leaves sucrose and that both are transported downward in the stem.

Major steps forward in the research on adventitious organ formation were the discovery of auxin and cytokinin and the use of plant tissue culture. Auxin was discovered early in the 1930s and used briefly after that to achieve adventitious root formation from cuttings. In 1939, White obtained shoots on genetic tumors of tobacco cultured in vitro and Nobécourt roots on in vitro grown carrot callus. Cytokinins were discovered in tissue-culture experiments in the 1950s, and it was found that excised tissues that are cultured in vitro on nutrient medium with auxin and/or cytokinin may form adventitious roots, shoots, or embryos with very high frequency. The enhanced incidence of regeneration in vitro is caused, among other things, by the presence of plant hormones in the medium that are taken up by the explants and increase the endogenous hormone levels. Other promotive factors include the high levels of organic and inorganic nutrients in the tissue culture medium, the absence of bacteria and fungi, and the high humidity.

Because in the case of adventitious embryo and shoot formation complete plants are formed from somatic cells (with shoot formation in two steps: from the adventitious shoots roots are regenerated to obtain a complete plant), these cells are called *totipotent*. It has been maintained that all cells in a plant body are totipotent. A systematic study of this question, though, is not practical to undertake. To describe the formation of new shoots, roots, or embryos from somatic cells, various terms are being used: in particular, adventitious root/shoot/embryo formation or root/shoot/embryo regeneration. The adventitious formation of shoots and roots is also often referred to as *organogenesis*. In this case, the adventitious structure has only one pole. When a somatic embryo has been formed, the regenerated structure has two poles (a root and a shoot pole) connected by vascular tissue. The use of the term organogenesis in plant tissue culture does not correspond with its general meaning, which refers to all formation of new organs, whether adventitious or not, (e.g., also to the formation of a side branch from an axillary bud). Therefore, the term *adventitious organogenesis* is more appropriate. Adventitious formation of embryos is called *somatic embryogenesis*. It is tempting to speculate that all adventitious regeneration involves the formation of an embryonic meristem, and that in

adventitious organogenesis either the shoot or the root pole is blocked at a very early stage, resulting in root or shoot formation, respectively. When somatic embryos, adventitious roots, and adventitious shoots are formed from the same explant, they originate from different types of cells (see the following). This is an indication, but not a proof, that organogenesis is not embryogenesis in which the development of one of the poles is blocked early.

A major drawback in the study of adventitious organogenesis in plants is the absence of broadly studied model systems, such as *Xenopus* and *Drosophila* in animal and *Arabidopsis* in plant developmental biology. Unfortunately, *Arabidopsis* is relatively difficult in tissue culture. For rooting, derooted mung bean seedlings have been studied by many authors, but have the disadvantage that they are relatively large, are not convenient in tissue culture and that small explants do not survive in vitro. A tissue culture system that is being used by some researchers consists of 1-mm stem slices cut from apple microcuttings (Fig. 2a–e). For adventitous shoot formation, regeneration from leaves of begonia has been studied frequently. In tissue culture, epidermal strips excised from tobacco stems consisting of 3–5 cell layers of epidermal and subepidermal cells are very useful explants (Fig. 2f). It should be noted that both the stem slices and the epidermal strips are simple systems in which interference from other parts of the plant is avoided as much as possible and in which the distance between the site of regeneration and the medium is short, so that gradients within the tissue are as small as possible.

Organogenesis and Regeneration in Animals

In embryonic development in animals, cell–cell interactions play a major role, and in transplant experiments,

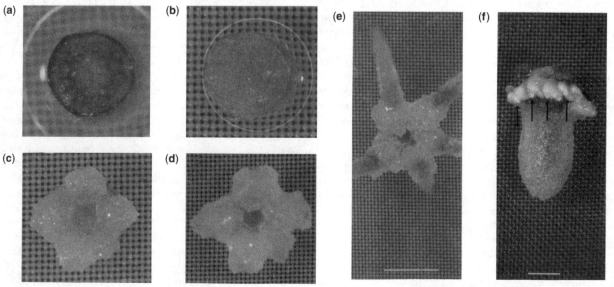

Figure 2. Two model systems to study adventitious organogenesis. Adventitious root formation in 1-mm slices excised from stems of apple microcuttings at 0, 5, 7, 9, and 15 days, respectively (**a–e**) and adventitious shoot formation from thin epidermal cell layers excised from tobacco stems (**f**); bar = 1 mm; photographs **a–e** have the same scale. (Photograph **f** kindly supplied by Dr. M. Smulders, Wageningen.)

numerous instances have been observed where a group of cells influences the differentiation of another group of cells. This phenomenon is referred to as *induction*. A spectacular example of induction is the grafting experiment of Mangold and Spemann published in 1924. They transferred a piece of blastopore lip from a newt early gastrula to another embryo (also early gastrula). The graft was placed at a different region in the host and induced the formation of an almost complete secondary embryo in the host embryo. This shows that the grafted piece of lip regulated the direction of development of the adjacent tissues. It was also found that only cells in a particular state are able to show the proper response to the signal. This particular state of reactivity is referred to as *competence*. In animal embryos, this state occurs only during an early, short period of time. The term *pluripotent* is used to describe that commitment is still only slight and that cells may develop along various pathways. By the signal of the inductor, cells become *determined* with respect to their future development. This is a gradual process, occurring over various cell generations. When the fate of a cell has been determined, it will follow that fate when grafted from one region of an embryo to another region. The formation of an organ from determined cells is often referred to as *differentiation*. It should be noted that the term differentiation is ambiguous because it is also used in a broad sense for all the process in which parts of an organism become different from one another or from their previous condition.

In higher animals, pluripotency is almost completely lost during the progress of development: Morphogenetic processes that have been terminated can only be reawakened to some extent after loss of an organ to renew the lost parts. The damage repaired may be a wound (that is replaced by new skin) or a lost organ. The renewal of a limb in salamanders and of a tail in lizards are well-known examples of the latter. The repair of lost parts in a full-grown organism is called *regeneration*. In the case of vertebrate limb regeneration, it is believed that the first step in regeneration is the loss of the differentiated state in cells beneath the wound epidermis. These cells start to divide and form a blastema consisting of cells that are *dedifferentiated* to some extent. The marked difference in regenerative capacity between the invertebrates and the vertebrates is well known. The former show so much regenerative power that some worms, for example, can be cut in two and thereafter restore all the missing parts in each half. At the other end of the scale, in higher animals, regeneration is restricted.

Figure 3 summarizes the succesive phases in differentiation. During the first phase, cells acquire the competence to respond to an organogenic stimulus. This may occur as part of the normal ontogenetic development (Fig. 3a) or in a process of dedifferentiation (Fig. 3b). The latter occurs when the organism has been wounded and previously differentiated cells start to repeat developmental processes. Then, during an induction period, cells become determined toward a specific developmental pathway. After that, they no longer require the stimulus and grow out to the new organ.

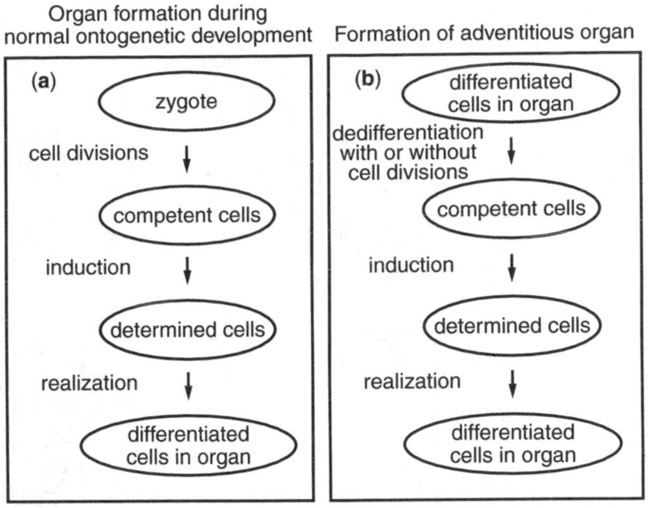

Figure 3. Successive steps in the formation of organs during ontogenesis (**a**) and during adventitious organogenesis (**b**).

Regeneration of Roots and Shoots is a Process Consisting of Distinct Steps

In plants, research on the different steps in regeneration has focused on histological and biochemical examinations and on the differential hormone requirements during the successive phases. Analogous to the stages that are distinguished in flowering, adventitious organogenesis has been divided into three phases. During the induction phase, the cells that will form the new organ do not show perceptible histological changes, but at the biochemical and molecular level events are believed to occur that are related to regeneration. After that, during the initiation phase, cells divide, leading to the formation of a new meristem and a new primordium. Finally, during the expression phase, a new shoot, root, or embryo is formed. Microscopic observations on root and shoot formation are shown in Figure 4. This characterization of three phases is used by many researchers, but is not suitable for two reasons. First, the term *induction* has a very different meaning in developmental biology. Furthermore, and more important, microscopic observations give only limited information about the developmental state of cells.

In animal developmental biology, the successive phases in ontogenetic processes and in regeneration have been defined according to the reaction of cells to the inductive stimuli (see previous section). Some researchers in plant developmental biology have analyzed adventitious organogenesis in a similar way. As a matter of fact, adventitious regeneration of shoots and roots in vitro is very suitable for such experiments, since the major inductive stimuli, auxin and cytokinin, can be added via the tissue culture medium and as explants can easily be transferred to a medium with different hormonal composition. For shoot regeneration, an extensive study has been carried out by Christianson and Warnick (4) in leaf explants of *Convolvulus*. They transferred explants at various times after the start of tissue culture from one medium to another, using three types of media, *viz.*, shoot-, root-, and callus-inducing media. These media had

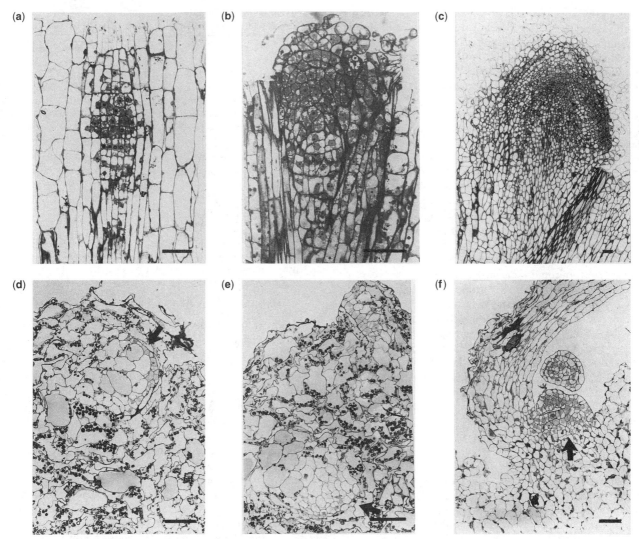

Figure 4. Microscopic observations on adventitious root formation from apple stem slices (**a–c**) and adventitious shoot formation from long-term kiwi callus (**d–f**). (**a**) Root meristemoid at 72 h; (**b**) early root primordium with the tendency to form a dome-like structure at 96 h; (**c**) root primordium at 144 h; (**d**) formation of a dome-like structure with tunica-like surface layer of cytoplasmic cells and corpus-like center of vacuolated cells (arrow); (**e**) kiwi callus with meristematic cells (arrow); (**f**) apical shoot meristem (arrow) leaf primordia and leaf. Bar = 25 μm in **a–c** and 50 μm in **d–f**. (Photographs kindly supplied by Dr. J. Jasik, Bratislava.)

very different auxin–cytokinin ratios. During the initial period after the start of culture, the hormonal composition could be varied over a wide range (all three media gave essentially the same results). After that, a phase occurred during which auxin and cytokinin should be applied in the proper concentrations. In the final, third phase, the hormonal composition could again be varied over a wide range. These results indicate the occurrence of three phases:

1. *Dedifferentiation*. Cells are at first not competent to respond to the organogenic stimulus, but acquire this competence during an initial phase of dedifferentiation.
2. *Induction*. After dedifferentiation, in the induction phase, cells are responsive to the organogenic

stimulus and become determined to form a specific organ, such as a shoot. Only during this phase, is the hormonal composition of the medium critical.

3. *Realization* (Christiansson and Warnick use the term differentiation). When the cells are determined, the new program of differentiation is initiated to produce a shoot. There is evidence that the first meristems that have been formed on an explant somehow inhibit the formation of additional meristems.

For shoot formation, this scheme has been confirmed in various species (5). The same scheme can be applied in adventitious root formation (6). In experiments on root regeneration from apple microcuttings, a different type of transfer experiments was carried out: During the rooting

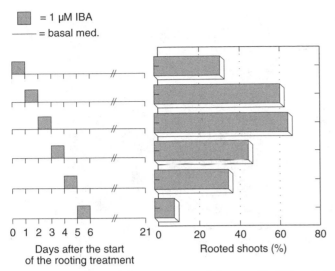

■ = 1 μM IBA
—— = basal med.

0 1 2 3 4 5 6   21
Days after the start
of the rooting treatment

0   20   40   60   80
Rooted shoots (%)

Figure 5. Evidence of the occurrence of phases with differential hormonal requirements during the rooting process. The effect of 24-h pulses with IBA on rooting of apple microcuttings was studied. The percentage of rooted shoots was determined after 10 d. Note that pulses at days 2, 3 and 4 resulted in the highest rooting percentage. (Data redrawn from Ref. 6.)

treatment, 24-h pulses with either auxin or cytokinin were given. There was a strongly enhanced responsiveness (inhibition by cytokinin and promotion by auxin) from 24 to 96 h after the start of the rooting treatment, indicating that during this time induction occurs (Fig. 5). During this period there is also an enhanced sensitivity to the antiauxin p-chlorophenoxyisobutyric acid and to salicylic acid, a phenolic compound that enhances oxidative breakdown of auxins. The occurrence of a lag period before stems become sensitive to auxin has already been reported by Went in 1939 (7).

Experiments that involve transfer from one medium to another may suffer from various pitfalls. (1) Hormones taken up by plant tissues are usually rapidly deactivated, either by conjugation or by oxidation. Thus it is supposed that during the period in which the hormones are applied, they are present at a significantly increased level and that after the pulse the endogenous concentration soon (within a few hours) reaches "normal" levels. However, a carry over effect may occur because inactivation may be slow, and it may be slow specifically in the target cells or free hormone may be released from conjugated hormones. (2) Added hormones may alter the metabolism of endogenous hormones during and after the pulse (8,9). (3) In a sample of explants or in the various meristematic centers within an explant, the regeneration process may be very asynchronous. This may obscure phases with distinct hormonal sensitivities. (4) It should also be noted that inappropriate hormonal conditions may have two distinct effects. The process may be arrested, and the tissue "waits" for the adequate hormonal stimulus. It may also be that the process is diverted into another direction. We presume that when a pulse with 6-benzlaminopurine (BAP) is given during the induction phase in adventitious root formation, BAP redirects the process and instead of organized growth in a root meristem nonorganized callus

growth is induced (6). As the distinct developmental states of cells and tissues are characterized by differential gene expression, ultimately, the successive phases have to be identified by the expression of different sets of genes.

The results discussed in this section indicate that the succession of steps in adventitious root and shoot formation regeneration is similar to regeneration in animals (Fig. 3b). The formation of somatic embryos can also be dissected into these phases (5). The first phase may involve a period of callus growth (indirect regeneration). Often, though, cells already existing in the explant become competent to respond to the organogenic/embryogenic stimulus after a (short) lag phase without any cell division or without cell division at a large scale (direct regeneration).

FACTORS INFLUENCING ADVENTITIOUS ORGANOGENESIS

Explant Effects

When explants are taken from plants growing ex vitro, they require surface sterilization with hypochlorite or alcohol. The portion of the explant that has died because the disinfectant has entered the tissue (tissue adjacent to the cut surface) should be removed. Endogenous contaminants are not killed by the disinfectants and often constitute a serious problem. Antibiotics may be used, but because they usually do not reach the site of contamination at a sufficiently high concentration, they do not kill the contaminants within the tissue. However, because antibiotics may block the growth of contaminants in the medium completely, they obscure the incidence of endogenous contamination. To remove endogenous contamination, excised meristems (that are often free of endogenous contaminants) can be grown in vitro to form complete plants that may or may not be free of contaminants. Alternatively, before taking the explants, endogenously contaminated plants may be given a warm-water treatment (10). Such treatment often kills all endogenous contaminants that are detectable during tissue culture. The duration of the treatment (30–180 min) and the temperature (between 42 and 55 °C) depend on the plant species, the condition of the plant material, and the type of contaminant. A high-temperature treatment may only be given to tissue that is resistant to the severe stress of the high temperature. In particular, dormant tissues are suitable, for example, dormant buds or bulb scales. The preparation of scale explants from a lily bulb is shown in Figure 6.

Adventitious shoots and roots may be induced from virtually all types of tissues that are excised from plants, but the ease of regeneration varies considerably. Rules of thumb are that the capability to regenerate decreases with the extent of differentiation (meristematic cells regenerate more easily than fully differentiated cells), the ontogenetic age (juvenile tissues are more capable of regeneration than adult tissues), the physiological age (recently formed tissue has the highest regeneration performance), and dormancy state (nondormant tissues have a higher regeneration capability than dormant ones). A major effect is exerted by (a)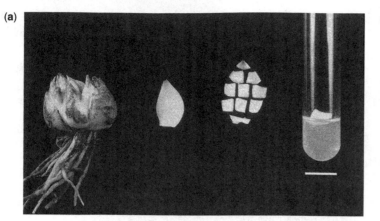

(b)

Figure 6. Preparation of scale explants from lily (**a**) and adventitious lily bulblets after 11 weeks of culture (**b**). Bar in (**a**) = 2.5 cm; bar in (**b**) = 1 cm. (Photographs kindly supplied by Ir. M.M. Langens, Lisse.)

the genotype. The capability to regenerate is probably determined by only few genes. The function of these genes is unknown. Experiments in tomato suggest that the genetic component associated with the capability to regenerate concerns not the sensitivity to hormones but the maintenance of morphogenetic competence (11).

When adventitious shoots and roots develop from the same organ, they originate from different tissues. In cultured carrot petioles, for example, application of indole-3-acetic acid (IAA) results in the formation of adventitious shoots, roots, and embryos, but from different tissues within the petiole (12). Roots originate from vacuolized cells near the vascular bundles. Vacuolized subepidermal cells give rise to somatic embryos and shoots are formed from large parenchyma cells. These data suggest that cells in the explant may be capable of only one of the regenerative pathways. It is not known whether this is because of inherent incapabilities of cells (the cells are not totipotent), or the position of a cell within the explant results in differences in the microenvironment that are favorable for one of the pathways (e.g., because of a hormonal gradient within the explant). The dependence of regeneration on the type of tissue indicates that, to achieve regeneration in recalcitrant species, it may be more important to have the right types of cells in an explant than to develop highly refined regeneration conditions (nutrients, hormones, etc.).

It has been suggested that one of the requirements for cells to exhibit their potential for regeneration is physical isolation from the maternal tissue: The excision of tissues from the native environment supposedly removes the restrictions that are necessary for proper functioning of cells in the whole plant. It is difficult to determine whether this is indeed a major factor, because excision of tissue from a plant and its culture in vitro influences regeneration in several other ways. First, wounding evokes a repair reaction triggered by compounds that are released when cells are damaged, and somatic cells dedifferentiate and form new tissue to cover the wound. These dedifferentiated cells or adjacent cells that also have been activated may be targets for organogenic signals taken up from the medium. Second, in plant tissues auxin is actively transported in a polar direction. Because of this, auxin taken up from the medium or synthesized by the explants accumulates at the basal cut surface. Usually regeneration occurs at this site. The involvement of polar auxin transport in determining the site of regeneration has been shown in experiments in which auxin transport inhibitors like 2,3,5-triiodobenzoic acid (TIBA) have been applied. In the presence of TIBA, regeneration occurs scattered all over the explant. Third, it should be noted that the epidermis of plants is relatively impermeable. Thus, when tissue is excised and cultured in vitro, nutritional and hormonal actors may easily enter the explant.

General Remarks about Plant Hormones

Plant hormones play a major role in adventitious organogenesis. In animal physiology, hormones denote substances that are synthesized in low amounts in one part of an organism and transported to target tissues in other parts where they exert an effect. In plants, chemical messengers have also been found. A classic example occurs in germinating barley seeds: Gibberellin synthesized and released by the embryo diffuses to the aleurone layer, where it induces the synthesis and secretion of hydrolytic enzymes. These enzymes degrade macromolecular reserves to small fragments that are used by the embryo for initial growth. Another notable example is the inhibition of the outgrowth of axillary buds by auxin synthesized in the apex and transported downward in the stem. In contrast to animal hormones, the synthesis of plant hormones is usually not restricted to a specific tissue, but may occur in many different tissues. Furthermore, plant hormones may be transported and act in distant tissues, but often they have their action at the site of synthesis. Another distinctive property of plant hormones is their lack of specificity: Each of them influences a wide range of processes. Auxin, for example, has been found to influence cell elongation, cell division, induction of primary vascular tissue, adventitious root, shoot and embryo formation, senescence, fruit growth, outgrowth of axillary buds, and sex expression. Because of the differences between animal and plant hormones, many researchers deny that the latter are genuine hormones and prefer to use terms like *plant growth substance* or *plant growth regulator*. Nevertheless, the term *plant hormone* is still widely used.

Most knowledge about the role of plant hormones is derived from studies in which hormones have been applied to plant tissues. Experimentation in vitro has several advantages: Tissue culture facilitates application of hormones, avoids possible microbial degradation of applied hormones, and allows study of the effect of hormones on isolated plant organs. Instead of the hormones themselves, compounds that affect their metabolism, transport, or action may be added. In many studies, the levels of endogenous hormones have been determined. More recently, researchers have used hormone mutants or plants transformed with cytokinin or auxin biosynthetic genes from *Agrobacterium tumefaciens* or with *rol* genes from *A. rhizogenes* (the latter influence among other things the signal transduction pathway). A promising new approach (13) is the transformation of protoplasts with *A. tumefaciens* containing a T-DNA derived vector with multiple enhancer sequences near the right border. When integrated, this construct activates transcription of adjacent genes. By culturing transformed protoplasts

at hormone levels that are either too high or too low, protoplasts with altered hormone metabolism or sensitivity may be isolated, and the genes affected by the transformation event may be isolated and studied. In tissue culture, two classes of plant hormones, cytokinins and auxins, are of major importance. They are required for growth of callus and cell suspensions, outgrowth of axillary buds, and regeneration of adventitious roots, shoots, and embryos. Ethylene has been studied frequently in relation with regeneration and likely plays a major role. Table 1 summarizes essential information about auxin, cytokinin, and ethylene with respect to regeneration. Other hormones, in particular, gibberellins, abscisic acid, polyamines, or jasmonates, are being used but only occasionally.

Plant hormones added to plant tissue culture media are taken up and increase the level within the tissue. After uptake, plant hormones are rapidly inactivated by conjugation or in some cases by oxidation. Ethylene is an exception, but this gaseous compound is rapidly released from the plant into the air. Usually only very small amounts of the applied hormones remain in the free form. It has been shown for applied auxin that an equilibrium exists between the free and the conjugated

Table 1. The Characteristics of the Three Plant Hormones That Play a Major Role in Regeneration

	Main effects in regeneration	Modulators of metabolism, action, or transport
Auxins: indole-3-acetic acid (IAA) indole-3-butyric acid (IBA) 1-naphthaleneacetic acid (NAA) phenylacetic acid (PAA) 2,4-dichlorophenoxyacetic acid (2,4-D) 2,4,5-trichlorophenoxyacetic acid (2,4,5-T) picloram dicamba p-chlorophenoxyacetic acid (CPA)	Adventitious root formation (at high conc.) Adventitious shoot formation (at low conc.) Induction of somatic embryos (in part. 2,4-D) Cell division Callus formation and growth Inhibition of root growth	2,3,4 Triiodobenzoic acid (TIBA) and 1-N-naphthylphthalamic acid (NPA) inhibit polar auxin transport p-Chlorophenoxyisobutyric acid (PCIB) inhibits auxin action as a genuine antiauxin by binding to the auxin receptor Phenolic compounds (e.g., ferulic acid or phloroglucinol) inhibit auxin oxidation Riboflavin strongly promotes photooxidation of IBA and IAA Transformation of plants with the auxin biosynthetic genes of *Agrobacterium tumefaciens* may increase endogenous auxin levels
Cytokinins: zeatin (Z) zeatinriboside (ZR) isopentenyladenine (iP) isopentenyladenosine (iPA) 6-benzylaminopurine (BAP) kinetin thidiazuron (TDZ) N-(2-chloro-4-pyridyl)-N'-phenylurea (CPPU or 4PU-30)	Adventitious shoot formation (at high conc.) Promotion of adventitious root formation (at very low conc.) Inhibition of adventitious root formation at higher concentrations Cell division Callus formation and growth	Compounds have been described that inhibit cytokinin synthesis (lovastatin), degradation, and action; usually, the effects are small and/or nonspecific Transformation of plants with the cytokinin biosynthetic genes of *A. tumefaciens* may increase endogenous cytokinin levels
Ethylene	Promotion or inhibition of adventitious regeneration depending on the time of application and on the genotype	1-Aminocyclopropane-1-carboxylic acid (ACC) is a direct precursor of ethylene and is metabolized by plant tissues to ethylene Aminoethoxyvinylglycine (AVG) inhibits ethylene synthesis; Co^{2+}, α-aminooxy-acetic acid and α-aminoisobutyric acid also inhibit ethylene synthesis but have a lower efficiency Silver inhibits ethylene action; silver is applied preferably as silver thiosulfate (STS). Auxin strongly increases ethylene synthesis

form, less than 1% being present in the free form. The effect of hormones not only depends on the rate of uptake from the medium and on the stability in the medium and in the tissue, but also on the sensitivity of the target tissue as cells may not recognize the hormonal signal, or are incapable of carrying out the desirable response. Applied hormones influence the synthesis or degradation of endogenous hormones belonging to the same class as the applied hormone or to other classes. All this results in a very complex situation in which it is sometimes difficult to reveal how the observed effect has been brought about.

Auxin, Cytokinin, and Regeneration

In 1944, Skoog and co-workers found that auxin stimulated root formation but inhibited shoot formation from callus. By that time, auxin was the only known plant hormone. Soon after, Skoog and co-workers demonstrated the requirement for a cell division promoting substance that could be satisfied by addition of autoclaved DNA. Quickly the active component was identified and named *kinetin*, the first known cytokinin. In a classic paper, Skoog and Miller (14) reported that root, shoot, and callus formation in tobacco explants cultured in vitro are brought about by high, low and intermediate auxin–cytokinin ratios, respectively. This concept has been repeatedly verified in many species. Recent confirmation comes from observations on transgenic plants that have increased cytokinin or auxin synthesis. However, since the process of regeneration can be dissected into a series of successive phases, each with its own hormonal requirements, the original concept requires some adjustments.

The dedifferentiation phase during which competence is acquired to respond to the organogenic stimulus depends upon auxin. This emerges from the activity of phenylacetic acid. This auxin is not effective in the second phase (induction) of rooting (3) and embryogenesis (15), but is nevertheless effective during the first phase (see Ref. 16 for a possible explanation). During the first phase, cytokinin is also required. Lovastatin, a compound that blocks cytokinin synthesis, specifically inhibits adventitious root formation when applied during the first phase, and not after that. The inhibition by lovastatin is reversed by adding a low dose of the cytokinin zeatin. The involvement of both plant hormones is not surprising, since both are probably required for cell division. In the second phase, induction, auxin and cytokinin should be supplemented at the appropriate concentrations. The transfer experiments of Christianson and Warnick (4) indicate that for adventitious shoot formation low levels of auxin and high levels of cytokinin are required. For root regeneration the situation should be reversed (6). In the third phase, realization, the hormonal composition of the medium is less critical. However, the high concentrations of exogenous hormones required for induction are often inhibitory. Here, other hormones may also play a role, among others abscisic acid in somatic embryogenesis.

With respect to the hormonal composition of the nutrient medium, some additional remarks should be made. First, the effect of hormonal supplements to nutrient media can be modified by inorganic nutrients (17). Second, the uptake of various hormones by explants from the medium may be very different. For example, in tobacco explants, NAA is taken up six and ten times faster than IAA or BAP, respectively (18). Third, since hormones are extensively metabolized, the actual concentration of hormones in the tissue has to be considered. Finally, applied hormones may alter the metabolism of endogenous hormones. For example, auxin increases both inactivation of applied cytokinin (8) and ethylene synthesis (9).

Other Hormones and Hormone-Like Factors

In regeneration, the hormone ethylene has received the most attention. Its effect is not well understood: It has been reported that ethylene stimulates adventitious organogenesis in some species, but is inhibitory in others (for rooting, see Ref. 19). Recent research on the action of ethylene in root regeneration (20) has indicated unambiguous effects for ethylene.

To understand the contradictory effects reported in the literature, some unique characteristics of ethylene should be taken into account (Table 2) (21–23). Differences in experimental designs related to these characteristics are often not considered and may explain some of the contradictory results. For example, in apple microcuttings silver thiosulfate (STS, a compound that blocks the effect of ethylene) has a promotive effect at high auxin concentration when the portion of the stem from where the roots develop is submerged in the solidified medium and not in slices that are cultured on top of the medium. This indicates that auxin increases ethylene production and

Table 2. Distinctive Characteristics of the Gaseous Plant Hormone Ethylene; Other Plant Hormones are Nonvolatile Compounds

Ethylene is a gaseous compound. Whereas plants reduce the endogenous concentrations of other hormones by enzymatic conversion (oxidation or conjugation), ethylene may simply diffuse from the tissue into the atmosphere. In Petri dishes, ethylene may accumulate in the headspace depending on how tightly the dish has been closed.

Since ethylene diffusion in water is ca. 10,000 times less than in air (21), ethylene cannot diffuse easily from submerged tissue. Thus, when (micro)cuttings are submerged with the basal part of the stem (from which the roots develop) in aqueous solution or in agar, ethylene may accumulate in the basal part. Plant tissues submerged in liquid medium may also accumulate ethylene.

Because of submerged conditions, anaerobiosis may occur, which prevents the formation of ethylene from ACC (21)

Knowledge about long-distance transport of ethylene in plants is scarce. Obviously, transport will be much faster in tissues with gas-filled intercellular spaces. Thus, when the tissue is submerged only partially, ethylene may escape via intercellular spaces.

Endogenous synthesis of ethylene is enhanced by auxin and wounding. It also depends on the orientation of the cuttings (22).

Secondary effects of applied compounds should be kept in mind. For example, ethylene may have a general senescing effect and it may also interfere with auxin, in particular, by inhibiting the polar auxin transport (23). STS may induce ethylene formation because Ag is a heavy metal and damages the tissue.

that, because ethylene cannot escape from the submerged, basal part of the stem, it accumulates to a high level and becomes inhibitory. In slices the ethylene levels in the tissue remain low, because ethylene produced by the slices quickly diffuses out of the tissue into the headspace of the Petri dish. Another possible reason why authors have obtained contradictory results is that the effect of ethylene depends on the timing of the application (20). Ethylene promotes rooting during the dedifferentiation phase. Possibly it acts as a wounding-related compound and enhances the sensitivity of the tissue to auxin. In this respect, it should be noted that wounding-related compounds may strongly promote regeneration (24,25). During the induction phase, ethylene inhibits rooting, as shown by pulses with STS. It depends on the genotype whether promotion during the dedifferentiation phase or inhibition during the induction phase prevails.

Phenolic compounds have been repeatedly suggested, but never experimentally proven, to be involved in plant development (26). Most research on the role of phenolic compounds in regeneration concerns adventitious root formation. Various authors have examined whether endogenous phenolic compounds are related with rootability. Such correlations have been found (among others, Ref. 27). It has also been reported that application of phenolic compounds enhances rooting, but few critical and extensive studies have been carried out. In apple stem slices, a very active phenolic compound is ferulic acid. It enhances rooting in the presence of IAA strongly, but with NAA only slightly. Because IAA can be oxidized but NAA not, ferulic acid likely acts as an inhibitor of auxin oxidation. Indeed, ferulic acid completely blocked enzymatic oxidation of IAA (Fig. 7). It should be noted that ferulic acid is a general antioxidant and also inhibits photochemical oxidation of IAA in the absence of plant tissues (Fig. 7). In general, tri- and most diphenolic compounds inhibit, and monophenolic compounds promote, IAA oxidation.

To study the interaction between phenolic compounds and auxin, we added some phenolic compounds at a wide range of IAA concentrations. Figure 8 shows results of an experiment with ferulic and salicylic acid. Salicylic acid is a monophenol and its addition results in increased instability of IAA. As expected, salicylic acid shifted the dose–response curve of IAA to the right and left the shape unaltered. Ferulic acid protects IAA and should shift the dose–response curve of IAA to the left. Indeed, ferulic acid shifted the optimum concentration of IAA somewhat to the left, but, just as other phenolic compounds (e.g., phloroglucinol), it also widened the dose–response curve of IAA (Fig. 8). This indicates that ferulic acid acts also in a way other than just by protecting IAA. Possibly phenolic compounds protect the tissue from the oxidative stress that is imposed by wounding when the explant is excised, resulting in a dose–response curve with a different shape.

As discussed previously, the action of auxin, cytokinin, ethylene, and phenolic compounds on regeneration has been revealed to some extent. All other hormones and many hormone-like compounds have been examined, and in some cases marked effects have been observed with polyamines, brassinosteroids, and jasmonates, and, occasionally, gibberellins and abscisic acid. However,

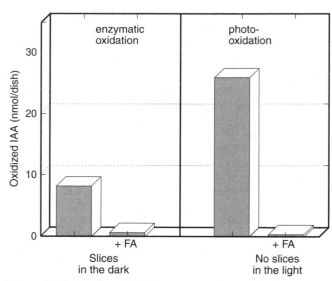

Figure 7. Oxidation of IAA in Petri dishes in the dark in the presence of explants (30 apple stem slices) or in the light without explants. Each Petri dish contained 20 ml solidified medium with 10 μM IAA. Note that addition of ferulic acid (FA) blocked oxidation almost completely.

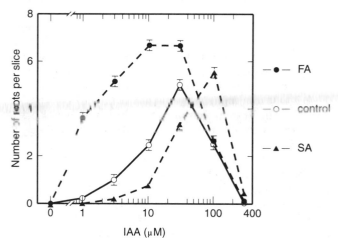

Figure 8. A method to determine the type of interaction between auxin and other compounds is to study the effect on a dose–response curve of IAA. Apple stem slices were treated for 5 d with a range IAA concentrations in the presence or absence of phenolic compounds. Note that salicylic acid (SA), a monophenol that enhances IAA oxidation, shifted the dose—response curve—as expected—to the right but did not alter the shape of the curve. Ferulic acid (FA) shifted the curve—as expected—to the left but also had a major effect on the shape; the width of the curve strongly increased. For an explanation see the text.

because the experiments hardly deal with the mechanism of action, these compounds are not discussed here. Relevant reviews are Ref. 28 for polyamines, Ref. 29 for brassinosteroids, and Ref. 30 for jasmonates.

Nutritional and Physical Factors

For mineral nutrition, most researchers use Murashige and Skoog's (MS) formulation (31) or derivatives, such as Linsmaier and Skoog's, Eriksson's, B5 and Schenk and Hildebrandt's. MS medium has been developed for callus growth in tobacco and is satisfactory for shoot and root growth in many species. The mineral composition of MS is similar to the mineral composition of well-growing healthy plants (cf. Ref. 32). It might be, though, that meristematic tissues have a different mineral composition and thus require another mineral composition of the medium. Reduction of the strength (e.g., half the concentration of all nutrients) is believed to be promotive in many species. As a rule, for adventitious rooting in vitro, the MS concentration is halved. Furthermore, the ratio of NH_4 to NO_3 and the amount of nitrogen may play a role. From the organic components, sucrose is usually added at a concentration of 2–4%. Occasionally, the type of carbohydrate is decisive for regeneration. Carbohydrates also play an osmotic role. Casein hydrolysate is occasionally added as organic nitrogen source and may enhance regeneration.

Explants can be cultured on solidified medium or in liquid medium. For gelation, agar, agarose (highly purified agar), or gelrite is used. All are complex polysaccharides. There are at least four major differences between culture in liquid and solidified media. (1) It is now well known that agar contains many impurities, including nutrients and hormone-like compounds. Which compounds occur in agar and at what concentration depends on the brand and on the batch. (2) Because of the relatively slow movement in solidified medium and because compounds may bind to agar, solidified medium may act as a slow-release source for medium components. (3) In liquid medium, all surfaces of the explants are exposed to the medium. In solidified medium, only the tissue that has contact with the medium may take up components. In this respect, it is important to note again that the epidermis of plants is relatively impermeable and that most uptake occurs via the cut surface, and possibly via the stomata and via young apical and axillary buds. (4) Finally, the rate of diffusion of gases out of submerged and emerged tissues is very different. The relatively poor solubility of gases in water leads in submerged tissues to partial anaerobiosis and accumulation of gases as CO_2 and ethylene.

The pH of media influences regeneration, but this is likely an indirect effect. The pH in the medium determines among other things the uptake of many compounds, solubility, the activity of enzymes secreted into the medium, and the activity of antibiotics. A large change of the pH within the tissues has a devastating effect. The buffering capacity of tissue culture media is low. Thus it has been observed that the pH of media, although set at pH 5.5, usually changes considerably. A buffer like tris-(hydroxymethyl)aminomethane (Tris) is very toxic. The only buffer that is suitable is 2-(N-morpholino)ethanesulfonic acid (MES), although at 10 mM (this is the commonly used concentration) MES is to some extent toxic (33; J. Hanecakova, personal communication).

The effect of light in tissue culture has received little attention. It results from a combination of several components: length (short day vs. long day), intensity, and quality. Often, but not always, low light intensity enhances regeneration. A complicating factor is that medium components may be photooxidized. This holds, for example, for the auxins IAA and IBA. With respect to light quality, both phytochrome and blue light effects have been found.

Temperature may have very marked effects during plant development, among other things with respect to the development and breaking of dormancy. In regeneration studies, the effect of temperature treatment before excision of explants has been examined. Cold-treated apple cotyledons and lily scales showed strongly enhanced regenerative capacity and decreased hormone requirements, respectively.

The involvement of electrical currents in development and regeneration has aroused interest by the study of the first cleavage of the zygote of the brown alga *Fucus*, which results in a large apical cell and a small basal cell from which the rhizoid develops. The cleavage of the zygote is preceded by polarization, which is triggered by environmental factors such as light and gravity. An electrical current develops through the zygote partly carried by Ca^{2+}. In somatic globular embryos of *Daucus* (a very early stage in which the embryo has a globular shape) an inward current has been detected in the region from which the apical meristems will develop and an outward current from the region of the future root meristem. An electrical current of ca. 1 µA increases root and shoot regeneration from callus (34).

BIOCHEMICAL AND MOLECULAR ASPECTS

Many researchers have carried out biochemical and molecular examinations in explants from which roots or shoots are regenerating. It is doubtful, however, whether these studies tell much about regeneration. For example, in a model system to study root regeneration, 1-mm stem slices cut from apple microcuttings, small meristemoids have been formed at 72 h. At this time, they make up far less than 1% of the volume of a slice. During the initial phase of regeneration, the number of cells involved in the regeneration process is much smaller. Obviously, biochemical and molecular characteristics of the meristemoids and in the cells from which the meristemoids will develop are swamped out by those of the surrounding cells. Furthermore, biochemical and molecular studies should be accompanied by microscopic studies, to establish whether the response occurs specifically in the cells from which the new organs are being formed. Unfortunately, such studies are scarce. Differences in enzyme activity can also be detected by the occurrence of the endproducts of the enzyme reaction. An example is short-lived starch grains that are formed during the very early stages of regeneration of shoots, roots, and embryos. In slices cut from apple microshoots and treated with auxin, short-lived starch grains appear during first day of culture in a ring consisting of cells of the vascular bundles and primary rays. During the next day, these starch grains are broken down and the cells enter division. From those in the primary rays, root primordia may develop (35). When auxin is not supplied, starch grains are also formed but at a much later time.

An involvement of methylation in developmental processes in plants can be deduced from the finding that plants that have substantially reduced levels of DNA methylation display many phenotypic abnormalities. Methylation also plays a major role in the development of mammals. For example, early in mammalian embryogenesis global demethylation occurs and after that remethylation. In *Arabidopsis* and tomato, methylation levels of DNA from young seedlings are approximately 20% lower than in mature leaves, whereas the highest methylation level occurs in seeds. Possibly hypermethylation during embryogenesis is followed by demethylation during germination and subsequent slow remethylation (36). Methylation also depends on the type of tissue. Cells of different tissues of the carrot plant, for example, are characterized by specific DNA methylation patterns. An increase in genome methylation was suggested to be necessary to disorganize a foregoing program (37). Fractionation of cell types of an embryogenic carrot cell line indicated a characteristic low genome methylation level of a fraction enriched in precursor cells of somatic embryos. In adventitious organ formation, azacytidin, a drug that causes hypomethylation, leads to reduced shoot and root regeneration. In apple stem slices, azacytidin acted specifically during the induction phase. It is, however, difficult to determine whether the drug acts by blocking methylation or otherwise (38; S. Marinova, personal communication).

A more recent possibility is to examine the expression of genes known to be expressed in root and shoot meristems. In *Arabidopsis*, the shoot meristemless (*STM*) gene has been isolated. In wild-type embryos, the *STM* transcript is found earliest in one or two cells of the late globular stage embryo. Seeds that are homozygous recessive (*stm/stm*) form after germination only a root, hypocotyl, and cotyledons, whereas the apical meristem is absent. Calluses from these incomplete seedlings are unable to regenerate shoots, showing that the *STM* gene is required in adventitious regeneration (39). In contrast, another gene required for the formation of an apical meristem in embryogenesis in *Arabidopsis*, pinhead (*PNH*), is not required for shoot regeneration. Other genes pivotal to shoot meristem development include Wuschel (*WUS*) and Zwille (*ZLL*), which seem to be required for the proper functioning of cells in the apical meristem of *Arabidopsis*, and no apical meristem (*NAM*), which defines the boundary between the meristem and organ primordia in petunia. These genes have not been examined with respect to adventitious shoot formation. A mutation that reduces the formation of axillary meristems, *lateral suppressor* in tomato, does not seem to affect adventitious shoot and root regeneration. The effect of other mutations that are involved in the formation of axillary meristems, *torosa-2* mutants in tomato and *barren stalk* mutants in maize, is not known. In *Arabidopsis*, mutants have been identified that are incapable of forming lateral roots. One of the genes involved, the *LRP1*, is also expressed early in adventitious root formation (40).

It should be remembered that the expression of gene families that are characteristic for certain developmental stages in cells are regulated by transcription factors produced by the expression of regulatory genes. The *STM* gene is such a regulatory gene. It encodes a *Knotted1* type of homeodomain protein. Tobacco plants constitutively expressing maize *Knotted1* form adventitious shoots on leaves of the most severe transformants (41), but in the monocot species maize, barley, and rice no ectopic shoots were observed on transgenic leaves. The overall appearance of plants overexpressing the *STM* gene resembles that of plants transformed with cytokinin synthesizing genes, and indeed the *STM*-overexpressing plants have increased levels of cytokinins.

Other molecular approaches are in situ hybridization studies with known genes such as transcription factors, auxin or cytokinin up-regulated genes, or cell cycle genes, or studies in transgenic plants transformed with the luciferase gene after selected promoters. Direct screening of mutant seeds for regenerative capacity is practically difficult.

CONCLUDING REMARKS

In plants, particularly during culture in vitro, somatic cells may be capable of regeneration. Adventitious regeneration of roots has been used for many ages in vegetative plant propagation via cuttings. Adventitious regeneration of shoots, roots, and embryos is an indispensable part of biotechnological breeding and propagation techniques and therefore one of the cornerstones of plant biotechnology. From a practical point of view, there are two main problems. First, plants that are produced via regeneration may be genetically different from the motherplant, a phenomenon referred to as *somaclonal variation* (42). It is not known whether this is a general problem or only occurs in a limited number of cases. A major reason for this uncertainty is that it is difficult to assess the extent of somaclonal variation in a population of regenerated plants (43). A second major problem is that regeneration does occur not at all, only infrequently, or only from cells that are not susceptible to genetic engineering. Until now, much of the knowledge about regeneration is empirical and research on basic mechanisms is still relatively scarce. Knowledge about the underlying mechanisms in regeneration expands rapidly, in particular because of the introduction of molecular techniques, and may solve this problem.

BIBLIOGRAPHY

1. A. Hagemann, *Gartenbauwisenschaft* **6**, 109–155 (1931).
2. R. Aloni, *Annu. Rev. Plant Physiol. Plant Mol. Biol.* **38**, 179–204 (1987).
3. R.H. Wetmore and J.P. Rier, *Am. J. Bot.* **50**, 418–430 (1963).
4. M.L. Christianson and D.A. Warnick, *Dev. Biol.* **95**, 288–293 (1983).
5. G.J. de Klerk, B. Arnholdt-Schmitt, R. Lieberei, and K.-H. Neumann, *Biol. Plant.* **39**, 53–66 (1997).
6. G.J. de Klerk, M. Keppel, J. Ter Brugge, and H. Meekes, *J. Exp. Bot.* **46**, 965–972 (1995).
7. F.W. Went, *Am. J. Bot.* **26**, 24–29 (1939).
8. L.M.S. Palni, L. Burch, and R. Horgan, *Planta* **174**, 231–234 (1988).

9. E.M. Meyer, P.W. Morgan, and S.F. Yang, in M.B. Wilkins, ed., *Advanced Plant Physiology*, Pitman, London, 1984, pp. 111–126.

10. M. Langens-Gerrits, M. Albers, and G.-J. de Klerk, *Plant Cell Tissue Organ Cult.* **52**, 75–77 (1998).

11. M. Koornneef, J. Bade, R. Verkerk, and P. Zabel, *Plant J.* **3**, 131–141 (1993).

12. F. Schäfer, B. Grieb, and K.H. Neumann, *Bot. Acta* **101**, 362–365, (1988).

13. H. Hayashi, I. Czaja, J. Schell, and R. Walden, *Science* **258**, 1350–1353 (1993).

14. F. Skoog and C.O. Miller, *Symp. Soc. Exp. Bioi.* **11**, 118–130 (1957).

15. K. Finstad, D.C.W. Brown, and K. Joy, *Plant Cell Tissue Organ Cult.* **34**, 125–132 (1993).

16. G.-J. de Klerk, J. Ter Brugge, J. Jasik, and S. Marinova, in A. Altman and Y. Waisel, eds., *Biology of Root Formation and Development*, Plenum, New York and London, 1997, pp. 111–116.

17. J.E. Preece, *Plant Tissue. Cult. Biotechnol.* **1**, 26–37 (1995).

18. A.J.M. Peeters, W. Gerads, G.W.M. Barendse, and G.J. Wullems, *Plant Physiol.* **97**, 402–408 (1991).

19. K.W. Mudge, in T.D. Davis, B. Haissig, and N. Sankhla, eds., *Adventitious Root Formation by Cuttings*, Dioscorides Press, Portland, Oreg., 1988, pp. 150–161.

20. G.-J. de Klerk, A. Paffen, J. Jasik, and V. Haralampieva, in A. Altman, M. Ziv, and S. Jzhar, eds., *Plant Biotechnology and In Vitro Biology in the 21st Century*, Kluwer Academic Press, Dordrecht, The Netherlands, 1999, pp. 41–44.

21. M.B. Jackson, *Annu. Rev. Plant Physiol.* **36**, 146–174 (1995).

22. L. De Wit, J.-H. Liu, and D.M. Reid, *Plant Cell Environ.* **13**, 237–242 (1990).

23. J. Suttle, *Plant Physiol.* **88**, 795–799 (1988).

24. P.J. Wilson and J. Van Staden, *Ann. Bot.* (London) **66**, 479–490 (1990).

25. W.M. Van der Krieken, J. Kodde, and M.H.M. Visser, in A. Altman and Y. Waisel, eds., *Biology of Root Formation and Development*, Plenum, New York and London, 1997, pp. 95–105.

26. B.V. Milborrow, in M.B. Wilkins, ed., *Advanced Plant Physiology*, Pitman, London, 1984, pp. 76–110.

27. P. Curir et al., *Plant Physiol.* **92**, 1148–1153 (1990).

28. A.W. Galston and R. Kaur-Sawhney, in P.J. Davies, ed., *Plant Hormones*, Kluwer Academic Press, Dordrecht, The Netherlands, 1995, pp. 158–178.

29. S.D. Clouse and J.M. Sasse, *Annu. Rev. Plant Physiol. Plant Mol. Biol.* **49**, 427–451 (1998).

30. G. Sembdner and B. Parthier, *Annu. Rev. Plant Physiol. Plant Mol. Biol.* **44**, 459–489 (1993).

31. T. Murashige and F. Skoog, *Physiol. Plant.* **15**, 473–497 (1962).

32. E. Epstein, in J. Bonner and J. Varner, eds., *Plant Biochemistry*, Academic Press, London, 1965, pp. 438–466.

33. D.E. Parfitt, A.A. Almehdi, and L.N. Bloksberg, *Sci. Hortic.* **36**, 157–163 (1988).

34. A. Goldsworthy, in A. Mizrahi, ed., *Biotechnology in Agriculture*, R. Liss, New York, 1988, pp. 35–52.

35. J. Jasik and G.-J. de Klerk, *Biol. Plant.* **39**, 79–90 (1997).

36. E.J. Finnegan, R.K. Genger, W.J. Peacock, and E.S. Dennis, *Annu. Rev. Plant Physiol. Plant Mol. Biol.* **49**, 223–247 (1998).

37. F. LoSchiavo et al., *Theor. Appl. Genet.* **77**, 325–331 (1989).

38. A.P. Prakash and P.P. Kumar, *Plant Cell Rep.* **16**, 719–724 (1997).

39. M.K. Barton and S. Poethig, *Development* (Cambridge, UK) **119**, 823–831 (1993).

40. D.L. Smith and N.V. Fedoroff, *Plant Cell* **7**, 735–745 (1995).

41. N. Sinha, R. Williams, and S. Hake, *Genes Dev.* **7**, 787–795 (1993).

42. P.J. Larkin and W.R. Scowcroft, *Theor. Appl. Genet.* **60**, 197–214 (1981).

43. G.-J. de Klerk, *Acta Bot. Neerl.* **39**, 129–144 (1990).

See also BIOREACTOR CULTURE OF PLANT ORGANS; MICROPROPAGATION OF PLANTS, PRINCIPLES AND PRACTICES; PHYSIOLOGY OF PLANT CELLS IN CULTURE; PLANT CELL CULTURES, SELECTION AND SCREENING FOR SECONDARY PRODUCT ACCUMULATION; SOMACLONAL VARIATION.

ANATOMY OF PLANT CELLS

HAZEL Y. WETZSTEIN
YI HE
University of Georgia
Athens, Georgia

OUTLINE

Introduction

Leaf Anatomy of *In Vitro*-Grown Plants

Hyperhydricity

Anatomy of Somatic Embryogenic Cultures

Anatomy of Organogenic Cultures

Anatomy of Plant Protoplast

Anatomy of Cell Suspension Cultures

Anatomy of Callus Cultures

Bibliography

INTRODUCTION

The cytology and anatomy of plant cells can exhibit considerable plasticity which is a means for plants to adapt and survive changing environmental conditions. These changes occur at the subcellular level via modifications in cell ultrastructure and at the cell and tissue level where differences in cell number and histological organization may arise. Modifications in plant anatomy can be marked. Unlike animals, plants can maintain sustained growth through meristematic regions. Plant cells also exhibit totipotency, that is, cells retain the capacity to develop into any structure of the mature plant. Totipotency is most apparent in meristematic and young tissues but can also be exhibited in differentiated cells. Thus, plant cells may differentiate (become progressively more specialized), dedifferentiate (revert to an undifferentiated, meristematic state), and then differentiate in a potentially different direction. Plant cells essentially can change their patterns of development. Controlled manipulation of cell totipotency is a foundation of *in vitro* culture upon which plants or plant parts are regenerated from cell and tissue cultures. Large numbers of individuals of similar genetic make up can be cloned by using this technology.

In vitro culture conditions are radically different from normal *ex vitro* conditions and vary depending on the type of culture system and its use. Applications of *in vitro* culture include micropropagation, secondary product production, gene transformation, mutation selection, and somatic hybridization. Modifications include the contributions of culture media such as plant growth regulators, nutrients, pH, gelling agents, and organic constituents. Environment conditions such as low light, high humidity, and limited gas movement often prevail. Development *in vitro* also occurs in the absence of *in situ* whole plant influences such as in somatic embryogenesis where development occurs in the absence of maternal tissues. Thus, the anatomy of cultured cells and tissues may exhibit unique characteristics divergent from development *in vivo*.

The objectives of this section are to describe the anatomical and cytological characteristics of plant cells and tissues grown *in vitro*. Plant culture systems are varied and include protoplasts, cell suspensions, callus cultures, organogenic cultures, and somatic embryogenic cultures. The different culture systems vary considerably in cell composition and culture conditions. Rather than a general review of plant anatomy, this work highlights some of the unique differences in cell structure and anatomical characteristics of the different culture systems. Some atypical or "abnormal" types of cell organization are described. Because plant form and function are

intimately related, the potential impact of these structural divergences on cell and tissue function are discussed. How specific factors (such as the culture environment, growth regulators, and/or medium) affect cell differentiation and development are covered in other sections of this encyclopedia.

LEAF ANATOMY OF *IN VITRO*-GROWN PLANTS

Leaves of plants grown *in vitro* can exhibit very divergent anatomy compared to plants grown under field or greenhouse conditions. Leaves grown *in vitro* characteristically are thinner, have a poorly differentiated mesophyll, large intercellular spaces, modified epicuticular wax formation, and altered stomatal numbers and configurations (for reviews, see Refs. 1–3). Anatomical differences observed between leaves grown under *in vitro* and *ex vitro* conditions are summarized in Table 1. Because of these divergences, cultured plants usually need to be gradually acclimatized to survive outside conditions of culture. Enhanced cuticular transpiration (resulting from lack of epicuticular wax development) and lack of guard cell functioning have been suggested as causes of mortality in cultured plants transferred to *ex vitro* conditions.

Unlike leaves grown in the field, those grown *in vitro* often lack or have a poorly differentiated palisade parenchyma. Mesophyllic cells characteristically appear spongy and are interspersed with numerous and large intercellular spaces. The limited differentiation of cells characteristically found in leaves developed *in vitro* is illustrated in Figure 1. The cytoplasm is frequently

Table 1. Summary of Anatomical Differences Between Leaves Grown under *Ex Vitro* Versus *In Vitro* Culture Conditions

Characteristic	*Ex vitro*	*In vitro*
Cuticular wax	Well developed	Markedly reduced
Stomatal orientation	More depressed	Raised and/or bulging
Stomatal shape and function	Ellipsoid; functional	Rounded; ± functional
Mesophyll	Differentiated into palisade and spongy regions	Limited or no palisade Development
Intercellular spaces	Small; fewer in number	Large; more extensive; Lacunae
Chloroplasts	Well-developed internal membranes; grana	Irregular internal membrane System with few grana
Cytoplasm	More extensive	Less extensive; limited to a parietal area

(a)

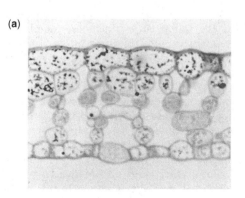

(b)

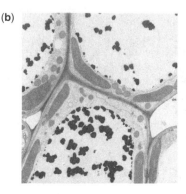

Figure 1. Structure of leaves developed *in vitro*. (**a**) Leaf cross section showing poorly differentiated palisade and spongy layers; mesophyllic cells are interspersed with large intercellular spaces. (**b**) A transmission electron micrograph of mesophyllic cells. Shown are the cytoplasm located in the cell periphery, large central vacuoles, and chloroplasts lacking internal membrane stacking

restricted to parietal areas. In some species, chloroplasts in leaves differentiated *in vitro* are flattened, and the internal membrane system is irregularly arranged and lacks organization into grana and stroma lamellae. Stomata may exhibit higher or lower densities in culture compared to field conditions. Guard cells frequently have higher surface topographies and rounder orientations compared to the more elliptical and sunken guard cells found in noncultured leaves. Reduced or defective stomatal functioning has been reported in a number of studies.

Epicuticular wax in many species is reportedly diminished or structurally divergent in leaves developed *in vitro*. Grout and Aston (4) related excessively high rates of water loss to the formation of much reduced quantities of structured epicuticular wax during culture. Gilly et al. (5) evaluated leaf cuticles from ivy plants grown *in vitro* by using transmission electron microscopy and radioactive labeling. They found that cuticle formation was effective and exhibited developmental and quantitative progressions from young to expanded leaves under both *in vitro* and *ex vitro* conditions. However, transfer of plants from *in vitro* to *ex vitro* conditions activated cuticle biosynthesis.

Thus, tissue-culture-regenerated plants may exhibit severe water loss and desiccation problems when transferred *ex vitro*. Photosynthetic capacity of plants may also be affected as a result of differences in chloroplast ultrastructure and mesophyllic differentiation. More normal leaf morphology resumes in new growth formed *ex vitro*, which suggests that divergent structure is a result of the *in vitro* environment. Environment factors *in vitro* can be regulated to promote photoautotrophic culture conditions and produce plantlets with fewer physiological and morphological disorders (see Refs. 2 and 6).

HYPERHYDRICITY

Hyperhydricity is severe physiological disorder that may affect micropropagated plants. Hyperhydricity is characterized by a series of severe anatomical abnormalities (Table 2), which in extreme cases may result in loss of propagative ability, culture decline, and mortality. Because affected tissues often exhibit a translucent, water-soaked appearance, the condition has also been referred to as "vitrification" or "glassiness." The rationale for using the term hyperhydricity is throughly discussed by Debergh et al. (7). A detailed discussion of hyperhydricity can be found in Ziv (1), George (8) and also see article on this subject.

Hyperhydricity primarily affects leaves and shoots in culture. Shoots often have shorter internodes, thickened stems, and leaves that are thickened, brittle, wrinkled, curled, and elongated. Stems of hyperhydric shoots have xylem and sclerenchyma tissues that are poorly lignified and less differentiated than in normal shoots. The ring of vascular bundles may have reduced sclerenchyma or be interrupted with parenchymatic areas. In some cases, reduced shoot apical dominance may be evident. Vietez et al. (9) found that apical meristems were smaller and composed of fewer cell layers in hyperhydric versus normal shoots. Likewise, maturation of cells in carnation shoot

Table 2. Morphological and Anatomical characteristics of Hyperhydric Shoots and Leaves

1. Swollen, thick shoots with shorter internodes or rosetted habit
2. Fasciated shoot apices and/or reduction of shoot apical dominance
3. Leaves which are translucent, glass-like, turgid, thickened and/or curled
4. Poor or lack of differentiated palisade cells; large intercellular spaces in the mesophyll
5. Reduced lignification; smaller vascular bundles with abnormal or fewer vessels, tracheids, and/or sclerencyma fibers
6. Cuticle thin and discontinuous; limited or no epicuticular wax
7. Epidermis discontinuous; epidermal holes
8. Stomata abnormal with deformed guard cells
9. Decreased cellulose and more limited cell wall development
10. Chloroplasts fewer in number; chloroplasts with abnormal organization in the grana and stroma, some with invaginations along the plastid envelope that fuses with the tonoplast; presence of membrane residues
11. Hypertrophy in the cortical and pith parenchyma

apices occurred sooner (10). Cells of the rib and peripheral meristematic regions became large, parenchymalike and vacuolate, and only the mother cells of the corpus and two tunica layers remained meristematic.

The leaves of hyperhydric shoots are characterized by lack of differentiation between palisade and spongy parenchyma with extensive intercellular spaces and lacunae dispersed throughout the mesophyll. Cells expansion is isodiametric with loss of a clear axis of elongation. Hyperhydric leaves also have fewer and smaller vascular bundles than nonhyperhydric leaves. Abnormal, protruding stomata are prevalent which may be nonfunctional with divergences, including grossly misshapen or twisted guard cells. In *Datura*, abnormal stomata represented 90% of the total number of stomata in hyperhydric plantlets (11).

Hyperhydric tissues may appear chlorophyll deficient and have fewer chloroplasts than corresponding normal tissue. Divergences in chloroplast ultrastructure have also been detected. In *Prunus avium* leaves, parenchymal cells of both vitrified and normal tissue-cultured leaves had a large central vacuole, limited cytoplasmic content, and chloroplasts with an irregularly arranged internal membrane system. However, in vitrified leaves, invaginations were observed along the plastid envelope that appeared to be fused with the tonoplast of vacuoles; membrane residues were frequently noted in intercellular spaces. Photosynthetic capacity appears to be reduced in some situations.

Decreased levels of cellulose and lignin were detected in vitrified carnation internodes which were consistent with increased deformational capabilities of cell walls. In accord, hypertrophy is commonly observed in cortical and pith cells of hyperhydric plants. This may be associated with the formation of large intercellular spaces as a result of separation of cells (schizogeny) and disintegration of cell walls (lysigeny). Abnormalities in both the pectic substances of the middle lamellae and the cellulose of cell walls were observed. Weakened cell wall structure may contribute to the appearance of epidermal discontinuities and perforations observed in some hyperhydric leaves. Other divergences of epidermal cell development include diminished or no epicuticular wax development.

Hyperhydric tissues characteristically have a water content higher than normal tissues. Liquid water in the intercellular spaces was observed through epidermal holes by using environmental scanning electron microscopy of *Gypsophila* (12). Nuclear magnetic resonance (NMR) imaging evaluated the distribution of water in the same species (13). Normal tissue-cultured and glasshouse-grown leaves had a high concentration of water within leaf vascular bundles. In contrast, vitrified leaves accumulated water in intercellular spaces. The presence of numerous hypertrophied cells that accumulate high levels of water contribute to the succulent appearance of hyperhydric tissues.

The degree of susceptibility varies with the kind of plant being cultured, the genotype, the specific conditions, and nature of the culture. A number of contributing causes for hyperhydricity have been proposed, many of which may overlap or interact. These include low irradiance, high relative humidity, water potential of the medium, nutrient composition, transpiration rate, growth regulators, temperature, low culture growth rate, pH of the medium, type of gelling agent, carbohydrate source, and poor gas exchange. Regardless of the exact cause, the phenomenon of hyperhydricity can be of important economic concern.

ANATOMY OF SOMATIC EMBRYOGENIC CULTURES

Somatic embryogenesis is the production of embryos from somatic cells in culture. Because fusion of gametes is not involved, large numbers of clonal embryos can be produced if embryogenic rates are high. Under ideal conditions, the morphology of somatic embryos mimics that of zygotic embryos with the development of a well formed apical meristem, radicle, and cotyledons. Both progress through similar stages: globular, heart, torpedo, and dicotyledonary stages in dicots; globular, scutellar, and coleoptilar stages in monocots; and embryonal suspensor mass, globular, torpedo, and multicotyledonary stages in conifers. The close similarity in somatic and zygotic embryo development has been cited as justification for using embryogenic cultures as models for studying plant embryogenesis. However, developmental conditions *in vitro* are very divergent from the *in vivo* situations where maternal tissues contribute biochemical, environmental, and physical inputs. Frequently, abnormalities in somatic embryo morphology and anatomy are observed that contribute to problems in embryo germination, plant regeneration, and survival.

In embryogenesis *in vivo*, the fertilized egg or zygote undergoes a set developmental sequences within the embryo sac that culminates in the production of the mature embryo. In contrast, somatic embryogenesis requires the induction of embryogenically competent somatic cells. Even when predetermined cells are used to initiate

cultures (as with meristematic or embryogenic cells), hormonal or environmental signals may be necessary to promote embryogenesis. Induction of embryogenic cultures is frequently accomplished by exposing explant tissue to an auxin-containing medium. Typically the medium is later changed to one with low or zero auxin for subsequent embryo development. Huang et al. (14) suggested that there is a shift in the polarity of cells and a change in cell function associated with inductive processes. In observations of cortical microtubules during induction in a rice embryogenic system, they found a change in terminally distributed microtubules to a more even distribution pattern after 12 hours.

Frequently, embryogenic induction is confined to specific responsive regions of the explant, for example, epidermal cells or specific mesophyllic layers. Tissues placed on an induction medium characteristically become mitotically active and form sectors composed of cells that are small, isodiametric, and characterized by a large nucleus and prominent nucleolus, a dense cytoplasm, small vacuoles, and high starch content. Continued meristematic activity can lead to the development of prominent embryogenic sectors and protrusions. These regions which give rise to embryos may be composed primarily of embryogenic meristematic cells or may be interspersed with nonembryogenic callus proliferations that characteristically have larger and more vacuolate cells (Fig. 2). In a number of embryogenic systems, higher incidences of abnormalities in embryo morphology are associated with the use of certain auxins (e.g., 2,4-dichlorophenoxyacetic acid versus naphthaleneacetic acid). Rodriguez and Wetzstein (15) found that auxin induction in *Carya* cultures promoted enhanced cell division, particularly in subepidermal cell layers. However, the type of auxin used caused notable differences in mitotic activity, location of embryogenic cell proliferation, epidermal cell continuity, callus growth, and embryo morphology. Cultures induced with naphthaleneacetic acid had embryogenic regions composed of homogeneous, isodiametric, and meristematic cells and produced embryos with normal morphology and a discrete shoot apex. In contrast, tissues induced on a medium that

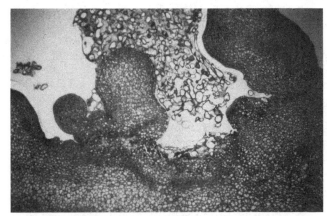

Figure 2. Section from a somatic embryogenic pecan culture. Developing somatic embryos are evident within proliferating proembryogenic regions that are intermixed with nonembryogenic callus

(a) (b)

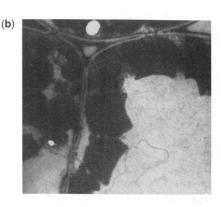

Figure 3. Transmission electron micrographs of zygotic and somatic embryonic cotyledons of pecan. (**a**) Zygotic embryos have cotyledonary cells that have extensive oil deposits, starch, and protein. (**b**) Somatic cells have a large central vacuolar space and limited oil reserves

contained 2,4-dichlorophenoxyacetic acid exhibited more intense and heterogeneous areas of cell division. Proliferating cell regions were composed of meristematic cells interspersed with callus.

Early concepts proposed that somatic embryogenesis required a single-cell origin and no vascular connections with the maternal tissue. However, a number of recent critical anatomical evaluations describe both unicellular and multicellular origin of embryos. Histological evaluations of somatic embryo development are reviewed extensively elsewhere (16,17). In many cases, somatic embryos originate from a complex of cells or from several morphologically competent adjacent cells. Embryogenesis occurs within discrete meristematic regions that have been referred to as proembryogenic masses, embryogenic masses, meristematic units, proembryonal cell complexes, proembryoids, and proembryogenic aggregates. From one to many somatic embryos may develop from these compact masses of embryogenic cells.

Globular embryo differentiation is characterized by cells that have numerous Golgi bodies, abundant endoplasmic reticulum (ER), thin cell walls, and tightly packed cells with only small intercellular spaces. During the development of globular and heart-shaped somatic embryos, the starch content of cells declines within the embryogenic regions from which embryos are derived. Starch accumulation and utilization are thought to be a source of energy for cell proliferation and growth. Somatic embryo development may occur with the formation of a well-defined suspensor, as exemplified in coniferous somatic embryos that characteristically have a substantial suspensor of highly vacuolate cells. In contrast, many studies report a suspensor that is reduced or absent. Isodiametric growth contributes to globular embryonic development and is followed by elongated growth, cotyledonal development, bilateral symmetry, hypocotyl elongation, and radicle development.

In some cases somatic embryos lack a well-developed shoot apex and exhibit poor polarity with varying degrees of embryo fasciation or fusion. Cotyledons may be underdeveloped, malformed, and/or exhibit a horn-shaped (i.e., collar-like) or a fan-shaped configuration. Failure to establish a functional shoot meristem may be the cause of conversion failure in somatic embryos (18). Wetzstein and Baker (19) found a poorly defined shoot apical region in horn-shaped somatic embryos, where mitotic cells were restricted to a few cell layers. In some abnormal embryos,

delimitation of an apical area may be lacking which is associated with epidermal rupture and disorganization of shoot apices caused by extensive callus proliferation at the apex (15). Defective polarization during early embryo development may contribute to the development of abnormally joined, multiple embryos. During embryonic development, endogenous auxin synthesis occurs in the shoot meristem, and polar transport is proposed as necessary for normal cotyledonal formation. Liu et al. (20) reported that inhibition of polar auxin transport from shoot apices caused malformed collar-like cotyledons suggesting that high exogenous auxin levels in the *in vitro* medium may contribute to abnormalities in somatic embryo development.

During the maturation period of embryogenesis, deposition of storage reserves occurs when carbohydrates, lipids, and protein components accumulate in cotyledonary cells. Somatic embryos characteristically have poor reserve accumulation. This is illustrated in Figure 3 which contrasts cotyledonary cells from zygotic and somatic embryos of pecan. Zygotic cells have extensive oil deposits, starch, and protein. In contrast, somatic cells have a large central vacuolar space and much more limited oil reserves. A general lack of storage reserves adversely affects later stages of development and subsequent germination of somatic embryos. A number of medium amendments and treatments (such as abscissic acid, osmoticums, cold treatment, and amino acids) have been incorporated into embryogenic protocols to enhance storage accumulation.

ANATOMY OF ORGANOGENIC CULTURES

In organogenic culture systems, plant regeneration occurs through the de novo formation of organs such as leaves, shoots, buds, or roots. Organogenesis is of great economic interest because of its role in plant micropropagation. De novo organ formation can be obtained from a variety of tissues that range from highly organized explants such as stems and leaves to less differentiated sources such as protoplasts, calli, or thin cell layers. When placed on appropriate media, specific cells of the explant may proliferate and directly give rise to organogenic precursor cells. Alternatively, organogenesis may occur indirectly through an intermediate proliferation of callus. Cells associated with stomata have been reported to be especially regenerative in a number of culture systems, including those that use stems, needles, and leaves as explants. Subepidermal cells (i.e., cortical or mesophyllic cells) exhibit mitotic activity and form cellular protrusions from which shoots regenerate.

Cell proliferation and the formation of meristemoids or meristematic centers is common to most organogenic systems. These regions are spherical masses of small meristematic cells that have large nuclei and densely staining cytoplasm. The formulation of the culture medium, particularly the type and concentration of plant growth regulators, can have a pronounced effect on their development. The initiation of meristematic centers has been traced to single cells and to small cell clusters or precursor cells. Meristemoids are the sites from which organs differentiate.

Plastid structure can change conspicuously in shoot regenerative cultures and frequently includes starch accumulation in regions where shoot differentiation will occur. In tobacco internode cultures, typical chloroplasts with extensive granal stacks were observed at culture initiation (21). Proplastids formed with prominent starch granules, few lamellar structures, and electron-dense materials within developing meristematic centers. Later, proplastids were observed in the apical dome, and chloroplasts were seen in developing leaf primordia.

The structures of leaves and shoots differentiated *in vitro* may vary significantly from those observed in plants grown *ex vitro*. Aspects of this are discussed under the sections on the anatomy of leaves and hyperhydricity.

ANATOMY OF PLANT PROTOPLAST

Plant protoplasts are the living parts of plant cells that contain the nucleus and cytoplasm when the cell wall is removed, usually by enzymatic digestion or mechanical means (Fig. 4). Protoplast cultures often serve as model systems for investigating cell wall regeneration and plant viral infections. Their lack of cell walls (but not semipermeable membranes) enables protoplasts to be used in a number of manipulations including genetic engineering (because protoplasts can absorb DNA, proteins, and other macromolecules) and fusion to create plant cell hybrids.

Protoplasts exhibit numerous structural variations depending on the tissue of origin and the preparative methods. Freshly isolated protoplasts are spherically shaped and generally have cytoplasmic strands that contain microtubules (Mts) radiating from the nuclear area and extending toward the cortical regions. Although protoplasts are usually uninucleate, multinucleate forms are observed that likely are formed via spontaneous protoplast fusion. Because they lack a cell wall and are surrounded by only a semipermeable membrane, protoplasts are extremely fragile and liable to physical or chemical damage.

It is widely held that development of a normal cell wall around a protoplast is necessary for subsequent cell division. Protoplasts can undergo dramatic structural changes to regenerate their cell walls after culturing on a suitable medium. Cell wall regeneration is initiated rapidly after protoplasts are removed from isolation media. For example, cellulose microfibril (CMF) deposition in tobacco protoplasts was noticeable within 30 minutes after removal of the original cell wall (22). CMF deposition seems to initiate randomly from distinct sites on the circumference of the protoplast surface, and the subsequent deposition exhibits an unevenly, parallel distribution. Further deposits occur in different orientations underneath earlier ones, indicating the beginning of wall layering. Although freshly prepared protoplasts are usually considered devoid of callose deposits (also a cell wall component), they may show threadlike spots within the first hours after isolation. These deposits may be a wound response or alternatively are incorporated as a newly synthesized cell wall. By forming a new cell wall, a protoplast becomes a regenerated cell. With subsequent enlargement and expansion, cells usually lose their spherical shape. In most cases, the regenerated cells become somewhat elongated and oval.

Microtubules (Mts) are believed to play important roles during plant growth and development. Their possible involvement in protoplast regeneration is comparatively well documented. Numerous studies suggest that the presence of a cortical Mt network in newly isolated protoplasts (Fig. 4) is a prerequisite for subsequent cell division. Freshly isolated protoplasts are characterized by markedly fewer cortical Mts than those in the donor tissues. In addition, the initial number of cortical Mts in protoplasts is inversely related to the degree of differentiation in their corresponding cells. Protoplasts originated from more mature cells have far fewer Mts

(a)

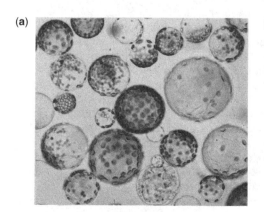

(b)

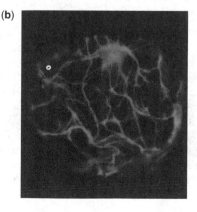

Figure 4. Protoplasts. (**a**) Light micrograph of protoplasts isolated from *Brassica napus* leaves. With the removal of the cell wall, the plasma membrane and protoplasts are spherical. Numerous chloroplasts are evident. (**b**) An immunocytochemical localization for microtubules in a protoplast from grape leaf mesophyll. A random orientation of Mts is shown

than less differentiated cells. Protoplasts from fully dedifferentiated suspension cultures usually contain an extensive network of cortical Mts. The Mts in freshly isolated protoplasts are randomly arranged. During the first days of culture, the cortical Mts gradually become more numerous and more organized. Some investigations indicate that reestablishing a cortical Mt network during the early stages of protoplast culture may be critical for orienting CMFs during cell wall formation and regaining mitotic activity.

The cells in plant tissues are usually interconnected by plasmodesmata to become a functional unit. Plamodesmata are assumed to play an important role in intercellular transport, cell-to-cell communication, cell differentiation, and growth. Plasmodesmata are usually established by protoplasmic strands retained between Golgi vesicles that aggregate to form the cell plate. However, the plasmodesmata connections are abruptly lost during protoplast preparations. In regenerating protoplasts, an aggregation of ER cisternae and Golgi vesicles usually occurs close to the cell membrane. Constrictions of these cisternae in the plane of the plasma membrane are presumed to form half-plasmodesmata which are usually not interconnected with other protoplasts (23). The formation of half-plasmodesmata during the very early stages of protoplast culture might be regarded as an attempt to reestablish intimate contact with neighboring cells. Most of the half-plasmodesmata fail to establish contact with adjacent cells and disappear from the surfaces of the cells during further culture. However, contact between adjacent cells is noticed where fusion of matching half-plasmodesmata is formed by partner cells.

Leaves are a convenient and thus commonly used source of tissue for protoplast isolation. The majority of mesophyllic protoplasts contain numerous chloroplasts of various sizes. Small spherical mitochondria are usually clustered around the nucleus and the chloroplasts. A considerable part of the ER accumulates around the chloroplasts. The remaining part of the ER network is present in the cortical cytoplasm and characteristically is partially fragmented and has reduced complexity. This may result from breakage of the anchoring points of the ER through or with the plasma membrane to the cell wall networks that exist in complete cells. During the first days of culture, the mitochondria are generally transformed into long vermiform organelles that distribute evenly throughout the cytoplasm. The clustered appearance of the ER in the vicinity of the chloroplasts is lost. The ER network seemingly repairs the broken ends and becomes more complex.

ANATOMY OF CELL SUSPENSION CULTURES

Plant suspensions are cultures composed of unorganized cell aggregates grown in a liquid medium. The cultures are usually initiated by inoculating a friable callus into a agitated liquid medium. During subsequent cell divisions, clumps of cells are continuously sloughed off. Single-cell cultures are seldom achieved because the dividing cells have a tendency to adhere to each other. Modifications of the culture medium can affect culture friability and the size and number of cells that compose cell clumps. Greater dissolution of the middle lamella has been observed in suspensions culture cells grown on media with high auxin concentrations. Environmental and culture conditions can affect the production of cell wall extracellular polysaccharides that may contribute to cell clumping.

The anatomy and morphology of a cell suspension culture can vary extensively depending on the species, explant, media composition, subculture method, and culture conditions. Cultures may consist of mitotically active, undifferentiated cells. Alternatively, cells can be highly differentiated with reduced mitotic activity. In suspension cultures that have large clumps composed of many cells, internal and external sectors can be divergent in their degree of differentiation and organization. Culture lines with desirable characteristics can be selected depending on the use of the culture. Suspension culture can be used as part of organogenic and embryogenic systems or to produce biosynthetic metabolites. Thus, the types of cell and their structural attributes are extremely divergent. A general discussion of all the cell forms is not feasible in this section.

ANATOMY OF CALLUS CULTURES

Callus is a proliferating mass of unorganized, generally undifferentiated cells, that can be initiated from a wide range of explant sources. The anatomical organization and structural characteristics of a callus depend in part upon the type and condition of cells used to initiate cultures and on the culture medium and conditions. The significance of callus cultures is that under appropriate circumstances, cells can differentiate and be induced to regenerate adventitious tissues and organs such as leaves, buds, shoots, or roots in organogenic cultures. Embryogenic callus can be produced and used to obtain somatic embryos. Callus also can be used to initiate cell suspensions or can be used to produce biosynthetic metabolites.

The initial stages of callus culture development include the induction, activation, and subsequent mitotic activity of cells (24). The smaller cells which result from this active phase of cell division have small vacuoles and an arrangement of mitochondria and plastids that suggests high metabolic activity. Cells generally are parenchymatous and can exhibit highly proliferative rates of growth. Varying degrees of cellular differentiation occur within the callus mass. Although a callus may appear amorphous, regions within a callus mass may exhibit significant differences in physiology and cytology. After initial establishment of callus cultures, cell division often becomes concentrated in cells located in the outer periphery of the proliferating clumps. This gives rise to an organization of surface meristematic layers surrounding a nondividing core. Divisions in the peripheral layers may persist, or in some cases decline, and may be supplemented by internally located, actively mitotic, growth centers. These mitotic structures have been called meristematic nodules or centers of morphogenesis. Nodules may remain stable and continue to produce parenchymatous cells or alternatively may have high regenerative capacity and give rise to plant organs. Internal cells may expand and differentiate into larger parenchymal cells. Differentiation of some cells into phloem elements or tracheids may proceed over time.

Callus cultures differ greatly in their texture, physical properties, and morphogenic capacity or competence. In some cases, callus lines have been isolated that differ in appearance and regenerative potential even when derived from the same explant. These differences may reflect the epigenetic potential of the cells or be caused by the appearance of genetic variability amongst the cells in the culture (25). Explants used to initiate callus cultures are composed of a heterogeneous mixture of cell types and thus may give rise to divergent types of callus even within a single culture. Critical selection and separation of callus types are important during routine subculturing so that sectors that have desirable characteristics can be maintained and optimized. For example, calli derived from immature embryos in maize have been classified into various types. Nonregenerable callus was of soft, granular tissues that was composed of elongated, vacuolated cells (26). Even within rapidly growing, highly embryogenic callus cultures, different morphotypes were identified. One form had embryos arising from a mantle of densely cytoplasmic cells that overlay a core of elongated vacuolate cells. Another granular type was composed of a thick mantle of meristematic cells over a lacunar core of elongated cells. The morphotypes exhibited divergent developmental patterns. Preferential selection could be used to maintain cultures enriched for a specific morphotype.

Morphogenic capacity is an important attribute of cal lus cultures. Thus, there has been interest in identifying cytological characteristic associated with regenerability. The ultrastructure of sunflower microcalli that differ in their regenerative potential was compared by Keller et al. (27) who found that nonregenerative cultures exhibited incomplete cellular divisions. In maize, buckwheat, and strawberry callus cultures, the cell walls of non-regenerable callus stained more extensively in the middle lamella with $KMnO_4$ than in regenerable tissues (28) which indicated alterations in the phenolic components of the cell walls.

Staining patterns correlated with the developmental capacity of the tissues. In a characterization of Aesculus callus culture cells (29), both embryogenic and nonembryogenic cells had organelle-rich cytoplasm indicative of cells that have high metabolic rates, that is, ribosomes grouped in polysomes, mitochondria with long electron-dense cristae, active dictyosomes, and frequent ER profiles. However, a feature of regenerable callus was diffuse cell wall degradation resulting from middle lamella digestion. This degradation accounted for the friability of the regenerative callus and led to the detachment of small embryogenic cell aggregates. In some systems, starch accumulation is a prerequisite to morphogenesis and accumulates preferentially in cells which will give rise to shoot primordia and somatic embryos. Starch is proposed as a direct source of energy reserves required for morphogenic processes.

BIBLIOGRAPHY

1. M. Ziv, in P.C. Derbergh and R.H. Zimmerman, eds., Micropropagation, Technology and Application, Academic Press, New York, 1991, pp. 45–69.
2. J. Pospíšilová, J. Solárová, and J. Čatský, Photosynthetica 26, 3–18 (1992).
3. M. Ziv, in J. Aitken, T. Kozai, and M. Lila Smith, eds., Automation and Environmental Control in Plant Tissue Culture, Kluwer Academic Publishers, Dordrecht, The Netherlands, 1995, pp. 493–516.
4. B.W.W. Grout and M.J. Aston, Hortic. Res. 17, 1–7 (1977).
5. C. Gilly, R. Rohr, and A. Chamel, Ann. Bot. (London) [N.S.] 80, 139–145 (1997).
6. T. Kozai, C. Kubota, and B.R. Jeong, Plant Cell, Tissue Organ Cult. 51, 49–56 (1997).
7. P. Derbergh et al., Plant Cell, Tissue Organ Cult. 30, 135–140 (1992).
8. E.F. George, Plant Propagation by Tissue Culture, Part 2, In Practice, Exegetics, Edington, England, 1996.
9. A.M. Vietez, A. Ballester, M.C. San-Jose, and E. Vietez, Physiol. Plant. 65, 177–184 (1985).
10. M.E. Welter, D.S. Clayton, M.A. Miller, and J.F. Petolino, Plant Cell Rep. 14, 725–729 (1995).
11. F.C. Miguens, R.P. Louro, and R.D. Machado, Plant Cell, Tissue Organ Cult. 32, 109–113 (1993).
12. K. Gribble et al., Plant Cell Rep. 15, 771–776 (1996).
13. K. Gribble et al., Protoplasma 201, 110–114 (1998).
14. B.Q. Huang, X.L. Ye, S.Y. Zee, and E.C. Yeung, In Vitro Cell Dev. Biol. Plant 33, 275–279 (1997).
15. A.P.M. Rodriguez and H.Y. Wetzstein, Protoplasma 204, 71–83 (1998).
16. D.W. Meinke, Annu. Rev. Plant Physiol. Plant Mol. Biol. 46, 369–394 (1995).
17. V.L. Dodeman, G. Ducreux, and M. Kreis, J. Exp. Bot. 48, 1493–1509 (1997).
18. T.C. Nickle and E.C. Yeung, Am. J. Bot. 80, 1284–1291 (1993).
19. H.Y. Wetzstein and C.M. Baker, Plant Sci. 92, 81–89 (1993).
20. C.-M. Liu, Z.-H. Xu, and N.-H. Chua, Plant Cell 5, 621–630 (1993).
21. M. Arai, T. Saito, Y. Kaneko, and H. Matsushima, Physiol. Plant. 99, 523–528 (1997).
22. T.N.M. van Amstel and H.M.P. Kengen, Can. J. Bot. 74, 1040–1049 (1996).
23. J. Monzer, Protoplasma 165, 86–95 (1991).
24. A.P. Aitchison, A.J. Macleod, and M.M. Yeoman, in H.E. Street, ed., Plant Tissue and Cell Culture, University of California Press, Berkeley, 1977, pp. 267–306.
25. E.F. George, Plant Propagation by Tissue Culture, Part 1, The Technology, Exegetics, Edington, England, 1993.
26. E. Werker and B. Leshem, Anna. Bot. (London) [N.S.] 59, 377–385 (1987).
27. A.V. Keller, N. Frey-Koonen, R. Wingender, and H. Schnabl, Plant Cell, Tissue Organ Cult. 37, 277–285 (1994).
28. V. Lozovaya et al., J. Plant Physiol. 148, 711–717 (1996).
29. P. Profumo, P. Gastaldo, and N. Rascio, Protoplasma 138, 89–97 (1987).

See also Cell and cell line characterization; Characterization of cells, microscopic; Measurement of cell viability; Plant cell cultures, selection and screening for secondary product accumulation.

ANGIOSPERMS

Margaret M. Ramsay
Royal Botanic Gardens, Kew
Richmond, Surrey
United Kingdom

OUTLINE

INTRODUCTION

Angiosperms are the dominant group of vascular plants on earth today with more than 250,000 species (1), characterized by possession of true flowers which are more advanced and complex than the reproductive structures of the gymnosperms.

The total numbers of species of economic value is difficult to estimate — 6000 species are known to be used in agriculture, forestry, fruit and vegetable growing, and pharmacognosy although only 100 to 200 are of major importance in world trade. Many thousand of species and cultivars are grown purely as ornamental garden plants.

Biologically the flower's prime function is the production of seed which may be produced as a result of self-pollination and fertilization, by cross pollination, or in some cases by apomixis (a nonsexual process).

Angiosperms have developed highly complex and diverse reproductive organs and they also possess an advanced level of cell structure and differentiation. Their high level of physiological efficiency and their wide range of vegetative plasticity and floral diversity have allowed the angiosperms to occupy a broad range of habitats that express a wide array of growth habits. This diversity of structure and function affects the uses that are made of the plant and also dictate the methods that are suitable for their propagation.

ADVANTAGES OF *IN VITRO* TECHNIQUES

The extent to which tissue culture methods can be used for genetic manipulation and for propagation is changing continuously. *In vitro* techniques have advantages over conventional methods, principally in their potential to produce large numbers of plants from small quantities of initial material by multiplication under controlled pathogen-free conditions. These techniques can be carried out on material that may be slow or difficult to propagate conventionally, and methods may be used to free plants from virus disease. Disadvantages are that advanced skills and specialized facilities are required, and because of this, costs of propagules are relatively high. The chances of producing genetically aberrant plants may also be increased. The term *in vitro* culture covers a wide range of techniques under sterile conditions including seed germination, micropropagation (shoot culture), meristem culture, and callus (tissue) and cell culture. In many cases more than one technique will be used to produce propagules.

IN VITRO SEED GERMINATION AND EMBRYO CULTURE

Seeds have several advantages as a means of propagation: they are often produced in large numbers, can normally be stored without loss of viability for long periods, and are usually free of pest and disease. As seeds arise through sexual reproduction, a wider genetic base can be maintained. This is an important consideration and benefit for material of conservation interest (2) but not if uniform offspring are required. Genetically uniform populations may be obtained using seed through the use of F1 seeds produced by crossing two homozygous parents, from inbred homozygous lines from autogamous species (e.g., wheat, barley, rice), and for those few genera that produce apomictic seed (seeds formed without fertilization and therefore genetically identical to the female parent) Although especially useful for plant breeders, it does not result in rapid and large-scale rates of propagation.

The main advantage of using plant tissue culture techniques for seeds are that dormancy or other germination requirements can be overcome *in vitro* by physical or chemical removal of testa and extended soaks in sterilants and/or water to remove possible inhibitors. Seedling material can be grown until of a suitable size for transfer to *in vivo* conditions or can be used as starting material for shoot culture. In some cases it is possible to initiate multiple shoot cultures directly from seed with the use of cytokinins: clusters of axillary and/or adventitious shoots form on germination and may be serially subcultured producing high rates of multiplication.

Culture of zygotic embryos dissected from seeds can also assist in the rapid production of seedlings from seeds that have a protracted dormancy period or when the genotype conveys a low embryo or seed viability resulting from some interspecific crosses.

However, dissection requires great skill and in some cases complex media with additional vitamins, growth regulators, amino acids, and coconut or other endosperm extracts. Ovule culture may be used as an alternative to embryo culture, often requiring a less complex medium for growth. Many orchids are grown both commercially and for conservation using immature seeds from green capsules (3).

PROPAGATION FROM VEGETATIVE MATERIAL

The most suitable explants for tissue culture are those in which there is a large proportion of either meristematic tissue or cells that retain an ability to express totipotency, that is, retain a latent capacity to produce a whole plant. It can be exhibited by some differentiated cells but not by terminally differentiated structures. Senescent tissue rarely result in successful cultures.

Methods available for propagation of plants from vegetative material are

1. by the multiplication of shoots from axillary buds
2. by the formation of adventitious shoots
3. by the formation of adventitious somatic embryos.

Cultures can be formed either directly on pieces of tissue or organs (explants) removed from the mother plant or indirectly from unorganized cells (in suspension cultures) or tissues (in callus cultures) established by the proliferation of cells within explants. These techniques are reviewed and described in detail by George (4).

Shoot Culture

The apices of lateral or main shoots, dissected from actively growing shoots or dormant buds are the usual starting material for shoot culture. Large explants have advantages over smaller ones for initiation because they contain more axillary buds and usually commence growth more rapidly. However it may be more difficult to establish aseptic conditions. Meristem tip cultures are initiated from much smaller explants and are used for virus elimination.

Growth regulators (usually cytokinins) are incorporated into the growth medium to promote growth and proliferation of axillary shoots in shoot culture. These remove the dominance of apical meristems so that axillary shoots are produced which serve as explants for repeated proliferation.

Vegetative shoots can sometimes be induced from meristems that would normally produce flowers or floral parts where young inflorescences are used (mature inflorescences give rise to floral structures). In sugar beet these arise from floral axillary buds whereas with cauliflower they are thought to arise from flower meristems. In most cases, however, shoots formed do not arise directly from flower meristems but adventitiously, often on the receptacle.

Adventitious Shoot Culture

Adventitious shoots arise directly from the tissues of an explant (but not from a meristem) rather than through a callus phase, and initiation largely depends on plant genotype. Shoots may be formed *in vitro* on explants from various organs (leaves, stems, petioles, flower petals, bulb scales, roots) of responsive plants. Several ornamental plants are propagated *in vitro* by direct shoot regeneration, particularly *Saintpaulia* and other members of the Gesneriaceae. In these cases, highly prolific cultures are initiated from leaf and petiole explants, and subculturing of shoot clumps provides efficient large-scale production without the formation of callus.

The use of growth regulators may sometimes result in proliferation of unorganized cells (callus) alongside direct shoot formation. This is not normally used for propagation because the disorganized cells may result in change in the genetic identity. Growth regulators may also cause proliferation of shoot primordia that require transfer to growth regulator-free media to develop further.

Although shoot regeneration from root apices and segments has been reported, root material grown in soil can be difficult to decontaminate to supply aseptic cultures. They can, however, be used as an initial source of shoots that can then be multiplied.

Adventitious Somatic Embryogenesis

Somatic embryos can be produced indirectly from disorganized cell and tissue culture and can be distinguished from adventitious shoots because they are bipolar (with both shoot and root poles), a coleoptile and scuttelum in monocotyledons and a shoot axis and cotyledons for dicotyledons. There is also no attachment to the originating tissue through vascular material, and most somatic embryos are detached easily from the surrounding cells. The stages in the development of somatic embryogenesis for dicotyledons are described as proembryo, globular, heart, torpedo, and plantlet stages which are similar to those that occur in differentiating zygotes after fertilization. Proembryos from monocotyledons give rise to discrete globular bodies with structures resembling the scuttelum and coleoptile.

Somatic embryogenesis was first observed in suspension cultures of carrot but has now been reported for more than 30 plant families. Somatic embryos normally arise in cultures of unorganized tissues, from callus, suspension cell, protoplast, or anther (haploid) cultures.

Callus Culture

Callus can be initiated from many tissues of higher plants, including plant organs and specific tissue types and cells. But callus is more easily established from young meristematic material, and the choice is much greater for dicotyledonous species. Primary callus forms on the original explant and can vary enormously in quantity, color, and friability in response to genotype and media used. Pieces of tissue removed from this material form the basis of secondary callus cultures which can be maintained for many years through subculture, although there may be problems in stability due to selection pressures if environmental conditions change. Callus can supply a source of uniform cells for regeneration, isolation of protoplasts, and for initiating suspension cultures.

Suspension Cell Culture

Although it is possible to initiate cell-suspension cultures without a preliminary callus phase, cultures are normally initiated using an inoculum of friable callus in a liquid medium which breaks up under agitation, divides, and forms clumps and chains of cells which then further fragment and multiply. Unlike large-scale culture of bacteria, culture of dispersed individual cells is not possible, and suspensions consist of cellular aggregates of varying size and composition. To ensure cell dispersion and good gaseous exchange, it is necessary to agitate the medium. Two main types of culture are carried out:

1. batch culture where cells are grown in a fixed volume of medium until there is an accumulation of inhibitors or nutrients are depleted. Usually on a small-scale, flasks and orbital shakers or stirred vessels are used.

2. continuous culture where a continuous supply of nutrient media maintains cell growth (particularly important for producing secondary metabolites).

In large-scale, continuous culture, agitation is achieved by using turbines and air-lift reactors in preference to mechanical stirring which can damage plant cells. Suspension cultures result in more rapid growth of plant cells compared to callus cultures because media and environmental conditions are more easily controlled. Somatic embryos and organs can be induced and cells/aggregates transferred to solid media to produce regenerative callus.

Protoplast Culture

The main use of protoplast culture is in plant breeding for specific gene transfer into crop species. To transfer large molecules or organelles into a plant cell to modify the genome or cytoplasm, it is necessary to overcome the obstacle of the protection afforded by the cellulose cell wall. This requires removing the wall to give an isolated protoplast which can then be penetrated directly using a syringe needle or by electroporation. Using appropriate media and conditions, the protoplasts will reform cell walls and multiply to form callus which through regeneration produces plants containing the desired modifications. The stages involved in protoplast isolation include selection of material and pretreatment to adjust osmotic pressure, enzymatic dissolution of the cell walls, and isolation and culture of the protoplasts. Fusion between the same or different species can take place to produce hybridization not possible by normal means, although this often results in instability and has limited application.

Anther Cultures

Immature pollen or microspores can be induced to form vegetative cells, instead of pollen grain when the appropriate medium is used, for producing haploid proembryos or callus tissue. This is of particular interest in plant breeding where homozygous progeny can be derived in one generation from an F1 heterozygote, substantially reducing the time required to check characteristics introduced through the traditional methods of back-crossing and self-pollination. Anther culture using microspores is usually more successful than using isolated pollen because there appears to be some stimulatory effect associated with the anther wall. Overall, however, anther culture has been fairly limited in its successful application, and Solonaceae species account for many of these and a strong genotypic influence.

SOMACLONAL VARIATION

Variation arising as a result of tissue culture (somaclonal variation) occurs widely and can be due to several causes (5). Some changes may be temporary or transitory, including phenotypic variation brought about by physiological modifications in response to culture conditions. This type of change normally disappears when the plants are transferred to *ex vitro* cultural conditions. Genetic change may also occur *in vitro*, and this may result from using differentiated cell cultures, media constituents, temperature of culture, or may be deliberately created in cultured cells using genetic engineering techniques. The choice of material may affect levels of genetic variation in cultures, and it is thought that the use of meristematic tissues is less likely to induce variation than organogenesis resulting from callus (2).

Many of the mutations produced during *in vitro* culture are deleterious, and their expression can result in reduced vigor or abnormal appearance. However, some can be useful for plant improvement, either immediately or when incorporated in a plant breeding program, and new genotypes may be of value in selecting cell lines to produce secondary metabolites. Many mutations remain undetected *in vitro* and may only become apparent on genetic analysis. Variability is particularly liable to occur during the induction and growth of a culture and during plant regeneration and in many instances result from breakage and structural alteration of chromosomes or chromatids which occur during the division of unorganized cells *in vitro*. Why this occurs is still unknown although many hypotheses have been advanced, in particular that these changes are related either directly or indirectly to alterations in the state of DNA methylation (6).

SECONDARY METABOLITE PRODUCTION

Higher plants contain a variety of substances that are useful medicines, food additives, perfumes, etc. Problems in obtaining these high-value-added substances from natural plants have stimulated research in the area of plant cell cultures for producing these useful metabolites, principally because this method is not affected by changes in environmental conditions such as climate or natural predation. Studies have concentrated on applying microbial fermentation technology to plant cell culture (7).

However, there are significant differences between microbial and plant-cell cultures which must be considered. Plant cells are very sensitive to shear forces due to the large size of the cells and the relatively inflexible cellulose cell wall. The large size of the plant cell also contributes to high doubling time (12 hours to several days) which prolongs the time for a successful fermentation run. The vacuole is the major site of product accumulation, unlike microorganisms where the product is secreted into the medium. Research on membrane permeabilization of plant cells is being carried out to address this, along with means of partially recycling of biomass, for instance, through immobilization. Problems remain with the low yield of plant metabolites, the unstable producing ability of cultured cells, and their slow growth rate.

Therefore, at present this process has not become a cost-effective technology, particularly if the metabolites can be easily manufactured by chemical or microbial fermentation methods. Now, Japanese firms are manufacturing a plant pigment (shikonin) from cultures of *Lithospermum erythrorhizon* and ginseng cell biomass on a commercial scale. Several other products including anticancer drugs may be close to commercialization. Investigation into the production of the antitumour compound taxol are also progressing.

CRYOPRESERVATION

Seed storage is the most common method of storing plant germplasm because of the low cost and low-tech equipment required. However seed storage is not applicable to all plant species. They may not tolerate the desiccation and low temperature requirements or may not produce seed naturally, only periodically, or in low numbers. All germplasm which has been genetically modified and all clonally propagated cultivars must be stored in a vegetative form.

Material that has been propagated *in vitro* can be viewed as a collection parallel to a seed bank and has been termed an *in vitro* active gene bank (8). In the interests of reducing labor and minimizing the risk of genetic drift or somaclonal variation, techniques are being developed for the long-term storage of this material. This storage can be of two types, reduced growth (*in vitro* slow-growth gene banks) and zero growth or cryopreservation (*in vitro* base gene bank).

Because mass propagative culture conditions are designed for rapid production, it is necessary to modify culture conditions to obtain slow growth. This usually takes the form of reducing the temperature to between 6–10 °C for temperate material and to 15–25 °C for tropical material which will typically extend the subculturing interval to between 1 and 2 years. However, this means that cultures are placed under conditions of stress and potential selection leading to deterioration and loss of clonal homogeneity.

Therefore the suspension of growth achieved though cryopreservation (freeze preservation in liquid nitrogen at −196 °C) offers a potentially more secure storage system. Cryopreservation is used routinely to maintain type cultures in microbiology and animal-cell culture and to store semen and embryos in the livestock industry and human medicine. Plant cryopreservation has received less attention largely due to the greater complexity and heterogeneity of plant material which varies enormously in response to culture requirements and response to freezing and thawing. However, plant-cell suspension cultures can now be routinely cryopreserved (homogeneity of cell cultures is an important factor). This procedure involves pregrowth (cells should be small at the exponential phase of growth) cryoprotection (usually chemical), cooling (slow cooling using a programmable freezer), storage, rapid thawing, post-thaw treatment to prevent injury by deplasmolysis, and recovery growth.

Cryopreservation of shoot cultures is far more complex and difficult because a shoot is much larger than a cell aggregate, has many different cell types, and thus is more likely to suffer structural damage as a result of freezing and thawing. This in turn is more likely to lead to adventitious regeneration through a callus phase rather than direct regeneration with the consequent risk of genetic instability. A wide variety of protocols have now been developed for cryopreserving vegetative material, particularly alginate encapsulation which was developed from artificial seed technologies. This technique does not require using potentially toxic cryoprotectants or controlled-rate freezing apparatus.

BIBLIOGRAPHY

1. V.H. Heywood, *Flowering Plants of the World*, B.T. Batsford Ltd., London, 1993.
2. M.F. Fay, *In Vitro Cell Dev. Biol.* **28P**, 1–4 (1992).
3. M.M. Ramsay and J. Stewart, *Bot. J. Linn. Soc.* **126**, 173–181 (1998).
4. E.F. George, *Plant Propagation by Tissue Culture*, 2nd ed., Exegetics Ltd., Westbury, U.K., 1993.
5. P.J. Larkin and W.R. Scowcroft, *Theor. Appl. Genet.* **60**, 197–214 (1981).
6. S.M. Kaeppler and R.L. Phillips, *In Vitro Cell Dev. Biol.* **29P**, 125–130 (1993).
7. M. Misawa, *Plant Tissue culture: An Alternative for Production of Useful Metabolites*, Agric. Serv. Bull. 108, FAO, Rome, 1994.
8. L.A. Withers, *Biol. J. Linn. Soc.* **43**, 31–42 (1991).

See also BRYOPHYTE IN VITRO CULTURES, SECONDARY PRODUCTS; CULTURE OF CONIFERS, DICOTYLEDONS, MONOCOT CELL CULTURE.

ANIMAL CELL CULTURE MEDIA

YUNG-SHYENG TSAO
Schering-Plough Research Institute
Union, New Jersey

SANDRA L. GOULD
DAVID K. ROBINSON
Merck & Co., Inc.
Rahway, New Jersey

OUTLINE

Serum

Chemically Defined Basal Media and Minimal Essential Media (MEM)

Serum-Free Media

Protein-Free Media

Medium Stability and Storage

Medium Development for Large-Scale Production

Development of Serum-Free Media: A Case Study

Bibliography

At the start of the twentieth century, researchers were trying to grow mammalian cells derived from biopsies, including cancer cells, fetal cells, and primary cells from various tissues. Since that time, medium development has followed one of two approaches; researchers have made use of biological fluids, including serum, embryo extracts, and protein digests, or they have studied those fluids and tried to find simpler substitutes for them. The following sections describe the basic components of cell culture media discovered by both of these approaches. These are (1) **serum** and its components, functions and processing; (2) **chemically defined basal media** that supplement serum for cell growth; (3) **serum-free media** (or serum-reconstitution); (4) **protein-free media** (or serum-protein replacement); (5) **medium stability and storage**; (6) medium development for **large-scale production**; and (7) **a case study** for developing a serum-free medium.

SERUM

Serum is the biological fluid most commonly used to support cell growth. The choice of serum, instead of other biological fluids, was initially due to its ability to minimize the carryover of other cells and later due to its availability, low cost, and ease of storage. Serum is typically added at concentrations of 1 to 10% (v/v) to supplement the common basal media described below. Serum contains a variety of factors that facilitate cell attachment and growth, including carrier proteins, attachment regulators, defense molecules, enzymes and their regulators, growth factors, and hormones. Serum also acts to protect cells against stresses induced by shear.

The major carrier protein, **albumin**, provides 80% of the colloidal osmotic pressure of blood (1). It contains many high-affinity binding sites for a wide variety of molecules with low solubility, including trace metals, hemin, cobalamin, hydrophobic molecules (e.g., long-chain fatty acids), amphiphilic nonionic detergent-like molecules (e.g., tryptophan, thyroxine, steroids, medium-chain fatty acids), and negatively charged organic molecules (e.g., bilirubin, iopanoate, and many dye-like compounds). The binding of these molecules is highly regulated by the allosteric binding of fatty acids (2). Albumin may deliver these molecules, especially fatty acids, as well as itself (as an amino acid source), directly into cells. It may also redistribute its "cargo," including vitamins such as **thyroxine, cobalamin, and retinol** to their specialized **binding proteins** for targeting to their own destinations (3). Albumin also carries a multitude of positive and negative charges, disulfide bonds, and binding sites that serve to control pH, redox potential, and osmotic pressure.

Lipoproteins HDL, LDL, and VLDL carry very hydrophobic molecules, such as phospholipids, cholesterol and triglycerides. HDL provides phospholipids to cells and removes cholesterol from cells (except liver cells). LDL does just the opposite. HDL also induces the secretion of endothelin from endothelial cells, epithial cells, and macrophages to promote the constriction and growth of surrounding smooth muscle cells, myocytes, and fibroblasts. LDL and VLDL carry endothelin-converting enzyme (ECE), which can convert endothelin into an active paracrine factor (local hormone) (4).

Other carrier proteins include fetuin, α-2 macroglobulin, and transferrin. **Fetuin's** serves as a fetal albumin and is paramount in delivering fatty acids and phospholipids to and removing cholesterol from the cells (5). Although cholesterol retrieving efficiency is about the same as that of HDL, its fatty acid and phospholipid-delivering efficiency is about three- to fourfold that of albumin (6). **α-2 macroglobulin** carries growth hormone, insulin, and PDGF to promote cell growth. This fraction of proteins is usually difficult to separate from fetuin and may in part be responsible for the mitogenic effects of fetal protein preparations (7). **Transferrin** can deliver iron (Fe) ions directly into cells. Human transferrin is about 40 times as effective as bovine transferrin (8).

Serum also contains a variety of other factors. **Attachment factors** (fibronectin, vitronectin, gelatin- or PAI-binding proteins, heparin, and FGF-2), proteases, and their inhibitors regulate cell spreading, growth, and cell–cell interactions. *In vivo*, **defense molecules** (antibodies, complement components, mannan-binding protein, and inflammation regulators) bind to microorganisms and signal for defense. In addition, there are **free-radical scavengers**: catalase, superoxide dismutase, and glutathione peroxidase.

Commercially, sera are obtained from a number of animal species. **Calf serum** is widely used due to its availability and low cost. Although calf serum contains three- to fourfold more transferrin than other sera, because of malnutrition and other stress conditions of the calves, it often requires supplementation with iron to physiological levels (9). **Fetal bovine serum** (FBS) provides the greatest growth potential because of the high level of proteins, for example, fetuin, α-2 macroglobulin, and other associated growth factors. **Horse serum** is occasionally used in place of bovine sera. **Human serum** is sometimes preferred for the growth of human cell lines. The growth potential of serum from these various species, as well as individual lots of serum from the same species, must often be tested to select sera with optimal quality. Serum lot-to-lot variations are due not only to the differences in donor animals, but also to various treatments during and after collection (10,11).

Blood is usually collected from slaughterhouses and coagulated to remove hemoglobin, fibrin, and some complements. The resulting serum is typically filtered with 0.1- or 0.2-μm filters to remove bacteria, fungi, and mycoplasmas and then stored frozen. Some sera are additionally filtered through 0.04-μm filters to reduce the potential risk of contamination with viruses. Sometimes γ-radiation is used to further reduce the potential for viral contamination, as a number of viruses are inactivated by such treatment (12–14). These filtration and treatment processes cannot remove complement or endotoxin and are unlikely to remove or inactivate prions, which are the putative causative agents of transmissible spongiform encephalopathies (TSE). Appropriate serum collection and storage practices are critical in minimizing contamination with endotoxins. Complement, which may cause lysis of certain cell types, can be inactivated by mild heat treatment, typically by incubating serum at 56 °C for 30 min. Each of these treatment methods can reduce the growth promoting properties of the sera. Because TSE is more prevalent in certain countries, the potential for contamination of sera can be reduced by sourcing bovine serum from countries with a low incidence of TSE among livestock herds, such as the United States, Canada, Australia, and New Zealand. Traceability of donor animals is necessary to ensure approriate sourcing. Human sera can contain a number of potentially infectious agents, and appropriate testing and donor tracing is critical in these cases (15). In general, sera are tested by the vendor for the ability to support the growth of test cell lines, as well as for sterility and endotoxin and hemoglobin levels as indicators of purity. Additional testing for common adventitious agents and protein, hormone, and vitamin composition can be provided upon request by vendors (16–20).

Sometimes sera are further modified to fit particular demands. Sera are treated with charcoal/dextran to remove hormones from serum used for hormone studies. Sera are fortified with nutrients or growth hormones to reduce the serum concentration, the protein load, and sometimes the cost, without affecting cell growth (21).

CHEMICALLY DEFINED BASAL MEDIA AND MINIMAL ESSENTIAL MEDIA (MEM)

Early attempts in developing chemically defined cell culture media yielded only minimal cell growth until the addition of serum by Earle in 1954 (22). In the same period, scientists analyzed the composition of various biological fluids and understood the nutritional requirement of some mammalian cell lines. Minimal essential media (MEM) supply the minimal requirements of the basal media—the nonprotein or peptide supplements that are added to serum growing cells. In 1955, Eagle developed the first MEM that could supplement serum for growing several mammalian cell lines. In the next 10 years, more and more cell metabolites and vitamin-related compounds were identified and added to media to improve the growth of a wide variety of different cell lines. This led to various modifications (Glasgow's, Dulbecco's, etc.) of Eagle's MEM and also to the development of serum-free and protein-free media (23).

Basal media and MEM supply the basic needs of cellular metabolism, which include ions, vitamins, metabolic regulators, and metabolites. Ions are important to maintain osmotic pressure, control membrane potential and transporting activities (Na^+, K^+), coordinate intra- and intercellular activities (Mg^+, Ca^+), participate in oxidation-reduction activities (SO_4^{-2}) and energy production (PO_4^{-3}), as well as balance the H^+ ions (HPO_4^{-2}, $H_2PO_4^-$, HCO_3^-, Cl^-) (24–27). B vitamins serve as functional-group carriers of enzymes in various metabolic pathways for all cell types. Other vitamins (A, C, D, E, K) regulate cell cycle, redox potential, and differentiation of some specific cell lines. Metabolites include glucose or other sugars (as major carbon sources), and amino acids (as major nitrogen sources), as well as lipid precursors (choline, inositol, ethanolamine, etc.) and nucleic acid precursors. Sometimes, other metabolic intermediates (pyruvate, succinate, etc.) are included to facilitate cell metabolism. Chemical buffers (sodium bicarbonate,

HEPES, etc.) and pH indicators (phenol red) can be also included in MEM. Often the various basal media must be tested in combination with different sera to find the most appropriate media for growing individual cell lines.

SERUM-FREE MEDIA

To minimize complications from serum in hormone-responsive cell culture studies, in 1975–1980 Sato replaced serum with specific hormones, growth factors, fetuin, serum albumin, transferrin, and mercaptoethanol (28,29). As described above, these replacements should fulfill most of the known fundamental functions of serum. In 1981–1987, Barnes revealed that serum contains many other components that are important for cell growth: binding proteins, adhesion proteins, enzymes and inhibitors, lipid-carrying proteins, etc. (30,31). Thereafter, it became possible to grow additional cell lines in serum-free media. Because the protein concentration was also greatly reduced, some novel protein products were easily identified and isolated from the culture for characterization (28–33). Extraction from secretory glands also yielded a rich discovery of growth factors (34). Although progress in serum-free medium development was relatively slow because of the need to determine the different nutrient and protection requirements of each cell line and because of difficulty in identifying and purifying active ingredients from serum, various media have been developed for a number of individual cell lines. The concentrations of hormones or growth factors included in typical serum-free media can be found in the literature (see Ref. 35 for a review).

The advent of hybridoma culture prompted development of serum-free media for antibody production. In 1978, Iscove reported the first such serum-free medium, in which serum was replaced by albumin, transferrin, and soybean lipids, and the basal medium (DMEM) was fortified with additional amino acids and vitamins, that is, Iscove's MEM or IMEM (36). Additional serum-free media were developed for a number of other specific hybridoma or lymphoid cell lines (37–41). In 1984, Kovar and Franek developed a serum-free medium containing transferrin, insulin, ethanolamine, linoleic acid, serum albumin and trace elements that could support the growth of four hybridoma and two myeloma cell lines (42). Other components of serum-free media include mercaptoethanol, casein, catalase, cholesterol, and steroid hormones, among others.

Although growth factors, such as IGF and EGF, can activate cells to grow, these and other growth factors are usually critical only for some subsets of cells to grow. However, in cases such as the regeneration of skin, nerve, skeletal muscle, kidney, liver, or hematopoiesis, where differentiation and coordination among cells are needed, the interrelationships among these growth factors become very critical. The culture of primitive and mature hematopoietic cells in serum-free media also requires erythropoietin (EPO), interleukins and colony-stimulating factors (CSFs) (43). The culture of marrow stromal (nonhematopoietic) cells also requires adding some interleukins and CSFs. Hormones are sometimes included for growing of target tissues and cells. Some proteases (e.g., thrombin), and inhibitors (e.g., α-1 antitrypsin) may be included (44,45). In growing attachment-dependent cells into tissue-like structures, for example, nerve, or skeletal muscle, the properties of the substrata are also important (46).

The methods for culturing endocrine, epithelial, fibroblastic, neuronal, and lymphoid cells in **serum-free media** have been extensively described by D.W. Barnes, et al. (47). An **updated review** was provided by K. Kitano (35). **Strategies** for optimizing serum-free media were described by G. Hewlett (48). A **guideline for making informed choices** of serum-free medium is presented by J.P. Mather (49). Serum-free media for **hematopoietic stem cells** were reviewed by J.S. Lebkowski et al. (50).

PROTEIN-FREE MEDIA

When scientists started to develop chemically defined, serum-free cell culture media, a protein-free approach was already under consideration. There was no success until 1965, when R.G. Ham developed the protein-free medium F-12, a mildly enriched basal medium, to support the clonal growth of Chinese hamster cell lines (51). This work, however, could not be reproduced after Ham moved his laboratory to Colorado. He thought that this was due to the removal of impurities from the higher grade chemical ingredients, such as thyroxine. In 1977, he added 19 trace elements, as well as additional calcium chloride, glutamine, and cysteine. This medium, MCDB 301, supplemented with either insulin or methylcellulose, could again support growth of the cell line (52). Ham thought that the methylcellulose might have also contained some impurities. In the same period, there were many other protein-free media developed by adding a variety of nonprotein supplements (53). An adaptation period of one week to several months is typically required for cells to grow readily in these media, suggesting that factor-independent subpopulations are being selected (54,55).

Individual serum functions can also be replaced by nonproteinaceous compounds. In 1986, serum's function as a **cholesterol carrier** was replaced by a cholesterol–cyclodextrin complex (56). In 1987–1989, transferrin as an **iron carrier** was replaced by ferric citrate (57,58). In 1990, albumin's function as a **metal carrier** was replaced by EDTA, transferrin was replaced by the iron-containing dye nitroprusside (59), and lipoprotein's function as a **lipid carrier** was replaced by lipid emulsions (60). In 1995, M.J. Keen developed a protein-free medium using β-cyclodextrin to replace HDL, LDL, and albumin to deliver cholesterol from cholesterol: phosphatidyl choline liposomes to cholesterol-auxotrophic NS0 myeloma cells (61,62).

Free radicals in the cell culture media are usually generated by xanthine oxidase which is activated by heat, anoxia, inferferon, or SH oxidation in the presence of cells (63). Free radicals can also be generated by excited riboflavin through UV-activated glass or plastic surfaces, including microcarriers (64). Free radicals and the resulting peroxidation can be controlled by cystine, cysteine, glutathione, Se- or Fe-containing enzymes, α-keto

acids, pyruvate, trace elements, and vitamin E. Although vitamin C can eliminate some free radicals, it can also produce O_2^- under hyperoxia (65). Damage from free radicals may also be controlled by reducing reagents (e.g., dithiothretol), as well as ferric cyanide and metal chelators which reduce peroxide production. However, because peroxides and free radicals may play a role in regulating the cell cycle, a delicate redox balance should be maintained.

Other roles of serum have also been addressed. **Osmolarity** is usually maintained by sodium chloride, but sometimes sucrose and osmoprotective compounds (proline, glycine, sarcosine, and glycine betaine) are used (66). **Damage from excessive shear** stress is usually prevented by adding of Pluronic F-68, polyethylene glycol (PEG), dextran, polyvinyl alcohol, polyvinyl pyrrolidone, and/or Methocel (67–69). **Metal ions** are usually delivered as a simple ionizable salts. However, EDTA, citrate, or gluconate may be used to increase their solubility (57,59). **Sparingly soluble lipids and steroids** are usually delivered as an emulsions or liposomes (59,70). Cyclodextrins may alternatively be used to increase their solubility (61,62). **Cell attachment** may be modulated by adding Arg-His-Asp (RGD) peptides, heparin or polylysine (71). **Cell growth promotion** may be enhanced by adding polyamines, such as putrescine, spermidine, spermine or their precursor ornithine (72).

MEDIUM STABILITY AND STORAGE

As described in the last section, UV radiation, heat, and high oxygen tension in the medium can generate free radicals, which can damage cells. Heat can also lead to the degradation of proteins, glutamine, and reducing agents. At $37\,^\circ$C, the half life of glutamine is on the order of 1 to 4 weeks dependent on the type and pH of the medium (73). Because the degradation of glutamine also generates ammonia, the addition of glutamine to compensate for losses may not be appropriate. However, media should not be frozen because some medium components such as tyrosine and tryptophan may precipitate and remain insoluble upon freezing and thawing (74). Thus, liquid media should be kept at low temperatures (4 to $8\,^\circ$C) and away from light during storage. Ascorbic acid (vitamin C) is another labile medium component, because it can be readily oxidized (74). Serum, proteins, glutamine, and ascorbic acid can be prepared separately and added to the medium just before use.

MEDIUM DEVELOPMENT FOR LARGE-SCALE PRODUCTION

In addition to the requirement to support cell growth, cell culture media used in commercial-scale production of biologicals should possess the following properties: the media and all of its components should be readily available in sufficient quantities, preferably from multiple vendors; the media must provide reproducible lot-to-lot performance; and, as is increasingly preferred by regulatory agencies, media used to produce human therapeutics should not contain any components, particulary proteins, of animal origin. These requirements are more readily satisfied by the chemically defined, serum- and protein-free media, described before. Additionally, for cells grown in serum-free media in high-shear environments, such as found in sparged stirred tanks, the media must provide factors that protect cells from shear damage. Finally, the medium composition should be optimized to maximize culture productivity.

For most suspension cells, lethal shear damage occurs in the areas of high energy dissipation associated with the rupture of bubbles introduced by either sparging or surface vortexing (75). For attachment-dependent cells in microcarrier culture, lethal shear damage can occur from mechanical agitation as well (for a review, see Ref. 76). The addition of certain polymers, such as methylcellulose, carboxy methylcellulose, hydroxyethyl starch, polyvinyl alcohol, polyethylene glycol, and Pluronic F68, to the media can at least partially protect cells from both bubble- and agitation-induced shear damage (for reviews, see Refs. 75, 77). In addition, cells can be adapted to high-shear environments by sequentially passaging cells in small shaker or spinner flasks, for example, while gradually increasing the agitation rate (55).

In many media, the growth of cells is limited by either the depletion of essential nutrients or the accumulation of waste products to inhibitory levels. Early attempts to improve culture productivity focused on analyzing the residual levels of nutrients in spent culture media. These depleted nutrients were then added back to the culture media to increase cell growth and culture longevity. In many cases, glucose and glutamine must be controlled at low levels to avoid excessive production of lactate and ammonia, which could otherwise limit cell growth. Careful attention to the composition of media has resulted in up to tenfold increases in monoclonal antibody production, yielding final titers of up to 2 g/L (see Ref. 78 for a review). More recently, stoichiometric analysis (79) and process control (80) have been applied, resulting in final antibody titers exceeding 2 g/L.

DEVELOPMENT OF SERUM-FREE MEDIA: A CASE STUDY

In many cases, a serum-free medium may not be available commercially or may not be adequate for a particular application. Developing a serum-free medium can be time- and labor-intensive, but the following steps may help streamline the process: (1) Perform a thorough literature search of media that support the growth of the particular cell line in question or of similar cell lines (e.g., same tissue origin, same species origin) to help identify a starting point and classes of additives to test. (2) Set specific goals relevant to the process. For example, set targets for the total medium protein content, the required cell growth rate, and/or the required product titer if any. (3) Define endpoints for evaluating the effects of changes to the medium (i.e., cell growth rate, maximum cell density, growth over several passages, and growth on particular substrates such as microcarriers). The case study described here used these tools to develop a serum-free, low-protein medium, called LPKM-1, to grow Vero cells and produce rotavirus (81).

Efficient production of rotavirus requires a tryptic cleavage of one of the outer coat proteins, VP4, for efficient infection of cells in vitro (82). The presence of serum required for growing of Vero host cells quenches trypsin activity. Therefore, it was necessary to remove serum before infecting the cells. Development of a serum-free, low-protein medium eliminated the need for extra steps in the rotavirus production process. Additionally, it was required that the serum-free medium also support sequential passages of Vero cells at reproducible growth rates similar to the growth rate in serum-containing medium and product titers comparable to a serum-containing process.

Vero cells are an extensively characterized African green monkey kidney cell line which are near diploid (i.e., normal DNA content or ploidy) (83). An initial literature search revealed that Taub and Livingston (84) published a serum-free medium developed to grow both primary kidney cells and some established kidney cell lines. This formulation, called K-1, is a 1:1 mixture of Dulbecco's Modified Eagle's Medium and Nutrient Mixture F-12 (Ham) with serum components replaced by insulin, transferrin, and a mixture of three hormones (prostaglandin E1, triiodothyronine, and hydrocortisone). However, K-1 could not support Vero cell growth over multiple passages. Therefore, K-1 was supplemented with a variety of additives. These supplemented media were then tested in both single- and multifactorial experiments (85) for their ability to support the growth of Vero cells for six passages in the absence of serum. By supplementing K-1 with a single recombinant growth factor, human EGF, Vero cells could be passaged repeatedly and maintain a constant growth rate. Moreover, the addition of dexamethasone further improved maximum cell density and doubling time. Dexamethasone and EGF have been shown to act synergistically to stimulate proliferation of primary human fibroblasts (86) and to support repeated passages of human diploid fibroblast cell lines (70).

A further literature search revealed that Medium 199 and its derivative, SFRE-199-1, had been used to grow Vero cells to high cell densities in serum-free formulations (87,88). By changing the K-1 basal medium mixture from 1:1 Dulbecco's Modified Eagle's Medium and Nutrient Mixture F-12 (Ham) to 1:1 Medium 199: Nutrient Mixture F-12 (Ham), a small but consistent improvement in maximum cell density was achieved.

Although the protein concentration in K-1 was already relatively low because most of the serum functions had been replaced by adding only a low level of a few defined proteins (i.e., insulin, transferrin, rEGF), the protein concentration was further decreased by determining the necessity of each component for cell growing. It was found that transferrin was not necessary for growing of Vero cells in this medium, while insulin and rEGF were essential. The combination of both ferrous and ferric salts present in the 1:1 mixtures of F-12 and DMEM or Medium 199 presumably provided sufficient iron to support Vero cell growth in the absence of transferrin. The concentration of bovine insulin could not be decreased but was replaced by recombinantly derived human insulin.

The final medium formulation, called LPKM, contained approximately 5 mg/L total protein. Vero cells grew at 60% of the rate for serum-containing medium and grew to 60% of the maximum cell density of cells in serum-containing medium. Cell growth was maintained for 20 passages in this medium (81).

BIBLIOGRAPHY

1. U. Kragh-Hansen, *Dan. Med. Bull.* **37**, 57–84, (1990).

2. J.H.M. Droge, L.H.M. Janssen, and J. Wilting, *Biochem. J.* **250**, 443–446 (1988).

3. L. Tragardh et al., *J. Biol. Chem.* **255**, 9243–9248, (1980).

4. B. Battistini et al., *Peptides (N.Y.)* **14**, 385–399 (1993).

5. W.S.L. Liao, R.W. Hamilton, and J.M. Taylor, *J. Biol. Chem.* **255**, 8046–8049, (1980).

6. L. Kumbla, S. Bhadra, and M.T.R. Subbiah, *FASEB J.* **5**, 2971–2975, (1991).

7. P. Libby, E.W. Raines, P.M. Cullinane, and R. Ross, *J. Cell. Physiol.* **125**(3), 357–366 (1985).

8. T.O. Messmer, *Biochim. Biophys. Acta* **320**, 663–670 (1973).

9. Hyclone Catalog (1997–1999), p. 7.

10. H.K. Naito and Y.S. Kwak, *J. Am. Coll. Nutr.* **11**, 8S–15S (1992).

11. C. Charcosset, M.Y. Jaffrin, and L. Ding, *ASAIO Trans.* **36**(3), M594–597, (1990).

12. C. House, J.A. House, and R.J. Yedloutchnig, *Can. J. Microbiol.* **36**, 737–740, (1990).

13. G.A. Erickson et al., *Dev. Biol. Stand.* **70**, 59–66 (1989).

14. G. Hanson, R. Wilkinson, and J. Black, *Art Sci.* **12**, 1–4, (1993).

15. H. Vrielink et al., *Lancet* **345**, 95–96, (1995).

16. G. Hanson, MT(ASCP), and L. Foster, *Art Sci.* **16**, 1–7, (1997).

17. *Gibco BRL Products & Reference Guide* (1997/1998), pp. 4–1 to 4–5.

18. *Bio Whittaker Catalog* (1996/1997), pp. 125–133.

19. *Irvine Scientific Catalog* (1996/1997), pp. 9–11.

20. *Sigma Catalog* (1998), pp. 22–23.

21. *Hyclone Catalog* (1997–1999), pp. 6–21.

22. W.R. Earle, E.L. Schilling, J.C. Bryant, and V.I. Evans, *J. Natl. Cancer Inst. (U.S.)* **14**, 1159–1163 (1954).

23. K. Higuchi, *Adv. Appl. Microbiol.* **16**, 111–136, (1973).

24. D.E. Metzler, *Biochemistry: The Chemical Reactions of Living Cells*, Academic Press, New York, 1977.

25. J.B. Finean, R. Coleman, and R.H. Mitchell, *Membranes and their Cellular Functions*, 2nd ed., Wiley, New York, (1979).

26. L. Stryer, *Biochemistry*, 4th ed., Freeman, New York, (1995).

27. F.J. Alvarez-Leefmans, F. Giraldez, and S.M. Gamino, *Can. J. Physiol. Pharmacol.* **65**, 915–925, (1987).

28. G. Sato and L. Reid, *Int. Rev. Biochem.* **20**, 219–251, (1978).

29. D. Barnes and G. Sato, *Cell* **22**, 649–655 (1980).

30. D. Barnes, *Bio Techniques* **5**, 534–542 (1987).

31. M.C. Glassy, J.P. Tharakan, and P.C. Chau, *Biotechnol. Bioeng.* **32**, 1015–1028, (1988).

32. H. Murakami, in A. Mizrahi, ed., *Advances in Biotechnological Processes*, Alan R. Liss, New York, 1989, pp. 107–141.

33. T. Ikeda, D. Danielpour, P.R. Galle, and D.A. Sirbasku, in D.W. Barnes, D.A. Sirbasku, and G.H. Sato, eds., *Cell Culture Methods for Molecular and Cell Biology*, **2**, pp. 217–241 (1984).

34. S.D. Elson, C.A. Browne, and G.D. Thorburn, *Biochem. Int.* **8**, 427–435, (1984).

35. K. Kitano, *Bio Technology* **17**, 73–106 (1991).

36. N. Iscove and F. Melchers, *J. Exp. Med.* **147**, 923–933 (1978).

37. F.J. Darfler and P.A. Insel, *J. Cell. Physiol.* **115**, 31–36, (1979).

38. T.H. Chang, Z. Steplewski, and H. Koprowski, *J. Immunol. Methods* **39**, 369–375, (1980).

39. H. Murakami et al., *Proc. Natl. Acad. Sci. U.S.A.* **79**, 1158–1162, (1982).

40. H. Murakami, in D.W. Barnes, D.A. Sirbasku, and G.H. Sato, eds., *Methods for Serum-free Culture of Neuronal and Lymphoid Cells*, Alan R. Liss, New York, pp. 197–205, (1984).

41. W.L. Cleveland, I. Wood, and B.F. Erlanger, *J. Immunol. Methods* **56**, 221–234, (1983).

42. J. Kovar and F. Franek, *Immunol Lett.* **7**, 339–345, (1984).

43. C.E. Sandstrom, W.M. Miller, and E.T. Papoutsakis, *Biotechno. Bioeng.* **43**, 706–733, (1994).

44. M.H. Simonian and M.L. White, in eds., D.W. Barnes, D.A. Sirbasku, and G.H. Sato, *Methods for Serum-Free Culture of Cells of the Endocrine Systems*, Alan R. Liss, New York, 1984, pp. 15–27.

45. J.K. Chen, *Life Sci.* **51**, 375–380, (1992).

46. R. Singhvi, G. Stephanopoulos, and D.I.C. Wang, *Biotechnol. Bioeng.* **43**, 764–771, (1994).

47. D.W. Barnes, D.A. Sirbasku, and G.H. Sato, eds., *Cell Culture Methods for Molecular and Cell Biology*, Vols. 1–4, Alan R. Liss, New York, 1986.

48. G. Hewlett, *Cytotechnology* **5**, 3–14, (1991).

49. J.P. Mather, *Methods Cell Biol.* **57**, 19–30, (1998).

50. J.S. Lebkowski, L.R. Schain, and T.B. Okarma, *Stem Cells* **13**, 607–612 (1995).

51. R.G. Ham, *Proc. Natl. Acad. Sci. U.S.A.* **53**, 288–293, (1965).

52. R.G. Ham, *In Vitro* **13**, 537–547, (1977).

53. H. Katsuta and T. Takoako, *Methods in Cell Biol.* **6**, 1–42, (1973).

54. R.L. Tarleton and A.M. Beyer, *Bio Techniques* 590–593 (1991).

55. D.K. Robinson, V. Yabannavar, and Y. Deo, in W.J. Harris and J.R. Adair, eds., *Antibody Therapeutics*, CRC Press, New York, pp. 184–219, (1997).

56. M. Mammami, G. Maume, and B.F. Maume, *Cell Biol. Toxicol.* **2**, 41–52 (1986).

57. J. Kovar and F. Franek, *Biotechnol. Lett.* **9**, 259–264, (1987).

58. Y.-J. Schneider, *J. Immunol. Methods* **116**, 65–77, (1989).

59. F.J. Darfler, *In Vitro Cell Dev. Biol.* **26**, 769–778, (1990).

60. F.J. Darfler, *In Vitro Cell Dev. Biol.* **26**, 779–783 (1990).

61. M.J. Keen and T.W. Steward, *Cytotechnology* **17**, 203–211 (1995).

62. M.J. Keen and C. Hale, *Cytotechnology* **18**, 207–217 (1996).

63. E. Cadenas, *Annu. Rev. Biochem.* **58**, 79–110 (1989).

64. L.S. Terada et al., *In Vitro Inflammation* **14**, 217–221, (1990).

65. P.L. Gutierrez, *Drug Metab. Rev.* **19**, 319–343, (1988).

66. K. Oyaas, T.E. Ellingsen, N. Dyrset, and D.W. Levine, *Biotechnol. Bioeng.* **44**, 991–998, (1994).

67. J.D. Michaels, J.F. Petersen, L.V. McIntire, and E.T. Papoutsakis, *Biotechnol. Bioeng.* **38**, 169–180 (1991).

68. L.A. van der Pol, I. Paijens, and J. Tramper, *J. Biotechnol.* **43**, 103–110 (1995).

69. J.D. Michaels et al., *Biotechnol. Bioeng.* **47**, 407–419 (1995).

70. W.J. Bettger, S.T. Boyce, B.J. Walthall, and R.G. Ham, *Proc. Natl. Acad. Sci. U.S.A.* **78**, 5588–5592 (1981).

71. C.H. Damsky and M. Bernfield, *Curr. Opin. Cell Biol.* **3**, 777–868 (1991).

72. K.J. Gawel-Thompson and R.M. Greene, *J. Cell. Physiol.* **140**, 359–370, (1989).

73. S.S. Ozturk and B.O. Palsson, *Biotechnol. Prog.* **6**, 121–128, (1990).

74. C. Waymouth, in (D.W. Barnes, D.A. Sirbasku, and G.H. Sato, eds., *Cell Culture Methods for Molecular and Cell Biology* Vol. 1, Alan R. Liss, New York, 1986, pp. 23–68.

75. J.J. Chalmers and F. Bavarian, *Biotechnol. Prog.* **7**, 151–158 (1991).

76. J.G. Aunins and H-J. Henzler, in H.-J. Rehm and G. Reed eds., *Biotechnology, A Multi-Volume Comprehensive Treatise*, 2nd ed., VCH,Weinheim, 1993, pp. 105–126.

77. D. Chattopadhyay, J.F. Rathman, and J.J. Chalmers *Biotechnol. Bioeng.* **45**, 473–480 (1995).

78. T. Bibila and D.K. Robinson, *Biotechnol. Prog.* **11**, 1–13 (1995).

79. L. Xie and D.I.C. Wang, *Biotechnol. Bioeng.* **43**, 1164–1174 (1994).

80. W. Zhou, J. Rehm, and W.S. Hu, *Biotechnol. Bioeng.* **46**, 579–587 (1995).

81. S.L. Gould, D.J. DiStefano, and D.K. Robinson, *In Vitro*, submitted (1999).

82. M.K. Estes, D.Y. Graham, E.M. Smith, and C.P. Gerba, *J. Gen. Virol.* **43**, 403–409 (1979).

83. S.K. Swanson et al., *J. Biol. Stand.*, **16**, 311–320 (1988).

84. M. Taub and D. Livingston, *Ann. N.Y. Acad. Sci.* 406–421 (1981).

85. G.E.P. Box, W.G. Hunter, and J.S. Hunter, *Statistics for Experimenters*, Wiley, New York, (1978).

86. J.D. Bulbar, W.H. Walsh, D.H. Garney, and D.D. Cunningham, *Proc. Natl. Acad. Sci. U.S.A.* **75**, 1882–1886 (1978).

87. J.M. Clark, C. Gebb, and M.D. Hirtenstein, *Dev. Biol. Stand.* **30**, 81–91 (1982).

88. J. Litwin, *Cytotechnology*, **10**, 169–174 (1992).

See also ANIMAL CELL CULTURE, PHYSIOCHEMICAL EFFECTS OF DISSOLVED OXYGEN AND REDOX POTENTIAL; ANIMAL CELL CULTURE, PHYSIOCHEMICAL EFFECTS OF OSMOLALITY AND TEMPERATURE; ANIMAL CELL CULTURE, PHYSIOCHEMICAL EFFECTS OF pH; ASEPTIC TECHNIQUES IN CELL CULTURE; CELL CYCLE SYNCHRONIZATION; CELL METABOLISM, ANIMAL; CONTAMINATION DETECTION IN ANIMAL CELL CULTURE; ENRICHMENT AND ISOLATION TECHNIQUES FOR ANIMAL CELL TYPES; MEASUREMENT OF CELL VIABILITY; OFF-LINE ANALYSIS IN ANIMAL CELL CULTURE, METHODS; OFF-LINE IMMUNOASSAYS IN BIOPROCESS CONTROL; ON-LINE ANALYSIS IN ANIMAL CELL CULTURE; STERILIZATION AND DECONTAMINATION.

ANIMAL CELL CULTURE, EFFECTS OF AGITATION AND AERATION ON CELL ADAPTATION

JEFF CHALMERS
The Ohio State University
Columbus, Ohio

OUTLINE

INTRODUCTION

Few subjects in animal cell culture technology have elicited as much debate, discussion, and controversy as the "shear sensitivity" of animal cells. In the early days of commercial, large-scale animal cell culture, many researchers believed that such cultures were not possible. However, this did not stop a number of major biotechnology companies from pursuing the use of suspended and anchorage-dependent animal cells to produce recombinant proteins for human use. As of 1993, three of the top ten biotechnology drugs on the market (based on gross sales) were produced in animal cells. This does not include all of the vaccines made in animal cells. Needless to say, animal cells are not too "shear sensitive" to be used in large-scale animal cell culture. However, if proper procedures and techniques are not used, one can very rapidly destroy the viability of, and in some cases rupture, most of the animal cells in a bioreactor. This article discusses the current understanding of the effects of hydrodynamic forces on cells, and the methods currently used to prevent the adverse effects of these forces.

Before this discussion begins, however, the terms used to characterize hydrodynamic conditions are presented. The term "shear sensitivity" itself is somewhat misleading in that it implies that the effect of hydrodynamic forces on cells can be characterized by the familiar shear stress-strain rate relationship:

$$\tau_{yx} = -\mu \left[\frac{\partial v_x}{\partial y} + \frac{\partial v_y}{\partial x} \right] \tag{1}$$

where τ_{yx} is the shear stress with typical units of dynes/cm^2, μ is the fluid viscosity, and $\partial v_x/\partial y$ and $\partial v_y/\partial x$ are the specific shear rates which are typically reported in reciprocal seconds. However, this relationship is applicable only to well-defined laminar flow. The fluid flow in most bioprocesses is not characterized as well-defined, laminar flow but instead is characterized as turbulent flow. Depending on the length scales of interest, turbulent flow conditions can be considered to range from somewhat defined to poorly defined. In complex turbulent flow, it is more appropriate to speak of the stress tensor, which is defined by

$$\bar{\bar{\tau}} = \mu(\nabla \bar{v} + (\nabla \bar{v})^T)$$

$$= \mu \begin{vmatrix} 2\dfrac{\partial v_x}{\partial x} & \dfrac{\partial v_y}{\partial x} + \dfrac{\partial v_x}{\partial y} & \dfrac{\partial v_z}{\partial x} + \dfrac{\partial v_x}{\partial z} \\ \dfrac{\partial v_x}{\partial y} + \dfrac{\partial v_y}{\partial x} & 2\dfrac{\partial v_y}{\partial y} & \dfrac{\partial v_z}{\partial y} + \dfrac{\partial v_y}{\partial z} \\ \dfrac{\partial v_x}{\partial z} + \dfrac{\partial v_z}{\partial x} & \dfrac{\partial v_y}{\partial z} + \dfrac{\partial v_z}{\partial y} & 2\dfrac{\partial v_z}{\partial z} \end{vmatrix} \quad (2)$$

To fully measure the "state of stress" in this type of system, the values of all nine terms in the matrix in equation 2 are needed. In addition, the flow is highly transient; consequently, each of these nine terms is rapidly changing. It should also be noted that the gradient in equation 1 corresponds to one of the nine gradient terms in the matrix in equation 2.

THE EFFECT OF HYDRODYNAMIC FORCES ON CELLS

During the last 20 years, a significant amount of research has been conducted on the nonlethal and lethal effects of hydrodynamic forces on cells. These studies can be broadly divided into two general categories: the effect of hydrodynamic forces on cells of medical interest and the effect of hydrodynamic forces on cells of biotechnological interest.

Cells of Medical Interest

Although it is beyond the scope of this article to discuss all of the nonlethal, metabolic, and gene expression studies on cells of medical interest, a brief overview will be given. A 1997 review article (1) lists 18 shear-stress-mediated cellular responses in endothelial cells which line arteries and veins (In this case, equation 1 is appropriate for defining the hydrodynamic forces acting on the cells). Although not as extensively studied, seven shear-stress-mediated cellular responses have been reported in the smooth muscle cells which are associated with arteries and veins (in a layer under the endothelial cells). These responses include activation of ion channels in the membrane and activation of specific proteins associated with the membrane, decreases in intracellular pH, and stimulation of specific cellular pathways to stimulate or repress the synthesis of specific mRNA molecules. Much of this work has been motivated by a desire to understand the pathology of atherosclerosis. However, recent reports on the effect of shear stress on other types of cells in the human body, such as bone, have also been published.

Cells of Biotechnological Interest

Although far fewer studies have been conducted on the nonlethal effects of hydrodynamic forces cells of biotechnological interest, compared to cells of medical interest, significant effects have been observed.

In typical turbulently mixed bioreactors these reported hydrodynamic effects include changes in cell viability,

size, and growth (2–5); metabolism (2,6); and the surface concentration of cell surface markers (receptors) (6–8). Although it is not possible from these reports to quantify the hydrodynamic forces at a level comparable to that in studies conducted on attached cells, the studies of Al-Rubeai et al. (4) and Lakhotia et al. (3) clearly indicate that suspended murine hybridoma and Chinese hamster ovary cells are damaged, at very high levels of agitation. In addition, a viable subpopulation is "selected" which is distinct from viable cells in low-agitation cultures. This distinction was indicated by a reduction in size and a "higher proliferative state relative to the control" which recovered to the level of the control if the agitation was reduced. More recently, McDowell and Papoutsakis (6) reported that increased levels of agitation in spinner vessels resulted in a significant increase in the number of CD13 receptors on the surface of suspended human promyelocytic leukemia cells, and a concurrent increase in mRNA levels for CD13. An example of a stress-associated response was described by Mufti and Shuler (9), who established that increasing agitation rates in spinner flasks increased levels of cytochrome p450 1A1 (CYP1A1) activity in various rodent and human hepatomas attached to microcarriers.

In contrast to these studies, which were conducted in bioreactors where the hydrodynamic forces are, relatively speaking, undefined, Ranjan et al. (10) subjected four cell lines to various levels of well-defined shear stress (equation 1). Two of these four cell lines were from endothelial tissues [primary human umbilical vein endothelial cells, (HUVEC) and bovine aortic endothelial cells, (BEAC)] and two cell lines from non-endothelial tissue (HeLa cells, which originated from a malignant tumor of the human cervix, and CHO cells). All four cell lines responded to moderate levels of laminar shear stress (25 dyn/cm^2) by expressing the c-fos protein. The c-fos protein is a product of the *c-fos* gene, a protooncogene which is a member of the AP-1 family of transcriptional cofactors that mediate transcriptional stimulation. The protein kinase C pathway, an important signaling pathway, is involved in the transcriptional activation of the *c-fos* gene. The previous finding is significant, because it demonstrates that at least one important shear stress response is conserved, i.e. it is not specific to cells of hemodynamically active origin.

ULTIMATE GOAL OF SCALE-UP

The ultimate goal in scaling up animal cell culturing processes is to maintain a homogeneous environment, preferably identical to that in the bench-scale system where the animal cell culturing process was most probably developed. This is a complex problem for two main reasons: (*1*) the need to add and remove gases and (*2*) the turbulent nature of the fluid flow that is typically needed to maintain a homogeneous environment. The first problem is compounded by the low solubility of oxygen in water, and the second by the general lack of understanding of turbulence at small scales.

Relatively speaking, oxygen is poorly soluble in water. Compared to other nutrients, such as sugars and amino acids, the saturating oxygen concentration is several orders of magnitude lower. This low solubility, combined with the increasingly high cell densities becoming possible in cell culture, leads to a situation in which the cells can consume all of the oxygen in a saturated cell culture medium in less than an hour and in some cases on the order of minutes. Although this rate of consumption is much less than that in bacterial cultures, where high-density cultures can consume all of the oxygen on the order of seconds, animal cell cultures greater than several liters in scale must have a means of introducing oxygen into the system other than by simply passing air over the medium–air interface at the top of the bioreactor. In addition to supplying oxygen, it is becoming more apparent through experience that in large-scale, high-density cultures, CO_2 must be removed to prevent inhibitory effects. A number of different techniques have been proposed and used to exchange gases with cell culture media, but these methods are usually problematic compared to the classical and typically used method of gas sparging directly into the vessel.

VARIOUS SEMIEMPIRICAL SCALE-UP METHODS

Establishing homogeneous conditions in a bioreactor requires fluid mixing. Even for the most soluble nutrients or inorganic salts, it is desirable and in some cases absolutely essential that the additives to the system be rapidly mixed to prevent any substantial gradients. The most extreme case is the addition of base to large (>1000 L) bioreactors. Without relatively rapid mixing, cells can remain suspended in very high pH conditions, which is unacceptable (11).

The typical and most straightforward method of mixing in a bioreactor (≥1 liter) is by using an impeller, either of the Rushton, pitched blade, or marine impeller design (12). Each of these designs has been used in the chemical process industry for many decades, and a substantial amount of empirical knowledge has been developed (13). But the majority of this knowledge is based on mixing compounds that are not damaged by hydrodynamic forces. In addition, the flow created by these impellers is turbulent, so that from a fundamental, first principle point of view one can predict only bulk flow conditions. This lack of fundamental understanding results from the complexity of fluid turbulence. Even though it has been studied for more than a century, only slow progress has been made in the understanding and the ability to predict flow conditions (velocity fields, energy dissipation, etc.) at very small scales under turbulent conditions. This lack of knowledge prevents scale-up of bioreactors for animal, insect, and plant cell culture using purely fundamental, predictive "first principles." Consequently, scale-up is typically accomplished by using "rules of thumb" and correlational approaches, along with the integration of our slowly increasing knowledge of "first principles." Some of these "rules of thumb" and correlational approaches are presented below and the remainder of the article is devoted to the state of our current fundamental "first principles" knowledge.

Rules of Thumb for Scale-Up of Bioreactors for Animal Cell Culture

First Rule of Thumb. Generally speaking, anchorage-dependent animal cells are more "shear-sensitive" than suspended cells. This observation is based on numerous comments by industrial and academic researchers and on reports in the literature. More on this concept, from experimental (qualitative observations) and correlational viewpoints is presented later.

Second Rule of Thumb. Without the use of "shear-protective" additives, sparging will kill all of the suspended cells in a bioreactor within a relatively short time. Again, this concept will be expanded upon later.

Third Rule of Thumb. Marine impellers create less shear damage to cells than a Rushton type impeller. Less experimental data exist to justify this assumption.

Fourth Rule of Thumb. Although gas exchange can be accomplished with membranes or silicone rubber tubing, the simplest and most straightforward method is to sparge gas directly into the system.

Fifth Rule of Thumb. The "shear sensitivity" of animal cells is cell-type-, and in some cases, cell-clone-specific. Although some cell lines, such as Chinese hamster ovary cells, have proven relatively tough, other cell lines are alleged to be much less robust.

Empirical and Correlational Approaches for Scale-up

As indicated directly or indirectly in the "Rules of Thumb" listed previously, a great deal of the confusion and seemingly contradictory information about the sensitivity of animal cells to hydrodynamic forces is related to (1) whether the cells are attached to microcarriers or in suspension and (2) the interaction of animal and insect cells with gas–liquid interfaces. In particular, it is when suspended cells attached to these interfaces are subjected to high hydrodynamic forces that one observes the "shear sensitivity" so commonly discussed for suspended cells.

Bioreactor Studies. Two complementary bioreactor studies were conducted by Oh et al. (14) and Kunas and Papoutsakis (15a) which significantly support the "Rules of Thumb." Before these studies, the prevailing belief was that animal and insect cells were damaged as a result of the hydrodynamic forces which arise from the action of an impeller. This belief was well justified by the results of Croughan et al. (16) and Cherry and Papoutsakis (17) and others which clearly demonstrated that animal cells could be removed from microcarriers in stirred bioreactors. However, this prevailing opinion was challenged when Oh et al. (14) and Kunas and Papoutsakis (15) demonstrated that suspended cells can withstand hydrodynamic forces generated by impellers at significantly higher levels than are typically used in bioreactors as long as care is taken to limit cell–bubble interactions. Kunas and Papoutsakis (18) carried this work one step further by developing a vessel in which air sparging and the upper air–medium interface were removed. The removal of this interface prevented the formation of a central vortex around the impeller shaft and the associated bubble entrainment. With this system, agitation rates up to 600 rpm were achieved without major cell damage, leading the authors to state, "Only when entrained bubbles interact with a freely moving gas–liquid interface, such as exists between the culture medium and the gas headspace, does significant cell damage occur."

This observation that suspended cell damage is not caused by mechanical mixing, but is the result of gas sparging (bubbles), was confirmed by several other research groups (19–23). In summary, it was shown that it is nearly impossible to grow suspended animal or insect cells in bubble columns, airlift bioreactors, or sparged, mechanically mixed bioreactors without using protective additives. More will be said later on the use and mechanism of protection of these additives.

Correlational Approach. A correlational approach was taken by researchers in Dr. H. Trampers' laboratory to relate cell damage in gas sparged bioreactors to bubbles in vessels (24). This correlational approach takes the following mathematical form:

$$k_d = \frac{24FX}{(\pi^2 d_b^3 D^2 H)} \tag{3}$$

where k_d = first order death rate constant (h^{-1})
  F = air flow rate into vessel (m^3/s)
  X = hypothetical killing volume (m^3)
  d_b = air bubble diameter (m)
  D = column diameter (m)
  H = height of column

The key concept in this relationship is the "hypothetical killing volume," X(m^3). Through a variety of different experimental conditions including different cell types and different height-to-diameter ratios of bubble columns, it was demonstrated that the "hypothetical killing volume" is correlated to a specific volume associated with each bubble (25,26).

A second correlational approach associated with gas sparging was developed by Wang et al. (27). This correlation was based on the hypothesis that cells are damaged by breakup and/or coalescence of bubbles in bioreactors which may take place in the sparger/agitation region or at the air–medium interface. Similar to the relationship presented by Tramper et al. (24), this correlation predicts that the local, specific death rate of suspended cells can be related to a specific volume and saturation constant associated with bubbles. This relationship takes the form,

$$p = \left(\frac{k_2 s}{K}\right) a \tag{4}$$

where p = local specific cell death rate (h^{-1})
  s = equivalent thickness of the inactivation region around a deformed bubble
  a = local specific bubble interfacial area (bubble surface area/ medium volume, m^{-1})
  k_2 = intrinsic cell inactivation rate constant (cells m^{-3}h^{-1})
  K = Michaelis–Menton saturation constant for cells absorbed into an "inactivation zone" around a bubble

An implicit assumption in this relationship is that this volume, $s \bullet a$, is sufficiently close to the bubble that it is directly proportional to the surface area of the bubble. When averaged over the entire vessel, this model predicts that the first-order death rate is linearly proportional to the specific interfacial area.

Even though it does not provide a mechanism for cell damage, equations 3 and 4 provide insight into cell damage, namely, (1) it is proportional to the bubble–medium interfacial area, and (2) it is independent of the height and diameter of the vessel. Although not initially obvious, this second point proved most insightful. Tramper et al. (24) experimentally demonstrated that at a constant gas sparge rate, the total death rate in a bubble column decreases as the column height increases. This indicated that the cell damage is located in either the bubble injection region or the bubble disengagement region at the air–medium interface.

Industrial Correlational Approaches. The observation that by using appropriate surface-active additives, such as Pluronic F-68, suspended cells can be grown without apparent damage in various sizes of bioreactors, led to several contributions to the scale-up literature by industrial researchers (11,28,29). These contributions focused primarily on the proper addition of O_2 to and removal of CO_2 from a culture as the primary criteria for scale-up and operation. However, like any correlation, they apply only to systems where they have been well studied. Care needs to be taken when new cell lines or new clones are used.

Fundamental, or First Principle Approaches to Scale-Up

The ultimate method for scaling up and operating a bioreactor is to base the scale-up on fundamental, or first principle approaches. Although significant progress has been made in this direction, it is debatable whether such a goal can ever be truly achieved because most, if not all, large-scale bioreactors operate under turbulent flow conditions and it is questionable whether turbulence can ever be understood and predicted at a scale relevant to animal cell culture. Nevertheless, significant progress has been and continues to be made in both the understanding and predictability of turbulence and complex flows.

To begin this discussion, several important points must be made. First, although the typical flow pattern in an animal cell bioreactor can be characterized as turbulent, it is far from random. Specifically, "flow structures" exist which can be associated with either specific events or mechanical structures in the vessel. This concept of flow structures, or "coherent structures," is well-known in turbulence research. Secondly, although absolute knowledge of the hydrodynamic forces acting on an animal cell in a bioreactor requires solving equation 2, on a micro scale, for all locations and times in a bioreactor (which is obviously impossible), a scalar quantity, the specific energy dissipation rate ϵ, is typically used in the mixing and fluid mechanics community to quantify the magnitude of the hydrodynamic forces in the flow. In nonmathematical terms, ϵ is defined as the rate of work done on a fluid element and has units of energy per unit volume per unit time (i.e. erg/cm^3-s or J/m^3-s). Theoretically, if one were to sum specific energy dissipation over the whole mixing vessel, it would equal the energy added through the impeller (if no energy is added to the system by any other means). Other scalar measures also exist, such as the scalar deformation rate. However, ϵ is a commonly used and convenient way to compare hydrodynamic forces in many different types of flow. The remainder of this section discusses the known flow structure in bioreactors and estimates of the energy dissipation associated with these flow structures. Finally, what little is known of the relationship between specific energy dissipation rate and cell damage is presented.

Flow Structures Associated with Gas Sparging. Because it has been well documented, as outlined before,

that sparging of gas into bioreactors damages cells, Handa-Corrigan et al. (30) and later Chalmers (31) used microscopic-video imaging systems to observe cell–bubble interactions. Although no images were provided, Handa et al. (30) suggested that the majority of cell damage takes place in the bubble disengagement region at the medium–air interface (the top of the culture). This suggestion was based on observations of cells experiencing violent, turbulent oscillations and surface deformations. They also observed cells entrained in the moving bubble surface interface and transported at high velocities through the draining bubble film. Summarizing these visual observations, they suggested three mechanisms of damage: (1) damage due to shearing in draining liquid films in foams, (2) rapid oscillations caused by bursting bubbles; and (3) physical loss of the cells in the foam.

Using higher magnification and specifically designed columns, Chalmers' group (31) presented photographic images of insect cells attached to rising bubbles and cells trapped in the foam layer. Figure 1 presents images of each of these observations.

In addition to attaching to rising bubbles and becoming trapped in the foam layer, cells can also be retained in the thin film of a bubble at the gas–medium interface. Figure 2 presents images of the bubble film just after a gas bubble came to rest at the air–medium interface. In the absence of Pluronic F-68 (Fig. 2a) a large number

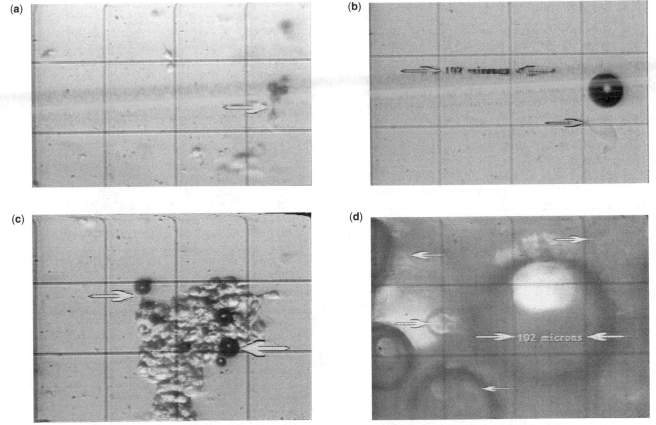

Figure 1. Microscopic video images of individual cells and clumps of insect cells attached to rising gas bubbles (**a–c**) and cells trapped in the foam layer (**d**). The bubbles appear as black spheres, and the cells are lighter spheres. Arrows indicate cell–bubble attachments (**a–c**) and cells (**d**). The distance between opposing arrowheads indicates the length scale in (**b**) and (**d**).

(a)

(b)

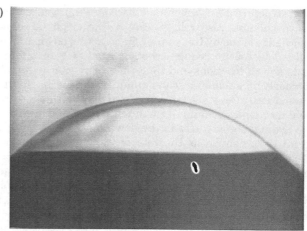

Figure 2. Microscopic video images of a gas bubble at an air–medium interface just after the bubble arrived. In (**a**) a large number of insect cells ($>10^3$) can be observed (white dots). In (**b**) Pluronic F-68 is present, and no cells can be observed on the film.

of cells ($>10^3$) can be seen attached to the bubble film. However, when Pluronic F-68 is present (Fig. 2b), no cells are present. Using visualization techniques including cell viability dyes, a third research group (32) demonstrated that cell damage is associated with the top, air–medium interface in sparged bioreactors without impellers.

Bubble Ruptures. Two fates await gas bubbles that approach the top air–medium interface: they can either become part of a previously present foam layer, or the bubbles can rupture. Both events are routinely observed in bioreactors. As shown earlier, cells can accumulate in the foam layer. It has been reported that by the end of a batch growth culture a significant amount of cells can be removed in this manner, and cell material in this foam contributes to the "bathtub ring" in a bioreactor.

To determine if enough cells are killed when a bubble ruptures to account for cell damage due to sparging when no noticeable foam forms, Trinh et al. (33) quantified the number of cells killed per bubble rupture. On average, 10^3 suspended insect cells were killed per 3.5-mm bubble rupture in the absence of Pluronic F-68 in a cell suspension of approximately 10^6 cells/mL. When Pluronic F-68 was present, no cell death was observed.

To further discern the mechanism by which cells are killed when a bubble ruptures, Trinh et al. (33) captured the upward jet that results when a bubble ruptures. The concentration of cells in this upward jet was approximately twice that in the bulk medium and all of the cells were dead. Again, when Pluronic F-68 was present in the medium, no cells were found in the upward jet. These observations led Trinh et al. (33) to suggest that the hypothetical killing volume suggested by the correlation of Tramper et al. (24) is in fact a thin layer surrounding the gas bubble that includes the adsorbed cells. It was also suggested that Pluronic F-68 prevents cells from adsorbing to the bubble, thereby preventing cell death.

Mechanism and Quantification of Bubble Rupture. The hydrodynamics of bubble rupture at a gas-liquid interface, though probably not turbulent, are complex and cannot be solved analytically. However, two research groups

conducted computer simulations, each using different methods, and obtained similar results (34,35). Figure 3 presents the predicted gas–liquid interface position and the regions of high energy dissipation as a function of specific time increments after a 0.77-mm bubble ruptures at a gas-liquid interface. Table 1 presents the total elapsed time and maximum energy dissipation rate for the rupture of three differently sized bubbles from these two simulations. For a comparison, Table 2 presents approximate rates of energy dissipation in which cell damage was reported in well-defined flow devices. As can be observed, the maximum energy dissipation rate associated with a bubble rupture is two to three orders

Table 1. The Rates of Energy Dissipation from Computer Simulations for the Rupture of Differently Sized Bubbles at a Gas–Medium Interface (34)

Bubble Diameter (mm)	Total elapsed Time for Bubble Rupture (s)	Maximum Energy Dissipation Rate (Ref. 34)	(erg/cm^3-s) (Ref. 35)
0.77	5.5×10^{-4}	9.52×10^8	—
1.77	2.0×10^{-3}	1.66×10^8	4.0×10^9
6.32	1.0×10^{-2}	9.4×10^5	8.0×10^4

Source: From Refs. 34 and 35.

Table 2. Rates of Energy Dissipation for Which Cell Damage Was Reported in Well-Defined Flow Devices

Cell Type	Instrument	Rate of Cell Damage (% min^{-1})	Rate of Dissipation (ergs/cm^3-s)
Insect	Cone and plate	33.5	3.15×10^5
Hydridoma	Concentric cylinder	3.4	2.20×10^5
Hydriboma	Double cup and bob	A	5.81×10^3
Mammalian	Capillary	16,900	4.80×10^8

[a]At 15 h cell viability was 73% (78% at time = 0) compared with 85% for a control culture.
Source: Ref. 34.

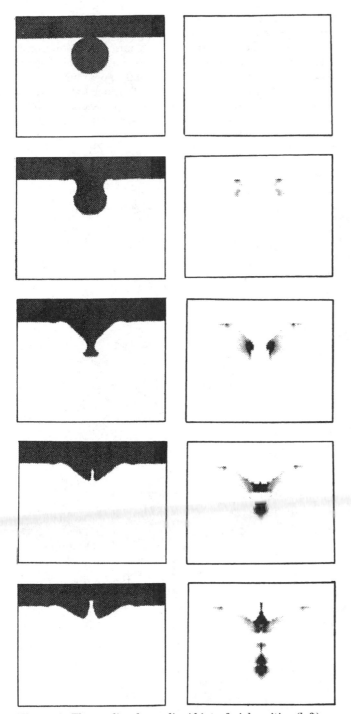

Figure 3. The predicted gas–liquid interfacial position (left), as a function of specific time increments after a 0.77-mm bubble at a gas–liquid interface ruptures, and the regions of high energy dissipation (right) associated with each time increment. The range of the gray scale is from 0 to 1×10^6 ergs/cm^3-s. From top to bottom, the time increments are 0.0, 2.0, 4.0, 4.5, and 5.5×10^{-4}s (from Ref. 34).

of magnitude higher than what has been reported as damaging cells.

Flow Structures Associated with the Impeller in Mixing Vessels. During the last 20 years, a great deal of progress has been made in understanding the flow structures in

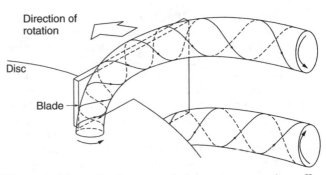

Figure 4. Schematic diagram of the vortexes coming off a Rushton impeller. Reprinted from Van't Riet et al., the trailing vortex system produced by Rushton turbine agitators, copyright 1975, p. 1093, with permission from Elsevier Science.

mixed vessels. Despite the high Reynold's number in these vessels, the flow is far from random. This was dramatically demonstrated by the work of van't Riet and Smith in the mid 1970's (36,37). Using a video camera which rotated at the same speed as the impeller in the vessel, van't Riet and Smith observed that two stable, standing vortices are convected away from the blade. A diagram (from the original publication) of these vortices coming off the Rushton blade is shown in Figure 4.

Continuing these vortex studies, detailed measurements using laser Doppler velocimetry (LDV) have been made by Yianneskis et al. (38,39), Wu and Patterson (40), Tatterson et al. (41), Stoots and Calabrese (42), and Zhou and Kresta (13,43), among others. These studies show that the flow associated with these vortices is characterized by steep mean-velocity gradients (anisotropic flow). As the vortices move outward from the blade they lose their identities and break down in a region of high turbulence which in some cases can be considered isotropic. These studies also point out that even within the impeller stream, the energy is not dissipated uniformly; rather, it can be considered to have localized regions of high fluid deformation and strong hydrodynamic processes.

In their study of the impeller stream of Rushton turbine impellers, Stoots and Calabrese (42) point out that even if the turbulent nature of the flow is not considered, one arrives at isolated flow regions where the local energy dissipation rates (calculated from mean-velocity data) exceed those based on impeller power draw per unit mass by almost an order of magnitude. Further, Wu and Patterson (40) estimated that the turbulent *energy dissipation rates* in the trailing vortices were about 20 times greater than the dissipation rate in the bulk of the tank. Similar results were also obtained by Ranade and Joshi (44) and Kresta and Wood (45) in their studies of pitched blade turbines.

Additionally, Stoots and Calabrese (42) generated detailed, three-dimensional maps of the flow in and around the vortices generated by a Rushton impeller. To generate these maps for one particular rotational speed, they recorded time-varying flow data at 19,260 spatial locations. These maps, some on the order of 1 mm in resolution, provide time-averaged velocity data. Although absolute values of energy dissipation cannot be calculated from these maps, (because they are based on time-averaged data) significant estimates can be made of hydrodynamic properties, including energy dissipation.

These LDV studies are near the maximum resolution, practically speaking, that one can obtain using single-point measurements, such as LDV. An alternative to LDV is to use three-dimensional particle tracking velocimetry (3-D PTV), which allows full-field measurements of turbulent flow conditions, as opposed to single-point measurements obtained with LDV. Using this technique, Venkat et al. (46,47) reported on the flow structures in a 250-ml spinner vessel and 2-liter and 20-liter bioreactors. Although these initial studies were at relatively low resolution, clear flow structures were observed in the vessels. Figure 5 presents cartoons of the flow structure observed in the spinner vessel, a 2-liter bioreactor, and a replica of a 20-liter bioreactor.

Attempts to Quantify the Energy Dissipation in Impeller-Associated Flow Structures. A number of different correlations have been proposed to predict the conditions (size, location, and operating rpm of impeller) at which animal cell damage occurs as a result of mixing. The most highly quoted and well known is the Kolmogoroff eddy length hypothesis put forward by Croughan et al. (16,48) and Cherry and Papoutsakis (17,49). This correlation, based on the Kolmogoroff theory of isotropic turbulence (50), was the first attempt to relate energy dissipation to cell damage. From Kolmogoroff's theory, an eddy length scale can be determined for which it is assumed that a majority of the energy associated with the turbulence is dissipated. If this length scale is on the order of a cell diameter (10 microns) or that of a microcarrier with cells attached (200 microns), cell damage is expected. The actual length scale is given by

$$\eta = \left(\frac{\upsilon^3}{\varepsilon}\right)^{1/4} \tag{5}$$

$$\varepsilon = \frac{P}{\rho_f V} \tag{6}$$

$$P = N_p \rho_f n^3 d_i^5 \tag{7}$$

In these equations η is the Kolmogoroff length scale (m), υ the kinematic viscosity (m²/s), ε the specific power dissipation (m²/s³), P the power consumed (W), V the dissipation volume (m³), N_p the power number (dimensionless), ρ_f the fluid density (kg/m³), n the impeller rpm (rev/s), and d_i the impeller diameter (m).

To use these relationships in bioreactor design, Croughan et al. (16,48) and Cherry and Papoutsakis (17,49) suggested that the designer choose operating conditions, so that the characteristic length scale η is greater than the cell diameter in suspended cell culture or microcarrier diameter in suspended microcarrier cultures.

In small-scale cultures (several liters and smaller), good agreement was observed between this approach and experimental studies. However, for larger scale systems this approach becomes problematic because of several limitations. First, central to this correlation is the volume V in which the turbulent energy is dissipated. Unfortunately, one can make only educated guesses as

to the value of this volume. Secondly, it has been shown that the maximum rate of turbulent energy dissipation occurs at length scales greater than the Kolmogoroff microscale (46). A third limitation is that the Kolmogoroff theory of isotropic turbulence, strictly speaking, applies only to isotropic turbulence. By definition, true isotropy requires that there be no directional preference in flow. However, as shown before with respect to flow structures in mixing vessels and bioreactors, there is significant three-dimensional flow, or anisotropic conditions. Despite

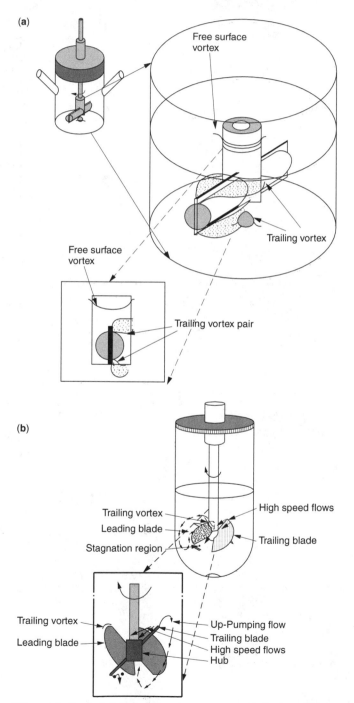

Figure 5. Cartoons of the flow structure around the impeller in a 250-mL spinner vessel (**a**), a 2-liter bioreactor (**b**), and a replica of a 20-L bioreactor (**c**) (from Ref. Chalmers, 47).

(c)

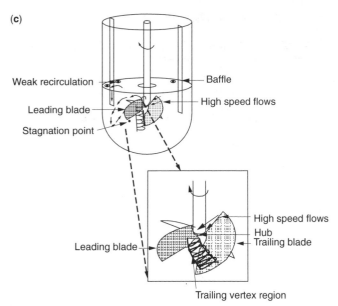

Figure 5. *Continued.*

this limitation, local isotropy can be assumed under the appropriate conditions. A more complete discussion of isotropy in mixing vessels can be found in a review article on turbulence in stirred tanks by Kresta (51)

Conservation of Energy Dissipation Around Impellers upon Scale-Up or Scale-Down. Based on the understanding that has been gained about the distribution of energy dissipation in mixing vessels, especially around the impeller, research is being conducted to compare the energy dissipation distribution around different types of impellers and to determine how this distribution changes upon scale-up. Specifically, Zhou and Kresta (13,43) compared the energy dissipation magnitude and distribution around three commonly used impellers (Rushton, pitched blade, and airfoil) for the same tank geometry and power input. In addition, nondimensional groups were developed which allow one to scale-up or scale-down a mixing vessel, so that the magnitude and distribution of energy dissipation can be conserved. Complementing these studies, Kresta (51) has written a review article that discusses the relationship of the local rate of dissipation of turbulent kinetic energy to various mixing processes and reviews various methods to estimate ε.

Use of Energy Dissipation as a Parameter to Quantify the Hydrodynamic Forces Acting on a Cell. A fundamental assumption of the last several sections has been that animal cell damage can be associated with the transient exposure of a cell to high levels of energy dissipation. This is a reasonable assumption, nevertheless but it has not been experimentally proven. This lack of a well-proven model relating a hydrodynamic property (ies) to cell damage/death was also recognized and discussed in a review article by Thomas and Zhang (52).

Despite the lack of a clear understanding of the hydrodynamics in bioreactors, researchers have been subjecting suspended animal cells to various types of well-defined, or pseudo well-defined hydrodynamic forces

in an attempt to relate those forces to cell damage/death. Table 2, referred to earlier in the section on bubble rupture, is a summary of some of those studies. However, all of those studies, except for the study using a capillary, were conducted over relatively long periods of time (order of minutes), and it is assumed that the "local" regions of high energy dissipation in bioreactors are present on the order of fractions of a second. For example, with respect to bubble rupture, the high levels of energy dissipation take place on the order of 10^{-3} seconds.

On the most fundamental basis, the rate of energy dissipation is only a scalar value used in an attempt to quantify highly complex, three-dimensional flow. Garcia-Briones and Chalmers (53) presented a model in which they concluded that two parameters should be used to determine the potential of a particular hydrodynamic condition to damage cells. The first parameter, the *state of stress of the fluid*, is similar to the energy dissipation rate, and the second parameter, the flow classification R_D, is a measure of the possibility of stress relaxation of the fluid element. This model was partially motivated by the work of Taylor (54), who showed that the breakup of oil drops in water is a function of the extensional characteristics of the flow. A flow with strong extensional characteristics broke up the drops at a lower level of the *state of stress in the fluid* than did a purely shear flow. This model proposes interesting possibilities, but it has yet to be experimentally tested.

Use of Medium Additives to Prevent Cell Damage/Death

A number of reasons exist for adding nonnutritive additives to cell culture media. Three of the most common purposes are (1) to deliver hydrophobic compounds, (such as lipids), to the cells; (2) to prevent or reduce foam formation; and (3) to prevent cell damage. As discussed before, at least three distinct mechanisms have been identified for cell alteration/damage/death in bioreactors: (1) the death of suspended cells attached to rupturing gas bubbles; (2) the actual entrapment of cells in a permanent foam layer at the air–medium interface; and (3) the alteration/damage of suspended cells and the actual removal of cells attached to microcarriers by hydrodynamic forces.

A number of additives have been used to address each of these purposes and to offset the various types of cell damage. However, addition of these additives, historically and currently, tends to be empirical with respect to the type and concentration. This empiricism is the result of a past lack of knowledge of cell damage mechanisms and our lack of understanding of the complex interfacial phenomena within bioreactors (a majority of additives are surface active).

A kinetic and thermodynamic approach to bubble rupture, has been taken by two different research groups to begin to understand and rank the effectiveness of different additives in preventing cell damage (55,56). Both groups concluded that the additives which rapidly (on the order of a second) and significantly (>10 ergs/cm^2) lowered the medium interfacial tension prevented cell-to-bubble attachment most effectively. In terms of specific additives, both groups agreed that Pluronic F-68, polyvinyl alcohol (PVA) and Methocel were the most effective.

With respect to cells trapped in the foam layer, Michaels et al. (57) used an experimental technique similar to foam floatation and ranked additives in terms of their effectiveness in preventing suspended cells from becoming trapped in the foam layer. This ranking from least effective to most is polyvinyl pyrrolidone, polyethylene glycol (PEG), serum-free medium with no additives, medium with 3% serum, Pluronic F-68, and Methocel A15LV.

The third type of cell damage (hydrodynamic forces without gas–liquid interfaces present), and the use of additives to prevent such damage, is much more difficult to quantify. Nevertheless, significant observations have been made. In particular, McDowell et al. (58) reported that adding Methocel A15LV, polyethylene glycol, and polyvinyl alcohol reduced quantifiable cellular responses to substantial increases in rotational speed in bioreactors. These quantifiable cellular responses included levels of specific cell-surface markers and glucose consumption and lactate production rates. Other research groups have also indicated that serum protects cells in studies which are not as clearly defined as the example cited.

CONCLUSIONS

In summary, a great deal of experience has been obtained over the years in culturing animal cells in large-scale bioreactors. Most of this experience has been obtained by empirical observations which have led to "rules-of-thumb" approaches to scale-up and operation of animal cell bioreactors. This approach and the complex nature of turbulent flow and interfacial phenomena, led to numerous misconceptions and seemingly contradictory reports. However, partially due to the commercial interest in animal cell culture and improvements in analytical approaches and technology, well controlled, quantitative studies have been and continue to be applied to these questions, and answers are beginning to emerge. With these continued studies, practitioners will be able to rely more on sound scientific principles and less on "rules-of thumb."

BIBLIOGRAPHY

1. M. Papadaki, S. Eskin, and *Biotechnol. Prog.* **13**, 209–221 (1997).

2. M. Al-Rubeai, S.K. Oh, R. Musaheb, and A.N. Emery, *Biotechnol. Lett.* **12**, 323–328 (1990).

3. S.K. Lakhotia, K.D. Bauer, and E.T. Papoutsakis, *Biotechnol. Bioeng.* **40**, 978–990 (1992).

4. M. Al-Rubeai, R.P. Singh, M.H. Goldman, and A.N. Emery, *Biotechnol. Bioeng.* **45**, 463–472 (1995).

5. M. Al-Rubeai, R.P. Singh, A.N. Emery, and Z. Zhang, *Biotechnol. Bioeng.* **46**, 88–92 (1995).

6. C. McDowell and E.T. Papoutsakis, *Biotechnol. Bioeng.* **60**, 239–250 (1998).

7. M. Al-Rubeai, A.N. Emery, S. Chalder, and M.H. Goldman, *J. Biotechnol.* **31**, 161–177 (1993).

8. S. Lakhotia, K.D. Bauer, and E.T. Papoutsakis, *Biotechnol. Bioeng.* **41**, 868–877 (1993).

9. N.A. Mufti and M.L. Shuler, *Biotechnol. Prog.* **11**, 659–663 (1995).

10. V. Ranjan, R. Waterbury, Z. Xiao, and S.L. Diamond, *Biotechnol. Bioeng.* **49**, 383–390 (1996).

11. S. Ozturk, *Cytotechnology* **22**, 3–16 (1996).

12. A. Nienow, *Appl. Mech. Rev.* **51**, 3–32 (1998).

13. G. Zhou and S.M. Kresta, *Trans. Inst. Chem. Eng.* **74**, 379–389 (1996).

14. S.K.W. Oh, A.W. Nienow, M. Al-Rubeai, and A.N. Emery, *J. Biotechnol.* **12**, 45–62 (1989).

15. K.T. Kunas and E.T. Papoutsakis, *J. Biotechnol.* **15**, 57–70 (1990).

16. M.S. Croughan, J.F. Hamel, and D.I.C. Wang, *Biotechnol. Bioeng.* **29**, 130–141 (1987).

17. R.S. Cherry and E.T. Papoutsakis, *Bioprocess. Eng.* **1**, 29–41 (1986).

18. K.T. Kunas and E.T. Papoutsakis, *Biotechnol. Bioeng.* **36**, 476–483 (1990).

19. A. Handa, A. Emery, and R.E. Spier, *Dev. Biol. Stand.* **66**, 241–252 (1987).

20. J. Tramper, J. Williams and D. Joustra, *Enzyme Microb. Technol.* **8**, 33–36 (1986).

21. B. Maiorella, D. Inlow, A. Shauger, and D. Harano, *Bio/Technology.* **6**, 1406–1410 (1988).

22. D.W. Murhammer and C.F. Goochee, *Bio/Technology.* **6**, 1411–1415 (1988).

23. D.W. Murhammer and C.F. Goochee, *Biotechnol. Prog.* **6**, 391–397 (1990).

24. J. Tramper, J.D. Smit, and J. Straatman, *Bioprocess. Eng.* **3**, 37–41 (1988).

25. D.E. Martens, C.D. de Gooijer, E.C. Beuvery, and J. Tramper, *Biotechnol. Bioeng.* **39**, 891–897 (1992).

26. I. Jobses, D. Martens, and J. Tramper, *Biotechnol. Bioeng.* **10**, 801–814 (1990).

27. N.S. Wang, J.D. Yang, and R.V. Calabrese, *J. Biotechnol.* **33**, 107–122 (1994).

28. W. Zhu et al., *Cytotechnology* **22**, 239–250 (1996).

29. D.R. Gray et al., *Cytotechnology* **22**, 65–78 (1996).

30. A. Handa-Corrigan, A.N. Emery, and R.E. Spier, *Enzyme Microb. Technol.* **11**, 230–235 (1989).

31. J.J. Chalmers, *Cytotechnology* **15**, 311–320 (1994).

32. D. Orton and D.I.C. Wang, Ph.D. Thesis, Massachusetts Institute of Technology, Cambridge, 1993.

33. K. Trinh, M. Garcia-Briones, F. Hink, and J.J. Chalmers, *Biotechnol. Bioeng.* **43**, 37–45 (1994).

34. M.A. Garcia-Briones, R.S. Brodkey, and J.J. Chalmers, *Chem. Eng. Sci.* **49**, 2301–2320 (1994).

35. J.M. Boulton-Stone and J.R. Blake, *J. Fluid Mech.* **154**, 437–466 (1993).

36. K. van't Riet, W. Bruijn, and J.M. Smith, *Chem. Eng. Sci.* **30**, 1093 (1975).

37. K. van't Riet, W. Bruijn, and J.M. Smith, *Chem. Eng. Sci.* **31**, 407–412 (1976).

38. M. Yianneskis, Z. Popiolek, and J.H. Whitelaw, *J. Fluid Mech.* **175**, 537–555 (1987).

39. M. Yianneskis and J.H. Whitelaw, *Trans. Inst. Chem. Eng.* **A71**, 543–550 (1993).

40. H. Wu and G.K. Patterson, *Chem. Eng. Sci.* **44**, 2207–2221 (1989).

41. G.B. Tatterson, H.S. Yuan, and R.S. Brodkey, *Chem. Eng. Sci.* **35**, 1369–1375 (1980).

42. C.M. Stoots and R.V. Calabrese, *AIChE J.* **41**, 1–11 (1995).

43. G. Zhou and S. Kresta, *AIChE J.* **42**, 2476–2490 (1996).

44. V.V. Ranade and J.B. Joshi, *Trans. Inst. Chem. Eng.* **A68**, 19–32 (1990).

45. S.M. Kresta and P.E. Wood, *Chem. Eng. Sci.* **48**, 1771–1794 (1993).

46. R. Venkat, L.R. Stock, and J.J. Chalmers, *Biotechnol. Bioeng.* **49**, 456–466 (1996).

47. R.V. Venkat and J.J. Chalmers, *Cytotechnology* **22**, 95–102 (1996).

48. M.S. Croughan, J.F. Hamel, and D.I.C. Wang, *Biotechnol. Bioeng.* **32**, 975–982 (1987).

49. R.S. Cherry and E.T. Papoutsakis, *Biotechnol. Bioeng.* **32**, 1001–1014 (1988).

50. A.N. Kolmogoroff, *Dokl. Acad. Sci.* USSR **30**, 301 (1941).

51. S. Kresta, *Can. J. Chem Eng.* **76**, 563–576 (1998).

52. C.R. Thomas and C.R. Zhang, in E. Galins and O.T. Ramirez, eds., *Advances in Bioprocess Engineering II*, Kluwer Academic Publishers, Dordrecht, The Netherlands, 1998, pp. 137–170.

53. M.A. Garcia-Briones and J.J. Chalmers, *Biotechnol. Bioeng.* **44**, 1089–1098 (1994).

54. G.I. Taylor, *Proc. R. Soc. London, A Ser.* **146**, 501–523 (1934).

55. D. Chattopadhyay, J. Rathman, and J.J. Chalmers, *Biotechnol. Bioeng.* **48**, 649–658 (1995).

56. J.D. Michaels et al., *Biotechnol. Bioeng.* **47**, 407–419 (1995).

57. J.D. Michaels et al., *Biotechnol. Bioeng.* **47**, 420–430 (1995).

58. C.L. McDowell, R.T. Carver, and E.T. Papoutsakis, *Biotechnol. Bioeng.* **60**, 251–258 (1998).

See also ANIMAL CELL CULTURE, PHYSIOCHEMICAL EFFECTS OF DISSOLVED OXYGEN AND REDOX POTENTIAL; BIOREACTOR CULTURE OF PLANT ORGANS; BIOREACTOR SCALE-DOWN; BIOREACTOR SCALE-UP; BIOREACTORS, AIRLIFT; BIOREACTORS, CONTINUOUS CULTURE OF PLANT CELLS; BIOREACTORS, MIST; ON-LINE ANALYSIS IN ANIMAL CELL CULTURE.

ANIMAL CELL CULTURE, PHYSIOCHEMICAL EFFECTS OF DISSOLVED OXYGEN AND REDOX POTENTIAL

GOVIND RAO
Medical Biotechnology Center and
Department of Chemical and
Biochemical Engineering
Baltimore, Maryland

MARCO A. CACCIUTTOLO
CYNTHIA OLIVER
MedImmune, Inc.
Gaithersburg, Maryland

OUTLINE

OXYGEN IN ANIMAL CELL CULTURE

Oxygen Requirements and Supply

Molecular oxygen is used by aerobic cells primarily to produce energy via oxidative phosphorylation. Oxygen is incorporated into the cells and reaches the mitochondria by diffusion. The rate of oxygen transfer into the cells depends on the concentration gradient across the cell membrane. Concentrations of dissolved oxygen (DO) in the cytoplasm of hepatocytes have been measured in vitro at 2 to 5 μM (1), which is equivalent to 1 to 2% of air saturation at 37 °C. (Air saturation refers to the amount of oxygen dissolved in the liquid phase at equilibrium with air at the given conditions of temperature and pressure. Table 1 lists the values for oxygen partial pressure in units commonly encountered in the literature over a range of oxygen pressures, and may be used to convert from one unit to another by simple linear interpolation.) This value of cytoplasmic dissolved oxygen is very similar to the reported critical concentration of oxygen required for mammalian cell survival (2).

Because of the relatively low solubility of oxygen in culture media at the conditions used to grow animal cells, efficient aeration mechanisms are required for the large-scale culture of animal cells in the commercial production of therapeutic products. In general, animal cells are more fragile than bacteria because of their lack of a cell wall, thereby necessitating gentler means of mixing. This challenge has prompted the development of very sophisticated agitation and gas delivery systems (3). By the introduction of surface-active agents, such as Pluronic F-68, damage to animal cells due to shear and bubble rupture has been minimized (4). In general, the physicochemical characteristics of the liquid–gas interface, the temperature, and the overall rate of mass transfer strongly influence the efficiency of oxygen delivery (5). Devices, such as spargers, baffles, and low-shear impellers, are commonly employed to supply oxygen to cultures in stirred tank bioreactors. Other bioreactor systems, such as air-lift bioreactors, employ direct gas sparging into the culture, thereby simultaneously accomplishing gas delivery and mixing.

Measurement of Oxygen Concentration and Uptake in Cell Cultures

For industrial applications, the level of oxygen in a culture is determined by the measurement of dissolved oxygen in the bulk phase of the culture fluid. The most common method for measuring DO in cultures is using polarographic oxygen sensors. Oxygen diffuses

Table 1. Equivalency Between Percentage of Dissolved Oxygen and Oxygen Partial Pressure

% DO (Air sat.)	% DO (Oxygen sat.)	Torr (mmHg)	Millibar	Atm.
0.00	0.00	0.00	0.00	0.00
20.00	4.18	31.77	42.35	0.04
50.00	10.45	79.42	105.88	0.10
100.00	20.90	158.84	211.77	0.21
478.47	100.00	760.00	1013.25	1.00

across a Teflon membrane and is reduced at a platinum cathode. The amount of current generated at the cathode is proportional to the oxygen diffusion rate across the membrane and is the basis for the DO measurement. These sensors are placed in the culture medium and can withstand steam sterilization. They measure the partial pressure of oxygen, which is related to the concentration of oxygen in the liquid phase through Henry's law:

$$pO_2 = H^*x$$

where H is Henry's solubility constant (atm* mol H_2O/mol O_2), pO_2 is the partial pressure of oxygen in the gas phase (atm), and x is the mole fraction of dissolved oxygen in the liquid phase (mol O_2/mol H_2O).

This relationship is valid only for low dissolved oxygen concentrations, and the values of H for different temperatures can be found in the literature (6). In addition, one should note that the presence of other solutes in water can significantly decrease the solubility of oxygen in water (5).

It is important to appreciate exactly what a DO sensor measures. The measurement is partial pressure not absolute concentration. In other words, a DO sensor calibrated in distilled water with nitrogen and air to read 0% and 100%, respectively, will read the same values in cell culture medium in equilibrium with nitrogen and air (at the same temperature and pressure), despite the differing absolute solubility of oxygen in the two liquids.

Polarographic oxygen sensors are prone to drift with time during the cultivation because of membrane fouling and electrolyte consumption. Recent developments in sensor technology may obviate these difficulties and provide the basis for more robust DO measurement techniques. For instance, optical sensors capable of noninvasive measurements are currently under investigation (7). These probes may eliminate the need for in situ steam sterilization, reduce the number of penetrations in the bioreactor vessel, provide off-line capability to troubleshoot the probes, and increase sensor lifetime. A picture of a system for noninvasive measurements is shown in Figure 1. Here, as described by Randers-Eichhorn et al. (7), an optical fiber bundle is held in line with an oxygen-sensing patch. The patch consists of an oxygen-sensitive ruthenium complex immobilized in silicone rubber. DO is measured by the change in fluorescent lifetime of the ruthenium complex in response to the dissolved oxygen pressure. Both headspace and dissolved oxygen pressures can be monitored and measured continuously and noninvasively. Figure 2 shows the setup for monitoring DO in a T-flask. DO profiles in a

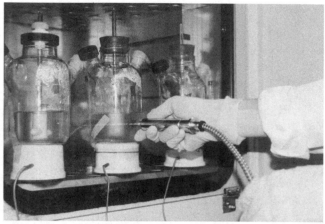

Figure 1. An optical sensor can be used to monitor multiple bioreactor units. The fiber bundle simply has to be placed in front of the oxygen-sensing patch to read the oxygen level in the spinner flask. Two patches are shown, one for the headspace and one for the liquid phase. Note that the readings can be made through the glass door of the incubator. See also Ref. 9 for more information.

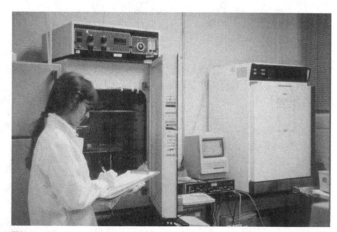

Figure 2. Picture showing oxygen sensor reading oxygen levels in a T-flask. The fiber bundle has two arms, one for excitation light from a modulated blue LED and the other to gather the emission for detection on a PMT. The lock-in amplifier detects the phase angle difference between the excitation and emission signals, and the data are viewed on a host computer. For further details see Ref. 7.

hybridoma culture in a T-flask with the cap open are shown in Figure 3. Evidently, the culture becomes hypoxic in about 40 hours and remains severely oxygen-limited through most of the culture period. This occurs even though the cap of the T-flask remains cracked open and

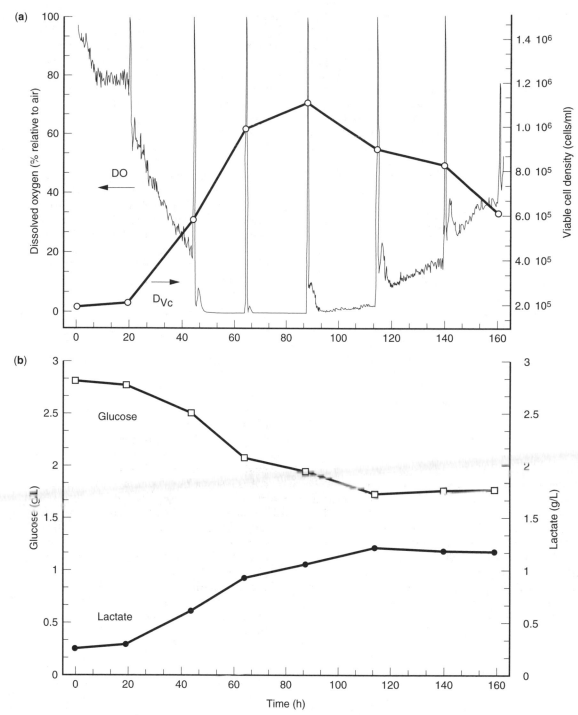

Figure 3. Data from T-flask oxygen measurement experiments. In this experiment the T-flask was operated with the cap cracked open. (**a**) The dissolved oxygen and viable cell count. Spikes in DO values are caused by mixing during sampling. (**b**) Glucose and lactate concentrations.

results from the liquid film resistance to mass transfer. Oxygen limitation apparently results in active anaerobic metabolization, as evidenced by the significant quantities of lactate formed (Fig. 3).

As can be seen in Figure 1, spinner flasks (and virtually any other transluscent system) can be equipped with sensor patches that can be scanned from the outside. Thus, one instrument could be multiplexed,

and a large number of spinners/roller bottles/T-flasks could be monitored at modest cost. In addition to DO, dissolved CO_2 measurements using the same technology have been recently reported (8). pH, glucose, and other analytes are also amenable to measurements based on this technology (9). Such sensing technologies are likely to be commercially available shortly and should revolutionize the current state of the art. Now, on-line measurements are generally made only in stirred tank bioreactors, not in other systems.

The measurement of oxygen uptake rate (OUR) provides a good estimate of the intracellular oxygen demand in cases where the oxygen supply rate is constant. Active animal cell cultures deplete oxygen from the medium at a rate that is proportional to the cell mass, provided that no other metabolite is limiting. With this premise, OUR has been utilized as a tool to estimate cell mass and to control cell culture processes (10–12).

Optimization of Oxygen Concentrations for Animal Cell Culture

Animal cells used to express protein products should be cultivated within an optimal range of pO_2. The partial pressure of oxygen in the blood stream ranges between 20 torr in venous blood and 70 torr in arterial blood (1). This is approximately equivalent to 15 and 45% of air saturation, respectively. Indeed, values between 10 and 50% of air saturation have been shown as optimal for the in vitro culture of mammalian cells (13,14). A range between 20 and 70% has been reported as optimum for maximum volumetric antibody accumulation by hybridomas (13,14). Under normal physiological oxygen levels, or *normoxia*, cell cultures can tolerate transient oxidative stress for short periods of time (a few hours) (15).

The effects of oxygen tensions outside the range considered as normoxia (10 to 100% of air saturation) are discussed in the following sections.

Hyperoxia

Elevated oxygen concentrations (usually above 100% of air saturation), or *hyperoxia*, have a negative effect on cell animal growth (15–18) and bacterial metabolism (19). Hyperoxia can damage cellular macromolecules, such as DNA (20,21), enzymes (22), other proteins (23–25), lipids, and cell membranes (20,26). The toxicity of oxygen is associated with the production of intracellular reactive oxygen species (ROS) (16,17,27,28), that damage a variety of cellular targets by a variety of mechanisms. When the levels of ROS exceed the levels of intracellular antioxidants, the result is a condition called oxidative

stress (29). The importance of oxidative stress in biological systems and its consequences have been extensively studied (30,31). The effects of oxidative stress become apparent after extended exposure (weeks) (32,33). Little information is available on the effects of hyperoxic oxygen levels on cell physiology in continuous cultures or perfusion cultures, that may run for months. There have been observations that some cell lines adapt to increasing levels of dissolved oxygen (34,35), but there is little information on the effect of hyperoxia-adapted cells on product expression or product quality. One may conjecture that genetic instability would be a problem under these conditions. In a recent report, Jan et al. (35) examined steady-state continuous cultures of hybridomas at elevated DO levels (up to 150%) and found lower steady-state cell concentrations at higher DO levels.

The following cascade of ROS formation and conversion results from single electron steps in the reduction of molecular oxygen to water:

$$O_2 \xrightarrow{-0.33\,\text{V (36, 37)}} O_2^{\bullet-} \xrightarrow{+0.94\,\text{V (37)}} H_2O_2 + Fe(II)$$
$$\text{oxygen} \qquad\qquad \text{superoxide}$$
$$\text{mol } O_2$$

$$\xrightarrow{+0.38\,\text{V (37)}} HO^{\bullet} + H_2O \xrightarrow{+2.33\,\text{V (37)}} H_2O$$
$$\text{hydroxyl radical}$$

The formation of these deleterious ROS, the rule rather than the exception, is a consequence of the unyielding second law of thermodynamics even under normoxic conditions, especially around actively respiring mitochondria and other microsomal fractions (27). Reactive oxygen species have properties that make them particularly harmful. These are listed in Table 2. Therefore, it is recommended that the cell culture practitioner make every attempt to avoid culture conditions that exacerbate oxidative stress. One should also pay careful attention to the medium formulation to keep an adequate antioxidant level and minimize levels of pro-oxidant species. Care should also be taken to avoid combinations of metal ions and reducing agents that could produce ROS through a redox cycling mechanism. For example, ferrous ions and ascorbate promote hydroxyl radical formation (30).

Table 2. Characteristics of Reactive Oxygen Species

	Lifetimes (s)	Physiological concentration	Diffusion distances (Å)	Diffusion constant (cm/min) (M)
$O_2^{\bullet-}$	10^{-6} (38)	3×10^{-10} (38)	It cannot cross membranes (37)	10^{-11}–10^{-12} (27)
$HO^{\bullet}$	3.7×10^{-9} (39)	?	$60^{(39)}$	?
H_2O_2	Enough to diffuse (27) intramitochondrial	2×10^{-5} (40) human epithelial lens cells	It can cross membranes$^{(30)}$	0.04 Erythrocyte plasma membranes$^{(27)}$
		2×10^{-4} (40) human cataract lens cells		0.20 Peroxisomal membrane$^{(27)}$
		1×10^{-6} (41) rat lives		0.02–0.42 water and other cell membranes$^{(27)}$
		10^{-7}–10^{-9} (27) intramitochondrial		

Table 3. Cellular Defenses for Removing Dioxygen Reduction Products

Product	Removal system and location in cells
$O_2^{\bullet-}$	Cu/Zn superoxide dismutase (SOD) in cytoplasm (eukaryotes)
	MnSOD in cytoplasm (prokaryotes)
	FeSOD and MnSOD in various organelles
H_2O_2 ($R-O_2H$)	*Catalase* in vesicles
	Peroxidases in vesicles
	Glutathione peroxidase found in cytoplasm (two-thirds) and mitochondrial matrix (one-third)
	Vitamin E in membranes
$HO^{\bullet}$	Almost impossible to eliminate once formed, but several compounds react very rapidly with it.

Source: Ref. 30.

Ascorbate readily reduces Fe(III) to Fe(II), which reacts with hydrogen peroxide to produce $HO^{\bullet}$ and Fe(III). The Fe(III) undergoes re-reduction to Fe(II), and the cycle continues.

As a consequence of the constant assault by ROS, evolution has provided defense mechanisms to cells in the form of enzymes and antioxidants, listed in Table 3 (38–41). These defense systems are induced in a matter of minutes (40) and are part of a broad stress response system. In continuous culture at 100% DO, Jan et al. (35) reported and adaptive response, where cells showed dramatically higher activities of several antioxidant enzymes.

Hypoxia

Even though there are beneficial effects on cell viability at oxygen levels below 10% of air saturation, or *hypoxia*, (32,42), oxygen may become the limiting nutrient for cell metabolism. Mitochondrial activity depends strongly on oxygen levels below the range of 2 to 10% of air saturation in the bulk phase, depending on the cell line (13,14). This may reduce the specific throughput of secreted proteins, thus decreasing the volumetric accumulation of product, and/or may extend the production phase by slowing cell metabolism. Thus, although hypoxia is considered beneficial in delaying cell senescence and minimizing chromosomal damage (43), these effects may not be as critical as ensuring appropriate delivery of oxygen to all cells in a bioreactor to avoid starvation or low product yields. Large-scale bioreactors are commonly operated at 50% of air saturation to avoid anaerobic conditions in the system. However, poor mixing in bioreactors generally ensures that uniform oxygen concentration is unlikely and that cells are much more likely to encounter zones of widely varying oxygen concentration. Scale-down studies, where the large-scale bioreactor environment is replicated at the laboratory scale, may provide more insights into better operating strategies. In contrast to the fermentation literature (44), not much published work exists on scale-down studies or regime analysis in animal cell culture.

Intriguing details are emerging on cellular responses to hypoxia at the level of gene expression (45). Although the responses of cells to hyperoxic stress have long been known, a unique transcription factor that is expressed only in hypoxic cells has recently been identified. This transcription factor is called hypoxia inducible factor 1 (or HIF-1). Its levels have been shown to respond quantitatively and inversely to the ambient oxygen tension. HIF-1 is thought to be turned on by an oxygen sensor that is believed to be a hemoprotein. Among the targets of HIF-1 expression are the genes for anaerobic metabolism. In addition, genes regulating angiogenesis, vasodilation, and erythropoetin and tyrosine hydroxylase synthesis (to increase breathing) are affected. For cell culture, it is likely that the switching on of anaerobic metabolism is the most significant consequence of hypoxia. Indeed, the widespread observation of increased lactate formation under oxygen-limiting conditions supports this contention. In addition, these findings also call for greater care during cell culture operations that result in oxygen deprivation, and also avoiding exposing cells to great variations in DO that may result in undesirable induction of hypoxic or hyperoxic stress responses.

Effects of Oxygen Concentration on Product Expression

Product expression levels are not dramatically affected by high levels of DO. For example, specific antibody production is maximal over a broad range of dissolved oxygen concentrations between 20 and 70% of air saturation (13,14). Although there is not a significant effect of elevated oxygen concentrations on the levels of secreted protein even at 100% of air saturation (15), the integrity of the molecule may be affected. In general, negative effects of hyperoxia on proteins may not be easily detected unless oxidation affects functional activity or physicochemical properties, such as cleavage, denaturation, or aggregation of the molecule (24,25). Therefore, it is likely that little or no benefit would be obtained for product expression at oxygen concentrations greater than 100% of air saturation. Jan et al. (35) observed that a higher DO in a chemostat resulted in a higher specific antibody production rate in hybridomas, but a concomitant lower cell concentration resulted in no significant change in volumetric productivity.

Hypoxic pO_2, on the other hand, has a marked effect on cell metabolism at oxygen pressures below 10% of air saturation (14). Lactic acid production per cell increases dramatically due to increased contribution of glycolysis to the total intracellular pool of ATP (14). The ratio of glucose to glutamine consumption rates increases with lower pO_2 (14), also indicating a stronger dependency on glycolysis to support cell functions.

In general, it is believed that normoxic levels of pO_2 have more beneficial effects on cell physiology (46) and on product expression in hybridomas (13,14).

Although it has been shown that the use of normoxic pO_2 consistently gives the best results in most animal cell culture applications, it is very difficult to predict the long-term effects of elevated oxygen concentrations for different cell lines because of the complex nature of the toxic effects of oxygen and its consequences, as well as the varying degree of cellular defenses in different cell lines. The origin of many of the observed spontaneous mutations (32,47) and loss of productivity in hybridomas are not easily explained and may be associated with the continuous exposure to reactive by-products of oxygen metabolism (16,20). At this stage of the technology, empirical observations are still necessary to fine-tune optimal oxygen levels in cell culture processes.

REDOX POTENTIAL IN CULTURES OF ANIMAL CELLS

Culture Redox Potential (CRP)

The culture redox environment has been of particular interest in cell culture applications because of its potential utility for monitoring cell-related events. Every time a chemical reaction involves the transfer of electrons, the species that donates the electrons is oxidized, and the species that receives the electrons is reduced. Because redox reactions involve transferring electrons, the change in redox potential is expressed in units of electrical potential, usually millivolts. In general, low CRP values indicate the net accumulation of reduced species, and high CRP values indicate the net accumulation of more oxidized species.

However, in many cases the interpretation of this measurement is rather difficult because it represents the net ratio of all of the reduced and oxidized species in the system and is also affected by DO and pH changes. In most cases, the individual species are not always clearly identifiable. In addition, the CRP does not necessarily represent the intracellular redox potential (IRP) of the cells, which may be the key parameter for assessing culture performance.

Measurement and Use of CRP in Cell Cultures

CRP can be measured directly in cell cultures by using a steam-sterilizable redox probe that is simple in design — a platinum electrode with an internal reference electrode is used and connected to a pH/mV meter. CRP measurements are prone to interference mainly by pH and dissolved oxygen (48,49). In some cases, redox electrodes have provided better results than oxygen electrodes for controlling oxygen levels in bioreactors (50,51). Differences in medium composition and/or culture conditions may also affect the absolute values of CRP obtained.

Attempts have been made to relate the measurements of CRP with hybridoma cell density and oxygen uptake rate (51,52). These follow several similar studies with bacterial and fungal cultures (53–57). The CRP profile from batch cultures of hybridomas correlate negatively with cell density (52). Changes in CRP occurred even when dissolved oxygen was controlled at a fixed value throughout the run, indicating that this change in CRP was possibly the result of the net accumulation of reduced species in the culture during the active cell growth phase. This correlation has been used to the estimate cell density in the culture on-line (52).

The direct relationship between measured extracellular CRP and the intracellular redox potential (IRP) is difficult to ascertain. Measurements of IRP using intracellular radiolabeled probes show clear differences among the different subcellular compartments, and the cytoplasm is in a more reducing state of IRP than the endoplasmic reticulum (58). This compartmental difference in the redox state within the cells may affect the proper folding of complex proteins, which occurs in the more oxidizing environment of intracellular organelles in the secretory pathway (58).

Not much information is available on the effects of CRP on cultured mammalian cells or on product expression. Typical values reported for cell cultures range from $+40$ mV (52) to -60 mV (58). On the other hand, intracellular values of redox potential for hybridomas range from -225 mV in the cytosol to between -145 mV and -178 mV in the secretory pathway (58). These values of IRP suggest the possibility that extremely low values of intracellular pO_2 are present in hybridomas, according to the linear relationship between CRP values and the logarithm of dissolved oxygen concentration (48). Intracellular values of pO_2 of about 1 to 2% of air saturation have been measured in hepatocytes (1). However, a more likely explanation is that the low IRP values are caused by high concentrations of reduced glutathione. This large difference between CRP and IRP may reflect large differences in the type and concentration of redox species between the medium and the cells. It also suggests that cells can maintain values of IRP within a very narrow range. Little is known of the relationship between IRP and CRP, how CRP affects IRP, or what types of molecules added to the culture will change the value of IRP. Levels of CRP outside of these reported values have not been investigated systematically, and therefore it is not possible to determine whether there is an optimum CRP for cell growth or product expression. However, recent work has shown that reducing agents added to the culture medium reverse the apoptosis-mediated cell death of sinusoidal mononucleocytes that occurred after isolation from human liver (59). This may indicate that the benefit of culturing cells at low CRP by adding reducing agents in the culture medium is possibly due to retarded oxidation of medium components. This observation also agrees with the beneficial effects of low pO_2 on cell growth and viability, as discussed in the previous section. CRP can be used to monitor low values of pO_2 more accurately than using an oxygen probe (50). Furthermore, it appears that using the information from both OUR and CRP profiles provides more robust control strategies by allowing discrimination between real metabolic events and operational failures (51). It is recommended that more cell culture practioners utilize any extra port on their bioreactors for monitoring CRP to gather further insights into its relationship with cellular metabolic activity and to exploit its apparent utility as a process control tool.

ACKNOWLEDGMENTS

Financial support from NSF, NIH, GENENTECH, MERCK and PFIZER to GR is gratefully acknowledged.

BIBLIOGRAPHY

1. D.P. Jones, *Am. J. Physiol.* **250**, C663–C675 (1986).

2. B. Chance, *Fed. Proc., Fed. Am. Soc. Exp. Biol.* **16**, 671–680 (1957).

3. R. Wagner and J. Lehmann, *Trends Biotechnol.* **6**, 1011–1014 (1988).

4. J.J. Chalmers, in M.L. Shuler, H.A. Wood, R.R. Granados, and D.A. Hammer eds. *Baculovirus Expression Systems and Biopesticides*, Wiley-Liss, New York, 1995, pp. 175–204.

5. J.E. Bailey and D.F. Ollis, *Biochemical Engineering Fundamentals*, 2nd ed., McGraw-Hill, New York, 1986.

6. P.E. Liley, R.C. Reid, and E. Buck, in R.H. Perry and D.W. Green, eds., *Perry's Chemical Engineer's Handbook*, McGraw-Hill, New York, 1984, pp. 3–103.

7. L. Randers-Eichhorn, R.A. Bartlett, D.D. Frey, and G. Rao, *Biotechnol. Bioeng.* **51**(4), 466–478 (1996).

8. Q. Chang, L. Randers-Eichhorn, J.R. Lakowicz, and G. Rao, *Biotechnol. Prog.* **14**, 326–331 (1998).

9. S.B. Bambot, J.R. Lakowicz, and Govind Rao, *Trends Biotechnol.* **13**, 106–115 (1995).

10. R. Dorresteijn et al., *Biotechnol. Bioeng.* **50**, 206–214 (1996).

11. K.T.K. Wong, L.K. Nielsen, P.F. Greenfield, and S. Reid, *Cytotechnology* **15**, 157–167 (1994).

12. D.W. Zabriskie. in M. Moo-Young ed., *Comprehensive Biotechnology*, Vol. 2, Pergamon, Oxford, 1985, pp. 175–190.

13. W.M. Miller, C.R. Wilke, and H.W. Blanch, *J. Cell. Physiol.* **132**, 524–530 (1987).

14. S.S. Ozturk and B.O. Palsson, *Biotechnol. Prog.* **6**, 437–446 (1990).

15. M.A. Cacciuttolo, L. Trinh, J.A. Lumpkin, and G. Rao, *Free Radical Biol. Med.* **14**, 167–170 (1993).

16. A.K. Bolin, A.J. Fisher, and D.M. Carter, *J. Exp. Med.* **160**, 152–166 (1984).

17. B. Halliwell and J.M.C. Gutteridge, *Biochem. J.* **219**, 1–14 (1984).

18. E. Meilhoc, K.D. Wittrup, and J.E. Bailey, *Bioprocess. Eng.* **5**, 263–274 (1990).

19. R.G. Forage, D.E.F. Harrison, and D.E. Pitt, in M. Moo-Young, ed., *Comprehensive Biotechnology*, Vol. 1, Pergamon, Oxford, 1986, pp. 256–257.

20. B.N. Ames, in Castellani ed., *DNA Damage and Repair*, Plenum, New York and London, 1989, pp. 291–298.

21. H. Joenje, *Mutat. Res.* **219**, 193–208 (1989).

22. C.G. Cochrane, I.U. Schraufstatter, P.A. Hyslop, and J.H. Jackson, in P.A. Cerutti, I. Fridovich, and J.M. McCord, eds., *Oxy-radicals in Molecular Biology and Pathology*, Liss, New York, 1988, pp. 125–136.

23. A. Levy, L. Zhang, and J.M. Rifkind, in P.A. Cerutti, I. Fridovich, and J.M. McCord, eds., *Oxy-radicals in Molecular Biology and Pathology*, Liss, New York, 1988, pp. 11–25.

24. K.J.A. Davies, *J. Biol. Chem.* **262**, 9895–9901 (1987).

25. C.N. Oliver, *Arch. Biochem. Biophys.* **253**, 62–72 (1987).

26. C. Richter, *Chem. Phys. Lipids* **44**, 175–189 (1987).

27. B. Chance, H. Sies, and A. Boveris, *Physiol. Rev.* **59**, 527–605 (1979).

28. T. Paraidathathu, H. De Groot, and J.P. Kehrer, *Free Radical. Biol. Med.* **13**, 289–297 (1992).

29. H. Sies, in H. Sies, ed., *Oxidative Stress*, Academic Press, Orlando, Fla., 1986, pp. 1–8.

30. B. Halliwell and J.M.C. Gutteridge, *Free Radicals in Biology and Medicine*, Clarendon Press, Oxford, 1991.

31. M. Martinez-Cayuela, *Biochimie* **77**, 147–161 (1995).

32. M. Meuth, *Biochim. Biophys. Acta* **1032**, 1–17 (1990).

33. G.E. Holmes, C. Bernstein, and H. Bernstein, *Mutat. Res.* **275**, 305–315 (1992).

34. H. Joenje, J.P. Gille, A.B. Oostra, and P. van der Valk, *Lab. Invest.* **53**, 420–428 (1985).

35. D.C.H. Jan, D.A. Petch, N. Huzel, and M. Butler, *Biotechnol. Bioeng.* **54**, 153–164 (1997).

36. A.U. Khan, *Photochem. Photobiol.* **28**, 615–627 (1978).

37. W. Pryor, *Annu. Rev. Physiol.* **48**, 657–667 (1986).

38. Fridovich, *Annu. Rev. Biochem.* **44**, 147–159 (1975).

39. R. Roots and S. Okada, *Radiat. Res.* **64**, 306–320 (1975).

40. Spector, N.J. Kleiman, R.-R. Huang, and R.-R. Wang, *Exp. Eye Res.* **49**, 685–698 (1989).

41. M. Hoffman, A. Mello-Filho, and R. Meneghini, *Biochim. Biophys. Acta* **781**, 234–238 (1984).

42. A.A. Lin and W.M. Miller, *Biotechnol. Bioeng.* **40**, 505–516 (1992).

43. L. Packer and K. Fuehr, *Nature (London)* **267**, 423 (1977).

44. A.P.J. Sweere, K.C.A.M. Luyben, and N.W.F. Kossen, *Enzyme Microb. Technol.* **9**, 386–398 (1987).

45. K. Guillemin and M.A. Krasnow, *Cell (Cambridge, Mass.)* **89**, 9–12 (1997).

46. W. Taylor, R. Camalier, and K. Sanford, *J. Cell. Physiol.* **95**, 33 (1978).

47. O.W. Merten, in A.O. Miller, ed., *Advanced Research on Animal Cell Culture Technology*, Kluwer Academic Publishers, Dordrecht, 'The Netherlands' 1987, pp. 367–100.

48. S.P. Ozturk, R. Mutharasan, and Govind Rao, *Handbook of Anaerobic Fermentation*, Dekker, New York, pp. 1187–1206.

49. A. Ishizaki, H. Shibai, and Y. Hirose, *Agric. Biol. Chem.* **38**, 2399–2406 (1974).

50. J.R. Birch and R. Arathoon, in A.S. Lubiniecki, ed., *Large-scale Mammalian Cell Culture Technology*, Dekker, New York, 1990, p. 263.

51. A.E. Higareda, L.D. Possani, and O.T. Ramirez, *Biotechnol. Bioeng.* **56**, 555–563 (1997).

52. K. Eyer and E. Heinzle, *Biotechnol. Bioeng.* **49**, 277–283 (1996).

53. S.K. Dahod, *Biotechnol. Bioeng.* **24**, 2123–2125 (1982).

54. W.B. Armiger, in M. Moo-Young ed., *Comprehensive Biotechnology*, Vol. 2, Pergamon, Oxford, 1985, pp. 133–148.

55. S.C.W. Kwong and Govind Rao, *Biotechnol. Bioeng.* **38**, 1034–1040 (1991).

56. S.C.W. Kwong, L. Randers, and Govind Rao, *Biotechnol. Prog.* **8**, 576–579 (1992)

57. T.H. Lee, Y.K. Chang, B.H. Chung, and Y.H. Park, *Biotechnol. Prog.* **14**, 959–962 (1998)

58. C. Hwang, A. Sinskey, and H.F. Lodish, *Science* **257**, 1496–1502 (1992).

59. K. Kinoshita et al., *J. Hepatol.* **26**(1), 103–110 (1997).

See also ANIMAL CELL CULTURE, EFFECTS OF AGITATION AND AERATION ON CELL ADAPTATION; ANIMAL CELL CULTURE, PHYSIOCHEMICAL EFFECTS OF pH; BIOREACTOR SCALE-DOWN; BIOREACTOR SCALE-UP; BIOREACTORS, AIRLIFT; BIOREACTORS, MIST; BIOREACTORS, STIRRED TANK; CELL METABOLISM, ANIMAL; ON-LINE ANALYSIS IN ANIMAL CELL CULTURE.

ANIMAL CELL CULTURE, PHYSIOCHEMICAL EFFECTS OF OSMOLALITY AND TEMPERATURE

KARIN ØYAAS
The Norwegian Pulp and Paper Research Institute
Trondheim
Norway

OUTLINE

INTRODUCTION

Today mammalian cell cultures are being used to produce several important health care products, including vaccines, hormones, growth factors, immunoregulators, and monoclonal antibodies. The very large quantities of these complex mammalian products needed has brought about a search for effective production methods. Research has concentrated on two main strategies, one of which has focused on increasing the cell density and thus the final product titer of the culture broth. Another strategy for improving the production processes has focused on increasing the specific productivity, thus forcing each cell to make products at high rates.

Optimization of product formation in a reactor requires a knowledge of the reaction system. Although animal cells are of considerable commercial importance, cell growth and product formation are not well understood; it is unclear what environmental factors affect these processes and over what range of levels these factors are important. Thus an understanding of the reaction system and a knowledge of the dependence of the cellular reaction rates on the local environmental conditions should enable optimization of reactor operating conditions to achieve maximum product formation.

Several reports have indicated that the rate of product formation by mammalian cells may be increased when the cells are exposed to conditions that are not optimal for cell growth. The application of several simple types of environmental stress, among them osmotic and temperature stress, have been shown to increase the specific rates of product formation by several mammalian cell lines. In the following discussion regarding osmolality and temperature effects, our focus will be on the growth and product formation of animal cells in culture (i.e., in vitro).

OSMOLALITY

Animal cells are sensitive to changes in medium ionic strength and osmolality (1). In animal cell processes, medium osmolality is a largely uncontrolled and ignored process parameter, which was defined mainly by a medium composition optimized for cell growth in the 1950s. Medium osmolality may change in the course of the growth cycle as a result of several factors, including the accumulation of metabolic products, or pH control with the addition of acid or base. Furthermore, in designing processes for high cell density culture, changes in growth medium formulations may involve increased levels of essential nutrients, and the corresponding increase in metabolites. Thus medium osmolality may be an important process variable in designing and improving animal cell processes.

Tolerance and Stress

All metabolically active cells contain 85–95% water, and any environmental factor that affects the activity, structure, or physical state of water poses a threat to life (2). Most cell culture media used in animal cell processes are designed to have an osmolality in the range of 270–330 mOsmol/kg, which is known to be quite acceptable for most cells (3). Osmolalities exceeding this level are designated hyperosmotic, while lower osmolalities are designated hypoosmotic.

Cell Growth. Hyperosmotic stress may be imposed as a sudden shock or be introduced gradually, allowing the cells to adapt to higher osmolalities. For a number of mammalian cell lines, abrupt hyperosmotic stress, caused by the addition of compounds such as inorganic salts or sugars, has been shown to suppress growth rate and maximum cell density, while extending culture longevity (4–6). Furthermore, when hybridoma cells were repeatedly grown in a hyperosmotic medium, the specific growth rate improved gradually during the first batches, stabilizing at a higher level than that obtained during abrupt osmotic stress at the same osmolality (7).

The responses of animal cells to hypoosmotic stress have been less extensively studied. However, reduced growth rates and cell densities have been reported for hybridoma cell lines grown in hypoosmotic media (8).

Osmoprotection—Osmoprotective Compounds. Osmoprotective processes have been shown to take place in prokaryotes, simple eukaryotes (e.g., yeast), and whole plants and animals exposed to osmotic stress (9–11). However, despite the diversity of organisms, and the varied and complex stresses that may be experienced, only a small number of fundamental adaptive strategies are followed in virtually all cases (2,10).

Exposure of cells to osmotic stress drives water into or out of the cell, causing an immediate swelling or shrinking. This initial volume perturbation is usually succeeded by a volume regulatory phase in which cells tend to return toward the volume they had in the isotonic medium. The volume regulatory phase is accomplished by modulation of membrane transport and/or metabolic pathways that alter the concentration of intracellular inorganic and organic solutes, collectively referred to as osmolytes (12–14). Little is known about the coordination between inorganic and organic osmolyte regulatory mechanisms. However, some results on unicellular algae suggest that readily available inorganic ions mediate short-term volume regulation while long-term maintenance depends on the accumulation of organic solutes (13). In the following discussion, the focus will be on the contributions of organic osmolytes in the cell volume regulatory response.

Since cell membranes are incapable of sustaining any sizeable osmotic-pressure differences, the osmotic pressure inside cells is close to that of the surroundings. Through homoisosmotic regulation, an animal's excretory system (kidneys in mammals) keeps the osmolality of the body fluid within certain limits. Therefore, most mammalian cells are not normally exposed to osmotic variation or high salt concentrations, and relatively little attention has been given to organic osmolytes in mammals. However, the cells of the mammalian excretory system experience high and varying extracellular concentrations of salt and urea. Through isosmotic intracellular regulation, the volume of these cells can be kept constant by regulation of the total number of moles of intracellular solutes (15,16).

Sugars (e.g., glucose, mannose, sucrose, trehalose), polyhydric alcohols (e.g., glycerol, mannitol, myo inositol), free amino acids (e.g., proline, glycine, alanine), methylamines (e.g., glycine betaine, sarcosine, glycerophosphorylcholine), and urea may be involved in the cell volume regulation process in both procaryotic and eucaryotic organisms exposed to water stress; see Table 1 (2,10,13,17).

When the osmoprotecive compound glycine betaine was included in the hyperosmotic culture medium, abruptly stressed mouse hybridoma cells were able to survive and grow at significantly higher medium osmolalities than cells exposed to the same medium in the absence of osmoprotective compounds (see Fig. 1). Similar, although less pronounced, effects were observed if sarcosine, proline, or glycine was added to the hyperosmotic growth medium (17). The inclusion of osmoprotective compounds affected both growth rates and cell densities.

Osmoprotective compounds may be synthesized in vivo by certain mammalian tissues. Glycerophosphorylcholine (GPC), glycine betaine, sorbitol, and myo-inositol have been found in rabbit and rat kidney cortexes and medullae (18,19).

Product Formation. The application of several simple types of environmental stress, among them osmotic stress, has been shown to affect product formation by animal cells. Abrupt hyperosmotic stress has been shown to increase the specific monoclonal antibody production rate, q_{MAb}, for a number of hybridoma cell lines (4,7,20–22). Furthermore, hyperosmotic stress caused by the addition of KI, NaI, or NaCl was found to stimulate the production of IL-8 by

Table 1. Distribution of Intracellular Organic Solutes Found in Various Organisms

Organic solute	Organism or cell type
Sugars; Monosaccharides (e.g., glucose, mannose, fructose)	Cyanobacteria, diatoms, fungi
Disaccharides (e.g., sucrose, trehalose)	Cyanobacteria, eubacteria, algae, plants, insects, crustaceans
Polyols (e.g., sorbitol, inositol, arabitol, mannitol, glycerol, glucosylglycerol)	Cyanobacteria, algae, fungi, plants, insects, crustaceans, mammalian renal cells, mammalian brain cells, mouse hybridoma cell
Amino acids and amino acid derivatives [e.g., proline, glutamate, taurine, ectoine, γ-aminobutyric acid (GABA)]	Eubacteria, plants, marine invertebrates, elasmobranch and teleost red blood cells, cyclostome fishes, Ehrlich ascites cells, mouse hybridoma cells
Methylamines [e.g., glycine betaine, sarcosine, trimethylamineoxide (TMAO), glycerophosphorylcholine]	Cyanobacteria, eubacteria, plants, marine invertebrates, marine cartilaginous fishes, coelacanth, mouse hybridoma cells
Urea	Mammalian renal cells, marine cartilaginous fishes, coelacanth, amphibians, lungfishes, snails
β-Dimethylsulfoniopropionate	Unicellular marine algae, marine macroalgae

Sources: Refs. 2,10,13, and 17.

cultured human peripheral blood mononuclear cells (23). The specific tPA production rate by recombinant CHO cells was also increased at elevated osmolalities and high pCO_2 levels in the growth medium (24).

However, enhanced specific productivities resulting from hyperosmotic stress is cell line specific (25,26). Furthermore, as cell growth is depressed at higher osmolalities, the enhanced productivities may not result in a substantial increase in the final product concentration during batch culture. However, when osmoprotective compounds were simultaneously included in a hyperosmotic hybridoma growth medium, high growth rates and specific antibody productivities were maintained at high medium osmolalities, giving significantly increased antibody titers as compared to those obtained in control culture (Fig. 2) (5).

Enhanced volumetric monoclonal antibody production resulting from hyperosmotic stress was also demonstrated in a perfusion system using calcium alginate-immobilized hybridoma cells (7). Through a gradual increase in osmolality, high cell densities were obtained while maintaining enhanced q_{MAb} of the immobilized cells. Furthermore, hybridoma cells were also demonstrated to maintain enhanced q_{MAb} as well as improved growth

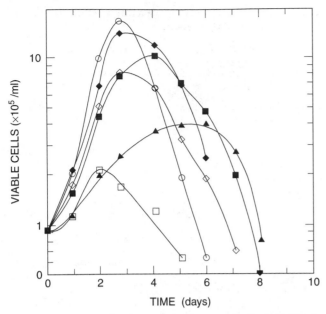

Figure 1. Growth of mouse hybridoma cell line 6H11 in NaCl-stressed growth media in the presence and absence of glycine betaine (15 mM). (○): Control, 330 mOsmol/kg, (◇): 60 mM NaCl added, 450 mOsmol/kg, (◆): 60 mM NaCl and 15 mM glycine betaine added, 465 mOsmol/kg, (□): 100 mM NaCl added, 510 mOsmol/kg, (■): 100 mM NaCl and 15 mM glycine betaine added, 525 mOsmol/kg, (▲): 140 mM NaCl and 15 mM glycine betaine added, 610 mOsmol/kg (Ref. 5).

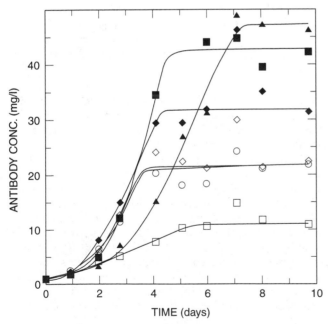

Figure 2. Monoclonal antibody production during growth of mouse hybridoma cell line 6H11 in NaCl-stressed growth media in the presence and absence of glycine betaine (15 mM). (○): Control, 330 mOsmol/kg, (◇): 60 mM NaCl added, 450 mOsmol/kg, (◆): 60 mM NaCl and 15 mM glycine betaine added, 465 mOsmol/kg, (□): 100 mM NaCl added, 510 mOsmol/kg, (■): 100 mM NaCl and 15 mM glycine betaine added, 525 mOsmol/kg, (▲): 140 mM NaCl and 15 mM glycine betaine added, 610 mOsmol/kg (Ref. 5).

rate after adaptation to hyperosmotic stress (27). Results reported by Yang et al. (6), indicate that, compared to the control, hybridoma cells respond to hyperosmotic media by maintaining a higher q_{MAb} for a prolonged period throughout the stationary phase.

While much work has been performed on the responses of animal cells to hyperosmotic stress, there are fewer reported studies on the responses to hypoosmotic stress. However, the effect of hypoosmotic stress resulting from NaCl subtraction on antibody production in two murine hybridoma cell lines has been studied (8). These cells, which had previously been shown to give increased cell specific antibody productivity, q_{MAb}, during hyperosmotic stress, did not display enhanced q_{MAb} when subjected to hypoosmotic stress.

Product Glycosylation

A variety of culture conditions may alter the glycosylation pattern of proteins (28). Kimura and Miller (29) studied the glycosylation of CHO-derived recombinant tissue plasminogen activator (tPA) produced at increased osmolality caused by the buildup of bicarbonate ions due to elevated pCO_2. Their results indicated that few, if any, changes occured in glycosylation site occupancy, expression of high-mannose oligosaccharides, or the distribution of surface charges on tPA in response to elevated pCO_2 and thus osmolality. However, both cell growth and tPA production were affected. In contrast to this, Ingrosso et al. (30) investigated the effects of repeated resealing of erythrocytes on the methyl acceptability of endogenous membrane proteins. Their results suggest that osmotic stress to the membrane of resealed erythrocytes may be responsible for increased protein methylation due to the appearance of new binding sites or an increased accessibility of existing sites.

Gene Expression

The molecular mechanisms involved in the responses of mammalian cells to osmotic stress have been most extensively studied in renal cell lines, since the renal medulla is the only tissue routinely exposed to a greatly hyperosmotic extracellular fluid. Here extracellular hypertonicity has been shown to increase the expression of genes involved in a generalized stress response, including the early response genes (Erg-1 and c-fos) (31), heat shock protein genes (α-B crystallin and HSP70) (31–33), and genes that encode proteins (specific enzymes and transporters) directly involved in the metabolism and transport of organic osmolytes (34–39). More specifically, hypertonicity elevates the abundance of aldose reductase, which is used for synthesis of the osmolyte sorbitol from glucose and increases the transcription of betaine, inositol, and taurine transporters used for the uptake of these osmolytes (40). The regulation and role of the heat shock proteins in osmotic regulation seems to be the subject of considerable ongoing research.

Programmed Cell Death — Apoptosis

The cause of culture death in bioreactors could be either apoptosis, in which the cells actively participate in their own death process, or necrosis, in which extreme environmental stress leads to catastrophic cell injury. In a study on SH-SY5Y human neuroblastoma cells, it was shown that these cells die by apoptosis under conditions of hyperosmotic stress (41).

TEMPERATURE

Temperature is a key environmental variable that may affect a large variety of animal cell parameters, ranging from growth and product formation to shear sensitivity and product quality. Fermentor temperature during animal cell cultivation is largely controlled at 37°C in order to simulate normal body temperature. Thus temperature has, until recently, received little attention as a process optimization variable (42). However, recent findings indicate that temperature manipulation may have several advantages in the optimization of animal cell processes.

Tolerance and Stress

Temperature stress has been reported to affect a large number of cellular functions in animal cells grown in culture. Table 2 attempts to give a short overview of some temperature effects reported in the literature (42–55).

Cell Growth and Cell Cycle. In a study using HeLa cells, Rao and Engelberg (43) found that exponential growth could be maintained only from 33 through 40°C. A downshift of temperature from 37 to 31–33°C prolonged the total generation time of cultured mammalian cells. In another study using human amnion cells, temperature was shown to affect all parts of the cell cycle and in ways that automatic of each period (52). However, the G_1 phase was the most severely affected of the four phases of the cell cycle.

In a more recent study on the effects of temperature shift on cell cycle, apoptosis, and nucleotide pools in CHO cell batch cultures, a temperature shift to 30°C after 48 h was found to cause a rapid decrease in the percent of cells in the S phase and accumulation of cells in the G_1 phase (45). Furthermore, culture viability was extended following temperature shift, as a result of delayed onset of apoptosis, while the overall rate of metabolism was reduced. Similarly, in a study of temperature effects on hybridoma cells, it was demonstrated that temperatures

Table 2. Effects of Temperature on Cellular Functions in Animal Cells

Effect	References
Viability	43–45
Metabolic rates	45–48
Specific rate of product formation/product titer	46, 49
Cell density and growth rate	43, 46, 49–51
Cell cycle	45, 52
Shear sensitivity	53
Cell morphology	53
Proteolysis	54
Product quality and integrity	42, 55

lower than 37°C increased the time that cells remained viable (44). Bloemkolk et al. (49) demonstrated that also for hybridoma cells lower temperatures caused cells to stay longer in the G_1 phase of the cell cycle. In batch cultures, the adverse effect on cell propagation was avoided, when temperature reduction was delayed until the first half of the exponential phase had been passed. Under these conditions, the maximum cell number was reached no later than in cultures performed at 37°C (47).

Metabolic Rates and Product Formation. In a study by Freitag et al. (48) the influence of the process temperature on the production of the anticoagulant rh-AT III, by cultivation of recombinant BHK21c13 cells in continous culture, was examined. By reducing the bioreactor temperature from 37 to 30°C during the later part of the exponential growth phase, the accumulation of toxic metabolites, such as lactate, could be curtailed. The glucose consumption/lactate production rate of the cells considered increased with process temperature between 30 and 37°C. The study concluded that in order to have a positive influence, temperature reduction must take place before the cell number stabilizes.

Reduction of fermenter temperature to 34°C was also found to have several advantages in the control of high-density perfusion culture of recombinant hamster cells (42). Low temperature reduced cell metabolic activities, reflected by the specific glucose uptake rate, specific lactate production rate, specific glutamine uptake rate, specific ammonia production rate, specific O_2 uptake rate, and the specific CO_2 evolution rate. Most pronounced was the change in the glucose/lactate metabolism. A consistent product quality and the improved molecule integrity were important advantages. However, although low temperature reduces the rate of proteolysis, thus favoring product integrity, it will not eliminate the problem (54).

Van der Pol et al. (51) monitored the monoclonal antibody production of an immobilized hybridoma cell line cultivated in a fluidized-bed reactor for nearly 900 h. The cultivation temperature was varied between 34 and 40°C. Raising the temperature from 34 up to 37°C resulted in a simultaneous increase of growth and specific antibody production rate. Specific metabolic rates of glucose, lactate, glutamine, and ammonium stayed constant in this temperature range. However, a further enhancement of temperature up to 40°C had a negative effect on growth rate, whereas the specific monoclonal antibody production rate showed a small increase. The other specific metabolic rates also increased in the temperature range between 38 to 40°C. The maximum specific MAb production rate and the maximum growth rate of the immobilized hybridoma cell line were reached at 40 and 37°C, respectively. The maximum cell-specific metabolic rates for glucose, glutamine, lactate, and ammonium were reached at 40°C, whereas $Y_{lac/glu}$ and $Y_{amm/gln}$ stayed constant for the investigated temperature range. The influence of temperature on the specific metabolic rates of immobilized cells and suspension cells was similar.

Reduced growth rates, glucose consumption rate, and lactate production were also demonstrated during low-temperature cultivations of adherent recombinant baby hamster kidney (BHK) cells (47). However, for these cells the maximum cell density and productivity seemed not to be affected by the temperature reduction.

Interspecies hybridomas, such as human x mouse hybrid cells, may show a high degree of instability for monoclonal antibody production. A reduction of cultivation temperature decreases the growth rate and thus the number of mitoses, where the segregation and loss of chromosomes occur. Reduced cultivation temperature was also found to stabilize the overall IgM-production rate over the total cultivation period for interspecies hybridomas (56).

Finally, the cellular responses to temperature stress were shown to be affected by the metabolic state of an EPO-producing CHO cell line, as cells maintained in serum-free medium were more sensitive than cells growing exponentially in the presence of serum (55). Furthermore, the EPO glycosylation patterns were not affected by exposure to a 42 °C/1-h heat shock. In contrast, a 45 °C/1-h heat shock terminated RNA and protein synthesis immediately and caused culture death in 12 h.

Shear Sensitivity

Ludwig et al. (53) studied the influence of temperature on the shear stress sensitivity of adherent BHK 21 cells. Decreasing the temperature lowered the growth rate and increased the ability of the BHK cells to withstand shear stress. Also, cell morphology was changed at low temperature; at 28 °C the cells were mostly spherical or triangular, as opposed to a confluent monolayer at 37 °C. These effects were ascribed to a change in the fluidity of the cell membrane, as the membrane congeals to a semicrystalline gel at low temperature.

Heat Shock Proteins/Stress Proteins

All cell types studied so far, from *E. coli* to human cells, possess a stress-response system. This is often referred to as the "heat shock" response due to the original observation that exposure to superoptimal heat resulted in production of a number of new proteins or drastically increased production of other proteins. In parallel with this, synthesis of most other proteins is inhibited. Furthermore, exposure to other agents such as nutrient starvation, osmotic stress, or toxic substances may also induce some of the heat shock proteins and some other proteins as well (54). We now know that proteins induced by heat exist in almost all living organisms, and that many of these proteins are related. Several of the genes for these proteins have been cloned, providing important data on their degree of similarity. This work has shown that two groups of stress proteins, in particular, the hsp 70 and hsp 90 families, are among the most highly conserved proteins in nature (57). Also, it has been shown that prior induction of heat shock proteins (e.g., by exposure to 39 °C) will protect cells from subsequent exposure to elevated temperature or other stresses.

Nishiyama et al. (58) have identified an RNA-binding protein induced in response to cold stress and designated as cold-inducible RNA-binding protein (CIRP). This protein was expressed in mouse testis (BALB/3T3 mouse fibroblasts). When the culture temperature was lowered from 37 to 32 °C, expression of CIRP was induced and growth of BALB/3T3 cells was impaired as compared to that at 37 °C. The results indicate that CIRP plays an essential role in cold-induced growth suppression of mouse fibroblasts. Overexpression of cold-inducible RNA-binding protein resulted in reduced growth rate and prolongation of the G_1 phase.

Programmed Cell Death — Apoptosis

The cause of culture death in bioreactors could be either apoptosis, in which the cells actively participate in their own death process, or necrosis, in which extreme environmental stress leads to catastrophic cell injury. Recently, Rajan Nagarathnamma and Sureshkumar (59) have shown that heat shock causes apoptosis enhancement in CC9C10 hybridoma cultures. Exposure to a 42 °C shock for 1 h increased the apoptosis extent (DNA fragmentation) by 32%.

ACKNOWLEDGMENT

The author wishes to thank Professor David W. Levine (Dept. of Biotechnology, Norwegian Institute of Technology, University of Trondheim) for critically reading this manuscript.

BIBLIOGRAPHY

1. C. Waymouth, in P.F. Kruse and M.K. Patterson, eds., *Tissue Culture: Methods and Applications*, Academic Press, New York, 1973, pp. 703–709.
2. G.N. Somero, in G.N. Somero, C.B. Osmond, and C.L.Bolis, eds., *Water and Life*, Springer-Verlag, Berlin, 1992, pp. 3–18.
3. R.I. Freshney, *Culture of Animal Cells*, 2nd ed., Alan R. Liss, New York, 1987.
4. S.K.W. Oh et al., *Biotechnol. Bioeng.* **42**, 601–610 (1993).
5. K. Øyaas et al., *Biotechnol. Bioeng.* **44**, 991–998 (1994).
6. X. Yang, G.W. Oehlert, and M.C. Flickinger, *Biotechnol. Bioeng.* **50**, 184–196 (1996).
7. S.Y. Park and G.M. Lee, *Biotechnol. Bioeng.* **48**, 699–705 (1995).
8. J.S. Ryu and G.M. Lee, *Biotechnol. Bioeng.* **55**, 565–570 (1997).
9. R.G. Wyn Jones and J. Gorham, in R. Gilles, ed., *Animals and Environmental Fitness*, Pergamon, New York, 1992, pp. 9–10.
10. P.H. Yancey et al., *Science* **217**, 1214–1222 (1982).
11. D. Le Rudulier et al., *Science* **224**, 1064–1068 (1984).
12. J.L. Eveloff and D.G. Warnock, *Am. J. Physiol.* **252**, F1–F10 (1987).
13. M.E. Chamberlin and K. Strange, *Am. J. Physiol.* **257**, C159–C173 (1989).
14. J.C. Parker, *Am. J. Physiol.* **265**, C1191–C1200 (1993).
15. R. Lange, *Oceanogr. Mar. Biol.* **10**, 97–136 (1972).
16. K. Fugelli, in R. Gilles, ed., *Animals and Environmental Fitness*, Vol. 1, Pergamon, New York, 1980, pp. 27–41.
17. K. Øyaas et al., *Biotechnol. Bioeng.* **43**, 77–89 (1994).
18. S. Bagnasco et al., *J. Biol. Chem.* **261**, 5872–5877 (1986).
19. S.R. Gullans et al., *Am. J. Physiol.* **255**, F626–F634 (1988).

20. K. Øyaas et al., in R.E. Spier, J.B. Griffiths, J. Stephanne, and P.J. Crooy, eds., *Advances in Animal Cell Biology and Technology for Bioprocesses*, Butterworth, London, 1989, pp. 212–220.

21. S.S. Ozturk and B.O. Palsson, *Biotechnol. Bioeng.* **37**, 989–993 (1991).

22. S. Reddy, K.D. Bauer, and W.M. Miller, *Biotechnol. Bioeng.* **40**, 947–964 (1992).

23. L. Shapiro and C.A. Dinarello, *Proc. Natl. Acad. Sci. U.S.A.* **92**, 12230–12234 (1995).

24. R. Kimura and W.M. Miller, *Biotechnol. Bioeng.* **52**, 152–160 (1996).

25. S. Reddy and W.M. Miller, *Biotechnol. Prog.* **10**, 165–173 (1994).

26. G.M. Lee and S.Y. Park, *Biotechnol. Lett.* **17**, 145–150 (1995).

27. S.Y. Park and G.M. Lee, *Bioprocess. Eng.* **13**, 79–86 (1995).

28. C.F. Goochee et al., in P. Todd, S.K. Sikdar, and M. Bier, eds., *Frontiers in Bioprocessing II*, American Chemical Society, Washington, D.C., 1992, pp. 199–240.

29. R. Kimura and W.M. Miller, *Biotechnol. Prog.* **13**, 311–317 (1997).

30. D. Ingrosso et al., *Eur. J. Biochem.* **244**, 918–922 (1997).

31. D. Cohen, J. Wasserman, and S. Gullans, *Am. J. Physiol.* **261**, C594–C601 (1991).

32. S. Dasgupta, T. Hohman, and D. Carper, *Exp. Eye Res.* **54**, 461–470 (1992).

33. D. Sheik-Hamad et al., *Am. J. Physiol.* **267**, F28–F34 (1994).

34. A. Garcia-Perez et al., *J. Biol. Chem.* **264**, 16815–16821 (1989).

35. F.L. Smardo, Jr., M.B. Burg, and A. Garcia-Perez, *Am. J. Physiol.* **262**, C776–C782 (1992).

36. S. Uchida et al., *J. Clin. Invest.* **91**, 1604–1607 (1993).

37. A. Yamauchi et al., *Am. J. Physiol.* **264**, F20–F23 (1993).

38. J.D. Ferraris et al., *Proc. Natl. Acad. Sci. U.S.A.* **91**, 10742–10746 (1994).

39. D. Sheik-Hamad et al., *Am. J. Physiol.* **270**, C253–C258 (1996).

40. M.B. Burg, *Am. J. Physiol.* **268**, F983–F996 (1995).

41. C.C. Matthews and E.L. Feldman, *J. Cell. Physiol.* **166**, 323–331 (1996).

42. S. Chuppa et al., *Biotechnol. Bioeng.* **55**, 328–338 (1997).

43. P.N. Rao and J. Engelberg, *Science* **148**, 1092–1094 (1965).

44. S. Reuveny, D. Velez, J.D. Macmillian, and L. Miller, *J. Immunol. Methods* **86**, 53–59 (1986).

45. A. Moore et al., *Cytotechnology* **23**, 47–54 (1997).

46. N. Barnabé and M. Butler, *Biotechnol. Bioeng.* **44**, 1235–1245 (1994).

47. R. Weidemann, A. Ludwig, and G. Kretzmer, *Cytotechnology* **15**, 111–116 (1994).

48. R. Freitag et al., *Cytotechnology* **21**, 205–215 (1996).

49. J.-W. Bloemkolk et al., *Biotechnol. Bioeng.* **40**, 427–431 (1992).

50. J. Gaertner and P. Dhurjati, *Biotechnol. Prog.* **9**, 298–308 (1993).

51. J.J. van der Pol et al., *Cytotechnology* **24**, 19–30 (1997).

52. J.E. Sisken, L. Morasca, and S. Kibby, *Exp. Cell Res.* **39**, 103–116 (1965).

53. A. Ludwig, J. Tomeczkowski, and G. Kretzmer, *Appl. Microbiol. Biotechnol.* **38**, 323–327 (1992).

54. S.-E. Enfors, *Trends Biotechnol.* **10**, 310–315 (1992).

55. E.I. Tsao et al., *Biotechnol. Bioeng.* **40**, 1190–1196 (1992).

56. O.-W. Merten, S. Reiter, and H. Katinger, *Dev. Biol. Stand.* **60**, 509–512 (1984).

57. D. Miller, *New Sci.* April 1, pp. 47–50 (1989).

58. H. Nishiyama et al., *J. Cell Biol.* **137**, 899–908 (1997).

59. M.M. Rajan Nagarathnamma and G.K. Sureshkumar, *Biotechnol. Lett.* **19**, 669–673 (1997).

See also ANIMAL CELL CULTURE MEDIA; BIOREACTORS, CONTINUOUS CULTURE OF PLANT CELLS; BIOREACTORS, RECIRCULATION BIOREACTOR IN PLANT CELL CULTURE; CELL CYCLE SYNCHRONIZATION; CELL GROWTH AND PROTEIN EXPRESSION KINETICS; CRYOPRESERVATION OF PLANT CELLS, TISSUES AND ORGANS; MICROPROPAGATION, HYPERHYDRICITY; PHYSIOLOGY OF PLANT CELLS IN CULTURE; PLANT CELL CULTURES, PHYSIOCHEMICAL EFFECTS ON GROWTH; PLANT PROTOPLASTS; HISTORY OF ANIMAL CELL TECHNOLOGY.

ANIMAL CELL CULTURE, PHYSIOCHEMICAL EFFECTS OF pH

CHRISTI L. MCDOWELL
ELEFTHERIOS T. PAPOUTSAKIS
Northwestern University
Evanston, Illinois

MICHAEL C. BORYS
Abbott Laboratories
Chemical and Agricultural Products Division
North Chicago, Illinois

OUTLINE

pH Homeostasis

pH Gradients in Tissues

pH Gradients in Cell Culture Systems

    pH Effects on Receptor Concentration

    pH Effects on Cell Differentiation

    pH Effects on Cell Metabolism

pH Effects on Cell Growth, Protein Production, and Glycosylation

Complex Effects of pH

Bibliography

pH affects the growth, metabolism, differentiation, and biophysics of animal cells in various ways. This section reviews a select number of these effects that are particularly relevant to animal cell culture: pH homeostasis; pH gradients in tissues and cell culture systems; pH effects on receptor concentration, cell differentiation, cell metabolism, protein production and glycosylation; and complex effects of pH.

pH HOMEOSTASIS

Modulators of pH homeostasis include Na^+/H^+ antiports, Cl^-/HCO_3^- antiports, and Na^+-(HCO_3^-) symports (or cotransporters). The most widely studied regulators of pH homeostasis are the Na^+/H^+ antiports or exchangers (NHEs). NHEs catalyze the electroneutral exchange of extracellular Na^+ (influx) for intracellular H^+ (efflux) with a stoichiometry of 1:1 (1). They are quiescent at normal physiological intracellular pH but become activated upon acidification of the cytosol or in response to mitogens, such as serum and granulocyte macrophage colony-stimulating factor (GM-CSF). In addition to a role in pH homeostasis, NHEs also play roles in regulating cell volume and osmolarity; the transcellular absorption of Na^+, acid/base equivalents, and water; and possibly cell proliferation (1,2). Mammalian NHEs are integral plasma membrane phospho(glyco)proteins with 10–12 putative transmembrane domains at the amino terminus and a long cytoplasmic region at the carboxyl terminus (3,4). To date, six members of the NHE family have been identified. NHE1, which is expressed in nearly all mammalian tissues, is localized to both the basolateral and apical (brush border) surfaces of several epithelial cells (5). It serves to regulate cystolic pH, cell volume, and possibly cell proliferation (3,6). NHEs 2 through 4 exhibit a more restricted tissue distribution and are found primarily in the kidney and gastrointestinal tract. NHE2 has been reported to be localized to both the basolateral and apical (brush border) membranes of epithelial cells (4). Although its exact physiological function is unclear, NHE2 may play a role in regulating cytosolic pH, cell volume, and cell proliferation in a manner similar to NHE1. NHE3, which is localized to the apical (brush border) membrane of renal proximal tubule and intestinal epithelia, is thought to play a role in the transepithelial reabsportion of Na^+ (3). NHE4, which is localized to the basolateral membrane of renal inner medullary tubules (an area of high osmolarity), is thought to play a role in regulating volume homeostasis (3,4). NHE5 is found in several nonepithelial tissues, including brain, testis, spleen, and skeletal muscle (7). Its function is not yet fully characterized. NHE6 exhibits ubiquitous tissue localization like NHE1, but is most abundant in mitochondrion-rich tissues, such as brain, skeletal muscle, and heart (8). It is believed to play a role in extruding of Na^+ from the alkaline matrix of respiring mitochondria, and therefore may regulate organellar volume (4).

There are two types of Cl^-/HCO_3^- antiports (or exchangers), Na^+-independent and Na^+-dependent. The Na^+-independent Cl^-/HCO_3^- exchanger is responsible for the electroneutral exchange of extracellular Cl^- (influx) for intracellular HCO_3^- (efflux) with a stoichiometry of 1:1 (1). In this manner, the Na^+-independent Cl^-/HCO_3^- exchanger acidifies the cytoplasm should alkalinization occur. Aside from a role in pH homeostasis, these exchangers may also play roles in regulating cell volume and osmolarity, in transepithelial transport of acids and bases, and in the transport of CO_2 between the peripheral tissues and the lungs (2). The Na^+-dependent Cl^-/HCO_3^- exchanger is responsible for the electroneutral exchange of extracellular Na^+ and HCO_3^- for intracellular Cl^- (and possibly H^+), with a stoichiometry of 1:1:2 for Na^+/Cl^-/acid–base equivalents, respectively (1). For example, in barnacle muscle fibers, 1 Na^+ and 1 HCO_3^-

are exchanged for 1 Cl^- and 1 H^+ (9). In this manner, the Na^+-dependent Cl^-/HCO_3^- exchanger is thought to play a role in extruding of acid from the cell. Both types of Cl^-/HCO_3^- exchangers have been reported to be present in fibroblasts, hepatocytes, erythrocytes, epithelial cells, and Vero cells, as well as in barnacle muscle fibers and squid giant axons.

The Na^+-(HCO_3^-) symporter is responsible for the coupled transport of Na^+ and HCO_3^- into the cell. Unlike the Na^+/H^+ and Cl^-/HCO_3^- exchangers, the Na^+-(HCO_3^-) symporter operates in an electrogenic fashion (i.e., the stoichiometry of Na^+ and HCO_3^- exchange is not 1:1, resulting in the creation of an electrical potential across the cell membrane). The Na^+-(HCO_3^-) symporter transports one Na^+ along with two or more HCO_3^-, resulting in net acid extrusion from the cell (10). Like the Na^+/H^+ and Cl^-/HCO_3^- exchangers, Na^+-(HCO_3^-) symporters also play roles in cellular functions other than pH homeostasis, such as regulating the transepithelial transport of Na^+, acid/base equivalents, and water (2). Na^+-(HCO_3^-) symporters have been reported to be present in mammalian hepatocytes and several types of epithelial cells, including frog retinal pigment epithelium, alveolar epithelial cells, and renal proximal tubules.

An interesting example of the way changes in antiporter expression can mediate cell adaptation to alterations in culture pH is the role of the band 3 protein in erythroid differentiation. Band 3, which is upregulated in the latter stages of erythrocyte differentiation, is structurally and functionally similar to the epithelial Na^+-independent Cl^-/HCO_3^- exchanger (11). As such, band 3 exchanges HCO_3^- ions for Cl^- ions, and acidifies the erythrocyte cytoplasm. Work done by McAdams et al. (12) demonstrated that in response to culture at pH 7.6, peripheral blood mononuclear CD34$^+$ cells exhibited a higher percentage of band 3 — positive cells (and acquired this erythroid differentiation marker sooner) compared to cells cultured at pH 7.35 or 7.1. They concluded from their experiments that pH plays an important role in erythroid differentiation. Since band 3 is a Cl^-/HCO_3^- exchanger, it is logical that cells exposed to pH 7.6 exhibit a greater amount of band 3 protein (than cells exposed to pH 7.35 or pH 7.1) in an attempt to acidify the cytoplasm under slightly alkaline conditions.

pH GRADIENTS IN TISSUES

Several studies in the literature have reported on the measurement of pH gradients in normal and tumor tissues. Examining the interstitial pH of tumor tissue is important in understanding tumor growth and also the response of tumors to various cancer treatments. Acidic extracellular pH has been shown to inhibit tumor cell proliferation and metabolism *in vitro* (13). Evidence in the literature also suggests that acidic pH may enhance tumor cell metastasis (14). In addition to these effects, pH plays an important role in determining the efficacy of various cancer treatments because acidic pH may reduce the proliferation of IL-2 stimulated lymphocytes (15) and increase the cellular resistance to radiation (16). Acidic pH may also enhance the cytotoxicity of certain alkylating drugs *in vitro* (17), increase the sensitivity of cells to hyperthermia (18), and serve as a prognostic indicator of tumor response to thermoradiotherapy (a combination of hyperthermia and radiation therapy) (19). Because pH has such a profound effect on tumor response to treatment, the ability to measure pH gradients in tumors noninvasively may prove useful in determining the most appropriate method of cancer therapy.

Martin and Jain (20) reported on the use of fluorescence ratio imaging microscopy (FRIM) to measure macroscopic pH gradients in normal and tumor tissues. The ability of FRIM to measure pH gradients was verified by *in vitro* calibration studies, in which the pH gradient of an isoelectrically focused polyacrylamide gel was measured by two independent pH measurements. The first measurement, using a flat membrane surface pH electrode, directly determined the spatial pH gradient along a track. The second measurement involved soaking the gel in BCECF (2′,7′-bis-(2-carboxyethyl)-5,6-carboxyfluorescein), a fluorochrome that exhibits pH-dependent spectral characteristics. Then the gel was excited with 495-nm light (maximum pH sensitivity of BCECF) followed by 440-nm light (at which BCECF is pH insensitive), and the fluorescent intensities along a track were recorded. The relationship between fluorescence and pH was determined by soaking sections of the gel in BCECF solutions at various pH and recording the fluorescence intensities. With this calibration curve, along with the fluorescent intensity information, the spatial pH gradient along a track could be determined. The average difference between the two independent measurements was less than 0.05 pH units, verifying the accuracy of FRIM for measuring spatial pH gradients. In vivo studies using FRIM were then done to measure the pH in VX2 carcinoma tumor tissue and its surrounding normal tissue grown in the rabbit ear chamber (tissue thickness limited to 50 μm). They found that the average pH of normal tissue was 7.18 ± 0.11 whereas that of tumor tissue was 6.75 ± 0.10.

In subsequent studies, Martin and Jain (21) examined the interstitial pH profiles of normal and tumor (VX2 carcinoma) tissues using FRIM at the microcirculatory level. They found that the interstitial pH, in normal tissue decreased 0.32 pH units over a distance of 50 μm away from a blood vessel, whereas the interstitial pH in tumor tissue decreased by 0.13 units over the same distance. Even though the pH gradient near the blood vessel in normal tissue was greater than that in tumor tissue, the tumor tissue had a greater proton concentration gradient (5.7×10^{-8} M over a distance of 0 to 50 μm away from the blood vessel) as compared to the normal tissue (4.5×10^{-8} M over the same distance). It is also important to note that the pH at the blood vessel wall in normal tissue was 7.38 ± 0.09 whereas that in tumor tissue was 6.79 ± 0.13. In temporal studies on the effects of hyperglycemia (high glucose concentration), they found that following a systemic injection of glucose (6 g glucose/kg of animal i.v.), the pH of tumor tissue dropped more than 0.2 units in 90 minutes, whereas the pH of normal tissue remained unchanged.

Dellian et al. (22) used an adapted method of FRIM (pinhole illumination-optical sectioning) to measure the pH in thick (2 mm as opposed to 50 μm) tumor tissues. They observed sharp pH gradients with variable spatial patterns between tumor blood vessels. On average, the pH decreased by 0.10 pH units over a distance of 40 μm away from the blood vessel wall and by 0.33 pH units over a 70 μm distance. Helmlinger et al. (23) reported the first combined measurement of pH gradients (using the pinhole illumination-optical sectioning method with FRIM) and pO_2 gradients (using phosphorescence quenching) in LS174T human adenocarcinoma tumors grown in SCID mice. They found that greater than 90% of the pH gradients between two tumor blood vessels had profiles with the highest pH nearest the blood vessel and the lowest pH furthest away from the blood vessel.

pH GRADIENTS IN CELL CULTURE SYSTEMS

Because variations in culture pH can affect several cellular properties and functions, effective pH control is necessary in mammalian cell culture for the production of diagnostic and therapeutic proteins, viral vaccines, and cells for somatic therapies. Unless pH is controlled properly, problems with cellular/product quality and/or uniformity may arise. Effective pH control is not always obtainable, however, as spatial inhomogeneities in the culture system may result in pH gradients. In addition, the extracellular pH may vary dramatically as the cell density increases, regardless of the culture system. Studies in the literature have reported on pH gradients in several different culture systems, including T-flasks, hollow fiber reactors, and stirred tank reactors. In T-flasks, Akatov et al. (24) found that within 6 hours of feeding, Chinese hamster fibroblasts had a local pH of 6.5, even though the bulk pH remained at 7.6. In studies with a hollow-fiber reactor system, Heath and Belfort (25) found gradients in nutrient concentrations, which could very well lead to gradients in pH and dissolved oxygen. Concerning pH gradients in a stirred tank reactor, Borys et al. (26) studied the effects of the degree of aggregation of recombinant CHO cells (cultured on microcarriers) on the production and glycosylation of the mouse placental lactogen-1 (mPL-1) protein. They found that alterations in the degree of CHO cell aggregation (i.e., from a single cell monolayer to large cellular aggregates) had an effect on mPL-1 protein expression and glycosylation. Borys et al. (26) proposed that these effects resulted from changes in the physiological conditions inside the cell aggregates (including changes in pH). Because effective pH control is necessary to obtain culture homogeneity but is not always obtainable, it is important to understand the effects of pH gradients on cellular properties and functions in several types of culture systems.

pH Effects on Receptor Concentration

Surface receptors mediate several functions including cell adhesion, catalysis of surface associated reactions, signal transduction, and recognition of cancerous or virally infected cells. Therefore, receptor surface content is an important issue in the production of whole cells for use in somatic therapies, including hematopoietic cells used for bone marrow or peripheral blood progenitor cell transplantation, and immune cells such as natural killer (NK) cells, tumor infiltrating lymphocytes (TIL), and antigen-specific cytotoxic T lymphocytes (CTL). Receptor expression is also an important issue in the production of viral vaccines. In light of this, cells used for somatic and gene therapies, and for viral vaccine production, must be cultivated in a way that maintains the expression of all necessary surface receptors in order to maintain biological functionality, as indicated by the treatment effectiveness or the amount and activity of vaccine produced. Thus, studying the effects of extracellular pH as a bioprocessing parameter on cellular receptor levels has applications in the large-scale culture of mammalian cells for somatic and gene therapies, and for viral vaccine production.

There are few studies in the literature on the effects of pH on receptor expression. Katafuchi et al. (27) examined the expression of the atrial natriuretic peptide (ANP) receptor on bovine aortic endothelial cells cultured in three different pH environments (pH 7.0, 7.4 and 7.7). Cells cultured in pH 7.0 medium exhibited approximately twice as many ANP receptors as compared to cells cultured in pH 7.4 medium. Furthermore, cells cultured in pH 7.7 medium exhibited only trace amounts of the ANP receptor. It is important to note that, although pH had dramatic effects on ANP receptor surface concentration, it did not affect the ANP receptor affinity for its ligand (ANP).

McDowell and Papoutsakis (28) examined the effects of extracellular pH on the surface content and mRNA level of the CD13 receptor on human-derived promyelocytic leukemia (HL60) cells cultured at pH 7.0, 7.2, and 7.4 in stirred tank bioreactors. Decreasing the pH from 7.4 to 7.0 and from 7.4 to 7.2 increased the CD13 receptor surface content of HL60 cells by approximately 70% and 60%, respectively. These changes in CD13 receptor surface content in response to extracellular pH were not correlated with changes in CD13 mRNA levels, as demonstrated by Northern blot analysis. Since extracellular pH affected CD13 receptor surface content, but did not affect CD13 mRNA levels, pH must affect another receptor processing step. It is possible that extracellular pH may affect CD13 receptor protein synthesis, trafficking to the cell membrane, internalization, degradation, or recycling. Additional experiments are needed in order to ascertain the mechanism through which extracellular pH affects CD13 receptor surface content.

pH Effects on Cell Differentiation

Human hematopoietic cells have great promise for use in bone marrow or peripheral blood progenitor cell transplantation therapies following high-dose chemotherapy and/or radiation treatments. In such therapies, these cells reconstitute the hematopoietic system, which is depleted during the eradication of cancer cells from the body. Once peripheral blood progenitor cells are administered to a patient, it generally takes between eight and twelve days for platelet and neutrophil counts to reach acceptable levels (29). During this period, the patient is in danger of internal bleeding and very susceptible to infections. Theoretically, these periods of thrombocytopenia (low platelet counts) and neutropenia (low neutrophil counts) could be

shortened by supplementing bone marrow or peripheral blood progenitor cells with mature megakaryocytes (precursors of platelets) (30) and neutrophils (31). In the *ex vivo* expansion of human hematopoietic cells for somatic therapies, extracellular pH is an important bioprocessing parameter because pH is proving to have a profound effect on cell differentiation. Depending upon the culture conditions (including extracellular pH and cytokine combinations), several types of cells may be produced from the same initial population. Therefore pH is an important culture parameter in determining the desired cellular end product, whether it be platelets for treating thrombocytopenia, neutrophils for treating neutropenia, or red blood cells and platelets for use in blood transfusions.

Concerning the effects of extracellular pH on primary hematopoietic cell differentiation, Zipori and Sasson (32) found that murine bone marrow cells cultured over the pH range of 6.5 to 8.0 exhibited maximum development of granulocyte/macrophage colonies at pH 7.4. This is consistent with the work of McAdams et al. (33) who found that in pH-adjusted methylcellulose cultures of human PB MNC (peripheral blood mononuclear cells) and CB MNC (cord blood mononuclear cells) over a range of pH 6.95 to 7.6, the differentiation of granulocyte/macrophage colony-forming units into colonies was maximum at pH 7.4. Two interesting observations were made from these experiments. First, PB MNC's were more sensitive to extreme pH values compared to CB MNC's. Second, granulocytic progenitors were more resistant to acidic pH compared to macrophage progenitors, because at pH 6.7 (the lowest pH value at which colonies were detected) the only colonies remaining were CFU-G.

McAdams et al. (33) also found that the differentiation of BFU-E (erythrocyteburst-forming units) was maximum at pH 7.4. In pH-adjusted liquid cultures of PB MNCs and CB MNCs, McAdams et al. (12,33) found that the number of BFU-E remaining in culture was not affected by low pH (7.15), but was reduced by 80% at high pH (7.6). This suggests that at pH 7.6, BFU-E are being depleted to produce mature erythroid cells. This was confirmed by a benzidine assay (benzidine detects the presence of hemoglobin), which showed that compared to the intermediate pH (7.35) cultures, less mature erythroid cells were present at low pH (7.15), whereas more mature erythroid cells were present at high pH (7.6). From these experiments, they concluded that low pH inhibits erythroid differentiation and high pH accelerats erythroid differentiation. Further studies by McAdams et al. (12), including morphological studies (Wright−Giemsa staining of cells; Fig. 1), flow cytometric analysis (using the antigens CD45RA and CD71), determination of the amount of band 3 (an antigen upregulated in the later stages of erythroid differentiation), and Western blot analysis of both band 3 and hemoglobin, support the conclusion that erythroid differentiation is accelerated at high pH (7.6) and inhibited or arrested at low pH (7.15).

Concerning the effects of extracellular pH on the differentiation of other types of cells, Fischkoff et al. (34) found that human-derived promyelocytic leukemia (HL60) cells cultured at pH 7.6−7.8 for one week differentiated into eosinophils and eosinophilic precursors. In subsequent

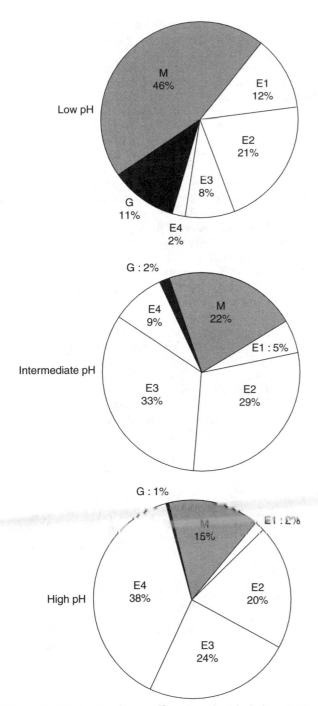

Figure 1. Effect of culture pH on erythroid differentiation: morphological analysis of cultured human hematopoietic cells under conditions favoring erythroid differentiation. G: granulocyte; M: monocyte; E1: pronormoblast (least mature erythroid); E2: basophilic normoblast; E3: polychromatic normoblast; E4: orthochromatic normoblast (most mature erythroid). Reprinted by permission from Ref. 12.

studies, Fischkoff and Rossi (35) identified stable sublines of the HL60 cell line which selectively differentiated into either neutrophils, monocytes, or eosinophils when treated with butyric acid. Consistent with earlier work, they found that the neutrophil and monocyte-directed sublines differentiated into eosinophils when cultured at pH 7.6 for two

months. In studies with the human erythroid cell lines K562 and KU-812, Endo et al. (36) observed an increase in spontaneous differentiation when the cells were cultured at pH 7.6, as opposed to pH 7.4. Fitzgerald et al. (37) examined the effects of pH on the *ex vivo* culture of Barrett's esophagus (BE), a premalignant epithelium associated with an increased risk for adenocarcinoma. As a determinant of cell differentiation, they measured the expression of villin, a binding protein that serves to link actin filaments in microvilli. Villin expression was not affected in BE samples exposed to a one-hour pulse of acidified media (pH 3–5) followed by culture at pH 7.4. In contrast, the number of cells exhibiting detectable amounts of villin increased from 25 to 50 or 83% in BE samples exposed to acidified media for 6 and 24 hours, respectively.

pH Effects on Cell Metabolism

Studies examining the effects of pH on cell metabolism have been done with HeLa cells (38), lymphoblastoid cells (39), hybridomas (40,41), and HL60 cells (28). In each instance, as the culture pH decreased, the cells exhibited a decrease in glucose consumption and lactate production rates. It is important to note that in addition to the decrease in glucose consumption and lactate production rates with decreasing culture pH, McQueen and Bailey (41) also observed a decrease in the hybridoma cell growth rate. This behavior is in contrast to the results of McDowell and Papoutsakis (28), who found that with decreasing culture pH, HL60 cells exhibited a decrease in glucose consumption and lactate production rates, and the cell growth rate remained largely unaffected. It is possible that the overall better yields of HL60 cells with decreasing pH (i.e., more cells per mole of glucose) could result from a decrease in metabolic energy requirements, utilization of a higher percentage of other nutrients at lower pH, or more likely, complex events involving several metabolic processes.

pH EFFECTS ON CELL GROWTH, PROTEIN PRODUCTION, AND GLYCOSYLATION

This section will review reported effects of culture pH on the expression rates and glycosylation of proteins by mammalian cells. The pH used for the culturing of mammalian cells normally ranges from pH 7.0–7.6. If not externally controlled, the pH will decrease to 6.6–6.9 due to the formation of lactic acid, which is a by-product of glucose metabolism. The optimal extracellular pH (pH_e) for mammalian cell culture has traditionally been determined by cell growth experiments. However, the optimal pH_e for cell growth may not always coincide with the optimal pH_e for protein production. For example, the maximum specific antibody expression rate of hybridoma cells has been associated with a pH stress response (40,42) at the acidic pH_e of ca. 6.8, which is considerably lower than the optimum pH_e reported for hybridoma cell growth (ca. pH_e 7.2) (40,42,43). There is no single pH_e value optimal for cell growth for all cell types. Rather, the optimal pH_e for cell growth for several different cell types ranges between pH 6.8–7.8 (44; Table 1).

Table 1. Optimal pH for Mammalian Cell Growth as Summarized by Waymouth (44)

Cell Line	Optimal pH for Cell Growth
HeLa	6.9
Rabbit lens	6.8–6.9
SV-40 transformed WI-38	7.4–7.5
Two human skin fibroblast cell lines	7.5–7.6 and 7.6–7.8

Table 2. Results of pH_e Optimization Studies for Two Murine Hybridomas and a Murine Myeloma Transfected to Express Human Antibody (45)

Cell Type	Parameter	Value	
		pH 7.0	pH 7.1
Hybridoma 1	Approx. max. cell den. (cells/ml)	1.6×10^6	1.6×10^6
	Spec. prod. rate (mg/10^{6-} cell/h)	240	260
	Time integral of cell conc.[a]	160	125
	Total MAb conc. (mg/L)	43	29
		pH 7.1	pH 7.2
Hybridoma 2	Approx. max. cell den. (cells/ml)	3×10^6	6×10^6
	Spec. prod. rate (mg/10^{6-} cell/h)	120	60
	Time integral of cell conc.	322	414
	Total MAb conc. (mg/L)	34	23
		pH 7.1	pH 7.4
GS-NS0	Approx. max. cell den. (cells/ml)	2.5×10^6	2.0×10^6
	Spec. prod. rate (mg/10^{6-} cell/h)	660	540
	Time Integral of Cell conc.	270	210
	Total MAb conc. (mg/L)	194	119

[a] 10^9 cell hours/L.

It is certain that there is also no single optimal pH_e for protein expression for all mammalian cell types. This was demonstrated by Wayte et al. (45) who reported the results of pH_e optimization studies for two murine hybridomas and a murine myeloma transfected to express human antibody (Table 2). The maximum cell density and specific MAb expression rate were similar for Hybridoma 1 at pH_e 7.0 and 7.1. However, the final MAb concentration was 50% greater at pH_e 7.0 because cell viability was maintained longer at the lower pH value. Despite a significant reduction in hybridoma 2 cell growth at pH_e 7.1, there was still a 50% increase in final MAb concentration compared to that at pH_e 7.2 due to a higher specific MAb expression rate. The small changes in pH_e affected the

three cell lines of Table 2 differently. However in each case, the pH_e optimization studies resulted in a 50–60% increase in MAb concentration.

There may be cases where it is advantageous to operate without pH control to allow the culture pH to naturally decrease as the culture progresses. Schmid et al. (46) compared MAb production from a mouse hybridoma in batch culture with pH control (pH_e 7.2) to a culture without pH control (7.2 initial pH_e, 6.6 final pH_e). The culture lifetime was approximately 2 days longer without pH control, and the final antibody concentration was ca. 60% greater than with pH control. This increase without pH control was attributed to a high initial cell growth rate at pH 7.2, and as the pH naturally decreased, to a decrease in both glucose and amino acid consumption rates along with an increase in specific antibody production.

Changes in pH_e have also been reported to affect the recombinant protein production rate of mouse placental lactogen I (mPL-I) expressed in Chinese hamster ovary (CHO) cells (47). In this study, recombinant mPL-I expression was carried out under non-growth conditions in serum-free medium. The maximal rates for mPL-I secretion, total protein secretion, and glucose consumption all occurred between pH_e 7.6 and 8.0. The dependency of the specific mPL-I expression rate on pH_e (Fig. 2) was found to be similar to the previously reported CHO cell growth rate dependence on pH_e (48). Kurano et al. (48) examined the effect of pH_e on CHO cells from pH_e 7.0–7.9,

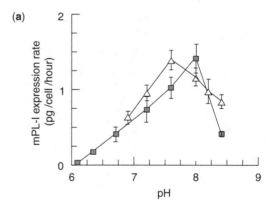

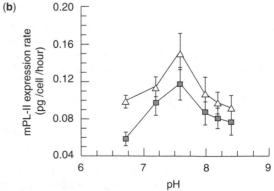

Figure 2. Extracellular pH affects the expression rates of recombinant proteins (**a**: mPL-I; **b**: mPL-II) by CHO cells. Reprinted by permission from Ref. 47

and found that the maximum specific glucose uptake rate occurred at pH_e 7.6. It is not unreasonable that in cells where protein expression is constitutive, such as with recombinant mPL-I expression, the effect of pH_e on protein production may be related to the overall effect of pH_e on cell metabolism and health. This would be in contrast to the optimum pH_e for specific MAb production by hybridomas, which was related to a pH_e stress response and found to correlate neither with the optimum pH_e for specific glucose uptake nor the optimum pH for cell growth (40).

Glycosylation is a post-translational modification to proteins in which an oligosaccharide (sugar) chain is attached and modified in the endoplasmic reticulum and Golgi organelles in eukaryotic cells (reviewed in Ref. 49). Structural changes in oligosaccharide chains have been reported to affect biological activity, antigenicity, circulatory lifetime and solubility of glycoproteins (49). Extracellular pH was found to affect the glycoform distributions of IgG from hybridomas (50), and mPL-I expressed in CHO cells (47). Glycoforms of mPL-I with similar molecular sizes were expressed between pH_e values from 7.2 to 8.0. However, decreased glycosylation of mPL-I occurred at both lower pH_e values ($pH_e \leq 6.9$) and at higher pH_e values ($pH_e \geq 8.22$).

Ammonia, which is a by product of glutamine metabolism, has been associated with decreases in cell growth rates and cell viability. Ammonia has been shown to affect protein glycosylation and specifically sialylation (51). It is now well established that the inhibitory effects of ammonia on mammalian cells increase with pH_e. It is believed that the unprotonated ammonia molecule (NH_3, see equation 1) is responsible for the inhibitory effects. In fact, direct correlations have been shown between increasing NH_3 concentrations, as calculated by the Henderson-Hasselbach equation (equation 2), and cell growth inhibition in hybridoma cultures (13), as well as inhibition of N-linked glycosylation by CHO cells (52).

$$NH_3 + H^+ \leftrightarrow NH_4^+ \tag{1}$$

$$pH = pK_a + \log(NH_3/NH_4^+) \tag{2}$$

In summary, there is no single pH_e value that can be specified as optimal for cell culture. There are reports that the optimal pH_e for cell growth may differ from that for protein expression and that the optimal pH_e will vary for different cell lines. Ultimately the decision on culture pH_e will take into account the effects of pH_e on the specific protein production rate, cell growth, nutrient consumption, and protein quality (e.g., glycosylation) to deliver the highest final protein concentration with acceptable quality. However, as the above references demonstrate, pH_e is a critical variable to optimize during the development of any cell culture process.

COMPLEX EFFECTS OF pH

An interesting example of pH regulation of a complex cellular process is the modulation of nitric oxide synthase (NOS) induction in macrophages. NOS is an enzyme that releases nitric oxide (NO) upon conversion of L-arginine to L-citrulline. NO participates in several mammalian organ functions including vasodilation, neurotransmission, inhibition of platelet aggregation, and mediation of the ability of macrophages to kill microbes and tumor cells. Three isoforms of NOS have been identified to date, endothelial, neuronal, and inducible. Endothelial and neuronal NOS are constitutively expressed and dependent on Ca^{2+} for activation. Inducible NOS (iNOS), so-called because it is expressed only after transcriptional induction, is the most common form and is found in several different cell types. Bellocq et al. (53) studied the effects of reduced medium pH on iNOS activity in rat peritoneal macrophages. They found that decreasing the medium pH from 7.4 to 7.0 resulted in a 2.5-fold increase in nitrate accumulation that was associated with an increase in iNOS mRNA levels. This increase in iNOS mRNA levels was found to result from an increase in the rate of iNOS mRNA transcription, as opposed to an increase in iNOS mRNA stability. The mechanism through which reduced pH led to the induction of iNOS transcription involved the transcription factor nuclear factor kappa B (NF-κB); exposing rat peritoneal macrophages to pH 7.0 medium (as opposed to pH 7.4 medium) resulted in an increase in NF-κB binding activity in the nucleus and in NF-κB-driven reporter gene expression. Treatment with drugs that interfered with NF-κB translocation to the nucleus resulted in no pH-induced nitrate accumulation, verifying the involvement of NF-κB in this pH-induced effect. Overall, the process by which reduced pH led to an increase in nitrate accumulation required the presence of TNFα. TNFα permitted NF-κB, which is initially bound to the inhibitory cytoplasmic protein IκB, to dissociate from IκB and translocate to the nucleus, where it bound to an appropriate recognition sequence, and initiate transcription of the iNOS gene.

BIBLIOGRAPHY

1. S. Grinstein, D. Rotin, and M.J. Mason, *Biochim. Biophys. Acta* **988**, 73–97 (1989).

2. R.L. Lubman and E.D. Crandall, *Am. J. Phys.* **262**, L1–L14 (1992).

3. C.H.C. Yun et al., *Am. J. Phys.* **269**, G1-G11 (1995).

4. J. Orlowski and S. Grinstein, *J. Biol. Chem.* **272**, 22373–22376 (1998).

5. S. Wakabayashi, M. Shigekawa, and J. Pouyssegur, *Phys. Revs.* **77**, 51–74 (1997).

6. S.D. D'Souza et al., *J. Biol. Chem.* **273**, 2035–2043 (1998).

7. C.A. Klanke et al., *Genomics* **25**, 615–622 (1995).

8. M. Numata, K. Petrecca, N. Lake, and J. Orlowski, *J. Biol. Chem.* **273**, 6951–6959 (1998).

9. W.F. Boron, W.C. McCormick, and A. Roos, *Am. J. Phys.* **240**, C80–C89 (1981).

10. D. Forestal, J. Haimovici, and P. Haddad, *Am. J. Phys.* **272**, G638–G645 (1997).

11. S. Fuerstenberg et al., *EMBO J.* **11**, 3355–3365 (1992).

12. T.A. McAdams, W.M. Miller, and E.T. Papoutsakis, *Br. J. Haematol.* **103**, 317–325 (1998).

13. J.J. Casciari, S.T. Sotirchos, and R.M. Sutherland, *J. Cell. Phys.* **151**, 386–394 (1992).

14. P.R. Young and S.M. Spevacek, *Biochim. Biophys. Acta* **1139**, 163–166 (1992).

15. D.A. Loeffler, P.L. Juneau, and S. Masserant, *Br. J. Cancer* **66**, 619–622 (1992).

16. I.F. Tannock and D. Rotin, *Cancer Res.* **49**, 4373–4384 (1989).

17. E. Jähde et al., *Cancer Res.* **49**, 2965–2972 (1989).

18. K.A. Ward and R.K. Jain, *Int. J. Hypertherm.* **4**, 223–250 (1988).

19. K. Engin et al., *Int. J. Radiat. Oncol., Biol., Phys.* **29**, 125–132 (1994).

20. G.R. Martin and R.K. Jain, *Microvasc. Res.* **46**, 216–230 (1993).

21. G.R. Martin and R.K. Jain, *Cancer Res.* **54**, 5670–5671 (1994).

22. M. Dellian, G. Helmlinger, F. Yuan, and R.K. Jain, *Br. J. Cancer* **74**, 1206–1215 (1996).

23. G. Helmlinger, F. Yuan, M. Dellian, and R.K. Jain, *Nat. Med.* **3**, 177–182 (1997).

24. V.S. Akatov, E.I. Lezhnev, A.M. Vexler, and L.N. Kublik, *Exp. Cell Res.* **160**, 412–418 (1985).

25. C, Heath and G. Belfort, *Adv. Biochem. Eng. Biotechnol.* **34**, 1–31 (1987).

26. M.C. Borys, D.I.H. Linzer, and E.T. Papoutsakis, *Ann. N.Y. Acad. Sci.* **745**, 360–371 (1994).

27. T. Katafuchi, H. Hagiwara, T. Ito, and S. Hirose, *Am. J. Phys.* **264**, C1345–C1349 (1993).

28. C.L. McDowell and E.T. Papoutsakis, *Biotechnol. Prog.* **14**, 567–572 (1998).

29. L.B. To et al., *Bone Marrow Transplant* **6**, 109–114 (1990).

30. J.A. LaIuppa, E.T. Papoutsakis, and W.M. Miller, *Stem Cells* **15**, 198–206 (1997).

31. S. Scheding et al., *Blood* **86**, 224a (1995).

32. D. Zipori and T. Sasson, *Exp. Hematol.* **9**, 663–674 (1981).

33. T.A. McAdams, W.M. Miller, and E.T. Papoutsakis, *Br. J. Haematol.* **97**, 889–895 (1997).

34. S.A. Fischkoff et al., *J. Exp. Med.* **160**, 179–196 (1984).

35. S.A. Fischkoff and R.M. Rossi, *Leuk. Res.* **14**, 979–998 (1990).

36. T. Endo, Y. Ishibashi, H. Okana, and Y. Fukumaki, *Leuk. Res.* **18**, 49–54 (1994).

37. R.C. Fitzgerald, M.B. Omary, and G. Triadafilopoulos, *J. Clin. Invest.* **98**, 2120–2128 (1996).

38. M.E. Barton, *Biotechnol. Bioeng.* **13**, 471–492 (1971).

39. J.R. Birch and D.J. Edwards, *Dev. Biol. Stand.* **46**, 59–63 (1980).

40. W.M. Miller, H.W. Blanch, and C.R. Wilke, *Biotechnol. Bioeng.* **32**, 947–965 (1988).

41. A. McQueen and J. Bailey, *Biotechnol. Bioeng.* **35**, 1067–1077 (1990).

42. P.M. Hayter, N.F. Kirby, and R.E. Spier, *Enzyme Microb. Technol.* **4**, 454–461 (1992).

43. C. Doyle and M. Butler, *J. Biotechnol.* **15**, 91–100 (1990).

44. C. Waymouth. in C. Waymouth, R.G. Ham, and P.C. Chapple, eds., *The Growth Requirements of Vertebrate Cells in Vitro*, Cambridge University Press, New York, (1981), pp. 105–117.

45. J. Wayte et al., *Genet. Eng. Biotechnol.* **17**, 125–132 (1997).

46. G. Schmid, H.W. Blanch, and C.R. Wilke, *Biotechnol. Lett.* **12**, 633–638 (1990).

47. M.C. Borys, D.I.H. Linzer, and E.T. Papoutsakis, *Bio/Technology* **11**, 720–724 (1993).

48. N. Kurano et al., *J. Biotechnol.* **15**, 101–112 (1990).

49. C.F. Goochee et al., in P. Todd, S.K. Sikdar, and M. Bier eds., Frontiers in Bioprocessing II, American Chemical Society, Washington DC., (1992), pp. 199–240.

50. R.J. Rothman, L. Warren, F.G. Vliegenhart, and K.J. Hard, *Biochemistry* **28**, 1377–1384 (1989).

51. D.C. Andersen and C.F. Goochee, *Biotechnol. Bioeng.* **47**, 96–105 (1995).

52. M.C. Borys, D.I.H. Linzer, and E.T. Papoutsakis, *Biotechnol. Bioeng.* **43**, 505–514 (1994).

53. A. Bellocq et al., *J. Biol. Chem.* **273**, 5086–5092 (1998).

See also ANIMAL CELL CULTURE MEDIA; ANIMAL CELL CULTURE, PHYSIOCHEMICAL EFFECTS OF DISSOLVED OXYGEN AND REDOX POTENTIAL; BIOREACTORS, PERFUSION; BIOREACTORS, STIRRED TANK; CELL METABOLISM, ANIMAL; ON-LINE ANALYSIS IN ANIMAL CELL CULTURE.

ANIMAL CELL PRODUCTS, OVERVIEW

BRYAN GRIFFITHS
SC& P
Porton, Salisbury
United Kingdom

OUTLINE

Introduction

Principles of the Production Process

Cell Products

　Viral Vaccines

　Immunoregulators

　Antibodies

　Recombinant Products

　Cell and Tissue Therapy

　Other Products

Conclusions

Bibliography

INTRODUCTION

The initial aims of cell and tissue culture were to study specialized cell behavior and function in vitro. However these aims could not be realized because only the ubiquitous dedifferentiated cell, or cells transformed by carcinogens, survived in culture. The discovery by Enders (1) in 1949 that human pathogenic viruses could be grown in cultured cells was both a revelation and the starting point for animal cell technology. Prior to this viruses could only be grown in living tissue; thus vaccine production used living organisms such as the embryonic chicken. The use of cultured cells to grow viruses opened up the possibility of a cheaper, easier, biologically safer, more controllable (and reproducible), and a larger-scale method for vaccine manufacture. It took only 5 years from Ender's discovery before the first cell-based vaccine was licenced for clinical use (Salk polio vaccine in primary monkey kidney cells in 1954).

Thus the first commercial product from animal cells was a vaccine, and these were the dominant products for the next 20–25 years. During this period new and biologically safer cell lines were introduced [WI-38 (2) and MRC-5 (3)] to replace the use of primary cells from monkey kidney or chick embryo, and a whole new range of human vaccines were licenced (measles, rabies, mumps, rubella) (4). Parallel development of veterinary vaccines (5) led to unit processes of thousands of liters in fermentation bioreactors (6). This was achieved through being able to use a continuous cell line capable of suspension growth, the BHK cell (7), which was considered biologically unsafe for human vaccines (potential presence of tumorigenic agents and viruses).

The development of animal cell biotechnology from 1954 to the present day has obviously been driven by technological advances, but the main influencing factor has been safety of the end product (8,9). Regulatory bodies such as WHO, FDA, etc. have set down at all stages of the process acceptable standards for cell products, and these have had to safeguard against both known and perceived hazards such as transforming viruses, disease agents, carcinogenic and immunologically damaging molecules, and, more recently, prions.

The key steps for manufacturing human products from animal cells can be summarized as follows;

1. Use of primary cells (to replace eggs and other living tissue/organs or animals) (1954)
2. Use of human diploid cells (HDC), such as WI-38 as a safe virus-free and noncarcinogenic substrate (1962)
3. Acceptance of cell lines derived from cancer tissue to produce human biologicals (1980, human interferon from Namalwa cells)
4. Production of monoclonal antibodies (1986) and their subsequent exploitation initially in the diagnostic field and subsequently as therapeutics

Table 1. 'Native' Animal Cell Products

Human vaccines	Polio (1954), measles (1963), rabies (1964), mumps (1969), rubella (1969), varicella (1976)
Veterinary vaccines	FMDV, rabies, Marek's, pseudorabies, BVD, Louping ill, bluetongue, avian influenza, canine distemper
Interferon	alpha-Interferon
Antibodies	Monoclonal antibodies

5. Acceptance of genetically engineered (recombinant) cell lines (1987, tPA from rCHO cells)
6. Use of cells as products for tissue replacement and gene therapy (1990s) and the future exploitation of stem cells

From the initial "native" cell products (Table 1) technological advances have not only led to the development of many new products (Table 2), but have also enabled more effective quality control tests to be introduced that can assess the final product for safety. Previously all ingredients of the process had to be proven to be completely safe in case any component got through with the final product; hence the requirement to use HDC for so long. However, if the dangers are known and can be detected, then less safe materials can be used if their absence in the final product can be determined. This, together with the increased scientific knowledge that has shown that many perceived dangers were not in fact real dangers, has opened up the use of cancer, recombinant, and other cell lines for the production of a very wide range of human biologicals (Tables 3 and 4).

PRINCIPLES OF THE PRODUCTION PROCESS

The basic principles common to most processes are summarized in Figure 1. The components are;

Table 2. Animal Cell Products—Licenced, in Trial, and Possible

Product range	Target diseases
Vaccines — native, recombinant and DNA	Viral infections, arthritis, MS, cancer
Immunoregulators — interferons, interleukins	Cancer, HIV, transplantation, tissue regeneration
Blood clotting factors — Factors VII, VIII, IX	Hemophilia A and B
Hormones — hGH, FSH (Gonal-F)	Dwarfism, fertility, contraception
Antibodies (monoclonal)	Diagnostics in vitro and in vivo, cancer, vascular remodeling
Tumor necrosis factors	Cancer
Colony stimulating factors	Neutophena, sepsis, infectious disease
Growth factors	MS, ulcers, diabetes, tissue repair
Gene therapy	Cystic fibrosis, cancer (colon, melanoma, renal cell, neuroblastoma, ovarian, breast, lung), HIV
CAMs (cell adhesion molecules)	Cancer, atherosclerosis, infections
Others — tPA	Myocardial infarction, thrombolytic occlusion
Erythropoietins	Anemia
Dismutases	Oxygen toxicity
Soluble receptors	Asthma, arthritis, septic shock
Antisense	Viral (e.g., HIV), cancer, inflammatory disease
Stem cells	Tissue engineering, cell therapy, cancer

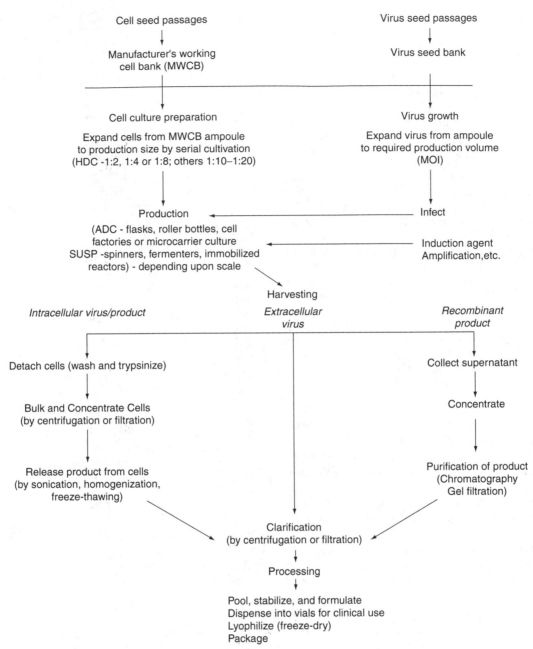

Figure 1. Steps in a production process using animal cells for vaccines, antibodies, and recombinant products shown schematically generalized.

1. *Seed banks (10).* Both the cell line and virus (for vaccines) have to be laid down in a fully tested and characterized bank. Thus each production batch will be initiated from identical cells (and virus) known to be viable and contamination-free.

2. *Cell seed expansion (10).* A series of culture steps is needed to expand the cell seed ampoule (e.g., 5 million cells) to production size (range $10^9 - 10^{13}$ cells). For HDC this is accomplished in steps of a split level of 1:2 or more usually 1:4 through a series of flasks, roller bottles, and possibly cell factories (A/S Nunc) (11). Other cell types are split at a 1:5 to 1:20 ratio. A similar buildup is needed for the virus seed.

3. *Production.* The production culture may be a batch of several hundred roller bottles (12), 30–50 cell factories, or a single bioreactor for suspension (100–10,000 L) (13) or microcarrier (50–500 L) cells (14–16). Although batch-type production is still the most common process, continuous processes where the product is harvested daily over a long period (20–100 d) are becoming increasingly used. Culture systems based on hollow fibers (17,18), porous microcarriers (19,20), or other immobilization techniques (21) are used for continuous perfusion processes. During the production phase the virus seed, or a promoter (e.g., for interferon) may be added.

Table 3. Licenced Engineered Animal Cell Products

Product	Trade name (year licenced)
Monoclonal antibodies	OKT3/Orthoclone (1987), Centoxin (1990), Reopro (1994), Myoscint (1989), Oncoscint (1990)
tPA	Activase/Actilyse (1987)
EPO	Epogen/Procrit/Eprex (1989), Epogin/Recormon (1990)
hGH	Saizen (1989)
HBsAg	GenHevac B Pasteur (1989), HBGamma (1990)
Interferon	Roferon (1991)
G-CSF	Granocyte (1991), Neupogen (1991)
Blood Factor VIII	Recombinate (1992), Kogenate (1993)
Dnase I	Pulmozyme (1993)
Glucocerebrosidase	Cerezyme (1994)
FSH	Gonal-F (1995)

Table 4. Engineered Products in Development (Examples)

Vaccines	HIV (gp 120, gp 160, CD4), HSV (gB, gD)
Interferons	Actimmune, Alferon, Betaferon, Infergen, Intron A, REBIF
Interleukins	Neumega, Proleukin, Sigosix
Colony stim. factors	Leukine, Neupogen
Growth factors	Betakine, Fiblast, Myotrophin, Somatokine
Erythropoietins	Procrit
tPA	Retaplase, nPA
Soluble receptors	Tenefuse, TNF
Antibodies	Leukoscan, Oncolym, Oncolysin B, Reopro, Zenopax

4. *Harvesting.* If the product is intracellular then the cells have to be harvested (trypsin and/or EDTA), washed and concentrated by centrifugation. Extracellular (secreted) products just need the culture supernatant to be collected.

5. *Downstream processing (22).* Intracellular products have to be extracted from the cells (by sonication, freeze-thawing, and/or homogenisation), and separated from the cells (centrifugation or filtration). Extracellular products need concentration and separation from the bulk supernatant.

6. *Formulation.* The product is added to a medium with protective and stabilising agents, and then usually freeze-dried (22).

7. *Quality control.* Throughout the process prescribed samples are taken for a range of QC tests to show safety, efficacy, and consistency in the process and product (9,23).

CELL PRODUCTS

Viral Vaccines

The first cell-based vaccine was against polio and was produced in monkey kidney cells. Then a series of HDC vaccines were licenced during the 1960s (24). The first recombinant vaccine was against hepatitis B, and currently the dominant target is HIV. Other viral diseases with trial vaccines are HSV, RSV, CMV, influenza, and rotavirus. The vaccine field has expanded from infectious diseases to include cancer (particularly melanoma, but also breast, colorectal, ovarian, and B-cell cancers), rheumatoid arthritis and multiple sclerosis, and contraception. The vaccine itself is showing a dynamic evolution from the original whole virus (dead or attenuated), through subunits (native and recombinant), to genetically deleted viruses, with the future aim of DNA vaccines and mucosal immunity. (see Viral Vaccine Production in Cell Culture)

Immunoregulators

The production of alpha interferon (Wellferon) by Wellcome (13) in Namalwa cells was the first licenced use of a cancer cell substrate for a human biological. The unrestricted growth in suspension culture of these cells allowed the first multi-thousand-liter (8000-L) unit process to be developed for human products. The knowledge gained from producing FMDV vaccine in suspension BHK cells (5,6) was of great benefit in developing this process.

There are a wide range of both interferons and interleukins occurring naturally (24), and to date alpha and gamma interferons and interleukin 2, 3, 4, 6, 11, and 12 have been manufactured in culture. (see Cell Products—Immunoregulators)

Antibodies

The production of monoclonal antibodies (25), first at the research level, then for diagnostics (including in vivo imaging), from the mid-1980s gave a huge impetus to industrial animal cell biotechnology (26). It stimulated the development of many novel "turnkey" culture units that could be used by relatively inexperienced (in cell culture) staff in the immunology laboratory, or as a production method in many startup biotechnology companies (17–19,27,28). The use of Mabs expanded from small-requirement (dose) diagnostics to large-dose therapeutics for HIV, cancer, allergic diseases, arthritis, renal prophylaxis, septic shock, transplantation, asthma, CMV, and anti-idiotype vaccines. The development of recombinant Mabs was largely driven by the need to "humanize" (29) the product to prevent immunological incompatibilities, leading to very short half-lives and making only single-dose treatment possible. The field has moved on to the use of adoptive immunotherapy, where the patient's cells are altered and grown in vitro and perfused back into the patient. Many novel products have been developed, such as the CD-4 receptor, which is a combination of the genes coding for the soluble form of the CD-4 receptor with the gene sequence for IgG molecules, which results in a soluble receptor for HIV. (see Cell Products—Antibodies)

Recombinant Products

The fact that cells with specialized in vivo functions, such as endocrine cells secreting hormones, could not be grown and replicated in culture with retention of their specialized properties has always been a great disappointment not only for advancing medical studies but also for using cells to manufacture naturally occurring biologicals. Thus genetic engineering techniques that allow the gene(s) responsible for production of required biologicals in a highly differentiated (nonculturable) cell to be inserted into a fast-growing robust cell line (30) opened up numerous possibilities for exploitation by animal cell technology (Tables 3 and 4).

Tissue Plasminogen Activator. The pioneering work to bring the first recombinant product from animal cells to the clinic was carried out by Genentech with tissue plasminogen activator (tPA) (31). This is a product necessary for dissolving blood clots for the treatment of myocardial infarction and thrombolytic occlusions. Alternative products, urokinase and streptokinase, were less specific and could cause general internal bleeding and other side effects. Attempts to develop a tPA process had been going on for many years. To put the problem in perspective, endothelial cells and other in vivo rich sources (e.g., human uterus) contain only 1 mg tPA/5 kg uterus (0.01 mg purified tPA /uterus) (32,33). Some tumor cell lines such as Bowes melanoma (33,34) secrete tPA at a higher rate (0.1 mg/L), but this was still uneconomical for a production process and (at that time) was considered unsafe, as coming from a human melanoma.

tPA was therefore an ideal product for cell technology and an example of a high-activity/low-concentration product that was clinically in demand. Genetic engineering not only allowed the product to be produced in a relatively safe cell line, but was used to amplify cell production $(50 \text{ mg}/10^9 \text{ CHO cells/day})$ from the low native secretion rates (34). This product was licenced as Activase/Actilyse in 1987.

Genentech engineered a CHO cell line able to express high levels of tPA, develop a large-scale (10,000-L) fermenter-based process, and provide the safety and efficacy data for the FDA to licence a recombinant product. This opened the way for a succession of new products, EPO being the next most notable one.

Erythropoietin. Erythropoietin (EPO) is a hormone produced by the kidney that controls the maturation of red blood (erythroid) cells, with clinical applications in anemia due to chronic renal failure. It is produced in a CHO cell transfected with the pSSVL-gHu Epo plasmid and is enhanced by gene amplification (35,36) using a roller bottle process. Purification is by a series of chromatography and gel filtration steps (37). The product was licenced in 1989 (by Amgen) as Epogen and in 1990 as Epogin.

Cell and Tissue Therapy

Cell Therapy. Cell therapy is the replacement, repair, or enhancement of biological function of damaged tissue or organs. This is achieved by transplantation of cells to a target organ by injection (e.g., fetal cells into the brain of patients with Parkinson's or Alzheimer's disease), or by implantation of cells selected/engineered to secrete missing gene products.

The first recorded application was to grow keratinocytes from a small skin biopsy into large cell sheets, which were then grafted onto burns patients (38,39). This has now advanced to commercial production of dermal replacement products (e.g., Dermagraft, 40). To avoid destruction of implants by the host's immune system, encapsulation of the transplant cells in semipermeable devices is widely used (41). Examples include pancreatic islet cells for diabetes (42), chromaffin cells for chronic pain (43), and genetically engineered BHK cells secreting neurotrophic factors for neurodegenerative diseases (44,45). It has not yet been possible to replace the liver or kidney, but artificial organs situated outside the patient containing primary or recombinant cells through which the patient's blood is perfused have been developed (46,47). Dialysis techniques only remove the toxic products, whereas the cells in the artificial organs perform biotransformations; that is, as well as degrading toxic products, they additionally regenerate many essential metabolites that are returned to the body.

The future of cell therapy is expected to be based on stem cells (self-renewing cells that give rise to phenotypically and genotypically identical daughter cells). Stem cells develop via a "committed progenitor stage" to a terminally differentiated cell. They are multipotent, that is, able to develop into a wide range of tissues and organs, but only fertilized germ cells are totipotent, that is, able to give rise to all cell tissues in the body. Control of the development of stem cells into the required tissue, or to stimulate quiescent "committed progenitor cells" of the required tissue, with the relevant growth factors and hormones would allow the most effective cell therapy possible. This approach is causing some ethical controversy, as the most suitable method of producing stem cells is to clone from the human embryo. The technique is to extract the genetic material from an adult patient needing transplantation, introduce it into a human egg with its nucleus removed, and grow the embryo in vitro for 8 divisions until stem cells can be treated with growth factors to form the required tissue (e.g., pancreas, nerve etc.).

The applications of cell therapy include: (reviewed by Gage, 48)

- Diabetes (pancreatic islet cells)
- Parkinson's disease (fetal dopamine cells)
- Duchenne's muscular dystrophy (myoblasts)
- Liver disease (parenchymal hepatocytes)
- Burns patients (keratinocytes and fibroblasts)
- Cartilage damage (chrondocytes)
- Pain (chromaffin cells)
- Cardiovascular disease (endothelial cells)
- Brain and spinal cord (neurotrophic factor secreting cells)
- Cancer (haemopoitic cells, bone marrow, adoptive cellular therapy)
- Retinal pigmented epithelium
- Huntingdon's disease

Gene Therapy. Gene therapy has the potential for treating a very wide range of human diseases, but since the first somatic gene therapy product (T-lymphocyte-directed gene therapy of ADA-SCID) went into trial in 1990, progress has been disappointingly slow, although there are over 300 clinical products at some stage of clinical trial. One problem is the development of safe and efficient gene-delivery systems, which needs careful regulation for safety (49). This has largely centered on engineering viruses as vectors of therapeutic genes.

Three modes of gene delivery are possible;

1. *ex vivo*. Remove the cells from the body, incubate with a vector, and then return engineered cells to the body (mainly applicable to blood cells).
2. *in situ*. Vector is placed directly into the affected tissues (e.g., infusion of adenoviral vectors into trachea and bronchi for cystic fibrosis patients).
3. *in vivo*. Vector is injected directly into the blood stream (this method is a goal but has not yet been used clinically).

The target diseases currently undergoing clinical trial are:

- Cancer (melanoma, colon, renal cell, neuroblastoma, ovarian, breast, lung)
- Genetic diseases (cystic fibrosis, Gaucher's, SCID)
- Viral (AIDS)

For a full review, see Anderson (50).

Other Products

The use of cells in toxicology, pharmacology, and testing (51) is an important area, as it allows more controlled experiments and significantly reduces the need for using animals (52).

A product area of increasing interest and potential is the cell-adhesion molecule (CAMs). These are molecules that mediate cell–cell and cell–matrix interactions and are being developed as drugs against inflammatory diseases. They also have the potential to treat metastatic diseases, atherosclerosis, and microbial infection. Chemokines present at sites of inflammation and disease bind leukocyte receptors and activate a family of CAMs known as integrins.

There is a sequential activation and interaction of multiple CAMs in the inflammatory process, which offers many targets for drug intercession. Target products include antisense and antagonists. Examples of such drugs undergoing trial are Cylexin (reperfusion injury), Integretin (arterial thrombosis, angina), and Celadin (inflammatory diseases). However, there are over 30 companies developing CAMs that are in various stages of preclinical and clinical trial, giving a high expectancy that some will soon make the clinic.

Other products not reviewed, such as hormones, blood factors (53), and growth factors (54) are given in the tables and are self-explanatory. For more information, see reviews by Griffiths (24,55), and Butler (56).

CONCLUSIONS

Since the first product (polio vaccine) in 1954, animal cell biotechnology has played a very significant role in producing a vast range of human and animal health products. The technology to produce cells in bulk and thus make an economical product has seen multiple small-flask cultures expand to 10,000-L unit processes and the ability to increase unit cell density from 2 to 100 million/mL. Manufacturers now have a choice between large batch cultures and smaller high-density perfusion cultures giving a daily harvest over 100 d. The ability to genetically engineer cells has had a great impact not only in making possible the production of many new biologicals that previously were unculturable, but also in enhancing cell productivity. The subject is still evolving fast, as it is no longer solely the province of the engineers running large bioreactors but now includes medical practitioners developing cell and gene therapy and tissue engineering solutions to disease. In fact, the day has arrived where the cell itself is the important product, rather than just being a vehicle or factory for producing proteins.

BIBLIOGRAPHY

1. J.F. Enders, T.H. Weller, and F.C. Robbins, *Science* **109**, 85–87 (1949).
2. L. Hayflick and P.S. Moorhead, *Exp. Cell Res.* **25**, 585–621 (1961).
3. J.P. Jacobs, *Nature (London)* **227**, 168–170 (1970).
4. A.J. Beale, *Animal Cell Biotechnology*, Vol. 5, Academic Press, London, 1992, pp. 189–200.
5. R.E. Spier, *Adv. Biotechnol. Processes* **2**, 33–59 (1983).
6. P.J. Radlett, T.W.F. Pay, and A.J.M. Garland, *Dev. Biol. Stand.* **60**, 160–170 (1985).
7. P.J. Radlett, *Adv. Biochem. Eng./Biotechnol.* **34**, 129–146 (1987).
8. J.C. Petricciani, *Animal Cell Biotechnology*, Vol. 3, Academic Press, London, 1988, pp. 13–25.
9. J.B. Griffiths and W. Noe, *Safety in Cell and Tissue Culture*, Kluwer Academic Publishers, Dordrecht, The Netherlands, 1998, pp. 135–154.
10. A. Doyle and J.B. Griffiths, *Animal Cell Biotechnology*, Vol. 5, Academic Press, London, 1992, pp. 75–96.
11. M. Mazur-Melnyk, *Cell and Tissue Culture: Laboratory Procedures in Biotechnology*, 1998, pp. 254–261.
12. G.F. Panina, *Animal Cell Biotechnology*, Vol. 1, Academic Press, London, 1985, pp. 211–242.
13. K.F. Pullen et al., *Dev. Biol. Stand.* **60**, 175–178 (1985).
14. A.L. vanWezel, *Nature (London)* **216**, 64–65 (1967).
15. B. Montagnon, J.C. Vincent-Falquet, and B. Fanget, *Dev. Biol. Stand.* **55**, 3742 (1984).
16. J.B. Griffiths, *Cell and Tissue Culture: Laboratory Procedures in Biotechnology*, 1998, pp. 262–267.
17. M.L. Wolf and M.D. Hirschel, *Animal Cell Biotechnology*, Vol. 6, Academic Press, London, 1994, pp. 237–258.
18. J.A.J. Hanek and J.M. Davis, *Cell and Tissue Culture: Laboratory Procedures*, Wiley, Chichester, 1993, pp. 28D:3.1–16.
19. P.W. Runstadler and S.R. Cernek, *Animal Cell Biotechnology*, Vol. 3, Academic Press, London, 1988, pp. 305–320.

20. D. Looby, *Cell and Tissue Culture: Laboratory Procedures in Biotechnology*, 1998, pp. 268–281.

21. J.B. Griffiths, *Animal Cell Biotechnology*, Vol. 4, Academic Press, London, 1990, pp. 149–166.

22. A. Mizrahi, *Adv. Biotechnol. Processes* **8**, (1988).

23. A. Doyle, *Cell and Tissue Culture: Laboratory Procedures in Biotechnology*, 1998, pp. 295–303.

24. J.B. Griffiths, *Mammalian Cell Biotechnology: A Practical Approach*, IRL Press, Oxford, U.K., 1991, pp. 207–235.

25. G. Kohler and C. Milstein, *Nature (London)* **256**, 495–497 (1975).

26. A. Mizrahi, *Adv. Biotechnol. Processes* **11**, (1989).

27. G.J. Berg, *Dev. Biol. Stand.* **66**, 195–209 (1987).

28. *Tecnomouse*, Integra Biosciences, AG, CH-8304 Wallisellen.

29. W.J. Harris, *Animal Cell Biotechnology*, Vol. 6, Academic Press, London, 1990, pp. 259–279.

30. P.G. Sanders, *Animal Cell Biotechnology*, Vol. 4, Academic Press, London, 1990, pp. 15–70.

31. Lubiniecki et al., *Animal Cell Biology and Technology for Bioprocesses*, Butterworth, Guildford, U.K., 1989, pp. 442–451.

32. J.B. Griffiths and A. Electricwala, *Adv. Biochem. Eng./Biotechnol.* **34**, 147–166 (1987).

33. T. Cartwright, *Animal Cell Biotechnology*, Vol. 5, Academic Press, London, 1992, pp. 218–246.

34. C. Kluft et al., *Adv. Biotechnol. Processes* **2**, 97–110 (1983).

35. S. Eridani, *Animal Cell Biotechnology*, Vol. 4, Academic Press, London, 1990, pp. 469–488.

36. Pat. Appl. WO 85/02610 (1986) Kirin -Amgen Inc.

37. Pat. Appl. WO 86/07594 (1986) Kirin -Amgen Inc.

38. J.G. Rheinwald and H. Green, *Cell* **6**, 331–344 (1975).

39. A.M. Burt and D.A. McGrouther, *Animal Cell Biotechnology*, Vol. 5, Academic Press, London, 1992, pp. 151–168.

40. G. Naughton, J. Mansbridge, and G. Gentzkow, *Artif. Organs* **21**, 11 (1997).

41. R.P. Lanza, J.L. Hayes, and W.L. Chick, *Nat. Biotechnol.* **14**, 1107 (1996).

42. D.W. Scharp et al., *Diabetes* **43**, 1167–1170 (1994).

43. J.M. Joseph et al., *Cell Transplant.* **3**, 355–364 (1994).

44. P. Aebischer et al., *Trans. Am. Soc. Artif. Intern. Organs* **32**, 134–137 (1986).

45. P. Aebischer, *Animal Cell Technology: From Vaccines to Genetic Medicine*, Kluwer Academic Publishers, Dordrecht, The Netherlands, 1997, pp. 29–31.

46. M.V. Peshwa et al., *Animal Cell Biotechnology: Products of Today, Prospects for Tomorrow*, Butterworth-Heinemann, Oxford, U.K., 1994, pp. 273–277.

47. H. Ijima et al., *Animal Cell Technology: From Vaccines to Genetic Medicine*, Kluwer Academic Publishers, Dordrecht, The Netherlands, 1997, pp. 577–583.

48. F.H. Gage, *Nature (London), Suppl.* **392**, 18–24 (1998).

49. O. Cohen-Haugenauer, *Hum. Gene Ther.* **6**, 773–785 (1995).

50. W.F. Anderson, *Nature (London), Suppl.* **392**, 25–30 (1998).

51. D.J. Benford, *Animal Cell Biotechnology*, Vol. 5, Academic Press, London, 1992, pp. 97–121.

52. B. Coulomb and L. Dubertret, *Animal Cell Biotechnology*, Vol. 5, Academic Press, London, 1992, pp. 123–168.

53. A.J. MacLeod, *Animal Cell Biotechnology*, Vol. 5, Academic Press, London, 1992, pp. 201–217.

54. P. Collodi and D.W. Barnes, *Animal Cell Biotechnology*, Vol. 5, Academic Press, London, 1992, pp. 247–278.

55. J.B. Griffiths, *Animal Cell Biotechnology*, Vol. 2, Academic Press, London, 1990, pp. 3–12.

56. M. Butler, *Animal Cell Technology, Principles and Products*, Open University Press, Milton Keynes, U.K., 1987.

See also CELL PRODUCTS—ANTIBODIES; CELL PRODUCTS—IMMUNOREGULATORS; CELL PRODUCTS—VIRAL GENE THERAPY VECTORS; ETHICAL ISSUES IN ANIMAL AND PLANT CELL TECHNOLOGY; MICROPROPAGATION OF PLANTS, PRINCIPLES AND PRACTICES; PLANT CELL CULTURES, SECONDARY PRODUCT ACCUMULATION; PLANT CELL CULTURES, SELECTION AND SCREENING FOR SECONDARY PRODUCT ACCUMULATION; TRANSCRIPTION, TRANSLATION AND THE CONTROL OF GENE EXPRESSION.

ANIMAL CELL TYPES, B CELL LINEAGE

K.B. WALKER
National Institute for Biological Standards and Control
South Mimms, England

OUTLINE

The B-Cell Lineage: Introduction

B Cells In Vivo

B Cells In Vitro

Bibliography

THE B-CELL LINEAGE: INTRODUCTION

In animal immune systems B cells provide humoral immunity through the secretion of antibodies in response to antigens (e.g., bacterial polysaccharides, virus glycoproteins) processed by the cell-mediated arm of the immune system, including T lymphocytes and cells of the monocyte–macrophage lineage (see sections on T lymphocytes, thymus cells, and macrophages). This section deals primarily with B cells of human origin, and there are some indicators that B-cell development in mouse is very similar to human (1). Among other species, the B cells of representatives from domestic livestock tend to be more deeply investigated, such as pig (2,3), chicken (4,5), and cow (6).

B CELLS IN VIVO

B lymphocytes—the name originating from "Bursa derived"—are a subpopulation of circulating lymphoid cells. In mammals these cells arise in the bone marrow from pluripotent stem cells under the influence of various cytokines (7,8). This process involves the commitment of the stem cell to the B-cell lineage, followed by differentiation through a number of recognized phenotypic and functional changes. Human bone marrow produces on the order of 10^{10} to 10^{11} B cells per day; however, only a small proportion of these cells manage to pass through the many developmental checkpoints before selection into the mature recirculating B-cell pool (9).

Table 1. Expression of Key Cell Markers in Different B-Cell Populations

B Cell Type	Level of Expression of Cell Markers					
Mature	SIgD	sIgM	CD21	CD22	CD23	Bcl-2
Immature	Low	Medium	Low	Low	Low	Low
Recirculating	High	Medium	Medium	High	Medium	High
Marginal zone	Low	High	High	High	Low	High

Table 2. Representative Examples of B-Cell Lines

Species	Cell Name	Characteristics/Applications	Collection Codes[a]
Bovine	BL-3	Established from neoplastic lymphode of a cow. Used in studies of bovine viruses.	ATCC CCL 240 ECACC 88112501
Human	BRISTOL 8	B lymphoblastoid cell line of defined human leucoyte antigen (HLA) type. Used in tissue typing studies.	ATCC TIB202 ECACC 88081201
	DAUDI	B lymphoid cell line derived from a case of Burkitt's lymphoma. Extensively used in studies of the mechanisms of leukemogenesis.	ATCC CCL213 ECACC 85011437
	HS-SULTAN	Plasmacytoma cells expressing IgG Kappa antibody. Used in tumorigenicity studies.	ATCC CRL 1484 DSMZ ACC78 ECACC 87012701
	IM-9	B lymphoblastoid cells isolated from a case of multiple myeloma. Reported to have receptors for insulin and calcitonin.	ATCC CCL 159 DSMZ ACC117 ECACC 86051302
	Mo-B	Epstein Barr virus (EBV) transformed B lymphoblast cells. Produces EBV antigens and is also infected with HTLV-II virus.	ATCC CCL 245 ECACC 90021503
	NAMALWA	Burkitt's lymphoma cell line. Used for the manufacture of human interferon.	ATCC CRL 1432 DSMZ ACC24 ECACC 87060801
	NC-37	EBV-transformed B lymphocytes. Used for chemical induction of EBV and EBV superinfection studies.	ATCC CCL 214 ECACC 89111414
Monkey/primate	26 CB-7	B lymphoblastoid cells expressing EBV-like antigen and are infected with *Herpes paio*.	ATCC CRL 1495 ECACC 89072101
	B95-8	Marmoset lymphoblastoid cells. Produces high titers of EBV. Also infected with a type D retrovirus.	ATCC CRL 1612 DSMZ ACC100 ECACC 85011419
	FR (JC)	EBV-transformed lymphoblastoid cell line derived from a female gorilla. One of a range of similar cell lines derived from high primates by in vitro EBV transformation.	ECACC 89072703
	LCL 8664	B-cell lymphoma cell line from a Rhesus monkey. Infected with B-lymphotrophic herpes virus (RhEBV).	ATCC CRL 1805 ECACC 91030709
Mouse	BCL 1 Clone CW13.20-3B3	A clone derived from BCL 1 tumor cells by limiting dilution. Used in the study of B-cell differentiation.	ATCC CRL 1669 ECACC 90061904
	MOPC 31C	Plasmacytoma cell line derived from a mineral oil induced tumor in Balb/C mice. Secretes IgG1.	ATCC CRL 130 ECACC 90110707
	MPC II	Myeloma cells derived from the Merwin plasma cell tumor-II in Balb/C mice. Produces fully assembled IgG 2b.	ATCC CCL 167 ECACC 91031103
	NFS-70 C10	Lymphoma cell produced by CaS-NS-7 ecotropic muvine leukaemia virus infection of an NFS/N mouse. Exhibits early pre-B-cell markers.	ATCC CRL 1694 ECACC 88041905
	NS0	Myeloma cell line used in the production of hybridomas (see hybridoma cell section).	ECACC 85110503
Porcine	L14	One of a series of cell lines established from peripheral blood mononuclear cells of a domestic boar infected with Shimozuma cells producing porcine retrovirus (Tsukuba-1). L14 cells contain retrovirus particles and express reverse transcriptase. The cells also express membrane IgM.	ECACC 91012317
Rat	Y3.Ag.1.2.3	A subclone of an azaguanine-resistant mutant of a LOU rat myeloma. Used as a fusion partner in the production of rat hybridomas (see hybridoma cell section).	ATCC CRL 1631 ECACC 85110502

[a]For further information on the cell lines available from resource centers, see the article on cell banks.

The average life span for a circulating B cell in a mouse is of the order of 6 weeks (10). The maturation of B cells upon interaction with antigen involves a number of changes, including uptake, processing, and presentation of antigen, an increased capacity to interact with T cells, and interaction with antigen-primed T cells, leading to proliferation and differentiation into plasma (antibody-secreting) cells (11,12). The continued development of B cells along this pathway to antibody-secreting cells is assured only by the timely exposure to "continuation signals" that ensure that the developing B cell does not undergo apoptosis (13,14).

B cells are responsible for the production of antibodies and the generation of the humoral immune response. Antibodies (or immunoglobulins) are proteins that have specificity to an antigen and can create complexes of the antigen and thereby facilitate a range of other immune responses. Phagocytosis of antigen–antibody complexes and complement activation are just two such processes initiated or facilitated by antibody–antigen binding.

There are a range of surface antigens (markers) that are used to identify and characterize B cells, including secretary IgD (SIgD), surface IgM (sIgM), CD21, CD22, CD23, Bcl-2, CD5, and CD116. Together these provide a phenotypic profile that enables characterization of particular subsets of the B-cell lineage, as shown in Table 1. In addition, such markers allow identification of the B-1a subset of B lymphocytes, which are rare in the periphery and have been associated with some autoimmune processes.

Antigen recognition and antigen binding to surface Ig (or the B-cell receptor, BCR), act as a signal to the B lymphocyte to undergo activation, proliferation, and subsequent differentiation. The presence of certain cytokines (from T helper cells) is crucial to this process, and modulation of cytokine levels can have profound effects on the maturation and nature of the B-cell response (11,12,15). One of the most complex processes involved in B-cell maturation is the maturation of the antibody response and immunoglobulin isotype switching (16,17). B-cell activation and differentiation may also be initiated by large non-protein molecules in a manner that does not require the presence of T helper cells—the so-called T-independent activation pathway. These activation signals tend to be polyclonal or nonspecific, whereas the T-dependent activation pathway is focused on antigen-specific B cells.

B CELLS IN VITRO

The development and maturation of B cells in vivo is a complex process with a number of distinct stages. These different stages are represented by a number of B-cell lines that have acted as excellent research tools to dissect particular aspects of B-cell development and antibody maturation (Table 2).

Many B-cell lines have been isolated from B-cell leukemias in humans and animals. B-cell lines can also be obtained from normal peripheral or spleen-derived B lymphocytes in vitro by transformation into immortal lines—one of the most common methods being Epstein Barr virus (EBV) transformation (18) (also see Cell Banking article). A useful source of EBV for this immortalization process is also a B-cell line B95-8, which releases EBV at high titer when incubated under low serum conditions at room temperature (see Table 2). Repeated intraperitoneal injection of irritants such as mineral oil or pristane have resulted in the derivation of lymphomas and a range of myeloma cell lines used in the production of hybridoma cells (see section on hybridoma cells). Clinical material derived from Hodgkin's lymphoma, myeloid leukemias, and lymphoblastoid leukemias have also allowed the derivation of a range of B-cell lines. Exposure to carcinogens (e.g., methyl nitrosourea or methycolanthrene) have also been used in mice and rats to induce leukemias or to produce ascites.

Since the development of hybridoma technology for the production of monoclonal antibodies, the number of antibody secreting lines has increased exponentially, and there are now whole chapters of the cell line catalogues devoted to these B-cell hybridoma lines. The technology of the production of these lines and their importance is covered in the section on hybridoma cells.

BIBLIOGRAPHY

1. P. Ghia, E. Ten Boekel, A.G. Rolink, and F. Melchers, *Immunol. Today* **19**, 480–485 (1998).
2. A.T. Bianchi, R.J. Zwart, S.H. Jeurissen, and H.W. Moonen-Leusen, *Vet. Immunol. Immunolpathol.* **33**, 201–221 (1992).
3. H. Yang and R.M. Parkhouse, *Immunology* **89**, 76–83 (1996).
4. M. Gomez Del Moral et al., *Anat. Rec.* **250**, 182–189 (1998).
5. D.M. Maslak and D.L. Reynolds, *Avian Dis.* **39**, 736–742 (1995).
6. Morrison et al., *Prog. Vet. Microbiol. Immunol.* **4**, 134–164 (1988).
7. T.W. Le Bien, *Curr. Opin. Immunol.* **10**, 188–195 (1998).
8. U. von Freeden Jeffrey et al., *J. Exp. Med.* **181**, 1519–1526 (1995).
9. M. Hertz and D. Nemazee, *Curr. Opin. Immunol.* **10**, 208–213 (1998).
10. I.C.M. MacLennan, *Curr. Opin. Immunol.* **10**, 220–225 (1998).
11. Y.J. Liu, S. Oldfield, and I.C.M. MacLennan, *Eur. J. Immunol.* **18**, 355–362 (1988).
12. J. Jacob, R. Kassir, and G.J. Kelsoe, *J. Exp. Med.* **173**, 1165–1175 (1991).
13. J.G. Cyster, S.B. Hartley, and C.C. Goodnow, *Nature (London)* **371**, 398–395 (1994).
14. J.G. Cyster and C.C. Goodnow, *Immunity* **3**, 691–701 (1995).
15. T. Reya and R. Grosschedl, *Curr. Opin. Immunol.* **10**, 158–165 (1998).
16. K. Andersson, J. Wrammert, and T. Leanderson, *Immunol. Rev.* **162**, 172–182 (1998).
17. I.C.M. MacLennan et al., *Immunol. Rev.* **156**, 53–66 (1997).
18. E.V. Walls and D.H. Crawford, in G.G.B. Klaus, ed., *Lymphocytes: A Practical Approach*, IRL Press, Oxford, U.K., pp. 149–162.

ANIMAL CELL TYPES, CELL LINES USED IN MANUFACTURE OF BIOLOGICAL PRODUCTS

G. STACEY
National Institute of Biological Standards and Control
South Mimms, England

OUTLINE

Animal cell lines are increasingly used in the manufacture of biologics intended for use as diagnostic and therapeutic reagents. Although monoclonal antibodies are used in both areas, hybridoma cell lines are dealt with elsewhere in the articles on hybridoma cells. In this article other cell lines used in the manufacture of natural cell products such as interferon, viral vaccines, and recombinant proteins will be discussed specifically. However, much of the general discussion applies equally to hybridoma cells used in production processes.

HISTORICAL PERSPECTIVE: ANIMAL CELLS IN THE MANUFACTURE OF BIOLOGICALS

Primary animal cells have been used for many years for vaccine production. These have generally proven acceptable and safe, but there are notable exceptions (see below), which have directed manufacturers and regulatory bodies to be very cautious in assessing new cell substrates. In addition, there has been a progressive move toward the validation and use of cell lines for which cell banks of fully authenticated and safety-tested cells can be established. The earliest cell lines used in production were human diploid fibroblasts (HDFs). MRC-5 and WI-38 are two of the best known and have been used in the manufacture of a number of licensed products (see below).

The next key development was the acceptance of the CHO cell line to produce "Activase" (tissue plasminogen activator), the first therapeutic protein manufactured from a transfected mammalian cell line to be marketed. Today, a wide range of potential diagnostic and therapeutic products are being developed in CHO cell-expression systems, and an ever expanding range of cell substrates is being worked with as candidate production cells. However, all will be subject to rigorous safety testing (see below) and validation before licensed products for which they are used as production substrates can be accepted for marketing.

THE NEED FOR ANIMAL CELLS AS SUBSTRATES FOR MANUFACTURING PROCESSES

For products such as vaccines (and especially live attenuated vaccines), the only way to produce the whole organism in its native form as a raw material is to replicate it by infecting animal cells that have the necessary complex biochemical machinery to generate virus particles. In addition, many complex recombinant products under development require the particular post-translational modifications that are best achieved in animal cells. In particular, the glycosylation that can be achieved in animal cells is closer to that which is found in the natural equivalents *in vivo* of recombinant therapeutic glycoproteins. This is of importance to provide therapeutic reagents that are not cleared rapidly by the liver *in vivo* and that have some resistance to degrading enzymes *in vivo* and during the *in vitro* production process (1). Bacterial and yeast host cells for recombinant DNA are extremely efficient and can produce high concentrations of product in culture supernatant but fail to glycosylate or produce inappropriate glycosylation (2,3).

THE CELL SUBSTRATES

Primary Cells

Primary cells are those isolated and cultured directly from tissue or organs. Such cells are still used in some cases for producing certain vaccines. Primary cells have been used for manufacturing a range of efficacious and safe vaccines. However, there have been some notable problems with cells as a source of contaminating organisms. Primary monkey kidney cells were identified in the 1960s as the source of the contamination of polio vaccine with SV40 virus (4). Although this contamination did not have an effect on vaccinees, it has required long term monitoring, and disease associations have been claimed more recently regarding the discovery of SV40 DNA in mesotheliomas (5–7). In an infamous case unrelated to vaccine production, the maintenance of rodent tumor cells *in vivo* led to the transmission of lethal infections to laboratory workers due to hantavirus in the laboratory (8).

As a result of these rare yet worrying examples and the international initiatives to reduce the use of animals in research and testing (9,10), there is a progressive movement to replace the use of animals, and thus primary cells, with continuous cell lines. However, the identification of cell lines with appropriate virus susceptibility and growth characteristics for scale-up of viral vaccines is a significant challenge. Even when promising candidate cell substrates are identified, they must undergo rigorous safety testing (see below) and validation for scale-up of cultures to the levels required for production.

HUMAN DIPLOID FIBROBLASTS

Human diploid fibroblasts (HDFs) were among the early cell types established as cell lines (11) and were quickly considered potential production cell substrates (12), although there were early concerns regarding their safety (13). Thousands of HDF cultures have been established and have been used widely in pharmacology, toxicology, biochemistry, genetics, and microbiology. However, relatively few have been adequately characterized to enable their use as substrates for manufacturing biologics. Examples of such cells that have been accepted for manufacturing are MRC-5 (14) and WI-38 (15). These particular cells have been rigorously tested for their stability in scale-up and for the presence of adventitious agents. In addition, they have a safe history of use over several decades which enhances their acceptability, although this cannot be used as evidence of the absence of adventitious organisms.

HDF cells are the predominant representative of a culture type designated as "finite cell lines" (16) because they have reproducibly limited capacity for replication, often quoted in the range of 50 passages. Thus, although they have been used successfully in vaccine production, ultimately they represent a limited resource in terms of population doublings. They also undergo genotypic and phenotypic changes that occur progressively as each culture of the cell line approaches senescence (17). Because of this, strict limits are set on the population doubling number at which the cells can still be used for production (18).

HDF cells and WI-38 and MRC-5, in particular, are valuable cell substrates for vaccine production because they are not tumor-derived and have long been recognized for their wide range of susceptibility to viruses (19). Scale-up of diploid fibroblast cultures is based on the growth of the cells as adherent monolayers in flat flasks or roller bottles, although attempts have been made to grow them in suspension culture. A commonly used means of establishing large-scale pseudosuspension culture bioreactor systems has been to grow the fibroblasts on microcarriers in a stirred vessel (20). It is also possible to subculture HDFs on microcarriers to rapidly achieve the cell numbers required for production runs (21). Work with MRC-5 has shown that the type of microcarrier substrate used for the growth of HDFs can significantly affect their virus productivity and therefore requires careful validation (22). An alternative form of culture to stirred microcarrier systems is the culture of cells at high density on a fixed matrix, often a bed of porous beads or hollow fibers, whereby the cell bed is perfused with growth medium from a reservoir. Such perfusion systems have been tested for producing polio vaccine from MRC-5 cells (23).

A range of HDF cultures including WI-38, MRC-5, IMR90, and HE2299, has been used in microcarrier systems with similar success in the experimental production of polio vaccine (24), although WI-38 and MRC-5 were the first HDFs to be set up for polio vaccine trials (25,26). HDF cells have also been used to produce live attenuated and inactivated hepatitis A vaccines (27,28), recombinant hepatitis B (29), and herpes simplex virus (30). Interestingly, in microcarrier systems for the growth of herpes simplex, it was shown that, although MRC-5 cultures showed a significantly slower growth rate than the Vero continuous cell line (see below), MRC-5 was much more efficient in producing virus and performed equivalently to Vero overall as a production cell line (31). For some viruses such as measles and rubella, multiple harvests from the same batch of inoculated HDFs can be achieved (32), although the quality of such pooled material requires careful validation. In the future, HDFs will continue to provide safe and well-characterized substrates for vaccine production. Should stocks of WI-38 and MRC-5 eventually become depleted, then a range of other HDF strains such as IMR90 and MRC-9 are available to take their place.

Continuous Cell Lines

Continuous cell lines have the advantage that they are capable of indefinite replication, but the means by which immortalization is achieved (e.g., tumor origin, viral transformation) raised questions early in their development regarding their acceptability as cell substrates for the manufacture of biologics (33). A significant concern has been the tumorigenic nature of these cells despite their history of safe use (34).

CHO Cells. CHO cells have been used since the 1950s in a wide range of studies (35) and have been well characterized in gene transfer experiments (36). CHO cell lines have been established, and the large amount of validation data on their use, characteristics, and resistance to infection with many human viruses (37) has led to their acceptance as safe substrates for producing biologics.

CHO cells are readily adapted to serum-free culture and have been successfully grown as suspension cultures for the producing recombinant protein in bioreactors of up to 10,000 L (38). They are now widely used to produce recombinant putative therapeutic proteins, and some, notably human factor VIII (39), and human erythropoietin (40), are now in routine clinical use.

C127I. The C127I cell line was derived from the mouse RIII mammary tumor (41). It has been used widely as a host for the expression of recombinant proteins, including human interferon (42), human erythropoietin (43), human growth hormone (44), and hepatitis surface antigen vaccine (45). It has frequently been used with recombinant bovine papilloma virus vectors and in a variety of approaches to inducing protein expression, including heat shock (46,47).

Myeloma Cells. More recently, myeloma cell lines, notably NS0 (ECACC 85110503) and SP2/0-Ag14 (ATCC CRL 1581, CRL 8287, ECACC 85072408), have been used as host cells for recombinant DNA products. A variety of expression systems have been used and two popular systems are dihydrofolate reductase (DHFR)-mediated gene amplification (48) and the glutamine synthetase (GS) selection marker that allows only survival of cells that can synthesize glutamine (49). Myeloma cells are deficient in glutamine synthetase and therefore are ideal candidates for the GS system. Thus the GS system has the advantage that much lower levels of the toxic selection agent methionine sulfoximine are required in myeloma cells compared to CHO. This also means that myeloma-GS systems involve lower levels of toxicity than are required with methotrexate selection used in myeloma-DHFR selection. The GS-myeloma system has been used for a wide range of recombinant proteins, including recombinant antibodies and interferon.

Myeloma cells are favored due to their growth as suspensions of single cells and the wealth of characterization data established in their use in generating hybridomas for manufacturing monoclonal antibodies (see hybridoma articles). They are also amenable to genetic modification to enhance their growth characteristics, such as the expression of recombinant IL-6 by Sp2/mIL-6 cells that is reported to promote survival of nascent hybrid cells (50).

Namalwa Cells. Namalwa, a human B-lymphoblastoid cell line (51) (ATCC CRL 1432, ECACC 87060801), was established as a production cell line for the expression of natural interferon. Its characteristic growth as a suspension culture makes it a good candidate for the traditional stirred vessel approaches to scale-up. Early attempts to increase the productivity of interferon from this cell line were attempted by a variety of means, including treatment with sodium butyrate (52) and low-temperature production (53). Subsequently, subclones of Namalwa, called Namalwa NJM-1, have been selected for production (54). Namalwa NJM-1 produces no endogenous proteases which obviously promotes the achievement of high levels of intact recombinant proteins. This clone has been used to produce a wide range of recombinant proteins (55–57).

Vero Cells. The Vero cell line was derived from the kidney of an African green monkey (*C. aethiops*) in 1962 and has proved a useful vaccine production substrate due to its susceptibility to a wide range of viruses and its rapid growth rate (58). Vero cells have been accepted for the manufacture of vaccines, including rabies and polio (59,60), and they have been adapted to microcarrier culture for high cell density bioreactor culture of up to 10,000 L for polio vaccine production (61). More recently Vero cells have been established as a production substrate for influenza vaccines using bioreactors of up to 12,000 L (62).

A number of different strains of the original Vero cell line have been established. Vero 76 (ATCC CRL 1587, ECACC 85020205) is a subclone of Vero that has lower maximum cell density in monolayer culture and is particularly susceptible to haemorrhagic fever viruses (unpublished data, P. Jacobs, NIBSC). Another subclone of Vero 76, Vero 1008 (ATCC CRL 1586, ECACC 85020206), shows a similar range of susceptibility to viruses, but the cells show some degree of contact inhibition and are suitable for the growth of slowly replicating viruses. The cell line Vero 317 (ECACC 89070502) developed at the RIKEN Cell Bank in Japan was derived from a Vero subclone Vero 303 and is adapted to grow serum-free in a glutamate medium.

When Vero cells were identified as likely vaccine substrates, the WHO initiated the production of a large cell bank which was subjected to rigorous safety testing. Ampoules from this bank are distributed by a number of international cell banks, including ATCC and ECACC. This bank provides a well-characterized seed stock from which manufacturers may generate their own cell banks.

New Candidate Production Cells, BHK-21, MDCK, HeLa. The BHK-21 cell line is probably the most popular of the newer cell substrates proposed for producing biologics. It was derived from the kidney of baby Syrian hamsters (63), and, in common with CHO, is readily adapted to serum-free culture conditions and suspension or adherent culture in a range of bioreactor types. BHK cells are well characterized in terms of the glycosylation of recombinant proteins that they can achieve. The two primary clones, BHK-21A and BHK-21B, differ in their glycosylating capability, and bi-, tri- and tetra-antennary structures are sialylated only in the B clone (64). It is likely that the BHK-21 cell lines will become well established as production substrates as more data are gathered on their use and effectiveness.

The production of influenza vaccines is still carried out exclusively in embryonated hens eggs which struggle to meet the demand for vaccines to strains that change annually. In the event of a pandemic due to a strain not included in the current vaccine, manufacturers would be hard pressed to supply the necessary new vaccine. Because of the requirement for a modern, rapidly responding manufacturing process, much work has been devoted to establishing *in vitro* culture systems. The canine kidney cell line MDCK (see kidney cell section) is a lead candidate production cell line and its acceptability is under consideration for vaccine preparation (65). Production systems based on Vero cells have also been developed to a level similar to MDCK with a number of successfully completed clinical trials.

HeLa cells were used to produce experimental adenoviral vaccines in the 1950s, but these were aborted because of the potential oncogenicity of the product. However, HeLa cells are again being proposed for manufacturing recombinant products, and their safety will require very careful consideration.

CHALLENGES FOR THE USE OF CELL SUBSTRATES

The use of animal cells as production substrates also introduces a number of challenges for biotechnology. The efficiency and stability of mammalian production systems are critical to their commercial viability. Genetic stability is an important issue because loss of expression of the product or unstable expression during scale-up will severely reduce the likelihood that a product reaches the market. Very careful consideration must therefore be given to coordinating the following elements:

- establishment of authenticated and safety-tested cell banks (see cell bank article)
- selection and validation of stable cell clones
- development of the scale-up process.

Furthermore, appropriate glycosylation may improve the efficacy of a therapeutic agent, and therefore its quality and reproducibility must be addressed in relevant cases.

Appropriate testing to ensure the authenticity of the cells and their freedom from adventitious agents is vital for each cell substrate. Use of a production cell line that has not been adequately authenticated and had appropriate safety testing may cause the failure of the product at an early stage of development and will prevent approval of the product at the product licensing stage. The safety testing of a proposed production cell line is governed initially by regulatory guidelines on cell substrates such as publications from the World Health Organisation (WHO) (66), the Center for Biologics Evaluation and Research (CBER, Rockville, Maryland) and the International Conference on Harmonisation (ICH) (67). These give only a guide to the safety testing that should be performed to ensure authenticity, genetic stability, and the absence of mycoplasma, bacteria, fungi, and viruses. A full and very careful scientific analysis of each new cell substrate is essential because not all contraindicatory outcomes can be addressed by adhering to the guidelines alone.

Although the current production cell lines provide a battery of successful systems for producing a variety of vaccines and recombinant proteins, it is likely that many more novel cell substrates will be presented for regulatory approval in production processes. It will be important to make careful scientific assessments of each on a case by case basis to assure the safety of biologics.

Continued research into new cell substrates and approaches to the scale-up of culture systems is vital. Current production systems are generally based on a high cell growth rate, and this diverts cell energy from production and may lead to genetic instability in large or prolonged culture systems. There have been some approaches to reduce the growth rate of standard cell substrates such as CHO to increase productivity. In the future, however, it may be better to look for other cell types that are better adapted to slow replication but high protein production.

BIBLIOGRAPHY

1. D.A. Cummings, *Glycobiology* 1, 115–135 (1991).
2. A. Hooker and D. James, *J. Interferon Cytokine Res.* 18, 287–295 (1998).
3. M. Fussenegger, J.E. Bailey, H. Hauser, and P.P. Mueller, *Trends Biotechnol.* 17, 35–41 (1999).
4. J.F. Fraumeni, C.R. Stark, E. Gold, and M.L. Lepow, *Science* 167, 59–60 (1970).
5. J.R. Testa et al., *Cancer Res.* 58, 4505–4509 (1998).
6. C. Mulatero, T. Surentheran, J. Breuar, and R.M. Rudd, *Thorax* 54, 60–61 (1999).
7. M. Carbone et al., *J. Cell. Physiol.* 180, 167–172 (1999).
8. G. Lloyd and N. Jones, *J. Infect.* 12, 117–125 (1984).
9. G. Zbinden, *Biomed. Environ. Sci.* 1, 90–100 (1988).
10. M. Bally and D.W Straughan, *Dev. Biol. Stand.* 86, 11–18 (1996).
11. H.E. Swimm and R.F. Parker, *Am. J. Hyg.* 66, 235–243 (1957).
12. L. Hayflick *Progress in Immuno Biological Standards*, Karger, Berlin and New York, 1969.
13. L. Hayflick, S.A. Plotkin, T.W. Norton, and H. Koprowski, *Am. J. Hyg.* 75, 240–258 (1962).
14. J.P. Jacobs, C.M. Jones, and J.P. Baille, *Nature* 227, 585–621 (1961).
15. L. Hayflick and P.S. Moorhead, *Exp. Cell Res.* 25, 585–621 (1961).
16. W.I. Scaeffer, *In Vitro Cell Dev. Biol.* 26, 97–101 (1990).
17. V.J. Cristofalo et al., *Exp. Gerontol.* 27, 429–432 (1992).
18. D.T. Wood and P.D. Minor, *Biologicals* 18, 143–146 (1990).
19. L. Hayflick and P.S. Moorhead, *Exp. Cell Res.* 25, 585–621 (1962).
20. J.M. Clark and M.D. Hirtenstein, *Ann. N.Y. Acad. Sci.* 369, 33–46 (1981).
21. E. Lindner, A.C. Arvidsson, I. Wergeland, and D. Billig, *Dev. Biol. Stand.* 66, 299–305 (1987).
22. J. Varani et al., *In Vitro Cell Dev. Biol.* 22, 575–582 (1986).
23. G.F. Mann and J. de Mucha, *Dev. Biol. Stand.* 37, 255–258 (1976).
24. A. von Seefried and J.H. Chun, *Dev. Biol. Stand.* 47, 25–33 (1981).
25. B. Larsson and J. Litwin, *Dev. Biol. Stand.* 46, 241–247 (1980).
26. P.B. Stones, *Dev. Biol. Stand.* 37, 251–253 (1976).
27. J. Peetermans, *Vaccine* 10(Suppl. 1), S99–S101 (1992).
28. A.W. Funkhouser et al., *J. Virol.* 70(11), 7948–7957 (1996).
29. L. Kutinova et al., *Arch. Virol.* 112(3–4), 181–193 (1990).
30. B. Thornton, I.D. McEntee, and B. Griffiths, *Dev. Biol. Stand.* 60, 475–481 (1985).
31. J.B. Griffiths, B. Thornton, and I.D. McEntee, *Dev. Biol. Stand.* 50, 103–110 (1981).
32. H. Mirchamsy et al., *Dev. Biol. Stand.* 37, 297–300 (1976).
33. J.C. Petricciani et al., *J. Natl. Cancer Inst. (U.S.)* 57(4), 915–919 (1976).
34. A.S. Lubiniecki, *Bioprocess Technol.* 10, 495–513 (1990).
35. T.T. Puck, Development of the Chinese Hamster Ovary for use in Somatic Cell Genetics in: ed., M.M. Gottesman, *Mol. Cell Genet.* 37–64 (1985).
36. R.J. Kaufman and P. Sharp, *J. Mol. Biol.* 159, 601–621 (1982).
37. M.E. Wiebe et al., in R. Spier, J.B. Griffiths, J.S. Stephenne, and S. Croy, eds., *Advances in Animal Cell Biotechnology and Technology for Bioprocesses*, Butterworth-Heinemann, Oxford, U.K., 1989, pp. 68–71.
38. W.R. Arathoon and J.R. Birch, *Science* 232, 1390–1392 (1986).
39. E. Gomperts, R. Lundblad, and R. Adamson, *Transfus. Med. Rev.* 6, 247–251 (1992).
40. J.K. Faudrey and W.E. Jelkman, *Z. Gesamte Inn. Med. Thre Grenzgeb.* 47, 231–238 (1992).
41. D.R. Lowry, E. Rands, and E.M. Seolnick, *J. Virol.* 26, 291–298 (1978).
42. R. Fukunaga, Y. Sokawa, and S. Nagata, *Proc. Natl. Acad. Sci. U.S.A.* 81(16), 5086–5090 (1984).
43. H. Link et al., *Blood* 84(10), 3327–3335 (1994).
44. M.J. Carter, T.J. Facklam, P.C. Long, and R.A. Scotland, *Dev. Biol. Stand.* 70, 101–107 (1989).
45. Y.H. Lee, J.Y. Lin, C.C. Pao, and J.S. Lo, *Biochem. Int.* 16(1), 101–109 (1988).
46. V.B. Reddy et al., *DNA* 6(5), 461–472 (1987).

47. M.M. Bendig, P.E. Stephens, M.I. Cockett, and C.C. Hentschel, *DNA* **6**(4), 343–352 (1987).

48. H. Dorai and G.P. Moore, *J. Immunol.* **139**, 4232–4241 (1987).

49. J.R. Birch et al., *Cytotechnology* **15**, 11–16 (1994).

50. J.F. Harris, R.G. Hawley, T.S. Hawley, and G.C. Crawford-Sharpe, *J. Immunol. Methods* **148**, 199–207 (1992).

51. G. Klein, L. Dombos, and B. Gothoskar, *Int. J. Cancer* **10**, 44–57 (1972).

52. M.D. Johnston, *J. Gen. Virol.* **50**, 191–194 (1980).

53. J. Morser and J. Shuttleworth, *J. Gen. Virol.* **56**, 163–174 (1981).

54. H. Miyaji et al., *Cytotechnology* **3**(2), 133–140 (1990).

55. S. Hosoi et al., *Cytotechnology* **19**(1), 1–10 (1995–1996).

56. S. Hosoi et al., *Cytotechnology* **7**(1), 25–32 (1991).

57. S. Hosoi et al., *Cytotechnology* **5**(Suppl. 2), S17–S34 (1991).

58. Y. Tasumura and Y. Kawakita, *Nippon Rinsho* (translated version) **21**, 1201–1215 (1963).

59. S. Seghal, D. Bhattacharya, and M. Bhardwaj, *J. Commun. Disord.* **29**(1), 23–28 (1997).

60. E. Vidor, C. Meschievitz, and S. Plotkin, *Pediatr. Infect. Dis. J.* **16**(3), 312–322 (1997).

61. B. Montagnon, J.C. Vincent-Falquet, and B. Fanget, *Dev. Biol. Stand.* **55**, 37–42 (1983).

62. O. Kistner et al., *Vaccine* **16**, 960–968 (1998).

63. I. MacPherson and M. Stoker, *Virology* **16**, 147 (1962).

64. A. Savage, in H. Hauser and R. Wagner, eds., *Glycosylation: A Posttranslational Modification*, deGruyter, Berlin and New York, 233–276 (1997).

65. F. Brown, J.S. Robertson, G.C. Schild, and J.M. Wood, *Dev. Biol. Stand.* **98** (1999).

66. WHO Expert Committee on Biological Standardization, *Requirements for the Use of Animal Cells as in vitro Substrates for the Production of Biologicals*, WHO Tech. Rep. Ser. 878, WHO, Geneva, 1998, pp. 19–56.

67. *Human Medicines Evaluation Unit: ICH Topic Q 5D – Quality of Biotechnological Products: Derivation and Characterization of Cell Substrates for Production of Biotechnological/Biological Products*, European Agency for the Evaluation of Medicinal Products, ICH Technical Coordination, ICH, London, 1997.

ANIMAL CELL TYPES, HYBRIDOMA CELLS

ALISON STACEY
Barley Hertfordshire
United Kingdom

OUTLINE

INTRODUCTION

In 1975 Kohler and Milstein first described the production of "continuous cultures of fused cells secreting antibody of predefined specificity," or monoclonal antibodies, as they are known today (1). Since then the generation of B-cell hybridomas and the production of monoclonal antibodies have become major activities in the biotechnology industry. The major uses of monoclonal antibodies are in diagnosis and immunotherapy. The first experiments produced mouse monoclonal antibodies; however, with improvements in technique and the development of new fusion partners, monoclonal antibodies from other species, including rat, hamster, human, and a variety of interspecies hybrids, can now be produced. Technologies have also been developed to allow the production of T-cell hybridomas with important immunological properties, including the expression of single cytokines (2).

The generation of hybridomas per se is straightforward; however, difficulties may arise in obtaining clones and antibodies of the desired specificity and biological characteristics. The process is time consuming (several months) and labor intensive, therefore, planning a coordinated hybridization and screening program is essential in order to maximize the chances of being successful. Figure 1 gives a schematic representation of the procedure for hybridoma production. Each of the stages is discussed in more detail in the following.

ANTIBODY-PRODUCING HYBRIDOMAS

In order to obtain an antibody producing hybridoma antibody producing lymphocytes, which have a limited life-span in vivo and in vitro, are fused with myeloma cells which are considered to be immortalized. Thus the hybrid cells theoretically have an infinite life-span and thus, a potentially unlimited the capacity for the production of antibody. The basic steps of the process have changed little since the time of Kohler and Milstein, however the critical steps of the process are the achievement of efficient and effective cell fusion and then to select clones with the appropriate characteristics.

Immunization

In order to maximize the chances of producing an antibody of the desired specificity, lymphocytes are hyper-immunized using either in vivo or in vitro techniques. Antigen presentation should be optimized to yield a high number of specifically stimulated B cells. This may require several attempts to develop the most effective protocol. It should also be borne in mind that the responses obtained may vary with the strain of animal used. The use of inbred strains of animal will help to reduce this. The development of some of the biological characteristics such

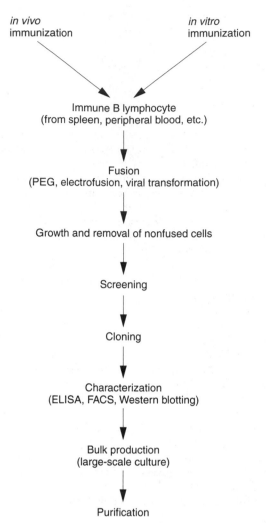

Figure 1. Procedure for monoclonal antibody production

as the isotype may be selected for during the immunization period. IgM antibodies are typically produced if the cells are stimulated once only or then stimulated with new antigens, such as different species of the same organism in order to produce cross-reactive antibodies, whereas IgG antibodies are produced following repeated immunization (one or more boosts) with the same antigen. The generation of IgA monoclonal antibodies can be achieved by gastric intubation (3).

The best results are obtained by using purified antigens, which should be well characterized before being used in immunization experiments. Use of a partially purified cocktail of antigens will lead to the majority of antibodies being directed against the immunodominant and often the major antigens.

In Vivo Immunization. In vivo immunization is used primarily in preparation of monoclonal antibodies from rodent species (i.e., rat and mouse). Immunization aims to elicit an immune response in the host organism. Immune lymphocytes are then taken from lymph nodes or spleen tissue and fused with myeloma cells, as indicated previously.

Use of Adjuvants. In general, particulate antigens are more immunogenic than soluble antigens. However, often antigens are prepared in soluble form, therefore requiring either precipitation using alum or suspension in adjuvant prior to injection (4). The most commonly used adjuvant is Freund's. This is available as a complete adjuvant comprising a mixture of oil, detergent, and heat-killed *Mycobacterium tuberculosis* cells or as an incomplete adjuvant comprising oil and detergent only. Other adjuvants include *Bordetella pertussis* cells which are added to alum-precipitated antigens. However, whole cell antigens such as bacteria do not require the use of adjuvants and may be administered in suspension in saline.

For some antigens where irritating adjuvants such as Freund's cannot be used or where the antigen has to be immobilized at the site of injection, for example, in the spleen, antigens may be immobilized onto inert carrier beads such as agarose or dextran. Proteins such as bovine serum albumin may also be used as carriers for small antigens.

Route of Injection. Most antigens are administered subcutaneously; however, in some cases antigens are delivered by intraperitoneal or even intravenous injection. As mentioned previously, in rare cases the specialist techniques of intrasplenic injection or gastric intubation may be required, but these are highly invasive.

In Vitro Immunization. More recently in vitro methods of immunization have been developed. While it is claimed that this approach was first described by Carrel and Ingebrichtsen in 1912 (5), the technique was established by Hengarter et al. (6). The main advantages of this technique compared to the conventional methods is the reduced amount of antigen and time required to produce hybridomas. Moreover, it enables weak immunogens to be used more successfully, as well as toxins and viruses that cannot be used with animals. A further advantage of this route of immunization is that both particulate and soluble antigens can be used in this technique without the use of adjuvants.

Mitogenic Stimulation. Mitogenic stimulation in vitro is a technique most commonly used in the production of human monoclonal antibodies (7). Cells are recovered from lymph nodes and stimulated with a mitogen such as pokeweed mitogen (PMW) (8,9). The stimulated cells are then fused with either mouse or human myeloma cells in order to produce a hybridoma; however, the number of stable clones obtained is greater with the use of mouse myeloma cells such as NS-1 cells.

This technology has been reported for the production of antitumor antibodies (9) and anti-HLA antigen antibodies. One of the advantages of this technique is that the majority of clones produced secrete IgG antibodies compared to IgM antibodies produced by in vitro immunization (see the preceding) or Epstein Barr virus transformation (see the following).

Epstein Barr Transformation. This method is again used primarily in the production of human monoclonal antibodies in which lymphocytes, derived from spleen, tonsil, or peripheral blood are transformed by incubation with Epstein Barr virus (EBV) to produce predominantly low-titer IgM antibody-producing B-lymphoblastoid cells. EBV can be isolated from the supernatant of cell lines B95-8 (10) or 1A2 (11) by filtration through a 0.45-μm filter. The transformed cells are then fused with mouse myeloma cells. This additional fusion step is necessary in order to stabilize antibody secretion, which may cease after a few weeks in culture and to enable cloning to be undertaken more effectively, since although B-lymphoblastoid cells grow easily, they are difficult to clone (12). A variety of antibodies against human leukocyte antigens have been generated in this way (13,14), and clones secreting IgM and IgG have been described.

Selection of Myeloma

There are many myeloma lines available that are suitable for the production of hybridomas. Most laboratories use well characterized lines that have proven provenance and give good fusion results. The most commonly used lines are given in Table 1 (15–22). The establishment of human hybridomas may be enhanced by fusion of mouse x human hybrid myelomas with human B cells, and stable hybrids expressing human IgG have been successfully produced in this way (23). In most instances, and in particular if hybridomas are to be used for human applications, myelomas that do not secrete endogenous antibody or light chain are recommended. All myelomas must be regularly checked to ensure that they retain their sensitivity to the selective agent hypoxanthine/aminopterin/thymidine (HAT) (24). HAT-sensitive lines should die in HAT-containing medium but

survive in the presence of 8-azaguanine or 6-thioguanine, typically at 30 μg/mL. HAT sensitivity is essential so that unfused myeloma cells will perish and not overgrow the newly fused hybrid cells (24).

Fusion

Cells can be fused using chemical polyethylene glycol (1,25) or by electrofusion. Other techniques have also been used, including recombinant DNA approaches. However, before a method is selected, it is important to ensure prior to fusion that the myeloma fusion partner has good viability and has been in exponential growth phase for several days. Moreover, wherever possible serum should be taken from the immunized animal 2–3 days prior to the scheduled fusion date in order to establish the presence of a circulating antibody response. If this is not possible, serum should be taken at the same time as the immune lymphocytes are recovered from the immunized animal. Cells from tissues should be brought into suspension by gently pushing them through sterile metal gauze.

The final prefusion task is preparation of either a mouse macrophage feeder layer or conditioned medium, which can be obtained from several suppliers of tissue culture reagents.

Fusion by Polyethylene Glycol. During chemical fusion of immune lymphocytes and myeloma cells typically the cells are incubated in a ratio of between 1:1 and 4:1 in the presence of polyethylene glycol 4000 (PEG 4000). PEG causes the aggregation of the integral proteins of the cell membrane, giving rise to protein-denuded areas (26).

Table 1. Most Commonly Used Myeloma Cells in the Production of Antibody-Producing Hybridomas

Species, strain	Full designation	Characteristics	Reference/collection code*
Mouse, BALB/c	P3X63Ag8	Derived from the Psk cell line that was established from the MOPC-21 plasmacytoma of BALB/c origin; secretes IgG_1, kappa	(1) ATCC CRL597 (U.1 clone of P3Xag8) ECACC 8501140
Mouse, BALB/c	P3X63Ag8.653	Non-Ig-secreting subclone of P3 × 63Ag8e	(15) ATCC CRL1580 ECACC 85011420
Mouse, BALB/c	P3/NS-1/1.Ag4.1	A further subclone of the P3 lineage that synthesizes kappa chains but does not secrete them; commonly called NS-1	(16) ECACC 85011427
Mouse, BALB/c	NS0/1	Non-Ig-secreting subclone of NS-1 cells	(17) ECACC 85110503
Mouse, BALB/c	SP2/0-Ag14	Non-Ig-secreting myeloma	(18,19) ATCC CRL8287 ECACC 85072401
Rat, LOU	210-RCY3-Ag1.2.3	Azaguanine-resistant subclone of the LOU rat myeloma; secretes Ig kappa	(20)
Rat, LOU	IR983F	Non-Ig-secreting LOU myeloma clone	(21)
Rat, AO	YB2-0	Non-Ig-secreting AO rat myeloma; a range of subclones have been isolated, including YB2/3HL and its derivative YB2/3.0Ag.YO (YO)	(22) ATCC CRL1662 ECACC 95110501
Human	LICR-2	Non-Ig secreting	(9)
Human × mouse	HF2 × 653	Non-Ig-secreting hybrid of a human lymphoblastoid cell line (WI-L 2–729-HF2) and the murine myeloma line P3XAg8.653; forms hybrids, by electrofusion, with human spleen cells and human lymphoblastoid cells	ECACC 90012609

*a*Further details of cell resource centers are given in the chapter on cell banks.

Where such areas come in contact between adjacent cells the membranes will merge and cells will fuse (26). The molecular weight of the PEG used is important (27). PEGs of low molecular weight will enter cells readily and may cause cytotoxicity, whereas PEGs of higher molecular weight, although less toxic, are more viscous (28). A weight of 4000 is a good compromise. Since all PEGs are toxic for mammalian cells, they should be removed by washing immediately after a short incubation period with the cells.

Electrofusion. Electrofusion is a technique more commonly used in the production of human monoclonal antibodies, since it is more readily adapted for use with sensitive cells (29,30). It also gives good fusion rates for small numbers of cells, which are difficult to handle using the PEG technique. During electrofusion pronase- and DNase-treated immune lymphocytes and myeloma cells are incubated in a ratio of 2:1 and are subjected to electrical pulses. Electrofusion is typically achieved by first applying an alternating electric field (10 kHz–1MHz) which causes cells to polarize and form "pearl chains" with their membranes in close contact (31). Short pulses of a higher voltage trigger cell fusion by causing membrane instabilities, which are the starting points for fusion of membranes of adjacent cells.

Growth of Hybridomas

Following fusion the cell preparations are cultured in conditioned medium or on feeder layers in the presence of selective medium to remove any unfused myeloma cells. For rat–rat, rat–mouse, or mouse–mouse hybridomas the selective agent is HAT. In addition, if a human–mouse heterohybridoma has been used as a fusion partner, ouabain will also be required in the medium in order to eliminate unfused heterohybrid fusion partner cells. All cultures are maintained in selective medium for 14 days, by which time all the unfused cells will have perished. The selective agents can then be removed from the growth medium. By this time small colonies should be visible and be screened for antibody production.

Screening, Characterization, and Cloning

Screening. There are several standard methods available for screening for antibody production (32). The method chosen will ultimately depend upon the envisaged use of the antibody; for example, for cell purification a cell binding assay using target cells will be the method of choice (flow cytometry), whereas for diagnosis an ELISA-based system may be more appropriate (33,34). After an initial culture period of up to 10 days after fusion, screening must be done frequently during the next 2–3 weeks in order to identify clones consistently producing antibodies of potential interest and to ensure detection of slower-growing clones that often appear to secrete high levels of antibody.

Once clones have been selected, further characterization should be undertaken. The assays used will again depend upon the final use of the antibody, but such tests will generally include isotype analysis, Western blotting,

isoelectric focusing, and affinity and avidity studies (35). Cells from these clones should then be cyropreserved (see the following) to safeguard against later loss from bacterial infection or loss of antibody-producing capacity, since all hybridomas are naturally unstable.

Cloning. Once potential clones have been identified, they should be cloned to ensure that they are truly clonal (36). Cloning may also increase antibody titer. This may be done by

- Dilution cloning [a small volume of cells (10 μL) is removed from the center of the clone and doubly diluted across a fresh cloning plate containing feeder cells or conditioned medium].
- End point cloning (36). [Cells from the clone are counted and resuspended at a concentration of 1000 cells/mL. 100 μL (100 cells) is then added to one well of fresh cloning plate containing feeder cells or conditioned medium and doubly diluted across the plate].
- Cloning in agar (36). [Cells from the clone are counted and suspended in agar at a concentration of 1000 cells/mL. After 4–5 days incubation the agar plates are overlaid with antispecies immunoglobulin. Precipitation occurs where clones are secreting antibody. These clones are then placed back into liquid medium and cultured in the normal manner].

A method for cloning in agar using FITC-labeled antibodies to reveal secretion of antibody can be used to identify the percentage of cells in a culture secreting antibody and also permits selection of viable clones secreting the highest levels of antibody (37).

Cloning should be carried out 2–5 times at this early stage until the cells can be expected to represent a single clone. Cells from each selected clone should be frozen after each cloning step. Cloning should also be carried out on a routine basis after periods of prolonged culture (i.e., 2–3 months), or before antibody production runs in order to eliminate the presence of any nonantibody producing cells that may overgrow the cells of interest.

Cryopreservation

It is essential to cryopreserve cells from potentially useful clones at all the stages of selection, growth, and production to safeguard against:

- Loss of cultures due to infection (bacterial, fungal, or mycoplasma)
- Loss of antibody secretion
- Death of cultures due to equipment failure

Standard techniques and equipment developed for cryopreserving mammalian cells are suitable for use with hybridoma cells. These include programmable rate-controlled freezers and isopropanol baths. Both systems aim to cool the cells at a controlled rate (typically 1 °C/min). A cryoprotectant, such as glycerol (10–15% v/v) or dimethyl sulfoxide (10% v/v), should be incorporated into the medium just prior to freezing (38) to protect cells from the damage caused by the formation of intracellular ice as the medium freezes.

The most important factors in successful cryopreservation are that the cultures to be frozen should be in the exponential growth phase and have high percentage viability in excess of 90%. In addition, hybridoma cells should be cryopreserved at cell concentrations in the range $(4-6) \times 10^6$/mL since levels of over 10^7/mL often result in low viability on recovery of cryopreserved cells. In some cultures viability may appear to be 90–100% by vital dye exclusion, but upon resuscitation the viability may fall to 50% or less. This may be due to a high proportion of apoptotic cells and large cell fragments that may retain membrane integrity but will ultimately die (39). Although suboptimally cryopreserved cultures may recover on thawing, the resulting cellular stress and loss of viability may result in a high proportion of surviving cells that do not secrete antibody, thus necessitating recloning.

Bulk Culture

Monoclonal antibodies can be obtained in quantity by either large-scale culture or from ascitic fluid. The antibody titer is higher in ascitic fluid (1–10 mg/mL compared to 20 µg–1 mg/mL), but the volume produced is limited. In addition, in some countries, such as the United Kingdom, the production of ascitic fluid is generally not permitted for ethical reasons. Therefore, increasingly antibody production is carried out in vitro. Cultures can be scaled up into larger tissue culture flasks or roller bottles (40) if relatively small amounts are required (1–10 mg antibody from up to 5 L of culture supernatant).

However, for larger quantities of antibody (i.e., culture volumes of 100–10,000 L) more specialized equipment is used and special considerations must be involved in bioreactor design (41). Typically stirred vessel fermentors are used where the cells are grown in batches in suspension (42), where productivity is strongly influenced by the characteristics of the culture at the initial point of inoculation (43,44). However, there are also continuous culture systems, such as hollow-fiber systems where cells are trapped in hollow fibers (45–47) and perfusion systems in which hybridoma cells are immobilized in a fixed or fluidized bed and perfused with culture medium (48). In these systems the replenishment of growth medium, temperature, partial pressures of oxygen (pO₂) and carbon dioxide (pCO₂), and pH are carefully controlled. In batch and continuous culture reactors viability, glucose levels, and the levels of certain metabolic waste products (e.g., ammonia, lactate) are also monitored to establish the general status of a culture. Scale up of a bench-scale reactor system is not always a simple process of increasing the dimensions of the culture vessel. A range of physical and hydrodynamic parameters will alter as scale increases and careful development of the culture environment is required to minimize the physical and biochemical stresses experienced by the hybridoma cells and to optimize their nutrition and hence maximize the yield of antibody (49,50).

Purification

Monoclonal antibodies can be purified by a variety of methods, the choice of which will depend upon the intended use of the antibody. For laboratory use a single step may be sufficient. Culture supernatants containing antibodies intended for use as diagnostic reagents and therapeutic reagents will require higher degrees of purification in multistep processes. The purification steps will also need to be validated for their capacity to eliminate or inactivate toxins and potential virus contamination including murine endogenous retroviruses (51,52). Although monoclonal antibodies have a high degree of homogeneity within species and subclass, each antibody is unique, and therefore purification systems may need to be adapted for each antibody purified. However, some generic methods are available such as Protein A/Protein G binding of immunoglobulin, cation or anion exchange, and gel permeation (53).

Purification of antibodies can be made more straightforward if the hybridoma culture is grown in serum low in immunoglobulins. However, more significant advantages in this respect can be achieved if the culture can be adapted to growth in a serum-free growth medium (54), which also reduces significantly the risk of virus contamination from the culture medium.

T-CELL HYBRIDOMAS

T–T cell hybridomas can serve to immortalize by fusion certain functions of T cells such as the secretion of a single cytokine, other immunological function, or expression of a surface receptor. As with monoclonal-antibody-producing hybridomas, it is important to select the correct fusion partner, that is, those known to fuse to activated T cells with a high efficiency, which proliferate rapidly and are sensitive to reagents selective for the growth of immortal hybrid cells (55). Most fusion partners are deficient in HGPRT or TK and are sensitive to HAT (55), and examples are given in Table 2. These lines have been used successfully to produce human (56, 2) and mouse (57,58) hybrids. In addition, a range of interspecies hybrids have been generated. Of particular interest is the AKR lymphoma T-cell line BW5147, which exhibits high fusion efficiencies and good growth characteristics and has enabled the production of stable T–T-cell hybrids (57).

Despite these successes, no human HAT-sensitive line with a high fusion efficiency and good growth has been described to date. Therefore, other methods have been developed to overcome the need to use HAT for the removal of unfused fusion partner cells and selection of clones of interest:

Table 2. Most Commonly Used Fusion Partners in the Production of T–T-cell Hybrids

Species	Full designation	Characteristics	References
Mouse	BW5147 AKR	T cell	(57)
Rat	C58 (NT)D	T cell	

- Soft agar cloning, since T–T-cell hybrids will form colonies, unlike the T-lymphoblastoid parent cells (2,59)
- Selection in azaserine–hypoxanthine (AH), which eliminates HGPRT⁻ cells (56)
- Treatment of T-cell lines with metabolic inhibitors (e.g., actinomycin-D, emetine) prior to fusion (60)
- By screening for antigenic determinants not present on the parent T-cell line

Despite the difficulties with fusion, hybrid stability, and cloning, the remaining steps are the same as those used to generate antibody-producing hybridomas. As with B-cell hybridomas, it is essential to plan and have adapted screening assays and cryopreservation protocols prior to fusion.

BIBLIOGRAPHY

1. G. Kohler and C. Milstein, *Nature (London)* **256**, 495–497 (1975).
2. C.D. Platsoucas, T.A. Cavelli, and J.A. Kunicka, *Hybridoma* **6**, 589–603 (1987).
3. D.E. Colwell, S.M. Michalek, and J.R. McGhee, *Methods Enzymol.* **121**, 42–51 (1986).
4. M.J. Dennis, in A. Doyle, J.B. Griffiths, and D.G. Newell, eds., *Cell & Tissue Culture: Laboratory Procedures*, Wiley, Chichester, U.K., 1994, pp. 25A: 1.1–1.5.
5. A. Carrel and R. Ingebrichtsen, *J. Exp. Med.* **15**, 287–291 (1912).
6. H. Hengarter, A.L. Luzzati, and M. Schreier, *Curr. Top. Microbiol. Immunol.* **81**, 92–99 (1978).
7. P.G. Abrams, J.L. Rossio, H.C. Stevenson, and K.A. Foon, *Methods Enzymol.* **121**, 107–119 (1986).
8. G. Wetzel and J.R. Kettman, *J. Immunol.* **126**, 723–728 (1981).
9. R.F. Martin et al., *Hum. Antibodies Hybridomas* **1**, 154–158 (1990).
10. J.H. Pope, in M.A. Epstein and B.G. Achong, eds., *The Epstein Barr Virus*, Springer-Verlag, Berlin, 1979, pp. 205–223.
11. A.W. Sadiak and M.E. Lostrum, in E.G. Engelman, S.K.H. Foung, J. Larrick, and A. Raubitscheck, eds., *Human Hybridomas and Monoclonal Antibodies*, Plenum, New York, 1985, pp. 167–185.
12. B.M. Kumpel and A. Martin, in A. Doyle, J.B. Griffiths, and D.G. Newell, eds., *Cell & Tissue Culture: Laboratory Procedures*, Wiley, Chichester, U.K., 1994, pp. 25E: 6.1–6.7.
13. K.M. Thompson et al., *Immunology* **58**, 157–160 (1986).
14. R.J.T. Hancock et al., *Br. J. Haematol.*, **71**, 125–129 (1988).
15. J.F. Kearney, Radbruch, B. Liesegang, and K. Rajewsky, *J. Immunol.* **123**, 1548–1550 (1979).
16. G. Kohler and C. Milstein, *Eur. J. Immunol.* **6**, 511–519 (1976).
17. G. Galfre and C. Milstein, *Methods Enzymol.* **73** (Part B), 3–46 (1981).
18. M. Schulman, C.D. Wilde, and G.A. Kohler, *Nature (London)* **276**, 269–270 (1978).
19. R. Laskov and M.D. Schraff, *J. Exp. Med.* **131**, 515–541 (1970).
20. G. Galfre, C. Milstein, and B. Wright, *Nature (London)* **277**, 131–132 (1979).
21. H. Bazin, in H. Peters, ed., *Proteins of the Biological Fluids*, Pergamon, Oxford, 1982, pp. 615–618.
22. H. Bazin, W.S. Pear, G. Klein, and J. Sumegi, in H. Bazin, ed., *Rat Hybridomas and Rat Monoclonal Antibodies*, CRC Press, Boca Raton, Fla., 1990, pp. 53–66.
23. E.B. Austin et al., *Immunology* **67**, 525–530 (1989).
24. H.M. McBride, in A. Doyle, J.B. Griffiths, and D.G. Newell, eds., *Cell & Tissue Culture: Laboratory Procedures*, Wiley, Chichester, U.K., 1994, pp. 25B: 1.1–1.4.
25. M.C. Glassy and M.I. Mally, in S.K. Balal, ed., *SAAS Bulletin of Biochemistry and Biotechnology*, Vol. 2, Southern Association of Agricultural Scientists, Nashville, Tenn., 1989, pp. 17–21.
26. M. Mally and M.C. Glassy, in C. Borrebaeck and I. Hagen, eds., *Electromanipulation in Hybridoma Technology*, Stockton Press, New York, 1990, pp. 71–88.
27. Q.F. Ahkong, D. Fisher, W. Tampien, and J.A. Lucy, *Nature (London)* **253**, 194–195 (1975).
28. S. Fazekas de St. Groth and D. Schneidegger, *J. Immunol. Methods* **35**, 1–21 (1980).
29. B.W. McBride, in A. Doyle, J.B. Griffiths, and D.G. Newell, eds., *Cell & Tissue Culture: Laboratory Procedures*, Wiley, Chichester, U.K., 1994, pp. 25B: 2.1–2.5.
30. R.W. Glaser, in A. Doyle, J.B. Griffiths, and D.G. Newell, eds., *Cell & Tissue Culture: Laboratory Procedures*, Wiley, Chichester, U.K., 1984, pp. 25E: 4.1–4.4.
31. E.W. Pascoe and S.J.Y. Saxby, in A. Doyle, J.B. Griffiths, and D.G. Newell, eds., *Cell & Tissue Culture: Laboratory Procedures*, Wiley, Chichester, U.K., 1984, pp. 26A: 1.1–1.10.
32. D. Catty, *Antibodies I and II: A Practical Approach*, IRL Press at Oxford University Press, Oxford, 1989.
33. H.M. McBride, in A. Doyle, J.B. Griffiths, and D.G. Newell, eds., *Cell & Tissue Culture: Laboratory Procedures*, Wiley, Chichester, U.K., 1984, pp. 25B: 4.1–4.13.
34. H.M. McBride, in A. Doyle, J.B. Griffths, and D.G. Newell, eds., *Cell & Tissue Culture: Laboratory Procedures*, Wiley, Chichester, U.K., 1984, pp. 25B: 5.1–5.4.
35. A. Johnstone and R. Thorpe, *Immunochemistry in Practice*, 3rd ed., Blackwell, New York, 1996.
36. S.A. Clark, in A. Doyle, J.B. Griffiths, and D.G. Newell, eds., *Cell & Tissue Culture: Laboratory Procedures*, Wiley, Chichester, U.K., 1984, pp. 25B: 6.1–6.12.
37. J.M. Davis, *Methods Enzymol*, **121**, 307–322 (1986).
38. A. Doyle, C.B. Morris, and W.J. Armitage, in A. Mizrahi, *Advances in Biotechnological Processes*, Vol. 7, Alan R. Liss, New York, 1988 pp. 1–17.
39. M. Solis-Recendez et al., *J. Biotechnol.* **38**, 117–127 (1994).
40. G.F. Panina, in R.E. Spier and J.B. Griffiths, eds., *Animal Cell Biotechnology*, Vol. 1, Academic Press, London, 1985 p. 211–242.
41. J. Varley and J. Birch, *Cytotechnology* **29**, 177–205 (1999).
42. J.P. Barford, C. Harbour, C.P. Marquis, and P.J. Phillipps, eds., *Fundamental and Applied Aspects of Animal Cell Cultivation*, Singapore University Press, Singapore, 1995.
43. R.L. Dutton, J.M. Scharer, and M. Moo-Young, *Cytotechnology* **29**, 1–10 (1999).
44. G.M. Lee, M.A. Kaminski, and B.O. Palsson, *Biotechnol. Lett.* **14**, 257–262 (1992).
45. E. Sjogren-Jansson and S. Jeansson, *J. Immunol. Methods* **84**, 359–364 (1985).
46. F.W. Falkenberg et al., *J. Immunol. Methods* **165**, 193–206 (1993).

47. E. Unterluggauer et al., *BioTechniques* **16**, 140–147 (1994).
48. D. de la Boise, M. Noiseaux, B. Massiex, and R. Lemieux, *Biotechnol. Bioeng.* **40**, 25–32 (1992).
49. Y. Christi, *Process Biochem.* **28**, 511–517 (1994).
50. A. Handa-Corrigan et al., *Enzyme Microb. Technol.* **14**, 58–63 (1992).
51. Working Party on Clinical Use of Antibodies, *Br. J. Cancer* **54**, 557–568 (1986).
52. S.R. Adamson, *Dev. Biol. Stand.* **93**, 89–96 (1998).
53. M.A. Kerr, L.M. Looms, and S.J. Thorpe, in M.A. Kerr and R. Thorpe, eds., *Immunochemistry Labfax*, BIOS Scientific Publishers (Blackwell Scientific Publications), Oxford, 1994 pp. 83–114.
54. E. Mariani et al., *J. Immunol.* **145**, 175–183 (1991).
55. F. Fox and C.D. Platsoucas, *Hum. Antibodies Hybridomas* **1**, 3–9 (1990).
56. S.K. Foung, D.T. Sasaki, F.C. Grumet, and E.G. Engelman, *Proc. Natl. Acad. Sci., U.S.A.* **79**, 7484–7488 (1982).
57. R.A. Goldsby, B.A. Osborne, E. Simpson, and L.A. Herzenberg, *Nature (London)*, **267**, 707–709 (1977).
58. P. Marrack et al., *Immunol. Rev.* **76**, 131–145 (1983).
59. J.E. Kunicka et al., *Hybridoma* **8**, 127–151 (1989).
60. Y. Kobayashi, M. Asada, M. Higuchi, and T. Osawa, *J. Immunol.* **128**, 2714–2718 (1982).

ANIMAL CELL TYPES, KIDNEY CELLS

G. HODGES
London, United Kingdom

OUTLINE

Kidney Cells *In Vitro*
Glomerular Cell Cultures
Tubular Cell Cultures
Bibliography

The vital excretory, homeostatic, and endocrine functions of the urinary system are performed by the kidney, and these highly vascular organs are responsible for a wide range of biochemical and physiological tasks (1,2). By a complex process involving filtration, active and passive absorption, and secretion, the kidneys retrieve essential electrolytes and metabolites; remove metabolic wastes and foreign substances; regulate ion, salt, and water concentrations of the body tissues; and secrete or respond to a wide variety of growth factors, hormones, and cytokines.

The kidney has a complex tissue organization with an internal structure divided into two distinct (outer cortex and inner medulla) regions that demonstrate striking structural and cellular heterogeneity (3,4). The basic morphofunctional unit of the kidney is represented by the nephron which contains some 15 types of renal epithelia organized segmentally along its length. The main elements of each nephron are the renal corpuscle (which represents the beginning of the nephron), the renal tubule,

and the associated vascular and urinary poles. The renal corpuscle that forms the blood-filtering unit of the nephron is composed of a glomerulus surrounded by a double-layered epithelial cup, the renal or Bowman's capsule. This capsule consists of an inner glomerular (or visceral) and an outer parietal (or capsular) layer. The glomerulus is made up of a small tuft of fenestrated capillaries with mesangial cells and their extracellular matrix (the mesangium) lying between the capillary loops (5). Wrapped around the glomerular capillary loops are the very elaborate and highly branched epithelial cells (podocytes) of the inner glomerular epithelium of Bowman's capsule (6). These cellular elements, together with the fused basal laminae of the capillary endothelial cells and podocytes, form the glomerular filtration apparatus of the kidney which is enclosed by the parietal layer of Bowman's capsule. Therefore, the glomerulus constitutes, a complex structure with each of the four main cell components (glomerular epithelial, parietal epithelial, glomerular endothelial, and mesangial) interacting in a specific way with the three spatially distinct extracellular matrices in the glomerulus (mesangial extracellular matrix and basement membranes of the capillary loops and Bowman's capsule). These all have basement membrane constituents as their major matrix type, but each is distinct in organization and composition (7,8). The mesangial cells resemble smooth muscle cells and are of considerable biological complexity (9–11). Several functions are attributed to them: vasoreactivity of the glomerular tuft; production of extracellular components to the mesangial matrix which provides structural support to the glomerular tuft; and, through their phagocytic properties, uptake and clearance of macromolecules such as immune complexes from the glomeruli with resident mesangial macrophages also contributing to this process.

The renal tubule arises as a direct extension of the parietal epithelium that lines Bowman's capsule and forms a continuous tube with varied histophysiological features characterizing its different segments (3). These perform functions of osmoregulation and excretion through filtration, selective reabsorption, and secretory processes, creating and maintaining osmotic and urea gradients (12). The tubular epithelial cells reflect this multiplicity of functions with unique and characteristic sets of enzymes, transport processes, and hormonal responses, depending on their cell type, tubule location, and environment (1,2,13). The major tubular segments comprise (1) the proximal thick segment, consisting of a proximal convoluted tubule and the straight descending thick limb of Henle's loop: the proximal tubule is lined with simple cuboidal epithelium that shows elaborate surface specializations associated with absorption and fluid transport functions and demonstrates distinct functional subregions (S_1, S_2, S_3) (14); (2) the thin descending and ascending segments of Henle's loop: the type of epithelial cell along the thin segment shows variation, and four regions of the epithelium are identified by different types of epithelial cells (14) which are thought to reflect specific active or passive roles in the countercurrent system involved in producing of hypertonic urine; and (3) the distal thick segment, consisting of a thick ascending limb of Henle, the macula densa (specialized epithelial cells located within the thick ascending limb; their apical surface is in contact with tubular fluid, and their basilar region is in contact with the glomerulus), and the distal convoluted tubule: the distal tubule is lined by a simple cuboidal epithelium devoid of a brush border and is involved in active reabsorption of sodium chloride; adiuretin (vasopressin) and mineralocorticoids play important roles in the function of the distal tubule. The juxtaglomerular apparatus is a specialized structure linking the distal end of the thick ascending limb at the macula densa with the glomerular vascular pole. It is thought to function in the control of glomerular hemodynamics and is composed of juxtaglomerular cells (renin-producing modified smooth muscle cells), the macula densa of the distal tubule (thought to serve as an osmoreceptor transmitting changes in sodium levels in the distal tubule to juxtaglomerular cells and regulating the release of renin by the juxtaglomerular cells in a paracrine manner), and the extraglomerular mesangium (15–17).

The collecting tubules and ducts of the kidney differ from the nephron in their embryonic origin (1). They form an intrarenal canalicular system that serves as a conduit for urine and are also involved in the reabsorption of water under the influence of the antidiuretic hormone, vasopressin. This canalicular system shows structurally and functionally distinct populations of epithelial cells with at least three cell types identified, principal, alpha-intercalated, and beta-intercalated cells, that are responsible for the multiple physiological functions of this kidney compartment (18). Striking interspecies differences exist in the prevalence and distribution of the different intercalated cells (19).

KIDNEY CELLS *IN VITRO*

Various approaches have been developed to study the kidney *in vitro*. Model systems include isolated perfused whole kidneys; isolated perfused nephron segments; renal tissue slices; isolated glomeruli, tubular and renal cells; primary and established cultures; and more recently, immortalization of glomerular and tubular cells using transfection technology. Although most renal cell culture studies use two-dimensional matrix support systems, culture models providing three-dimensional matrix support have resulted in cell phenotypes that more closely reflect the *in vivo* situation. Whole-organ metanephric cultures, in which organotypic renal proximal tubular and epithelial glomerular differentiation occur, have been also used to study various aspects of renal epithelial growth and differentiation, proximal tubular cystic disease, and drug toxicity (20), and individual microdissected cysts from the kidneys of patients with polycystic kidney disease have been successfully cultured (21). Optimal growth and differentiation conditions for each renal cell type are most likely quite different, and considerable influence is effected by different substrates and by perfusion and coculture systems. These remain areas which, although addressed in the past, need to be further defined.

The techniques, culture conditions, applications, and relative merits of these different *in vitro* approaches are selectively discussed in reviews of renal cell culture (see Refs. 13,22–30).

To circumvent the problems associated with primary cultures, a large number of established cell lines have been developed including transformed or tumor-derived renal epithelial cell lines and transfected epithelial cell lines of glomerular or tubular origin. Representative renal cell lines are listed in Table 1, and the most widely used lines are outlined following.

GLOMERULAR CELL CULTURES

Culture of all four glomerular cell types—glomerular (visceral) epithelial (or podocyte), parietal epithelial, glomerular endothelial, and mesangial—from several different animal species, including human, has been reported in the literature (5,20,25,31–34). Continuous cell lines have been successfully developed from the various glomerular cell types and are presented in Table 1.

Cultures have been generally established from glomeruli obtained from renal tissue using sieving or centrifugation techniques or a combination of both, followed by culture for outgrowth of cells. To isolate glomeruli, cortices are separated from the medullae of excised kidneys, and small fragments of cortical tissue are sequentially sieved through a series of stainless steel meshes of calibrated pore size resulting in a tissue suspension of mainly decapsulated glomeruli. Small numbers of encapsulated glomeruli and tubular fragments may be contained. Glomerular cells can be grown by (1) direct plating of the isolated glomeruli for outgrowth of cells or (2) enzymatic dissociation of the glomeruli and plating out of the dissociated cells. Glomeruli subjected to vigorous proteinase digestion of the basement membrane can result in the culture of a wider variety of cells (35). Culture of podocytes under standard conditions leads to dedifferentiation, including the loss of foot processes and of markers for differentiated podocytes. Partial differentiation of cultured podocytes can be achieved through confluent growth and minimal subcultivation (36). The outgrowth of mesangial cells is favored where whole glomeruli are plated out on plastic culture flasks and maintained in a medium containing relatively high amounts of serum. This method depends on the differential growth capacities of the intrinsic glomerular cells. Whole glomeruli, preincubated with a cocktail of enzymes, usually proteinases, can generate glomerular "cores" devoid of glomerular epithelial and endothelial cells, consisting mostly of mesangium and capillary loops from which mesangial cells can rapidly grow as relatively homogeneous populations. Prolonged culture of mesangial cells forms a multifocal nodular structure, or hillocks, composed of cells and extracellular matrix, which may mimic the situation in the glomerular mesangium (37).

TUBULAR CELL CULTURES

The *in vitro* culture of virtually every segment of the nephron and intrarenal canicular system has been successfully undertaken from a variety of mammalian species,

Table 1. Selected Renal Cell Lines: Principal Characteristics and Applications

Species/cell line	Characteristics and applications	References/collection code
Cat CrFK	• Established in 1964 from cortical portion of the kidneys of a 10- to 12-week-old normal female domestic cat. • Used extensively in feline-virus research, viral infectivity assays, and for study of the biology of various retroviruses and derived vectors.	*In Vitro* **9**, 176 (1973) ATCC CCL 94 ECACC 86093002
FL74-UCD-1	• Established from lymphomatous kidney of cat inoculated with virus obtained from a cat with spontaneous leukemia. The cells are persistently infected with KT-FeLV-UCD-1 feline leukemia virus. • Cells capable of producing large amounts of feline leukemia of attenuated virulence for cats: used for vaccine preparation.	*Nature (London)* **222**, 589–590 (1990) ATCC CRL 8012 U.S.Pat. 4,264,587
Dog MDCK	• MDCK is easily cloned. Confluent MDCK cells form polarized monolayers with domes. MDCK cells are nontumorigenic but are transformed by virus or chemical mutagenesis. • Used as model of distal nephron epithelium. Used in studies of growth regulation, tubule biogenesis, cell polarity, and differentiated functions and as a model system for natural transporting epithelia to study drug transport interactions and/or interactions with drugs and excipients. Grown on semipermeable supports, forms a model barrier epithelium • Transfected MDCK cells expressing the transgene MUC1 provide a model for analyzing the effect of mucins on epithelial permeabilities. • Supports growth of a wide range of animal viruses.	*Proc. Soc. Exp. Biol. Med.* **98**, 574 (1958) ATCC CCL 34 ECACC 85011435 ECACC 841219903 Note: The two ECACC lines originate from different depositors. Although their common identity has been confirmed by DNA fingerprinting, ECACC 84121903 is reported to have a wider virus host range.
Hamster[a] BHK-21 (clone 13)	• Subclone (clone 13) of the parental line established in 1961 from the pooled kidneys of five unsexed, one-day old Syrian hamsters. • Many genetically engineered variants reported in the literature. • BHK-21 cl 13 and variants used extensively for virus replication studies and, as hosts for transfection studies.	*Virology*, **16**, 147 (1962) ATCC CCL 10 DSMZ ACC 61 ECACC 85011433
Human[a] 293 (HEK293)	• Derived from a primary culture of human fetal kidney transformed by sheared human adenovirus type 5 (Ad 5) DNA. Particularly sensitive to human adenovirus and adenovirus DNA, contain and express the transforming genes of Ad 5. Many transfected subclones developed from the 293 cell line. • Used in the isolation of transformation defective, host-range mutants of Ad 5, and excellent for titrating human adenoviruses.	*J. Gen. Virol.* **36**, 59–72 (1977) ATCC CRL 1573 DSMZ ACC 305 ECACC 85120602
ACHN	• Cell line initiated in 1979 from malignant pleural effusion of a 22-year-old Caucasian male with widely metastatic renal adenocarcinoma. Cells are growth-inhibited by human interferons. • Used in cytotoxicity studies; cloning of cDNA encoding endothelin-2 precursor; study of cell growth inhibition; regulation of protein expression; and in interferon/interferon-induction studies.	ATCC CRL 1611 ECACC 88100508
G-401	• Derived from a renal Wilm's tumor. Highly undifferentiated, epithelial-like cell type; highly transformed; produces nephroblast growth factor (NB-GF). • Used in tumorigenicity studies; post-transcriptional control of N-myc gene expression, and in isolation and localization of renin-binding protein gene.	*Science* **236**, 175 (1987) ATCC CRL 1441 ECACC 87042204
HK-2	• Proximal tubular cell (PTC) line derived from normal kidney. Immortalized by transduction with human papilloma virus 16 (HPV-16) E6/E7 genes. Cells retain phenotype indicative of well-differentiated PTCs and functional characteristics of proximal tubular epithelium such as Na dependant/phlorizin sensitive sugar transport and adenylate cyclase reponsiveness to parathyroid, but not to antidiuretic hormone. Cells capable of gluconeogenesis. • Used as *in vitro* model system for proximal tubule studies; cells can reproduce experimental results obtained with freshly isolated PTCs.	*Kidney Int.* **45**, 48 (1994) ATCC CRL 2190
Monkey[a] BS-C-1	• Epithelial cell line established in 1961 from primary cell cultures of African green monkey kidney. • Used for viral diagnostic studies; as a model for *Cryptosporidium parvum* infectivity assay, and as a transfection host	*J. Immunol.* **91**, 416 (1963) ATCC CCL 26 ECACC 85011422

(continued)

Table 1. *Continued*

Species/cell line	Characteristics and applications	References/collection code
CV-1/EBNA-1	• Derived from CV-1 by transfection with DNA encoding Epstein–Barr virus nuclear antigen (EBNA-1) and with a vector containing cytomegalovirus regulatory sequences. • CV-1 is a fibroblast cell line, established in 1964 from the kidney of a male adult African green monkey. It is used as host for transfected DNA expression.	*EMBO J.* **10**, 2821 (1991) ATCC CRL 10478 U.S.Pat. 5,262,522
JTC-12	• Derived from the cynomolgus monkey: probably of proximal tubule origin. • Exhibits sodium-dependent glucose and phosphate transport and responsiveness to parathyroid hormone, but not to vasopressin or calcitonin.	*Biochim. Biophys. Acta* **541**, 467 (1978)
Vero	• Established from the kidney of a normal adult African green monkey. • Cells are susceptible to several viruses. Used in virus replication studies and plaque assays and as indicator line for mycoplasma testing.	ATCC CCL 18 ECACC 84113001 WHO cell bank: ECACC 88020401
Mouse As4.1	• Derived from a renin-expressing kidney tumor induced by tissue-specific oncogene-mediated tumorigenesis in transgenic mice (strain ATCC CRL 2193 [C57BL/10Ros-pd x C3H/HeRos]F1). • Express high levels of renin mRNA and synthesize prorenin and renin. Produce high levels of interleukin-6 • Used as a model for molecular biology studies of renin-producing kidney cells.	Sigmund et al. (1990)
M-1A	Cortical collecting duct (CCD) cell line established from normal renal tissue taken from an SV40 [Tg(SV40E)Bri/7] transgenic mouse. • M-1 cells retain characteristics of CCD. Most lines cloned from M-1. Characteristics of either intercalated cells (ICC) or principal cells (PC) of the CCD. • When grown on permeable supports, cells show a high transepithelial resistance and develop a lumen-negative trans-epithelial potential difference.	*Kidney Int.* **39**, 1168 (1991) ATCC CRL 2038 ECACC 95092201
mIMCD-3	mIMCD-3 is an inner medullary collecting (IMCD) cell line derived in 1991 from a mouse transgenic for the early region of SV4O [Tg(SV40E)bri/7]. Polarized cell line with characteristics of the terminal IMCD.	ATCC CRL 2123
SV40-MES 13	Glomerular mesangium C57BL/6J x SJL/J mouse transgenic for early region of SV40.	*Kidney Int.* **33**, 677–684 (1988) ATCC CRL 1927
Opussum OK	• Proximal tubular cell line derived from the kidney cortex of a normal adult female American opussum. Retains several properties of proximal tubular epithelial cells in culture and expresses a low level of angiotensinogen gene. Cells display a variety of receptors in culture. Considered an excellent *in vitro* model for the kidney proximal tubule epithelium. • Used to study mechanisms of hormonal regulation of transport processes and intracellular processing of biologically active peptides. Widely used as a model for studies of receptors. • Stable transformants of OK cells developed with the ANG-GH fusion gene providing useful *in vitro* systems to study the regulation of expression of the renal angiotensinogen gene.	*In Vitro* **14**, 239 (1978) ATCC 1840 ECACC 91021202
Pig LLC-PK$_1$	• A renal tubular cell line established in 1958 from a cortical mince of kidney from a normal 3–4-week-old male Hampshire pig. • Cells are not clonally derived and may be heterogeneous; clonal variants can be selected. • Express a mixture of distal and proximal tubule characteristics. Plasminogen activator but not renin; do not express distal tubular marker protein or calbindin D-28K. • Express some of the differentiated transport functions of proximal tubule cells. Domelike structures form in confluent monolayers. • Develop rapid resistance *in vitro* to the antiproliferative effect of TGF-β_1. Cells survive in suspension but show no significant growth. Form multicellular spheroids. • Used in transepithelial transport studies and as a convenient model for the *in vitro* study of proximal tubule function although, this cell line responds to arginine vasopressin, a distal tubule marker. • Used in the production of plasminogen activator.	*In Vitro* **12**, 670 (1976) ATCC CL 101 (formerly ATCC CRL 1392) ECACC 86121112 U.S.Pat. 3,904,480

Table 1. *Continued*

Species/cell line	Characteristics and applications	References/collection code
Potoroo PtK1 (NBL-3)	• Established in 1962 from a kidney of normal adult female Potoroo. First permanent cell line of marsupial origin to be established. • Cell line important because of low number, large size, and distinct morphology of the chromosomes. Used in cytogenetic studies.	*Nature (London)* **194**, 406 (1962). ATCC CCL 35 [ATCC CRL 6493] ECACC 91013163
Rabbit Vept	• Proximal tubule epithelial cell line established in 1989 from primary cultures of proximal tubules (S1 segment) microdissected from superficial slices of cortex from the kidney of a normal male New Zealand rabbit • Cells retain electrolyte transport characteristics of the proximal tubule and receptor and signaling mechanisms for angiotensin II.	Romero et al. (1992) ATCC CRL 2087
Toad A6	• Established in 1965 from a mince of whole kidney from a normal adult male toad. Resemble collecting duct epithelium; exhibit a high transepithelial electrical resistance; respond to aldosterone and increase cAMP levels in response to vasopressin. The cells are morphologically homogeneous. • A6 cells support the replication of Granoff's frog viruses, but no inclusion-tumor associated virus. • A6 cells provide a useful model system for studies of transport function in kidney, electrophysiology, and ion transport.	*Am. J. Physiol.* **241**, C154 (1981) ATCC CCL 102 ECACC 89072613

[a]The Vero monkey and BHK-21 fibroblast-like cell lines epithelial-like cell line are prominent in the biopharmaceutical field and are discussed in the section on cell lines used in manufacturing.

including humans (13,25,38). Methods have included the following.

- Primary culture of microdissected specific renal tubule segments with monolayer growth of individual epithelial cell types (39). Confluent primary cultures of such microdissected segments express differentiated functions of the tubule epithelial cell of origin and can be maintained in their differentiated state in culture for varying periods of time. The major limitations to this approach are notably the labor-intensive nature of tubule isolation; the relatively small numbers of cells generated from any one donor; variability among sequential tubule preparations; the limited life span and/or period of expression of differentiated characteristics, and the need to continuously prove that specific tubular cells have, indeed, been obtained. The choice of assay can circumvent, to some extent, these limitations by using of high specific activity radiotracers, by the scaling down of biochemical assays to the microlevel, and by the application of polymerase chain reaction (PCR) technology.

- Use of gravity sedimentation enriched populations of enzyme-digested tubule fragments (40) or size and density separation of enzymatically treated tissue using centrifugation (41).

- Immunodissection involving dispersion of renal tissue preparations into single cells or into mixtures of cells and tubule fragments, and the separation of cells based on the reactivity of cell surface determinants with specific antibodies coated on polystyrene dishes (42). Immunodissection protocols have been developed for isolating cells from proximal tubule, thick ascending limb, distal tubule, and cortical and medullary collecting tubules. The technique, however, is technically rigorous and demands careful immunologic and cell type characterization to be successful.

- Fluorescence-activated cell sorting in combination with immunodissection has been developed for isolating near homogeneous populations of the principal and intercalated cells of the cortical collecting duct (43).

- Magnetic separation and primary culture of specific tubular cells, based on physical dissociation of cortical tissue followed by sieving and magnetic removal of iron oxide-laden glomeruli (44). An immunomagnetic procedure has been developed where antibody-coated ferrous particles permit magnetic separation of specific cell populations (45).

- Use of colloidal silica as a substratum: there is evidence that proximal tubule cells can grow to confluence more rapidly on colloidal silica-coated tissue culture dishes than on tissue culture polystyrene, show a polarized morphology, and retain differentiated markers after passaging (41).

Various renal tubular cell culture models and immortalized cell lines have been established (46), and immortalization of primary tubular cell cultures is achieved in several ways (47). More recently, transfection of primary tubular cell cultures has provided a route for developing well-differentiated cell lines that are more representative of the different nephronic segments (13). Proximal nephron cell lines have been produced by targeted oncogenesis in transgenic mice using pyruvate kinase-SV40 (T) antigen hybrid gene (48); proximal and distal cell functions have been maintained in SV40-transformed tubular cell lines derived from rabbit kidney cortex (49);

and SV40-transformed lines have been derived from proximal tubule cells of hypertensive and normotensive rats (50) and from the thick ascending limb of Henle's loop of rabbit and mouse (51,52). Immortalized cell lines have further been generated from human proximal tubular cell cultures transduced with human papilloma virus (HPV) *E6/E7* genes (53) or using a hybrid adeno 12-SV40 vector (54). Renal cell lines have also been produced from proximal tubule (S_3 segment), distal tubule, cortical, and outer medullary collecting duct microdissected from kidneys of transgenic C57BL/6 mice that harbor the large T-antigen gene of temperature-sensitive mutant simian virus, pSVtsA58(ori-) (55).

Extensively used established tubular cell lines for renal physiological and toxicological studies are the canine-derived MDCK (56), the porcine-derived LLC-PK$_1$ (57), the opossum-derived OK (58), and the Xenopus-derived A6 (59). These established cell lines are not derived by cloning, however, and therefore do not represent a single cell type, nor do they completely represent the functional characteristics of their purported tubular origins. The MDCK cell line has been extremely well characterized and shows some features consistent with a distal tubular origin (60). The original cell line, however, is made up of cells with different morphology and varying degrees of junctional resistance, as well as cells that do not appear to be derived from the distal tubule (61). More recently, MDCK has been proposed as a model for studying the renal cortical collecting duct, and cell subtypes were cloned that resemble principal and intercalated cells (62). Dome, cluster, and tubular formations of MDCK cells are substrate-dependent (63), and, grown on semipermeable supports, MDCK cells develop apical membranes of very low permeability that provide a model of barrier epithelium (64).

The LLC-PK resembles predominantly, but not entirely, proximal tubular epithelium; two other less extensively studied cell lines that share many characteristics of the renal proximal tubule are the cynomolgus monkey-derived JTC-12 and OK cell lines, although the Xenopus A6 cell line resembles collecting duct epithelium. In substrate-dependent culture, LLC-PK$_1$ and OK cells grow as a polarized epithelium where the apical cell membrane faces the culture membrane.

BIBLIOGRAPHY

1. D.W. Seldin and G. Giebisch, eds., *The Kidney: Physiology and Pathophysiology*, 2nd ed., Raven Press, New York, 1992.
2. G.B. Haycock, in A.R. Mundy, J.M. Fitzpatrick, D.E. Neal, and N.J.R. George, eds., *The Scientific Basis of Urology*, ISIS Medical Media, Oxford, U.K., 1999, pp. 53–90.
3. W. Kriz and L. Bankir, *Am. J. Physiol.* **254LF**, F1–F8 (1988).
4. J.M. Fitzpatrick and P.G. Horgan, in A.R. Mundy, J.M. Fitzpatrick, D.E. Neal, and N.J.R. George, eds., *The Scientific Basis of Urology*, ISIS Medical Media, Oxford, U.K., 1999, pp. 53–90.
5. M. Davies, *Kidney Int.* **45**, 320–327, (1994).
6. P. Mundel and W. Kriz, *Anat. Embryol.* **192**, 385–397 (1995).
7. J.R. Couchman, L.A. Beavan, and K.J. McCarthy, *Kidney Int.* **45**, 328–335 (1994).
8. S. Gauer, J. Yao, H.O. Schoecklmann, and R.B. Sterzel, *Kidney Int.* **51**, 1447–1453 (1997).
9. D. Schlondorff, *FASEB J.* **1**, 272–281 (1987).
10. P. Mene and M.J. Dunn, *Physiol Rev.* **69**, 1347–1424 (1989).
11. F. Prols, A. Hartner, H.O. Schoecklmann, and R.B. Sterzel, *Exp. Nephrol.* **7**, 137–146 (1999).
12. V. Zeidel, *Am. J. Physiol.* **271**, (Renal Fluid Electrolyte Physiol., 40), F243–F245 (1996).
13. W. Pfaller and G. Gstraunthaaler, *Environ. Health Perspect.* **106**(Suppl. 2), 559–569 (1998).
14. K.M. Madsen and C.C. Tisher, *Am. J. Physiol.* **250**, F1–F15 (1986).
15. P.D. Bell and J.Y. Lapointe, *Clin. Exp. Pharmacol. Physiol.* **24**, 541–547 (1997).
16. H. Osswald, B. Muhlbauer, and V. Vallon, *Blood Purif.* **15**, 243–252 (1997).
17. S. Bachmann and I. Oberbaumer, *Kidney Int., Suppl.* **67**, S29–S33 (1998).
18. S. Kloth et al., *Differentiation* **63**, 21–32 (1998).
19. Y.H. Kim et al., *J. Am. Soc. Nephrol.* **10**, 1–12 (1999).
20. W.E. Sweeney and E.D. Avner, *J. Tissue Cult. Methods* **13**, 163–168 (1991).
21. P.D. Wilson, R.W. Schrier, R.D. Breckon, and P.A. Gabow, *Kidney Int.* **30**, 371–378 (1986).
22. M.F. Horster, *Klin Wochenschr.* **58**, 965–973, (1980).
23. M.A. Smith, W.R. Hewitt, and J.B. Hook, in C.K. Atterwill and C.E. Steele, eds., *In Vitro Methods in Toxicology*, Cambridge University Press, Cambridge, U.K., pp. 13–35 (1987).
24. J.I. Kreisberg and P.D. Wilson, *J. Electron Microsc. Technol.* **9**, 235–263 (1998).
25. L.J. Striker, R.L. Tannen, M.A. Lange, and G.E. Striker, *Int. Rev. Exp. Pathol.* **30**, 55–105 (1988).
26. P.D. Williams, *In Vitro Cell Dev Biol.* **25**, 800–805 (1989).
27. M.F. Horster and M. Sone, *Methods Enzymol.* **191**, 409–427 (1990).
28. N.L. Simmons, *Methods Enzymol.* **191**, 426–436 (1990).
29. R. Della Bruna and A. Kurtz, *Exp. Nephrol.* **3**, 219–222 (1995).
30. M. Taub, *Methods Mol. Biol.* **75**, 153–161 (1997).
31. J.I. Kreisberg and M.J. Karnovsky, *Kidney Int.* **23**, 439–447 (1983).
32. D.A. Troyer and J.I. Kreisberg, *Methods Enzymol.* **191**, 141–152 (1990).
33. P. Mundel and W. Kriz, *Exp. Nephrol.* **4**, 263–266 (1996).
34. R. Ardaillou, *Cell Biol. Toxicol.* **12**, 257–261 (1996).
35. S. Ringstead and G.B. Robinson, *Virchows Arch. A: Pathol. Anat. Histol.* **425**, 391–398 (1994).
36. P. Mundel, J. Reiser, and W. Kriz, *J. Am. Soc. Nephrol.* **8**, 697–705 (1997).
37. M. Kitamura and Y. Ishikawa, *Kidney Int.* **53**, 690–697 (1998).
38. P.D. Wilson, *Miner. Electrolyte Metab.* **12**, 71–83 (1986).
39. P.D. Wilson, *J. Tissue Cult. Methods* **13**, 137–142 (1991).
40. M.R. Rosenberg and G. Michalopoulos, *J. Cell. Physiol.* **131**, 107–113 (1987).
41. D.P. Rodeheaver, M.D. Aleo, and R.G. Schnellmann, *In Vitro Cell Dev. Biol.* **26**, 898–904 (1990).
42. W.L. Smith and W.S. Spielman, in *Functional Epithelial Cells in Culture*, K.S. Martin and J.D. Valentick, eds., Alan R. Liss, New York, 1989, pp. 303–323.

43. A. Naray-Fejes-Toth and G. Fejes-Toth, *J. Tissue Cult. Methods* **13**, 179–184 (1991).

44. J.H. Pizzonia et al., *In Vitro Cell Dev. Biol.* **27A**, 409–416 (1991).

45. M. Taub, *BioTechniques* **25**, 990–994 (1998).

46. G. Gstraunthaaler and W. Pfaller, in R.R. Watson, ed., *In Vitro Methods of Toxicology*, CRC Press, Baton Rouge, La., 1992, pp. 94–106.

47. U. Hopfer et al., *Am. J. Physiol.* **270**, C1–C11 (1996).

48. N. Cartier et al., *J. Cell Sci.* **104**, 695–704 (1993).

49. A. Vandewalle et al., *J. Cell Physiol.* **141**, 203–221 (1989).

50. P.G. Woost et al., *Kidney Int.* **50**, 125–134 (1996).

51. D.M. Scott, *Differentiation* **36**, 35–46 (1987).

52. G. Wolf et al., *Am. J. Physiol.* **268**, F940–F947 (1995).

53. M.J. Ryan et al., *Kidney Int.* **45**, 48–57 (1994).

54. L.C. Racusen et al., *J. Lab. Clin. Med.* **129**, 318–329 (1997).

55. M. Hosoyamada, M. Obbinata, M. Suzuki, and H. Endou, *Arch. Toxicol.* **70**, 284–292 (1996).

56. C.R. Gaush, W.L. Hard, and T.F. Smith, *Proc. Soc. Exp. Biol. Med.* **122**, 931–935 (1966).

57. R.N. Hull, W.R. Cherry, and G.W. Weaver, *In Vitro* **12**, 670–677 (1976).

58. H. Koyama et al., *In Vitro* **14**, 239–246 (1978).

59. K.A. Rafferty, in M. Mizell, ed., *Biology of Amphibian Tumors*, Springer-Verlag, Berlin, pp. 52–81 (1969).

60. J.E. Lever, *Miner. Electrolyte Metab.* **12**, 14–19 (1986).

61. C. Yeaman, K.K. Grindstaff, and W.J. Nelson, *Physiol Rev.* **79**, 73–98 (1999).

62. M. Gekle, S. Wunsch, H. Oberleithner, and S. Silbernagl, *Pfluegers Arch.* **428**, 157–162 (1994).

63. W.W. Minuth, S. Kloth, V. Maier, and R. Dermietzel, *In Vitro Cell Dev. Biol.* **30A**, 12–14 (1994).

64. J.P. Lavelle et al., *Am. J. Physiol.* (Renal Physiol 42) **273**, F67–F75 (1997).

ANIMAL CELL TYPES, LIVER CELLS

G. STACEY
National Institute of Biological Standards and Control
South Mimms, England

OUTLINE

Hepatocytes *In Vivo*
Hepatocytes *In Vitro*
Hepatic Cell Lines
Conclusion
Bibliography

The liver comprises 2–5% of the body weight of adult mammals, and is an extremely powerful organ that is involved in controlling the quality of the blood. It elicits the removal of unwanted materials, the detoxification of hazardous chemicals, and the synthesis of a range of important plasma proteins such as albumin. It is also a unique organ in its capacity to regenerate itself to full capacity from residual functional tissue. However, the liver is surprisingly uncomplicated in structure and is based on a single histological unit called the acinus (1). The acinus comprises a parenchymal cell mass around a terminal portal venule and a hepatic arteriole that provides the blood supply to the acinar venous sinusoids (Fig. 1). The venous sinusoids are lined with endothelial cells with extremely large cellular pores (up to 1 μ) and kupffer cells that are reticular-endothelial cells with the capacity to phagocytose bacteria and other foreign matter. At the basal region of this endothelial layer is a physical space called the space of Diss between the endothelium and the hepatocytes that is organized in hepatic plates with the bile-collecting canaliculi located on the side of each hepatic plate opposite to the space of Diss. The space of Diss drains to a lymphatic vessel. The hepatic plates with the endothelial cell and Kupffer cell layers radiate from the central core of the acinus to give it a wheel-like structure.

Despite its structural homogeneity the biochemical functions are polarized to some extent. Notably, glutamine synthesis and ketogenesis are most prominent in the periportal vein and gluconeogenesis, urea synthesis, and oxidative phosphorylation are concentrated in the central venous region (2–4). The drug metabolizing activities are also more abundant in the centri-lobular regions (5,6).

Many cell types are represented in the liver, including hepatocytes, Kupffer cells, endothelial cells, bile duct cells, mesenchymal cells, and nerve cells. A cell type referred to as "oval cells" have been proposed as a precursor cell component of the liver (7) and may be derived from the bile duct. However, the overwhelmingly predominant cell type performing the majority of liver function is the hepatocyte which occurs at a density of approximately 10^8 cells/g liver tissue in humans.

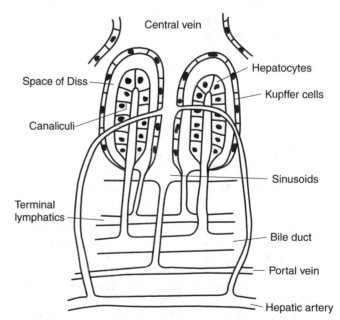

Figure 1. Simplified diagram of the structural organization within the human liver acinus. It shows the hepatic plates that radiate from the central venule and the flow of lymph and bile produced.

HEPATOCYTES *IN VIVO*

The proportion of liver cells that are hepatocytes varies among species and is higher in human (80%) (8) than in rat (60%) (9). The size of hepatocytes also varies among species, and human hepatocytes are generally larger than those of rat and dog (10). Within an individual liver, hepatocyte size exhibits a gradient across the structure of the acinus. In adult human liver, the hepatocytes are mono- or binuclear and predominantly tetraploid. In adult animals, hepatocytes divide very infrequently and mitotic indexes are very low, for example, 0.03% in the adult rat (11).

The hepatocyte is the center of active chemical biotransformation in the liver although some metabolizing enzymes are found in the endothelial, Kupffer, and bile duct cells (12). Hepatocytes perform a combination of enzyme-mediated activation and detoxification reactions divided into two groups called phase I and phase II. Phase I processes are generally hydrolytic and "redox" reactions that provide the functional groups to which phase II reactions conjugate ligands to enable excretion via membrane transporters (13). Cytochrome P-450 (CYP) mono-oxigenase is a key Phase I enzyme system that is representative of a supergene family of at least 74 different groups of which 14 are common to all mammals (14). UDP-glucuronyl-transferases are the most abundant phase II enzymes which like CYP are located in the hepatocyte endoplasmic reticulum. The three UDP-glucuronyl-transfereses are responsible for conjugation of UDP-glucuronic acid to endogenous substrates such as bilirubin, steroids, estradiol, and xenobiotics. The most important of the other phase II enzymes are the sulfotransferases, which effect sulfation in the hepatocyte cytosol, and the glutathione-*S*-transferases (GST) (15) which mediate conjugation of reduced glutathione to a range of electrophilic compounds. The four main classes of these phase II enzymes (alpha, mu, pi, and theta) are found in a number of mammalian species (16) and are important in protecting against the toxic effects of carcinogens (17). As already indicated, numerous factors influence the relative enzyme activities, and these include species and strain (18), endocrine status (19), age (20), disease (21), and nutrition (22). Drug metabolizing activities among individual humans can be very variable (23,24) and in the case of CYPs and GSTs, may have a genetic basis (23,24).

HEPATOCYTES *IN VITRO*

The liver has been studied *in vitro* by using isolated perfused livers (25), liver slices (26), isolated cells, subcellular fractions, and cell lines. Technological developments have led to renewed interest in the liver slice models whereby tissue slices of 250 µm can be prepared reliably and provide metabolic data over 2–3 days. However, the most common *in vitro* systems involve isolated hepatocytes.

Hepatocyte cell suspensions are usually prepared by a two-step collagenase perfusion method first described by Berry and Friend (27) for rat liver and now has been applied to various species, including rabbit, dog, pig, and humans (28–30). The first stage of perfusion releases desmosome and hemidesmosome junctions between adjacent cells and the extracellular matrix by a divalent cation-free buffer, whose action may be assisted by a chelator such as EGTA or EDTA (31). Then, perfusion with collagenase digests the liver matrix to yield a single-cell suspension. The whole liver is usually dissociated for small mammals whereas wedge biopsies are more frequently used for large species. Kupffer and endothelial cells can be prepared by collagenase–pronase treatment (32).

Suspensions of hepatocytes allowed to grow on glass or plastic surfaces reestablish their polarity and form an epithelial monolayer (33). Numerous hepatocyte functions are retained *in vitro* such as Na dependent transport functions for bile acids (34), amino acids (35), and sugars (36), although a general decrease in liver specific functions is observed and had been ascribed to a dedifferentiation process (37). Up to 50% CYP in rat hepatocytes its lost during the first 24–48h *in vitro* (38–41) and, although inducers may be used to regain CYP activity, this response is not directly comparable to that *in vivo* (42). Similar changes are also observed for UDP-glucuronyl transferases (43), glutathione transferases (44), and other liver-specific genes such as albumin and transferrin, which may fall to less than 10% of their transcription rate *in vivo* after only 24h *in vitro* (45,46). These changes differ among species although human hepatocytes *in vitro* retain the activity of all of the main CYP groups (47).

The culture medium has a marked influence on the performance of hepatocytes *in vitro* (48), and the levels of certain amino acids have very specific effects on hepatocytic function (49). Differentiation of hepatocytes *in vitro* may be enhanced by additives such as dexamethazone which enhances transcription of matrix proteins such as collagen. Insulin and epidermal growth factor stimulate DNA synthesis in hepatocytes, but the most powerful mitogen for stimulating hepatocytic growth is hepatocytic growth factor (50,51).

Coculture of hepatocytes with other cell types such as sinusoidal cells (52) and fibroblasts (53) have shown some improvement in hepatocyte survival. However, the most successful system of coculture uses rat epithelial liver cells derived from biliary epithelium, and this has been successful in producing cultures that survive for several weeks and retain the synthesis of plasma proteins and phase I and II enzymes (54–56).

Generation of spheroid cultures using PolyHEMA (57) encourages the long-term expression of plasma proteins such a albumin and transferrin (57,58) and the deposition of extracellular matrix proteins (59).

The diversity of preparative methods, culture media, and substrates means that there may be some difficulty in comparing data using these different approaches. The problem of standardizing hepatocyte cultures has been reviewed by Skett and Bayliss (60).

The growth of hepatocytes on plastic surfaces or membranes coated with collagen has provided some improvement in hepatocytic survival (61). However, a novel collagen sandwich culture method has shown significant improvements in function, as well as prolonged viability of the hepatocytes (62). Bioreactor systems using hepatocytes in alginate beads have also been beneficial in maintaining hepatocytic functions (63). Further devices called rotating wall vessels, which avoid the turbulence created at the gas–growth medium interface to yield low shear-stress culture systems (so-called microgravity), are also being used now (64). These systems have enabled the *in vitro* production of relatively large pieces of tissue with recognizable liver histology (64). Rat hepatocytes cultured in real microgravity conditions have shown changes in Kupffer cells and glycogen storage (65).

HEPATIC CELL LINES

Hepatocyte cell lines have been obtained from hepatomas and transfection of normal hepatocytes with recombinant constructs that express oncogenes (Table 1). However, although some of the hepatoma cell lines such as HepG2 and C3A retain a number of the synthetic properties of hepatocytes, no cell lines have been produced that express the full range of phase I and phase II enzymes found in primary cultures. In addition enzyme activities that are

Table 1. Examples of Hepatic Cell Lines

Species	Cell Name	Characteristics/Applications[a]	Catalogue Codes[b]
Human	WRL68	Derived from embryo cells. See U.S. Patent 3,935,066	ATCC CCL48 ECACC89121403
	PLC/PRF/5	Contains hepatitis B sequences. Used for isolating a wide range of viruses and has some significance in toxicology	ATCC CRL 8024 ECACC85061113
	Hep3B	Produces a variety of plasma proteins, e.g., fibrinogen, alpha-fetprotein, transferrin, albumin, complement C3, and alpha-2-macroglobulin	ATCC HB8064 DSMZ ACC93 ECACC86062703
	HepG2	Produces a variety of plasma proteins, e.g., prothrombin, antithrombin III, alpha-fetprotein, complement C3, and fibrinogen	ATCC HB8065 DSMZ ACC180 ECACC85011430
	Chang Liver (HeLa)	Widely used in virology, biochemistry, and transplantation science but now well documented to be a subclone of HeLa cells	ATCC CCL13 ECACC88111403
	SK-HEP-1	Isolated from liver adenocarcinoma and forms large cell carcinomas in nude mice. Reported to secrete alpha-1 antitrypsin	ATCC HTB52 DSMZ ACC141 ECACC91091816
Mouse	MC/9	Fetal cells that secrete histamine and leukotrienes. Also express IgE receptors that permit sensitization to antigens following exposure to antigen-specific IgE	ATCC CRL 8306 ECACC90112405
	H2.35	SV40 (tsA255) transformed hepatocytes. Temperature-sensitive expression of serum albumin mRNA. Undergoes morphological changes and induction of albumin secretion in response to extracellular matrix gel substrata	ECACC94050407 ATCC CRL1995
	NCTC clone 1469	Derived from NCTC 721. Used in biochemical and nutritional studies and is susceptible to vesicular stomatitis virus (Indiana)	ATCC CCL9.1, ECACC88111403
	HEPA 1-6	Mouse hepatoma culture derived from the BW7756 tumor described as secreting several liver-specific products	DSMZ ACC175 ECACC92110305
Rat	Anr4	Derived from the ARL J301-3 cell line by transformation with the EJ-ras oncogene	ECACC90071808
	ARL-6	Provides the substrate for an *in vitro* test for tumorigenicity of rat tumor cells that is more sensitive than nude mouse inoculation	ECACC87050701
	BRL 3A	Generated by primary cloning of untransformed (normal) cells. Noted for secretion of mitogen-like peptides under serum-free culture conditions	ATCC CRL1442 ECACC85111503
	Clone 9	Used for *in vitro* studies of carcinogenesis and nutrition	ATCC CRL1439 ECACC88072203
	MH1C1	Derived from "Morris hepatoma" cells but has distinct characteristics. Cells synthesize serum albumin and tyrosine asparaginase. Also reported to express a variety of dehydrogenase, synthetase, transferase enzymes, and cytochrome P450 of significance in toxicology studies. Some sources may be HeLa contaminated	ATCC CCL144 DSMZ ACC106 ECACC85112702
	Phi1	Derived from Fu5 cells, this cell line is resistant to thioguanine (6 µg/mL) and is reported to be highly differentiated	ECACC85061101
Fish	R1	Rainbow trout liver. Has been subcloned to yield cell line D11	DSMZ ACC 56
Bird	LMH/2A	Chemically induced hepatocarcinoma cells derived from a leghorn chicken. These cells are transfected with the chicken estrogen receptor	ATCC CRL2118

[a]Further details of biochemical activities of cell lines of importance are available through Forschungszentrum, the CCLTOPS catalogue, Prof. F.J. Weibel, GSF- fur Umwelt and Gesundheit, Neuherberg, Germany.
[b]Descriptions of the different sources of cells and their addresses can be found in the chapter on cell banking.

retained in these cell lines may show variability in serial passage over time. One well-known "hepatocyte" cell line, "Chang Liver," is cross-contaminated with HeLa cells (66). Nonparenchymal cell lines have been described, and rat liver epithelial cell lines, probably derived from biliary epithelium, can transform and propagate indefinitely *in vitro*.

CONCLUSION

Although a number of cell types are found in the liver, the hepatocyte is by far the most predominant and important cell type. Liver function can be studied by whole-organ perfusion or tissue slice analysis but these suffer significantly from variation among organs. No cell line can yet provide the full range of activities identified in primary hepatocytic cultures. However, the diversity of methods for the preparation and culture of hepatocytes means that standardization is a major problem.

BIBLIOGRAPHY

1. A.M. Rappaport, Z.J. Borowy, W.M. Longheed, and W.N. Lotto, *Anat. Rec.* **119**, 11–33 (1954).
2. J.J. Gumico et al., *Hepatology* **6**, 932–944 (1986).
3. K. Jungermann and N. Katz, *Physiol. Rev.* **69**, 708–751 (1989).
4. R. Gerbhardt, *Pharmacol. Ther.* **53**, 275–354 (1992).
5. J. Baron et al., in J. Caldwell and G.D. Paulson, eds., *Foreign Compound Metabolism*, Taylor & Francis, London, 1984, pp. 17–36.
6. D. Ratanasavanh et al., *Hepatology* **13**, 1142–1151 (1991).
7. W.F. Grigioni et al., *Hepatology* **21**, 1543–1546 (1995).
8. G.A. Gates et al., *J. Lab. Clin. Med.* **57**, 182–184 (1961).
9. R. Daoust, *Ann. Inst. Biol. Sci. Publ.* **4**, 3–10 (1958).
10. H.P. Rohr et al., in H. Popper and F. Schaffner, eds., *Progress of Liver Diseases*, Vol. 5, Grune & Stratton, New York, 1976, pp. 24–34.
11. R. Carriere, *Int. Rev. Cytol.* **25**, 201–277 (1969).
12. P. Steinberg et al., *Mol. Pharmacol.* **32**, 463–470 (1987)
13. G. Jedlitschky et al., *Cancer Res.* **56**, 988–994 (1996).
14. D.R. Nelson et al., *Pharmacogenetics* **6**, 1–42 (1996).
15. B. Ketterer and L.G. Christodoulides, *Adv. Pharmacol.* **27**, 37–69 (1994).
16. B. Mannervick et al., *Biochem. J.* **282**, 305–308 (1992).
17. J.D. Hayes and D.J. Pulford, *Crit. Rev. Biochem. Mol. Biol.* **30**, 445–600 (1995).
18. C.H. Walker, in J.W. Bridges and L.F.N. Chausseaud, eds., *Progress in Drug Metabolism*, Vol. 5, New York, Wiley, 1980, pp. 113–164.
19. J.A. Gustafsson, A. Mode, G. Norstedt, and P. Skett, *Annu. Rev. Physiol.* **4**, 51–60 (1983).
20. C.N. Falany, *Trends. Pharmacol. Sci.* **12**, 254–259 (1991).
21. F.P. Guengerich, *Annu. Rev. Nutr.* **4**, 207–231 (1984).
22. W.R. Bidlack, R.C. Brown, and C. Mohan, *Fed. Proc. Fed. Am. Soc. Exp. Biol.* **45**, 142–148 (1986).
23. B. Van Ommen et al., *Biochem. J.* **269**, 609–613 (1990).
24. B. Ketterer et al., *Biochem. Pharmacol.* **41**, 635–638 (1991).
25. S.A. Belinsky et al., *Mol. Pharmacol.* **25**, 158–164 (1984).
26. P.F. Smith et al., *In Vitro Cell Dev. Biol.* **22**, 706–712 (1986).
27. M.N. Berry and D.S. Friend, *J. Cell Biol.* **43**, 506–520 (1969).
28. G.A.E. Van't Klooster et al., *Xenobiotica* **22**, 523–534 (1992).
29. C.A. McQueen, in C.A. Tyson and J.M. Frazier, eds., *Methods in Toxicology*, Vol. 1, Academic Press, San Diego, Calif., 1993, pp. 255–270.
30. R.G. Ulrich et al., *Toxicol. Lett.* **82/83**, 107–115 (1995).
31. T.D. Seilaff et al., *Transplantation* **59**, 1459–1463 (1995).
32. D.L. Knook, N. Blansjaar, and E. Sleyster, *Exp. Cell Res.* **109**, 317–329 (1977).
33. M. Maurice, E. Rogier, D. Cassio, and G. Feldmann, *J. Cell Sci.* **90**, 79–92 (1988).
34. L.R. Schwarz et al., *Eur. J. Biochem.* **55**, 617–623 (1975).
35. M.S. Kilberg, *J. Membr. Biol.* **69**, 1–12 (1982).
36. H. Baur and H.W. Heldt, *Eur. J. Biochem.* **74**, 397–403 (1977).
37. R.W. Van Dyke, J.E. Stephens, and B.F. Scharschmidt, *Am. J. Physiol.* **244**, G484–G492, (1982).
38. P.S. Guzelian, D.M. Bissell, and U.A. Meyer, *Gastroenterology* **72**, 1232–1239 (1977).
39. W.E. Fahl et al., *Arch. Biochem. Biophys.* **192**, 61–72 (1979).
40. C.J. Maslansky and G.M. Williams, *In Vitro Cell Dev. Biol.* **18**, 663–693 (1982).
41. J.M. Bégué, C. Guguen-Guillouzo, N. Pasdeloup, and A. Guillouzo, *Hepatology* **4**, 839–842 (1984).
42. P.B. Watkins et al., *J. Biol. Chem.* **261**, 6264–6271 (1986).
43. J. Forster, G. Luippold, and L.R. Schwartz, *Drug Metab. Dispos.* **4**, 353–360 (1986).
44. Y. Vandenberghe et al., *Biochem. Pharmacol.* **37**, 2482–2485 (1988).
45. D.F. Clayton and J.E. Darnell, *Mol. Cell Biol.* **3**, 1552–1561 (1983).
46. J.M. Fraslin et al., *EMBO J.* **4**, 2487–2491 (1985).
47. F. Morel et al., *Eur. J. Biochem.* **191**, 437–444 (1990).
48. C.M. Ryan et al., *Surgery* **113**, 48–54 (1993).
49. K. Hasegawa, Y. Mijata, and B. Carr, *J. Cell. Physiol.* **158**, 365–373 (1994).
50. T. Nakamura et al., *Nature (London)* **342**, 440–443 (1989).
51. A. Srain, *Gut.* **35**, 433–436 (1994).
52. J.C. Wanson, R. Mosselmans, A. Brouwer, and D.L. Knook, *Biol. Cell* **56**, 7–16 (1979).
53. G. Michalopoulos, F. Russel, and C. Biles, *In Vitro Cell Dev. Biol.* **15**, 796–806 (1979).
54. J.P. Lebreton et al., *Biochem. J.* **235**, 421–427 (1986).
55. J. Conner et al., *Biochem. J.* **266**, 683–688 (1990).
56. C. Lerche et al., *Eur. J. Biochem.* **244**, 98–106 (1997).
57. J.Z. Tong et al., *Exp. Cell Res.* **200**, 326–332 (1992).
58. Y. Sakai et al., *Cell Transplant.* **5**, 505–511 (1996).
59. L. Knook, N. Blansjaar, and E. Sleyster, *Exp. Cell Res.* **109**, 317–329 (1977).
60. P. Skett and M. Bayliss, *Xenobiotica* **26**, 1–7 (1996).
61. C. Guery et al., *Toxicol. In Vitro* **7**, 453–459 (1993).
62. E. Knop et al., *Anat. Rec.* **242**, 337–349 (1995).
63. B. Femond et al., *Cell Biol. Toxicol.* **12**, 325–329 (1996).
64. T. Battle, T. Maguire, H.J. Moulsdale, and A. Doyle, *Cell Biol. Toxicol.* **15**, 3–12 (1999).
65. R.N. Racine and S.M. Cournier, *J. Appl. Physiol.* **73**, 136S–141S, (1992).
66. G.N. Stacey, B. Bolton, and A. Doyle, *In Vitro Cell Dev. Biol.* **29A**, 123A, (1993).

ANIMAL CELL TYPES, MONOCYTE, MACROPHAGE LINEAGE

K.B. WALKER
National Institute of Biological Standards and Control
South Mimms, England

OUTLINE

THE MONOCYTE LINEAGE

Cells of the monocytic lineage arise from stem cells in the bone marrow and are released into the blood, where they are observed predominantly as cells of monocytic morphology expressing high levels of CD14 (Gram-negative bacterial lipopolysaccharide receptor) and the monocyte-specific cytochemical marker sodium fluoride sensitive esterase (1). Monocytes are usually negative for CD16 (Fc gamma receptor type III) expression, but a subset of CD14-positive, CD16-positive monocytes has been identified (2) that appear to have distinct functional properties (3). The monocyte lineage of leukocytes gives rise to macrophages and other cells, including dendritic cells (4) and osteoclasts, which are an essential cellular component in the biology of bone (5).

MACROPHAGES IN VIVO

While macrophages have a role to play in the immune response (6,7), they are best known for their capacity to ingest and degrade or phagocytose a range of materials. The other major group of phagocytes are neutrophils (polymorphonuclear cells) of the granulocytic lineage of leukocytes (8). Macrophages exist as freely circulating mononuclear phagocytic cells in the blood and lymph and as resident macrophages in various tissues and organs. Resident macrophages include Kupffer cells in the liver (9), microglial cells in the brain (10,11), and the mesangial phagocytes of the kidney, to name a few.

The materials that macrophages phagocytose range from antibody–antigen complexes through to viruses, bacteria, dead cells, and membrane fragments. Macrophages play an important role in wound healing by phagocytosing dead cells and debris (12) and secreting chemoattractants and growth factors (13,14). Macrophages are also professional antigen-presenting cells (15). This term refers to their capacity to ingest, process, and present protein antigens to T helper cells via the MHC class I or II pathways (15).

Macrophages mature and differentiate under the influence of cytokines and growth factors and show marked phenotypic and functional variation (16,17). Circulating macrophages passage between the blood and lymph via trafficking through the peripheral lymph nodes. Macrophages tend to be nonproliferating cells and may survive for many years, unlike neutrophil phagocytes, which are produced for short-term responses. Some of the tissue resident macrophages are believed to remain resident for the life of the host. Macrophages are a pleomorphic and phenotypically varied population due to the profound effect that the microenvironment has upon them (18).

In response to injury or infection, the release of soluble factors creates a chemotactic gradient that initiates the infiltration of macrophages (among other cell types) into the site of injury or infection (19). Macrophages are one of the first in the waves of cells entering such an area. These cells phagocytose bacteria and tissue debris and can migrate to the draining lymph nodes to process and present antigens to T helper cells in order to prompt cell-mediated and humoral immunity (16). At latter stages, macrophages can be stimulated by exposure to particular cytokines (macrophage activating factors, MAFs) (20) resulting in a number of changes in the macrophage. These changes include increased expression of MHC-II, induction of biochemical pathways that mediate killing of bacteria, and the increased expression of some surface markers. Macrophages are relatively nondiscriminatory in their functions, as they do not respond in an antigen-specific manner. Thus amplification of macrophage activity by recall responses does not occur directly by macrophages but is a secondary consequence of antigen-specific lymphocyte activity with release of macrophage activating factors such as interferon gamma and granulocyte macrophage colony-stimulating factor, to name two.

Macrophages are crucial to the immune system and are central to protection against bacterial infection. They are central players in antigen presentation and have a major part to play in certain aspects of antiviral immunity, including the expression of interferon gamma (21).

MACROPHAGES IN VITRO

Macrophages have been difficult to transform into stable continuous cell lines of a well-defined phenotype. Because of the innate plasticity of this cell type, such cell lines that have been derived have tended to be unstable and change phenotype over time. However, some relatively stable monocyte-like or macrophage-like cell lines have been isolated and characterized (Table 1). They have been extremely useful in dissecting the details of macrophage activation and maturation and the mechanisms of macrophage-mediated microbicidal function.

The distinction between monocyte-like and macrophage-like cells can sometimes be rather subtle, but in general it can be said that monocyte-like are the less differentiated being progenitor cells. The monocyte-like cell lines often require exposure to macrophage activating factors and other cytokines to evoke a broader range of

Table 1. Examples of Cell Lines of This Myeloid Lineage

Species	Cell Name	Characteristics/Applications	Collection Codes[a]
Human	HL-60	Derived from promyelocytic leukemia cells isolated by lymphophoresis. Differentiation is enhanced by treatment with compounds such as dimethyl sulfoxide and sodium butyrate. A series of subclones have been isolated with subtly different characteristics in terms of their capacity to differentiate.	ATCC CCL 240 ECACC 88112501
	THP	Derived from a case of acute monocytic leukemia. Reported to have antibody (Fc) and complement (3b) receptors and to be capable of differentiation into macrophage–like cells.	ATCC TIB202 ECACC 88081201
	U937	Cell line derived from a case of histiocytic lymphoma. Expresses many monocyte characteristics typical of cells of histiocytic origin.	ATCC CRL 1593 ECACC 87010802
Mouse	416B	Mouse (BDF) myelomonocytic leukemia cell line. Forms myelomonocytic, megakaryocytic, and erythropoietic colonies in vivo.	ECACC 85061103
	J774.2	Cell line of monocyte–macrophage lineage cloned from the J774.1 tumor. Synthesizes lymphocyte activating factor and interleukin-1.	ATCC TIB 67 ECACC 85011428
	P388.D1	Isolated from P388 macrophage cells for high production of interleukin-1. Useful in antibody-dependent cell-mediated cytotoxicity studies.	ATCC TIB 63 ECACC 85011439
	RAW 264	Monocyte–macrophage cell line isolated from a tumor induced by intraperitoneal injection of Abelson leukemia virus (A-MuLV). Reported to exhibit phagocytosis and lysozyme secretion. Mediates antibody-dependent lysis of sheep erythrocytes and tumor target cells.	ATCC TIB 71 ECACC 85062803
	Wehi 3b	Myelomonocytic cells induced by mineral oil injections of Balb/C mice. Expresses interleukin-3.	ECACC 86013003

[a]For further information on cell lines available from resource centers, see the article on cell banks.

macrophage-like functions and markers. One particularly valuable macrophage precursor cell line is the human prolmyelocyitc leukemia line HL60 (22) (Table 1), which can be stimulated by a variety of compounds (typically dimethyl sulfoxide or sodium butyrate) to differentiate into cells representative of the monocyte or neutrophil lineages. Some of these differentiated cell types have been established as stable cell lines including HL60 15-12 (23) and HL60 Ast.4 (24). However, some similarly derived cell lines, such as HL60 M2 and HL60 Ast.3 (24), are reported to differentiate solely in the neutrophil lineage. A second histiocyte cell line, U937, has also been used for a wide range of research purposes (Table 1) and responds to differentiating agents (25).

Much work has been done using these and other monocytic cell lines in identifying the factors critical to macrophage activation. Aspects of phagocytosis and intracellular trafficking of bacteria and antigens have also been fruitfully investigated using these lines.

DENDRITIC CELLS

Categories and Characterization of Dendritic Cells

Dendritic cells are unequaled in their potency for initiating the primary immune response through their characteristically powerful capability for presenting antigen to T cells. Dendritic cells are derived from CD34-positive leukocytes either directly from CD34-positive bone marrow progenitors (26) or from dendritic cell/macrophage precursors in the peripheral circulation (27). These cells appear in almost all tissues and have slightly different

characteristics in different microenvironments. These differences have led to identification of different categories of dendritic cells, including interstitial dendritic cells (gut, kidney, heart, and lung), interdigitating dendritic cells (thymic medulla and secondary lymphoid tissue), Langerhans cells (skin epidermis and mucosa), lymph dendritic cells (veiled cells), and blood dendritic cells. Follicular dendritic cells may not to be of the leukocyte lineage and are primarily involved in maintaining humoral immunity by presenting antigens to B cells, immune memory, and maturation of antibody affinity (28).

Dendritic cells at a particular site will alter phenotype over their lifespan; however, there are a number of features that are characteristic of this cell type (28):

- Dendritic morphology
- Low phagocytic activity
- Induction of proliferation in allogeneic naive T cells
- CD1a positive with high levels of major histocompatibility complex II (MHC II) and accessory molecules such as CD54, CD58, CD80, and CD86
- Absence of CD14 and nonspecific esterase
- Intracytoplasmic Birkbeck granules seen in Langerhans cells of the epidermis

The capability of dendritic cells for antigen presentation has notably implicated them in responses to allergens (29) and to virus infection (30). Dendritic cells are also known to be involved in the selection and survival of T cells (31). The power of their antigen-presenting activity is illustrated by reports that T-cell responses to extremely low levels of bacterial superantigens are elicited solely through dendritic cells (32). In addition, their significance in immunity means that they are important factors in transplantation immunology and appear to migrate from allografts and stimulate T cells in the host lymphoid tissue (33,34). The cell biology of dendritic cells is obviously complex (35), and the wider potential of these cells has been indicated by the observation of the close association between Langerhans cells and nerve processes (36) and the potential role of dendritic cells as an interface with the nervous system has been discussed by Ibrahim et al. (37).

Dendritic Cells in Vitro

Dendritic cells are difficult to isolate due to their scarcity in tissue. The most efficient methods for their preparation have been from blood using granulocyte macrophage colony stimulating factor (GM-CSF) to cause proliferation of dendritic cells and addition of IL-4 to suppress the growth of macrophages (38). In vivo the microenvironment is critical to the growth and phenotype of dendritic cells, and this is illustrated in vitro by the promotion of full maturation when hepatic dendritic cells are grown on collagen I (39). However, the level of activity in stem cell research means that the developments in the culture of dendritic cells is rapidly progressing (40,41).

Dendritic cell lines have been derived from mouse epidermis and spleen cells (42,43). These cell lines express co-stimulatory molecules, show efficient antigen-presenting activity, and are proving valuable in the study of maturation of dendritic cells. Another dendritic cell line from fetal murine skin is reported to be MHC class I positive and MHC class II negative (44). The cells resemble fetal dendritic cells and are potent stimulators of CD8+ T cells, therefore, they have an important role in evoking a cytotoxic T-cell response. A range of cells have been isolated more recently (45,46), but there is still a need for more continuous cell lines to represent the full range of dendritic cells.

The importance of dendritic cells in quality and level of the immune response (47) and their involvement in both primary and secondary phases of the immune response means that they are being actively investigated in new approaches to vaccination (48,49).

The ability of dendritic cells to damp down the response to auto- and allo-antigens and allergens means that they remain important in transplant science (50,51), allergy, and autoimmune diseases (52). The antitumor activities of denritic cells are well known (53), and some cancer therapies have also been developed commonly based on the presentation of tumor-associated antigens by in vitro dendritic cells. Novel genetic engineering approaches are also opening up new models for research (54) and therapy, and it seems that, as the cell biology of this vital cell type is resolved, numerous clinical benefits will follow.

BIBLIOGRAPHY

1. B. Passlick, D. Fleiger, and H.W.L. Zeigler-Heitbrock, *Blood* **74**, 2527–2534 (1989).
2. H.W.L. Zeigler-Heitbrock, B. Passlick, and D. Fleiger, *Hybridoma* **7**, 521–527
3. H.W.L. Zeigler-Heitbrock, *Immunol. Today* **9**, 424–428 (1996).
4. J.H. Peters, R. Gieseler, B. Thiele, and F. Steinbach, *Immunol. Today* **6**, 273–278 (1996).
5. S. Hayashi et al., *Biochem. Cell Biol.* **76**, 911–922 (1998).
6. S. Gordon, *Res. Immunol.* **149**, 685–688 (1998).
7. G. Mazarella et al., *Arch. Chest Dis.* **53**, 92–96 (1998).
8. S.W. Edwards and F. Watson, *Immunol. Today* **16**, 508–510 (1995).
9. M.L. Rose et al., *Drug Metab. Rev.* **31**, 87–116 (1999).
10. G. Stoll and S. Jander, *Prog. Neurobiol.* **58**, 233–247 (1999).
11. K. Dobrenis, *Methods* **16**, 320–344 (1998).
12. R.E. Ellis, J.Y. Yuan, and H.R. Horvitz, *Annu. Rev. Cell Biol.* **7**, 663–698 (1991).
13. K. Wilson, *Nurs. Crit. Care* **2**, 291–296 (1997).
14. D. Chantry et al., *J. Leukocyte Biol.* **64**, 49–54 (1998).
15. E.R. Unanue, *Annu. Rev. Immunol.* **2**, 395–428 (1984).
16. S. Gordon et al., *J. Cell Sci. Suppl.* **9**, 1–26 (1988).
17. A.F. Valledor, F.E. Borras, M. Cullen-Young, and A. Caleda, *J. Leukocyte Biol.* **63**, 405–17 (1998).
18. R. Andreesen, ed., *The Macrophage*, Karger, Berlin and New York, 1992.
19. M.A. Horton, ed., *Macrophages and Related Cells*, Kluwer Academic Press, Dordrecht, The Netherlands, 1993.
20. A.A. te Velde, in C. Bruijnzeel, ed., *Immunopharmacology of Macrophages and other Antigen-presenting Cells*, Blackwell, London, 1994.
21. S. Gessani and F. Belardelli, *Cytokine Growth Factor Rev.* **9**, 117–123 (1998).
22. G.D. Birnie, *Br. J. Cancer, Suppl.* **9**, 41–45 (1988).
23. C.M. Bunce, J.M. Lord, A.K.-Y. Wong, and G. Brown, *Br. J. Cancer* **57**, 559–562 (1988).
24. C.M. Bunce, A.G. Fisher, D. Toksoz, and D. Brown, *Exp. Haematol.* **11**, 828 (1983).
25. J.W. Larrick, D.G. Fischer, S.J. Anderson, and H.S. Koren, *J. Immunol.* **125**, 6–12 (1980).
26. J.W. Young, P. Szabolcs, and M.A.S. Moore, *J. Exp. Med.* **182**, 1111–1120 (1995).
27. C.D.L. Reid, S. Stackpoole, A. Meager, and J. Tikerpae, *J. Immunol.* **149**, 2681–2688 (1992).
28. C. Caux, Y.-J. Liu, and J. Banchereau, *Immunol. Today* **16**, 2–4 (1992).
29. E. Sprecher and Y. Becker, *Arch. Virol.* **132**, 1–28 (1993).
30. S. Nair et al., *J. Virol.* **67**, 4062–4069 (1993).
31. T. Broker, *J. Leukocyte Biol.* **66**, 331–335 (1999).
32. N. Bhardwaj et al., *J. Exp. Med.* **178**, 633–642 (1993).
33. C.P. Larsen et al., *J. Exp. Med.* **172**, 1483–1493 (1990).
34. C.P. Larsen, P.J. Morris, and J.M. Austyn, *J. Exp. Med.* **171**, 307–314 (1990).
35. M.T. Lotze, *J. Leukocyte Biol.* **66**, 195–200 (1999).
36. J. Hosoi et al., *Nature (London)* **363**, 159–163 (1993).
37. M.A.A. Ibrahim, B.M. Chain, and D.R. Katz, *Immunol. Today* **16**, 181–186 (1995).
38. N. Romani et al., *J. Exp. Med.* **180**, 83–93 (1994).
39. L. Lu et al., *J. Exp. Med.* **179**, 182–1834 (1994).
40. C. Gasperi et al., *J. Luekocyte Biol.* **66**, 263–237 (1999).
41. M.E. Hill et al., *Immunology* **97**, 325–332 (1999).
42. S. Xu et al., *J. Immunol.* **154**, 2697–2705 (1995).
43. P. Paglia et al., *J. Exp. Med.* **178**, 1893–1901 (1993).

44. A. Elbe, S. Schleischitz, D. Strunk, and D. Stingl, *J. Immunol.* **153**, 2878–2889 (1994).

45. R. Nunez, *Immunol. Lett.* **68**, 173–186 (1999).

46. U.S. Pat. 5,648,219 (1997), ATCC ref CRL 11904.

47. C.R. De Sousa, A. Sher, and P. Kaye, *Curr. Opin. Immunol.* **11**, 592–599 (1999).

48. A. Takashima and A. Morita, *J. Leukocyte Biol.* **66**, 350–356 (1999).

49. H. Takahashi, Y. Nakagawa, K. Yokomuro, and J.A. Berzofsky, *Int. Immunol.* **5**, 849–857 (1993).

50. A.W. Thompson and L. Lu, *Transplantation* **68**, 1–8 (1999).

51. A.W. Thompson et al., *J. Leukocyte Biol.* **66**, 322–330 (1999).

52. B. Ludwig et al., *Immunol. Rev.* **169**, 45–54 (1999).

53. P. Bjork, *Clin. Immunol.* **92**, 119–127 (1999).

54. L. Lu et al., *J. Leukocyte Biol.* **66**, 233–236 (1999).

ANIMAL CELL TYPES, OVARIAN CELLS

G. HODGES
London, United Kingdom

OUTLINE

Ovarian Cells *In Vitro*

Ovarian Cell Culture

    Ovarian Surface Epithelium-Derived Cell Cultures

    Ovarian Granulosa-Derived Cell Cultures

    Ovarian Theca-Derived Cell Cultures

Bibliography

The ovary is among the more complex organs of the body and has two interrelated functions, mature oocyte production and steroid hormone biosynthesis that produces two major groups of steroid hormones (estrogens and progestrogens) which stimulate the development and function of secondary sex organs, placental and mammary (1). Its functions are achieved by numerous cell types (1), and all of these cells have some tendency to undergo malignant transformation (2).

The ovary is structurally organized into a peripheral cortex, which contains ovarian follicles in various stages of development, and a smaller inner medullary zone of highly vascularized loose connective tissue. Surrounding the ovary is a mesodermally derived outer covering of simple cuboidal epithelium, the ovarian surface epithelium (OSE). This is a complex and physiologically versatile structure that plays an important role in normal ovarian physiology (3). OSE cells continually proliferate and recolonize the ovarian surface following ovulation (4). Sited between the surface epithelium and the ovarian cortex is an inner avascular dense connective tissue capsule, the tunica albuginea. This capsule is continuous with the richly cellular cortex that harbors the oocyte-containing ovarian follicles embedded in a loose connective tissue (stroma) of fibroblasts, fibrocytes, and scattered smooth muscle fibers.

The ovarian follicles form discrete morphofunctional units and basic types are identified on the basis of their developmental state: primordial, growing, mature or graafian, and atretic follicles. Follicle development represents a complex process that is regulated by endocrine, as well as paracrine and autocrine factors (6). The primordial (inactive) follicle consists of a primary oocyte surrounded by a single layer of squamous follicular cells. In the growing follicle, the oocyte enlarges, and the follicle epithelium becomes a several-layered cuboidal stratified epithelium (stratum granulosum), where the follicular cells are then identified as granulosa cells. Ovarian granulosa cells are the site of the conversion of thecal cell-derived androgenic precursors into estrogens, and they play an essential role in the maturation of the developing ovum. As the follicle matures, the granulosa cells form a thickened mound, the cumulus oophorus, in the region associated with the oocyte. Stromal connective tissue immediately surrounding the follicle differentiates into the steroid hormone-producing theca folliculi. This differentiates into two layers, the theca interna—an inner, highly vascularized layer of cuboidal steroid-secreting cells (thecal androgens play fundamental roles in the ovary, secrete as estrogen precursors, and have local regulatory actions) and the theca externa—an outer layer consisting mainly of vascular connective tissue. Interactions between mesenchymal-derived thecal cells and epithelial-derived granulosa cells are essential for follicular development in the ovary, mediate gonadotropin actions (6), and lead to the formation of a mature tertiary (or graafian) follicle that is a vesicular formation bulging from the ovarian surface. After ovulation and oocyte release, the OSE cells undergo several rounds of division to repair the wound created by follicular rupture. During postovulatory repair, OSE cells reversibly modulate to a more fibroblast-like form and can modulate *in vitro* from an epithelial to a mesenchymal morphology in response to a variety of environmental cues (7).

Following ovulation, remnants of the follicle form the corpus luteum, a temporary estrogen and progesterone-producing endocrine organ that arises through differentiation of granulosa and theca interna cells into granulosa lutein and theca lutein cells. Granulosa lutein cells secrete high levels of estrogen, but small amounts of progesterone, and in pregnancy secrete the hormone relaxin. Ovarian thecal and theca lutein cells are the major source of 19-carbon (C_{19}) steroids (i.e., dehydroepiandrosterone, androstenedione, and testosterone) produced by the ovary and necessary as a substrate for granulosa cell biosynthesis of estrogen (8).

OVARIAN CELLS *IN VITRO*

Human ovarian cancer is a leading cause of death from gynecologic malignancy (9), and the vast majority of ovarian tumors originate from cells of the ovarian surface epithelium (7,10). Interest in ovarian cancer is reflected by a literature containing descriptions of more than 100 human ovarian tumor cell lines. The majority of these are epithelial cell lines derived mostly from serous, mucinous, endometrioid, and clear cell carcinomas of the ovary; a few lines have been established from teratomas and granulosa cancers (11). A greater number of lines have been established from ascitic fluids rather than from solid tumors, and from patients with poor prognosis. Relatively few cell lines are from well-differentiated or benign tumors, and the development of ovarian cell lines from normal tissue has on the whole been disappointing. Human ovarian carcinoma cell lines have been genetically engineered to produce therapeutic proteins, for example, retroviral-mediated gene transduction of cell lines for isolating cytokine interleukin-4-secreting clones (12). Established ovarian cell lines have been used extensively in identifying growth factors, growth factor receptors, and oncogene expression; in studying the mechanisms of the action of steroids and antisteroids; in developing predictive tests of steroid and chemotherapeutic drug sensitivity; in characterizing tumor markers; in generating specific monoclonal antibodies with potential applications in tumor diagnosis; in monitoring disease progression; and in developing novel treatment strategies (for selected reviews, see Refs. 2,7,8). Cell lines from ovarian tumors of untreated patients and of patients with disease refractory to combination chemotherapy have been developed as an approach to the study of multiple drug resistance in human ovarian cancer. Drug-resistant cell lines have also, been established by exposing cell lines from untreated patients to individual drugs of interest.

Representative ovarian cell lines are listed in Table 1 (13–21). A number of these cell lines have been transfected with a variety of vector systems and used to express recombinant proteins of commercial and therapeutic importance. CHO and Sf9 ovarian cell lines provide important culture systems for biopharmaceutical manufacturing processes (see the section on Cell Lines Used in Manufacturing).

OVARIAN CELL CULTURE

The ovary has been studied *in vitro* using a number of different culture systems, and panels of ovarian cancer cell lines have been established for use in fundamental and applied research programs (selectively reviewed in Refs. 11 and 22). The most general approach, to date, has been the initiation of primary monolayer cultures leading to the establishment of continuous cell lines. Many of the human cell lines have been derived from clinical material. Solid ovarian tumours are minced into 1- to 2-mm^3 pieces and dissociated enzymatically into small clusters of cells are prior to culture. Ascitic fluids, when used as the source material, are centrifuged and the cells pelleted and extensively washed prior to culture; cell lines can be established by cloning from the ascitic fluid (23). In some cases, cultures may be transplanted into athymic nude mice, the solid tumor excised from the animals, and *in vitro* cultures reestablished. Human ovarian tumor cell lines have been established from human tumor xenografts in athymic mice. Ovarian tumor cells have been grown directly in soft agar systems, which favor the growth of tumor cells. A review of the literature indicates that there are no particular growth

factors or media that result in cell line development. Other three-dimensional culture systems have been used to provide better agreement with *in vivo* characteristics and better predictive models than conventionally used monolayer cultures for selective screening of drugs, such as the use of multilayered cell cultures (24), collagenous sponge matrix (25), multicellular spheroid models grown in spinner culture (26), or culture of cells on microcarrier beads in low-turbulence rotating wall vessels (27).

Ovarian Surface Epithelium-Derived Cell Cultures

Tissue culture studies of normal ovarian surface epithelium are discussed in Auersperg et al. (7). Cultured human OSE cells are reported to produce growth factors, cytokines, cell adhesion molecules, and protetolytic enzymes, and contain 17β-hydroxysteroid dehydrogenase activity. The phenotype of normal OSE cells in culture is profoundly influenced by the extracellular matrix (ECM).

Human OSE cells may be biopsied from normal human ovaries *in situ* or from oophorectomized specimens (28,29). The ovarian surface is scraped with a silicone rubber spatula, and fragments of detached OSE are rinsed from the spatula into culture medium, centrifuged, resuspended in fresh medium, and transferred to plastic tissue culture dishes. This procedure allows setting up successful primary monolayer cell cultures. However, the number of viable cells obtained from surgical specimens is low (in the 10^4 range), and cultures have a generally limited life span ranging from 15 to 25 population doublings. OSE in low-passage culture has been successfully transfected with SV40 large T antigen, and the resulting immortalized OSE lines (IOSE) provide large cell numbers. Rat ovarian surface OSE cell lines have been initiated from aseptically removed ovaries subjected *in vitro* to trypsin treatment for selective removal of ovarian surface epithelium (30).

Human OSE-derived cell lines include Caov-3 (16), HEY (16), IOSE-Van (31), MLS (32), NIH:OVCAR-3 (17), OW-1, SAU, and SKA (33). Other animal cell lines include spontaneously transformed tumorigenic rat ovarian surface epithelial cell lines (34), spontaneously immortalized rat ovarian surface epithelial cell ROSE 199 and ROSE 239 lines and sublines (35–37), and mouse ovarian surface epithelial (MOSE) cell lines p53-def-MOSE (a p53-deficient MOSE), and T-Ag-MOSE (a SV40 large T antigen transfected MOSE) (38).

Ovarian Granulosa-Derived Cell Cultures

Granulosa cell cultures have been prepared from whole individual follicles dissected intact from the ovarian stroma, incised to reveal the granulosa cell layer, and the granulosa cells were removed with a platinum loop and transferred to a culture medium. Dissected follicles from oophorectomized specimens can provide nonluteinized granulosa cells, and granulosa lutein cells can be obtained as by-products of oocyte aspirations from *in vitro* fertilization (IVF) procedures and embryo transfer (29).

Granulosa cell lines include HTOG, an estrogen-producing human ovarian granulosa tumour cell line (39); HGL5 established by transformation of human luteinized

Table 1. Selected List of Commonly Used Ovarian Cell Lines: Principal Characteristics and Applications

Species/Cell name reference	Characteristics/Applications	Collection code/Literature
Human		
A2780	Derived from the ovarian carcinoma of an untreated patient. Cells grow as monolayers, as multilayers, and in suspension in spinner cultures. The cell line is cisplatin-sensitive and is parent line to numerous sublines developed for use in drug studies. A2780cis (ECACC No. 93112517): a cisplatin-resistant line, cross-A2780adr (ECACC No. 93112520): an adriamycin-resistant line, cross-resistant to melphalan and vinblastine. AG6000: highly gemcitabine-resistant variant. The A2780 series of cell lines is widely used in combination chemotherapy studies; in studying drug resistance mechanisms; in screening new drugs (especially against drug resistant tumors).	ECACC No. 93112519 Ref. 13
BG-1	Derived from ovarian adenocarcinoma. Has functional estrogen and progesterone receptors in clinically significant levels. Highly estrogen-responsive *in vitro*. Highly resistant to cisplatin; sensitive to paclitaxel and radiation. Good model for study of hormone responsiveness in ovarian tissue and design of combination chemotherapy regimens.	Refs. 14,15
Caov-3	Established in 1976 from ovarian adenocarcinoma of surface epithelial origin taken from patient treated with cytoxan, adriamycin, 5-fluorouracil, Fur IV. Overexpresses a mutant p53. Radiosensitive. Growth inhibited by all-*trans*-retinoic acid. Used in cytokine, retinoid, anticancer therapy studies.	ATCC HTB 75 Ref. 16
HEY	Derived from a nude mouse xenograft (HX-62) of a peritoneal deposit of a patient with a moderately differentiated ovarian papillary cystadenocarcinoma of surface epithelial origin. Cells show functional TGF beta receptors. Sublines include cells trans-fected with the murine interferon beta gene. Used in cytokine and therapy studies.	Ref. 16
NIH:OVCAR 3	Established from the malignant ascites of a patient with progressive adenocarcinoma of the ovary after combination chemotherapy with cyclophosphamide, adriamycin, and cisplatin. Resistant to clinically relevant concentrations of adriamycin, melphalan, and cisplatin. Shows androgen and estrogen receptors and high-affinity receptors for interleukin-1. Forms multicell spheroids. Useful model for cytotoxic drug resistance studies and experimental evaluation of hormonal therapy in ovarian cancer. Multicell spheroids used as an *in vitro* model of micrometastases of ovarian carcinoma.	ATCC HTB 161 Ref. 17
OAW 42	Derived from ascites of patient with serous cystadenocarcinoma. Has retained ability to form free floating cysts *in vitro* and produces extracellular matrix. Has defined chemosensitivity pattern. Number of drug-resistant variant cell sublines. Valuable for studies on biology of human ovarian cancer and for multiple drug resistance studies.	ECACC No. 85073102 Ref. 11
PA-1	Derived from ascitic fluid of patient with malignant ovarian teratoma. Secretes large quantities of bioactive follistatin. Expresses wild-type p53. Forms multicell tumor spheroids. Useful model for studies on some developmental mechanisms in human cells and drug-mediated destruction of micrometastases.	ATCC CRL 1572 ECACC No. 90013101 Ref. 18
SK-OV-3	Derived from ascites of patient with moderately well differentiated adenocarcinoma (no prior platinum therapy). Contains a p53 deletion mutation; shows high degree of radioresistance; resistant to tumour necrosis factor and to growth inhibitory effects of all-*trans*-retinoic acid. Multidrug resistant sublines developed. Useful model for chemotherapy studies including evaluation of broad-spectrum platinum drugs.	ATCC HTB 77 ECACC No. 91091004 Ref. 19
SW 626	Derived from well-differentiated papillary cystadenocarcinoma. Expresses a temperature-sensitive (ts) p53 mutant. Useful model to define regulation and expression of both the gastrin gene and peptide in ectopic (nongastrointestinal) tissues and for cytotoxicity and biological activity evaluation of gold and tin compounds in ovarian cancer cells.	ATCC HTB 78 Ref. 19
Hamster[a]		
CHO-K1	Subclone of parental CHO cell line initiated from ovary biopsy of normal adult Chinese hamster Requires proline due to absence of gene for proline synthesis. Histological identity of CHO not confirmed. Great number of CHO sublines developed; widely used in recombinant protein production; CHO expression systems dominant in manufacture of biopharmaceuticals.	ATCC CCL 61 DSMZ ACC 110 ECACC No. 85051005 Ref. 20
Insect[a]		
Sf9	Clonal derivative of parent pupal ovarian line IPLB-SF-21-AE of the Fall Army *Spodoptera frugiperda*; highly susceptible to baculovirus infection. Used in production of protein products genetically manipulated into baculovirus vector systems.	ATCC CRL 1711 DSMZ ACC 125 ECACC No. 89070101 Ref. 21

[a]CHO and Sf9 cells are also described in the section on cells used in manufacturing.

granulosa cells with the E6 and E7 regions of human papillomavirus (40); SIGC, a spontaneously immortalized clonal granulosa cell line derived from primary rat ovarian granulosa cell cultures, SV-SIGC (a pSV3 neotransfected clonal derivative), and T-SV-SIGC (a nude mouse tumor-derived cell line) (41,42); OV312, a nonmetastatic radiation-induced murine ovarian granulosa cell tumor line (43); and AIMS/GRXII, (a goat cell line derived from granulosa cells subjected to luteinizing hormone stress (44).

Ovarian Theca-Derived Cell Cultures

Thecal cultures have been prepared by removing the theca layer, devoid of adhering granulosa cells, and placing enzymatically treating minced fragments dispersed into single cells into a culture medium (41,45).

Cell lines include HTOT established from a theca cell tumor (39). A human ovarian thecal-like tumor (HOTT) cell culture system that produces excessive amounts of C_{19} steroids has been developed from an ovarian tumour. This may serve as an appropriate model to study regulation of human ovarian thecal C_{19} steroidogenesis and the expression of steroid-metabolizing enzymes (46).

BIBLIOGRAPHY

1. G.F. Whitman, T.E. Nolan, and D.G. Gallup, in *Cancer of the Ovary*, Raven Press, New York, 1993, pp. 1–20.
2. T.C. Hamilton, *Curr. Probl. Cancer* **16**, 1–57 (1992).
3. N. Auersperg, S.L. Maines-Bandiera, and P.A. Kruck in *Ovarian Cancer Vol. 3*, Chapman & Hall Medical, London, 1995, pp. 167–189.
4. W.J. Murdoch, *Biol. Rev. Cambridge Philos. Soc.* **71**, 529–543 (1996).
5. J.F. Roche, *Rev. Reprod.* **1**, 19–27 (1996).
6. J.A. Parrott and M.K. Skinner, *Endocrinology* **139**, 2240–2245 (1998).
7. N. Auersperg et al., *Semin. Oncol.* **25**, 2281–2304 (1998).
8. S.G. Hillier and M. Tetsuka, *Baillière's Clin. Obstet. Gynaecol.* **11**, 249–260 (1997).
9. M. Daley and G.I. Obrams, *Semin. Oncol.* **25**, 255–264 (1998).
10. H. Salazar et al., in *Ovarian Cancer Vol. 3*, Chapman & Hall Medical, London, 1995, pp. 145–156.
11. A. Wilson, in A. Doyle, J.B. Griffiths, and D.G. Newell, eds., *Cell & Tissue Culture: Laboratory Procedures*, Wiley, Chichester, U.K., 1996, pp. 15C:1.1.
12. A.D. Santin et al., *Gynecol. Oncol.* **58**, 230–239 (1995).
13. B.C. Behrens et al., *Cancer Res.* **47**, 414–418 (1987).
14. K.R. Geisinger et al., *Cancer*, **63**, 280–288 (1989).
15. W.S. Baldwin et al., *In Vitro Cell Dev. Biol. Anim.* **34**, 649–654 (1998).
16. R.N. Buick, R. Pullano, and J.M. Trent, *Cancer Res.* **45**, 3668–3676 (1985).
17. T.C. Hamilton et al., *Cancer Res.* **43**, 5379–5389 (1983).
18. J. Zeuthen et al., *Int. J. Cancer.* **25**, 19–32 (1980).
19. J. Fogh, J.M. Fogh, and T. Orfeo, *J. Natl. Cancer Inst.* **59**, 221–226 (1977).
20. F.T. Kao and T.T. Puck, *Proc. Natl. Acad. Sci. U.S.A.* **60**, 1275–1281 (1968).
21. J.L. Vaughn and S.A. Weiss, *Bioprocess Technol.* **10**, 597–618 (1990).
22. R.D.H. Whelan, L.K. Hosking, S.A. Shellard, B.T. Hill, and J.R.W. Masters, ed., in *Human Cancer in Primary Culture: A Handbook*, Kluwer Academic Publishers, Dordrecht, The Netherlands, 1991, pp. 253–260.
23. L. Wasserman et al., *Eur. J. Cancer* **28**, 22–27 (1992).
24. E. Smitskamp-Wilms et al., *Eur. J. Cancer* **34**, 921–926 (1998).
25. H.G. Dyck et al., *Int. J. Cancer* **69**, 429–436 (1996).
26. E.K. Rofstad, in *Human Cancer in Primary Culture: A Handbook*, J.R.W. Masters, ed., Kluwer Academic Publishers, Dordrecht, The Netherlands, 1991, pp. 81–101.
27. T.J. Goodwin, T.L. Prewett, G.F. Spaulding, and J.L. Becker, *In Vitro Cell Dev. Biol. Anim.* **33**, 366–374 (1997).
28. N. Auersperg, S.L. Maines-Bandiera, H.G. Dyck and P.A. Kruk, *Lab Invest.* **71**, 510–518 (1994).
29. S.G. Hillier, R.A. Anderson, A.R.W. Williams, and M. Tetsuka, *Mol. Hum. Reprod.* **4**, 811–815 (1998).
30. A.K. Godwin et al., *J. Natl. Cancer Inst.* **84**, 592–601 (1992).
31. S.L. Maines-Bandiera, P.A. Kruk, and N. Auersperg, *Am. J. Obstet. Gynecol.* **167**, 729–735 (1992).
32. E.K. Rofstad and R.M. Sutherland, *Br. J. Cancer* **59**, 28–35 (1989).
33. S.A. Khan et al., *Gynecol. Oncol.* **66**, 501–508 (1997).
34. J.R. Testa et al., *Cancer Res.* **54**, 2778–2784 (1994).
35. A.T. Adams and N. Auersperg, *Exp. Cell Biol.* **53**, 181–188 (1985).
36. P.A. Kruk and N. Auersper, *In Vitro Cell Dev. Biol. Anim.* **30A**, 217–225 (1994).
37. R.P. Perez et al., *Gynecol. Oncol.* **58**, 312–319 (1996).
38. M. Kudo and M. Uhibuya, *Pathol. Res. Pract.* **194**, 705–730 (1998).
39. I. Ishiwata et al., *J. Natl. Cancer Inst.* **72**, 789–800 (1984).
40. W.E. Rainey et al., *J. Clin. Endocrinol. Metab.* **78**, 705–710 (1994).
41. L.S. Stein, G. Stoica, R. Tilley, and R.C. Burghardt, *Cancer Res.* **51**, 96–706 (1991).
42. L.S. Stein, D.W. Stein, J. Echols, and R.C. Burghardt, *Exp. Cell Res.* **207**, 19–32 (1993).
43. K. Yanagihara et al., *Jpn. J. Cancer Res.* **86**, 347–356 (1995).
44. T.N. Chapekar and A.K. Malik, *Pathobiology* **59**, 345–350 (1991).
45. J.M. McAllister, W. Byrd, and E.R. Simpson, *J. Clin. Endocrinol. Metab.* **79**, 106–112 (1996).
46. W.E. Rainey et al., *J. Clin. Endocrinol. Metab.* **81**, 257–263 (1996).

ANIMAL CELL TYPES, T LYMPHOCYTES

RICHARD STEBBINGS
National Institute of Biological Standards and Control
South Mimms, England

OUTLINE

Introduction

T-Lymphocyte Subpopulations and Function

INTRODUCTION

T lymphocytes play an essential role in initiating and regulating antigen-specific immune responses that clear pathogens and protect against reinfection. The ability of T lymphocytes specifically to recognize and respond to infection is determined by their antigen-specific T-cell receptor (TCR). The DNA sequences that confer antigen specificity to these TCR are assembled at four different loci (α and β or γ and δ) from V (variable-region), J (joining-region), and in some cases D (diversity-region) gene segments (1). This hypermutable V(D)J recombination generates a wide repertoire of recognition that allows the immune system to field a gamut of individual T lymphocytes, each expressing a different TCR specific for individual antigens (for a description of the development of T cells, see the section on thymus cells). In the absence of foreign antigen, T lymphocytes are small resting cells in the G_0 phase of the cell cycle. Although essentially inactive, they do express their TCR plus ancillary molecules and traffic between the blood and secondary lymphoid tissues, where trapped antigen is presented, by antigen-presenting cells (APC), for immune surveillance. Activation, following interaction with specific antigen, is accompanied by an increase in size (blastogenesis) and entry into G_1. Activated T lymphocytes proliferate to generate a larger population of specific effector cells (clonal expansion) as well as memory cells that can produce a faster and more efficient "secondary" response to reinfection (2).

T-LYMPHOCYTE SUBPOPULATIONS AND FUNCTION

Antigen is presented to T lymphocytes as peptide antigens bound to the highly polymorphic major histocompatibility complex (MHC) I or II molecules on the surface of APC. Recognition of foreign peptide sequences by T lymphocytes is restricted by initial recognition of self-MHC, such that only self-APC can trigger T-cell activation. The T lymphocyte ancillary molecules CD4 and CD8 mediate binding to MHC I and II, respectively, enhancing adhesion between T cells and APC (3). Mutually exclusive expression of CD4 or CD8, by subpopulations of mature peripheral T lymphocytes, serves as a phenotypic marker of T helper (T_H) or T cytotoxic (T_C) cell function, respectively.

The cells of the body display a representative sample of their intracellular contents at their cell surface to prevent their interiors from being a privileged site for pathogen replication. Short peptide fragments generated in the cytoplasm by multicatalytic proteolysis of free polypeptides are bound to freshly synthesized MHC class I molecules and presented at the cell surface for perusal by $CD8^+$ T_C (4,5). This MHC I class antigen presentation pathway is primarily aimed at the detection of viral infections and tumor cells. Activated T_C cells recognizing cells presenting non-self-peptides, on MHC class I molecules, kill them by releasing lytic perforin/granzyme granules onto the target cell's surface or trigger target cell suicide via the FAS–FAS ligand pathway (6). Activated T_C cells can also secrete cytokines to suppress viral replication (7).

Exogenous antigen, captured by professional APC, enters the endocytic pathway, where it is subjected to proteolysis to form long peptide fragments. These peptides are loaded onto the exposed binding site of newly synthesized MHC class II molecules and transported to the cell surface for scrutiny by $CD4^+$ T_H cells. T_H cells are regulatory cells that control the development and enhance the function of effector cells as well as control viral infections (8). Activated T_H cells secrete cytokines that augment T_C and B lymphocyte responses, enhance APC function, and directly inhibit viral replication (9). Additionally, a subset of $CD4^+$ T cells are capable of lysing virally infected cells in an MHC class II restricted fashion (10). The T_H subset has been further subdivided into two functionally different populations based on the range of cytokines they secrete in response to activation (11). This dichotomy is based on T_{H1} cells promoting cell-mediated immunity, through secretion of the cytokines interferon γ, interleukin 2, and tumor necrosis factor β, and T_{H2} cells promoting humoral immunity through secretion of interleukins 4, 5, 10, and 13. T lymphocytes can be still further subdivided into naive and antigen-primed memory cells, distinguished by their expression of the CD45RA or CD45RO isoforms, respectively (12). Naive T_H cells are believed to produce mainly interleukin 2, while memory cells secrete either of the polarized T_{H1} or T_{H2} range of cytokines (13).

PRIMARY T-LYMPHOCYTE CULTURES

Primary lymphoid cell cultures containing T lymphocytes are routinely prepared from peripheral blood mononuclear cells (PBMC) separated from whole blood on a Percoll density gradient and resuspended in complete medium (RPMI 1640 culture medium supplemented with 10% fetal calf serum, 10 mmol/L L-glutamine, 100 U/mL penicillin, 100 µg/mL streptomycin, and 5×10^{-5} mol/L 2-mercaptoethanol). Alternatively, T-lymphocyte-containing primary cultures can be prepared from various lymphoid tissues disaggregated to give a single cell suspension. However, these cultures can only be successfully cultured for short periods of time and require supplementation with T-cell growth factors such as interleukin 2 (IL2). Furthermore, the heterogeneous nature of these primary cultures and the different subpopulations of T lymphocytes within can limit interpretation of data from experiments employing them.

GENERATION OF T-CELL CLONES

Since T lymphocytes with defined specificities form only a very minor proportion of the total T-cell pool, selective enrichment for particular subsets is carried before cloning.

For human studies PBMC derived from individuals vaccinated against or infected with a particular pathogen under investigation are employed. With mice, antigen-reactive T cells are derived from draining lymph nodes, prepared as a single cell suspension, 7–10 days after immunization with antigen emulsified in complete Freund's adjuvant. Harvested lymphoid cells, washed and resuspended in complete medium, are then incubated at $(3–5) \times 10^3$ cells/mL in 96-well round-bottom plates in the presence of 6×10^4 gamma-irradiated (2000R) allogeneic APC and 20–100 µg antigen, in a final volume of 100 µL. For human T cells autologous irradiated PBMC are used as a source of APC. With murine lymph node cells (LNC), irradiated syngeneic spleen cells are used as APC. After 7 days, cultures should be fed with 100 µL of fresh complete medium supplemented with 20 U/mL IL2. At day 14 proliferating cultures, identified as clusters of blast cells, can be transferred into 1 mL of complete medium containing 20 U/mL IL2 and cultured in 24-well plates. After several restimulations with fresh antigen and irradiated APC, T-cell cultures should be subcloned at 10, 3, and 1 cell per well in 96-well round-bottom plates in the presence of irradiated APC, phytohemagglutinin (PHA) 1 µg/mL, and IL2 20 U/mL. These cultures can then be maintained in complete medium supplemented with IL2 and only require restimulation with irradiated autologous APC, PHA, and IL2 every 2–3 weeks. These T-cell clones should be rested for 7 days prior to in vitro testing.

This T-cell cloning technique of repeated antigen stimulation in the presence of IL2 can result in the selection of T lymphocytes responsive to IL2 but not responsive for antigen. In order to exclude such cells, it is necessary to test the antigen specificity of expanded clones generated after the limiting dilution subcloning step. To test specificity, T-cell clones at 2×10^4 cells/well should be incubated in triplicate in 96-well plates in the presence of 1×10^5 irradiated autologous APC and various dilutions of antigen (e.g., 0.2–20 µg/mL) in a final volume of 100 µL of complete medium. After 3 days, cultures should be pulsed overnight with [^{3}H] TdR (1 µCi/well). The level [^{3}H] TdR incorporation can be measured by liquid scintillation, using a beta counter, to determine which cultures are proliferating specifically to antigen.

T-CELL LINES

Herpes virus saimiri (HVS), a T-cell tropic γ-herpes virus of nonhuman primates, has been shown to immortalize efficiently human T lymphocytes (14). HVS-immortalized human T-cell lines can be maintained in culture supplemented with IL2, but without the need for antigen or mitogen stimulation (15). Unlike human T-cell lines derived from spontaneous tumors or those transformed with human T-cell leukemia virus (16), HVS-immortalized T cells retain important properties of conventionally cultured T cells, including: phenotypic cell surface markers, antigen specificity, intact signal transduction, and the cytokine profile of activated mature T lymphocytes (17,18).

A number of T-cell lines (Table 1) have been derived from the blood of patients with various T-cell leukemias

Table 1. T-cell Lines Used Widely in Immunological Research

Cell name	Origin	Characteristics and applications	Catalogue codes[a]
JURKAT E6-1	Human—ALL	Produces large amounts of IL2 in response to stimulation with phorbol-12-myristate-13-acetate (PMA) and anti-CD3 antibody	ATCC. TIB-152 ECACC. 88042803 ARP. ARP027
HUT-78	Human—SS	Secrete IL2 and tumor necrosis factor α; growth enhanced by added IL2	ATCC. HTB-161 ECACC. 88041901 ARP. ARP002
HUT-102	Human—SS	Growth rate enhanced by added IL2; sheds HTLV-I in culture that can be used to transform primary T lymphocytes	ATCC. TIB-162
MOLT-4	Human—ALL	Stable T-cell line that forms rosettes with sheep erythrocytes	ATCC. CRL-1582 ECACC. 85011413 ARP. ARP011
CEM-CCRF	Human—ALL	Supports HIV-1 replication in vitro; subclone CEM-4 (ARP006) expresses high levels of CD4	ATCC. CCL-119 ECACC. 85112105 ARP. ARP005
SUP-T1	Human—TLL	Expresses high levels of CD4 and supports HIV replication in vitro	ATCC. CRL-1942 ARP. ARP024
C8166	Human—HTLV	Used in coculture with PBMC to detect replicating HIV-1; forms syncitia after HIV-1 infection	ECACC. 88051601 ARP. ARP013
EL4	Murine—TL	Produces high levels of IL2 in response to stimulation with phorbol-12-myristate-13-acetate	ATCC. TIB-39 ECACC. 85023105
CTLL	Murine—CC	Cytotoxic T-cell clone; growth rate is dependent on added IL2; often used in bioassays to measure IL2 activity	ECACC. 87031904

[a]Further information on resource centers is in the chapter on cell banks.

(19–22) and spontaneous lymphomas (23). Unlike primary T-lymphocyte cultures, these cell lines survive indefinitely without the need for exogenous growth factors, but demonstrate various abnormal growth properties, chromosome number, and function.

ABBREVIATIONS

ALL, acute lymphoblastic leukemia; ARP, AIDS Reagent Project, NIBSC, UK; ATCC American Type Culture Collection, USA; CC, adapted to IL2-dependent continuous culture; ECACC, European Collection of Cell Cultures, CAMR, UK; SS, Sezary syndrome; TLL, T-lymphoblastic leukemia; TL, thymic lymphoma.

BIBLIOGRAPHY

1. S.A. Lewis, *Adv. Immunol.* **56**, 27–150 (1994).
2. R.W. Dutton, L.M. Bradley, and S.L. Swain, *Annu. Rev. Immunol.* 201–223 (1998).
3. C. Doyle and J.L. Strominger, *Nature (London)* **330**, 256–259 (1987).
4. A.R.M. Townsend et al., *Cell* **44**, 959–968 (1986).
5. Y.S. Hahn, B. Yang, and T.J. Braciale, *Immunol. Rev.* **151**, 31–49 (1996).
6. D. Kagi et al., *Science* **265**, 528–530 (1994).
7. O.O. Yang et al., *J. Virol.* **71**, 3120–3128 (1997).
8. S. Jonjic et al., *J. Exp. Med.* **169**, 1199–1212 (1989).
9. J.L. Davignon et al., *J. Virol.* **70**, 2162–2169 (1996).
10. M.D. Lindsley, D.J. Torpey, 3rd, and C.R. Rinaldo Jr., *J. Immunol.* **136**, 3045–3051 (1986).
11. T. Mossman et al., *J. Immunol.* **136**, 2348–2355 (1996).
12. M. Sanders, M.W. Makgoba, and S. Shaw, *Immunol. Today* **9**, 195–199 (1988).
13. S. Swain et al., *Immunol. Rev.* **150**, 143–167 (1996).
14. B. Biesinger et al., *Proc. Natl. Acad. Sci. U.S.A.* **89**, 3116–3119 (1992).
15. E. Meinl, R. Hohlfeld, H. Wekerle, and B. Fleckenstein, *Immunol. Today* **16**, 55–58 (1995).
16. N. Yamamoto et al., *Science* **217**, 737–739 (1982).
17. C. Vella et al., *J. Gen. Virol.* **78**, 1405–1409 (1997).
18. K. Saha, G. McKinley, and D.J. Volsky, *J. Immunol. Methods* **206**, 21–23 (1997).
19. G.E. Foley et al., *Cancer* **18**, 522–529 (1965).
20. J. Minowada, T. Ohnuma, and G.E. Moore, *J. Nat. Cancer Inst. (U.S.)* **49**, 891–895 (1972).
21. U. Schneider, H.-U. Schwenk, and G. Bornkamm, *Int. J. Cancer* **19**, 621–626 (1977).
22. S.D. Smith et al., *Cancer Res.* **44**, 5657–5660 (1984).
23. A.F. Gazdar et al., *Blood* **55**, 409–417 (1980).

ASEPTIC TECHNIQUES IN CELL CULTURE

JOHN M. DAVIS
KEVIN L. SHADE
Bio-Products Laboratory
Elstree, Herts.
United Kingdom

OUTLINE

INTRODUCTION

Microbial Contamination

We live in a world in which we are constantly surrounded by microbes in our environment, as well as upon and within our own bodies. Yet the successful performance of cell culture demands that we maintain and manipulate our cultures free of all microbial contamination. Because we cannot sterilize our cultures after manipulating them, we are completely dependent on *aseptic technique* to ensure that, in handling our reagents, cells, and all the associated equipment, we maintain the sterile environment that our cultures require.

Aseptic technique is a combination of many procedures all designed with the single goal of *minimizing the probability* that a microbe gains access to the cell culture environment. It is important to appreciate this concept of minimizing the probability of contamination because even the best aseptic technique cannot absolutely guarantee the maintenance of sterility (indeed, all sterilizing techniques work on this same principle) (1). Thus a methodical and fastidious approach is required with attention to each element of each procedure. Dropping an element or cutting corners may not necessarily result in actual contamination the first time but will *increase the probability* that contamination will occur. Consequently, it is essential that all the elements of an aseptic procedure are carried out consistently every time, making sterility breakdowns rare events.

Cellular Cross-Contamination

Cell culture requires more, however, than just the type of aseptic technique used to exclude microbes. Our work becomes at best meaningless and at worst potentially dangerous if we are not culturing the cells we think we are. Thus aseptic technique for cell culture must include procedures to minimize the possibility of contaminating one cell line with another. Evidence that such inadvertent cross-contamination could occur was first presented in the 1970s when various cell lines from a number of different laboratories were all shown by isoenzyme analysis and/or karyotyping to be HeLa cells, and these findings were later extended to include contamination involving cells other than HeLa (2,3). Thus a second aim of all aseptic technique used for cell culture is to minimize the probability of contaminating our pure characterized cultures with other cells.

Biohazards

So far we have concentrated on the role of aseptic technique in protecting the cells that we are culturing. However, another element in almost all aseptic technique used for cell culture is protection of the operator from hazards posed by the cells or (re)agents used in their culture.

Both human and nonhuman cells may carry viruses pathogenic to humans, and at least one actual fatality has been documented which was caused by inadequacies in the handling of such cells (4). Thus any new or poorly characterized cell line, or any cell line new to a laboratory, *must* be handled as if it harbored a potential pathogen. Indeed, although the risk may be less when handling cell lines which have been extensively screened for microbiological contamination, the assays for such contaminants have limitations in both the level of contamination and the range of organisms they can detect. Furthermore, the continual discovery

of previously unidentified human viral pathogens — for example, in the last 20 years we have seen the discovery of human retroviruses and various "new" human herpes and hepatitis viruses — suggests that however extensive our microbiological screening, we can never be certain that a cell line might not contain a potentially dangerous viral "passenger". Moreover, the risk is not limited to viruses because other contaminating microbes (such as mycoplasma) could be hazardous, and it may be that the cells themselves could pose a threat if inoculated into an operator via a puncture wound (5).

The commonest cell culture *reagent* that could be biohazardous is animal serum, of which the most widely used is fetal calf serum. This can contain a variety of microorganisms, but mycoplasma and bovine viruses are probably the most important. The agent which causes bovine spongiform encephalopathy (BSE) is also of concern, particularly because it appears to be the cause of new-variant Creutzfeld–Jakob disease (CJD) in humans (6,7). The chances of such agents being present in a batch of serum can be minimized by purchasing only from the most reputable suppliers, who can document the source and health of the animals slaughtered and the standard of abattoir procedures and who process and test the serum to minimize the chance that such agents appear in the final product. Sourcing from a country where BSE is not endemic is important, but as there was (and probably still is) a lot of "fake" Australasian and U.S. serum on the market (much of it originating in countries where bluetongue, foot-and-mouth disease and other viruses are a problem) (8), purchasing from a reputable supplier is doubly important.

Other reagents which may pose a hazard are those intentionally added to cultures, such as viruses which are being propagated or assayed. (Radioisotopes and other harmful chemicals may also pose a risk, but these hazards cannot be addressed by aseptic technique.)

Because of all the potential hazards outlined above, it is essential to carry out a full risk assessment before starting the culture of any cell line. The results of this will have a major impact on the aseptic techniques used and most importantly on the environment in which they are carried out. Reference must always be made to the relevant local and national regulations governing the handling and containment of biological organisms (see, for example, Ref. 9).

Cell Culture Environments

The environments in which cell culture may be carried out vary both in the degree of protection given to the cells against external contamination and the degree of protection afforded the operator against potential hazards posed by the culture and its manipulation. In the early days of cell culture, procedures were carried out on the open laboratory bench, an environment which did little to protect the culture and nothing to protect the operator. As laminar flow cabinets or hoods (LAFs) and microbiological safety cabinets (MSCs) became available, and our understanding of the possible hidden hazards associated with cell culture increased, so the move has been away from the open bench toward a more protective environment. However, cell culture can still be carried out successfully on the open bench under circumstances where the risk assessment indicates a minimal risk to the operator, and, where using an appropriate MSC is not possible, doing so may still be justified. The vast majority of the principles and aseptic techniques used for culture are the same whether they are to be performed on the open bench or carried out within the environment of a LAF or MSC. Thus, the general principles of aseptic technique for cell culture will be described first, as applied to working on the open bench. This will be followed by an examination of high-efficiency particulate air (HEPA) filtration, the range of sophisticated equipment now available which employs HEPA filtration and provides protection to the operator and/or the cell cultures, and the adaptations in technique required to use this equipment safely and successfully. Finally, clean rooms for use in cell culture will be examined.

ASEPTIC TECHNIQUE: GENERAL CONSIDERATIONS

Culture Equipment

All materials which come into direct contact with a cell culture *must* be sterile—aseptic technique only aims to maintain that sterility. Consequently all equipment such as flasks, bottles, dishes, pipettes, etc. must be treated by a sterilizing technique that gives a very high probability of sterility and is appropriate for the individual piece of equipment and the materials from which it is manufactured. The conditions of subsequent storage before use are also important. All sterile cell culture equipment should be stored in a clean, dry environment away from unnecessary air flow and protected from possible sources of physical, chemical, or other damage to itself or any of the packaging which maintains its sterile integrity. The integrity of such packaging (e.g., the bags containing single-use pipettes or the autoclave bags containing autoclaved materials) must be confirmed before the equipment is used: any equipment in damaged or inadequate packaging must be discarded or repackaged and resterilized before use. Sometimes damaged equipment comes in perfectly good packaging; for example, crushed and cracked cell culture flasks can arrive inside a plastic bag which is still airtight. *Never* be tempted to use equipment over which you have the slightest doubt concerning sterility. It may be expensive to discard or resterilize but not as expensive as your wasted time and the other materials you will waste working on an infected culture.

All equipment to be used directly in or indirectly associated with aseptic techniques must be carefully chosen with the three aims of aseptic technique constantly in mind:

- protection of the cells from microbes;
- protection of the cells from cross-contamination with other cells; and
- protection of the operator from possible associated hazards.

Illustrations could be given that relate to almost any piece of equipment, but two are given here:

1. Pipetting aids must always be used to operate pipettes, thereby protecting both the operator and the cells from each other. There are many designs, and the most comfortable and efficient for you and your type of work should be selected; one that is too awkward to use or takes too long to pipette the volumes you use will lead to fatigue, and a tired operator is a sloppy operator with poor aseptic technique.

2. An incubator is an essential piece of equipment but must be kept clean to minimize the growth of microbes within it; so choose one that is easy to clean and has no inaccessible areas that will harbor microorganisms. If humidified, the incubator should also have very good control of the humidity because condensation on flasks and dishes can be a major source of microbial contamination.

Working Area

Whether working in a microbiological safety cabinet or not, a suitable working area must be chosen where the environmental flow of potentially contaminating particles can be minimized. A separate room is preferable, which is not a thoroughfare and which is designated only for use when employing aseptic techniques. This should be kept as clean as reasonably possible with the minimum amount of equipment necessary and preferably without shelving (which can harbor dust) above the working area. Air movement should be kept to a minimum commensurate with adequate ventilation and temperature control. If a separate room is not available, find a quiet corner of the laboratory where you can get as near as possible to these conditions. (Siting of MSCs is dealt with further in the later section on MSCs.)

The work surface *must* be kept clean and tidy. Start with a clear surface, and wash it down with 70% ethanol or other liquid disinfectant. Then introduce to this clean area *only* those items required for a particular procedure. If these are many, it is better to break the procedure down into a series of steps and change the items in your clean area at suitable points, rather than work in a clutter. Arrange your equipment so that you have easy access to all items without having to reach across one to get to another, and make sure there is a large open space in the center on which to work. Having too many items in your clean area will inevitably lead, sooner or later, to you touching the sterile surface of (for example) a pipette against a nonsterile item.

Always work within the central range of your vision, and cultivate a sensitivity to possible undesired contacts. The same principle applies here as mentioned before: if you are in any doubt about the sterility of an item, do not use it—it is just not worth it.

Should you spill a liquid, mop it up immediately, and swab the surface down with your liquid disinfectant. Finally, when the work is done, remove all items from the work surface and swab it down again.

Personal Hygiene

Before Work. Wash your hands. If you have long hair, tie it back, or wear a suitable cap to keep it away from your face and the work. (Hair is a good source of microorganisms, particularly yeast, and when working with a Bunsen flame it is not unknown for individuals with long hair worn loose to set it on fire.)

During Work. Never eat, drink, smoke, chew, or apply cosmetics in the laboratory. This is to avoid bringing the hands (which may have hazardous material on them) into contact with the mouth or other mucous membranes which offer easy access to the body. Similarly, always cover any cuts or cracks in the skin to avoid the ingress of unwanted material. While working in a LAF or MSC or with open bottles and cultures, do not talk, cough, or sneeze

At the End of Work. Always wash your hands, irrespective of whether you have been wearing gloves.

Clothing

Before starting any aseptic handling, make sure you are correctly clothed. A clean laboratory coat with long sleeves and close-fitting cuffs is the minimum requirement. Gloves can be useful because they contain any flakes of skin and loosely adherent microbes which may be present on the hands. Furthermore, if pulled up over the cuffs of the laboratory coat, they will contain any flakes of skin or microbes which might otherwise be expelled from the cuffs by currents of air forced down the sleeves by your movements. Latex surgical gloves without a powder coating are the best type to use. Gloves should be sprayed or swabbed regularly with a liquid disinfectant during use, but (particularly if you are using a Bunsen) do not forget that 70% ethanol is flammable. Gloves also offer you a degree of protection from your cells. Keep an eye on the condition of your gloves, and if they get holed, discard them immediately, and replace them with fresh ones. The use of a face mask is also worth considering, particularly if you have a mustache and/or beard (see previous comments on hair).

Swabbing/Spraying

As far as possible, the surfaces of all nonsterile items should be dried (if necessary), then sprayed or swabbed with a liquid disinfectant before being introduced into your clean working area. This is particularly important for items coming from refrigerators, water baths, or humidified incubators, where microbial growth can be rife. Again, do not forget that 70% ethanol is flammable!

Capping

Deep screw caps are preferable to other types of closures for all bottles used routinely in aseptic techniques. They generally offer a good seal when screwed down and offer a significant degree of protection to a bottle's contents even when only placed loosely in position, a situation which often occurs during aseptic handling. Bottles with ISO threads can be used with a variety of caps of different materials, as well as other fittings designed for liquid handling.

Pouring

Pouring from one sterile container to another should be avoided if at all possible. Pouring generates aerosols, and they can carry cells or infectious biological reagents to other cultures or to the operator. However the most common risk is to sterility maintenance because a bridge of liquid can be formed between the nonsterile outside and the sterile inside of a vessel and act as a conduit to introduce microorganisms. It follows that, if pouring *MUST* be done, then it must be done as a single delivery in one tip; even then it still carries a significant risk.

Flaming

When working on an open bench, it is common to use the flame of a Bunsen or similar burner to flame the necks of bottles and screw caps before and after use and glass pipettes only before use. This practice comes from microbiological technique, and its purported mechanism of action, and indeed it usefulness in cell culture, is open to some discussion. Any sterilizing or fixing effect on microbes is restricted to dry surfaces in direct contact with the flame. Some workers claim that if one works close to the flame, an updraft is created that prevents particles settling onto the work; others claim that the convection currents produced are not unidirectional and cause more problems than they solve. Either way, *flaming is not an essential part of sterile technique*, even on the open bench, and should be avoided if at all possible when working in laminar flow or microbiological safety cabinets because it disturbs the air flow and can be a fire hazard.

Pipetting

Pipettes commonly used in cell culture handle volumes from around 1 to 100 mL. Below 1 mL, Gilson- or Eppendorf-type pipettes would normally be used along with single-use, sterile, disposable plastic tips. From 1 mL upward, available standard glass or disposable plastic pipettes probably represent the easiest way of manipulating measured volumes of liquids. Syringes can be used, but the hypodermic needles which fit them are not long enough to reach the bottom of most cell culture vessels and carry the risk of causing needle-stick and associated injuries to operators. Mixing needles — basically a length of plastic tubing connected to a suitable fitting for a syringe — are a better option, will reach the bottom of all except some of the larger vessels, and are available presterilized (e.g., from Henleys Medical Supplies, Welwyn Garden City, Herts, United Kingdom). However syringes may still prove unsuitable for manipulating some cells, because of the high shear forces created when liquid passes through the relatively small aperture where the needle is mounted.

As mentioned before, pipettes must be used only in cell culture applications in association with a pipetting aid, which must be suitable for the range of pipettes to be used. All pipettes should be fitted with a cotton (or similar) plug to maintain the sterility of the inside of the pipette when used with the (nonsterile) pipette aid. Suitable pipettes may be conveniently purchased as preplugged sterile single-use plastic items which are discarded after use. Despite this, many laboratories still use glass pipettes which appear to be a cheaper option in the long run. However, when the cost of unplugging, cleaning, drying, replugging, and resterilization are factored in, any financial advantage of glass pipettes is less evident.

When using a pipette, a number of precautions must be taken. The first is when inserting the pipette (particularly if it is a glass pipette) into the pipette aid. The pipette must be held at a point close to the pipette aid and excessive force must be avoided; it is very easy to break a pipette, and the broken end may then lacerate the hand or arm of the operator. Then, when sucking up liquid into the pipette, it is important that the cotton plug is not wetted, otherwise microbes can be introduced into the sterile liquid as it is being expelled from the pipette. Finally, to reduce the formation of aerosols, liquid should be expelled from the pipette as gently as possible commensurate with the technique being employed. Bubbles should not be blown through cell culture solutions, and the tip of a pipette should never be held above the lip of a vessel while expelling a liquid.

ASEPTIC TECHNIQUE: BASIC PROCEDURES

The range of aseptic manipulations that may be employed during cell culture is almost infinite, so two basic procedures will be used to illustrate the way in which the principles discussed previously are brought together to form an aseptic procedure.

Manipulation of Cells and Liquid Reagents

This is illustrated by a procedure for supplementing a bottle of medium with fetal calf serum and then passaging a suspension cell line on the open bench.

Protocol

1. Bring medium and serum from the refrigerator or freezer, check that the caps are secure, and then warm/thaw them in a water bath set at 37 °C.

   The water bath should be sited away from the clean working area, because it can be a source of contamination, particularly bacteria and algae.

2. Prepare your clean area by removing all unnecessary items, spray/swab the bench with liquid disinfectant, and allow the surface to dry.

3. Bring to your working area only the equipment you need for the first step (supplementing the bottle of medium). The can of pipettes and the pipetting aid should be sprayed down with liquid disinfectant before being placed in an easily accessible position within the clean working area. The Bunsen burner should also be introduced to the area; it should be clean, but should *never* be sprayed with ethanol (in case some has not evaporated by the time it is lit). Finally the medium and serum are removed from the water bath, dried, sprayed with liquid disinfectant, and placed in the clean working area.

4. If ethanol or another flammable disinfectant has been used, wait until it has all evaporated before

lighting the Bunsen. Open the pipette can, and place the lid out of the way but still within the clean working area, (i.e., with its mouth neither facing the ceiling nor in contact with the bench).

   The lid can be placed under the open end of the pipette can. This tilts the can upwards at a convenient angle for removing pipettes and ensures that the lid is out of the way.

5. Loosen the caps of the medium and serum bottles.

6. Remove a pipette carefully from the pipette can, touching the top of the pipette as little as possible with the fingers, and allowing the selected pipette to touch the other pipettes as little as possible, particularly near their tops.

   The aim here is not to touch any part of the pipette which will come into contact with your sterile solutions or equipment on the tops of the other pipettes which may have been touched by nonsterile hands or gloves.

7. Insert the pipette carefully into the pipetting aid, then flame the pipette by pushing it lengthwise through the Bunsen flame, rotating it through 180°, then pulling it back through the flame. This should take no more than 3 seconds.

   Remember to hold the pipette close to its point of insertion into the pipetting aid and not to use excessive force, to avoid the dangers of breaking a pipette.

8. Pick up the bottle of serum in your free hand.

9. Grasp the lid of the bottle in the crook formed between the little finger and the palm of the hand holding the pipetting aid, and unscrew the bottle from the lid.

10. Flame the neck of the bottle by rotating it briefly in the Bunsen flame, then holding the pipette and bottle at an angle such that the pipetting aid and hand never get positioned vertically above the open mouth of the bottle, insert the pipette into the bottle

    On the open bench, particles may fall downward from nonsterile items such as hands or pipetting aids onto whatever is underneath. Thus, the open necks of sterile bottles or flasks should never be below such nonsterile items, and this is the reason for working at an angle. This also holds when working in a device with a vertical (top to bottom) airflow, such as a Class II MSC (see later section).

11. Draw the required amount of serum into the pipette, then withdraw the pipette from the bottle.

12. Flame the neck of the serum bottle, then bring it to its lid, and place the lid securely on the bottle. Put down the serum bottle and pick up the bottle of medium, and repeat steps 9 and 10.

13. Expel the contents of the pipette into the medium, withdraw the pipette, reflame the neck of the bottle, and replace the lid. Finally, discard the pipette into a container of disinfectant.

To continue the process and use the freshly prepared medium to passage a suspension cell line,

14. After screwing down its lid, remove the bottle which contained the serum from the clean working area, as it is no longer required.

This is to keep the working area as uncluttered as possible.

15. Take a new, sterile, plastic cell culture flask out of its sterile wrapping, ensure that it is not damaged and that its cap is secure, and place it in the clean working area.

When first removed from its wrapping, the flask should be sterile both inside and out, and thus should not need spraying with disinfectant.

16. Pipette an appropriate amount of the freshly prepared medium into the flask using the technique detailed in steps 5 to 13, but here of course the liquid is withdrawn from the bottle of medium and expelled into the culture flask. It is recommended that the neck of the culture flask is *not* flamed, as the plastic could easily melt or catch fire.

17. Remove from the incubator the flask containing the cells to be passaged, and make sure its cap is secure. If there is any wetness on the outside of the flask, dry it immediately because such moisture is an excellent source of microbes. Spray or swab the outside of the flask with liquid disinfectant if this will not affect the culture (i.e., if the flask is sealed and there is no chance of the disinfectant reaching the cells). Once the flask has dried, introduce the flask to the clean working area.

18. Resuspend the cells, and pipette an appropriate volume from the cell-containing flask into the flask containing fresh medium, again using the technique described in steps 5 to 13, and *not* flaming the necks of the plastic flasks. When expelling the cell suspension into the fresh medium, place the tip of the pipette below the surface of the medium, and once the cells are expelled do not blow bubbles through the medium. Both of these precautions are to avoid the formation of aerosols.

19. Place the new flask of cells in the incubator, and discard the old flask of cells (to be destroyed by incineration or autoclaving). If the work is now finished, turn off the Bunsen, replace and/or secure any closures (lids on pipette cans, caps on bottles etc.), remove all items from the clean working area, and spray and wipe down with liquid disinfectant.

General Note. Ethanol, including 70% ethanol solution, poses a serious fire hazard when used with techniques that employ a Bunsen burner. It is recommended that a nonflammable disinfectant is used if at all possible.

As far as possible, only one cell line should be handled at a time, and each cell line must have a bottle(s) of medium dedicated for use only with that cell line. All cultures and bottles of medium for one cell line must be removed from the working area, the Bunsen turned off, and the area sprayed/swabbed down with disinfectant before introducing another cell line to the area. This should only be done after a gap of at least 5 minutes to allow any possible cell-containing aerosols to dissipate. These precautions are to avoid cellular cross-contamination.

Remember that working on the bench offers no protection to the operator from any biohazards present in the cell culture and should only be undertaken if the risk assessment permits.

Working Without a Bunsen. When working without a Bunsen, one must be even more fastidious with regard to maintaining the initial sterility of one's equipment. The advantage is that the fire hazard is eliminated, but one loses the comfort that if the outside of a pipette (for example) were to be nonsterile, then passing it through the Bunsen flame *might* sterilize it. Thus for example, it may be worth using individually wrapped single-use rather than reusable pipettes to avoid the potential for contamination when removing the pipette from the pipette can. Individually wrapped pipettes should be opened at the end furthest from the tip, and the open end turned inside out over the wrapping still in place, so that when the pipette is withdrawn, it can touch only the inner surface of the wrapping, which should still be sterile. Similar precautions should be taken when removing other sorts of wrapping from sterile equipment.

Provided the maximum precautions are taken to maintain the initial sterility of all equipment, aseptic technique performed without a Bunsen can give at least as good results in sterility maintenance as using a Bunsen. Indeed, the decreased time of exposure to the environment caused by not having to flame equipment may be advantageous in this respect.

Aseptic Manipulation of Equipment

The previous handling guidelines should be sufficient to cover most cell culture operations carried out on a small scale. However, as the scale of operation increases, it may be necessary to aseptically assemble sterile pieces of equipment, for example, connect segments of tubing or connect tubing to culture or media vessels. This is best achieved by completely covering the equipment to be sterilized in a wrapping which microbes cannot penetrate. Then, once sterilized, the equipment remains sterile until it is unwrapped immediately before aseptic assembly. In many cases it may be worth covering the actual assembly points separately to help maintain sterility. Then assembly is simply achieved after discarding the final wrapping on all assembly points. As with the aseptic techniques already described, manipulations are best performed wearing (sterile) gloves in an environment which will protect the equipment from airborne contaminants. In this case that might be a Class II MSC but could also be a vertical or horizontal LAF because sterile equipment poses no microbiological risk to the operator.

However, although the actual points of connection must always be sterile, it is not always possible for the whole of each piece of equipment to be sterile. An example of this might be when replacing an empty media vessel with a full one on a pilot-scale culture system or when sampling from such a system via a septum. Such manipulations may be facilitated by using a surface sterilizing reagent, such as beta iodine or (as used in our laboratory) 0.2% chlorhexidine gluconate ("Hibitane") in 70% ethanol. As an example, when disconnecting one piece of tubing connected to another by a Luer or similar fitting and replacing it with a new one, one would proceed as follows:

1. Douse both halves of the existing connection and both the new connector half and its cover, plug, or sheath with the liquid surface sterilizing agent. This can be done conveniently by squirting the liquid disinfectant over the area from a squeeze bottle.

2. Using swabs made of segments of sterile butter muslin (or similar absorbent and autoclavable cloth) soaked in the surface sterilizing agent, wrap the four half-connectors separately. This should be done so that the swab on each half-connector butts up against, but does not overlap, that on the other half. Leave the swabs in place for at least 3 minutes to sterilize the covered surface.

3. Rapidly remove the cover, plug, or sheath from the new connector, break the old connection, and make the new connection. This is done with the swabs still in place, and they are only discarded after the connection has been completed.

Wearing (sterile) gloves is strongly recommended. Also, note that if any of the tubes already contain liquid, then they must be clamped before attempting this procedure.

It should be noted that the swabs soaked in surface sterilizing agent can fulfill two purposes. The first is to sterilize the area surrounding a connection. Thus, if there is any slight inaccuracy in making a connection, then the surrounding area which might be touched will be sterile and will not compromise the sterility of the system. The second role is to act as a barrier between one's gloves (which may not be sterile) and the sterile surfaces.

Sampling from a septum is achieved in a similar way, but only the surface of the septum is sterilized using a soaked swab, and in this case the swab is discarded before penetrating the septum with a sterile needle. (By the term septum here, we mean a small unit often marketed as a sterile injection site, not the large industrial type fitted to fermenters and similar equipment.)

As with all aseptic techniques, the surface sterilizing agent/swab technique gives no total assurance of sterility, and the relatively short exposure times and nature of the chemicals involved means that some organisms may not be inactivated. However, it appears to be an effective technique when used against those organisms commonly found in laboratories. We have run cell cultures in bioreactors using antibiotic-free medium for more than 6 months in an ordinary laboratory with hundreds of aseptic manipulations carried out as described. The cultures tested negative for bacteria, fungi, yeast, and mycoplasma at the end of this period.

HEPA FILTRATION

Introduction

Although aseptic techniques performed on the open bench can be successful, they offer no protection to the operator from any hazards posed by the work, and the high particulate load in the air of the average laboratory means that the chances of contamination of the work are always significant. These problems can be greatly reduced by performing the work in a suitable environment which incorporates a controlled flow of air from which the vast majority of both inert particles and associated viable contamination has been removed. This is achieved by using HEPA (high-efficiency particulate air) filtration.

HEPA filters are defined by their particle removal efficiency and their flow rate. They have a removal efficiency of at least 99.97%, and they attain this efficiency when the air velocity through the filter is approximately 2 cm/s. (4 ft/min). Higher grade filters have been developed recently (e.g., ULPA [ultra low penetration air] filters), that are intended for the microelectronics industry where contamination of electronic components with very small particles can be a significant problem. This level of filtration is not normally required for pharmaceutical and biological (including cell culture) applications. Indeed, providing a level of filtration greater than that required can have significant cost penalties, and the associated working practices may make the work unnecessarily difficult and time-consuming.

Mechanisms of Filtration

Filtration of a gaseous medium is not a simple process of sieve retention but rather the sum effect of several different mechanisms. The relative contributions of these various mechanisms can vary considerably, depending on the size of the particles and the air velocity passing through the filter.

Sieve retention, the simplest mechanism of particle retention, occurs when the particle is larger than the spaces between the fibers of the filter medium. This sieving effect is of significance only for relatively large particles (>2 μm), and these would normally be captured by the coarser and less expensive prefilter placed upstream of a HEPA filter. It would not be cost-effective to use HEPA filters to remove large particles.

The three most important particle removal mechanisms of HEPA filters are inertial impaction, diffusive retention, and interception (Fig. 1).

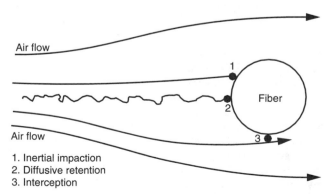

Figure 1. Schematic representation of mechanisms of filtration.

Inertial Impaction. As the air flows around the fibers of the filter medium (see section on the construction of HEPA filters), because of their inertia, the larger particles do not follow the air flow but continue in their original direction and become embedded in the fibers. This mechanism is a major factor for particles >1 μm. As the air velocity through the filter increases, this mechanism becomes more effective.

Diffusive Retention. Small particles (<0.1 μm) are retained primarily by impaction resulting from Brownian motion (i.e., randomized movement caused by these particles being hit by other small particles or the molecules of the gas in which they are suspended) which causes them to deviate from the direction of the air flow. This random motion results in the impaction of the particles on the fibers of the filter.

Interception. Particles of an intermediate size (ca. 0.1–0.5 μm) have less inertia than the larger particles and tend to follow the flow of air around the fibers. If a particle strikes a fiber as it passes it tangentially or if it approaches one closer than half its diameter, it will be captured and retained by it. This mechanism is interception.

Once a particle has come into contact with the fiber of the filter, it is held by electrostatic forces and will become difficult to dislodge. It is the combination of these mechanisms that explains the efficacy of HEPA filters.

Filter Efficiency

Figure 2 shows the classical efficiency curve of a HEPA filter. The minimum efficiency is for a particle of about 0.2–0.3 μm, and particles both larger and smaller are removed more efficiently. However, the actual minimum efficiency is variable and will depend on the following factors (10):

the density of the individual particles;

the velocity and mean free path of the particles;

the thickness of the medium;

the velocity, pressure, and temperature of the air; and

the sizing and distribution of the fibers within the filter medium.

Because penetration is a function of so many variables, the actual shape of the curve and the minimum efficiency will vary. Rather than consider a single most penetrating particle size, it has been suggested that it would be better to think in terms of the most penetrating particle range, for example, 0.1–1.0 μm (10).

Filter efficiency decreases with increasing flow rate. Conversely the lower the flow rate, the more efficiently a filter will retain particles. Air flow velocities of around 90 ft/min, frequently found in clean rooms, should not result in problems due to particle penetration of intact HEPA filters because any upstream particles are likely to be either too small or too large to penetrate to any significant degree. The large surface area of HEPA filters is designed to provide an adequate volume of filtered air at the flow rate required to maintain suitable conditions within the clean room or device (e.g., laminar flow cabinet). By increasing the amount of filter medium in a filter it is possible to decrease the pressure drop across it and also to increase its efficiency. If the air velocity were reduced sufficiently, it could be possible to achieve 99.97% removal of particles in the range 0.1–1.0 μm. with a relatively coarse filter. However, the dynamics of filtration is but one factor necessary to achieve suitable air quality in any particular situation. The filtered air flow volume needs to be sufficient to allow adequate flushing of the critical area to cope with the particulate and microbiological contamination that may be introduced by the process equipment or personnel.

Construction of HEPA Filters

The filter medium is a paper formed from fine glass fibrils. The higher the efficiency of the filter, the higher the proportion of smaller fibers. The filters can be constructed in a number of ways, each method intended to produce a relatively large surface area of filter medium which is sealed within a rigid frame and with a suitable gasket to prevent the leakage of unfiltered air into the working environment. The filter medium is mounted in the frame, normally in one of the following ways (11):

Standard Filters. The sheets of filter medium are folded in parallel pleats, and the pleats are separated by corrugated aluminum or kraft paper separators. The filter medium and separators are assembled and bonded to the frame with a resin. The frame may be constructed of metal, wood/chipboard, or plastic.

For most biological or pharmaceutical applications, filters with wooden/chipboard frames would not be considered suitable because these materials can themselves shed particles and because of the potential for microbiological growth on the frame, particularly if they became damp. However, in some applications for example, in certain

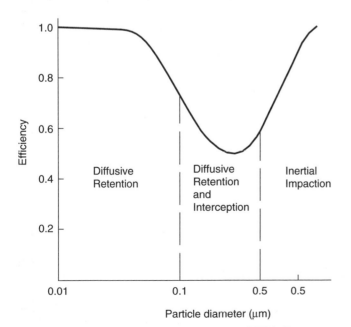

Figure 2. Classical efficiency curve for a HEPA filter.

chemical or biological processes, the use of wooden frames may be indicated to facilitate incineration after use.

Minipleat Filters. This format is a modification of the standard filter but without the use of separators. The filter medium is folded in tighter pleats (approximately 2–3 pleats/cm compared to approximately 1 pleat/cm for the standard filters). To allow the air to flow satisfactorily, the paper has built-in nylon cords or molded media which protrude slightly from the paper folds to hold them apart. These filters can provide greater volume flow capacity and more uniform face velocities compared to standard filters, but they are more expensive. However, because they hold approximately twice the surface area of filter medium compared to the same size of standard filter unit, they may have advantages where space constraints are a problem or where extended use between filter changes is required.

Flanders Filters. This type of filter is constructed from crenellated corrugated filter paper, and the surface 'texture' acts in lieu of separators to keep the pleats apart. The crenellated surface and the absence of separators result in approximately 40% more available filtration area compared to standard filters. These filters, however, fit only patented frames. The filter-holding frames have a slot filled with thixotropic gel which accepts a tongue on the filter edge, achieving an air tight seal without using a compressed gasket (the gaskets on both standard and minipleat filters use compressible gaskets made from closed-cell neoprene, PTFE or molded polyurethane).

Testing HEPA Filters

Filters are tested by the manufacturers to assign or confirm their particle removal efficiency. In addition, on-site testing after delivery and installation needs to be carried out to verify that no damage has occurred during storage, transportation, or installation. At times confusion has arisen regarding the purpose and intention of this on-site testing.

The tests carried out by the filter manufacturers use relatively sophisticated equipment to control challenge aerosol particle size, air flow rates, etc. and would normally consist of one of the following.

Hot DOP (dioctylphthalate) Test. Thermally generated particles of DOP, of an essentially monodisperse aerosol with a mean diameter of approximately 0.3 μm, is generated upstream of the filter and the percentage penetration determined by comparing the downstream concentration of particles to the upstream (100%) concentration.

Sodium Flame Test. An aerosol of particles of sodium chloride, mainly within the range of 0.02–2.0 μm with a median of 0.6 μm, is generated by atomizing a sodium chloride solution and evaporating the water. By comparing the upstream and downstream concentrations photometrically, the percentage penetration can be determined. The hot DOP and sodium flame tests produce substantially similar size distribution aerosols, and the tests give comparable results.

Oil Mist Test. This is the basis of the German (DIN 24184) test. The principle of the test is the same as the DOP and sodium flame tests, but the test aerosol is an oil mist with a particle size between 0.3 and 0.5 μm.

The test normally carried out on-site is a cold version of the DOP test or a variant of it using, for instance, a different challenge aerosol such as mineral oil. An aerosol of polydisperse particles in the range 0.1–3.0 μm is generated, usually at ambient temperature, and introduced upstream of the filter while the downstream filter face is scanned with a photometer to detect any leaks, again by comparing the upstream and downstream concentrations of the aerosol. It is normally accepted that a photometer reading of >0.01% is indicative of an unacceptable leak. There are a number of fundamental differences between this test and the more sophisticated tests used by the filter manufacturers to verify particle removal efficiency (e.g., particle size distribution, concentration of the aerosol challenge, temperature control, air flow velocity control, etc.). This *in situ* test is intended to check the integrity of the filter installation and is not suitable to verify the particle removal efficiency of the filters.

HOODS AND CABINETS EMPLOYING HEPA FILTRATION

These can be divided into two categories: laminar flow hoods, where a flow of HEPA-filtered air is used to protect the work from contamination by particulates but which offer little or no protection to the operator from hazards posed by the work; and microbiological safety cabinets, all of which offer protection to the operator but which may or may not offer protection to the work.

Laminar Flow Hoods

Horizontal Flow. The flow of HEPA filtered air in these hoods is directed from the back of the hood across the work surface toward the operator (Fig. 3). Thus these hoods should be used only for manipulating clean, preferably sterile, nonhazardous equipment and solutions, NEVER FOR CELL CULTURE.

Vertical Flow. The airflow in these hoods is from top to bottom (Fig. 4). Thus air is not blown directly from the work at the operator as in a horizontal flow hood (see above), but the absence of a front screen and of filtration of the exhaust air means that there is still no significant protection offered to the operator. Thus it is recommended that these hoods too should not be used for cell culture.

Microbiological Safety Cabinets

A microbiological safety cabinet is defined as a "cabinet intended to offer protection to the user and environment from the aerosol hazards of handling infected and other hazardous biological material, but excluding radioactive, toxic and corrosive substances, with air discharged to the atmosphere being filtered" (12). Some, *but not all* of these cabinets will also offer protection to the work from environmental contaminants.

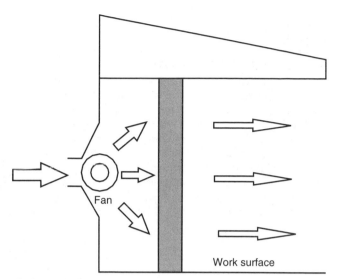

Figure 3. Schematic cross section of a horizontal laminar flow hood (arrows indicate airflow; hatched area = HEPA filter).

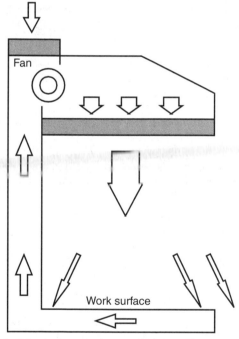

Figure 4. Schematic cross section of a vertical laminar flow hood (arrows indicate airflow; hatched area = HEPA filters).

Class I MSCs. These cabinets operate by pulling in a constant stream of air from the room through the working aperture and passing it through a HEPA filter to exhaust (Fig. 5). The inward air flow through the working aperture protects the operator from particulates generated within the cabinet, and these particulates are prevented from escaping to the environment by the HEPA filtration of the exhaust. However, this type of cabinet offers no protection to the work from particles generated outside the cabinet. Class I MSCs are generally used for handling viruses and other biological agents which pose a moderate risk to the operator.

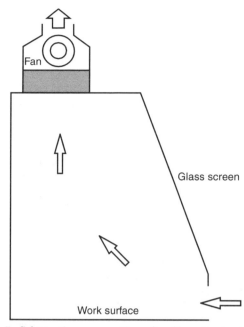

Figure 5. Schematic cross section of a Class I microbiological safety cabinet (arrows indicate airflow; hatched area = HEPA filter).

Class II MSCs. These cabinets are most commonly used for cell culture because they are easy to use but offer protection both to the operator from particulates generated inside the cabinet and to the work from particulates generated outside the cabinet (Fig. 6).

The work is protected by the vertical flow of HEPA filtered air from the top of the cabinet and by the fact that air entering the front of the cabinet is drawn directly away beneath the work surface without passing over the work.

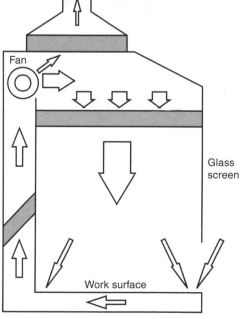

Figure 6. Schematic cross section of a Class II microbiological safety cabinet (arrows indicate airflow; hatched areas = HEPA filters).

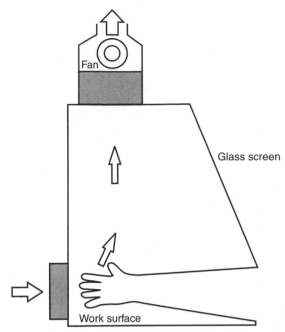

Figure 7. Schematic cross section of a Class III microbiological safety cabinet (arrows indicate airflow; hatched areas = HEPA filters).

The operator is protected by the inflow of air at the working aperture which prevents the outflow of particles from inside the cabinet and by the fact that all the exhaust air is HEPA filtered before being released to the environment. Only a percentage of the total airflow within the cabinet is sent to exhaust. The amount required is that needed to compensate for the air constantly flowing into the cabinet at the front aperture; the balance of the air is recycled within the cabinet.

Class III MSCs. These cabinets offer maximum protection to both the work and the operator and environment but are extremely cumbersome to use and are normally employed only when manipulating dangerous (Hazard Group 3 or 4) (Ref. 9) pathogens (Fig. 7). In these cabinets, the operator is separated from the work by gloves mechanically attached to the cabinet, and both the inlet and exhaust air have airborne particulates removed by HEPA filtration systems.

WORKING WITHIN LAMINAR FLOW HOODS AND MICROBIOLOGICAL SAFETY CABINETS

General

All cabinets/hoods should be switched on before use, and allowed to run for at least 15 minutes to establish the correct air flow and clear any airborne particles. When using cabinets that employ a blanking plate for the front aperture (Class I and II MSCs), the blanking plate should be removed immediately before starting the fan and placed on a *clean* surface (not the floor!). It is replaced once the fan has been stopped after use (although many cabinets are designed to run continuously). In all hoods/cabinets, the use of Bunsen and similar burners should be avoided

if at all possible because they disturb the airflow pattern and may be a fire hazard.

All hoods/cabinets must be kept as clean as possible, and the working surface should be cleaned with a suitable disinfectant before and after every use. Spills should be mopped up immediately and the working surface wiped with disinfectant. If anything should be spilt down the perforations in the working surface of a vertical laminar flow hood or Class II microbiological safety cabinet, then the working surface must be removed and the underlying area cleaned and treated with a suitable disinfectant. In any case this should be done at least weekly, even if no known spills have occurred.

To operate efficiently, the hood/cabinet must be tested regularly (at intervals of not more than 1 year) for HEPA filter integrity, airflow direction and rates, and particulate containment and elimination of external contamination, if appropriate (see section on Performance Tests). If used for handling pathogens, testing must be performed more frequently (i.e., at least every 6 months). All hoods/cabinets must be tested when first installed and whenever they are relocated within or between laboratories (see section on Performance Tests).

Laminar Flow Hoods

Horizontal Laminar Flow Hoods. Because the airflow is directed at the operator these hoods must *never* be used for cell culture or when manipulating any other potentially hazardous material.

The type of aseptic technique described previously should be adopted when working in such hoods. However, it must be modified so that any items that may not be sterile (e.g., hands, pipetting aids) should, instead of not being directly above any sterile openings (e.g., open bottles), not be directly *behind* them.

Vertical Laminar Flow Hoods. General aseptic technique is carried out as described for use on the open bench without a Bunsen burner.

Microbiological Safety Cabinets

Class I Microbiological Safety Cabinets. As stated before, these cabinets are designed to protect the operator from the material being handled but offer no protection to that material or the work being performed with it. Thus such a cabinet must be positioned where the incoming air is as clean as possible. In addition, so as not to compromise the containment of particulates within the cabinet, it must be sited where there is minimum air turbulence outside the cabinet—see the next section (on Class II cabinets) for further details.

The lack of protection offered to the work means that the amount of work carried out in a Class I cabinet should be kept to a minimum. For example, if performing a virus assay by plaque formation on a cell monolayer, the handling of the (uninfected) cells when preparing the monolayer should be carried out in a Class II cabinet. Only the preparation, dilution, and addition of the virus should be performed in the Class I cabinet.

Performing manipulations within the Class I cabinet is rather different from that described for all of the other types of hoods or cabinets because the only way to protect the work from the incoming unfiltered air is to position equipment, for example open bottles or culture dishes, so that the airflow is directed away from any sterile areas. Thus when removing material from a bottle, for example, the base of the bottle should be pointed into the airflow with the neck pointing toward the extract filter. In cases where equipment (e.g., petri dishes or multiwell plates) must remain horizontal when opened, it should be positioned well to the rear of the cabinet away from the front aperture. The length of time that vessels are left open must in all cases be kept to a minimum. In other respects, aseptic technique used in Class I cabinets is the same as that described previously.

Class II Microbiological Safety Cabinets. Protection of both the work and the operator depends on the integrity of the inward airflow at the working aperture and its efficient and immediate removal through the perforations/slits in the front of the working surface. Eddies caused by (1) turbulent air outside the cabinet or (2) around the arms of the worker can direct nonsterile incoming air across the top of the work surface, or cause air from within the cabinet to escape, unfiltered, to the environment. To avoid this:

1. Steps must be taken to reduce to a minimum the turbulence of the air in the vicinity of the front of the cabinet. This means that very careful consideration must be given to the siting of the cabinet in terms of its proximity to architechtural features (walls, columns, doors), work benches, other cabinets, and routes of movement by other workers — see Figures 8 and 9. Furthermore, in the vicinity of the cabinet, the velocity of the air due to the room's ventilation system must be kept to a minimum. It is particularly important that air entering the room via grilles, diffusers, etc. should not discharge directly across or toward the aperture of the cabinet, and every attempt should be made to keep all air velocities in the room below 0.3 m/s. Clearly, when designing a new laboratory, the siting of the MSCs must be decided early and incorporated into the design process. Positioning of cabinets in an existing laboratory similarly requires the most careful consideration, and all cabinets must be performance tested once in their new position, even if they have been moved only a short distance within the same room. These requirements apply equally to Class I and Class II cabinets.

2. Operators should avoid rapid movement of their arms in the working aperture and should avoid coughing, sneezing, singing, or whistling while working.

   It should always be remembered that the front perforations/slots are a nonsterile area, and thus all work should be performed at least 5 cm away from them. Similarly the perforations/slots at the rear of the working surface should be kept clear so

as not to impede the downward flow of air in the cabinet.

In all other respects, aseptic technique for use in a Class II MSC is the same as for work on the open bench without a Bunsen. Care must be taken, however, not to place any piece of apparatus in a position where it will obstruct the airflow within the cabinet (e.g., over a grille) or to inadvertently allow a piece of sterile apparatus to come into the nonsterile airflow at the front of the cabinet, for example, when removing a pipette from a pipette can.

Class III Microbiological Safety Cabinets. These are such specialized pieces of equipment for handling such hazardous materials that no advice will be offered here. They must be used only after the most thorough and rigorous training.

TESTING CLASS I AND CLASS II MICROBIOLOGICAL SAFETY CABINETS

Class I and Class II MSCs require some common basic tests of function to demonstrate satisfactory performance, but a Class II cabinet also needs additional testing due to its design and method of operation.

The full details of the tests to be carried out and detailed test methods should be obtained from the relevant official standards (e.g., Ref. 12). However, the following is an overview of the testing requirements for these two types of safety cabinets, together with an indication of their relevance.

Performance Tests

Table 1 provides a quick reference to the performance tests needed for Class I and Class II cabinets.

Operator Protection Tests. Ideally there should be no escape of aerosols from inside the cabinet to the external environment. However, with open-fronted cabinets, some escape will be inevitable, but this should be within "acceptable" limits to minimize any risk to the operator.

This test is intended to quantify the degree of protection offered by the cabinet and is defined in British Standard 5726 as "the ratio of exposure to airborne contamination generated on the open bench to the exposure resulting from the same dispersal of airborne contamination generated within the cabinet" (12).

Table 1. Performance Tests for Class I and Class II MSCs

Test	Class I	Class II
Operator protection test	Yes	Yes
External contamination test	No	Yes
Cross-contamination test (where applicable)	No	Yes
Airflow checks	Yes	Yes
Temperature checks	No	Yes
Tests for leakage	Yes	Yes

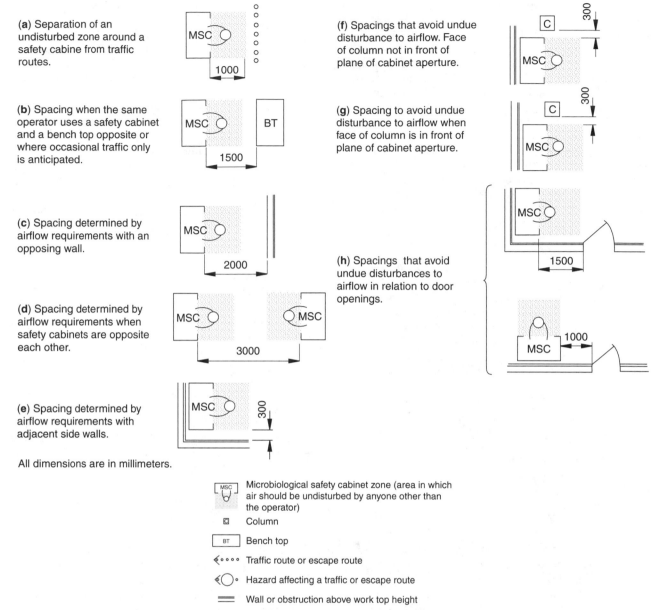

Figure 8. Siting of microbiological safety cabinets: Recommendations for minimum distances to avoid disturbance to the safety cabinet and its operator. (Note: Siting arrangements which should be avoided are overlaid with a cross.) (Reproduced from Ref. 12 with permission).

Basically, the test consists of generating a detectable aerosol within the cabinet and determining the proportion of the aerosol which escapes to the outside of the cabinet through the front aperture. During the test the airflow entering the cabinet is disturbed in order to simulate the effect of an operator's arm, by introducing a cylinder through the front aperture. A number of different aerosols have been described, some of which have been incorporated into official standards.

Biological Method. This test uses a bacterial spore suspension (e.g. *Bacillus subtilis var globigii*), which is generated inside the cabinet using a nebulizer, and determines the number of spores escaping by using microbiological air samplers situated outside the cabinet but close to the open aperture. The protection factor is

calculated by comparing the numbers generated inside to those detected outside.

Spraying spore suspensions can be potentially problematical because of the risk of inhalation by the operators in addition to the problem of contaminating a cabinet which is intended for cell culture work where bacteria-free conditions are needed. Another potential problem sometimes overlooked is that the precision of microbiological enumeration methods is not particularly good and can vary according to the type of air sampler used. Any variability in the counting method may well have an effect on the accuracy of the calculated protection factor. Another disadvantage to this method is that the agar plates exposed in the air samplers need to be incubated to allow the bacteria to grow and the result

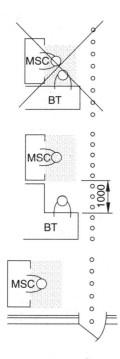

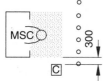

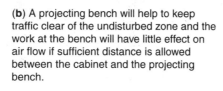

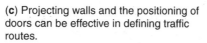

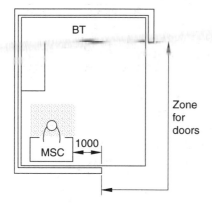

(a) A bench at right angles to a safety cabinet may keep traffic away from the undisturbed zone but work at the bench will cause disturbances to the air flow.

(b) A projecting bench will help to keep traffic clear of the undisturbed zone and the work at the bench will have little effect on air flow if sufficient distance is allowed between the cabinet and the projecting bench.

(c) Projecting walls and the positioning of doors can be effective in defining traffic routes.

(d) Columns can assist the definition of traffic routes.

(e) In a small laboratory, the safety cabinet should be clear of personnel entering through the doors.

All dimensions are in millimeters.

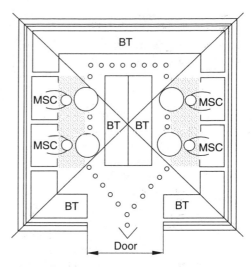

(f) Danger of too much air movement in front of safety cabinets should be alleviated by allowing more space between the apertures of the safety cabinets and the bench tops.

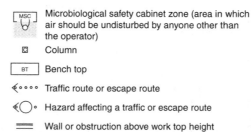

(g) Danger of too much movement in front of safety cabinets should be avoided by allowing more space between the apertures of the safety cabinets and the bench tops.

MSC	Microbiological safety cabinet zone (area in which air should be undisturbed by anyone other than the operator)
C	Column
BT	Bench top
◄°°°°	Traffic route or escape route
◄◯°	Hazard affecting a traffic or escape route
═══	Wall or obstruction above work top height

Figure 9. Siting of microbiological safety cabinets: Avoiding disturbance due to other personnel. (Note: Siting arrangements which should be avoided are overlaid with a cross.) (Reproduced from Ref. 12, with permission).

will not, therefore, be available until several days after testing.

The Potassium Iodide (KI Discus) Test. The arrangement of aerosol generator, cylinder, and samplers is similar to that used in the biological test, but in this case the challenge aerosol is produced by dropping a solution of potassium iodide under controlled conditions onto a rapidly rotating disc (13). The aerosol dries rapidly and produces particles in the size range of 3–10 μm. The particles are collected in membrane filter-based air samplers. Then the filters are placed in a solution of palladium chloride enabling the particles to be seen and counted under a low magnification microscope.

The biological and potassium iodide methods give comparable results.

Other aerosol materials have been used, such as polystyrene microspheres and an optical brightener (14,15), using the same kind of nebulizer as that used for spores. These seem to be suitable alternatives to the biological and KI discus tests.

The requirement for this test is that no individual protection factor obtained should be less than 1.0×10^5. This means that an operator working in a safety cabinet should be exposed to less than 1/100,000 the level of airborne contamination within the cabinet. Similarly, it means that the operator is exposed to less than 1/100,000 the level of airborne contamination compared to working on an open bench (assuming the same level of aerosol generation).

External Contamination Test. This test is intended to demonstrate whether the curtain of air descending at the front of a Class II cabinet prevents contamination from the room from entering the working area of the cabinet where it might compromise the work being carried out. A spore suspension and nebulizer, as used in the Biological Operator Protection Method, is used to generate a large aerosol of spores outside the cabinet but close to the open aperture for a period of not less than 4 minutes, and the number of spores entering the cabinet during this period and for a further 5 minutes afterward is detected using exposed agar plates which are distributed over the working surface. The challenge aerosol should contain at least 3×10^6 spores.

Both the British and U.S. National Sanitation Foundation standards permit no more than five colonies per test (in total, not per agar plate), and a control test, where the test is repeated but with the cabinet switched off, should have more than 300 colonies present on the agar plates.

This test carries similar problems of potential contamination with bacterial spores, as mentioned in the previous section, but the potassium iodide test could be used as a quicker and safer alternative. Another alternative would be to use an electronic particle counter to compare the particle counts both inside and outside the cabinet. However, unless a sufficiently large room particle count is obtained, then test sensitivity may be a problem, in which case a particulate aerosol would need to be generated. A disadvantage of employing an alternative method to that specified in the relevant standard is the problem of setting acceptable limits for the test method employed.

Cross-Contamination Test. This test is intended to demonstrate whether aerosols generated on one side of the cabinet will contaminate materials at the other side. There are differences in the test methods described in the U.S. and British standards, although the principle is the same. A nebulizer is used to generate a spore aerosol on one side of the cabinet, and agar plates are used to detect the degree of contamination on the other side of the cabinet. The problems of bacterial contamination mentioned earlier are equally applicable with this test. However, the relevance of the test needs to be considered carefully because it may not be applicable for most safety cabinet applications. It was originally introduced to test relatively large cabinet installations where more than one operator was working within a single cabinet, and to provide some evidence that cross-contamination did not occur between one operation and the other. Where a single operator is working within a cabinet, this type of test is probably not necessary.

Airflow Checks. To function properly, air must enter Class I and Class II cabinets with a velocity sufficient to retain within the working area any aerosols or particles that may be generated. However, this flow rate should not be so high that turbulence is created within the cabinet, otherwise there is an increased risk that air will spill out of the open aperture. Air movements into the cabinet can also be influenced by air currents from within the room (e.g., from the ventilation system, other equipment operating in the vicinity, opening and closing of doors, movement of people, etc.). The control of the inward flow of air in Class II cabinets is further complicated by the gain in energy (in the form of heat) of the recirculating air which can disturb the balance between room air and cabinet air at their interface (16).

For Class I and Class II cabinets, it is necessary to measure the inflow, and for Class II cabinets also the downward flow inside the cabinet. The expected airflows should be within the limits given in chart 1.

Chart 1. Airflow Values

Cabinet Type	Inward Velocity	Downward Velocity
Class I	0.7–1.0 m/s[a]	N/A
Class II	≥0.4 m/s[b]	0.25–0.5 m/s[a]

[a]No individual measurement should differ from the mean by more than 20%.
[b]With Class II cabinets, the downward flow of air within the cabinet cannot be so easily measured because the inward flow of air at the open aperture varies in the vertical plane and normally has a higher velocity at the bottom. This also makes it difficult to make meaningful measurements of the inward airflow using anemometer readings at the aperture (or to set limits for a range of readings across this open face). To overcome this problem, it is customary to measure the velocity and then calculate the volume of air being extracted in the exhaust dust; this volume is equal to that drawn in through the front opening. Then the mean inward velocity can be calculated by dividing the volume of discharge air by the cross-sectional area of the front aperture (16).

In addition to velocity measurements, air visualization tests can be useful for checking that air flows inward over the whole of the working aperture. This can be particularly useful for demonstrating any effects caused by air currents within the room under different operating conditions. This can be done relatively simply using commercially available smoke pencils or cotton wool swabs soaked in titanium tetrachloride.

Although this test is simple to perform and can provide evidence of satisfactory air movement, it is not particularly sensitive because smoke streams are transient and sometimes difficult to see (17). The use of schlieren photography has been successfully used to visualize air flow patterns and provide more meaningful data on air movements in safety cabinets (18). However, although this is an elegant technique, it is not a method available to most testing facilities.

Temperature Checks. This is a particular requirement for Class II cabinets because of the potential heat gain referred to before. After four hours of continuous running with the fan(s) working and the lights on, the air temperature inside the cabinet measured 100 mm above the center of the working space should not rise by more than 8 °C above ambient laboratory temperature (12).

Test for Leakage. This test is to demonstrate the absence of any significant leaks through filters, seals, and construction joints. The same principle is used as that described earlier for the *in situ* testing of HEPA filters, using cold DOP or a suitable mineral oil. A test aerosol is generated on the dirty side of the filter, seals, and joints, and the clean sides are scanned with a photometer probe.

Again it should be recognized that this test is not suitable for confirming the efficiency rating of the installed HEPA filters but is only intended to detect leaks which may affect the integrity of the cabinet.

Testing Program

The following is not meant to be a comprehensive list of all the checks and tests to be carried out, but is given as broad guidance.

Testing after Installation. After a cabinet has been delivered or moved, a number of tests need to be carried out. Checks should be performed on the integrity of filters, gaskets, and construction joints to confirm that no damage or movement that might affect performance has occurred during transportation or installation. Performance tests for airflow, operator protection and, ideally, external contamination should be performed. The test for cross-contamination may be appropriate in some cases. If the cabinet was dismantled for transportation and then reassembled, checks must also be made to detect any leakage of the cabinet body.

Periodic Testing. A suitable regular testing program will have to be put in place but the level and frequency of testing will be a matter of judgment, depending on the particular application and the type of work being carried out and also on the need to satisfy health and safety requirements and whether full compliance with the relevant published standard is required.

To confirm continual satisfactory operation, the checks performed ought to include airflow measurements (and possibly air visualization checks), tests of filter and seal integrity, and operator protection tests. As a minimum, these tests should be carried out annually, but more frequent testing will be required if known pathogens are handled.

CLEAN ROOMS FOR CELL CULTURE USE

A number of definitions of a clean room have been published. That described in BS 5295 (19) is appropriate and defines a clean room as "A room with control of particulate contamination, constructed and used in such a way as to minimise the introduction, generation and retention of particles inside the room and in which the temperature, humidity and pressure shall be controlled as necessary". In cell culture and pharmaceutical clean rooms, the purpose of the facility is to provide an environment that limits the risk of introducing adventitious contamination into the product or process.

Modern clean room technology was originally developed for the military and space industries and was quickly utilized in the electronics industry. This technology was then embraced by the pharmaceutical and biological industries for aseptic operations. The basic design, construction, and operation of clean rooms for the different industries and different applications followed similar principles.

A significant advance was the development of HEPA filters (20). Following the introduction of conventional turbulent flow facilities using HEPA filtered air, the laminar flow room concept developed, although this probably had greater application in the micro-electronics industry, and generally it was not seen as cost-effective or directly applicable for many pharmaceutical clean room operations. This basic clean room technology was undoubtedly the best available at the time, but its limitations for pharmaceutical and similar aseptic operations have become apparent. The direct involvement of people within the clean rooms introduces a potentially significant particulate and microbiological bioburden to the process (21,22). The problems associated with the control of the personnel activities is becoming increasingly realized and we are now seeing reference to clean zones and the introduction of isolator systems into clean room activities. These can give a greater level of protection to the process and can also provide added protection to the staff when the materials they are handling may be toxic or a potential source of infection.

Clean Room Standards

Following the advent of clean rooms, the need for suitable formal standards, against which they could be designed, built, and tested, became apparent, Over the years a number of different standards have been produced, but the basis of the classification methods has resulted in considerable confusion.

The first published clean room standard was the U.S. Federal Standard 209 that has been amended

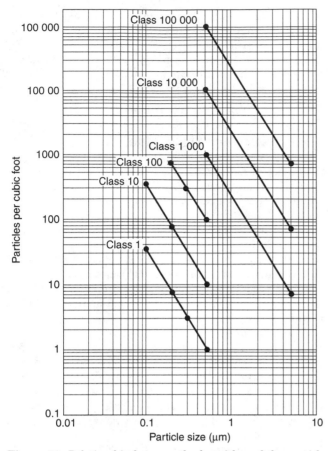

Figure 10. Relationship between the logarithm of the particle concentration (particles per cubic foot) and the logarithm of the particle size (in micrometers). It shows the maximum number of particles of a given size for the designated room class.

Table 2. Example of the Notation Used in Different Standards for the Same Class of Room (in This Case a Federal Standard 209, Class 100 room)

Standard	Class
US Federal Standard 209D	100
US Federal Standard 209E	M3.5
BS 5295 (1976)	1
BS 5295 (1989)	E or F
U.K. GMP Guide (1983)	1/A or 1/B
E.C. GMP Guide	A or B
French AFNOR	4000
German VDI	3
Australian 1386	3.5
Proposed ISO Std 209	5

each of the systems used, but the differences, in part, reflect difficulties in producing universally acceptable classification systems and nomenclature.

Table 3 provides some guidance on the rationale behind the notation used for some of the various national standards.

Table 3. Basis of the Notation Used for Various Room Classification Standards

Standard	Basis of Notation
Fed. Std. 209 (A–D)	Number of particles $\geq 0.5\ \mu m/ft^3$. A room with 100 particles/ft^3 has Class 100 conditions.
Fed. Std. 209 E	Incorporates metric units and designates a room with 3530 particles/m^3 (equivalent to 100/ft^3) as M3.5 (3.5 is the log of 3530).
BS 5295 (1976)	Rounded down metric version of Fed. Std. 209, Class 100 converted to 3000 particles/m^3 and called Class 1
BS 5295 (1989)	Based on Fed. Std. 209 but more accurately in line with stated values. Class 100 converted to 3500 particles/m^3, called Class E or F.
U.K. GMP Guide (1983)	Based on BS 5295 (1976) but Class 1 (Class 100 equivalent) became Grade 1/A or 1/B.
E.C. GMP Guide	Based on U.K. GMP Guide and a Class 100 equivalent became Grade A or B.
French AFNOR	Metric version used (counts/m^3) but defined it in a rounded up version; Class 100 equivalent became Class 4000.
German VDI	Based on particles/m^3 but at the $>1.0\ \mu m$ size. Class number derived from the logarithm of the number of particles/m^3. A room with 1000 particles $> 1.0\ \mu m/m^3$ is Class 3.
Australian Std. 1386	Based on Fed. Std. 209 but using the number of particles $>0.5\ \mu m/dm^3$. Class 100 (3530 particles/m^3) equates to 3.5 particles/dm^3 and is called Class 3.5.
Proposed ISO Std. 209	Based on Japanese approach using a method similar to the German VDI standard, i.e., the log of the number of particles/m^3 but utilizing the 0.1 μm particle size.

and revised a number of times. Subsequently, other national standards have been produced, but the basis of the particulate classification system adopted for each of them was the same as that for the original Federal Standard 209 (see Fig. 10). Some of these standards have introduced different terminology, a different basis for classification, differing levels of requirements, etc. There have also been differences in the scope of some of these standards. For instance, those intended only as a specification and method for classifying clean rooms include U.S. Federal Standard 209, French AFNOR, and Dutch VCTN-1, and those that encompass aspects of design practice and guidance for creating and monitoring controlled environments include British Standard 5295 and German VDI 2083.

The differences among the various clean room standards has undoubtedly combined to create confusion and uncertainty. Problems can also be compounded where companies operate or trade in different countries and where compliance with different standards is required.

To emphasize some of the differences and highlight the confusion that can easily be caused, Table 2 shows the notation used in different standards for the same class of room (in this case a Federal Standard 209 Class 100 room).

In these examples, ten different terms have been used to describe essentially the same thing. There are merits to

Table 4. Comparison of the Classification of Rooms with Various Particulate Levels Using Different Standards

Max. count/m³ at 0.5 µm and above	Fed. Std. 209D	Fed. Std. 209E	E.C. GMP Guide	B.S. 5295 1989	ISO 209
3.5					2
10		M1			
35.3	1	M1.5		C	3
100		M2			
353	10	M2.5		D	4
1000		M3			
3530	100	M3.5	A or B	E or F	5
10,000		M4			
35,300	1000	M4.5		G or H	6
100,000		M5			
353,000	10,000	M5.5	C	J	7
1,000,000		M6			
3,350,000	100,000	M6.5	D	K	8
10,000,000		M7			

To put some of these classifications into context, Table 4 provides direct comparisons for some of the standards.

No single definitive standard or guide can be applied in designing and operating clean rooms (23). There are a number of such documents, and the supplier and user will need to develop precise specifications based on the relevant standard and the exact requirements of the facility.

Now we are seeing a concerted effort to harmonize clean room standards with the new ISO clean room standards that are being developed. However, due to the involvement of so many organizations worldwide where broad agreement from so many participating organizations is needed, harmonization is unlikely to be a simple process nor is it likely to show rapid progress.

Design and Operation of Clean Rooms

Control of Airborne-Contamination. The control of airborne contamination in a clean room is accomplished by the following means:

1. Preventing Entry: This is achieved by filtering the air entering the clean room and preventing the ingress of external air by using positive air pressure relative to the external environment.
2. Purging: The air handling system changes the air in the room at a sufficient rate to remove or dilute particulate matter generated within the room from either the process or the personnel.
3. Minimizing the generation of particulate matter: Room appointments (floors, walls, equipment etc.) are chosen for their resistance to particle generation. Clean room clothing is made of nonlinting material designed to minimize the release into the room of particles shed by personnel. (e.g., skin flakes, hair, etc.).
4. Providing localized protection for the product or process from the settling of particulate matter, from product-to-product contamination, and for the protection of staff from potential hazards.

5. Providing designated areas for personnel entry, for the entry of equipment and parts, etc., and for the cleaning of equipment and parts.
6. Controlling the flow of materials through the process steps and controlling the personnel activities.

Critical Design Parameters. A number of critical parameters that can be summarized as follows need to be addressed in any clean room facility:

total particulate cleanliness
microbiological cleanliness (where necessary)
HEPA filtration specification
room pressurization
temperature and humidity
air change rates
hazards from products and processes
disinfection and cleaning procedures

Total particulate cleanliness: This will be dictated by the nature of the work to be carried out. Conventional aseptic filling operations will require a Class 100/M3.5 room, but in most instances cell culture work will probably require only a Class 10,000/M5.5 room. Overspecifying a room needs to be avoided to prevent unnecessary construction and maintenance costs.

Microbiological cleanliness: Again this will depend on the nature of the work and the need to control this parameter. For aseptic filling operations, a comprehensive microbiological monitoring program would be needed to demonstrate compliance with appropriate standards. The nature of cell culture work probably makes this requirement unnecessary in most cases.

HEPA filtration specification: This aspect was dealt with separately earlier in the chapter. However, it should be recognized that the type of grille or diffuser on the filtered air inlet will influence air movement in a conventionally ventilated room (23). Careful consideration will need to be given to the siting of safety cabinets to avoid the disruption of air movement into the cabinet (see previous section on MSCs).

Room pressurization: Establishing pressure levels is an important requirement to prevent the ingress of external, unfiltered air into the clean room or clean zone. Depending on the size and complexity of the facility, it may be necessary to provide a cascade of differential pressures across rooms or zones within the clean room. Room pressures can also be an indicator of a number of performance characteristics such as air balance stability, HEPA filter blockage, control and fan faults, etc. It is normal to provide a pressure differential of at least 15 Pa between adjacent areas (the highest pressure is in the most critical area), although some reduction in this figure may have to be accomodated in relatively complex facilities with multiple pressure differentials to avoid excessive overpressure relative to the outside (24).

Temperature and humidity: These parameters need to be controlled to create comfortable conditions for the staff and avoid excessive generation of particles (bearing in mind that under normal conditions the personnel are usually the major source of particulate contamination in clean rooms). As a general guide, the temperature should be around 20 °C and the humidity levels not more than 50% RH although there may at times be differences to cater for specific requirements (e.g., low RH for moisture-sensitive materials).

Air change rates: This determines the quantity of air moving through the classified clean space. It is needed to provide an adequate degree of flushing of the room or zone to avoid buildup of particulate matter. In trying to determine the required air change rate, due allowance will also have to be given to compensate for internal heat gain, process exhaust (e.g., loss of air through safety cabinets), for loss due to leakage (e.g., through doors), and for the air volume required to pressurize the room. Minimum air change rates of 20 room volumes per hour are often quoted but rates of 30–35/hour are not uncommon to compensate for the parameters mentioned.

Hazards from products and processes: Many of the potential hazards that may be encountered during cell culture have been dealt with in earlier sections. Additional hazards may be posed by large-scale operations or specialized techniques, particularly those capable of producing aerosols. It is essential that a full risk assessment is carried out and that appropriate measures are put in place in the clean room to deal with identified hazards.

Disinfection and cleaning procedures: There are many factors that influence the efficacy of disinfectant procedures. The choice of cleaning agents and disinfectants, how they are prepared and used, and the environment in which they are used can have a bearing on their efficacy. There is no compound that has all the characteristics of an ideal disinfectant, so the choice for any particular situation will need to be made carefully, taking into account health and safety considerations, the risk of corrosion of treated surfaces, and the potential for residues, as well as the factors that influence efficacy (1). Choices must be made specifically for each working environment.

Process and Personnel Influence. Although the design and maintenance of a clean room is undoubtedly important, probably even more important are the activities of the personnel working within it. The clean room only provides a suitable environment within which the processes can be operated. The design and control of the processes and the personnel activities need to be given very careful consideration. Poor practices within a well designed and maintained clean room will inevitably result in contamination problems. Good techniques are crucially important, otherwise all the effort and cost expended in providing the clean room facility and associated equipment will be wasted. In some of the earlier sections we have attempted to describe some aspects of good technique as applicable to cell culture. Further details can be found in Refs. 25 and 26.

APPENDIX 1

BASIS OF PARTICULATE CLASSIFICATION SYSTEM

The basis of the original U.S. Federal Standard 209 particulate classification is the straight line relationship between the logarithm of the particle concentration plotted against the logarithm of the particle size. This can be seen on the graph shown in Figure 10 which is based on that illustrated in Federal Standard 209D.

The 0.5 μm size is the reference particle size. Extrapolating the line allows other particle sizes in the same class to be determined. Drawing the lines parallel to each other creates a logarithmic relationship for each particle size among the various classes. This provides an elegant means of constructing a particle classification system.

In Federal Standard 209E the class limits are given in counts per cubic feet, as well as counts per cubic meter, which makes interpretation easier. The standard also allows concentration limits to be calculated for intermediate classes using the following equations:

1.
$$\text{Particles/m}^3 = 10^M \left(\frac{0.5}{d}\right)^{2.2}$$

where d is the particle size in μm and M is the numerical designation of the class based on SI units (i.e., for a Class $M3.5$ room, $M = 3.5$).

2.
$$\text{Particles/ft}^3 = N_c \left(\frac{0.5}{d}\right)^{2.2}$$

where d is the particle size in μm and N_c is the numerical designation of the class based on English (U.S. customary) units (i.e., for a class 100 room, $N_c = 100$).

However, it should be recognized that this type of particle size distribution is defined for classification purposes, and it will not necessarily represent the actual size distributions found in any particular cleanroom situation. The source, type and amount of particulate matter will vary considerably from one facility to another and within an individual facility when different activities take place. Simple idealized particle size distributions with straight line relationships will not always occur in practice, a factor not always recognized by users or by regulatory authorities.

ACKNOWLEDGMENT

Extracts from B.S.5726: 1992 are reproduced with the permission of BSI under licence no. PD\ 1998 1079. Complete editions of the standard can be obtained by post from BSI Customer Services, 389 Chiswick High Road, London W4 4AL, United Kingdom.

BIBLIOGRAPHY

1. P.L. Roberts, in J.M. Davis, ed., *Basic Cell Culture; A Practical Approach*, Oxford University Press, Oxford, U.K., 1994, pp. 27–55.

2. W.A. Nelson-Rees and R.R. Flandermeyer, *Science* **191**, 96–98 (1976).

3. W.A. Nelson-Rees, D.W. Daniels, and R.R. Flandermeyer, *Science* **212**, 446–452 (1981).

4. K. Hummeller et al., *N. Eng. J. Med.* **261**, 64–68 (1959).

5. E.A. Gugel and M.E. Sanders, *N. Eng. J. Med.* **315**, 1487 (1986).

6. A.F. Hill et al., *Nature (London)* **389**, 448–450 (1997).

7. M.E. Bruce et al., *Nature (London)* **389**, 498–501 (1997).

8. J. Hodgson, *Bio Technology* **9**, 1320–1324 (1991).

9. Advisory Committee on Dangerous Pathogens, *Categorization of Biological Agents According to Hazard and Categories of Containment*, 4th edn., HM Stationery Office, London, 1995.

10. G.H. Caldwell and W. Whyte, in W. Whyte, ed., *Cleanroom Design*, Wiley, London, 1991, pp. 181–204.

11. J.A. Diamond and M. Wrighton, *Parenteral Society Tutorial No. 3, Parenteral Quality*, U.K., 1986.

12. British Standards Institute, *Microbiological Safety Cabinets*, Parts 1–4, BS 5726, British Standards Institute, London, 1992.

13. N. Foord and O.M. Lidwell, *J. Hyg.* **75**, 15–56 (1975).

14. J.A. Matthews, Thesis, Council for National Academic Awards, London, 1985.

15. D.A. Kennedy, *B. Health Saf. Soc. News.* **15**, 22–30 (1987).

16. C.H. Collins, *Laboratory-acquired Infections*, Butterworth, Sevenoaks, U.K., 1988.

17. R.P. Clark, *Lab. Prac.*, September, pp. 926–929 (1980).

18. R.P. Clark and B.J. Mullan, *J. Appl. Bacteriol.* **45**, 131–135 (1978).

19. British Standards Institute, *Environmental Cleanliness in Enclosed Spaces*, Parts 0–4, BS 5295, British Standards Institute, London, 1989.

20. R. Tetzlaff, in W.P. Olson and M.J. Groves, eds., *Aseptic Pharmaceutical Manufacturing: Technology for the 1990's*, Interpharm Press, 1987, pp. 367–401.

21. P.R. Austin, *Contam. Control* **5**, 26–32 (1966).

22. H. Heuring, *Contam. Control* **9**, 18–20 (1970).

23. G.J. Farquharson and W. Whyte, in W. Whyte, ed., *Cleanroom Design*, Wiley, London, 1991, pp. 57–84.

24. G.J. Farquharson, *Clean Rooms Int.* March/April, pp. 18–20 (1992).

25. J.M. Davis, ed., *Basic Cell Culture; A Practical Approach*, Oxford University Press, Oxford, U.K., 1994.

26. R.I. Freshney, *Culture of Animal Cells: A Manual of Basic Technique*, 3rd ed., Alan R. Liss, New York, 1994.

See also BIOREACTOR OPERATIONS — PREPARATION, STERILIZATION, CHARGING, CULTURE INITIATION AND HARVESTING; cGMP COMPLIANCE FOR PRODUCTION ROOMS IN BIOTECHNOLOGY: A CASE STUDY; CONTAMINATION DETECTION AND ELIMINATION IN PLANT CELL CULTURE; CONTAMINATION DETECTION IN ANIMAL CELL CULTURE; CONTAMINATION OF CELL CULTURES, MYCOPLASMA; STERILIZATION AND DECONTAMINATION.

BIOREACTOR CULTURE OF PLANT ORGANS

MOTOMU AKITA
Kinki University
Wakayama, Japan

OUTLINE

INTRODUCTION

Plant organ culture techniques are necessary for producing valuable compounds, such as medicines, or providing valuable seedlings, such as virus-free plants on a commercial scale. The technology for bacterial culture is also applicable to plant organ culture. However, because plant organs have several unique properties, bioreactor systems specialized for plant organ culture must be established. In this article the properties of plant organs are described from the viewpoint of applying bioreactor techniques. The possibility and potential of scale-up culture of plant organs is also described with several examples.

AIMS OF BIOREACTER CULTURE OF PLANT ORGANS

Dedifferentiated and suspended cells are easier to culture using bioreactors compared with organ cells, but plant cells lose the capability of synthesizing many kinds of metabolites under undifferentiated conditions. Although all kinds of metabolites are not always synthesized even if whole plantlets are cultured, in many cases a series of metabolic activities appear and/or increase after differentiation. Plant organs are selected as the material for bioreactor culture to produce such kinds of metabolites. If the metabolite is synthesized in the roots, normal roots or hairy roots are used because they have many advantages for bioreactor culture compared with shoots. The shape of a root is much simpler than that of a shoots. Mechanical agitation is often applicable to root culture because roots have relatively high tolerance to physical stress compared with shoots. In addition, roots have relatively high potential for regeneration from homogenates produced by vigorous agitation (see later). The metabolite content of roots possibly decreases when the shoot is regenerated (1). However many metabolites are produced only in leaf tissues, not in roots.

Bioreactors are also used for clonal propagation of valuable seedlings. Seedlings are provided mainly by shoot culture. When a root has high potential in regenerating shoots, the techniques of root culture can also be applied to mass produce of plant seedlings (2). Cultured shoots are used after acclimatization except for several kinds of plants that can be directly transplanted to the soil, for example, potato microtubers (3). In most cases, the growth and quality of shoots varies significantly in a bioreactor, but such variation can be reduced during acclimatization if the cultured plants are not severely damaged. In addition, variation during the culture can be negligible when the seedlings are commercially produced after several years of field cultivation. Because most *in vitro* derived shoots have to be acclimatized immediately after culture, the number of products is restricted by the space and equipment for acclimatization (e.g., a greenhouse). By contrast, several kinds of *in vitro* derived storage organs are free from such limitations because they can be easily stored and directly used without acclimatization.

CHARACTERISTICS OF PLANT ORGANS IN BIOREACTER SYSTEMS

Growth Limitations

Because their growth rate is so slow, long-term culture is characteristic of plant organs. In the case of hairy roots, the doubling time (T_d) is about two days. This T_d value is large compared with microorganisms, such as *Escherichia coli* (T_d = approximately 21 min at 40 °C) and *Asperigillus niger* (T_d = approximately 2 h at 30 °C). In the longest cases, up to six months of culture are required to develop storage organs.

Because shoots and roots commonly contain approximately 6 to 7% of dry matter, 60 to 70 gDW/L of cells can be packed in the vessel if the bioreactor is completely filled with organs. However, 50% of the volume in a bioreactor may have to be occupied by the medium for supplying nutrients within the culture. Therefore, 30 to 35 gDW/L can be expected as a maximum density of cultures when calculations are based on medium volume. Actually, roots can be grown in excess of 20 gDW/L in a shaking flask, but further growth is difficult. This growth limitation is caused mainly by formation of highly dense clumps and severely limited mass transfer within the culture (see clump formation). The density of the culture is expected to increase if the culture contains more dry matter such as with storage organs (dry matter content is approximately 10 to 20%).

Sensitivity to Physical Stress

Plant organs are sensitive to physical stress. The characteristic response to stress is medium browning and growth suppression. Although biosynthesis of stress-related compounds is stimulated by increased physical stress in some cases, culture efficiency is usually decreased by shear damage. For example, callus is induced on hairy roots by vigorous stirring, and secondary metabolite production declines (4). The sensitivity varies significantly between plant species and their organs. Normally the root is relatively insensitive to stress compared with shoots.

Because mechanical agitation is simple and effective in reducing mass transfer resistance in submerged systems, establishing cell lines that have high tolerance to shear stress is important in improving culture efficiency.

Clump Formation and Limitations of Mass Transfer

Clump or mat formation is one of the characteristic phenomena of plant organ culture using bioreactors. Growth is inhibited by clump formation resulting from restriction of nutrient supply, including oxygen. In submerged systems, for example, channeling of sparged air and impaired liquid mixing due to clump formation are typical and important causes of reduced growth of hairy roots (5). Resistance to oxygen transport is especially important because the oxygen requirement of cells is high, but oxygen solubility is low in the medium (1.38 mmol oxygen dissolves in 1 liter of water at 20 °C at 1 atm). Stimulating the oxygen supply has also demonstrated that growth is enhanced by aeration using oxygen-enriched air, though growth can be inhibited by toxicity when pure oxygen is used.

There are several kind of resistance to nutrient transport from the medium to cell cultures (6). The tissue itself has a series of resistances to mass transfer from the epidermis to the stele (internal mass transfer resistance). This resistance varies with the plant species and the cell line. Other types of resistance arise from clump formation. One resistance is in the liquid–solid boundary layer at the surface of each cultured organ inside the clump, and another is the resistance to convective mass transfer.

Mass transfer resistance in the boundary layer is influenced by many factors, such as the surface characteristics and size of the organs, the distance between organs, the oxygen demand of the organs, the external oxygen concentration, and the liquid properties, including the excretions from the cells. This resistance can be reduced by increasing the liquid velocity in the vicinity of the organs with sufficient agitation using a system, such as an isolated impeller (see root growth).

Clump formation also reduces the bulk liquid velocities outside the clumps (resistance to convective mass transfer). The bulk velocity of the external medium flow influences the internal liquid velocity of the clumps, and sufficient external liquid velocity is required to supply enough nutrients against the internal resistance to mass transfer. For example, Prince et al. investigated the effect of external flow velocity on internal convection using an *Allium cepa* root ball which was 3.6 cm in diameter and had 0.32 gFW/mL of packing density (7). The oxygen transfer restriction was eliminated in this small root ball when the linear liquid velocity outside the root ball was about 1 cm/s.

Air bubbles are often trapped in clumps. Liquid movement is inhibited and, in addition, the air–liquid mass transfer resistance is rises in such trapped air areas. Channeling of the air is also observed.

Resistance to mass transfer is increased by clump formation in any type of bioreactor system including nonsubmerged systems.

Heterogeneous Quality of Cultures

Plant organs, especially shoots, are commonly cultured in nonmechanically agitated bioreactors to avoid physical stress. Roots and shoots are not distributed homogeneously in the medium in such types of bioreactors (8). The specific gravity of explant depends on their species and the conditions of culture. Cultures which have high

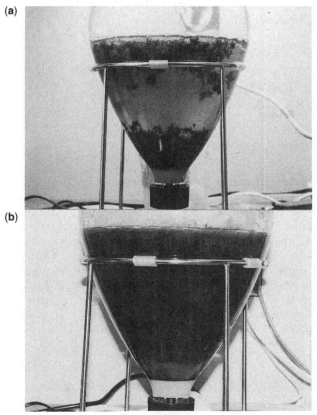

Figure 1. Growth of *Thymus vulgaris* shoots in an air-sparged bioreactor. (**a**) Distribution of shoots in the early period of culture. (**b**) The last period of culture. Vessel volume is 3-L.

specific gravity sink in the medium, but their fine shoots and/or roots float on the medium surface by trapping air bubbles. When the air flow rate is sufficiently low to prevent physical stress, two groups of the cultures are formed in the reactor (Fig. 1a). One is at the surface of the medium and another is at the bottom of the reactor. The cultures that sink to the bottom cling to each other and then develop as a large clump. Floating cultures also develop clumps, and finally the reactor is completely filled with one large clump (Fig. 1b). Cultured plants never move around and/or mix with each other after that. This indicates that the history and quality of the culture may be not homogeneous in a bioreactor even if the shapes and sizes of the cultures are similar. The quality of each root or shoot also varies within the same clump. Sufficient nutrients are supplied but physical stress is relatively high at the surface area of the clumps. In contrast, nutrient supply is limited but shear stress is low inside the clump.

When cultures are dispersed by mechanical agitation, it is possible to achieve homogeneous distribution of organs in a culture. However, the agitation must be carefully controlled to prevent damage from physical stress.

PROBLEMS OF LARGE-SCALE CULTURE OF PLANT ORGANS

Techniques for Inoculation and Transfer Between Bioreactors

Plant organs are very large, and a large mass of explants is necessary for inoculating a bioreactor because of the slow growth rate. Therefore, eliminating bacterial contamination during inoculation is quite important especially in large-scale bioreactor culture. For example, Kawamura et al. reported on an apparatus for inoculating plant organs into large-scale bioreactors that consists of an autoclavable bag and a connector that fits tightly to the hole for inoculating the bioreactor (9) (Fig. 2). More than 5 **kg** FW of plant organs cultured in flasks can be transferred to large-scale bioreactors aseptically (9–11; see also later sections).

Until now, there has been no system for transferring large cultures between bioreactors. If a cultures homogenate has sufficient potential for regeneration, it can be used as an inoculant for a bioreactor and/or for transferring cultures between bioreactors. Metabolic activities can be maintained even if the tissues are completely homogenized by strong agitation (12). The roots of some plants can be regenerated from their brief homogenates without special treatment. Briefly homogenized root was used to inoculate a trickle-bed reactor (13). Root homogenates may also be transferred between bioreactors through pipes (14). These results indicate the possibility of developing a large system for root culture using root homogenates. However, it is usually impossible to use shoots in such homogenized systems. Suspended cells which have high potential for differentiating and embryo or embryogenic callus can be used as materials for inoculation and/or transfer between bioreactors for shoot and root culture.

Mano and Matsuhashi reported that transgenic plants of horseradish have significantly high potential to be regenerated from leaf segments (15). This may

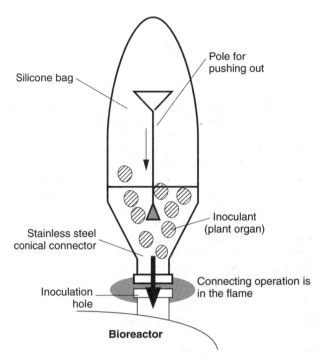

Figure 2. Schematic diagram of the apparatus for inoculating plant organs into a large scale bioreactor (modified from Ref. (10).

indicate the possibility of developing a system in which shoots proliferate after serious injuries, such as brief homogenization.

Measuring the Growth of Roots and Shoots in Bioreactors

Measuring and monitoring of the growth of organs in a bioreactor is quite important in controling the culture, but direct and representative sampling is difficult. The total weight of culture can be monitored in nonsubmerged types of bioreactors by measuring the changes in the total weight of the bioreactor, including cultured roots and shoots. This direct measurement system can also be installed on large-scale bioreactors. However, it is difficult to measure the total weight of cultures in submerged types of bioreactors. If the medium is totally removed from the culture vessel, the net weight can be measured in the same way as for nonsubmerged types. However, special equipment and techniques (e.g., a medium transfer mechanism or a reservoir) are required. The weight of the medium held in the culture clumps has to be estimated correctly in both cases.

Several methods for indirectly monitoring growth have been proposed, and they are commonly used for the bioreactor culture of plant organs. For example, Taya et al. reported that the growth of hairy root can be estimated by measuring the change of the electron conductivity (EC) of the medium as for suspended cells (16). Plants absorb ionic compounds from the medium during their growth. Ion uptake activities are highly regulated in plants, and uptake of ions depends on the demand rather than on the concentration of the medium (17). Because a decrease in the EC value of the medium reflects the absorption of the ions by the plants, the EC value is proportional to the growth of plants. This methodology is very effective, and many authors have used it for monitoring growth. For example, Muranaka et al. successfully maintained a high-density culture of the hairy root of *Duboisia leichhardtii* by using this method, and they succeeded in efficiently producing alkaloids during a long culture period (18). The EC value represents the total electrical resistance of the medium, and it can be affected by ionic compounds excreted and/or secreted from the cells, especially in prolonged culture or under stressed conditions. Measuring the consumption of specific compounds, such as ammonium ion, may be an alternative parameter for estimating growth (see also Fig. 4). Osmotic pressure of the medium may also be useful if a suitable carbon source, such as monosaccharides, is used instead of sucrose (19). Automated systems for monitoring consumption of several compounds from the medium are also proposed. Such systems are useful for selecting the most suitable parameter for estimating and controlling culture growth.

ROOT GROWING IN BIOREACTORS

Because the density of root clumps is higher than that of shoot clumps, resistance to mass transfer is significantly higher for root clumps compared with shoots. The environmental condition of the root between the surface and inner area of such dense clumps is completely different. Mass transfer is severely limited to the inner area of the clumps (see previous). Restriction of the oxygen supply influences growth and root development which are completely inhibited at low concentrations of oxygen (20). Because metabolic activity also decreases in the inner area of the clumps, metabolite productivity decreases. Therefore, one of the key techniques for bioreactor culture of roots is maintaining mass transfer and homogeneously supplying substrates, especially oxygen, to each root. Inhibition of mass transfer is thought to affect normal roots more severely than hairy roots (see growth of hairy roots).

Root Growth in Submerged Types of Bioreactors

Roots can be grown using nonagitated air-sparged types of bioreactors, but growth and metabolite productivity are decreased by clump formation. Because the root shape is simple and roots have relatively higher tolerance

to mechanical stress, they can be cultured using agitated bioreactors equipped with appropriate impeller shapes, depending on their sensitivity to the physical stress.

Mats and/or clumps of roots can be easily formed at projections, such as electrodes, sampling tubes, and ringed air spargers, even if agitated types of bioreactors are used. Once small clumps are formed, their size increases quickly because roots cling to one another. Eliminating such projections from the inner space of the bioreactor is strongly recommended to prevent clump formation. In addition, intermittent and/or reversed agitation sometimes stimulates growth because clump formation is efficiently prevented (Fig. 3). Figure 4 shows the effect of intermittent and reversed agitation on root culture using an air-sparged bioreactor. Consumption of substances from the medium was stimulated by root dispersion especially in the late stage of the culture when large root clumps were observed in the control bioreactor. Root growth was clearly stimulated by the dispersion (about 1.5 times higher than the control). Dispersion also increased the alkaloid content of the root more than twice the control (21).

Figures 5a and 6 show an example of the large-scale culture of the normal *Atropa belladonna* root (9). A 500-L bioreactor (Fig. 5a) filled with 300 L of medium was used in this experiment. Projections were totally eliminated from the inner area of the bioreactor. After the roots were transferred from shaking flasks, the cultures were slowly (7 rpm), reversely, and intermittently agitated (1 h of agitation per week) using large impellers. Whereas small root clumps (less than 5 cm in diameter) were formed, enlargement of size was completely prevented by agitation. About 35.6 kg FW of roots were produced from 2.7 kg FW of the inoculants during 56 days of culture. Table 1 shows the alkaloid content and the respiratory activity of the cultured roots. Alkaloid content and respiratory activity decreased in clumps. This result clearly indicates that roots can be cultured in a large scale using submerged systems, but root dispersion is necessary for successful culture.

Medium flow resulting from strong agitation can stimulate mass transfer inside of the root mat or clumps. When roots are protected from severe physical stress, growth is stimulated by such strong agitation. The most efficient examples reported used an isolated

(a) 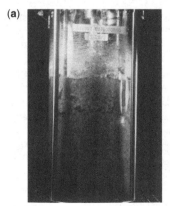 **(b)**

Figure 3. The effect of root dispersion on the growth of root cultures in a bioreactor. *Atropa belladonna* root was cultured using a newly designed bioreactor in which roots are dispersed by slow agitation. (**a**) Cultures were intermittently agitated every 3 days for 40 min at 30 rpm. (**b**) A bioreactor without agitation and baffles was used as the control. Medium volume was 8-L. Approximately 120 g roots were inoculated and cultured at 30 °C.

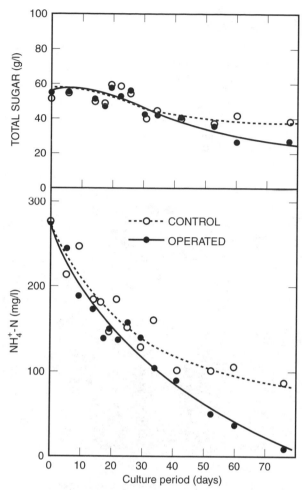

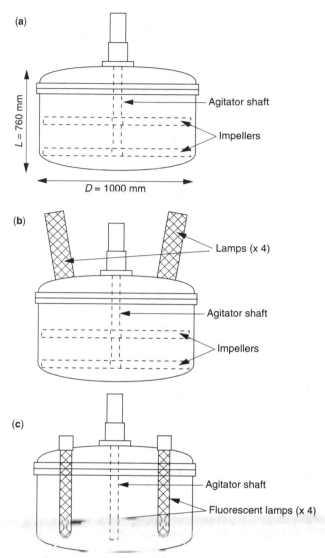

Figure 4. The effect of root dispersion on sugar and ammonium consumption in bioreactor culture. *Atropa belladonna* root was cultured using bioreactors as shown in Figure 3. Closed circles indicate the results when roots are well dispersed by slow agitation (modified from Ref. 22).

Figure 5. Diagram of a 500-L bioreactor used for plant organ culture. A porous, disk type of sparger (300 mm in diameter) was set in the bottom of the bioreactor. (**a**) For *Atropa belladonna* roots, impellers were operated at 7 rpm for 1 hr every week. (**b**) For *Stevia rebaudiana* shoots, impellers were used only for medium preparation. Four lamps were on the bioreactor. (**c**) For *Stevia rebaudiana* shoots, impellers were removed and four fluorescent lamps were inserted.

impeller system (e.g. Model TBR-2, Sakura Seiki Co., Tokyo, Japan). This system is a turbine-blade type of bioreactor in which the cultivating space is isolated from the agitation space by a mesh. The root mat is entrapped on the mesh, and then medium is passed through it by forced flow. Medium movement inside the root mat is stimulated significantly, and the mass transfer limitations are completely overcome. Though the scale-up is difficult because it is hard to provide sufficient uniform flow of medium on a large scale, the root density can be increased significantly by using this system. Kondo et al. reported on the growth stimulation in this system by comparing it with several types of bioreactors (22). Uozumi et al. successfully cultured a hairy root up to 30 gDW/L using this bioreactor (23). The capability to produce secondary metabolites is retained in a high density root mat (14 gDW/L) by using this system (18).

Mass transfer within each root in the clump was also increased by entrapping roots on a matrix made from fibers which have large pores (e.g., hollow fibers).

Root Growth in Nonsubmerged Types of Bioreactors

Several types of culture systems were proposed in which roots are not continuously submerged in the medium, for example, trickle bed (13), rotating drum (22), droplet (24), ebb and flow (25), and nutrient mist (26). Stimulation of root growth rate is sometimes expected using these systems. These systems can also be used for continuous product recovery from the medium or for experiments in plant–microbe interactions.

Growth enhancement by these types of bioreactors is caused mainly by increased oxygen supply. The root surface is in direct contact with atmospheric oxygen (approximately 20% v/v) in these systems. Although there is some mass transfer resistance at the medium boundary

Table 1. Clump Formation and Its Influence on Alkaloid Content and Respiratory Activity of *Atropa belladonna* Root Cultured in a Large Scale Bioreactor[a]

	Existence (% of total DW)	Alkaloid Content(mg/gDW)		Respiratory Activity (mg O_2/gDW/hr)
		Hyoscyamine	Scopolamine	
Dispersed	74	1.64 ± 0.006	0.400 ± 0.032	2.75
Clump	26	1.43 ± 0.054	0.233 ± 0.006	3.55([b]1.95)

[a] A 500-L scale bioreacter with 300-L of medium was used. Roots were cultured for 56 days. The total weight of the root at the end of the culture was 35.6 kgFW (about 13 times the amount of inoculant)

[b] Inner area of the clump.

Figure 6. Large-scale culture of *Atropa belladonna* roots using a 500-L bioreactor.

layer at the root surface, oxygen supply is easily stimulated in these systems, compared with submerged systems, because of the large gradient of oxygen concentration. By providing gravity flow, convective mass transfer limitations are also overcome in root clumps. However, because the density of the root mats or clumps increases, mass transfer resistance within roots also gradually increases during culture. Root dispersion is difficult in the systems, so some kind of "spacer", such as fibers of a porous material, may be necessary within the roots to stimulate mass transfer inside the clumps.

Several authors tried comparing culture efficiency with types of bioreactors (22,27,28). Their results suggest that root growth is not necessarily enhanced in the nonsubmerged systems compared with submerged and agitated systems, such as impeller isolated bioreactors. The final density of the roots in a bioreactor, for example, is clearly lower in most of the nonsubmerged systems except for ebb-and-flow or trickle bed systems (27). Growth is severely decreased by uneven wetting especially when medium dispersion is insufficient.

Nutrient supply is controlled by the medium exchange rate in this type of culture. Therefore, holdup volume in the root mats or clumps is an important factor in controlling growth. Cuello et al. analyzed the duration for gas- and liquid-phase exchange in root clumps using an ebb-and-flow system and succeeded in designing suitable operating conditions (29).

Metabolite productivity is sometimes increased using these systems. The stimulation may be caused by suitable gas conditions in the bioreactor. In submerged systems, the synthesis of some kinds of metabolites is inhibited by continuous bubbling because dissolved carbon dioxide is purged from the medium (30). Productivity of these metabolites is increased when carbon dioxide enhanced air is supplied to the bioreactor, but special equipment and/or cost are required (31). Nonsubmerged types of bioreactors may be used in such cases to improve productivity and lower the costs.

Rapid Growth of Hairy Roots in Bioreactors

Transgenic roots produced by *Agrobacterium rhizogenes* are called hairy roots because of their shape. Agrobacteria provide a valuable system for gene manipulation but the simple shape and rapid growth of the transgenic roots are also great advantages in bioengineering.

Normal roots of some plants, which are induced by treatment with auxins, can also be grown rapidly like hairy roots when their shape is maintained as "hairy". This suggests that the rapid growth of hairy roots may result from their characteristic thin and branched shape, though the transferred genes can also affect the growth. The surface area per unit of root biomass is increased in such thin and branched roots, so that nutrients are absorbed efficiently. Calluslike tissues are not observed in well-established hairy root cultures, and this is also important in maintaining a higher growth rate. When calluslike tissues are induced among the roots, mass transfer is severely inhibited. In addition, callus formation tends to be stimulated in the inner area of the root clumps where the oxygen supply is limited. Hairy roots may maintain relatively low resistance to mass transfer through the culture period, compared with the normal, callus-forming roots.

GROWTH OF SHOOTS IN BIOREACTORS

Mass propagation of shoots using bioreactors was first reported by Takayama, Misawa, and co-researchers in the 1980s using air-sparged systems. Now, the possibility of shoot mass propagation using bioreactor culture is suggested for many kinds of plants (32). Shoots have a complex shape compared with roots. They easily cling to each other and form clumps as large as the roots, though the density of the clump is usually low. The resistance to mass transfer inside of the clumps is also thought to be low. Although an increase in dissolved oxygen stimulates growth and proliferation of the shoots, the oxygen required to maintain their metabolic activities in a bioreactor may be smaller than that for the root. This may be the reason that shoots can be successfully propagated using simple systems, such as bubble column type bioreactors which have relatively low $K_L a$ values, if the shear stress is prevented (33).

Shoot Growth in Submerged Types of Bioreactors

Air-sparged, nonagitated types of submerged bioreactors are commonly used to culture shoots. If a suitable culture system is established on a small scale, scale-up to a large bioreactor is relatively easy. The same medium composition as in shake culture can be used for large-scale bioreactors. Table 2 shows an example of the effect of scale-up on the growth of *Stevia rebaudiana* shoots. When shoots were cultured in the same medium, about 90% of the shoot density was obtained using a large-scale bioreactor (500 L) compared with the shake flask.

Figures 7, 8 and 9 show examples of large-scale culture of shoots (11). A 500-L bioreactor (Figs. 5b and 7) was used to culture *Stevia rebaudiana* shoots. The aeration rate was 0.05 vvm (15 L/min of air flow for 300-L of medium). The inner surface of the bioreactor was completely smooth, and there are no space or objects to trap the shoots. When 3 kg FW of shoot primordia were transferred to the bioreactor and cultured for 1 month, 47.9 kg FW of shoots were produced. When the illuminating system was changed, as shown in Figure 5c, and sucrose was fed during culture, growth was stimulated, and 64.6 kg FW of shoots were finally propagated from 460 kg FW of primordia after 4 weeks of culture (10). In either case, although the bioreactor was completely filled with the shoots except for the region just above the sparger (Fig. 9), there was no area in which shoots died. This indicates that nutrient transfer was sufficient in every area of the

Table 2. Effect of Scale Up on the Growth of *Stevia rebaudiana* Shoots[a]

Culture Scale	Medium Volume (L)	Total Weight (g FW / culture)	Final Density (g FW/L)[b]	DW/FW (%)
Shaking Flask	0.2	35.7	179	3.7
Bioreactor (Small)	5	877	175	3.3
Bioreactor (Large)	300	47.9×10^3	160	3.5

[a] The same composition of the medium was used at every culture scale. About 10 gFW/L of shoots were inoculated and cultured for about 1 month.
[b] Total weight of shoot/medium volume.

Figure 7. Large-scale bioreactor used for culturing *Stevia rebaudiana* shoots (see also Fig. 5b).

Figure 8. Large-scale culture of *Stevia rebaudiana* shoots using a 500-L bioreactor. The bioreactor shown in Figure 7 was used.

Figure 9. Inner area of the shoot clump formed in the large-scale culture of *Stevia rebaudiana*. Shoots were removed and the inner area was shown. Shoots distributed in the inner area were etiolated.

bioreactor, even though air sparging was the only force for medium movement. There was no callus formation in this culture condition.

Variation in quality is a problem for the bioreactor culture of shoots (see also quality of cultures). Shoot quality depends on their position in the bioreactor. For example, shoots cultured just above the sparger were injured by air bubbles and acclimatization of such shoots was significantly more difficult (11). Because sparging air is the main source of physical stress in this type of bioreactor, the type of sparger and air flow rate have to be carefully designed.

Illumination influences shoot morphology and quality. When the bioreactor is completely filled with shoots, no shoots move around in the bioreactor. Light strength decreases exponentially with the distance from the lamps. Figure 10 shows an example of the effect of the variation in light strength on *Stevia rebaudiana*. When fluorescent lamps were used as the illuminator in a bioreactor, shoots distributed far from the lumps were etiolated, and leaf development was significantly suppressed (Fig. 10a). Light strength also influences the metabolite content. In this example, stevioside content varied significantly and clearly decreased in etiolated shoots (Fig. 10b).

There are two types of storage organs used in culturing. One is the storage organs, such as potato microtubers and propagules of yam, that is induced on the adventitious buds of stems. A two-step culture method which consists of a shoot proliferation step and a subsequent step for storage organ formation is commonly used for this type of plant. Another type is the organ that is formed directly on the basal tissues of the storage organ, such as lily bulblets and corms of taro. A one-step culture method can be used for this type of plant, but the two-step method often increases culture efficiency.

For the former type of plant, the physiological conditions of the shoots in the shoot proliferation step significantly influence culture efficiency. Thus, there are similar problems, such as the positional effect, as described before. For example, potato microtuber formation is severely inhibited on shoots that are previously cultured under completely submerged conditions (34). On the other hand, positioning in a bioreactor has a relatively small influence on the latter type of plant. The shapes and properties of these type of plants are not clearly changed through the culture period. An efficient illuminating system is not required because illumination has potentially inhibitory effects on the development of

(a)

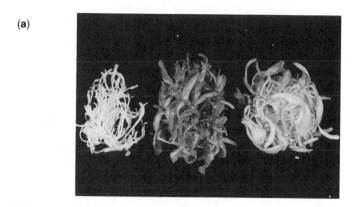

(b)

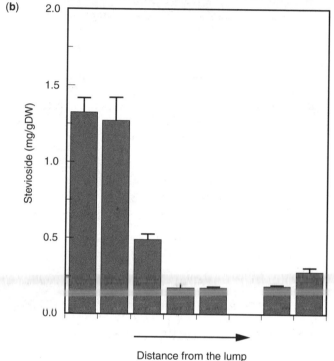

Distance from the lump

Figure 10. The effect of illumination on the development of leaves and stevioside content of *Stevia revaudiana* shoots cultured in a large-scale bioreactor. (**a**) Variation of the shape and leaf development of the shoots. (**b**) Content of stevioside. Stevioside content was drastically decreased in etiolated shoots.

storage organs. Thus, if an appropriate culture conditions are established, commercially valuable storage organs, such as virus-free bulbs, can be efficiently produced on a large scale. Takahashi et al. reported on a method for mass propagation of lily bulbs using a 2000-L bioreactor (35).

Shoot Growth in Nonsubmerged Types of Bioreactors

Recent reports indicate that the growth of shoots can also be stimulated by preventing their continuous submersion in a liquid medium. The most important factor for growth stimulation in these systems is thought to be enhanced oxygen supply, as in the case of root culture. In addition, some kinds of plants show abnormal growth in liquid medium. Medium composition affects such phenomena, but abnormal metabolites or ethylene are also probably produced when plants are continuously submerged or when the medium is continuously aerated. The strength and quality of light can be improved in nonsubmerged systems.

Weathers et al. reported that shoots of several plants are efficiently propagated in a nutrient mist culture system (36). Improved and simplified systems of nutrient mist culture have been proposed too (37). Growth can be stimulated in droplet culture systems and temporary immersion systems, such as ebb-and-flow type of bioreactors. Temporary immersion in the medium stimulated storage organ formation of potatoes (38).

Nonsubmerged systems are also useful for simplifying culture systems, and simplification is quite important for commercial propagation of plant organs.

FUTURE PROSPECTS

Research using large-scale bioreactors indicates the high potential of plant organ culture techniques for establishing efficient systems to produce valuable products. However, long-term culture is required for plant organ cultures. Because the equipment for culturing is monopolized during this period, the economical efficiency of cultures is low. If a product release system is developed, plant organs can be used as biocatalysts, and the efficiency for producing chemicals can be improved. In addition, several approaches, such as those described following, are thought necessary for improving culture efficiency.

1. Enhancement and acceleration of the growth rate by gene engineering.
2. Optimization of culture systems by improving bioreactor.
3. Simplification of the system and using low energy alternatives.
4. Utilization of characteristic plant activities, such as photosynthesis may need special attention in practical applications.

BIBLIOGRAPHY

1. T. Aoki et al., *Plant Cell Rep.* **16**, 282–286 (1997).
2. N. Uozumi, Y. Nakashimada, Y. Kato, and T. Kobayashi, *J. Ferment. Bioeng.* **74**, 21–26 (1992).
3. M. Akita and S. Takayama, *Plant Tissue Cult. Lett.* **10**, 255–259 (1993).
4. J.D. Hamill et al., *Bio/Technology* **5**, 800–804 (1987).
5. S.A. McKelvey, J.A. Gehrig, K.A. Hollar, and W.R. Curtis, *Biotechnol. Prog.* **9**, 317–322 (1993).
6. S. Yu et al., in P.M. Doran, ed., *Hairy Roots: Culture and Applications*, Harwood Academic Publishers, Amsterdam, 1997, pp. 139–150.
7. C.L. Prince, V. Bringi, and M.L. Shuler, *Biotechnol. Prog.* **7**, 195–199 (1991).
8. S. Takayama, *Abstr., 6th Int. Cong. Plant Tissue Cell Cult.*, Minnesota Univ., Minneapolis, 1996, p. 449.
9. M. Kawamura, T. Shigeoka, M. Akita, and Y. Kobayashi, *J. Ferment. Bioeng.* **82**, 618–619 (1996).
10. M. Akita, T. Shigeoka, Y. Koizumi, and M. Kawamura, *Plant Cell Rep.* **13**, 180–183 (1994).
11. M. Akita, T. Shigeoka, Y. Koizumi, and M. Kawamura, *J. SHITA* **6**, 113–121 (1994).
12. GB Pat. Appli. 2,162,537 (February 5, 1986), F. Mavitune and P.D. Williams (to Albright & Wilson Ltd.).
13. H.E. Flores and W.R. Curtis, *Proc. N.Y. Acad. Sci.* **655**, 188–209 (1992).
14. S. Takayama and N. Saiki, *Abstr., Annu. Mee. SHITA*, Mie Univ., 1993, p. 48.
15. Y. Mano and M. Matsuhashi, *Plant Cell Rep.* **14**, 370–374 (1995).
16. M. Taya et al., *J. Chem. Eng. Jpn.* **22**, 84–89 (1989).
17. J. Imsande and B. Touraine, *Plant Physiol.* **105**, 3–7 (1994).
18. T. Muranaka, H. Ohkawa, and Y. Yamada, *Appl. Microbiol. Biotechnol.* **40**, 219–223 (1993).
19. H. Tanaka, M. Uemura, Y. Kaneko, and H. Aoyagi, *J. Ferment. Bioeng.* **76**, 501–504 (1993).
20. V.P. Repunte, M. Taya, and S. Tone, *J. Chem. Eng. Jpn.* **29**, 874–880 (1996).
21. M. Akita et al., *Mem. Sch. B.O.S.T. Kinki Univ.* **1**, 30–39 (1997).
22. O. Kondo, H. Honda, M. Taya, and T. Kobayashi, *Appl. Microbiol. Biotechnol.* **32**, 291–294 (1989).
23. N. Uozumi et al., *J. Ferment. Bioeng.* **72**, 457–460 (1991).
24. P.D.G. Wilson et al., in H.J.J. Nijkamp, L.H.W. Van Der Plas, and J. Van Aartrijk, eds., *Progress in Plant Cellular and Molecular Biology*, Kluwer Academic Publishers, Dordrecht, The Netherlands, 1990, pp. 700–705.
25. J.L. Cuello, P.N. Walker, and W.R. Curtis, *Am. Soc. Agric. Eng. Int. Winter Meet.* Chicago, 1991, Pap. No. 917528.
26. A. DiIorio, R.D. Cheetham, and P.J. Weathers, *Appl. Microbiol. Biotechnol.* **37**, 457–462 (1992).
27. W.R. Curtis, *Cur. Opin. Biotechnol.* **4**, 205–210 (1993).
28. A.M. Nuutila, A.-S. Lindqvist, and V. Kauppinen, *Biotechnol. Tech.* **11**, 363–366 (1997).
29. J.L. Cuello, P.N. Walker, and W.R. Curtis, *Book Abst., Int. Symp. Plant Prod. Closed Ecosyst.*, Narita, 1996, p. 132.
30. Y. Kobayashi, H. Fukui, and M. Tabata, *Plant Cell Rep.* **9**, 496–499 (1991).
31. Y. Kobayashi et al., *Appl. Microbiol. Biotechnol.* **40**, 215–218 (1993).
32. S. Takayama, in Y.P.S. Bajaj, ed., *Biotechnology in Agriculture and Forestry*, Vol. 17 Springer-Verlag, Berlin, 1991, pp. 495–515.
33. S. Takayama and M. Misawa, *Plant Cell Physiol.* **48**, 121–125 (1981).
34. M. Akita and S. Takayama, *Plant Cell, Tissue Organ Cult.* **36**, 177–182 (1993).
35. S. Takahashi, K. Matsubara, H. Yamagata, and T. Morimoto, *Acta Hortic.* **319**, 83–88 (1992).
36. P.J. Weathers, R.D. Cheetham, and K.L. Giles, *Acta Hortic.* **230**, 39–44 (1988).
37. C. Chatterjee, M.J. Correll, P.J. Weathers, and B.E. Wyslouzil, *Biotechnol. Tech.* **11**, 155–158 (1997).
38. M. Akita and S. Takayama, *Plant Cell Rep.* **13**, 184–187 (1994).

See also ADVENTITIOUS ORGANOGENESIS; MICROPROPAGATION OF PLANTS, PRINCIPLES AND PRACTICES.

BIOREACTOR OPERATIONS—PREPARATION, STERILIZATION, CHARGING, CULTURE INITIATION AND HARVESTING

DAVID R. GRAY
Chiron Corporation
Emeryville, California

OUTLINE

Preparation
  Cleaning
  Instrumentation
  Maintenance of Valves and Changing of O-Rings and Gaskets
  Connections to Receive Charging Fluids
Sterilization
  Theory
  Principles (Unambiguous Flow of Steam)
  Monitoring and Control
Charging
  Fluid Transfers
  Volume Control (Weight, Level)
  Foam/Tangential Inlets
  Sterile Connections/Disconnections

PREPARATION

Bioreactor operations may be segregated into two different areas that constitute the overall production cycle. The first area includes operations involved with the preparation of the bioreactor system before the production phase. The second area deals with the operation and control of the bioreactor when cells are present, the production phase. This section describes operations in the first of these categories, the activities of harvesting, cleaning, sterilization and instrumenting of the reactor before inoculation. These operations encompass the physical and chemical processes that govern the cleaning, sanitization of the process contacting surfaces, sterilization of equipment and media, instrumentation, aseptic addition of sterile media components, calibration of measurement devices, preinoculation preparation, system checks for sterile operation (contamination control), and some aspects of fundamental maintenance. The sequences of events that must occur in starting up a bioreactor system are shown in Figure 1. Following cleaning to remove residual cell product and media components from the system, the equipment is sterilized using pressurized, saturated, clean steam at temperatures that inactivate any residual live bacteria. Media components, which may be degraded by exposure to high temperature, must be sterilized by filtration followed by aseptic transfer to the reactor. All sensors and probes used in monitoring or controlling the production phase must be installed, sterilized, and calibrated.

The precise methods used in bioreactor operations are a function of the scale of the reactor and the objectives of the process. The process may be required for manufacturing, development, proof of concept, or research. The scale of operation of a bioreactor is a function of the product demand (kg/yr.), process productivity (kg/m³), and the facility occupancy (fractional yr.). In process development, the scale of operation may be 1–10% of the putative manufacturing scale and may feature a flexible, highly instrumented reactor. Cell culture bioreactors range from 10 L for simple development scale operation, 100–500 L for pilot- scale operation, and may be as large as 10,000 L for manufacturing (1). Plant-cell culture reactors of 20,000 L and greater are used in tobacco cell suspension culture (2,3). Process economics strongly affects the choice of scale in commercial manufacturing and can dictate the use of high volume bioreactors for the industrial production of low-value market products.

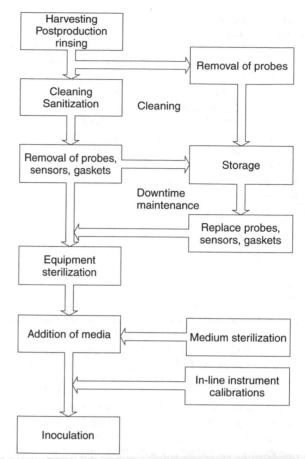

Figure 1. Bioreactor operations. Preparation of the bioreactor and associated skid components following the production phase through to inoculation. The sequence of activities are shown with reference to their location section numbers in this discussion of bioreactor operations.

Bioreactors and process vessels up to 500 L in volume can be mobile and autoclavable and are amenable to manual cleaning procedures (referred to as cleaning out of place or COP). Tanks of 1000 L and greater are normally fixed-in-place, immobile, with hard-piped systems to supply nutrients, gases, and utilities to the bioreactor skid. (Note: A skid refers to the integrated bioreactor system including the associated piping, instrumentation, and point of use utility connections). Associated process recovery equipment and utility equipment would be also fixed-in-place. Fixed piping facilities are operated without long lengths of hoses on the floor, and spills and leakage from connections are minimized or avoided. A fixed piping layout requires installation of a large number of valves and components to direct the flow of fluids. These systems require a large initial capital expense. Cleaning-in-place (CIP) procedures are required to ensure that all process components are effectively cleaned.

Consistent optimal performance in bioreactor processes requires operation within narrow ranges of environmental parameters. In this respect, bioreactors are often equipped for automatic process control. Feedback control systems require output signals from the bioreactor sensors to provide information for determining controller outputs to actuate valves, pumps, and other devices remotely. Consequently, instrumentation of the bioreactor, calibration of probes, maintenance, and procedures for testing important sensors are important set-up operations. Sterilization of the equipment to provide asepsis is a key requirement. This is achieved by using pressurized (15 psig) steam at 121 °C. The sterilization procedure is usually designed as an automatic process that uses sequentially programmed actuation of valves. For small and simple bioreactors, sterilization can be a manual or a semiautomatic process. Heat-sensitive media components are sterilized by submicron filtration. Sterilization of heat-sensitive plant tissue may be accomplished by chemical sterilization. Maintenance of asepsis requires careful control procedures following the SIP process, for example, good sterile (aseptic) technique, avoidance of low internal pressure relative to the environment, and the use of aseptic connections.

Cleaning

Cleaning is an important unit operation in the process industry. In the chemical process industries, the term cleaning refers to the removal of deposits in the form of scale or sediment following the prolonged use of such equipment. This is considered part of a scheduled maintenance program. These cleaning procedures are important in preserving and maintaining process surfaces, but may be equally important in many cases to prevent corrosion or scale buildup, which could decrease the efficiency of operation. Such cleaning may not be obligatory after every process batch and may be driven by economic factors. In the food processing industry, cleaning is absolutely required for producing clean and sterile equipment surfaces to avoid product spoilage. In the biotechnology industry, cleaning is a mandatory process which must achieve high standards of cleaning, sanitization, and reproducibility. Large-scale systems require cleaning in place (CIP) methods.

A simple cleaning procedure, after termination of a development bioreactor process, may be used when strict cleaning is not demanded. The basic procedure first involves disabling the controller alarms and information-gathering processors, such as chart recorders. Addition pumps are set to manual and output set to zero, or else switched to the off mode. The pH control is set to manual or off, and the sparger airflow rate lowered to a minimum flow to prevent debris from entering the sparger holes. The agitator should be set for general mixing during harvesting of the broth and during the rinsing and cleaning procedure. After the evacuation of the cells and product from the reactor, the inside surfaces are rinsed with DI water. After rinsing, the pH and DOT probes are removed from the vessel ports and replaced with blind plugs. The vessel is filled to operational volume with DI water, and concentrated sodium hydroxide is added to a final concentration of 0.5 N. The reactor is sanitized by using caustic for 1–4 hours at 40–60 °C. During caustic cleaning, the agitator is operated at maximum rpm, and the spargers should be operated at a low airflow rate.

For small-scale vessels, the agitator motor and external disposable tubing connections would be removed. The top

dish is opened and the internal surfaces rinsed with DI water. All O-rings are removed and discarded. Probes are removed and replaced with blind plugs. pH probes are placed in buffer. The empty vessel is rinsed with DI water several times through the top ports. The equipment can be washed with 7X detergent and finally rinsed with DI water. Spargers and small bore SS tubing are cleaned by circulating of 0.5 N NaOH for 30–60 minutes. After final DI water rinsing, the vessels and components are dried by using air.

Large-scale bioreactor systems are cleaned in place (CIP). In a CIP process all product-contacting surfaces are cleaned by circulating cleaning agents within the process equipment. Some of the equipment components may be removed and cleaned out of place. This may be required for hard to clean equipment where adequate flow of CIP cleaning reagents is not possible, (e.g., internal settlers, spargers, and threaded components).

In clinical product manufacturing processes, the cleaning procedures must be validated to demonstrate their ability to remove cellular components, products, and solutes from the equipment surfaces. This becomes increasingly stringent for biotechnological manufacturing process equipment in which multiple products are produced. Absence of batch-to-batch and product-to-product carryover must be demonstrated in clinical product production. Thus CIP is also concerned with minimizing the risk of batch carryover.

The principles and standards used for cleaning and sanitization in the biotechnology and pharmaceutical industry originated in the dairy industry (4). The fundamental principles of CIP unit operation are based on the 3-A dairy standards (5). The important keys to successful CIP are given here:

- Construct equipment with cleanable smooth surfaces. In biotechnology, 316L stainless steel (SS316L) has become the standard. Seamless tubing is preferred.
- Design piping and other ancillary process equipment to ensure direct contact with the cleaning reagents. This requires excluding hard to clean valves, dead legs, and instruments from the bioreactor skid design.
- Select cleaning reagents capable of dissolving organic residues (proteins, lipids, carbohydrates).

Figure 2 provides a view of the important elements in CIP processes.

Typical CIP cleaning agents (shown in Table 1) are caustic soda (organic dissolving power and bactericidal power), sodium carbonate (limited organic dissolving power but compatible with brass), sodium metasilicate (dispersing and rinsing), trisodium phosphate (dispersing power), nonionic detergents (wetting agents, dispersing and rinsing, organic dissolving power), sodium tripolyphosphate (sequestering power), and EDTA and sodium gluconate (sequestering power and calcium dissolving power). A discussion of CIP reagents is given in Ref. 6 for the difficult cleaning challenges experienced in the brewing industry. The choice of reagent depends on the nature of the process residue and soil. In biotechnology applications, circulation of sodium hydroxide (0.1 N) at

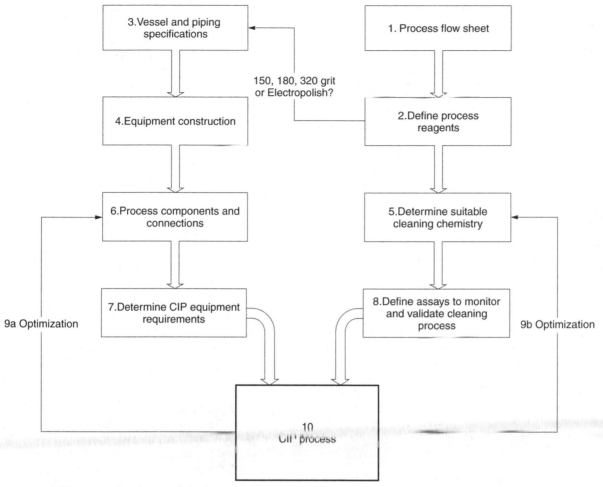

Figure 2. Fundamental elements for developing a CIP process for effective cleaning of bioreactor systems. The process flow sheet provides information for specifying of the process equipment and cleaning chemistry for the likely process residuals. Specification of electropolished surfaces may be based on the potential difficulty of cleaning the process, equipment. Numbers reflect the time sequence of the activities.

Table 1. Composition of CIP Reagents Used in Large-Scale Cell Culture Equipment Cleaning

Category	Compound	Comments
Prerinsing	Sodium phosphate buffer (50–100 mM, pH 6–8) DI water	Removal of loose protein and cellular material before cleaning
Caustic	0.1–1 N Sodium hydroxide 30–60 °C	Strong organic dissolving power
Caustic detergents	0.1–1 N Sodium hydroxide + 0.04% Sodium dodecyl benzene sulfonate (SDBS)	SDBS helps solubilize oil/lipid material and is easier to remove than SDS
Alkaline detergents	Trisodium phosphate and sodium hydroxide	Improved dispersion of soils; not used often in fermenters
Chlorinated caustic and alkaline detergents	Sodium hypochlorite (50–100 ppm Cl$_2$) with) trisodium phosphate and sodium hydroxide	Inceased cleaning and sanitization; has been used in filration membrane cleaning; requires extensive flushing with DI water to remove chlorine (can destroy oxide layer and lead to pitting)
Neutral detergents	0.1–1% Triton X-100, 0.1% Tween 80	Occasionally used as a orthogonal choice to caustic for hard to clean protein precipitates
Acids	Phosphoric acid	Used in neutralization step

50 °C is normally suitable for cleaning and may be coupled in series with a detergent (1% Triton X100 or 1% SDBS) to enhance the organic dissolving power.

CIP. CIP methods are commonplace in the food, brewing, and pharmaceutical industries. In the food industry and in particular the dairy sector, a number of guidelines were established many years ago. The foremost of these is the document entitled "3-A Accepted Practices for Permanently Installed Sanitary Products, Pipelines, and Cleaning Systems (5), which provides a basic guide for designing process systems and cleaning procedures used to clean such equipment. Discussions of CIP relating to bioreactor skid systems have described the importance of selecting effective cleaning chemistry (7), materials of construction (8,9), and sanitary valves and components (10,11).

CIP is a process for cleaning process equipment without dismantling it. CIP is the procedure by which flushing, washing, rinsing, and sanitizing solutions are brought into immediate contact with all soiled surfaces and continuously replenished. The cleaning process is essentially chemical, and in large-scale generally requires recirculating cleaning reagents to minimize water and chemical costs. CIP is the expedient choice for large-scale processing equipment because it is a reproducible, labor-saving process that provides consistency with minimal equipment damage. CIP is achieved by contacting all process surfaces with the cleaning solution at a correct flow rate. For piping, adequate flow rate and exposure to the cleaning agent is needed. If piping of several different diameters is present in the system, the larger tubing will be the path of least resistance, and special design considerations may be required.

In designing a CIP process for cleaning cell-culture equipment, some basic guidelines can be followed. The equipment and piping should be arranged to physically circulate the liquid across product contacting surfaces at velocities that promote liquid turbulence in piping and sufficient flow across all of the internal surfaces of vessels. Suitable cleaning reagents should be chosen for effectiveness in solubilizing the components of the cell membranes and product and in eliminating any potential process solutes that may reside in equipment.

The general process equipment design requirements for efficient cleaning are as follows:

- Use 316L stainless steel for all process contacting equipment surfaces (see Table 2).
- The equipment should confine the cleaning solutions.

- Use seamless SS316L tubing.
- Pitch the bottoms of flat vessels not less than 1/4″ per ft from rear to front and 1/2″ per ft from side to center to provide reasonable flow for moving of suspended solids.
- Pitch flat top surfaces approximately 1/2″ per ft from center to sidewalls to encourage the continual flow of water sprayed on the surfaces toward the sidewalls.
- A minimum radius of 1″ is desirable at all tank corners, whether vertical or horizontal.
- Equip the vessel with an adequate permanent vent to protect against all changes in pressure or vacuum resulting from the heating and cooling associated with the cleaning process or design it with appropriate vacuum and pressure ratings.
- Inert-gas fusion-welded joints are the most suitable for all permanent connections in transfer systems constructed of 316L stainless steel.
- Clamp-type joints of CIP design are acceptable for semipermanent connections. Such joints should maintain the alignment of the interconnecting fittings, have gaskets which remain flush with interior surfaces, and be designed to ensure that pressure exists on each side of the gasket to avoid product buildup that might occur in joints that are otherwise watertight.
- Pitch piping at 1/16″ to 1/8″ per ft sloping to the drain ports. The pitch must be continuous.
- Avoid dead-ends and branches in piping. Locate any required branches or tees in a horizontal position and limit them to not more than three pipe diameters long. Vertical dead-ends are undesirable because trapped air prevents cleaning solution from reaching the upper portion of the fitting.
- Use valves of diaphragm, pinch, or metal bellows sealed stem designs (12,13). Other designs such as ball valves are prone to accumulate particles and debris in the vicinity of the gasket (14).
- Construct equipment of components and materials which meet 3-A standards. Use appropriate materials such as EPDM, Teflon EPDM-backed silicone, and Teflon for gaskets.
- Polish contact surfaces to a #4 finish (RMS surface roughness factor of Ra 30–35 μinch) or better finish. This may also be specified as 150 grit (See Table 3) and for a Ra of 15–30 μinch 180 grit. Electropolishing, although not mandatory, may be a good choice to facilitate easy cleaning. Electropolishing

Table 2. Composition of Common AISI Grades of Austenitic Stainless Steels

Grade	% Carbon by weight	% Chromium by weight	% Nickel by weight	% Titanium by weight	% Molybdenum by weight
304	0.08	18–20	8–11		
304L	0.03	18–20	8–12		
321	0.08	17–19	9–12	$\geq 5\times$ Carbon	
316	0.08	16–18	10–14		2–3
316L	0.03	16–18	10–14		2–3

Table 3. Specified Stainless Steel Surface Finishes

Finish/Polish	Grit	Ra[a] (μ inch)	Appearance	Process	Uses	Relative Cost
No. 1 finish			Rough, dull, Imperfections	Hot rolled, annealed, and descaled	Not used in biotechnology	
No. 2D finish			Dull	Cold rolled, annealed, and descaled		
No. 2B finish			Bright gray	Cold rolled, annealed, descaled, and rolled through polish roll	A preliminary step to higher quality finishes; nonproduct contacting areas; media preparation tanks	1.00
No. 3 polish	80–120	40–50	Semipolished			1.04
No. 4 polish	150–180	15–30	Bright finish with grain marks; no surface imperfections		Standard 3-A polish Common dairy finish; bioreactors	1.05
No. 6 polish	240	10–15	Dull satin	Brushed in a medium of abrasives and oil	ms	1.07
No. 7 polish	320	<10	Highly reflective but some grit lines	Buffed, finely ground surfaces; a high-quality SS	Tissue cultures and QWFI system	1.08
No. 8 polish	320	<10	Mirror finish	Successive stage polishing with finer abrasives; extensive buffing with rouges to remove grit lines; high-quality SS	Bioreactors	
Electropolished	320	<10	Smooth, bright finish	Removal of a thin layer 0.01″ to 0.0001″ deep	WFI process equipment; bioreactors	1.10

[a]The average Ra is derived from readings taken at four cross sections approximately 90° apart. Ra readings taken across grain

involves polishing to at least a 180-grit finish followed by anodic electrochemical dissolution of vessel surface irregularities. The electropolishing process produces a smoother surface finish, reducing the surface roughness (Ra) to 0.18–0.22 μm (<10 μinch), and improves cleanability by removing pockets, that might retain solutes and microbes (15). Electropolishing passivates the stainless steel surfaces by forming a stable oxide layer that renders the surface resistant to corrosion. Chemical repassivation is necessary periodically using nitric acid or chelating solutions (16).

- For large-scale systems (>1000-L bioreactor) the CIP unit should operate reliably with minimal quantities of solutions in the total system to reduce water, chemical, and steam requirements and the cost of treating aqueous waste.

The process of CIP may also be assigned some basic operational guidelines to ensure success. These are as follows:

- During CIP, the equipment remains assembled.
- Piping systems can be effectively cleaned via recirculation of flush, wash, and rinse solutions at flow rates that produce a velocity of ≥5 ft.s^{-1} in the largest diameter piping in a CIP circuit. A minimum Reynolds number of 10,000 leads to good radial mixing of solutes and soil, efficient heat transfer, effective mass transfer for removal of solutes, and a level of momentum that scours the equipment surfaces.

- Tanks can be effectively cleaned by distributing flush, wash and rinse solutions on the upper surfaces at pumping rates equivalent to 2.0–2.5 gpm/ft^2 of circumference for vertical vessels or at 0.2–0.3 gpm ft^2 of internal surface for horizontal and rectangular tanks.
- Use of one or more spray balls inside vessels, to distribute the cleaning fluids evenly across all the inside surfaces of the vessel and to avoid producing a "bathtub ring" that results from simply filling and draining a vessel.
- Use multiple spray balls in tanks that contain baffles and other internals that may interfere with the CIP fluid flow.

The essential cleaning problem is typical for all vessels. Vertical sight glasses, side ports for pH, dO_2 and temperature probes are present in the vessel sidewalls and present horizontal ledges. Inlet ports exist above the liquid level for acid, alkali, and antifoam addition, that may be vulnerable to aerosol deposition and drying. Spargers for efficient oxygenation and CO_2 and ammonia addition must be CIP cleanable. Agitators are top or bottom entry designs requiring seals at the point of vessel entry. Magnetic drive agitators with simple internal ceramic seals can be used for mixing in small-scale suspension bioreactors. These mixers should be removed periodically to check and renew worn seal surfaces. Additional bearings within the vessel zone are needed for large-scale reactors with long agitator shafts. Foam at the liquid surface as a result of bubble aeration may lead to a dry biomass ring on the vessel surface. Pressure vessels are designed to operate at a maximum of 40–45 psig (150–180 °C) and are equipped with pressure rupture disks. These disks may be in a remote location inside a nozzle outlet and require periodic maintenance and cleaning if foaming occurs. The vessel top-dish may also contain entry ports for sensors such as foam probes and pressure sensors, which will be fouled by foam and aerosol-entrained debris. Such problems may require spray-ball installations to access such locations with CIP reagents.

The degree of difficulty of cleaning is a function of the quality of the equipment surfaces, sanitary design, and the type of organism or biological material that causes the soiling. A mycelial polysaccharide-type soil will be harder to clean than a yeast or bacterial soil. Caustic, detergent, and water (Reverse Osmosis (RO water), Water For Injection (WFI)) rinses are generally suitable reagents for cleaning in cell culture reactors. Occasional nitric acid rinses for repassivation are performed. Rinsing of hydroxide can be monitored by pH, and rinsing of detergent can be monitored by a shake/foam test. From the standpoint of water usage, it is more efficient to use repeated low-volume rinses versus a few large-volume rinses.

Sanitary design of equipment is a fundamental requirement. The surface texture of equipment should be 180 grit or higher for vessels and 150 grit or higher for piping. In constructing the bioreactor, the surface quality should be preserved. Following mechanical polishing, the surfaces are washed and dried, oxides are removed by passivation, and surfaces maybe electropolished. Avoidance of dead legs in piping, use of high-quality welding to avoid hard to clean pits in cracks, and the use of diaphragm valves, and regular maintenance are fundamental to CIP. With electropolished vessels, CIP prevents damage to the surface that may be inflicted during COP operations. In addition to the aforementioned principle related to process equipment and CIP flow parameters, it is necessary to consider the design of the CIP recirculation equipment. The CIP systems consist of a centrifugal recirculation pump, heat exchanger, piping, and valves and tanks for storing CIP reagents. A waste collection vessel may be present for spent cleaning reagents enabling pH adjustment prior to evacuation to a sewer. In CIP systems, the multiple tank approach involves using a single tank per CIP solution. A single tank for each reagent utilizes more floor space but enables faster cleaning of many pieces of equipment. The single-tank approach, where one tank is used for all solutions, takes up less space, provides more flexibility, and will cost less but is less economical in the use of cleaning solutions (17). In medium-scale systems, it is usual to use the bioprocess tanks as CIP storage tanks. In this mode of operation, the bioreactor is rinsed and cleaned first and is then used as the CIP storage vessel for the external system cleaning. A vessel that is used for storing CIP reagent should possess a suitably sized drain port to allow an adequate flow rate from the supply tank to the inlet of the CIP pump. An NPSH calculation indicates the maximum frictional losses tolerable.

$$\mathrm{NPSH} = P_a + P_z + P_{vp} + P_f$$

where NPSH is the net positive suction head

P_a is the atmospheric pressure
P_z is the difference in vertical elevation between the supply tank liquid level and the pump inlet
P_{vp} is the vapor pressure of the liquid
P_f is the pressure drop due to frictional losses

Minimizing frictional losses by oversizing the supply tank outlet and adequately elevating of the supply tank are important. Tanks should be fitted with a vortex breaker above the tank outlet to avoid air entrainment and pump cavitation. Avoidance of cavitation and vortex formation will enable effective tank drainage and emptying. This is most important in the final water rinsing steps.

Validation of the CIP process is required in clinical product manufacturing processes but is a useful exercise in most situations. IQ and OQ Validation involves inspecting the installation of the valves and equipment to eliminate dead legs in the assemblies and to ensure that the cleaning fluid circulates at appropriate velocity of through the system. The PQ involves measuring the process residuals after cleaning. A minimum volume of sequestering solution is circulated through the equipment to contact and solubilize any residuals. The detection of residual is by a Lowry Petersen protein assay (18), which can detect ≥ 2 µg/mL protein. A total organic carbon (TOC) assay measures low quantities (0.5 ppm) of organic carbon containing residuals and has been applied in measuring residuals following CIP (19). Direct surface swabbing, in conjunction with TOC, can be used to investigate specific harder to clean surfaces such as gaskets. Acceptance criteria for process residuals depend on the potential level likely to associate with the flow of product through the process. Typically CIP should reduce the residual protein level to <100 ppm.

SIP. The steam-in-place (SIP) process has developed from the requirement to sterilize equipment that is too bulky or heavy to be moved into autoclaves or ovens. Although sterilization may be accomplished by wet heat, dry heat or chemicals, SIP has become synonymous with "sterilize-in-place." The biotechnology industry has adopted saturated steam as the effective sterilizing agent for fixed-in-place equipment and autoclaving bioreactors and other tanks involved in parenteral filling operations (20).

Feed lines, inlet gas, exhaust lines, sensors, liquid feed filters, gas filters and vessels must be SIP compatible. A typical SIP cycle for a vessel involves several stages:

• Heat to purge (raising the temperature from ambient to ~105 °C)

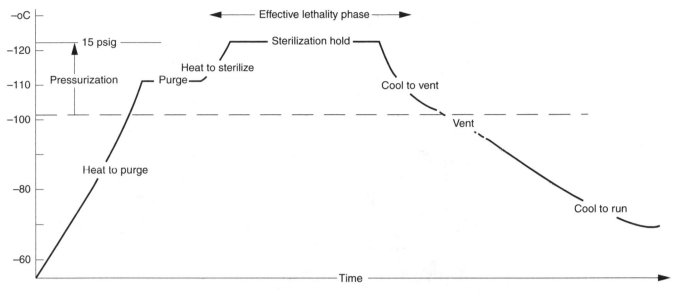

Figure 3. Typical sterilization temperature–time profile.

- Purge (eliminate air by steam at ~105 °C). Air purging in SIP must be achieved by displacing the air with steam. In the air purge and initial heating stage, steam is introduced into the top of the vessel ports and the top port of the vessel heating/cooling jacket. In the example of a simple skid in Figure 4, the vessel drain port BRHV-1 on the base of the vessel, the steam solenoid valves SV-1 through SV-4, diaphragm valves D8, D10, D12 and D14, and exhaust line pressure regulating valve NV-1 are opened. The steam line to the vessel jacket, SV6, and the steam trap solenoid valve SV-7 are open and external steam/steam condensate heats the vessel walls and minimizes condensate buildup on the interior walls. The steam displaces the air from the vessel through the harvest drain line for several minutes, and the harvest line diaphragm valve D17 is closed to direct the steam condensate to the steam trap through the open diaphragm valve 18. If the air is not effectively removed by the air purge, the vessel will develop an air barrier, which increases the heat transfer resistance from the jacket. If this occurs, the mixture of air and steam will lower the saturation pressure of the steam resulting in a sterilization temperature below the targeted value of 121 °C for a 15-psig sterilization pressure.

- Heat to sterilize (elevate temperature from 105 °C to 121 °C). In the heating period, the pressure in the system is allowed to approach the supply steam pressure (15 psig) by closing exit valves D14. Condensation is produced during the heating phase and is removed from the system through steam traps placed in drains, vents or the harvest line at the base of the vessel. Removal of condensate is very important because it may have the opportunity to supercool on the walls of the equipment.

- As the temperature increases, the steam flow through the system is reduced to the amount consumed by the traps and orifices plus the amount needed for heating. A manual steam pressure reduction valve (PRV-S) regulates steam pressure (18–25 psig).

- 'Sterilization (hold at 121 °C for a designated duration of between 15 and 60 minutes). When the sterilization temperature is reached, the time and temperature are monitored. When the sterilization hold period is completed, steam valves SV1-SV-4 are closed, shutting off the steam. Vent valve NV-1 is used to reduce the system pressure after opening valve D14.

- Cool to vent (reduce temperature from 121 °C to 100–105 °C). After shutting off the steam supply, the pressure in the system decreases as a result of steam condensation and flow through the vessel vent filter.

- Vent sterile air (open D15) is used to displace the condensing steam vapor to avoid vacuum formation and to maintain a slight internal positive pressure to prevent possible ingress of gaskets, probe components, and nonsterile air. Allow the temperature to reach <100 °C. The vessel drain valve BRHV-1 is closed, and the steam trace is initiated through solenoid valve SV-5. Residual heat in the vessel evaporates the cooling condensate as the pressure is reduced.

- Cool to run (reduce temperature to run temperature of 20–37 °C, and maintain positive pressure inside the vessel).

Dead legs are most likely to be present somewhere in hard-piped systems. They may be present in tee connections for instrument housings, valves, or short pipe lengths before isolation valves. Unlike conditions in a straight pipe or a tank, air cannot be displaced by convection or a washout process. In the worst-case analysis, we assume that the air would be eliminated by diffusion from the dead leg into the main path. Orientation

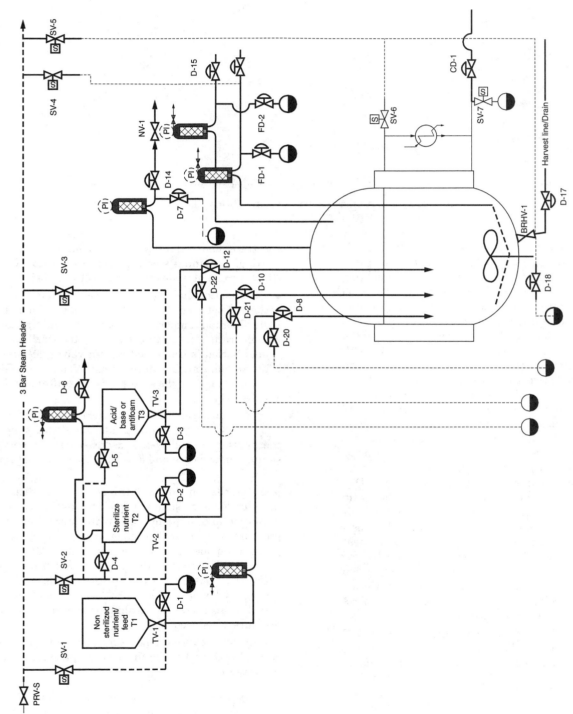

Figure 4. Bioreactor-skid pipe and valve arrangements. Representation of a fixed-in-place bioreactor showing interconnections with nutrient supply, inlet gases, and exhaust gas supplies. T1 piping is an example of filter sterilization of a nutrient to be fed to the bioreactor. T2 and T3 piping layouts are typical examples of direct addition of sterile components. TV-1, TV-2, and TV-3 are flush-mounted, steam-sterilizable, piston-design, tank harvest valves.

of the dead leg should allow condensate to drain into the sloped main path.

Autoclaves purge air through a slow vacuum cycle prior to the heating phase. Use of vacuum purging of air for large-scale process equipment is limited by the expense required to install a plant-wide vacuum system and the vacuum pressure rating of the process vessels and equipment. Bioreactors are typically built as pressure vessels rated for 3-bar internal pressure and for full internal vacuum at 150 °C. In SIP, the relative positive pressure in the vessel is always maintained. In autoclaving, it is necessary to ensure that non-pressure-rated vessels are fitted with large enough vent filters to accommodate the rate of pressure change during vacuum/steam cycling and slow vent cool down.

Very few detailed engineering descriptions of the SIP process are provided in the literature. However, a singular detailed analysis has been provided by Noble (21). The following discussion of steam-sterilization design is taken from Noble's topical paper.

Air purging. During the purge phase, air is mixed with steam by convection and is removed as an air/steam mixture. The mixing process is poorly defined because forced circulation is not present. In practice, the momentum of the in-flowing steam provides the mixing energy. The fastest removal of air can be anticipated when the agitation and flow approximate to an ideal Continuous Stirred Tank Reactor (CSTR). For this to occur, the mixing time (t_H) must be much less than the space-time (τ). For design purposes assume that

$$t_H = 0.1\tau \qquad (1)$$

For a turbulent steam jet (Re_i value of 10000–90000), t_H can be estimated from an experimental correlation given by Henzler (22):

$$t_H = \frac{C_H d_i}{v_i} \qquad (2)$$

where the subscript (i) refers to the diameter d and velocity v of the inflowing jet. The dimensionless mixing number C_H is (in the turbulent range) a function only of the geometry:

$$C_H = 1.1 \left(\frac{D}{d_i}\right)^{2.75} \qquad (3)$$

for $L = D$.

For this geometry, the space-time is defined by

$$\tau = \frac{D^3}{(v_i)d_i^2} \qquad (4)$$

In this case it is assumed that the steam is not condensing. In actual reality, the value is then larger because condensation reduces the flow. Substituting these relations equation 6, there is no ratio of d_I/D that also ensures turbulent jet flow. The relationship in equation 1 is not feasible and this infers that, in practice, good mixing, provided by steam inflow, cannot be created in tanks unless special nozzle developments can improve jet-mixing performance. In reality, the mixing time is less than the space-time so that some mixing is to be expected. From a design perspective, d_i should be minimized and v_i should be maximized to reduce processing time and create a turbulent steam jet.

At maximum steam flow, the operation can be characterized by

$$\tau = \frac{4V}{\pi d_i^2 (v_i)} = 0.0364 \frac{V}{d_i^2} \qquad (5)$$

and

$$w_s = \frac{\rho V}{\tau} \qquad (6)$$

For a 1-inch steam inlet supplying a 1000 L bioreactor

$$\tau = 56.4 \text{ S}$$

and the steam flow is given as

$$w_s = 0.1213 \text{ kg} \cdot \text{s}^{-1} = 76.6 \text{ kg} \cdot \text{hr}^{-1}$$

for saturated steam at 1-bar gauge pressure.

It is recommended in practice that the air-purge step should be allowed a factor of 4 times τ.

Air is transported down to the harvest valve at the base of the vessel air and radially to the vessel wall where it may form a boundary layer. Cold air is denser than steam (ρ air is 1.225 kg m^{-3} at STP, steam 0.013 kg m^{-3} at 121 °C). Trapped air may be found predominantly in the base of the vessel. This raises the importance of bottom vent purging.

In the equation of motion, the driving force for forced convection was described in Bird (20) as ∇P. The corresponding driving force for free convection is $\Delta\rho g$, which is related to the Grashof number. For the extreme combination of air at 25 °C and steam at 121 °C,

$$\Delta\rho g \approx 12 \text{ Nm}^3$$

whereas the average pressure gradient, assuming a 2-atm steam supply and a drain at 1 atm is

$$\nabla P \approx \frac{\Delta P}{L} \approx \frac{10^5}{5} = 2 \cdot 10^4 \text{ Nm}^3$$

Providing that the steam flow is not largely throttled during the air purge, free convection should be of minor importance, and an intact air body should not remain in the vessel. When the steam and air become mixed, a gravity-induced polarization will not be a measurable phenomenon. Gradients will be insignificant, as predicted using the relationship from Bird et al. (23, p. 576) employed for a similar situation.

$$\left(\frac{X_s}{X_{s0}}\right)^{\overline{V_a}} \left(\frac{X_{a0}}{X_a}\right)^{\overline{V_s}} = \exp\left[(\overline{V}_s M_a - \overline{V}_a M_s)\frac{gL}{RT}\right] \qquad (7)$$

After mixing, the air purging will occur when steam mixes with the air and exits through both the drain and vent lines.

Heating. Heating the vessel space by displacing colder gas and by sending heat through the vessel wall will increase the temperature of the internal surfaces of the equipment. In this situation, heat transfer resistance becomes relevant. Insulation of vessels and pipe will reduce condensate load.

Thermal response time is estimated from

$$t_T = \frac{\Delta x^2}{2\alpha} \qquad (8)$$

where α is the thermal diffusivity and the thermal resistance is calculated from

$$R_S = \frac{\Delta x}{k} \qquad (9)$$

The temperature changes are very rapid in the wall and condensate film compared to the insulation. The condensate film plays a critical role, and a change in its thickness can have a large impact on the response of the system. This is a result of its relatively low thermal diffusivity.

Sterilization Hold. In this stage, a pseudosteady state can be assumed. Temperature is constant, and concentrations do not change significantly. Because temperature nonuniformities are minimal, normal convection will be unimportant. Transport of heat and mass will depend on diffusive driving forces.

Under these conditions the heat transport can be estimated. In a pipe it is required that the pipe wall temperature be at 121 °C. For a well-insulated pipe, the heat flux becomes

$$\left(\frac{Q}{A}\right)_{INS} = \frac{(121-25)2k_{INS}}{D\ln(D+2\Delta x_{INS}/D)} \qquad (10)$$

Ignoring curvature, equation 15 gives

$$\left(\frac{Q}{A}\right)_{INS} = 344 \; W \cdot m^{-2}$$

which is calculated when $\Delta x = D$ and $D = 2.54$ cm, $K_{INS} = 0.05 W \cdot m^{-1} \cdot °C^{-1}$, $\Delta x_{INS} = 0.0254$ m.

A correlation is available from Perry and Green (24) for convective and radiative heat transfer coefficient for noninsulated and insulated steel pipe in a room at 80 °F. The result for a 1-inch diameter pipe is $16.1 \; W \cdot m^{-2} \cdot °C^{-1}$ which can be used to calculate Q/A below

$$\left(\frac{Q}{A}\right)_{bare} = h\delta T = 16.1(121-25) = 1550 \; W \cdot m^{-2} \qquad (11)$$

The fluxes correspond to the required minimum steam flow from which the expected condensate flow can be estimated. The minimum flow w_s is the entering steam flow and exiting condensate flow:

$$w_s = \left(\frac{Q}{A}\right) \cdot \frac{\pi D L}{H} \qquad (12)$$

Laminar flow will prevail at the pipe exit. At the entrance of the pipe, turbulent flow will arise if the pipe exceeds a critical length, corresponding to a value of Re = 2000 at the entrance:

$$L_{crit} = \frac{500\mu_s H}{(Q/A)} \qquad (13)$$

L_{crit} is in the range of 10–50 m for the heat flux determined in equation 10. Assuming a worst case of laminar steam flow and horizontal stratified flow of condensate, the depth of liquid (Δx_c) condensate in a round pipe is hard to calculate. Rectangular channel models have been described (25) and may provide a relevant solution, considering the low depth of condensate likely in SIP where curvature effects are not so important:

$$w_s = \frac{4a\Delta x_c^3 N}{K} \qquad (14)$$

where a is the channel width, K is a geometric constant,

$$N = \frac{\rho^2 g(\sin\beta)}{\mu c} \qquad (15)$$

and β is the angle between the pipe and the horizontal. For pipes with a diameter of 1 inch, a good approximation is $a = 10\Delta x_c$, and substituting equation 12 gives

$$\Delta x_c \propto \left[\frac{(Q/A)DL}{\sin\beta}\right]^{1/4} \qquad (16)$$

Steady-State Models and Analysis. A one-dimensional model is depicted in Figure 5 for a region at a horizontal wall. During the sterilization hold period, the largest nonuniformities in properties would be expected in this region.

Heat transfer in the condensate layer occurs only by conduction. This is described by

$$\left(\frac{Q}{A}\right) = k_c \frac{(T_{INT} - T_W)}{\delta_C} \qquad (17)$$

Assuming that the maximum allowable temperature difference is limited to 0.5 °C, a maximum condensate layer may be derived using the heat fluxes defined earlier.

$$(\delta_{C,max})_{INS} = 1.0 \cdot 10^{-3} m$$

$$(\delta_{C,max})_{bare} = 2.2 \cdot 10^{-4} m$$

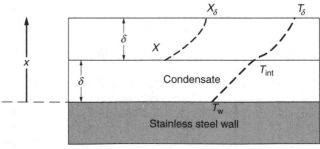

Figure 5. One-dimensional model of temperature profile for condensate layer at pipe wall. Adapted from Ref. 21.

Relating equation 17 to equation 16, a design aid for placing steam traps or orifices can be determined as

$$\Delta L = \frac{(9.28 \cdot 10^{16})k_c \sin\beta}{(Q/A)^5 D \Delta T^4} \tag{18}$$

Using a 1″ pipe sloped to 1/8″/ft as an example, the length at which the buildup of condensate creates a 0.5 °C drop is given below:

$$(\Delta L_{max})_{INS} = 110 \text{ m}$$

and

$$(\Delta L_{max})_{bare} = 0.057 \text{ m}$$

Clearly, the result shows the importance of insulating piping. A similar result is obtained by using flow relationships for round channels instead of equation 12. Because liquid flow in this model is laminar, it will not thermally mix. A temperature probe in the drain will measure the mixed condensate stream. From this average temperature measurement, it will be difficult to measure the extent of subcooling. This average temperature can be computed from

$$\langle T \rangle = \frac{\int_0^{\delta_c} T v_z dx}{\int_0^{\delta_c} v_z dx} \tag{19}$$

In a rectangular flow model, the velocity profile is well known and is given as

$$\langle T \rangle = T_{INT} - (T_{INT} - T_w)/6 \tag{20}$$

Where the cold spot is associated with T_w, the equation indicates that the temperature deviations are damped severely in the resultant T.

Gas Phase. A model for one-dimensional condensation of vapor in the presence of a noncondensable gas is well-known and used in psychrometric calculations (see Ref. 23, example 18.5-1). The basic equations to be solved are heat transport, mass transport, air properties, (assuming ideal gas), and steam properties (obtained from curve fits to the steam tables). The sum of mole fractions is limited to one, and the pressure is constrained to be constant over the boundary layer. The heat transport equation is then given by

$$-\left(\frac{Q}{A}\right) = -k\frac{dT}{dx} + N_s M_s [C_p(T - T_{INT}) + H] \tag{21}$$

where the constant N_s is the molar flux of steam to the wall. The molar flux is constant in this model and is the solution for diffusion through the stagnant film:

$$N_s = \frac{c_{ln}\wp_{as}}{\delta}\ln\left(\frac{1 - X_{s,\delta}}{1 - X_{s,INT}}\right) \tag{22}$$

Constant concentration has been assumed and set to the log mean average. The variation in concentration is normally insignificant. There are two degrees of freedom in this model. The wall temperature can be fixed at a suitable sterilization temperature (121 °C). The other variable is the amount of air in the boundary layer. A

simple analytical solution may be obtained if a value of N_s is assumed. The value is constrained between zero and $N_{s,max}$:

$$N_{s,max} = -\frac{(Q/A)}{HM_s} \tag{23}$$

which would exist when no resistance to mass transfer in the gas phase exists.

With the constant B defined as

$$B = \frac{(Q/A)}{N_s M_s C_p} + \frac{H - C_p T_{INT}}{C_p} \tag{24}$$

the integration of equation 21 can be expressed as

$$T_\delta = (T_{INT} + B)\exp\left(\frac{N_s M_s C_p \delta}{k}\right) - B \tag{25}$$

Given the temperatures at the boundaries, the partial pressures of steam can be evaluated from the following quadratic fit to the steam table over the range of 120–141 °C:

$$P_s = (8.91534 \cdot 10^5) - (1.76864 \cdot 10^4)T + 99.28T^2 [P_a] \tag{26}$$

The molar concentrations of steam can also be determined by a similar curve fit:

$$c_x = \frac{0.8892}{134.829 - 1.60466T + 0.004999T^2} (\text{kg} \cdot \text{mol} \cdot \text{m}^{-3})$$

The concentration of air can be expressed from the ideal gas law, yielding the following relationship:

$$c_{a,\delta} = \frac{(P_{s,INT} - P_{s,\delta}) + c_{a,INT}R(T_{INT} + 273.16)}{R(T_\delta + 273.16)} \tag{28}$$

Converting mole fractions to concentrations with

$$X_s = \frac{c_s}{(c_s + c_a)} \tag{29}$$

these relations can be solved simultaneously with equation 27 to find the concentrations at the two limits.

The total system pressure is then

$$P_{tot} = P_{s,INT} + 8313.7c_{a,INT}(T_{INT} + 272.16) \tag{30}$$

which is an approximate measure of the air trapped in the system. The exact fraction of the original air that is trapped can be expressed as

$$f_a = \frac{\int_0^\delta c_a dx}{\delta(P/RT)} \tag{31}$$

From equation 22, this can be derived as

$$f_a = c_{a,INT}\left[\exp\left(\frac{\delta N_s}{c_{ln}\wp_{as}}\right) - 1\right] \cdot \frac{c_{ln}\wp_{as}}{\delta N_s} \cdot \frac{298R}{1.015 \cdot 10^5} \tag{32}$$

Examining the equations, the solution space can be defined by two parameters δN_s and $(Q/A)/N_s$. As these parameters increase, the gas-phase resistance increases. Because N_s is not of particular interest, the space is shown in terms of f_a, (Q/A), and D. Under the stated assumptions, the boundary layer thickness is related simply to D by

$$\delta = \frac{D}{2} \tag{33}$$

The parameters used in the analysis are shown in the list of symbols at the end of the section and reflect steam properties at 121 °C. The diffusivity is estimated from a prediction developed by Slattery and Bird for low-pressure mixtures of steam and nonpolar gases, available in Bird et al. (23).

Figure 6 shows the magnitude of the gas-phase resistance. It is concluded that the amount of air trapped in a pipe would be low and $f_a < 0.1$. In Figure 7, it is shown that the gas-phase resistance in smaller pipes can be neglected because the air removal requirements are not large. In tanks it may be considered that gas-phase resistances may be significant, thereby resulting in nonhomogeneity of temperature. In reality, the gas layer would be close to the vertical walls. The lower temperature would be measured by an RTD thermometer at the vessel wall or else could be mapped using thermocouples attached to the vessel surfaces.

Dead legs in ports, nozzles and piping may harbor air which will cause significant gas-phase heat transfer resistance. When condensate can drain from dead legs, it can increase the mixing of steam and air through convection and aid in purging. In the worst case, dead-leg pockets will eventually reach temperature through diffusive heat transfer. Minimization of the L/D ratio at ≤ 6 is recommended to facilitate air purging and the ability to reach sterilization temperature in the dead-leg zone.

Use of double steam blocks on vessel inlet lines will enable sterilization through valves and longer lengths of piping.

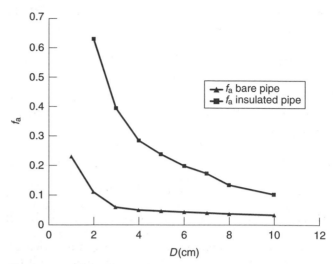

Figure 7. Limits of trapped air as a function of pipe diameter for a temperature drop of 0.5 °C. For the gas-phase temperature span of 0.5 °C, the interface temperature was fixed at 121 °C. The trapped air is expressed as f_a, which is calculated from equation 31. Adapted from Ref. 21.

Steam traps are important components in SIP and should be used in all low-point locations where condensate can accumulate.

In the final analysis steam-sterilization cycles are designed to provide significant overkill, providing greater than 30 logs of kill for a spore-forming organism. Through adherence to high standards for cleaning equipment, maintenance, and diligence in operation, the probability of contamination is virtually negligible. Small discrepancies in temperature (0.5 °C) will have no significant effect on increasing or decreasing the very low probability that a contaminant survives the SIP process. Contamination is most frequently a result of operational errors during sampling and production. Equipment related contamination is usually through waterborne sources, for example, as a result of leaking cooling coils or faulty seals and gaskets.

Instrumentation

Instruments measure the parameters in the bioreactor and provide a means for adjusting the conditions. Adjustments are through manual intervention or by automated control of the bioreactor. A variety of sensors and probes are available (see Table 4).

Instrumentation has been an important part of bioreactor operation for over thirty years. At the start of the era, in-line instruments were confined to temperature control and simple indications of DOT and pH, enabling manual adjustments of aeration and acid/base addition. At that time, sensors were leading the control envelope. Computer-assisted bioreactor control applications emerged in the mid 1960s and then increased rapidly during the 1970s. Currently, powerful, compact computer systems provide an ideal partnership for the wide variety of the instruments that are available for use in bioreactors. The early concerns for the increased risk of contamination or failure through the use of multiple

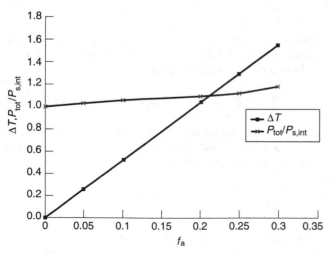

Figure 6. Magnitude of the gas-phase heat-transfer resistance. Conditions for pipe diameter $D = 1$ inch; $T_{\text{int}} = 121\,°C$; $Q/A = 1550\ \text{W·m}^{-2}$ (bare pipe). The air fraction f_a is calculated from equation 31. ΔT is the temperature span in the gas phase, and $P_{\text{tot}}/P_{\text{a,int}}$ shows the contribution of air to the total pressure. Adapted from Ref. 21.

Table 4. Sensors and Measurement Devices

Physical Parameter	Sensor	Principle	Features
Gas flowrate	Rotameter	Bouyancy position	Simple but limited use
	Mass flowmeters (Porter)	Temperature difference	Automatic, accurate, requires calibration; gas specific (C_p)
Liquid Flow Rate	Turbine, gear, and helical flowmeters (e.g., Hoffer,)	Positive displacement volume rpm	Common, simple, low–high flow rates
	Mass flowmeters (e.g., Porter)	Temperature difference	Automatic, accurate, requires calibration
	Load cells/Scales	Δ weight/time	
	Peristaltic pumps	Volume delivered/rpm	Requires calibration; tubing creep and wear can occur in prolonged use
	Platinum resistance RTD	Change in resistance with T (°C)	Reliable linear response of resistance and temperature
	Thermocouple		
Temperature	Hg-glass thermometer	Expansion of fluid with increasing temperature	Standard for cross-checking using thermowell in vessel
	Galvanic DOT in-tank probe	Electrochemical reaction at cathode	Requires maintenance, post-SIP calibration, and checking
	In-tank probe	Change of pH in bicarbonate electrolyte (combined with pH probe)	Slow response; requires post-SIP calibration and checking
DOT (dO_2)	In-tank probe	Electrochemical	Requires pre-SIP calibration; check during run
pCO_2	In-tank probe	CO_2 diffusion and pH change in bicarbonate buffer	Calibration using known pCO_2
pH	In-tank probe	Hydrogen ion passage across ceramic pore	Requires calibration before sterilization Checking required off-line

in-line devices are compensated for by the increases in productivity gained with reliable and reproducible process control. Nowadays, the use of control loops using mass spectrometer-based off-gas analysis is common in industrial manufacturing processes (26).

Data analysis, process control, and determination of physiological parameters in bioreactors depend on measuring chemical and physical variables. Measurements are obtained either by manual off-line measurements or else monitored by in-tank sensors. pH and DOT probes have been widely used in the industry for many years. Many new probe devices are available for in-tank sensing of pCO_2 and biomass, and new autosamplers can facilitate rapid in-line measurement of metabolites (glucose, glutamine) and waste products (lactate, ammonia).

Bioreactors include nozzles and ports for inserting probes and sensors and connecting lines for adding or removing liquids and gases. The number and type of ports are specified before the construction of the bioreactors. In jacketed bioreactors built as pressure vessels, the insertion of probes through the vessel wall results in a decrease in the jacketed surface area available for heat transfer. This is not a serious restriction for plant and cell culture with low metabolic heat generation compared to bacteria. Ports are placed to enable locating sensors and probes in appropriate positions in the vessel for monitoring the designated parameters. Ports designated for pH and DOT probes are located near the bottom of the tank so that the probes are submerged in media at all times during the process. This requires side entry ports on vessels 50 L or greater in volume. In fed-batch reactors, the initial volume

could be 0.2–0.6 of the total vessel volume, and the minimum liquid height should cover the probes. In small scale bioreactors (<20 L), it is possible to use a top-head plate probe insertion in combination with long probes to access the liquid media. Vessels with many internal features such as aeration tubing, baffles, agitators, spin filters, or internal settlers impose limitations on probe location.

For larger scale vessels, probes are inserted through the side of the vessel at the base of the cylindrical section. In large, tall vessels, additional probes are placed at the middle or higher section of the cylinder. The commonly used probes, pH, DOT, and newer probe types, are designed to fit into the Ingold 25-mm probe port. The guide tube is stainless steel (SS316L) and is specified as DIN 1 4435. The probes are inserted in a metal support cage that fits the 25-mm port. The probe tip is exposed to the bioreactor media and the probe stem is sealed by an O-ring. The metal sleeve is sealed in the Ingold probe holder by O-rings and is secured by a threaded, drilled cap at the end of the probe holder cylinder. A typical arrangement is shown in Figure 8.

pH. Growth and production rates may be strongly influenced by pH. Components of the media may have limited stability outside of the physiological pH range. Degradation of glutamine in cell culture media is influenced by pH with a rate constant of degradation of 2.2×10^{-4} hr^{-1} at pH 7.0 and 2.8×10^{-3} at pH 8.0 at 37 °C which is equivalent to a half-life of 7–8 days (27). Thus it is important to control the pH of the medium before inoculation, as well as during the production phase. Probes

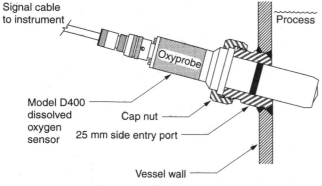

Figure 8. Vessel side port.

for measuring of pH were developed by Ingold in 1947. In-vessel, steam-sterilizable pH probes have been available for nearly 50 years. Use of an in-line pH electrode in a fermenter was reported in 1959 (28).

The glass probes used for insertion in bioreactors consist of combined reference and indicator electrodes. Probes are available 13, 16, 21, 26, 31 and 36 cm long. The range of measurement is 1–14 pH with a temperature range of −5 to 135 °C and a pressure range (in housing) of 100 psig at 25 °C and 30 psig at 135 °C. The reference electrolyte is a silver chloride/potassium chloride solution. Internal temperature correction is achieved through an integral thermistor temperature sensor. The outer glass body of the probe is constructed of potash soda lead glass. The terminal glass bulb is a proprietary silica glass made with small percentages of alkali metals and rare earth metals. The porous ceramic liquid junction, through which the process and probe are connected, is made of silica plus complex mineral compounds, fired at high temperature to form an inert, porous ceramic matrix. The internal electrolyte gel is 3.8 M potassium chloride, a cellulose thickening agent, and butylparaben (0.005%), an antifungal preservative.

A standard Ingold probe measures pH with an accuracy of 0.04 pH ±0.01 and a reproducibility of 0.01–0.02 pH units. A drift of the pH signal may occur in a solution of constant (H+) concentration but should be less than 0.02 pH units hr^{-1}. Gel electrolyte (containing cellulose in the electrolyte) probes are the current, preferred choice, when SIP or autoclaving is required. Nongel electrolyte probes require housing pressurization to avoid boiling of the electrolyte during sterilization.

Following the insertion of the probe in the vessel port, the probe is ready for SIP or autoclaving. Control of pH is accomplished through a PID controller. Ideally, pH controllers should be split-range PID loop controllers. Split-range controllers have range-specific set points and provide separate PID outputs for actuating acid and base peristaltic pumps or valves. Split-range loop controllers are suited to pH control because they enable separation of the PID constants for the acid and base additions. With standard PID controllers, it is possible to have the acid pump overshoot the set point leading to a demand for base addition. This often leads to overadjustment of pH with a consequence of increased osmolality and potentially decreased productivity. When a common PID controller is

used, a significant deadband and/or a decrease in the maximum addition rate of the acid and base can prevent the acid/base overcorrection. For large-scale reactors, acid and base may be added by using pneumatic tank pressure with actuating pulse-modulated solenoid valves.

DOT (dO$_2$ Probes). DOT is an important parameter to monitor and generally will be controlled in a bioreactor. Oxygen is required for cellular energy production and biosynthesis of product and cellular constituents. It is consumed at a rate of approximately 12×10^{-17} mol. cell^{-1}· s^{-1} (29). Oxygen is sparingly soluble in aqueous media (10 ppm or 260 μM) and may become limiting in high cell-density cultures. A high level of dissolved oxygen (hyperbaric DOT) is associated with oxygen-radical cell death (30). Two types of electrodes, galvanic and polarographic probes, are currently used for measuring dissolved oxygen. The major difference between polarographic and galvanic sensors is the source of the polarization voltage. The most commonly used probe is the polarographic probe, obtainable originally from Ingold, in which the voltage is supplied by a power supply in the amplifier. In the galvanic probe, the voltage is supplied by a separate electrochemical reaction. In the polarographic probe, the cathode is platinum and the anode is a silver/silver chloride half-cell.

Cathode reaction (Pt):

$$O_2 + 2H_2O + 2e^- \rightarrow H_2O_2 + 2OH^-$$

Anode reaction (Ag/AgCl):

$$4Ag + 4Cl^- \rightarrow 4AgCl + 4e^-$$

Overall reaction:

$$4Ag + O_2 + 2H_2O + 4Cl^- \rightarrow 4AgCl + 4OH^-$$

The overall reaction produces alkaline conditions and consumes chloride ions. Silver chloride accumulates at the anode but does not impair the electrode. The chloride ions have to be replenished through periodic addition of fresh KCl. The electrode current depends on the oxygen partial pressure, not on the oxygen solubility in the solution. For a specific solution, the O$_2$ concentration C is proportional to its partial pressure (pO$_2$) according to Henry's law:

$$C = KapO_2$$

where a = solubility factor of the solution and K is Henry's constant. The solubility factor is strongly influenced by the temperature and composition of the solution (illustrated in Table 5).

The cathode, anode, and the electrolyte of polarographic electrodes are separated from the measuring medium by a membrane permeable to oxygen but impermeable to most ions. The rate-controlling step is the diffusion of oxygen from the liquid medium through the membrane to the surface of the cathode. The current output of the probe is proportional to the activity or the partial pressure of the oxygen in the liquid medium.

Table 5. Oxygen Solubility and DOT Measurement

Medium Saturated with Air	Solubility of O_2 at 20 °C and 760 mmHg (mg O_2/L)	% DOT (100% in water at 20 °C and 760 mmHg)
Water	9.2	100
4 mol/L KCl	2.0	100
50% methanol/H_2O	21.9	100

DOT is controlled by using a staged control strategy. As the overall oxygen uptake rate increases, the DOT falls. In the first stage of DOT control, the agitation rate will be increased stepwise up to a maximum designated rpm. The next stage involves an increase in the airflow rate up to a designated maximum Lpm. The third stage involves decreasing the air flow rate and supplementing with oxygen gas up to a maximum Lpm. A fourth stage involves an increase in the bioreactor operating pressure. In cell culture, the rpm is generally fixed, and the DOT is controlled only through an increase in airflow rate and oxygen supplementation. Thus a DOT PID controller must provide output signals for rpm control and valve positioning for gas flow-rate control.

For lengthy perfusion mode operation, it is advisable to install a backup probe. In-line DOT probes may be readjusted by using off-line blood gas analyzer measurements. Daily off-line checks should be performed to monitor probe drift during culture processes.

Temperature Probes. Platinum 100-Ω resistance thermometers are used extensively in bioreactors. They can be inserted into an Ingold 25-mm standard probe holder. They may also be placed in the inlet and outlet of bioreactor cooling jackets to monitor heat removal/generation during culture operation. They need to be calibrated periodically to avoid large offsets. A temperature measurement accuracy of 0.1–0.2 °C is achievable with RTD probes. Platinum RTD temperature sensors exhibit a linear, stable resistance versus temperature relationship. They are available in sanitary designs that conform to 3-A Sanitary Standards and feature product contact surfaces designed for CIP applications. Temperature control requires PID control of cooling jacket valves. In general, cooling is more likely to be required for large-scale vessels with lower surface area to volume ratios. Small vessels operating with animal or plant cells have low metabolic heat production, and control of culture temperature is likely to require heating the jacket. Unjacketed tanks may be fitted with heating bands (31) which operate as on/off devices and can be used with a pulse-modulating on/off PID controller. (Pulse modulation involves variation of the on-time/off-time and number of events/unit time).

pCO_2. CO_2 is toxic to CHO cells at levels >105 mmHg which is evidenced by reduced growth rate and productivity (32). It is likely that aqueous dissolved CO_2 influences the internal pH of the cell (pH_i). CO_2 may accumulate to high levels in bioreactors at high cell density and where

removal of CO_2 from the broth is low due to low sparging rates and microbubble oxygenation. Absence of CO_2 or bicarbonate ions in cell culture media can also lead to repression of growth of animal cells (33). Therefore, it is important to monitor the CO_2 level in a bioreactor. Control of dCO_2 is usually not automated and is regulated through manipulation of sparging rates or vessel head-sweep airflow rate to reduce the headspace pCO_2. In sparged reactors, the total removal rate of CO_2 is limited by the maximum sparge rate that can be tolerated before foam or cell/bubble damage rates become limiting.

Cell culture media often use a bicarbonate buffer system. The reactions of the CO_2/bicarbonate buffer system are as follows:

$$CO_2 + H_2O \longleftrightarrow H_2CO_3 \text{ (at pH < 8.0 reversible, forward reaction rate is slow)}$$

$$H_2CO_3 \longleftrightarrow HCO_3^- + H^+ \text{(rapid rate)}$$

$$CO_2 + H_2O \longleftrightarrow HCO_3^- + H^+ \text{(overall reaction } K_a = 5.2 \times 10^{-7} \text{ M)}$$

If a bicarbonate-buffered medium is allowed to stand in air, the carbon dioxide will leave solution, and the pH of the medium will increase. To prevent this, media should be stored below a 5% v/v CO_2 air mixture.

CO_2 can been measured off-line using blood gas analyzers (34) or by in-line sterilizable pCO_2 probes. In-line probes were described in 1980 (35). The Severinghaus type pCO_2 probe is essentially a combination of a pH probe with a CO_2 permeable silicone membrane. Dissolved CO_2 passes through the membrane into the NaCl/NaHCO$_3$ electrolyte where it dissociates and increases the H$^+$ concentration and hence decreases the pH. Response time is relatively slow (1–2 minutes) but decreases at higher pCO_2 levels. The probes are compatible with the standard 25-mm Ingold port. The probe is SIP and autoclave compatible.

Another approach has been the use of in-liquid silicone tubing through which N_2 is purged at a low flow rate (36). A mass spectrometer is used to determine the CO_2 concentration. This approach requires inserting silicone tubing into the liquid medium by using a modified 19-mm or 25-mm DIN 1 4435 Ingold-style fitting which is drilled and fitted to provide a SS316L tube connection inside the vessel and a sanitary connector externally. The tubing is connected on the inside and sterilized with the reactor.

Biomass Probes. Biomass is most often measured off-line using turbidity or direct visual counting to determine cell concentration. In-line measurement devices are now available and have been used in mammalian-cell bioreactors (37). The back-scattering in-line probes show a linear response to cell density in the range of 2–15 million cells per mL (38). Transmission laser probe types such as the Cerex (Cerex Corp., USA), Wedgewood (Wedgewood Technologies, CA, USA) and Monitek (Monitek Technologies, CA, USA), perform well in hybridoma suspension culture (39). The probes are Ingold 25-mm type design and are SIP compatible. Another in-line laser turbidimeter (ASR Co. Ltd., Tokyo) has been used to continuously monitor microbial culture (40) and can also be applied in cell-culture reactors.

Capacitance sensors have been examined for on-line biomass determination of attachment-dependent HeLa cells on Cytodex carrier beads (41). The biomass determination is based on the capacitance imparted by cell membranes. This type of sensor can distinguish biomass from the inert carrier matrix. Concentrations in the range of 2–3 million cells per mL were measured.

Biomass probes are calibrated outside of the bioreactor by using a high-density suspension of cells. The medium is used to zero (offset) the probe and a high concentration cell sample provides the span measurement. Calibration information is cell-specific and can be stored in the controller memory.

Flow-Rate Sensors. Flow measurement devices are required to control gas and liquid flow rates. In manual systems, gas flow rate can be adjusted and monitored by using rotameters. In automated control, gas mass flowmeters are used. These meters use a thermal sensing device, which is influenced by the gas density and specific heat capacity, to determine flow rate. Mass flowmeters require periodic calibration.

Precise liquid flowmeters are needed for perfusion culture feeds. Feed rates may be as low as 5 Lpd and cell removal rates in continuous culture may be as little as 1 Lpd. Turbine positive displacement meters and mass flowmeters can be used for measuring and controlling of flow rate.

The gas flowmeters are placed before the sterilizing filters and are not sterilized. The liquid mass flowmeters may have to be placed in the sterile fluid circuit and would have to be SIP or autoclave compatible.

Nutrient/Metabolite Sensing. External off-line analyzers such as the YSI Biolyzer are used for determining of glucose, glutamine, and ammonia. In-line detection requires using an external low-volume recirculation loop with a microporous tangential cross-flow filter. External loop recirculation rates of 50–100 mL/min are used to effect permeation of the soluble media across the membrane. The clarified filtrate is sent to an aseptic sampling module and then sequentially loaded to an off-line YSI type analyzer. A 12-L perfusion system was operated with a Millipore 0.22-μm cross-flow filter in an Applikon minimum holdup volume external cross-flow holder (38). At a laboratory scale (20 L), the device can be operated aseptically in-line. The vessel/device connection is made by using autoclaved tubing which is connected aseptically to the sterilized membrane device. Connection with a large-scale SIP-based bioreactor requires producing SIP compatible sample loop devices.

Pressure. Pressure influences the DOT and pCO_2 in the medium. Bioreactors are operated with pressures in the range of nominal 0–10 psig. Inlet gas flowrate and outlet pipe diameter and filtration sizing must be compatible to prevent pressure buildup and thus enable control of pressure using a sterilizable, needle design control valve (42). Pressure should be measured in the

vessel headspace and before the exhaust filter. Pressure sensors (43) are available in Ingold style fittings or in sanitary flush-mounted Triclamp designs. A simple pressure-indicating gauge can be installed at a vessel top-plate nozzle where simple pressure indication is desired. This may be useful in addition to a sensor for local and immediate indication for operators. Sanitary 3-A design gauges with flush-mounted designs are available (44).

Liquid Level. Liquid Level probes are used to measure liquid level and provide the sensor to control the bioreactor medium level. Control of medium level is important in perfusion-based cultures where a constant bioreactor volume is desired. Long batch runs should also be monitored for liquid level to provide a measurement for replacing evaporative loss or for monitoring vessel volume during fed-batch bioreactor processes.

The following approaches are available:

- determination of bioreactor weight using a load cell (45) for large reactors or floor-mounted scales for smaller reactors (46)
- magnetostrictive sensors (47)
- capacitance sensors (48)

Capacitance sensors measure the change in dielectric constant between the vessel wall and the probe and detect the liquid/air interface. They are fitted through the top-plate of the vessel through a 19-mm Ingold type or sanitary clamp fitting. They are steam-sterilizable and are fitted before SIP.

Others. Process control loops require actuators to respond to output signals from the controller. Peristaltic pumps can be activated and regulated by external signals. The flow rate must be calibrated with rpm to enable precise tuning of the control loop. Remote actuating valves and metering pumps play some part in large-scale process control loops. In regulating the gas flow rate, proportional valves are placed outside the sterile boundary with a sterile air filter between the nonsterilizable device and the bioreactor. A representation of a basic controlled-perfusion bioreactor is shown in Figure 9.

The level of sophistication of bioreactor sensors and instrumentation depends on goal. In manufacturing, the trend is more toward conservatism and minimal failure risk whereas in research, maximization of the information accrual rate dominates the strategy.

Maintenance of Valves and Changing of O-Rings and Gaskets

Valves are an essential element of any bioreactor system. They divert fluids through fixed piping manifolds and delivery lines in bioreactor skids. They isolate fluids from product and equipment in the production phase and are fundamental in ensuring the direction of steam, nutrients, gases, and product during the phases of the production cycle. Valves that are part of the primary product contacting system and the sterilizable zones of the bioreactor should comply with sanitary design standards. Several valve designs are used in a bioreactor system.

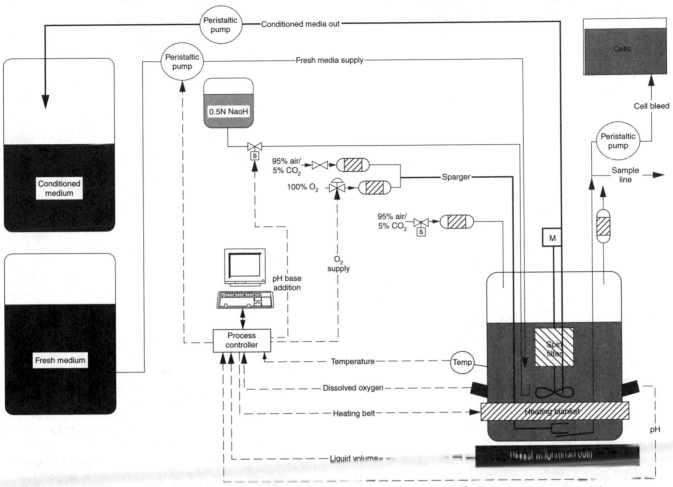

Figure 9. A perfusion bioreactor system. Liquid level control using a vessel load cell and regulation of a medium inlet pump.

Diaphragm valves (49) are the most usual choice for on/off selection in the sanitary flow path. Ball valves are available in SS316L three-piece designs (50) and are used as absolute shut-off valves for utilities and to a smaller extent in the sanitary flow path. Simple plug valves (51) are useful in small diameter utility line applications and can operate with steam but are not regarded as a sanitary design. Several specialized valves are used for sampling, harvesting, and in automated CIP skids. Aseptic-designed tank harvest valves (52) have a steamable piston design and provide a flush-mounted, drainable face inside the vessel. The piston design produces a steam barrier while the valve is closed. When the valve is opened, the steam supply to the valve stem is cut off, and samples or product can flow out of the vessel. Sample valves are similar in design to harvest valves but are located in the vessel sidewall.

Diaphragm valves feature a smooth, crevice-free flow path free of any entrapment areas, and a self-draining design which is the key choice for SIP/CIP applications. The isolating diaphragm protects flow medium from external contamination while containing the medium within the valve bore.

All moving parts are isolated from contacting the flow medium. In addition to the standard two-way forged and cast valve bodies, the manufacturers of sanitary diaphragm valves offer many important process-related features:

- Multiport diverter valves divert or combine process streams.
- Close-coupled branch valves minimize the dead leg on sampling valve lines.
- Tee pattern valves provide excellent drainability and minimal deadleg for process sampling.
- The standard forged valve body provides the highest available surface finish and the optimum surface for electropolishing. Because these bodies are forged from wrought stainless steel bar (ASTM A-182 grade 316L), they have minimum porosity and low ferrite content (0.5% maximum). As a result, the migration of oxides throughout the system is minimized. These bodies comply with FDA guidelines as well as cGMP principles and are ideal for high purity applications in the bioprocessing industry.

In addition to the stainless steel valve body, the diaphragm contacts the process stream and/or utilities. A variety of elastomers are available for bioreactor use in valves and as gaskets. The major elastomers used are given in Table 6. In bioproduct processes it may be important to pick an elastomer that possesses the right physical properties for

Table 6. O-Ring and Gasket Elastomers

Elastomer Valve Diaphragms and Gaskets	Color	Temperature Range (°F)		Regulatory Acceptance	
		Lower	Upper	FDA	USP
Butyl rubber	Black	−22	266	✓	✓
Black EPDM Peroxide-cured	Black	−40	266	✓	✓
Black EPDM Post peroxide-cured	Black	−40	266	✓	✓
Clack EPDM sulfur-cured	Black	−40	266	✓	
White EPDM sulfur-cured	White	−40	248	✓	
TFE butyl-backed	White face Black backing	−4	302	✓	✓
TFE Viton-backed	White face Black backing	23	347		
Steam grade TFE Viton-backed	White face Black backing	23	347		
TFE EPDM-backed	White face Black backing	−4	320	✓	✓
Sream grade TFE EPDM-Backed	White face Black backing	−4	320	✓	✓
Silicone FDA	White	−40	302	✓	
White Butyl FDA	White	−40	230	✓	

the application and is also acceptable to the FDA for safety and nonleachables.

- EPDM is available for valve diaphragms and flat gaskets, but not in an O-ring format without the use of additives that render it unacceptable by the FDA for use in product-contacting applications in biopharmaceutical production. In this case EPR O-rings are used. An important consideration is the potential for toxicity of the materials to the cells in the bioreactor. Some materials such as PVC may be toxic to the cells in the reactor. Tests can be performed in shake flasks to evaluate the toxicity of particular polymeric elastomers. It has been shown that silicone, C-flex, and Teflon are noninhibitory to the growth of hybridomas (53).

- Teflon/EPDM-backed diaphragms are used in Saunders-type diaphragm valves for product-contacting vessels and lines.

- Buna-N material will handle most food, dairy, pharmaceutical, and sanitary services. It is the backbone of elastomer for the food and edibles industries and has excellent resistance to compression set, tear, and abrasion. It has good acid and mild alkali resistance.

- EPDM (ethylene propylene rubber) is excellent for hot water and steam service up to 325 °F, and in addition it is very abrasion resistant and has excellent resistance to ozone, sunlight, weather, and deionized water. EPDM also has good tensile strength and good resistance to mild acids, alkalis, and alcohols (rated −65 °F to 350 °F and short term to 400 °F).

- Viton (fluorocarbon rubber) material has excellent mechanical, chemical, and heat resistance. It is particularly well suited for hot, fatty, and oil products. Viton has poor serviceability in steam contact application. EPDM is favored, for steam service. Viton is especially good for hard vacuum service because of its high molecular weight and low gas permeability. It has been used to −65 °F in some static seals due to its high flexibility (temperature ranges 0 °F to 400 °F under continuous duty and will take 600 °F for short periods of time).

- Silicone is known for its purity and nonleaching characteristics. Its ability to withstand many chemicals and combination of chemicals makes it a good choice in the pharmaceutical industry. Silicone has excellent low-temperature flexibility.

Harvest valves should be steam-sterilizable, piston-type valves with O-ring seals and should be flush-mounted to provide a smooth join with the vessel interior surface to eliminate dead pockets. In several designs, the valve housing can be pressurized with steam between operations to reduce the risk of contamination. A harvest valve should be sized to allow harvesting of the bioreactor and recirculation of CIP fluids.

In skid systems, there are many valves and gaskets which must be checked and installed before each batch run. EPDM-type gaskets, present in utility lines, are likely to be stable for months and are replaced during planned maintenance activities. The vessel and cell-culture-contacting gasket materials should be carefully inspected after each run. Flat Teflon gaskets are normally replaced after each process because they are prone to cold flow and deformation. Silicone rubber O-rings can be used for several cycles without problems and simple inspection and replacement are usual rather than regular replacement. After reinstalling gaskets, a low-pressure leak test will detect poorly seated gaskets and O-rings.

When replacing O-rings, it is important to clean the surfaces to eliminate small debris such as gasket particulates. Where circular O-rings are required to fit inside a tight sleeve, it is necessary to place a small amount of silicone gel on the O-ring to prevent any damage. If O-ring damage is persistent, then steel surfaces should be inspected. Any burrs or aberrations should be carefully machined away. Probe holders and other O-ring sealing systems are engineered to provide suitable seats and grooves. For O-rings, straight-sided grooves are best to prevent extrusion or nibbling, but 5° sloping sides are easier to machine and are suitable for very high pressure (15,000 psig.). The sides should be finished to a 32 μinch RMS (root mean square) with no burrs, nicks, or scratches. Any rubbing surfaces should be 8–16 μinch RMS.

When replacing flat gaskets, for example, sight glass gaskets, take care to prevent crimping the gasket. Tighten faces evenly by working with geometrically diapositioned nuts in a sequenced tightening procedure. Use a torque wrench that is correctly adjusted when it is stipulated that nuts are to be tightened to a set torque value.

Connections to Receive Charging Fluids

Several line connections are required for autoclaved and mobile vessels, after sterilization. For acid, base, antifoam, nutrients, inlet gases, and inoculation. Pilot-scale bioreactors may contain components that are disposable or are assembled using sterilized-silicone tubing. Assembly of components is specific to the vessel's particular design. In general, flexible silicone tubing can be used to link CO_2, air, oxygen, nutrient, and some utility lines to the reactor. These components need to be sterilizable by autoclaving because they will not tolerate steam under pressure in an SIP process. Tubing connections can be made to attach vessel stainless steel ferrules or dip tubes. These connections are secured by attaching external Tygon ties. The tubing is assembled and connected to the vessel before autoclaving. Sterilizing gas filters and liquid filters used in operating the bioreactor are also attached to the appropriate tubing and autoclaved. Final connections are made to the utility lines outside of the sterile boundary, for example, between the air filter inlet and the air utility line. When direct sterile tubing connections are to be made, a hot tubing weld must be used. This is accomplished by a sterile-tubing welder (Sterile Connection Devices-SCD) (54). The SCD device makes a sterile heat-fused connection of C-flex tubing. A vessel can be autoclaved with tubing, fitted with a terminal piece of C-flex tubing and a guarding sterile vent filter, and attached to various connections. A remote, sterile, C-flex capped line may be fused to the vessel C-flex line to form a complete assembly connection. This can be used for small-scale vessels on small tubing (up to size 16).

Larger bioreactor skid designs are made up of numerous connections for transferring the liquid medium, alkali and acid for pH control, antifoam, sterile quality air, carbon dioxide, and oxygen. Peripheral hard-piped connections, to be made within the sterile zone of the skid, must be from a sanitary design and must be steam-sterilizable. Instrument air, indirect utility lines, heating/cooling jacket water, and other non-process-contacting materials are also associated with the skid. Non-process-contacting connections are usually made with typical threaded pipe connections, swaged connections, or with plastic PVC glued pipe of ferrule or threaded design. Deionized water lines are usually constructed of schedule 80 plastic PVC (1–2″ pipe) which delivers the deionized water to tanks where it is batch-sterilized or else transferred to a stainless steel piped sterilizable filter unit to be sterile-filtered into aseptic equipment. Teflon-lined flexible hose with sanitary connectors can be used to transfer products, nutrients and other sterile components between equipment. Utility lines may be joined to process piping through quick-coupling fittings.

Sterile connectors are available in various forms, the most common of which is the Tri-clamp fitting. Pressure gauges, valves, pumps and other associated important skid equipment are available with sanitary design connections. The connectors most commonly used are described here and are illustrated in Figure 10.

Tri-Clamp. The Tri-clamp (TC) type fitting consist of a grooved, 316L stainless flange, an annular elastomeric gasket with a built-in O-ring that fits the flange groove, and a three-piece hinged 304 or 316 stainless clamp that

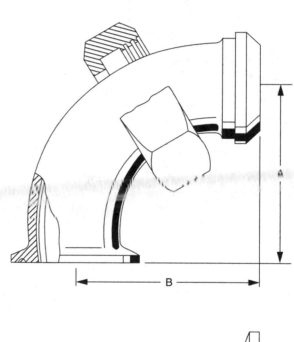

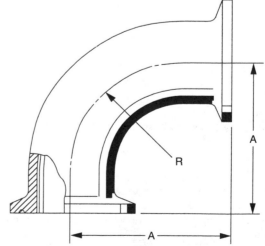

Figure 10. Sanitary design connectors.

fits around the assembled flanges and gasket. The fitting is self-aligning and easy to assemble or disassemble and in the 1.5–4 inch pipe diameter range provides a flush inside surface suitable for CIP. The fractional 0.5 and 0.75 inch fittings are useful in assembling small, fixed, stainless tubing. However, it is advisable to fit the 0.75-inch gasket into a 0.5-inch ferrule to facilitate CIP reagent access to the flat face of the ferrule (similarly fit a 1-inch ferrule with a 1.5-inch gasket).

Bevel Seat. Bevel seat fittings employ union nut assembly and Acme sanitary threads for connection (55). This fitting is more difficult to align and assemble than the Tri-clamp, and it is important to check the tightening of the exterior nut in lengths of pipe subject to vibration. The fitting is suitable for CIP, and the alignment and vibrational problems may be alleviated through correct use of pipe supports.

STERILIZATION

Sterilization inactivates potential contaminants and provides asepsis (the absence of bacterial contamination). Sterilization processes are designed to reduce the probability of survival of a single bacterial spore to less than 10^{-3}. The ability to achieve the low probability ($p \leq 0.001$) of contamination is met by the steam-sterilization procedure. In addition to wet heat sterilization with pressurized steam, dry heat (1800 °C) for small equipment components, hot air (170 °C for 2 h), radiation (γ rays, X rays, high-energy electron beams,) for serum medium, and nonionizing radiation (UV rays) for surface sterilization are potentially useful. Where high temperature sterilization is not feasible, such as where the vegetative organism surfaces must be sterilized (seeds, plant materials), chemical sterilization methods are required.

Empty vessel sterilization is used when cell culture media, containing temperature-sensitive nutrients and potentially cross-reactive components (glucose and amino acids). In this situation, the presterilization microbial loads are lower than situations in which an unsterilized medium is present in the bioreactor before sterilization.

Vessels which are in the 10 to 100-L range may fit into autoclaves and hence undergo programmed sterilization using a slow vacuum/steam cycle heat-up, a 60-minute sterilization cycle at 121 °C, and a slow venting cool-down cycle. The vessel probes, addition line connectors, or silicone tube lines are unclamped and open but protected by a terminal, 47-mm, 0.45-μm gas-sterilizing grade disc filter.

SIP is the method for sterilizing vessels that are skid-mounted.

Theory

The destruction of microorganisms by heat implies loss of viability. The destruction of organisms by heat at a constant temperature follows a first-order rate equation given by

$$\frac{dN}{dt} = -kN$$

where k = Reaction rate constant, min^{-1}, $k = f(T)$ and N = number of viable organisms.

$$N = N_0 e^{-kt}$$

t = time (min)

Decimal Reduction Time (D). D is the time (min) of exposure to heat to reduce viable microbes by 90% at temperature T.

$$\frac{N}{N_0} = 0.1 = e^{-kD}$$

Thus

$$D = \frac{2.303}{k}$$

Spores are more resistant to thermal inactivation than vegetative organisms. The k value for spores is less than for vegetative cells. The rate constant k is a strong function of temperature. At 121 °C, the value of k for *Bacillus stearothermophilus* is around 0.77 min^{-1} (56). The D value would then be calculated as 3.0 minutes. In other words, a log reduction in viable *Bacillus stearothermophilus* cell number would be obtained every 3 minutes during the sterilization hold period at 121 °C. By contrast, the k values for heat intolerant organisms like *Escherichia coli* are very large at 121 °C. The D values for *E. coli* RV308 in the temperature range of 56–62 °C are shown in Table 7. The k values from this data are 0.75 min^{-1} at 56 °C and 12.6 min^{-1} at 62 °C, indicating a very steep curve for k versus temperature and clearly showing that such organisms are undetectable at 121 °C.

The rate constant k is a function of temperature, and an Arhenius type expression can be used as follows:

$$k = ae^{-E/RT}$$

where a = empirical constant, min^{-1}
  T = absolute temperature, K
  E = activation energy, $\text{cal} \cdot \text{g}^{-1} \cdot \text{mole}^{-1}$.
  R is the gas constant = 1.98 $\text{cal} \cdot \text{g}^{-1} \cdot \text{mole}^{-1} \cdot \text{K}^{-1}$

On the basis of Eyring's theory of absolute reaction rate;

$$k = gTe^{\frac{-\Delta H}{RT}} e^{R\frac{\Delta S}{}}$$

where

  g = factor including Boltzmann constant and Planck's constant
  ΔH = heat of reaction activation
  ΔS = entropy change of activation

Table 7. Heat Kill D values for *E. Coli* RV308

Temperature (°C)	D Value (s)
56	185
58	63
60	26
62	11

In 1921 Bieglow published the Q_{10} theory (57) which said that

$$D = ae^{-b'}$$

where a' and b' are empirical constants.
Thus the reaction rate constant can be calculated.

Z-Value. The Z-value represents the number of degrees of temperature required to change the D-value by a factor of 10. The Z-value is defined as the slope of log D versus temperature. This is a straight-line graph.

$$\frac{\Delta \log_{10} D}{\Delta T} = \frac{1}{Z}$$

The D values determined for the various temperatures can then be used to calculate the Z-value for a sterilization process. Knowing the Z-value allows determining lethality at a particular temperature compared to a known lethality delivered at a reference temperature:

$$F_R = [\Delta T] \cdot 10^{\frac{(T-R)}{Z}}$$

where

R = reference temperature having a known D-value (lethality)
T = temperature
F_R = the equivalent time at the reference temperature required to produce the level of lethality delivered at temperature T.

Dividing F_R by D_R gives the log reduction achieved at temperature T per unit time.

Del Value (∇). The Del value is frequently used in analyzing sterilization. The heating up and cooling down phases provide a finite kill effect of contaminant in addition to the sterilization hold phase. (The Del value accounts for this effect and can be described as the $\log_e$ of the population survival ratio throughout the whole sterilization cycle.)

$$\ln \frac{N}{N_0} = -Ae^{-E/RT \cdot t}$$

$$\ln \frac{N}{N_0} = \nabla$$

and,

$$\nabla = A \int_0^T e^{-E/RT} dT$$

Note that

$$N = \frac{N_0}{e^\nabla}$$

For a complete cycle, the overall Del value is determined according to the following equations:

$$\nabla_{TOTAL} = \ln \frac{N_0}{N} = \nabla_{HEATING} + \nabla_{HOLDING} + \nabla_{COOLING}$$

$$\nabla_{TOTAL} = a \int_T^{121} e^{-E/RT} dT + ae^{-E/RT} T_{121} + a \int_{121}^T e^{-E/RT} dT$$

In practice, the heating and cooling curves are integrated, and Del may be determined. Del factors represent the number of $\log_e$ (or $\log_{10} / 2.303$) reductions in population throughout the whole sterilization cycle.

Principles (Unambiguous Flow of Steam)

The multiple addition lines and integral components that must be sterilized to ensure aseptic operation during the production phase complicate sterilization of fixed-in-place systems. It is important to design a sterilization cycle to ensure that saturated steam at 121 °C contacts all of the equipment surfaces and areas for a specific duration to guarantee achieving design ∇ factor at all locations. It is not possible to sterilize all skid components concurrently because steam would be required to flow in multiple directions and would be difficult to direct. Sequential sterilization of systems is likely to be needed. Many valves will be needed to enable the flow of steam in the correct direction. The commonly used steam block valve arrangement is a key element in SIP. Steam blocks may be unidirectional or bidirectional (Fig. 11).

During the heat-up phase, condensate is produced and must be eliminated from the system to prevent cool spots and hindering steam flow. Steam traps are used at suitable points in the system, and piping is arranged with slopes to enable condensate drainage to the steam traps. Vents are placed at equipment high points to accelerate air displacement by steam. Steam supply pressure should be regulated to 20–30 psig to ensure that steam can enter sterilizing areas at 15-psig pressure when pulse control valves are opened.

Manual. Manual sterilization involves sequential operation of steam supply, steam trap bypass, and vessel-associated valves by an operator. The following simplified operations are the basic elements for achieving empty vessel sterilization.

- Open vessel drain valve
- Open gas exhaust line valves (if pressure-regulating valve is in-line, open the valve manually at the controller).
- Open all steam trap valves, and open steam trap bypass lines.
- Open up steam lines to vessel inlet nozzles (gas inlets, nutrient, acid/base) and the vessel jacket top valve.
- Close the exhaust line valve and steam trap bypass lines when steam is exiting the steam trap bypass lines and the vent lines.
- Ensure that steam is flowing into the vessel and is exiting through the drain-line steam trap and the exhaust filter line steam trap. Check that the individual TD steam traps are functioning correctly.
- Monitor the temperature during heat-up (use an external thermowell thermocouple or thermometer).

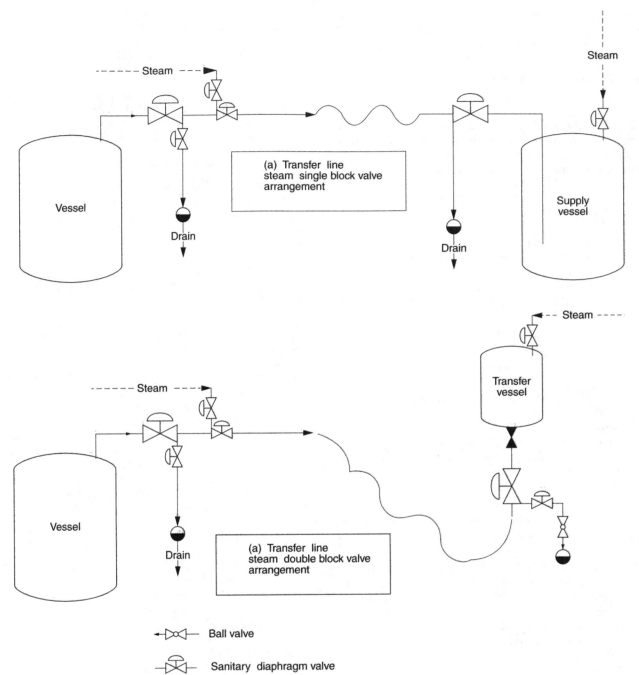

Figure 11. Steam block assemblies. Single block assembly (**a**) for sterilization of a media addition line. Double block assembly (**b**) for sterilization of a transfer line pre transfer and tank and transfer line post-transfer.

- After 5 minutes at 100 °C close the exhaust line valve, and regulate the inlet steam supply pressure at 15–20 psig at the steam PRV valve.
- As the vessel temperature approaches 121 °C regulate the supply steam pressure to 15 psig.
- Sterilize for 30–60 minutes for empty vessel sterilization.
- After sterilization shut off the steam supply, and allow the vessel to cool to 110 °C. If there is a cooling jacket start the flow of cooling water through the

jacket. Open the exhaust valve, and close the harvest valves.
- When the vessel temperature reaches 105–110 °C, flow sterile air into the vessel to maintain 5–10 psig pressure during cool down.
- Cool the vessel to the desired run temperature using the jacket temperature control loop.

Automatic. Automatic SIP is most commonly used on skid systems where many valves are in the sterilizable zone. A sequential program that automatically opens and closes valves during the progressive stages of the SIP cycle drives the sterilization cycle. Purging of air from the system can be accelerated by incorporating steam trap bypass lines that include a pulse open/closed solenoid valve. This is a suitable feature for automated systems. The pulsed valves open in the early heat-up phase and then close as the air and condensate are removed. A temperature sensor in the purge line could provide the threshold signal for opening/closing the solenoid valve. The valves are of the fail-closed design option, that is in the absence of electrical or pneumatic actuation they are closed.

A sequential program cycle for a bioreactor may be divided into the following nine stages.

Idle. This represents the vessel downtime configuration when SIP is inactive.

Heat to Purge. Vessel temperature is raised from ambient to a purge temperature (100–105 °C). The next stage is initiated when the temperature reaches the purge temperature.

Purge. The vessel temperature is maintained at the purge temperature for a specified period of time. The trigger to move to the next stage is when the elapsed time equals the purge time entered in the controller. Purge time is in the range of 2–10 minutes.

Heat to Sterilize. In this stage, the air exhaust system is closed, and the vessel temperature is increased from the purge temperature up to the sterilization temperature. Steam passes into the vessel and vessel jacket, and condensate is removed through the steam traps. When the sterilization temperature (121 °C) is reached, the next stage is initiated.

Sterilization. The vessel temperature is maintained at the sterilization temperature for a designated period of time. When the timer clock indicates that the designated duration has elapsed, the next stage is initiated.

Cool to Vent. The steam supply to the jacket of the vessel is shut off, and chilled water enters the jacket. The circulation loop is initiated, and the vessel temperature decreases to the stipulated vent temperature (~105 °C).

Vent. The air inlet valve opens to allow air to enter the vessel thereby preventing the possibility of a vacuum forming on further cooling.

Cool to Run. Temperature is decreased to the stipulated operating temperature.

Run. Run is the initial fermentation configuration, and the valves are positioned to allow the control loops to function. The system-specific valve-sequencing program is established, and a step logic program is entered into the sequence controller.

Steam. Some basic definitions for steam follow:

- Live steam is steam at temperatures greater than 100 °C or greater than 1 atmosphere pressure.
- Saturated steam is steam at the dew point.
- Superheated steam is steam above the dew point.
- Service steam is steam used for general purposes; it may be contaminated with dissolved components of the feed water, rust, scale, water treatment

chemicals, feed-water additives used to purify the boiler water, and volatile organic additives, typically amines and ammonia derivatives that help prevent corrosion. Examples of boiler water additives are sodium hydroxide (pH 13.2), sodium bisulfite, sodium sulfite, morpholine (pH 11.4), cyclohexylamide, and diethylaminoethanol (pH 11.2). The impurities present can be pyrogenic, (i.e., cause increased temperature) if present in pharmaceutical drug substances that are injected.

- Clean steam is steam free of pyrogens and chemicals, that is produced from demineralized and prepurified feed waters. Clean steam should be generated from stainless steel stills with stainless steel condensers and tubing.

Clean steam is characterized as dry saturated steam, having no additives, relatively low pH, and equal in purity (when condensed) to the water purity acceptable for final rinsing of drug contact surfaces.

When water changes phase from liquid to vapor in the form of steam at a given pressure, heat is added without changing the temperature of the fluid. By increasing the heat input into the water, the amount of steam produced is increased. Under constant pressure, the enthalpy value increases in proportion to the temperature until the boiling point is reached. While the water is changing phase, the temperature remains constant, but the enthalpy is increasing.

For a given pressure, enthalpy of water at the boiling point is commonly referred to as h_f. This is the value of enthalpy for water that is on the verge of boiling, also known as saturated water.

The temperature at which the phase change occurs is known as the saturation temperature.

At this point, heat can be added until all of the water is changed into steam without changing temperature. When this happens, the value of enthalpy is referred to as h_g, and this is known as saturated steam. The values of h_f and h_g are tabulated in standard steam tables (Table 8).

Additionally, there is a value h_{fg} that indicates the amount of steam necessary to go from saturated water to saturated steam. This value relates to the quality of steam. When the amount of steam is half of the total weight, the steam quality is said to be 50%. When the amount of steam is equal to the total weight, the steam quality is 100%. Because this quality is generally not achievable in real circumstances, the accepted industry standard for steam quality is 99.5%.

Steam Traps and Drains. Steam traps are a critical component in SIP systems. A steam trap is a self-actuating valve that opens in the presence of condensate and/or noncondensable gases (air, CO_2, H_2 present in steam) and closes in the presence of steam (58). Steam traps are present in equipment that uses steam for sterilization. They are placed in lines draining condensate from the bioreactor. Several types of steam traps are available. Mechanical traps operate by the difference in density between steam and condensate. This category includes float and thermostatic traps and inverted bucket traps.

Table 8. Steam Table

Gauge Pressure (psig)	Absolute Pressure (psia)	Temperature (°C)	Heat Content			Volume Steam V_g (m³/Kg)
			Sensible h_f (kJ/Kg)	Latent $h_f g$ (kJ/Kg)	Total h_f (kJ/Kg)	
0	14.7	100.00	419.0	2257.1	2676.1	1.673
1	15.7	101.89	427.0	2252.0	2678.9	1.573
2	16.7	103.61	434.4	2247.3	2681.7	1.486
3	17.7	105.28	441.4	2242.9	2684.3	1.405
4	18.7	106.94	448.1	2238.5	2686.6	1.336
5	19.7	108.56	454.6	2234.3	2688.9	1.274
6	20.7	110.00	460.7	2230.6	2691.2	1.211
7	21.7	111.33	466.5	2226.8	2693.3	1.161
8	22.7	112.67	472.3	2223.1	2695.4	1.117
9	23.7	113.94	477.9	2219.6	2697.5	1.074
10	24.7	115.22	483.5	2215.9	2699.4	1.030
11	25.7	116.44	488.6	2212.7	2701.2	0.993
12	26.7	117.61	493.7	2209.4	2702.9	0.955
13	27.7	118.78	498.6	2205.9	2704.5	0.924
14	28.7	119.94	503.2	2202.9	2706.1	0.893
15	29.7	121.00	507.9	2199.9	2707.8	0.868
16	30.7	122.06	512.3	2197.1	2709.4	0.837
17	31.7	123.11	516.7	2194.1	2710.8	0.812
18	32.7	124.11	520.9	2191.5	2712.4	0.793
19	33.7	125.11	525.1	2188.7	2713.8	0.768
20	34.7	126.00	529.0	2186.2	2715.2	0.749

Thermostatic traps operate on the principle that saturated process steam is hotter than either its condensate or steam mixed with noncondensable gas. When separated from steam, condensate cools below the steam temperature. A thermostatic trap opens its valve to discharge condensate when it detects a lower temperature. Balanced pressure and bimetallic traps and liquid and wax expansion thermostatic traps are included in this category.

Thermodynamic traps (TD traps) are the most commonly used traps in biotechnology applications. The TD trap uses the velocity and pressure of flash steam to operate the condensate discharge valve. The discharge valve is a simple flat disc (see Fig. 12). When closed, the flat disc is lying on the flat seat covering both the inlet and discharge orifices. Cool condensate at process pressure readily flows through the orifice, lifts the disk, and flows over the seat to the discharge orifices. When condensate reaches steam temperature, it flashes, which causes a decrease in pressure below the disk as the gas velocity increases, and the disk is forced down onto the seat which then seals the orifices. A TD trap can remove air from the system at startup but not if the system pressure increases very quickly because high velocity air, like flash steam, can force the disc closed and seal the orifice. Thus TD traps often require a separate air vent or bypass when used on batch applications that shut down and start up frequently and rapidly, for example, batch bioreactors. Because the TD trap has only one moving part, the disc, it is extremely rugged. TD disc traps are available in SS316L and with NPT or tube ends for $\frac{1}{4}''$, $\frac{3}{8}''$, $\frac{1}{2}''$, $\frac{3}{4}''$ and 1″ tubing (59). They may be installed in any position and are compact and easy to maintain.

Steam condensate is collected in the drain pan. A drain is essentially a fluid collection device that is separated from the process piping by an air gap which prevents pressure backup into the system. Drains should have sufficient capacity to handle full flow of fluids. Drains are connected to the facility floor drains. The air gap in a drain also prevents back-siphoning of waste into process equipment.

Ethylene Oxide. Ethylene oxide (EtO) can be used to sterilize materials that cannot be sterilized by heat. This may be performed by direct batch exposure to EtO or in a suitably designed gas-diffusion system (60). In a gas-diffusion system, a package is wrapped in a permeable material typically of low-density polyethylene (LDPE) film.

The direct batch gas sterilization process loses efficiency as product lot size increases as a result of the progressive increase in the load density. By comparison, product

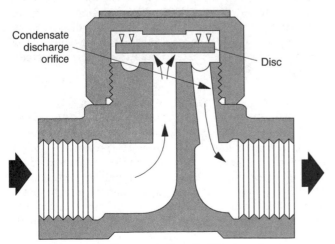

Figure 12. Representation of a TD steam trap.

sterilized in a biological indicator evaluation resistometer (BIER vessel) has only one layer of protective barrier that separates the product from the critical process elements (temperature, humidity, and EtO concentration).

Ethylene oxide sterilization is effective only in defined humidity ranges. Moisture levels of 20–40% were used to inactivate spore samples (61). These levels are rarely employed by the sterilization industry. The original Kaye and Phillips experiment was performed directly on a microbial site that did not include multiple layers of packaging to hinder gas permeation.

A modern, industrial-size vessel typically injects steam into the headspace until a relative humidity of 50–80% is achieved. Additional steam is used to maintain the high level of humidity throughout the conditioning dwell. The same holds true for EtO concentration. Although a BIER vessel may inactivate spores with a sterilant concentration of 350 mg/L, industrial vessels use 600–1000 mg/L.

EtO gas-diffusion technology is a safety-conscious, effective, and efficient way for routine terminal EtO sterilization of small components. By controlling temperature and humidity and selecting packaging materials with compatible EtO permeation properties, a gas-diffusion sterilization system can be installed, validated, and operated to deliver six logs of lethality (SAL) in 10–12 hours of exposure.

Hypochlorite. Hypochlorite has frequently been used for surface disinfection. It has broad antimicrobial activity, rapid bacteriacidal action, is stable in water. Other advantages are high solubility and nontoxicity at use concentration. The active species is hypochlorous acid (HOCl). Spore-forming organisms are more resistant than vegetative organisms and viruses. It should be noted that heavy metals can reduce its sterilizing ability. Hypochlorite is used in water to prevent bacterial colonization. Hypochlorite contains chloride ions that can cause pitting of stainless steel or stress corrosion cracking around welds at temperatures greater than 50 °C. In general, bleach is used reluctantly for the cleaning and sanitizing of stainless steel equipment. If fermenter cooling water is chlorinated, care should be exercised to drain and flush the cooling system before sterilization. In large vessels, the coils may be left filled with water during sterilization.

Vapor-phase sterilization of seeds can be accomplished by overnight chlorination. Seeds are incubated in the vapor phase of acid-treated bleach in a fume hood. Such procedures are limited to very small-scale operation due to the toxicity of chlorine.

Monitoring and Control

A bioreactor is a complex system that requires preventive maintenance to assure smooth operation. Leaky seals resulting from faulty O-rings and gaskets may result in loss of the process through contamination. Contamination control in production is based on initial good design and high standards of fabrication for the equipment and its support systems, piping, valves and seals. Operator training, adequate sterilization procedures, and ongoing preventive maintenance programs go a long way toward minimizing the risk of contamination. A laboratory logbook

is useful for recording maintenance activity. Bioreactor skid supply lines contain numerous connections that depend on gaskets for correct functioning and maintenance of asepsis. Careful selection and checking of gaskets is important. All flanges and fittings should be checked routinely to ensure correct sealing and operation.

Once the basic procedures for SIP are in place, it is important to investigate the procedures through a testing method that will detect any potential flaws in the equipment setup or method. Flaws may exist in the components in the system, aberrant procedures, and operator errors, or in utility deficiencies.

Thermocouples can be placed inside equipment to monitor the SIP process. Special gaskets are available with features that enable inserting a thermocouple through a sanitary standard flange (62). For example, it is possible to insert four probes through a single gasket-clamp connector. These devices, along with the temperature probes and data logging equipment, enable thermal mapping and validation of the sterilization process. Special gaskets, which accommodate biological indicator (BI) devices such as *Bacillus stearothermophilus* bacterial spore strips, can be used to investigate the SAL values at various places during the SIP cycle.

Dummy Runs with Bacterial Culture Media. System integrity may be tested using a sterile run method. This involves incubating the bioreactor system with a suitable bacterial growth medium to allow investigating system integrity. A carefully planned sterility test will enable the user to investigate the equipment and associated operational procedures. Growing bacteria in doubling times range from 20 minutes for wild type *E. coli* to 6 hours for some yeasts and fungi in nonoptimal conditions.

Cell-culture media provide a rich nutritional environment for potential contaminating microbes. Using cell-culture medium is an expensive proposition for large-scale sterility testing at a cost of $10 L^{-1}. Sterility can be checked economically by using simple microbial growth media in the bioreactor. Lauria Broth (LB) requires few components, involves simple preparation, is cheap, and will provide adequate nutrition for a range of bacteria. The LB medium consists of

10 g · L^{-1} Bacto tryptone
5 g · L^{-1} Bacto yeast extract
10 g · L^{-1} NaCl

To prepare the LB medium, fill the vessel to approximately 70% of the final operating volume, and add the solid media components with mixing. Heat the medium to 50 °C to dissolve the components.

Adjust the pH with NaOH to 7.0, and add further DI water to final operating volume. The bioreactor should be set up with probes and associated systems in the same way as the intended culture process. The LB medium can be sterilized in place in the vessel at 121 °C for 20 minutes. After sterilization the LB medium is cooled to 30 °C. Add sterile glucose to the vessel to give a final 10 g/L concentration. The agitation is set to provide surface aeration, and the vessel is incubated for 1–2 days. A

Table 9. Media and Conditions for Sterile Test

Medium	Temperature		Aerobic	Anaerobic
	20°C	37°C		
Thioglycollate broth	+	+	+	
Sabouraud's medium	+	+	+	
Blood agar plate (streaked sample)		+	+	+
Deoxycholate plate (streaked sample)		+	+	+
Cell-culture medium (250-mL shake flask)		+	+	

250–1000 mL sample of medium can be removed after each stage of the test. This sample can be further incubated in a shaker flask to increase the sensitivity of detecting of contaminants in the bioreactor. In the subsequent step of the bioreactor sterility test, the aeration is activated (if it is part of the process being tested), and the vessel is incubated for a further 1–2 days. In the third stage the nutrient additions are activated, and a defined volume (0.1% v/V–1% v/V) of nutrient is added to the LB medium. Contamination is evident by growth in the shake flask samples followed by decreasing DOT in the bioreactor because O_2 is taken up by contaminating organisms. The sensitivity of the detection method depends on the contamination level that may be present after sterilization or after initiating the subsequent stages of the test.
Given:

- SOUR is approximately 1.5 mM. hr^{-1}. A 650^{-1} for bacterial contaminant (63).
- 1.0 OD 650 is ~2×10^6 cell·mL^{-1}.
- Specific growth rate μ of contaminant of 1.0 hr^{-1} (i.e., a doubling time of 0.693 hr).
- DOT of 100% air saturation is 240 µM O_2 for air-saturated water.

We calculate that an A 650 of 0.02 (4×10^4 cells mL^{-1}) is required to give a 10% decrease of DOT per hour. This is a sensitive test of contamination in conditions where aeration is not provided.

Incubating samples from the reactor in shakers may increase contaminant detection resolution. The bigger the proportion of the sample relative to the bioreactor, the higher the sensitivity of the assay. If a single viable contaminant is present in the sample, it will be evidenced by the development of turbidity over the 7-day duration of the test. The most sensitive test would be to filter the entire test medium through a 0.2 µm sterile filter and then incubate the filter in nutrient agar to detect a single contaminant colony. This would be problematic because it may not be possible to filter the large volume of media aseptically and with typical test filters, and the test would be limited to one time. Removing a large volume into a shaker or sterile bag may increase sensitivity to 1–2 contaminants/L.

The LB medium does not provide a universal growth medium that supports a wide range of contaminants. This may be addressed by incubating the LB medium sample in small volumes of cell culture media after sampling or else into other selective media. Aliquots removed from the sterility test can be incubated under different conditions in other media such as indicated in Table 9. Tests should be incubated for at least 7 days.

Contamination can be detected by visual checking (A_{650}) or with a microscope at 400× magnification. Contaminants detected should be typed using a commercial microbial identification kit. Typically, *Pseudomonas* sp. are of aqueous origin, *Bacillus* sp. airborne, and *Staphylococcus* sp. are typical of handling contamination.

Leak Testing. After completing the assembly of the empty clean bioreactor vessel, it is wise to perform a simple pressure test before the sterilization process to ensure that seals and gaskets are providing the necessary sealing. Internal vessel surfaces should be dry before a pressure test because small deformities may be filled with water, which is held by capillary action and will not be displaced by the 10–15 psig overpressure used in the test. First, make sure that all vessel fittings are securely fastened, caps are tightened, and silicone tubing is clamped shut. Attach a pressure gauge to the exit exhaust gas filter. Connect a 12-psig, regulated air supply line to the inlet air filter, and pressurize the vessel to 10 psig. When the vessel pressure has reached 10 psig, clamp the inlet gas line, and hold for 30 minutes. The pressure should not drop by more than 1 psig for a vessel of approximately 100 L in volume. For larger vessels, a longer period of time will be necessary to detect the drop in pressure caused by a small leak. During the pressure test, connections and probe holders can be inspected using a soapy-liquid squirt bottle. Dispense the soap around potentially suspect components, for example, probe insert ports, top port connections, and silicone tubing connections. A leak is detected by observing a foam layer around the leak. After pressure testing, slowly release the pressure in the reactor through the exhaust line.

A pressure test will indicate a poorly seated gasket and other faulty leaking components before sterilization, and so is worthwhile in leading to greater certainity of achieving successful asepsis after sterilization. For large-scale skid systems of high volume, a pressure hold test may not be entirely successful in identifying problems. Very sensitive testing is possible using tracer gases such as Freon or sulfur hexafluoride. Indeed such methods have been used for integrity testing of large-scale fermenters in the pharmaceutical industry and can detect the presence of microscopic leaks. The cooling coils in large industrial fermenters are likely to contain many welded sites that may develop microscopic leaks which can cause contamination of the bioreactor through ingress of cooling water contaminants. The coils can be pressurized (20–30 pisig) with a Freon 12 gas mixture. Leaks are indicated by the accumulation of Freon in the fermenter which is detected by a halogen gas detector (64).

Table 10. Postinoculation Sampling Matrix

Sample	Medium (additions present)	Inoculation Cells	Comments
1. Sterile sampling of medium	Sterile medium from supply vessel or bag pretransfer to bioreactor	None	Observe for any contaminant growth over 7 days
2. Sterile sample of medium from bioreactor	Preinoculation medium from bioreactor	None	Observe for any contaminant growth over 7 days
3. Test of growth medium	Sterile medium from supply vessel or bag pretransfer to bioreactor	Healthy growing stock cells (from culture deposit or research source)	Test medium for growth (specific growth rate μ and capacity — cell yield)
4. Test of growth medium after transfer manipulations and residence in bioreactor	Preinoculation medium from bioreactor	Healthy growing stock cells (from culture deposit or research source)	Test for possible medium transfer effect on medium and bioreactor/medium interactions; any decrease in μ and yield from test 3 may indicate that changes to medium have occured
5. Test of growth medium after inoculation manipulations	Postinoculation medium; aseptically centrifuge out cells, and decant and save medium	Healthy growing stock cells (from culture deposit or research source)	Test for possible inoculation effect on medium; any decrease in μ and yield from test 3 and 4 may indicate that changes to the medium have occured
6. Test cells preinoculation	Sterile medium from supply vessel or bag pretransfer to bioreactor	Cells from inoculum vessel sample diluted to original inoculation density	Examine μ of cells to determine any decrease in cell function; tests health of inoculum
7. Test cells postinoculation	Sterile medium from supply vessel or bag pretransfer to bioreactor	Cells from bioreactor postinoculation aseptically centrifuged to concentrate and then diluted to original inoculation density	Examine μ of cells to determine any decrease in cell function; compare with (7) to determine any effect on cell health during the inoculation process

CHARGING

Bioreactors and ancillary feed tanks are charged with liquid media components. Typical bacterial, yeast, and plant cell culture media can be autoclaved or sterilized in place. Some components, for example, vitamins, are heat-sensitive and would be added as poststerilization sterile additions. The cell-culture medium is not autoclaved and must be sterilized by filtration. Sterile media may be stored in commercially available (65) γ-irradiated polyethylene-vinyl acetate sterile bags or transferred to the bioreactor. Sterile media would normally be stored at cold room temperature (2–10 °C). Sera supplied by various manufacturers are triple-filtered sterile solutions. Other manufacturers may supply γ-irradiated sterile serum. When serum is required in the medium, it is added to the bioreactor as a sterile addition using a sterile addition vessel through a sterile connection. Filter-sterilized protein additions, for example, growth factors, are similarly added as sterile additions.

Ancillary tanks are charged with base for pH control, specific nutrients for batch-fed and perfusion bioreactors, and may be antifoam. Addition tanks may be empty vessel-sterilized or autoclaved, and the component is added aseptically to the sterile tanks or vessel. Sodium hydroxide (0.1–1 N) is filter-sterilized into the storage vessel using a polysulfone-based 0.2-μm sterilizing filter. Potassium hydroxide is sterilized in place or autoclaved if it is to be used (rarely used in pH control). Antifoams such as PPG 2000 are diluted and sterilized in place or are autoclaved. If ammonium hydroxide is used, it is obtained as a sterile concentrated ammonia solution that can be diluted with sterile water and aseptically transferred to a presterilized vessel. Fluid is transferred by various simple methods, which are described in the following sections. The equipment layouts are illustrated in Figure 13.

Fluid Transfers

Following SIP or autoclaving, media can be transferred to the bioreactor. Small-scale bioreactors are assembled with sterile tubing connected to the inlet ports. This tubing will consist of a length of silicone tubing (size 15, 16, or 24) attached to a length of Pharmed tubing (size 15 or 24) and finally a short length of C-flex tubing (size 16) which is fitted with a terminal sterile 47-mm disposable vent-type filter that prevents ingress of contaminants after sterilization. This standard setup is present on all lines that use a peristaltic pump with a size 15, 16, or 24 head and require sterile connection to media or base or other fed tanks. Pharmed tubing is robust and suited for use in peristaltic pump heads. The soft C-Flex tubing can be sterilely welded to other tubes fitted with terminal C-Flex. In small-scale and pilot-scale operation, medium may be stored in sterile bags thereby saving capital for the purchase of tanks. The sterile gamma irradiated bags have found much use in recent years for small-scale cell culture operation in conjunction with the sterile-welding technology. In larger scale systems, the media would be stored in sterilizable tanks or made fresh and 0.2-μm filter-sterilized into the bioreactor.

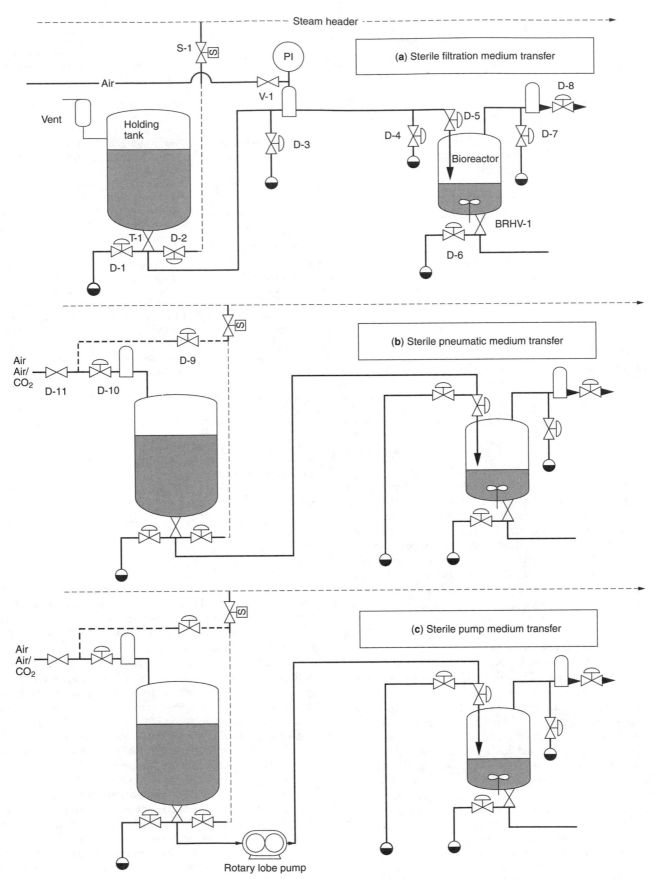

Figure 13. Charging media to bioreactor. Tank overpressure, pumping, and filtration are used for transferring media to the sterilized vessel. Valve arrangements provide a means to sterilize addition lines pretransfer.

Overpressure or Sterile Air. Sterile media can be transferred from the sterile storage tank to the bioreactor using gas pressure as the driving force. This requires that the storage tank be designed as a pressure vessel. Because the storage tank would most likely possess a cooling jacket to preserve media at 2–12 °C, it would already be rated for external pressure and be capable of meeting the design criteria for internal positive pressure. Bicarbonate-containing cell-culture media are stored under 95% v/v/air, 5% v/v CO_2 to maintain the bicarbonate buffer system at a stable pH (7.2–7.4). The gas mixture is delivered through a steam-sterilized 0.2-μm hydrophobic gas filter to the top dish port. A pressure of 10–15 psig is applied to the vessel headspace. A steam-sterilizable transfer line is present between the medium vessel and the bioreactor. The fundamental operations for sterile transfer are as follows. (Fig. 13a):

- Charge the medium vessel with sterile medium, and close the drain port valve. The bioreactor media transfer port inlet valve D-5 is closed.
- Open steam supply valves S-1 and D-2 to medium vessel drain port and transfer line. Open the steam trap valve D-1 on the bioreactor transfer port steam block assembly and D-4 at the bioreactor inlet port.
- Steam the transfer line for 30 minutes at 121 °C.
- Shut off the steam supply to transfer line valves S-1 and D-2, and allow the line to cool to ambient temperature. Pressurize the empty bioreactor with sterile air at 10 psig and open valve D-5, close D4, and allow air to push condensate out through D-1 until the temperature falls below 100 °C.
- Close the steam trap isolation valve D-1 at the medium vessel.
- Maintain positive pressure (5–10 psig) in the transfer line during cooling.
- Pressurize the medium vessel to 15 psig with a sterile air/CO_2 mixture (if not, bicarbonate-based medium system air can used).
- When the line is cooled to ambient temperature, begin the medium transfer. Open the medium vessel drain port T-1 and the bioreactor inlet port D-5. Open the bioreactor exhaust line valve D-8.
- When the prescribed volume has transferred (indicated by bioreactor or media vessel load cell or level sensors), close the medium vessel drain port valve and allow the line pressure to dissipate into the bioreactor.
- When the transfer is complete, close both the bioreactor inlet port valve D-5 and the medium tank valve T-1.

Pump (Sterilized). Small volumes of sterile media may be transferred between vessels or from sterile media bags through sterilized silicone/Pharmed tubing. A peristaltic pump is used for the transfer. Flow rates are relatively slow, and this method may not be suitable for large scale operations. In small-scale operation, the method allows precise volume transfers. The highest flow rate possible with a peristaltic pump is 1000 Lph with large tubing

sizes. More typically, flow rates for common peristaltic pumps are in the range of 0.1–1 Lpm.

In larger systems, the pump must be sterilizable in place. A steam-sterilizable rotary lobe pump can be used for this application (66). Transfer line steam-sterilization is as described in the example of pneumatic transfer.

Through Sterilizing Filter. Nonparticulate liquid media may be sterilized by filtration through an absolute 0.2-μm liquid filter. Manufacturers supply presterilized liquid filters. For small-scale operations, these filters may be setup with sterilized silicone/Pharmed tubing and used in a transfer line to the bioreactor. In large-scale operations, it is usual to insert the filter in a stainless steel housing located in the sterilizable transfer line. The filter is steam-sterilized in the forward direction and cooled with concomitant air pressurization on the upstream side of the filter cartridge to prevent backpressure on the cartridge. In clinical manufacturing, sterilizing filters must be validated by an integrity test after use (67). The integrity test is performed by either a bubble-point or diffusion-forward-flow measurement (68). In a bubble-point test, the filter is removed and placed in a test housing and wet. The micropores of the filter contain water held by strong capillary forces. Upstream air pressure is increased stagewise and held for a short period of time. At some pressure, the water will be pushed out of the capillaries by the air pressure, and the upstream pressure will collapse as air bubbles pass through the filter. The pressure at which this occurs is known as the bubble-point pressure and is a function of the size of the pores:

$$P = \frac{4k \cos \theta \sigma}{d}$$

where P = bubble-point pressure
d = pore diameter
k = shape factor corection
$\cos \theta$ = liquid-solid contact angle
σ = surface tension

A 0.2-μm filter has a bubble point between 40–50 psig depending on the specific materials and manufacturer. In another test, the decrease in pressure as oxygen diffuses across a wetted membrane is measured over a fixed time period.

$$\Delta P = \frac{D \cdot (t) \cdot (P_a)}{V_h}$$

where D = diffusion rate (ml.min^{-1})
t = time (mins)
P_a = atmospheric pressure (psi)
V_h = upstream reservoir volume of apparatus (ml)
ΔP = pressure drop (psig)

Filter integrity is validated by a suitable bubble-point result or by a diffusion test pressure drop within the manufacturer's specification.

Filters are available as 4″ and 10″–30″ pleated cartridges which provide approximately 5 to 15 ft^2 of surface area (69–71). Low protein binding filter materials should be used when filtering media that contain low levels of growth factors or other medium proteins. Bio-inert type materials are available from manufacturers (e.g., Pall with nylon 6, 6 polyamide.) Clean water flow rates of these cartridge style membranes are in the range of $0.2-0.5 \, \text{L} \cdot \text{min}^{-1} \cdot \text{ft}^{-2} \cdot \text{psig}^{-1}$. Sanitary 3-A design housings are available with a double O-ring cartridge seal.

Volume Control (Weight, Level)

During transfer of media to the bioreactor, it is desirable to monitor the progress of the transfer by measuring the volume or mass transferred. As liquid enters the bioreactor, the weight increases as mass is added. A load cell can measure the mass of the reactor. Alternatively, a capacitance level probe (72) could be used to indicate when the liquid has reached a predetermined level and provide a signal to stop the transfer pump or close valves in the transfer circuit.

A floor load scale can be used for mobile vessels up to 500 L in operating volume. Fixed-in-place tanks require skid-mounted load cells to measure the vessel weight. Volume control can be achieved by activating of a feed pump to maintain a fixed volume. A common method of volume control in a perfusion system is to set the tank dilution rate by controlling the medium outlet continuous pump at a set flow rate and using a level sensor or load cell to regulate the flow rate from the fresh medium inlet pump. The flow control may be by PID pulsed on/off control or by PID regulated continuous flow rate. Continuous flow control is recommended in high-density cell culture.

Typical arrangements for simple liquid level control are shown in Figure 9.

Foam/Tangential Inlets

Foam results when bubbles persist at the liquid surface. Because the medium will contain proteins and other foam-generating materials (high surface tension), it is usual to experience foam in bioreactors that are sparged. When frit spargers are used, a stable fine textured foam may persist throughout a run (73). The foam layer will provide a gas mass transfer resistance between the headspace and the liquid surface. This may lead to a small increase of the pCO_2 of the culture during the production phase. Larger diameter bubbles do not tend to form stable foam packs and can be used to break down the foam layer to a certain extent. Foam may be entrained in the exhaust lines where it may coat and block exhaust filters and lead to a failure to control the bioreactor at optimal conditions. To avoid catastrophic foam-out problems, a break tank can be installed between the vessel and the exhaust filter. Foam may lead to specific CIP complications when the foam transports cellular material and media into the filter housing and top-plate nozzles.

Foam can be detected by a conductivity probe that can send a signal to alert operators or to initiate addition of antifoam in standard type bacterial culture. Antifoam agents are potentially toxic to cell culture and are rarely used in foam control. Mechanical devices are most often the choice for cell culture foam separation during the production phase. Defoaming has been achieved by inserting a preformed polysiloxane hydrophobic screen at the top of the liquid surface in an ascites suspension culture (74).

During transfer of the medium in the preparative phase, it useful to add the liquid components through vessel side ports. Ideally, these ports would be arranged tangentially on the side of the vessel. This minimizes the degree of splashing and air entrainment during addition and the formation of a foam layer on the surface of the media. There are also designs in the patent literature that describe abatement of foam using tangential inlets coupled with a low rpm centrifugal separator (75). Foam layers may impede the ability to adjust pH correctly if base is retained in the foam layer after addition through a top-plate nozzle. It is advisable to set the pH controller on manual or the integral function at a low value in attempting to adjust the pH when a foam pack exists.

Sterile Connections/Disconnections

Following sterilization, it is necessary to connect feed tanks, base tanks, sample lines, and perhaps gas lines to the bioreactor. Small autoclaved systems are usually set up with flexible tubing for later connections. Sterile tubing may be connected by using a sterile tubing welder. These devices are supplied as medical devices for connecting tubes aseptically (Terumo Medical Corporation Elkton, MD 21921 USA SCD IIB Sterile Tubing Welder). The type of tubing used to make a sterile connection can be PVC with a nominal OD of 0.21 inches and a wall thickness of 0.030 inches. Users of this equipment have had successful results in fusing thermoplastic tubing such as C-Flex (size 16). The equipment is portable and can be operated inside a class 100 hood if required. The SCD welder makes a sterile connection between the two lengths of tubing by thermal welding. Two pieces of tubing are placed in the tubing holder on the machine. The welding wafer heats to approximately $500 \, ^\circ\text{F}$ ensuring sterility and melts through both pieces of tubing. With the wafer maintaining a functionally closed system, one tubing holder moves back to align the two pieces of tubing being connected. The instrument then simultaneously retracts the wafer and welds the tubes together. The two pieces of tubing to be welded together should be previously autoclaved and have a terminal 0.2-μm 47-mm disc air filter. Because the tubing size is limited to size 16 with a nominal ID of 0.21 inches the flow rate of fluids is restricted. Exhaust gas flow rates of >5 Lpm may cause some pressure buildup in this small diameter tubing. If a line is to be severed or removed, place two tubing clamps over the line, and cut distally from the fermenter. If the line is to be reused, use a tubing welder to join a short C-flex tubing length to the vessel C-flex line.

For larger vessels, larger diameter tubing connections must be made. It is not possible to make a strict sterile connection between two hoses without steaming afterward. Where aseptic connections are required to link equipment, such as tanks, it is important to include steam-block arrangements in the interconnecting lines (Fig. 11). Steam-block arrangements enable sterilization

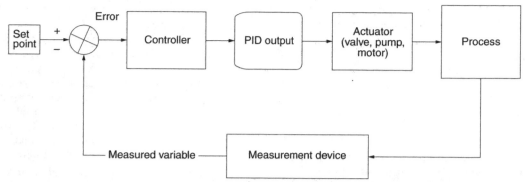

Figure 14. Process control loop. Feedback control loop showing PID controller. Comparison of the process pH to the set point value generates an error value. This electronic value feeds into a PID algorithmic calculation which generates a PID output to open a valve or to activate a pump motor resulting in the addition of acid (CO_2) or base to the process tank.

of the transfer lines independently of the vessel sterilization cycle.

pH Priming

Before initiating automatic pH control, it is important to prime the acid and base lines. This involves filling the lines between the supply tank and bioreactor with acid or base, setting the controller to manual mode, and manually pulsing the addition pump or regulating valve to allow the base or acid to flow up to the vessel. An operator would have to visually observe the process of priming to assure that the lines are filled. In very large scale, use of a closed circulation loop or of an in-line conductivity probe could automate systems priming. A simplified example of a pH control loop is shown in Figure 14.

If priming is not done, it is possible that the initial pH control will overshoot and may cause the cells to experience wide pH fluctuations. Process control of pH is best achieved using PID process loop controllers. In PID control, the difference between the set-point pH and the actual measured pH would be used to generate a controller output signal for actuating of the base or acid pump. In a split PID control loop, the acid and base systems have individual PID controllers. In a standard PID control loop, the output signal is calculated from a control algorithm PID equation.

$$\text{Output} = K_\text{c} \cdot E + \frac{1}{t_\text{i}} \int_0 E \, dt + t_\text{d} \frac{dE}{dt}$$

where E = set-point measurement value
K_c = controller gain (proportional)
t_i = integral time (integral)
t_d = derivative time (derivative)

This expression explains that an unprimed line will lead to an artificial delay in the change of the pH because base or acid is not present through the line. This will lead to an increase in the influence of the integral term, and a high output value will be calculated, which will likely lead to an overshoot in pH. To prevent this, the controller should be set in manual mode during priming and installation.

Instrumentation Checks

Sterilization may affect the calibration of various probes. Simple poststerilization checks can be made and any necessary readjustments can be made.

Temperature. The RTD probe can be checked by measuring the temperature of the medium with a mercury in glass thermometer in a thermowell port filled with a nonvolatile liquid such as glycerol.

pH. As mentioned earlier, a standard Ingold probe measures pH with an accuracy of 0.04 pH $\pm$ 0.01 and a reproducibility of 0.01–0.02 pH units. A drift of the pH signal may occur in a solution of constant (H^+) concentration. However, the rate of drift should be less than 0.02 pH units hr^{-1}. Larger rates of drift are usually due to fouling of the liquid junction.

The response of a pH probe should be 59.19 mV/pH at 25 °C. The response time should be less than 30 seconds. Slow response time indicates thickening of the gel layer and accompanies increased electrical resistance, low slope, and zero-point drift. The aging of the probe is markedly increased by increased temperature. Normal probe life is 1–3 years at 20 °C and less than a week at 120 °C. Repeated dehydration, resulting from leaving a probe inside a dry vessel before or after a process, decreases the effective probe life. A dry probe gel layer may require days to rehydrate. The reference electrode is also adversely affected when the probe is allowed to dry out. Salt crystals may form in the reference electrode liquid junction causing an increase in the reference electrode impedance. Soaking in 3.8 M KCl may reverse this effect.

Because of potential problems due to poor storage and maintenance of probes, several points for maintaining and checking probes are provided here.

- During downtime, immerse the pH probe tips in 3.8 M KCl. Do not store at pH > 10 or in nonionic solutions for a prolonged time because pH sensitization of the glass may occur which renders the probe sluggish at low pH.

- When changing electrolyte or during cleaning, use an aspirator to remove the entire electrolyte from the inner and outer reservoirs.

- Clean diaphragm junctions regularly (2–3 process cycles) when immersion in high protein media occurs. Place the cleaning solution (supplied by manufacturer) in a probe cap over the tip, and invert the probe overnight. The inversion causes the small hydrostatic pressure to push the cleaning solution through the diaphragm ceramic frit. Rinse thoroughly and replenish the probe electrolyte. Allow 1–2 hours for stabilization before calibration and use. Preferably, leave the probe overnight in 3.8 M KCl.

- Never clean the probe with tissue. It can create a static charge that will render the probe very sluggish.

Probes are calibrated by a two-point calibration method. The probe is immersed in a pH 7.0 standard. The standard electrolyte is pH 7.0 ± 0.2 pH units at 20 °C. The probe reading is adjusted to this value by adjusting the meter zero. The probe is rinsed and placed in pH 4.0 ± 0.2 pH unit buffer, and the meter span is adjusted to read pH 4.0. The process should be repeated once. At this point, the probe is inserted into the 25-mm Ingold-type vessel port. In lengthy perfusion-mode bioreactor processes, it is usual to install more than one probe as backup sensors. Probes are steam-sterilizable in place.

Several useful diagnostic tests can indicate the state of a before deciding whether it is reliable and suitable for in-line duty. Some practical tests are given here:

- A probe that gives a stable reading at pH 7.0 ± 0.2 pH should indicate a value of pH 4.0–4.15 within 30 seconds after placement in a pH 4.0 calibration buffer.

- An mV slope of <35mV/pH unit indicates that a probe has an internal short circuit and should be discarded. This may be indicated by an unusually high span value (>140% normal new probe).

- Place a high impedance ohmmeter (10 MΩ) in parallel with the probe, which is immersed in 3.8 M KCl, by placing one electrode in the beaker of 3.8 M KCl and locating the other electrode in the probe electrolyte through the fill hole. Use only Ag/AgCl wires to prevent half-cell formation. The resistance should typically be 3–15 kΩ. If the reading is 20 kΩ or greater, the junction is blocked and needs cleaning.

- Electrical resistance between the reference element and the lead out signal wire should be too great to measure. However, any detected measurable resistance indicates cracking of the isolating glass and electrolyte leakage. This results in an inoperative probe.

Other indications and cautions for nongel electrolyte probes are as follows:

- Ensure that the sealing plug is removed from the probe before use to equalize the pressure of the probe within the housing. During sterilization at 121 °C, the pressure will increase in the probe as the equilibrium vapor pressure increases.

- Maintain positive pressure on the probe housing to create a flow of electrolyte out of the electrode junction. The probe housing pressure should be approximately 6–8 psig greater than the sum of the positional hydrostatic plus vessel head pressure during operation. Note: A drop of electrolyte should form at the probe junction at positive housing pressure if the junction is not blocked. The electrolyte leak rate is 0.5–1.0 mL/day at 2-bar overpressure.

- For all pH probes, a normal drift of 0.05 pH/day can occur and is usually corrected by off-line grab sample and calibration measurement. It is important in this method that the sampling technique does not impart an error to the pH correction. This may be a problem in high cell-density cultures, high pCO$_2$ and changing sample temperature. To minimize errors of this nature, a contained sampling device such as a siliconized syringe is used.

DO$_2$. Ingold DOT probes give a current of 60 nA at air-saturation DOT. In N$_2$ saturated solution, the current should be less than 1% of the air saturation value. Response time from saturation to N$_2$ should be 45–90 seconds. Probes are stable at constant temperature and pressure, and the change in output in air saturated solution is less than 2%/week. The dissolved oxygen (DO) probe should be plugged in to the amplifier with the power to the control box on. If the probe has been unplugged from the amplifier or the power to the control box is off for more than one hour, the probe must be repolarized for 12 to 24 hours. Probes that have been unplugged or have had power interrupted for less than one hour must be repolarized for approximately two times the power off/unplugged time period.

Place the probe in air-saturated water, and record the dissolved oxygen value. This can be set-up using a 2000-mL graduated cylinder filled with deionized water and sparging the flask with air at approximately 5 Lpm for about 10 minutes. Next, place the probe in an oxygen-free atmosphere, that is, pure nitrogen. Use an empty 500-ml Erlenmeyer flask sparged with nitrogen at approximately 1 Lpm. The dissolved oxygen reading should drop to less than 2 to 5% of the saturated air value.

Leave the probe in the nitrogen atmosphere, and record the minimum steady-state dissolved oxygen value and the output current or amplifier voltage. If the voltage is greater than 0.04 volts, the probe must be rejected. If the probe passes the test, set the DO to zero. Placing the probe in air-saturated water can check the response time of the probe. On changing from nitrogen to air-saturated water, the probe should attain 98% of the air value within 45 to 90 seconds.

The DOT probe is calibrated in the vessel after sterilization and addition of sterile media. The bioreactor should be at the run temperature and sufficient media should be in the vessel to cover the probe. The rpm is set at a level that gives good bulk mixing. Attach a 95% v/v N$_2$/5% CO$_2$ gas line to the inlet sparger and headspace inlet filters. Sparge the mixture until the output signal is asymptotic for >30 minutes. Then adjust the probe zero on the DO controller amplifier. After the zero adjustment is completed, disconnect the N_2/CO_2 line, and flow 95% air/5% CO_2 through the gas inlet lines. When an asymptotic maximum is achieved, set the span to 95% airsaturation.

CULTURE INITIATION

Transfer of Cell Inoculum

Inoculum transfer essentially involves transferring an actively growing culture of cells from one growth vessel to another larger growth vessel containing fresh medium. The operation of transferring the culture should impose little or no physiological stress on the cells. The important aspects of inoculum transfer are as follows:

- Transfer in minimum time.
- Avoid DOT <20%.
- Avoid foam generation in transfer equipment.
- Preset receiving environmental conditions to run conditions.
  - Set pH at set-point value.
  - Set DOT at 95% saturation.
  - Set temperature at set point.
  - Set an initial low agitation rate (for bulk mixing).
- Sterilize transfer lines and allow them to cool ahead of time.

Ensure that the vessel valves are in the correct open/closed positions for the culture run conditions. Particular attention should be given to ensuring that the receiving vessel airflow rate is set at a low rate and that the gas vent (exhaust) line is open to avoid any elevation of backpressure toward the inoculum vessel or container. Any backflow of gas could result in foam production in the transfer lines or overpressurization of the inoculum vessel and subsequent breakage of connecting joints.

In small-scale vessels, it is possible to locate the inoculum inlet line below the initial liquid level line. This avoids the possibility that vessel headspace internal devices will hinder the inoculum. For larger vessels, a subsurface transfer line may impart hydrostatic pressure which must be overcome by pressurizing the inoculum sufficiently.

Overpressurizing an inoculum vessel can lead to transient high pCO_2 and O_2 concentrations and potential physiological shock to the cells. Transfer is achieved by using gravity feed or a peristaltic pump to avoid this problem. Small-scale cultures may be grown in small spinner jars or flasks that are set up with internal stainless tubing to facilitate transfer (Fig. 15). The inoculum can be transferred by gravity feed or by using a peristaltic pump. The spinner flask should be agitated by gentle swirling during the procedure. It is generally not possible to maintain the flasks at temperature in an incubator during the procedure when the vessels are separated by distance. The flask could be wrapped in insulating foam for the short inoculation duration if desired. Larger scale inoculum

transfer may require using a slight gas overpressure to convey the inoculum. Using a minimum pressure, avoiding any large pressure gradients, and avoiding foaming are key design parameters.

Initial Mixing/Planting Conditions

After inoculation, a number of checks and controls are performed. The control and monitoring systems are switched from manual to automatic remote control mode. Initial aeration conditions are set for minimal flow rate to maintain the DOT and pH (CO_2). Overaeration during and just after inoculation can lead to production of a foam layer at the liquid surface which can trap cells and potentially inactivate them. In general, bubbles can inflict serious forces as they disengage at the liquid/gas surface, and through interfacial tension forces. These forces may irreversibly damage cells (76). In cell culture adding a low quantity of Pluronic Polyol F68 to the medium may very significantly decrease the rate of bubble damage (77). When cell density is low, this can lead to a significant lag phase through the reduction in viable cell concentration following an increased specific death rate. The agitation rate should be set to the lowest Rpm that keeps the cells suspended. The DOT will usually remain unchanged because the overall OUR is low relative to the volumetric OTR (Oxygen Transfer Rate).

The pCO_2 is a function of the % CO_2 in the aeration mixture and can be verified by checking the mixture component flow rates and the liquid pCO_2 using a BGA. In batch bioreactor operation, the volume will change only by a small amount, which would be the balance of loss through evaporation and entrainment and gain through the addition of base and other reagents. Nutrient feed should be initiated soon after the inoculation. The pump and hardware are checked to ensure that the lines are set up for the flow of nutrient feed to the vessel.

Initial Sampling

Samples are removed for sterility testing and off-line chemistry checks. A 100–200 mL sample of preinoculation medium is taken and stored at 2°–10°C before inoculation with postinoculation cells from the bioreactor. A small aliquot of the fresh sampled bioreactor medium is used to determine preinoculation glucose, glutamine, ammonia, lactate, pCO_2 (BGA), DOT (BGA), and pH (BGA) values.

After completing the inoculation, a small sample of the inoculum is retained to determine viable cell concentration. The bioreactor is sampled to provide measure postinoculation viable cell count, pH (BGA), and pCO_2 (BGA). The culture is visually examined under a phase-contrast microscope to determine cell count and viability and to inspect the morphology of the cells.

Control Cultures Set Up

Various shake flask or small-scale incubation tests are setup. The tests are designed to detect any source of contamination that occurs during the inoculation and to compare and possibly troubleshoot differences in the cell

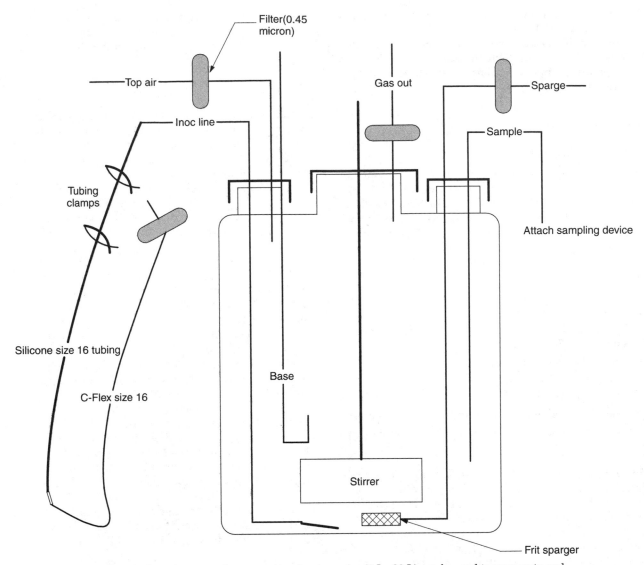

Figure 15. Inoculum transfer apparatus. A spinner jar (1 L–20 L) can be used to propagate and transfer an inoculum. The use of C-Flex tubing facilitates attachment to the bioreactor by using a SCD tubing welder.

growth rate in the bioreactor and the shake flask controls. The tests are depicted in Table 10 (see page 165).

HARVESTING

In cell culture, where the product of interest is secreted into the medium, the harvesting process should not result in cell damage or lysis. Operation at low temperature (4–12 °C) is used to minimize enzymatic degradation of glycoproteins. In cell culture, sialadase may be active and cause degradation. It may be possible to minimize the degradation process by using inhibitors such as 2,3-dehydro-2-deoxy-N-acetyl neuramic acid (78). Prevention of mechanical damage and lysis involves chilling and using low-shear transfer systems. Peristaltic pumps with silicone tubing, pneumatic transfer through overpressure of the reactor, or gravity transfer in small-scale vessels are recommended. Additionally, static

laboratory centrifugation is a gentle approach useful for clarifying relatively small volumes of culture liquid. The largest laboratory rotor capacities are 5 liters. The researcher can gauge the time needed to process the material with such an approach. It would be feasible to process up to 30 liters of culture broth in ~2.5 h using a laboratory batch centrifuge.

Filtration techniques may be used to separate the cells from the product. When cross-flow filtration is employed to separate cells and medium, it is important to minimize cell lysis, which could result in a decrease in yield and purity of the product, as well as decreased performance of the membrane system. Avoidance of average wall shear rates greater than 3000 s^{-1}, using membrane pore sizes of 0.2 um, and minimizing of shear rates (less than 25 dyn $\cdot$ cm^{-2}) minimize the potential cell damage in cross-flow filtration (79). Addition of Pluronic polyol F68 (0.1%w/v) prevents damage from gas bubbles generated by mixing and shearing. If small membrane systems of 1 ft^2 are used, then

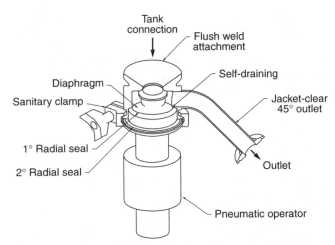

Figure 16. Typical flush-mounted tank harvest valve design.

peristaltic pumps (Watson-Marlow or Cole-Palmer) are adequate. For larger systems, larger gap rotary positive-displacement pumps are employed (Flowtech, Waukesha). Cross-linked cellulosic membranes are available (Sartorius Hydrosart) which are useful in minimizing surface fouling and for facilitating high protein flux.

Discharge

The simplest methods for transferring the bioreactor contents from the bioreactor are mentioned in the next sections.

Gravity. In a facility in which tanks are arranged at multiple levels, gravitational force may be used to drain the bioreactor. The harvest valve is designed to allow sufficient flow from the tank. Harvest valve designs are flush-mounted with the vessel internal dish surface and use piston-type operation to open the closed positioning (Fig. 16). The flow rate from the tank will drop as the liquid height decreases. The following equation relates the time for reducing of the liquid height from the original height (H) to the empty vessel height designated as the reference height of zero.

$$T = \frac{1}{C_D} \sqrt{\frac{2}{g}} \cdot \left(\frac{D}{d}\right)^{0.5} \cdot (H^{0.5})$$

where D is the tank diameter, d is the outlet valve and piping diameter, and C_D is the discharge coefficient of the harvest orifice. Time restrictions limit this operation to small-volume systems (<500 L).

Sterile Air Overpressure. Sterile air overpressure at the bioreactor is frequently used to make liquid transfers. An air supply feeds a sterilizing inlet filter. The exhaust gas lines, inlet nutrient, and base lines are closed, and the vessel head pressure is increased to between 10–20 psig. Transfer is initiated by opening the harvest valve and the receiving vessel inlet and vent valves. When processing fragile cells, a minimum overpressure is chosen to minimize potential lysis caused by rapid pressure decreases through valves and fittings.

Pump. To transfer small-scale bioreactor contents, it is possible to use a peristaltic pump with silicone tubing. Although peristaltic pumps are available up to 4″ in diameter and with deliverable flow rates of 1000 Lph it is generally not feasible to operate such systems smoothly. Placing persistaltic pumps at a distance from the reactor outlet may cause pressure losses that result in restriction of flow to the pump and subsequent tubing collapse. Rotary lobe pumps have been used for liquid transfer in cell culture. Recirculation of hybridoma cells through a recirculation loop of 0.75″ diameter and 8 ft long for 1 hour at flow rates up to 16.7 gpm did not cause any decrease in cell viability in the presence of Pluronic F68. The turbulent shear associated with flow through the system did not damage cells over a range of Reynolds numbers up to 71000, corresponding to a Kolomogorov eddy length of 12 µm (79). Cell damage at Kolomogorov eddy lengths of 3.5 µm or less at high power dissipation rates have been reported (80). Circulation of hybridoma cells through a rotary lobe pump showed cell damage at tip speeds of 350 cm·s⁻¹ (79). When cell damage is to be avoided, the choice of pump should enable operation at low rpm with maximum tip speed in the region of 250 cm·s⁻¹.

BIBLIOGRAPHY

1. W.R. Arathoon and J.R. Birch, *Science* **232**, 1390–1395 (1986).
2. A.E. Humphrey, E. Galindo, and O.T. Ramirez, eds., *Advances in Bioprocess Engineering*, Kluwer Academic Publishers, Dordrecht, The Netherlands, 1994, pp. 103–107.
3. T. Hashimoto, in S. Azechi, S. Sugita, and K. Suziki, *Proc. 5th Int. Plant Tissue Cell Culture*, Tokyo, July 1982, p. 403.
4. D.A. Vulborling, *Spec. Tech. Publ. ASTM STP 538*, 196–209 (1973).
5. International Association of Milk, Food and Environmental Sanitarians, The United States Public Health Service, The Dairy Industry Committee, *J. Milk Food Technol.* **37**(1), 605–02 1974.
6. D. Parry, *Process Biochem.*, July/August, pp. 27–29 (1974).
7. D.G. Adams and D. Agarwal, *Pharm. Eng.* **10**, 9–15 (1990).
8. C.T. Cowan and C.R. Thomas, *Process Biochem.* February, pp. 5–11 (1988).
9. E. Grave, *Biopharmacology* pp. 22–28 (1988).
10. A.K. Lloyd, *Food Manuf.* February, pp. 37–42 (1972).
11. D.A. Seiberling and W.J. Harper, *Am. Milk Rev.* **19**(1), 30–31, 34 (1957).
12. Y. Christi, *Chem Eng. Prog.* **88**(9), 80–85 (1992).
13. G. Hauser, *Inst. Chem. Eng. Symp. Ser.* **126**, 435–445 (1992).
14. A. Grasshoff, *Trans. Inst. Chem Eng.* **70**(C2), 69–77 (1992).
15. J. Villafranca and E.M. Zambrano, *Pharm. Eng.* **5**(6), 28–30 (1985).
16. D.C. Coleman and R.W. Evans, *Pharm. Eng.* **10**(2), 43–49 (1990).
17. T. Myers, T. Kasica, and S. Chrai, *J. Parenter. Sci. Technol.* **41**, 9–15 (1987).
18. G.L. Petersen, *Anal. Biochem.* **100**, 201–220 (1979).
19. R. Baffi et al., *J. Parenter. Sci. Technol.* **45**(1), 13–19 (1991).
20. J. Agalloco, *J. Parenter. Sci. Technol.* **44**(5), 253–256 (1990).
21. P.T. Noble, *Biotechnol. Prog.* **8**(4), 275–284 (1992).

22. H.J. Henzler, Untersuchungen Zum Homogenisieren Von Fluesigkeiten Oder Gasen *VDI-Forschungsh.* **587**, VDI Verslag, Dusseldort (1978).

23. R.B. Bird, W.E. Stewart, and E.N. Lightfoot, *Transport Phenomena*, Wiley, New York, 1960.

24. R.H. Perry and D.W. Green, *Chemical Engineers Handbook*, McGraw-Hill, New York, 1984.

25. L.G. Straub, E.Silberman, and H.C. Nelson, *Trans. Am. Soc. Civ. Eng.* **123**, 685–717 (1958).

26. J. Alford et al., *Ann. N.Y. Acad. Sci.* **721**, 326–336 (1994).

27. S.S. Seaver, J.L. Rudolph, and J.E. Gabriels, *Bio Techniques* **2**(4), 254–260 (1984).

28. G. Gualandi, E. Caldorola, and E.B. Chain, *Sci. Rep. Ist. Super. Sanita* **2**, 50, (1959).

29. H.P.J. Bonarius, C.D. de Gooijer, J. Tramper, and G. Schmid, in R.E. Spiers, ed., *Animal Cell Technology*, Proc. 13th ESACT Meet., Butterworth-Heinemann, Wiltshire, U.K., 1995.

30. L. Packer, ed., *Methods in Enzymology*, Vol. 105, Academic Press, New York, 1984.

31. Heating Belt systems available from Watlow, St. Louis, MO.

32. D.R. Gray et al., *Cytotechnology* **22**, 65–78 (1996).

33. A. Itagaki and G. Kimura, *Exp. Cell Res.* **83**(2), 351–361 (1974).

34. Blood Gas Analyzer available from Ciba Corning Diagnostics, Corning, NY.

35. E. Puhar, A. Einele, H. Buhler, and W. Ingold, *Botechnol. Bioeng.* **22**, 2411–2416 (1980).

36. S.K. Dahod, *Biotechnol. Prog.* **9**, 655–660 (1993).

37. Available from Aquasant-Messtechnik AG, Switzerland.

38. K.B. Konstantinov, Y.-S. Tsai, D. Moles, and R. Metanguihan, *Biotechnol. Prog.*, **12**, 100–109 (1996).

39. P. Wu et al., *Biotechnol. Bioeng.* **45**, 495–502 (1994).

40. T. Yamane, *Biotechnol. Prog.* **9**, 81–85 (1993).

41. R.M. Mantanguihan, *Bioprocess Eng.* **11**, 213–222 (1994).

42. Supplied by Badger Meter Inc., Research Control Valve (Orion), Tulsa, Okla.

43. Supplied by Sensortronics, Covina, Calif.

44. Pressure gauges supplied by Andersen Instruments Co. Inc., Fultonville, N.Y.

45. Load Cells supplied by Sensortronics, Covina, Calif.

46. Floor scales supplied by Fairbanks Scales, Kansas City, Mo.

47. Mag-Flow meters available from Yokogawa Corporation, Newnan, Ga.

48. Capacitance probes from Endress and Hauser, Greenwood, Ind.

49. Diaphragm valves supplied by Tri-Clover Inc., Kenosha, Wis.

50. Sanitary design ball valves from Worcester Controls, Marlborough, Mass.

51. Plug valves available from Nupro, Willoughby, Ohio.

52. Sanitary design valves supplied by Triclover Inc., Kenosha, Wis.

53. W.M. Miller, A.A. Lin, C.R. Wilke, and H.W. Blanch, *Biotechnol. Letts.* **8**(7), 463–468 (1986).

54. SCD IIB Sterile Tubing Welder available from Terumo Medical Corporation Elkton, Md.

55. Clamp and Bevel seat fittings available from Tri-Clover Inc., Kenosha, Wis.

56. S. Aiba, A.E. Humphrey, and N.F. Millis, *Biochemical Engineering*, Academic Press, New York, 1965, Chapter 8, p. 194, Table 8.1.

57. W.E. Bigelow, *J. Infect. Dis.* **29**, 528 (1921).

58. J. Radle, L.R. O'Dell, and B. Mackay, *Chem. Eng. Prog.* January pp. 30–47 (1992).

59. Steam Traps supplied by Spirax Sarco Inc., Blythewood, S.C.

60. Ethylene Oxide Sterijet System supplied by Andersen Products, Inc., Haw River, N.C.

61. S. Kaye and C.R. Phillips, *Am. J Hyg.* **50**, 296–306 (1949).

62. Tri-Clamp fitting validation gaskets available from Rubberfab Mold and Gasket, Andover, N.J.

63. S. Chen, Chiron Corporation, Calif. (personal communication) (1999).

64. C.A. Perkowski, G.R. Daransky, and J. Williams, *Biotechnol. Bioeng.* **26**, 857–859 (1984).

65. Sterilized polyethylene vinyl acetate clear Bio-Pharm bags supplied by Stedim Laboratories, Concord, Calif.

66. Steam Sterilizable Positive displacement pumps supplied by Waukesha Cherry -Burrell, Charlotte, N.C.

67. S.S. Chrai, 1988, *Pharm. Technol.*, October, pp. 62–71 (1988).

68. A. Kononov, *Pharm. Eng.* **13**, 30–35 (1993).

69. Sartorius Corporation, Edgewood, N.Y.

70. Millipore Corporaton, Bedford, Mass.

71. Pall Corporation, East Hills, N.Y.

72. 3-A Sanitary Capacitance probes supplied by Garner Industries Inc., Lincoln, Neb.

73. S. Zhang, A. Handa-Corrigan, and R.E. Spier, *Biotechnol. Bioeng.* **41**, 685–692 (1993).

74. M. Ishida et al., *Cytotechnology* **4**, 215–225 (1990).

75. U.S. Pat. 4,553,990 (November 19, 1985), A. Hofmann (to, Linde AG).

76. A. Handa-Corrigan, A.N. Emery, and R.E. Spier, *Enzyme Microb. Technol.* **11**, 230–235 (1989).

77. B.L. Maioella, D. Inlow, A. Shauger, and D. Harano, *Bio/-Technology* **6**, 1406–1410 (1988).

78. M.J. Gramer et al., *Bio/Technology* **13**, 692–698 (1995).

79. B. Maiorella, G. Dorin, A. Carion, and D. Harano, *Biotechnol. Bioeng.* **37**, 121–126 (1991).

80. A. McQuen, E. Meilhoc, and J.E. Bailey, *Biotechnol. Lett.* **9**(12), 831–836 (1987).

See also BIOREACTORS, CONTINUOUS CULTURE OF PLANT CELLS; BIOREACTORS, PERFUSION; BIOREACTORS, RECIRCULATION BIOREACTOR IN PLANT CELL CULTURE; BIOREACTORS, STIRRED TANK; CONTAMINATION DETECTION AND ELIMINATION IN PLANT CELL CULTURE; CONTAMINATION DETECTION IN ANIMAL CELL CULTURE; CONTAMINATION OF CELL CULTURES, MYCOPLASMA; CULTURE ESTABLISHMENT, PLANT CELL CULTURE; EQUIPMENT AND LABORATORY DESIGN FOR CELL CULTURE; IMMUNO FLOW INJECTION ANALYSIS IN BIOPROCESS CONTROL; MEASUREMENT OF CELL VIABILITY; OFF-LINE ANALYSIS IN ANIMAL CELL CULTURE, METHODS; OFF-LINE IMMUNOASSAYS IN BIOPROCESS CONTROL; ON-LINE ANALYSIS IN ANIMAL CELL CULTURE; STERILIZATION AND DECONTAMINATION.

BIOREACTOR SCALE-DOWN

LAURA A. PALOMARES
OCTAVIO T. RAMÍREZ
Instituto de Biotecnología
Universidad Nacional Autónoma de México
Cuernavaca, México

OUTLINE

Ideally, a bioreactor should operate under controlled and homogeneous conditions. However, concentration gradients due to deficient mixing can develop in inhomogeneous systems, such as hollow-fiber, packed-bed, and microcarrier reactors, and also in the so-called homogeneous systems such as stirred-tanks, airlifts, and bubble columns. Environmental heterogeneities are the result of bioreactor design and operating conditions, and in general, lead to poor culture performance. Diverse environmental heterogeneities have been documented for suspended cultures of animal and plant cells, including gradients in carbon dioxide, pH, microcarrier concentration, and culture segregation due to formation of cell aggregates and clumps. Gradients in dissolved oxygen or substrate concentration can also occur in large-scale animal and plant-cell culture as predicted from analyzing the characteristic times of a particular process/bioreactor combination. A full description and analysis of environmental heterogeneities, their causes, and their consequences, both in homogeneous and heterogeneous cell culture systems, have been presented in the article "Bioreactor Scale-Up" of this Encyclopedia. Here, scale-down is described as a method of approaching the problem of heterogeneity in large-scale cultures.

The classical approach for scaling-up a fermentation process on the basis of geometrical similarity has the long recognized drawback that only one fundamental variable can be maintained constant whereas the rest will vary as the process is translated to the larger scale. Constant power input per unit volume, constant volumetric oxygen transfer coefficient, or constant speed at the impeller tip are among the most commonly used scale-up criteria. However, as seen in Table 1 (1), using any of these criteria will result in a lower agitation rate in the large-scale vessel compared to the small-scale bioreactor. Lower agitation rates will, in turn, favor the development of a nonuniform environment. If the desired scale-up goal is to maintain the same homogeneity as in the small-scale bioreactor, then mixing time (defined in the article "Bioreactor Scale-Up") should be maintained constant. Nonetheless, this would require a substantial increase in agitation rate on the larger scale, a task that is usually not possible due to equipment limitations and/or the fragile nature of most cell cultures. Accordingly, the well-mixed condition obtained in laboratory-scale vessels cannot be maintained

on larger scales. This is evidenced in Figure 1 (2), where it can be seen that the mean circulation time t_c of Newtonian and pseudoplastic fluids increases with reactor volume.

Circulation time is defined as the time necessary for a fluid element to return to a fixed reference position after circulating through a stirred tank. As a rule of thumb, circulation time t_c is (3)

$$t_c = \frac{t_M}{4} \tag{1}$$

where t_M is mixing time.

Several correlations exist to estimate t_c, some of which are listed here. However, correct estimation of t_c demands

Table 1. Comparison of Scale-Up Methods

Method	N for the 10,000-L Vessel
Constant power/unit volume, P/V	
Nongassed power	107
Gassed power	85
Constant volumetric O_2 transfer rate, $k_L a$	79
Constant shear, ND	50
Constant mixing time	1260
Constant momentum factor	5

Source: Scaling-up of a 10-L vessel at 500 rmp and 1 vvm to a geometrically similar 10,000-L vessel operated at the same volumetric aeration rate. Reprinted by permission from Ref. 1.

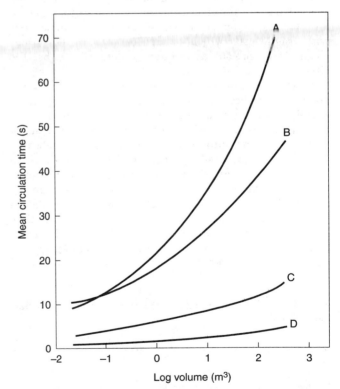

Figure 1. Variation of mean circulation time with vessel volume. Tank configuration $H/T_v = 1$, single disc turbine $D/T_v = 0.33$, power 1.67 k W/m^3. A. pseudoplastic fluid. B. Newtonian fluid (1 Pa s). C. Newtonian fluid (0.1 Pa s). D. Newtonian fluid (0.01 Pa s). Reprinted by permission from Ref.2.

a case by case analysis. Circulation time is considered to be the total tank volume divided by the impeller pumping capacity Q_p (4,5). The impeller pumping capacity is defined as

$$Q_p = KND^3 \qquad (2)$$

where K is the discharge coefficient, N is the impeller speed, and D is the impeller diameter. The discharge coefficient is a function of tank and impeller geometries. Some discharge coefficients for baffled tanks with standard configuration are listed in Table 2. Accordingly, the general form of t_c is

$$t_c = \frac{V}{KND^3} \qquad (3)$$

Tank geometry is explicitly included in other correlations for calculating t_c, for example (3),

$$t_c = \frac{V}{2HN\pi^2 T_v^2} \qquad (4)$$

where H is liquid height and T_v is tank diameter.

In a culture where concentration gradients are present, a cell will be exposed to a fluctuating environment during its circulation through the bioreactor. If severe gradients develop, then localized zones will exist in the bioreactor which can adversely affect the culture. For instance, dissolved oxygen or substrate can be limiting in stagnant zones, whereas in point-of-addition zones, the cells can be inhibited by high substrate or acid/base concentrations. Special attention should also be placed on shear stress gradients, particularly with fragile cells which will be exposed to deleterious shear stresses on a frequency determined by the circulation time.

Circulation time is a fundamental parameter in scale-down studies because it determines the time spent by a cell in the various bioreactor regions and provides the framework for scaling the fluctuations in a large-scale fermenter into a test system (6). As indicated by Yegneswaran et al. (6), actual circulation times in large fermenters are distributed over a range of values, which can be described by a lognormal probability distribution (see Fig. 2). Therefore, using a circulation time distribution rather than a mean circulation time should better mimic the real situation. As seen in Figure 1, mean circulation times for typical microbial pilot and small-scale fermenters (less than 100 liters) do not exceed

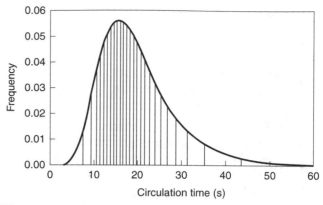

Figure 2. Frequency of circulation times for a lognormal distribution. Reprinted by permission from Ref. 6.

10 s. In 10,000-L fermenters, t_c can range from 5 to 30 s for Newtonian fluids and can be as high as 40 s for pseudoplastic fluids. In comparison, t_c for animal- and plant-cell cultures are severalfold higher than those for microbial fermentations at equivalent scales. As reviewed in detail in the article "Bioreactor Scale-Up", mixing times for typical large scale stirred-tank cell-culture vessels can range between 40 to 200 s, whereas for airlift and bubble-column bioreactors, t_M can range between 200 to 1,000 s. Accordingly circulation times for large animal- and plant-cell cultures will be between 10 to 250 s.

THE SCALE-DOWN APPROACH

Scaling-down is a semiempirical method, based on regime analysis, which consists of reproducing on a small scale the conditions prevailing (or those that will be possible to attain) on the large scale (3,7). Scaling-down can be applied to two general situations. It can be used as a method for scaling a new process or operation to new larger scale equipment or installation. Alternatively, scale-down can be used for diagnosing problems (poor yields, cell death, etc.) and predicting the outcome of modifications to existing large-scale processes or equipment (changes in cell line, media composition, operating conditions, or bioreactor fittings). In practice, the second general situation is most commonly encountered (3). In both cases, small-scale experiments are designed from knowledge of the fundamental parameters limiting the process on the large scale. The experimental results can then be used to optimize and predict the performance of the new (or modified) process or operation on the large scale. An underlying value of downscaling experiments is generating data under conditions resembling real conditions but without actual experimentation on the real scale, a task that is impractical and economically unfeasible. A schematic representation of the scale-down method is presented in Figure 3, which shows that it consists of four main steps (8,9): regime analysis, simulation, optimization, and application. A more detailed review of the scale-down method can be found in Ref. 9.

Table 2. Typical Discharge Coefficients for Standard Baffled Tanks[a]

Impeller	Discharge coefficient
Disk style	0.95 ± 0.28
Propeller	0.55 ± 0.1
Pitched-blade turbine, six-blade 45°	0.92
Pitched-blade turbine, three-blade 45°	0.76
Reciprocating impellers	$0.15-0.30^a$

Source: From Ref. 5.
Note: [a] Various D/T_v ratios.

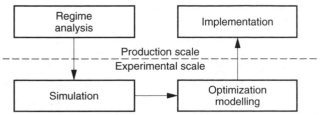

Figure 3. Scale-down procedure. Adapted from Refs. 8,9.

Regime Analysis

Regime analysis is based on identifying the limiting mechanisms that determine the bioprocess on a large scale (10). This is acheived by obtaining the characteristic times (also called relaxation times or time constants) of the subprocesses comprising the system (3). Characteristic time is defined as capacity divided by flow of the particular subprocess (9,10). Three types of information are needed for determining characteristic time: the bioreactor configuration, technological models (for instance, models that describe mixing and heat and mass transfer), and microbiological models (for instance, models that describe the relationship between production rate and growth rate or between growth rate and substrate concentration) (11). Any model has inherent limitations and thus, must be used with caution. Accordingly, analysis of subprocesses should be based only on comparison of the order of magnitude of the various characteristic times (11). When characteristic times of intracellular mechanisms are of the same order of magnitude as those of changes in the extracellular microenvironment (such as mixing and mass transfer), then both subprocesses will interact and will affect the behavior of the cell (3). In Figure 4 (12), the relative magnitudes of characteristic times of important cellular and environmental processes are compared. Characteristic times of various subprocesses relevant to cell culture are summarized in Table 3. A more detailed discussion of characteristic times, including some examples and their derivation, as well as indications for selecting significant subprocesses can be found in Refs. 3,7, and 9.

Small characteristic times are indicative of a fast process. In contrast, a slow process is distinguished by a large characteristic time and will be the potentially limiting step (13). For instance, Nienow et al. (13) reported oxygen uptake, mass transfer, and mixing characteristic times for recombinant CHO cells and NSO cells in an 8-m^3 agitated

Table 3. Characteristic Times

Process	Formula
Mixing	t_M (experimentally determined)
Mass transfer	$t_{mt} = \dfrac{1}{k_L a}$
Circulation	$t_c = \dfrac{t_M}{4}$
Growth	$t_\mu = \dfrac{1}{\mu}$
Heat transfer	$t_{ht} = \dfrac{V_\rho C_P}{UA}$
Diffusion	$t_{diff} = \dfrac{\lambda_k^2}{D_{o/w}}$
Reaction	$t_{rxn} = \dfrac{c}{q_{O_2} X}$

Source: Adapted from Ref. 11.

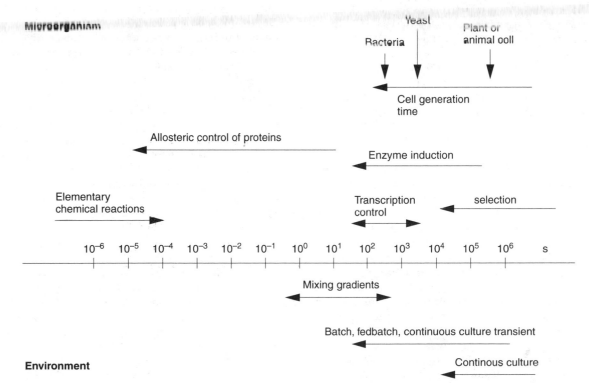

Figure 4. Characteristic times of important cellular and environmental processes (Adapted from Refs. 3,12).

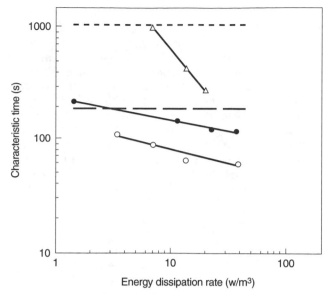

Figure 5. Characteristic times in an agitated tank at the 8 m^3 scale with a 2/9T_v Rushton turbine. Mass transfer time ($\triangle$) and t_M ($\bigcirc$) at a superficial gas velocity of 2×10^{-4} m/s. t_M ($\bullet$) without aeration. Short dashed line indicates reaction time for 4×10^5 CHO 320 cells/mL. Long dashed line indicates reaction time for 2×10^6 NSO cells/mL. Reprinted by permission from Ref. 13.

bioreactor (Fig. 5). As seen in Figure 5, oxygen uptake characteristic time for CHO cells is larger than the mass transfer and mixing characteristic times at all energy dissipation rates studied. Thus, it can be anticipated that no oxygen limitation or dissolved oxygen gradients will form under these conditions. In contrast, for higher cell concentrations and/or specific oxygen uptake rates, as is the case for the NSO cells, characteristic time for mass transfer will be longer than for oxygen uptake. Accordingly, it can be predicted that oxygen supply will be insufficient. A similar analysis based on characteristic times for plant-cell cultures is presented in the article "Bioreactor Scale-Up". In such a case, oxygen gradients will develop because the characteristic mixing time is larger than the oxygen uptake and mass transfer characteristic times.

Simulation

In this step, the rate limiting mechanism (or mechanisms), identified from regime analysis of the production scale, is simulated on a laboratory scale. It is not necessary to maintain geometric similarity during scaling-down to have conditions representative of the full scale. What is important is to maintain the relevant characteristic times constant between both scales (7). A fundamental difference between large and laboratory-scale bioreactors is that the latter are inherently well mixed. Because phenomena resulting from mixing deficiencies on a large scale are among the most common problems analyzed during scaling-down, ingenious experimental setups have to be designed to create a nonuniform environment during laboratory-scale simulations. Furthermore, spatial gradients (i.e., those that are a function of the location in the reactor) that originate

from mixing deficiencies have a low characteristic time (on the order of seconds to minutes). Therefore, spatial gradients are difficult to simulate because very fast fluctuations are required in the downscaled experiments (14). In contrast, temporal gradients (those that depend on process time) have high characteristic time (on the order of minutes to hours) and are relatively simple to simulate (14). The commonly employed experimental configurations for scale-down studies are described in detail later.

Optimization and Modeling

After the limiting mechanisms on the large scale have been identified and reliable methods have been established to simulate them in a laboratory bioreactor, then an optimization study can be pursued. A broad range of conditions can be readily tested in scale-down experiments, including the effect of frequency, amplitude, and axis of oscillations of a particular culture variable on cell physiology and metabolism. The period, amplitude, and oscillation axis will correspond to circulation time, magnitude of gradients, and mean bulk condition, respectively, on the full scale. In this way, the influence of changes in physical subprocesses on cellular subprocesses can be assessed. Unfortunately, very little information exists on bioprocess optimization with respect to gradients (14). Furthermore, equipment, economic, and operating constraints on the large scale will prevent the implementation of all possible optimization results (3). For instance, an optimization result can dictate the best configuration for distributing an ideal number of base addition or substrate feeding ports in a fed-batch or perfusion operation, as well as the best concentration of the corresponding feeding solutions. However, on the full scale, it may be impractical to implement the exact results of such optimization exercise due to concerns for contamination risks and/or increased operating and equipment complexity. In any case, the optimization results should consider only the regimes attainable in full scale.

The results of the optimization step can be used to generate improved microbiological models. This is important because the vast majority of microbiological models are based on balanced growth, (pseudo) steady-state phenomena, or temporal gradients, but almost no models exist describing cell metabolism and physiology under transient oscillatory conditions resulting from spatial gradients (3,14). Furthermore, computational fluid dynamics (CFD) can be a powerful tool for complementing the simulation and optimization studies through estimates of concentration, fluid velocity, and shear rate profiles. In particular, CFD can yield important information on the maximum and minimum values of relevant variables such as pH, shear stress, temperature, and concentrations of substrate, dissolved oxygen, carbon dioxide and by-products. In addition, the fraction of culture volume exposed to a particular detrimental condition (information that is necessary for representative downscaling experiments) can be derived from CFD. Nonetheless, only limited studies of CFD for plant or animal cell cultures exist (15).

Application

The final step in the downscaling method is translating the laboratory experimental results to the production scale (see Fig. 3). Here again, quantitative published information is very scarce for the general area of microbial fermentations and is almost nonexistent for animal- and plant-cell culture. An interesting case study is reported by Geraats (16) for the production of a lipase by a *Pseudomonas alcaligenes*. Although this example deals with a bacterial process, it is briefly described here because it is rare to find such published information and it clearly illustrates the power of the scale-down approach for any type of culture. In this case study, a 65% reduction in maximum lipase concentration occurred when the process was scaled-up from 10 L to an existing 100-m^3 fermenter. To investigate the possible causes of the production loss, downscaling experiments at 10 L and 100 L were performed. The aspects studied included gradients of soy oil, pH, and oxygen; raw materials, medium preparation, and sterilization and inoculation procedures; dissolved carbon dioxide concentration; and shear and air-broth interfaces. Through such an approach, dissolved carbon dioxide was identified as the main cause of the scaling-up problem. By increasing the ventilation rate and decreasing the head pressure combined with a lower pH, lipase production at full scale finally reached the values obtained on the 10-L scale. It should be noted that culture parameters on a larger scale will not necessarily be inferior to those on smaller scale. For instance, although antibody production by CHO cells was worst in a 1,000-L production tank compared to a 2-L bioreactor, cell growth in the larger vessel was almost doubled (17). In this case, analysis of the hydrodynamic parameters for the two vessels revealed that cell damage due to disengagement of small bubbles was more important than mechanical stresses. Therefore, cell death was higher in the smaller bioreactor because superficial gas velocity was also higher compared to the larger vessel. Furthermore, if scale-down studies reveal an absence of detrimental effects caused by environmental gradients, there could be situations where an oscillatory operation could be beneficial on the large scale. For instance, when a cell culture is insensitive to particular dissolved oxygen gradients, then energy savings and reduction of deleterious shear stresses could be achieved by an intermittent mixing and/or gassing operation. Examples of the beneficial effects of an oscillatory environment can be found for various microbial fermentations (18–20).

SCALE-DOWN STUDIES IN CELL CULTURE

The development of improved operating strategies aimed at attaining cell concentrations higher than in conventional cultures will result in increased oxygen consumption and heat and carbon dioxide evolution. Moreover, a move toward the use of highly concentrated feeding solutions should be expected in improved fed-batch and perfusion strategies. Undoubtedly, all of this will worsen the problems, of bioreactor heterogeneity, observed until now. Diverse culture heterogeneities and problems due to scale-up or transient environmental changes have been reported for plant- and animal-cell culture (described in Refs. 17 and 21). For instance, culture segregation due to cell clumping is common in plant-cell culture, changes in protein glycosylation patterns can be caused by transient periods of glucose excess, heat-shock proteins are induced by transient changes in temperature, and protein structural changes can occur during scale-up. A detailed description of other environmental heterogeneities and their adverse effects on cell culture is presented in the article "Bioreactor Scale-Up". Nonetheless, scale-down has scarcely been used as a tool for simulating gradients or predicting their effect on cell culture. The reports by Slater (17), discussed previously, and by Nienow et al. (13) are among the very few available studies of scale-down in cell culture. In the latter study, Nienow et al. (13) used scaled-down vessels to simulate the mixing times of an 8-m^3 animal-cell bioreactor and concluded that severe pH gradients are likely to develop in the full-scale system. To our knowledge, only the work of Rhiel and Murhammer (22) has assessed the effect of oxygen oscillations in animal-cell culture, specifically on the physiology of insect cells. They reported an adverse effect of dissolved oxygen oscillations between 0 and 15% in cell growth and recombinant protein production. In their study, the cultures remained at each DO concentration for 30 min, and it took from 6 to 7 min to shift between both DO set points. Nonetheless, although interesting, this work does not represent the behavior of actual large-scale cultures because the oscillation frequency tested was too slow.

The apparent lack of interest in scaling-down studies can be a consequence of the relatively small scale at which plants and animal-cell cultures have been commonly performed, as well as their high metabolic characteristic times. However, it can be expected that, as operation of high cell-density, large-scale cultures increases, the application of downscaling experiments will increase. The characteristics of animal and plant cells and their culture systems will demand the design of novel experimental configurations to adequately simulate large-scale conditions in laboratories. However, there is a lack of published information about systems expressly designed for cell cultures. Accordingly, the description of experimental configurations for scale-down studies, presented in the next section, is based on information generated from the microbial fermentation area. Most of the experimental configurations described in the next section can be readily applied to animal and plant cells, and only particular attention must be placed on the fact that circulation times and shear sensitivities of cell cultures are severalfold higher than those of microbial fermentation.

EXPERIMENTAL CONFIGURATIONS FOR SCALE-DOWN STUDIES

Special experimental configurations are needed for laboratory simulation of the oscillatory environment to which a cell is exposed when spatial gradients are formed in a large-scale vessel. Scale-down systems can be classified into two general groups depending on the component studied: those for simulating gradients in dissolved oxygen or carbon dioxide and those for simulating substrate or pH gradients. A detailed description of these two general scale-down systems and a comparison of the advantages and disadvantages of the various configurations in cell culture applications follow. Designs of novel experimental configurations, not discussed here, will be necessary to simulate gradients of other variables such as temperature and shear stress, as well as gradients in inhomogeneous systems.

Simulation of Dissolved Oxygen or Carbon Dioxide Gradients

The most commonly used experimental arrangements for oscillating the concentration of a dissolved gas are schematically summarized in Figure 6. Such general designs have been applied to oscillating dissolved oxygen concentration but can also be readily applied to oscillating dissolved carbon dioxide concentration, which is of particular importance in cell culture. These designs can be divided into two general groups: one- (Fig. 6: a, b, c, and d) and two-compartment (Fig. 6: e, and f) systems.

In one-compartment systems, the cells are maintained in a single vessel, and the dissolved gas concentration is oscillated by various means (6,20,22–24). In the simplest arrangement (Fig. 6a), two different gas streams are intermittently sparged (or alternatively transferred through a membrane or liquid surface) to a stirred-tank, airlift, or bubble-column bioreactor. Various combinations

of gas streams can be used, such as carbon dioxide and air, air and nitrogen, oxygen-enriched air and air, or on/off aeration. Cyclic shifts between the two gases are generated by following a predetermined pattern that is independent of the actual DO of the culture (i.e., open-loop control). By assigning unequal sparging times to each gas, it is possible to simulate different culture volume fractions subjected to different conditions (for instance, the volumes of stagnant and well-mixed zones). In this way, the time spent by a cell in each condition can be simulated. Likewise, switching frequency can also be manipulated to simulate circulation time. A mean constant circulation time can be simulated by choosing a constant switching frequency. However, as described before, a real situation is better represented by a circulation time distribution (see Fig. 2), which can also be simulated by establishing a random switching frequency. For instance, in a laboratory-scale reactor, Yegneswaran et al. (6) utilized the Monte Carlo method to control random on/off cycles of aeration to reproduce the lognormal distribution of circulation times shown in Figure 2. A simulation of the expected DO profile using such a random method, compared to a constant switching frequency, is shown in Figure 7. It can be seen that the Monte Carlo strategy, compared to constant oscillations, resulted in occasional longer periods of low DO which could have caused the reported reduction in antibiotic productivity. These results clearly indicate that the strategy chosen for simulating spatial gradients can strongly affect the experimental outcome.

Single-compartment systems with open-loop control, as described previously (Fig. 6a), have the advantages of simple operation and reduced equipment and instrumentation requirements. However, actual control of DO (or dissolved

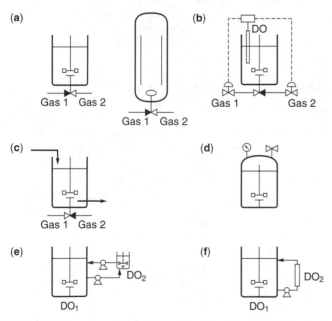

Figure 6. Experimental configurations for simulating dissolved oxygen or carbon dioxide gradients. One compartment systems: (**a**) Batch stirred and air-lift bioreactors with open-loop gas control. (**b**) With closed-loop gas control. (**c**) Chemostat. (**d**) Pressure manipulated reactor. Two compartment systems: (**e**) Combination of STR with different oxygen concentrations. (**f**) STR and PFR combination. DO_1 and DO_2 refer to different dissolved oxygen or carbon dioxide concentrations. DO refers to a dissolved oxygen electrode.

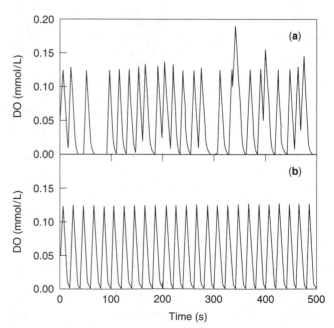

Figure 7. Simulated dissolved oxygen profiles for Monte Carlo and periodic aeration experiments in bacterial cultures. Aeration was on for 5 s and off for 15 s, on average. Reprinted by permission from Ref. 6.

carbon dioxide) cannot be attained in such systems. In the case of DO, this problem can be overcome if the flows of the various gases used are controlled through a closed-loop algorithm based on the signal of a DO electrode (Fig. 6b). For instance, a closed-loop PID (proportional-integral-derivative) control, where oxygen partial pressure was varied according to the difference between DO measurements and a square-type oscillating DO set point, is shown in Figure 8 (18). In this case, it was possible to obtain periodic sinusoidal DO profiles of predetermined periods, amplitudes, and oscillation axes in actual fermentations (18). Only few other closed-loop algorithms for oscillating DO have been reported (20,23). A disadvantage of the closed-loop systems, besides their complexity, is the need for fast response electrodes when short oscillation periods are required.

Many experimental scale-down systems are based on batch operation, and thus, spatial gradients are simulated along with temporal gradients. In some instances, it is desirable to discriminate between the effect of both types of gradients and/or assess the effect of gradients under steady-state conditions. This can be achieved in experimental arrangements similar to those described before, but where a continuous or perfusion operation mode is established (see Fig. 6c). An example of this configuration, applied in a microbial fermentation, is reported by Buse et al. (24). A fluctuating dissolved gas concentration can also be attained in one-compartment configurations simply by oscillating the total pressure of the system (Fig. 6d). An advantage of these configurations is that pressure gradients, which can occur in large-scale vessels, can also be simulated (23). Nonetheless, special pressure-proof vessels are required and foaming problems can originate during decompression.

A general disadvantage of single-compartment (open-loop and closed-loop) systems is their slow dynamic response. This is a drawback when short oscillation periods are desired but can be partially solved by increasing agitation and gassing rates. However, this solution cannot be applied to fragile cell cultures. A questionable aspect of single-compartment systems

is that all cells are simultaneously exposed to the same condition, potentially inducing a synchronization phenomena, something that is unlikely to occur in a real large-scale culture. Some limitations of the experimental arrangements described before can be overcome by two-compartment systems.

The most commonly used two-compartment configurations are composed of two interconnected stirred-tank bioreactors, each maintained at a different dissolved gas concentration, and where culture broth is circulated between them via a pump system (Fig. 6e) (25,26). In this case, mean circulation time is equal to the total volume of the two vessels divided by the circulation flow rate (25). A large-scale vessel can be considered to be composed of two main zones, a small well-mixed zone in the impeller vicinity and a poorly mixed region in the rest of the tank. Thus, vessels of different sizes are commonly used to simulate these two different culture volume fractions (25). Accordingly, for DO, the large vessel maintained at a low DO will simulate the large poorly mixed zone, whereas the small vessel maintained at a high DO will represent the well-mixed zone. An advantage of these systems is that the residence time distribution in each vessel will naturally simulate the circulation time distribution in a large-scale culture. In addition, algorithms for controlling dissolved gas concentration in each vessel are simpler than for one-vessel configurations because complex dynamic behavior of mass transfer and electrode response is not involved. Furthermore, short oscillation periods can be attained in these systems simply by increasing the flow rate between the two vessels. Nonetheless, shear stresses at the pumps will limit their use for fragile cells. Disadvantages of two-vessel configurations also include the need for duplicate equipment, increased contamination risk, and operating failure at the circulation loops/pumps. It could also be questionable that cells in the two-stirred-tank configuration experience a step change between two different dissolved gas concentrations, contrary to what happens in a real situation where cells gradually move between both conditions. This sudden step change can be prevented if a two-compartment configuration, as shown in Figure 6f, is used. In this case, cells are circulated from a stirred vessel to a plug-flow reactor, where a gradual change in dissolved gas concentration due to cell metabolism occurs (27). Such types of systems have been successfully used for microbial cultures with high metabolic demands but might be inadequate for animal and plant cells with relatively lower metabolic demands. Finally, it is interesting to note that for cultures where pH is controlled by a carbonate buffer system, as in most mammalian-cell cultures, the configuration shown in Figure 6e can also be used to simulate pH gradients simply by maintaining a different dissolved carbon dioxide concentration in each vessel.

Simulation of Substrate or pH Gradients

Substrate gradients can occur in large-scale cultures at feeding points during fed-batch and perfusion operation modes, and pH gradients can occur in acid/base addition zones. In both cases, cells will be transiently exposed to potentially deleterious high substrate or acid/base

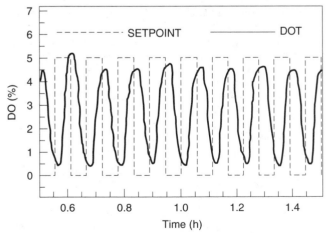

Figure 8. Closed-loop control strategy for oscillating dissolved oxygen at a predetermined frequency, amplitude, and oscillation axis. Reprinted from Ref. 18.

concentrations. Simulating such gradients in a small-scale bioreactor can be a challenge because, in contrast to the experimental designs discussed in the previous section, recovery of the basal condition after pH or substrate pulses strongly depends on metabolic activity. A notable exception is the case mentioned before for generating pH oscillations by fluctuations of dissolved carbon dioxide concentration. Various experimental configurations for simulating substrate concentration or pH gradients are schematically summarized in Figure 9. It should be noted that only very few reports exist describing these experimental arrangements, even for microbial fermentation systems. Accordingly, their utility in animal- and plant-cell culture, which present much lower metabolic demands than microbial fermentations, remains to be proven.

The simplest experimental design for simulating substrate or pH gradients consists of a chemostat with pulsed additions of substrate or acid/base solutions (Fig. 9a) (28). In this case, the circulation time is simulated by the frequency of the pulses, whereas the magnitude of the gradient is simulated by the concentration of the pulse. Alternatively, two-compartment systems have also been proposed for simulating nutrient and pH gradients (29,30). These usually consist of a stirred tank and a plug-flow reactor (PFR) connected in parallel in which medium is circulated between the two compartments via a pump system (Fig. 9b, c). The stirred-tank reactor represents the well-mixed zone of a large bioreactor, whereas the PFR represents the stagnant or feed injection zone (31). Depending on where substrate is added, zones of low or high nutrient concentration or pH can be simulated. If substrate is added in the stirred-tank reactor, as shown in Figure 9b, then zones of low substrate concentration will develop in the PFR. Likewise, zones of high substrate concentration will be formed in the PFR if the substrate is added directly into it. In this case, t_c will be represented by the residence time in each reactor, as already discussed in the previous section. Finally, experimental arrangements shown in Figure 9d can be designed for simulating gradients of dissolved gas concentration and pH or substrate concentration simultaneously (19). It should be stressed that the practical feasibility of applying such experimental designs in animal- and plant-cell culture still needs to be demonstrated. In particular, it should be tested if simulation of real cell-culture circulation times can be attained under practical operating conditions.

NOMENCLATURE

A	Area
C_p	Heat capacity
D	Impeller diameter
$D_{o/w}$	Diffusion coefficient of oxygen in water
H	Liquid height
K	Discharge coefficient
N	Impeller speed
P	Power
q	Specific production rate
Q_p	Pumping capacity
STR	Stirred tank reactor
t_c	Circulation time
t_M	Mixing time
t_{mt}	Time constant for mass transfer
t_{rxn}	Time constant for oxygen consumption
T_v	Vessel diameter
U	Overall heat transfer coefficient
V	Liquid volume
vvm	Volume of air per volume of liquid per minute
λ_k	Size of the smallest turbulence-generated eddy
μ	Specific growth rate
ρ	Liquid density

BIBLIOGRAPHY

1. D.I.C. Wang et al., *Fermentation and Enzyme Technology*, Wiley-Interscience, New York, 1979, pp. 208.
2. C. Anderson, G.A. LeGrys, and G.L. Solomons, *The Chem. Eng.* February, 43–49 (1982).
3. K.C.A.M. Luyben, in U. Mortensen and H.J. Noorman, eds., *Proceedings of the International Symposium on Bioreactor Performance*, Nordic Programme on Bioprocess Engineering, Denmark, 1993, pp. 159–169.
4. I.J. Dunn and E. Heinzle, in U. Mortensen and H.J. Noorman, eds., *Proceedings of the International Symposium on Bioreactor Performance*, Nordic Programme on Bioprocess Engineering, Denmark, 1993, pp. 189–206.
5. G.B. Tatterson, *Fluid Mixing and Gas Dispersion in Agitated Tanks*, McGraw-Hill, New York, 1991, pp. 121–323.
6. P.K. Yegneswaran, M.R. Gray, and B.G. Thompson, *Biotechnol. Bioeng.* **38**, 1203–1209 (1991).
7. N.W.F. Kossen, in E. Galindo and O.T. Ramírez, eds., *Advances in Bioprocess Engineering*, Kluwer Academic Publishers, Dordrecht, The Netherlands, 1994, pp. 53–65.
8. N.M.G. Oosterhuis, Ph.D. Thesis, Delft University of Technology, Delft, 1984.

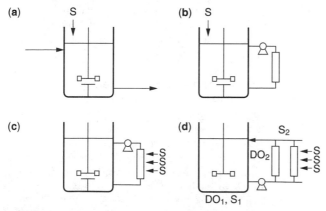

Figure 9. Experimental configurations for simulating substrate or pH gradients. (**a**) Chemostat. (**b**) STR and PFR. Substrate added to the STR to simulate substrate limited regions. (**c**) STR and PFR. Substrate added to the PFR to simulate high substrate concentration zones. (**d**) Combination of a stirred tank with two PFRs for simultaneous simulation of substrate and oxygen gradients. S, S_1 and S_2 refer to substrate, acid, or base feeding solutions.

9. A.P.J. Sweere, K.Ch.A.M. Luyben, and N.W.F. Kossen, *Enzyme Microb. Technol.* **9**, 386–398 (1987).

10. S. Frandsen, J. Nielsen, and J. Villadsen, in U. Mortensen and H.J. Noorman, eds., *Proceedings of the International Symposium on Bioreactor Performance*, Nordic Programme on Bioprocess Engineering, Denmark, 1993, pp. 171–179.

11. J. Feijen and J.J. Homeester, in M. Berovic and T. Koloini, eds., *Bioreactor Engineering Course Workshop Notes*, Boris Kidric Institute of Chemistry, Ljubljana, 1991, pp. 93–116.

12. J.E. Bailey and D.F. Ollis, *Biochemical Engineering Fundamental*, 2nd ed., McGraw-Hill, New York, 1986, pp. 208–214.

13. A.W. Nienow et al., *Cytotechnology* **22**, 87–94 (1996).

14. J. Feijen, J.J.M. Hofmeester, and D. Groen, in L. Alberghina, L. Frontali, and P. Sensi, eds., *ECB6: Proceedings of the Sixth European Congress on Biotechnology*, Elsevier Science, London, 1994, pp. 919–926.

15. F. Boysan, K.R. Cliffe, F. Leckie, and A.S. Scragg, in R. King, ed., *Bioreactor Fluid Dynamics*, Elsevier Applied Science, London, 1988, pp. 245–258.

16. S.M.G. Geraats, in E. Galindo and O.T. Ramírez, eds., *Advances in Bioprocess Engineering*, Kluwer Academic Publishers, Dordrecht, The Netherlands, 1994, pp. 41–46.

17. N.K.H. Slater, in U. Mortensen and H.J. Noorman, eds., *Proceedings of the International Symposium on Bioreactor Performance*, Nordic Programme on Bioprocess Engineering, Denmark, 1993, pp. 15–22.

18. A. De León, E. Galindo, and O.T. Ramírez, in R.D. Schmid, ed., *Biochemical Engineering 3*, Universität Stuttgart, Germany, 1995, pp. 200–202.

19. G. Larsson, S. George, M. Törnkvist, and S.O. Enfors, in U. Mortensen and H.J. Noorman, eds., *Proceedings of the International Symposium on Bioreactor Performance*, Nordic Programme on Bioprocess Engineering, Denmark, 1993, pp. 15–22.

20. M. Träger, G.N. Qazi, R. Buse, and U. Onken, *J. Ferment. Bioeng.* **74**, 282–287 (1992).

21. R.A. Taticek, M. Moo-Young, and R.L. Legge, *Plant Cell, Tissue Organ Cult.* **24**, 139–158 (1991).

22. M. Rhiel and D.W. Murhammer, *Biotechnol. Bioeng.* **47**, 640–650 (1995).

23. H. Takebe et al., *J. Ferment. Bioeng.* **78**, 93–99 (1994).

24. R. Buse, G.N. Qazi, and U. Onken, *J. Biotechnol.* **26**, 231–244 (1992).

25. N.M.G. Osterhuis, N.W.F. Kossen, A.P.C. Olivier, and E.S. Schenk, *Biotechnol. Bioeng.* **27**, 711–720 (1985).

26. G. Larsson and S.O. Enfors, *Bioprocess Eng.* **3**, 123–127 (1988).

27. C. Abel, U. Hübner, and K. Schügerl, *J. Biotechnol.* **32**, 45–57 (1994).

28. E. Heinzle, J. Moes, and I.J. Dunn, *Biotechnol. Lett.* **7**, 235–240 (1985).

29. P. Neubauer, L. Häggström, and S.O. Enfors, *Biotechnol. Bioeng.* **47**, 139–146 (1995).

30. F. Bylund et al., *Bioprocess Eng.* **20**, 377–389 (1999).

31. S. George, G. Larsson, and S.O. Enfors, in L. Alberghina, L. Frontali, and P. Sensi, eds., *ECB6: Proceedings of the Sixth European Congress on Biotechnology*, Elsevier Science, London, 1994, pp. 883–886.

See also BIOREACTOR SCALE-UP; BIOREACTORS, STIRRED TANK; CELL CYCLE SYNCHRONIZATION; CELL GROWTH AND PROTEIN EXPRESSION KINETICS; OFF-LINE ANALYSIS IN ANIMAL CELL CULTURE, METHODS; ON-LINE ANALYSIS IN ANIMAL CELL CULTURE.

BIOREACTOR SCALE-UP

LAURA A. PALOMARES
OCTAVIO T. RAMÍREZ
Instituto de Biotecnología
Universidad Nacional Autónoma de México

OUTLINE

Geometric Considerations
    Aspect Ratio, Homogeneity, and Gradients
    Heating and Cooling
Nomenclature
Bibliography

Bioprocess and bioreactor scale-up depends strictly on the substance being produced. For many biotechnology products, such as antibiotics, alcohols, organic acids, amino acids, and enzymes, a commercially viable process is attained only if a final fermenter scale on the order of hundreds of thousands of liters is reached. Such large volumes are needed to satisfy markets requiring very large amounts (10^2 to 10^7 ton/year) of products with very low added value (10^{-1} to 10^3 \$/kg). In contrast, products derived from animal and plant-cell culture are generally costly therapeutics or fine chemicals whose price can be as high as 10^4 to 10^9 \$/kg, but their worldwide demand is only 10^{-1} to 10^3 kg/year. Such behavior is depicted in Figure 1 showing an inverse logarithmic relationship between product costs and market size (1). Accordingly, to produce many substances derived from tissue culture,

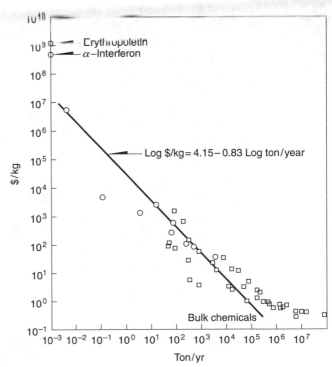

Figure 1. Relationship of chemical prices to annual production. Cell culture derived products (○). Adapted and reprinted by permission from Ref. 1.

at most only hundreds to thousands of liters per year are required to satisfy the demand, and thus, the largest bioreactors needed hardly come close to the scales of other biotechnology products. This can be exemplified by two extreme cases, human erythropoietin (EPO) and taxol. EPO has the largest market for any therapeutic recombinant protein produced by mammalian-cell culture. At an EPO selling price of $0.012/U (130,000 U/mg), the total estimated 1998 U.S. demand ($1,600 \times 10^6$/yr) could be satisfied with only ca. 1 kg/yr. This production could be generated through a very modest hypothetical bioprocess (batch operation, overall yield 500 U/mL-cycle, 10-day cycles) in a single bioreactor of less than 10,000 L. Similarly, the estimated world requirements of taxol (200 kg/yr at a bulk price of $500,000/kg) could be satisfied by a single 70,000-L plant cell bioreactor yielding 0.25 g/L of product and operated in 15-day cycles (assuming a 50% purification yield) (1,2). These examples assume that the markets are satisfied by a single manufacturer using a single vessel, which is an improbable scenario. Furthermore, the development of improved cell lines and bioreactor operation modes, such as perfusion culture, has resulted in substantial increases in cell and product concentrations, which would further reduce the bioreactor volumes calculated.

In general, although the largest homogeneous cultures of suspended animal cells range between 4,000 to 20,000 L, inhomogeneous systems (microcarrier, packed-beds, etc.) for anchorage-dependent animal cells usually do not exceed the 1000-L scale. Problems, such as mass transfer limitations and concentration gradients inherent in inhomogeneous systems, can explain their lower scales. Published data on the industrial production of products derived from cell culture are limited. Some of the largest cultures reported in the literature are summarized in Table 1 (2–10). For example, the foot-and-mouth disease vaccine obtained from baby hamster kidney cells (BHK21) has been produced by Wellcome for almost 30 years in a 1,000-L reactor, and more recently at a 10,000-L scale (3).

Cells from human origin (lymphoblastoid Namalwa cells) have been cultured for interferon production in reactors up to 8,000 L (6), and hybridomas for Mab production have been cultured in 2,000-L air-lift bioreactors (7). Some of the largest reported scales for microcarrier cultures, packed-beds and ceramic cartridges, are 7,000, 100 and 240 L, respectively (5,8). Compared to animal cells, the scales of plant-cell cultures tend to be larger. For instance, tobacco plant cells have been cultured in 20,000-L reactors (3), and a 75,000-L bioreactor is in operation for taxol production (2). Nevertheless, despite these successful experiences, large-scale culture of higher eukaryotic cells is still a challenge, particularly for anchorage-dependent cell lines or fragile cells, such as insect cells.

The first step in translating a cell-culture process to commercial scale is to define the product requirements, the overall production and the downstream processing economics. A decision has to be made whether a single large-scale batch reactor or smaller multiple continuous reactors are used. This decision can be made only after considering the whole process, the operating and capital costs, available facilities, and even assessment of the consequences of a possible contamination during the culture. From a technical standpoint, one of the major issues for scaling up a cell-culture bioreactor is the oxygen transfer problem. In general, the fragile nature of most higher eukaryotic cells and foaming problems due to the proteinaceous composition of many medium formulations limit the use of high power inputs and intense submerged aeration. As a process is scaled up, bioreactors must be more reliable, safer, and cheaper and must comply with regulations. In particular, the tight regulation imposed on the production of cell-culture-derived therapeutics has significantly constrained the free development of improved and novel bioprocesses on a commercial scale. The dogma that the "process determines the product," mostly hatched from our analytical limitations in fully characterizing cell-culture products, has resulted in a

Table 1. Some of the Largest Scale Cultures Reported for Higher Eukaryotic Cells

Cell Line	Scale (L)	Reactor	Product	Reference
Taxus sp.	75,000	Agitated-tank	Taxol	2
Tobacco cells	20,000	Agitated-tank	Nicotine	3
BHK-21	10,000	Agitated-tank	Foot-and-mouth disease vaccine	4
CHO	10,000	Agitated-tank	tPA	5
Namalwa cells	8,000	Agitated-tank	Lymphoblastoid interferon	6
Bowes melanoma	7,000	Agitated-tank/ microcarriers	tPA	5
Murine hybridomas	2,000	Air-lift	Mab	7
Vero cells	1,000	Agitated-tank/ microcarriers	Killed polio virus vaccine	8
Murine hybridomas	1,000	Stirred-tank	Mab vs. cell-surface antigens of adenocarcinomas	9
BHK cells	500	Agitated-tank/ perfusion	Factor VIII	10

costly and treacherous path for any company trying to modify its approved process. Thus, many therapeutics are produced by suboptimal processes and scales. However, this should change as regulation moves toward focusing on the characterization of products rather than on the bioprocess.

GEOMETRIC CONSIDERATIONS

Aspect Ratio, Homogeneity, and Gradients

Stirred Tanks. The purpose of a bioreactor is to provide the appropriate ambient conditions required for maximal growth and/or product production in a contained system. Ideally, the culture environment should also be homogeneous and controlled. These conditions can be effectively achieved in agitated tanks, which are the preferred design for scaling up anchorage-independent cell cultures. Stirred-tank bioreactors for cell culture are cylindrical vessels with an aspect ratio (liquid height to diameter ratio) usually in the range of $1:1$ to $3:1$ (3,4). Nonetheless, the aspect ratio should be kept under $2:1$, and impellers should be spaced to avoid top-to-bottom heterogeneity (1.0 to 1.5 impeller diameters apart) (11). Furthermore, oxygen transfer from the liquid surface, which is particularly important for animal-cell culture, can be improved by keeping the aspect ratio close to unity. Special attention should be given to the headspace dimensions if foaming problems are anticipated. In general, foam and liquid height increase from gas holdup are handled by leaving 20–30% of the vessel empty (11). A common feature of stirred-tank bioreactors is a hemispherical bottom, which prevents stagnant zones even at low agitation rates. Nonetheless, as pointed out by Charles and Wilson (11), the cost of a hemispherical bottom vessel can increase total costs for the vessel by as much as 50%, yet, no clear evidence of its superior performance over a standard dished-bottom vessel exists. Probably the feature most distinguishing between different stirred-tank bioreactor designs is the impeller configuration. Many different types of impellers have been used to culture animal and plant cells in stirred tanks, including marine propellers, sails, pitched blades, Vibromixers (perforated discs placed on a vertically reciprocating shaft), anchors, paddles, helical screws, cell lift and centrifugal impellers (for a more detailed discussion (see Biorectors, Airlift) Ref. 12). Typical geometric characteristics of commonly used impellers are shown in Figure 2 (12–14). Many impellers have been designed to prevent mechanical damage to the fragile cells and still maintain a homogeneous environment and high oxygen transfer rates. Likewise, mechanical damage to the cells can be reduced by excluding baffles, minimizing other inserts, and placing the impeller shaft eccentrically to avoid vortex formation. Finally, vessel design should consider the best use of available materials. However, excessive costs are usually reduced by simply selecting a vessel from reactors already available from specialized vendors.

Tank geometry must be carefully selected because it influences the homogeneity of the bioreactor, especially as scale increases. Furthermore, heat and mass transfer are directly influenced by mixing, which in turn depends on the operating variables and geometric characteristics of the bioreactor. For instance, the mixing time (the time required for the system to reach a specified degree of uniformity) for a nonaerated Newtonian fluid in a baffled stirred tank with a flat-blade turbine impeller can be obtained from Figure 3 and equation 1 (14–16).

$$N^* = N t_M \left(\frac{g}{N^2 D} \right)^{1/6} \frac{D^2}{T_v^{1.5} H^{0.5}} \tag{1}$$

where N^* is dimensionless mixing time, N is the impeller speed, t_M is the mixing time, g is gravitational acceleration, D is the impeller diameter, T_v is the vessel diameter, and H is the liquid height.

Only scarce information exists for homogenization and concentration gradients in large-scale animal or plant-cell bioreactors. Among such studies, Langheinrich et al. (17) obtained mixing time data for typical operating conditions (maximum agitation speed of $1 \, s^{-1}$, maximum air sparging rate of 0.005 VVM, superficial air velocity of $2 \times 10^{-4} \, m/s$) of an 8,000-L stirred-tank bioreactor (Rushton turbine $D = 2/9 T_v$) used for animal-cell culture at Glaxo Wellcome Research and Development. Because a draw and fill operation is employed during production, aspect ratios of 0.3, 1.0, and 1.3 were studied. It was shown that mixing time, obtained from pH tracer and decolorization experiments, could be correlated with the total energy dissipation rate, ε_T (see Figs. 4 and 5), defined by equation 2:

$$\varepsilon_T = \frac{P_o \rho N^3 D^5}{V} + u_G g \rho \tag{2}$$

where P_o is the power number (see eq. 27), ρ is the liquid density, V is the liquid volume, and u_G is the superficial gas velocity. The first term on the right-hand side of Equation 2 is the contribution from agitation, whereas the second term is from aeration. Interestingly, experimental data of Langheinrich et al. (17) fitted literature correlations well (equations 3 and 4) obtained under agitation and aeration conditions two orders of magnitude above those studied in the animal-cell-culture bioreactor.

$$t_M N P_o^{1/3} = 5.3 \left(\frac{T_v}{D} \right)^2 \tag{3}$$

for aspect ratios of 1 (18), and

$$t_M N P_o^{1/3} = 3.3 \left(\frac{H}{D} \right)^{2.43} \tag{4}$$

for aspect ratios of 1 to 3 (19).

Several important conclusions for the particular system used are drawn from the study of Langheinrich et al. (17). First, gradients in pH and dissolved oxygen can occur in large-scale animal-cell culture because long mixing times in the range of 40 to 200 s were obtained under typical operating conditions. Second, homogenization is dramatically affected by geometric characteristics. A fourfold increase in mixing times can occur as aspect ratios

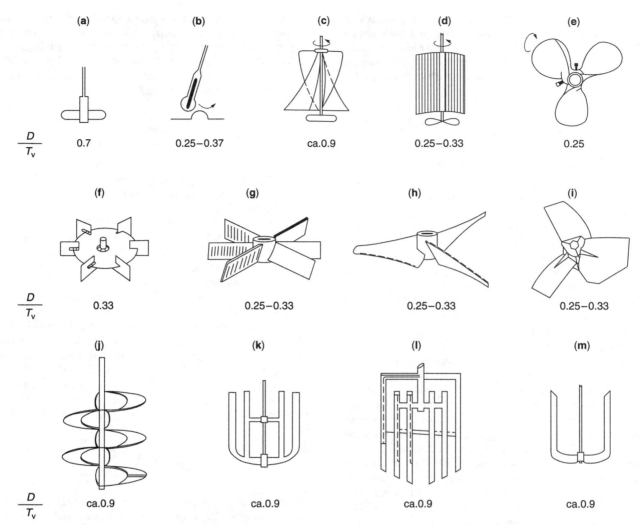

$\dfrac{D}{T_v}$

(a) 0.7   (b) 0.25–0.37   (c) ca.0.9   (d) 0.25–0.33   (e) 0.25

(f) 0.33   (g) 0.25–0.33   (h) 0.25–0.33   (i) 0.25–0.33

(j) ca.0.9   (k) ca.0.9   (l) ca.0.9   (m) ca.0.9

Figure 2. Commonly used impellers and their geometric characteristics, used in animal and plant cell culture. D = impeller diameter; T_v = vessel diameter; (**a**) suspended magnetic bar; (**b**) spherical rotating flex stirrer; (**c**) sail impeller; (**d**) membrane basket agitator; (**e**) marine propeller; (**f**) disc turbine; (**g**) angled blade; (**h**) and (**i**) profiled impellers; (**j**) helical ribbon impeller; (**k**) gate-anchor impeller; (**l**) gate impeller; (**m**) anchor impeller; Adapted from (Ref. 12–14).

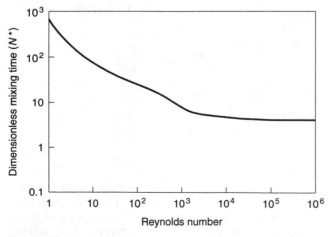

Figure 3. Dimensionless mixing time as a function of Reynolds number for turbine impellers in baffled vessels as determined from pH tracer experiments. Reprinted by permission from (Ref. 16).

increase from 0.3 to 1.3 (Fig. 4). And third, sparging substantially enhances mixing, particularly in the upper parts of the vessel and at aspect ratios greater than 1 (Fig. 5).

It has been reported that poor mixing in stirred-tank reactors causes significant problems in myeloma, hybridoma, CHO, and insect-cell cultures at high cell density (20–22). Undesirable carbon dioxide accumulation occured in 110 to 500-L reactors as a result of low mass transfer characteristics of the systems. Furthermore, inhomogeneities in microcarrier concentration and pH gradients have been determined in large-scale stirred vessels (23,24). For instance, severe localized cell lysis problems at the point of base addition, allegedly caused by long mixing times, were observed in a 15-L reactor with cell retention (22). Another problem aggravated by deficient mixing is culture segregation, which originates from the formation of cell aggregates and clumps during high cell-density cultures (22). Problems related to mixing can negatively affect overall culture performance. However, due to the fragile nature of animal cells, such problems

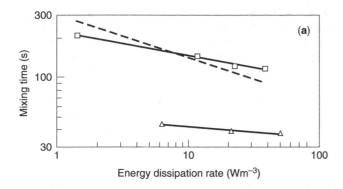

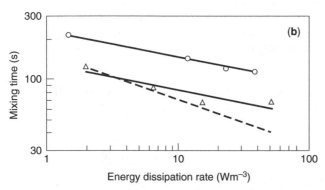

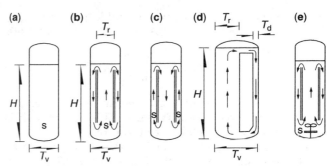

Figure 6. Typical geometric characteristics of common air-lift and bubble-column bioreactors used for cell culture. For all the vessels shown, $H/T_v > 2$. For internal loop air-lifts, $T_r/T_v > 0.6$. For external loop air-lifts, $T_r/T_d \approx 2$. (**a**) bubble column; (**b**) and (**c**) internal loop air-lifts; (**d**) external loop air-lift; (**e**) air-lift with impeller. Arrows indicate direction of liquid flow and "s" indicates sparger position. Adapted from (Ref. 13).

Figure 4. Effect of specific energy dissipation rate from agitation on mixing time. Determinations based on pH tracer experiments for various aspect ratios in an unaerated large-scale bioreactor. (**a**) $H = 0.3\, T_v(\triangle)$, $H = 1.3\, T_v(\square)$, dashed line = prediction by Cooke et al. (19); (**b**) $H = T_v(\triangle)$, $H = 1.3\, T_v(\bigcirc)$, dashed line = prediction by Ruszkowski (18). Reprinted by permission from (Ref. 17).

Air-Lift and Bubble-Column Bioreactors. Air-lift and bubble column reactors are also commonly used configurations for culturing suspended cells (typical geometric characteristics are shown in Fig. 6). The height to diameter ratio of these reactors can range between 6 : 1 to as high as 12 : 1, but is typically 10 : 1. Nonetheless, air-lift reactors with aspect ratios even below 2 have been reported for animal- and plant-cell cultures. A more detailed discussion of air-lift and bubble-column bioreactors can be found in the article Bioreactors, Airlift. As in stirred-tank vessels, geometric characteristics directly influence homogeneity, mass, and heat transfer in air-lift and bubble-column bioreactors. For instance, the relationship between geometric characteristics and mixing time is shown in equations 5 and 6 (14,25):

$$t_M = Bt_c \left(\frac{A_d}{A_r}\right)^{0.5} \tag{5}$$

where B is a constant equal to 3.5 and to 5.2 for internal- and external-loop air-lift reactors, respectively, A_d and A_r are the downcomer and riser cross-sectional areas, respectively, and t_c is the circulation time or time taken for a liquid element to complete a circulation cycle in the reactor. Circulation time of air-lift reactors (either internal-loop or external-loop with short horizontal top and bottom sections) can be determined from the superficial liquid velocity and geometric parameters according to

$$t_c = \frac{H_r}{u_{Lr}} + \frac{H_d A_d}{u_{Lr} A_r} \tag{6}$$

where H_r and H_d are the riser and downcomer heights, respectively, and u_{Lr} is the superficial liquid velocity in the riser. u_{Lr} is affected, among other variables, by liquid density, gas holdups, frictional loss coefficients, and frictional pressure drops in riser and downcomer through complex relationships usually solved by iterative algorithms. However, simplified correlations between u_{Lr} and geometric characteristics and operating parameters have been developed for particular conditions. For instance, Popovic and Robinson (26) proposed the following equation

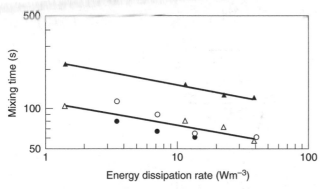

Figure 5. Effect of specific energy dissipation rate from agitation and aeration on mixing time. Determinations based on pH tracer experiments for various aspect ratios in a large-scale bioreactor. Closed symbols; determinations at the liquid surface. Open symbols; determinations at the end of impeller discharge stream. Unaerated ($\triangle$) and aerated at $u_G = 2 \times 10^{-4}$ m/s ($\bigcirc$). Adapted and reprinted by permission from (Ref. 17).

cannot be solved by increasing agitation intensity. Therefore, other solutions must be found. For instance, pH gradients can be eliminated by adding base at multiple points properly localized in the reactor (22), and carbon dioxide removal can be enhanced by manipulating air bubble size by suitable design of the sparging system (21).

for a pseudoplastic fluid in an external-loop air-lift ($u_{Gr} \geq$ 0.04 m/s, $H = 1.88$ m, and A_d/A_r of 0.111, 0.25, and 0.44):

$$u_{Lr} = 0.23\, u_{Gr}^{0.32} \left(\frac{A_d}{A_r}\right)^{0.97} \mu_{app}^{-0.39} \qquad (7)$$

where u_{Gr} is the superficial gas velocity in the riser and μ_{app} is the apparent viscosity (Pa s). This last variable is important only in some plant-cell cultures.

Using equations 1, 5, 6, and 7, Doran (14) compared the time constants of hypothetical 10,000-L stirred-tank and external-loop air-lift plant-cell reactors of typical dimensions (details of geometric characteristics are given in Ref. 14). Several conclusions can be drawn from the results of such simulation, shown in Figure 7. First, under typical operating conditions, the mixing time for a large-scale air-lift can be substantially longer than that for a stirred vessel of the same volume. For instance, whereas mixing times from 200 s to as long as 1000 s were predicted for the air-lift, the range of mixing times for the stirred vessel was only between 20 to 200 s. A similar conclusion

was reported by Bello et al. (25) where the mixing time per unit volume of stirred tanks was three to five times shorter than in air-lift reactors (either internal- or external-loop) at the same total power input per unit volume. It is interesting to note that the calculations shown in Figure 7 agree closely with experimental determinations by Langheinrich et al. (17), discussed in the previous section, for an animal-cell stirred-tank reactor of similar size. Doran (14) also calculated the time constants t_{rxn} for oxygen consumption (ratio of equilibrium oxygen concentration in the culture liquid to the volumetric oxygen consumption rate), and for mass transfer, t_{mt} (inverse of the volumetric mass transfer coefficient). Mixing problems will be encountered if $t_M > t_{rxn}$, and oxygen limitation will occur if $t_{mt} > t_{rxn}$. Accordingly, it was concluded that even for the long mixing times of the air-lift reactor, oxygen limitation or dissolved oxygen gradients will not occur at low plant-cell concentrations (5 kg/m^3, Fig. 7a). Nonetheless, as the cell concentration increases to 30 kg/m^3, mixing becomes limiting in the air-lift reactor at gas velocities below 0.5 m/s, and thus,

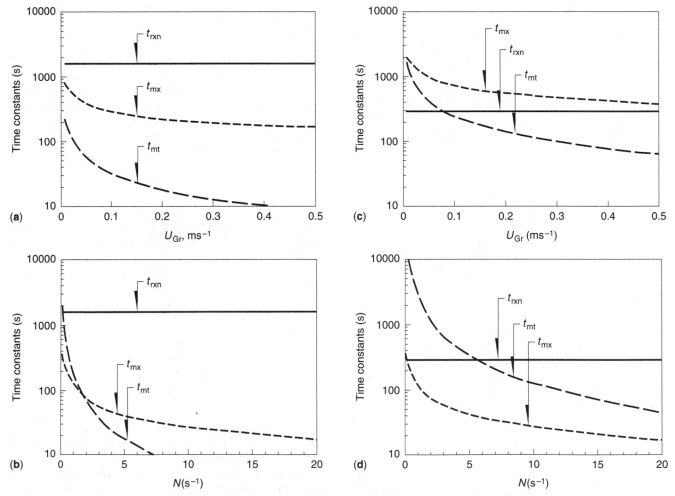

Figure 7. Comparison of time constants for 10,000-L hypothetical plant-cell bioreactors. (**a**) external loop air-lift reactor, cell concentration 5 kg/m^3 dry weight; (**b**) stirred-tank reactor, cell concentration 5 kg/m^3 dry weight, (**c**) external loop air-lift reactor, cell concentration 30 kg/m^3 dry weight; (**d**) stirred-tank reactor, cell concentration 30 kg/m^3 dry weight. Reprinted by permission from (Ref. 14).

oxygen gradients will develop (Fig. 7c). Furthermore, for high cell concentration, oxygen limitation will occur in the air-lift reactor below gas velocities of 0.1 m/s and in the stirred-tank reactor below an agitation rate of 6 s^{-1} (Fig. 7c and 7d). Notice that, due to cell fragility, most bioreactors are operated at gas velocity and agitation rates well below 0.5 m/s and 6 s^{-1}, respectively.

Mixing problems in air-lift vessels have been documented experimentally (27). As for stirred-tank vessels, the fragile nature of most higher eukaryotic cells limits the options for improving homogeneity in air-lift and bubble-column reactors. Accordingly, some solutions to mixing problems, such as adding impellers to the draft tube of air-lift reactors have not been widely applied and appear to contradict the design essence of an air-lift reactor. Suitable design of reactor geometry can be an efficient alternative to improving homogeneity. For instance, minimal mixing times of internal loop air-lift reactors have been observed for riser to column diameter ratios (T_r/T_v) above 0.6 (14). Accordingly, a riser to downcomer cross-sectional area equal to unity ($T_r/T_v = 0.71$) has been proposed as an effective design criterion (28). Other geometric characteristics that can considerably affect homogeneity of air-lift reactors are the draft tube and sparger positions, the draft tube to liquid height ratio, and the curvature of the bottom sections.

Two types of flow regimes have been described in bubble column reactors, homogeneous and heterogeneous (29). The homogeneous flow regime, characterized by the absence of a circulatory flow, occurs at low gas superficial velocities (<0.04 m/s) and when sparger holes are uniformly distributed on the bottom of the vessel. Typically bubble columns operate under a heterogeneous flow regime where circulatory flow is present as a result of an uneven distribution of sparger holes or high superficial gas velocities. Mixing time has been correlated with geometric characteristics and operation parameters for bubble columns under a heterogeneous flow regime according to (29)

$$t_M = 11\frac{H}{T_v} + (gu_{Gs}T_v^{-2})^{-0.33} \tag{8}$$

where u_{Gs} is the superficial gas velocity corrected for local pressure.

A comparison of calculated mixing times for air-lift and bubble-column bioreactors has been presented by van't Riet and Tramper (29). This comparison revealed that for low superficial gas velocities (0.001 m/s), mixing times of bubble column reactors were shorter than for air-lifts. The opposite occurs at gas velocities above 0.001 m/s. Nonetheless, for all conditions calculated and an aspect ratio of 10, differences in mixing times between both reactors were small and never exceeded 30%. Furthermore, such differences were even smaller as the reactor volume increased. Accordingly, the behavior of large scale air-lift and bubble-column reactors in terms of homogeneity and the presence of concentration gradients should be similar.

Although high superficial velocities improve homogeneity and prevent formation of concentration gradients in air-lift and bubble-column reactors, foaming problems and bubble-associated cell damage limit the use of high aeration rates. Tramper et al. (30) proposed a killing volume hypothesis where cells are killed within a hypothetical volume associated with each air bubble. The specific death rate constant k_d was then correlated with the operating and geometric characteristics of bubble columns according to

$$k_d = \frac{24F_gV_k}{\pi^2 d_b^3 T_v^2 H} \tag{9}$$

where F_g is the gas flow rate, d_b is the bubble diameter, and V_k is the killing volume which is constant for a particular system and can be determined experimentally. Equation 9 shows that the aspect ratio of bubble-column bioreactors can have important consequences on animal-cell survival. For instance, a decrease in aspect ratio will result in an increase in the death-rate constant for a given reactor volume and superficial gas velocity. Thus, a high aspect ratio should be the design criterion.

Inhomogeneous Bioreactors. Inhomogenous systems can be defined as those where cells grow either attached to or entrapped by a solid substratum arranged within a particular device or bioreactor configuration. Many different types of substrata and geometric configurations exist, however inhomogeneous systems can be simply classified according to Table 2 (31). In almost all inhomogeneous systems, oxygen transfer will be the limiting factor, and thus, as discussed below, dissolved oxygen concentration gradients will be present. Accordingly, predicting dissolved oxygen profiles constitutes the basis for a rational design of many inhomogeneous systems. The simplest designs include static units, such as dishes, flasks, and trays, where oxygen is transferred through the liquid surface only by diffusional mechanisms. As described by Murdin et al. (32), the dissolved oxygen profile can then be predicted from a steady-state mass balance according to

$$c = \frac{OUR}{2D_{o/w}}\left(\left[\frac{2D_{o/w}c_m}{OUR}\right]^{0.5} - y\right)^2 \tag{10}$$

where c and c_m are dissolved oxygen concentration in the liquid and at saturation, respectively; OUR is oxygen

Table 2. Clasification of Inhomogeneous Systems[a]

Plate devices	
Static	T-flasks, plate units, trays, multiwell plates, petri dishes
Dynamic	Plate units, roller bottles
Films	Bags, spiral wound
Tubes	Macroforms, hollow fibers
Packed beds	Ceramic, glass spheres, helixes, springs, diatomeceous earth, polystyrene jacks, sponges
Monolithic	Ceramic matrix
Microcarriers	Simple solid, collagen-coated, porous
Microencapsulation	Alginate, agarose, polylysine

[a]Adapted from (Ref. 31).

uptake rate; y is liquid depth, and $D_{o/w}$ is the diffusion coefficient of oxygen in water (2.6×10^{-5} cm^2/s). For a typical maximum cell concentration (3×10^6 cell/mL) and typical specific oxygen uptake rate (4×10^{-10} mmol/cell-h), equation 10 predicts that medium height must be maintained below 2 mm to avoid oxygen limitation. This result indicates that high superficial area to culture volume ratios A/V, usually in the range of 2.5 to 5 cm^{-1}, are required, that is, two to three orders of magnitude higher than for large-scale homogeneous bioreactors. A modest increase (two- to three-fold) in substratum area without a corresponding increase in A/V ratios can be achieved by circulating medium through multistacked static units or by rotating vessels and flasks containing multiple internal surfaces. Scalability of high A/V culture units is limited because their cumbersome configurations require labor-intensive operations. Yet, volumes as large as 200 L have been reported (31). High substratum areas to culture volume ratios and concomitantly high cell concentrations ($>1 \times 10^8$ cell/mL), can be attained in some inhomogeneous bioreactors, such as packed-beds (4 to 100 cm^{-1}) and hollow-fiber (30 to 200 cm^{-1}). High gas/liquid areas needed for oxygen transfer are eliminated by perfusing an oxygen saturated medium. This results in compact bioreactor designs, as well as in the possibility of culturing anchorage-independent cells in a perfusion mode at low shear stresses and close to *in vivo* cell concentrations.

Although inhomogeneous systems are extensively used in lab-scale applications and commercial size operations, a strictly controlled and uniform environment cannot be maintained. Thus undesirable heterogeneities and existence of concentration gradients will inevitably occur during scale-up. Accordingly inhomogeneous systems are usually scaled up by increasing the number of small scale units, rather than increasing the size of the equipment. Many disadvantages exist for such an approach, including higher capital and operating costs, poorly controlled and monitored operation, and unit-to-unit variability. Therefore, the challenge is to design inhomogeneous bioreactors that approach a more uniform environment. This can be achieved through an appropriate design of geometric characteristics and operating conditions. Because many different types of inhomogenous systems exist, each case deserves a unique analysis. Nonetheless, geometric and homogeneity considerations of microcarrier systems, hollow-fiber, and packed-bed reactors are analyzed below because these are among the most important systems for commercial-scale operations. A more detailed discussion of inhomogeneous bioreactors can be found in the article Bioreactors, Continuous Culture of Plant Cells.

Hollow Fiber Reactors. Conventional hollow-fiber (HF) reactors are composed of a bundle of hollow fibers sealed in an outer casing, that form a shell-and-tube configuration 3.2 to 9 cm in diameter and 6 to 30 cm long. HF are anisotropic structures composed of a thin (0.1 to 2.0 µm) inner or "active" skin and a thick (50 to 85 µm) spongy support layer. The internal fiber diameter commonly ranges from 40 to 200 µm, and the molecular weight cutoffs (molecular weight of species retained) vary from 300 to 300,000 Da. The medium is perfused from the

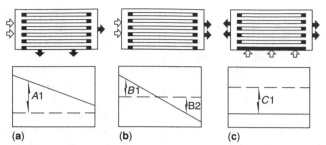

Figure 8. Schematic representation of common operating modes and pressure distribution in hollow-fiber reactors. Open arrows indicate feed inlet, and closed arrows indicate effluent streams. The tube-side pressure is represented by a solid line and shell-side pressure by a dashed line. (**a**) Open-shell ultrafiltration and transmembrane pressure difference, A1. (**b**) Closed-shell ultrafiltration and transmembrane pressure difference B1 and B2. (**c**) Cross-flow ultrafiltration and transmembrane pressure difference, C1. Reprinted by permission from (Ref. 33).

lumen side and transported through the porous membrane to the shell side, where the cells are immobilized. As illustrated in Figure 8, there are three common operating modes (open shell, closed shell, and cross-flow), each resulting in a different pressure distribution throughout the HF cartridge (33). The axial pressure drop inside the fibers can be expressed, to a good approximation, by equation 11, obtained from the Hagen–Poiseuille equation for flow in a pipe (34):

$$\frac{dp_c}{dz} = -\frac{8\mu_b}{r_c^2}\overline{u}_c \qquad (11)$$

where p_c and $\overline{u}_c$ are pressure and average axial velocity inside the capillary, respectively; Z is the axial position; and r_c is the inside capillary radius. The transmembrane flux depends on the pressure difference between the shell and lumen sides of the fiber. For the open and closed-shell operating modes, the pressure gradient and thus medium flux decreases along the axial direction, causing undesirable concentration gradients. The result is a higher cell concentration at the inlet section and only negligible growth after a certain cartridge length. Most commercial HF devices operate in the closed mode, resulting in reversed flux at the downstream section of the reactor. Thus, whereas fresh medium is supplied to the cells in the inlet section, exhausted medium and toxic wastes are removed in the outlet segment, creating severe environmental heterogeneities. Accordingly, scaleup of HF reactors in the axial direction is limited and explains the short length of commercially available devices. In conventional HF reactors, only diffusional transport exists on the shell side. Thus, fibers must be tightly packed to avoid nutrient limitation and/or by-product accumulation at the interstitial zones. Even so, necrosis usually occurs because the distance from the interstices to the inner capillary skin is 150 µm at best. Necrotic zones will still occur even if convective transport is forced through the shell side, because the tissue-like cell concentrations prevent a uniform flow distribution. Accordingly, severe radial gradients are also present in many HF reactors. Concentration polarization and fouling of the inner lumen surface is an additional problem that can lead to reactor inhomogeneities. Experimental determinations of axial and radial heterogeneities are well documented (33,35,36).

Many solutions have been proposed to improve homogeneity and prevent nutrient limitation and by-product buildup in HF reactors, most of them based on modifying geometric characteristics and operating conditions (35,36), for instance, combination of fibers and membranes in flat-bed HF arrangements, introduction of porous distribution tubes to perfuse either fresh medium or oxygen, mixing aeration and medium supply fibers, combination of different fiber types, selection of molecular weight cutoffs, and flow alternation between shell and lumen sides. Still scale-up of HF reactors is performed primarily by increasing the number of devices modularly. Predicting dissolved oxygen profiles in the radial direction can help design more efficient HF reactors (fiber size, distance between fibers, etc). A simplified approach is to perform a steady-state mass balance around a cylindrical capillary where only diffusion is considered. Dissolved oxygen profiles are then obtained, as detailed by Murdin et al. (32), from the following equation:

$$c = c_m + \frac{OUR}{4D_{o/w}}\left[r^2 - r_c^2 - 2r_0^2 \log\frac{r}{r_c}\right] \quad (12)$$

where r is the radius and r_0 is the radius at which the oxygen concentration falls to zero.

Packed-Bed Reactors. Packed-bed reactors are also widely used to culture anchorage-dependent and independent cells (37). In general, an aspect ratio of 3 is suitable for packed-beds. As shown in Table 2, many different materials have been used as packing matrixes. The diameter of the various matrix particles commonly ranges from 0.1 to 5 mm, resulting in typical void fractions of 0.3 to 0.6. Depending on the packing material used, very high substratum area to culture volume ratios can be reached (see above), which approximate those of hollow-fiber reactors. However, in contrast to hollow-fiber reactors, properly designed packed-bed reactors should not present radial gradients, and thus, a straightforward scale-up in this direction is possible. Still, axial concentration gradients, are inherently present and will limit reactor length. Again, dissolved oxygen is almost always the limiting substrate, and its concentration profile along the axial direction can be determined using a macroscopic mass balance (38):

$$c = c_0 - \left[\frac{OUR}{u_L}\right]z \quad (13)$$

where c_0 is the inlet dissolved oxygen concentration and u_L is the superficial liquid velocity based on an empty bed. Concentration gradients will decrease as u_L increases (see equation 13). Nonetheless, the maximum u_L will be limited by the permissible maximum pressure drop, which can be determined from the Ergun equation (39):

$$\Delta_p = \frac{150z\mu_b(1-\varepsilon)^2 u_L}{T_p^2\varepsilon^3} + \frac{1.75z\rho u_L^2(1-\varepsilon)}{T_p\varepsilon^3} \quad (14)$$

where ε is the void fraction, T_p is the particle diameter, and μ_b is the culture medium viscosity. Maximum shear

stresses in the bed must also be considered to determine the maximum allowable u_L.

Microcarriers. Stirred vessels and air-lifts are the reactor configurations used for cultivating cells attached to microcarriers. Furthermore, the density of microcarriers is only slightly above that of the culture medium (1.03 to 1.05 g/mL). Thus, geometric considerations and homogeneity characteristics of the bulk liquid phase will be the same as those for homogeneous systems (detailed in previous sections). Nonetheless, concentration gradients within porous microcarriers (typical diameter range between 100 and 500 μm) can still exist. Dissolved oxygen profiles within porous microcarriers can be modeled from a steady-state mass balance of oxygen diffusion into a sphere, as described by Murdin et al. (32). Alternatively, mass transfer limitations and thus formation of undesirable gradients within a porous microcarrier can be determined from plots of the observable module (obtained from the Thiele modulus) Φ versus the effectiveness factor η, shown in Figure 9 (40). For spherical particles and OUR as the rate of interest, Φ and η are defined as

$$\eta = \frac{observed\ OUR}{OUR\ without\ diffusional\ limitation} \quad (15)$$

$$\Phi = \frac{OUR_0}{D_e c_m}\left(\frac{V_p}{A_p}\right)^2 \quad (16)$$

where A_p and V_p are the external surface area and volume of a microcarrier, respectively, and OUR_0 is the observed oxygen uptake rate. D_e is the effective diffusion coefficient, which in turn is a function of the microcarrier porosity and tortuosity factor. From Figure 9 and equations 15 and 16, it can be seen that oxygen limitation will occur at an Observable Module above 0.3. Accordingly, oxygen limitation can be avoided ($\Phi < 0.3$) through proper design of microcarrier geometry (see Fig. 9 and equations 15 and 16). This approach can also be applied to other immobilized systems such as hollow fibers and microencapsulation.

Heating and Cooling

Maintaining a constant and homogeneous temperature is an essential requirement in cell culture. Temperature

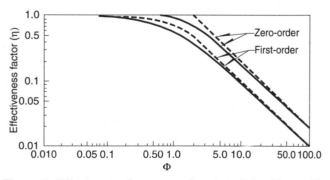

Figure 9. Effectiveness factor as a function of the Observable Modulus. Michaelis–Menten kinetics. Continuous line; spherical geometry; dashedline; slab geometry. Reprinted by permission from (Ref. 40).

variations of 1 °C can reduce cell growth, viability, and/or product production (41). It has also been reported that temperature affects cell cycle phases, metabolism, and cell resistance to shear (42), whereas its control is particularly relevant when temperature-inducible genes are used for recombinant protein production (43). High temperatures may also increase proteolytic activity and alter protein post-translational modifications (42). Moreover, culture susceptibility to viral infection can be modified by temperature (44). Although viral infection is an undesirable event for some processes, in others, such as in virus production for vaccines or bioinsecticides, viral progeny are the desired product. As shown in Table 3 (45–51), a wide range of optimum temperatures exists for tissue culture. Mammalian-cell lines are usually cultured at or near 37 °C, whereas cells from other origins, such as insect, amphibian, fish, or plant, require lower temperatures, usually between 10 and 30 °C. The optimum temperature for growth is not always the most effective for product production. In an attempt to provide the most favorable conditions for growth and product production, several groups have proposed production strategies that demand fine control of temperature (42,45). Moreover, temperature monitoring has been proposed as a tool for designing control strategies (42,52). Altogether, it is clear that adequate temperature control and monitoring are imperative, especially when applications of cell culture are scaled up to a commercial level.

For fermentation of lower eukaryotes or prokaryotes, heat transfer becomes increasingly problematic as the fermenter is scaled up. This is due to the high cell concentrations reached, the high heat generated from metabolic activity, and the large volumes of fermenters employed. The importance of heat transfer in such a process can be illustrated by the production of single cell protein, where operating costs from heat removal can be higher or at least equal to those from oxygen transfer (13).

A particular problem that occurs at large scales is that heat transfer from the vessel walls is severely limited because reactor volume increases by the cube whereas reactor area only increases by the square. A contrasting situation exists for cell culture, where heat transfer is not considered an important problem and has been widely overlooked, whereas other aspects such as oxygen transfer have received much more attention. The reasons for this are the lower cell concentrations, smaller power inputs, lower sizes of commercial reactors, and lower generation of metabolic heat of animal- and plant-cell culture compared to microbial fermentation. Nonetheless, careful attention must still be placed on the design of the heat transfer and temperature control systems of large-scale cell cultures, particularly if temperatures near ambient are required. Heat transfer can become important as higher cell density cultures are developed and larger reactor volumes are needed. Moreover, the fragile nature of most cells limits the use of highly turbulent mixing that is necessary if high heat transfer rates are needed. Accordingly, heat transfer phenomena must be considered during the design of a large-scale cell-culture process.

Effective heat transfer is required to maintain constant temperature in the bioreactor and also to sterilize and cool the vessel. The latter is important to assure product stability at the end of the culture. In this section, a general introduction to heat transfer in bioreactors will be given with special emphasis on cell culture. Only maintaining a desired temperature during the culture phase will be discussed because sterilization and product recovery are discussed elsewhere. Finally, the relevance of heat transfer will be evaluated from calculations for hypothetical large-scale cultures.

Heat Transfer in Bioreactors: The Basics. The main heat flows (heat per unit time per unit volume) in a stirred-tank bioreactor are shown in Figure 10. The heat accumulation

Table 3. Optimum Temperatures for Cell Culture

Cell Line	Optimum Temperature (°C)	Parameter (maximum value)	Reference
Murine hybridoma HB-32	33	Viability index	45
Murine hybridoma HB-32	39	q_{Mab}	45
Rat–mouse–mouse trioma	37	X_t and q_{Mab}	46
CHO cells	37	Growth rate	41
Insect cells	27–29	X_t	47
Insect cell–baculovirus expression system	22–29	Recombinant protein production[a]	47
Catharantus roseus plant cells	30	X_t	48
Catharantus roseus plant cells	21.5	Product synthesis	48
Perilla frutescens plant cells	28	X_t	49
Perilla frutescens plant cells	25	Anthocyanin productivity	49
Chlorella vulgaris SO-26 green alga cells	20	Specific growth rate	50
Chlorella vulgaris SO-26 green alga cells	14	Cellular sugar content	50
Salmonid fish cell lines	15–20	X_t	51
Nonsalmonid fish cell lines	20–30	X_t	51

[a]Optimum temperature should be determined for each recombinant protein produced.

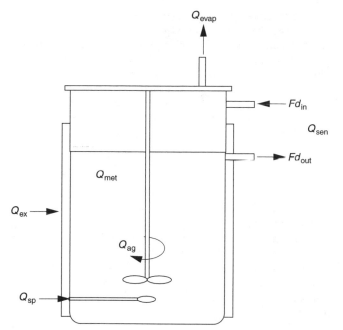

Figure 10. Main heat flows in a stirred-tank bioreactor.

rate Q_{acc} in such a reactor at constant pressure is given by

$$Q_{acc} = Q_{met} - Q_{sen} + Q_{sp} + Q_{ag} - Q_{evap} + Q_{ex} \quad (17)$$

where Q_{met} is heat produced by the metabolism of the cells, Q_{sen} is the sensible enthalpy gain by flow streams (outlet-inlet), Q_{ag} is the heat generated by power input from agitation, Q_{sp} is the heat generated by sparging, Q_{evap} is the heat removed by evaporation, and Q_{ex} is the heat transferred by the heat exchanger. Heat transferred between the environment and the reactor can be important for small-scale vessels but becomes negligible as volume increases. Heat transferred from the environment can be included in Q_{ex}. If the temperature of the reactor is maintained constant, then

$$Q_{acc} = 0 \quad (18)$$

Each of the terms in equation 17 is individually analyzed below.

Metabolic Activity. Among other things, heat generated by metabolic activity depends on the type of cell, the growth phase, the type and concentration of substrates, the concentration of by-products, the shear and osmotic stresses, and the physicochemical properties of the culture medium (53). This has been extensively analyzed for cultures of prokaryotes or lower eukaryotes, but only very scarce information exists for animal- and plant-cell cultures. Heat generated from aerobic growth can be calculated from the oxygen consumption using the oxycaloric equivalent, which is defined as the heat yield from oxygen consumed. For any cell type, fully aerobic metabolism, and all biologically useful organic substrates, the oxycaloric equivalent has been determined to be -450 ($\pm5\%$) kJ/mol O_2 consumed (53). However,

compared to *in vivo*, the oxycaloric equivalent is always more negative for cells grown *in vitro*. This is caused by nutritional deficiencies present in the *in vitro* environment, which must be fulfilled through anaerobic processes by cells rapidly growing (53). Under such conditions, use of the oxycaloric equivalent to calculate the heat produced from cell metabolism can lead to an underestimation of almost 40%. Heat flow rates for several biological models have been determined under controlled pH and dissolved oxygen using microcalorimeters or bench-scale calorimeters (52–54). Heat fluxes (heat flow rate per cell) for some mammalian-cell lines are shown in Table 4. However, caution should be taken when using such values because the heat generated by metabolism is affected by the various parameters mentioned previously.

If metabolic pathways are lumped into one global reaction, then the heat produced by cellular metabolism can also be calculated from the standard enthalpies of substrates, biomass, and products. Again, caution should be taken because metabolic reactions are variable and depend on the particular system studied, including substrate availability, physicochemical conditions, and toxicity of by-products. Even when catabolism of higher eukaryotic cells is very complex, Kemp and Guan (53) have concluded that, for heat balance considerations, the metabolic processes of animal cells can be described essentially by glycolysis, oxidative phosphorylation, the pentose phosphate

Table 4. Specific Heat Flux for Some Cell Types[a]

Cell Type	Heat flux, pW/cell
Human erythrocytes	0.01
Human platelets	0.06
Bovine sperm	1.3 ± 0.1
Human neutrophils	2.5 ± 0.3
Human lymphocytes	5
Horse lymphocytes	8
Human T-lymphoma	8 ± 1
3T3 mouse fibroblasts	17
Chinese hamster ovary 320 (recombinant)	~ 23
kB	25
Vero	27 ± 2
Mouse lymphocyte hybridoma	30–50
HeLa-53G	31.2
Mouse macrophage hybridoma, 2C11-12	32 ± 2
Chinese hamster ovary K1	38
Human foreskin fibroblast	40 ± 10
Rat white adipocytes	40
Human white adipocytes	49 ± 15
Human melanoma, H1477	80
Human keranocytes	83 ± 12
Hamster brown adipocytes	110
SV-K14 (transformed) keranocytes	134 ± 35
Rat hepatocytes	329 ± 13
Kluyveromyces fragilis	1.2^b

[a]Reprinted by permission from Ref. 53.
[b]In g/L, calculated from data of Ref. 54.

pathway, and glutaminolysis. Accordingly, they summarized animal-cell growth and product production as

$$\text{Glucose} + \text{glutamine} + O_2 \rightarrow \text{biomass} + \text{product}$$
$$+ \text{lactate} + CO_2 + NH_3 + H_2O \quad (19)$$

Then Q_{met} can be calculated from the standard enthalpies of formation of each compound, if reaction stoichiometry is known, according to (13)

$$Q_{\text{met}} = \sum_{i=1}^{m} b_i H_{p,i} - \sum_{j=1}^{n} a_j H_{s,j} \quad (20)$$

where m and n are the number of products and substrates, respectively; b_i and a_j are the stoichiometric coefficients of product i and substrate j, respectively; and $H_{p,i}$ and $H_{s,j}$ are the enthalpies of product i and substrate j, respectively. Note that the contribution of anabolism to heat generation is generally negligible (52). Some standard enthalpies of formation for common cell culture substrates and byproducts are shown in Table 5 (55,56). Von Stockar and Marison (54) have determined that neither temperature

Table 5. Standard Enthalpies of Formation of Various Compounds at 25 °C, 1 atm[a]

Compound	Formula[b]	ΔH_f°, kcal/mol
Acetic acid	$C_2H_4O_2$ (aq)	−94.5
Acetate ion	$C_2H_3O_2^-$ (aq)	−116.1
Alanine	$C_3H_7O_2$ N(s)	−88.9
Ammonia	NH_3 (g)	−41.4
	NH_3 (aq)	−19.4
Ammonium chloride	NH_4Cl (s)	−75.4
Ammonium ion	NH_4^+ (aq)	−31.8
Bicarbonate ion	HCO_3^+ (aq)	−165.5
Biomass	$CH_{1.8}O_{0.5}N_{0.2}$ (aq)	−21.9
Carbon dioxide	CO_2 (g)	−94.0
	CO_2 (l)	−98.9
Carbonic acid	H_2CO_3	−167.2
Ethanol	C_2H_5OH (l)	−66.3
Formic acid	CH_2O_2 (aq)	−85.1
α-β-D-Glucose	$C_6H_{12}O_6$ (aq)	−302.0
Glutamic acid	$C_5H_8O_4N$ (s)	−170.4
Hydrochloric acid	HCl (g)	−22.1
	HCl (aq)	−18.0
Hydrogen ion	H^+ (aq)	0
Lactic acid	$C_3H_6O_3$ (l)	−124
Methane	CH_4 (g)	−17.9
Methanol	CH_3OH (l)	−57.0
Oxygen	O_2 (g)	0
	O_2 (aq)	−2.9
Potassium chloride	KCl (s)	−104.2
Pyruvic acid	$C_3H_3O_3$ (s)	−114.1
Sodium chloride	$NaCl$ (s)	−98.2
Sodium hydroxide	$NaOH$ (s)	−102.0
Succinic acid	$C_4H_6O_4$ (l)	−178.8
Sucrose	$C_{12}H_{24}O_{12}$ (s)	−371.6
Sulfuric acid	H_2SO_4 (l)	−193.9
Water	H_2O (g)	−57.8
	H_2O (l)	−68.3

[a] Adapted from Refs. 13, 55, 56.
[b] (s), (l), (g), and (aq) refer to the thermodynamic state of compounds.

nor the thermodynamic state of compounds are significant for energy balance calculations of aerobic (or mainly aerobic) cell growth. Therefore, the values in Table 5 can be utilized for energy balances for culture temperatures different from 25 °C without any further corrections.

Inlet and Outlet Flows. The heat flow Q_{sen} contributed by substances fed to or eliminated from the reactor, if no phase change occurs and if inlet and outlet flows are equal, is given by

$$Q_{\text{sen}} = F_{\text{fd}} C_p (T_{\text{out}} - T_{\text{in}}) \quad (21)$$

where F_{fd} is flow rate, C_p is the heat capacity of the stream, and T_{out} and T_{in} are the temperature of the outlet and inlet flows, respectively. Heat generation can also originate from the temperature difference between the gas and the culture broth. For instance, if gases are compressed adiabatically, their temperature can be higher than the liquid, particularly for cultures controlled at low temperatures (29). Accordingly, the sensible enthalpy gain by the gaseous stream is given by

$$Q_{\text{sen}_g} = F_g C_{P_g} \rho_g (T_g - T_b) \quad (22)$$

where T_b is the temperature of the culture broth, ρ_g is the gas density, and the subindex g refers to the gas phase.

Other reactions that occur in the system simultaneously with cell growth, such as neutralization of buffers in culture media, also contribute to the energy balance (56). The enthalpies of neutralization of physiological buffers, shown in Table 6, are usually exothermic in culture conditions (56).

Power Input by Agitation and Sparging. An important source of heat in a bioreactor is the power input from agitation and sparging. The power P_g, transferred to the reactor by sparging, can be calculated from the following equation (13,57):

$$P_g = \rho_g F_g \left[\frac{RT_b}{MW} \ln \frac{p_1}{p_2} + \alpha \frac{u_{G_0}^2}{2} \right] \quad (23)$$

where R is the gas constant, MW is the gas molecular weight, p_1 is the pressure at the sparger, p_2 is pressure at the vessel top, α is the fraction of gas kinetic energy transferred to the liquid (typically 0.06), and u_{G_0} is the gas velocity at the sparger orifice. For well-designed spargers,

Table 6. Enthalpies of Neutralization of Physiological Buffer Systems for the Reaction $A^- + H^+ \rightarrow AH^a$

Buffer System	T (°C)	$\Delta_b H_{H^+}$ (kcal/mol H^+)	pK_a
Bicarbonate	25	−2 to −1.8	6.4
HCO_3/CO_2 (aq)	37	−1.2	6.3
Carboxyl groups			
Succinate	25	−0.07	5.6
Citrate	25	+0.8	6.4
Phosphate	25	−1	6.9
$HPO_4^{2-}/H_2PO_4^-$	37	−0.86	pH 6.6

[a] Adapted from Ref. 56.

the term $\alpha \dfrac{u_{G_0}^2}{2}$, which represents the jet kinetic energy developed at the sparger holes, is small and can be neglected (13). Equation 23 can be used to calculate power dissipation due to sparging into stirred-tank, bubble-column, and air-lift bioreactors. Notice that, equation 23, neglecting the jet kinetic energy at the sparger holes, corresponds to the term of the energy dissipation rate from aeration in equation 2. This is obtained by substituting in equation 23 the molar gas flow rate and pressure difference between the sparger and top of the vessel, given by equations 24 and 25, respectively:

$$\dot{n} = \frac{F_g (p_1 - p_2)}{\ln \left(\dfrac{p_1}{p_2} \right) R T_b} \tag{24}$$

$$(p_1 - p_2) = \rho g H \tag{25}$$

where $\dot{n}$ is the molar gas flow rate.

The power P transferred to a Newtonian fluid by agitation depends on the fluid density, viscosity, agitation speed, impeller type, impeller diameter, vessel diameter, liquid height, and other geometric characteristics and parameters of the system (16). Then, using dimensional analysis, the energy dissipated to the liquid by agitation is a function of

$$\frac{P}{\rho N^3 D^5} = \text{fn} \left\{ \frac{\rho N D^2}{\mu_b}, \frac{N^2 D}{g}, \frac{T}{D}, \frac{W}{D}, \frac{H}{D}, \text{etc} \right\} \tag{26}$$

where W is the impeller width. P_o, Re, and Fr are the Power, Reynolds, and Froude numbers, respectively, defined as

$$P_o = \frac{P}{\rho N^3 D^5} \tag{27}$$

$$\text{Re} = \frac{\rho N D^2}{\mu_b} \tag{28}$$

$$\text{Fr} = \frac{N^2 D}{g} \tag{29}$$

Fr is important only if gross vortexing exists in the reactor. If vortexing is prevented by baffling or off-center stirring, then the Power number is only a function of Re and geometrical ratios (16). As shown in Figure 11, this is true for the laminar flow regime (Re $< 3 \times 10^5$), where P_o is related to Re through the proportionality constant K_p, which depends only on the system geometry:

$$P_o = \frac{K_p}{\text{Re}} \tag{30}$$

In the turbulent flow regime, the power input is independent of Re. Figure 11 can be used to predict P_o for various system geometries. For other geometries, P_o must be determined experimentally. For vessels equipped with multiple impellers, the interaction between them is not significant if the distance between impellers exceeds their diameter (29). Power consumed will then be the addition of the power consumed by each impeller. Furthermore, if an agitated vessel is also aerated, then power input from

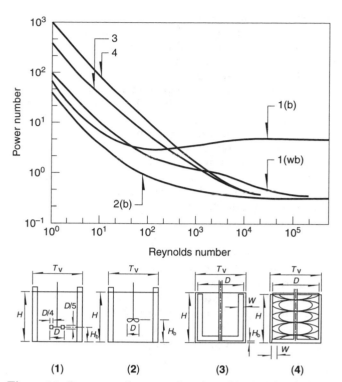

Figure 11. Power number as a function of Reynolds number. (1) turbine stirrer, $T_v/D = 3.33$, $H_b/D = 1$, wb = without baffles, b = with baffles; (2) marine propeller, angle $25°$, $T_v/D = 3.33$, $H_b/D = 1.5$; (3) anchor impeller, $T_v/D = 1.02$, $H_b/D = 0.01$, $W/D = 0.1$; (4) helical ribbon impeller, $T_v/D = 1.02$, $H_b/D = 0.1$, $W/D = 0.1$. Reprinted by permission from (Ref. 29).

agitation will decrease with respect to the unaerated case. Van't Riet and Tramper (29) recommend the correlation by Hughmark for turbine stirrers in water for calculating aerated power inputs P_a:

$$\frac{P_a}{P} = 0.10 \left(\frac{F_g}{NV} \right)^{-0.25} \left(\frac{N^2 D^4}{g W V^{2/3}} \right)^{-0.20} \tag{31}$$

Evaporation. If the inlet gas is at a lower temperature than the bioreactor, heat loss due to evaporation can be significant. Furthermore, evaporation can contribute to substantial water loss in the fermenter, which has to be replaced or avoided by using effective condensers. As the gas temperature increases, the water content of saturated air increases. The latent heat of evaporation for water λ is 41 kJ/mol, and the heat loss due to evaporation can be calculated as (13)

$$Q_{\text{evap}} \approx \lambda F'_g (w_{\text{out}} - w_{\text{in}}) \tag{32}$$

where F'_g is dry air flow and w_{out} and w_{in} are the water contents of the inlet and outlet flows, respectively. Water contents can be calculated from psychrometric charts.

Heat Exchangers. Heat exchangers have to be included in vessels to control the culture temperature. Vessels can be jacketed, have internal or external coils, and even an external heat exchanger can be used. Jacketed vessels are used in small-scale reactors, but as scale increases, the heat transfer area per unit volume decreases, and internal or external coils may be required for efficient control of temperatures. As illustrated in the practical cases shown below, a jacket is more than sufficient to transfer the necessary heat in typical large-scale cell cultures. Furthermore, shear damage of cells in pumps and risk of contamination limit the use of external heat exchangers.

To maintain a constant culture temperature, the heat generated (or removed) by all of the processes analyzed previously must equal the heat transferred by the exchanger. Heat transfer depends on the overall heat flow resistance R', and the temperature gradient ΔT between the exchanger and the culture broth:

$$Q_{ex} = \frac{\Delta T}{R'} \qquad (33)$$

Calculation of ΔT depends on the system geometry and on the direction of motion of the fluids. For heat transfer from jackets in agitated-tank reactors, ΔT is given by the log mean temperature difference (LMTD):

$$\text{LMTD} = \frac{(T_i - T_o)}{\ln\left[\dfrac{T_b - T_o}{T_b - T_i}\right]} \qquad (34)$$

where T_i and T_o are the temperatures of the coolant entering and leaving the reactor, respectively. T_b is constant if a perfectly mixed vessel is assumed. In turn, R' is defined as

$$R' = \frac{1}{UA} \qquad (35)$$

where A is the area for heat transfer and U is the overall heat transfer coefficient. U depends on the nature of the material across which heat is transferred, the system geometry, and the nature of the fluid flows involved. As a rule of thumb, U for a clean vessel usually ranges between 280 to 1,137 W/m^2 °C (11). However, a wider range exists for various practical situations (58). Lower values correspond to very viscous non-Newtonian fluids, whereas larger values correspond to nonviscous Newtonian fluids. The overall heat transfer resistance is the sum of the individual resistances in series in the path of heat transfer. Therefore, for the jacket of thin-walled vessels, where the area normal to the direction of heat

transfer can be considered constant, U is given by the following expression:

$$\frac{1}{U} = \frac{1}{h_i} + \frac{t}{k_m} + \frac{1}{h_o} + \frac{1}{h_{f_i}} + \frac{1}{h_{f_o}} \qquad (36)$$

where h_i and h_o are the broth and coolant side heat-transfer coefficients, respectively; t is the vessel wall thickness; k_m is the metal thermal conductivity; and h_{f_i} and h_{f_o} are the inside and outside fouling heat transfer coefficients, respectively. Fouling is caused by deposits (microbial film, scale, dirt, etc.) which depend on operating conditions and should be kept to a minimum by maintaining vessels and jackets or coils clean. Fouling resistances for various systems can be found elsewhere (58). Individual heat-transfer coefficients depend on the fluid properties, geometric characteristics, and operating conditions of the system through complex relationships. Thus, from a practical standpoint, individual heat-transfer coefficients are obtained from empirical correlations, some provided by vessel vendors, relating the Nusselt number (Nu) with the Reynolds (Re) and Prandtl (Pr) numbers. For instance, h_i for stirred vessels can be determined as

$$\text{Nu} = a\text{Re}^b\text{Pr}^c\left(\frac{\mu_b}{\mu_w}\right)^d \qquad (37)$$

where a, b, c, and d are constants (listed in Table 7 for various reactor configurations), μ_w is the broth viscosity at the wall temperature, and Nu and Pr are defined as

$$\text{Nu} = \frac{h_i T_v}{k} \qquad (38)$$

$$\text{Pr} = \frac{C_p \mu_b}{k} \qquad (39)$$

where k is the broth thermal conductivity. Equation 37 is acceptable for unaerated fluids. However, no correlations have been published for aerated conditions (11). Depending on stirring speed and flooding phenomena, aeration can improve or deteriorate h_i. Nonetheless, the influence of gassing is minor in relation to other variables (29).

Liquid circulation in bubble columns is determined by the gas flow rate, and thus, h_i can be determined as (29)

$$h_i = 9391(u_{G_s})^{0.25}\left(\frac{\mu_{wt}}{\mu_b}\right)^{0.35} \qquad (40)$$

Table 7. Values of Constants in Equation 37 for Jacketed Vessels[a]

Impeller	a	b	c	d	Re
Jacketed vessels					
Paddle	0.36	0.66	0.33	0.21	$300 - 3 \times 10^5$
Pitched-blade turbine	0.53	0.66	0.33	0.24	$80 - 200$
Disk flat-blade turbine	0.54	0.66	0.33	0.14	$4 - 3$
Propeller	0.54	0.66	0.33	0.14	2×10^5
Anchor	1	0.5	0.33	0.18	$10 - 300$
Anchor	0.36	0.66	0.33	0.18	$300 - 40,000$
Helical ribbon	0.63	0.5	0.33	0.18	$8 - 10^5$
Helical coils					
Paddle	0.87	0.62	0.33	0.14	$300 - 4 \times 10^5$

[a]Adapted from Ref. 58.

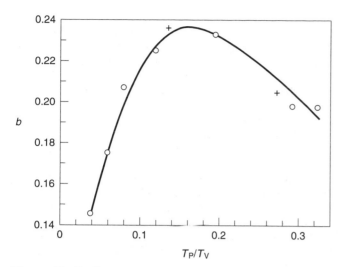

Figure 12. Coefficient b of equation 42, T_p/T_v in inches per foot. Reprinted by permission from (Ref. 58).

where μ_{wt} is the water viscosity and μ_b is the broth viscosity. Equation 40 is applicable to vessels with diameters ranging from 0.1 to 1 m and for broths viscosities between 10^{-3} to 5×10^{-2} N s/m^2.

The liquid velocity has to be known to perform heat transfer calculations in air-lift bioreactors. This is rather complex and, as exemplified by van't Riet and Tramper (29), an average liquid velocity can only be estimated from numerical simulations. Alternatively, simplified correlations, such as the one shown in equation 7, exist. However, liquid velocity gradients are present, and thus, a single value for heat transfer cannot be given. Flow in the downcomer can be considered a single phase. Thus, h_i can be determined from correlations of turbulent flow in pipes (29):

$$\frac{h_i T_d}{k} = 0.027 \left(\frac{\rho \overline{u_{L_d}} T_d}{\mu_b} \right)^{0.8} \mathrm{Pr}^{0.33} \tag{41}$$

where $\overline{u_{L_d}}$ is the average superficial liquid velocity in the downcomer and T_d is diameter of the downcomer. The flow regime in the riser can be described in the same manner as in a bubble column, that is, either homogeneous or heterogeneous flow (see previous section) (29). The flow regime is determined mainly by the total hydrodynamic resistance of the bioreactor. If the riser acts as a bubble column in the heterogeneous flow regime, equation 40 can be used to calculate the heat-transfer coefficient. For "true" air-lift flow (homogeneous flow), equation 41 can be used.

For packed-bed reactors, the heat-transfer coefficient between the inner container surface and the fluid stream can be calculated from the following correlation (58):

$$\mathrm{Nu} = b_1 b T_v^{0.17} \left(\frac{T_v G}{\mu_b} \right)^{0.83} \mathrm{Pr} \tag{42}$$

where T_p is the diameter of the particle, G is the superficial mass velocity, k and C_p refer to the liquid phase, b is a coefficient dependent on T_p/T_v (see Fig. 12), and b_1 is equal to 1.22 (SI) or 1 (U.S. customary).

Heat-transfer coefficients for other inhomogeneous systems, such as hollow-fiber and fluidized-bed, have to be evaluated for each individual case because the packing, liquid velocity, and the aspect ratio have to be considered. Various correlations for such systems can be found elsewhere (58). Finally, once all of the terms of equation 36 are known, the heat transferred by the exchanger can be calculated by combining equations 33–35:

$$Q_{ex} = UA \left(\frac{T_1 - T_o}{\ln \left[\dfrac{T_b - T_o}{T_b - T_i} \right]} \right) \tag{43}$$

and the cooling water flow rate F_w can be calculated from

$$F_w = \frac{Q_{ex}}{C_P(T_o - T_i)} \tag{44}$$

Many industrial operations still rely on simply multiplying the number of small-scale units (such as T-flasks or roller bottles) and placing them in contained suites. For such cases a different approach must be taken to solve heat-transfer problems. These are usually solved through suitable design of the HVAC (heating, ventilation, and air conditioning) systems. A "mega roll" industrial unit has been described by Panina (59). This unit has a capacity of 7,200 roller bottles of 1 L in cylindrical wire racks. Racks are positioned in an incubator of 269.8 m^3 heated with electric heaters that generate 14 kW. Heat is uniformly distributed by forced air circulation, and temperature is controlled by a heat exchanger fed with cold water.

Practical Cases. To assess the magnitude of the various heat flows and thus the importance of heat transfer in a cell-culture process, heat balances for an hybridoma culture in an hypothetical 10,000-L water-jacketed stirred-tank reactor are presented. Comparison is made with a hypothetical baculovirus-infected insect-cell culture in the same vessel. The behavior of the large-scale hybridoma culture is assumed to follow actual kinetic data obtained from a 1-L agitated batch reactor under the experimental conditions described by Higareda et al. (60). Such an assumption is valid because the data at the 1-L scale, shown in Figure 13, are similar to data reported by Backer et al. (9) at the 1000-L scale. The calculations are performed for a hypothetical vessel, detailed in Table 8, which closely resembles the reactor used for CHO cell culture at Glaxo Wellcome and described by Nienow et al. (61).

Viable cell concentration and oxygen uptake rate (OUR) are shown in Figure 13a. The horizontal line represents the time of glutamine depletion which corresponds to the onset of respiration cessation and the end of exponential growth. Similar behavior has been described by Higareda et al. (60). Heat flow during the lag and exponential growth phases was calculated using 50 pW/cell, which corresponds to the highest heat flux reported for a mouse hybridoma (see Table 4). By choosing such a high value, the calculated results should yield the worst case scenario in terms of heat generation. Guan et al. (52) showed

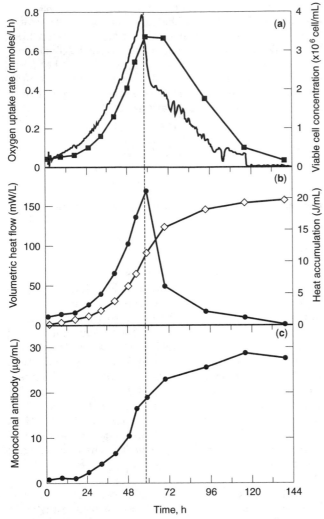

Figure 13. Kinetic data of a hybridoma batch culture in a 1-L stirred-tank reactor. (a) oxygen uptake rate (—), viable cell concentration (■); (b) heat flow (●), heat accumulation (◇); (c) monoclonal antibody. Dottedline represents the time of Gln depletion.

Table 8. Details of the Hypothetical 10,000-L Bioreactor[a]

Working volume	8,000 L
Aspect ratio	1.3
Liquid height	2.54 m
Tank diameter	2 m
Vessel wall thickness	6×10^{-3} m
Baffles	4
Turbine	1 Rushton
Turbine diameter	0.44 m
Distance of turbine from vessel bottom	$2/9\ T_v$
Heat exchanger	Water jacket
Area of water jacket	12 m²
Aeration	Superficial and submerged
Superficial gas velocity	2×10^{-4} m/s
Temperature	37 °C
Dissolved oxygen concentration	20% w.r.t. air sat.
Metal conductivity	80 W/m² °C

[a]Some values are based on Ref. 61.

that, as the metabolic activity of a culture decreases, the heat generated approaches the oxycaloric equivalent. Accordingly, during the stationary and death phases, the heat flow was calculated using the oxycaloric equivalent, 450 J/mmol O_2 consumed. The calculated profiles of volumetric heat flow and accumulated heat (integral of heat flow with time) are shown in Figure 13b. It can be seen that volumetric heat flow and accumulated heat have a maximum of 0.17 W/L and 20 J/mL. Thus, for the 10,000-L (8,000-L working volume) reactor, a maximum heat flow of 1.36 kW is expected. In comparison, maximum heat flows of 7.4 W/L and 26 W/L have been reported for typical aerobic growth of yeast (at a relatively low maximum biomass concentration of 6.2 g/L) and unicellular protein production in a 2.3×10^6-L reactor, respectively (54), that is, heat generated from the metabolism of the hypothetical hybridoma culture will be between one to two orders of magnitude lower than that for other microbial fermentation.

To satisfy the oxygen requirements of the culture shown in Figure 13, the oxygen transfer rate (OTR) must equal OUR. In this example, the maximum OUR was 0.8 mmoles/L h. Thus, an oxygen transfer coefficient, k_La, of at least 5.47 h^{-1} is required (assuming $c_m = 0.183$ mM). As determined by Nienow et al. (61), such a k_La can be obtained if 50 W/m³ are supplied to the reactor by agitation and sparging. Accordingly the total heat dissipated by the reactor would then be 1.76 kW (i.e., 1.36 kW + (8 m³ × 0.05 kW/m³)). Interestingly, under such conditions 80% of the heat flux will originate from cellular metabolism, whereas only 23% results from aeration and agitation. For a superficial gas velocity of 2×10^{-4} m/s (see Table 8) and using equations 23 or 2, the power supplied by sparging will be 1.96 W/m³. Thus, 48.04 W/m³ are needed for agitation. An impeller speed of 1.67 s^{-1} can be calculated from equation 27 if a Power number of 5 is assumed for the reactor (61). This results in a fully turbulent flow regime with a Reynolds number of 3×10^5 (assuming that the culture broth behaves like water), but where shear damage is not expected. A broth side heat-transfer coefficient of 1,694 W/m² °C can then be calculated using equation 37. Finally, if fouling layers in the tank or jacket are not formed and if the heat transfer coefficient on the coolant side is negligible, an overall heat-transfer coefficient of 1,503 W/m² °C results from equation 36 and data from Table 8 (vessel wall thickness and metal thermal conductivity). This value agrees with overall heat-transfer coefficients of jacketed vessels using brine as the coolant and water as the fluid in vessel, which range from 230 to 2,625 W/m² °C (58).

To maintain a constant temperature in the reactor, heat generated from cellular metabolism, agitation, and sparging has to be removed by the water jacket, dissipation to the environment, and evaporation. Again, for the worst case scenario, maximum heat removal from the exchanger can be determined if heat losses to the environment and from evaporation are neglected. This is possible if the reactor is well insulated and a water-saturated gas stream at the culture temperature is used for aeration. Accordingly, solving equation 43 for a jacket area of 12 m² yields a LMTD of 0.09 °C. Multiple combinations of inlet and outlet water temperatures below 37 °C but above 36 °C, can satisfy the calculated LMTD. For instance, water entering the jacket at a temperature of 36 °C and at a flow rate of 22 L/min (see equation 44) would be necessary to control the culture temperature at 37 °C. Interestingly, such a result indicates that for the hypothetical hybridoma culture, the heat transfer problem actually becomes a heating task, because jacket inlet water is required at a temperature above ambient.

A different situation will result for another model culture, for instance, protein production by the insect-cell baculovirus expression vector system (47,62). An insect-cell culture, typically maintained at 27 °C, can reach a concentration of 8×10^6 cell/mL and an OUR of 17 mmol/Lh during infection with a baculovirus. Using the oxycaloric equivalent, a heat flow from metabolic activity of 17.2 kW will result for the same hypothetical vessel described above, that is, about 13 times higher than for the hybridoma culture. For comparison, the same agitation and aeration rates used for the hybridoma culture are also used in this case. Because the OUR is severalfold higher, it is assumed that the OTR can be increased by oxygen enrichment, additional membrane aeration, and/or pressure increase, or other means. It should be cautioned that for an actual situation, the practical utility of such alternatives must be carefully evaluated. The fragile nature of insect cells also rules out the use of higher agitation and aeration rates. Accordingly, the total heat generated will be 17.6 kW. Following the same procedure as before, a LMTD of 0.98 °C can be calculated. Such a LMTD could be satisfied, at least theoretically, for inlet water temperatures ranging from 6 to 26 °C. Thus, more options exist than for the hybridoma culture for designing the heat transfer system. In conclusion, the cases illustrated show that jacketed vessels are more than sufficient to satisfy heat-transfer requirements for cell culture at the typical largest scales performed to date. Moreover, compared to microbial systems or to oxygen transfer requirements in cell culture, the heat transfer problem is not a major challenge but cannot be overlooked.

NOMENCLATURE

A	Area
A_d	Downcomer cross-sectional area
a_j	Stoichiometric coefficient of substrate j
A_p	Area of microcarrier
A_r	Riser cross-sectional area
b_i	Stoichiometric coefficient of product i
c	Dissolved oxygen concentration
c_m	Dissolved oxygen concentration at saturation
c_0	Inlet dissolved oxygen concentration
C_p	Heat capacity
D	Impeller diameter
d_b	Bubble diameter
D_e	Effective diffusion coefficient
$D_{o/w}$	Diffusion coefficient of oxygen in water
F_g	Gas flow rate
F'_g	Dry gas flow rate
F_{fd}	Flow rate (feed)
Fr	Froude number
Fw	Cooling water flow rate
g	Gravitational acceleration
G	Superficial mass velocity
H	Liquid height
H_b	Height of stirrer above vessel bottom
H_d	Downcomer height
h_{fi}	Inside fouling heat-transfer coefficient
h_{fo}	Outside fouling heat-transfer coefficient
h_i	Broth side heat-transfer coefficient
h_o	Coolant side heat-transfer coefficient
$H_{p,i}$	Enthalpy of product i
H_r	Riser height
$H_{s,j}$	Enthalpy of substrate j
k	Thermal conductivity
k_d	Specific death rate constant
k_m	Metal thermal conductivity
K_P	Proportionality constant
LMTD	Log mean temperature difference
m	Number of products
Mab	Monoclonal antibody
MW	Molecular weight
n	Number of substrates
$\dot{n}$	Molar gas flow rate
N	Impeller speed
N^*	Dimensionless mixing time
Nu	Nusselt number
OUR	Oxygen uptake rate
OTR	Oxygen transfer rate
P	Power
p_1	Pressure at the sparger
p_2	Pressure at top of vessel
p_a	Aerated power input
p_c	Pressure in capillary
P_g	Power transferred to liquid from sparging
P_o	Power number
Pr	Prandtl number
q	Specific production rate
Q_{acc}	Heat accumulation rate
Q_{ag}	Heat produced by agitation
Q_{evap}	Heat removed by evaporation
Q_{ex}	Heat transferred by the heat exchangers
Q_{met}	Heat produced by metabolism
Q_{sen}	Sensible enthalpy gain by flow streams
Q_{sen_g}	Sensible enthalpy gain by gas stream
Q_{sp}	Heat produced by sparging
R	Gas constant
R_l	Resistance to heat flow
r	Radius
r_0	Radius at which oxygen concentration falls to zero
r_c	Capillary radius
Re	Reynolds number
t	Vessel wall thickness or time
T	Temperature
T_b	Broth temperature
t_c	Circulation time
T_d	Downcomer diameter
T_g	Temperature of gas
T_i	Jacket fluid inlet temperature

T_{in}	Temperature of the inlet flow
t_M	Mixing time
t_{mt}	Time constant for mass transfer
T_o	Jacket fluid outlet temperature
T_{out}	Temperature of the outlet flow
T_p	Diameter of particle
T_r	Riser diameter
t_{rxn}	Time constant for oxygen consumption
T_v	Vessel diameter
U	Overall heat transfer coefficient
$\overline{u_c}$	Average axial velocity inside capillary
u_G	Superficial gas velocity
u_{G_0}	Gas velocity at the sparger orifice
u_{G_r}	Superficial gas velocity in the riser
u_{G_s}	Pressure corrected gas velocity
u_L	Superficial liquid velocity
u_{L_r}	Superficial liquid velocity in the riser
$\overline{u_{L_d}}$	Average superficial liquid velocity in downcomer
V	Liquid volume
V_k	Killing volume
V_p	Volume of a microcarrier
VVM	Volume of air per volume of liquid per minute
W	Width of impeller
w_{in}	Water content of inlet gas stream
w_{out}	Water content of outlet gas stream
X_t	Total cell concentration
y	Liquid depth
z	Axial position
α	Fraction of gas kinetic energy transferred to the liquid
Δ_p	Pressure drop
ε	Void fraction
ε_T	Total energy dissipation rate
Φ	Observable module
η	Effectiveness factor
λ	Latent heat of evaporation of water
μ	Viscosity
μ_{app}	Apparent viscosity
μ_b	Broth viscosity
μ_w	Broth viscosity at the wall temperature
μ_{wt}	Water viscosity
ρ	Liquid density
ρ_g	Gas density

BIBLIOGRAPHY

1. A.E. Humphrey, in E. Galindo and O.T. Ramírez, eds., *Advances in Bioprocess Engineering*, Kluwer Academic Publishers, Dordrecht, The Netherlands, 1994, pp. 103–107.

2. K. Venkat, *International Course on Medicinal Plants*, Instituto Mexicano del Seguro Social, Mexico, December 8–12, 1997.

3. W.R. Arathoon and J.R. Birch, *Science* **232**, 1390–1395 (1986).

4. G.L. Smith in B.K. Lydersen, N.A. D'Elia, and K.L. Nelson, eds., *Bioprocess Engineering: Systems, Equipment and Facilities*, Wiley, New York, 1994, pp. 69–84.

5. T. Cartwright, in R.E. Spier and J.B. Griffiths, eds., *Animal Cell Biotechnology*, Vol. 5, Academic Press, San Diego, Calif., 1992, pp. 217–245.

6. A. Handa-Corrigan, in R.E. Spier and J.B. Griffiths, eds., *Animal Cell Biotechnology*, Vol. 4, Academic Press, San Diego, Calif., 1990, pp. 123–132.

7. N.B. Finter, A.J.M. Garland, and R.C. Telling, in A.S. Lubiniecki, ed., *Large Scale Mammalian Cell Culture Technology*, Dekker, New York, 1990, pp. 1–14.

8. R.E. Spier, in R.E. Spier and J.B. Griffiths, eds., *Animal Cell Biotechnology*, Vol. 5, Academic Press, San Diego, Calif., 1992, pp. 1–46.

9. M.P. Backer et al., *Biotechnol. Bioeng.* **32**, 993–1000 (1988).

10. B.G.D. Bödeker et al., in R.E. Spier, J.B. Griffiths, and W. Berthold, eds., *Animal Cell Technology: Products of Today, Prospects for Tomorrow*, Butterworth-Heinemann, Oxford, 1994, pp. 580–583.

11. M. Charles and J. Wilson, in B.K. Lydersen, N.A. D'Elia, and K.L. Nelson, eds., *Bioprocess Engineering: Systems, Equipment and Facilities*, Wiley, New York, 1994, pp. 3–68.

12. J.B. Griffiths, in R.E. Spier and J.B. Griffiths, eds., *Animal Cell Biotechnology*, Vol. 3, Academic Press, San Diego, Calif., 1988, pp. 179–220.

13. B. Atkinson and F. Mavituna, *Biochemical Engineering and Biotechnology Handbook*, 2nd ed., Stockton Press, New York, 1991, pp. 476–485.

14. P.M. Doran, *Adv. in Biochem. Eng./Biotechnol.*, **48**, 115–168 (1993).

15. K.W. Norwood and A.B. Metzner, *AIChE J.* **6**, 432–440 (1960).

16. M.F. Edwards in N. Harnby, M.F. Edwards, and A.W. Nienow, eds., *Mixing in the Process Industries*, Butterworth, London, 1985, pp. 131–144.

17. C. Langheinrich et al., *Trans. Inst. Chem. Eng.* **76**, 107–116 (1998).

18. S. Ruszkowski, *Inst. Chem. Eng. Symp. Seri.* **136**, 283–292 (1994).

19. M. Cooke, J.C. Middleton, and J. Bush, in R. King, ed., *Second international Conference on Bioreactor Fluid Dynamics*, Elsevier Applied Science, New York, 1988, pp. 37–64.

20. A. Garnier et al., *Cytotechnology* **22**, 53–63 (1996).

21. D.R. Gray et al., *Cytotechnology* **22**, 65–78 (1996).

22. S. Ozturk, *Cytotechnology* **22**, 3–6 (1996).

23. J.G. Aunins et al., *Cell Culture Engineering IV*, San Diego, Calif., March 7–12, 1994.

24. M.E. Brown and J.R. Birch, *Cell Culture Engineering V*, San Diego, Calif., January 28–February 2, 1996.

25. R.A. Bello, C.W. Robinson, and M. Moo-Young, *Can. J. Chem. Eng.* **62**, 573–577 (1984).

26. M. Popovic and C.W. Robinson, *Biotechnol. Bioeng.* **32**, 301–312 (1988).

27. H. Tanaka, *Process Biochem.* **22**, 106–113 (1987).

28. I. Rousseau and J.D. Bu'Lock, *Biotechnol. Lett.* **2**, 475–480 (1980).

29. K. van't Riet and J. Tramper, *Basic Bioreactor Design*, Dekker, New York, 1991.

30. J. Tramper, D. Smit, J. Straatman, and J.M. Vlak, *Bioprocess Eng.* **3**, 37–41 (1998).

31. R.E. Spier, in R.E. Spier and J.B. Griffiths, eds., *Animal Cell Biotechnology*, Vol. 1, Academic Press, London, 1985, pp. 243–263.

32. A.D. Murdin, N.F. Kirkby, R. Wilson, and R.E. Spier, in R.E. Spier and J.B. Griffiths, eds., *Animal Cell Biotechnology*, Vol. 3, Academic Press, San Diego, Calif., 1988, pp. 55–74.

33. J.P. Tharakan and P.C. Chau, *Biotechnol. Bioeng.* **28**, 1064–1071 (1986).

34. M.S. Dandavati, M.R. Doshi, and W.N. Gill, *Chem. Eng. Sci.* **30**, 877–886 (1975).

35. B. Griffiths, in A.S. Lubiniecki, ed., *Large-Scale Mammalian Cell Culture Technology*, Dekker, New York, 1990, pp. 217–250.

36. O.T. Schonherr and P.J.T.A. Van Gelder, in R.E. Spier and J.B. Griffiths, eds., *Animal Cell Biotechnology*, Vol. 3, Academic Press, San Diego, Calif., 1988, pp. 337–355.

37. B. Griffiths, in R.E. Spier and J.B. Griffiths, eds., *Animal Cell Biotechnology*, Vol. 4, Academic Press, San Diego, Calif., 1990, pp. 149–166.

38. J.M. Smith, *Chemical Engineering Kinetics*, 3rd ed., McGraw-Hill, New York, 1981.

39. R.B. Bird, W.E. Stewart, and E.N. Lightfoot, *Transport Phenomena*, Wiley, New York, 1960, pp. 180–207.

40. J.E. Bailey and D.F. Ollis, *Biochemical Engineering Fundamentals*, 2nd ed., McGraw-Hill, New York, 1986, pp. 208–214.

41. N. Kurano et al., *J. Biotechnol.* **15**, 101–112 (1990).

42. S. Chuppa et al., *Biotechnol. Bioeng.* **55**, 328–338 (1997).

43. A.S. Fiorino et al., *In Vitro Cell Dev. Biol. Anim.* **34**, 247–258 (1998).

44. T. David-Pfeuty and Y. Nouvian-Dooghe, *Oncogene* **7**, 1611–1623 (1992).

45. G.K. Sureshkumar and R. Mutharasan, *Biotechnol. Bioeng.* **37**, 292–295 (1992).

46. J.W. Bloemkolk, M.R. Gray, F. Merchant, and T.R. Mosmann, *Biotechnol. Bioeng.* **40**, 427–431 (1992).

47. L.A. Palomares and O.T. Ramírez, in E. Galindo and O.T. Ramírez, eds., *Advances in Bioprocess Engineering II.*, Kluwer Academic Publishers, Dordrecht, The Netherlands, 1998, pp. 25–52.

48. C.M. Bailey and H. Nicholson, *Biotechnol. Bioeng.* **35**, 252–259 (1990).

49. J.J. Zhong and T. Yoshida, *J. Ferment. Bioeng.* **76**, 530–531 (1993).

50. H. Hosono et al., *J. Ferment. Bioeng.* **78**, 235–240 (1994).

51. R.D. Fernandez, M. Yoshimizu, T. Kimura, and Y. Ezura, *J. Aquat. Anim. Health* **5**, 137–147 (1993).

52. Y. Guan, P.T. Evans, and R.B. Kemp, *Biotechnol. Bioeng.* **58**, 464–477 (1998).

53. R.B. Kemp and Y. Guan, *Thermochim. Acta* **300**, 199–211 (1997).

54. U. von Stockar and I.W. Marison, *Thermochim. Acta* **193**, 215–242 (1991).

55. G.W. Castellan, *Physical Chemistry*, Adison-Wesley, Reading, MSS., 1971.

56. U. von Stockar et al., *Biochim. Biophys. Acta* **1183**, 221–240 (1993).

57. G.B. Tatterson, *Fluid Mixing and Gas Dispersion in Agitated Tanks*, McGraw-Hill, New York, 1991, pp. 302–304.

58. J.G. Knudsen et al., in R.H. Perry and D. Green, eds., *Perry's Chemical Engineer's Handbook*, McGraw-Hill, New York, 1984, pp. 10-1–10-68.

59. G.F. Panina, in R.E. Spier and J.B. Griffiths, eds., *Animal Cell Biotechnology*, Vol. 1, Academic Press, San Diego, Calif., 1985, pp. 211–242.

60. A.E. Higareda, L.D. Possani, and O.T. Ramírez, *Biotechnol. Bioeng.* **56**, 555–563 (1997).

61. A.W. Nienow et al., *Cytotechnology*, **22**, 87–94 (1996).

62. L.A. Palomares and O.T. Ramírez, *Cytotechnology* **22**, 225–237 (1996).

See also ANIMAL CELL CULTURE MEDIA; ANIMAL CELL CULTURE, PHYSIOCHEMICAL EFFECTS OF DISSOLVED OXYGEN AND REDOX POTENTIAL; ANIMAL CELL CULTURE, PHYSIOCHEMICAL EFFECTS OF OSMOLALITY AND TEMPERATURE; ANIMAL CELL CULTURE, PHYSIOCHEMICAL EFFECTS OF pH; ASEPTIC TECHNIQUES IN CELL CULTURE; BIOREACTOR OPERATIONS — PREPARATION, STERILIZATION, CHARGING, CULTURE INITIATION AND HARVESTING; BIOREACTORS, STIRRED TANK; CELL GROWTH AND PROTEIN EXPRESSION KINETICS; FLUX ANALYSIS OF MAMMALIAN CELL CULTURE: METHODS AND APPLICATIONS; MEASUREMENT OF CELL VIABILITY; OFF-LINE ANALYSIS IN ANIMAL CELL CULTURE, METHODS; OFF-LINE IMMUNOASSAYS IN BIOPROCESS CONTROL; ON-LINE ANALYSIS IN ANIMAL CELL CULTURE; STERILIZATION AND DECONTAMINATION; VIRUS PRODUCTION FROM CELL CULTURE, KINETICS; IMMUNO FLOW INJECTION ANALYSIS IN BIOPROCESS CONTROL.

BIOREACTORS, AIRLIFT

SHINSAKU TAKAYAMA
Tokai University
Shizuoka, Japan

OUTLINE

INTRODUCTION

Industrially important organisms, such as microbes, and plant and animal cells, are aerobically cultured in mechanically agitated and pneumatically driven bioreactors. Mechanically agitated bioreactors, the most standardized bioreactor system in industrial processes, have high construction costs, complicated structures, considerable power input, higher process cost, and higher risks of microbial contamination than equivalent preumatic reactors. Pneumatically driven bioreactors, such as bubble column and airlift bioreactors, are preferable as plant culture bioreactors because they lack the drawbacks of mechanically agitated bioreactors.

The use of airlift bioreactors is generally accepted as the standard procedure for growing plant cell and tissue suspension cultures. Airlift bioreactors use air to circulate the medium (1), and the medium is driven in one direction because aeration occurs within a draft tube or external loop and rising bubbles make the medium flow (2). These types of bioreactors are excellent because they supply sufficient oxygen and substrate and create low shear stress (2).

Various types of airlift bioreactors have been developed and their characteristics have been evaluated (for culturing plant cells and root culture (3–18), yeast (5,19,20), fungus (21), bacterial (19,22,23), and animal cells (24)). The type of airlift bioreactor affects growth and metabolite production by its effects on physical and chemical properties.

This article reviews the basic construction of airlift bioreactors, their characteristics, types, and applications in plant cell and tissue cultures.

GENERAL CHARACTERISTICS

Fundamental Design

Airlift bioreactors are a modification of the bubble column bioreactor which generates medium flow by aeration and circulates medium in one direction. Fundamentally, airlift bioreactors should be composed of at least two columns, a riser column, a downcomer column, and connections between them at the top and the bottom of the vessel to circulate the medium. Various types of bioreactors conform to this requirement and are recognized as airlift bioreactors. An airlift bioreactor is either a draft-tube airlift bioreactor or an external-loop bioreactor, as in Figure 1. The bubbles sparged into a draft tube from an air sparger rise in the riser column (draft tube) and generate upward medium flow. The medium which reaches the surface pours into the annular space between the draft tube and the enclosing bioreactor vessel (downcomer) and flows downward; thus the medium circulates.

The fundamental design feature to consider is the bioreactor's ratio of height to the diameter. Sometimes a ratio of only about 1.5 to 3.0 was used in designing the airlift bioreactor but these values are insufficient for efficient mixing. Values of 5 or more are necessary.

In practice, the bioreactor vessel is generally made of a glass column up to a size where it cannot be autoclaved. These bioreactors are equipped with several ports or tubes,

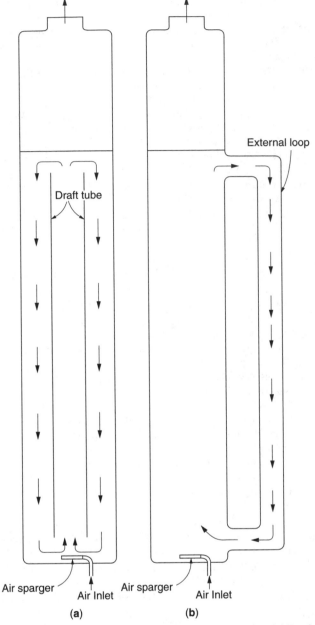

Figure 1. Basic configurations of airlift bioreactors. (**a**) Draft-tube airlift bioreactor. (**b**) External-loop airlift bioreactor.

such as an inoculation port, sensor port (pH, DO, ORP, etc.), sampling tubes, feeding tubes, and control systems.

Aeration and Medium Flow Characteristics

Bubble Generation and Holdup (25). Aiba et al. described the characteristics of bubble generation and holdup (25). The size of bubbles sparged from the orifice of a sparger at a low aeration rate was calculated from equation 1;

$$\frac{\pi}{6} \cdot d_B{}^3 \Delta \rho g = \pi d \sigma \tag{1}$$

where d_B is the diameter of the bubbles (mm), d is the diameter of the orifice (m), $\Delta \rho$ is the difference between air and liquid density (g/cm³), g is the acceleration of gravity (m/s^2), and σ is the surface tension of the liquid (dyn/cm). In equation 1, the left-hand side refers to the buoyancy of bubbles, and the right-hand side is the power equivalent to the retention of bubbles. This equation was experimentally consistent when aeration rate Q was within the limits of 0.02 to 0.5 cm^3/s. Within this limit, the diameter of bubbles d_B (mm) correlated with $d^{1/3}$ and did not depend on the aeration rate Q (cm^3/s). Above the limit of $Q = 0.5$cm^3/s, equation 1 was in consistent, and so, an experimental equation 2 was used to estimate d_B (26).

$$d_B \propto Qn', \quad n' = 0.2-1.0 \qquad (2)$$

The relationship between d_B (mm) and the superficial gas velocity V_B (m/s) is presented in Figure 2 (27). The values are split into two graphs plotted in the same figure. The bubbles were mostly spherical at $d_B \leq 1.5$, and in this range, V_B correlated with d_B^{1-2} (27).

The bubble shape begins to transform in the d_B range of 1.5 to 6 mm, and when d_B exceeded 6 mm, the bubble shape became mushroom-like in appearance.

The generation of bubbles and their hold-up are important parameters relative to oxygenation, shear generation, and flow and agitation of the medium in airlift bioreactors.

Medium Flow Characteristics. The direction and velocity of the medium flow fluctuate severely in a simple aeration bioreactor (Fig. 3a). In general, the inhibition of cell growth in simply aerated bioreactors especially at higher aeration rates, is attributed to the generation of severe shear stress caused by turbulent and fluctuating flow of the medium

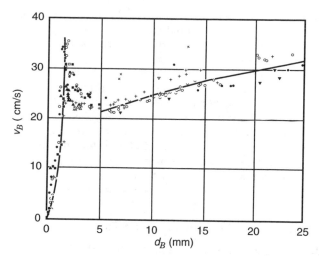

Figure 2. Superficial velocity of single bubble in water (27). Ordinate: superficial bubble velocity, V_β(cm/s). Abscissa: diameter of bubble, d_B (mm).

in the bioreactor. Such fluctuation was reduced by the using an airlift bioreactor (Fig. 3b). These results indicate the importance of the basic design and construction of in scaling up a bioreactor.

Medium flow direction in the riser column is mostly unidirectional without stagnation, but after the medium reaches the top of the vessel and is released into the free space, stagnation occurs (S. Takayama, unpublished result). These flow characteristics of a airlift bioreactor can be analysed by measuring medium circulation time as

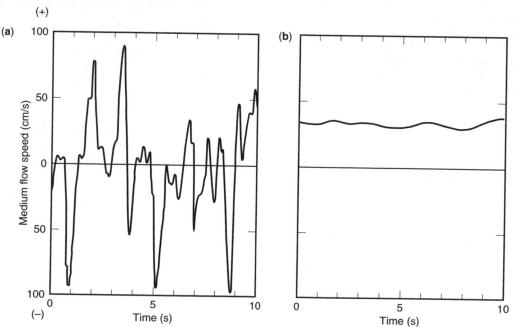

Figure 3. Medium flow characteristics in two types of 50-L bioreactors. (**a**) Simple aeration bioreactor. (**b**) airlift bioreactor. An electromagnetic flow meter was installed 10 cm below the medium surface in the culture vessel in the center of a simple aeration bioeactor and in the downcomer portion of an airlift bioreactor, and the medium flow velocity was recorded. The plus and minus symbols in the y axis indicate the direction of medium flow. The aeration rate was 20 L/min (2).

described in the next section where the difference between the observed and calculated circulation time is correlated with stagnation of the medium in the superficial and downcomer areas.

Medium flow in an airlift bioreactor is also affected by the shape and type of sparger. The straight bar or ring-shaped brass sparger with several openings (0.5 to 1 mm in diameter) generates rather large bubbles and induces turbulent flow but fine bubbles generated from a sintered sparger (plate or pipe) induce mild and slow medium flow. A plate sparger made of sintered material placed at the tapered bottom of the bioreactor is effective in preventing cell sedimentation in areas of poor mixing.

Medium Circulation Time and Fluid Velocity

Medium circulation time in an airlift bioreactor is affected by the aeration rate and by the volume of the medium. Thomas and Janes (28) reported the circulation time measured for air fluxes of 400 and 1000 mL/min, as shown in Figure 4. Circulation time was longes at lower gas flux which can be attributed to cycle-to-cycle variations in the number of close encounters between the stagnant particles and the bubbles rising through the draft tube. The curves of mean-value circulation time obtained with volumes of water ranging from 2000 to 2250 mL at the two air fluxes exhibited well-defined minima when the volume of water was about 2150 mL, irrespective of the gas flux.

Medium circulation was also affected by the internal diameter (or cross-sectional area) of the draft tube. Although the medium flow velocity was high in a narrow draft tube and slow in a wide draft tube, the flow characteristics measured in volume (mL/s) were the reverse (Fig. 5). Because the medium flow velocity in the narrow draft tube was high, the transit time required for flow through the draft tube (riser column) was quite short, but the the time required for recirculation in the downcomer was long (Fig. 6a). As the diameter of draft tube increases, the transit time in the draft tube lengthens, and the recirculation time shortens (Fig. 6a).

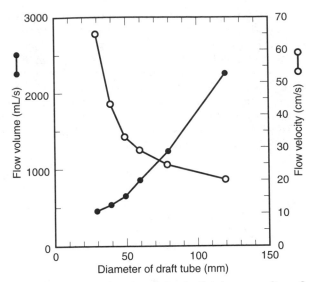

Figure 5. Effects of the size of the draft tube on medium flow velocity and flow volume of the medium in the transit of an airlift tube in a 10-L draft-tube airlift bioreactor (working volume, 8 L) (S. Takayama, unpublished result).

The difference between the observed and calculated circulation time ($a-b$ in Fig. 6b) reveals the stagnation time in the recirculation at the top of the vessel and in the downcomer portion of the column. The stagnation time was almost the same with all draft-tube diameters (Fig. 6b).

Although the oxygen transfer characteristics of airlift bioreactors are generally inferior to those of bubble column or aeration-agitation bioreactors, the liquid circulation of the airlift bioreactor makes this system superior to the bubble column or aeration-agitation bioreactors with respect to mixing (29,30).

Medium Mixing and Axial Biomass Distribution

The mixing behaviour of 4.0-liter airlift bioreactors was analyzed by Kiese et al. using a pulse technique, to measure the response of the water-filled bioreactor to a pH signal (22). The results shown in Figure 7 indicate that both circulation and mixing times as a function of air flow rate decrease sharply with increasing air flow up to superficial gas velocities of about 1 cm/s; and are nearly constant at higher air flow rates (22).

The medium circulation velocities in the riser U_{lr} at numerous superficial gas velocities U_{sgr} based on the riser cross section are represented by the simple equation (5),

$$U_{lr} = aU_{sgr}$$

where a and b were determined to be 0.96 and 0.39 when both velocities are expressed in m/s. The mixing times t_m depend on the air flow rate, as shown in Figure 8, and the sharpest decline of t_m occurs up to $U_{sgr} = 0.0175$ m/s (ca. 1.0 vvm), above which t_m slowly decreases from 22 s to a constant value of 17 s at 0.045 m/s.

In aeration bioreactors including the airlift bioreactor, the level of agitation which is necessary to limit shear, may be insufficient for mixing, expecially when

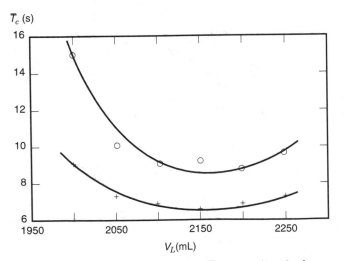

Figure 4. Mean-value circulation time, T_C, versus liquid volume, V_L, at air flow rates (mL/min) of 400 (○) and 1000 (+) (28).

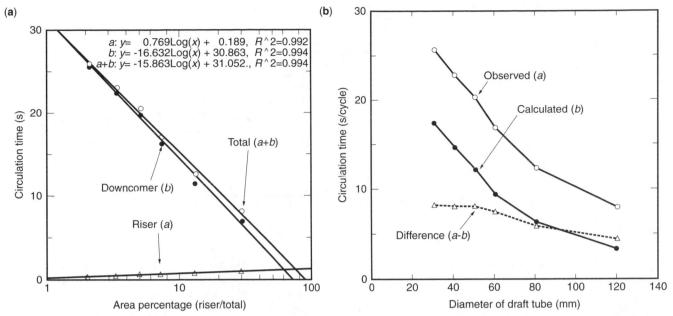

Figure 6. Effects of the size of the draft tube on medium circulation time of a draft-tube airlift bioreactor (S. Takayama, unpublished results). (**a**) Effect of the cross-sectional area percentage of the riser column to the total area on medium circulation time in a draft-tube airlift bioreactor. Bioreactor vessel size: 180 (i.d.) × 370 mm (length), 10 liter. Working volume: 8 liters Draft tube: 30–120 mm (i.d.) × 220 mm (length). (**b**) Comparison of calculated and observed medium circulation times.

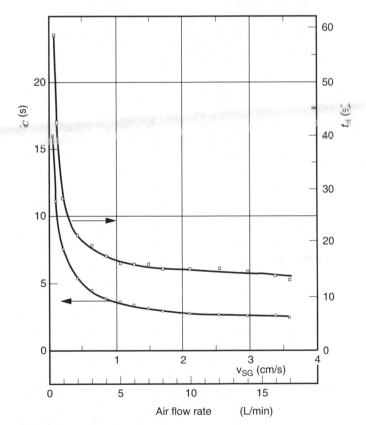

Figure 7. Effect of superficial gas velocity on mixing (t_m) and circulation times t_c (22).

cell concentrations exceed about 20 g DW/L (30) (dry weight/L).

The axial biomass distribution of the cell suspension was examined by Assa and Bar using *Phaseolus vulgaris* at a cell density of 12.5 g DW/L, and the result is shown in Figure 9 (5). In the denser and clump-forming plant cells, a clear trend of gradual biomass accumulation (the local concentrations increase from top to bottom) occurs

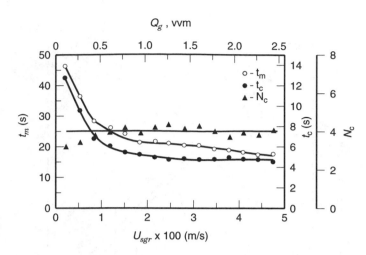

Figure 8. Mixing time, t_m, circulation time, t_c, and circulation number, N_c, as a function of the superficial riser gas velocity u_{sgr} for water (5).

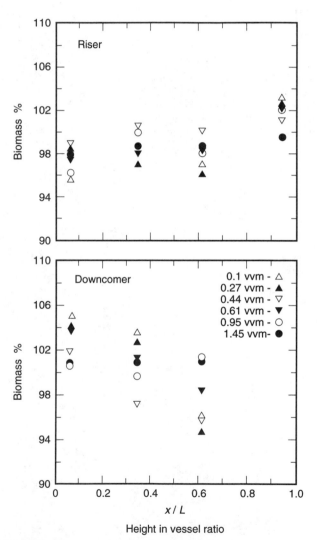

Height in vessel ratio

Figure 9. Biomass axial distribution expressed as a percentage of the overall concentration of 12.5 g DW/L of a cell suspension of *Phaseolus vulgaris* in water at various aeration rates (5). Aeration rate VVM, volume gas/volume liquid/minute.

in the downcomer, a slightly lower but uniform biomass level in the riser, and a somewhat higher level at the top of the vessel. These density gradients in the downcomer

could be attributable to the the gravity effect which leads to settling of cell aggregates. A more uniform distribution in the riser may be attributed to turbulent convection currents induced by gas sparging. The slightly higher concentration at the top of the vessel may result from a flotation effect induced by the rising bubbles.

Volumetric Oxygen Transfer Coefficient (K_La)

Plant cells are aerobically cultured in the bioreactor, so the supply of the oxygen into the bioreactor should be taken into account as a limiting factor in improving the overall efficiency of cell cultures. The supply of oxygen into the bioreactor is given by the volumetric oxygen transfer coefficient K_La values, expressed as

$$\ln \frac{C_s - C}{C_s - C_0} = -K_L a \cdot t$$

where C_s is the saturated dissolved oxygen (mg/L), C_o is the dissolved oxygen at time 0 (mg/L), C is the dissolved oxygen (mg/L) at definite time t (h^{-1}), and t is time (h). The oxygen requirement of plant cells is quite low compared to microorganisms, K_La values of about 10 h^{-1} are sufficient for biomass production of *Nicotiana tabacum* cells. The final biomass concentration becomes constant. But when K_La was set below 10 h^{-1}, cell yield depended on K_La values (31). The factors that affect K_L and a are the mixing conditions in the bulk liquid, the diffusion coefficient, the viscosity and the surface tension of the medium, the air-flow rate, gas hold up, and the bubble size (32).

The specific interfacial mass transfer coefficient K_L is constant for a fixed medium and temperature and is relatively insensitive to fluid dynamics in the bioreactor (33), but the specific interfacial area a is difficult to measure, and so the two parameters are combined and referred to as the volumetric mass transfer coefficient, K_La. The difference in K_La is mainly attributed to differences in the specific interfacial area a which is affected by aeration rate, bubble size, and mixing. K_La values are also affected by types of bioreactors and the draft-tube diameter. In the scale-up of an airlift bioreactor, the long residence time of small air bubbles in tall columns may lead to the depletion of oxygen from these bubbles which results in a decline of K_La (30).

Effect of Various Types of Bioreactors. Although differences between the airlift and bubble column bioreactors are subtle, the differences with respect to oxygen transfer and mixing are significant (30). Bello et al. demonstrated that the oxygen transfer characteristics of airlift bioreactors are generally inferior to those of the bubble columns because of the liquid circulation and limited oxygen transfer in the downcomer (29). This is partly explained by the fact that the dissolved oxygen levels were higher, in the riser section of the airlift bioreactor, whereas in the downcomer section they were lower than those realized in an aerated stirred tank bioreactor (23).

The effect of aeration rate on the volumetric oxygen transfer coefficient (K_La), in a 10-liter airlift bioreactor (working volume of 8 liters) with a draft tube was compared with those of a simple aeration and aeration-agitation bioreactors with and without baffle plates (Fig. 10). K_La values in airlift bioreactors were almost the same as in an simple aeration bioreactor, but these values were inferior to those of aeration-agitation bioreactors (with and without baffle plates). The K_La value for the aeration rate is expressed by the following simple correlations:

$$\text{airlift:} K_La = 0.669 + 2.925x, \quad r^2 = 0.997$$
$$\text{aeration:} K_La = 0.499 + 2.965x, \quad r^2 = 0.999$$
$$\text{agitation:} K_La = 0.140 + 5.438x, \quad r^2 = 0.996$$
$$\text{agitation + baffle:} K_La = 0.447 + 7.879x, \quad r^2 = 0.998$$

where x is the aeration rate (liter/min) (S. Takayama, unpublished results).

In airlift bioreactors, an optimal aeration rate or superficial air velocity is required for maximum cell growth and secondary metabolite production, as shown in Figure 11 (34).

Diameter of Draft Tube. K_La values in the airlift bioreactor are almost the same as those in simple aeration bioreactors, but the K_La values are affected by the

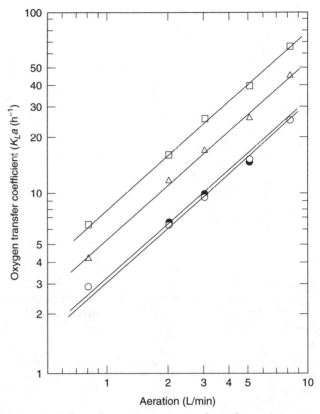

Figure 10. Effect of aeration rate on the volumetric oxygen transfer coefficient in various types of biorectors (S. Takayama, unpublished results). (○) airlift; (●) simple aeration; (△) agitation without baffle plate; (□) agitation (U_{gr}) with baffle plate.

size of the draft tube, as shown in Figure 12. The K_La values slightly decrease when the diameter of draft tube decreases. This phenomenon can be explained by the fact that bubble velocity in the riser column was stimulated

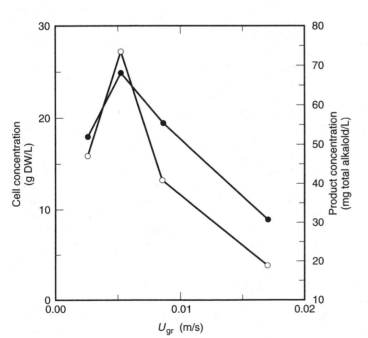

Figure 11. Cell and product concentrations versus superficial air velocity in the riser (34). (○): cell concentration, (●): product concentration.

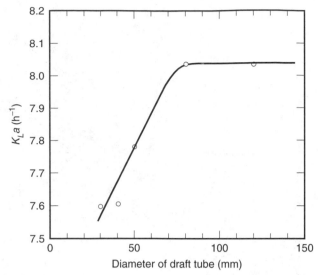

Figure 12. Effect of the diameter of the draft tube on the volumetric oxygen transfer coefficient ($K_L a$) in a draft-tube airlift bioreactor (S. Takayama, unpublished result). Aeration: 2 L/min; working volume: 8 L in a 10-liter vessel.

by decreasing the diameter of the draft tube, resulting in the reduction of the bubble surface area contacting the medium per unit aeration rate. In these situations, the a value (interfacial area of bubbles) of $K_L a$ at unit gas flux decreases because the bubbles rise rapidly without stagnation in narrow draft tubes.

Cell Density and Viscosity Change. The favorable effect of using the airlift bioreactor may result from reduced shear stress. Ballica and Ryu analyzed the effects of rheological properties and mass transfer on cell mass and tropane alkaloid production (34). When plant cells grow well, they can occupy 40 to 60% (about 15 to 20 g DW/L) of the entire culture volume, and the apparent viscosity becomes high. Tanaka examined the relationship between the apparent viscosity and the concentration of solids in supension and concluded that when the cell density exceeds 10 g/L, the slope of the apparent viscosity curve increases rapidly. When cell density reaches 30 g/L (DW), the culture medium becomes difficult to agitate and supply with oxygen (Fig. 13) (35). Under such high cell density conditions, the viscosity of the broth increased up to 100 cP (centpoise). The culture broth exhibited pseudoplastic behavior and resulted in a marked decrease in $K_L a$ values and cell growth rate (36).

The dependence of $K_L a$ on aeration rate and apparent viscosity (Fig. 14a) is given by the following empirical correlation (34):

$$K_L a = A(U_{gr})^{0.3}(\eta_{eff})^{-0.4}$$

where η_{eff} is the effective viscosity (cP), U_{gr} is the superficial air velocity in the riser (m/s), and A is a parameter related to the power input (34). This correlation is effective for plant cell culture and for satisfactory bioreactor operation. The $K_L a$ depends on the apparent viscosity of the cell suspension to the power of -0.4 over a cell concentration range of 0.3 to 0.8 mL/mL (Fig. 14b) (34).

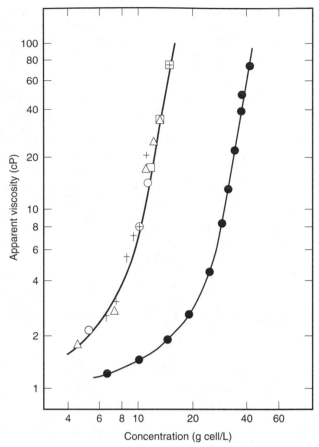

Figure 13. Relationship between apparent viscosity and concentration of cells and pseudocells in culture media (35). □: *C. roseus*; ○: *C. tricupsidata*; △: *N. tabucum*, +: granulated agar; ●: sepharose 6B beads.

Shear Stress (Reynolds Stress)

Because plant cells usually have high tensile strength but low shear resistance (32,37,38), the cells are more susceptible to shear under turbulence (34).

Airlift bioreactors are superior for yield and productivity of cell mass and secondary metabolites because of low shear stress and an adequate mass transfer coefficient compared with other bioreactor systems (6,9,36).

Wagner and Vogelmann compared different types of bioreactors and found that the yield of anthraquinones in the airlift bioreactor was about double that found in those bioreactors with flat blade turbine impellers, perforated disk impellers, or draft-tube bioreactors with Kaplan turbine impellers (9). The yield was also about 30% higher than that of a shake flask culture. Similarly, Asaka et al. reported that the production of ginsenoside in embryogenic tissue cultures of *Panax ginseng* was more than double the amount in an airlift bioreactor compared with a paddle impeller and internal turbine bioreactor, though the ginsenoside content on a gram dry cell basis was not so different (6).

Ballica and Ryu analyzed the effect of rheological properties and mass transfer on cell mass and tropane alkaloid production in a 1.2-liter draft-tube airlift bioreactor (34). The result of the the experiment on the

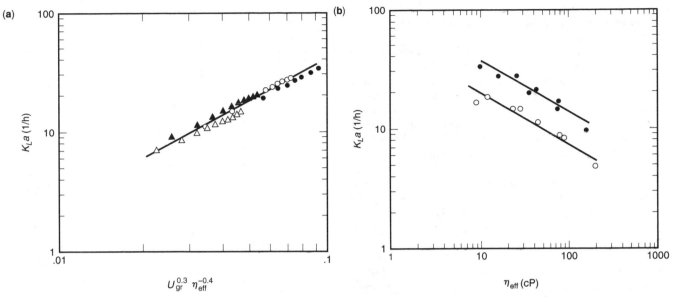

Figure 14. Effects of the physical properties in the culture on the volumetric oxygen transfer coefficient in a draft-tube airlift bioreactor (34). (**a**) K_La versus $(U_{gr})^{0.3}$ $(\eta_{eff})^{-0.4}$ for various cell concentrations: (**b**) Changes in the volumetric oxygen transfer coefficient with apparent viscosity at aeration rates of (●) 0.85 cm/s (96.16 s-1) and (○) 0.17 cm/s.

effect of superficial air velocity in the riser on the Reynolds stress (Fig. 15) reveals that the Reynolds stress increases at low cell concentrations and these stresses for higher cell concentrations are smaller (34). The Reynolds stress, τRe can be calculated from the following equation (35):

$$\tau\,\mathrm{Re} = 0.37\rho_I \left(\frac{\varepsilon}{\nu}\right) dp^2$$

where ε is the power dissipation per unit mass (cm²/s³), ν is the kinematic viscosity of the suspension (cm²/s), ρ_I is the fluid density (g/mL), and dp is the particle diameter (mm).

Destruction of cell aggregates is stimulated at higher Reyolds stress (Fig. 16). A greater size of cell aggregates or older cells also resulted in increased damage and susceptibility to Reynolds stress (34,39).

Power Requirements

In general, the requirement for oxygen in plant cell cultures is low compared with microbial and animal cells,

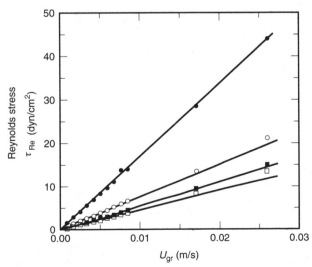

Figure 15. Changes in the estimated Reynolds stress (τ_{Re}) with superficial air velocity (U_{gr}) in the riser for various cell suspension concentrations (34). (●) PCV = 0.3 mL/mL; (○) PCV = 0.4 mL/mL; (□) PCV = 0.5 mL/mL; (▫); PCV = 0.6 mL/mL.

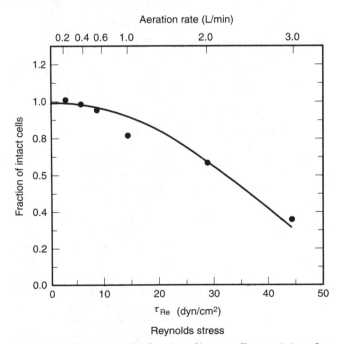

Figure 16. Changes in the fraction of intact cells remaining after hydrodynamic shearing. The fraction of intact cells is the ratio of dry weight of intact cells at the start to that at the end of the shear test (at 12 h) (34).

resulting in lower energy input in plant cell cultures and thus reducing the costs required for aeration and agitation (40).

Power requirements of the airlift bioreactor were analyzed by Gavrilescu and Roman in producing bacitracin in microbial cell cultures (23). They considered the total power (operating cost) involved in the fermentation process in the stirred vessel at an aeration rate of 0.2 vvm as 100% and found that the energy saving in the airlift bioreactor would be 17% at 1.0 vvm, 33% at 0.5 vvm, and 64% at 0.2 vvm respectively, which led to important energy savings (23).

Similar results were also reported by Malfait et al. for filamentous mold cultivation. The results revealed that the power requirements of the airlift bioreactor were approximately 50% of those of the mechanically agitated system (21).

The cost reductions are not limited to operating energy costs. The simplicity of the airlift bioreactor reduces construction and maintenance costs tremendously, which can result in an overall cost reduction in airlift bioreactors.

TYPES OF AIRLIFT BIOREACTORS

Various types of airlift bioreactors have been developed and used for plant cell and root cultures. These bioreactors are fundamentally classified by their structures.

Draft Tube, Internal Loop

The draft tube airlift bioreactor is the standard among various airlift bioreactors and is used for laboratory, pilot, and industrial-scale cell suspension cultures. The term "internal loop" is also used for this type of bioreactor. The construction of this type of bioreactor is simple, as illustrated in Figure 17. (41). A cylindrical tube situated vertically in the center of culture vessel is used as a draft tube. The air sparged into the draft tube from the air sparger flows up in the draft tube and circulates the medium within the bioreactor. Fundamentally, only a few revisions of the aeration or bubble column bioreactor, make the draft-tube airlift bioreactors possible. A modification of the internal loop results in the partition plate airlift bioreactor (Fig. 18) (16).

To operate a draft-tube airlift biroeactors efficiently, a ratio of height to diameter of 5 or more is necessary (22). The efficiency of medium circulation depends on the aeration rate and on the ratio of the cross-sectional area of the draft tube to the total cross-sectional area of the culture vessel. As described in the previous sections, the draft tube bioreactor is superior to bubble column or aeration-agitation bioreactors because of low shear stress, good mixing, and sufficient oxygen supply. In spite of these excellent characteristics, use of the draft-tube airlift bioreactor is limited compared with simple air-sparged bioreactors, bubble column bioreactors or aeration-agitation bioreactors. The uses of airlift bioreactors which have appeared in scientific reports are listed in Table 1 (42–45). The main objective of the research was to produce secondary metabolites and

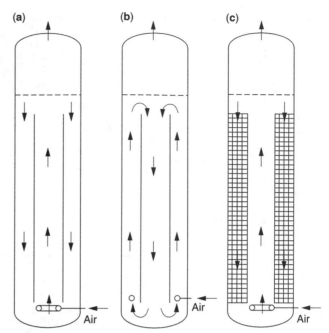

Figure 17. Types of draft tube airlift bioreactors (1,41). (**a**) Sparged into draft tube. (**b**) Sparged into annular space. (**c**) Mesh airlift for hairy root culture.

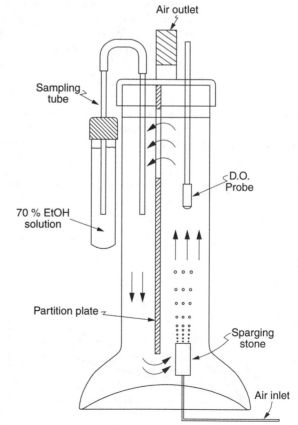

Figure 18. Partition-plate internal-loop airlift bioreactor (16).

analyze the biochemical engineering dimensions needed for scale-up.

Table 1. Research Using a Draft-Tube Airlift Bioreactor for Plant Cell and Tissue Cultures

Plant Name	Research Objective	Volume (L)	Ref.
Digitalis lanata	Biotransformation/β-methyldigitoxin	200	3
Digitalis lanata	Two-stage culture/deacetyllanatoside C diosgenin	20	4
Dioscorea deltoidea		2.4	42
Eschscholtzia californica	Benzophenanthridine alkaloids/two-phase	2.3	43
Datura stramonium	Tropane alkaloids	1.2	35
Thalictrum rugosum	Berberine	2.3	44
Phaseolus vulgaris	Biomass production	5.0	5
Panax ginseng	Embryogenic tissue/ginsenoside	3.0	6
Nicotiana tabacum	Simulation model/growth and product	2.0	7
Tripterygium wilfordii	Cone-shaped draft tube/tripdiolide	10	8
Molinda citrofolia	Types of bioreactors/anthraquinone	10	9
—	Review article	—	37
Catharanthus roseus	Large-scale cultivation/ajmalicine	20	10
Trigonella foenum-graceum	Hairy root/column mesh	9	11
Atropa belladonna	Hairy root/tropane alkalod	2.5	12
—	Review/hairy root	—	41
—	Review/scale-up	—	30
—	Review/scale-up	—	38
—	Review/scale-up	—	36
—	Review/scale-up	—	32
—	Review /polymodal pilot-plant fermentor	—	45
—	Review /fluid dynamics	—	28
Catharanthus roseus	Surface immobilization	6	13
Catharanthus roseus	Surface immobilization	20	14
Papaver somniferum	Surface immobilization/sanguinarine	6	15
Fragaria ananassa	Internal logitudinal partition panel airlift	0.55	16
Chenopodium rubrum	Semicontinuous/photoautotrophic	20	17
Chenopodium rubrum	Photoautotrophic	20	18

The various types of draft-tube airlift bioreactors reported (1,41) and partly scaled-up to industry scale are little used as laboratory bioreactors (22). Draft-tube airlift bioreactors are the most promising among various bioreactors because of their simple design and ease of construction.

External-Loop

The external-loop airlift bioreactor, known as the tower cycling bioreactor (46) or airlift tower-loop reactor (19), consists of two separate columns, a riser column into which air is sparged through an air sparger, a downcomer column (external recirculation tube) for liquid circulation, a connection between the tops and the bottoms of the columns to circulate the medium between the two columns of the vessel. The medium flow is generated by aeration sparged into the riser column. Operation at high biomass levels is possible in these systems. Fowler reported that cultures of *Catharanthus roseus* at biomass levels up to 30 g DW/L were grown without developing a major stagnant zone in 10- and 100-liter vessels (39). A laboratory-scale external-loop airlift bioreactor whose vessel consisted of a 100-mm bore Pyrex tube, 1.27 m long, as was constructed by Smart and Fowler (47), as shown in Figure 19. Such bioreactor systems are quite difficult to sterilize by autoclaving, so Smart and Fowler (47) sterilized the vessel by washing first with a 1% solution (V/V) of Steriguard bacterial detergent and then repeatedly steaming with live steam for 1.5 h on three successive days.

This type of airlift bioreactor is used on a large scale in industrial microbial fermentation (19), so application to plant cell culture is also feasible.

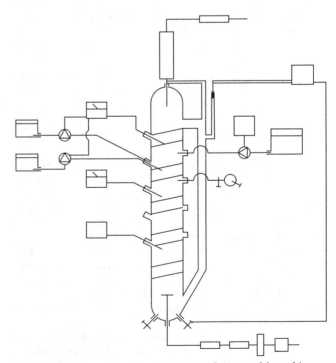

Figure 19. Schematic representation of external-loop bioreactor (48).

Sidelong Airlift Bioreactor (48)

Airlift bioreactors require bubbling to generate medium circulation and mass transfer. Although a tower shape in most airlift bioreactors, is considered the fundamental prerequisite for efficient operation, the author invented a new type of simple airlift bioreactor (48). The configuration is Sidelong, as shown in Figure 20 (sidelong) A-left. An air sparger was lifted at one side of the vessel wall, and this configuration is fundamental for generating medium flow. The medium flow characteristics of the sidelong airlift bioreactor are quite different from the simple bubbling bioreactor (Fig. 19), as indicated in Figure 20b,c. In the 50-liter sidelong bioreactor, the trajectory of the medium flow revealed clear circles, and the circulation time without cells was 6.5 s (Fig. 20c, left). The flow velocity, measured by a electromagnetic hydrocurrent meter, was highest at the periphery of the vessel and lowest in the central region (Fig. 20b, left). The trajectory of a particle suspended in the medium revealed mixing of the medium within several circulations.

The characteristics of this type of airlift have not yet been fully analyzed, but this type of bioreactor was invented to meet the need for cultivating regenerated plant tissues and organs and, of course, cell suspension cultures.

Modification of Airlift Bioreactors for Organ Cultures

Airlift bioreactors are used mostly for cell suspension cultures. Their characteristics are not suitable for regenerated tissue and organ cultures. In plant tissues, suspension cultures of regenerated tissues and organs are damaged severely, as indicated by Takayama and Misawa (49), because of severe shear stress, which increases by increasing the size of the culture (34). Even in turbulent conditions, the tissues attached to the coils of the stainless steel pipes which circulate the water used for temperature control can grow fairly. For example, in the bubble column fermenter, most of the shoots attached to the coil of the bioreactor, which prevented excessive injury to the tissues from aeration, and consequently, growth was luxuriant (49). This result suggests that immobilization of the tissues is the fundamental prerequisite for designing bioreactors for cultivating regenerated tissues and organ cultures.

Modifications of airlift bioreactors, equipped with a biomass separation meshes, were applied to regenerated tissue and organ cultures, especially bulbs, corms, embryos, shoot cultures, and root cultures, as shown in Figure 21a, b (50). Bubbling in the aeration vessel circulates the medium, and the cultured tissue segments are retained by stainless steel or nylon mesh located at the bottom of the culture vessel. The medium was sparged with bubbles in a riser column to generate the circulation. Tissues grew fairly because they suffered no damage from severe shear stress caused by aeration.

Meshes have also been used in hairy root cultures (11,41). Modified columns (Fig. 21c,d) (41) and column mesh bioreactor (Fig. 21d) (11) were proposed in which biomass separation takes place on the stainless steel mesh, perforated plate(s) or mesh basket(s) located in the bioreactor space over the air sparger and the liquid–gas mixing zone (41).

Surface Immobilized Plant Cell (SICP) Bioreactor

Plant cell suspension cultures should always be supplied with ample oxygen and nutrients. These conditions are far removed from the conditions of cells in intact tissues. The cells in suspension culture are sensitive to shear stress generated by aeration and mixing. As examined in this article, although the airlift bioreactor is considered suitable for plant cell cultures because of lower shear stress compared with other bioreactor systems, such as mechanically design agitation bioreactors or bubble column bioreactors, the problems of shear stress still remained to be resolved on scale-up.

Cell immobilization techniques have been suggested as a more appropriate way to culture these cells which grow naturally in tissue (14). A surface-immobilized plant cell (SIPC) bioreactor (13–15) was built as an airlift bioreactor, as shown in Figure 22 (13). A nonwoven short-fiber polyester material that attached readily and efficiently to plant cells by natural adhesion was used as the immobilization support. The immobilization material was cut into strips which were wound around a cage made of welded stainless steel rods in a hexagonal or square spiral configuration, and the spacing between the immobilizing layers was 1.2 cm. The immobilization material cage was placed vertically into the bioreactor vessel, and the riser tube of an airlift bioreactor was located at the center of the cage.

Alternative Bioreactor (Airlift Plus Impeller)

The airlift bioreactor is excellent for plant cell cultures up to a cell density of 20 g cell dry weight per liter (9), but the problems encountered in using airlift bioreactors change due to the rheological properties at higher cell densities. High cell density decreases oxygenation and mixing due to stagnation of the cell suspension. The results of these phenomena are decreased cell growth and metabolite production. To solve these problems airlift bioreactor systems equipped with Kaplan turbine impellers at the bottom of the draft tube were used to stimulate axial medium flow (Fig. 23). The Kaplan turbine is well known as an axial hydraulic action type turbine with adjustable runner blades operated with a high flow rate. Wagner and Vogelmann studied airlift reactors with Kaplan turbines as a hydromechanical mixing system which should achieve increased cell densities (9). The effect of using this type of bioreactor on yield and productivity for cell mass and anthraquinones in *Morinda citrifolia* was examined. The cell mass concentration was practically identical with that in the airlift bioreactor, but the yield of anthraquinones was only about one-third that compared to the airlift bioreactor (9). The decline of anthraquinone production may be the result of increased shear stress as the tip velocity of the Kaplan turbine exceeded a defined critical value (9). Such results indicate that this type of bioreactor is insuitable for cell strains sensitive to shear stress, but plant cells reveal a wide range of sensitivity to shear

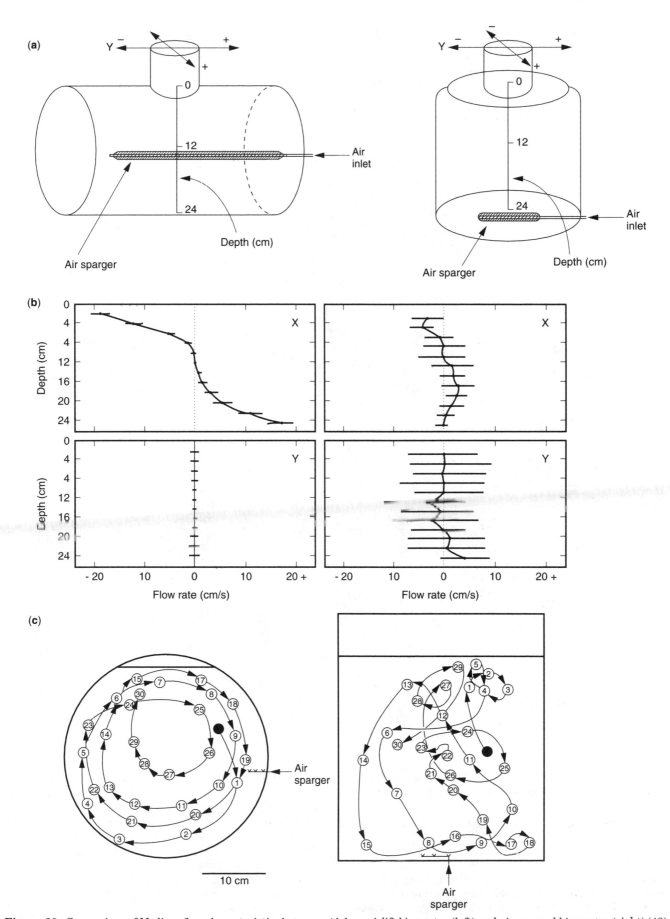

Figure 20. Comparison of Medium flow characteristics between sidelong airlift bioreactor (left) and air-sparged bioreactor (right) (48).

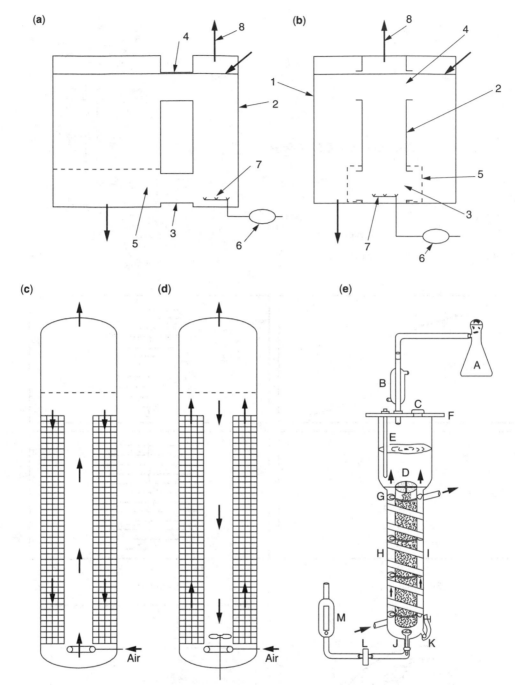

Figure 21. Types of airlift bioreactors for organ culture (1,41). (**a**) External airlift bioreactor with separator mesh. (**b**) Draft-tube airlift bioreactor with separator mesh. (**c**) Mesh airlift for hairy root culture. (**d**) Mesh airlift with turbine for hairy root culture. (**e**) Column mesh bioreactor for hairy root culture.

stress, and selected species or strains resistant to severe shear stress may be cultured in this type of bioreactor.

APPLICATIONS OF AIRLIFT BIOREACTORS

Anthraquinone Production

Wagner and Vogelmann used *Morinda citrofolia* as a model cell culture to compare four different fermenter designs for both growth and secondary metabolite production (Fig. 24) (9,51). In comparison with shake flasks the airlift bioreactor produced similar maximum dry weight yields, but a 30% increase in maximum anthraquinone yield. By contrast, in all the mechanically stirred bioreactors, the maximum dry weight showed a small reduction and, significantly, anthraquinone yields were markedly diminished. The primary reason for these differences was cell breakage caused by the shear stress generated by mechanical stirring. Cell breakage was confirmed by microscopic observation and also by an

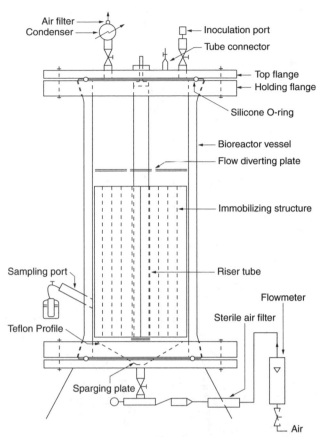

Figure 22. Surface-immobilized plant cell (SICP) bioreactor (13).

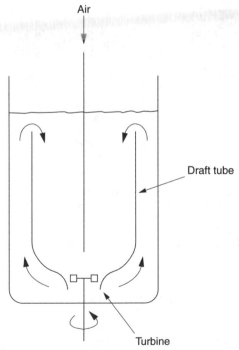

Figure 23. Draft-tube bioreactor with Kaplan turbine (9).

associated increase in culture pH resulting from the release of cell contents. Judging from these medium pH changes, cell breakage becomes noticeable at the end of

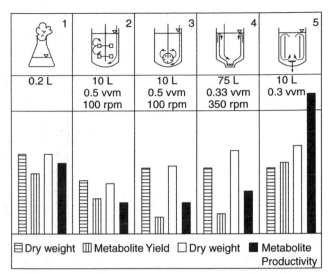

Figure 24. Comparison of yield and productivity for cell mass and anthraquinones in various bioreactor systems (9). 1. Shake flask; 2. flat blade turbine, 3. perforated disk impeller; 4. draft-tube reactor with Kaplan turbine; 5. airlift reactor.

the growth phase, well before anthraquinone accumulation reaches a maximum value in the stationary phase.

Diosgenin Production (42)

Diosgenin was produced by freely suspended cell cultures of *Dioscorea deltoidea* using a 2.4-liter draft-tube airlift bioreactor containing 2.0 liters of medium. At low oxygen supply rates (K_La value of 3.0 h^{-1}) diosgenin was not produced but the bioreactor successfully produced cell mass when the phosphate concentration in the medium was less than ca. 2 mM. At high oxygen supply rates (K_La values of 17.1 h^{-1}), both cell mass and diosgenin were produced at all phosphate concentrations. A high oxygen supply rate (K_La of 17.1 h^{-1}) led to the greatest formation of diosgenin in 30 g/L sucrose when the sucrose-to-phosphorus mole ratio was 42.5 : 1. Cells immobilized by entrapment in reticulated polyurethane foam were compared with cell suspension in the airlift bioreactor but led to delayed development of a cell suspension culture. The diosgenin formation in this case was significantly higher, but all of this increase was attributable to the suspended cells.

Biotransformation of β-Methyldigitoxin to β-Methyldigitoxin (3)

Alfermann et al. reported using a 200-L draft-tube airlift bioreactor to produce β-methyldigitoxin by hydroxylation of β-methyldigitoxin by *Digitalis lanana* cell suspension cultures (3). They converted a normal commercial bioreactor into an airlift biorector with an inner draft tube. Cell cultures were scaled-up from 1-L Erlenmeyer flasks to the 200-L airlift biorector (Fig. 25). In the 200-L bioreactor, a typical biotransformation of β-methyldigitoxin to β-methyldigitoxin was produced and 430 mg of β-methyldigitoxin resulted. This was about 70% of the substrate added. About 10% of the substrate remained untransformed. The amount of

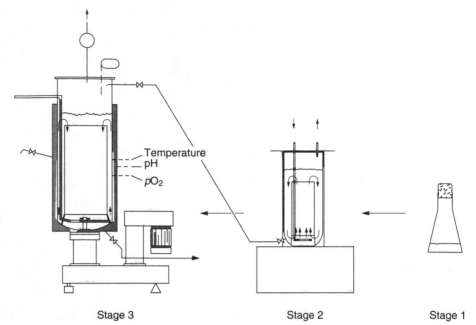

Figure 25. Scaling up of *Digitalis lanata* cell cultures to a working volume of 200 L. For stages 2 and 3, conventional bioreactors converted into airlifts with inner draft tubes are used (3). Stage 1. 1-L Erlenmeyer flask. Stage 2. 30-L airlift. Stage 3. 200-L airlift.

β-methyldigitoxin produced in a 200-L reactor operated for 23 days would be sufficient to prepare more than 800,000 cardiovascular tablets for treating 1,000 patients who have heart disease, for more than one year.

Two-stage Cultivation of Digitalis Lanata Cells: Semicontinuous Production of Deacetyllanatoside C in 20-Liter Airlift Bioreactors (4)

Two bioreactors were connected for efficient production of deacetyllanatoside C, a cardenolide of the important digoxin series. Digitoxin was used as the substrate for the biotransformation. The process was established at a 20-L scale using two airlift bioreactors. One bioreactor (Fig. 26-left: working volume 12 liters in a 20-liter airlift bioreactor) was used for cell growth, and another (Fig. 26-right: working volume 18 liters in a 20-liter airlift bioreactor) for deacetyllanatoside C production. Growth and production phases were synchronized, and the process finally ran semicontinuously in a 7-day cycle. Six consecutive production runs were carried out yielding a total of 43.8 g d'eacetyllanatoside C.

Two-Phase Airlift Bioreactor (43)

The-phase airlift bioreactor is a technique for enhancing the productivity of plant metabolites. Byun and Pedersen reported using compounded silicone fluid to make two-phase cultures to increase the productivity of bezophenanthridine alkaloids in cell suspension cultures of *Eschscholtzia calfornica* in an airlift bioreactor (43). Compounded silicone fluid solubilizes various organic compounds. Its specific gravity is 0.97, and it is immiscible with culture media. These characteristics are excellent in making an accumulation phase clearly separated from the culture medium. In the experiment, the volume of the accumulation phase was 200 mL and that of culture medium was 2.1 liters. The accumulation phase

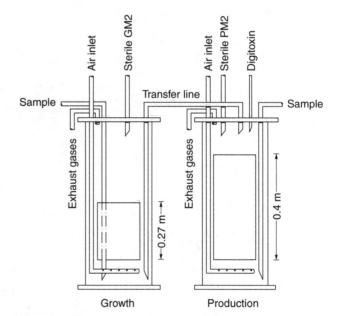

Figure 26. Bioreactor configuration for two-stage culture by semicontinuous growth for digitoxin biotransformation using *Digitalis lanata* cells. Cells were grown in a 20-L airlift bioreactor (working volume 12 L) and inoculated into a second 20-L airlift bioreactor (working volume 18 L) for deacetyllanatoside C production (4).

always remained at the top of the culture medium, and the contact area between the culture medium and the accumulation phase for mass transfer remained constant throughout the operation. In the two-phase airlift bioreactor, the major part of the alkaloid was in the accumulation phase (153 mg/L) compared with the alkaloid found in the aqueous cellular phase (cell and medium) (8.2 mg/L). The net production of the alkaloid in a two-phase airlift bioreactor was almost 1.5 times higher than that in a normal airlift bioreactor operation.

The addition of elicitors (yeast elicitor) combined with the two-phase technique further stimulated the productivity, and the net production of the alkaloid was almost 3.5 times higher than that of a normal airlift bioreactor operation.

Airlift Bioreactors for Plant Roots (11)

Plant cell culture is a promising technology for large-scale production of useful quantities of secondary metabolities and some industrial-scale bioreactors have been constructed, but a problem frequently encountered in developing of an industrial process for producing various plant metabolites is the instability of secondary metabolism which leads to a decline in the secondary metabolite content in serial subcultures. Once the content declines, it is difficult to recover productivity. Some examples are the tropane alkaloids, vinca alkaloids, steviol glycosides, and harringtonine. Alternatives to cell cultures, regenerated tissue, and organ cultures (especially transformed root (hairy root) cultures) are considered promising because their genetic ability to produce various secondary metabolites is stable. Airlift bioreactors are examples among various culture vessels. Rodrigues-Mandiola et al. reported using airlift bioreactors to produce diosgenin by hairy root cultures of *Trigonella foenum-graceum* (11). They designed two airlift bioreactors, a draft-tube bioreactor and a column-mesh bioreactor (Fig. 21e), both of them which had a 9-liter nominal capacity and an aspect ratio (height/diameter) of 6.4:1. The ratio of both the draft tube and cylindrical mesh to the bioreactor diameter was 3:2. The bioreactors were aerated at 0.8 liter/min (0.1 vvm) using a sintered sparger 4 cm in diameter. At first, hairy roots of *Trigonella foenum-graceum* were cultured in a draft tube bioreactor where the roots progressively attached to the centralizing support arms. Most of the roots grew in all directions, distributed throughout the bioreactor, and gradually formed a cylindrical ring net. The medium flow through the rootnet without any limitation even after 8 weeks of culture indicated that nutrients, including oxygen, were available to the roots. Based on the results of *Trigonella* hairy root cultivation in the draft-tube bioreactor, a column mesh bioreactor was designed, and hairy roots of *Nicotiana rustica* were cultured. The hairy roots gradually became attached to the mesh (which provides the immobilization sites), and grew in all directions, finally forming a cylinder of roots which completely filled the bioreactor. Harvesting the roots was quite easy because the roots were packed into a cylindrical ring. Production of diosgenin was observed in these systems. These designs can be easily adopted by any competent plant cell culture laboratory interested in developing pilot-scale root systems.

BIBLIOGRAPHY

1. B. Atkinson and F. Mavituna, *Biochemical Engineering and Biotechnology Handbook*, 2nd ed., Macmillan, New York, 1991.
2. S. Takayama and M. Akita, *Plant Cell Tiss. Org. Cult.* **39**, 147–156 (1994).
3. A.W. Alfermann et al., in S.H. Mantell and H. Smith, eds., *Plant Biotechnology*, Cambridge University Press, Cambridge, U.K., 1983, pp. 67–74.
4. W. Krcis and E. Reinhard, *J. Biotechnol.* **16**, 123–136 (1990).
5. A. Assa and R. Bar, *Biotechnol. Bioeng.* **38**, 1325–1330 (1991).
6. I. Asaka et al., *Biotechnol. Lett.* **15**, 1259–1264 (1993).
7. N. Shibasaki, T. Yonemoto, and T. Tadaki, *J. Chem. Technol. Biotechnol.* **58**, 151–157 (1993).
8. P.M. Townsley and F. Webster, *Biotechnol. Lett.* **5**, 13–18 (1983).
9. F. Wagner and H. Vogelmann, in W. Barz, E. Reinhard, and M.H. Zenk, eds., *Plant Tissue Culture and Its Biotechnological Application*, Springer-Verlag, Berlin, 1977, pp. 245–252.
10. D.P. Fulzele and M.R. Heble, *J. Biotechnol.* **35**, 1–7 (1994).
11. M.A. Rodriguez-Mandiola, A. Staffod, R. Cresswell, and C. Arias-Castro, *Enzyme Microb. Technol.* **13**, 697–702 (1991).
12. J.M. Sharp and M.P. Doran, *J. Biotechnol.* **16**, 171–186 (1990).
13. J. Archambault, B. Volesky, and W.G.W. Kurz, *Biotechnol. Bioeng.* **35**, 702–711 (1990).
14. J. Archambault, *Enzyme Microb. Technol.* **13**, 882–892 (1991).
15. J. Archambault, R.D. Williams, M. Perrier, and C. Chavarie, *J. Biotechnol.* **46**, 121–129 (1996).
16. Y.C. Hong, T.P. Labuza, and S.K. Harlander, *Biotechnol. Prog.* **5**, 137–143 (1989).
17. U. Fischer et al., *Plant Cell Tiss. Organ Cult.* **38**, 123–134 (1994).
18. U. Fischer and A.W. Alfermann, *J. Biotechnol.* **41**, 19–28 (1995).
19. R. Luttmann et al., *Biotechnol. Bioeng.* **24**, 1851–1870 (1982).
20. S. Frohlich et al., *Biotechnol. Bioeng.* **38**, 56–64 (1991).
21. J.L. Malfait, D.J. Wilcox, D.G. Mercer, and L.D. Barker, *Biotechnol. Bioeng.* **23**, 863–878 (1981).
22. S. Kiese, H.G. Ebner, and U. Onken, *Biotechnol. Lett.* **8**, 345–350 (1980).
23. M. Gavrilescu and R.-V. Roman, *Acta Biotechnol.* **13**, 161–175 (1993).
24. F. Unterluggauer et al., *BioTechniques* **16**, 140–144, 146–147 (1994).
25. S. Aiba, A.E. Humphrey, and N.F. Millis, *Biochemical Engineering*, University of Tokyo Press, Tokyo, 1965, pp. 1–345.
26. W.W. Eckenfelder, Jr., *Chem. Eng. Prog.* **52**, 286 (1956).
27. D.W. Van Klevelen and P.J. Hoftijze, *Chem. Eng. Prog.* **46**, 29 (1950).
28. N.H. Thomas and D.A. Janes, ed., *Ann. N.Y. Acad. Sci.* **506**, 171–189 (1987).
29. R.A. Bello, C.W. Robinson, and M. Moo-Young, *Biotechnol. Bioeng.* **27**, 369 (1985).
30. G.F. Payne, M.L. Shuler, and P. Brodelius, in B.J. Lydersen, ed., *Large Scale Cell Culture Technology*, Carl Hanser Verlag, Munich, 1987, pp. 193–229.
31. A. Kato, Y. Shimizu, and S. Noguchi, *J. Ferment. Technol.* **53**, 744–751 (1975).
32. M.M.R. Fonseca, F. Mavituna, and P. Brodelius, in M.S.S. Pais, F. Mavituna, and J.M. Novais, eds., *Plant Cell Biotechnology*, Springer-Verlag, Berlin, 1988, pp. 389–401.
33. H. Blenke, *Adv. Biochem. Eng.* **13**, 121 (1979).

34. R. Ballica and D.D.Y. Ryu, *Biotechnol. Bioeng.* **42**, 1181–1189 (1993).

35. H. Tanaka, *Biotechnol. Bioeng.* **24**, 2591–2596 (1982).

36. A. Pareilleux and N. Chaubet, *Biotechnol. Lett.* **2**, 291–296 (1980).

37. S.H. Mantell, J.A. Matthews, and R.A. McKee, *Principles of Plant Biotechnology: An Introduction to Genetic Engineering in Plants*, Blackwell, London, 1985.

38. M.L. Schuler, in M.S.S. Pais, F. Mavituna, and J.M. Novais, eds., *Plant Cell Biotechnology*, Springer-Verlag, Berlin, 1988 pp. 329–342.

39. M.W. Fowler, in S.H. Mantell and H. Smith, eds., *Plant Biotechnology*, Cambridge University Press, Cambridge, U.K., 1983, pp. 4–37.

40. R.G.S. Bidwell, *Plant Physiology*, 2nd ed., Macmillan, New York, 1979 pp. 1–726.

41. H. Wysokinska and A. Chimel, *Acta Biotechnol.* **17**, 131–159 (1997).

42. G.H. Robertson et al., *Biotechnol. Bioeng.* **34**, 1114–1125 (1989).

43. S.Y. Byun and H. Pedersen, *Biotechnol. Bioeng.* **44**, 14–20 (1994).

44. D.-I. Kim, H. Pedersen, and C.-K. Chin, *Biotechnol. Bioeng.* **38**, 331–339 (1991).

45. P. Hambleton, J.B. Griffiths, D.R. Cameron, and J. Melling, *J. Chem. Technol. Biotechnol.* **50**, 167–180 (1991).

46. C.H. Lin et al., *Biotechnol. Bioeng.* **18**, 1557–1572 (1976).

47. N.J. Smart and M.W. Fowler, *J. Exp. Bot.* **35**, 531–537 (1984).

48. Jpn. Pat. Appl 63–192382 (August 9, 1988), S. Takayama (to Kyowa Hakko Kogyo, Co. Ltd.).

49. S. Takayama and M. Misawa, *Plant Cell Physiol.* **22**, 461–467 (1981).

50. Jpn. Pat. Appl 62–179383 (August 6, 1987), S. Takayama (to Kyowa Hakko Kogyo, Co. Ltd.).

51. G. Wilson, in T.A. Thorpe, ed., *Frontiers of Plant Tissue Culture*, 1978, pp. 169–177.

52. P. Morris, N.J. Smart, and M.W. Fowler, *Plant Cell Tissue Organ Cult.* **2**, 207–216 (1983).

See also ANIMAL CELL CULTURE, EFFECTS OF AGITATION AND AERATION ON CELL ADAPTATION; ANIMAL CELL CULTURE, PHYSIOCHEMICAL EFFECTS OF DISSOLVED OXYGEN AND REDOX POTENTIAL; ASEPTIC TECHNIQUES IN CELL CULTURE; BIOREACTORS, PERFUSION; BIOREACTORS, RECIRCULATION BIOREACTOR IN PLANT CELL CULTURE; CULTURE ESTABLISHMENT, PLANT CELL CULTURE; EQUIPMENT AND LABORATORY DESIGN FOR CELL CULTURE; OFF-LINE ANALYSIS IN ANIMAL CELL CULTURE, METHODS; OFF-LINE IMMUNOASSAYS IN BIOPROCESS CONTROL; ON-LINE ANALYSIS IN ANIMAL CELL CULTURE; PLANT CELL CULTURES, SECONDARY PRODUCT ACCUMULATION; STERILIZATION AND DECONTAMINATION.

BIOREACTORS, CONTINUOUS CULTURE OF PLANT CELLS

H.J.G. TEN HOOPEN
Delft University for Technology
The Netherlands

OUTLINE

Introduction

Assumptions in Continuous Culture Theory and Their Pitfalls

The Practical Setup of a Continuous Culture

Mathematical Description of Continuous Culture

Applications of the Continuous Cultivation of Plant Cells

Nomenclature

Bibliography

INTRODUCTION

The first suspension culture of plant cells in a nutrient solution was reported in 1954 by Muir (1). Only two years later cultures of plant cells in bioreactors were carried out by Melchers and Engelmann (2). Bioreactors can be employed in batch, fed batch, or continuous mode. In a continuous culture, the nutrients consumed by the organisms are continuously replenished by an inflow of fresh medium. In most cases the culture volume is kept constant by an equal outflow of the cell suspension (Fig. 1). Stable steady states can be achieved at dilution rates (flow rate/volume) smaller than the maximum specific growth rate of the organism, as predicted by mathematical derivation. Intuitively, it is understandable that steady states occur, because the limiting nutrient concentration and the biomass growth are coupled in a stable control cycle: Increasing nutrient concentration stimulates growth resulting in increased biomass concentration, but increased biomass concentration decreases the limiting nutrient concentration. In a steady state, all concentrations (biomass, biomass composition, and nutrients) are constant. The biomass concentration in the culture

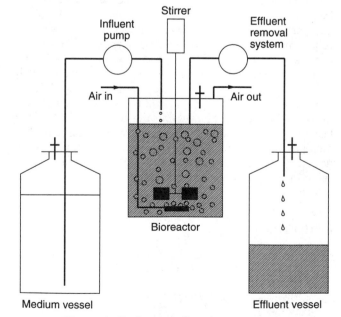

Figure 1. Basic setup of continuous culture.

is controlled by the concentration of the growth-limiting nutrient in the medium; the average specific growth rate in the culture is identical with the dilution rate. This "balanced growth" in the chemostat is a very attractive tool in physiological studies or in research on growth and production kinetics. It provides an experimental system with completely controlled conditions. In fact, it is opposite to batch culture, where all conditions are continuously changing. (In specific cases this may be advantageous; for example, nongrowing cells in the stationary stage can only be studied in a batch culture.) Furthermore, in continuous culture it is possible to study a cell culture under various limitations (carbon, nitrogen, or phosphate source). Continuous culture could be attractive for industrial fermentations too because of the relatively long run times compared to short down times in the production process. However, the number of current applications in the fermentation industry is rather small. The most common type of continuous culture, controlled by the influent flow, is more specifically called a *chemostat*.

A different approach of running a continuous culture is by monitoring the biomass concentration and controlling the influent/effluent rate by this parameter. The monitoring device used is a turbidimeter, and the system is called *turbidostat*. A disadvantage of this system is that the tubidimetric biomass determination is not very reliable, because of its sensitivity for changes in cell or aggregate size. Phototrophic biomass uses light as an energy source. Therefore, phototrophic cultures can be grown in continuous culture with light as the limiting "nutrient." The concept of this photostat has several specific features, such as the correlation between cell density and light absorbance. In a low-density culture part of the "nutrient" light is lost through the glass wall or absorbed by the wall of the vessel. This makes the theoretical and practical basis considerably different from the chemostat. Phototrophic plant cell cultures are not very common. Therefore, only the chemostat will be discussed in detail. Some publications refer to continuous culture systems, if the medium flow is continuous but the biomass is retained in the culture vessel (perfusion cultures, hairy root cultures, immobilized cells). These systems are not discussed here.

The number of applications of continuous culture of plant cells is small compared to continuous cultures of bacteria and yeasts. There are several reasons for this difference. Plant cells in suspension culture have some characteristics that complicate the performance of a continuous culture. The low specific growth rate of plant cells (generation times usually between 20 and 60 h) means that it takes a long time to reach a steady state; the total duration of an experiment can extend from some weeks up to more than a year. The risk of calamities (contamination, equipment failure) is then, of course, considerable. On a volume basis, plant cells are up to 200,000 times larger than bacteria and may have a relatively high specific density because of the formation of intracellular starch. Furthermore, in a plant cell culture there will be always some aggregate formation. The size of these aggregates can vary widely with the cell line and the environmental conditions, but it can be up to several

millimeters in diameter. The large and "heavy" cells and aggregates tend to settle in the reactor. Another problem is the sticky character of several cell lines and the excretion of sticky and viscosity-increasing polysaccharides. This makes it difficult to maintain an ideally mixed suspension, a prerequisite for continuous culture.

In subsequent sections, practical problems and pitfalls in the setup of the continuous culture, the design of continuous culture equipment, the mathematical description of the continuous culture, and a literature survey will be discussed, all with an emphasis on plant cells in suspension culture.

ASSUMPTIONS IN CONTINUOUS CULTURE THEORY AND THEIR PITFALLS

Continuous culture theory is built on a set of assumptions (Table 1). The inflow of fresh medium into the well-mixed culture should be continuous. All cells in the culture should face the same environmental conditions, particularly the concentration of the growth-limiting component. However, in most cases influent is added dropwise at one place in the culture, hence discontinuously in time and place. After addition to the culture, nutrients are diluted by mixing, taken up by the cells, and metabolized. The problem of the discontinuous addition can be analyzed by a comparison of the characteristic times of the involved mechanisms: addition, mixing, uptake, and metabolic conversion. When the characteristic time for uptake is short compared to the mixing time, a drop of medium serves only a part of the culture. This effect is smoothed if the mixing time is short compared to the metabolic conversion time. All cells will pass through a relatively nutrient-rich zone before starvation will occur. However, a fast uptake of the limiting nutrient and a slow metabolic conversion causes a relatively high nutrient concentration in the cell; this could trigger alternative metabolic pathways like formation of storage products or production of secondary metabolites, especially in the case of plant cells.

Another assumption in continuous culture theory is that the biomass suspension is ideally mixed. Then the biomass concentration in the vessel and in the effluent are equal. As already mentioned in the introduction, this is a critical point with plant cells. This assumption can be checked by estimating the *effluent efficiency*. This efficiency is 100% if the biomass concentration in the vessel and in the effluent are equal. Inadequate mixing in the culture vessel is caused by three phenomena: wall growth, settling, and buoyancy. The effect (decrease) on the

Table 1. Assumptions in the Operation of Continuous Culture

Assumption	Points of concern
Continuous feed	Addition by drops
All biomass in suspension and ideally mixed	Wall growth, settling, buoyancy
Biomass in reactor and effluent equal (100% efficiency)	Effluent removal device
Biomass in steady state	Intracellular adaptation
Limiting factor defined	Defined media, validation

effluent efficiency is increased by the use of an inadequate effluent removal system. It is more common than realized, not only in plant cell cultures. Noorman et al. (3) analyzed the kinetic consequences of nonideal mixing.

The biomass in a continuous culture is assumed to be in steady state. However, multicompartment plant cells may require a relatively long intracellular adaptation time to reach balanced growth and steady-state biomass composition compared to bacteria. After a change in conditions it takes roughly 5 residence times [= 5/dilution rate, where dilution rate = flow (m^3/s)/volume (m^3)] to reach a steady state if hydrodynamics is the ruling mechanism. In plant cell cultures, where the intracellular adaptation might be controlling, there are examples known where as many as 12 residence times were needed to attain a constant biomass composition in a phosphate-limited continuous culture.

The growth-limiting factor in continuous culture studies is always assumed to be known, but analysis of such experiments may reveal scepticism about this. A completely defined medium is advantageous, and medium components such as coconut milk should be avoided. A distinct test of the limiting factor is to decrease the concentration of the apparent limiting component; if this results in a proportional change in the biomass concentration, the assumption is correct.

THE PRACTICAL SETUP OF A CONTINUOUS CULTURE

The concept of the continuous culture as described in the introduction can be translated into the basic design (Fig. 1): a culture vessel with an ideally mixed cell culture, a medium vessel for the addition of fresh medium through the influent pump, and an effluent system (a pump or an overflow device) connected to a waste vessel. The influent pump controls the dilution rate; the effluent system is designed in such a way that the volume remains constant for example through the level of the outlet tube. Many additions to and modifications of this setup are possible, dependent on the particular application. A culture of aerobic organisms needs an aeration system. Probes for online measurements of volume, temperature, pH, redox potential, and foam development can be attached. Control systems based on the sensor signals to control volume, temperature, pH, and foam level may be installed. Online analysis equipment to determine the influent and effluent gas composition and essential medium components are very useful. A sampling device for off-line analysis of the broth is in fact essential. A condenser to avoid loss of water from the culture vessel by evaporation is especially recommended for slow-growing organisms, such as plant cells. In these cultures the evaporation can be considerable in comparison with the medium flow through the system. Several modifications have been reported on the aeration and stirring systems. In the case of plant cells the objective is to achieve sufficient mixing of the culture without damaging the cells. Aeration and stirring generate hydrodynamic (shear) stress in the reactor. Therefore, the applicability of stirred tank reactors have been disputed for several years. However, Scragg et al. (4) and Meijer et al. (5,6) demonstrated that cells from various plant

species are relatively shear tolerant. They concluded that large differences in shear sensitivity exist among plant cell lines, and that stirred tank reactors in most cases are the best choice for plant cell cultures. An option to reduce the negative effects of shear stress imposed by the impeller in a stirred tank can be achieved by modifying the design of the routinely used Rushton turbine impeller. Among those modified impellers are the hollow paddle (7,8), the marine (9), the cell lift (9,10), the large flat blade (11), and the helical-ribbon (12) impeller.

The effluent system needs special attention in continuous cultures of plant cells. As mentioned before, plant cells settle and stick to themselves and the culture vessel relatively easy. In combination with the low flow rates in a continuous culture of plant cells (because of the low growth rate), this makes the common effluent systems fail. These systems remove effluent from the surface of the culture, either by an overflow device, or an effluent pump through a level tube. Application of this system in a plant cell culture causes a low effluent efficiency, because the surface of the culture shows a difference in composition with the average of the culture due to settling or buoyancy. Furthermore, the cells will stick and eventually block the effluent tubing, especially because of the low flow rates in continuous culture of plant cells. Van Gulik et al. (13) developed an outflow device avoiding these problems (Fig. 2). The objective was to design a system that removes effluent from an optimally mixed section of the reactor (close to the stirrer) and at a high flow rate to avoid sticking or settling in the tubes. To achieve this the outflow device operates discontinuously. At regular time intervals a pneumatic valve system forces the effluent under air pressure into the waste vessel at a high flow rate. The timing is such that the culture volume stays within 1% of the set point. The effluent efficiency of this system was tested and approached the theoretical maximum of 100%.

The sticky and foamy character of some cell cultures makes it difficult to keep the cells inside the broth: They tend to form a layer above the suspension. Bertola and Klis (14) attacked this problem by the use of short periods of increased stirrer speed at regular intervals. Cells were taken up in the suspension again, and no cell damage was observed. Another way to avoid this problem is to leave out as much as possible objects protruding through the culture surface, such as baffles and probes. A double-wall bioreactor with sensor ports on the bottom is advantageous from this viewpoint. The wall can be cleaned either by shaking the culture vessel daily, or with a magnetically activated cleaner. Foam can be controlled by addition of antifoam substances, either at regular time intervals based on observation of the foam development, or controlled by a foam sensor. Bond et al. (15) published a study on the effect of antifoam on plant cells. The tested antifoam substances appeared not to harm the plant cells under normal experimental conditions. A point of general concern in continuous culture is of course the maintenance of aseptic conditions. High-quality equipment is a first prerequisite, particularly with the long run times of the continuous culture. A thorough leak test before sterilization is essential. In general, membrane filters are excellent to prevent contamination. However, it should be

(a)

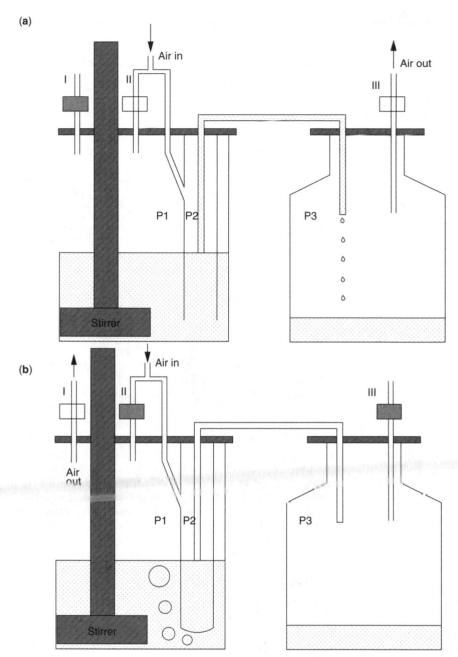

(b)

Figure 2. Principle of the effluent removal system. P1 = pressure inside the headspace of the reactor, P2 = pressure inside the effluent removal tube, and P3 = pressure inside the headspace of the effluent vessel. (**a**) When valve I is closed while at the same time valves II and III are opened, this will cause pressure P2 being equilibrated with pressure P1, resulting in a rise of the culture fluid in the effluent removal tube. Because of the aeration of the reactor, P1 (= P2) will exceed P3, resulting in withdrawal of all culture fluid above the inner tube. (**b**) If valve I is open and valves II and III are closed, a small stream of air keeps the effluent removal tube empty to avoid wall growth and clogging. In this case pressure P2 (= P3) exceeds P1 (13).

realized that the amount of aeration gas passing through the bioreactor during the culture period is huge. Therefore, it is advisable to extend the contamination prevention in the air inlet with a large cottonwool depth filter. Membrane filters may be clogged by moisture. To avoid blowing up the equipment in such a case, because of the pressure of the air inlet line, it is essential to build in some safety device. An inexpensive safety measure is to provide a relatively loosely connected tube in the setup.

MATHEMATICAL DESCRIPTION OF CONTINUOUS CULTURE

The formulation of unstructured models of fermentation processes is comprehensively treated in textbooks on biotechnology (16,17). Here, only a few very general aspects are presented. The basis of these models is a set of balance equations for extensive properties. The general form of the balance equation is:

rate of accumulation = rate of inflow − rate of outflow
+ formation rate
− consumption rate

The extensive properties considered to be important in the case of continuous culture of plant cells will be viable biomass X, nonviable biomass, limiting substrate S, and product. Nonviable biomass and product can be left out if the concentrations are negligible. This set of equations is completed with kinetic rate equations for biomass growth, biomass decay, substrate consumption, and product formation. As an example the general balance for biomass read (See end of article for nomenclature):

$$d\frac{VX}{dt} = F_iX_i - F_oX_o + Vr_X - Vr_d \qquad (1)$$

In a standard chemostat the biomass concentration in the influent is zero. With the assumption that the death rate of the biomass may be neglected, and realizing that the accumulation term in a steady-state situation is zero, Eq. (1) becomes:

$$0 = -F_oX_o + Vr_x \qquad (2)$$

Introduction of the dilution rate D and the specific growth rate μ in Eq. (2):

$$D = \frac{F}{V} \qquad (3)$$

$$\mu = \frac{r_x}{X} \qquad (4)$$

gives the important correlation:

$$\mu = D \qquad (5)$$

This illustrates the characteristic of the chemostat that the average specific growth rate can be controlled by setting the dilution rate.

By introduction of relations for r_X and r_s, the system is completely described. The commonly used growth expression is the Monod equation:

$$r_X = \mu X = \mu_{max}\frac{S}{S + K_s}X \qquad (6)$$

where μ_{max} is the maximum specific growth rate and K_s the saturation constant. Substrate is assumed to be used for growth and for maintenance purposes of the biomass according to the expression:

$$r_s = q_sX = \left(\frac{\mu}{Y_{SX}} + m_s\right)X \qquad (7)$$

Y_{sx} is the true yield of biomass on substrate and m_s gives the substrate consumption for maintenance processes. Worked examples of Eqs. (6) and (7) are depicted in Figure 3. The parameters used are typical of plant cells. On the curve of r_x it is shown that at a substrate concentration equal to the saturation constant K_s the growth rate has half the value of the maximum growth rate. The curve of r_s shows a shortcoming of the model used at substrate concentration 0; there is still substrate consumption although there is no substrate available. This situation is better described by a different concept, where the maintenance energy is provided by endogenous respiration of biomass. With the introduction of the Eqs. (6) and (7) the set of equations describing the growth of an organism in continuous culture is complete. The biomass and the substrate concentration can be solved as a function of the dilution rate D, resulting in:

$$X = \frac{D(S_i - S_o)}{D/Y_{SX} + m_s} \qquad (8)$$

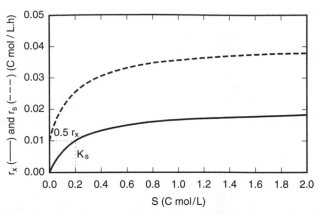

Figure 3. Growh rate r_x (——) and consumption rate r_s (- - -) versus limiting substrate concentration S [Eqs. (6) and (7)]. Parameters used $\mu_{max} = 0.02h^{-1}$, $K_s = 0.2$ C mol/L, $Y_{sx} = 0.65$ C mol/C mol, $m_s = 0.01h^{-1}$, $X = 1$ C mol/L.

$$S = \frac{k_sD}{\mu_{max} - D} \qquad (9)$$

These solutions still contain the organism-dependent parameters Y_{sx}, m_s, and K_s.

The yield Y_{sx} and the maintenance coefficient m_s can be determined from a series of continuous culture experiments at various dilution rates. An example of Eqs. (8) and (9) is plotted in Figure 4. At low dilution rates (= specific growth rates), the substrate consumption for maintenance is relatively important, resulting in a low apparent yield of biomass on substrate. At high dilution rates near the maximum specific growth rate, the culture starts to wash out, resulting in biomass concentration zero and maximal unconsumed substrate concentration. Van Gulik et al. (18,19) used this model extended with balance and rate equation for product formation and validated this model in a comprehensive continuous culture study. They could fit their experimental data roughly with this model and could estimate the organism-specific parameters. However, statistical analysis revealed that this model was certainly not a perfect description of a plant cell culture in continuous culture. Important observations were that although the carbon source was presumed to be the limiting nutrient, the phosphate concentration was always almost zero in the broth, and that the cells formed internal storage polysaccharides ("starch") in a dilution-rate-dependent quantity. On the basis of these observations van Gulik et al. (19) developed a structured model as depicted in Figure 5. The biomass was divided into two compartments: the active biomass part where all the biochemical activities take place, and the storage compartment containing the storage polysaccharides. The flow of material between medium and active biomass compartment, and between active biomass and the storage compartment, was controlled by the external carbon source and the internal phosphate concentration. This model is considerably more complex than the basic model mentioned previously. It includes 7 balance equations, 9 rate equations, and 15 unknown parameters. The number and structure of the equations is not a problem; it can

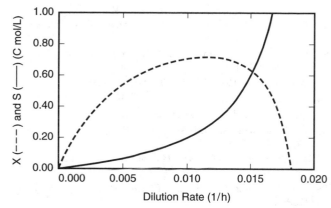

Figure 4. Biomass concentration X (- - -) and limiting substrate concentration S (——) versus dilution rate D in a chemostat [Eqs. (8) and (9)]. Parameters used $S_i = 2$ C mol/L; for other parameters see Figure 3.

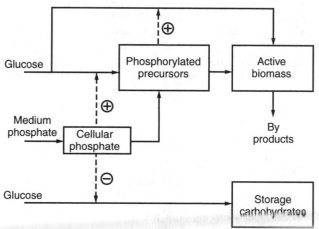

Figure 5. Schematic representation of the structured growth model (19).

be simulated by a computer. However, for a validation of the model it is necessary to have estimates of the real values of the parameters. A sensitivity analysis of the parameters revealed that only 7 of them were important. These parameters were either determined or obtained from the literature. Three parameters were optimized by fitting the experimental data with the model equations. The results of continuous culture and batch experiments were used to validate the model. It presented a satisfactory description of both type of experiments.

APPLICATIONS OF THE CONTINUOUS CULTIVATION OF PLANT CELLS

The earliest attempts to set up a continuous culture of plant cells were published by Tulecke (20) in 1966 and by Miller et al. (21) in 1968. These studies have only historical value, because the authors could not achieve any steady states. The systems developed by Wilson et al. in 1971 (22), Kurz (23), and Wilson (24) are more important, because these designs have been used in several other studies. The chemostat system of Wilson (22) consisted of a round-bottomed Pyrex flask with a volume of 4 L. The

culture was aerated through a sintered glass sparger, and mixing of the culture broth was accomplished by means of a magnetic stirrer. Effluent was withdrawn using a specially designed system consisting of a valve, controlled by an constant-level device. They achieved steady-state growth of *Acer pseudoplatanus* cells at several dilution rates. Kurz (23) designed a chemostat that was rather unique in its mixing and aeration system. A small and long cylindrical vessel is mixed and aerated by large air bubbles filling up the diameter of the tube rising at regular intervals from the bottom of the cylinder. Influent was introduced at the bottom, and effluent was removed through an overflow device at the top. A different air-mixed chemostat culture system was used by Wilson (24). This relatively simple chemostat was made of an inverted erlenmeyer flask. Mixing and oxygenation of the culture were accomplished by forced aeration. Also in this design the effluent was withdrawn by means of an overflow tube. Unfortunately, Sahai and Shuler (25) concluded in a detailed study that the outflow efficiency of these systems is below 100%. This makes the quantitative results of these investigations relatively useless. However, several studies using these setups showed at least that continuous cultures of plant cells could be carried out under carbon, nitrogen, and phosphate limitation. The work done by de Gucht and van der Plas (26) is an exception. They used the Kurz chemostat, but were aware of the problem of the outflow efficiency and took it into account in their calculations. They achieved six steady states with a culture of *Petunia hybrida* cells. From their experimental data they estimated the maintenance and true yield of biomass on glucose and ATP.

Bertola and Klis (14) reported the use of a continuous culture system that was based on a 2-L standard turbine-stirred fermenter and the outflow system designed by Wilson (22). These authors published glucose-limited chemostat cultures of *Phaseolus vulgaris* cells. They determined the outflow efficiency to be 53%. After some modifications the outflow efficiency was improved to 80% (27). Their results were corrected for the outflow efficiency of the chemostat system. Van Gulik et al. (18,19) used a comparable system, but with a more sophisticated outflow device and coupled gas analysis equipment. Combination with off-line analysis of the spent medium and the biomass provided the opportunity to check their experimental data with elemental balances. Results of these investigations have been mentioned.

The aim of these studies was mainly the formulation and validation of models describing growth kinetics. The application of continuous culture for production of secondary metabolites has not been investigated extensively. In most cases secondary product formation is considerably depressed under growth conditions. Sahai and Shuler (28) proposed a multistage continuous culture system with variation in growth rates to create optimal conditions for both growth of the biomass and product formation. Lima Costa (29) investigated the production of an extracellular protease by a cell culture of *Cynara cardunculus* in continuous culture. The production of this protease was growth coupled, and continuous culture appeared superior to batch culture in this particular case.

NOMENCLATURE

D	dilution rate (1/s)
F	flow (m³/s)
S	substrate concentration (kg/m³)
X	biomass concentration (kg/m³)
V	volume (m³)
K_s	saturation coefficient (kg/m³)
Y_{sx}	true yield (−)
m_s	maintenance coefficient (1/s)
r_d	death rate (kg/m³/s)
r_s	substrate consumption rate (kg/m³/s)
r_x	growth rate (kg/m³/s)
q_s	specific substrate consumption rate (1/s)
μ	specific growth rate (1/s)
μ_{max}	maximum specific growth rate (1/s)

Subscripts

i	inflow
o	outflow

BIBLIOGRAPHY

1. W.H. Muir, A.C. Hildebrandt, and A.J. Riker, *Science* **119**, 877–878 (1954).

2. G. Melchers and U. Engelmann, *Naturwissenschaften* **20**, 564–565 (1955).

3. H.J. Noorman, J. Baksteen, and J.J. Heijnen, *J. Gen. Microbiol.* **137**, 2171–2177 (1991).

4. A.H. Scragg, E.J. Allan, and F. Leckie, *Enzyme Microb. Technol.* **10**, 361–367 (1988).

5. J.J. Meijer et al., *Enzyme Microb. Technol.* **16**, 467–477 (1994).

6. J.J. Meijer, H.J.G. ten Hoopen, K.Ch.A.M. Luyben, and K.R. Libbenga, *Enzyme Microb. Technol.* **15**, 234–238 (1993).

7. K. Matsubara et al., *J. Chem. Technol. Biotechnol.* **46**, 61–69, (1989).

8. H. Tanaka, *Process Biochem.* **22**, 106–113 (1987).

9. W.J. Treat, C.R. Engler, and E.J. Soltes, *Biotechnol. Bioeng.* **34**, 1191–1202 (1989).

10. D.I. Kim, G.H. Cho, H. Pedersen, and C.K. Chin, *Appl. Microbiol. Biotechnol.* **34**, 726–729 (1991).

11. B.S. Hooker, J.M. Lee, and G. An, *Biotechnol. Bioeng.* **35**, 296–304 (1990).

12. M. Jolicoeur, C. Chavarie, P.J. Carreau, and J. Archambault, *Biotechnol. Bioeng.* **39**, 511–521 (1992).

13. W.M. van Gulik et al., *Appl. Microbiol. Biotechnol.* **30**, 270–275 (1989).

14. M.A. Bertola and F.M. Klis, *J. Exp. Bot.* **30**, 1223–1231 (1979).

15. P.A. Bond, P. Hegarty, and A.H. Scragg, in O.M. Neijssel, R.R. van der Meer, and K.Ch.A.M. Luyben, eds., *Proceedings of the 4th European Congress on Biotechnology 1987, Vol. 2*, Elsevier, Amsterdam, 1987, pp. 440–443.

16. J.A. Roels, *Energetics and Kinetics in Biotechnology*, Elsevier, New York, 1983, pp. 223–237.

17. J. Bu'lock and B. Kristiansen, *Basic Biotechnology*, Academic Press, London, 1987, pp. 75–131.

18. W.M. van Gulik, H.J.G. ten Hoopen, and J.J. Heijnen, *Biotechnol. Bioeng.* **40**, 863–874 (1992).

19. W.M. van Gulik, H.J.G. ten Hoopen, and J.J. Heijnen, *Biotechnol. Bioeng.* **41**, 771–780 (1993).

20. W. Tulecke, *Ann. N.Y. Acad. Sci.* **139**, 162–175 (1966).

21. R.A. Miller, J.P. Shyluk, O.L. Gamborg, and J.W. Kirkpatrick *Science* **159**, 540 (1968).

22. S.B. Wilson, P.J. King, and H.E. Street, *J. Exp. Bot.* **22**, 177–207 (1971).

23. W.G.W. Kurz, *Exp. Cell Res.* **64**, 476–479 (1971).

24. G. Wilson, *Ann. Bot. (London)* [N.S.] **40**, 919–932 (1976).

25. O.P. Sahai, and M.L. Shuler, *Can. J. Bot.* **60**, 692–700 (1982).

26. L.P.E. de Gucht and L.H.W. van der Plas, *Biotechnol. Bioeng.* **47**, 42–52 (1995).

27. M.A. Bertola, Ph.D. thesis, University of Amsterdam, 1980, pp. 45–60.

28. O.P. Sahai and M.L. Shuler, *Biotechnol. Bioeng.* **26**, 27–36 (1984).

29. M.E. Lima Costa et al., *Enzyme Microb. Technol.* **19**, 493–500 (1996).

See also BIOREACTORS, PERFUSION; BIOREACTORS, RECIRCULATION BIOREACTOR IN PLANT CELL CULTURE; BIOREACTORS, STIRRED TANK; CELL GROWTH AND PROTEIN EXPRESSION KINETICS; CONTAMINATION DETECTION AND ELIMINATION IN PLANT CELL CULTURE; OFF-LINE IMMUNOASSAYS IN BIOPROCESS CONTROL; ON-LINE ANALYSIS IN ANIMAL CELL CULTURE; PHYSIOLOGY OF PLANT CELLS IN CULTURE; STERILIZATION AND DECONTAMINATION.

BIOREACTORS, MIST

PAMELA J. WEATHERS
Biology and Biotechnology Department

BARBARA E. WYSLOUZIL
Chemical Engineering Department
Worcester Polytechnic Institute
Worcester, Massachusetts

OUTLINE

Introduction
    Plants as Gas Exchangers
    Gas Solubility Issues
    Spray, Mist, and Fog
Nutrient Mist Reactor Design and Operation
Nutrient Mist Bioreactors for Micropropagation
Nutrient Mist Bioreactors for Hairy Root Culture
    Lab-Scale Studies
    Pilot-Scale Studies
    Engineering Considerations
Summary and Conclusions
Acknowledgments
Bibliography

INTRODUCTION

Reports describing and discussing successful growth of differentiated tissues in specialized bioreactors cover a wide array of reactors including bubble columns, a rotating drum, dispersed liquid reactors, and mechanical agitation with impeller isolation. All these reactors try to minimize tissue damage while providing adequate nutrients and gases to the growing biomass. Recent studies show that reactors using gas as the continuous phase and media as the disperse phase may offer the best conditions for growth and productivity of transformed root cultures, especially at high root densities, because of the high level of O_2 available to the growing roots. Nutrient mist reactors have been developed to provide media dispersed as a fine mist ($D_p < {\sim}30\ \mu m$) to the growing tissues. Specific reactors have been designed for growing whole plants, for use in micropropagation, and for growth of differentiated plant organs, especially hairy roots.

Plants as Gas Exchangers

Plants have evolved as extremely efficient gas exchangers, especially for the primary gases of respiration and photosynthesis, CO_2 and O_2, and the phytohormone, ethylene. Other gases are also readily exchanged, as evinced by NASA's work on the use of plants for cleaning the atmosphere in closed living environments. In the light, green shoots photosynthesize as well as respire, resulting in a net production of O_2 and consumption of CO_2; in the dark, depending on the ambient O_2 concentration, plants produce CO_2 via either aerobic respiration or fermentation. Furthermore, the process of photorespiration in some plants results in a loss of CO_2 from cells normally fixing CO_2, yielding a net loss of carbon. Most of this gas exchange occurs through the stomata found in the leaves of higher plants. In woody plants, lenticels, located on stems, facilitate gas exchange across the bark and into the living tissues of the secondary phloem and xylem.

Roots are also excellent gas exchangers, although less is known about the biochemistry and dynamics of their processes. Since they are primarily underground organs, roots have until recently proved difficult to study in vivo. Because they are not exposed to significant amounts of light, roots primarily respire and do not photosynthesize, resulting mainly in O_2 uptake and CO_2 evolution. When exposed to light, however, roots can form chloroplasts and perform photosynthesis, thus altering their net metabolic status. Furthermore, the roots of some plant species show increased growth in the presence of CO_2 levels that are significantly elevated from ambient values.

All plant tissues can both produce and absorb the gaseous phytohormone ethylene (C_2H_4). Ethylene has profound effects on growth, development, and even the production of secondary metabolites. Furthermore, CO_2 inhibits the last step in ethylene biosynthesis, which can affect overall plant metabolism and development. Thus CO_2 interacts both with ethylene and O_2 to regulate plant growth and metabolism. Clearly, there is a need to be able to study plant and plant tissue responses to changes in the gaseous environment. The nutrient mist bioreactor meets this need and also offers an alternative reactor for growing differentiated plant tissues, especially those sensitive to changes in their gaseous environment.

Gas Solubility Issues

Gases have limited solubility in water, making it difficult to obtain efficient gas transfer to roots and other plant tissues grown in liquid. The solubilities are strong functions of temperature, pressure, and the presence of other solutes such as sugars and salts (1,2). For the three gases most important to plant cultures, O_2, CO_2, and C_2H_4, and under typical culture conditions, the solubility decreases as temperature and the concentration of salts and sugars increase. In water at $25\,°C$, oxygen is the least soluble of the three gases ($1.26\ mM\ L^{-1}$) followed by ethylene (~4 times more soluble) and then carbon dioxide (~25 times more soluble). Thus in liquid-phase reactors solubility limits the amount of gas available to a growing tissue culture and severely restricts the range of gas concentrations that can be investigated experimentally.

Spray, Mist, and Fog

To obtain good gas transfer into the liquid media, either the liquid is dispersed into a gas phase or gas is dispersed into a liquid phase. In bubble column and other liquid-based reactors, gas is dispersed into the liquid phase. In a mist reactor, liquid is dispersed into the gas phase using a number of different methods. The most common methods for dispersing the droplets in a plant tissue reactor include a spinning disk, an atomizer using compressed gas, or an ultrasonic generator. The spinning disk has been primarily used with whole plant aeroponic systems (3). The drawback of the spinning disk is that it produces a wide range of droplet sizes unless the disk spins very rapidly. At high velocity the droplets formed are small and offer a large surface-to-volume ratio, yielding increased gas transport into the liquid. Unfortunately, the high speed of the droplets being thrown at the plant tissue can result in tissue damage, and so a tradeoff must be made between good gas exchange and tissue damage. Atomizers using compressed gas have also been used extensively with whole plants (3) and to a lesser degree for in vitro cultures (4). The use of compressed gas for production of aerosols through nozzles can yield very small droplet sizes, ensuring good gas transfer into the liquid and, thus, to the plant tissue. However, the presence of the salts in plant culture media results in salt build up at the nozzle orifice and eventual clogging. Furthermore, use of compressed gases is costly.

The use of ultrasonics has, by far, proved to be the most effective method of generating nutrient aerosols for in vitro plant cultures and will be the main focus of this discussion. In most mist reactors a submerged ultrasonic transducer powered by electrical energy produces sound waves in the 1.7 MHz range. The energy, partially lost as heat, is transferred to the liquid–gas interface where capillary waves are produced. The aerosol is created by the release of very small particles of liquid being propelled into the air from the crest of the waves. For transducers that operate below 800 kHz, the number mean diameter D_n of the droplets depends on the surface tension of the liquid σ, the density ρ, and the transducer frequency ν as $D_n = 0.34$ $(8\,\pi\sigma/\rho\nu^2)^{1/3}$ (5). For the synthetic wax used by Lang (5), where $\sigma = 21.7$ dyn/cm, $\rho = 0.82$ g/cm^3, and $\nu = 400$ kHz, the predicted particle diameter $D_n = 5.5 \times 10^{-4}$ cm. At higher frequencies the droplet size also depends on the power density at the liquid surface (6). Droplet density (number of droplet particles in the air stream), on the other hand, depends primarily on the rate at which the droplets are removed from the production region. An air stream is usually used to move the droplets from the production zone to the plants. Increasing the gas flow rate initially increases both the number and the size of droplets that can be transferred away from the surface of the liquid. Extremely high gas velocities can, however, result in a loss of the largest particles due to impingement onto the surfaces of the mist generator. This effect obviously depends on the particular reactor geometry. At high gas flow rates, it is important to humidify the carrier gas stream before it enters the mist generator to avoid excessive media loss and changes in the media composition due to evaporation.

Droplet size is an important factor in the mist reactor and can be characterized by several different parameters. The most common are the number mean particle diameter D_n and the volume mean diameter D_v. Distinguishing between these terms is particularly important when measured droplet particle sizes are reported. The terms used to describe dispersions of liquids in a carrier gas have rather broad definitions according to Perry and Green (7), leading to some confusion in the literature. Generally, mists are characterized by droplet sizes in the range of 0.01–10 μm, while fogs have droplets in the range of 1–100 μm, and finally, sprays have droplets in the range of 10–10,000 μm (7). Droplets formed using compressed gas atomizers range from 1 to 100 μm in diameter, while ultrasonic systems typically produce droplets in the 1–35 μm diameter range, with $D_v \sim 7$–10 μm (3,8). The advantages of working with smaller droplets include their long settling times, which means it is easy to transport and distribute the dispersed liquid, the closely related ability of the small droplets to penetrate dense

tissue cultures and thus distribute the nutrients more evenly, and the large surface area of the mist that ensures good mass transfer of gases between the two phases.

Droplet size is particularly important for ensuring adequate transfer of both liquids and solubilized gases from the culture medium to the plant tissue. Very little work has been done to develop operational limits for plants or plant cultures. For whole plants undergoing high rates of transpiration, droplets that are too small will not supply enough water to make up for water loss. Yet if the droplet size is too large, there may be inadequate gas transfer to the roots because of the formation of a layer of water along the root surface; the system essentially reverts to a liquid- instead of a gas-phase reactor. For whole plants, the minimum acceptable droplet size is not known precisely, but it is thought to be about 1 μm (R.D. Zobel and P.D. Weathers, unpublished results). Droplet size limitations for growth of in vitro cultures have not been studied other than to measure droplet deposition in dense beds of hairy roots (9) and our current studies of the role that droplet size plays in hairy root growth (B.E. Wyslouzil, unpublished).

NUTRIENT MIST REACTOR DESIGN AND OPERATION

The basic design of a nutrient mist reactor always includes the following components: a mist generator, a growth chamber, a mist carrier-gas supply, a nutrient reservoir, a liquid feed pump, a timer for turning the mist feed on and off, and the associated tubing (Fig. 1). In early versions of the mist reactor, the ultrasonic transducer was in direct contact with the nutrient medium. The life of the transducer was significantly diminished by repeated autoclaving, and experiments often ended prematurely because electrical components failed (10). The use of an acoustic window fabricated from a polypropylene Rubbermaid container to separate the

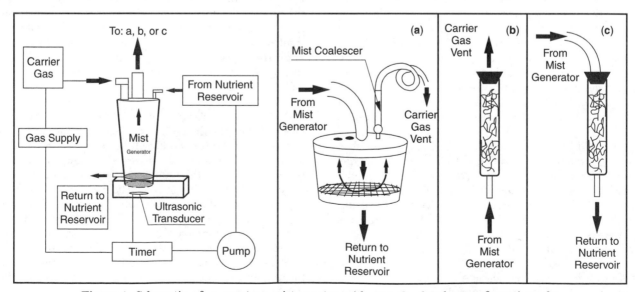

Figure 1. Schematic of a nutrient mist reactor with growth chamber configurations for (**a**) micropropagation, (**b**) hairy root culture with mist input in an upflow mode, and (**c**) hairy root culture with mist input in a downflow mode. (Figure courtesy of M. Correll.)

transducer from the nutrient media made it unnecessary to autoclave the transducer and led to the development of a simple, easy-to-construct mist reactor (11). A thin sheet of Teflon also works well as an acoustic window, and this material has the added advantages of higher temperature tolerance than polypropylene as well as being easily incorporated into a reactor of almost any shape or dimension (R.G. Waterbury and B.E. Wyslouzil, unpublished results). With these improvements, together with the use of a home humidifier base as the transducer source, it is now easy to produce a reliable mist reactor cheaply and in almost any configuration.

To accommodate the special growth characteristics of each species of plant or the type of tissue to be cultured, for example, hairy roots versus plantlets, the growth chamber of a nutrient mist reactor must be properly configured. For plantlets, callus, and general purpose micropropagation, the growth chamber [Figure 1(a)] must provide a large surface area, covered with some absorptive material such as cheesecloth or filter paper to maintain adequate moisture just beneath the tissue. The tissue support is situated about 1 in. from the bottom of the culture chamber, so that a small amount of culture media can collect underneath the support and maintain a constant level of relative humidity within the chamber. Adequate headspace above the tissue support is also important so that shoots are able to emerge fully. The chamber is preferably constructed out of an autoclavable plastic, such as glass-clear polycarbonate or polypropylene, with a gas-tight lid into which ports for tubing connections can be inserted. Inoculation of the growth chamber in micropropagation studies is still performed by hand. Since the reactors are larger than standard micropropagation boxes and the gas phase composition and humidity can be controlled, it should be possible to grow up plant clones with only one transfer step. Furthermore, acclimatization of the plantlets can occur in the reactor itself, thus reducing the amount of manual labor significantly.

For transformed roots the growth chamber must contain a trellis or other packing material on which the roots can be immobilized. To encourage even mist distribution, the growth chambers tend to be tall and narrow rather than the short, wide chambers used in micropropagation studies. Inoculation of the reactor is currently accomplished by flooding the growth chamber with media, adding the roots, and running the reactor as a bubble column until the roots immobilize. Other more labor-intensive methods have also been used, such as hand loading individual roots onto a trellis as well as growing roots in shake flasks with packing rings until the roots attached and then loading the rings into the growth chamber. A detailed description for how to construct a complete bioreactor is given in Ref. 11.

Typical operating conditions for the nutrient mist reactor include air flow rates in the range of 1–6 LPM, and mist duty cycles ranging from 1 min on/15 min off to 5 min on/5 min off. Room air filtered through a 0.2-μm filter is usually used as the carrier gas, but other gases, for example, CO_2 or C_2H_4, can also be added. The reactor can be operated in batch mode by collecting and recycling the mist back to the reservoir after it exits the culture chamber. It can also be run in continuous mode by discarding the mist after it passes over the tissue, and adding fresh media to the reservoir. In many of the studies to date, the nutrient media used in the experiments has been "conditioned." Conditioned medium is produced by growing roots in shake flasks containing fresh, autoclaved media for up to 2 weeks. The roots adjust the distribution of sugars between sucrose, glucose, and fructose, the levels of salts, and exude other, unidentified compounds that have a positive effect on root growth. Conditioned medium is introduced into the autoclaved media reservoir by filtering it through a sterile filter.

NUTRIENT MIST BIOREACTORS FOR MICROPROPAGATION

Species reported to have been micropropagated in experimental mist reactors are listed in Table 1 (12–18) For a discussion of reports prior to 1991, see the review by Weathers and Zobel (3). Work with *Solanum* by Kurata et al. (17) suggested that mist culture was more effective if provided from the root zone (bottom up) of differentiated plantlets. We tested this concept using Boston fern (*Nephrylepis*) and found that misting from the bottom was significantly better than from the top of the growth chamber even with different misting cycles (Table 2). However, at present it is much easier to work with the mist inlet port on the top of the growth chamber compared to the bottom; so most research in micropropagation is conducted using a top-down configuration.

One big advantage of the mist system is that the problem of hyperhydration can be reduced or greatly eliminated. Woo and Park (16) found that *Dianthus* shoots grown in flask cultures were seriously hyperhydrated even if the shoots were only partially submerged; mist-grown cultures showed about 25% less hyperhydration.

Table 1. Plants Propagated in Vitro Using Nutrient Mist Reactors

Plant Species	Inoculum Source	Ref.
Asparagus	shoots, callus	12
Brassica	anthers	4
Cinchona	nodal explants	4
Cordyline	shooting tissue	13
Daucus	callus, shoots	14
	leaf, petiole	15
Dianthus	nodal explants	11
	shoots	16
Ficus	callus + shooting meristems	4
Lycopersicon	nodal explants	4
Musa	shooting tissue	13
Nephrylepis	shooting tissue	13
	shooting tissue	10
Pinus	shoots	P. Weathers, unpublished
Saintpaulia	shoots, callus	K.L. Giles, unpublished
Solanum	nodal explants	17
Taxus	shoots for rooting	18

Table 2. Bottom vs. Top Misting of Ferns in a Nutrient Mist Bioreactor after 7 d

Mist Feed Point	Increase in Fresh Weight (g)
A:	
Top	0.75± 0.29
Bottom	0.97± 0.06
Agar control	0.83± 0.22
B:	
Top	0.18± 0.08
Bottom	0.34± 0.09
Agar control	0.33± 0.08

A: Mist cycle: 5 min on/15 min off for 3 d, then 5 min on/ 20 min off for 4 days. B: Mist cycle: continuous misting for first 30 min, then 5 min on/20 min off for 7 d.

Hyperhydration can be fully controlled, as shown in experiments with *Dianthus*, by altering the mist cycle, and the CO_2 and humidity levels in the growth chamber (19).

Other interesting observations were made by Tisserat et al. (15) with *Daucus* shootlets grown for up to 4 weeks in a bottom-fed mist reactor. Besides a large increase in growth compared to controls, there was also direct induction of asexual embryoids from the enlarged leaflet surfaces. These were not observed in any of the controls grown in liquid or on agar. When *Daucus* callus was grown in the Tisserat system, more biomass was produced than in any of the liquid or agar controls. Embryogenic callus produced significantly more somatic embryos than on agar; no embryos were produced in the liquid systems. Clearly the mist reactor offers great potential for culturing a wide array of plant tissues used in micropropagation.

NUTRIENT MIST BIOREACTORS FOR HAIRY ROOT CULTURE

Lab-Scale Studies

Earlier work using mist reactors for growing transformed roots was recently reviewed by Weathers et al. (20). Species of transformed roots that have been successfully grown in mist reactors include *Beta vulgaris*, *Artemisia annua*, *Carthamus tinctorius*, and *Nicotiana* sp. The reactors were operated in both batch and continuous mode. In general, the overall biomass density achieved in the mist reactors (g/L) was comparable to that achieved in shake flasks. Like the roots of intact plants, transformed roots also respond to changes in gas composition with the level of the response varying with plant species. To date, the gas studies have focused mainly on CO_2, whose effect was to decrease the lag time in some root cultures (21). Systematic studies changing the levels of O_2 or C_2H_4 have not been conducted to date. The choice of mist cycle is also a variable that must be investigated in developing a growth protocol for roots in the mist reactor (22).

More recently, the simplified acoustic window reactor developed by Chatterjee et al. (11,23) with the growth chamber configured as in Figure 1(b) was used to measure the growth kinetics of *A. annua* hairy roots. The reactor

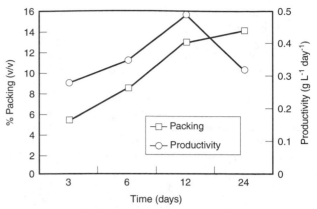

Figure 2. Productivity and packing density of hairy roots of *A. annua* grown in a nutrient mist reactor for up to 24 d. (Figure courtesy of T. Smith.)

was inoculated with roots grown for 1 week in shake flasks containing plastic mesh rings. At this point the roots were firmly attached to the rings and were manually loaded into the reactor. Good growth was observed even at a sparse mist feed rate of 4 min of mist per hour (1 min in every 15 min = about 5.52 mL/h). Three reactors were run simultaneously and harvested after 3, 6, 12, and 24 d of growth (3 replicates for each experiment). As root bed densities increased from about 5% to 15%, the productivity ($gDWL^{-1}$ d^{-1}) gradually increased from days 3 to 12 (Fig. 2). From day 12 to 24 the average productivity decreased from about 0.49 to 0.32 g DWL^{-1} d^{-1}, suggesting that some factor was beginning to limit growth. At the low feed rates used, media may have been limiting, and thus to achieve the desired productivities of 3–4 $gDWL^{-1}$ d^{-1}, increasing the mist duty cycle may be required.

Recently the mist reactor was redesigned and constructed with the growth chamber configured as in Figure 1(c) and using a mist generator with a teflon acoustic window. The reactor is inoculated and first runs as a bubble column until the roots attach to the stainless steel support (requiring 6–9 days). Preliminary results show that for three-week runs done in triplicate, biomass accumulation (gFW/L) is significantly lower than in a parallel bubble column reactor or in flasks. A nutrient feed rate of only ~8 mL/h (5 min of mist fed 3 times per hour) again led us to believe that growth may be nutrient limited. However, in one experiment with uneven inoculum distribution, a packing density of 806 gFW/L (59.6 g DW/L) was achieved in the lower 25% of the growth chamber. Although overall reactor productivity was low, local productivity was extremely high and mist was still exiting the root bed. This suggests that mist reactors have the potential to achieve higher biomass densities than bubble column reactors, where oxygen would most likely be limiting at such a high density.

Pilot-Scale Studies

There has been only one pilot-scale (500-L) study of hairy roots in a mist reactor. Using a unique root immobilization matrix, *Datura* roots were grown for 40 days to a final density of 8 gDW L^{-1} with an estimated productivity of 1.3 gDW L^{-1} d^{-1} (24). The final packing density was about 8% (FW/V), which is considerably less than the laboratory-scale studies described previously.

Engineering Considerations

In a gas-phase reactor, adequate media distribution rather than gas control may limit reactor performance. To understand this problem, the applicability of the standard models for aerosol deposition in randomly packed fibrous filter beds to mist deposition across a bed of hairy roots was analyzed (9). The analysis showed that on a local level the root–droplet system meets the assumptions for the standard models of particle capture by impaction, interception, and diffusion used in industrial filters. However, the overall structure of the root bed introduces uncertainty into the correct choice of root packing fraction and gas velocity required by the model. For reasonable parameter values, the minimum in the deposition efficiency curves is close to the peak in the mist number and mass distributions, implying that mist can penetrate deeply even into a dense bed of roots. Furthermore, a simplified treatment of the root hairs showed that they should contribute significantly to mist droplet capture (9).

To verify the model, changes in the mist size distribution were measured across manually packed beds of *A. annua* transformed roots as a function of droplet size, bed length, and gas flow rate at a root packing fraction $\alpha = 0.5$ (9). There was good agreement between the measured and predicted values for the droplet diameter where the deposition efficiency across the bed is 50%, $D_{0.5}$, as long as the Reynolds number (Re) was not too high (Re <10). Here the Reynolds number $Re = D_r V_g \rho_g / \mu_g$ is based on the root diameter D_r, the velocity of the gas V_g, the density of the gas ρ_g, and the viscosity of the gas μ_g. When $D_r = 0.1$ cm, $V_g = 10$ cm/s, $\rho_g = 1.2 \times 10^{-3}$ g/cm^3, and $\mu_g = 1.8 \times 10^{-4}$ g/(cm s), Re = 6.7. Agreement between the model and the experiments broke down when the flow rate was increased to the point where the creeping flow assumptions were no longer valid (10 < Re < 20). At the lower velocities typically present in the mist bioreactor, the standard aerosol models should, therefore, provide a good description of particle deposition at these high packing densities.

Deposition of mist in situ onto growing root beds of *A. annua* was also measured using the simplified acoustic window mist reactor described by Chatterjee et al. (11). The experiments measured deposition by tracking the total volume of liquid leaving the mist reactor as a function of time and comparing it to the liquid being fed into the mist generating chamber. Although these experiments are less informative than measuring complete size distributions, they did show an initial rapid increase in the bed capture efficiency immediately after inoculation, later leveling off to a value of roughly 50% for a 20-cm-thick root bed despite packing density increases from 4 to 15% (23). This suggests that root bed depths of 25–40 cm are quite realistic in a mist reactor for each mist input port and that it is possible to achieve a balance between mist droplet deposition and penetration throughout the bed volume.

SUMMARY AND CONCLUSIONS

The nutrient mist reactor represents one extreme of the spectrum of potential bioreactor environments. With gas as the continuous phase rather than liquid, the reactor is particularly suited to systematically studying the response of plants and plant organs to changes in the gas-phase environment. It has proven to be a versatile design that can be successfully used in micropropagation and hairy root culture. Further studies with the reactor will continue to improve our understanding of the interaction of plants and plant tissues with their environment.

ACKNOWLEDGMENTS

This work was supported in part by grants from the National Science Foundation, BES-9414858, the National Institutes of Health, 1 R21 AI39170-01, and the United States Department of Agriculture, 93-38420-8804.

BIBLIOGRAPHY

1. B. Atkinson and F. Mavituna, *Biochemical Engineering and Biotechnology Handbook*, 2nd ed., Macmillan, Basingstoke, U.K., 1991, pp. 703–705.

2. C.J. Geankopolis, *Transport Processes and Unit Operations*, 3rd ed., Prentice-Hall, Englewood Cliffs, N.J., 1993, pp. 586–587, 884.

3. P.J. Weathers and R.D. Zobel, *Biotechnol. Adva.* **10**, 93–115 (1992).

4. P.J. Weathers and K.L. Giles, *In Vitro Cell Dev. Biol.* **24**, 727–732 (1988).

5. R.J. Lang, *J. Acoust. Soc. Am.* **34**, 6–8 (1962).

6. M.A. Tarr, G. Zhu, and R.F. Browner, *Appl. Spectrosc.* **45**, 1424–1432 (1991).

7. R.H. Perry and D.W. Green, *Perry's Chemical Engineer's Handbook*, 7th ed., McGraw-Hill, New York, 1997, pp. 14-62–14-69, 14-81–14-82.

8. M. Whipple, M.S. Thesis, Worcester Polytechnic Institute, Worcester, Mass., 1995.

9. B.E. Wyslouzil et al., *Biotechnol. Prog.* **13**, 185–194 (1997).

10. C.S. Buer et al., *In Vitro Plant* **32**, 299–304 (1996).

11. C. Chatterjee et al., *Biotechnol. Tech.* **11**, 155–158 (1997).

12. R.D. Cheetham, C. Mikloiche, M. Glubiak, and P. Weathers, *Plant Cell, Tissue Organ Cult.* **31**, 15–19 (1992).

13. P.J. Weathers, R.D. Cheetham, and K.L. Giles, *Acta Hortic.* **230**, 39–43 (1988).

14. P.J. Weathers, A. DiIorio, and R.D. Cheetham, in *Proceedings of the Biotech USA Conference*, Conf. Management Corp., Norwalk, Conn., 1989 pp. 247–256.

15. B. Tisserat, D. Jones, and P. Galletta, *HortTechnology* **3**, 75–78 (1993).

16. S.H. Woo and J.M. Park, *Biotechno. Tech.* **7**, 697–702 (1993).

17. K. Kurata, Y. Ibaraki, and E. Goto, *Am. Soc. Agric. Eng.* **34**(2), 621–624 (1991).

18. L.J.W. Butterfield, P.J. Weathers, and H.E. Flores, *HortScience* (1999) (in press).

19. M. Correll and P.J. Weathers, *In Vitro Plant* **34**, 64-A, (Abstr. No. 1030) (1998).

20. P. Weathers, B.E. Wyslouzil, and M. Whipple, in P.M. Doran, ed., *Hairy Roots*, Gordon & Breach/Harwood Academic, U.K., 1997, pp. 191–199.

21. A.A. DiIorio, R.D. Cheetham, and P.J. Weathers, *Appl. Microbiol. Biotechnol.* **37**, 463–467 (1992).

22. A.A. DiIorio, R.D. Cheetham, and P.J. Weathers, *Appl. Microbiol. Biotechnol.* **37**, 457–462 (1992).

23. C. Chatterjee, Ph.D. Thesis, Worcester Polytechnic Institute, Worcester, Mass., 1996.

24. D.G. Wilson, in P.M. Doran, ed., *Hairy Roots*, Gordon & Breach/Harwood Academic, U.K., 1997, pp. 179–190.

See also ASEPTIC TECHNIQUES IN CELL CULTURE; BIOREACTOR OPERATIONS—PREPARATION, STERILIZATION, CHARGING, CULTURE INITIATION AND HARVESTING; HAIRY ROOTS, BIOREACTOR GROWTH; PLANT CELL CULTURES, SECONDARY PRODUCT ACCUMULATION.

BIOREACTORS, PERFUSION

WEI WEN SU
University of Missouri-Columbia
Columbia, Missouri

OUTLINE

Introduction

When to Use Perfusion?

Overview of Plant Cell Perfusion Bioreactor Designs

  Separation by Sedimentation

  Separation by Filtration

Analysis of Design and Operating Parameters

  Design of Cell Retention Devices

  Design Considerations for Mixing and Oxygen Transfer

  Operating Parameters

  Scale-up

Process Monitoring, Optimization, and Control

Conclusions and Future Outlook

Bibliography

INTRODUCTION

Perfusion is continuous or semicontinuous flow of a physiological nutrient solution through a population of cells. It implies that the cells are retained within the culture unit as opposed to continuous-flow culture (chemostat) which washes the cells out with the withdrawn medium. By using a perfusion system, one can easily accomplish in situ medium exchange and thus achieve better control of the culture environment (dissolved oxygen, pH, substrate and hormone concentrations), remove toxic extracellular by-products, and constantly recover secreted products to reduce feedback inhibition. As a result, improved reactor productivity can usually be attained.

Culture perfusion was initially employed mainly in medical applications. The technique was inspired by the observation that human cells in vivo are continuously supplied with blood, lymph, or other body fluids to keep them in a constant physiological environment. The use of perfusion has also become a common practice in animal cell cultivation, mainly in response to the quest to attain high cell concentration for improving cell culture volumetric productivity. Application of the perfusion technique in plant cell and tissue cultures is not as common as in animal cell cultures. Nonetheless, it is an emerging technology that deserves greater attention. The use of perfusion (or similar concept) in plant cell cultures was first reported in the mid-80s, when Pareilleux and Vinas (1) and Ammirato and Styer (2) proposed using perfusion techniques in studying alkaloid production in a steady-state culture with cell retention and in somatic embryo production, respectively. Matsubara and Fujita (3) also demonstrated the potential of perfusion culture in the production of berberine by *Coptis japonica*. Kim and co-workers (4) reported their study of using a cell-lift impeller in the perfusion culture of *Thalictrum rugosum* cell suspension for berberine production. Our lab has developed and studied a number of plant cell perfusion bioreactors for secondary metabolite and secreted protein production (5–12). Moorhouse et al. (13) reported using a dual-membrane stirrer system in a bioreactor, similar to that of Lehmann et al. (14), for cultivating of plant somatic embryos in suspension cultures.

The scope of this atricle is to review current developments in perfusion bioreactor systems for plant cell and tissue cultures. Several key subjects relating to plant cell perfusion bioreactors are discussed:

(1) applications of perfusion systems;

(2) perfusion bioreactor designs;

(3) analysis of design and operating parameters; and

(4) process monitoring, optimization, and control.

WHEN TO USE PERFUSION?

An effective bioreactor operating strategy should provide high productivity, high product yield (product formed per substrate consumed), and high product content (product/cell weight). Higher productivity means more product can be formed per unit time per unit reactor volume. High product content makes downstream separation and purification easier and more efficient. High product yield reduces the cost of substrates. Generally, the operating strategy is determined on the basis of the kinetic pattern of product formation and the way the product is translocated following its synthesis (i.e., whether the product is excreted into the medium or retained in the cell). It is very common to relate the pattern of product synthesis to cell growth, that is, product synthesis is frequently characterized as growth associated or nongrowth associated.

The most suitable type of products to be manufactured by using perfusion bioreactors are the nongrowth-associated, secreted products. In this case, the reactor is operated at high cell density without rapid cell division for a long period. The product can be continuously harvested from the spent medium. One of the major limitations that prevents wider usage of plant cell cultures for industrial production of biologics is slow cell growth. Use of perfusion culture operating at a growth-arrest state and high cell density will circumvent this problem. For growth-associated, secreted products, perfusion culture with a bleed stream can also be considered. To increase the product output, perfusion culture should be operated at a high perfusion rate with the bleed stream adjusted to give a highly specific cell growth rate. For growth-associated, intracellular products, high productivity can be achieved by increasing the cell growth rate in a single-stage bioreactor. Because the product is stored within the cell, a culture strategy that gives a high biomass output rate with a high product content is most desirable. Besides semicontinuous culture, perfusion culture with a bleed stream may be considered for producing this type of product. In this case, a much higher cell density can be obtained in perfusion cultures compared to continuous or semicontinuous cultures, because cells are retained within the reactor via a cell retention device. By incorporating a bleed stream, the perfusion reactor can be operated at steady state at a very high cell concentration. For a culture system that follows simple Monod kinetics, the maximum biomass output rate in a perfusion reactor with a bleed stream is higher than that in a chemostat by a factor of $1/\beta$, where β is the bleed ratio (the ratio between the flow rates of the bleed stream and the feed stream). Perfusion culture has been used in the commercial production of berberine, an intracellular, growth-associated metabolite, by Mitsui Petrochemical Industries, Ltd. (3). Kim and co-workers also used perfusion cultivation in the production of berberine (4). In our laboratory, a high cell density of 35 g dry weight/L and a rosmarinic acid concentration of 4 g/L were achieved in the perfusion culture of *Anchusa officinalis* (9,10).

In general, perfusion culture has the following advantages over batch culture:

1. Reduced plant down time
2. Reduced unproductive growth phase as a proportion of the total process time
3. More consistent product quality because conditions in the bioreactor are held stable
4. More extensive automation with improved control
5. Higher cell density and productivity

Protagonists of batch processes point to the simplicity of the plant required, greater process flexibility (i.e. the capacity to produce different products in different production runs), and reduced quantities of product at risk because a process failure is limited to a single batch (15). When using perfusion systems, it is also important to determine the long-term genetic stability of the cells employed.

OVERVIEW OF PLANT CELL PERFUSION BIOREACTOR DESIGNS

What sets a perfusion bioreactor apart from a regular bioreactor is its ability to achieve medium exchange with cell retention. Therefore, a means for separating cell and medium needs to be incorporated into the perfusion bioreactor. In designing a cell retention device for plant

cell perfusion cultures, factors to be considered include high culture biotic phase volume and viscosity, non-Newtonian fluid behavior, cell shear sensitivity, cell and aggregate size distribution and morphology, as well as scalability. Currently, two major classes of techniques are practiced for separating cells from the medium in perfusion reactors, gravitational or centrifugal sedimentation and tangential filtration (e.g., axial rotating filtration or cross-flow filtration) or vortex-flow filtration.

Separation by Sedimentation

Cells can be retained by sedimentation by incorporating an in situ or ex situ settler (or an external continuous centrifuge) into a bioreactor. Gravitational sedimentation is a gentle process for cell separation. It is mechanically simple, and exploits the inertial properties of the cells, instead of using physical barriers as in filtration, to accomplish the separation. Sedimentation devices, however, have a large holdup volume. Considering the large particle size of plant cells, sedimentation is an attractive method for cell/medium separation. An in situ settler, in the form of a settling (decanting) column, has been used in perfusion cultures of *Catheranthus roseus* (1) and *T. rugosum* (4) (Fig. 1). Several other designs based on gravitational sedimentation have also been reported, although they are not tested in plant cell cultures. Sato et al. (16) describe an internal conical cell separator consisting of a tapered conical sedimentation chamber affixed to the headplate of a stirred bioreactor. Spent medium is removed through the top of the cone, and cells settle and slide down the steep wall of the cone to the bioreactor (Fig. 2a). All of the above designs take advantage of the reduced culture fluid superficial upward velocity in a settling zone that has an enlarged cross-sectional area.

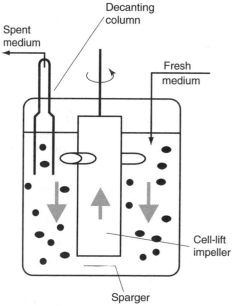

Figure 1. A schematic diagram of the perfusion bioreactor used by Kim et al. (4); cells and medium were separated by using a decanting column.

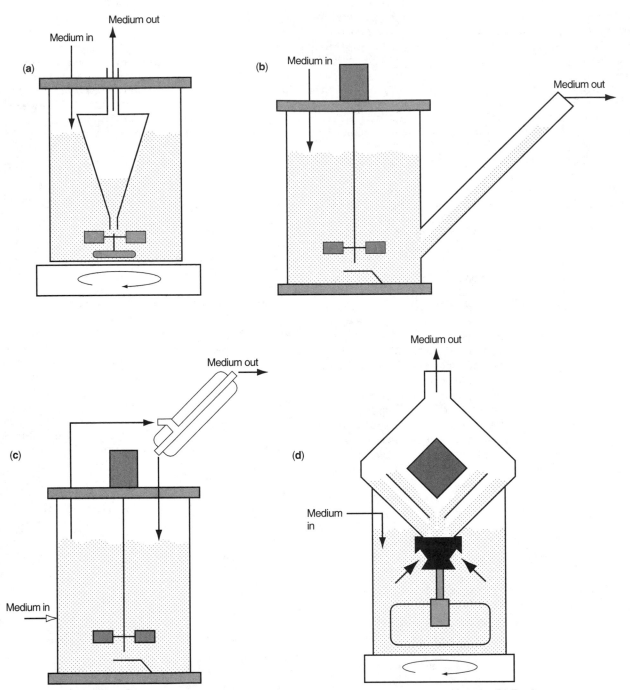

Figure 2. Various settler designs for perfusion bioreactors: (**a**) Sato et al. (16); (**b**) Davison et al. (21); (**c**) Searls et al. (22); (**d**) Tyo and Thilly (23).

Inclined sedimentation is an alternative to vertical sedimentation. Cell-settling velocity can be increased by using inclined sedimentation. This phenomenon was first observed by Boycott (17) and analyzed, among others, in the work of Leung and Probstein (18). For a dilute, monodisperse particle suspension, the particle sedimentation rate can be expressed by

$$w_{si} = w_{sv} \left(1 + \frac{h}{H} \sin \alpha \right) \qquad (1)$$

where w_{si} is the settling rate in inclined sedimentation, w_{sv} is the settling rate in vertical sedimentation, h is the vertical height of the solid suspension, H is the plate distance of the settler, and α is the angle of inclination from the vertical axis. To take full advantage of inclined sedimentation, a shallow settler (small H) with a large angle of inclination should be used. The large inclination angle, however, needs to be counterbalanced by the ease with which cells slide back to the fermenter. Typically, an inclination angle from 30 to 60° is used. Inclined sedimentation has been applied to the lamella settlers used in wastewater treatment processes (19), a yeast cultivation system (20), a mixed culture fermentation of yeast and bacterium (21), and animal cell cultures (22,23). In the reactor used by Davison et al. (21), a large inclined shallow settling chamber is integrated into a stirred fermenter as a side arm (Fig. 2b). As the culture is withdrawn through the settling chamber, cells undergo countercurrent inclined sedimentation, settle and slide down the lower surface of the settler to the fermenter. Searls et al. (22) modified the inclined settler used by Davison et al. by placing it on the top of the bioreactor and therefore allowing cells to undergo crosscurrent sedimentation (Fig. 2c). To alleviate cell attachment to the settler wall, Searls et al. (22) also tried bubbling to clean the settler periodically and using a vibrator to shake the settler. To increase sedimentation capacity, Tyo and Thilly (23) describe a system consisting of a series of nested truncated cones with an innermost solid cone. This system is basically a modified version of that used by Sato et al. (16). In this reactor, cells and medium are pumped up through the cones while cells are collected on the lower face of the cones, thereby resulting in a virtually cell-free fluid (Fig. 2d). For plant cell cultures, a countercurrent inclined settler similar to that used by Davison et al. (21) was tested by Su et al. (9) to retain *A. officinalis* cells in a stirred-tank, perfusion culture. Although it works well generally with low cell densities, a large number of cells agglomerate and become immobilized in the settler

at high cell density. This problem is augmented because inclined sedimentation is enhanced by using a shallow settler, as indicated in equation 1. For countercurrent sedimentation, another problem is the lack of means to drive the concentrated cell sediment back to the bulk of the reactor. In all of the designs mentioned above, the settler is in direct contact with the turbulent region of the bioreactor, and therefore the sedimentation process is vulnerable to disturbance by fluid turbulence. In addition, the entrance into these settlers is small and hence prone to clogging.

Our lab has developed and tested a series of perfusion bioreactors designed by incorporating an internal settling zone into air-lift bioreactors. For high density plant cell suspensions, problems in circulating viscous and shear sensitive cell cultures through an external loop, where anoxic condition and shear damage are likely (24), make the use of internal separation devices preferable. Our initial design incorporated a conical "skirt baffle" with an internal-loop, air-lift bioreactor (Fig. 3). Although developed independently, our design is similar to the Applikon® "multipurpose tower bioreactor (MTB)" (Applikon Dependable Instruments BV, the Netherlands). With the internal-loop design, gas bubble recirculation into the downcomer section is inevitable. Despite the downward liquid flow in the region, some of these bubbles would rise into the settling zone and hence disturb cell sedimentation. In addition, we observed a thick layer of

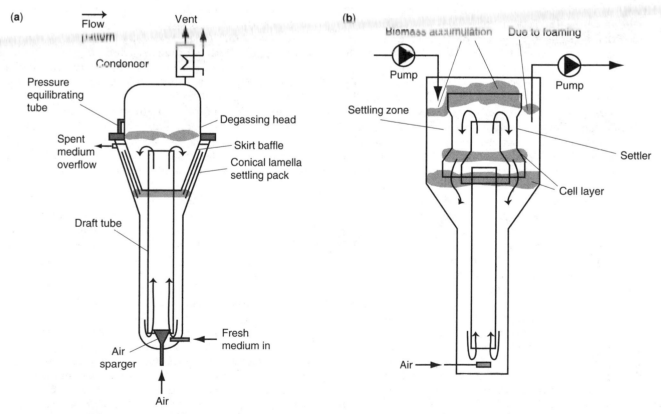

Figure 3. Accumulation of plant cell flocs in various regions of the internal-loop, air-lift perfusion bioreactor (W.W. Su, unpublished) and the Applikon® MTB (redrawn form Ref. 25, courtesy of Applikon Dependable Instruments BV, the Netherlands).

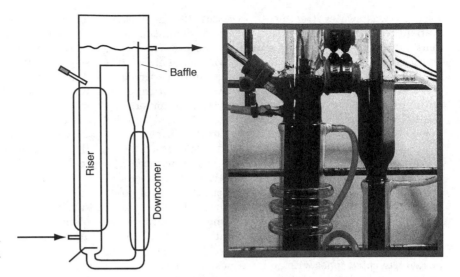

Figure 4. A schematic diagram of an external-loop, air-lift perfusion bioreactor with a basic settler design.

biomass accumulated at the entrance to the settling zone (W.W. Su, unpublished). In a study of the MTB, Hellinga and Luyben also noted accumulation of plant cell flocs in the settling zone due primarily to the escape of recycled air bubbles (25) (Fig. 3). To overcome the problem caused by the recycled bubbles, the internal cell settling zone was integrated with an external-loop, air-lift bioreactor. In its simplest form, the internal settling zone in the external-loop, air-lift perfusion bioreactor is created by inserting a baffle plate into the upper portion of the downcomer (Fig. 4). Clarified spent medium containing secreted metabolic products is continuously removed via the overflow from the settling zone, and fresh medium is fed into the well-mixed riser to replenish nutrients. The improvement comes from the much lower downcomer gas holdup in the external-loop, air-lift reactor due to more efficient gas disengagement in the reactor headspace. By taking advantage of the well-defined flow pattern in the air-lift reactor, the downward bulk liquid flow in the downcomer can guide the cell particles away from the settling region, so that few cell particles are present in the settler. During a 14-day perfusion culture of *A. officinalis* in this bioreactor, a totally clear settling zone was noted for the first 8 days of the culture during which the culture packed cell volume (PCV) was below 60%. It should be pointed out, however, that noticeable eddies provoked some degree of mixing at the entrance of the settling region. The eddies carried some of the cells into the inlet region of the settler. Nonetheless, these cells were carried back to the bulk of the culture by the downward liquid flow, and hence there was a constant turnover of cells at the settler entrance. With increased gas velocity and thus faster culture circulation, mixing at the settler entrance was augmented. The effect of aeration on cell retention efficiency, however, was minor compared with perfusion rate and cell loading (5) due to the fact that the settling zone for the most part was not stirred by the localized mixing at its entrance. As the cell loading exceeds 60% PCV, cell retention in the perfusion, air-lift bioreactor becomes very sensitive to the increase in perfusion rate. It was evident that a cell-free settling zone could no longer be achieved, as indicated in the later stage of perfusion culture. In

this case, countercurrent sedimentation took place in the settler. To improve the sedimentation capacity, a conical settling zone can be incorporated into the upper portion of the downcomer (Fig. 5a). A series of nested truncated cones can be inserted into the settling zone to convert it into a lamella settler (Fig. 5b). To improve oxygen supply in the settler, membrane aeration tubing (Fig. 5c) or a surface aerator (Fig. 5d) may be installed to provide bubble-free aeration without disturbing cell sedimentation (8).

The perfusion bioreactor design of Hamamoto et al. (26) that incorporates an annular settling zone into a stirred-tank bioreactor has been tested in our laboratory for perfusion culture of a plant cell suspension. The settling zone in this reactor is created by inserting a cylindrical baffle into a stirred-tank reactor (Fig. 6). When *A. officinalis* cells are grown in this reactor and perfused at 0.2 to 0.4 per day, cell dry weight reaches over 20 g/L after two-week cultivation, the maximum PCV exceeds 75%, and secreted protein productivity reaches 0.3 g/L/day (W.W. Su, unpublished). Similar to that which occurred in the perfusion, air-lift bioreactor, plant cells start being "pushed" into the annular settling zone once culture PCV exceeds 60%. At a PCV of 75%, the settling zone is practically filled with cells. Therefore, it is necessary to include a bleed stream when the PCV exceeds 60%. At such a high biotic phase volume, the concentrated plant cell suspension usually does not readily behave as a fluid and thus is very difficult to pump, especially at low pumping speeds. As a result, intermittent pumping at a higher pump rate is done.

From the standpoint of system versatility, it is easier to retrofit an existing reactor with an external separator than with an internal separation device. The major drawback of external separators, however, involves the problems noted earlier that are associated with continuous circulation of the culture suspension through an external loop. W.W. Su (unpublished) recently examined the use of an external conical settler integrated with a stirred-tank bioreactor (Fig. 7) for perfusion culture of *A. officinalis*. Pressure equilibration between the settler and the reactor is very important for proper operation of the system. The settler was essentially cell-free for the most part when the PCV

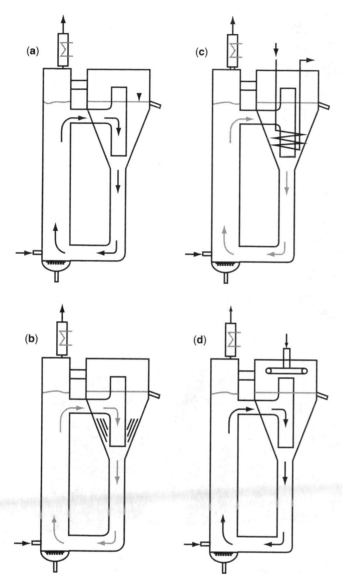

Figure 5. External-loop, air-lift perfusion bioreactor with a conical settling zone: (**a**) basic design; (**b**) with conical lamella settling plates incorporated in the settling zone; (**c**) with membrane tubing in the settling zone for bubble-free aeration; (**d**) with a surface aerator above the settling zone.

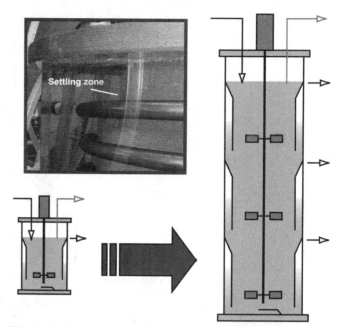

Figure 6. Stirred-tank perfusion bioreactor with an annular settling zone.

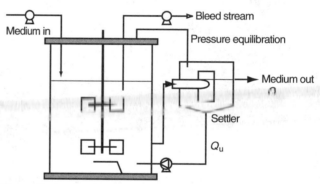

Figure 7. A stirred-tank bioreactor integrated with an external settler.

was below ca. 50%. As a result, virtually cell-free effluent can be withdrawn through the settler. At a PCV >60%, it became difficult to recirculate the culture, and cells started to accumulate in the settler. At this point, further increase of the underflow rate was ineffective in removing the cells accumulated in the settler because it merely sucked more culture directly from the bioreactor through a channel in the T-shaped conduit (Fig. 7). Installation of a slow-moving scraper to sweep the cells accumulated in the settler into the underflow drain may be necessary to overcome this problem. At such a high cell density, an intermittent bleed stream has to be initiated to stabilize the culture.

External centrifugal separators can be used to enhance cell sedimentation. A continuous centrifuge has been used in the *C. japonica* perfusion culture for berberine

production in a 6 m³ reactor (3). Conventional centrifugal separators involve rotary seals and are constructed to provide good clarification of the supernatant but usually generate too much mechanical stress on the cells. Recently, a new semicontinuous centrifuge design (Sorvall Centritech Lab) was reported for hybridoma perfusion cultures (27). The centrifuge (Fig. 8) uses a so-called "inverted question mark" design which makes it possible to rotate one end of a flexible tube while the other stands still, thus eliminating the need for rotating seals (Fig. 9). In this system, separation takes place in a conical slot (Fig. 8), where the suspension enters at an inner radius through a tube at one end of the rotor, flows around the circumference, and exits through the tube at the other end. Because cells are slightly heavier than the liquid, they move outward and are discharged by a pneumatic system through a tube connected to the separation chamber at its outer radius. As the suspension flows around the rotor's circumference, it is subject to low centrifugal forces for enough time for separation to occur. Use of

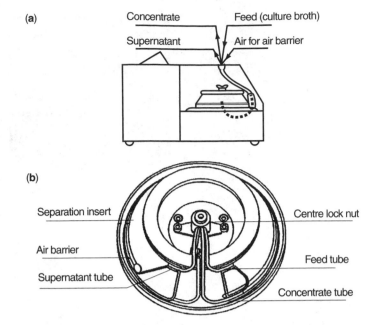

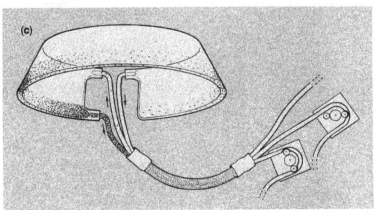

Figure 8. The Centritech® LAB centrifuge: (**a**) profile of apparatus with rotor and "inverted question mark" tubing bundle; (**b**) rotor assembly; (**c**) separation insert; (**d**) the entire system. Reproduced from the Centritech® Separation Systems product bulletin, courtesy of Sorvall, Inc., Newtown, CT.

the centrifuge intermittently decreased the daily cell residence time outside the bioreactor, the daily pelleted-cell residence time in the centrifuge, and the frequency of cell passage to the centrifuge. It was hypothesized by Johnson et al. (27) that having cells periodically packed at the bottom of the centrifuge insert is deleterious to the culture by exposing the pelleted cells to prolonged nutrient limitations. At this time, there are no data available on the application of the Centritech centrifuge to plant cell cultures. In general, although centrifugal separators, are more suitable for large-scale perfusion operations, they are complex and are more prone to mechanical failure than simple gravitational sedimentation devices. These devices also have the disadvantages of external separators.

Separation by Filtration

Cell and medium can be separated by using stationary or moving filters. A stirred tank reactor with insitu filtration by a stationary stainless steel filter (Fig. 10) was used to culture *A. officinalis* to high cell density (6). In this system, a stainless steel composite disc filter with a 70 μm pore opening and an effective filter area of 44 cm^2 (Fuji Filter Manufacturing Co., Japan) was used as the filter medium in a 2.5 L bioreactor. To examine the effect of agitation on the in situ filtration process in the reactor, increases in the suction head of the pump that drove the filtration process were measured at impeller speeds of 100 and 200 rpm. Surprisingly, the suction head attained at 200 rpm was about 40% higher than at 100 rpm. Direct observation of cake deposition in the reactor from a dilute cell suspension revealed that the filter cake formed at 100 rpm was thicker but less compact. Aeration at up to 0.4 vvm had little effect on the filtration rate because the bulk fluid flow was dominated by the impeller hydrodynamics (7). According to Mackley and Sherman (28) in their study of cross-flow cake filtration with 125 to 180 μm polyethylene particles,

Rotating a tube fixed at one end
provides no flexibility.

When one end rotates, the other end
must rotate in the same direction..

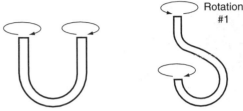
If the tube is flexible, it can be bent into
a "U" or an inverted question mark, but now the ends
rotate in opposite directions when viewed from the top.

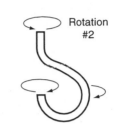

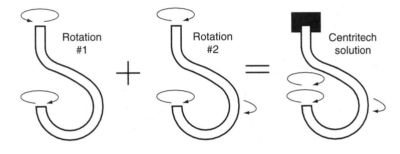

Figure 9. Working principle of the "inverted question mark" tubing bundle in the Centritech® centrifuge. Reproduced from the Centritech® Separation Systems product bulletin, courtesy of Sorvall, Inc., Newtown, CT.

when the tangential fluid flow dominated, particles rolled along the cake surface until captured at a stable site, and this packing process is very selective. Accordingly, a very compact cake with high specific cake resistance would formed by tangential flow.

Flux decay due to increased mass transport resistances is inevitable in any filtration process. It is a general practice to use backflushing to clean the filter. The success of backflushing depends on the flux decay mechanism which is determined mostly by the characteristics of the filter and the cell culture. A two-compartment filtration/backwash chamber (Fig. 10) may be employed to improve filter backflushing (7). While spent medium is removed through one of the compartments, the other compartment is purged with air for backwash and culture aeration. This process alternates between the two compartments by controlling the on/off cycle of the solenoid pinch valves with a programmable timer.

The most common moving filter is the spin filter, a cylindrical filter attached to the agitator shaft (Fig. 11). Spin filter technology has been used in somatic plant embryo cultures (2). The centrifugal force generated via the rotation of the cylindrical filter is expected to hinder filter clogging. However, a study of Yabannavar et al. (29) showed that high spinning rates also promote fluid exchange across the filter and hence reduces particle retention efficiency. Therefore, there is a trade-off in using high rotation speed to prevent filter clogging. The spin filter method is not particularly suitable for viscous plant cell suspensions due to the difficulty of preventing filter fouling (9). No detailed study has been published on the mechanism of spin filter fouling by plant cell

cultures. It is speculated that secreted polysaccharides, proteins, and DNA fragments liberated following cell death may be responsible for filter fouling. In addition to spin filters, Lehmann et al. (14) installed hydrophilic microporous polypropylene membrane fibers inside the reactor as a microfilter/membrane stirrer to retain and mix animal cells in a bioreactor (Fig. 12). A similar reactor was used by Moorhouse et al. (13) for plant somatic embryo cultivation. Vortex-flow filtration (VFF) has been found superior to cross-flow filtration for perfusion animal cell cultures (30). VFF is based upon Taylor vortices, established by rotating a cylindrical filter inside a second cylinder. The sample is fed under pressure between these two cylindrical surfaces, forcing filtrate across the filter and into the inner cylinder for collection. The vortices keep the filter surface constantly clean, thus preventing clogging. No information is available on using this type of device in perfusion plant cell cultures.

ANALYSIS OF DESIGN AND OPERATING PARAMETERS

The main design objective for a perfusion bioreactor is to retain cell particles effectively, so that a high cell concentration can be achieved, and to operate functionally at that high cell loading. Therefore, The key design parameters include efficient cell and medium separation and effective mixing and oxygen transfer in high-density cultures. The key operating parameters include perfusion and bleed rates, mode of perfusion (close vs. open), and medium composition.

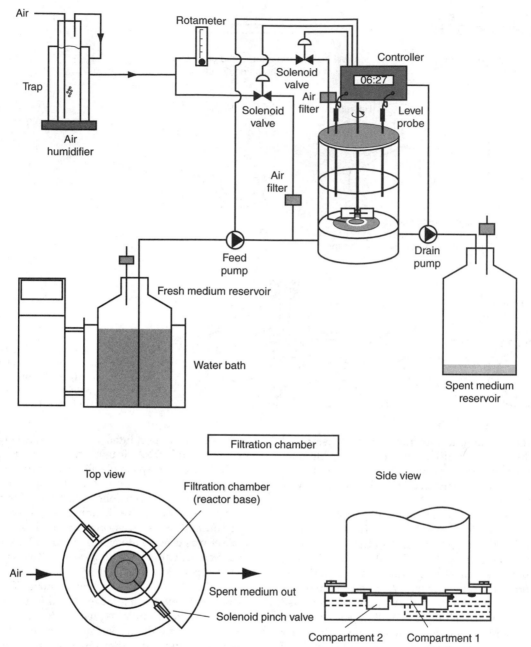

Figure 10. Perfusion bioreactor with insitu filtration. Reproduced form Ref. 9, courtesy of Springer-verlag, Berlin.

Design of Cell Retention Devices

For external gravitational settlers, solid-flux analysis (19) used in designing settlers for wastewater treatment may be used. W.W. Su (unpublished) has shown that hindered sedimentation due to high cell concentration in plant suspension culture can be modeled by a simple exponential relationship,

$$\nu_i = \nu_{io} e^{-KX} \qquad (2)$$

where ν_i is the particle settling rate at a specific cell concentration X; ν_{io} is the settling rate in an infinitely dilute cell suspension, which can be estimated from Stokes' law; and K is an empirical constant related to the cell suspension's surface properties, size distribution,

morphology, and viscosity. When X is expressed as PCV, the K values range from 5 to 15 for 1 to 2-week-old tobacco and *Anchusa* cell cultures (W.W. Su, unpublished). Solid flux is expressed by the following equation:

$$SF = \nu_{io} \left[e^{-KX} \cdot X + \left(\frac{U_B}{\nu_{io}} \right) X \right] \qquad (3)$$

The limiting solid flux (SF_L) occurs at a cell loading X, that can be calculated from the following equation:

$$\begin{cases} (1 - K \cdot X) e^{-KX} + \dfrac{U_B}{\nu_{io}} = 0 \\ X > \dfrac{2}{K} \end{cases} \qquad (4)$$

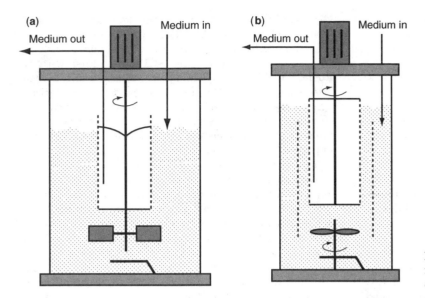

Figure 11. (a) Basic spin-filter biopreactor; (b) Modified spin-filter bioreactor with a perforated draft tube and separate motors for the folter and the impeller.

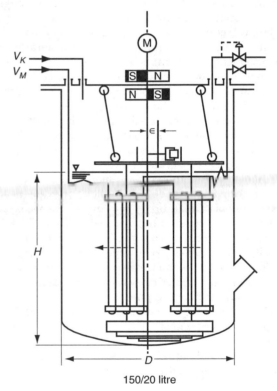

150/20 litre

Figure 12. Perfusion bioreactor with a dual-membrane stirrer (14). Reproduced from Ref. 14, courtesy of Academic Press, New York.

where U_B is the settler underflow rate. Then, the area requirement A of the external settler can be determined as

$$A = \left(\frac{Q \cdot X_o}{SF_L - U_B \cdot X_o} \right) \qquad (5)$$

where Q is the volumetric effluent flow rate (equal to the perfusion rate times reactor working volume), and X_o is the cell concentration in the perfusion bioreactor. For internal settling devices, the settling zone may be sized

by equating the upward superficial liquid velocity (equal to the volumetric nutrient feed rate/reactor cross-sectional area, in the case of the annular settling zone) and the terminal sedimentation rate of the cell particles. Here the hindered settling effect should be considered in calculating the particle-settling rate using, for instance, equation 2. The use of equation 5 is demonstrated in the following example. For a culture having a K value of 10, a ν_{io} value of 1 mm/s, and the settler underflow rate U_B set at 0.0055 mm/s, from equation 4 the limiting solid flux occurs at a cell loading of 70% PCV, and the corresponding SF_L is 0.0045 mm/s. For a 500 L bioreactor with a perfusion rate of 0.5 per day, the calculated settler area is 0.82 m² (a settler diameter of about 1 m).

For continuous centrifugal separators, the sigma factor concept ($\Sigma = Q'/\nu_{io}$, where Q' is the volumetric throughput of the centrifuge) (19) may be used to determine the centrifuge size. The sigma factor for certain newer centrifuge types, such as the Centritech centrifuge, has yet to be derived. The residence time in the centrifuge, frequency of cell passage to the centrifuge (during intermittent operations), the magnitude of centrifugal force applied, and the rate of external culture circulation (which may lead to shear and anoxic stresses) will need to be optimized.

For filtration-type separators, several reports have been published dealing with engineering analysis of spin filters for animal cell cultures (29,31,32). Both filter material and rotation speed, as well as culture viscosity, are important factors. For animal cells, cell colonization is prevented, provided that filter rotation exceeds around 0.6 m/s tip speed (31). Addition of deoxyribonuclease significantly reduces membrane fouling during animal cell perfusion culture using a cross-flow membrane module (30), suggesting that DNA fragments liberated following cell death are partly responsible for membrane fouling.

Design Considerations for Mixing and Oxygen Transfer

Perfusion cultures are operated at high cell densities. Most plant cell cultures are viscous at high cell concentrations.

Generally this results from the high biotic phase volume and the elongated cell morphology, whereas the spent culture media usually have low viscosities. High-density plant cell suspension cultures behave as non-Newtonian fluids. Power-law models, including Bingham plastics, pseudoplastics, and Casson fluids, have been used to describe the rheological characteristics of plant cell suspensions (33). Culture rheological properties can affect mixing and oxygen transport substantially. For instance, in pseudoplastic fluids, apparent viscosity is lower at higher shear. In this case, mixing and bubble dispersion will be better in the impeller region where high shear exists, and areas away from the impeller will experience a higher apparent viscosity, leading to poor mixing and oxygen transfer.

Culture rheological characteristics may be altered by changes in medium osmolarity. Cell enlargement resulted in increased PCV and led to a significant increase in apparent culture viscosity in the later stage of batch cultivation, in which medium osmolarity approached zero. The addition of 5% mannitol reduced the cultures apparent viscosity by almost 84% without affecting cell growth (9). In perfused A. officinalis cultures, a high sucrose concentration (6%) in the perfusion medium maintains medium osmotic pressure at the low perfusion rate. No significant increase in cellular water content was observed during perfusion culture. The ability to control cell enlargement or, more so, to reduce cell size becomes very crucial, especially in the light of the impact of biotic phase volume on cell retention efficiency in perfusion bioreactors. Further, mixing and oxygen transfer problems in the bioreactor are expected to be minimized as a result of the culture's improved rheological characteristics.

Cell sedimentation is hindered in dense cell cultures, which have a high solid fraction. Because the culture viscosity is high, however, suspension homogeneity becomes a primary concern in mixing. Effective impeller design is of particular importance in stirred tank reactors. The standard Rushton turbine provides efficient bubble dispersion with good mixing. It does have a few disadvantages, including high power requirement, highly localized shear, and a drastic drop in power upon aeration (33). This latter phenomenon is especially profound in viscous culture fluids, leading to poor pumping capacity (ability to circulate fluid elements). These limitations have encouraged the use of other impeller types. Large-diameter impellers including helical ribbon impellers and helical screw (spiral) stirrers generally provide high homogenization efficiency at low agitation speeds. By adding three vertical surface baffles, Jolicoeur et al. reported improved mixing in a plant cell bioreactor agitated by a helical ribbon impeller (34). A helical screw (spiral) stirrer was used in the cultivating of *Coleus blumei* (35). These impellers, however, suffer from inefficiency in inducing bubble dispersion (33). Recently introduced axial-flow hydrofoil impellers with large solidity ratios (the horizontally projected area of the blades divided by the horizontal area swept out by the blades) have been reported to give efficient gas dispersion and bulk blending in highly viscous mycelial and gum fermentations (36,37). The use of

large-solidity-ratio hydrofoil impellers in high-density tobacco cell cultures has been tested recently in our laboratory. Although mixing time is reduced with the hydrofoil impeller (Lightnin A315) compared with the Rushton turbine at the same power input, oxygen transfer rate is lower (W.W. Su, unpublished). For pneumatically agitated vessels, superficial gas velocity (and hence liquid circulation velocity), as well as reactor geometry, can affect the suspension efficiency. Particle suspension is generally improved with a high reactor aspect (height to diameter) ratio, inclusion of a draft tube, or a contoured bottom (33).

Oxygen demand has to be met, so that the dissolved oxygen level can be maintained above a certain critical concentration. For a cell culture with a typical oxygen demand of 1 μmole/g dry weight/min and a cell density of 10 g dry weight/L, a volumetric oxygen transfer coefficient (k_La) of 3 h^{-1} is sufficient to maintain the dissolved oxygen concentration at 20% air saturation. Even for cultures with 10 times higher oxygen demand, the oxygen transfer requirement is still moderate. However, due to the high culture viscosity, cell shear sensitivity, and cell aggregation discussed in the preceding sections, oxygen transfer in plant cell cultures becomes a challenging problem.

Stirred tank bioreactors with sparged aeration have been used widely in plant cell cultures. In viscous culture suspensions, impeller configuration has a strong impact on k_La especially at high rotational speeds (33). Although the Rushton turbine has certain drawbacks as noted above, it is still considered one of the most efficient impellers for bubble dispersion and oxygen transfer in fermentation. However, because of concern with excess shear at increased agitation rates, low impeller speeds are commonly used in plant cell bioreactors. In this case, sparger design to provide finer bubbles may be desirable because bubble break-up at low impeller speeds is not efficient. In pneumatically agitated reactors, the main factors that affect k_La are superficial gas velocity, culture viscosity, and bubble coalescence, sparger design and location, as well as geometric parameters, such as the ratio of the downcomer and riser cross-sectional areas in air-lift reactors (33). Although higher oxygen transfer rates can be achieved by increasing the superficial gas velocity, this may lead to desorption of volatile compounds, foaming, and excess shear. In perfusion, air-lift bioreactors, increased downcomer gas holdup at increased gas velocity may disturb cell sedimentation in the settling zone.

Operating Parameters

One of the most important operating parameters is the culture perfusion rate. Because of the generally slow metabolic rates of plant cells, it is believed that low perfusion rates (less than 1/day) are sufficient to accomplish high-density cultivation, as demonstrated in the perfusion culture of the A. officinalis system (with a maximum specific growth rate of 0.45/day), where a low perfusion rate of 0.2 to 0.4/day was found satisfactory (6,10). Moreover, by operating at low perfusion rates, product in the spent medium is more concentrated, which is beneficial to downstream processing. It should be noted, however, that perfusion serves two main purposes, supplement of nutrient and removal of extracellular inhibitory metabolites. The dominant factor should be identified, and the perfusion rate and perfusion medium should be designed accordingly. To reduce the medium cost, partial reuse of the spent medim by replenishing selected nutrient may be used (38). Bleed rate is another important operating parameter in perfusion cultures. It can be used to manipulate the culture's specific growth rate and to reduce the dead cell concentration.

Scale-up

Cell retention by either sedimentation or filtration is governed by the mass fluxes (solid flux and filtrate flux, respectively), that is when the perfusion rate (vol/vol/time) is to be kept constant during scale-up, the fluxes will increase with (reactor volume)$^{1/3}$. For gravitational sedimentation, multiple settling zones (Fig. 6) or use of lamella settling pack (Fig. 3) may be a viable solution. For centrifugal separators, scale-up may be based on the sigma factor. For filtration devices, procedures for scaling up cross-flow microfilters (14) and vortex-flow filters (39) can be followed. For spin filters, Yabannavar et al. (32) have shown that the filter can be kept unclogged by maintaining the ratio of permeation drag to lift drag constant and that $N_{spinfilter} \propto (V)^{-1/6}$, where $N_{spinfilter}$ is the rotational speed of the filter and V is the reactor volume. They also reported that the rotational speed of the agitator should be proportional to $V^{-2/9}$ to maintain a constant power per unit volume during scale-up. Under these conditions, fluid exchange per unit volume decreases slightly during scale up ($\propto V^{-1/18}$). Note that increased fluid exchange may lead to increased cell leakage through the perfusate.

PROCESS MONITORING, OPTIMIZATION, AND CONTROL

Development of optimal perfusion strategy is of great importance because it affects process performance and also influences the medium cost, which represents a significant portion of total operating cost. Kyung et al. (40) reported a strategy for automated adjustment of perfusion rates in animal cell cultures based on on-line oxygen uptake rate (OUR) measurement to keep the medium glucose concentration around a set point. Using the stoichiometric ratio of oxygen/glucose consumption derived from batch cultures, the amount of glucose consumed can be estimated on-line by integrating OUR over time. As the amount of glucose consumed reaches a set point, the reactor is switched to the perfusion mode. After perfusion commences, the perfusion rate required to maintain a constant glucose concentration is calculated on-line from the OUR integral. Ozturk et al. (41) controlled the glucose concentration in a perfusion animal cell culture by directly monitoring glucose concentration in the culture with a glucose sensor and automatically adjusting the perfusion rate. These strategies are aimed at controlling nutrient supply. When toxic removal is the limiting factor, accumulation of the inhibitory substances may be controlled directly by diluting them using a high perfusion rate (which also dilutes the desired product)

or by controlling nutrient supply to reduce the production of the inhibitory waste metabolites. Büntemeyer et al. (38) discussed a strategy to optimize medium use for perfusion animal cell cultures. They used a stirred tank dialysis system to generate defined limitation and inhibition states by decoupling nutrient supply and inhibitor accumulation at constant cell densities in a continuous culture. It was shown that reusing a portion of an ultrafiltered perfusate by recombining it with fresh medium is suitable for sustaining cell viability, where ultrafiltering is used to remove low molecular weight inhibitory substances.

Adjustment of perfusion rate presents several unique challenges for perfused plant cell cultures. For instance, growth of cultured plant cells is often regulated by the so-called "conservative substrates", substrates that are taken up by cells but not metabolized immediately and, instead are, stored in an intracellular pool. Phosphate is a typical example of conservative substrate. This means that control of extracellular nutrient concentration may not be an effective strategy for perfusion plant cell cultures. Rather, a strategy that allows controlling an intracellular pool of conservative substrates may be necessary to maximize productivity. Our laboratory has recently developed an intracellular phosphate estimater based on on-line OUR measurement and the extended Kalman filter technology (42) for controlling intracellular phosphate level in a phosphate-limited perfusion plant cell culture (W.W. Su, unpublished). Another complication for controlling perfused plant cell cultures is that unknown "conditioning factors" are often present in the spent media of plant cell cultures, especially with respect to secondary metabolite production. It is necessary to analyze the medium usage to identify the limiting factor involved in the perfusion process, whether that is nutrient limitation, toxic buildup, or dilution of conditioning factors.

CONCLUSIONS AND FUTURE OUTLOOK

Perfusion culture represents a versatile system for achieving continuous, high-density plant cell cultures. Effective and relatively simple cell retention devices, such as those based on gravitational sedimentation, have demonstrated good potential with long-term operational stability. Very little information is available, however, on the cellular kinetics, physiology, and genetic stability of high-density perfusion plant cell cultures. Also, data are yet to be collected to determine whether high cell density creates an "allelopathic" stress, seen in other biological and ecological systems (43).

BIBLIOGRAPHY

1. A. Pareilleux and R.A. Vinas, *Appl. Microbiol. Biotechnol.* **19**, 316–320 (1984).

2. P.V. Ammirato and D.J. Styer, in M. Zaitlin, P. Day, and A. Hollaender, eds., *Biotechnology in Plant Science*, Academic Press, New York, 1985, pp. 161–178.

3. K. Matsubara and Y. Fujita, in A. Komamine, M. Misawa, and F. DiCosmo, eds., *Plant Cell Culture in Japan*, CMC, Tokyo, 1991, pp. 39–44.

4. D.I. Kim, G.H. Cho, H. Pedersen, and C.K. Chin, *Appl. Microbiol. Biotechnol.* **34**, 726–729 (1991).

5. W.W. Su, B.J. He, H. Liang, and S. Sun, *J. Biotechnol.* **50**, 225–233 (1996).

6. W.W. Su, F.Lei, and N.P. Kao, *Appl. Microbiol. Biotechnol.* **44**, 293–299 (1995).

7. W.W. Su, *Biotechnol. Tech.* **9**, 259–264 (1995).

8. U. S. Pat. 5,342,781 (August. 30, 1994), W.W. Su (to University of Hawaii, Honolulu, Hawaii).

9. W.W. Su, E.C. Asali, and A.E. Humphrey, in Y.P.S. Bajaj, ed., *Biotechnology in Agriculture and Forestry*, Vol. 26, Springer-Verlag, Berlin, 1994, pp. 1–20.

10. W.W. Su, F. Lei, and L.Y. Su, *Biotechnol. Bioeng.* **42**, 884–890 (1993).

11. W.W. Su and A.E. Humphrey, *Biotechnol. Lett.* **13**, 889–892 (1991).

12. W.W. Su and A.E. Humphrey, *Biotechnol. Lett.* **12**, 793–798 (1990).

13. S.D. Moorhouse, et al. *Plant Growth Regul.* **20**, 53–56 (1996).

14. J. Lehmann, J. Vorlop, and H. Büntemeyer, in R.E. Spier and J.B. Griffiths, eds., *Animal Cell Biotechnology*, Vol. 3, Academic Press, New York, 1988, pp. 221–237.

15. M. Rhodes and J.R. Birch, *Bio/Technology* **6**, 518–523 (1988).

16. S. Sato, K. Kawamura, and N. Fujiyoshi, *J. Tissue Cult. Methods* **8**, 167–171 (1983).

17. A.E. Boycott, *Nature, (Londan)* **104**, 532 (1920).

18. W.-F. Leung and R.F. Probstein, *Ind. Eng. Chem.* **22**, 58–67 (1983).

19. U. Wiesmann and H. Binder, *Adv. Biochem. Eng./Biotechnol.*, **24**, 119–171 (1982).

20. J. Tabera and M.A. Iznaola, *Biotechnol. Bioeng.* **33**, 1296–1305 (1989).

21. B.H. Davison, K.Y. San, and G. Stephanopoulos, *Biotechnol. Prog.* **1**, 260–268 (1985)

22. J.A. Searls, P. Todd, and D.S. Kompala, *Biotechnol. Prog.* **10**, 198–206 (1994).

23. M.A. Tyo and W.G. Thilly, *Paper AIChE Annu. Meet.* San Francisco, 1989, Pap. 306.

24. O. Holst and B. Mattiasson, in B. Mattiasson and O. Holst eds., *Extractive Bioconversions*, Dekker, New York, 1991, pp. 11–26.

25. C. Hellinga and K.C.A.M. Luyben, *Report on a Study of the Multipurpose Tower Bioreactor*, Applikon Dependable Instruments, 1998.

26. U. S. Pat. 4,814,278 (March 21, 1989), K. Hamamoto et al. (to Teijin Limited, Osaka, Japan).

27. M. Johnson et al. *Biotechnol. Prog.* **12**, 855–864 (1996).

28. M.R. Mackley and N.E. Sherman, *Chem. Eng. Sci.* **47**, 3067–3084 (1992).

29. V.M. Yabannavar, V. Singh, and N.V. Connelly, *Biotechnol. Bioeng.* **40**, 925–933 (1992).

30. S. Mercille etal., in R.E. Spier, J.B. Griffiths, and W. Berthold, eds., *Animal Cell Technology: Products of Today, Prospects for Tomorrow*, Butterworth-Heinemann, Oxford, 1994, pp. 532–538.

31. E. Favre and T. Thaler, *Cytotechnology* **9**, 11–19 (1992).

32. V.M. Yabannavar, V. Singh, and N.V. Connelly, *Biotechnol. Bioeng.* **43**, 159–164 (1994).

33. W.W. Su, *Appl. Biochem. Biotechnol.* **50**, 189–230 (1995).

34. M. Jolicoeur, C. Chavarie, P.J. Carreau, and J. Archambault, *Biotechnol. Bioeng.* **39**, 511–521 (1992).

35. B. Ulbrich, W. Wiesner, and H. Arens, in K.-H. Neumann, W. Barz, and E. Reinhard, eds., *Primary and Secondary Metabolism of Plant Cell Cultures*, Springer, Berlin, 1985, pp. 293–303.

36. E. Galindo and A.W. Nienow, *Biotechnol. Prog.* **8**, 233–239 (1992).

37. K. Gbewonyo, D. Dimasi, and B.C. Buckland., in C.S. Ho and J.Y. Oldshue, eds., *Biotechnology Processes Scale-up and Mixing*, AIChE, New York, 1987, pp. 128–134.

38. H. Büntemeyer, C. Wallerius, and J. Lehmann, *Cytotechnology* **9**, 59–67 (1992).

39. M. Mateus and J.M.S. Cabral, *AIChE J.* **41**, 764–769 (1995).

40. Y.-S. Kyung, M.V. Peshwa, and W.-S. Hu, *Cytotechnololy* **14**, 183–190 (1994).

41. S.S. Ozturk, J.C. Thrift, J.D. Blackie, and D. Naveh, *Biotechnol. Bioeng.* **53**, 372–378 (1997).

42. J. Albiol, J. Robusté, C. Casas, and M. Poch, *Biotechnol. Prog.* **9**, 174–178 (1993).

43. M. Javanmardian and B.O. Palsson, *Biotechnol. Bioeng.* **38**, *1182–1189 (1991).*

See also ASEPTIC TECHNIQUES IN CELL CULTURE; BIOREACTOR OPERATIONS—PREPARATION, STERILIZATION, CHARGING, CULTURE INITIATION AND HARVESTING; BIOREACTORS, RECIRCULATION BIOREACTOR IN PLANT CELL CULTURE; IMMUNO FLOW INJECTION ANALYSIS IN BIOPROCESS CONTROL; OFF-LINE IMMUNOASSAYS IN BIOPROCESS CONTROL; ON-LINE ANALYSIS IN ANIMAL CELL CULTURE; PHYSIOLOGY OF PLANT CELLS IN CULTURE.

BIOREACTORS, RECIRCULATION BIOREACTOR IN PLANT CELL CULTURE

H.J.G. TEN HOOPEN
J.E. SCHLATMANN
Delft University for Technology
The Netherlands

OUTLINE

INTRODUCTION

A very commonly used "bioreactor" in plant cell biotechnology is the shake flask. It is employed in all kinds of fundamental studies of physiology, plant genetics, metabolic pathways, etc. Also in preliminary studies on the application of plant cells, such as the commercial production of secondary metabolites or the use of plant cells for biotransformation purposes, the shake flask is mainly used. There are some major drawbacks to this approach. The environmental conditions for the cells in a shake flask are not well controlled, as explained more in detail later. The reproducibility of shake flask experiments is rather poor, particularly if secondary metabolism is studied because of the subtle control of these biosynthetic pathways. This lack of reproducibility results from each sample in an experimental series being one single culture with its own environmental conditions.

In commercial applications of plant cells the use of large-scale bioreactors will be necessary. Scaling-up the preliminary shake-flask experiments to the bioreactor conditions is very complex. The two systems differ in almost every aspect: geometry, mixing, and ventilation. This causes large differences in the conditions, especially in the composition of the gaseous phase and the concentrations of the dissolved gases. These differences influence all primary and secondary metabolic processes of the cell culture.

The preferable alternative for the shake flask is a well-mixed and aerated bioreactor: The environmental conditions are much better controlled by stirring and aeration, and multiple samples can be drawn from one culture in a reproducible way instead of sacrificing one culture flask for each sample during an experiment. On the other hand, the bioreactor also has its typical problems. A disadvantage of the bioreactor could be hydrodynamic shear forces. Air flow rate and impeller speed (power input) generate hydrodynamic (shear) stress in the reactor, which can be considerably higher than in the shake flask. For many years plant cells have been presumed to be intrinsically shear sensitive. Therefore, stirred tank reactors have been suspect for several years. However, Scragg et al. (1) and Meijer et al. (2) demonstrated that cells from various plant species are shear tolerant. Meijer et al. (3) reviewed literature data on the effects of hydrodynamic stress on cultured plant cells. They concluded that large differences in hydrodynamic stress sensitivity exist among various plant cell lines, and that the opinion that plant cells are all sensitive to hydrodynamic stress has to be revised. An option to reduce the negative effects of shear stress imposed by the impeller in a stirred tank can be achieved by modifying the design of the routinely used Rushton turbine impeller. Among those modified impellers are the hollow paddle (4,5), the marine (6), the cell lift (6,7), the large flat blade (8), and the helical-ribbon (9) impeller.

A second important difference between a shake flask and a bioreactor is the air flow. This leads to a completely different gas regime in the bioreactor. Whether this is a problem depends on the specific case considered. Various effects of a high (inital) air flow rate on biomass growth has been described. In some studies the detrimental effects of a high air flow rate were attributed to increased shear stress. But in the main effects of aeration have been attributed to the removal of CO_2 or other key gaseous metabolites. These effects will be discussed more in detail in the following. Here the principal differences in gas exchange between shake flask and bioreactor, as illustrated in Figure 1, are discussed. In the cell suspension oxygen is consumed and carbon dioxide and other volatile compounds are produced. In the case of a shake flask oxygen has to

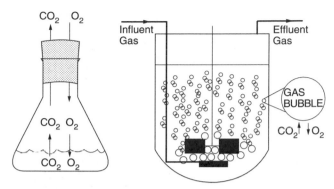

Figure 1. Comparison between the gas transfer in a shake flask and a stirred and aerated bioreactor.

be transferred from the surrounding gas phase through the flask closure into the internal gas phase and from the internal gas phase through the liquid–gas interphase into the culture. Volatiles produced by the cells follow the opposite route. The transfer rates through the flask closure depend on the structure of the particular closure, which explains partly the low reproducibility of shake flasks. The gas transfer through the gas–liquid surface depends on the shaking conditions, the composition of the medium (salts, detergents, etc.) and the shape of the flask. A quantitative analysis of this mass transfer system was given by van Suijdam et al. (10). In Table 1 experimental data are shown of the composition of the gas phase in a shake flask during growth of a plant cell culture. It is clear that the dissolved gas concentrations can reach extreme values in the flask.

A third point of concern in bioreactor studies is the interrelation among mixing, mass transfer, and shear. Increasing the aeration rate to enhance the ventilation or increasing the stirrer speed to enhance the mass transfer also changes the hydrodynamic shear forces in the reactor. In studies on the effects of gaseous compounds or the effects of shear, it is essential to separate these effects carefully. A typical example of the confusion that can be caused by the interpretation of the difference between shake flask and bioreactor experiments is shown in results reported by Moo-Young and Chisti (11). They compared the growth of *Eschscholtzia californica* in shake flasks and in slanted-bottom bubble columns. They used this reactor type because of the presumed shear sensitivity of the plant cell culture. The biomass growth in the different reactor systems presented by the authors are shown

Table 1. Gas-Phase Composition in a 300-mL Shake Flask with 60 mL Suspension of Plant Cells during Growth (Flask Closure Cotton Wool and Aluminum Foil)

Age of the culture (days)	Oxygen (%)	Carbon dioxide (%)	Dry weight (g/L)
0	21	0	3
1	20	3	3
3	16	9	6
5	14	11	11.5
7	13	11	16.5

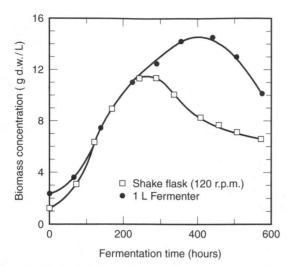

Figure 2. *Eschscholtzia californica* biomass concentration versus fermentation time for 250-mL shake flasks agitated at 120 rpm and for a 1-L slanted-bottom fermentor (11).

Table 2. Typical Oxygen Uptake Rates for Plant Cell Suspensions in 10^{-5} mol/kg dry weight/s

Growing cells, batch culture	3–15
Growing cells, continuous culture	9–15
Nongrowing cells in production medium	0.9–3
Nongrowing cells in production medium (active biomass)	3–15

in Figure 2 and explained as an effect of hydrodynamic shear. Taking into account the data in Table 1 and realizing that the shake flask has a relatively low mass transfer rate, it is probable that the shown difference in biomass productivity results not from a shear effect, but from the low dissolved oxygen concentration in a shake flask with a high biomass concentration.

The choice of the most suitable bioreactor type for plant cell cultures has been an issue for several years. The discussion concentrates on the possibilities of the air lift and stirred bioreactor, respectively. Mass transfer, mixing, and oxygen requirements are met by stirring and aeration in stirred vessels, and by aeration alone in air lift bioreactors. Both the mass transfer coefficient and the mixing time are dependent on the air flow rate, the fraction air in the liquid (gas holdup), the stirrer speed (usually expressed as the power input), the reactor geometry, and the rheological properties of the culture. Air-driven bioreactors are completely dependent on the aeration for transport processes. These bioreactors are less suitable for cultivating cells at a high density, due to insufficient mixing and the development of anaerobic zones (12). A study of the biomass axial distribution in an air lift bioreactor with internal loop indicated that already at biomass concentrations of 12.5 g/L the *Phaseolus vulgaris* cells were not homogeneously distributed over the bioreactor (13). Therefore, the stirred and aerated bioreactor is the preferred reactor type, in spite of the possibility of a shear problem with shear-sensitive cells, as discussed before.

To fully understand the behavior of the cells in the bioreactor it is necessary to further examine the effect of the typical gas regime. In the bioreactor the culture is almost in equilibrium with the aeration gas, and gaseous metabolites produced are discharged almost completely. Therefore, it is essential to deal in more detail with the effects of the gaseous compounds on growth and production.

Heterotrophic growing plant cells need oxygen for their growth. Some data on the oxygen consumption by plant cells are given in Table 2. The oxygen uptake is dependent on the activity of the cells. Growing cells consume more oxygen than stationary-phase cells. Therefore, the uptake data for batch cultures show a larger variation than the data for continuous culture. The low uptake values of the batch cultures are from the stationary-phase cells. Cells in production media, which do not support growth, are comparable with the stationary-phase cells. The lower oxygen uptake values can be explained partly by the higher amount of storage products (e.g., starch) in the nongrowing cells. Those starchlike products increase the dry weight, but do not contribute to the respiration. Therefore, it is clearer to express the oxygen uptake rate per gram of active biomass. Active biomass is defined as the dry weight minus the storage products. The oxygen uptake per cell may vary a great deal, because the size of the cells is very variable. However, the larger cells contain a lot of inactive material, such as storage carbohydrates and vacuoles. Therefore, a rough estimate could be made by calculating the uptake rate for a young and small plant cell, which contains a large amount of active biomass. Assuming an average weight of 6×10^{-12} kg per cell, the oxygen uptake rate for a growing cell should be between 5 and 10×10^{-16} mol/cell/s. The effect of the oxygen concentration on growth can be roughly described by a Monod correlation. Growth limitation is observed mainly at dissolved oxygen concentrations below 10% saturation. In shake flask cultures this concentration can be reached easily because of the low gas exchange, especially at the end of exponential growth phase, when the cell density is high, as shown before. In a bioreactor the oxygen concentration can be maintained above the critical value by a high air flow rate and a high stirrer speed. However, negative effects of high air flow rate on biomass growth has been reported by several authors (14–17). In some studies the detrimental effects were attributed to an increased hydrodynamic stress (18,19). In most studies the negative effects of aeration have been attributed to the removal of carbon dioxide or other key gaseous metabolites (15–17,20). Literature data on the effect of carbon dioxide enrichment are not consistent. Enrichment sometimes diminishes the negative effects (14,20), but in several investigations an effect was not found. In at least one study a negative effect of a high carbon dioxide concentration was reported (15). Carbon dioxide is not the only metabolite of plant cells that is exchanged between liquid and gas phase. The plant hormone ethylene is a prominent example of a gaseous compound with a strong effect upon various metabolic processes. Other active compounds may play a less well-documented role here, too. The effects of gaseous compounds on growth of plant cells is comparable for different species, because of the general character of the primary metabolism. However, the variety of secondary pathways in plant species, with various functions and various control strategies, makes the effect of the concentration of gaseous metabolites on plant secondary metabolism much more unpredictable. There are only a few studies on the effect of the gaseous metabolite concentration on the biosynthesis of secondary metabolites by plant cells. These are not systematic studies, but the effects of gaseous metabolites may explain the results of such research. Lee and Shuler (21) investigated the effect of different flask closures, and thus the effect of gas exchange, on ajmalicine production by *Catharanthus roseus* cell cultures. They reported a lower ajmalicine production in shake flasks with a reduced gas exchange. Scragg et al. (22) investigated the effect of three different concentrations CO_2 in the incoming gas flow (1, 2, and 4%) on serpentine production by *C. roseus* in a 30-L air lift reactor. They found a positive correlation between the CO_2 concentration and serpentine production, but the final serpentine content of the cells was always lower than in the cultures for inoculation of the bioreactor. Fujita (23) and Matsubara et al. (4) found a reduced berberine production when increasing cell densities of *Coptis japonica* cultures were used. Matsubara et al. (4) concluded that an accumulation of metabolites was one of the two main reasons for the reduced berberine production.

In conclusion, it can be stated that a more appropriate control of the composition of the gaseous phase should enlarge the value of the bioreactor in plant cell biotechnology.

THE RECIRCULATION BIOREACTOR

The challenge is to set up a bioreactor system in which the concentration of oxygen and all gaseous metabolites of the cell culture can be controlled independently of the hydrodynamic shear regime in the reactor. The conventional method to control the oxygen concentration in a bioreactor is to manipulate the aeration rate and the stirrer speed. In that case shear conditions change and gaseous metabolites are removed from the culture. A more sophisticated approach was used by Smith et al. (24). In their concept, on-line measurements of dissolved oxygen and carbon dioxide concentrations were used in an adaptive feed-forward control strategy based on a mathematical model for mass transfer of both gases. The gas composition of the influent gas was controlled by a timed open/shutoff procedure on the In-line gas valves. A comparable system was described by Jay et al. (25) and Schlatmann et al. (26). They controlled the influent gas composition by controlling the flow through the mass flow meter devices (Fig. 3). The concentration of oxygen and carbon dioxide can be controlled in such systems, but unknown, biologicaly active gaseous metabolites of the cells are lost by ventilation of the culture. Schlatmann et al. (27) designed the recirculation bioreactor to keep these compounds in the culture under constant hydrodynamic conditions.

Experimental Setup

The setup is shown in Figure 4. A standard stirred and aerated bioreactor was adapted in such a way that a part of the exhaust gas was recirculated. Fresh air was introduced into the system between the recirculation pump and the sparger. The composition of the gas phase can be manipulated by the ratio between fresh air flow and total gas flow. In the next section the calculation to achieve a required gas regime in the bioreactor in shown. The system as depicted here is restricted in its possibilities by the fixed composition of air. Enlargement of the applications can be achieved by controlling the composition of the influent gas, as mentioned (24–26).

Calculation of the Flow of Fresh Air in the Recirculation Bioreactor

The recirculation bioreactor may serve various purposes. It could be the purpose to keep as much gaseous metabolites in the system as possible and at the same time keep the oxygen concentration high enough to support growth and/or production. Another option is to maintain a certain carbon dioxide concentration in the bioreactor. An interesting possibility is the comparison of shake flask and bioreactor experiments, keeping the concentration of gaseous metabolites comparable. This example is worked out here, but the calculation of the recirculation rate follows comparable lines for the various options. The starting point for the calculation of the flow of fresh air, ϕ_v, is the balances over the gas and liquid phase for the accumulated gaseous compound. The gas-phase balance of the recirculation bioreactor is given by

$$V_{GF}\frac{dC_G}{dt} = \phi_V(C_{OG} - C_G) - k_L a_F\left(\frac{C_G}{H_G} - C_L\right)V_{LF} \quad (1)$$

The variables used herein are defined as follows:

ϕ_V	flow of fresh air in the recirculation bioreactor (L/h)
C_{OG}	environmental concentration of the gaseous compound (mol/L)
C_G	gas-phase concentration of the gaseous compound (mol/L)
C_L	liquid-phase concentration of the gaseous compound (mol/L)
H_G	Henry coefficient for the gaseous compound (−)
$k_L a_F$	mass transfer coefficient in the bioreactor (h^{-1})
$k_L a_S$	mass transfer coefficient in the shake flask (h^{-1})
K_W	transfer coefficient of the shake flask closure (L/h)
r_G	production rate of the gaseous compound (mol/L/h)
t	time (h)
V_{GF}	gas volume in the bioreactor (L)
V_{LF}	liquid volume in the bioreactor (L)
V_{LS}	liquid volume in the shake flask (L)

When the pressure is assumed to be constant, the concentrations of the gaseous compounds can be expressed in moles per liter. The amount of recirculated air (recirculation rate) is not important for the calculation;

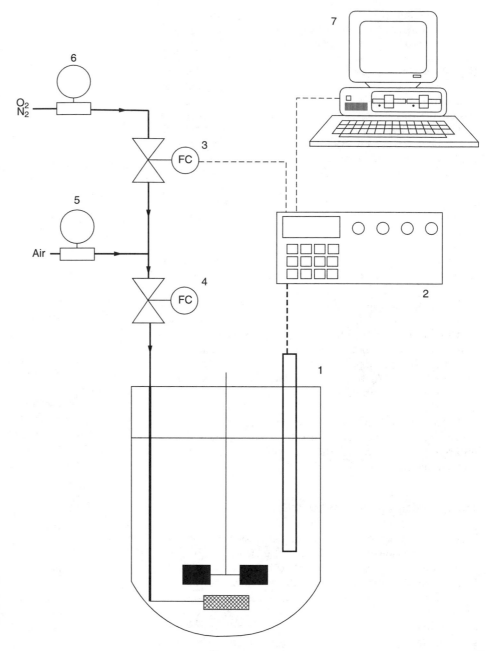

Figure 3. Setup for DO control in a stirred bioreactor. (1) O_2 electrode, (2) PID controller, (3,4) mass flow controller, (5,6) manometer, and (7) data acquisition system (26).

it only influences the k_La of the reactor. When (pseudo) steady state is assumed, Eq. (1) simplifies to

$$\phi_V(C_{OG} - C_G) = k_L a_F \left(\frac{C_G}{H_G} - C_L \right) V_{LF} \qquad (2)$$

The balance for the liquid phase is denoted by

$$V_{LF} \frac{dC_L}{dt} = k_L a_F \left(\frac{C_G}{H_G} - C_L \right) V_{LF} - r_G V_{LF} \qquad (3)$$

In (pseudo) steady state Eq. (3) is reduced to

$$k_L a_F \left(\frac{C_G}{H_G} - C_L \right) = r_G \qquad (4)$$

Combining Eqs. (2) and (4) yields the gas-phase concentration C_G

$$C_G = C_{OG} - \frac{r_G V_{LF}}{\phi_V} \qquad (5)$$

The liquid-phase concentration C_L follows directly from Eq. (4)

$$C_L = \frac{C_G}{H_G} - \frac{r_G}{k_L a_F} \qquad (6)$$

Finally, when Eqs. (5) and (6) are combined, the liquid-phase concentration C_L can be expressed as

$$C_L = \frac{C_{OG}}{H_G} - \frac{r_G V_{LF}}{\phi_V H_G} - \frac{r_G}{k_L a_F} \qquad (7)$$

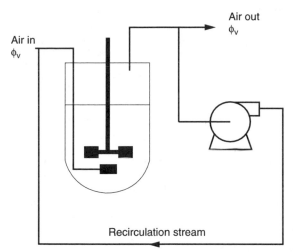

Figure 4. Recirculation bioreactor (26).

The same can be derived for the shake flask. The liquid-phase balance is the same. In gas-phase balance ϕ_V, $k_L a_F$, and V_F are substituted by K_W, $k_L a_S$, and V_S, respectively. The liquid-phase concentration C_L in a shake flask is then given by

$$C_L = \frac{C_{OG}}{H_G} - \frac{r_G V_{LS}}{K_W H_G} - \frac{r_G}{k_L a_S} \qquad (8)$$

To obtain the same situation for the cells in both culture systems, the liquid-phase concentrations C_L should be equal. As carbon dioxide is the most important accumulated gaseous compound (in absolute amounts), it was selected for the calculation. So, assuming that the gas-phase concentration of the environment C_{OG} and the CO_2 production rate r_G are the same in shake flask and bioreactor, combination of Eqs. (7) and (8) and rearrangement gives

$$\frac{V_{LF}}{\phi_V} = \frac{V_{LS}}{K_W} + \frac{H_G}{k_L a_S} - \frac{H_G}{k_L a_F} \qquad (9)$$

For the calculation of the flow of fresh air K_W, $k_L a_F$, and $k_L a_S$ had to be determined. The $k_L a$'s for oxygen were determined with both static measurement and the gassing-out method (with diluted apple sauce as a model suspension). The K_W of the silicon foam stoppers for oxygen was determined. The values of K_W and $k_L a$ for carbon dioxide can be calculated from the K_W and $k_L a$ for oxygen using the diffusion coefficients for the gas and the liquid phase, respectively (28). In a typical case a shake flask with a liquid volume V_{LS} of 0.06 L was compared with a bioreactor with a liquid volume V_{LF} of 1.8 L. The $K_L a$'s for the shake flask and the bioreactor, respectively, were estimated as 41 and 36 h^{-1}, the K_W of the silicon foam stoppers as 0.064 L/h. With a Henry coefficient H_G of 1.21, it can be calculated that a flow of fresh air of 1.9 L/h through the recirculation bioreactor will provide an optimal similarity with the dissolved gas composition of the shake flask. (In these calculations the reactions of CO_2 in water have been neglected, because in the plant cell culture the pH remains close to 5 throughout the

experiment. At this pH the amount of CO_2 that reacts with water is low.)

Applications of the Recirculation Bioreactor

The design of an industrial fermentation process proceeds through the following steps: preliminary laboratory experiments in shake flasks, optimization of the production parameters in a laboratory-scale fermenter, and scaleup to the production size in one or more steps. The production parameters can be divided into scale dependent and scale independent. If we look into the production of a secondary metabolite by a plant cell culture, the concentration of dissolved gaseous metabolites is a very important parameter, as pointed out in the introduction. Furthermore, the effect of this parameter is specific for each different product. Although the influence of a controlling factor on a metabolic pathway in itself is scale independent, the concentration of these specific controlling factors is scale dependent, because the concentration depends on mass transfer, mixing, and biomass concentration. A complicating factor is that the individual gaseous metabolites might be undefined but nevertheless very active. The recirculation bioreactor provides a solution for these problems and is an essential tool in the scaleup procedure. The gas regime in the shake flask can be mimicked in a bioreactor, the optimal concentration of the gaseous metabolites can be estimated for the production process, and the large-scale conditions can be set to achieve this optimal concentration in the production process. Two examples are briefly presented: the comparison between shake flask and bioreactor and the estimation of the optimal concentration of the solved gaseous metabolites for production of a secondary metabolite.

Example 1: Comparison between a Shake Flask and a Bioreactor Experiment. In a study (26) on the production of the secondary metabolite ajmalicine by the suspension culture of *C. roseus*, the production was optimized in shake flasks. It appeared that the production was almost completely lost, if the process was carried out in an aerated and stirred bioreactor. Application of the recirculation bioreactor, in which under equal shear conditions the gas-phase composition of the shake flask was mimicked, restored the productivity (Fig. 5). It proves that gaseous compounds produced by the cell suspension were essential for the ajmalicine production and shows the particular possibilities of this bioreactor design.

Example 2: Optimization of the Secondary Metabolite Production. As shown in Example 1, dissolved gaseous metabolites are essential for the production of ajmalicine by cell cultures of *C. roseus*. For the design of a production-scale process the optimum concentration has to be estimated. Schlatmann et al. (27) investigated this in a 3-L recirculation bioreactor with a working volume of 2 L. The recirculation flow was 60 L/h. The flow of fresh air introduced in the head space of the reactor was varied between 1.8 and 120 L/h. It was known from previous experiments that the oxygen concentration should be kept above 80% air saturation to optimize the ajmalicine production. If the oxygen concentration went below 80%

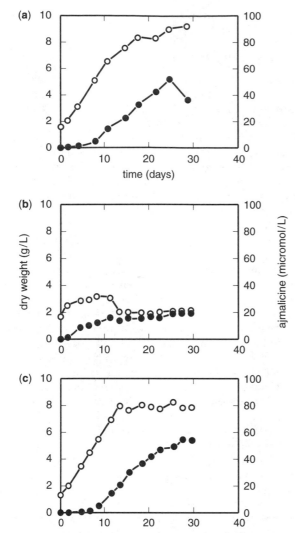

Figure 5. Time courses of biomass dry weight (○) and ajmalicine (●) production by *Catharanthus roseus* in (**a**) a shake flask, (**b**) a standard stirred and aerated bioreactor, and (**c**) a recirculation bioreactor (26).

Table 3. Dissolved Gaseous Metabolite Concentrations Expressed as Carbon Dioxide Concentration in the Recirculation Bioreactor and the Corresponding Ajmalicine Production by *Catharanthus roseus* Cells

Flow rate of fresh air introduced in the recirculation system	Dissolved gaseous metabolites concentration (mmol CO_2/L)	Specific production rate of ajmalicine by *C. roseus* cells (µ mol/g AB/d)
1.8	0.93–1.24	0.49
7.5	0.22–0.33	1.50
15	0.13–0.19	1.58
30	0.09–0.12	3.36
60	0.08–0.10	4.39
120	0.06–0.07	2.25

storage polysaccharides. From this result it is relatively easy to calculate the refreshment rate in the recirculation bioreactor at each scale of production.

BIBLIOGRAPHY

1. A.H. Scragg, E.J. Allan, and F. Leckie, *Enzyme Microb. Technol.* **10**, 361–367 (1988).

2. J.J. Meijer et al. *Enzyme Microb. Technol.* **16**, 467–477 (1994).

3. J.J. Meijer, H.J.G. ten Hoopen, K.Ch.A.M. Luyben, and K.R. Libbenga, *Enzyme Microb. Technol.* **15**, 234–238 (1993).

4. K. Matsubara et al. *J. Chem. Technol. Biotechnol.* **46**, 61–69 (1989).

5. H. Tanaka, *Process Biochem.* **22**, 106–113 (1987).

6. W.J. Treat, C.R. Engler, and E.J. Soltes, *Biotechnol. Bioeng.* **34**, 1191–1202 (1989).

7. D.I. Kim, G.H. Cho, H. Pedersen, and C.K. Chin, *Appl. Microbiol. Biotechnol.* **34**, 726–729 (1991).

8. B.S. Hooker, J.M. Lee, and G. An, *Biotechnol. Bioeng.* **35**, 296–304 (1990).

9. M. Jolicoeur, C. Chavarie, P.J. Carreau, and J. Archambault, *Biotechnol. Bioeng.* **39**, 511–521 (1992).

10. J.C. van Suijdam, N.W.F. Kossen, and A.C. Joha, *Biotechnol. Bioeng.* **20**, 1695–1709 (1978).

11. M. Moo-Young and Y. Chisti, *Bio/Technology* **6**, pp. 1291–1296 (1988).

12. A.K. Panda, S. Mishra, V.S. Bisaria, and S.S. Bhojwani, *Enzyme Microb. Technol.* **11**, 386–397 (1989).

13. A. Assa and R. Bar, *Biotechnol. Bioeng.* **38**, 1325–1330 (1991).

14. J.P. Ducos, G. Feron, and A. Pareilleux, *Plant Cell, Tissue Organ Cult.* **13**, 167–177 (1988).

15. P.K. Hegarty, N.J. Smart, A.H. Scragg, and M.W. Fowler, *J. Exp. Bot.* **37**, 1911–1920 (1986).

16. B. Maurel and A. Pareilleux, *Biotechnol. Lett.* **7**, 313–318 (1985).

17. N.J. Smart and M.W. Fowler, *Biotechnol. Lett.* **3**, 171–176 (1981).

18. H. Tanaka, *Biotechnol. Bioeng.* **23**, 1203–1218 (1981).

19. H. Tanaka, H. Semba, T. Jitsufuchi, and H. Harada, *Biotechnol. Lett.* **7**, 485–490 (1988).

20. J.P. Ducos and A. Pareilleux, *Appl. Microbiol. Biotechnol.* **25**, 101–105 (1986).

(as established by DO measurement) oxygen-enriched air was introduced without changing the total flow (to keep the shear conditions unchanged). Carbon dioxide was used as a marker for the total gaseous metabolite concentration, because it can be measured easily in the gas flow. The dissolved gaseous metabolite concentrations, expressed as dissolved CO_2 in these experiments, are shown in Table 3. The variation in the concentrations was caused by variation in the amount of biomass by sampling during the experiment. Less biomass means less production of gaseous metabolites. This has an effect on the concentrations, especially at lower flow rates of fresh air. The time courses of ajmalicine production were determined, and from these data the specific production rate was estimated. The results in Table 3 expressed as µmol ajmalicine/g active biomass/day, show a pronounced maximum at a relatively low concentration of dissolved gaseous metabolites. Active biomass (AB) means the metabolically active part of the biomass and is calculated as total dry weight minus intracellular

21. C.W.T. Lee and M.L. Shuler, *Biotechnol. Tech.* **5**, 173–178 (1991).

22. A.H. Scragg et al., in C. Webb and F. Mavituna, eds., *Plant and Animal Cells: Process possibilities*, Ellis Horwood, Chichester, England, 1987, pp. 77–91.

23. Y. Fujita, *Ciba Found. Symp.* **137**, 228–235 (1988).

24. J.M. Smith, S.W. Dawison, and G.F. Payne, *Biotechnol. Bioeng.* **35**, 1088–1101 (1990).

25. V. Jay, S. Genestier and J.-C. Courduroux, *Plant Cell Rep.* **11**, 605–608 (1992).

26. J.E. Schlatmann et al., *Biotechnol. Bioeng.* **41**, 253–262 (1993).

27. J.E. Schlatmann et al., *Biotechnol. Bioeng.* **44**, 461–468 (1994).

28. C.N. Satterfield *Mass Transfer in Heterogenous Catalysis*, Krieger Publ. Co., Huntington, N.Y., 1981.

See also ASEPTIC TECHNIQUES IN CELL CULTURE; BIOREACTOR OPERATIONS—PREPARATION, STERILIZATION, CHARGING, CULTURE INITIATION AND HARVESTING; BIOREACTORS, PERFUSION; OFF-LINE IMMUNOASSAYS IN BIOPROCESS CONTROL; ON-LINE ANALYSIS IN ANIMAL CELL CULTURE; PHYSIOLOGY OF PLANT CELLS IN CULTURE.

BIOREACTORS, STIRRED TANK

PAULINE M. DORAN
University of New South Wales
Sydney, Australia

OUTLINE

Bioreactors with mechanical stirring offer important advantages for large-scale culture of suspended plant cells. Greater power can be input to the culture compared with air-driven reactors; for viscous broths containing high cell densities, this translates into better mixing and oxygen transfer. The availability of two independent means of manipulating the hydrodynamic environment, adjusting the stirring speed and gas flow rate, also allows greater flexibility of reactor operation than in air-driven systems. Yet, these potential advantages may never be realized in full in any particular plant-cell application because the large shear forces generated by conventional impellers and the shear sensitivity of the cells limit the operating conditions that can be employed. Reactor engineering and innovation in this area are aimed at finding an appropriate balance between the beneficial and destructive effects of energy dissipation and hydrodynamic shear in plant-cell suspensions. To this end, a wide range of vessel, impeller, and sparger configurations has been applied in experimental studies. This article outlines the function of stirring equipment with reference to plant-cell systems and the effect of reactor operating conditions on culture performance.

PLANT-CELL PROPERTIES AND REACTOR ENGINEERING

The principal functions of a bioreactor used for aerobic cell culture are to provide adequate oxygen transfer, mixing, and heat transfer without the deleterious effects of excessive hydrodynamic shear. The properties of the biomass can impose significant technical constraints on the design of appropriate bioreactors, and this is particularly true for suspended plant cells. Unfortunately, it is difficult to generalize about several of the important engineering characteristics of plant suspensions. For example, some cultures produce viscous non-Newtonian broths (1–3), while others exhibit essentially Newtonian behavior (4) and only moderate viscosities (5,6). The tendency of plant cells to clump together varies considerably depending on the species and culture conditions (3,7–12), and there are differing reports about the extent to which plant cells are shear sensitive (12–16). Despite this variability, however, plant-cell cultures offer a potentially highly challenging combination of features unlike those encountered in microbial or animal-cell systems. When complex broth rheology and cell shear sensitivity are coupled together, providing adequate mixing, solids suspension, and oxygen transfer can be difficult. Although plant cultures also possess some ameliorating characteristics, such as relatively low oxygen demand and, consequently, cooling requirements, mixing and oxygen transfer are still of major concern. Developing technology to meet these challenges is critical for applications involving secondary metabolite production, because the low yields often associated with dedifferentiated plant cells mean that very high biomass densities are required for economic feasibility.

Stirred-tank bioreactors offer several advantages for culturing suspended plant cells. Although stirred, air-lift and bubble-column reactors achieve similarly adequate rates of mixing and mass transfer in low viscosity fluids, air-driven reactors do not perform well for high-density cultures of viscosity greater than about 100 mPa s (17). Stirred vessels are more suitable for viscous broths because greater power can be input by mechanical agitation. Experimental (5,18,19) and theoretical (20) studies have demonstrated that mixing becomes limiting in airlift reactors when plant-cell densities reach $20-30 \text{ kg m}^{-3}$ dry weight. At these concentrations, airlift performance deteriorates markedly due to poor gas disengagement, impeded liquid circulation, development of unmixed zones, and gravity settling of the cells. Increasing the aeration rate is the only operating response available to improve the hydrodynamic functioning of air-driven reactors; however, overventilation of plant-cell cultures has been shown to reduce growth and product formation because carbon dioxide and other necessary volatile components such as ethylene are stripped from the broth at high gas throughputs (21). Development of dense foam layers and cell flotation are also features of plant-cell cultures that are exacerbated at high aeration rates. Foaming should be avoided because its treatment ultimately requires adding chemical antifoam agents that can significantly reduce $k_L a$ values for oxygen transfer (22). Mechanical agitation and the ability to increase the power input independent of gas flow rate help reduce these problems in stirred-tank reactors.

As well as providing important technical benefits for viscous plant-cell culture, stirred bioreactors have a proven record of performance and reliability in the fermentation industry and have been well studied and characterized in large-scale operations. These features account for stirred tanks being the reactor configuration of choice for commercial scale-up of suspended plant-cell processes, as indicated in Table 1. Although application of stirred bioreactors is standard fermentation technology, conventional high-shear agitation is inappropriate for plant-cell culture. Modifications in equipment configuration and/or operating practice are usually required to avoid mechanical damage of the cells or other detrimental metabolic effects associated with high levels of hydrodynamic shear. The main parameters that affect shear levels in stirred vessels are the impeller geometry and stirrer speed; however, although many different styles of impeller have been tested experimentally, there are few general guidelines for optimizing this aspect of stirred-reactor design for suspended plant-cell culture (33). Because the generation of shear forces is intimately related to the effectiveness of mixing and mass transfer in bioreactors, development of systems for low-shear agitation which still provide adequate circulation and gas dispersion remains a key challenge.

EQUIPMENT AND OPERATING CHARACTERISTICS

The functions of stirring in a bioreactor are (1) to eliminate concentration and temperature gradients by homogenizing the vessel contents, (2) to disperse gas in the fermentation broth to aerate the culture, (3) to promote heat transfer between the broth and heat exchange surfaces, and (4) to suspend the cells. The equipment chosen for stirring has a significant influence on how well these functions can be performed. The principal mechanical features and operating details of a range of mixing hardware are outlined briefly in the following paragraphs. This description provides a background for the more detailed discussion of plant-cell applications presented in subsequent sections.

Stirred Vessels

Figure 1 shows a typical arrangement of equipment in a stirred tank (34). Not shown is the equipment for heat transfer: in large vessels this may be a helical cooling coil immersed in the broth; for small fermenters an external water jacket is often sufficient. Standard stirred-tank reactors are cylindrical in cross section. The base profile of the tank may be varied as illustrated in Figure 2; rounded rather than angled corners discourage the formation of stagnant pockets into which fluid currents cannot penetrate. The energy dissipation rate required to suspend solids in stirred tanks is very sensitive to the shape of the vessel base, and the modified geometries shown in Figure 2 can all significantly enhance the suspension of particles compared with flat-bottom tanks (35,36). Because the solubility of oxygen in aqueous liquids is limited to only about 8 ppm under typical fermentation conditions, the oxygen content of the culture broth must be continuously replenished as oxygen is consumed by the cells. In stirred vessels, this is most commonly achieved using a sparger which releases air bubbles into the culture below the impeller. When headplate access is at a premium or when mechanical stresses on the stirrer can be alleviated by using a shorter shaft, for example, in viscous fluids, the stirrer shaft may enter through the base of the vessel rather than the top, as shown in Figure 3. For either top-entering or bottom-entering stirrers, the point at which the stirrer shaft enters the tank is a potential contamination site, and several types of stirrer seal have been developed to combat this problem. Large fermenters are usually equipped with mechanical seals; for smaller vessels, magnetically coupled drives may be used so that the stirrer shaft does not actually pass through the fermenter body. In this case, a magnet attached to the end of the shaft inside the fermenter is driven by a rotating magnet in a housing outside. Use of magnetic coupling is limited to relatively small fermenters ($\leq \sim 800$ L) and low viscosity broths. In vessels with a single impeller of diameter $\frac{1}{4} - \frac{1}{2}$ the tank diameter, the depth of liquid is usually limited to $1.0-1.25$ times the tank diameter to facilitate good mixing. In tall fermenters with large aspect ratios, multiple impellers mounted on the stirrer shaft are commonly used to improve oxygen transfer conditions; an example is shown in Figure 4. In multiple impeller systems, the distance between the agitators should be $1.0-1.5$ impeller diameters. If the impellers are too far apart, unagitated zones develop between them; conversely, impellers located too close together produce flow streams that interfere with each other and disrupt circulation to the far reaches of the vessel.

Baffles

Baffles are vertical strips of metal mounted against the walls of stirred vessels (Fig. 1) to prevent gross vortexing of the liquid. Typically, four baffles of width $\frac{1}{10} - \frac{1}{12}$ the tank diameter are equally spaced around the tank circumference. Baffles may be attached to the wall (Fig. 4) or mounted slightly away (Fig. 1) to prevent sedimentation or development of stagnant zones at the inner edge. Baffles create turbulence in the fluid by breaking up the

Table 1. Stirred-Tank Reactors for Scale-Up of Suspended Plant-Cell Culture

Reactor volume (L)	Equipment and Operating Conditions	Plant Species	Company	Reference
75, 750, 7,500, 15,000 and 75,000	Cascade of five stirred tanks Semicontinuous operation	*Echinacea purpurea, Panax ginseng, Rauwolfia serpentina, Chenopodium* sp, and tobacco	DIVERSA Gesellschaft für Bio- und Verfahrenstechnik mbH, Germany	23
		Taxus spp	Phyton, Inc. and Bristol-Meyers Squibb Co, USA	B. Bringi (personal communication)
20,000 and 25,000	Two-stage operation Culture period of 4 weeks	*Panax ginseng*	Nitto Denko Corp., Japan	24,25
20,000	Batch, semicontinuous and continuous operation Dual four-blade paddles, 45° blade angle	*Nicotiana tabacum*	Japan Tobacco Inc., Japan	26
10,000	Semicontinuous operation Axial flow impeller	Various, for polysaccharide production	Cooperative Research Centre for Industrial Plant Biopolymers, Australia	D. McManus (personal communication)
6,000	High-density fed-batch culture	*Coptis japonica*	Mitsui Petrochemical Industries Ltd, Japan	27
5,000	Batch operation Triple six-flat-blade turbines	*Catharanthus roseus*	Gesellschaft für Biotechnologische Forschung mbH, Germany	28
4,000	Axial flow turbine with draft tube	*Polianthes tuberosa*	KAO Corp., Japan	H. Sawada (personal communication)
200 and 2,000	Batch operation Dual disc turbines	*Nicotiana tabacum*	Japan Tobacco Inc., Japan	26
800	Batch operation Dual six-flat-blade turbines	*Solanum aviculare*	Gesellschaft für Biotechnologische Forschung mbH, Germany	29
200 and 750	Two-stage culture	*Lithospermum erythrorhizon*	Mitsui Petrochemical Industries Ltd., Japan	30,31
100 and 630	Culture cycle of 14 days	Ginseng	Omutninsk Chemical Plant, Russia	32

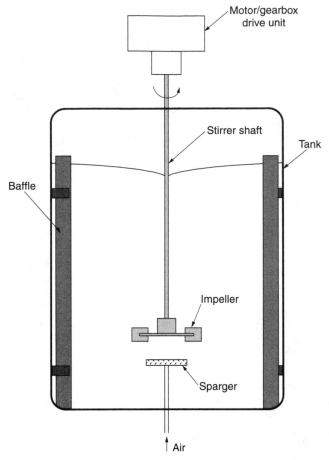

Figure 1. Typical equipment configuration for a stirred-tank bioreactor with single impeller used for aerobic cell culture.

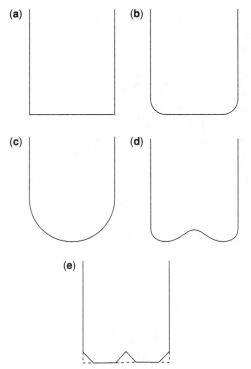

Figure 2. Bottom variations in stirred-tank bioreactors: (a) flat, (b) dished, (c) round, (d) contoured, and (e) cone and fillet.

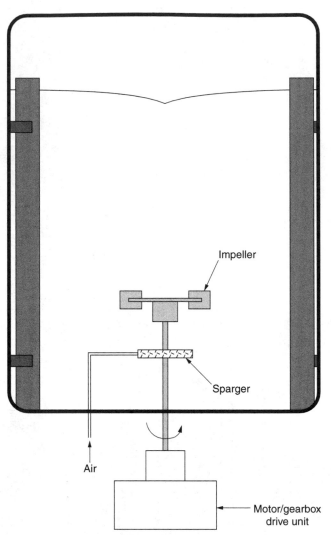

Figure 3. Stirred-tank bioreactor with bottom-entry stirrer.

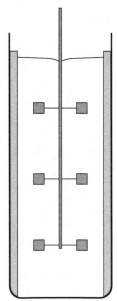

Figure 4. Stirred-tank bioreactor with multiple impellers.

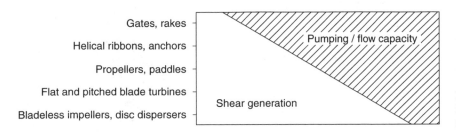

Figure 5. Characteristics of different impellers in terms of the ratio of shear generation to flow capacity.

circular flow generated by rotation of the stirrer. Circular flow is relatively ineffective for mixing, which is better facilitated by axial and radial currents. Circular flow can also lead to vortex development and gas entrainment. If allowed to reach down to the impeller, deep vortices impose extraordinarily high mechanical stresses on the stirrer shaft, bearings and seals, which must be avoided to prevent equipment damage.

Impellers

Impellers can be broadly classified depending on whether they generate high local shear rates, or whether they have strong pumping capacity for development of large-scale flow currents. Both functions are required for good mixing, but they do not usually work together. The characteristics of several impellers in terms of the relative prominence of their shear and flow generating functions are indicated in Figure 5.

Propellers, paddles, and turbines are remote-clearance impellers with typical diameters of $\frac{1}{4} - \frac{2}{3}$ the tank diameter and are operated at high speeds to generate tip velocities of the order of $3\ \mathrm{m\ s^{-1}}$. When remote-clearance agitators are used in low-viscosity fluids, a turbulent region of high shear is generated near the impeller where the essential mixing process occurs, as material is exchanged between different liquid streams. High shear levels close to the impeller are also responsible for bubble breakup in sparged systems. Because the mixing process should involve fluid from all parts of the vessel, the impeller must be able to generate circulatory currents with sufficient velocity to carry material from the furthermost reaches of the tank to the impeller. In viscous fluids, it is often not possible for mechanical and economic reasons to rotate the impeller fast enough to generate turbulence; impellers used for viscous mixing are designed to provide maximum bulk movement of material. Anchor and helical impellers are large agitators that are installed with small wall clearances and operated at low speeds to generate bulk fluid currents. For viscous fluids or when high shear rates must be avoided, slow-speed, low-turbulence, high-flow impellers are preferred to high-speed, high-turbulence, low-flow impellers.

There is a wide range of impellers available commercially, some of which have been applied to plant-cell culture. The general characteristics of agitators used in plant-cell reactors are described following.

Rushton Turbine. As shown in Figure 6, the Rushton turbine is a disc turbine with six flat blades. This impeller has been the agitator of choice in the fermentation industry since the 1950s, largely because it has been well studied

Figure 6. Rushton turbine.

and characterized. Rushton impellers generate predominantly radial or horizontal flow, but have relatively low pumping efficiency resulting in low circulatory flows per unit power consumed (37,38). Rushton turbines are often chosen for their gas handling capacity and perform well for gas dispersion. Bubble breakup is very effective near the impeller blades as air is forced into these high shear regions by the turbine disc. Changes in gas dispersion conditions with increasing stirrer speed or gas flow rate are illustrated in Figure 7 (39). At high gassing rates or low stirrer speeds (Fig. 7a), the gas-handling capacity of the impeller is exceeded, the impeller ceases to pump the two-phase fluid, and the flow pattern is dominated by buoyant gas-liquid flow up the middle of the vessel as in a bubble column. This condition corresponds to impeller flooding. As the stirrer speed is increased or the gas flow rate reduced, the impeller starts to disperse the gas towards the walls of the tank (Fig. 7b). Complete gas dispersion

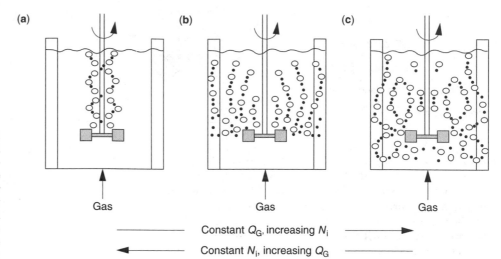

Figure 7. Flow patterns developed by a Rushton turbine under gassed conditions as a function of stirrer speed N_i and gas flow rate Q_G: (**a**) impeller flooding; (**b**) impeller loading, where the gas is dispersed by the impeller but poorly distributed throughout the vessel; (**c**) complete gas dispersion below as well as above the impeller. (Adapted from Ref. 39.)

below as well as above the impeller (Fig. 7c) is achieved at even higher stirrer speeds or lower gas flow rates; this is the desirable operating condition because it results in homogeneous dispersion of gas to all parts of the vessel. The ability to handle high gas velocities without flooding is an advantage for improving oxygen transfer conditions in the reactor. The gas handling capacity of Rushton turbines can be improved by increasing the impeller to tank diameter ratio (40).

Pairs of low-pressure, high-speed trailing vortices are generated behind the flat blades of Rushton turbines (41,42). This hydrodynamic feature is responsible for the relatively high level of power drawn by Rushton turbines under nonaerated conditions. With sparging, gas readily accumulates in these low-pressure regions, producing ventilated cavities behind the blades and causing a significant drop in power. The extent of this power reduction is a complex function of the stirrer speed, air flow rate, vessel size and fluid properties, and cannot yet be accurately predicted (43). The rheological properties of the fluid exert considerable influence: practical operating conditions usually give a 40–50% loss of power with sparging in low-viscosity fluids (44), but this can increase to 80% or more in viscous or non-Newtonian fluids (40).

Rushton turbines are effective for solids suspension in three-phase systems. An impeller clearance above the floor of the tank of $\frac{1}{4}$ the tank diameter is recommended because this allows gas dispersion under the impeller as well as adequate agitation in the upper parts of the vessel (35,45). Suspension of solids is generally improved by reducing the impeller off-bottom clearance, but this can cause flow instabilities in aerated systems (46).

Single and multiple Rushton turbines have been applied to suspended plant-cell culture in several experimental investigations, as indicated in Table 2. The suitability of Rushton turbines for large-scale plant-cell culture has also been analyzed in a theoretical engineering study based on a reactor size of 10 m³ (33).

Other Disc Turbines. Disc turbines may have more or fewer blades than the Rushton turbine's standard six. Increasing the number of blades up to eighteen reduces

the rate at which the power drops with increasing gas flow and significantly improves the gas-handling capacity of the impeller (95). The unaerated power number also increases with the number of blades at constant impeller diameter (96); even so, a twelve-blade disc turbine has been shown to disperse about three times more gas than a larger Rushton impeller operating with the same power consumption and stirrer speed (95). Although turbine impellers with more than six blades have not been tested for plant-cell culture, single and dual agitators with four flat blades have been used in several studies (Table 2). As an extrapolation of the above findings, decreasing the number of blades from six to four can be expected to reduce the gas-handling capacity of the impeller, as well as the unaerated power number.

The blades on disc turbines may be pitched or angled rather than flat. This reduces the unaerated power number by about 50% and generates impeller discharge currents with a more pronounced downward velocity component than the radial flows from a Rushton turbine (46). At a given gas flow rate, lower stirrer speeds can be used without flooding compared with Rushton turbines of similar geometry; this has been attributed to the greater projected area of the pitched blades (38). The impeller discharge efficiency and gas holdup produced using pitched blade disc turbines are also improved relative to Rushton impellers (38). Disc turbines with angled blades have been applied to plant-cell culture (Table 2).

Propellers. Propellers are usually of the three-blade marine type shown in Figure 8 (see page 261). They generate a flow pattern with mainly axial velocity components. Propellers can be operated in either downward or upward pumping modes; downward is more common. They are generally installed with diameters around $\frac{1}{3}$ the tank diameter.

The unaerated power number for marine propellers is about $\frac{1}{10}$ that of Rushton turbines, depending on the blade pitch and other geometric parameters (96). With gassing, significant power losses may occur; aerated downward pumping propellers can also exhibit flow and power draw instabilities (46). Propellers are very effective for solids

Table 2. Examples of Stirred-Reactor Configurations Applied to Suspended Plant-Cell Culture[a]

Impeller	Reactor Specifications	Impeller Specifications	Stirrer Speed (rpm)	Sparger Specifications	Aeration Rate (vvm)	Reactor Operation	Maximum Cell Density	Species	Reference
Rushton turbine	$V_L = 2$ L $D_T = 15$ cm Concave sides, flat bottom No baffles	$D_i/D_T = 0.50$ $W_B/D_i = 0.32$ $L_B/D_i = 0.32$ $D_D/D_i = 0.60$	1,000	—	—	Batch	6.4 g L^{-1} dry weight	*Catharanthus roseus*	47
Rushton turbine	$V_L = 1.45$ L	$D_i = 9$ cm	100–300	—	0.16	Chemostat, dilution rate = 0.35 d^{-1}	1.36 g L^{-1} dry weight (at 150 rpm)	*Daucus carota*	48
Dual Rushton turbines	$V_L = 8.5$ L $D_T = 23$ cm Concave sides, flat bottom	$D_i/D_T = 0.37$ $W_B/D_i = 0.22$ $C_S/D_T = 0.33$ $C_{i2}/H_L = 1$	100	Cross-shaped glass, 5-cm arms, two rows of 2-mm holes per arm; or 8-cm sintered glass disc, pore size 3	0.24	Batch	14 g L^{-1} dry weight (with cross-shaped sparger)	*Catharanthus roseus*	8
Dual Rushton turbines	$V_L = 11$ L $D_T = 23$ cm Concave sides, flat bottom No baffles	$D_i/D_T = 0.35$ or 0.52	100	Crucifix or sintered disc	0.009–0.2	Batch	14 g L^{-1} dry weight	*Catharanthus roseus*	49
Dual six-flat-blade disc turbines	Volume = 12 L $D_T = 23$ cm Concave sides, flat bottom No baffles, or three linked internals or baffles $W_{BF}/D_T = 0.08$	$D_i/D_T = 0.37$ $W_B/D_i = 0.035$ above disc $W_B/D_i = 0.21$ below disc $C_{i2}/H_L = 1$	300	In unbaffled vessel, sintered glass, 3-cm diameter; in baffled vessel, one of the three internals supplied air	—	Batch	13.6 g L^{-1} dry weight (unbaffled)	*Catharanthus roseus*	50
Dual disc turbines: six blades lower, four blades upper	$V_L = 20$ L $D_T = 28.4$ cm Three baffles Foam breaker	$D_i/D_T = 0.50$	50	Cross-shaped stainless steel pipe with 21 × 1.0–1.5 mm holes	0.75–1.1	Semicontinuous	18.1 g L^{-1} dry weight	*Nicotiana tabacum*	51
Dual disc turbines	Volume = 2,000 L $D_T = 125$ cm	$D_i/D_T = 0.34$	10–100	—	—	Batch	12 g L^{-1} dry weight	*Nicotiana tabacum*	26
Six-blade disc turbine with inclined vanes	$V_L = 2$ L $D_T = 15$ cm Concave sides, flat bottom No baffles	$D_i/D_T = 0.50$ Blades inclined 60° from the vertical	1,000	—	—	Batch	7.7 g L^{-1} dry weight	*Catharanthus roseus*	47

(continued)

Table 2. Continued

Impeller	Reactor Specifications	Impeller Specifications	Stirrer Speed (rpm)	Sparger Specifications	Aeration Rate (vvm)	Reactor Operation	Maximum Cell Density	Species	Reference
Angled disc turbine	$V_L = 25$ L	—	100	—	0.25	Fed-batch	17 g L^{-1} dry weight	*Panax ginseng*	52
Six-flat-blade turbine	$V_L = 2.5$ L, $D_T = 15$ cm, $H_L/D_T = 0.93$, Four baffles	$D_i/D_T = 0.56$, $W_B/D_i = 0.20$	115	Ring, 8 holes	1	Batch	17 g L^{-1} dry weight	*Vinca rosea*	53
Six-flat-blade turbine	$V_L = 1.8$ L, $D_T = 13$ cm, $H_L/D_T = 1.2$, Round bottom, Four baffles, $W_{BF}/D_T = 0.11$	$D_i/D_T = 0.35$, $W_B/D_i = 0.24$, $C_i/D_T = 0.23$	300	—	0.11	Batch	12.3 g L^{-1} dry weight	*Atropa belladonna*	54
Six-blade impeller	$V_L = 3$ L, $D_T = 15$ cm, Concave sides, flat bottom	$D_i/D_T = 0.49$	480	—	0.033	Batch	8.2 g L^{-1} dry weight	*Helianthus annuus*	55
Six-blade turbine with magnet stirring plate	$V_L = 0.55$ L, $D_T = 8.5$ cm, $H_L/D_T = 1.14$, Flat bottom, Three baffles, $W_{BF}/D_T = 0.09$	$D_i/D_T = 0.59$, $W_B/D_i = 0.16$, $C_i/D_T = 0.47$	200–500	Surface aeration	0.18	—	—	*Carthamus tinctorius*	56
Dual six-flat-blade turbines	$V_L = 800$ L	—	60–80	—	0.063	Batch	135 g L^{-1} fresh weight	*Solanum demissum*	29
Dual six-flat-blade turbines	Volume = 13.5 L	—	200	Ring, 6 × 1-mm pores	0.2	Batch	15 g L^{-1} dry weight	*Perilla frutescens*	57
Dual six-blade turbines	$V_L = 1.8$ L, Three baffles	—	250	Sintered steel	0.33	Batch with gas recirculation	8.1 ± 0.6 g L^{-1} dry weight	*Catharanthus roseus*	58
Dual six-blade turbines	$V_L = 2$ L, Three baffles, $H_{BF} = 14$ cm, $W_{BF} = 1.4$ cm	$D_i = 4.5$ cm	250	Sintered steel	0.2	Two-stage batch	14 g L^{-1} dry weight	*Catharanthus roseus*	59
Dual six-blade turbines	$V_L = 1.8$ L	—	125	Sintered steel	0.28	Chemostat, dilution rate = 0.19 d^{-1}	2.8 g L^{-1} dry weight	*Catharanthus roseus*	60
Triple six-flat-blade turbines	$V_L = 5,000$ L	—	30	—	0.010–0.017	Batch	11 g L^{-1} dry weight	*Catharanthus roseus*	28
Four-blade turbine	$V_L = 3$ L, $D_T = 14$ cm	$D_i/D_T = 0.56$, $W_B/D_i = 0.19$, $C_i/D_T = 0.36$	150	—	0.5	Batch	26 g L^{-1} dry weight	*Nicotiana tabacum*	61

Table 2. Continued

Impeller	Reactor Specifications	Impeller Specifications	Stirrer Speed (rpm)	Sparger Specifications	Aeration Rate (vvm)	Reactor Operation	Maximum Cell Density	Species	Reference
Four-flat-blade impeller	$V_L = 3$ L $D_T = 14$ cm	$D_i/D_T = 0.54$ $W_B/D_i = 0.67-1.8$ $L_B/D_i = 0.33$ $C_i/D_T = 0.27$	50–200	—	0.2	Batch	580 g L^{-1} fresh weight (at 150 rpm and $W_B/D_i = 1.8$)	*Nicotiana tabacum*	13
Four-flat-blade impeller	$V_L = 3$ L	$D_i = 5.6$ cm	150	One-hole nozzle	0.7	Batch	18.5 g L^{-1} dry weight	*Nicotiana tabacum*	62
Four-blade impellers	$V_L = 1.7$ L	—	100	—	0.12	Batch	4.6 g L^{-1} dry weight	*Daucus carota* (embryogenic)	63
Dual four-flat-blade turbines	$V_L = 3$ L $D_T = 14.8$ cm	$D_i/D_T = 0.51$	150–400	—	0.2	Batch	13 g L^{-1} dry weight	*Perilla frutescens*	64
Dual four-blade impellers	$V_L = 1.8-3.6$ L	—	80	—	0.3	Two-stage chemostat	6.0 g L^{-1} dry weight (at dilution rate = 0.10 d^{-1})	*Dioscorea deltoidea*	65
Dual flat-blade turbines: six blades lower, four blades upper	$V_L = 13$ L	$D_i = 10$ cm	100	27 × 1.0–1.8-mm holes	—	Batch	15 g L^{-1} dry weight	*Nicotiana tabacum*	66
Dual flat-blade turbines: six blades lower, four blades upper	$V_L = 34$ L $D_T = 35$ cm Magnetically coupled stirrer shaft	$D_i/D_T = 0.50$ $C_{i1}/D_T = 0.37$ $C_{i2}/D_T = 0.77$	100	—	0.7	Chemostat, dilution rate = 0.42 d^{-1}	15 g L^{-1} dry weight	*Nicotiana tabacum*	67
Flat-blade and 45° angle-blade turbines	$V_L = 10$ L	—	80	Three extra-coarse glass frits	Up to 0.6 (air) and up to 0.2 (pure oxygen)	Batch	8.2 g L^{-1} dry weight	*Dioscorea deltoidea*	68
Triple turbines	Volume = 5 L	—	90	—	1.0	Batch	6.7 g L^{-1} dry weight	*Digitalis lanata*	69
Sail impeller	$V_L = 3$ L $D_T = 14$ cm	$D_i/D_T = 0.54$ $W_B/D_i = 1.8$ $L_B/D_i = 0.33$ $C_i/D_T = 0.27$	150	—	0.2	Batch	590 g L^{-1} fresh weight	*Nicotiana tabacum*	13
Dual six-flat-blade impellers connected by crossed Teflon ribbons	$V_L = 4$ L Baffles Bottom-entry stirrer shaft	$W_R = 2.5$ cm $L_R = 22$ cm	40	—	0.075	Batch	2.7 g L^{-1} dry weight	*Fragaria × ananassa* cv. Brighton	70

(continued)

Table 2. Continued

Impeller	Reactor Specifications	Impeller Specifications	Stirrer Speed (rpm)	Sparger Specifications	Aeration Rate (vvm)	Reactor Operation	Maximum Cell Density	Species	Reference
Marine blade impeller	$V_L = 0.5$ L Magnetically coupled stirrer shaft	—	150–180	—	1.2	Batch	9 g L^{-1} dry weight	*Medicago sativa*	71
Marine-type impeller	$V_L = 50$ L	$D_i = 15$ cm	150–420	—	0.20–0.45	Multicycle draw-fill	29 g L^{-1} dry weight	*Panax ginseng*	72
Marine three-blade impeller	$V_L = 2$ L $D_T = 13$ cm $H_L/D_T = 1.15$ Round bottom	$D_i/D_T = 0.35$ $H_i/D_i = 1.33$ $C_i/D_T = 0.23$ Blade pitch = 45°	100	Pipe with row of 7×1-mm holes	0.43	Batch	11.8 g L^{-1} dry weight	*Nicotiana tabacum*	73
Marine impeller	$V_L = 3$ L $D_T = 14.3$ cm	$D_i/D_T = 0.59$	120–170	—	0.033	Batch	15.5 g L^{-1} dry weight	*Perilla frutescens*	16
Propeller	$V_L = 2.3$ L		80–140	Bubble-free aeration	—	Semicontinuous perfusion	26 g L^{-1} dry weight	*Anchusa officinalis*	74
Kaplan turbine	$V_L = 75$ L Draft-tube reactor		350	—	0.33	Batch	15.3 g L^{-1} dry weight	*Morinda citrifolia*	12
Perforated disc	Volume = 10 L		100	—	0.5	Batch	—	*Morinda citrifolia*	12
Six-blade paddles	$V_L = 4$ L Round bottom No baffles		90–160	Hollow stirrer shaft	0.25–0.5	Batch	8.4 g L^{-1} dry weight	*Daucus carota* (embryogenic)	75
Modified (gate-type) paddle	$V_L = 5$ L $D_T = 20$ cm $H_L/D_T = 0.80$ No baffles	$D_i/D_T = 0.67$ $W_B/D_i = 0.60$	60	Ring, 8 holes	0.5–1.5	Fed-batch	30 g L^{-1} dry weight	*Cudrania tricuspidata*	18
Modified (gate-type) paddle	$V_L = 2$ L $D_T = 13$ cm $H_L/D_T = 1.15$ Three baffles $W_{BF}/D_T = 0.12$ $H_{BF}/H_L = 0.20$	$D_i/D_T = 0.54$ $H_i/D_i = 1.0$ $C_i/H_L = 0.20$	175 or 280	Ring	0.5 or 0.25	Batch	9 g L^{-1} dry weight (at 175 rpm and 0.5 vvm)	*Catharanthus roseus*	76
Modified (gate-type) paddle	$V_L = 2$ L	—	50	—	1.0	Batch	11.6 g L^{-1} dry weight	*Panax ginseng* (embryogenic)	77
Modified paddle (hollow stirring wing)	Volume = 2.5 L	—	—	—	0.05 (oxygen)	Fed-batch	75 g L^{-1} dry weight	*Coptis japonica*	78
Two-blade grid paddle	$V_L = 2$ L $D_T = 13$ cm $H_L/D_T = 1.23$ Four baffles	$D_i/D_T = 0.53$ $H_i/D_i = 2.1$	50	Spiral, 16 holes $D_S/D_T = 0.46$	0.3	Fed-batch	30 g L^{-1} dry weight	*Oryza sativa*	6

Table 2. *Continued*

Impeller	Reactor Specifications	Impeller Specifications	Stirrer Speed (rpm)	Sparger Specifications	Aeration Rate (vvm)	Reactor Operation	Maximum Cell Density	Species	Reference
Dual angled four-blade paddles	Volume = 20,000 L D_T = 250 cm	D_i/D_T = 0.50 Blade angle = 45°	10–40	—	—	Batch	17 g L^{-1} dry weight	*Nicotiana tabacum*	26
Anchor impeller	Volume = 15 L D_T = 22 cm	D_i/D_T = 0.50 H_i/D_i = 2.5	—	—	—		—	*Fragaria ananassa*	79
Anchor turbine	V_L = 25 L	—	100	—	0.25	Fed-batch	15.3 g L^{-1} dry weight	*Panax ginseng*	52
Anchor impeller	V_L = 32 L D_T = 25.9 cm H_L/D_T =~ 3	D_i/D_T = 0.74	40	—	0.6	Batch	—	*Coleus blumei*	19,80
Double helical ribbon impeller (upward pumping)	V_L = 10 L D_T = 21 cm Profiled base Three vertical surface baffles, 8 × 10 × 0.16 cm	D_i/D_T = 0.90 W_R/D_i = 0.195 L_R/D_i = 4.12	120	Surface aeration (air and oxygen-enriched air)	0.1	Batch	27.6 g L^{-1} dry weight	*Catharanthus roseus*	2
Double helical ribbon impeller (upward pumping)	V_L = 10 L D_T = 21 cm Profiled base Three vertical surface baffles, 8 × 10 × 0.16 cm	D_i/D_T = 0.90 W_R/D_i = 0.195 L_R/D_i = 4.12	60	Surface aeration (air and oxygen-enriched air)	0.05	Batch	10.4 g L^{-1} dry weight	*Eschscholtzia californica* (embryogenic)	81
Helical agitator	V_L = 32 L D_T = 25.9 cm H_L/D_T =~ 3	D_i/D_T = 0.89 Seven helices Helix spacing = 0.32 D_i	50	Ring, 6 × 1.5-mm holes D_S/L = 0.39	0.6	Batch	—	*Coleus blumei*	80
Helical stirrer (downward pumping)	—	—	100–300	—	—	Batch	—	*Berberis wilsoniae*	82
Helical stirrer	V_L = 27 L	—	100	—	0.15–0.74	Batch	27 g L^{-1} dry weight	*Digitalis lanata*	83
Spin stirrer	V_L = 27 L	—	100	—	0.15–0.55	Batch	27 g L^{-1} dry weight	*Digitalis lanata*	83

(continued)

Table 2. *Continued*

Impeller	Reactor Specifications	Impeller Specifications	Stirrer Speed (rpm)	Sparger Specifications	Aeration Rate (vvm)	Reactor Operation	Maximum Cell Density	Species	Reference
Cell-lift impeller	V_L = 1.25 or 2.5 L	—	60	Ceramic stone, off-center	—	Batch	6.5 g L^{-1} dry weight	*Glycine max* (photomixotrophic)	84
Cell-lift impeller	Volume = 5 L Dished base	—	—	Sintered stainless steel	—	Perfusion	27.6 g L^{-1} dry weight	*Thalictrum rugosum*	85
Membrane stirrer	V_L = 19 L	Membrane tubing wound into a basket configuration	30–50	Bubble-free aeration with pulses of bubbling through the membrane tubing (air and oxygen-enriched air)	—	Fed-batch	50 g L^{-1} dry weight	*Thalictrum rugosum*	86
Taylor–Couette membrane stirrer	V_L = 2.2 L Magnetically coupled stirrer	D_i/D_T = 0.53	200	Bubble-free aeration	0.95	Batch	93 g L^{-1} fresh weight	*Beta vulgaris*	87
Spin filter	Magnetically coupled stirrer	—	—	—	—	Batch	10.3 g L^{-1} dry weight	*Daucus carota*	88
Magnetic stirrer bar	V_L = 3 L Dual carboys	D_i = 7.5 cm	90	Stainless steel, porosity 10 μm	1.5	Batch	5.2 g L^{-1} dry weight	*Mentha spicata*	89
Magnetic stirrer bar	V_L = 2 L Round-bottom flask Stirrer stabilizer rod	—	125	Fritted glass	0.025	Chemostat, dilution rate = 0.36 d^{-1}	—	*Glycine max*	90
Magnetic stirrer bar	V_L = 20 L	C_i = 1 cm	200	—	0.05	Batch, turbidostat and chemostat	7.3 g L^{-1} dry weight (batch)	*Acer pseudoplatanus*	91
Magnetic stirrer bar	V_L = 4 L Round-bottom flask	D_i = 6 cm	260	Pipe (1-mm diameter) or sintered glass, porosity 2	0.125	Turbidostat	3 g L^{-1} dry weight	*Acer pseudoplatanus*	92
Magnetic stirrer bar	V_L = 3.5 L	—	—	—	—	Multicycle draw-fill	1.7 g L^{-1} dry weight	*Haplopappus gracilis*	93
Double-bar magnetic stirrer	V_L = 2 L V-shaped flask	—	200–300	—	—	Multicycle draw-fill	8.4 g L^{-1} dry weight	*Phaseolus vulgaris*	94

^{a}Symbols are defined in the Nomenclature section.

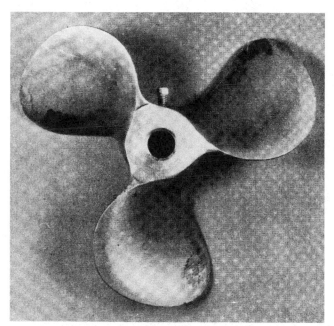

Figure 8. Marine propeller.

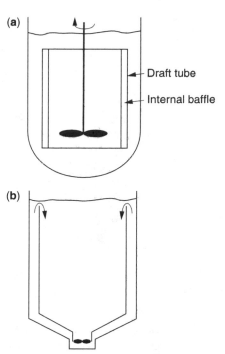

Figure 9. Axial flow impellers with draft tube. (**a**) Vessel with draft tube and internal baffles. The internal baffles prevent vortexing and liquid draw-down into the draft tube. (Adapted from Ref. (98).) (**b**) Draft tube reactor with Kaplan turbine used for culture of *Morinda citrifolia* plant cells. (Adapted from Ref. 12.)

suspension with low power consumption and outperform disc turbines and paddles in this respect (97). Several investigators have used marine propellers for suspended plant-cell culture (Table 2).

Axial Flow Impellers with Draft Tube. The use of a draft tube with an axial flow impeller such as a marine propeller can significantly improve particle suspension in systems containing solids (97,98). Two configurations for this type of stirring system are shown in Figure 9. The position and size of the draft tube have a strong influence on the hydrodynamic conditions generated. For example, if the bottom of the tube restricts the impeller outflow, the loss of local pressure causes a marked increase in the energy dissipation rate required for solids suspension (97). The power saved by installing draft tubes to assist particle suspension is reported to be greatest at low solids concentrations (98). An axial-flow Kaplan turbine in a 75-L vessel with a draft tube performed satisfactorily for plant-cell culture (Table 2) and compared well with other impeller configurations (12).

Perforated Discs. Rotating perforated discs are most commonly employed in two-phase liquid extraction systems, including aqueous extraction of biological molecules such as proteins (99,100). This type of contactor is very effective for high-shear liquid droplet dispersion and is associated with high mass transfer rates and separation efficiencies. Application of perforated discs for stirring in bioreactors is not well characterized in the literature; however, a perforated disc impeller operated at a speed of 100 rpm was tested by Wagner and Vogelmann (12) for culturing *Morinda citrifolia* plant cells in a 10-L vessel. Its performance in terms of biomass and secondary metabolite production was similar to that of dual flat-blade turbines and a Kaplan turbine with draft tube.

Intermig Impellers. The Intermig is an impeller design of Ekato GmbH, Germany. As shown in Figure 10, Intermig impellers are used in dual configuration, orthogonal to each other and positioned on the stirrer shaft about one impeller diameter apart. Each impeller has two large blades connected at the hub; each large blade supports smaller blades at its ends. The smaller blades may be either unslotted (Fig. 10a) or slotted to form four wing tips (Fig. 10b). The large and small blades pump in opposite directions; the recommended arrangement is for the small outer blades to pump downwards while the central blades pump upwards. Because the unaerated power number for each impeller is very low (about 0.35; Ref. 101), at least two Intermigs with diameter 0.6–0.7 times the tank diameter would be used instead of a single Rushton turbine.

The flow produced by ungassed Intermig impellers is very chaotic (102), and energy dissipation rates are reported to be more uniformly distributed throughout the vessel than with Rushton turbines (103). With aeration, the flow becomes radial due to the development of large ventilated cavities at the tips of each blade. These cavities extend back into the liquid, increase in size with increasing gas flow rate, and reduce the power draw by up to about 50% (40,43). Mixing of non-Newtonian fluids has been found to be equally satisfactory with Intermig and Rushton impellers under a range of operating conditions, with the presence or absence of slots on the blades having little effect (104). The energy dissipation rates required for particle suspension by dual Intermigs are higher than for Rushton turbines and significantly greater than for downward pumping axial flow impellers (36).

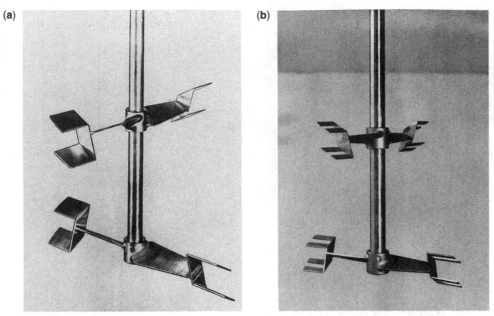

Figure 10. Dual Intermig agitators with (**a**) unslotted wing tips and (**b**) slotted wing tips. (Photographs provided courtesy of Ekato Rühr- und Mischtechnik GmbH, Germany.)

Unslotted Intermig impellers have been tested for pilot- and large-scale suspended plant-cell culture (P. Forschner, Ekato GmbH, personal communication; Ref. 23). Advantages of low shearing force, good mixing, good gas dispersion, and low energy consumption have been cited (23), but more detailed quantitative information was not provided.

Paddles. As shown in Figure 11a, flat-blade paddles are simple in design and construction and commonly have two or four blades. Paddles generate radial flow patterns similar to those produced by Rushton turbines (105,106), and are intermediate between circulation- and dispersive-type impellers in their function and operating characteristics. They are often applied as large-blade-area, low-speed agitators that push or carry liquid around the vessel. Under these conditions, no high-speed liquid streams are generated, and little top to bottom mixing takes place (107). Flat-blade paddles perform poorly compared with other impellers in terms of gas distribution, gas holdup, power consumption, and discharge efficiency, and have been reported to be unsuitable for use in gas-liquid systems (38). Their small projected blade area means that they flood relatively easily. Particle suspension efficiency in large-scale tanks is also lower with two-flat-blade paddles than with Rushton turbines (106).

A wide variety of modified paddle geometries is used in industry for mixing applications; some examples are illustrated in Figures 11b and 11c. These impellers are usually employed at low speeds for mixing viscous liquids. Several semiconventional paddle configurations such as modified gate and hollow stirring wing paddles have been investigated for use in plant-cell reactors (Table 2). In these studies, the stirring speeds were higher (up to 630 rpm; Ref. 18) than those commonly used in viscous fluids, and cell densities of 75 g L^{-1} dry weight were supported without sedimentation (78). Oxygen transfer

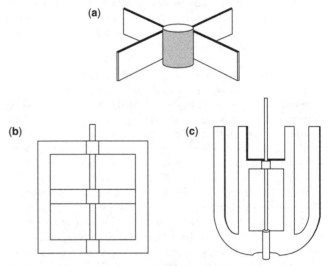

Figure 11. Flat-blade paddle and examples of modified paddle geometries: (**a**) four-flat-blade paddle, (**b**) frame- or wing-type paddle, (**c**) gate- or grid-type paddle.

rates with a wing-type paddle were increased by enriching the air supply with oxygen because raising the gas flow rate reduced biomass levels (78).

Anchors. Anchor impellers such as that shown in Figure 12 are applied when circulatory rather than dispersive mixing is required. Often used with high viscosity fluids for laminar blending and heat transfer applications, anchors are typically large-diameter agitators operated with small wall clearance (around 1–5% of the tank diameter) and low speed (5–20 rpm). While producing highest shear levels near the wall, anchors generate a relatively uniform distribution of energy dissipation rates compared

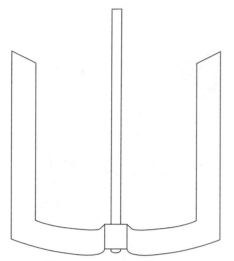

Figure 12. Anchor impeller.

mixing is required (110); breakup of bubbles sparged into the vessel is also not very effective. Wall clearance exerts a strong influence on the power requirements for mixing with anchors (111).

Anchors with relatively small impeller to tank diameter ratios of 0.50–0.74 have been tested for plant-cell culture at speeds up to 100 rpm (Table 2), but have not performed as well as other agitators for secondary metabolite production (52,80).

Helical Ribbon Impellers. Helical ribbon impellers are large-diameter agitators with spiral-like motion operated at small clearance from the vessel wall to promote fluid mixing and minimize the formation of stagnant zones. Depending on the direction of rotation, upward or downward pumping can be achieved. As shown in Figure 13, helical impellers may be constructed with either single or double helices. Like anchor impellers, helical ribbons are usually operated at low speeds and shear rates and have lower gas dispersing capacity than turbines and other high-speed agitators. In non-Newtonian fluids, oxygen transfer depends more strongly on gas flow rate than power input or stirrer speed, reflecting the limited ability of helical impellers to disperse air bubbles (112). Aeration may be facilitated, however, by installing surface baffles (113). Circumferential flows are dominant in stirred tanks fitted with helical agitators; however, upward flows are also generated by the outer helix and downward flows are prominent along the

with small-diameter impellers (105). However, by creating primarily tangential flows, anchor impellers do not provide strong top to bottom mixing, although secondary axial flows also develop at higher rotational speeds (108). Fluid near the stirrer shaft may be relatively stagnant during anchor operation because there is little mixing from near the wall to the center of the vessel (109). Therefore, anchor impellers are not recommended if completely uniform

(a)

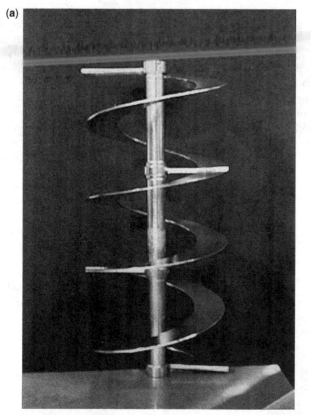

(b)

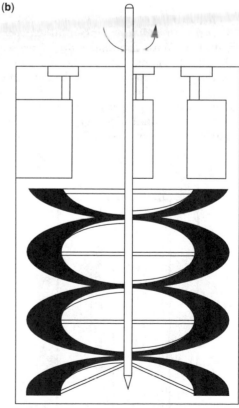

Figure 13. (**a**) Helical ribbon impeller. (Photograph provided courtesy of B. Braun Biotech International, Germany.) (**b**) Bioreactor with double helical ribbon impeller and three surface baffles. (Adapted from Ref. 2.)

stirrer shaft (114). Changes in impeller geometry can have a significant effect on mixing (115); for example, the mixing time may be doubled by increasing the blade width at constant impeller diameter (116). Overall, helical impellers give a positive top to bottom turnover without unmixed regions, thus representing a significant advantage over anchor agitators. Power requirements are similar to those for anchor impellers (117).

Helical impellers have been used for plant-cell culture in several studies (Table 2). Biomass concentrations of 27–28 g L^{-1} dry weight were supported using either air sparging below the impeller (83) or surface aeration only with oxygen-enriched air and surface baffles (2). For large-scale applications, oxygen transfer may present some difficulties. The success of helical impellers for plant-cell culture has been attributed to their high pumping capacity and low mechanical shear forces; however, the investment costs associated with this type of stirrer are greater than with conventional equipment (2).

Spin Stirrers. As illustrated in Figure 14, spin stirrers consist of vertical stirrer blades that rotate independently on a stirring assembly. The length of the blades ensures that as much fluid as possible is kept in motion by the action of the stirrer. There is little general information in the literature about the performance characteristics of spin stirrers; however, they have been used for suspended plant-cell culture (Table 2) at biomass concentrations up to 27 g L^{-1} dry weight (83). When applied for culture of *Berberis wilsoniae* cells, production of protoberberine alkaloids was lower than with a bladed turbine (82).

Cell-Lift Impellers. These novel impellers comprise a draft tube topped with radial or tangential exit ports, as shown in Figure 15. With rotation, reduced pressure is created at the tips of the exit ports, drawing liquid from the bottom of the reactor up the draft tube and out to the sides of the vessel for low-shear mixing. Mixing may be enhanced by installing a sparger at the base of the draft tube to generate additional airlift-type circulation (85). Cell-lift impellers have been tested for plant-cell culture

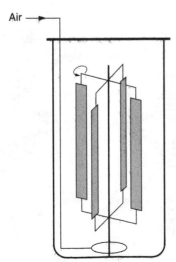

Figure 14. Bioreactor with spin stirrer, each blade capable of individual rotation. (Adapted from Ref. 83.)

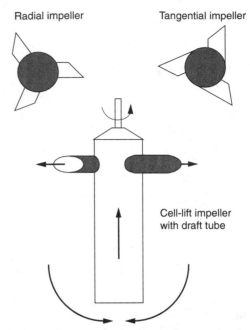

Figure 15. Cell-lift impeller, showing radial and tangential exit port configurations. (Adapted from Ref. 84.)

(Table 2). It was found in these studies that the tangential port configuration provided better lift than the radial design (84).

Membrane Stirrers. Membrane stirrers are used for bubble-free aeration when the effects of gas sparging, such as cell shear damage or foaming, must be avoided. To date, the major application has been in animal-cell culture. Aeration is achieved by gas exchange through silicone or microporous polypropylene tubing immersed in the culture broth. For bubble-free aeration, the gas pressure inside the tubing must not exceed the bubble point, at which bubbles appear on the outside of the tube walls. Several different membrane stirrer devices have been developed (118,119). To prevent the cells from settling and to promote mixing, the tubing may be kept in motion as an effective stirrer; this also enhances mass transfer of oxygen from the air or oxygen-enriched air in the tubing to the culture fluid, and facilitates penetration of liquid into the inner sections of the tubing coil. Reactors with membrane stirrers have been used to culture plant cells to high density (50 g L^{-1} dry weight) with little foaming and no cell flotation (Table 2).

An alternative reactor configuration using gas-permeable membranes and based on the Taylor-Couette flow principle has also been designed for suspended plant-cell culture (Table 2). In this reactor, a perforated cylindrical glass tube covered with silicone membrane sheeting was rotated as a primitive type of stirrer generating vortex flow in the annulus between the tube and the walls of the fermenter. The inside of the glass tube was maintained at high pressure relative to the annulus by using an air pump, so that bubble-free aeration of the culture was achieved. Some sedimentation of plant cells occurred in this reactor, prompting the investigators to suggest that the vessel could be operated with its axis horizontal (87).

Spin Filters. Spin filters are used in bioreactors for continuous perfusion culture at high cell densities. They have been applied mainly for intensive culture of animal cells. A simple spin filter system is shown in Figure 16. Culture fluid is removed from inside the spin filter with minimal cell loss in the spent medium, while fouling is reduced by rotating the filter unit. Because spin filters are not designed to perform the normal mixing and gas dispersion functions of an impeller, problems with oxygen transfer may occur unless other devices, such as turbine agitators or membrane tubing aeration, are used to enhance mass transfer rates (120). As indicated in Table 2, spin filters have been used for culture of plant cells (88).

Magnetic Stirrers. Many of the first attempts to cultivate suspended plant cells in bioreactors involved the use of Teflon-coated magnetic stirrer bars driven by external drives at the base of the culture vessel (Table 2). However, because only limited power can be input by magnetic stirrers and the liquid currents generated are relatively weak, mixing was most likely inadequate in these reactors except at low cell densities. Magnetic stirrer bars are usually placed on the bottom of the vessel and rotate in contact with the reactor floor, so that plant-cell damage by abrasion in this region is also highly probable. This problem was avoided in one study by suspending the stirrer bar 1 cm from the bottom of the vessel using a flexible Teflon stalk (91). The usual position of magnetic stirrers also means that gas must be sparged into the liquid from above the stirrer. Because contact between the bubbles and agitator is minimized by this arrangement, gas dispersion and therefore oxygen transfer may be severely restricted. This is reflected in the relatively low biomass densities supported by plant-cell reactors with magnetic stirrer bars, as indicated in Table 2.

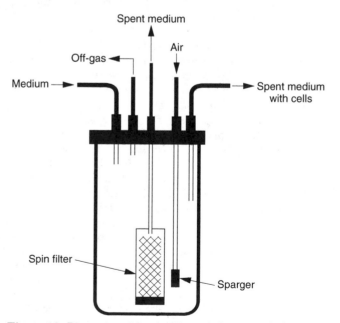

Figure 16. Bioreactor with spin filter unit for removal of medium in perfusion culture. The reactor also has provision for continuous whole broth removal. (Adapted with permission from Ref. 88.)

Spargers

Several types of sparger are in common use. Porous spargers of sintered metal, glass, or ceramic are applied mainly in small-scale applications; gas throughput is limited because the sparger poses a high resistance to flow. Cells growing through the fine holes and blocking the sparger can also be a problem. Orifice spargers, also known as perforated pipes, are constructed by making small holes in piping which is then fashioned into a ring or cross and placed at the base of the reactor. The individual holes on orifice spargers should be large enough to minimize blockages. Point or nozzle spargers are used in many agitated fermenters from laboratory to production scale. These spargers consist of a single open or partially closed pipe providing a point-source stream of air bubbles. Advantages compared with other sparger designs include low resistance to gas flow and small risk of blockage.

The design and location of the sparger have an important influence on gas-liquid hydrodynamics and power consumption in stirred vessels. With Rushton turbines, the fall in aerated power can be reduced by installing a large sparger, for example, a ring sparger with diameter 1.2 times that of the impeller, rather than a small sparger (121). Use of a large ring sparger affects the formation of ventilated cavities behind the agitator blades, thus allowing a 50% increase in the aeration rate that could be used without flooding compared with a point or small ring sparger. The effect of sparger geometry is even more pronounced with downward pumping axial flow impellers: the use of point spargers increases the likelihood and severity of torque and power instabilities that can occur with these impellers in gassed systems (40). The gas-handling ability of downward pumping agitators can be improved and the stirrer speed required for complete solids suspension reduced by replacing point spargers with ring spargers of diameter about 0.8 times the impeller diameter, and increasing the separation between the sparger and impeller from 0.1 to up to 0.2 times the tank diameter (40,122).

PLANT-CELL CULTURE IN STIRRED BIOREACTORS

Studies of suspended plant-cell culture in stirred vessels have been carried out for at least 35–40 years. A wide variety of reactor, impeller and sparger geometries has been tested in batch, fed-batch, perfusion, draw-fill, chemostat, and turbidostat operations. A summary of the equipment and operating conditions used and the maximum cell densities achieved is shown in Table 2. The list includes results from vessels with working volumes of 0.5–20,000 L equipped with single, dual, or a stack of three impellers on the stirrer shaft.

Stirrer systems for plant-cell culture have usually been evaluated in terms of biological rather than physical parameters, that is, cell growth and alkaloid levels are measured in preference to more direct indicators of stirrer performance, such as mixing time, power consumption, or gas holdup. Most papers that describe reactor culture of plant cells do not provide complete details of the vessel internals; for example, the number and width of the baffles, the diameter and position of the impeller, the type and position of the sparger, the vessel diameter, and the liquid height are often not specified. Yet, all of these geometric variables influence the effectiveness of fluid circulation and bubble dispersion in aerobic fermenters. Despite the large number of experimental studies carried out, few general-purpose guidelines have emerged for designing and operating stirrers for plant-cell culture (33). The lack of widely accepted design principles reflects in part the inherent complexity of stirring hydrodynamics, especially for three-phase systems such as aerobic cell suspensions, as well as the variation in critical properties, for example, rheology, cell shear sensitivity, and level of aggregation, among different plant-cell suspensions.

Comparative Studies of Reactor Hardware: Impellers, Baffles, and Spargers

Several comparative studies have been carried out using the same cell line and vessel configuration to determine which impeller design produces the best outcome in terms of plant-cell growth and/or production of valuable metabolites. These investigations are summarized in Table 3. The results highlight the difficulty associated with drawing general conclusions about impeller performance in plant-cell systems because the outcomes obtained using the same or similar impellers varied considerably in different studies.

Baffles in stirred tanks can be expected to increase levels of hydrodynamic shear as well as reduce vortexing and improve mixing. Experiments have been carried out to examine the effect of baffles on plant-cell growth and product formation. Total protoberberine levels (g L^{-1}) produced by *Berberis wilsoniae* cells in a reactor stirred with a flat-blade turbine were about 35% greater with baffles than without (82). In other work with a cell-lift impeller, plant cells tended to accumulate in the space between the baffles and vessel wall, resulting in poor local mixing on both sides of the baffles. Removal of the baffles alleviated these fouling problems but did not significantly affect growth; only slight differences in biomass were observed for reactors with and without baffles (84). For *Catharanthus roseus* cultured in a 12-L reactor with dual 8.5-cm flat-blade turbines, the presence or absence of baffles did not have a significant effect on cell growth at most stirrer speeds. Alkaloid levels measured as mg g^{-1} dry weight were similar in baffled and unbaffled vessels at 100 rpm; however at 700 rpm, specific alkaloid levels were 3.0–3.5 times higher in the baffled tank (50).

The effect of different sparger designs on *Catharanthus roseus* cultures was investigated using a 12.5-L bioreactor with dual Rushton turbines (8). A sintered glass disc sparger of diameter 8 cm gave higher k_La values than a glass cross-shaped sparger with two rows of 2-mm holes, when operated at the same aeration rate and stirrer speed. These spargers were employed at various air flow rates to produce initial k_La values ranging from 2.5 to 20 h^{-1} at a constant stirrer speed of 100 rpm. The highest biomass level was achieved with the sintered sparger at an intermediate initial k_La of 12.5 h^{-1}; k_La

values above 12.5 h^{-1} obtained by using the same sparger depressed growth. In a cell-lift reactor, replacing a single orifice sparger with a porous ceramic air stone resulted in the generation of smaller bubbles and an increase in k_La; changing the sparger position to slightly off-center also improved mixing in the vessel (84). A new type of spiral sparger with fine holes was used to produce small bubbles in a 2-L fermenter equipped with a two-blade grid paddle (6). High k_La values of 20 h^{-1} were obtained at a stirrer speed of 50 rpm, whereas 235 rpm was required to achieve the same k_La using a dual flat-blade turbine system and conventional ring sparger.

Mechanical Shear

Stirred-tank reactors have been used in several studies to assess the sensitivity of plant cells to mechanical forces, and to determine the restrictions this property imposes on reactor operating conditions (15,16,50,56,76). Plant-cell responses to short- (2–24 h) and long-term shear forces have been examined, and various criteria used to evaluate shear effects. In some studies, the problem was viewed in terms of threshold values of shear rate (15,16,123) or shear stress (123,124), above which the growth rate, cell viability, product yield, or some other critical culture parameter is significantly altered. For this approach to be useful, the shear conditions present during cell culture must be quantified. Several methods have been applied to estimate shear levels in stirred bioreactors used for plant-cell culture. Profiles of velocity and shear rate in an 11-L reactor stirred with dual Rushton turbines were determined using flow simulation methods; at a stirring rate of 100 rpm, shear rates varied from a maximum of 470 s^{-1} near the impeller to less than 5 s^{-1} in the bulk of the liquid (49). In other work, average and maximum shear rates were evaluated as a function of stirrer speed, impeller diameter, tank diameter, and blade width using correlations from the literature (47,50). However, because the hydrodynamic conditions in stirred reactors are ill defined and spatially variable, it is difficult to predict the actual levels of shear experienced by the cells and the durations of exposure. Another limitation associated with the concept of threshold shear rates or shear stresses for cell damage is that different threshold values may apply, depending on the period of exposure of the cells to the shear. If the response of plant cells to shear depends on the duration as well as the intensity of the shear forces, a parameter which takes the time of exposure into account would provide a better basis for predicting shear damage. Recently, the cumulative energy dissipated on the cells has been used as a correlating factor for shear effects in stirred plant-cell reactors (33,48,54). This parameter can be calculated as the integral of the average power dissipated on the cells with respect to time. Under turbulent conditions, growth of *Atropa belladonna* suspensions in a baffled bioreactor stirred with a six-flat-blade turbine was observed to decline rapidly at energy dissipation values of about 10^8 J m^{-3} (54). Sublytic effects such as changes in aggregate size, protein release, and mitochondrial activity may occur at lower dissipation energies (54,125).

Table 3. Results from Comparative Studies of Different Impellers Applied to Suspended Plant-Cell Culture[a]

Impeller and Operating Conditions	Performance Indicators	Reactor Specifications	Reactor Operation	Plant Species	Reference
Disc turbine (100 rpm, 0.25 vvm)	14.4 g L^{-1} dry weight cells and 50.1 mg L^{-1} saponins produced	V_L = 25 L	Fed-batch, 28 days	*Panax ginseng*	52
Anchor turbine (100 rpm, 0.25 vvm)	15.3 g L^{-1} dry weight cells and 43.5 mg L^{-1} saponins produced				
Angled disc turbine (100 rpm, 0.25 vvm)	17.0 g L^{-1} dry weight cells and 40.1 mg L^{-1} saponins produced				
Six-flat-blade turbine (D_i = 6.4 cm, W_B = 1.4 cm, 225 rpm, 1.0 vvm)	$k_L a$ = 42 h^{-1} at a cell concentration of 11 g L^{-1} dry weight. Mixing and gas dispersion poor at cell densities above 20 g L^{-1} dry weight	V_L = 1 L D_T = 11.3 cm H_L = 10 cm Ring sparger, 8 holes	—	*Cudrania tricuspidata*	18
Modified (gate-type) paddle (D_i = 7.0 cm, H_i = 5.0 cm, 400 rpm, 1.0 vvm)	$k_L a$ ≈ 52 h^{-1} at a cell concentration of 11 g L^{-1} dry weight. Mixing and gas dispersion satisfactory at cell densities above 20 g L^{-1} dry weight				
Rushton turbine (D_i = 7.5 cm, 1,000 rpm)	Biomass doubling time = 7.5 days. Maximum cell concentration = 6.4 g L^{-1} dry weight. Maximum alkaloid concentration = 10.4 mg L^{-1}	V_L = 2 L D_T = 15 cm Concave sides, flat bottom No baffles	Batch, up to 14 days	*Catharanthus roseus*	47
Vaned six-blade disc turbine, blades inclined 30° from the vertical (D_i = 7.5 cm, 1,000 rpm)	Biomass doubling time = 6.1 days. Maximum cell concentration = 6.8 g L^{-1} dry weight. Maximum alkaloid concentration = 5.5 mg L^{-1}				
Vaned six-blade disc turbine, blades inclined 60° from the vertical (D_i = 7.5 cm, 1,000 rpm)	Biomass doubling time = 4.5 days. Maximum cell concentration = 7.7 g L^{-1} dry weight. Maximum alkaloid concentration = 24.2 mg L^{-1}				
Flat-blade turbine (with baffles)	Concentration of protoberberine alkaloids produced = 3.0 g L^{-1}	Volume = 30 L	Batch	*Berberis wilsoniae*	82
Flat-blade turbine (without baffles)	Concentration of protoberberine alkaloids produced = 2.2 g L^{-1}				
Helical stirrer (upward pumping)	Concentration of protoberberine alkaloids produced = 2.3 g L^{-1}				
Spin stirrer	Lower maximum protoberberine alkaloid concentration than with the flat-blade turbine and baffles				

(continued)

Table 3. Continued

Impeller and Operating Conditions	Performance Indicators	Reactor Specifications	Reactor Operation	Plant Species	Reference
Four-flat-blade impeller ($D_i = 5.6$ cm, $W_B = 1.5$ cm, 100–200 rpm, 0.2 vvm)	No growth at 200 rpm due to shear damage Poor mixing at 100 rpm Maximum cell concentration = 520 g L^{-1} fresh weight after 12 days at 150 rpm	$V_L = 3$ L $D_T = 14$ cm	Batch	*Nicotiana tabacum*	13
Large four-flat-blade impellers ($D_i = 7.6$ cm, $W_B = 5.1$–14 cm, 50–200 rpm, 0.2 vvm)	Growth rate increased with increasing blade width Maximum cell concentration = 580 g L^{-1} fresh weight after 7–8 days at 150 rpm with $W_B = 14$ cm Some cell damage				
Sail impeller ($D_i = 7.6$ cm, $W_B = 14$ cm, nylon cloth blades, 150 rpm, 0.2 vvm)	Longer lag phase Maximum cell concentration = 590 g L^{-1} fresh weight after 9 days at 150 rpm No cell damage Higher power requirements than for the large flat-blade impellers				
Dual Rushton turbines (60 rpm)	Maximum cell concentration = 75 g L^{-1} fresh weight after 15 days	$V_L = 2.0$–2.5 L	Batch	*Glycine max*	84
Dual marine impellers (60 rpm)	Maximum cell concentration = 15 g L^{-1} fresh weight after 11 days				
Cell-lift impeller (60 rpm)	Maximum cell concentration = 82 g L^{-1} fresh weight after 13 days Reduction in average biomass aggregate size				
Dual turbines (100 rpm, 0.5 vvm)	Lower biomass yield and productivity but higher anthraquinone yield than perforated disc	Volume = 10 L	Batch	*Morinda citrifolia*	12
Perforated disc (100 rpm, 0.5 vvm)	Lower anthraquinone yield but higher biomass yield and productivity than dual turbines				
Dual flat-blade turbines: six blades lower, four blades upper ($D_i = 6.5$ cm, $W_B = 1.4$ cm, 235 rpm, ring sparger, 0.3 vvm)	Maximum cell concentration = 9.5 g L^{-1} dry weight after 29 days Biomass yield from sugar = 0.29 g g^{-1} Cell damage observed	$V_L = 2$ L $D_T = 13$ cm $H_L = 16$ cm Four baffles	Batch, 29 days	*Catharanthus roseus*	6
Two-blade grid paddle ($D_i = 6.9$ cm, $H_i = 14.8$ cm, 50 rpm, spiral sparger, 0.3 vvm)	Maximum cell concentration = 19 g L^{-1} dry weight after 25 days Biomass yield from sugar = 0.43 g g^{-1} No cell damage observed				
Internal turbine (400 rpm, 1.0 vvm)	14.5 g dry weight cells and 65.7 mg total ginsenoside produced	$V_L = 2$ L	Batch, 42 days	*Panax ginseng* (embryogenic)	77
Paddle (50 rpm, 1.0 vvm)	23.2 g dry weight cells and 77.7 mg total ginsenoside produced				

Table 3. Continued

Impeller and Operating Conditions	Performance Indicators	Reactor Specifications	Reactor Operation	Plant Species	Reference
Dual vaned six-flat-blade disc turbines (D_i = 12 cm, W_B = 1.8 cm, vane width = 0.3 cm, 100 rpm)	Biomass doubling time = 2.5 days Maximum cell concentration = 11.1 g L^{-1} dry weight Maximum alkaloid concentration = 21.8 mg L^{-1}	V_L = 7.5 L D_T = 23 cm H_L = 18.5 cm Concave sides, flat bottom No baffles C_{i1} = 7.5 cm C_{i2} = 18.5 cm	Batch	*Catharanthus roseus*	47
Dual vaned six-blade disc turbines, blades inclined 60° from the vertical (D_i = 8.5 cm, 100 rpm)	Biomass doubling time = 1.3 days Maximum cell concentration = 11.9 g L^{-1} dry weight Maximum alkaloid concentration = 25.8 mg L^{-1}				
Anchor (D_i = 19 cm, 40 rpm, 0.6 vvm)	Maximum rosmarinic acid concentration = 2.1 g L^{-1}	V_L = 32 L D_T = 25.9 cm H_L = ~ 78 cm	Batch	*Coleus blumei*	80
Helical agitator (D_i = 23 cm, 50 rpm, 0.6 vvm)	Maximum rosmarinic acid concentration = 3.3 g L^{-1}				
Spin stirrer (100 rpm, 0.15–0.55 vvm)	After 20 days, 25 g L^{-1} dry weight cells and 776 mg L^{-1} β-methyldigoxin produced	V_L = 27 L	Batch	*Digitalis lanata*	83
Helical stirrer (100 rpm, 0.15–0.74 vvm)	After 20 days, 23 g L^{-1} dry weight cells and 720 mg L^{-1} β-methyldigoxin produced				
Rushton turbine (D_i = 0.78 or 1.15 m, W_B = 0.16 or 0.23 m)	Reaction not limited by mixing and oxygen transfer as determined by analysis of characteristic times; complete gas dispersion; complete solids suspension; no cell shear damage	V_L = 10, 000 L D_T = 2.3 m H_L = 2.3 m Four baffles Ring sparger	Continuous, steady-state	Not specified — theoretical study	33
Pitched blade turbine, downward pumping (D_i = 0.92 or 1.15 m, W_B = 0.18 or 0.23 m)					
Pitched blade turbine, upward pumping (D_i = 0.92 or 1.15 m, W_B = 0.18 or 0.23 m)	The upward pumping pitched blade turbine showed greatest potential for effective impeller operation according to the above criteria; the larger Rushton turbine was also promising.				

[a]Symbols are defined in the Nomenclature section.

Varying degrees of shear damage are known to be inflicted on plant cells by different types of impeller (6,13). If cumulative energy dissipation, as outlined above, is a satisfactory correlating parameter for shear damage irrespective of impeller design, it can be expected that of the remote-clearance impellers, low-power-number agitators such as pitched blade turbines, propellers, and hydrofoil impellers would be less damaging at the same stirrer speed and impeller diameter than high-power-number devices such as Rushton turbines. However, recent work with microbial cultures and a range of different impellers has indicated that the hydrodynamic forces generated in fermenters and their shear effects on cells do not depend solely on overall energy dissipation levels. Additional factors, such as the distribution of energy dissipation rates within the vessel, the sweep volume of the impeller, and the trailing vortex structure behind the impeller blades, are responsible for generating local shear conditions close to the impeller where most cell damage takes place. In addition, impellers that generate strong circulation currents transport cells to the impeller region more frequently than agitators that do not promote rapid bulk liquid turnover, thus increasing the risk of shear effects. A correlating procedure for shear damage that takes all of these factors into account has been developed by Jüsten et al. (126,127) for mycelial cultures and was applied recently to plant-cell systems in a theoretical study (33). The utility of this approach for plant cultures remains to be tested experimentally.

The cell aggregate size distribution in plant-cell cultures can vary significantly with stirring conditions and has been used as an indicator of shear effects in stirred reactors (6,15,47,54,56). Changes in aggregate size due to action of the impeller can influence growth and productivity in the culture. Not only is mass transfer of oxygen and nutrients affected (128,129), but a range of plant-cell functions and properties including secondary metabolite synthesis, phenolic acid accumulation, enzyme activity, and levels of RNA, DNA, and protein has been found to depend on aggregate size (129–131). Forced breakup of aggregates, whether by shattering or attrition, may also affect viability if the plasmodesmata or cell walls are damaged. Although mechanisms of aggregate disruption have been investigated and predictive models developed for mycelial cultures in stirred reactors (132–134), similar analyses have not been carried out for plant-cell systems. The factors affecting plant aggregate size appear to be complex; for example, aggregates may be larger in the presence of high shear levels (8), and complex cycles of aggregate growth, plasmolysis, and necrosis have been described in *Nicotiana tabacum* cell cultures under different agitation and aeration conditions (26). Lack of understanding in this area means that it is difficult to control cell aggregation in agitated reactors; some models of cell damage predict that the effects of hydrodynamic shear are reduced with increasing scale (126). Consequently, it may not be possible to reproduce aggregate sizes at different equipment scales, thus affecting the success of plant reactor scale-up.

Mixing

Mixing is necessary in reactors to avoid zones of nutrient depletion. In vessels used for cell culture, the biomass represents a sink for substrates, and the sparger and medium inlet ports are nutrient sources. Unless the cells are brought into regular and frequent contact with nutrients through good mixing, they experience deleterious physiological effects that cause a decline in overall culture performance. In the worst case, if bulk liquid velocities are very low, the cells settle out of suspension. This creates a region of high nutrient demand, rapid depletion of substrates, and high local concentration gradients at the bottom of the reactor. Because of the larger size of plant cells compared with microorganisms and their tendency to form macroscopic aggregates, cell settling is of greater concern in plant-cell cultures than in most other fermentation processes.

Problems with incomplete mixing in stirred plant cell reactors have been reported in several studies (12,13,18,84). For continuous reactor operation, especially when culture overflow is used to maintain a constant liquid volume, unless special attention is given to the effluent removal, the cell concentration in the exit stream may be significantly different from that inside the vessel due to gradients of cell density developed in the broth (60,135–137). In most investigations of plant-cell reactors, the adequacy of mixing has been judged qualitatively rather than measured, based on visual observation of the absence of gross biomass settling or bulking. Although carried out for plant cells in airlift contactors (138), monitoring of spatial biomass distributions in stirred reactors has not been reported. However, even in plant-cell broths that appear homogeneous, rates of mixing may not be adequate relative to the rates of other processes in the reactor, such as oxygen uptake and oxygen transfer. Because oxygen transfer takes place mainly in the impeller region in large stirred vessels, the time taken by cells to travel in bulk flow currents from the impeller through the remainder of the tank and back to the oxygen supply is an important consideration in assessing mixing performance.

The rate of mixing is generally quantified experimentally using tracer injection techniques (17,34), whereby the response of the system to a chemical or physical stimulus is recorded. The mixing time is the time required to achieve a given degree of homogeneity after the stimulus is imposed; the greater the mixing time, the slower is bulk mixing in the vessel, and the greater the likelihood that zones of nutrient depletion can develop. Mixing times for stirred plant-cell reactors have been measured in a limited number of experimental investigations. Using a pulse of hot water as a tracer, Leckie et al. (50) estimated the mixing time in an unbaffled stirred tank of working volume 8.5 L used to culture *Catharanthus roseus* cells. With dual flat-blade turbines operated at 100 rpm, the mixing time determined from the measured temperature fluctuations was 3.6 s. In other work with an upward pumping double helical ribbon impeller, the mixing time was determined by using an iodine decolorization method (2). In a flat-bottom vessel with a working volume of 11 L without baffles mixing times of 18–25 s were measured as the stirrer speed was decreased from 150 to 120 rpm. Contouring the vessel base and installing surface baffles significantly reduced these mixing times to 4–9 s.

The value of the mixing time can be used to determine whether mixing in a reactor is adequate relative to the rate of oxygen transfer or the cellular oxygen demand. Regime analysis or comparison of characteristic times (17,139) has been used for this purpose for other types of cell culture using experimental data (140), but its use with plant-cell suspensions has been limited to theoretical analyses based on empirical correlations for mixing time (20,33).

Gas-Liquid Oxygen Transfer

An important function of reactors used for aerobic cell culture is providing adequate oxygen. Stirred plant-cell reactors are most commonly aerated by bubbling air from spargers; however, surface aeration only (2) or bubble-free aeration using membrane tubing (74,87) may also be used (Table 2). The effectiveness of gas-liquid oxygen transfer is represented by the volumetric oxygen transfer coefficient $k_L a$; this parameter allows comparison of the oxygen transfer capacities of different reactors irrespective of their size and shape, or the magnitude of the concentration driving force. When oxygen limits growth in cell cultures, the value of $k_L a$ determines the maximum biomass density that can be supported (34), so that increasing $k_L a$ is a common objective in reactor design and operation. Reports of higher final biomass concentrations with increasing initial $k_L a$ in batch cultures of tobacco cells (66) demonstrate this relationship. Oxygen transfer coefficients are measured experimentally using a variety of techniques (34); however, a common problem with all $k_L a$ measurements is that it is necessary to assume that the reactor contents are perfectly mixed. If the dissolved oxygen concentration is not uniform in the vessel, measured $k_L a$ results are essentially meaningless. Because complete gas dispersion below as well as above the impeller requires relatively high power inputs for most agitators, gas-phase uniformity is often compromised in plant-cell reactors to avoid the effects of shear damage (33). This means that dissolved oxygen levels and therefore experimental $k_L a$ values, will vary depending on where in the reactor the dissolved oxygen level is measured. Practical estimation of $k_L a$ is also subject to several other common errors that affect the accuracy and value of the final results (141).

The power input, gas flow rate, and liquid properties are the principal determinants for $k_L a$ in stirred vessels. Dimensional correlations that are independent of agitator type have been developed in the following form (43,141):

$$k_L a = A \left(\frac{P_T}{V_L}\right)^\alpha u_G^\beta \tag{1}$$

where $k_L a$ has units of s^{-1}, P_T has units of W, V_L has units of m^3, and u_G has units of $m\,s^{-1}$. The total power dissipation P_T is usually taken to be the sum of the aerated shaft power and the power

provided by isothermal expansion of the gas (140,142). The values of α and β are largely insensitive to the properties of the culture broth (143) and usually fall in the range of 0.2–1.0. In contrast, liquid properties strongly influence the value of A, which includes a dependence on viscosity raised to a negative power of roughly 0.5–1 (10,142,144,145). The effects of other broth properties on $k_L a$ are complex and impossible to predict, with cell concentration and morphology, surface active agents, and coalescence-inhibiting substances such as salts, having a major influence on bubble sizes and gas-liquid interfacial conditions. Because changes in medium composition can alter $k_L a$ values by orders of magnitude, measurements using water or fluids other than the actual cell broth do not indicate the oxygen transfer rates achievable during reactor culture. Oxygen mass transfer coefficients were measured in many studies of stirred plant-cell reactors; examples of the results are summarized in Table 4. Because plant cells have lower specific oxygen requirements than most microbial cells, $k_L a$ values of 5–100 h^{-1} are sufficient for plant-cell cultures, including those grown to high density (66,78,149). This range can be compared with 60–1000 h^{-1} for microbial cells and 0.1–25 h^{-1} for animal cells (150). At suboptimal $k_L a$ values, plant-cell cultures may exhibit linear rather than exponential growth kinetics (66,147). The minimum $k_L a$ necessary to achieve exponential growth can be calculated using stoichiometric analysis (148).

Although the effect of suspended particles such as cells on $k_L a$ is highly variable (39), it is often found that suspended solids promote bubble coalescence, lower the gas holdup, dampen turbulence, and interfere with gas-liquid contact, thereby reducing $k_L a$. In some cases, however, these effects are negligible below threshold particle concentrations (45,151). The data available for plant suspensions indicate that increasing the cell concentration to values as high as 20–60 g L^{-1} dry weight may have only a limited effect on $k_L a$ in stirred reactors (18,78). The influence of biomass concentration on the oxygen transfer coefficient can be expressed alternatively in terms of the relationship between $k_L a$ and the apparent viscosity of the broth (10). In other work, $k_L a$ at constant cell concentration was found to be sensitive to changes in osmotic pressure, apparent viscosity, and the biomass fresh weight to dry weight ratio. At a fixed cell density of 10.2 g L^{-1} dry weight, $k_L a$ increased from 4.8 to 5.7 h^{-1} as the fresh weight to dry weight ratio was reduced from 19.6 to 11.0 (152).

The efficiency of gas-liquid mass transfer in bioreactors depends to a large extent on the characteristics of the bubbles in the culture broth, in particular, their size. In stirred reactors with highly developed turbulent flow, the size of the bubbles produced by the sparger bears little relationship to those present in the vessel as bubble breakup and coalescence due to action of the impeller are the main influences on bubble size (153). However, because shear damage is a major concern in plant-cell culture, stirred reactors will in most cases be restricted to operation at slow stirrer speeds, with the bubbles rising freely through

Table 4. Values of the Oxygen Transfer Coefficient (k_La) Measured in Stirred Plant Cell Reactors

Impeller	Stirrer Speed (rpm)	Aeration Rate (vvm)	Reactor Working Volume (L)	Liquid	Cell Concentration (g L^{-1} dry weight)	Plant Species	k_La (h^{-1})	Reference
Six-flat-blade turbine	225	1.0	1	Culture broth	0 and 11	*Cudrania tricuspidata*	43 and 42	18
Dual six-blade turbines	250	0.33	1.8	Sterilized culture broth with antifoam agent	—	*Catharanthus roseus*	40 ± 2	58
Marine impeller	100–325	0.43	2	Culture broth	2.5–12	*Nicotiana tabacum*	4.5–22	73
Modified paddle (hollow stirring wing)	—	0.2–1.0	2.5	Culture broth treated with formaldehyde	20–80	*Coptis japonica*	30–95	78
Modified (gate-type) paddle	400	1.0	1	Culture broth	0 and 21.6	*Cudrania tricuspidata*	55 and 48	18
—	200	—	—	Culture broth	—	*Nicotiana tabacum*	20–85	26
Rushton turbine	60	0.1–0.3	1.25–2.5	—	—	—	4.5–9.8	84
Dual Rushton turbines	0–300	0.012–0.14	8.5	—	—	—	0.5–30	8
Turbine	250	0.5–1.5	2	Distilled water	—	—	29–51	146
Four-flat-blade impeller	150	0.2–5.6	3	—	—	—	5–50	62
Dual flat-blade turbines: six blades lower, four blades upper	0–200	0.25–1.0	12	Deionized water	—	—	5–35	66
Dual flat-blade turbines: six blades lower, four blades upper	100–400	0.3–0.5	2	—	—	—	5–130	6
Marine impeller	60	0.1–0.3	1.25–2.5	—	—	—	2.2–7.3	84
Two-blade grid paddles	0–200	0.3–0.5	2	—	—	—	15–60	6
Double helical ribbon (upward pumping)	60 and 100	0.05 (surface aeration)	10	—	—	—	1.4 and 5.0	81
Dual six-flat-blade impellers connected by Teflon ribbons	40	0.075	4	Water	—	—	6.7	70
Cell-lift impeller	60	0.1–0.3	1.25–2.5	Medium, nonbiological	—	—	4.6–20.1	84
—	80	0.054–0.35	3.5	—	—	—	$k_La = 31.07 \ (vvm)^{0.792}$	147,148

the liquid after minimal interaction with the impeller. The same is true for large-diameter agitators such as anchors and helical impellers, which either cannot be operated at high speed because of the excessive power required or do not generate the high-velocity turbulent flows needed for bubble disruption and dispersion. Because these conditions produce low gas holdups and k_La values, alternative low-shear strategies for improving the mass transfer coefficient may be used. Particular attention may be given to design of the sparger because the delivery of small bubbles is advantageous for mass transfer when the impeller does not impart sufficient energy for continual bubble breakup. Improvements in k_La were achieved in plant-cell reactors stirred with dual Rushton (8) or grid paddle (6) impellers by using spargers that produce a greater number of smaller bubbles at the same gas flow rate. There may be practical limitations associated with this approach, however, because plant cells have a high tendency for wall growth and will invade and block porous or small-orifice spargers. In addition, in large bioreactors affected by inadequate mixing due to stirrer speed restrictions, reducing the size of the bubbles cannot remedy the problem of poor gas distribution or the development of oxygen gradients in the vessel (33).

As indicated in equation 1, the value of k_La in stirred reactors can be improved by increasing the gas flow rate at constant stirrer speed. Numerous studies have shown, however, that there are limits to which this strategy can be pursued, because reduced growth and metabolite production occur at high gas throughputs (21). This phenomenon has been attributed to stripping of carbon dioxide and other volatile components such as ethylene from the culture broth. The importance of gaseous metabolites in plant-cell cultures was highlighted by Schlatmann et al. (68), who demonstrated that recirculating the reactor off-gas can be used to achieve a three- to four fold increase in alkaloid concentrations in a stirred bioreactor, to reach product levels similar to those found in shake flasks. Because the mass transfer coefficient for stripping volatile compounds increases with any increase in the k_La for oxygen transfer, losses of these compounds are greater when the oxygen transfer capacity of the reactor is improved by increasing the gas flow rate or stirrer speed. Plant-cell properties and metabolism may also be affected in other ways as a function of k_La. In a stirred bioreactor used for culturing *Catharanthus roseus* cells, ajmalicine was not produced at initial k_La values above 10 h^{-1}, whereas the rate of serpentine synthesis was enhanced with increasing k_La (8). The maximum specific oxygen uptake rate by cells grown at an initial k_La of 20 h^{-1} was higher than for cells grown at $k_La = 2.5$ h^{-1}. Dissolved oxygen concentrations in these reactors were not reported, although altered dissolved oxygen tension is the direct result of changes in k_La. The relationship between plant-cell metabolism and oxygen concentration may be complex, with growth, secondary metabolite synthesis, and differentiation processes such as embryogenesis affected in different ways (62,63,81,154,155). The various influences of k_La on the performance of plant-cell cultures reflect this complexity.

Gas-Liquid Hydrodynamics

For stirred bioreactors to perform their mixing and mass transfer functions satisfactorily, appropriate gas-liquid hydrodynamic conditions must be generated by the impeller. Neither bulk blending nor gas dispersion is effective when the impeller is flooded (Fig. 7a), and implicit in application of k_La correlations such as equation 1 is that the reactor is thoroughly mixed and of uniform composition. Impellers should be operated to achieve complete dispersion of gas bubbles above and below the impeller (Fig. 7c). For a given tank and agitator, the stirrer speed necessary to achieve this condition depends on the gas flow rate. The higher the gassing rate, the greater the stirring speed required.

Despite the importance of these hydrodynamic considerations, there are few reports in the literature that mention impeller flooding or are concerned with gas dispersion conditions in plant-cell reactors (33). At the restricted stirrer speeds employed in many applications, adequate bubble dispersion must often be compromised to minimize shear damage. Using visual observations and computer simulation of bubble trajectories, Boysan et al. (49) demonstrated that an 11-L plant-cell reactor stirred with dual Rushton turbines operated under flooded conditions at air flow rates as low as 0.009–0.2 vvm. Based on this result and especially as the volumetric power input required to prevent flooding increases with reactor scale (40), many of the larger stirred reactors listed in Table 2 and operated at slow stirrer speeds are likely not to have achieved effective dispersion of the gas phase. This situation would be considered unacceptable and impractical in most applications of aerated stirred tanks.

Foaming

Foaming is particularly pronounced in plant-cell suspensions, and its effects are exacerbated in reactor cultures by entrained cells attaching to the walls of the vessel and forming a thick meringue or crust in the reactor headspace. Plant-cell meringues usually extend across the entire diameter of the vessel and prevent adequate circulation of medium, resulting in severe loss of culture performance. Foam development in plant suspensions is due mainly to the cell-free broth; the presence or absence of cells has a relatively minor influence (22). The volume of foam produced depends on the gas flow rate in the reactor; however, foam formation and stability at any given time during the culture are also strongly affected by broth properties such as pH (22). There are many reports on the use of chemical antifoam agents in stirred plant-cell reactors (22,61,156), but a major disadvantage of antifoams is that they reduce the value of k_La and consequently, the ability of the reactor to sustain high cell densities (22). Mechanical foam destruction is preferable if it can be made sufficiently effective, and various approaches to mechanical foam control have been reported for stirred plant reactors. Leckie et al. (47,50) installed a flat-blade turbine impeller at the liquid surface of a 12-L stirred bioreactor to act as a foam breaker; alternatively, periodic increases in stirring speed have been applied to clean off layers of plant cells attached to the reactor walls (48,135). For example,

2 s of high speed agitation every hour was effective in controlling meringue formation in a 2-L fermenter used for continuous culture of *Phaseolus vulgaris* cells. In other plant-cell systems, foaming and wall growth were avoided by limiting the stirrer speed and air flow rate to relatively low levels (66).

Reactor Monitoring and Control

Few investigations have been carried out to develop or improve monitoring and control strategies in stirred plant-cell reactors. Real-time, on-line monitoring is usually limited to physical variables such as temperature, pH, agitation speed, air flow rate, culture volume, and dissolved oxygen tension. In some cases, concentrations of carbon dioxide and oxygen in the reactor off-gas have also been measured by either mass spectrometry or paramagnetic/infrared analysis (60,157,158). Knowledge of the off-gas composition allows calculation of parameters such as the respiratory quotient, oxygen uptake rate, and carbon dioxide evolution rate, which can be used to indicate changes in plant-cell activity, including secondary metabolite synthesis (64,154). The metabolic state of plant cells in stirred bioreactors has been indirectly monitored using measurements of NAD(P)H fluorescence (159). The oxidation-reduction state of the culture was reflected in changes in fluorescent levels, so that aerobic-anaerobic transitions could be analyzed *in situ* and on-line. These findings could be used to predict impending oxygen limitations in plant-cell reactors.

Development of methods for on-line monitoring of cell concentration is an important goal for large-scale reactor operation. Because culture conditions and properties can change during the time taken for samples removed from the reactor to be analyzed off-line in the laboratory, control actions based on such measurements are not as effective as those responding to on-line data. An additional problem with plant-cell cultures is the high risk of contamination if culture samples must be removed frequently from the vessel for analysis. Several methods have been described for indirect monitoring of plant-cell concentration, including measurement of medium conductivity, dielectric permittivity, and culture turbidity (57,160,161). Turbidimetric methods have been applied to plant-cell suspensions to control biomass density in continuously stirred turbidostat reactors (91,92), as well as for explicit cell concentration analysis in batch cultures (57). Use of a laser turbidimeter capable of producing a monochromatic light source has been found to overcome problems in plant-cell cultures with interference from pigmented intracellular components (57).

FUTURE INVESTIGATIONS

Although a large number of studies on plant-cell culture has been carried out in stirred reactors using a wide range of different impellers (Table 2), many of the newer agitators being developed commercially have not yet been tested. These modern impellers offer a range of technical features that may better satisfy the requirements of plant cells and improve culture performance. The main

engineering benefits are low power number, reduced loss of power with aeration, and greater gas handling capacity. Impellers with small power numbers have the advantage in plant-cell applications of being able to be operated at higher stirrer speeds without exceeding the energy dissipation rate that causes shear damage; alternatively, larger diameter impellers can be employed for the same power input. In either case, the mixing time is reduced, and bulk blending is enhanced. Impellers that retain most of their power draw with gassing have the advantage that actual energy dissipation rates and power requirements can be determined with greater reliability. The ability to disperse higher gas flow rates without flooding is also an important feature for plant-cell cultures. With shear damage a concern, this makes raising the velocity of the gas, whether recycled or fresh, a more feasible method for increasing $k_L a$.

Curved-blade disc turbines such as that shown in Figure 17a have been produced by several companies, including Scaba, Chemineer, and ICI. These impellers are similar to Rushton turbines, except that changing the shape of the blades has a dramatic effect on the power requirements and gas-handling characteristics. Rotation with the concave side forward prevents the formation of large trailing cavities behind the blades, thus reducing the unaerated power number to $\frac{1}{4} - \frac{1}{2}$ that of a Rushton turbine (162,163) and minimizing the power loss with gassing (163,164). The impeller is also much more difficult to flood; the six-blade Scaba SRGT turbine handles gas flow rates about three times higher than those that cause flooding of Rushton turbines (163).

Hydrofoil impellers such as those shown in Figures 17b and 17c are relatively new agitators with improved hydrodynamic characteristics. The pitch angle and blade width are varied along the length of hydrofoil blades, and the leading edges are rounded like an airplane wing to provide energy-efficient flow over the top. The shape of hydrofoil impellers promotes effective pumping with strong axial velocities at low power number. Most hydrofoils are operated for downward pumping; however, because downflow impellers are generally more prone to flooding than Rushton turbines and can generate hydrodynamic instabilities in aerated systems (43,165), upward pumping hydrofoils have greater promise for plant-cell applications (33). Still in the development stage (43), up-pumping hydrofoils have high gas-handling capacity, virtually no reduction in power draw with gassing over a wide range of air flow rates, and no flow instabilities. Neither downflow nor upflow hydrofoils have been tested in plant-cell reactors.

CONCLUSIONS

The ability to provide adequate mixing and oxygen transfer is the most important benefit of using stirred reactors for suspended plant-cell culture. However, whether these conditions can be achieved depends on the extent to which cell shear sensitivity limits the stirrer speeds that can be employed. Other factors that play a major role in determining levels of plant-cell growth and metabolite synthesis, such as aggregate formation and

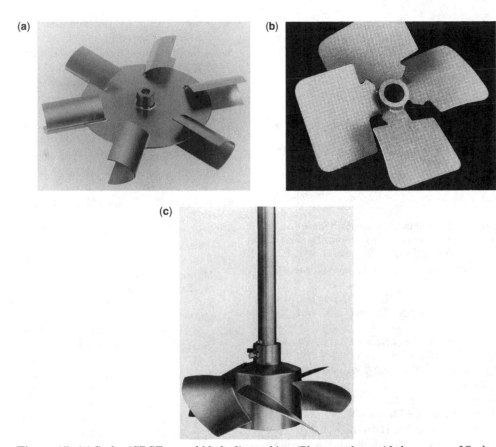

Figure 17. (**a**) Scaba 6SRGT curved-blade disc turbine. (Photograph provided courtesy of Scaba, Sweden.) (**b**) Lightnin A315 hydrofoil impeller. (Photograph provided courtesy of Lightnin Mixers Pty Ltd., Australia.) (**c**) Chemineer Maxflo T hydrofoil impeller. (Photograph provided courtesy of Chemineer Inc., USA.)

disruption, foam development, and stripping of volatile broth components, are also controlled by the operating conditions employed in stirred-tank reactors. To date, few general-purpose rules have been developed to assist in the design of stirred vessels for plant-cell culture. This reflects in large part the wide range of performance outcomes so far observed using different reactor hardware, operating regimes, and plant-cell systems.

NOMENCLATURE

A	Dimensional constant in equation 1	
C_i	Impeller clearance above the vessel floor	m (ft)
C_S	Clearance between the sparger and impeller	m (ft)
D_D	Impeller disc diameter	m (ft)
D_i	Impeller diameter	m (ft)
D_S	Sparger diameter (ring type sparger)	m (ft)
D_T	Tank diameter	m (ft)
H_{BF}	Baffle height	m (ft)
H_i	Impeller height	m (ft)
H_L	Height of liquid in reactor	m (ft)
$k_L a$	Oxygen mass transfer coefficient	s^{-1}
L_B	Impeller blade length	m (ft)
L_R	Ribbon length	m (ft)
N_i	Stirrer speed	s^{-1}
n_{BF}	Number of baffles	–
P_T	Total power input by agitation and gassing	W (Btu min^{-1})
Q_G	Volumetric gas flow rate	$m^3\ s^{-1}$ ($ft^3\ s^{-1}$)
u_G	Superficial gas velocity	$m\ s^{-1}$ ($ft\ s^{-1}$)
V_L	Liquid volume in reactor	m^3 (ft^3)
W_B	Impeller blade width	m (ft)
W_{BF}	Baffle width	m (ft)
W_R	Ribbon width	m (ft)
α, β	Exponents in Equation 1	–

Subscripts

1,2   Individual impellers in dual agitator systems

BIBLIOGRAPHY

1. R. Ballica and D.D.Y. Ryu, *Biotechnol. Bioeng.* **42**, 1181–1189 (1993).

2. M. Jolicoeur, C. Chavarie, P.J. Carreau, and J. Archambault, *Biotechnol. Bioeng.* **39**, 511–521 (1992).

3. A. Kato, S. Kawazoe, and Y. Soh, *J. Ferment. Technol.* **56**, 224–228 (1978).

4. W.R. Curtis and A.H. Emery, *Biotechnol. Bioeng.* **42**, 520–526 (1993).

5. A.H. Scragg, P.A. Bond, and M.W. Fowler, in *Proceedings of the Sixth European Conference on Mixing*, Pavia, Italy, May 24–26, 1988, BHRA, The Fluid Engineering Centre, Cranfield, U.K., 1988, pp. 457–464.

6. H. Yokoi et al., *J. Ferment. Bioeng.* **75**, 48–52 (1993).

7. S.S. Hoekstra, P.A.A. Harkes, R. Verpoorte, and K.R. Libbenga, *Plant Cell Rep.* **8**, 571–574 (1990).

8. F. Leckie, A.H. Scragg, and K.C. Cliffe, *Biotechnol. Bioeng.* **37**, 364–370 (1991).

9. J.E. Prenosil et al., *Enzyme Microb. Technol.* **9**, 450–458 (1987).

10. H. Tanaka, *Biotechnol. Bioeng.* **24**, 425–442 (1982).

11. P. Tsoulpha and P.M. Doran, *J. Biotechnol.* **19**, 99–110 (1991).

12. F. Wagner and H. Vogelmann, in W. Barz, E. Reinhard, and M. Zenk, eds., *Plant Tissue Culture and its Bio-technological Application*, Springer-Verlag, Berlin, 1977, pp. 245–252.

13. B.S. Hooker, J.M. Lee, and G. An, *Biotechnol. Bioeng.* **35**, 296–304 (1990).

14. J.J. Meijer, H.J.G. ten Hoopen, K.Ch.A.M. Luyben, and K.R. Libbenga, *Enzyme Microb. Technol.* **15**, 234–238 (1993).

15. A.H. Scragg, E.J. Allan, and F. Leckie, *Enzyme Microb. Technol.* **10**, 361–367 (1988).

16. J.-J. Zhong, K. Fujiyama, T. Seki, and T. Yoshida, *Biotechnol. Bioeng.* **44**, 649–654 (1994).

17. K. van't Riet and J. Tramper, *Basic Bioreactor Design*, Dekker, New York, 1991, pp. 8–12, 204–207, 401–418.

18. H. Tanaka, *Biotechnol. Bioeng.* **23**, 1203–1218 (1981).

19. B. Ulbrich, W. Wiesner, and H. Arens, in K.-H. Neumann, W. Barz, and E. Reinhard, eds., *Primary and Secondary Metabolism of Plant Cell Cultures*, Springer-Verlag, Berlin, 1985, pp. 293–303.

20. P.M. Doran, *Adv. Biochem. Eng./Biotechnol.* **48**, 115–168 (1993).

21. P.K. Hegarty, N.J. Smart, A.H. Scragg, and M.W. Fowler, *J. Exp. Bot.* **37**, 1911–1920 (1986).

22. R. Wongsamuth and P.M. Doran, *Biotechnol. Bioeng.* **44**, 481–488 (1994).

23. E. Rittershaus, J. Ulrich, and K. Westphal, *Int. Assoc. Plant Tissue Cult. Newsl.* **61**, 2–10 (1990).

24. K. Ushiyama, in A. Komamine, ed., *Plant Cell Culture in Japan*, CMC, Tokyo, 1991, pp. 92–98.

25. K. Ushiyama and K. Hibino, *Abstr. 213th Am. Chem. Soc. Nat. Meet.*, San Francisco, 1997, Abstr. No. AGFD057.

26. T. Hashimoto and S. Azechi, in Y.P.S. Bajaj, ed., *Biotechnology in Agriculture and Forestry*, Vol. 4, Springer-Verlag, Berlin, 1988, pp. 104–122.

27. K. Matsubara and Y. Fujita, in A. Komamine ed., *Plant Cell Culture in Japan*, CMC, Tokyo, 1991, pp. 39–44.

28. O. Schiel and J. Berlin, *Plant Cell Tissue Organ Cult.* **8**, 153–161 (1987).

29. J. Berlin et al., *J. Exp. Bot.* **36**, 1985–1995 (1985).

30. M.E. Curtin, *Bio/Technology* **1**, 649–657 (1983).

31. Y. Fujita, M. Tabata, A. Nishi, and Y. Yamada, in A. Fujiwara, ed., *Plant Tissue Culture 1982*, Japanese Association for Plant Tissue Culture, Tokyo, 1982, pp. 399–400.

32. N.V. Shamkov et al., *Biotekhnologiya* **1**, 32–34 (1991); *Chem. Abstr.* **115** 27653v.

33. P.M. Doran, *Biotechnol. Prog.* **15**, 319–335 (1999).

34. P.M. Doran, *Bioprocess Engineering Principles*, Academic Press, London, 1995, pp. 140–160, 201, 210–213.

35. J.J. Frijlink, A. Bakker, and J.M. Smith, *Chem. Eng. Sci.* **45**, 1703–1718 (1990).

36. S. Ibrahim and A.W. Nienow, *Trans. Inst. Chem. Eng., Part A* **74**, 679–688 (1996).

37. R.M. Hockey and J.M. Nouri, *Chem. Eng. Sci.* **51**, 4405–4421 (1996).

38. G.J. Xu et al., *Can. J. Chem. Eng.* **75**, 299–306 (1997).

39. A.W. Nienow, M. Konno, and W. Bujalski, *Chem. Eng. Res. Des.* **64**, 35–42 (1986).

40. A.W. Nienow, *Chem. Eng. Prog.* **86**(2), 61–71 (1990).

41. K. van't Riet and J.M. Smith, *Chem. Eng. Sci.* **28**, 1031–1037 (1973).

42. K. van't Riet and J.M. Smith, *Chem. Eng. Sci.* **30**, 1093–1105 (1975).

43. A.W. Nienow, *Trans. Inst. Chem. Eng., Part A* **74**, 417–423 (1996).

44. A.W. Nienow, M.M.C.G. Warmoeskerken, J.M. Smith, and M. Konno, in *Proceedings of the Fifth European Conference on Mixing*, Wurzburg, Germany, June 10–12, 1985, BHRA, The Fluid Engineering Centre, Cranfield, U.K., 1985, pp. 143–154.

45. C.M. Chapman, A.W. Nienow, M. Cooke, and J.C. Middleton, *Chem. Eng. Res. Des.* **61**, 167–181 (1983).

46. C.M. Chapman, A.W. Nienow, M. Cooke, and J.C. Middleton, *Chem. Eng. Res. Des.* **61**, 82–95 (1983).

47. F. Leckie, A.H. Scragg, and K.R. Cliffe, *Enzyme Microb. Technol.* **13**, 801–810 (1991).

48. E.H. Dunlop, P.K. Namdev, and M.Z. Rosenberg, *Chem. Eng. Sci.* **49**, 2263–2276 (1994).

49. F. Boysan, K.R. Cliffe, F. Leckie, and A.S. Scragg, in R. King, ed., *Bioreactor Fluid Dynamics*, Elsevier Applied Science, London, 1988, pp. 245–258.

50. F. Leckie, A.H. Scragg, and K.C. Cliffe, *Enzyme Microb. Technol.* **13**, 296–305 (1991).

51. K. Kato et al., *Agric. Biol. Chem.* **36**, 899–902 (1972).

52. T. Furuya, T. Yoshikawa, Y. Orihara, and H. Oda, *J. Nat. Prod.* **47**, 70–75 (1984).

53. H. Tanaka, F. Nishijima, M. Suwa, and T. Iwamoto, *Biotechnol. Bioeng.* **25**, 2359–2370 (1983).

54. R. Wongsamuth and P.M. Doran, *J. Chem. Tech. Biotechnol.* **69**, 15–26 (1997).

55. A.H. Scragg, *Enzyme Microb. Technol.* **12**, 82–85 (1990).

56. T. Takeda, M. Seki, and S. Furusaki, *J. Chem. Eng. Jpn.* **27**, 466–471 (1994).

57. J.-J. Zhong, K. Fujiyama, T. Seki, and T. Yoshida, *Biotechnol. Bioeng.* **42**, 542–546 (1993).

58. J.E. Schlatmann et al., *Biotechnol. Bioeng.* **41**, 253–262 (1993).

59. J.E. Schlatmann et al., *Biotechnol. Bioeng.* **47**, 53–59 (1995).

60. W.M. van Gulik et al., *Appl. Microbiol. Biotechnol.* **30**, 270–275 (1989).

61. G.-Q. Li, J.H. Shin, and J.M. Lee, *Biotechnol. Tech.* **9**, 713–718 (1995).

62. J. Gao and J.M. Lee, *Biotechnol. Prog.* **8**, 285–290 (1992).

63. V. Jay, S. Genestier, and J.-C. Courduroux, *Plant Cell Rep.* **11**, 605–608 (1992).

64. J.-J. Zhong, K.B. Konstantinov, and T. Yoshida, *J. Ferment. Bioeng.* **77**, 445–447 (1994).

65. B. Tal, J.S. Rokem, and I. Goldberg, *Plant Cell Rep.* **2**, 219–222 (1983).

66. A. Kato, Y. Shimizu, and S. Nagai, *J. Ferment. Technol.* **53**, 744–751 (1975).

67. A. Kato et al., *J. Ferment. Technol.* **58**, 373–382 (1980).

68. D. Drapeau, H.W. Blanch, and C.R. Wilke, *Biotechnol. Bioeng.* **28**, 1555–1563 (1986).

69. P. Markkanen, T. Idman, and V. Kauppinen, *Ann. N.Y. Acad. Sci.* **434**, 491–495 (1984).

70. Y.C. Hong, T.P. Labuza, and S.K. Harlander, *Biotechnol. Prog.* **5**, 137–143 (1989).

71. K.A. McDonald and A.P. Jackman, *Plant Cell Rep.* **8**, 455–458 (1989).

72. A.Kh. Lipsky, *J. Biotechnol.* **26**, 83–97 (1992).

73. C.-H. Ho, K.A. Henderson, and G.L. Rorrer, *Biotechnol. Prog.* **11**, 140–145 (1995).

74. W.W. Su and A.E. Humphrey, *Biotechnol. Lett.* **13**, 889–892 (1991).

75. R.H.J. Kessell and A.H. Carr, *J. Exp. Bot.* **23**, 996–1007 (1972).

76. H. Tanaka, H. Semba, T. Jitsufuchi, and H. Harada, *Biotechnol. Lett.* **10**, 485–490 (1988).

77. I. Asaka et al., *Biotechnol. Lett.* **15**, 1259–1264 (1993).

78. K. Matsubara et al., *J. Chem. Tech. Biotechnol.* **46**, 61–69 (1989).

79. M. Kessler, H.J.G. ten Hoopen, J.J. Heijnen, and S. Furusaki, *Biotechnol. Tech.* **11**, 507–510 (1997).

80. B. Ulbrich, *Found. Biotech. Ind.*, Ferment. Res. **4**, 147–164 (1986).

81. J. Archambault et al., *Biotechnol. Bioeng.* **44**, 930–943 (1994).

82. W. Kreis and E. Reinhard, *Planta Med.* **55**, 409–416 (1989).

83. H. Spieler, A.W. Alfermann, and E. Reinhard, *Appl. Microbiol. Biotechnol.* **23**, 1–4 (1985).

84. W.J. Treat, C.R. Engler, and E.J. Soltes, *Biotechnol. Bioeng.* **34**, 1191–1202 (1989).

85. D.-I. Kim, G.H. Cho, H. Pedersen, and C.-K. Chin, *Appl. Microbiol. Biotechnol.* **34**, 726–729 (1991).

86. G.W. Piehl, J. Berlin, C. Mollenschott, and J. Lehmann, *Appl. Microbiol. Biotechnol.* **29**, 456–461 (1988).

87. D.A. Janes, N.H. Thomas, and J.A. Callow, *Biotechnol. Tech.* **1**, 257–262 (1987).

88. D.J. Styer, in R.R. Henke, K.W. Hughes, M.J. Constantin, and A. Hollaender, eds., *Tissue Culture in Forestry and Agriculture*, Plenum, New York, 1985, pp. 117–130.

89. C.-J. Wang and E.J. Staba, *J. Pharm. Sci.* **52**, 1058–1062 (1963).

90. R.A. Miller, J.P. Shyluk, O.L. Gamborg, and J.W. Kirkpatrick, *Science* **159**, 540–542 (1968).

91. R. Bligny, *Plant Physiol.* **59**, 502–505 (1977).

92. S.B. Wilson, P.J. King, and H.E. Street, *J. Exp. Bot.* **22**, 177–207 (1971).

93. F. Constabel, J.P. Shyluk, and O.L. Gamborg, *Planta* **96**, 306–316 (1971).

94. I.A. Veliky and S.M. Martin, *Can. J. Microbiol.* **16**, 223–226 (1970).

95. W. Bruijn, K. van't Riet, and J.M. Smith, *Trans. Inst. Chem. Eng.* **52**, 88–104 (1974).

96. J.H. Rushton, E.W. Costich, and H.J. Everett, *Chem. Eng. Prog.* **46**, 467–476 (1950).

97. A.W. Nienow, in N. Harnby, M.F. Edwards, and A.W. Nienow, eds., *Mixing in the Process Industries*, 2nd ed., Butterworth, Oxford, 1992, pp. 364–393.

98. M.J. Cliff, M.F. Edwards, and I.N. Ohiaeri, *Inst. Chem. Eng. Symp. Ser.* **64**, M1–M11 (1981).

99. J.S.R. Coimbra, F. Mojola, and A.J.A. Meirelles, *J. Chem. Eng. Jpn.* **31**, 277–280 (1998).

100. A.L.F. Porto et al., *Biotechnol. Tech.* **11**, 641–643 (1997).

101. A.W. Nienow, *Trends Biotechnol.* **8**, 224–233 (1990).

102. S. Ibrahim and A.W. Nienow, *Trans. Inst. Chem. Eng., Part A* **73**, 485–491 (1995).

103. H.D. Laufhütte and A. Mersmann, *Chem. Eng. Technol.* **10**, 56–63 (1987).

104. E. Galindo et al., *Chem. Eng. Technol.* **19**, 315–323 (1996).

105. M. Kaminoyama, K. Arai, and M. Kamiwano, *J. Chem. Eng. Jpn.* **27**, 17–24 (1994).

106. A.W. Nienow and D. Miles, *Chem. Eng. J.* **15**, 13–24 (1978).

107. F.A. Holland and F.S. Chapman, *Liquid Mixing and Processing in Stirred Tanks*, Reinhold, London, 1966, pp. 21–22.

108. M. Ohta, M. Kuriyama, K. Arai, and S. Saito, *J. Chem. Eng. Jpn.* **18**, 81–84 (1985).

109. D.C. Peters and J.M. Smith, *Trans. Inst. Chem. Eng.* **45**, T360–T366 (1967).

110. S. Nagata, *Mixing: Principles and Applications*, Kodansha, Tokyo, 1975, p. 198.

111. J.L. Beckner and J.M. Smith, *Trans. Inst. Chem. Eng.* **44**, T224–T236 (1966).

112. A. Tecante and L. Choplin, *Can. J. Chem. Eng.* **71**, 859–865 (1993).

113. A.A. Kamen, C. Chavarie, G. André, and J. Archambault, *Chem. Eng. Sci.* **47**, 2375–2380 (1992).

114. M. Kaminoyama and M. Kamiwano, *Inst. Chem. Eng. Symp. Ser.* **136**, 541–548 (1994).

115. F. Rieger, V. Novák, and D. Havelková, *Chem. Eng. J.* **33**, 143–150 (1986).

116. P.J. Carreau, I. Patterson, and C.Y. Yap, *Can. J. Chem. Eng.* **54**, 135–142 (1976).

117. M. Zlokarnik and H. Judat, in W. Gerhartz, ed., *Ullmann's Encyclopedia of Industrial Chemistry*, 5th ed., VCH, Weinheim, 1988, pp. 25-1–25-33.

118. J. Lehmann, G.W. Piehl, and R. Schulz, *Dev. Biol. Stand.* **66**, 227–240 (1987).

119. J. Lehmann, J. Vorlop, and H. Büntemeyer, in R.E. Spier and J.B. Griffiths, eds., *Animal Cell Biotechnology*, Vol. 3, Academic Press, Orlando, Fla., 1988, pp. 221–237.

120. A.N. Emery, D.C.-H. Jan, and M. Al-Rubeai, *Appl. Microbiol. Biotechnol.* **43**, 1028–1033 (1995).

121. A.W. Nienow et al., in R. King, ed., *Bioreactor Fluid Dynamics*, Elsevier Applied Science, London, 1988, pp. 159–177.

122. J.J. Frijlink, M. Kolijn, and J.M. Smith, *Inst. Chem. Eng. Symp. Ser.* **89**, 49–58 (1984).

123. H. Chen, J.-J. Wang, and Y.-G. Liu, *Abstr. VIIth Int. Congr. Plant Tissue Cell Cult.*, (IAPTC), Amsterdam, June, 24–29 1990, p. 341.

124. A. Pareilleux, *NATO ASI Ser., H* **18**, 313–328 (1988).

125. P.K. Namdev and E.H. Dunlop, *Appl. Biochem. Biotechnol.* **54**, 109–131 (1995).

126. P. Jüsten, G.C. Paul, A.W. Nienow, and C.R. Thomas, *Biotechnol. Bioeng.* **52**, 672–684 (1996).

127. P. Jüsten, G.C. Paul, A.W. Nienow, and C.R. Thomas, *Bioprocess Eng.* **18**, 7–16 (1998).

128. I. Ananta, M.A. Subroto, and P.M. Doran, *Biotechnol. Bioeng.* **47**, 541–549 (1995).

129. A.C. Hulst, M.M.T. Meyer, H. Breteler, and J. Tramper, *Appl. Microbiol. Biotechnol.* **30**, 18–25 (1989).

130. A.M. Kinnersley and D.K. Dougall, *Planta* **149**, 200–204 (1980).

131. T. Kuboi and Y. Yamada, *Phytochemistry* **15**, 397–400 (1976).

132. H. Taguchi, T. Yoshida, Y. Tomita, and S. Teramoto, *J. Ferment. Technol.* **46**, 814–822 (1968).

133. H. Tanaka, J. Takahashi, and K. Ueda, *J. Ferment. Technol.* **53**, 18–26 (1975).

134. J.C. van Suijdam and B. Metz, *Biotechnol. Bioeng.* **23**, 111–148 (1981).

135. M.A. Bertola and F.M. Klis, *J. Exp. Bot.* **30**, 1223–1231 (1979).

136. O.P. Sahai and M.L. Shuler, *Can. J. Bot.* **60**, 692–700 (1982).

137. H.J.G. ten Hoopen, W.M. van Gulik, and J.J. Heijnen, *In Vitro Cell. Dev. Biol.* **28P**, 115–120 (1992).

138. A. Assa and R. Bar, *Biotechnol. Bioeng.* **38**, 1325–1330 (1991).

139. N.W.F. Kossen, in *Proceeding of the Third European Congress on Biotechnology*, Vol. 4, VCH, Weinheim, 1984, pp. IV-257–IV-282.

140. A.W. Nienow et al., *Cytotechnology* **22**, 87–94 (1996).

141. K. van't Riet, *Ind. Eng. Chem. Process Des. Dev.* **18**, 357–363 (1979).

142. M. Cooke, J.C. Middleton, and J.R. Bush, in R. King, ed., *Bioreactor Fluid Dynamics*, Elsevier Applied Science, London, 1988, pp. 37–64.

143. T. Martin, C.M. McFarlane, and A.W. Nienow, in *Proceedings of the Eighth European Conference on Mixing*, Cambridge, September 21–23, 1994, Institution of Chemical Engineers, Rugby, U.K., 1994, pp. 57–64.

144. B.C. Buckland et al., in R. King, ed., *Bioreactor Fluid Dynamics*, Elsevier Applied Science, London, 1988, pp. 1–15.

145. Z. Xueming et al., *Chin. J. Chem. Eng.* **2**, 198–209 (1994).

146. W.M. van Gulik et al., *Biotechnol. Prog.* **10**, 335–339 (1994).

147. A. Pareilleux and R. Vinas, *J. Ferment. Technol.* **61**, 429–433 (1983).

148. A. Pareilleux and N. Chaubet, *Eur. J. Appl. Microbiol. Biotechnol.* **11**, 222–225 (1981).

149. N.J. Smart and M.W. Fowler, *Biotechnol. Lett.* **3**, 171–176 (1981).

150. R.E. Spier and B. Griffiths, *Dev. Biol. Stand.* **55**, 81–92 (1984).

151. M. Greaves and V.Y. Loh, in *Proceedings of the Fifth European Conference on Mixing*, Wurzburg, Germany, June 10–12 1985, BHRA, The Fluid Engineering Centre, Cranfield, U.K., 1985, pp. 451–467.

152. Y.-H. Zhang et al., *Biotechnol. Lett.* **19**, 943–945 (1997).

153. S.P.S. Andrew, *Trans. Inst. Chem. Eng.* **60**, 3–13 (1982).

154. Y. Kobayashi, H. Fukui, and M. Tabata, *Plant Cell Rep.* **8**, 255–258 (1989).

155. J.E. Schlatmann, J.L. Vinke, H.J.G. ten Hoopen, and J.J. Heijnen, *Biotechnol. Bioeng.* **45**, 435–439 (1995).

156. J.-J. Zhong, T. Seki, S.-I. Kinoshita, and T. Yoshida, *World J. Microbiol. Biotechnol.* **8**, 106–109 (1992).

157. P. Nikolova, M. Moo-Young, and R.L. Legge, *Plant Cell Tissue Organ Cult.* **25**, 219–224 (1991).

158. D. Rho, C. Bédard, and J. Archambault, *Appl. Microbiol. Biotechnol.* **33**, 59–65 (1990).

159. E.C. Asali, R. Mutharasan, and A.E. Humphrey, *J. Biotechnol.* **23**, 83–94 (1992).

160. G.H. Markx, C.L. Davey, D.B. Kell, and P. Morris, *J. Biotechnol.* **20**, 279–290 (1991).

161. M. Taya, M. Hegglin, J.E. Prenosil, and J.R. Bourne, *Enzyme Microb. Technol.* **11**, 170–176 (1989).

162. E. Galindo and A.W. Nienow, *Chem. Eng. Technol.* **16**, 102–108 (1993).

163. F. Saito, A.W. Nienow, S. Chatwin, and I.P.T. Moore, *J. Chem. Eng. Jpn.* **25**, 281–287 (1992).

164. M.M.C.G. Warmoeskerken and J.M. Smith, *Chem. Eng. Res. Des.* **67**, 193–198 (1989).

165. C.M. McFarlane, X.-M. Zhao, and A.W. Nienow, *Biotechnol. Prog.* **11**, 608–618 (1995).

See also ASEPTIC TECHNIQUES IN CELL CULTURE; BIOREACTOR OPERATIONS—PREPARATION, STERILIZATION, CHARGING, CULTURE INITIATION AND HARVESTING; CELL CYCLE IN SUSPENSION CULTURED PLANT CELLS; IMMUNO FLOW INJECTION ANALYSIS IN BIOPROCESS CONTROL; MICROPROPAGATION OF PLANTS, PRINCIPLES AND PRACTICES; OFF-LINE ANALYSIS IN ANIMAL CELL CULTURE, METHODS; OFF-LINE IMMUNOASSAYS IN BIOPROCESS CONTROL; ON-LINE ANALYSIS IN ANIMAL CELL CULTURE; PLANT CELL CULTURES, PHYSIOCHEMICAL EFFECTS ON GROWTH; STERILIZATION AND DECONTAMINATION.

BRYOPHYTE IN VITRO CULTURES, SECONDARY PRODUCTS

HANS BECKER
FR 12.3 Fachrichtung Pharmakognosie und Analytische Phytochemie Universität des Saarlandes
Saarbrücken, Germany

OUTLINE

Introduction

Culture Conditions

Product Formation

Biotransformations

Biosynthetic Studies

Bibliography

INTRODUCTION

Bryophytes represent an independent branch of the plant kingdom with more than 11,000 species (1). Morphologically they take up a position intermediate between the thallophytes and the cormophytes. The bryophyte plant is the haploid gametophyte, the diploid sporophyte remains attached to the gametophyte during its life. From the taxonomic point of view they are divided into three classes:

- Hornworts (Anthoceratae)
- Liverworts (Hepaticae)
- Mosses (Musci or Bryatae)

The hornworts always adapt simple thallus forms. The liverworts comprise species with thallose vegetation as well as those with a small stem bearing simple leaves without a midrib. The mosses are characterized by a small stem and leaves with a midrib. From the chemical point of view liverworts are the richest group. Most of them contain oilbodies within the cells. These oilbodies are rich in various terpenoidic compounds. Besides those bibenzyl and bisbibenzyls are typical compounds for liverworts.

Compared to other plants, the chemical investigation of bryophytes is a rather young discipline. There are several reasons for this. It is difficult to collect larger amounts of material, and laborious workup procedures are needed for obtaining pure species. A stereomicroscope has to be used frequently to separate the respective species from unwanted contamination. Field cultivation of bryophytes similar to that of higher plants is not practicable. Even if the required ecological conditions were created, mixed populations would result. Despite these difficulties a large number of new natural products, some of them with novel and unique skeletons, have been isolated from bryophytes during the past thirty years (2,3).

The difficulty of obtaining larger amounts of plant material without contamination by other species may be overcome by in vitro cultures. Bryophyte cultures offer effectively the same advantages as in vitro cultures of higher plants. Plants from all climatic zones can be grown in cultures. A comparatively constant production of homogeneous material can be achieved through controlled culture conditions, giving sufficient amounts for subsequent analysis (4). Furthermore, liverwort cultures produce qualitatively and quantitatively almost the same compounds as field-collected plants.

CULTURE CONDITIONS

Initiation of aseptic plantlets is more difficult for bryophytes than for higher plants. There are two starting points. One possibility is to surface sterilize intact spore capsules and to open these capsules under sterile conditions (5). Capsules are prepared from the female gametophytes and soaked for 2 min in 2% NaOCl solution containing 1% Tween 80. The capsules are then washed with sterile water. The washed capsules are opened with sterile forceps, and the spores are spread out on MSK 2 (6) or B5 medium (7) containing 2% glucose or sucrose and 0.9% agar. The other way is to surface sterilize fragments of the thallus. As the thallus is very thin and as leaflets consist often of one single cell layer, it is very difficult to kill the attached microorganisms without affecting the cells of the plant (8). Apical notches of approximately 5 mm^2 in area (thallose liverworts) or apical tips of 5–10 mm length (folious liverworts and mosses) are cut and surface sterilized. This treatment can be carried out in the so-called washing machine by using a syringe with a modified syringe holder in which the plant parts are placed (9). After NaOCl treatment and washing with sterile water the plant tissue is transferred to MSK-2 or B5 solid medium. As the bryophytes are often only a few millimeters in size, they can be cultivated easily in differentiated form even in fermenters. Depending on the species, 0.1–5 kg of fresh plant material may be obtained within one year starting from a single spore of a leaflet.

Culture conditions for optimal growth of *Ricciocarpos natans* were studied in detail (10). Out of six basal mineral media (Benecke, Gamborg B 5, Knop b, Knudson, Lorenzen, MSK-2, White) fresh weight increase was best on Gamborg B 5 medium. Continous light (2000–6000 lux) and the addition of 2% sucrose or glucose stimulated growth. No growth appeared in the dark, even with the addition of sugars, and only poor growth was observed in light without the addition of sugars.

For the induction of callus 2,4-D in a concentration of 1 mg L^{-1} is added to the medium. Usually, after 6 weeks of growth under constant light at 1000 lux, callus formation can be observed. Friable callus tissue can then be transferred to liquid MSK-2 medium containing 1 mg L^{-1} 2,4-D and 2% glucose to start cell suspension cultures. Light is an essential requirement for growth (11).

PRODUCT FORMATION

The lipophilic oil bodies of the liverworts accumulate various mono-, sesqui-, and diterpenoids. The presence of blue-colored sesquiterpenoid azulenes can be recognized in the blue oil bodies of *Calypogeia azurea* (12). Takeda and Katoh (13) studied the production of total essential oil and individual products of *C. granulata* and compared the GC spectrum of intact plants, redifferentiated plants, and suspension-cultured cells (20 and 30 d old). The spectrum of compounds was quite similar for the various sesquiterpene-derived compounds. 1,4-Dimethyl-azulene was the main compound of the four samples analyzed. From in vitro cultures of differentiated *C. azurea* a variety of ten azulenes have been isolated (14). Nakagawa et al. (15) isolated 4-methylazulene-1-carbaldehyde (**1**) and 4-methylazulene-1-carboxylic acid from cultures of the same plant. Vanadate may act as an abiotic elicitor and stimulate the production of 1,4-dimethylazulene twofold.

OHC

1

The wealth of compounds that may be isolated from in vitro cultured liverworts has been demonstrated, for example, with *Ricciocarpos natans* (10,16–20), *Marchantia polymorpha* (21–23), and *Jamesoniella autumnalis* (8,24–27) and *Fossombronia* species (5,28,29). *R. natans* grows floating on the surface of stagnant waters in temperate regions of both the Northern and Southern Hemispheres. If the marshy pool dries out, it keeps growing on the moist soil. Cultures have been maintained in 200-mL Erlenmeyer flasks on liquid B 5 medium according to

Gamborg (7), supplemented with 1% sucrose. The fresh weight of the culture increased 12-fold in 24 days from an initial weight of 1.5 g (10). The lipophilic fraction of the extract yielded five novel sesquiterpenoids, three monocyclofarnesane derivatives, and two cuparane derivatives. In addition to those new compounds, (-)-limonene, cuparene, phytol, and lunularin were isolated (16). From the hydrophilic fraction of the extract three new bibenzyl glycosides were isolated: 5,4-dihydroxybibenzy-2-O-ß-D-glucopyranoisde, 5,3,4'-trihydroxybibenzyl-2-O-ß-D-glucopyranoside, and 2'-carboxy-4,3'-dihydroxybibenzyl--3-O-ß-D-glycopyranoside (2). In addition, phenylethanoid glycosides were detected [salidroside, ß-(3,4-dihydroxy-phenyl-ethyl-O-ß-glucopyranoside)], which have not been previously reported for bryophytes (17). It has been shown that lunularin, a bibenzyl common to all liverworts, is synthesized by the plants via prelunularic acid, lunularic acid, and decarboxylation of the latter (30,31). The existence of prelunularin in cultures of R. natans suggests that there is a second pathway leading from lunularic acid to lunularin (19). Many liverworts in the field as well as in the culture turn from green to red during aging. The pigment is attached to the cell wall. In vitro cultures allow to study the phenomenon in detail. We isolated a new anthocyanidin [riccionidin A (3)] and a dimer (riccionidin B) by means of multilayer counter-current chromatography (MLCC) (18). Under standard conditions the formation of riccionidin A and B started after two weeks of culture. After three weeks the content of both pigments increased greatly. At the same time nitrate and phosphate were depleted from the medium. In nitrogen- as well as phosphate-deficient medium pigment formation started immediately. Light intensity also had an effect on the riccionidins. The pigment formation increased when a higher illumination rate (5000 versus 2000 1x) was applied (20).

and protoplasts have been carried out (for a review, see 23). Suspension cultures have been maintained photomixotrophically (32) and photoautotrophically using 1% CO_2 in air as sole carbon source (31).

Bisbibenzyls are characteristic constituents of M. polymorpha. The main bisbibenzyl, marchantin A, is of particular interest because of its biological activities, which include cytotoxicity against tumor cells, antimicrobic, and muscle-relaxing activities (33). Therefore, bisbibenzyl formation in differentiated aseptic cultures and cell suspension cultures was investigated. Marchantin A (4) and H were the main bisbibenzyls isolated from an aseptic culture derived from collected material near Heidelberg, Germany (21,22). Whereas the marchantin H content remained almost constant — except for a short period after subculturing — marchantin A accumulation started after four weeks and continued even during the stationary phase. It could be shown that nitrate limitation was responsible for the marchantin accumulation. Marchantin A formation could be strongly induced by the addition of cupric sulfate to the medium in a final concentration of 1 mM. Chitosan and yeast extract were not effective as elicitors.

4

Bryophytes are capable of producing arachidonic acid (ARA) and eicapentaenoic acid (EPA) at levels that amount to up to 40% of their total fatty acids. A large production of ARA and EPA in cell cultures of M. polymorpha has been reported. The values of ARA and EPA were up to 7.1 and 3.9 mg/g dry weight of cells, respectively, grown under photomiotrophic conditions (34).

Jamesoniella autumnalis is a common liverwort found throughout most temperate forest regions in the Northern Hemisphere. Plants usually form green and somewhat opaque patches, becoming reddish brown in sunlight. Several cultures of different origin of J. autumnalis were generated from field-collected gametophytes and routinely subcultured on B5 medium (7) whose microelements had been replaced by those of the MSK 2 medium (6). The medium was solidified with 0.9% agar and contained 2% sucrose. The cultures were kept in 200-mL flasks at $20 \pm 1.5\,°C$ under constant light. J. autumnalis grew as a completely differentiated surface culture. The growth rate varied quite significantly among the cultures of different origin. Fresh and dry weight increase were about eightfold within 80 days.

2

3

Marchantia polymorpha is a worldwide-distributed thallose liverwort. It can usually be found on wet walls, among moist shores of creeks, and as a weed in greenhouses. Because of its wide distrubution, many studies with intact plant callus, suspension cultures,

Five sesquiterpene hydrocarbons (anastreptene, β-barbatene, bicyclogermacrene, α- and β-bisabolene) and one sesquiterpene alcohol (spathulenol) were detected in field-grown material as well as in the cultures (8).

A large variety of ent-labdanes and clerodane-derived furanoditerpenes has been isolated from the nonvolatile terpenoid fraction (25,27). The structure of a highly oxygenated diterpene (5) (24) has been elucidated by X-ray analysis. Furthermore five new lignan derivatives, 2, 3, 6'-tricarboxy-6,7-dihydroxy-1(3')-2'-pyranonyl-1,2-dihydronaphthalene (6), its two methylesters, 2,6'-dicarboxy-6,7-dihydroxy-1(3')-2'-pyranonyl-1,2-dihydronaphthalene and 2,3-di-carboxy-6,7-dihydroxy-1-(3', 4'-dihydroxy)phenylnaphthalene, were isolated from the methanol extract from *J. autumnalis* (26).

5

6

The liverwort genus *Fossombronia* Raddi consists of about 50 species, but until recently reports on secondary metabolites of this genus have been scarce (2). In vitro cultures from *F. pusilla* derived from spores produced the sacculatal-derived diterpenes perrottetinal A and B and 8-hydroxy-9-hydroperrottetianal A (7). α-(-)-Santonin, a well-known sesquiterpene from higher plants, has also been isolated from the cultures (5). A petrol ether extract and the isolated terpenes exerted antibacterial activity.

F. alaskana, a rare arctic liverwort, was axenically cultured (28). Phytochemical investigation of the gametophytes afforded five new *epi*-neoverrucosane-type diterpenoids (5-oxo-*epi*-neoverrucosane, 13-hydroxy-5-oxo-*epi*-neoverrucosane, 8α-acetoxy-13-hydroxy-5-oxo-*epi*-neoverrucosane, 8α-16-diacetoxy-13-hydroxy-5-oxo-*epi*-neoverrucosane (8), and 8α-13-dihydroxy-5-oxo-*epi*-neoverrucosane) together with the previously

7

known 5β-hydroxy-*epi*-neoverucosane. In addition, the new homoverrucosane-type diterpene 5,18-dihydroxy-*epi*-homoverrucosane was isolated (29).

8

F. pusilla and *F. alaskana* yielded three new hopane-type triterpenes [22-hydroxy-29-methyl-hopanoate, 20-hydroxy-22(29)-hopen (9) and 22(30)-hopen-29-oic acid] together with adianton, diploterol, tetrahymanol, and caryophyllene-6,7-epoxide (28).

9

BIOTRANSFORMATIONS

Many microorganisms are used for biotransformation studies. Cell suspension cultures of *M. polymorpha* were assayed for their capability to transform steroids. Testosterone was selectively converted to 6β-hydroxytestosterone, whereas the 17α-hydroxyl group of the isomeric epitestosterone was oxidized, yielding the appropriate ketone androst-4-ene-3,17-dione (35). In separate experiments with the same cell culture, androst-4-ene-3,17-dione was stereoselectively reduced at C-17, producing testosterone (36). 1,4-Androstadiene-3,17-dione was converted to 17ß-hydroxy-1,4-androstadiene-3-dione as the major metabolite and 4-androstene-3,17-dione and testosterone. 4-Androstene-3,11,17-trione was stereoselectively metabolized at C-17 to the corresponding 17α-hydroxy-4-androsten-3,11-dione as the sole product.

BIOSYNTHETIC STUDIES

In spite of the remarkable diversity of terpenoids produced by liverworts, investigations directed to the biosynthesis of these compounds in nonvascular plants are rare compared to the numerous studies on terpenoid metabolism in higher plants. Comparative investigations on such phylogenetically distant plants might provide useful insights into the evolutionary origins and structure–function relationships of the responsible biosynthetic enzymes.

A cell-free extract of in vitro cultured *R. natans* catalyzes the cyclization of geranyl diphosphate to 4S-(-)-limonene (37). The enzyme was partially purified by combination of anion-exchange and hydroxylapatite chromatography. In the stereochemistry of the coupled isomerization–cyclization reaction and in its general properties, the limonene synthase from this bryophyte resembles the corresponding monoterpene cyclases from gymnosperm and angiosperm species. The monoterpene cyclases sabinene synthase and bornyl diphosphate synthase are involved in (−)-sabinene and (+)-bornylacetate formation in different variants of the liverwort *Conocephalum conicum* cultured in vitro (38).

Aseptic cultures of bryophytes offer good experimental conditions for biosynthetic studies. The simple morphology supports the uptake of the respective compounds under physiological conditions, in contrast to undifferentiated cell cultures, where the dedifferentiated status of the cells might have a drastic impact on many biochemical processes. Therefore, ^{13}C-labeling experiments should give a realistic impression of the biosynthetic origin of different terpenoid classes.

The biosynthetic sequences of highly oxidized cadalenes (39,40) and irregular sesquiterpenes such as pinguisanes (41), ß-barbatene (42), and kelsoene (43) have been examined by GC-MS and NMR analyses of the biosynthetically labeled compounds, incorporating isotopically labeled precursors such as acetate and mevalonate. Geometrically specific cyclization of farnesyl diphosphate to cadinanes was examined using partially purified enzymes from cultured cells of the liverwort *Heteroscyphus planus* (44,45). Furthermore, preferential labeling of the farnesyl diphosphate-derived portion in chloroplastidic terpenoid biosynthesis was confirmed by ^{2}H- and ^{13}C-enriched NMR peaks (46,47).

In the past, the mevalonic acid pathway has been accepted as the only biosynthetic route leading to isopentenyldiphosphate (IPP), operating in all organisms. Recently, an alternative, nonmevalonoid pathway of IPP biosynthesis has been discovered in bacteria and plants (48,49). In this glyceraldehyde–pyruvate pathway the first putative C_5 intermediate, 1-deoxyxylulose 5-phosphate, is formed by condensation of an activated C_2 unit (derived from pyruvate) and glyceraldehyde-3-phosphate.

The incorporation of ^{13}C-labeled glucose into borneol, bornyl acetate, the sesquiterpenes cubebanol and ricciocarpin A, phytol, and stigmasterol has been studied in axenic cultures of the liverworts *R. natans* and *Conocephalum conicum*. Quantitative ^{13}C NMR spectroscopic analysis of the resulting labeling patterns showed that the isoprene building blocks of the sesquiterpenes and stigmasterol are built up via the mevalonic acid pathway, whereas the isoprene units of the monoterpenes and the diterpene phytol are exclusively derived from the glyceraldehyde–pyruvate pathway. These results indicate the involvement of both isopentenyl diphosphate biosynthetic pathways in different cellular compartments (50,51).

BIBLIOGRAPHY

1. W.B. Schofield, *Introduction to Bryology*, Macmillan, New York, 1985.
2. Y. Asakawa, in W. Herz et al., eds., *Progress in the Chemistry of Organic Natural Products*, Springer, New York 1995, pp. 1–618.
3. H.D. Zinsmeister, H. Becker, and T. Eicher, *Angew. Chem., Int. Ed. Engl.* **30**, 130–147 (1991).
4. H. Becker, *J. Hattori Bot. Lab.* **76**, 283–291 (1994).
5. M. Sauerwein and H. Becker, *Planta Med.* **56**, 365–367 (1990).
6. K. Katoh et al., *Physiol. Plant.* **49**, 241–247 (1980).
7. O.L. Gamborg, R.A. Miller, and K. Ojima, *Exp. Cell Res.* **50**, 151–158 (1968).
8. H. Becker and M. Blechschmidt, *Flavour Fragrance J.* **10**, 187–191 (1995).
9. D.V. Basile, *Bull. Torrey Bot. Club* **99**, 313–316 (1972).
10. G. Wurzel and H. Becker, *Z.Naturforsch.*, **45C**, 13–18 (1990).
11. K. Katoh, in J.M. Glime, ed., *Methods in Bryology*, Hattori Bot. Lab., Nichinan, 1998, pp. 99–105.
12. S. Huneck, *Z. Naturforsch.* **18B**, 1126 (1963).
13. R. Takeda and K. Katoh, *Planta* **151**, 525–530 (1981).
14. U. Siegel et al., *Phytochemistry* **31**, 1671–1678 (1992).
15. S. Nakagawa et al., *Plant Cell Physiol.* **34**, 421–429 (1993).
16. G. Wurzel and H. Becker, *Phytochemistry* **28**, 2565–2568 (1990).
17. S. Kunz and H. Becker, *Phytochemistry* **31**, 3981–3983 (1992).
18. S. Kunz and H. Becker, *Phytochemistry* **36**, 675–677 (1994).
19. S. Kunz, G. Burkhardt, and H. Becker, *Phytochemistry* **35**, 233–235 (1994).
20. S. Kunz and H. Becker, *Z. Naturforsch.* **50C**, 235–240 (1995).
21. K.P. Adam and H. Becker, *Z. Naturforsch.* **48C**, 839–842 (1993).
22. K.P. Adam and H. Becker, *Phytochemistry* **35**, 139–143 (1994).
23. K.P. Adam, in Y.P.S. Bajaj, ed., *Biotechnology in Agriculture and Forestry*, Vol. 37, Part IV, Springer-Verlag, Berlin, 1996, pp. 186–201.
24. H. Tazaki et al., *Phytochemistry* **37**, 491–494 (1994).
25. H. Tazaki, J. Zapp, and H. Becker, *Phytochemistry* **39**, 859–868 (1995).
26. H. Tazaki, K.P. Adam, and H. Becker, *Phytochemistry* **40**, 1671–1675 (1995).
27. H. Tazaki, K. Nabeta, and H. Becker, *Phytochemistry* **48**, 681–685 (1998).
28. C. Grammes, G. Burkhardt, and H. Becker, *Phytochemistry* **35**, 1293–1296 (1994).
29. C. Grammes et al., *Phytochemistry* **44**, 1492–1502 (1997).
30. Y. Ohta, S. Abe, H. Komura, and M. Kobayashi, *J. Hattori Bot. Lab.* **56**, 249–254 (1984).

31. K. Katoh, *Physiol. Plant.* **59**, 242–248 (1983).

32. K. Katoh, *Physiol. Plant.* **57**, 67–74 (1983).

33. Y. Asakawa, in H.D. Zinsmeister and R. Mues, eds., *Bryophytes, Their Chemistry and Chemical Taxonomy*, Clarendon Press, Oxford, 1990, pp. 369–410.

34. Y. Shinmen et al., *Phytochemistry* **30**, 3255–3260 (1991).

35. H. Hamada, H. Konishi, H.J. Williams, and A.I. Scott, *Phytochemistry* **30**, 2269–2270 (1991).

36. H. Hamada, S. Naka, and H. Kurban, *Chem. Lett.*, pp. 2111–2112 (1993).

37. K.P. Adam, J. Crock, and R. Croteau, *Arch. Biochem. Biophys.* **332**, 352–256 (1996).

38. K.P. Adam and R. Croteau, *Phytochemistry* **49**, 475–480 (1998).

39. K. Nabeta, Y. Mototani, H. Tazaki, and H. Okuyama, *Phytochemistry* **35**, 915–920 (1994).

40. K. Nabeta, T. Ishikawa, T. Kawae, and H. Okuyama, *J. Chem. Soc., Perkin Trans.* pp. 3277–3280 (1994).

41. H. Tazaki et al., *Chem. Commun.*, pp. 1101–1102 (1997).

42. K. Nabeta et al., *Chem. Commun.*, p. 169 (1998).

43. K. Nabeta et al., *Chem. Commun.*, p. 1485 (1998).

44. K. Nabeta et al., *J. Chem. Soc., Perkin Trans. I*, PP. 1935–1939 (1995).

45. K. Nabeta et al., *J. Chem. Soc., Perkin Trans. I*, 2065–2070 (1997).

46. K. Nabeta, T. Ishikawa, and H. Okuyama, *J. Chem. Soc., Perkin Trans. I*, pp. 3111–3115 (1995).

47. K. Nabeta, T. Saitoh, K. Adachi, and K. Komuro, *Chem. Commun.*, pp. 671–672 (1998).

48. M. Rohmer et al., *Biochem. J.* **295**, 517–524 (1993).

49. M.K. Schwarz and Ph.D. Thesis, ETH, Zürich, Switzerland, 1994.

50. R. Thiel, K.P. Adam, J. Zapp, and H. Becker, *Pharm. Pharmacol. Lett.* **7**, 103–105 (1997).

51. K.P. Adam, R. Thiel, J. Zapp, and H. Becker, *Arch. Biochem. Biophys.* **354**, 181–187 (1998).

See also Plant cell culture, effect on growth and secondary product accumulation of organic compounds and surfactants; Plant cell cultures, secondary product accumulation; Plant cell cultures, selection and screening for secondary product accumulation.

C

CELL AND CELL LINE CHARACTERIZATION

A. DOYLE
Wellcome Trust
London, England

G. STACEY
National Institute of Biological Standards and Control
South Mimms, England

OUTLINE

INTRODUCTION

The characterization of a new cell line is essential and should be carried out at both the earliest passage of cultures once established and at frequent intervals thereafter (R.J. Hay, 1988). The occurrence of cross-contamination is not merely anecdotal; documented cases have been widely reported (1,2). Some earlier reports indicated that levels of cross-contamination may exceed 30% of cultures tested (3). The preferential inclusion of suspect cultures in these reports does not detract from the fact that cross-contamination is a serious problem. The classic example is that of HeLa contamination (1), for which conventional cytogenetic analysis in association with isoenzyme studies was used to verify the species and, with human samples, the race of origin. This is particularly easy in the case of the HeLa cell line, since it has characteristic cytogenetic markers, and in isoenzyme analysis, the type B rather than the more usual type A glucose-6-phosphate dehydrogenase isoform is present. A more traditional technique is that of immunological characterization — in essence a known sample of cells or tissue is used to raise an antibody in the rabbit. The antibody is then used in a fluorescence study against the test cells and an anti-rabbit globulin conjugated with FITC (fluorescein isothiocyanate) is added and then can be visualized on a fluorescence microscope. In recent times molecular biology has had a significant influence on cell characterization and DNA fingerprinting provides a particularly useful tool for identity testing.

CHROMOSOME ANALYSIS

The determination of the chromosomal complement of a cell line provides a direct method of confirming the species of origin. It also allows the detection of gross aberrations in chromosome number and/or morphology. Cytogenetic analysis is very useful for specific identification of cell lines with unique chromosome markers. In one study of 47 cell lines reported by O'Brien et al. (4), two cell lines could not be differentiated by eight separate enzyme tests, but were readily distinguished by karyology. However, it should be borne in mind that very careful interpretation is required to differentiate cell lines of normal karyotype beyond the level of species. To follow is an outline of the standard technique for cytogenetic analysis for a monolayer cell culture.

Slide Preparation — Karyology — Materials

Hypotonic trypsin/versene solution (Hypo-TV) at 37 °C for harvesting adherent cells

Hypotonic KCl solution (Hypo-KCl) at room temperature

Heparin, 10 units per mL

Glacial acetic acid at room temperature (it is important to use this fresh)

Working colcemid solution (100 or 50 μg/mL) at room temperature

Slides, precleaned, wet, and chilled

MONOLAYER CULTURES

Day 1

Passage the cells 1 day before use. One 80–90% confluent 75-cm² culture is generally sufficient for a chromosome preparation.

Day 2

1. Add colcemid solution to the cells at a final concentration of 0.01–0.03 µg/mL and incubate at 37 °C for 1 h. The incubation time for colchicine treatment is 45 min for a fast-growing culture, 1–1.5 h for an "average" culture, 2–3 h for diploid cell cultures, and over 3 h for slow-growing cultures.

2. Label each centrifuge tube (15-mL size) and add 0.6 mL fetal bovine serum.

3. After incubation with colcemid, collect the medium in a centrifuge tube and centrifuge at 150 g for 5 min. After decanting supernatant, resuspend cells in Hypo-KCl and add to the harvest from the Hypo-TV treatment (Step 7).

4. Add 6 mL Hypo-TV per 75-cm² flask and incubate at 37 °C for 10 min.

5. Aspirate to suspend cells. If a large number of cells are still attached, treat the cultures with Hypo-TV again.

6. Transfer the cell suspension into the tube prefilled with serum (Step 2) and centrifuge 150 g for 5 min.

7. Decant supernatant without disturbing the cell pellet; add Hypo-KCl little by little with continuous agitation to a final volume of 4 mL. After centrifugation and removal of supernatant, add KCl gradually by increasing the volume with intermittent agitation. Finally, add the contents of one (or two) Pasteur pipettes to each. Between each addition, agitate cells by gentle pipetting.

8. Leave at room temperature for 10 min, then centrifuge and decant most of the supernatant, leaving an equal volume to the cell pellet for resuspending cells.

9. Resuspend cells gently but thoroughly to make a uniform cell suspension.

10. Add freshly prepared ice cold 1:3 glacial acetic acid/methanol fixative slowly as indicated for Hypo-KCl (Step 8) to a total of 4 mL and leave for 15 min at room temperature. The fixative must be made fresh just before use. After mixing, it must always be kept on ice.

11. Decant as much supernatant as possible without disturbing the cell pellet. Repeat Step 10.

12. Repeat Step 10; however, leave at room temperature for 10 min and then centrifuge.

13. Remove most of the supernatant, leaving behind a volume, which is about ten times the cell mass, and then resuspend the cells.

14. Pick up a wet, clean, chilled slide with a pair of forceps. Hold the frosted edge with your fingers and shake off excess water using a fanning motion. (Be sure the frosted side faces up.) Hold the slide slightly downward (at about a 30° angle), place a few drops of cell suspension onto the upper edge of the slide just below the frosted edge to let the suspension run down slowly, and at the same time blow gently and evenly over the surface to spread cells over the entire surface. Wipe off excess liquid from the edges and back of the slide, and air dry by leaving the slide on a paper towel.

15. When dried, examine the quality of the chromosomes under phase microscopy to assess metaphase spreads and cell density. Cells should be evenly distributed and chromosomes from each cell should be close together but without frequent overlapping. If the quality is good, make at least 10 slides. Experience suggests that slides prepared the same day as the cell harvest give the best quality spreading and staining. However, it may be advisable to also store the fixed cell suspension for repeat analyses. To make slides from this stored material, cells are again treated in fresh fixative at least three times with a 10-min incubation between each centrifugation.

16. Use an indelible marker pen to record on the slide: cell line, number of slide, date of preparation.

17. Leave slides overnight at room temperature for further drying and then store in a slide box.

Slides kept at room temperature may be used even after one month of storage and still give good staining. As slides get older, staining results become unpredictable. Keep slides in a cold and dry environment, or seal in a container filled with an inert gas (e.g., argon) and store at 4 °C.

Day 3

Stain slides.

GIEMSA STAINING

The most commonly used procedure for chromosome banding is Giemsa staining (5,6). The method allows for a simple chromosome count and provides an estimation of the rate of polyploidy, a key issue related to cell stability.

1. Immerse slides in 1% Giemsa solution in a Coplin jar, at room temperature, for 30 min.

2. Rinse thoroughly with distilled water. Use a squeezing bottle to stream water over the slide surface evenly. Examine under 63x and 100x water-emulsion objective lenses.

ISOENZYME ANALYSIS

Isoenzyme analysis is used for the speciation of cell lines and for the detection of contamination of one cell line with another, although a relatively high level of contamination is necessary (>10%). This method utilizes the property that isoenzymes have similar substrate specificity, but different molecular structures, which in turn affects their electrophoretic mobility. Each species therefore has a characteristic isoenzyme mobility pattern. While the species of origin of a cell line is usually indicated with only two isoenzyme tests (lactate dehydrogenase and glucose-6-phosphate dehydrogenase) (4), specific identification of a cell line requires a larger battery of tests (3). Generally the use of four isoenzymes can give adequate results (7).

To follow is the standard technique for isoenzyme analysis utilizing the "Authentikit" system supplied by Innovative Chemistry, Inc., Marshfield, MA.

PREPARATION OF REAGENTS

1. *Cell extraction buffer*

   Prepare a 50 mM solution of Tris (for final volume of 100 mL) in 80 ml water.
   Adjust to pH 7.5 (either 1 M HCl or 1 M NaOH).
   Add 1 mM EDTA and then 2% Triton X-100.
   Adjust the pH if necessary and make up the volume with deionized water. Store at 4 °C.

2. *Enzyme substrates*. The list of tabulated migration distances and ratios supplied by the kit manufacturer includes the following:

   Aspartate aminotransferase (AST)
   Glucose-6-phosphate dehydrogenase (G6PD)
   Lactate dehydrogenase (LDH)
   Malate dehydrogenase (MD)
   Mannose phosphate isomerase (MPI)
   Nucleoside phosphorylase (NP)
   Peptidase B (Pep B)

3. *Visualization*. Detection of the enzyme bands is provided by an insoluble purple formazan dye when 3-(4,5-dimethylthiazol-2-yl)-2,5-diphenyl-2H-tetrazolium bromide (MTT) is reduced in the presence of phenazine methosulfate (PMS) and the appropriate substrate.

4. *Standards*. Extracts of the mouse cell line NCTC Clone L929 (standard) and the human cell line HeLa (control) can be obtained from the kit manufacturer in lyophilized form. If required, they can be prepared from growing cultures, but these should be obtained from a documented and authenticated source (i.e., culture collection).

Materials

Agarose film—agarose gel on transparent polystyrene sheet (Authentikit, Innovative Laboratories, Marshfield, MA)
Barbital buffer 0.05 mM, pH 8.6
Hamilton syringe, 5–10 μL with tapered needle
Electrophoresis cell (Innovative Laboratories)
Power supply to provide 160 V DC, constant voltage (Innovative Laboratories)

A minimum of 10^7 cells is necessary for each analysis.

1. Prepare a cell pellet (centrifugation at 150 g for 5 min). Attached cells must be removed with the appropriate enzyme.
2. Decant the culture medium, resuspend the cell pellet in 15 ml Earle's balanced salt solution at 4 °C. Centrifuge the cells at 150 g for 5 min.
3. Mark the volume of the pellet on the side of the tube, and then drain the buffer by inverting the tube on absorbent paper. Dry the inside of the tube, taking care not to touch the pellet. Add an equal volume of cell extraction buffer to the pellet, mix with a micropipette tip, and stand on ice for 15 min.
4. Mix again and examine a drop sandwiched between a glass microscope slide and a coverslip. The cells should now be lysed. If necessary, mix again and leave for another 15 min on ice.
5. Centrifuge at 150 g at 4 °C for 10 min. Remove the supernatant with a micropipette to a microtube and place on ice.
6. Into each chamber of the electrophoresis cell base place 95 mL 0.05 M barbital buffer. Ensure that the liquid levels are equal. Then fill the chamber cover with 500 mL ice-cold water.
7. Remove the agarose gel (Authentikit) from the rigid mould by peeling the film back. Lay the gel face up with the plastic film against the work surface.
8. Using a Hamilton syringe with a Teflon tip, place 1 μL sample into each of the pre-formed slots.
9. Position the gels into the chamber cover with the agarose facing upward. Ensure that the positive and negative indicators on the gel match those of the chamber.
10. Connect the power and run the gel at 160 V DC for 25 min. The time must be precise.
11. Switch off the power, lift the chamber cover, and stand on paper towels to drain for 30 sec.
12. Remove the agarose film and place gel side up on the work surface. Pour the contents of the enzyme substrate vial down the center of the gel, and using the edge of a pipette, spread over the gel until fully covered.
13. Drain off excess substrate by holding the corner of the film on a paper towel. Place gel side up onto an incubator tray with a sheet of damp filter paper. Cover the tray with the lid and incubate at 37 °C until the enzyme bands appear (i.e., up to 20 min).
14. Stop the reaction by immersing in tap water and leave until fully destained.
15. Dry the gels at 60 °C in an oven for 60 min.

EVALUATION OF RESULTS

To complete this it is necessary to obtain a copy of the tabulated results from Authentikit (Fig. 1).

1. Measure the distance the enzyme bands have migrated.

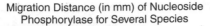

Migration Distance (in mm) of Nucleoside
Phosphorylase for Several Species

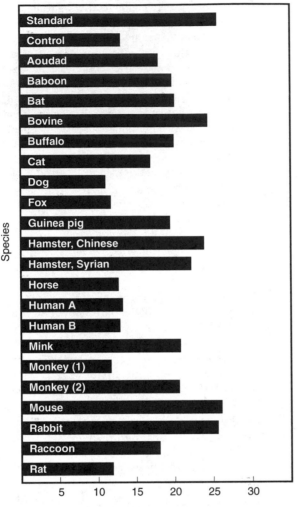

(1) Monkey, Cercopithecus
(2) Monkey, Rhesus

Figure 1. Migration distances for various species. *Source*: Cell and Tissue Culture: Laboratory procedures. A. Boyle and J.B. Griffiths (eds.) John Wiley & Sons, Chichester.

Migration distances (in mm) of nucleoside
phosphorylase for several species

Species	Migration Distance	Ratio
Standard	24.6	1.00
Control	12.8	0.52
1. Aoudad	17.2	0.70
2. Baboon	18.8	0.75
3. Bat	19.7	0.79
4. Bovine	23.1	0.94
5. Buffalo	19.7	0.80
6. Cat	16.4	0.64
7. Dog	10.6	0.42
8. Fox	11.4	0.46
9. Guinea pig	18.7	0.75
10. Hamster, Chinese	23.0	0.94
11. Hamster, Syrian	21.4	0.87
12. Horse	12.4	0.50
13. Human A	12.9	0.52
14. Human B	12.8	0.52
15. Mink	20.0	0.81
16. Monkey, Cercopithecus	11.3	0.46
17. Monkey, Rhesus	20.0	0.81
18. Mouse	25.2	1.02
19. Rabbit	25.1	1.01
20. Raccoon	17.5	0.71
21. Rat	11.5	0.46

2. Check that the values for the standard and control are within ±2 mm of the listed values.

3. If they are, proceed to Step 5. If not, apply the following correction factor:

$$\text{Factor} = \frac{\text{Actual value for standard (mm)}}{\text{Listed (tabular) value for standard}} \quad (1)$$

4. Multiply the actual values for the control and test samples by the factor and compare these with the listed values. They should now be within 1–2 mm of their listed values.

5. Calculate the migration ratio as follows:

$$\text{Factor} = \frac{\text{Migration distance of control or test (mm)}}{\text{Migration distance of standard (mm)}}$$

$$(2)$$

The values for the control and test samples are compared with those given in the tabulated results provided. Identify the species, giving similar ratios for each enzyme examined, and list them.

By a process of elimination, the species should be identified (Fig. 1). *Note*: If a cell line of human origin is being checked, it is necessary to include G6PD in the examination, as this enzyme can detect HeLa contamination. This cell line has the Type B G6PD, which is very unusual in cell lines derived from Caucasians.

The enzymes selected for analysis may initially depend on the supposed species of the cell lines being tested. However, it is recommended to use at least four enzymes to confirm such preliminary results. The routine use of glucose-6-phosphate dehydrogenase and lactate dehydrogenase will give an indication of species of origins, and nucleoside phosphorylase and malate dehydrogenase will provide confirmations of a particular species match.

DNA FINGERPRINTING AND DNA PROFILING

The term *DNA fingerprinting* has been applied to a variety of methods involving PCR or Southern blotting and a range of primer sequences or nucleic acid probes for different targets in the genome. All these methods are based on a determination of the size of DNA sequences, which are hypervariable between different individuals. However, the specific genetic identity of an individual can only be demonstrated by analyzing a number of loci. This can be achieved by analyzing a number of specific loci using a panel of DNA probes (8) or multiplex PCR. In addition, there are techniques that identify multiple genetic loci simultaneously by the use of random primers (9) or DNA probes that cross-hybridize with a range of related hypervariable DNA sequences. The latter technique is commonly called multilocus DNA fingerprinting and was the first DNA fingerprinting technique to be reported (10). This technique has been used widely for human paternity and forensic studies and in analysis of population genetics for a wide range of animals and plants. The original probes 33.15 and 33.6 reported by Jeffreys (11) have proven extremely valuable for authentication and quality control of animal cell cultures (12,13), and DNA fingerprinting is now regarded as a useful component of routine identity testing for cell lines used in the production of biological reagents (14).

The probe 33.15 has proved to be extremely useful for analyzing cell lines from a wide range of species and has been validated for routine use at the European Collection of Cell Cultures (ECACC, Salisbury, UK) (13,15). The current methodology uses an oligonucleotide for the consensus core sequence of the original 33.15 probe (11) and is described in the following.

PREPARATION AND RESTRICTION ENZYME DIGESTION OF HIGH-MOLECULAR-WEIGHT DNA FROM CELL LINES

Materials

Cell suspension solution: 0.2 M sodium acetate (pH 7.5), RNAse (0.34 mg/mL)

Chloroform/isoamyl alcohol (24 : 1)

Phenol/chloroform: Phenol (saturated solution of pH >7.6 in tris buffer) mixed 1 : 1 with chloroform/isoamylalcohol (24 : 1)

TE buffer: 10 mM Tris, 1 mM disodium EDTA, pH 7.5

Electrophoresis buffer: 10 fold dilution in distilled water of 10 × TBE (162 g/L Tris, 46.3 g/L ortho-boric acid, 9.5 g/L disodium EDTA, pH 7.5) with 0.5 mg/L ethidium bromide

Depurination solution: 0.25 M hydrochloric acid in distilled water

Denaturation solution: 1.5 M sodium chloride, 0.5 M sodium hydroxide

Neutralization solution: 3.0 M sodium chloride, 0.5 M Tris-HCl pH 7.5, 1 mM disodium EDTA

X20 SSC: 175 g/L sodium chloride, 88.2 g/L trisodium citrate at pH 7.4

Stop mix: 1 mL loading buffer (e.g., Sigma G2526), 0.25 mL ethidium bromide (10 mg/mL), 0.25 mL 10XTBE.

Sterile microtubes

Microfuge

10% (w/v) SDS pH 7.2

RNAse type 1A (10 mg/mL)

Proteinase K (10 mg/mL)

Micropipetter (1 mL) and phenol-resistant disposable tips

Incubator or infrared lamp

uv spectrophotometer

Restriction enzymes (HinfI, HaeIII) and digestion buffer (from enzyme manufacturer)

Sterile distilled water (0.22 μm filtered)

Submarine horizontal electrophoresis equipment

Agarose (low endo-osmosis)

Whatman 3 MM chromatography paper (or equivalent)

Nylon membrane (20 cm × 20 cm) (e.g., Hybond-N, Amersham)

Wash trays (approximately 28 cm × 22 cm × 4 cm)

1. Resuspend a pellet of approximately 5×10^6 cells in a microtube with 200-μL cell resuspension.

2. Add 25 μL 10% SDS and mix by inversion.

3. Incubate at 50 °C for 5 h.

4. Microfuge (10,000 g for 5 min) and transfer the upper aqueous phase to a fresh labeled microtube.

5. Mix by inversion with 200 μL phenol/chloroform.

6. Microfuge (10,000 g for 5 min) and transfer the aqueous phase to a fresh microtube.

7. Repeat steps 4, 5 and 6 twice.

8. Mix by inversion with 200 μL chloroform and repeat step 6.

9. Precipitate the DNA with 2 volumes of cold absolute ethanol.

10. The DNA is pelleted by microfuging (12,000 g for 5 min), the alcohol is aspirated, and the pellet partially dried in air (37 °C or under an infrared lamp).

11. The DNA pellet is resuspended and dissolved at 37 °C overnight with mixing) in 30–40 μL TE and quantified by uv spectrophotometry at 260 and 280 nm. An optical density (OD) value of 1 corresponds to a double-stranded DNA concentration of 50 μg/mL. The ratio A260/A280 gives an indication of nucleic acid purity, and this ratio should be 1.8 for pure DNA.

12. A 5 μg sample of the DNA is digested at 37 °C overnight after mixing with Hinf1 or HaeIII enzyme (5–10 units/μg DNA), 4 μL 10× enzyme reaction buffer and sterile water up to a total volume of 40 μL.

13. The enzyme digest is stopped by addition of 8 μL 6 × stop mix.

SOUTHERN BLOT

Materials

Lambda/HinDIII molecular weight marker

Electrophoresis buffer (as in the preceding)

Plastic wash trays (1 L capacity)

20xSSC (see above)

uv transilluminator (315 nm)

1. Quantify the DNA in each digest and run an analytical 0.8% agarose gel of at least 20 cm in length and electrophorese 5 µg digested DNA with 1/5 volume 6× stop mix. Allow the 2.3 kb fragment of a Lambda/HinDIII molecular weight marker to run the full length of the gel.

2. Gently agitate the gel consecutively, with intermediate distilled water washes, in the following buffers: neutralization solution (15 min), denaturation solution (30 min), neutralization solution (30 min).

3. Transfer the gel onto a supported paper platform (two sheets Whatman 3MM, Whatman-Labsales, or equivalent) drawing from a 20xSSC reservoir of at least 1 L.

4. Place a nylon membrane (20 cm × 20 cm, Hybond N, Genetic Research Instruments Ltd., or equivalent) over the top surface of the gel and then cover with four similar-sized sheets of Whatman 3MM soaked in the 20xSSC transfer buffer. Place two piles of absorbent paper towels (e.g., Kimberley Clark) over the soaked sheets and apply a 1 kg weight evenly on top of the towels.

5. Transfer of DNA fragments should be complete after blotting overnight, when the nylon membrane can be removed, dried, and fixed over a uv transilluminator (315 nm for 5 min).

VISUALIZATION OF DNA FINGERPRINTS BY CHEMILUMINESCENCE WITH THE MULTILOCUS PROBE 33.15

Details of this procedure were provided with permission of the probe manufacturer Cellmark Diagnostics (Abingdon, UK).

Materials

1xSSC: 1 : 20 20xSSC in distilled water

Wash solution 1: 160 mL/L 0.5 M Na_2HPO_4, pH 7.2, 10 mL/L 10% SDS

Wash solution 2: 13.8 g/L maleic acid and 8.7 g/L NaCl in distilled water at pH 7.5

Prehybridization solution: 990 mL/L 0.5 M $Na_2 HPO_4$, pH 7.2, 10 mL/L 10% SDS

Hybridization solution: 900 mL/L prehybridization buffer, 100 mL/L of 100 g/L casein

Hammarsten in stringency wash solution 2

0.1% SDS in distilled water (dehybridization solution)

Incubator (hybridization oven) at 50 °C

Trays or equivalent for hybridization

Probe (NICE™ 33.15 probe, Cellmark Diagnostics)

Lumiphos 530 (Cellmark Diagnostics)

X-ray casettes

Fuji-RX X-ray film

Developing chemicals and dark room facilities

1. Wet Southern membranes (up to 10 per hybridization in 1xSSC and place into 500 mL sterile prehybridization solution and agitate at 50 °C for 20 min.

2. Transfer the membranes individually to 160 ml hybridization solution, add the contents of one NICE™ probe vial, and gently agitate at 50 °C for 20 min.

3. Transfer the membranes individually to 500 mL prewarmed stringency wash solution 1 and gently agitate at 50 °C for 10 min.

4. Repeat step 3.

5. Transfer each membrane to 500 mL stringency wash solution 2 (see above) and gently agitate at room temperature for 10 min.

6. Repeat step 5.

7. Drain each membrane and place DNA side up on a transparent polyester sheet, apply 3 mL Lumiphos 530 luminescent reagent (*Note*: use a fume cabinet), place another polyester sheet on top of the gel, squeezing out excess Lumiphos, and seal the membrane between the sheets with tape. The sealed gels are then fixed into an X-ray development casette.

8. Fix two sheets of X-ray film over the sealed membranes and incubate at 37 °C for 3−5 h.

9. Develop the top X-ray film according to the manufacturer's instructions and check that all the expected bands in the standard HeLa DNA fingerprint are present and clear.

10. If increased exposure is required, reincubate the second X-ray film as before (*Note*: additional films can be exposed for up to 3 days after the Lumiphos has been added). *Note*: Membranes can be stripped with 0.1% SDS at 80 °C and rinsed in 1XSSC for reprobing.

INTERPRETATION OF DNA FINGERPRINT PATTERNS

For animal cell lines from species commonly encountered in cell culture (i.e., human, rodent, primate) 15−25 bands will be visualized using the described probe. The majority of other animal species will yield useful band patterns, although the number of bands may be outside the normal range indicated. High degrees of band sharing will be observed for mouse and rat cell lines from the same strain of origin. However, differences in pattern of 1−3 bands between different cell lines are usual (16), and even hybridoma clones isolated from the same fusion experiment can be differentiated (15).

Multilocus DNA fingerprinting can provide highly reproducible results (15), provided that certain precautions and quality control procedures are adopted. The most important of these factors are:

Use of a standard genomic DNA (e.g., HeLa, K562) on each gel. This will act as a control for all stages of the process.

Reproducible migration distance of molecular weight markers. This will greatly assist comparisons between gels.

Preblotting: Compare samples of high-molecular-weight DNA and digested DNA by agarose minigel electrophoresis to confirm absence of significant DNA degradation and completion of DNA digestion, respectively.

ANTIBODY STAINING FOR SPECIES VERIFICATION

The indirect fluorescent antibody staining technique has been used as a powerful technique for verifying the species of cell lines (17–19). This technique is a two-stage process. A species-specific antiserum, produced in rabbits, is used to label test cells in parallel with positive and negative controls. Next, goat antirabbit globulin, coupled to the fluorescent dye fluorescein isothiocyanate (FITC), is applied.

One particular advantage of this technique is sensitivity; mixtures containing as little as 1:1000 of cells of the wrong species can be visualized. This would not be possible with isoenzyme analysis or modern molecular approaches, although it does have the disadvantage of the use of animals.

PREPARATION OF ANTISERUM

Materials

Cell lines with authenticated species of origin

Trypan blue solution, 0.4% in Hanks balanced salt solution

FITC-conjugated, goat antirabbit antiserum

Microscope slides and coverslips

1. Prepare cells of verified species by scraping from a flask using a rubber policeman. Wash by suspending in HBSS with centrifugation at 150 g for 10 min and repeat three times.
2. Resuspend in HBSS to give a cell count of 5×10^5/mL (inoculation 1), 10^6/mL (inoculation 2), and 10^7/mL (inoculation 3).
3. Inoculate 1 mL in one ear vein of a healthy rabbit twice weekly for 3 weeks, increasing the dose each week.
4. Administer three additional booster injections at 10^7 cells/mL on a weekly basis thereafter.
5. After the third booster injection it is necessary to perform test bleedings, collect serum, and examine by serial dilution. Prepare a cell suspension at 10^6 cells/mL in growth medium and evaluate cytotoxicity with Trypan blue stain.
6. If the titres are satisfactory (1:8 or above), collect blood by cardiac puncture, allow to clot for 2 h

at room temperature, and centrifuge at 200 g for 15 min.
7. Remove serum and inactivate its complement component by incubating at 56 °C for 30 min, then dilute and distribute in aliquots and store at −70 °C.

Procedure for Test Cell Line

1. Harvest adherent cells by trypsinization and wash three times by suspending the cells in HBSS at pH 7.5 with subsequent centrifugation to form a cell pellet.
2. Resuspend the washed cells in HBSS to a final cell density of 5×10^6/mL.
3. Mix 0.1 mL cell suspension and 0.1 mL diluted antiserum and place in a humidified incubation chamber at room temperature for 30 min. The correct dilution of antisera will need to be determined beforehand for every preparation of antiserum with positive control cells. Nonspecific absorption can be avoided by diluting the antiserum to an appropriate level.
4. Cells are washed to remove nonabsorbed antiserum with three changes of HBSS and are incubated with occasional shaking for 30 min in the dark with 0.1 mL FITC-conjugated, goat antirabbit antiserum.
5. Following a further three washes with HBSS, a drop of cell suspension is sealed under a coverslip. This is then examined by fluorescence microscopy at 500x using appropriate illumination/optics for FITC.
6. A positive result is seen as bright fluorescence on the cell surface.

Normally, controls of cells of the suspected species, a related species, and an unrelated species are included.

VIABILITY TESTING USING THE MTT ASSAY

The MTT assay (20) is a sensitive, quantitative, and reliable colorimetric assay that provides information on viability and on cell proliferation. Mitochondrial dehydrogenase enzymes in cells can convert a yellow water-soluble substrate 3-(4,5-dimethylthiazol-2-yl)-2,5-diphenyl tetrazolium bromide (MTT) into a dark blue formazan product, which is insoluble. The amount of formazan is directly proportional to the cell number in a range of cell lines (21). The MTT assay is especially useful with cells, that are no longer dividing but remain active.

Reagents

MTT stain: MTT stock solution of 5 mg/mL (Sigma) prepared in phosphate-buffered saline (PBS) pH 7.5 and filtered through a 0.22-μm filter to sterilize and remove insoluble residue

Acidified propan-2-ol:0.04 M HCl in propan-2-ol

Materials

96-well microtitre plate

microELISA reader

Procedure

Caution: MTT is a mutagenic and toxic agent.

1. To 100 μL cell suspension or cell monolayer in each microtitre well add 10 μL MTT (5 mg/mL) and incubate in a humidified incubator at 37 °C for 3 h.

2. Add acidified propan-2-ol to each well and mix thoroughly to dissolve insoluble blue formazan crystals.

3. Read plate on a microELISA reader using a test wavelength of 570 nm and reference wavelength of 630 nm. Plates must be read within 30 min of adding acidified propan-2-ol.

The MTT assay can also be used to quantify cell activation by determining the maximal velocity (v) of the reaction (22).

A linear relation exists between cell count and MTT formazan up to 10^6 cells. For experiments with higher cell densities, it is recommended that a standard curve is constructed (23).

NEUTRAL RED ASSAY

Neutral red (3-amino-7-dimethyl-2-methylphenazine hydrochloride) is water-soluble, weakly basic, and a supravital dye. It has the valuable property of accumulating in the lyosomes of viable cells.

The neutral red (NR) assay has become a standard viability test for cytotoxicity determinations (24). Damaged or dead cells remain unstained by neutral red since it is no longer retained in cytoplasmic vacuoles and the plasma membrane does not act as a barrier to retain the dye (25).

Reagents

Neutral red. 4 mg/mL stock solution: Dilute 1:100 into medium, incubate overnight at 37 °C, and centrifuge before use. Protect from light.

1% CaCl$_2$/0.5% formaldehyde. Mix 6.5 mL 37% formaldehyde with 50 mL 10% CaCl$_2$ and 445 mL distilled water.

1% Acetic acid/50% ethanol. Mix 4.75 mL acetic acid with 250 mL 95% ethanol and 245 mL distilled water.

Materials

96-well tissue culture plates

ELISA reader

Microplate shaker

Eight-channel pipette

1. Prepare a cell suspension. Count cells and accurately adjust the cell concentration to achieve about 60–70% confluence at time of addition of test agent. The normal range is from 9×10^3 to 4×10^4 cells per well of a 96-well plate.

2. Seed 0.2 mL cells to each well in a 96-well plate and incubate at 37 °C for 24 h.

3. Aspirate to remove medium and add fresh medium containing graded dilutions of test agent. Incubate for desired interval.

4. Examine cultures with an inverted microscope to determine highest tolerated dose (HTD), which is about 90% cell survival, and those concentrations leading to total cell destruction. An appropriate concentration range of test agent can then be selected for a subsequent neutral red assay (26). It is important to examine at least 4–8 wells per concentration of test agent.

5. At the time of assay, remove medium with test agent and incubate cells with fresh medium containing 40 μg/mL neutral red. The medium should be incubated at 37 °C overnight to allow precipitation of dye crystals. Centrifuge medium at 1500 g for 10 min before use and add 0.2 mL decanted supernatant to each well. The first two wells on each plate should receive medium without neutral red and serve as blanks.

6. Incubate for 3 h to allow incorporation of neutral red into surviving cells.

7. Remove the medium by inverting the plate and wash cells. Rapid rinsing with a mixture of 1% calcium chloride and 0.5% formaldehyde is recommended.

8. Extract dye into the supernatant with 0.2 mL of a solution of 1% acetic acid/50% ethanol. After 10 min at room temperature and rapid agitation for a few seconds on a microtitre plate shaker, scan the plate with an ELISA reader with a 540-nm filter.

9. Compare absorbance of dye extracts with control cells against unknowns by determining the arithmetic mean (sum of replicate values divided by the number of replicates) for each set of concentrations of test agent. Calculate the percentage cell population viability.

$$\% \text{ viability} = \frac{\text{Mean absorbance of experimental cells}}{\text{Mean absorbance of control cells}}$$

(3)

10. Use individual data points for each experimental concentration presented as the arithmetic mean plus or minus the standard error of the mean to construct concentration–response toxicity curves. Such curves are used to calculate midpoint toxicities or NR$_{50}$ values.

CONCLUSIONS

Cell line characterization methods have developed from basic chromosome analysis through to isoenzyme analysis and finally reached the molecular biology age with DNA profiling techniques. Each has its own benefits and drawbacks, and the nature of the expertise of each laboratory can influence the ultimate choice of routine characterization method used. It is important to stress that characterization is a requirement; cases of cross-contamination of cell lines is not simply anecdotal, and years of research can be invalidated as a consequence of inadequate quality control applied both at the start and during the course of a cell-based project.

Quantification methods are part of everyday routine; the more they can be automated the better, and the key techniques of MTT and neutral red assay are provided here, as they are the more relevant to in-process testing in animal cell technology.

BIBLIOGRAPHY

1. W.A. Nelson-Rees, D.W. Daniels, and R.R. Flandermeyer, *Science* **212**, 446–452 (1981).

2. P.D. Van Helden et al., *Cancer Res.* **48**, 5660–5662 (1988).

3. D.M. Halton, W.D. Peterson, and B. Hukku, *In Vitro* **19**, 16–24 (1983).

4. S.J. O'Brien, J.E. Shannon, and M.H. Gail, *In Vitro* **16**, 119–135 (1980).

5. A.T. Sumner, H.J. Evans, and R.A. Buckland, *Nature New Bio.* **232**, 31–32 (1971).

6. M. Seabright, *Chromosoma* **36**, 204–210 (1972).

7. G.N. Stacey, H. Hoelzl, J.R. Stephenson, and A. Doyle, *Biologicals* **25**, 75–85 (1997).

8. M. Honma, G.N. Stacey, and H. Mizusawa, in A. Doyle, J.B. Griffiths, and D.G. Newell, eds., *Cell and Tissue Culture: Laboratory Procedures* **9A**, Wiley, Chichester, England, p. 1 (1994).

9. J.G.K. Williams et al., *Nucleic Acids Res.* **18**, 6531–6535 (1990).

10. A.J. Jeffreys, V. Wilson, and S-L. Thein, *Nature* **316**, 76–79 (1985).

11. A.J. Jeffreys, V. Wilson, and S.L. Thein, *Nature* (London) **314**, 67–73 (1985).

12. D.A. Gilbert et al., *Am. J. Hum. Genet* **47**, 400–517 (1990).

13. G.N. Stacey, B.J. Bolton, and A. Doyle, *Nature* **357**, 261–262 (1992).

14. World Health Organization Expert Committee on Biological Standardization and Executive Board, *Requirements for the Use of Animal Cells as in vitro Substrates for the Production of Biologicals*, WHO, Geneva (in press).

15. G.N. Stacey et al., *Cytotechnology* **8**, 13–20 (1992).

16. G.N. Stacey, B.J. Bolton, A. Doyle, and J.B. Griffiths, *Cytotechnology* **9**, 211–216 (1992).

17. C.S. Skulberg, in J. Fogh, ed., *Contamination In Cell Culture*, Academic Press, New York, 1973, pp. 2–23.

18. R.J. Hay, in R.I. Freshney, ed., *Animal Cell Culture: A Practical Approach*, IRL Press, Oxford, 1986, pp.71–112.

19. R.J. Hay and M.L. Macey, in A. Doyle and J.B. Griffiths, ed., *Cell and Tissue Culture: Laboratory procedures* **9A**, Wiley, Chichester, England, 1995, pp. 1–3.

20. T. Mossman, *J. Immunol. Methods* **65**, 55–63 (1983).

21. M. Al Rubeai and R. Spier, in R.E. Spier, J.B. Griffiths, J. Stephenne, and P.J. Croog, eds., *Advances in Animal Cell Biology and Technology for Bioprocesses*, Butterworth, London, pp. 143–155. 1989.

22. D. Gerlier and N. Thomasset, *J. Immuol. Methods* **94**, 57–63 (1986).

23. M. Al Rubeai, in A. Doyle and J.B. Giffiths, eds., *Cell and Tissue Culture: Laboratory Procedures* **40**, Wiley, Chichester, England, 1995, pp. 1–2.

24. H. Babich and E. Borenfreund, *Alternatives Lab. Anim.* pp. 129–144 (1990).

25. E. Borenfreund and H. Babich, in A. Doyle and J.B. Griffiths, eds., *Cell and Tissue Culture: Laboratory Procedures* **40**. Wiley, Chichester, England, 1995.

26. E. Borenfreund and Borrero, *Cell Biol. Toxicol* **1**, 55–65 (1984).

See also CELL STABILITY, ANIMAL; CELL STRUCTURE AND MOTION, CYTOSKELETON AND CELL MOVEMENT; CELL SURFACE RECEPTORS: STRUCTURE, ACTIVATION, AND SIGNALING; CELLULAR TRANSFORMATION, CHARACTERISTICS; ENRICHMENT AND ISOLATION TECHNIQUES FOR ANIMAL CELL TYPES.

CELL BANKS: A SERVICE TO ANIMAL CELL TECHNOLOGY

G. STACEY
National Institute of
Biological Standards and Control
South Mimms, England

A. DOYLE
Wellcome Trust
London, England

OUTLINE

Overview of Culture Collection Activities and Operations

Accessioning

Master and Distribution Cell Banking System

Cryopreservation

Ampoules

    Cryopreservation Method

Resuscitation of Cyropreserved Cultures

Shipping and Distribution

Quality Control of Cell Lines

    Testing for Bacteria and Fungi

    Mycoplasma Testing

Cell Identification and Characterization

    DNA Fingerprinting

    Isoenzyme Analysis

Quality-Control Services

    Depository Activities

Overview of Representative Culture Collections

American Type Culture Collection (ATCC) — An Historical Perspective

Current Facilities and Operations

Coriell Institute for Medical Research, the Home of the Coriell Cell Repositories

    Establishment and Development

From Hard Copy to Internet: The Coriell WWW Catalogues

Submission of Cell Lines

Since the first culture collection was established in Prague in 1889 by Frantisek Kral, culture collections have developed an important role as a service and support activity that aids the biotechnologist both in industry and academia. Service collections are repositories of cultures that promote accessibility of authentic quality-controlled cell cultures. Culture collections form a worldwide network that promotes awareness of new developments in cell culture, including advances in research and regulatory affairs. This article will give details on their activities and profiles of the major centers.

All culture collections are different; however, three general rules of operation can be recognized (Table 1) (1). At the initial level, repositories provide safe storage alone. At the next level, the collection acts as a centralized resource for an institution, and repository commitments must meet the needs of a number of departments and distribution to collaborators. Finally, at the level at which gene banks and service collections must operate, the demand from users becomes significant. Thus, in the case of service collections, customer service, sales, and marketing are essential activities, even for nonprofit service culture collections.

Each culture collection has a remit that may arise from its scientific origins or source of financial support. Some collections are very specialized, with an emphasis on research activity. Others adopt a more pragmatic service approach focused on improving availability of a comprehensive range of cultures and related services (which can aid income generation). Culture collections are not just safe repositories for important biological reference materials. Even the most commercially minded centers recognize the importance of continued development to ensure the relevance of its cultures to the current and future needs of the user community.

Table 1. Three Different Levels for the Operation of a Culture Collection

Activity	Safe depository	Stock facility	Service collection
Cell culture and quality control	$+^a$	+	+
Collection activities[b]	−	+	+
Extended activities[c]	−	−	+

[a] Quality control may only include viability testing.
[b] Including authentication, cataloguing, and distribution.
[c] Including training courses, technology development, conference exhibitions, and acquisition of new cultures.
Source: Published by permission of *Cryo-Letters* (1).

Most collections are actively involved in an international network mediated through organizations such as the European Culture Collection Organisation (ECCO Secretary, M-L. Suihko, VTT, Espoo, Finland; email *maija-liisa.suihko@vtt.fi*) and the World Federation for Culture Collections (WFFC Secretary, A. Doyle, Wellcome Trust, UK; email *a.doyle@wellcome.ac.uk*). The catalogues of some collections are accessible on the Internet (e.g., *http://www.ch.ic.ac.uk/medbact.index.html*). In addition, there are established systems such as WDCM (World Directory of Collections of Cultures of Micro-organisms, WFCC World Data Centre of Micro-organisms, Saitama, Japan) and a new EU-funded initiative called CABRI (Common Access to Biological Resource Information; email *http://www.cabri.org*), which aim to provide efficient access to the databases of many collections. A special feature of CABRI is that it is focused on harmonized protocols for culture banking and quality control such that a high minimum standard is set for the European member collections.

OVERVIEW OF CULTURE COLLECTION ACTIVITIES AND OPERATIONS

In order to be able reproducibily to prepare cell lines of a suitably high standard for distribution to the research community and the pharmaceutical industry alike, culture collection need expertise in a wide area, ranging from technical expertise required for the growth, maintenance, and preservation of materials, strict acceptance criteria developed over many years, and also administrative skills required for data handling, cataloguing, and sales/order processing systems. Obviously the precise details of data handling, cataloguing, the choice of software, and database design will vary with each culture collection, as will the sales order processing systems. However, there is a high degree of commonality regarding the growth, preservation, and quality-control procedures used by the major culture collections. All collections have high standards regarding the quality of material they supply to their customers. A minimum level of number of viable cells per sample in conjunction with a minimum percentage viability is applied. In addition, all cultures are guaranteed to be free of contaminants (i.e., adventitious agents and cross-contaminating cell lines). Thus to be able to maintain all these aspects, culture collections must have technical expertise in these areas:

- The culture of a diverse collection of cell types derived from different species and different tissues
- Cryopreservation of a diverse collection of cell types derived from different species and different tissues
- Optimum storage conditions and monitoring
- Rigorous cell banking procedures
- Rigorous quality-control procedures and standards
- Shipping and distribution

The remainder of this article is devoted to these topics, with particular emphasis upon the quality-control aspects. In addition, there are overviews of representative culture collections regarding their remit, their development in response to this remit, and also areas of specific interest.

ACCESSIONING

The means by which collections obtain and process new cell line accessions is critical to the value of each repository. This value can be measured in terms of the immediate quality of the cells and information that can be supplied to customers and in terms of the development of the scientific value of the collection. Each collection needs to have a rigorously applied accessioning that provides documented links between cell line information and authentication/quality-control data with specific cryopreserved stocks. This protocol should also provide controls to prevent distribution of cells before appropriate quality-control procedures have been completed.

Proactive collections may solicit new accessions by searching the current literature. This helps to ensure that cell banks are already prepared and tested as the research community requests them.

When accepting a new accession, adequate information must be obtained to identify the scientific characteristics and specific of each cell line. However, it is important in cell data records to categorize the information clearly which data are (*1*) generated within the collection, (*2*) generated by the depositor, and (*3*) taken from published research. Obviously the last section may contain data from cells that may not be authentic or altered through prolonged passage. In addition to scientific data, any risk that may be associated with the cells must be identified to ensure safe laboratory containment and operator protection. In many cases there are no evident infectious hazards associated with the cells. Experience in tissue culture indicates that the general risk of direct infection from cells is low and the recommended containment (e.g., European containment level 2)(2) provides operator protection against many unexpected pathogens. Nevertheless, the few serious instances of laboratory infection from in vitro cell cultures are sufficient to require a continuous approach that can be greatly assisted by classifying cells based on species and tissue of origin (3,4). Thus careful and documented accessioning procedures are central to the provision of reliable, reproducible, and safe cell cultures. Their establishment should be a priority for new and developing collections.

MASTER AND DISTRIBUTION CELL BANKING SYSTEM

The implementation of a master and working bank system (Fig. 1) is probably the most important procedure in the maintenance of cell cultures, and is essential for the provision of reproducible cultures. Moreover, the continuous propagation of cell cultures can lead to the loss of desirable characteristics present in low-passage cell cultures. A master and working cell banking system can help to overcome this by providing preserved material with minimized passaging. A master cell stock should be prepared and fully characterized in terms of cell

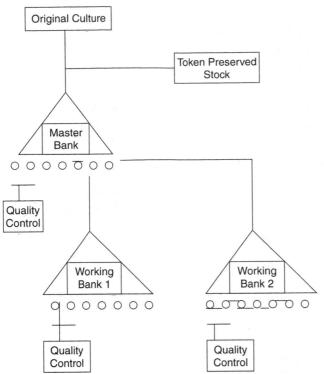

Figure 1. A typical master cell banking and working cell bank system (o, ampoule of cryopreserved cells). During the first few subcultures a 'token' preserved stock of 1–3 ampoules is prepared in case of contamination during preparation of the master cell bank. Preparation of master cell banks in this way ensures that all cultures provided for the working bank are identical in terms of subculture history (i.e. 'age'). This type of system in *bona fide* culture collections is a key component in promoting standardization in biological research and biotechnology. (Reproduced by permission of Cryo Letters) (2).

numbers, cell viability, presence/absence of contaminants, and presence of the characteristics of interest for that cell line, for example, the ability to support the growth of viruses, differentiated phenotype. Once characterized, this stock can now be used to prepare a series of working cell stocks, which should also undergo a quality-control screen to establish the absence of contaminants and conformity with the master cell stock. The major advantage of this system of cell banking is the high degree of standardization and reproducibility achieved, which ensures that all experiments performed at different times and in different laboratories are conducted on cells of a similar passage number and population doubling.

CRYOPRESERVATION

Before a culture collection can take on the long-term storage of an organism, appropriate preservation procedures must be in place. Each stored ampoule must contain a representative complement of cell types to ensure faithful recovery of a culture with the characteristics of the original. If the culture can be preserved (i.e., by cryopreservation at −196 °C), then it is essential to examine

post-thaw stability, since the preservation process may apply selection pressures leading to alteration of culture characteristics on recovery from liquid nitrogen storage. This is especially important when storing cultures that comprise diverse cell types. Suboptimal cryopreservation exposes a culture to the risk of overgrowth by genetically altered or even contaminant cells. While reliable procedures for cryopreservation are available for most bacteria, yeasts, and animal cells, such methodologies for other cell types, including protists, some fungi, plant cells, and animal embryos and gametes, remain the preserve of specialist laboratories.

It is impractical for most laboratories to maintain cell lines in culture indefinitely; moreover, cell cultures undergo genetic drift with continuous passage, which risks losing their differentiated characteristics. It is recommended practice in most cell banks that a limit be set for the number of population doublings that any cell lines should undergo before reinitiation from a fresh cryopreserved ampoule. This is of less importance for undifferentiated cell lines of infinite lifespan, although such cultures remain susceptible to variation and may "fail to perform."

Nearly all animal cell lines can be cryopreserved successfully in liquid nitrogen at −196 °C. A few simple criteria must be adhered to in order to provide guaranteed success with cryopreservation.

The essentials for efficient cryopreservation are slow freezing, at a rate of between −1 and −3 °C per minute, and fast thawing, which is achieved by placing ampoules in a water bath at 37 °C. The addition of a cryoprotectant, such as dimethyl sulfoxide (DMSO) or glycerol, is important for survival during cryopreservation. Nevertheless, DMSO does have toxic properties and, as a solvent, may facilitate the passage of other toxic chemicals through the skin of the operator. Therefore, care must be taken in handling the cryoprotectant, and some consideration may be necessary as to whether cells should be washed immediately on thawing.

AMPOULES

The decision on whether to use glass or plastic ampoules will depend upon the practice of each cell bank. Complete closure, eliminating nitrogen, can only be achieved with sealed glass ampoules. For most laboratories it is more practical to use the screw-cap plastic ampoule. These are made of a special grade of plastic designed to withstand extreme cold (cryotubes). Even though they are more convenient to use, precautions are still necessary in handling ampoules of this type. Ampoules with an external thread on the body of the cryotube are recommended to avoid contamination.

Cryopreservation Method

A generic cryopreservation protocol is outlined here:

1. It is best practice to cryopreserve cultures that have high levels of viability and in the exponential phase of growth.

2. A viable cell count is performed, such as by trypan blue exclusion. For most cultures over 90% of cells should be viable prior to preservation.

3. Cells are centrifuged and resuspended in sufficient cryopreservation medium to fill the required ampoules. In general each ampoule should contain at least 4×10^6 cells per ampoule but should not exceed 2×10^7 cells per ampoule.

4. A recommended medium is whole serum (e.g., newborn calf or foetal bovine serum) to which 7–10% (v/v) cryoprotectant is added (e.g., DMSO or glycerol). By eliminating culture medium, which usually contains bicarbonate, a pH closer to neutral can be maintained during the freezing process. It is possible to use 20% serum with culture medium and cryoprotectant. However, great care is necessary to avoid the pH becoming alkaline during dispensing of ampoules.

5. Resuspend the cell pellet in the appropriate volume of cryopreservation medium and dispense into the ampoules in 1 mL aliquots. The ampoules should be clearly labeled with the cell designation, passage number, and date of cryopreservation. A short equilibration period (upto 1 h) at 4 °C is recommended.

6. Cool the ampoules at between −1 to −3 °C per minute until at least −60 °C is reached. The most suitable rate should be determined by prior experimentation. For example, −1 °C per minute is used for most cell types. A programmable controlled-rate freezer can achieve this by means of an electronically controlled input valve on the nitrogen supply to a cooling chamber. When such equipment is not readily available, an alternative method involving freezing cells in a −70 °C freezer using an insulated container may be used. Ampoules are placed in a polystyrene box or block so that each ampoule is insulated by about 1.0–1.5 cm of material, including the top and bottom of the ampoule. The box is placed in a central position at a quarter to a third from the top of the freezer. The ampoules are left to cool for 16–24 h, then transferred to a liquid nitrogen storage vessel.

   A simpler method is to use the "two-stage freezer," where ampoules are placed at a predetermined depth in the neck of a dewar flask containing liquid nitrogen. After a period of 16–24 h the ampoules can be plunged directly into liquid nitrogen. Typical cooling profiles achieved by these methods are shown in Figure 2.

7. After freezing, the ampoules are immediately transferred to a nitrogen storage vessel and kept in the gas phase.

RESUSCITATION OF CYROPRESERVED CULTURES

In order to obtain the maximum level of viability in cultures recovered after cyropreservation, cells should be thawed quickly, in a waterbath set at 37 °C. In addition, for some cell types the cyroprotectant should be removed by centrifugation to minimize any potential toxic effects. A typical protocol follows:

1. Ampoules for resuscitation should be thawed rapidly by placing them in a water bath at 37 °C. To provide protection should the ampoule explode, place ampoules inside a screw-capped metal container with holes to allow free passage of water. To avoid possible explosion from trapped nitrogen, allow the contained ampoule to stand at room temperature for 2 min before placing in the water bath. During all stages of handling the ampoule, a full face mask and appropriate gloves should be worn.

2. Once thawed, the ampoule is wiped with a tissue soaked in disinfectant (e.g., 70% ethanol, glutaraldehyde). The contents are transferred with a pipette to a 15 mL centrifuge tube. Slowly add 0.5–1.0 mL of prewarmed growth medium to the tube, which is then either transferred to a prepared culture flask or a further 10 mL of prewarmed growth medium is added to the cells and the cells are centrifuged to remove the cryoprotectant. Centrifugation is important if it is considered necessary to remove the cryoprotectant due to toxicity or if, as in the case of DMSO, the cryoprotectant can alter the characteristics of the cells.

3. Cultures should be seeded into tissue culture flasks at between 30–50% of their final (maximum) cell density. This allows the cells rapidly to condition the medium and achieve exponential growth.

SHIPPING AND DISTRIBUTION

Material is supplied either in the form of frozen ampoules requiring dry ice for shipment or as growing cultures. The supply of growing cultures gives the packages longer shelf life in transit (i.e., up to 5 days) compared with a maximum of 3 days recommended for dry-ice packages. This is particularly useful in reducing the cost of distribution. Details on the requirements for packaging and shipping are given in the following. Cell cultures are categorized under the International Air Transport Association (IATA) Regulations as "Diagnostic Specimens" and will only be classified as "Dangerous Goods" where they are known to represent an infectious hazard to humans or animals. Growing cell cultures should be packaged as described in IATA packing instruction 650, and frozen ampoules are most readily transported in dry ice, for which there are additional packing regulations, given in IATA packing instruction 904. However, if it is required to ship frozen cells at liquid nitrogen temperature, it is advisable to use dry shippers where the liquid is absorbed into the container wall. The use of such containers avoids the need to conform to the requirements of IATA packing instruction 202. For the most up-to-date information on the IATA Regulations, the reader is directed to the IATA website (*http://www.iata.org/cargo/dg/dgr/htm*).

Senders of any biological material are now required to provide a material safety data sheet describing the contents of the package. Culture collections achieve this by means

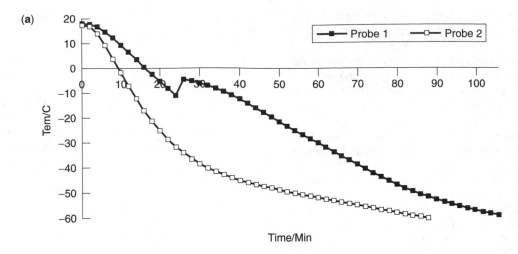

Probe 1: Temperature in ampoule
Probe 2: Temperature inside polystyrene box

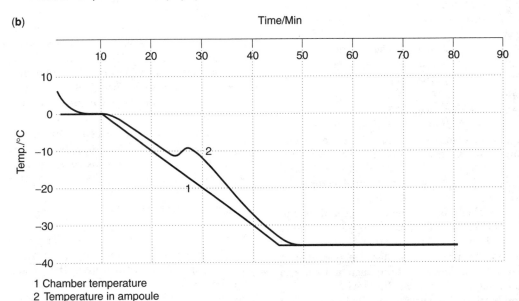

1 Chamber temperature
2 Temperature in ampoule

Figure 2. Cooling profiles in ampoules containing standard cryoprotectant using an insulated polystyrene box (**a**) and a controlled-cooling-rate freezer (Kryo10, Planer Products, Sunbury, UK) (**b**).

of a cell line data sheet (Fig. 3), which describes the cell culture, its culture condition, and characterization data.

QUALITY CONTROL OF CELL LINES

A key issue that is central to the role of cell banks is the routine testing that is carried out to ensure that cell lines accessioned into the public collections are free from the commonest forms of microbial contaminants. These fall onto three groups—bacteria and fungi, mycoplasma, and viruses. Protocols for the most commonly used methods by all the culture collections for the detection of bacteria and fungi and mycoplasma are given in the following. The testing for viruses is more specialized, and there is less commonality between the collections; therefore, the techniques used are discussed in connection with reference to each collection.

Testing for Bacteria and Fungi

A routine for many laboratories is the inclusion of antibiotics in culture media that are added as if they are a growth requirement like glucose or foetal bovine serum. However, their inclusion can simply mask low levels of contamination, and most cell banks do not include antibiotics in growth media unless isolating cells from primary material. Often when cells are cultured for the first time without antibiotics, contamination readily becomes all too apparent. There is usually an increase in pH, accompanied by turbidity in the medium. (Reagents used in the preparation of growth media may have contamination present; therefore, routine checks are obligatory before the newly prepared stocks are used.) The two methods generally used for the detection of bacteria and fungi are microbiological culture and direct

European Collection of Cell Cultures

Cell Line Data Sheet   Despatch Note: 214731

Cell Name: HEL92.1.7
ECACC No: 92111706

Description: Human   Caucasian erythroleukaemia

Culture Medium: RPMI 1640 + 2 mM Clutamine + 10% Foetal Bovine Serum (FBS).

Sub-culture routine: Maintain cultures between 3–9 × 100,000 cells/ml; 5% CO_2;
37 °C.

Morphology: Lymphoblast

Passage No: CB No: 98L052   Cells/mL:

Expected Viability: Resuscitate into: 15 mls

References: Science 1982;261:1233

- -
Important Safety Information for Handling Cell Lines

Cell Culture Hazards.

Cell line of human/primate origin. There is no evidence for the presence of infectious virus or toxic products. However ECACC recommends that cultures are handled at ACDP Category 2 containment.

Materials Safety Data

This information is provided for your own risk assesment procedures - within the UK if you have any general safety enquiries please consult the COSHH regulations 1988 from the HSE or contact the COSHH information service (Tel 0171 221 0870).

Frozen Ampoules

These are 1 ml plastic cryotubes containing cells (refer to the cell culture hazard) and FBS including 10% (v/v) dimethylsulphoxide (DMSO) which is toxic and readily penetrates the skin. The ampoules are packed in solid carbon dioxide pellets which will cause frost-bite on contact with skin. On very rare occasions residual liquid nitrogen may be present in ampoules. On warming, these ampoules may present an explosive hazard.

Growing Cultures

Cultures are despatched in 25 cm sq. tissue culture flasks filled with culture medium. Flasks are sealed with tape and placed in a closed plastic wrapper. The flasks are surrounded with both sufficient material to absorb the entire contents of the flask in case of accidents and also a layer of crushproof packaging.

- -

End of Data Sheet for Cell HEL92.1.7

Figure 3. A typical material safety data sheet (ECACC, UK) that incorporates general cell line data as well as a description of the materials supplied and any specific known hazards.

observations using Gram's stain. The methods are outlined in the following.

Detection of Bacteria and Fungi by Microbiological Culture. There are two types of culture medium recommended by the European Pharmacopoeia (5) and the United States Code of Federal Regulations (6). These relate to requirements for quality control of cell substrates utilized in vaccine manufacture.

1. Fluid thiogylcollate medium for the detection of aerobic and anaerobic organisms.
2. Soya bean-casein digest (tryptose soya broth) for the detection of aerobes, facultative anaerobes, and fungi.

Method:

1. Inoculate 1 mL of test sample into each of two pairs of 15 mL broth (thioglycollate or tryptose). This should include spent growth medium and cells. Adherent cells should be scraped from the cell sheet. Enzymes may kill contaminating organisms.
2. Incubate one pair of test samples at 25 °C and the other at 35 °C.
3. Examine daily. If bacteria are present, the broths become turbid or develop a sediment within 1–2 days. However, incubation for up to 2 weeks is recommended. Further examination can be carried out by agar culture or Gram's stain and, if necessary,

antibiotic sensitivity tests may be performed on any isolate.

4. Positive controls using reference organisms are recommended, although this depends upon the facilities available to conduct these tests (e.g., anaerobic—*Bacillus subtilis*; aerobic—*Clostridium sporogenes*).

Treatment of Contamination. In the event of bacterial or fungal infection, ideally cultures should be discarded and fresh cells recovered from frozen stocks following extensive decontamination of work surfaces and tissue culture cabinets. Fresh medium should also be used. However, in the case of irreplaceable stocks, this may not be possible, in which case treatment of the culture with antibiotics can be attempted. There are numerous antibiotics available for use with tissue cultures. The antibiotic selected will depend upon the Gram stain result of the contaminating organism; however, several antibiotics are active against both Gram positive and Gram negative bacteria, for example, streptomycin sulfate and kanamycin. A common approach is to use a combination, such as streptomycin sulfate and penicillin, which can be purchased as a cocktail from tissue culture suppliers.

Mycoplasma Testing

Mycoplasma infection of cell lines was first recorded in 1956 by Robinson et al. (7). Mycoplasmas are the smallest free-living, self-replicating organisms (0.2–2 μm in diameter). They lack a cell wall, are incapable of peptidoglycan synthesis, and are therefore not susceptible to some of the antibiotics commonly used in tissue culture, such as penicillin and its derivatives. Due to the lack of cell wall, mycoplasma are pleiomorphic, varying in shape from spherical to filamentous or helical cells. Mycoplasmas are extracellular parasites that usually attach to the external surface of the cell membrane. There are 50 species of mycoplasma. The most common ones in terms of tissue culture are *Mycoplasma arginini* (human origin), *M. fermentans* (human origin), *M. hyorhinis* (pig origin), *M. orale* (human origin), and *Acholeplasma laidlawii* (bovine origin) (8).

The incidence of this form of contamination (typically 5–35% of cultures), which is not normally visible on either routine macroscopic or microscopic examination of the cells in culture, varies considerably from laboratory to laboratory. The majority of cell repositories worldwide will not distribute cell banks contaminated with mycoplasma. The use of tissue culture products from recognized suppliers will reduce the risk of introducing mycoplasma into the tissue culture facility, since all suppliers now screen their reagents for the presence of mycoplasma. In addition, most culture collections batch test media and serum before use.

The effects of mycoplasma in culture are well known (9) but worth reviewing:

- Induction of morphological changes
- Chromosomal aberrations
- Affect the growth rate of cells
- Alter amino acid and nucleic acid metabolism
- Induce cell transformation

Their presence is not compatible with controlled in vitro systems and is of concern to the regulatory authorities with reference to in-cell substrates used in the manufacture of "biologicals" for both therapeutic and diagnostic use. In view of this, prescribed tests should be undertaken (10–12).

The methods available for detection are in some cases considered old fashioned and labor intensive, such as DNA staining (Hoechst stain) and microbiological culture, although PCR methods are increasingly popular (13). Other tests include uridine/uracil ratio analysis, electron microscopy, biochemical detection of adenosine phosphorylase, and DNA–RNA hybridization. Since each technique has disadvantages and limitations, at least two procedures should be used to minimize false-positive and false-negative results. Protocols for the most commonly used techniques of agar culture and DNA fluorescence staining are given in the following.

DNA Staining Method (Hoechst 33528). The fluorochrome dye Hoechst 33528 binds specifically to the adenine–thymidine regions of DNA, causing fluorescence when viewed under ultraviolet (UV) light. A fluorescence microscope equipped for epi-fluorescence with a 340–380 nm excitation filter and 430 nm suppression is required. Due to the fundamental requirement for mycoplasma-free cell lines distributed by cell banks, a detailed outline of this technique is provided, as it remains the most usual detection method in current use.

Method: All cell cultures should be passaged at least three times in antibiotic-free medium prior to testing. Failure to do this may lead to false negative results.

1. Harvest adherent cells by scraping the cell sheet with a sterile swab. Resuspend in the original culture medium at approximately 5×10^5 cells/mL. Suspension cell lines should be tested directly at a similar concentration.

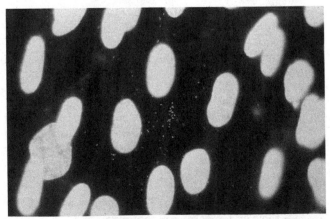

Figure 4. Cell preparation of the Vero cell lines with Hoechst 33528 DNA stain and viewed at 100× magnification using a fluorescence microscope. (*Source*: Photographs kindly supplied by D. Hockley, NIBSC, UK.)

2. Add 2–3 mL of test cells to each of two 22-mm sterile coverslips in 35-mm-diameter tissue culture dishes. Incubate at 37 °C in a humidified atmosphere of 5% CO_2 and 95% air. Examine one sample at 24 h and the other after 72 h.

3. Before fixing, examine the cells on an inverted microscope (10× objective magnification) for evidence of microbial contamination.

4. Working in a fume cupboard, to each dish add to the medium 2 mL of freshly prepared Carnoy's fixative (methanol:glacial acetic acid, 3:1 v/v) dropwise from the edge of the culture dish. Take care not to disturb the cells. Leave for 3 min at room temperature.

5. Pipette the fixative into a waste bottle and immediately add another 2 mL of fixative. Leave for a further 3 min at room temperature.

6. Air dry the coverslip in air on the inverted tissue culture dish lid for 30 min.

7. Make a stock solution of Hoechst 33528 stain (10 mg/100 mL) and store protected from light. Dilute this to the working concentration just before use. In a fume cupboard and wearing gloves, add 2 mL freshly prepared Hoechst 33528 stain (100 µg/L in distilled water) to each dish and leave for 5 min, shielding the coverslip from light.

8. Still working in the fume cupboard, decant the stain. Dispose of according to local safety procedures for toxic substances.

9. Add one drop of mountant (22.2 mL 0.1 M citric acid, 27.8 mL 0.2 disodium phosphate, adjusted to pH 5.5) to a glass slide and mount the coverslip, cell side down.

10. Examine under UV fluorescence at 100× magnification. Uninfected cultures observed under fluorescence microscopy are seen as fluorescing nuclei against a negative background; cultures contaminated with mycoplasma appear as brightly fluorescent, uniformly shaped bodies surrounding the cells, or as small clusters between the cells. Artifacts may be fluorescent and interfere with the interpretation of results, but they will appear larger in size than mycoplasma and irregular in shape.

Sometimes it may prove difficult to differentiate between contaminated and uncontaminated cultures of cells. To overcome this problem, and also for reasons of standardization, the assay may be performed using a monolayer culture as an indicator onto which the test sample is inoculated. This method also has the advantage of being able to screen serum, cell culture supernatants, and other reagents that do not contain cells. Indicator cells, which should have a high cytoplasm/nucleus ratio, such as Vero (African Green Monkey kidney) cells, are incubated in tissue culture dishes containing coverslips for 12–24 h prior to inoculation of the test sample. As with test cells, indicator cells are inoculated at a concentration such that a semiconfluent monolayer is formed by the time of staining (e.g., 5×10^4 cells/mL).

Some preparations may show extracellular fluorescence caused by disintegrating nuclei. Fluorescent debris is usually not of uniform size and is too large to be mycoplasma. Contaminating bacteria or fungi will also stain if present, but will appear much larger than mycoplasma. Compare with positive and negative slides. Positive slides using positive control strains (e.g., *M. hyorhinis* and *M. orale*) inoculated at a concentration of 100 c.f.u/dish can be an invaluable help to identification.

Positive controls should be handled in specialist facilities quite separate from the tissue culture laboratory. However, for a small tissue culture unit that is not able to set aside specialist facilities, positive control slides can be obtained from commercial suppliers.

The main advantage of the DNA staining method is the speed (less than 1 day) at which results are obtained, and that the noncultivable strains of *M. hyorhinis* can be detected. However, this method is not as sensitive as the culture method. It is generally considered that approximately 10^4 mycoplasma per mL are required to produce a clear positive result by the Hoechst stain technique.

Microbiological Culture. Most mycoplasma cell culture contaminants will grow on standardized agar and broth media, with the exception of certain strains of *M. hyorhinis*. As in the case with microbial positive controls, the culture methods to be described should be performed in a laboratory quite separate from the main tissue culture area.

Method:

1. Harvest adherent cells by scraping the cell sheet with a sterile swab. Resuspend the cells in the original culture medium at a concentration of approximately 5×10^5 cells/mL. Test suspension cells directly at the same concentration.

2. Inoculate an agar plate with 0.1 mL of the test sample.

3. Inoculate 2 mL mycoplasma broth with 0.2 mL of the test sample.

4. Incubate test plate under anaerobic conditions, for example, using the Gaspak anaerobic system (BBL) at 36 °C for 21 days and test broth at 36 °C.

5. After 7 and 14 days incubation subculture the broth (0.1 mL) onto agar plates. Incubate anaerobically at 36 °C as above.

6. Examine all agar plates after 7, 14, and 21 days incubation under 40× and 100× magnification using an inverted microscope for detection of mycoplasma colonies.

Both agar and broth media should be checked prior to use for their ability to grow species of mycoplasma known to contaminate cell lines, such as *A. laidlawii*, *M. arginini*, *M. fermentans*, *M. hominis*, *M hyorhinis*, and *M. orale*. Type strains are available from the National Collection of Type Cultures (U.K.) or the American Type Culture Collection (U.S.). The main advantage of the method is that, theoretically, one viable organism per inoculum can be detected, compared with 10^4 mL for the Hoechst staining method.

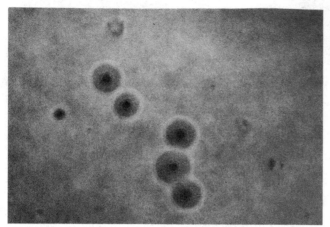

Figure 5. Mycoplasma colonies on agar viewed using an inverted microscope.

Treatment of Contamination. As with bacterially infected material, the best course of action is to discard the infected material and resuscitate the earlier passage material, which may be uninfected. However, for some cell lines this may not be possible. Therefore, in cases such as this it is possible to attempt to eradicate the infection. The eradication of mycoplasma is less straightforward and has less chance of success than the eradication of bacteria. However, by using one of the agents proven to be effective against mycoplasma, success rates of 50–60% may be obtained. Agents such as MRA (mycoplasma removal agent), Ciprofloxacin, Baytril, or BM Cyclin have been used with success by the culture collections. It should be borne in mind that even following apparently successful eradication cell lines should be monitored to check that removal is permanent.

CELL IDENTIFICATION AND CHARACTERIZATION

Cell lines are used because of their unique characteristics. Therefore, it is essential to ensure that these characteristics are still in evidence after undergoing cell banking and cyropreservation. In order to do this, culture collections undertake cell characterization and authentication tests to check the species of origin of a cell line and confirm the genetic makeup of cell in terms of their DNA fingerprint or karyotype. In addition, the presence of specific markers and cell surface receptors may be confirmed. However, no collection is in the position to verify and examine every aspect of a given cell line. The verification of a cell line is crucial since there are documented cases of some well-recognized cell lines, including KB, WISH, and Chang liver cells, being contaminated with HeLa cells (14–18). Many researchers are unaware of these problems. It should be borne in mind that the use of contaminated cells may yield anomalous results.

The identity tests to be described rely on the detection of genetic differences between individuals of the same species and of different species. There are several tests used throughout culture collections to detect these differences. These include DNA fingerprinting and isoenzyme analysis,

which are discussed in the following. Some culture collections perform supplementary tests to augment the standard techniques. These are discussed with reference to the specific collections.

DNA Fingerprinting

DNA fingerprinting visualizes numerous polymorphic loci spread throughout the genome on different chromosomes. These loci are known as variable number tandem repeats (VNTRs). The number and size of VNTRs present in the genome vary with the individual, and specific loci show differences between individuals that are inherited in a Mendelian manner. DNA fingerprints may be obtained from a large number of species, including mammals, rodents, insects, and fish. The process involves the preparation of Southern blots of genomic DNA digested with a restriction enzyme (e.g., HinfI, Hae III). A labeled VNTR probe such as Jeffrey's multilocus probe 33.15 (19) is then hybridized to the blot, and the bound probe visualized by "autoradiography" detecting the label (e.g., alkaline phosphatase or ^{32}P). The main uses of DNA fingerprinting are to identify cell lines from a wide range of species, to demonstrate cross-contamination between cell lines, and to gain information about genetic stability (20).

DNA profiles can also be obtained using probes to highly polymorphic single loci. Simultaneous analysis of numerous single loci by PCR generates a cell-line-specific DNA profile (e.g., GENEPRINT™, Promega, Madison, WI; AmpFLSTR Profiler Plus Amplification Kit, Perkin Elmer, Foster City, CA). In this system VNTR sequences at precise locations in the genome are amplified using specific primers. The PCR products are then separated on agarose gels and the size of resulting bands determined.

Isoenzyme Analysis

Isoenzyme analysis is based upon the existence of enzymes of similar or identical substrate specificity, but different molecular structures with different electrophoretic mobilities. Such enzymes are present in most types of animal cells, and enzymes can be selected for which the electrophoretic pattern varies between but not within species. Substrates are available for up to 20 isoenzymes, including glucose 6 phosphate dehydrogenase, lactate dehydrogenase, malate dehydrogenase, nucleoside phosphorylase, peptidase A, and peptidase B. However, in most cases it is only necessary to study 2–4 enzymes to obtain an identification for the species of origin. A variety of electrophoresis matrices may be used, but most collections now use preformed agarose gels and the associated kit (Authentikit™, Innovative Chemistry, Marshfield, MA). This system gives rapid (3 h), permanent results and a choice of up to 13 enzymes. The main use of isoenzyme analysis is to confirm the species of origin of the cell line in question and to exclude cross-contamination between human, mouse, and rat cell lines (21). Isoenzyme analysis may also be used to exclude or confirm HeLa-like contaminations, since this cell line has the unusual form of glucose 6 phosphate dehydrogenase.

Table 2. Quality-Control Tests[a] Performed Routinely on Cell Banks at Eight Major Service Culture Collections

| | QC testing | | | | | Identity testing | | | |
	Bacteria and Fungi	Mycoplasma DNA staining	Mycoplasma culture	Mycoplasma PCR	Other tests	Virus testing (PCR)	DNA fingerprinting/ profiling	Isoemzyme analysis	Other tests
ATCC	√	√	√	√		√	√	√	Cytogenetics
Coriell	√	√	√			√	√	√	Cytogenetics
DSMZ	√	√	√	√		√	√	√	Cytogenetics Immunophenotyping
ECACC	√	√	√		GenProbe	√	√	√	
ICLC	√	√		√	GenProbe Mycotect	√	√	√	
IFO	√	√	√	√		√	√	√	
JCRB	√	√	√	√		√	√	√	
RIKEN	√	√	√	√		√	√	√	

[a]These and other specialist techniques and facilities exist within these collections and may be accessible as services to research and industry. In particular, these services include training, patent depositories, and safe depositories.

QUALITY-CONTROL SERVICES

As indicated, quality is a vital aspect of the culture collection ethos. All the quality-control tests described are offered as services, including an eradication service for mycoplasma, bacteria, and fungi. Cell characterization tests are also offered. For most collections the techniques used are identical. However, in some cases additional tests are undertaken. In addition, many other facets of culture collection expertise are offered as services to scientists working with cell cultures. Special aspects of these activities are highlighted in the descriptions of these collections to follow. Table 2 summarizes the quality-control services offered by eight representative culture collections.

Depository Activities

Patent Depositories. Under the terms and conditions of the treaty animal and human cell lines and hybridomas fall within the definition of "microorganisms" for the purposes of patent applications. Certain cell banks have been authorized to act as International Depository Authorities (IDAs) under the Budapest Treaty to accept samples of microorganisms in support of patent applications.

The procedures leading to a valid deposit vary between centers; however, there are common features behind the operations of patent depositories.

The Budapest Treaty on the International Recognition of the Deposit of Microorganisms for the Purposes of Patent Procedure was developed in 1977 and came into force towards the end of 1980 (22). Its main effect is to remove the need for inventors to deposit their cell lines in a culture collection in every country in which they intend to seek patent protection. Thus a deposit made in any IDA is acceptable in all countries party to the Treaty as meeting the deposit requirements of its own national laws. Cell banks become an IDA through formal nomination by a contracting state, which must ensure that the collection can comply with the requirements of the Treaty. The Budapest Treaty thus provides a uniform system for patent deposits and lays down procedures that depositor and depository must follow, the duration of

deposit (at least 30 years or 5 years following the most recent request for a culture, whichever is the longer), and a mechanism for the release of samples.

Cell lines deposited to comply with patent disclosure requirements must be made available to the public during the patenting procedure. Unlike a written description, however, the cell line is the representation of the invention itself, and thus the exact conditions of release are extremely important.

General Requirements for an Effective Patent Depository. Each IDA has its own application procedures, deposit forms, and administrative requirements. However, there are general conditions applied by all:

- Multiple, replicate samples requested (ca. 10–20) prepared from a single batch of cells.
- A scientific description must be provided.
- A statement of pathogenicity is required (this is a special concern with human cell lines derived from patients with viral infections).
- A unique deposit number is given on completion of the administrative and quality control procedures. A preliminary deposit number may be allocated on receipt of the samples. This must *be confirmed* on completion of the IDA procedures.
- A viability test is conducted. This is often a trypan blue exclusion test and a study of cell line growth over a test period.
- Each cell lines is tested for microbial contamination. The presence of mycoplasma is a major problem. Most IDAs will not accept cell lines contaminated with mycoplasma as a valid deposit. Acceptance of such deposits as mixed cultures has been proposed but is not a satisfactory solution.
- Once all the laboratory tests and administrative procedures are complete and the IDA fee paid, the deposit is accepted and the international forms completed and released.

Safe Deposit Facilities. A low-profile but highly valuable function of cell banks is to offer a safe deposit facility for important cell lines used in research or in pharmaceutical manufacturing. This provides a safe, secure storage location away from the parent institution and thus protects important material from loss due to disaster. The stocks held can be simply transferred from existing banks to the cell bank or the cell bank may perform culture, quality control, and preservation. Whatever the origin of the material, it is only sensible to have as a minimum a viability test carried out on a sample of the material on deposit. Cell banks have secure storage facilities, normally with "state-of-the-art" alarm systems, with technical staff on call 24 hours a day.

Generation of New Cell Lines. A wide range of techniques have been utilized to generate cell lines, including irradiation, chemical mutagens, viruses, and more recently recombinant DNA technology. A common feature of many cell lines is a transformed phenotype that tends to reflect dedifferentiation and certainly loss of many of the characteristics of the cell types that cell lines are required to emulate. The development of rDNA approaches whereby proliferation of the culture is regulated may enable the establishment of cultures of more highly differentiated characteristics. However, a major problem is the reliability of the molecular switch between differentiated phenotype and proliferative phenotype.

New inducible expression systems such as those based on biochemical inducers active in insect cells may offer more reliable approaches. However, where the phenotype of the cells is not critical (e.g., genetic linkage studies) Epstein Barr virus (EBV) transformation of human mononuclear cell to generate B lymphoblastoid cultures is one of the most reliable and straightforward techniques available.

EBV Transformation. A reliable long-term source of human genetic variant DNA is provided by the availability of Epstein Barr virus transformed B-lymphoblastoid cell lines. Many collections are involved in the derivation of cell lines from patient-derived peripheral blood samples (i.e., ECACC, ICLC). The development of this technology has meant that B-lymphocytes can be isolated and stored untransformed and then subjected to EBV in culture at a later date to generate a cell line. This can be very useful in judging the value of material to a project or the wider scientific community before undergoing the time-consuming and expensive process of generating a cell line for future storage and distribution.

Such cell line resources have become an invaluable tool in genetic studies from HLA-defined cell lines used in quality control/technique development in tissue typing laboratories (23) to the determination of the genes involved in non-insulin-dependent diabetes mellitus (24).

Outlined in the following are three key techniques:

- Isolation of lymphocytes
- Preparation of virus
- EBV transformation

Isolation of Lymphocytes

1. Maintain the whole blood in lithium/sodium heparin, or ACD collection tubes at room temperature prior to use.

2. Dilute the blood with an equal volume in RPMI-1640 containing 10 units/mL heparin. Addition of heparin is necessary if the blood is not to be separated immediately.

3. Carefully layer the blood/RPMI-1640 mixture onto the separation medium (i.e., Ficoll-Isopaque) at a ratio of 2 volumes of blood to 1 volume of separation medium. Centrifuge at 500 g for 20 min at room temperature.

4. Carefully remove the mononuclear cell layer by using a wide-bore Pasteur pipette. If a good separation has not been obtained, the tubes can be centrifuged for a further 10–20 min.

5. The cells are washed by adding RPMI-1640 to volume of 15 mL and mixing gently. Centrifuge at 400 g for 10 min.

6. Check to see that a cell pellet has formed before discarding the supernatant. If a pellet is not visible, centrifuge for a further 10 min. If there is a very low lymphocyte count, the pellet may only just be visible to the naked eye.

7. Resuspend in a 15-mL tube and wash the cells as above.

8. Repeat this wash step once more. If desired, a viable cell count can be performed at this stage.

9. If the cells are to be frozen, they should be resuspended in freeze media such as RPMI-1640 supplemented with 20% FCS and 10% DMSO and frozen using the protocol given above. The lymphocytes should then be stored in vapor-phase liquid nitrogen until required. However, they can be stored temporarily for up to 1 month at −80 °C.

Preparation of EBV-Containing Supernatant from the Cell Line B95-8

1. Maintain a culture of the marmoset cell line B95-8 in RPMI-1640 supplemented with 5% FBS at concentration of $(3-9) \times 10^5$ cells/mL by diluting the cells 1 : 2 to 1 : 4 every 3–4 days.

2. When a sufficient number of cells has been obtained (dependent upon the amount of supernatant required), dilute the cells to 2×10^5 cells in RPMI-1640 supplemented with 2% FBS.

3. Gas the flasks with 5% CO_2 in air, screw on the caps tightly, and incubate at 33 °C for 2–3 weeks without any changes of medium.

4. On the day of harvest stand the flask up and allow the cells to settle.

5. Decant the supernatant into tubes and centrifuge at 400 g for 10 min at room temperature.

6. Filter the supernatant through a 0.45-μm sterile filter to remove cell debris and large particles.

7. Aliquot the supernatant at the required volume (usually 1 mL) and store at −80 °C or below, preferably in vapor-phase liquid nitrogen. Avoid freeze/thawing of the virus, as this will reduce the efficiency of EBV transformation.

8. Prior to use, thaw at 37 °C in a water bath and use immediately.

Establishment of Human B-Lymphoblastoid Cell Lines from Peripheral Blood Mononuclear Cells

1. Centrifuge the cell sample to be transformed. Discard the supernatant and resuspend the cell pellet in 1 mL EBV supernatant.

2. Incubate at 37 °C for 30 min. Gently agitate and incubate for a further 30 min.

3. Centrifuge at 150 g for 10 min.

4. Resuspend cells at a concentration of $(1-2) \times 10^6$ cells/mL in RPMI-1640 supplemented with 20% FBS, 1% phytohaemagglutinin (PHA), penicillin (100 units/mL), streptomycin (100 µg/mL), and neomycin (50 µg/mL). Pipette 1 mL of cells into each well of a 24-well flat-bottom culture dish and incubate at 37 °C in 5% CO_2 in air.

5. The cultures should be fed twice weekly by removing half the supernatant and replacing with fresh culture medium, taking care not to disturb the cells.

6. After 1–2 weeks, the foci of B cells should be visible under an inverted microscope, and the initial proliferation of T cells should regress.

7. The cultures may be expanded into 25-mL tissue culture flasks once established (usually after 2–3 weeks). This is often indicated by the medium becoming acidic and turning yellow and large clumps of cells becoming visible. The flasks should be incubated upright and the cultures split 1:3 every 2–4 days depending upon growth rate.

As soon as the cultures are established, an aliquot should be cryopreserved and tested for mycoplasma (see above).

Training and Educational Role of Culture Collections. A key role of culture collections is the provision of training and education. This varies from the basic introductory courses on culture, preservation, and quality-control methods to more advanced programs, including such diverse topics as cell-line immortalization and large-scale culture methods. This often represents a good opportunity to make contact with the specialists within the depository who can assist in developing a long-term source of guidance in techniques and new developments. This aspect of culture collection activity also opens up the international network of culture collections and the wide range of knowledge and expertise that this offers.

OVERVIEW OF REPRESENTATIVE CULTURE COLLECTIONS

There are a number of major service culture collections throughout the world. An overview of their development and activities is useful in that it highlights the varied specialist expertise and facilities from each center. As described, all the collections operate using a battery of standard techniques: master and working cell banking systems, cryopreservation and culture techniques, and quality-control procedures. Nevertheless, each collection has evolved differently depending on their scientific origins and source of funding. The international network of service culture collections therefore offers a rich resource of science and technology that is available to all.

AMERICAN TYPE CULTURE COLLECTION (ATCC) — AN HISTORICAL PERSPECTIVE

Contributed by R. Hay

For centuries mankind has recognized and reaped the benefits of microbial fermentations, establishing private collections to catalyze the effective processes required. The American Type Culture Collection (ATCC) is a global nonprofit resource developed over the years to acquire, preserve, authenticate, and distribute biologicals and related information. However, the growth has been controlled because ATCC's goal has always been to assure that the specimens most useful to the scientific community are added to the collection and secured for future use. The activity is offered for advancement, validation, and application of scientific knowledge. Most of the biological holdings have been available through the generosity of numerous donors, many of whom supply their material without charge and with minimal restrictions on distribution. Advisors have willingly given of their time and talents over the years to ensure that the various reference collections developed are complete and that the strains included are appropriately characterized. The curators, scientists, and support personnel in the scientific and service departments have contributed not only by expanding, characterizing, and distributing, but also in the compilation of historical information and data relating to seed and distribution stocks. Thus responsibility for both the problems and achievements associated with the development of such reference resources are shared by a great many individuals from within the scientific community (25). The perspective provided here illustrates the broad range of service required for development of the ATCC during the past century. The milestones in the development of ATCC are summarized in Table 3.

As previously stated, the first recorded public service collection was developed by Frantisek Kral at the Institute of Hygiene in The Faculty of Medicine in Prague. An initial catalogue of holdings appeared in 1902. Kral supplied cultures of bacteria, yeasts, filamentous fungi, and media from "Krals Bakteriologisches Laboratorium" in Czechoslovakia. A successor, E. Pribram, transferred the collection, first to the University of Vienna, Austria, then to Loyola University School of Medicine in Chicago.

Coincidentally, in the United States, charter members of the Society of American Bacteriologists (SAB), established in 1899, recognized the need for a central characterization and distribution agency for microbial cultures.

Table 3. Milestones in ATCC History

Dates	Occurrence/event
1899	Society of American Bacteriologists (now ASM) established
1911	Bureau for the Distribution of Bacterial Cultures established in New York
1922	SAB "Committee-in-Charge" plans collection with direction from Lore Rogers
1925	ATCC founded and housed at John McCormick Institute in Chicago
1937	ATCC relocates to Georgetown University in Washington, D.C.; M. Mollari, Curator
1947	ATCC incorporated as nonprofit scientific institution and moves to independent quarters; Ruth Gordon, Curator
1956	ATCC purchases and transfers to its own facility on M street; Freeman A. Weiss, Curator
1958	Viral and Rickettsial Registry added
1960	The first cellular cryopreservation is accomplished at ATCC; W. Clark, Curator
1962	NCI-sponsored program for developing a cell culture collection begins
1963–1964	ATCC's Rockville facility is constructed and occupied
1973–1980	ATCC strengthened under Richard Donovick with increase in grant and contract income; one new floor and a new wing are added to the existing building
1980	Collection of rDNA and vectors begins with fledging Molecular Biology Program
1981	ATCC named first International Depositing Authority for patents under Budapest treaty
1980–1993	With R.E. Stevenson as Director, three additional buildings are purchased to allow growth, and plans for relocation are explored
1984	Bioinformatics activities including hybridoma database development grow, and the Bioinformatics Program becomes firmly established
1994–1997	Decision is made for move to Manassas, VA, next to Prince William campus of George Mason University; R.H. Cypess, President and CEO, directs construction planning, reorganization, and move preparation
1995	An alliance is formed with The Institute for Genome Research to distribute cDNA clones
1998	ATCC moves entire collection with essential support equipment intact to Manassas without incident; largest contract ever awarded ATCC is received to develop reagent resource for study of malaria; ATCC is poised for further service and growth in the new millennium

Accordingly, "The Bureau for the Distribution of Bacterial Cultures" was established in 1911 at the Museum of Natural History in New York City with C. Winslow as curator. As news of the Bureau and its mission became known, deposits and exchanges occurred, including transfer of selected holdings from the then-famous Kral Collection in Vienna.

In 1922 SAB assumed responsibility for the collection, moving total stocks of 175 strains in a single suitcase to the Army Medical Museum in Washington, D.C. To expand the developing resource a "Committee-in-Charge" representing the SAB, the John McCormick Institute for Infectious Diseases, the National Research Council, the American Phytopathological Society, and the American Zoological Society was organized for action. Charter members included well-known scientists of the time, many of whom continued to interact closely with ATCC during most of its formative years. Lore Rogers, S.J. Nichols, R.E. Buchanan, C.A. Kofoid, C.L. Shear, and F.P. Gey are examples. They proposed to establish the institution "to preserve cultures of micro-organisms that have historic and scientific interest and to provide a center for obtaining cultures needed in education and research." With funding obtained from the Rockefeller Foundation, the institution was founded in 1925 and was named The American Type Culture Collection. The John McCormick Institute of Infectious Diseases in Chicago offered support, including housing for the fledging collection, and the 175 strains were transferred from New York accordingly. The ATCC remained in Chicago for 12 years, increasing holdings to about 2000 strains of bacteria, filamentous

fungi, and yeasts and initiating a policy of periodic catalogue publications plus addition of a nominal charge for cultures ($1–2 per strain). Resources at the McCormick Institute diminished during the depression of 1932–1937 forcing ATCC to relocate in 1937, this time to the Georgetown University Medical School in Washington, D.C. Under curatorship of M. Mollari and for the following 10 years, the organization grew in staff numbers and capability. Protozoa and algae were added to the collection. Application of freeze drying became prevalent and commercial sales increased due to the requirements of the antibiotics industry. At that point in history, ATCC also began services as the USPHS depository for cultures isolated during epidemics. In 1947 the ATCC was incorporated as a nonprofit scientific organization.

Space constraints prompted another relocation, and the "Committee-in-Charge" rented an apartment building still in Georgetown, Washington D.C. After initial renovation the collection of some 3000 cultures was moved in September 1947 to the first independent ATCC quarters with Ruth Gordon as curator.

Through collaborations with individuals and groups of laboratories in the Washington area (USDA, NIH, Army Medical School, and BBL) the collection continued to expand, but economics were still a constant concern. The "Committee-in-Charge" became the Board of Trustees, and representatives from the National Academy of Sciences, the Mycological Society of America, and the American Institute of Biological Sciences were added. Today 25 organizations are represented on the ATCC Board of Directors.

In 1956 The Collection, requiring additional space, was able to purchase and renovate a brownstone apartment building at 2112 M Street. This was the first facility purchase for the ATCC. When this property was sold in 1964, the equity became the core for a small Foundation Fund to play an important role in development of The ATCC's financial situation.

In 1958 the Viral and Rickettsial Registry was established at the ATCC with financial assistance from NIH. For several decades stocks were contributed in bulk by collaborators, with dispensing to ampoules, storage, and distribution accomplished by ATCC staff. Production and authentication of the majority of virus stocks are now performed exclusively by personnel in the ATCC Virology Program.

Late in the 1950s conferences at the National Research Council, NCI, and the Rockefeller Institute emphasized growing awareness on the part of the scientific community of microbial and cross-contamination in cell line and virus stocks. Rapid growth in biomedical research and the resultant need for reliable reagents stimulated Government and industry together to fund design and construction of a new ATCC in Rockville, Maryland. The NIH, NSF, and a variety of industrial sources provided in excess of one million dollars, with ATCC adding core matching funds. The Collection consisting of about 8500 cultures was moved to the new 35,000 square foot facility February 23, 1964. William Clark had succeeded Freeman Weiss as curator and, with reorganization and title changes, had become the Director. The collections were subdivided by organism type to form departments, more recently becoming "programs."

Currently ATCC has 7 scientific programs, namely, Bacteriology (bacteria, mycoplasma, bacteriophage); Bioinformatics (information databases on ATCC and other holdings); Cell Biology (cell cultures and transplantable tumors); Molecular Biology (molecular probes, vectors, rDNAs, cDNA clones); Mycology and Botany (filamentous fungi, yeasts, and plant tissue cultures); Protistology (protozoa and green algae); and Virology (viruses, chlamydia, rickettsiae and antisera). Somewhat concomitant with planning and construction of the ATCC building in Rockville was an NCI-sponsored program for development of the cell culture collection. Four cooperating national laboratories participated with advice from the Cell Culture Collection Co-ordinating (C4) committee, a group of experts selected by the NCI (26). The four core support laboratories included the Child Research Center of Michigan in Detroit, the South Jersey Medical Research Foundation (now the Coriell Institute for Medical Research, see following section) in Camden, the Naval Biological Laboratory (University of California, Berkeley), and the ATCC. The C4 committee met periodically with staff of the cooperating laboratories to recommend new accessions and review problems. Distribution of cell cultures was to become a financial boon for the organization. Research on the use of vapor-phase and liquid nitrogen for preservation and storage of microorganisms and cell lines documented broad utility. Freeze drying remained the technology of choice for microbial distribution stocks. Despite modest continued funding for growth of the collections, in 1971 a reduction

of support from NIH precipitated another financial crisis for the ATCC (27).

With the recruitment in 1973 of Richard Donovick as Director with years of experience in industry, a new era for the ATCC of expansion and stability had begun. Donovick was a staunch advocate for the ATCC, recognizing the community requirement for a stronger institution. He was skilled at negotiations with staff and funding agencies alike, motivating personnel to achieve needed growth in both collection and research activities. In the 7–8 years of his tenure grants and contracts totaling millions of dollars were attracted from numerous agencies by the staff. The Collection increased both in number of holdings and in distribution. An additional floor plus an entire new wing were constructed at the Rockville facility, once again with support from NCI, a group of industrial supporters, and the ATCC.

ATCC had been accepting deposits for patent purposes since 1949, when the World International Property Organization (WIPO) initiated negotiations to establish the International Budapest Treaty to simplify the process of patenting biologicals (see above). In 1981 the ATCC was named the first International Depository Authority (IDA) under that treaty. This continues not only as an important business activity for the ATCC, but also as a means of acquiring new, interesting, and useful holdings. When Donovick retired in 1980, another science administrator, Robert E. Stevenson, became the Director. Having background with industry (Union Carbide) and years of management experience at NIH, Stevenson continued to oversee the expansion of ATCC's service and facility. Cell line distribution, for example, increased 714%. Three new buildings were purchased, adjacent to the original red brick structure in Rockville. The educational commission of ATCC grew as scientific workshops were developed and offered. Research and development programs were added through appropriate external support.

In the latter half of the 1980s, once again, the need by ATCC for added space prompted evaluation of reconstruction and relocation options. Economic considerations ruled out onsite construction, and other locations such as College Park, Maryland, DeMoines, Iowa, and Lansing, Michigan were discussed. Design of a proposed new facility was initiated, and potential funding sources were explored (28). With Stevenson's announced retirement the Board and senior ATCC staff began interviews for his replacement. Candidate Raymond H. Cypess, immediately recognized as a visionary, suggested ATCC consider more fully a move to Virginia and affiliation with George Mason University. He spoke forcefully about ATCC's staff, reputation, and potential for the future. Cypess was offered and accepted the position, renamed ATCC President and CEO, in 1993.

In 1995 a decision by the CEO and Board was made to move to Manassas, Virginia, another city on the outskirts of Washington, D.C. The state of Virginia, George Mason University, Prince William County, and ATCC provided funding. Construction plans made earlier were modified as necessary, and the Manassas project was begun. The new, state-of-the-art facilities were completed for occupancy beginning in March 1998 in Manassas, Prince William

County. The address of this facility is 10801 University Blvd. All receiving, shipping, storage, and production-related lab work is now performed in this facility.

Concomitant with preparation for and execution of the move to Manassas, the ATCC experienced exciting program growth. Two contracts awarded recently are among the largest in ATCC's history. The first for 4.6 million dollars was awarded in 1997 for repository management services offered to CDC in Georgia. The second, with about 9.8 million dollars being provided by NIAID, is for a joint program developing a malaria resource. In addition, a series of smaller but nonetheless important programs have been secured competitively by ATCC scientists and collaborators. Clearly the ATCC, with expert staff, new facilities, and a well-defined mission, is poised for even more growth and achievement in the new millennium.

CURRENT FACILITIES AND OPERATIONS

The new headquarters is divided into three areas — administrative, laboratories and operations. The two-story laboratory wing provides housing 24 labs plus support areas. This structure, designed strictly for biological work, features a variable-air-volume HVAC system utilizing 100% makeup air. Laboratories surround a central core that provides support areas such as instrumentation rooms and environmental rooms (hot and cool). All laboratories are fully equipped and are designed to Biological Safety Level 2 containment standards. All work is done in biological safety cabinets or fume and chemical hoods as required. A Biological Safety Level 3 laboratory suite is also included. Each of the six collections has its own laboratory suites.

The operations area contains the media preparation laboratory, glassware facility, a preservation laboratory for lyophilization and programmed freezing, the repository, the packaging and shipping area, and support areas. The repository provides a storage area of 8,200 sq. ft. This includes a cold room for storage at 4 °C and space for 65 vapor-phase liquid nitrogen refrigerators (-170 °C). The LN$_2$ refrigerators are supplied from a 6,000-gallon liquid nitrogen tank through vacuum-insulated lines. The facility also supports 55 ultra-low-temperature mechanical freezers (-80 °C). A standby motor generator provides power for the freezers and most computer and lab equipment in the event of failure of commercial power.

A backup safe storage facility is maintained at another location remote from the Manassas buildings. There several ampoules of each strain or line are maintained for safety in the unlikely case of a catastrophic event occurring at the headquarters. Spare liquid nitrogen refrigerators and mechanical freezers are maintained in case of mechanical failure. Additionally, ATCC maintains a large supply of dry ice, which can be used for emergency cooling.

All freezers are equipped with dual, independent temperature monitoring systems. Both systems can trigger facility alarms. One system is to provide alarms such as temperature and power. The other system monitors and records freezer temperatures on a central computer system.

A packaging and shipping area is adjacent to the repository. The administrative wing houses customer service and corporate functions.

The ATCC also occupies some 50,000 square feet of modern laboratory and support space in a second new building close by. This structure is shared with George Mason University and will be used by the Bioinformatics group and extensively otherwise for research.

ATCC's holdings of metazoan and microbial cells, nucleic acids, and other macromolecular entities have now increased to more than 92,000 strains of microorganisms and cell isolates, represented by 2,000,000 specimens and more than 1,000,000 DNA sequences. Thus the scale and sophistication of the ATCC activities assures its position as the key international collection.

CORIELL INSTITUTE FOR MEDICAL RESEARCH, THE HOME OF THE CORIELL CELL REPOSITORIES

Contributed by R. Johnson and J. Beck

Establishment and Development

The people make the place, and the Coriell Institute for Medical Research in Camden, New Jersey, is no exception. Its founding father, Lewis Coriell M.D., PhD., after whom the Institute was renamed in 1995, provided the impetus and the vision needed to establish the laboratories. Lewis Coriell's creativity resulted in the development of the necessary technology to grow human and animal cells without antibiotics, and to freeze and successfully recover cells from liquid nitrogen. Lewis Coriell came to Camden in 1948 as Director of the Municipal Hospital for Infectious Diseases, bringing with him a background in pediatric medicine and immunology and a strong interest in polio. Many of the patients were victims of polio, requiring long periods of careful rehabilitation and hospital treatment. With colleagues at the University of Pennsylvania Hospital and elsewhere, Coriell embarked on a series of successful field trials of gamma globulin and, later, of Salk vaccine during outbreaks of polio in the early 1950s. Coriell's earlier contribution to the Salk vaccine lay in the adoption of his cell culture methodology for growing the virus. The laboratory confirmations of antibody titers and the presence of polio virus were largely carried out in the Camden Hospital, and the success of these activities brought considerable notice and with it the offer of resourcing Coriell's work. With three and a half acres of land donated by the city of Camden and financial support from the local business community, the laboratories were built in 1956.

Coriell continued his studies with viruses, but the emphasis was changing to a more rigorous analysis of cell culture methodologies as he and others developed new and important human and animal cell lines. Basic growth media were designed by Earle, Eagle, and others. Early cell culture procedures, though not hit or miss, were far from perfected in the 1950s and 1960s. Coriell, with a group of colleagues, most prominently Arthur Greene and Warren Nichols, and later Gerald McGarrity, developed many of the basic techniques in common use today in cell culture. Key technology, such as laminar flow hoods equipped with HEPA filters, liberated the researcher from the threat of culture contamination and permitted workers in the Institute to grow cultures routinely without antibiotics. Today the last of Lewis Coriell's original laminar flow hoods stands on display, guarding the cell culture laboratories nearby.

Special attention was paid to the storage of cell cultures so that they could be recovered from liquid nitrogen easily and reproducibly, and Coriell developed the programmed freezing schedule of 1 degree per minute, which has provided a standard reference technique for the preservation of animal cells. Cultures laid down over 30 years ago by this procedure can be recovered with similar success rates as those frozen down much more recently. These fundamental improvements in handling and storing cells were the starting point for the major cell repository activities that began at the Institute in 1964 with a National Cancer Institute collection and rapidly expanded in the early 1970s, the birth of the period of somatic cell genetics. In 1972 the NIH invited Lewis Coriell to establish the first national library of cells from individuals with inherited diseases. The collection, funded by the National Institute of General Medical Sciences (NIGMS), continuously since then, had gathered together cell lines, mostly fibroblasts in the early years and lymphoblastoid cultures subsequently, from individuals with inherited diseases. In this way NIH wished to create a rich cellular resource of well-characterized cell lines for the research community to use and identify the genetic basis of many diseases. The main reasons for creating a library of mutant human cell lines, together with those from obligate heterozygotes and normal controls, were readily apparent to the committee set up by NIH to address this issue.

The committee responses are summarized by the following quotations:

> Much time and research money can be saved if investigators in different laboratories are not required independently to develop and maintain cell lines.
>
> Use of identical cell lines for study in different laboratories should facilitate the comparison of data and the interpretation of research results.
>
> Some cell lines will represent rare diseases that otherwise would not be readily available to all interested investigators.

Since 1972, the NIGMS Human Mutant Cell Repository at Coriell has acquired more than 35,000 cell lines representing more than 1,000 of the 4,000 known genetic diseases. Many of these diseases are or are likely to be single-gene disorders, but many others will have complex genetic etiology, such as schizophrenia, diabetes, and rheumatoid arthritis. Also included in the NIGMS Human Mutant Cell Repository is a large number of chromosomally abnormal cell lines, many with informative translocations, or regions of monosomy, a renowned collection of human rodent somatic cell hybrids, widely used for mapping purposes and for the production of mapping and regional mapping panels. There is an increasingly large range of human diversity panels from many ethnic groups and a large number of

Table 4. Biological Materials Processed and Distributed at the Coriell Institute (figures are from inception of the Collection to February 1, 1998)

Repository	Preparation		Distribution	
	Specimens processed	2-mg DNA preparations	Cell cultures	DNA aliquots
NIGMS	16,214	2,507	75,173	26,711
NIA	4,563	n/a	19,679	n/a
NIMH	14,508	5,259	56	22,867
HBDI	4,392	2,347	545	12,736
ADA	4,070	597	2,736	490
NCI	2,150	53	n/a	n/a
Other	4,029	409	1,609	378
Totals	**49,926**	**11,172**	**99,879**	**63,182**

three-generation extended families (including 51 Centre d'Etude du Polymorphism Humain, CEPH, families) suitable for linkage studies. By 1998 75,000 cell lines from this collection had been distributed to more than 40 countries and also more than 26,000 aliquots of DNA (DNA isolation was initiated in 1990). Table 4 provides information about materials processed and the distribution from the different collections at Coriell since the repository was established.

From the 1970s, the Coriell Institute won contracts from the NIGMS and the National Institute on Aging (NIA) to establish and maintain what have become the world's largest cell repositories for the study of genetic and aging-related diseases, respectively. In 1980 the National Institute of Mental Health awarded the Coriell Institute a contract to establish a cell repository for the study of the genetic basis of Alzheimer's, manic depression, and schizophrenia, and in 1993 the American Diabetes Association (ADA) contracted the Coriell to develop a multiethnic family-based resource to identify the genes important in the etiology of type 2 (non-insulin-dependent diabetes). The Aging Cell Repository was initiated in 1974 by Dr. Warren Nichols. It contains a unique set of normal fibroblast cultures from the Gerontology Research Center, Baltimore. This material, part of the Baltimore Longitudinal Study, represents fibroblasts taken from individuals at different times during their lives. The collection is also rich in age-related diseases with strong genetic components, including Alzheimer's (many extended kindreds), Werner's syndrome, and progeria. Since 1974 over 20,000 cell cultures have been distributed form the NIA repository. In addition to these collections, in 1996 Dr. Jeanne Beck helped establish the National Cancer Institute (NCI) Co-operative Family Registry for Breast Cancer Studies (CFRBCS). The CFRBCS is a multisite cooperative consortium of investigators who collect biological specimens from participants with a family history of breast cancer, breast/ovarian cancer, or Li–Fraumeni syndrome and their relatives. Coriell participates in this consortium of the NCI by processing and maintaining the collection from several sites.

The CFRBCS also collects related family history, clinical, demographic, and basic epidemiological data on risk-factor exposures, as well as followup epidemiological data and data on recurrence, new morbidity, and mortality in the participating families. These repositories of specimens and related databases are particularly suited to support interdisciplinary and translational breast cancer research and are available to investigators who receive approval from the CFRBCS advisory committee.

FROM HARD COPY TO INTERNET: THE CORIELL WWW CATALOGUES

In 1994 Coriell published the last of its printed catalogues, a 351-page catalogue for the Aging Repository and 1085 pages for the Human Genetic Mutant Cell Repository. The Bioinformatics group has now developed WWW catalogues for three of the cell repository collections: The NIGMS Human Genetic Mutant Cell Repository, The NIA Aging Cell Repository, and The ADA Diabetes Repository.

These catalogues may be accessed through the Coriell Cell Repositories home page (*http://locus.umdnj.edu.ccr*). Each of these catalogues has unique features; however, certain search and navigational tools are standard. There are quick searches for specific gene mutations, chromosomal breakpoints, repository numbers, and diseases. The structured search groups relevant search fields into functional categories and contains pick lists of available search terms. Pedigrees identify family relationships and genetic status. All catalogues link to information about ordering materials. The NIGMS and NIA catalogues provide direct links to OMIM and reciprocal links from OMIM to specific diseases in the collection, a browseable list of diseases through a diagnosis list, and a section to highlight what is new in the collection. Each bibliographic reference listed in the catalogue is linked directly to PubMed, and there is also a link from PubMed to relevant samples in the collections. In addition, the NIGMS catalogue has a browseable list of genes that are linked to the Human Gene Mutation Database, Cardiff, UK. Each mutation identified is linked to allelic variants in OMIM and to all samples in the collection that carry that specific mutation. For the somatic cell hybrid collection, there are ideograms displaying which portion of the chromosome is included in each hybrid. All resources for each chromosome have been assembled by chromosome and include a gene map providing a listing of all genes for which samples are available, translocation breakpoint ideograms, an ideogram indicating regions of monosomy, trisomy, and additional polyploidy, regional mapping panels, if available, and all hybrids for that chromosome that are in the collection. Finally, a suggestion box enables users to provide feedback and to indicate additional samples that would be useful for their research. Each of these catalogues provides valuable information for the cell cultures and DNA samples that are available in a user-friendly, easily navigable format.

SUBMISSION OF CELL LINES

Each submission of human cells for inclusion in the NIGMS or NIA repositories requires clinical and laboratory documentation of the diagnosis and a copy of the Institutional Review Board (IRB) approved consent form used to obtain the specimen. For submission to the NIGMS

Human Genetic Mutant Cell Repository, a model informed consent form is available from the Repository. This model informed consent has been reviewed by the Office for Protection from Research Risks (OPRR) and approved by the NIGMS Human Genetic Mutant Cell Repository IRB. In addition, the OPRR has provided guidance on Protection for Human Subjects in the NIGMS Human Genetic Mutant Cell Repository and Submission of Non-Identifiable Material to the Repository. This guidance on the Protection for Human Subjects states "... research material may only be utilized in accordance with the conditions stipulated by the cell repository IRB. Any additional use of this material requires prior review and approval by the cell repository IRB and, where appropriate, by an IRB at the recipient site, which must be convened under an applicable OPRR-approved Assurance."

AVAILABILITY OF CELL LINES AND DNA

The major collections (NIGMS, NIA, and the ADA) all have restricted access. However, cell cultures and DNA samples are widely distributed to qualified professionals who are associated with recognized research, medical, educational, or industrial organizations engaged in health-related research or health delivery. For NIGMS and NIA collections, before cell cultures or DNA samples can be ordered, to ensure compliance with federal regulations for the protection of human subjects (45 CFR part 46) the principal investigator must provide the Repository with a description of the research to be done with the cell cultures of DNA samples ("Statement of Research Intent"). The principal investigator and the institutional official who can make legal commitments on behalf of the institution must also sign an "Assurance Form" detailing the terms and conditions of sale. Both the Assurance Form and Statement of Research Intent must accompany the order placed with the Repository. For the ADA collection, permission must be sought from Matt Peterson, Director, Research Programs (*mpeterson@diabetes.org*).

Despite the hurdle that meets every order, Coriell distributed over 3,000 cell cultures and more than 3,300 DNA samples in 1998 to researchers in 36 countries. In turn, Coriell receives feedback from many scientists in the form of bibliographic data that include the use of Repository materials—information that is then added to the database and abstracted to the WWW catalogues. Thus the Coriell is very significant in the international culture collections, with its roots in human genetics.

DSMZ: DEPARTMENT OF HUMAN AND ANIMAL CELL CULTURES

Contributed by R. MacLeod and D. Fritze

The DSMZ-Deutsche Sammlung von Mikroorganismen und Zellkulturen GmbH (German Collection of Microorganisms and Cell Cultures) is an independent, nonprofit organization dedicated to the acquisition, characterization, preservation, and distribution of bacteria, animal cell lines, fungi, phages, plasmids, plant cell cultures, plant viruses, and yeasts. Established in 1969 in Göttingen, the DSM originally belonged to the Gesellschaft fur Strahelnforschung (GSF, München), but was later moved to Braunschweig to become a department within the Gesellschaft für Biotechnologische Forschung (GBF). In 1987 the DSM was recognized as an independent institute in the form of a GmbH (Ltd.) that is owned by the State of Lower Saxony. In 1996, after becoming an "Institut der blauen Liste" (now designated "Wissenschaftsgemein-schaft Gottfried Wilhelm Leibnitz"), the abbreviation of the name was changed to DSMZ, acknowledging the grow-ing importance of the cell culture department established some 7 years earlier. The DSMZ is financed entirely by the German government (50% from the Federal Ministry of Research and Technology and 50% by the state gov-ernment). DSMZ is governed by a board of directors (in consultation with a scientific advisory board) that includes representatives of federal and state ministries, universi-ties, and biotechnology.

CELL CULTURE COLLECTION

The DSMZ collection of continuous, immortalized human and animal cell lines, established in 1989, covers the generality of those used in biomedicine and biotechnology. The greatest variety of cell lines are those demanded by biomedical workers representing the many different kinds of human tumors, and all major human tumors from which cell lines have been established are represented in the collection, including the largest number of leukemia–lymphoma cell lines publicly available. The DSMZ welcomes the deposit of new cell lines of proven or likely utility. The DSMZ is also recognized as an International patent repository for cell lines according to the Budapest Treaty.

Each cell line is distinguished by characteristic features that make them unique and biomedically or biotechnologically interesting. In order to maintain these characteristics, emphasis is placed upon a program of extensive quality and identity control and on the characterization of cell lines. The degree of testing necessary reflects the growing realization that general cell repositories are the facilities best equipped to tackle the problems of cell line cross-contamination (affecting one-in-six new cell lines) and mycoplasma contamination (affecting a third of all cell lines). Tests for contamination and adventitious agents used at DSMZ include the standard tests described previously. Mycoplasma is also detected by RNA hybridization and PCR (29,30). Added to this is the duty to demonstrate the potential utility to other researchers of newer human tumor cell lines that form a significant part of the DSMZ holdings.

CYTOGENETICS

Cytogenetic analysis of continuous cell lines serves two purposes: first, cytogenetics supplies a precise written means of identification that may be used either to confirm cell line identities (where karyotypes have been previously published) or to detect intraspecies or interspecies cross-contamination (along with DNA profiling or isoenzyme analysis, respectively); and second, recurrent cytogenetic rearrangements provide a key with which to check likely tumors of origin and predict oncogenic changes at the molecular level. Characterization includes classical karyotyping analysis using G-banding with the assistance of chromosome painting or single-locus FISH probes. The DSMZ provides ISCN karyotypes for almost all human cell lines held.

IMMUNOPHENOTYPING

Human cell lines are analyzed routinely for expres-sion of differentiation and activation marker proteins. Hematopoietic cell lines are screened for expression of cell surface markers with a set of monoclonal antibodies classified according to the CD system. This is a two-step staining procedure where cells are incubated with antigen-specific murine monoclonal antibodies. Binding of antibody is then assessed by an immunoflouresence technique using FITC-conjugated antimouse Ig antisera. The distribution of antigens is analyzed by flow cytometry or light microscopy. This technique allows determination of cell lineage and limited determination of differentia-tion status. Other human cell types are checked for the presence of tissue-specific intermediary filaments.

VIRUS DETECTION

At the DSMZ the major human pathogenic viruses are detected using either ELISA for the detection of reverse transcriptase activity, indicating the presence of retrovirsuses in cell culture supernatants, or PCR/RT-PCR for the detection of the respective DNA or RNA signatures of HIV-1, HTLV-1 and -2, EBV, hepatitis B (HBV), and hepatitis C (HCV).

In the ELISA, digoxigenin- and biotin-labeled deoxyu-ridine- triphosphates are incorporated in the presence of viral reverse transcriptase (RT) during the synthesis of a DNA strand along a synthetic target of single-stranded DNA molecules. The newly synthesized DNA is trapped on a streptavidin-coated microtiter plate and subsequently detected immunologically by binding of peroxidase-labeled antidigoxigenin antibodies.

The presence of DNA sequences integrated into the cellular genome from other viruses (HIV, HTLV, EBV, HBV, and HCV) are detected by PCR.

RESEARCH ACTIVITY

An active program of collection-based research is per-formed at the DSMZ attracting external funding. Current topics include: identification of novel oncogene fusion partners in lymphoma cell lines; signal transduction in leukemia cell lines in response to cytokine stimulation; cytokine secretion by human tumor cell lines; occurrence of tumor suppressor gene alterations in tumor cell lines; and cytogenetic characterization of tumorigenic chromosome changes.

WEB SITE ON THE INTERNET

The DSMZ catalogue of human and animal cell lines including characterization data is freely accessible at the DSMZ interactive database, which includes a search engine. The web site can be found at <*www.gbf.de/dsmz/dsmzhome.html*>. The electronic catalogue is fully searchable and is continuously updated with regard to stock levels.

EUROPEAN COLLECTION OF CELL CULTURES (ECACC)

Contributed by B. Bolton

The European Collection of Cell Cultures (ECACC), was established in 1984 at the Centre for Applied Microbiology and Research to provide a patent depository for animal cell cultures. Initial funding was realized by Dr. Bryan Griffiths through an application to the European Commission, which secured joint funding by the U.K. Department of Trade and Industry for the establishment of this new collection. The primary remit of this collection was to fill a gap in the existing network of European biological collections regarding patent deposits of animal cells. Finance was provided for an initial 6-year period with the remit for the Centre to become ultimately self-financing. From an initial staff of five in 1984, the Collection grew to 40 in 1996. During its evolution ECACC has developed to comprise several distinct collections, which are described in the following.

GENERAL COLLECTION

The general collection consists of approximately 500 cell lines deposited from researchers world-wide covering a diverse range of species (see Table 5). The predominant species within the collection is human, representing 607 from 49 different tissues (see Table 6).

In addition, there are over 300 hybridoma cell lines producing monoclonal antibodies to a wide range of determinants such as bacterial, viral, oncogene, plant, immunoglobulins, MHC, and CD antigens. Through the generosity of numerous laboratories throughout the world, ECACC is currently able to offer the largest and most comprehensive collection of human and animal cell lines in Europe.

THE COLLECTION CELL LINES OF HUMAN GENETIC AND CHROMOSOME ABNORMALITIES

This is a specific collection of over 22,000 accessions from individuals and whole families with over 600 different genetic disorders represented. The collection began as an EU-funded Biotechnology Action Programme project, which was jointly developed by Dr. Alan Doyle of ECACC and Prof. Galjaard of the Department of Clinical Genetics, Erasmus University, Rotterdam, The Netherlands, in 1986. The majority of these cell lines are EBV-transformed lymphoblastoid lines; however, the collection does include skin fibroblasts and amniotic fluid cell cultures. Since May 1990 substantial funding for this

Table 5. Species Variation in ECACC General Collection

Species	No. of cell lines	Species	No. of cell lines
Baboon	1	Insect	34
Bat	2	Lion	1
Bear	1	Lizard	2
Bovine	27	Mink	2
Buffalo	2	Marsupial mouse	1
Chick	1	Mollusc	1
Chicken	3	Monkey	38
Chimpanzee	1	Mouse	475
Deer	2	Muntjac	4
Dog	15	Owl monkey	1
Dolphin	1	Opossum	1
Donkey	1	Orangutan	1
Drosphilia	15	Ovine	12
Duck	1	Porcine	18
Feline	1	Potoroo	2
Ferret	1	Quail	2
Fish	8	Rabbit	9
Fox	1	Rat	220
Fog	3	Raccoon	1
Gerbil	2	Red fox	1
Gibbon	1	Red panda	1
Goat	1	Snake	2
Goose	1	Sun bear	1
Gorilla	1	Tick	10
Guinea Pig	3	Toad	1
Hamster	45	Trout	1
Heterohybrids	35	Turkey	1
Horse	2	Turtle	1
Human	607	Wolf	1

resource has been received from the MRC Human Genome Mapping Project (HGMP), which has enabled a subsidized EBV transformation service to be made available to U.K.-registered HGMP users through a scientific liaison committee. The grant also subsidizes distribution of cell lines to research laboratories throughout the international gene mapping community. Since 1989 ECACC has also received funding from the British Diabetic Association, for the establishment of specific repositories of cell lines derived from patient with types I and II diabetes. This repository now holds over 5,000 cell lines and DNA samples, which have been distributed world-wide for diabetes research. A similar repository has been established for research into Rheumatoid Arthritis by the U.K. Arthritis and Rheumatism Council.

HLA-DEFINED CELL LINE COLLECTION

In addition to the 9th and 10th ECACC now holds the complete 12th International Histocompatability Workshop cell line panel. This is available as cell lines and purified DNA. Complete panels have already been distributed to approximately 80 laboratories world-wide.

CULTURES FOR PATENT DEPOSIT PURPOSES

ECACC is recognized as an International Depository Authority (IDA) under the terms of the Budapest Treaty

Table 6. Human Tissue Variation in the ECACC General Collection

Tissue	No. of cell lines	Tissue	No. of cell lines
Adrenal	1	Lung	34
Amnion	3	Lymphoblast	9
B-lymphoblast	5	Lymphocyte	7
B-lymphocyte	7	Melanoma	2
Bladder	7	Leukemias (various)	14
Blood (miscellaneous)	16	Monocyte	1
Bone marrow	2	Muscle	1
Brain	4	Nerve	3
Breast	8	Nose (nasal septum)	1
Cervix	10	Oral (epidermal)	1
Colon	30	Ovary	9
Conjunctiva	2	Pancreas	6
Embryo (whole)	2	Pharynx	1
Embryo (lung)	2	Placenta	1
Embryo (palatal mesenchyme)	1	Prostate	1
Embryo (intestine)	1	Rectum	4
Endothelial	1	Retina	1
Fetal lung	13	Skin (various)	47
Foreskin (newborn)	1	Stomach	3
Heart	1	T-cell	8
Illeacaeum	1	Thyroid	5
Intestine (small)	1	Umbilical cord	2
Kidney	3	Uterus	2
Larynx	3	Vagina	1
Liver	6		

of 1977. This allows scientists, operating in countries that are signatories to the treaty, to deposit cell lines (including cells of human origin and halo inherent viruses (up to and including the Advisory Committee on Dangerous Pathogens category 4 organisms); bacteria, pathogenic protozoa; plant cells; DNA of eukaryotic origin; yeast and fungi to secure a patent on a process involving the microorganism.

MICROBIAL QUALITY CONTROL

It is ECACC policy only to distribute cells known to be free from contamination with mycoplasma, bacteria, and fungi. The tests routinely used at ECACC for detection of mycoplasma are the standard techniques of Hoechst Staining and Microbiological culture. ECACC also offers this service to regulatory protocols.

AUTHENTICITY OF CULTURES

Before a cell line becomes part of ECACC's collection, it undergoes characterization/authenticity tests to confirm the identity and species origin of the cell line. This is an essential requirement in the management of cell stocks, as cross-contamination between cell cultures is a serious risk for any user of cell cultures. Two basic methods are used at ECACC to verify a cell line; each method has its own advantages and disadvantages. Isoenzyme analysis relies on detecting polymorphic enzyme variants (see preceding section), and several different enzymes may be used for this type of analysis; however, in most cases, a good indication of the species of origin is often obtained with just two enzymes: glucose-6-phosphate dehydrogenase and lactate dehydrogenase.

The second technique used at ECACC, DNA fingerprinting, has become more common in recent years as laboratories have moved toward the use of molecular-based techniques. ECACC uses the Jeffrey's multilocus probes 33.15 and 33.6 (19), which cross-hybridize with a wide range of species and can be used to screen for intraspecies as well as interspecies contamination. This technology means that new deposits to ECACC will have their originality verified by comparison with the fingerprints of existing cell lines. In collaboration with the Human Genome Resource Centre Cambridge, ECACC is currently evaluating the use of multiplex PCR for human cell lines. Using up to ten primer sets, this produces a less complex fingerprint with fewer bands, which can be more easily digitized and stored on a database for comparison with other fingerprints.

The capabilities of the ECACC extend beyond standard cell culture activities such as immortalization; large-scale culture and other services can be found on the internet site (*www.camr.org.uk/ecacc.htm*). A CD-ROM catalogue suitable for PC and Applemac containing full details on all ECACC's collection accessions is available on request from ECACC.

THE INTERLAB CELL LINE COLLECTION (ICLC)

Contributed by B. Parodi, O. Aresu, P. Visconti, M. Cesaro, R. Tononini, and T. Russo

The quality and reproducibility of in vitro assays largely depends on the knowledge of the features of the cell lines used, and on the guarantee of operating with certified material. The cell line collections arose around the world in response to this need for well-characterized material, free of cellular and microbial contaminants.

In Italy, researchers are becoming more and more aware of the need for quality control of the cell lines in their experiments. In view of this, the Interlab Cell Line Collection (ICLC) was set up in November 1994 in Genoa. ICLC belongs to the National Institute for Cancer Research and is located at the Advanced Biotechnology Centre, the first nucleus of the Scientific Park for Biotechnology of Genoa. The personnel involved in the Collection had previously contributed to setting up the Cell Line Data Base, which is currently available through the World Wide Web as a hypertextual database of human and animal cell lines (*http://www.biotech.ist.unige.it/interlab/cldb/html*).

ICLC ACTIVITY

The activity of ICLC includes: collection, production, characterization, quality control, expansion, storage and distribution of human and animal cell lines and hybridomas.

Collection and Production of Cell Lines

A new cell line may be included in the Collection for different reasons: A researcher decides to deposit, usually to make the cell line available to the scientific community and for reasons of security; a researcher is obliged to deposit (by journal editors, for patent purposes); a customer asks for a cell line that is not yet available; the Collection itself performs bibliographic searches on newly established cell lines and sends a letter of request to the originator. ICLC collects the cell lines from the originators, by guaranteeing a service of safe deposit. Since February 1996, ICLC has also been recognized by the World Intellectual Property Organisation as an International Deposit Authority under the Budapest Treaty. Animal cell lines, including cells of human origin and hybridomas, are accepted for patent deposit. The service of production of Epstein Barr transformed human B lymphoblastoid cell lines is also available. All new lines for deposit into the collection are handled in a separate area until the absence of mycoplasma has been confirmed. For security reasons all master and distribution banks are stored in at least two separate liquid nitrogen containers.

Characterization

Isoenzyme analysis is used for the identification of the species of origin of animal cell lines. The AuthentiKit™ System (Innovative Chemistry, Inc., Marshfield, MA) provides a panel of enzymatic substrates (AST, G6PD, LD, MD, MPI, NP, PepB) that identify enzymes with similar catalytic activities but different electrophoretic mobilities in different species. When a cell line is tested, the species identity can be fully characterized, and cross-contamination of a cell lines with another can also be detected (21).

The species confirmation for human cell lines is performed by PCR amplification of a region of the human major histocompatibility complex, class II gene HLA-DPbeta1. This assay is based on the use of a pair of primers that amplify a 310-bp fragment of the nonpolymorphic region of the human gene HLA-DP beta.

The ploidy of cell lines is also examined, in collaboration with the Cytometry Unit of the Advanced Biotechnology Centre.

Distribution of the Cell Lines and Information on the Services Offered by ICLC

The cell lines are distributed to qualified researchers, as frozen ampoules or growing cultures. The cells are sent by courier the week following the order. The ICLC catalogue now contains 150 cell lines of human and animal origin. Among the 96 human cell lines, large panels of well-characterized tumor cell lines are available (neuroblastoma, breast carcinoma, lung carcinoma, etc.).

The description of the service offered, together with the catalogue of ICLC are available online through the Internet, and further information can be obtained by email (iclc@ist.unige.it). At the end of the CABRI project (described previously) the catalogue will also be available at the CABRI web site.

Mycoplasma Detection and Treatment

It has been reported that as many as 10–15% of all tissue culture lines are contaminated by mycoplasma (31). Detection of these mycoplasma-contaminated lines is needed for a good quality. At the Interlab Cell Line Collection (ICLC) three methods are used to detect mycoplasma infection: DNA staining by Hoechst 33258, nested double-step PCR amplification, and Myco Tect assay. Agar culture is being introduced.

Nested Double-Step PCR Amplification

The double-step PCR analysis employs a pool of primers (9 outers and 9 inners) that anneals to genes sequences coding for the evolutionary conserved 16S rRNA of some 25 different mycoplasma species, including the ones most commonly found in cell cultures (30).

The amplification product of the outer primers is 504–519 bp while the inner primers give a product of 318–333 bp. This mycoplasma detection method has several advantages: simplicity and speed, high sensitivity, and specificity.

Biochemical Assay with 5-MPDR (Mycotect Assay, Gibco BRL)

This assay is based on a defined biochemical difference between mycoplasma and their host mammalian cells. Mycoplasmas contain significant amounts of adenosine phosphorylase (32), an enzyme present in small amounts in mammalian cells that converts 6-methylpurine deoxyriboside (6-MPDR), a nontoxic analogue of adenosine into 6-methylpurine and 6-methylpurine riboside. These products are toxic to mammalian cells. A culture infected with mycoplasma can be detected by incubation with 6-MPDR and subsequent monitoring for mammalian cell toxicity

THE CABRI PROJECT

ICLC takes part in the Common Access to Biotechnological Resources and Information (CABRI), an EU-funded project that involves 21 European Collections of biological material and a number of bioinformatic centers. The project aims to standardize the operating procedures, and in particular the quality-control methods used in the different collections, and to set up a bioinformatic network that links the catalogue databases. At the end of the project, which is projected to the end of 1999, the user will have the possibility of easily accessing high-quality biological resources via the Internet, collecting information on human and animal cells, bacteria, fungi, yeasts, plasmids, animal and plant viruses, and DNA probes and directly ordering online the material of interest.

Through initiatives like the CABRI project, the ICLC looks forward to a future of active development and increasing interactions within the European and world-wide network of culture collections.

ANIMAL CELL BANK AT THE INTERNATIONAL FOR FERMENTATION, OSAKA, JAPAN

Contributed by M. Takeuchi

History of the IFO

Since the early part of the twentieth century, authentic cultures of microorganisms have been collected and preserved by several national Japanese organizations. The collections of the Department of Agricultural Chemistry at the University of Tokyo (FAT), the Central Laboratory of the South Manchuria Railway Co. (CLMR), and the Government Research Institute of Formosa (GRIF) are noteworthy examples.

The IFO collection was started in 1944 with cultures derived from the GRIF and FAT collections, and in 1946 the cultures from the CLMR collection were added with the generous help of Professor Hirosuke Nagamishi of Hiroshima University. IFO is greatly indebted to Takeda Chemical Industries, Ltd., for donations and to the Japanese government for grants and contracts to support maintenance of the collection. IFO is a nonprofit organization for the collection, preservation, and distribution of authentic cultures of microorganisms and animal cells useful for scientific and technological studies. The number of cultures in the collection reached 16,000 by the end of 1997.

ORGANIZATION OF IFO

The IFO Foundation has a board of 10 member trustees, including the chairman Dr. Katsura Morita, 15 councilors, and 2 auditors. The IFO institute employs a director, Dr. Masao Takeuchi, 3 business staff, and 11 research staff working in five sections: bacteria, actinomycetes, yeast, fungi, and animal cells. Table 7 indicates the level of activity of each section in 1998.

In the IFO culture collection 16,000 strains are stored and 11,300 of these are fully catalogued in the List of Cultures, Micro organisms 10th edition (1996), or Animal Cell Lines 5th edition (1998). In 1998 9,000 strains were distributed to researchers around the world, including corporate researchers (17%) and foreign researchers (10%). In each research section new microorganisms have been isolated from nature.

Table 7. The Level of Activity of Each Section at the IFO in 1998

Section	Strains preserved	Strains distributed	Strains specially collected
Bacteria	2900	4000	*Sphingomonas*
Actinonycetes	1700	800	*Streptomyces*
Yeast	3100	1500	*Saccharomyces*
Fungi	7900	2500	Marine fungi
Animal cells	800	300	Neural-related cells
Total	**16400**	**9100**	

In addition to the general collection, IFO also preserves safety deposits and patent deposits of microorganisms and cells. IFO has joint activities with the European Patent Organisation (EPO) based on the Budapest Treaty. IFO receives deposits of microorganisms and animal cells for patent applications. Quality control of safety deposit material is carried out depending on the preservation contract with the depositor. Patent deposit materials are preserved to ensure priority of patents. The IFO must check the viability of these materials.

ANIMAL CELL SECTION

Two animal cell banks, the ATCC and the Coriell Institute for Medical Research in the United States have for many years provided the majority of research resources for animal cells. During a 10-years period starting in the mid-1980s several other cell banks were established, for example, ECACC in the U.K. and DSMZ in Germany. In Japan RGB of RIKEN, JCRB of Japan Health and Welfare Ministry, and the IFO cell bank were also established.

The animal cell section of IFO was established in 1984 for preserving patent deposits of animal cells in Japan. It operates as a public service and nonprofit organization under Japanese Tissue Culture Association (JTCA), guidelines for good cell banking. These guidelines were drawn up in 1987 with reference to the World Federation of Culture Collection and the U.S. Federation of Culture Collection guidelines.

In 1998 approximately 800 cell lines were stored in this facility, and approximately 200 cell lines are listed as available for distribution. The section now focuses upon human cells and brain-derived neural cells

QUALITY CONTROL AT IFO CELL BANK

Accession to Distribution of Cell Lines

Cultures obtained from depositors are propagated according to instructions to prepare the first seed stock. Cultures derived from such seed stock material are then subjected to critical characterization, including a series of tests for microbial contamination, including by mycoplasma, and isoenzyme analysis to verify species.

If these tests suggest that the cell line is acceptable, it is expanded to produce seed and distribution stocks. Antibiotics are not used in the culture media at any stage of cultivation, to avoid masking infection.

Contamination

One of the most serious problems in tissue culture is mycoplasma contamination. The IFO investigated the contamination rates in 687 cell lines between May 1984 and October 1993. Of these 185 were contaminated with mycoplasma, 8 with bacteria, and 2 with yeast (Table 8). From the results of a contamination check at IFO it is common knowledge that KATO III cells were contaminated with mycoplasma; however, the original established line has not been contaminated.

Table 8. Contamination Levels Detected at IFO

Contaminant	No. of lines affected
Bacteria	8
Yeast	2
Mycoplasma	185
Other cell lines	15
Noncontaminated	477
Total No. of lines tested	**687**

Table 9. Incidence of Mycoplasma Contamination in Japan

	% of Contamination (no. of cell lines tested)			
Year	IFO	JCRB	RIKEN	Total
1984–1985	30% (171)	28% (196)		29% (368)
1986–1987	30% (143)	14% (342)	40% (218)	25% (703)
1988–1989	18% (80)	9% (209)	10% (125)	11% (411)
1990–1991	32% (191)	4% (128)	20% (323)	20% (642)
1992	15% (68)	0% (64)	36% (77)	18% (209)
Total	**27% (653)**	**14% (937)**	**26% (743)**	**21% (2332)**

Cross-contamination with other cell lines was also found in 15 cell lines, and these 15 lines were immediately destroyed.

Incidences of Mycoplasma Contamination

IFO uses two methods for the routine screening of cell lines for mycoplasma contamination. These are DNA staining and PCR amplification of the DNA sequence between the mycoplasma 16s and 23s rDNA genes. Mycoplasma contamination was examined by Hoechst staining in cell lines collected from within Japan and abroad by three cell banks in Japan (IFO, Japanese Cancer Resources Bank, JCRB, and RIKEN Institute of Physical and Chemical Research) (33). The results are given in Table 9. This survey showed that the incidence of contamination decreased by 1992. Our campaign against mycoplasma contamination may have caused this. Our samples were taken from a variety of institutions, suggesting that Japanese science operates with a contamination levels of almost 20%.

Virus Contamination Test

Many cell lines come into contact with viruses during cultivation. At IFO PCR-based assays have been developed to replace the conventional method of observing cytopathic effects on indicator cells. These include detection of human papilloma virus (HPV) by PCR using sequences homologous to the E6 open reading frame (34) and retroviruses by RT-PCR (35).

DATA MANAGEMENT OF THE IFO CELL BANK

Data management is one of the most important aspects of cell banking. The database, designed by the Cell Bank committee of the JTCA, is used by the IFO and other cell banks in Japan. The IFO database and cell line listings can be viewed on the IFO home page (*http://www.soc.nacsis.ac.jp/ifo/index.html*).

THE COLLECTION AND RESEARCH

The members of IFO cell banks are also engaged in research, including immunology, cellular biochemistry and developmental biology. The main research interest is in the mechanisms of differentiation of animal cells. Human megakaryoblastic cell lines and brain-related cells have been studied in terms of their biological properties. There are few neural-related cells that retain their biological function in vitro. A current research goal at IFO has been to isolate new cell lines from mouse brain and establish neural precursor cell lines from mouse brain or embryonic mouse olfactory neuron tissue. Two such lines have now been isolated and fully characterized and are now available from the IFO (36,37). These two lines should provide a good model to study the mechanisms of survival, proliferation, and differentiation of the precursor cells in the central nervous system or in the peripheral nervous system. Moreover, the animal cell bank of the IFO endeavors to establish a specialized bank to collect and maintain neurocytic brain-related cells.

JAPANESE COLLECTION OF RESEARCH BIORESOURCES (JCRB CELL BANK)

Contributed by H. Mizusawa, H. Tanabe, T, Masui, and T. Sofuni

Brief History of the JCRB Cell Bank (From 1984 to 1999)

The Japanese Collection of Research Bioresources (JCRB) was originally established as the Japanese Cancer Research Resources Bank program by the Ministry of Health and Welfare (H&W) in 1984. The JCRB included a gene bank as well as a cell bank, and they were located separately in two independent institutions. The cell bank was located at the Division of Genetics and Mutagenesis of the National Institute of Health Sciences (Yoga, Setagaya, Tokyo), and the Gene Bank was located at the National Institute of Infectious Diseases (Toyama, Shinjuku, Tokyo). Both institutes are members of the Ministry of Health and Welfare of Japan. Operation of the JCRB cell and gene banks was fully funded by the Japanese Foundation for the Promotion of Cancer Research from the data of inception in 1984 to 1995.

Being the first cell bank in Japan, the JCRB had to establish the entire cell bank systems in place today. The systems were based upon the cell banking model developed by Dr. Hay and his colleagues at ATCC.

Therefore, when we first started the cell bank, we required the assistance of outside research laboratories with experience in cell culture quickly to accelerate the activity of the JCRB cell bank. The following institutes joined the cell bank program as cooperative cell banks: Institute of Medical Research of Tokyo University; Kihara Institute of Biology of Yokohama City University; Tokyo University; Kihara Institute of Biology of Yokohama City

University; Tokyo Metropolitan Institute of Gerontology; Radiation Biology Research Centre of Kyoto University; Okayama University, and Radiation Effect Research Foundation at Hiroshima. This cooperative system was maintained until 1995 and played an important role in collecting many cell lines established in Japan. Approximately 40% of the cell lines in the JCRB cell bank were collected through activities of the cooperative cell banks. The JCRB cell bank listed 570 cell lines as of 1998. 449 of the 570 cell lines were collected from Japanese scientists.

The JCRB cell bank has also built a system to distribute cell materials. Distribution started immediately after starting the operation. From its inception to 1995, all cell materials were distributed free of charge, since full funding was provided by the Foundation for Promotion of Cancer Research. Although the depositors restricted JCRB from distributing materials to scientists not involved in the Cancer Research project of the H&W, there was soon much demand for these resources. In fact, the distribution from our cell bank at 1995 reached approximately 4500 ampoules per year.

Due to the growth in activity of the cell bank, the H&W decided to make JCRB permanent; however, a small handling charge for the distribution of cell materials was introduced to help maintain the economic condition of the bank. The Japan Health Sciences Foundation was brought in to manage the distribution of the cell lines. However, the involvement of the government in the distribution of materials and the collection of handling fees created a tremendous amount of paperwork and caused significant delays in shipping the materials to scientists.

Since the foundation is a nonprofit and a nongovernmental organization, it can manage the distribution more quickly and efficiently. Therefore, the Japan Health Sciences Foundation established the Human Science Research Resources Bank (HSRRB) in 1995 in the Osaka branch of the National Institute of Health Sciences. The JCRB now supplies seed ampoules to the HSRRB, and the HSRRB propagates cell lines for distribution and storage at their laboratory.

Now the JCRB cell bank is responsible for collecting cell lines and maintaining quality control of seed cultures. It is also involved in developing new methods for quality control and new materials for research. Since the quality controls for seed cultures are crucial for the distribution of ampoules, the JCRB has focused its efforts on developing an extensive quality control system at the cell bank of the H&W.

QUALITY CONTROL OF CELL LINES OF THE JCRB/HSRRB CELL BANK

Mycoplasma-Free Cell Culture

Until recently, there have been few strong arguments by Japanese scientists supporting the use mycoplasma-free cell lines. Therefore, numerous cell lines, up to 40% of lines collected from Japanese laboratories (33), were found to be contaminated with mycoplasma. In view of this, one of the first quality-control measures of the JCRB was to establish a system that would ensure that the cell lines would remain free of contamination, especially from mycoplasma.

The first system that we established was a system of detecting mycoplasma using the traditional staining method by Hoechst 33258 with amplification of Vero cells in culture (38). Currently this method is complemented by a nested PCR technique developed by Harasawa (39,40). Both methods are now used to confirm the absence of mycoplasma in every cell line that is cultured.

Many cell lines that are deposited at the JCRB are still found to be contaminated by mycoplasma. This stressed the importance of finding a good system to eliminate mycoplasmas from cell cultures. Initially, we had difficulty finding a good reagent. However, the antibiotic of choice now is MC210 (Dai Nippon Pharmaceutical Co., Osaka, Japan), which is effective in removing mycoplasma from greater than 90% of the cell cultures tested.

After treatment and confirmation of decontamination, the decontaminated cell cultures are then stored as seed stocks. When we prepare seed cultures as frozen ampoules, portions of the seed stock are applied to extensive cell culture for over 30 passages continuously. Portions of the cell cultures are then continuously withdrawn and examined for contaminants. At the end of the passages we determined that the ampoules prepared were negative for mycoplasma contamination.

Prevention of Cross-Culture Contamination

The JCRB cell bank uses chromosomal analysis and isoenzyme analysis to detect cross-culture contamination. Although these methods work well in identifying the species among a variety of animals, with many lines derived from the same species it is more important to identify individuality for the large number of human lines in the JCRB collection.

To achieve this the JCRB modified the Jeffrey DNA fingerprinting technique and established a DNA profiling system (41,42). Using this technique, all cell lines in the JCRB cell bank have been tested. Among all the human lines tested two were found to be identical. These lines have been removed from the catalogues. Since these experiments are done routinely, one of the major weaknesses with the DNA fingerprinting technique is the need to use radioactive chemicals for Southern blot and hybridization analysis. Therefore, even with the DNA profiling system, it has been difficult safely to identify the individuality of each cell culture.

Recently, a precise method using short tandem repeat sequences (STR) to identify the individuality of cell lines derived from humans has been introduced (43) and made available commercially (44). Since this technique is much easier to use than previous ones, the JCRB has decided to use the STR methods to identify cell lines routinely in the laboratory.

Detecting Virus Contamination in Cell Culture

Other contaminants that are important in the cell bank are viruses. However, this is a time consuming and expensive activity. In addition it required special facilities to ensure that when viruses replicate in co-culture methods there is no risk to staff and other cell lines.

JCRB currently uses a newly developed nested PCR system that normally detects mycoplasma contamination to detect bovine diarrhoea viruses. Using newly developed primer sequence sets (45), we are now able to test for BVDV in all cell lines using the nested PCR system.

JCRB/HSRRB CELL BANK SYSTEM

Since 1995, the HSRRB has taken over the responsibility of distributing cell materials from the JCRB. Thus the JCRB cell bank was restarted as a master cell bank, where it could focus on increasing the amount and variety of cell line collections and on developing quality-control methods to develop qualified cell lines for the future. The JCRB is now solely responsible for controlling the quality of seed cell cultures and eliminating cross-culture contamination. The cells for distribution are propagated at the HSRRB, and the HSRRB is responsible for controlling the quality of distributed ampoules. Currently, quality-control measures of the JCRB Cell Bank include tests for mycoplasma contamination, isoenzyme analysis to identify animal species, and chromosome analysis for further genomic characterization. In the near future, the JCRB will institute new tests that look for different viral contaminants and identify the individuality of the cell lines. Since the JCRB cell bank is located in the Tokyo area and the HSRRB is located in Osaka, some difficulties in operating the cell bank have been experienced. However, it is hoped that this isolation is only temporary and that the two laboratories will be reunited in the future.

WEB SITE ON THE INTERNET

Information about the JCRB cell bank is now available on the World Wide Web (WWW), including information about quality controls. In addition, the JCRB website includes a link to the HSRRB website, which contains information about the distribution process. Internet addresses for both the JCRB and the HSRRB websites are: JCRB Cell Bank URL: *http://cellbank.nihs. go.jp/*; email: *cellbank@nihs.go.jp*; HSRRB Cell Bank URL: *http://www.jhsf.or.jp/English/index_e.html*; email: *hsrrb@nihs.go.jp*.

RIKEN CELL BANK

Contributed by T. Ohno

RIKEN Cell Bank has been established as a nonprofit public collection for deposit, isolation, quality control, preservation, and distribution of cultured animal cell lines produced in the field of life science. Supported by the Science and Technology Agency of Japan, it was set up in the Tsukuba Life Science center of the Institute of Physical and Chemical research (RIKEN) in 1987. Subsequently a DNA Bank, BioInfo Bank, and Plant Cell Bank joined to form a consortium, the RIKEN Gene Bank.

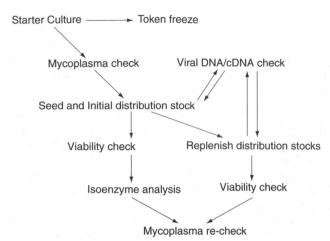

Figure 6. RIKEN cell bank accessioning scheme.

RIKEN Cell Bank has collected more than 1500 animal cell lines and had been distributing more than 2300 ampoules per year; this is the largest distribution from public cell line collections in Japan. The cell lines collected have been divided into two major groups, namely, the certified cell line (RCB-CCL) and the cell-repository lines (RCB-CRL).

RCB-CRL consists of established cell lines from 25 animal species; normal cell types from human, monkey, and suncus; genetic mutant cell types; 12 types of seminormal established cell lines such as NIH-3T3 cells; serum-free/protein-free cultured cell lines; temperature-sensitive cell lines; differentiation-inducible cell lines; fish cell lines; insect cell lines; recombinant DNA transformed cell lines; myelomas and hybridomas; other cell lines such as those from salamanders. Also, cell line libraries consisting of variants derived from a parent cell lines are installed in RIKEN Cell Bank (i.e., Kimura's 3Y1 library and Mitsuhashi's insect cell line library).

ACCESSIONING SCHEME

RIKEN Cell Bank accepts and distributes cell lines after the routine quality control applied to RCB-CRLs (Fig. 6). The accessioning scheme is based upon that of the ATCC, but additional viral DNA detection tests for human and monkey cell lines are included. Human immunodeficiency cirus-1 and hepatitis B–type virus are also tested for; however, to date no cell lines have been shown to be contaminated with these viruses.

CELL LINE DEPOSITION, RESTRICTION FOR DISTRIBUTION, AND CREDIT POINTS

Depositors of cultured animal cell lines are welcome if the cell line or a similar one has not been previously registered with RIKEN Cell Bank. The certificate of acceptance of the cell line is issued after the line is found to be free of mycoplasma, and one credit point is given to the depositor for each line accepted. The credit point can then be exchanged by the depositor for an ampoule of any other cell lines free of charge.

Cell lines are made available subject to any reasonable restrictions requested by the depositor/developer.

SAFETY DEPOSIT

RIKEN Cell Bank accepts safety deposit of cell lines in the form of prefrozen ampoules. These ampoules are kept in gas-phase liquid nitrogen storage tanks in the range -140 to $-170\,°C$ and are never thawed.

MYCOPLASMA DETECTION

Detection methods used at the RIKEN Cell Bank are Hoechst DNA staining and PCR amplification of mycoplasma DNA, but species identification is not performed. From 1985 to 1998 23.4% of 1083 cell lines tested were found to be contaminated with mycoplasma. For the researchers handling new cell lines that are in doubt regarding mycoplasma status, RIKEN Cell Bank offers inspection for mycoplasma.

RESEARCH ACTIVITIES

The management of RIKEN Cell Bank considers that research activities related to animal cell biology should be an important aspect of the bank and should be developed carefully from the base of the current banking service activities. RIKEN Cell Bank is accepting long-term collaboration with intra- and interinstitutional laboratories through use of particular cell lines and culture techniques and also through sharing students, research associates, research ideas, and funds. Current projects include:

- Induction and expansion of human killer lymphocytes for tumor immunotherapy
- Tissue engineering for development of hybrid-type artificial tissues and organs
- Development of primary culture techniques for human cells
- Development of hypersensitive detection methods for virus genes
- In situ freezing method for cultured cells

CATALOGUES

RIKEN Cell Bank Publishes its General Catalogue biannually, a Newsletter twice yearly and email news 7–8 times per year. All the cell lines available for distribution are searchable in the RIKEN Cell Bank home page (*http://www.rtc.riken.go.jp/*).

WORLD REFERENCE CELL LINES

Through a long-term cooperation RIKEN Cell Bank and the European Collection of Cell Cultures (ECACC) have established two World Reference Cell Lines for comparative operation in the field of cell culture. One

is a hybridoma cell line HyGPD-YK-1-1, which produces IgG. The line has no commercial value but is intended as a useful cell line for the comparative development of serum-free culture media. The second line is the Vero cell line (WHO-standardized cell line) held at ECACC.

The RIKEN Cell Bank is committed to the provision of authentic cell lines free of microbial contaminants. It is also active in other aspects of standardization in cell culture, which is an ongoing role for the RIKEN Cell Bank within the international network of culture collections.

BIBLIOGRAPHY

1. G.N. Stacey and A. Doyle, *Cryo-Letters, Suppl.* **1**, 31–39 (1998).
2. Advisory Committee on Dangerous Pathogens, *Categorization of Biological Agents According to Hazard and Categories of Containment*, 4th ed., HSE Books, Sudbury, U.K. 1995.
3. W. Frommer et al., *Appl. Microbiol. Biotechnol.* **39**, 141–147 (1993).
4. G. Stacey, A. Doyle, and D. Tyrrell, in G. Stacey, A. Doyle, and P. Hambleton, eds., *Safety in Cell and Tissue Culture*, Kluwer Academic Press, Dordrecht, The Netherlands, 1998, pp. 1–25.
5. European Pharmacopoeia, *Sterility Testing*, 2.6.1, Suppl. 1999, European Department for the Quality of Medicines, European Pharmacopoeia Secretariat, Strasbourg. Available at: *http://www.pheur.org*.
6. FDA Code of Federal Regulations, *Sterility Testing*, 21 610.12, FDA, Rockville, Md., 1993.
7. L.B. Robinson, R.B. Wichellausen, and B. Roizman, *Science* **124**, 1147–1148 (1956).
8. G.J. McGarrity and H. Kotani, in S. Razin and M.F. Barile, eds., *The Mycoplasma, Vol. IV*, Academic Press, New York, 1985, pp. 353–390.
9. R.A. Del Giudice and R.S. Gardella, in *In Vitro Monogr.* **5**, 104–115 (1983).
10. FDA Code of Federal Regulations, *Mycoplasma Testing*, 21 610.30, FDA, Rockville, Md., 1993.
11. *Human Medicines Evaluation Unit: ICH Topic Q 5D – Quality of Biotechnological Products: Derivation and Characterization of Cell Substrates for Production of Biotechnological/Biolocial Products*, European Agency for the Evaluation of Medicinal Products, ICH Technical Coordination, ICH, London, 1997.
12. European Pharmacopoeia, *Mycoplasma Testing*, 2.6.7, Suppl. 1999, European Department for the Quality of Medicines, European Pharmacopoeia Secretariat, Strasbourg. Available at: *http://www.pheur.org*
13. G.N. Stacey, B. Parodi, and A. Doyle, *Exp. Clin. Cancer Res.* **14**(Suppl. 1), 210–211 (1995).
14. S. Grimwade, *Nature (London)* **259**, 172 (1976).
15. B.J. Culliton, *Science* **184**, 1058–1059 (1974).
16. J. Fogh, N.B. Holmgren, and P.P. Ludovici, *In Vitro Cell Dev. Biol.* **7**, 26–41 (1971).
17. S.M. Gartler, *Nature (London)* **217**, 750–751 (1968).
18. G.N. Stacey, B.J. Bolton, and A. Doyle, *In Vitro Dev. Biol.* **29A** (Part II), 123A (1993).
19. A.J. Jeffreys, V. Wilson, and S.L. Thein, *Nature (London)* **316**, 76–79 (1985).
20. G.N. Stacey et al., *Cytotechnology* **8**, 5–13 (1992).
21. R.W. Nims et al., *In Vitro Dev. Biol.* **34**, 35–39 (1998).

22. Anonymous, *Guide to the Depositing of Microorganisms under the Budapest Treaty*, World Intellectual Property Organization, Geneva, 1995.

23. W.S. Sly et al., *Tissue Antigens, Seminal Ref.* **7**, 165–172 (1976).

24. P. Reed et al., *Hum. Mol. Genet.* **6**, 1011–1016 (1997).

25. R.J. Hay et al., *J. Cell. Biochem., Suppl.* **24**, 107–130 (1996).

26. W.F. Scherer, *NCI Monogr.* **7**, 3–5 (1962).

27. W.A. Clark and D.H. Geary, *Adv. Appl. Microbiol.* **17**, 295–309 (1974).

28. R.E. Stevenson and H. Hatt, *Encycl. Microbiol.* **1**, 615–619 (1992).

29. C.C. Uphoff, S.M. Gignac, and H.G. Drexler, *J. Immunol. Methods* **149**, 43–53 (1992).

30. A. Hopert et al., *In Vitro Cell Dev. Biol.* **29A**, 819–821 (1993).

31. C.R. Woese, in A. Balows, et al., eds., *The Prokaryotes*, Vol I, Springer-Verlag, Berlin, 1991, pp. 3–18.

32. G.J. McGarrity and D.A. Carson, *Exp. Cell Res.* **139**, 199–206 (1982).

33. M. Takeuchi et al., *Bull. JFCC* **9**, 3–18 (1993).

34. *Res. Commun.* **18**, 6–12 (1997).

35. *IFO Res. Commun.* **19**, in press (1999). Available at: *http://wwwsoc.nacsis.ac.jp/ifo/*

36. Y. Nakagaito et al., *In Vitro Cell Dev. Biol.* **34**, 585–592 (1998).

37. M. Satoh and M. Takeuchi, *Dev. Brain Res.* **87**, 111–119 (1995).

38. H. Mizusawa et al., *Tanpakushitsu Kakusan Koso* **31**, 1470–1480 (1986).

39. R. Harasawa, H. Mizusawa, and K. Koshimizu, *Microbiol. Immunol.* **30**, 919–921 (1986).

40. R. Harasawa et al., *Mol. Cell. Probes* **5**, 103–109 (1991).

41. M. Honma et al., *In Vitro Cell Dev. Biol.* **28A**, 24–28 (1992).

42. E. Kataoka et al., *In Vitro Cell Dev. Biol.* **28A**, 553–556 (1992).

43. A. Edwards, A. Civitello, H.A. Hammond, and C.T. Caskey, *Am. J. Hum. Genet.* **49**, 746–756 (1991).

44. M.D. Riccardone, A.M. Lins, J.W. Schumm, and M.M. Holland, *BioTechniques* **23**, 742–747 (1997).

45. R. Harasawa et al., *Tissue Cult. Res. Commun.* **12**, 215–220 (1993).

CELL CYCLE EVENTS AND CELL CYCLE-DEPENDENT PROCESSES

Nicholas R. Abu-Absi
Friedrich Srienc
The University of Minnesota
St. Paul, Minnesota

OUTLINE

INTRODUCTION

The growth of eukaryotic cells is governed by a complex array of biochemical reactions that are collectively known as the cell cycle. The cell cycle controls are intimately coupled to the internal and external states of the cell, and thus ultimately control when a cell devotes the majority of its energy to growth, DNA replication, successful cell division, or cell maintenance. A recent explosion in the amount of research devoted to cell cycle controls has revealed that the activity of enzymes known as cyclin-dependent kinases (CDKs) regulates the order of cellular events that culminate in cell division. Furthermore, these so-called cyclin-dependent kinases are themselves subject to strict controls that govern their expression, activation, and degradation, giving a cell the freedom to respond to a variety of situations. An overview of the findings from the numerous biological studies that have been conducted on cell cycle control is presented here. Additionally, since the cell cycle controls the most fundamental aspects of cell growth and division, it can be linked to the dynamics of many other cellular processes important to cell culture such as protein accumulation and secretion. Therefore, in order to achieve a more complete understanding of the cell cycle, a discussion of some of the investigations that have coupled the cell cycle to the dynamics of cellular process is presented. Indeed, the framework diagrammed here offers many opportunities for further research that can be used to engineer more efficient methods of cell culture.

THE BIOLOGY OF THE CELL CYCLE

The cell cycle is a group of complex interrelated events that exists to ensure that cells successfully divide to form two normal daughter cells. The eukaryotic cell cycle was originally divided into two stages, interphase and mitosis, as observed by light microscopy. Interphase was defined as the period between two successive rounds of cell division, whereas mitosis was identified as the period when the cell undergoes many structural changes, divides its chromosomes, and undergoes cell division. The advent of biochemical and genetic techniques later showed that interphase consists of many cellular processes necessary for the regulation of cell proliferation that cannot be discerned under the light microscope. The cell cycle was then partitioned into four separate phases: G_1 (or Gap 1), DNA synthesis (S phase), G_2 (Gap 2), and mitosis (M phase), where G_1 refers to the gap between mitosis and DNA synthesis and G_2 to the gap occurring after DNA synthesis and prior to mitosis.

Understanding the complex network of biochemical events that guide a cell through successful completion of cell division is of utmost importance in understanding how cells behave in response to different situations. There has been an explosion in the amount of cell cycle related research being done in the past ten years. This is due, in part, to the fact that a wide variety of cellular processes are thought to be somehow regulated by cell cycle events and proteins. Additionally, improper control of cell cycle processes in higher eukaryotes has been linked to the formation of cancerous tumors. This article discusses the biochemical events that occur in the cell cycle and the control mechanism that maintains temporal order of these events.

The G_1 and G_0 Phases

Upon initial inspection, it would seem that the G_1 phase of the cell cycle is a particularly uneventful time period (or gap) between the more spectacular events that occur during mitosis and DNA synthesis. Closer analysis of this interval, however, reveals that the cell is closely monitoring its environmental and internal state to determine whether or not the cell has sufficient resources and an intact genome to safely complete the rest of the cell cycle. In contrast to the other phases of the cell cycle, G_1 is highly variable between different cell lines and growth conditions. For example, if certain factors are not available, some cells do not progress towards the S phase, but instead may enter a resting state called the G_0 phase or quiescence. During this time, the cell undergoes a significant reduction in RNA and protein synthesis and will remain in this arrested state with unduplicated DNA content until the cell receives the proper stimuli.

The sequence of events that take place during the G_1 phase has been deduced by observing how cells respond when cultured in different environments. For instance, when cultured cells are deprived of serum containing growth factors, the synthesis of new proteins is drastically reduced and a portion of newborn cells cease to advance through the cell cycle. Through observations made using time-lapse cinematography on a culture of Swiss 3T3 cells, it was determined that only those cells that exited from mitosis less than four hours prior to growth factor deprivation were arrested in G_0, while all other cells continued to mitosis as normal (1). This observation led to the division of G_1 into two subphases. The first subphase consists of cells that are still dependent on growth factors for cell cycle progression and are referred to as G_1 postmitotic (G_{1pm}) cells, while cells which can proceed through the rest of the cycle are called G_1 presynthesis phase (G_{1ps}) cells. The transition point where cells pass from G_{1pm} to G_{1ps} is commonly called the restriction point in mammalian cells (2) and is analogous to "start" in yeast. During G_{1pm}, progression through the cycle is highly dependent on growth factors and amino acids (3) which

are necessary for the production of certain regulatory proteins (called cyclins) that are discussed later. It is thought that the purpose of the G_{1ps} period is to allow cells to grow, perhaps so that later in the cycle the cell can focus its energy on other cellular processes such as DNA replication in S phase and reorganization of the cellular infrastructure during mitosis. This also explains the fact that the G_{1ps} portion of G_1 accounts for nearly all the temporal variability in the cell cycle, both within a population and between cell lines (2,3). For example, after having passed the restriction point, larger cells may proceed almost immediately to the S phase, while smaller cells may linger in G_{1ps} for up to ten hours before beginning the transition to the S phase (1). Additionally, although growth factors are not required to promote S phase entry once past the restriction point, the length of G_{1ps} is significantly increased when factors such as insulin and insulinlike growth factor (IGF) are lacking in the environment, presumably due to a decrease in net protein synthesis. Under these conditions, cells undergo mitosis with a significantly reduced protein content (2), suggesting that different growth factors play an important role in growth and progression through the cell cycle.

A culture of 3T3 cells arrested in G_0 by serum deprivation can reenter the cell cycle and proceed through the restriction point only after stimulation by the addition of platelet-derived growth factor (PDGF) (1), and this process is referred to as *competence*. Factors that initiate competence, however, vary between cell lines (4), and several other growth factor/cell line combinations have been described (5). The biochemical effects of competency factors such as PDGF are discussed later. Time-lapse cinematography also revealed that cell cultures that lack serum do not contain cells arrested all at one point, but at numerous time points throughout G_{1pm}. Additionally, when stimulated to proliferate, arrested cells require an extra 8 hours in G_1 before proceeding to the S phase, presumably so the cells can recover from G_0 (2).

The DNA Synthesis Phase

The S phase is the period when the cell replicates its entire genome so that each daughter cell will receive a complete, intact set of chromosomes. Due to the immense size of mammalian genomes, multiple bidirectional replication sites are necessary in order for the synthesis of new DNA to be accomplished in such a short time (approximately 8 hours in most cell lines). In addition to chromosome duplication, the cell must also synthesize proteins such as histones for the formation of nucleosomes and the chromosomal scaffolding. The replication of DNA has been shown to be a highly ordered process. This was observed by culturing cells with a thymidine analog called bromodeoxyuridine (BrdU). Cells in the process of synthesizing DNA take up BrdU and integrate it into newly formed DNA. Cells can then be separated by their position in S phase, based on the amount of BrdU in the DNA, to determine the exact time of replication of certain genes (6–8). These studies revealed several aspects of the replication of certain genes. First, individual genes within a cell line are replicated at defined periods in the

S phase, with many of the more active genes replicated earlier than others. Furthermore, newly replicated DNA is packed into nucleosomes a short distance behind the replication fork (1).

It is also very important that the cells not replicate the same chromosome more than once; thus there is a mechanism that ensures that this does not happen and is known as the rereplication block (2). Early cell fusion experiments done in human HeLa cells showed that nuclei prior to S phase could be induced to undergo DNA synthesis early, whereas nuclei that have already undergone a round of DNA synthesis do not undergo another round of DNA synthesis (3). Later cell fusion experiments show that nuclei with replicated DNA can undergo another round of DNA synthesis before mitosis after the nuclear membrane has been permeablized and then repaired (4). A model involving a replication licensing factor (RLF) was then proposed to describe these phenomena. DNA would only be replicated if RLF was bound to the chromatin to be broken down by the replication machinery during the next S phase. During mitosis, the nuclear envelope breaks down, allowing new RLF to access the chromosomes and bind to them, thus licensing the chromosomes to undergo a single round of DNA replication. Support for this model was gained when two protein complexes, called RLF-M and RLF-B, were shown to be required to modify chromatin before replication is allowed to proceed (5). These protein complexes were shown to bind to chromatin in late mitosis and be gradually removed during the S phase (6,7). The rereplication block model described earlier has recently been modified to accommodate a more detailed understanding of the proteins and mechanisms involved (2). Chromosomes in G_1 must first be "primed" by the establishment of a prereplication complex (pre-RC) containing an origin recognition complex (ORC), an RLF complex, and other proteins. After initiation of DNA synthesis, the pre-RC is disrupted to become the post-RC, which still contains the ORC. The ORC may be some sort of "docking" factor, which allows other components of the pre-RC to assemble on the DNA.

The G_2 Phase and Mitosis

Cells enter the G_2 phase upon completion of DNA synthesis, and this period lasts between 2 and 4 hours, culminating with the onset of chromosome condensation. During this time, duplication of the centriole is completed, and cells produce many mRNAs and proteins that will be used during mitosis. Cells may also use this period of time before mitosis to ensure that the cell is sufficiently large and is ready to undergo cell division.

Mitosis is generally the shortest phase of the cell cycle, lasting only 15 min to 1 hr, depending on the cell line. However, visual inspection of a cell culture reveals that this is a very physically active period for the cell as it goes through many structural changes in preparation for cell division. Mitosis is traditionally divided into 6 stages that describe the mechanical events that occur during this period: prophase, prometaphase, metaphase, anaphase, telophase, and cytokinesis. The beginning of mitosis, known as prophase, is marked by condensation of

the chromosomes into compact structures, which allows sister chromatids to be more easily separated. The cytoskeletal microtubules begin to disassemble, and the mitotic spindle, a weblike array of microtubules that mediates the events in mitosis, begins to form. Also during this phase, the Golgi apparatus breaks up into tiny vesicles and pinocytosis and RNA synthesis stop. Prometaphase begins with the disruption of the nuclear envelope, followed by the attachment of the chromosomes to the microtubules that make up the mitotic spindle. The microtubules attach to protein structures on the DNA called kinetochores, which as we shall see are very important in regulating progression through mitosis. During metaphase, the mitotic spindle arranges all the chromosomes in a single plane across the middle of the cell called the metaphase plate. After alignment of the chromosomes, the cell enters anaphase, when sister chromatids are pulled to opposite sides of the cell by the microtubules so that there is a complete genome on either side of the cell. Microtubules not attached to kinetochores then elongate, pushing the spindle poles further apart. Telophase then begins, resulting in the formation of two nuclei, one around each set of chromosomes on opposite sides of the cell. The cell then begins the process of cell division, called cytokinesis, when a ring of actin and myosin that has formed around the center of the cell begins to contract. Upon completion of this process, two new daughter cells are formed, each containing a complete, intact genome and cytoplasmic organelles sufficient to support further growth and proliferation.

The temporal organization of the cell cycle phases is shown in Figure 1, where the numbers and arrows represent when the corresponding cyclin/CDK complexes listed in Table 1 are active.

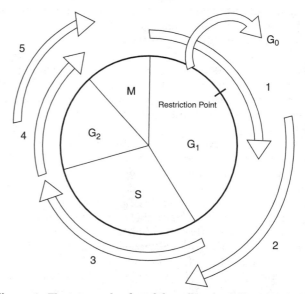

Figure 1. The temporal order of the cell cycle. Cells exit mitosis after cell division and are sensitive to the presence of growth factors until they pass the restriction point. The large block arrows depict the active period of the best-defined cyclin/CDK complexes. The numbers next to the arrows correspond with the numbers assigned to the various complexes as defined in Table 1.

Table 1. The Major Cyclin/CDK Complexes are shown, along with the Cell Cycle Phases where each Complex is Active and their Respective Functions

Complex #	Cyclin/CDK complex	Cell cycle phase	Function	Ref.
1	cyclin D/CDK4	G_0, early–mid G_1	Stimulates exit from G_0 and progression through the restriction point; phosphorylates pRb	22,27,30,77–79
2	cyclin E/CDK2	G_1/S transition	Triggers S phase entry; activates cdc25A	33,76,81,83
3	cyclin A/CDK2	S	Phosphorylates DNA synthesis machinery; inhibits E2F/DP1 transcription factor	18,36,37,91
4	cyclin A/CDK1	G_2, M	Triggers formation of mitotic spindle	39,93,97
5	cyclin B/CDK1	G_2/M transition, M	Triggers entry into M; responsible for many structural changes in mitosis	18,39,46,98–102
6	cyclin H/CDK7	All cell cycle phase?	Cyclin activating kinase (CAK); phosphorylates other cyclin/CDK complexes	49,50

The numbers listed with each complex correspond with the numbers listed in Figure 1. See the text or the references for more detail on the substrates of the complexes.

CONTROL OF THE CELL CYCLE

Successful completion of cell division requires that the cell reside in an environment capable of supporting growth, and that the cell cycle proceed as normal. As discussed previously, cells require the presence of growth factors and essential amino acids to support the synthesis of new proteins. In addition, the immediate environment must not be overcrowded such that daughter cells have room to grow and further divide. If any of these conditions are not met, cells enter the G_0 phase, and will eventually die if environmental conditions are not improved. Similarly, if the events in the cell cycle do not occur in the proper sequence, daughter cells may not inherit the proper genetic material or organelles to properly function, and thus die. Therefore, cells must possess some means of sensing whether or not the environment is able to support growth. There must also be a control mechanism in place that regulates the sequence of events that make up the cell cycle. An understanding of the mechanisms that control cell growth is of interest to many scientists including oncologists and biochemical engineers. Oncologists may be able to use the knowledge of the events regulating the cell cycle to develop therapies against certain types of cancers, while the control network of cells in culture can be engineered to reduce a culture's dependence on serum.

Central to the mechanisms that control the events of the cell cycle are checkpoints. Checkpoints refer to points in the cell cycle where certain criteria must be met before the cell cycle can proceed any further. For example, the restriction point in the G_1 phase is a checkpoint, since cells not cultured in the presence of growth factors are arrested in G_0 and not allowed to proceed through the cell cycle until growth factors are supplied. Indeed several other checkpoints that control the order of the cell cycle phases and of the events within each phase exist throughout the cell cycle. For instance, cells with unduplicated DNA do not enter mitosis, and chromosomes are only replicated once in each cycle. Additionally, cells with damaged DNA do not undergo mitosis; rather they become arrested in G_2 to allow the DNA repair machinery to correct any defects in the genetic material. Controls that mediate the order of events in mitosis also exist. For example, entry to metaphase is prevented until the chromosomes are lined up at the metaphase plate. As we will see, complexes made up of two classes of proteins are involved in regulating all these events throughout the cell cycle. These two classes of proteins are cyclins and cyclin-dependent kinases (CDKs).

Partners in Cell Cycle Regulation: Cyclins and CDKs

Cyclins were originally defined as molecules that accumulated throughout the cell cycle in marine invertebrate eggs and were then degraded during each mitosis (8). Several years later, the amino acid sequences of various cyclins were compared, and a consensus sequence of approximately 100 amino acids was found and dubbed the "cyclin box" (9). Cyclins are now defined on the basis of the presence of the cyclin box in the native sequence. Currently, 12 cyclins have been identified in mammalian cells (cyclins A–H, with 2 A, 3 B, and 3 D type cyclins) (10). Expression of the cyclin box domain genes facilitates binding to CDKs whose kinase activity is dependent on being bound to a cyclin molecule (11) and are thus defined as proteins that require binding to cyclins in order to attain activity (12). There are 7 known CDKs in animal cells (named CDK 1–7). Different cyclins may interact with different CDKs at specific points in the cell cycle, and each of these complexes is specific for substrates with differing functions. Since several of the complexes have similar substrate specificity in vitro, the change in specificity is thought to be due not only to changes in the affinity towards different substrates, but also to localization of complexes in distinct parts of the cell by various cyclins (13). We shall first briefly discuss the different cyclin/CDK complexes that form throughout the cell cycle, and then review the regulation of these complexes. The major cyclin/CDK complexes that are active throughout the cell cycle are listed in Table 1 along with their corresponding times of activity and their functions.

The major studies that resulted in the discovery of G_1 cyclins were performed in budding yeast (14). It was shown that three cyclin molecules, CLN1, CLN2, and CLN3, have important functions in advancing cells through the restriction point (called start in yeast). In addition, inactivation of all three molecules is required to cause G_1 arrest of the cell cycle. By searching for factors that could complement the function of CLNs in yeast, several mammalian cyclins were isolated: cyclin C, cyclin D, and cyclin E (15). Other studies revealed several other G_1 cyclins, cyclins G and F; however, neither the roles that these cyclins play nor their CDK partners are known (16). It has been shown that cyclin C mRNA synthesis reaches its peak in mid-G_1, and cyclin G mRNA synthesis starts soon after growth factor stimulation of quiescent cells (17). Conversely, cyclins D and E are fairly well characterized and have been the focus of much attention, since defects in control of these proteins have been implicated as a cause of cancer (18).

While D-type cyclins have been shown to associate with CDK2, CDK5, and CDK6, they seem to interact primarily with CDK4 (19). Transcription of various combinations of D-type cyclins (cyclins D1, D2, and D3) is stimulated by different mitogens (growth factors) depending on the cell type, and their synthesis throughout the rest of the cell cycle is maintained by the presence of growth factors (20). Removal of growth factors halts the synthesis of D cyclins, resulting in a rapid decrease in their levels, since they have very short half-lives (21). Furthermore, in cycling cells, the level of cyclin D does not vary, and only a small peak showing its accumulation is observed. As noted previously, growth factor deprivation has no effect on cells that have passed through the G1 restriction point, and these cells continue to cycle until their progeny become arrested in early G_1. Additionally, microinjecting cells with cyclin D1-specific antibodies results in their cell cycle arrest only during early to mid-G_1 (22), while overexpression of cyclin D1 results in a reduction in the length of G_1, the size of the cell, and growth-factor dependency (23,24). These findings indicate that the primary function of D cyclins is to link the cell cycle machinery with the extracellular environment, which permits the cell to become committed to DNA synthesis when in a favorable environment or become arrested in G_0 under unfavorable conditions.

Cyclin E is expressed soon after cyclin D expression in G_0 cells stimulated with growth factors, and its level peaks late in G_1. It has been shown to interact with a single catalytic subunit, CDK2 (25). In contrast to D cyclins, cyclin E is synthesized periodically and is rapidly degraded soon after entry into the S phase, thus adhering to the original definition of a cyclin (26). Similar to cyclin D, cyclin E function has been shown to be essential for progression through G_1, since microinjection of anti-cyclin E antibodies prevents cells from entering the S phase, while its overexpression results in acceleration towards the S phase (27). This acceleration is, however, balanced by an increase in the lengths of the S and G_2 phases so that the length of the entire cycle remains unchanged. The increase in the length of the S and G_2 phases following increased expression of cyclin E may be due to the fact that cells in which cyclin E is overexpressed are smaller in size than cells expressing normal levels of cyclin E. Since an increased amount of cyclin E advances progression to the S phase early, it is possible that the cyclin E/CDK2 complex is responsible for initiating the synthesis of proteins that are used during DNA synthesis. These findings, when taken together, suggest that the cyclin E/CDK2 complex is involved in regulating the passage from G_1 into the S phase (16).

Upon entry into S phase, cyclin E is degraded, and cyclin A takes its place as the partner of CDK2 (28). Transcription of cyclin A is begun immediately after entry into S phase, and its inhibition by microinjection of anti-cyclin A antibodies allows, at most, only 10% of the DNA to be replicated (29), and may also disrupt the temporal relationship between DNA synthesis and mitosis (30). As a cell progresses into the G_2 phase, cyclin A is no longer associated with CDK2. However, it forms a new complex with CDK1 (also known as cdc2 due to its homology with the budding yeast kinase of the same name) to regulate progression through G_2 and into mitosis. It has been shown that CDK1 forms complexes with both cyclin A and cyclin B near the G_2/M transition (31). These complexes are located in different parts of the cell, since cyclin A accumulates in the nucleus, while cyclin B resides in the cytoplasm until the breakdown of the nuclear envelope during mitosis (32). Once they have completed their respective functions, both cyclins are then rapidly degraded later in mitosis.

Experiments done in fission yeast suggest that the cell monitors the presence of cyclin B, since cells undergo another round of DNA replication when cyclin B/CDK1 complexes are not present (33). This implies that the level of cyclin B/CDK1 complexes, both active and inactive, are constantly monitored and the presence of these complexes maintains cells in G_2. Although this theory has not been investigated in other eukaryotes, it is not entirely unlikely given the high degree of conservation of cell cycle controls among eukaryotic cells.

Regulation of the Cyclin/CDK Complex Activity. Three different mechanisms allow tight control over the myriad of cyclin/CDK complexes: (*1*) cyclin transcription and degradation, (*2*) CDK phosphorylation and dephosphorylation, and (*3*) binding of the cyclin/CDK complexes to various inhibitory proteins. These three forms of cyclin/CDK regulation interact to form an intricate regulatory network that allows cells to control events temporally throughout the cell cycle and to ensure that these processes are successfully completed. These three mechanism of cyclin/CDK complex regulation are depicted in Figure 2.

Intracellular concentrations of different cyclins, and thus the activity of cyclin/CDK complexes, are regulated by synthesis and degradation. There are two different pathways by which cyclins can be degraded, depending on the type of the cyclin. Cyclins D and E, which function in early G_1 and the G_1/S transition, respectively, are inherently unstable and thus have short half-lives of

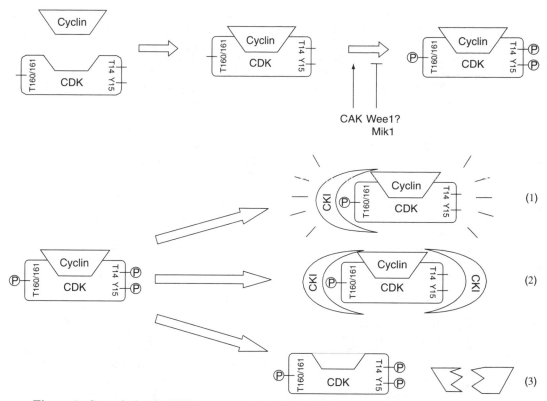

Figure 2. Control of cyclin/CDK complex activity. Cyclin/CDK activity is governed by three mechanisms: phosphorylation/dephosphorylation reactions, cyclin-dependent kinase inhibitors (CKIs), and cyclin expression and degradation. After assembly of the cyclin/CDK complex, the activity of CAK activates the complex, while phosphorylation by Mik1 and Wee1 inhibits complex activity. The phosphorylated complex is then activated by the phosphatase activity of cdc25 and is denoted as complex (1) in this figure. Complex (2) has the correct residues phosphorylated which are necessary for activation; however, it is bound by two CKIs and is inactivated. Complex (3) is shown inactivated by cyclin degradation.

approximately 30 min. The instability of these cyclins has been attributed to the presence of a PEST sequence near the C-terminus of the cyclin box (34). In contrast, cyclins A and B, which function in the S, G$_2$, and M phases, are stable throughout interphase, even when they are not bound with CDKs. However, these cyclins are degraded during mitosis in the proteosomes through a ubiquitin-mediated pathway (35). These proteins contain a sequence of amino acids near the amino terminus called the "destruction box" that marks them for destruction. Degradation of these proteins is also regulated so that cyclin/CDK complexes that are necessary for cell cycle progression are not inactivated before their specific tasks are successfully completed. For example, it has been shown that proteolysis of cyclin A and B does not occur in *Xenopus* egg extracts until purified cyclin B/CDK1 complexes are added (36). This suggests that the proteolysis pathway is initiated by active cyclin B/CDK1 complexes. Conversely, the addition of purified cyclinA/CDK1 complexes to the extracts had the opposite effect of delaying cyclin degradation triggered by cyclin B/CDK1 kinases (37). The fact that only cyclin B/CDK1 complexes can activate proteolysis may explain why cyclin degradation does not occur until correct assembly of the mitotic spindle is achieved in metaphase (38). The exact mechanism of

regulation of cyclin degradation is unknown, but it has been hypothesized that cyclin B/CDK1 activates a cyclin-specific enzyme that catalyzes the conjugation of ubiquitin to cyclins (39).

Cyclin/CDK complexes are also regulated by phosphorylation and dephosphorylation. Specifically, phosphorylation on T160/161 (threonine at amino acid 160/161) (depending on the CDK) is necessary for activation of the complex, while phosphorylation of T14 and Y15 (tyrosine 15) residues inhibits cyclin/CDK activity (Fig. 2). The T160/161 residue is located on a "T loop," which is conserved in all CDKs. When not phosphorylated, this loop prevents kinase activity by covering the substrate binding domain. Phosphorylation at this site serves two purposes: It increases the affinity between the cyclin and the CDK and therefore stabilizes the complex, and it alters the conformation of the T loop, allowing access to a substrate (40).

The kinase that is responsible for phosphorylation of this residue on several CDKs is also a cyclin/CDK complex. This complex, commonly referred to as CDK activating kinase (CAK), is made up of cyclin H, CDK 7, and a third 32-kDa protein (41). Although the level of CAK activity has not been shown to vary in a cell cycle dependent manner, it is possible that the level of cyclin H is regulated by its synthesis and degradation (38).

The kinase responsible for activating CAK has not been identified, but it has been shown not to be an autophosphorylation process (41,42). Further studies done with this complex have shown that it may play a role in activating the RNA polymerase II C-terminus domain kinase of transcription factor IIH (43,44). This suggests that cyclin H and CDK7 may have a more diverse role than merely activating other CDKs (38).

Cyclin/CDK complexes are also subject to negative regulation through phosphorylation of the T14 and Y15 residues. These residues are found within the ATP binding site of the enzyme. Phosphorylation of the Y15 residue inhibits kinase activity by interfering with the transfer of a phosphate group to a bound substrate (45), while phosphorylation of the T14 residue acts by interfering with binding of ATP (46). The regulation of phosphorylation on these residues is not yet defined in animal cells. In addition, only the process of the tyrosine phosphorylation is known in yeast, since phosphorylation of the threonine residue is not necessary in this organism (10). In yeast, the tyrosine residue is phosphorylated by the Wee1 and Mik1 kinases (47). The Wee1 kinase is in turn regulated by the Nim1 kinase, which inhibits Wee1 activity. The kinase responsible for phosphorylation of the Y15 residue in animal cells has been isolated (48) and is closely related to Mik1 (10). The dephosphorylation of these residues, which is necessary for cyclin/CDK activation, is performed by the phosphatases cdc25A, B, and C (49). Cdc25A has been shown to be activated by the cyclin E/CDK2 complex during transition to S phase. Furthermore, activated cdc25A has been shown to activate both cyclin E/CDK2 and cyclin A/CDK2 complexes (10). The cdc25C phosphatase activates cyclin B/CDK1 complexes at the transition from G_2 to mitosis (10). The function of cdc25B is as yet unknown.

The activity of cyclin/CDK complexes is also controlled by certain proteins, known as cyclin-dependent kinase inhibitors (CKIs or CDIs), that bind to these complexes and repress their activation. Several different inhibitors have been isolated from mammalian cells that interact with either different cyclin/CDK complexes or the CDKs themselves. These proteins are commonly named according to their molecular weights, such as the 21-kDa inhibitor known as p21. Other cyclin/CDK inhibitors are p27, p15, and p16. These inhibitors are intricately involved in regulating progression through the G_1 phase, and have thus been the subject of recent reviews (50–52).

Immunoprecipitation studies have revealed a very interesting fact: Cyclin D1 is commonly found among a quaternary complex of proteins that includes either CDK4 or CDK6, p21, and proliferating cell nuclear antigen (PCNA) (discussed below) (19). The p21 protein acts as an inhibitor by binding directly to cyclin B/CDK1, cyclin E/CDK2, cyclin A/CDK2, and cyclin D/CDK4 complexes. These cyclin/CDK complexes containing p21 are, oddly enough, still enzymatically active in vitro, which directly conflicts with studies showing that p21 is an inhibitor of CDK activity and is able to induce cell cycle arrest (53,54). It was later demonstrated that active cyclin/CDK/p21 complexes isolated from cycling cells could be inactivated by the addition of more p21 (55), suggesting that the stoichiometry of p21 binding to these complexes is very important. For example, in cycling cells, there is only enough p21 present so that each cyclin/CDK complex has a single p21 molecule and thus remains active. However, when the amount of p21 is increased or the cyclin/CDK pool depleted, the ratio of p21 to cyclin/CDK complex increases so that more than one p21 molecule binds to each complex, resulting in inhibition of the complex kinase activity. We will explore the circumstances under which this sort of inhibition becomes important in a later section.

A second inhibitor that has been identified is called p27. p27 and p21 have been shown to have highly similar amino acid sequences near their amino terminus domains. In addition, they also share the same range of specificity for cyclin/CDK complexes. The cell cycle arrest of certain cell lines in response to some environmental signals such as contact inhibition and transforming growth factor-β (TGF-β) has been shown to be due, in part, to inactivation of cyclin/CDK complexes by this inhibitor (56), and the exact mechanism of this type of arrest will be described shortly. A recent discovery has given some insight into one of the mechanisms by which this protein inhibits cyclin/CDK activity. Cyclin/CDK complexes that have not yet been activated by CAK phosphorylation can be bound by p27, which then blocks the CAK's access to the threonine residue found on the CDK subunit (57). Presumably another mechanism exists for inhibition of active complexes.

The main function of the p27 inhibitor seems to be to prevent the G_1/S transition from occurring, when the need arises, by inhibiting the activity of cyclin E/CDK2 complexes (58). In normal cycling cells, the p27 inhibitor seems to be dispersed among the different cyclin/CDK complexes that are active during G_1; therefore, it does not have a significant impact on any essential events. However, in response to stimulation by TGF-β it has been suggested that the expression of CDK4 is down regulated, resulting in an increased amount of p27 bound to cyclin E/CDK2 complexes (59).

The p16 inhibitor was originally found associated with CDK4 in a transformed cell line (60). Unlike the p21 and p27 inhibitors, p16 was found to associate specifically with the cyclin D CDKs CDK4 and CDK6. It competes with cyclin D for the CDK subunits and may even act to disrupt cyclin D/CDK complexes. This inhibitor is thought to play an important role in regulating growth, since the p16 gene has been shown to be deleted in many tumor cell lines (61).

Examination of TGF-β arrested cells led to the discovery of another inhibitor, p15, which was shown to be bound to CDK4 (62). Similar to p16, p15 binds to CDK4 and CDK6, competes with cyclin D, and may be involved in arresting cells treated with TGF-β in G_1. TGF-β stimulates the synthesis of p15 so that its level is increased approximately 30 times. This allows p15 to sequester the CDK4 complexes so that a much larger amount of the p27 inhibitor binds to and inactivates cyclin E/CDK2 complexes, resulting in G_1 arrest.

The Cyclin/CDK Complexes in Action

Now that we have discussed the events that occur in the cell cycle and defined the major regulatory proteins, we are in a position to examine the mechanisms that control cell cycle progression. Indeed, control of the activity of the cyclin/CDK complexes is at the heart of cell cycle regulation, since these complexes either directly or indirectly act on cellular constituents that are capable of bringing about changes in intracellular events. This implies that the proper regulation of these cyclin/CDK complexes is essential for maintaining the temporal integrity of cell cycle events. This is accomplished by repressing the activity of the appropriate kinase through different combinations of cyclin degradation, phosphorylation/dephosphorylation of specific CDK residues, and the action of CDK inhibitors. These controls not only maintain temporal stability, but also ensure that DNA synthesis is not carried out when DNA is damaged and that mitosis does not occur in the presence of damaged or misaligned chromosomes.

G_1 to S Phase Control. An important property in normal cells is their ability to sense the state of the extracellular environment, such as the presence of various growth or inhibitory factors or contact inhibition. This communication between the cell and its surroundings takes place in the G_0 and G_1 phases, and the cells respond by becoming quiescent, by undergoing programmed cell death (known as apoptosis), or by continuing with the cell cycle. The interaction with the environment ceases when the cell passes the restriction point, after which the cell is said to be committed to DNA synthesis. We will now consider the machinery responsible for directing the path that the cell will traverse in response to signals from the environment.

Several lines of evidence indicate that the product of the retinoblastoma gene, pRb, is responsible for linking signals from the extracellular domain (63). Specifically, mutation or deletion of this gene is found in many cancerous cell lines that have lost their ability to regulate proliferation (64). Additionally, pRb accumulation was found to coincide with suppression of growth in TGF-β induced arrest (65), while inhibition of pRb results in a resumption of cell cycle progression. Mammalian DNA viruses, whose main purpose is to activate quiescent cells so that they can use the cell's replication machinery to duplicate their own DNA, encode enzymes that phosphorylate the hypophosphorylated form of pRb (66). This led to the belief that pRb is in its active state when under phosphorylated, and that phosphorylation of this protein results in its inactivation. This is supported by the fact that hypophosphorylated pRb is present in G_0 cells and is phosphorylated later during the G_1 phase. In its active form, pRb binds to several proteins, including a heterodimer of proteins from the E2F and DP family of proteins. This heterodimer, called DRTF1/E2F (67), is a transcription factor, which stimulates the transcription of numerous proteins that are necessary for cell cycle progression such as CDK1, cyclin A, and a variety of proteins that are indispensable during DNA synthesis (68).

In order to further characterize G_1 events, it is necessary to find out what is responsible for inactivating pRb. A likely candidate seems to be the cyclin D/CDK4 complex. The presence of this active kinase is highly dependent on the presence of growth factors that stimulate the synthesis of cyclin D, and thus the activity of this complex appears during the time period when pRb is inactivated. Moreover, cells that lack functional pRb do not require the presence of cyclin D/CDK4 to proceed through the cell cycle, implying that the primary function of this complex is to inactivate pRb (69). Indeed, the belief that pRb is a substrate of this complex was confirmed, since cyclin D/CDK4 is capable of binding to and phosphorylating pRb in vitro (70,71). The following feedback loop, involving the levels of cyclin D and the functional state of pRb, has been proposed (72). Cells that lack functional pRb have drastically reduced levels of cyclin D. However, addition of pRb to these cells results in the synthesis of cyclin D. Active kinase complexes are then allowed to form, and act by phosphorylating pRb. The hyperphosphorylated form of pRb down-regulates the levels of cyclin D during late G_1, causing the inactivation of CDK4, which in turn allows pRb to be activated once again by the action of phosphatases, and the cycle begins anew upon completion of mitosis.

Since the CDK inhibitors p21 and p27 are constitutively expressed in cycling cells, these molecules are present throughout the G_1 phase. This indicates that in order for cells to advance past the restriction point, the level of cyclin D/CDK4 complexes must increase until it exceeds the number of inhibitor molecules present so that there is at most one CKI per cyclin/CDK complex to allow kinase activation (56). This model not only satisfies the timing of cyclin D/CDK4 activation, but suggests a dual role for this complex. Its primary function is to phosphorylate pRb to promote the transcription of important genes, while its secondary role is to sequester the CKIs so that the cyclin E/CDK2 complex, which is the next kinase to be active in the cell cycle, cannot be inhibited by these molecules, thus allowing the transition to the S phase to occur (50). A schematic representation of the control mechanism that regulates progression through G_1 is shown in Figure 3.

As mentioned previously, the activity of cyclin E/CDK2 complexes during the cell cycle has been narrowed down to the transition from G_1 to S. Some of the possible substrates for this complex include the so-called "pocket proteins," which include pRb, p107, and p130 (68). The term *pocket protein* refers to those proteins that are involved in binding to certain proteins, such as transcription factors, and releasing them after activation of the pocket protein by the action of a CDK kinase. The cyclin E/CDK2 complex is localized in the nucleus, where it is bound to E2F, p107, and p130 (although it has not been shown to phosphorylate these proteins) (73), suggesting that it too plays a role in regulation of transcription. The E2F subunit that interacts with cyclin E/CDK2 has been shown to differ from the E2F found associated with pRb, which hints that they have different properties (74). One confirmed substrate of cyclin E/CDK2 is the cdc25A phosphatase, which is involved in activating CDKs by cleaving the inhibitory phosphates on both a tyrosine and a threonine residue (75). The cdc25A phosphatase has been proposed to activate both the cyclin E/CDK2 and cyclin A/CDK2 complexes, creating a positive feedback loop between cyclin E/CDK2 and cdc25A [see Fig. 4(a)]. If this feedback loop does indeed exist, it has

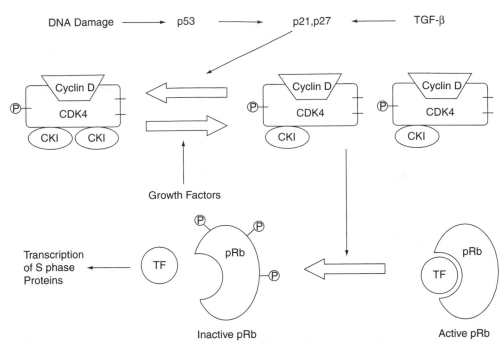

Figure 3. The G_1 control mechanism. Schematic of the control mechanism that governs the progression past the restriction point and through G_1. The presence of growth factors stimulates the expression of cyclin D to promote the formation of cyclin D/CDK4 complexes and increase the amount of complex to a level above the inhibitory threshold of CKIs. This increase in cyclin D/CDK4 complexes results in kinase activity, which then promotes the release of transcription factors. DNA damage results in the expression of p53, which increases the amount of CKIs present to arrest cells in G_1 and allow DNA repair to take place.

been hypothesized that the activation of cdc25A occurs in response to the completion of some G_1 event like the duplication of the centrosome or the accumulation of certain DNA synthesis proteins (76).

Experiments done in yeast suggest that proteolysis may be involved in activating the cyclin E/CDK2 homologue, which plays an important role in the G_1/S transition (77). It has been shown that a protein called cdc34 is necessary for the G_1/S transition to take place, possibly because it is responsible for initiating the degradation of certain CKIs. Functional homologues of the cdc34 protein have been isolated in vertebrate cells and are thought to stimulate the degradation of the p27 inhibitor (78). This, in addition to isolation of CKIs by the cyclin D/CDK4 complex, could possibly aid in beginning the onset of the synthesis phase.

In summary, a cell becomes arrested in G_0 in response to some extracellular signal that triggers a decrease in the level of cyclin D below a threshold level of CKIs. This allows p21 and p27 to bind to cyclin D/CDK4 complexes in a 2-to-1 ratio, resulting in an inhibition of kinase activity. Following stimulation of cyclin D synthesis due to the presence of growth factors, there is an increase in cyclin D/CDK4 kinase activity, since there is no longer enough inhibitor present to inactivate all the complexes. This active CDK kinase then serves two purposes: (1) to inactivate pRb, allowing the release of transcription factors that induce the synthesis of other cyclins and DNA synthesis proteins, and (2) to isolate the CKIs and allow activation of the cyclin E/CDK2 complex. This complex then stimulates the release of additional

transcription factors and initiates a positive feedback loop that triggers the onset of the DNA synthesis phase.

Once cells move beyond the restriction point, they are committed to entering the DNA synthesis phase. Cells do, however, have the power to delay the onset of later events to ensure the proper coordination of important processes. For example, failure to repair damaged DNA greatly increases the occurrence of genetic mutations, which can have disastrous consequences for the cell. Thus an additional checkpoint exists in late G_1 phase that ensures that the DNA is free from damage and ready to be replicated. We will now see that cells possess the means to sense when there is damage to the genetic material and then halt cell cycle progression to allow the DNA repair machinery time to repair any defects in the genetic material. It has been determined that there is a single protein, p53, which is held accountable for inducing G_1 arrest in response to DNA damage (79) (see Fig. 3 for a schematic representation).

By inducing DNA damage in normal cells, it was determined that cells responded by increasing the level of p53 and the CKI p21 (80). The increase in p21 then leads to the inhibition of both cyclin D/CDK4 and cyclin E/CDK2 complexes to prevent entry into the S phase. Additionally, p21 also functions to oversee the proper repair of DNA by interaction with PCNA. PCNA is required for both DNA synthesis and repair. When bound with p21, PCNA is no longer able to interact with polymerase-δ, thus preventing DNA synthesis; however, the DNA repair function remains unaffected by the inhibitor (81). Therefore, p21 must

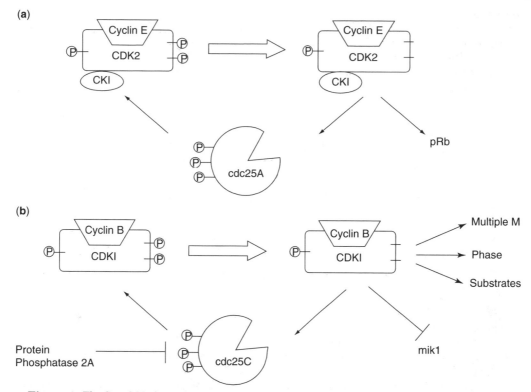

Figure 4. The S and M phase entry mechanisms. The positive feedback mechanisms that trigger the entry into S phase (**a**) and mitosis (**b**). The transition to S phase is triggered when an active cyclin E/CDK2 complex activates the cdc25A phosphatase, which, in turn, activates more cyclin E/CDK2 complexes, causing a sharp rise in active complexes. These complexes then phosphorylate pRb to stimulate the release of transcription factors. The M phase transition is triggered by a positive feedback mechanism between cyclin B/CDK1 complexes similar to that described for part (**a**). In addition, the active cyclin B/CDK1 complex inactivates the mik1 kinase that phosphorylates the complex on an inhibitory residue, thus contributing to the steep rise in cyclin B/CDK1 activity.

inhibit both CDK activity and the PCNA/polymerase-δ interaction in order for DNA repair to be carried out before advancing to the S phase.

S Phase and G$_2$ Progression. Shortly after the cyclin E/CDK2 complex triggers entry into the S phase, cyclin E is degraded and CDK2 interacts with cyclin A. There is evidence that cyclin E may be degraded in a ubiquitin-mediated pathway as in budding yeast, but this has not been confirmed (82). The cyclin A/CDK2 complex is localized in the nucleus where it associates with p107, p130, and E2F (83). Interacting with E2F allows the kinase to phosphorylate the DP1 subunit of the transcription factor complex, which inhibits its DNA binding capability (84). This suggests that when cyclin A synthesis begins in early S phase, it aids in halting the transcription of genes triggered by the E2F complex in the late G$_1$ phase. No other substrates of this cyclin/CDK complex have been identified except for the helicase RF-A, which is one component of the DNA synthesis machinery (10).

Upon completion of DNA synthesis, CDK2 is degraded and cyclin A forms a new complex with CDK1 (85). It is not known if this complex has any function in the G$_2$ phase, but its activity has been implicated in

triggering initial events that occur in mitosis. During G$_2$, cyclin B begins associating with CDK1 in the cytoplasm and is transported to the nucleus just prior to mitosis. Mitosis begins when the cyclin B/CDK1 complex is activated, and it has been suggested that a checkpoint mechanism exists that prevents activation of this complex until duplication of the centriole has been completed (86). One model that describes this transition involves regulating cdc25C activity by down-regulating protein phosphatase 2A (PP2A), which controls cdc25C activity (76). Signals could be initiated from the DNA replication machinery or from the proper organization of chromatin that would inhibit the action of PP2A. This allows cdc25C to become activated and dephosphorylate the inhibitory residues on CDK1, causing its activation. A positive feedback loop is then initiated [see Fig. 4(b)], whereby active cyclin B/CDK1 complexes activate more cdc25C and, in budding yeast, inactivate the wee1 kinase by phosphorylating a residue that inhibits its action (87). This mechanism brings about a sharp transition into mitosis almost immediately after a small amount of CDK1 is activated.

Another mechanism exists that allows cells with damaged DNA to become arrested in G$_2$. In response to DNA damage, transcription of the CHK1 kinase is initiated. This kinase targets cdc25C and phosphorylates the serine-216 residue, which allows formation of a binding site for the 14-3-3 protein that inhibits cdc25C activity (88). This then allows the DNA repair machinery to work, since the cyclin B/CDK1 complex cannot be activated by cdc25C.

Supervising the Order of Events in Mitosis. The point when the cyclin A/CDK1 complex is most active occurs just after the cell enters into mitosis, which coincides with nucleation of the centrosomes and the formation of the long microtubular extensions of the mitotic spindle. Indeed, it has been shown that cyclin A/CDK1 activity has a profound effect on the nucleating ability of the centrosomes in *Xenopus* extracts (89). Once spindle formation is complete, cyclin A begins to be degraded and the activity of cyclin B/CDK1 complexes takes over.

The cyclin B/CDK1 complex is either indirectly or directly involved in regulating almost all the structural reorganization that takes place during mitosis. Chromosome condensation begins in prophase, when cyclin B/CDK1 phosphorylates histone H1, transcription factors, and several other proteins (90). Phosphorylation of these proteins is thought to weaken the interaction between these proteins and the DNA, allowing the chromosomes to pack more tightly (91). In addition, after formation of the mitotic spindle is complete, the cyclin B/CDK1 complex reduces the nucleating ability of the centrosomes, which causes the microtubules to shorten to a length suitable for the alignment and separation of the chromosomes (10). The breakdown of the nuclear envelope, which marks the beginning of prometaphase, and the reorganization of the cytoskeleton are also brought about by phosphorylation. The nuclear lamina are present in a hypophosphorylated form in an intact nucleus, and when phosphorylated by the cyclin B/CDK1 complex, the lamina network breaks down, contributing to nuclear disassembly (92). Lamina phosphorylation itself is not sufficient to achieve disruption of the nucleus, however, and these mechanisms remain unclear (38). Caldesmon has also been shown to be a substrate of the mitotic CDK complex, and microtubular reorganization ensues after its phosphorylation. Caldesmon binds to actin and calmodulin, and its phosphorylation weakens its affinity for actin, causing the microfilaments to dissociate (93). This process causes the cells to take on a round shape, which is characteristic of anchorage-dependent cells undergoing mitosis. The activity of the cyclin B/CDK1 complex is also responsible for repressing the formation of the contractile bundle until the cell enters anaphase (94). The cyclin B/CDK1 complex phosphorylates myosin, which prevents its association with actin filaments; so the contractile bundle is unable to form. This inhibitory function of the complex stays in effect until cyclin B is destroyed in the ubiquitin degradation pathway. The activity of the cyclin B/CDK1 complex is, ironically, responsible for the initiation of its own destruction (95), which stays in effect until the pathway responsible for its destruction is turned off in late G_1, possibly by the action of the cyclin E/CDK2 complex (96).

The final regulatory mechanism that exists in the cell cycle is a checkpoint that prevents the onset of anaphase until several other structural events have taken place in the cell. Some of these events include the assembly of a bipolar spindle, attachment of the chromosomes to the spindles through the kinetochore, and the migration of the chromosomes to the metaphase plate. This suggests that proper completion of these processes is somehow monitored to ensure that anaphase does not begin prematurely (97). Several experiments (recently reviewed in, Ref. 98) suggest that one of two events signals that the cell is ready to undergo chromosome separation and cell division. These events are either the attachment of the chromosomes to the mitotic spindle (99) or the presence of tension on the kinetochores due to their attachment to a bipolar spindle (100). Since all the structural reorganization processes that occur in mitosis depend on the successful completion of earlier events, failure to complete any one of these processes could result in the failure of the signaling event to occur. This would then stimulate a cascade of phosphorylation events culminating in cell cycle arrest in mitosis.

Understanding Cell Cycle Regulation through Mathematical Modeling

Understanding the complex array of biochemical events outlined above can be difficult at times. Moreover, the controls responsible for cell cycle regulation are rapidly becoming more and more complicated. This complexity makes it extremely difficult to understand the mechanisms involved through logical thinking alone. Mathematical modeling has been used for the purpose of explaining the various physiological events that are commonly observed in different situations in *Xenopus* oocytes and yeast cells. The cell cycle controls discussed in this article have been described for animal cells. However, the majority of initial experiments on the cell cycle were carried out in yeast cells and in *Xenopus* oocyte extracts. The components that make up the control systems in these organisms are largely identical to those of animal cells, and many findings resulting from studies on these organisms have been extrapolated to apply to animal cells. An initial review of the literature would suggest that the control systems in these organisms are different, but this is not the case, since only the terminology used to represent the components differs. The cell cycle nomenclature used in these organisms has been reviewed elsewhere (10,38).

Modeling of the molecular controls that are thought to control the cell cycle in *Xenopus* oocytes (101) and yeast cells (102–104) has been done in order to determine the effect of changing the activity of various components of the regulatory network. We now believe that progression through the cell cycle is governed by the activity of cyclin/CDK complexes; that is, a sharp rise in a complex's activity results in a transition to the next phase. The aim of the molecular modeling done by Novak and Tyson was to determine whether or not the control mechanisms presented above could be used to describe the observed behavior of cells in a variety of situations, or more simply stated, to compare theory with experiment. In order to do this, a set of equations (one for each component of the system) was generated, based on the mechanisms described above, to describe the rate of change of each component as a function of time by translating the proposed control mechanisms into simple reaction rate equations. The resulting set of differential equations was then solved using numerical techniques, and the activity of the cyclin/CDK complexes could be followed with time, and the sharp oscillations in complex activity could be varied. Since sets of equations commonly contain more than one solution (depending on rate constants and initial conditions used), phase-plane plots were constructed, and observation of limit-cycle-type behavior in these plots indicates an oscillatory solution, which is characteristic of proposed cell cycle control (see Refs. 102,105 for details). This type of model is easily solvable by today's personal computers and has been successfully implemented to describe the observed behavior of a variety of yeast cell cycle mutants (103,104) and of embryo and oocyte maturation (101).

IMPLICATIONS FOR ANIMAL CELL CULTURE

The use of large-scale cultures of animal cells is becoming increasingly important as the demand for their biological products grows. Animal cell cultures are used for the production of many different products of diagnostic and therapeutic value, including vaccines, monoclonal antibodies, and in some cases, the cells themselves (for example, to replace the function of damaged tissue). Developing culture conditions to optimize the production and secretion of proteins is therefore of utmost importance in meeting these demands. Consequently, the key to designing such systems is not only determining the kinetics of growth, secretion, and proliferation, but also understanding how the cell cycle regulates certain cellular processes. The heterogeneous nature of cell populations has made it extremely difficult to obtain these data. Various mathematical models have been developed to describe cell behavior under different conditions. However, they are, for the most part, unable to account for the heterogeneous nature of cell populations and usually do not provide insight into variations introduced by the cell cycle. One model used to describe cell growth within a heterogeneous population of cells, known as population balances, has been developed (106–109). However, the implementation of this approach has been hampered by the difficulty in obtaining the single-cell parameters necessary for its use.

These problems are being tackled through the development and improvement of tools, such as flow cytometry and fluorescent markers, which allow the examination of single cell parameters within a heterogeneous population. Flow cytometry is a very sensitive tool that allows one to simultaneously obtain several single cell parameters related to growth physiology from a large number of cells in the population. Collecting these data allows one to determine the size, DNA content (and thus cell cycle position), and protein content (when the protein is conjugated to a fluorescent marker) of individual cells in a population. Many flow cytometric studies relating the cell cycle to optimization of cell culture performance are available in the literature. Several of these analyses and their findings are outlined below.

Cell Cycle Dependent Protein Accumulation

Hybridoma cells have received the most attention in this area since they are widely used in the production of monoclonal antibodies (MAbs). Numerous studies have suggested that lower rates of growth result in higher protein production (110–112, and papers referenced in Ref. 113). It has been argued that because populations growing at slow rates have a larger fraction of G_1 cells, the G_1 phase must be a period of enhanced protein production and secretion. However, few studies have been done to give insight into the mechanism for this behavior. Additionally, other studies have been done that show a positive correlation between growth rate and protein production (briefly discussed in Ref. 113). These variations in response observed by different researchers may be explained since the overall growth rate of a culture depends on both cell growth and cell death. This means that the same overall growth rate can be observed in two cultures with different rates of cell loss and cell growth. The conclusion that can be made from the examination of these different investigations is that it is important to be aware of the different methods used to quantify growth and protein production in cell cultures since they may result in conflicting results.

Al-Rubeai and colleagues used flow cytometric results obtained from synchronous cultures, and concluded that the rates of both Ig synthesis and secretion are at a maximum at the G_1/S transition (114). The synchronization of the culture was achieved by the addition of thymidine, which arrests cells prior to entry of the S phase. This type of analysis presents another problem, however, since factors used for cell cycle arrest have been shown to alter the cell metabolism (115), so that these results should be viewed with care. A more complete series of studies on rates of MAb production during the cell cycle was performed by Kromenaker and Srienc. The first study involved using flow cytometric data obtained in asynchronous cultures and applying these data to population balances to find the single cell rates of protein accumulation in various cell cycle phases for both antibody-producing and -nonproducing cell lines (116). It was found that during the early G_1 phase, producer cells with low protein content have a specific rate of accumulation of protein approximately 3–5 times greater than the specific rate observed for the remainder of G_1 phase cells. This indicates that the rate of protein synthesis is much greater than the rates of protein degradation and secretion in the G_1 phase. The S phase of the cell cycle was shown to a have a low specific rate of protein accumulation that remained relatively steady throughout this period. Extending this analysis to the $G_2 + M$ phase yielded a negative specific rate of protein accumulation, indicating that protein is lost due to secretion and degradation during this period of the cell cycle. From these findings, it was concluded that the disruption of the secretion system that occurs during the division of the cell accounts for the extremely high rate of accumulation observed in early G_1 since the transport system probably has not had time to recover from its disruption.

A later report by Kromenaker and Srienc described the relation between growth rate and secretion rates during the cell cycle as determined using single-cell data and population balances (117). Lactic acid was used as a growth suppresser, and the concentrations were changed to vary the specific growth rate. This study brought several previously unconsidered factors to light. By analysis of the DNA distribution of the populations at different growth rates, it was found that the lengths of all the cell cycle phases were increased by the same factor that the specific growth rate was decreased. This is in direct contrast to other studies that suggest that the increase in productivity is due to the increase in the length of the most productive phase of the cell cycle. Second, the application of population balances in this study revealed that the net specific rate of antibody synthesis is independent of growth rate. Additionally, the specific secretion rate of $G_2 + M$ cells is greater than the rate in the other two phases, and remains constant with decreasing growth rates. In contrast, the specific secretion rates for G_1 and S phases increase with decreasing growth rate. These findings are consistent with the hypothesis that the disruption of the secretion pathway during cell division results in lower secretion throughout G_1 until the Golgi apparatus has had sufficient time to reassemble. During this time, antibodies accumulate in the cell until they can be secreted during $G_2 + M$, before the Golgi apparatus is disrupted once again. Thus, at lower growth rates, more time is spent in G_1; so an increased secretion rate is observed since the cells have had time to recover from cell division. A subsequent study tracked the amount of cell surface antibodies as a function of DNA content (118). Since prior experiments determined that the amount of cell surface antibodies observed is a good measure of the rate of secretion, these data were used to determine the pattern of secretion during the cell cycle. The results obtained from this method also agreed with the hypothesized regulation of secretion described above.

Understanding the regulation of foreign genes in animal cells is also of great interest since recombinant DNA is commonly used to express proteins of interest in immortal cell lines such as Chinese hamster ovary cells (CHO cells). This provided the motivation for studies in which the expression of foreign genes in animal cells was tracked throughout the cell cycle. One study tracked the secretion of tissue-like plasminogen activator (t-PA) in recombinant CHO cell cultures synchronized by fluorescence-activated cell sorting (FACS) based upon DNA content (119). This study shows that secretion of t-PA reaches its maximum near the late S phase, while the increase in this protein is maximum in the G_1 phase. These results are fairly similar to those obtained by Kromenaker and Srienc using different methods. Gu and colleagues also measured the expression of foreign β-galactosidase during the cell cycle in recombinant CHO cells (120) and in recombinant mouse cells (121) and hypothesize that control of foreign gene expression is promoter dependent. These particular studies, however, showed that the use of either the cytomegalovirus (CMV) promoter or the mouse mammary tumor virus (MMTV) promoter causes the production of β-galactosidase mainly in the S phase.

Cell Cycle-Dependent Behavior of Culture Systems

There is a steady increase in the number of reports in the literature that look to the cell cycle to explain various phenomena, such as differing growth rates between cultures and cell death due to shear stress in bioreactors. For example, the effect of different inoculation sources was examined by Al-Rubeai et al. (122). Two cultures that were inoculated with cells from either the exponential or stationary growth phase were examined using flow cytometry. The culture inoculated with cells from the exponential phase grew at a much greater rate than the other culture, and therefore reached the death phase much quicker due to nutrient exhaustion and to buildup of toxins. Analysis of the cell cycle, cell size, and mitochondrial activity of the cultures revealed differences among the cultures. For instance, the slower-growing culture was found to consist of a larger portion of cells in G_1, while the culture with the greater growth rate had a larger fraction of cells in the S phase. Since G_1 cells are generally smaller than the S phase cells, the culture with a greater fraction of cells synthesizing new DNA will use more nutrients than the smaller cells. Additionally, cultures with elevated mitochondrial activity were also shown to have a greater capacity for proliferation. These findings led the researchers to conclude that examination of these parameters can provide a means of predicting how the culture will behave and a means of explaining variations in the growth patterns of cultures.

It is also of great interest to determine how cells behave in different culture systems. For cells in continuous culture (123), it has been found that the fraction of cells in the G_1 phase decreases with increasing growth rate, while the fraction of cells in the S phase increases. As expected, the G_1 phase accounts for most of the resulting variability in the lengths of the cell cycle phases; However, the S phase is affected slightly more than the G_2 and M phases. A more detailed analysis of the cell cycle was performed in cultures containing microcarriers, and the fraction of contact-inhibited cells was determined and compared to a model (124). Microcarriers are commonly employed in cultures of anchorage-dependent cells such as CHO cells, since these types of cells must be attached to a surface in order to proliferate. When the culture grows to the point when there is no longer any surface area available for newborn cells, contact inhibition occurs and a portion of the cells cease to proliferate even when provided with an ample supply of nutrients. Analysis of this system yielded the distribution of cells in the cell cycle during the lag phase, exponential growth, and confluency. During the lag phase, the culture is predominantly in the G_1/G_0 phases and rapidly enters the other cell cycle phases upon reaching exponential growth. As the culture reaches confluency, the culture gradually shifts back to predominantly the G_1/G_0 phases; however, a portion of the cells continue cycling. These findings are similar to those of suspension cultures. Furthermore, a method was developed to determine the fraction of cells affected by contact inhibition, and the results showed tight agreement with a model that had been developed.

Furthermore, when culturing animal cells in large-scale bioreactors, the cells can be injured due to the shear stress that results from agitation and sparging. There have been several studies examining the mechanism of cell damage and seeking to minimize injury to the cell through the use of various shear-protecting agents such as Pluronic F-68 and serum. However, there have been relatively few studies examining the effect that these forces may have on cell metabolism and growth. It is evident that shear forces have an effect on the overall growth rate of the culture due to an increase in cell death. Lakhotia and co-workers (125) analyzed the rate of DNA synthesis and the cell cycle phase distributions of cultures grown in small-scale bioreactors with gas sparging and varying agitation intensities to determine the effect that these forces may have on growth kinetics of the surviving cells. Measurement of the kinetics was accomplished by measuring the uptake of BrdU and the DNA content distributions of the surviving cells with the aid of flow cytometry. They determined that in cultures exposed to high agitation, the fraction of cells in the S phase increased by up to 45%, while the fraction of cells in G_1 decreased by up to 50% compared to control cultures. The simultaneous analysis of the DNA synthesis rate suggests that the change in the cell cycle phase distribution is due to an increase in the DNA synthesis rate. Additionally, when cultured in the absence of shear stress, the apparent specific growth rate of the cells previously exposed to the shear forces was initially higher than, and later the same as, that of control cells. It was also shown that an increase in the agitation intensity resulted in a further increase of the DNA synthesis rate. By subjecting cells to extreme, short-term stresses induced by turbulent capillary flow, it was determined that larger cells are more susceptible to damage than smaller cells (126). This conclusion was made based on the preferential disappearance of S and G_2 cells when passed through the capillary. Another possible explanation for these results is that an apoptotic response triggered by the shear forces caused a decrease in the size of the cells; however, verification of this hypothesis is difficult.

Engineering the Cell Cycle

Once a more complete picture of how the cell cycle regulates cell physiology and metabolism has been painted, the cell cycle can possibly be manipulated in culture in such a way that cells will perform desired processes at an optimal level. For example, controlling the regulatory mechanisms of cell growth and proliferation, or controlling the rate at which cells advance through the cell cycle, can lead to the enhancement of protein production, improved economic feasibility, and easier downstream processing. Indeed, there have been several studies published to date that address these issues and will undoubtedly lead to further examination of directing cellular operations through the engineering of the cell cycle in cells in culture.

Cells in culture are subjected to many different forms of stress that may ultimately lead to cell death. Cell death can be caused due to some external factors such as hydrodynamic forces that may cause the cell to rupture. Additionally, the response to certain stimuli, such as limitation of glutamine, glucose, or serum or the buildup of toxins, may result in programmed cell death, called *apoptosis*, which is regulated by some of the same pathways as the cell cycle (this topic is discussed in detail in another article). When in culture, cells reach their maximum cell density in a few days, after which the process of programmed death is induced in the majority of these cells (127). The discovery that the product of the protooncogene *Bcl-2* is able to suppress apoptosis led to its use in cell culture to prolong the productive period of the culture (128–131). Several recent studies show that cells expressing this gene increased MAb production by 40% (129) and prolonged the viable culture time by up to 4 days by delaying the onset of apoptosis for 2 days (131). Additionally, it was determined that the increase in MAb production was equally due to both the prolonged survival of cells and to the increase in protein production per cell (131). The expression of the *Bcl-2* gene also allows cells to be cultured with a significant decrease in serum supplementation (130). Specifically, *Bcl-2* overexpressing cells could maintain high viable cell concentrations for up to 4 days, while the control culture lasted only 1 day when supplemented with 0.5% serum. Moreover, cells cultures supplemented with 9% serum with an additional 2% serum feed after 4 days sustained productivity for 10 days, while the control culture underwent apoptosis after 4 days (130). These experiments show that the cell cycle controls can be bypassed when expression of the appropriate gene is induced in culture.

A similar approach has been used in CHO cells to bypass the cell cycle controls that require the presence of growth factors for efficient proliferation. The observation that cells stimulated by growth factors possess elevated cyclin E levels led researchers to transfect CHO cells with a cyclin E expression vector (132). The expression of this gene in cells cultured in protein-free media resulted in cell proliferation and levels of cyclin E, cell morphologies, and cell cycle phase distributions comparable to those obtained in growth-factor-stimulated cells. An approach related to the one used above employed expression of the transcription factor E2F-1 in protein-free media to bypass the growth factor requirements for proliferation (133). Although the desired effect of eliminating growth-factor dependency of the cells was achieved in both cases, the two approaches had different effects on several cellular properties. For instance, the cells expressing cyclin E rapidly round up and move into suspension, whereas the E2F-1 cells adhered more tightly to the surface. Additionally, the same doubling time was observed for both cell types; however, the cyclin E cells resided in G_1 for approximately 20 min longer, while the S phase was reduced by 2 hours as compared to the E2F-1 cells. Finally, the expression of certain proteins varied between the two cell types. The expression of cyclin A was shown to be 2 times higher in the E2-F–expressing cells, and these same cells expressed a greater number of proteins at higher levels, relative to wild-type cells, than did the cyclin E expressing cells. These results present two different ways of circumventing growth-factor-dependent controls that show different effects on cellular processes. Additionally, these findings may give some insight into the roles that these factors play in the regulation of the cell cycle.

Another recent study involved increasing protein production in cultured cells by arresting cells in G_1 (134). This was done by transfecting CHO cells with a gene coding for either p21, p27, or p53175P controlled by a tetracycline-regulated promoter. The p53 protein used in this study is a mutant that has lost its apoptotic function; so the culture would undergo G_1 arrest and not programmed cell death. The culture was allowed to proliferate until it reached the desired cell density, after which expression of the cell cycle arrest molecules was stimulated, thus blocking proliferation. This sort of cytostatic process has several advantages over other processes involving uncontrolled growth of cultures, including slower rates of nutrient exhaustion, higher protein production, and low genetic drift.

Gene Therapy

A firm grasp of the events that control the cell cycle will also yield significant advances towards the treatment of various illnesses, including cancer and genetic disorders. The connection between the cell cycle and some cancers has been reviewed elsewhere (18,51,135,136). Many genetic disorders that alter the way cells function result from either the absence of an enzyme or the presence of a flawed enzyme. This situation may be remedied by the introduction of the appropriate genes into a target cell, a procedure known as gene therapy. Although cells that have incorporated the desired gene into their genome have displayed the desired effects, this technique is limited by low rates of gene transfer (137). Numerous studies have attempted to uncover the factors that determine transfer efficiency, resulting in the hypothesis that gene transfer is cell cycle dependent (138,139). More specifically, since the most widely used method for gene transfer is using recombinant retroviruses, genes introduced into cells in this manner are not integrated into the host's genome until mitosis. It has been postulated that nuclear envelope breakdown is necessary for the retro-viral integration complexes to act. Based on these studies, Andreadis and Palsson suggested that transfer efficiency depends both on cell cycle and retroviral life cycle events and developed a mathematical model describing the interplay between the two (140). This model aided in the design of experiments that allow the direct measure of intracellular retroviral decay (138). Furthermore, they developed a method to monitor the cell cycle dependence of gene transfer by using the transduction of β-galactosidase as a marker. Thus it was determined that almost all transduced cells show up in the early G_1 phase, supporting the hypothesis that only mitotic cells are susceptible to gene transfer (139). These studies reveal the importance of the kinetics of both retroviral decay and cell cycle progression of the target cells in the efficiency of gene therapy.

CONCLUSIONS

An overview of the mechanisms by which cells respond to different environmental stimuli has been presented. Additionally, this summary has made it evident that even after the commitment to synthesize new DNA is made, the cell possesses the ability to monitor and direct internal events. For instance, the transcription and activation of key enzymes is regulated by the cell cycle controls to ensure that necessary cellular processes are carried out successfully. The existence of this type of control suggests that almost every cellular process is somehow linked to the cell cycle machinery to guarantee the proper coordination of events throughout the cell cycle. A failure to maintain the order of these events may result in cell death or a disruption of normal cell behavior. Thus a more complete understanding of factors that have an effect on cell cycle regulation would allow one to gain insight into better directing the cellular processes bound to cell cycle control. This knowledge is characterized by the recent explosion in the amount of studies that have focused on better understanding the components involved in directing cell division and their influence on fundamental cellular processes. Additionally, a brief overview of some of the studies devoted to understanding cell cycle dependent processes and optimizing cell behavior in culture is presented here. It seems that obtaining a better understanding of the cell cycle and understanding how the cell cycle directs fundamental components of cell growth, such as protein production and secretion, may provide the missing link that will allow researchers to not only improve cell culture, but may result in the development of treatments for many types of ailments and dysfunctions, such as cancer, which are indubitably cell cycle related.

BIBLIOGRAPHY

1. R.A. Laskey, M.P. Fairman and J.J. Blow, *Science* **246**, 609–613 (1989).
2. B. Stillman, *Science* **274**, 1659–1664 (1996).
3. P.N. Rao and R.T. Johnson, *Nature (London)* **159**, 159–164 (1970).
4. G.H. Leno, C.S. Downes, and R.A. Laskey, *Cell* **69**, 151–158 (1992).
5. J.P.J. Chong et al., *Nature (London)* **375**, 360–361 (1995).
6. P. Romanowski, M.A. Madine, and R.A. Laskey, *Proc. Natl. Acad. Sci., U.S.A.* **93**, 10189–10194 (1996).
7. Y. Kubota et al., *Cell* **81**, 601–609 (1995).
8. T. Evans et al., *Cell* **33**, 389–396 (1983).
9. T. Hunt, *Semin. Cell Biol.* **2**, 213–222 (1991).
10. J. Pines, *Adv. Cancer Res.* **66** (1995).
11. H. Kobayashi et al., *Mol. Biol. Cell* **3**, 1279–1294 (1992).
12. J. Pines and T. Hunter, *Trends Cell Biol.* **1**, 117–121 (1991).
13. D.S. Peeper et al., *EMBO J.* **12**, 1947–1954 (1993).
14. S.I. Reed, *Annu. Rev. Cell Biol.* **8**, 529–561 (1992).
15. C.J. Sherr, *Cell* **73**, 1059–1065 (1993).
16. X. Grana and E.P. Reddy, *Oncogene* **11**, 211–219 (1995).
17. G.F. Draetta, *Curr. Opin. Cell Biol.* **6**, 842–846 (1994).
18. C.C. Orlowski and R.W. Furlanetto, *Endocrinol. Metab. Clin. North Amer.* **25**, 491–502 (1996).
19. Y. Xiong, H. Zhang, and D. Beach, *Cell* **71**, 504–514 (1992).
20. C.J. Sherr, *Trends Biochem. Sci.* **20**, 187–190 (1995).
21. H. Matsushime, M. Roussel, R. Ashmun, and C.J. Sherr, *Cell* **65**, 701–713 (1991).
22. V. Baldin et al., *Genes Dev.* **7**, 812–821 (1993).

23. D.E. Quelle et al., *Genes Dev.* **7**, 1559–1571 (1993).

24. D. Resnitzky, M. Gossen, H. Bujard, and S.I. Reed, *Mol. Cell. Biol.* **14**, 1669–1679 (1994).

25. A. Koff et al., *Science* **257**, 1689–1694 (1992).

26. V. Dulic, E. Lees, and S.I. Reed, *Science* **257**, 1958–1961 (1992).

27. M. Ohtsubo and J.M. Roberts, *Science* **259**, 1908–1912 (1993).

28. L.H. Tsai, E. Harlow, and M. Meyerson, *Nature (London)* **353**, 174–177 (1991).

29. L.H. Tsai et al., *Oncogene* **8**, 1593–1602 (1993).

30. D.H. Walker and J.L. Maller, *Nature (London)* **354**, 314–317 (1991).

31. G. Draetta et al., *Cell* **56**, 829–838 (1989).

32. J. Pines and T. Hunter, *J. Cell Biol.* **115**, 1–17 (1991).

33. M.J. O'Conell and P. Nurse, *Curr. Opin. Cell Biol.* **6**, 867–871 (1994).

34. S.I. Reed et al., *Cold Spring Harbor Symp. Quant. Biol.* **56**, 61–67 (1991).

35. M. Glotzer, A.W. Murray, and M.W. Kirschner, *Nature (London)* **349**, 132–138 (1991).

36. M.A. Felix et al., *Nature (London)* **346**, 379–382 (1990).

37. T. Lorca et al., *J. Cell Sci.* **102**, 55–62 (1992).

38. J. Pines, *Biochem. J.* **308**, 697–711 (1995).

39. A. Hershko et al., *J. Biol. Chem.* **269**, 4940–4946 (1994).

40. J. Pines, *Cancer Bio.* **5**, 305–313 (1994).

41. R.P. Fisher and D.O. Morgan, *Cell* **78**, 713–724 (1994).

42. T.P. Makela et al., *Nature (London)* **371**, 254–257 (1994).

43. W.J. Feaver, J.Q. Svejstrup, N.L. Henry, and R.D. Kornberg, *Cell* **79**, 1103–1109 (1994).

44. R. Roy et al., *Cell* **79**, 1093–1101 (1994).

45. E. Abherton Wessler, J.L. Parker, R.L. Goahlon, and H. Piwnion-Worms, *Mol. Cell Biol.* **12**, 1675–1685 (1993).

46. J.A. Endicott, P. Nurse, and L. Johnson, *Protein Eng.* **7**, 243–253 (1994).

47. C. Featherstone and P. Russell, *Nature (London)* **349**, 808–811 (1991).

48. R.N. Booher, R.J. Deshaies, and M.W. Kirschner, *EMBO J.* **12**, 3417–3426 (1993).

49. K. Galaktianov and D. Beach, *Cell* **67**, 1181–1194 (1991).

50. C.J. Sherr and J.M. Roberts, *Genes Dev.* **9**, 1149–1163 (1995).

51. T.K. MacLachlan, N. Sang, and A. Giordano, *Crit. Rev. Eukaryotic Gene Expression* **5**, 127–156 (1995).

52. S.J. Elledge and J.W. Harper, *Curr. Opin. Cell Biol.* **6**, 847–852 (1994).

53. Y. Xiong et al., *Nature (London)* **366**, 701–704 (1993).

54. J.W. Harper et al., *Cell* **75**, 805–816 (1993).

55. H. Zhang, G.J. Hannon, and D. Beach, *Genes Dev.* **8**, 1750–1758 (1994).

56. K. Polyak et al., *Genes Dev.* **8**, 9–22 (1994).

57. K. Polyak et al., *Cell* **78**, 59–66 (1994).

58. E.J. Firpo, A. Koff, M.J. Solomon, and J.M. Roberts, *Mol. Cell. Biol.* **14**, 4889–4901 (1994).

59. M.E. Ewan, H.K. Sluss, L.L. Whitehouse, and D.M. Livingston, *Cell* **74**, 1009–1020 (1993).

60. M. Serrano, G.J. Hannon, and D. Beach, *Nature (London)* **366**, 704–707 (1993).

61. T. Nobori et al., *Nature (London)* **368**, 753–756 (1994).

62. G.J. Hannon and D. Beach, *Nature (London)* **371**, 257–261 (1994).

63. D.S. Peeper and R. Bernards, *FEBS Lett.* **410**, 11–16 (1997).

64. R.A. Weinberg, *Science* **254**, 1138–1146 (1991).

65. M. Laiho et al., *Cell* **62**, 175–185 (1990).

66. E. Moran *Curr. Opin. Genet Dev.* **3**, 63–70 (1993).

67. S.P. Chellappan et al., *Cell* **65**, 1053–1061 (1991).

68. E.W.-F. Lam and N.B.L. Thangue, *Curr. Opin. Cell Biol.* **6**, 859–866 (1994).

69. J. Lukas et al., *J. Cell Biol.* **125**, 625–638 (1994).

70. M.E. Ewan et al., *Cell* **73**, 487–497 (1993).

71. S.F. Dowdy et al., *Cell* **73**, 499–511 (1993).

72. H. Muller et al., *Proc. Natl. Acad. Sci., U.S.A.* **91**, 2945–2949 (1994).

73. E. Lees et al., *Genes Dev.* **6**, 1874–1885 (1992).

74. N. La-Thangue, *Curr. Opin. Cell Biol.* **6**, 443–450 (1994).

75. I. Hoffman, G. Draetta, and E. Karsenti, *EMBO J.* **13**, 4302–4310 (1994).

76. I. Hoffmann and E. Karsenti, *J. Cell Sci., Suppl.* **18**, 75–79 (1994).

77. E. Schwob, T. Bohm, M.D. Mendenhall, and K. Nasmyth, *Cell* **79**, 181–184 (1994).

78. M. Pagano et al., *Science* **269**, 682–685 (1993).

79. M.B. Kastan et al., *Cancer Res.* **51**, 587–597 (1991).

80. A. Di-Leonardo, S.P. Linke, K. Clarkin, and G.M. Wahl, *Genes Dev.* **8**, 2540–2551 (1994).

81. M.K.K. Shivji et al., *Curr. Biol.* **4**, 1062–1068 (1994).

82. R.W. King et al., *Science* **274**, 1652–1659 (1996).

83. M.E. Ewan, B. Faha, E. Harlow, and D.M. Livingston, *Science* **255**, 85–90 (1992).

84. W. Krek et al., *Cell* **78**, 161–172 (1994).

85. M. Pagano, R. Pepperkok, F. Verde, and G. Draetta, *EMBO J.* **11**, 961–971 (1992).

86. A. Bailley and M. Bornens, *Nature (London)* **355**, 300–301 (1992).

87. T.R. Coleman and W.G. Dunphy, *Curr. Opin. Cell Biol.* **6**, 877–882 (1994).

88. Y. Sanchez et al., *Science* **277**, 1497–1501 (1997).

89. F. Verde et al., *J. Cell Biol.* **118**, 1097–1108 (1992).

90. E.A. Nigg, *Curr. Opin. Cell Biol.* **5**, 187–193 (1993).

91. R. Reeves, *Curr. Opin. Cell Biol.* **4**, 413–423 (1992).

92. E.A. Nigg, *Curr. Opin. Cell Biol.* **4**, 105–109 (1992).

93. S. Yamashiro, Y. Yamakita, R. Ishikawa, and F. Matsumura, *Nature (London)* **344**, 675–678 (1990).

94. Y. Yamakita, S. Yamashiro, and F. Matsumura, *J. Cell Biol.* **124**, 129–137 (1994).

95. F.C. Luka, E.K. Shibuya, C.E. Dohrmann, and J.V. Ruderman, *EMBO J.* **10**, 4311–4320 (1990).

96. J.A. Knoblich et al., *Cell* **77**, 107–120 (1994).

97. W. Wells, *Trends Cell Biol.* **6**, 226–234 (1996).

98. S.J. Elledge, *Science* **274**, 664–1672 (1996).

99. X. Li and R.B. Nicklas, *Nature (London)* **373**, 630–632 (1995).

100. C.L. Reider, R.W. Cole, A. Khodjakov, and G. Sluder, *J. Cell Biol.* **130**, 941–948 (1995).

101. B. Novak and J.J. Tyson, *J. Cell Sci.* **106**, 1153–1168 (1993).

102. B. Novak and J.J. Tyson, *J. Theor. Biol.* **165**, 101–134 (1993).

103. B. Novak and J.J. Tyson, *J. Theor. Biol.* **173**, 283–305 (1995).

104. B. Novak and J.J. Tyson, *Proc. Natl. Acad. Sci. U.S.A.* **94**, 9147–9152 (1997).

105. J.J. Tyson et al., *Trends Biochem. Sci.* **21**, 89–96 (1996).

106. J.F. Collins and M.H. Richmond, *J. Gen. Microbiol.* **28**, 15–33 (1962).

107. R.J. Harvey, A.G. Marr, and P.R. Painter, *J. Bacteriol.* **93**, 605–617 (1967).

108. A.G. Fredrickson, D. Ramkrishna, and H.M. Tsuchiya, *Math. Biosci.* **1**, 327–374 (1967).

109. D. Ramkrishna, A.G. Fredrickson, and H.M. Tsuchiya, *Bull. Math. Biophys.* **30**, 319–323 (1968).

110. S. Terada, E. Suzuki, H. Ueda, and F. Makishima, *Cytokine* **8**, 889–894 (1996).

111. K. Takahashi et al., *Cytotechnology* **15**, 57–64 (1994).

112. F. Makishima, S. Terada, T. Mikami, and E. Suzuki, *Cytotechnology* **10**, 15–23 (1992).

113. M. Al-Rubeai, A.N. Emery, S. Chalder, and D.C. Jan, *Cytotechnology* **9**, 85–97 (1992).

114. M. Al-Rubeai and A.N. Emery, *J. Biotechnol.* **16**, 67–86 (1990).

115. N.J. Cowan and C. Milstein, *Biochem. J.* **128**, 445–454 (1972).

116. S.J. Kromenaker and F. Srienc, *Biotechnol. Bioeng.* **38**, 665–677 (1991).

117. S.J. Kromenaker and F. Srienc, *J. Biotechnol.* **34**, 13–34 (1994).

118. M. Cherlet, S.J. Kromenaker, and F. Srienc, *Biotechnol. Bioeng.* **47**, 535–540 (1995).

119. M. Kubbies and H. Stockinger, *Exp. Cell Res.* **188**, 267–271 (1990).

120. M.B. Gu, P. Todd, and D.S. Kompala, *Biotechnol. Bioeng.* **42**, 1113–1123 (1993).

121. M.B. Gu, P. Todd, and D.S. Kompala, *Biotechnol. Bioeng.* **50**, 229–237 (1996).

122. M. Al-Rubeai, M.S. Chalder, R. Bird, and A.N. Emery, *Cytotechnology* **7**, 179–186 (1991).

123. D.E. Martens et al., *Biotechnol. Bioeng.* **41**, 429–439 (1993).

124. K.A. Hawboldt, T.I. Linardos, N. Kalogerakis, and L.A. Behie, *J. Biotechnol.* **34**, 133–147 (1994).

125. S. Lakhotia, K.D. Bauer, and E.T. Papoutsakis, *Biotechnol. Bioeng.* **40**, 978–990 (1992).

126. M. Al-Rubeai, R.P. Singh, A.N. Emery, and Z. Zhang, *Biotechnol. Bioeng.* **46**, 88–92 (1995).

127. R.P. Singh, M. Al-Rubeai, C.D. Gregory, and A.N. Emery, *Biotechnol. Bioeng.* **44**, 720–726 (1994).

128. E. Suzuki et al., *Cytotechnology* **23**, 55–59 (1997).

129. N.H. Simpson, A.E. Milner, and M. Al-Rubeai, *Biotechnol. Bioeng.* **54**, 1–16 (1997).

130. S. Terada, Y. Itoh, H. Ueda, and E. Suzuki, *Cytotechnology* **24**, 135–141 (1997).

131. Y. Itoh, H. Ueda, and E. Suzuki, *Biotechnol. Bioeng.* **48**, 118–122 (1995).

132. W.A. Renner et al., *Biotechnol. Bioeng.* **47**, 476–482 (1995).

133. K.H. Lee, A. Sburlati, W.A. Renner, and J.E. Bailey, *Biotechnol. Bioeng.* **50**, 273–279 (1996).

134. M. Fussenegger, X. Mazur, and J.E. Bailey, *Biotechnol. Bioeng.* **55**, 927–939 (1997).

135. C. Cordon-Cardo, *Am. J. Pathol.* **147**, 545–559 (1995).

136. L.H. Hartwell and M.B. Kastan, *Science* **266**, 1821–1828 (1994).

137. B. Palsson and S. Andreadis, *Exp. Hematol.* **25**, 94–102 (1997).

138. S.T. Andreadis, D. Brott, A.O. Fuller, and B.O. Palsson, *J. Virol.* **71**, 7541–7548 (1997).

139. S. Andreadis, A.O. Fuller, and B.O. Palsson, *Biotechnol. Bioeng.* **58**, 272–281 (1998).

140. S. Andreadis and B.O. Palsson, *J. Theor. Biol.* **182**, 1–20 (1996).

See also CELL CYCLE IN SUSPENSION CULTURED PLANT CELLS; CELL CYCLE SYNCHRONIZATION; CELL DETACHMENT; CELL GROWTH AND PROTEIN EXPRESSION KINETICS; CHARACTERIZATION AND DETERMINATION OF CELL DEATH BY APOPTOSIS; PROGRAMMED CELL DEATH (APOPTOSIS) IN CULTURE, IMPACT AND MODULATION.

CELL CYCLE IN SUSPENSION CULTURED PLANT CELLS

M.R. FOWLER
De Montfort University
Leicester, United Kingdom

OUTLINE

INTRODUCTION

Cell division plays a crucial role in the development of all complex multicellular eukaryotes. This is particularly true of plants, which, unlike animals, can grow and alter form throughout their life cycles. Therefore, an understanding of the mechanisms that regulate the plant-cell division cycle is of central importance if a clear picture of plant development is to be arrived at. Such scientific advances also hold the promise of making significant advances in other fields such as agriculture. Many agriculturally desirable traits (such as yield) could be improved by manipulating the pattern and/or frequency of plant-cell division. Furthering our understanding of the plant-cell division cycle holds out the hope of increasing our "scientific" knowledge and also of making a positive contribution to the quality of life of many people.

THE CELL CYCLE

The basic function of the cell cycle is to replicate the genetic material and to partition this genetic material into each of two daughter cells. The now classical four-phase model of the cell cycle was first defined in the early 1950s when it was shown that DNA synthesis occurs during a discrete period in the interphase. The cell cycle is composed of the S (or synthesis) phase, during which the DNA content of the cell is doubled, to provide genetic material for subsequent segregation and partitioning into daughter cells. This segregation and partitioning occurs during mitosis (the M phase) and cytokinesis. The two remaining stages of the cell cycle, G_1 and G_2 separate the S and M phases. Originally "G" was used to designate "gap" because there was little cellular activity observable during these phases. Now it is clear, however, that despite the lack of easily observed physical changes in the structure of the cell, many of the controls of cell division occur during the G_1 and G_2 phases.

PLANT CELL DIVISION

The basic processes that occur in plant cell division (DNA replication and partitioning into daughter cells) are similar to those in other higher eukaryotes. Cell division in plants, though has some unique physical characteristics (the formation of the preprophase band, phragmoplast, and cellulose cell wall and the lack of some clearly defined components of the mitotic spindle apparatus), and the way that cell division is integrated into plant development is also unique. In animals the form of the organism is determined in the embryo, cell migration occurs and, in the mature organism, cell proliferation replaces lost or damaged cells. In contrast, plants are capable of growth and of altering form throughout their life cycles by cell division in specialized regions termed meristems, and subsequent tissue development and cell migration does not occur (1,2).

MODEL SYSTEMS FOR INVESTIGATING PLANT CELL DIVISION

The system of choice for investigations into plant-cell division is often a cell-suspension culture. Ideally, the cell suspension utilized should be composed of a homogenous population of single cells or very small cell clusters and for experimental ease should be fairly rapidly growing. It is not always possible to obtain a cell suspension that matches these ideals, necessitating the use of alternative cell-suspension cultures. Although cell suspensions cannot mimic all of the complex interactions that regulate cell division in planta, they have for some time been considered an ideal system in which to study the control of plant-cell division, especially at the biochemical and gene expression levels. In principle, cell-suspension cultures are uncomplicated by lack of uniformity in the material, allow the environmental conditions (such as light intensity, temperature, growth medium nutrient, and growth regulator composition) to be controlled and manipulated, and have the added advantage that cultures can be maintained indefinitely in a sterile condition.

The use of Synchronized Cells in Suspension Culture

The particular events that occur in specific phases of the cell cycle can be investigated by using cells in suspension culture that undergo a synchronous cell cycle. To achieve this, all of the cells in the suspension culture have to be in the same cell-cycle phase at the same time and progress through the cell cycle at precisely the same rate. In practice this ideal is impossible to achieve. Recent advances in technology and the availability of newly developed chemicals have, though, resulted in synchronization protocols that give a sufficiently high degree of synchronization to allow using cells in suspension culture to investigate cell-cycle phase-specific events.

Plant cells in suspension culture are usually synchronized by treatments that either inhibit DNA synthesis and prevent progression through the S phase (hydroxyurea, aphidicolin, thymidine) or prevent mitosis (colchicine, propyzamide, oryzalin), although other treatments (such as nutrient or plant growth regulator removal or cold shock) that result in arrest at other points in the cell cycle have also been used. Application of the treatment to the cells results in arresting the cell cycle at a particular phase. Removal of the treatment, usually by washing the cells, removes the inhibition and allows the cells to progress through the cell cycle, hopefully synchronously. By monitoring DNA synthesis and/or the mitotic index of the culture, the particular phases of the cell cycle can be identified. In practice, as soon as the inhibition is removed, the degree of synchrony in the culture begins to decline, so that useful synchrony is maintained only for one cell cycle or only a portion of a cell cycle. Therefore, it may be necessary to synchronize cells at more than one point to investigate the events in all four phases of the cell cycle.

The Tobacco BY-2 Cell Suspension. A cell-suspension culture derived from tobacco has become, what might be considered, the standard for investigations into fundamental aspects of the plant-cell-division cycle (3). It is by no means the only cell suspension culture used, but in the limited space available in this article it will be used to illustrate the features desirable in any cell-suspension culture used for investigations into the plant-cell cycle. The tobacco BY-2 cell line has a batch culture cycle of only seven days, during which the cells multiply 80 to 100-fold (a high growth rate). BY-2 cells can be synchronized to a high degree (70–80% by mitotic index) by treatment with the DNA polymerase inhibitor aphidicolin (plant cells in suspension culture are not easily synchronized to such a high degree; less than 40% is common). To investigate events that occur after mitosis, when the extent of synchrony from the aphidicolin treatment is much reduced, the synchronization of the BY-2 cells can be further improved by treating aphidicolin-synchronized cells with propyzamide (an antitubulin drug) to induce arrest at mitosis. This results in synchrony of more than 80% (by mitotic index) and allows investigating events following mitosis (3).

Cell-Cycle-Dependent Gene Expression

Specific phases in the cell cycle are often accompanied by an increase in the expression of a small number of genes involved in that particular phase of the cell cycle. The identification of genes expressed at particular points during the cell cycle can give an initial indication of the processes at that particular point. The way in which this expression is regulated is also of interest because the phase of the cell cycle must be sensed in some way by the transcriptional machinery to ensure that expression occurs at the correct time. Only a few genes are expressed so periodically during the cell cycle of eukaryotes. The classical examples are genes encoding proteins involved in DNA replication and cyclins (cyclins are described later in this article). To identify or map the expression of genes such as these, a cell suspension that can be synchronized to a very high degree is required. Therefore, the tobacco BY-2 cell line is widely used in such studies (4), although other cell lines, particularly a *Catharanthus roseus* (periwinkle) cell suspension (5), have been used.

These systems have been used to study the expression pattern of a variety of genes that are expressed periodically during the plant cell cycle, particularly those encoding proteins involved in DNA replication (see also Table 1) (6–12).

Table 1. Cell-Cycle, Phase-Specific Gene Expression Patterns of Selected Sequence. (This list is not exhaustive, and a more complete review of cell-cycle phase-dependent gene expression is given in Ref. 4.)

Gene/Gene Product	Expression Pattern	Reference
Histones[a]	Very late G_1/S	6,7
RNR (ss)[b]	S	8
PCNA[c]	S	9
dUTPase[d]	S	4
DNA polymerase α[e]	S	4
DHFR-TS[f]	S	4
Extensins[g]	S	10
HSP90[h]	S	4
cyc5Gm[i]	G_2/M	11
cyc3Gm[j]	S/G_2	11
cyc1Gm[k]	S	11
cdc2cAm[l]	S/G_2/M	12
cdc2dAm[m]	G_2/M	12

[a] Histones comprise a complex multigene family. The expression of H1, H2A, H2B, H3, and H4 appears to be temporally correlated with DNA synthesis.
[b] The small subunit of ribonucleotide reductase. Involved in DNA synthesis.
[c] Proliferating cell nuclear antigen. Involved in DNA synthesis.
[d] Deoxyuridine triphosphatase. Involved in DNA synthesis.
[e] Involved in nuclear DNA synthesis.
[f] Bifunctional dihydrofolate reductase-thymidylate synthase. Involved in DNA synthesis.
[g] Components of cell wall.
[h] Possible cell-cycle regulator via activity of Wee1 kinase homologue.
[i] cyc5Gm. soybean B-type cyclin.
[j] cyc3Gm. soybean A-type cyclin.
[k] cyc1Gm. soybean A-type cyclin.
[l,m] *Antirrhinum* non-PSTAIRE encoding cdc2-like sequences (see Table 2). cdc2aAm and cdc2bAm which encode perfectly conserved PSTAIRE motifs do not exhibit cell cycle periodicity.

Table 2. PSTAIRE Epitopes of Mammalian and Plant CDKs. (The PSTAIRE motif is shown in bold, as are the designations of plant sequences coding for unique PSTAIRE motifs.)

Cdk	PSTAIRE motif
Cdc2 (Cdk1), Cdk2, Cdk3, all plant homologues that complement yeast mutants	**PSTAIRE**
Cdk4	**PISTVRE**
Cdk5	**PSSALRE**
Cdk6	**PLSTIRE**
Cdk7	**NRTALRE**
Cdk8	**SMSACRE**
cdc2bAt[a], **cdc2cAm**[b], **cdc2dMs**[c]	**PPTALRE**
cdc2dAm[b], **cdc2fMs**[c]	**PPTTLRE**
R20s[d]	**NFTALRE**
cdc2eMs[c]	**SPTAIRE**
PCTAIRE	**PCTAIRE**
PITALRE	**PITALRE**
PITSLRE (p58-GTA)	**PITSLRE**
PITAIRE (CHED), **cdc2cMs**[c], **cdc2Ps2**[e]	**PITAIRE**
Bvcrk1[f]	**KFMA-RE**

Table compiled from data in Refs. 2, 14, and 24.
[a] *Arabidopsis thaliana*
[b] *Antirrhinum majus*
[c] *Medicago sativa*
[d] *Oryza sativa*
[e] *Pisum sativum*
[f] *Beta vulgaris* cdc2-related kinase.

Histones. Histone gene expression provides a paradigm for cell-cycle-specific gene expression. The expression of histone genes is usually closely correlated with the S phase of the cell cycle in plants, as is the case in other eukaryotes, although the initiation of histone gene transcription does in fact occur just before the onset of the S phase. However, differences in the expression pattern between plants and animals have been noted. In plants the amount of histone mRNA is not directly coupled to the DNA synthesis level, indeed DNA synthesis can be interrupted or slowed without unduly affecting transcription of histone genes (6,7).

Enzymes Involved in DNA Synthesis. Proliferating cell nuclear antigen (PCNA), a DNA polymerase δ auxiliary protein, the small subunit of ribonucleotide reductase, DNA polymerase α, and a bifunctional dihydrofolate reductase-thymidylate kinase have all been shown to be expressed in late G_1 and S in synchronized plant cells, in much the same pattern as observed in other eukaryotes (4,8,9).

Other Sequences. Several sequences have been identified by differential screening of cDNA libraries prepared from periwinkle cells at various phases of the cell-division cycle (5,13). This illustrates the use of synchronized plant cells to identify novel plant genes expressed in a cell-cycle-dependent manner. The sequences are designated by the prefix *cyc* and should not be confused with cyclin sequences. A few examples are described later. A more comprehensive description can be found in Ref. 4, 5, and 13.

cyc19. *cyc19* encodes a heat shock protein (HSP90), whose expression correlates with the rate of DNA synthesis. Although a role in cell cycle regulation is not immediately obvious, a member of the HSP90 family has, been identified in yeast, as a regulator of Wcc1 kinase (a key negative cell-cycle regulator described later in this article) activity.

cyc15 and cyc17. These sequences encode hydroxy-proline-rich glycoproteins known as extensins which are part of the plant cell wall structure. Both *cyc15* and *cyc17* are expressed during the S phase, but expression is also high in the stationary phase (nondividing) cells (10).

CONTROLLING PROGRESSION THROUGH THE CELL CYCLE

Because the basic processes of cell division are common to all eukaryotes, perhaps it should come as no surprise that many of the controls acting on the cell cycle appear to be conserved among all eukaryotes, including plants. Most of the information about these controls has originated from excellent work in yeast and mammalian cells, where potential cures for cancer have provided a considerable impetus for fine research. Research into plant-cell division lags behind and is often aimed at identifying controls homologous to those already identified in yeast and mammalian cells. It is now becoming apparent though, that the plant-cell-division cycle is likely to exhibit some unique features, even in these highly conserved control systems.

In this short section we will look at developments in a rapidly expanding area of plant-cell division research where the use of synchronized plant cell-suspension cultures has also made possible many advances in our understanding. By necessity, much of the information will, be given in the context of information from other systems, and readers are directed to other sources for some of this information (14).

Cyclin-Dependent Kinases and Cyclins

Evidence from a wide variety of eukaryotes has indicated for some time that there are two major control points in the cell cycle, one situated in G_1 and one situated in late G_2. These control points regulate progression through the cell cycle and are crucial in determining the timing of the S phase and M phase, and thus the rate of cell proliferation and the size of the cell at cytokinesis. The G_1 control point is termed "START" in yeast, the R or restriction point in mammalian cells and is probably one of the principal control points identified in plants (14–16). This G_1 control point regulates the entry of cells into the S phase. The G_2 control point, which is probably the second principal control point (16) identified in plants, controls the entry of cells into the M phase. Now a large body of evidence indicates that these control points involve the action of a group of serine/threonine protein kinases (Fig. 1) known as cyclin-dependent kinases (or Cdks) (14). The activity of these Cdks depends on the binding of a second, regulatory

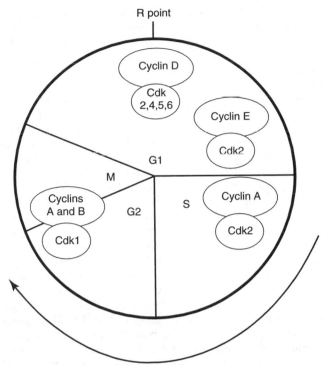

Figure 1. A simplified representation of the active Cdk/cyclin complexes formed during the cell cycle in mammalian cells. For clarity, only the well-characterized Cdk/cyclin complexes are shown. Cyclins D1, 2, and 3 in conjunction with various Cdks are thought to mediate the R-point (the mammalian equivalent of "START" in yeast cells) and integrate extracellular signals into the cell-cycle machinery. Cyclin E binds only to Cdk2 and is activated later in the cell cycle near the G_1 to S-phase transition. Both cyclin D and cyclin E associated kinases bring about S phase, at least in part, by controlling E2F mediated transcription via the phosphorylation of retinoblastoma (Rb) and other associated proteins (such as p107 and p130). The phosphorylation of Rb, p107, and p130 releases free E2F transcription factors which are postulated to initiate transcription from genes required for the S phase. Cyclin E may also be required during the initial stages of the S phase. In the S phase, Cdk2 then associates with cyclin A. Cdk2/cyclin A complexes may be required for continued DNA synthesis and may help to terminate S-phase specific transcription. The transition from G_2 to mitosis is brought about by cyclin A and cyclin B/Cdk1 (Cdc2) complexes. Cyclin B/Cdk1 complexes are known to be capable of phosphorylating many proteins (at least *in vitro*) that undergo phosphorylation at mitosis *in vivo*. It is proposed that these phosphorylation events bring about the physical changes necessary for mitosis.

protein termed a cyclin and on specific phosphorylation and dephosphorylation events.

The prototypic Cdk was identified in the fission yeast *Schizosaccharomyces pombe* (17) as the product of the *cdc2* gene in experiments with mutated cells that exhibited abnormal cell-cycle regulation (a budding yeast homologue termed *CDC28* was also rapidly identified). It was demonstrated that this single Cdk was sufficient to mediate both control points. Rapid progress in a quest for homologues of these sequences in higher eukaryotes soon led to the identification of a large number of Cdk encoding sequences in mammalian cells and the suggestion that different Cdks in higher eukaryotes may mediate the different control points (2,14,18–22). Cdks are often designated by reference to a particular motif (in subdomain III), the so-called PSTAIRE motif. Cdks can be divided into two groups, those with perfectly conserved PSTAIRE motifs and those with divergent PSTAIRE motifs. There is also a further group of sequences that encode divergent PSTAIRE motifs and that, although they belong to the same family of serine/threonine protein kinases as do Cdks, exhibit structural differences (14). These sequences are generally longer, encoding proteins with amino and/or carboxy terminal extensions, and there is also some speculation that some of these proteins may not require cyclin binding to be activated. These larger non-PSTAIRE sequences are not necessarily directly involved in controlling cell division but instead have roles in processes such as signal transduction (23). Therefore, they may be more closely involved in integrating cell division into development. A large number of Cdks in plants have also been identified and, as is the case with mammalian Cdk sequences, these sequences can be distinguished by their PSTAIRE motifs. In general, plant Cdks with perfectly conserved PSTAIRE regions are capable of complementing temperature-sensitive yeast mutants (although with different degrees of efficiency), although some cell cycle phase specificity may be implied from such experiments. Thus Cdc2a from alfalfa can complement only a G_2 control point defective yeast mutant, and Cdc2b from alfalfa can complement only a G_1 control point defective yeast mutant (24,25). Both of these sequences have perfectly conserved PSTAIRE regions and are more than 85% identical at the amino acid level. These PSTAIRE encoding sequences are considered to be constitutively expressed during the cell cycle. A variety of non-PSTAIRE encoding cdk sequences have been identified, some of which represent PSATIRE motifs as yet not identified in yeast or mammalian cells. These non-PSTAIRE encoding cdk sequences are generally expressed later in the cell cycle (from the S phase onward) and do not complement yeast temperature-sensitive mutants (12,24,25). These could then represent Cdks involved in some unique aspect of the plant-cell cycle, perhaps in some of the specific structural changes that occur in plant-cell division. As this catalogue of genes increases, it is now apparent that plants, like animals, also contain the larger non-PSTAIRE sequences. Some of these sequences are almost identical to previously isolated mammalian sequences, but others appear to be unique to plants (26,27).

Cdks, by definition, bind a second regulatory protein, termed a cyclin. At least nine types of cyclin have been identified in mammalian cells and are designated by the letters A to I. Some types of cyclin may interact with only one specific Cdk, whereas others may interact with several Cdks. Cyclins are involved in determining the substrate specificity and cellular location of the Cdk complex. As with Cdks, all cyclins share some common structural characteristics (such as the cyclin box, which is required for interaction with the Cdk), but it is possible to divide cyclins into two large groups on the basis of their cell-cycle phase-specific gene expression pattern. The first

group comprises the so-called mitotic cyclins (cyclins A, B, and F) that are required for progression into mitosis and are characterized by the presence of a "destruction box" which is responsible for protein instability; the second group comprises the so-called G_1 cyclins (cyclins C, D, and E) that are required for entry into and progression through the S phase (14). Homologues of cyclins A, B, and D have been identified in a wide variety of plant species (28–31), and in some cases functional equivalence has been demonstrated. Initially, it proved difficult to characterize plant mitotic cyclins as A-type or B-type on the basis of sequence similarity, a situation that has only recently been resolved by complex sequence analysis (28). Transcription patterns of plant A-type and B-type cyclins generally follow the pattern of their yeast and mammalian counterparts. Plant D-type cyclins contain the characteristic LxCxE motif (29–31) required for interaction with the retinoblastoma (Rb) protein, and like their mammalian counterparts, are expressed early in the cell cycle and respond to extracellular mitogens (29–31). The presence of the LxCxE motif and the recent identification of Rb homologues in plants (32,33) suggests that their role, like their mammalian counterparts, is, to mediate E2F-regulated transcription and therefore bring about the events necessary for the S phase.

This information suggests a similar picture of cell-cycle regulation in plants and animals, and given the pace at which the field is progressing, it may yet be asserted with confidence that the core components of the cell-cycle control points are indeed the same. There are, however, some data that suggest that fundamental differences between plants and animals may exist. In plants a particular group of A-type cyclins is known to be expressed very early in the cell cycle before D-type cyclin expression (34). In mammalian cells, A-type cyclin expression is initiated at the onset of the S phase after D-type cyclin expression. As yet, homologues of E-type cyclins have not been identified in plants, yet it is known in mammalian cells that cyclin E is required for Rb phosphorylation and progression through the S phase (35). It is possible though that plant D-type cyclins (of which there are more than the three present in mammalian cells) also fulfill the roles of mammalian cyclin E.

The activity of the cyclin/Cdk complexes is also mediated by other controls. The best characterized is by altering the phosphorylation status of key residues of the Cdk catalytic subunit, such as Thr14, Tyr15, and Thr161 (or their equivalents). Cdk activation depends on the phosphorylation of Thr161 and the dephosphorylation of Thr14 and Tyr15. The phosphorylation of Thr161 is brought about by Cdk activating kinase (CAK) which is itself a cyclin/Cdk complex (14). Inhibitory phosphorylations of Thr14 and Tyr15 (which prevent premature activation of the cyclin/Cdk complex) are carried out by the products of the *wee1* and *myt1* genes (14,36). Dephosphorylation of these two residues is carried out by the product of the *cdc25* gene(s). Plant Cdks have equivalents to these residues, but little data as to their phosphorylation status during the cell cycle is available. Whether similar phosphorylation and dephosphorylation events are involved in regulating plant Cdk activity is therefore a little difficult to judge. Evidence from experiments using mutated *cdc2* genes (in which the Thr14 and Tyr15 residues were changed to non-phosphorylatable amino acids) had little effect on plant development (37), suggesting that phosphorylation of these residues is not important in regulating plant Cdk activity (although the results are difficult to interpret in certain terms). Some evidence supports the presence of similar phosphorylation events in plant cells. Recently PCR fragments highly identical to *cdc25* sequences from various organisms have been isolated from several monocotyledonous plant species, and expression of yeast cdc25 in plants does lead, at least in some ways, to the expected phenotype (38,39). Plant sequences encoding CAK activity have also been recently reported (40). However, no homologues of *myt1* have yet been reported.

SUMMARY

Plant cell cultures provide a powerful system for investigating the molecular controls that act during the plant-cell-division control cycle and have enabled significant advances to be made. Plant cell division, although basically similar to cell division in other eukaryotes, displays some unique physical characteristics that may be reflected in some differences in the otherwise highly conserved control mechanisms of eukaryotic cell division. In addition to the information presented in this short article, plant-cell suspensions have made contributions to other areas of plant-cell-cycle research, for instance, enabling the roles of phytohormones in controlling cell division to be more clearly defined. Understanding the fundamental mechanisms controlling the plant-cell-division cycle via investigations which, at least in part, utilize plant-cell suspensions promises to make meaningful contributions to crop improvement in the not too distant future.

BIBLIOGRAPHY

1. O. Shaul, M. Van Montagu, and D. Inzé, *Crit. Rev. Plant Sci.* **15**, 97–112 (1996).
2. M.R. Fowler et al., *Mol. Biotechnol.* **10**, 123–153.
3. T. Nagata, Y. Nemoto, and S. Hasezawa, *Int. Rev. Cytol.* **132**, 1–30 (1992).
4. M. Ito, in D. Francis, D. Dudits, and D. Inzé, eds., *Plant Cell Division*, Portland Press, London, 1998, pp. 166–186.
5. H. Kodama et al., *Plant Physiol.* **95**, 406–411 (1991).
6. N. Chaubet and C. Gigot, in D. Francis, D. Dudits, and D. Inzé, eds., *Plant Cell Division*, Portland Press, London, 1998, pp. 269–283.
7. M. Iwabuchi, T. Nakayama, and T. Meshi, in D. Francis, D. Dudits, and D. Inzé, eds., *Plant Cell Division*, Portland Press, London, 1998, pp. 285–300.
8. G. Philipps, B. Clement, and C. Gigot, *FEBS Lett.* **358**, 67–70 (1995).
9. H. Kodama et al., *Eur. J. Biochem.* **197**, 495–503 (1991).
10. M. Ito, H. Kodama, A. Komamine, and A. Watanabe, *Plant Mol. Biol.* **36**, 343–351 (1998).
11. H. Kouchi, M. Sekine, and S. Hata, *Plant Cell.* **7**, 1143–1155 (1995).
12. P.R. Fobert et al., *Plant Cell* **8**, 1465–1476 (1996).
13. M. Ito, H. Kodama, and A. Komamine, *Plant J.* **1**, 141–148 (1991).
14. C. Hutchison and D. Glover, *Cell Cycle Control*, IRL Press, Oxford, U.K., 1995.
15. A.B. Pardee, *Science* **246**, 603–608 (1989).
16. J. Van't Hof, *Brookhaven Symp.* **25**, 125–165 (1973).
17. P. Nurse and Y. Bissett, *Nature (London)* **292**, 558–560 (1981).
18. M.G. Lee and P. Nurse, *Nature (London)* **327**, 31–35 (1987).
19. S.J. Ellidge and M.R. Spotswood, *EMBO J.* **10**, 2653–2659 (1991).
20. F. Fang and J.W. Newport, *Cell* **66**, 731–742 (1991).
21. L.-H. Tsai et al., *Oncogene* **8**, 1593–1602 (1993).
22. M. Meyerson et al., *EMBO J.* **11**, 2909–2917 (1992).
23. T. Okuda, J.L. Cleveland, and J.R. Downing, *Oncogene* **7**, 2249–2258 (1992).
24. D. Dudits et al., in D. Francis, D. Dudits, and D. Inzé, eds., *Plant Cell Division*, Portland Press, London, 1998, pp. 21–45.
25. G. Segers, P. Rouzé, M. Van Montagu, and D. Inzé, in J.A. Bryant and D. Chiatante, eds., *Plant Cell Proliferation and Its Regulation in Growth and Development*, Wiley, Chichester, 1997, pp. 1–19.
26. Z. Magyar et al., *Plant Cell* **9**, 223–235 (1997).
27. A. Atanassova et al., in preparation.
28. J.-P. Renaudin et al., *Plant Mol. Biol.* **32**, 1003–1018 (1996).
29. R. Soni, J.P. Carmichael, Z.H. Shah, and J.A.H. Murray, *Plant Cell* **7**, 85–103 (1995).
30. M. Dahl et al., *Plant Cell* **7**, 1847–1857 (1995).
31. L. De Veylder, M. Van Montagu, and D. Inzé, in D. Francis, D. Dudits, and D. Inzé, eds., *Plant Cell Division*, Portland Press, London, 1998, pp. 1–19.
32. G. Grafi et al., *Proc. Natl. Acad. Sci. U.S.A.* **93**, 8962–8967 (1996).
33. Q. Xe, P. Sanz-Burgos, G.J. Hannon, and C. Gutierrez, *EMBO J.* **15**, 4900–4908 (1996).
34. I. Meskiene et al., *Plant Cell* **7**, 759–771 (1995).
35. M. Hatakeyama and R.A. Weinberg, *Prog. Cell Cycle Res.* **1**, 9–19 (1995).
36. P.R. Mueller, T.R. Coleman, A. Kumagai, and W.G. Dunphy, *Science* **270**, 86–90 (1995).
37. A. Hemerly et al., *EMBO J.* **14**, 3925–3936 (1995).
38. P.A. Sabelli, S.R. Burgess, A.K. Kush, and P.R. Shewry, in D. Francis, D. Dudits, and D. Inzé, eds., *Plant Cell Division*, Portland Press, London, 1998, pp. 243–268.
39. M.H. Bell, N.G. Halford, J.C. Ormrod, and D. Francis, *Plant Mol. Biol.* **23**, 445–451 (1993).
40. M. Umeda et al., *Proc. Natl. Acad. Sci. U.S.A.* **95**, 5021–5026 (1998).

See also CELL CYCLE EVENTS AND CELL CYCLE-DEPENDENT PROCESSES; CELL CYCLE SYNCHRONIZATION; CELL GROWTH AND PROTEIN EXPRESSION KINETICS; PROGRAMMED CELL DEATH (APOPTOSIS) IN CULTURE, IMPACT AND MODULATION.

CELL CYCLE SYNCHRONIZATION

Paul Holmes
Mohamed Al-Rubeai
University of Birmingham
Edgbaston, Birmingham
United Kingdom

OUTLINE

INTRODUCTION

Central to cell proliferation is the cell cycle which is characterized by precisely regulated, both spatially and temporally, gene expression. Study of cells synchronized with respect to their cell-cycle phase is seen as a useful research tool for investigating cell-cycle-regulated gene expression and cellular physiology in both fundamental and applied cell biology. Potentially, understanding cell-cycle-related gene expression could be of use in developing strategies to control proliferative diseases such as cancer, or to improve bioprocesses for securing biologically active proteins by increasing cell growth rates, reducing cell death rates and enabling culture strategies to maintain cells in their most physiologically productive state. Methods for synchronizing cells in all phases of the cell cycle are described and their relative merits and shortcomings are discussed.

THE CELL CYCLE: A BRIEF OVERVIEW

During the culture of transformed mammalian cells, three distinct phases are seen, namely, lag, exponential growth, and a stationary phase. Initially after inoculation there is a lag phase characterized by no increase in cell number. This is the period before any significant occurrence of first mitotic cell divisions. During this time cells are progressing through the cell cycle prior to the first division. As a brief overview, the cell cycle consists of, in its most simple terms, four stages, G1, S, G2, and mitosis (Fig. 1). Preparation for replication of chromosomes occurs during G1. It is here that a checkpoint, the so-called restriction or R point (1) exists, at which cell cycle progression can be halted so that DNA damage can be corrected before nucleic acid is replicated, thus reducing the number of mutations carried over into subsequent progeny cells. Past the restriction point cells are committed to replication and there is no requirement for growth factors or high levels of protein synthesis for entry into the next stage of the cell cycle, the S phase. It is also from G1 that cells can exit and return to the cell cycle from a quiescent phase termed G0. The S phase is characterized by replication of nucleic acid and associated proteins, for example histones, and is followed by replication of cytoplasm during G2. G2 also has a checkpoint which can prevent entry into mitosis based on the physiological conditions of the cell. Cell-cycle progression is prevented if unreplicated or damaged DNA is detected, thus ensuring that all chromosomes are replicated and repaired before segregation at mitosis (2). Mitosis completes the cell cycle and results in two daughter cells in G1. In mitosis another checkpoint exists which monitors spindle fiber formation and attachment of spindle fibers to the kinetchores, preventing mitosis if such structures are not in place.

Progression through the cell cycle is controlled by cyclins. Nine cyclins have been identified to date and are designated A through I. As denoted by their name, cyclins are synthesized and degraded in a precisely timed sequence within the cell cycle. Levels of the particular cyclins are regulated at the level of transcription, as well as by targeted degradation via the ubiquitin pathway. So far, the best characterized cyclins are A, B, D, and

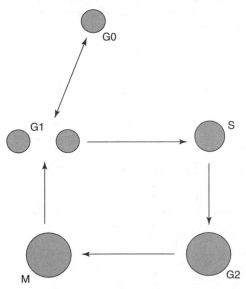

Figure 1. Diagrammatic representation of the cell cycle showing sequential progression from G1 through S to G2 then back to G1 via mitosis which results in two daughter cells. Cells may exit the cell cycle by entering a quiescent phase, denoted G0, or by cell death, most frequently by apoptosis.

E (3). D and E type cyclins are expressed during G0/G1 and are called START cyclins. START, also known as the restriction checkpoint in mammalian cells, is the point in late G1 at which the cell commits itself to another round of DNA replication, as mentioned before. The D-type cyclins (D1, D2, and D3) are expressed in response to growth factors or mitogens and rapidly degrade when mitogens are withdrawn. In cells that are proliferating continuously, their levels are less variable during the cell cycle than cyclins E, A, or B1. Expression of a particular D-type cyclin is tissue-specific. For example, T lymphocytes express more cyclin D3 than D2 and are cyclin D1-negative. D-type cyclins appear to promote G0 to G1 transitions and influence the rate of G1 progression. Absence of D-type cyclins in specific cell types may signal a switch between proliferation and differentiation. In general, cyclin E is induced later in G1 than the D-type cyclins and is likely to be involved in G1 to S phase transition. Cyclins A and B1 are mitotic cyclins (3). Cyclin A is synthesized during the S phase and degrades during anaphase. Cyclin B1, the first identified cyclin, is synthesized during late S, maximally expressed during the transition from G2 to M, and degraded during anaphase. Cyclins complex with cyclin-dependent kinases (cdks) via a 150 amino acid sequence termed the cyclin box. This process activates the kinase which in turn phosphorylates many proteins including transcription factors which in turn induce gene expression and allow continued progression through the cell cycle (4).

Cell-Cycle Monitoring

Because of the asynchronous nature of proliferating cell cultures, synchronization has for thirty years or more been seen as a requirement to study cell cycle phase specificity of gene expression and cellular physiology (5). A convenient method for assessing cell-cycle phase and hence degree of synchronicity is flow cytometry (FC) which allows analysis of a large number of cells (20,000 or more) rapidly, reliably, and consistently, thus yielding statistically valid data relating to the cell-cycle phases of a population or selected subpopulation of cells (6). Moreover, FC can be advantageous in removing nonviable or apoptotic cells from the analysis and can also be used for multiple staining to correlate other parameters, for example, intracellular protein content, to cell-cycle phase (6), and is discussed later. For cell-cycle analysis, cells are fixed in 70% ethanol at −20 °C for at least 30 minutes and stored at this temperature until required for analysis. Keeping the temperature as close to −20 °C as possible during fixation is critical for producing sharp peaks in the subsequent analysis. Cells can be stored in this state for up to 12 weeks. Before analysis, cells are pelleted from the ethanol solution at 375 g for five minutes and then washed in PBS. Then the cells are resuspended in RNAse A (Sigma R6513; 50 µg/mL made up in PBS: Note that this molecular biology grade RNAse A does not require boiling before use; most other grades, however, need to be boiled for 10 minutes before use to remove any contaminating DNAse activity and may also need to be used at higher concentrations, that is up to 250 µg/mL) and incubated at 37 °C for 30 minutes, a process which

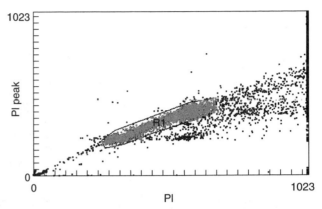

Figure 2. Gating a scatter plot PMT4 int vs. PMT4 peak scatter plot to remove debris and dead cells from subsequent flow cytometric analysis (PMT4 = 620 nm). Events enclosed in R1 will be included in subsequent analysis whereas events to the left of the gate are cell debris and is also where apoptotic cells, if present, will appear. To the right and beneath the gate are cell doublets and aggregates.

digests RNA and leaves DNA intact thereby removing the fluorescent signal induced by interchelating stain binding to the RNA secondary structure. Propidium iodide (PI) from a stock of 1 mg/mL in distilled water is added to a final concentration of 50 µg/mL, and the cells are stained for 15 minutes after which they are analyzed by FC.

For analysis a flow cytometer equipped with an argon laser (excitation wavelength 488 nm) is used. A 620 nm band-pass filter is used to collect red light emission which is plotted as integral signal against peak signal. Gating the linear data from this plot frees subsequent analysis of events due to debris, apoptotic cells, and doublets (Fig. 2). This protocol works well for cells of good viability. If the culture viability is below 90%, then these dead cells need to be removed from the analysis. This is done by incubating cells with DNAse I (Sigma DN-25; 0.5 mg/mL in SMT buffer: 2.42 g/L Tris, 85.6 g/L sucrose, 1.01 g/L MgCl$_2$) for 15 minutes, then thoroughly washing the cells three times in PBS before fixation.

Data obtained are deconvoluted mathematically to give the percentages of cells in the G1/G0, S, and G2/M phases of the cell cycle (Fig. 3). Many programs are now available for this task and provide rapid, consistent, and reliable data analysis, for example, Multicycle (Phoenix Flow Systems, San Diego, CA 92121), which is based upon a polynomial S-phase algorithm with an iterative, nonlinear least square fit.

Measurement of Cell Cycle and Cell-Cycle Phase Duration

Duration of cell-cycle phases and the whole cycle can vary considerably among different cell types. Measurement of cell-cycle phase duration is desirable for kinetic studies of the cell cycle and can also be helpful for phased synchronization protocols especially by methods which rely on S-phase blockade. One method for measuring cell-cycle phase duration relies on incorporating tritiated thymidine (HTdr) and a mitotic block using cholchicine. Then a time course of the number of isotope-incorporated cells is followed. Two experiments are performed to

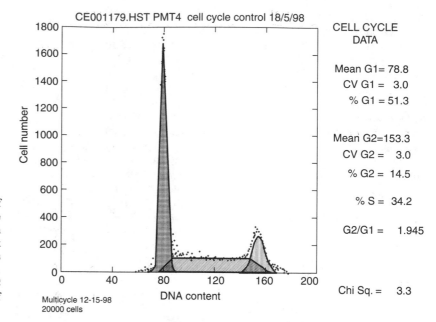

Figure 3. A typical flow cytometric analysis of exponentially growing cells showing cell-cycle phases for 20,000 cells. The x axis of the histogram shows DNA content, so consequently the first peak represents G1 cells, and the second peak represents G2 and mitotic cells. In between are S-phase cells. Data is deconvoluted mathematically by the FC system software and quantifies the percentage of cells in each cell-cycle phase.

estimate the duration of each cell-cycle phase. The first estimates the mean length of G1 and uses continuous HTdr labelling; the second, using a HTdr pulse and chase, estimates the durations of the G2 and S phase. Although this experiment takes a relatively long time, up to four weeks, it is versatile and can even be adapted to *in vivo* use. Additionally, this method allows for accurate determination of cell cycle duration even with quiescent cells present and also allows for reliable quantification of these dormant cells (7,8).

To exponentially growing cells is added 1 µCi/mL [3H]-Tdr (= 6.7 Ci/mM) and 0.01 µg/mL Colcemid, thereby achieving the continuous label experiment. After 30 minutes, half of the cells are removed, washed to remove the labelled thymidine, and resuspended in a Colcemid-containing medium without labeled Tdr so achieving a pulse label. Samples are then taken from each experiment after 30 minutes, 1 hour, $1\frac{1}{2}$ and 2 hours, and then at hourly intervals till seven hours, smeared onto a microscope slide, and fixed with methanol. Autoradiograms are prepared by the dip method using Sukara NR-M2 emulsion and after exposure for three to four weeks are developed with D 19b (Eastman Kodak) for 5 minutes at 18 °C. After fixing, washing, and drying, cells are stained with Giemsa and examined microscopically. Unlabeled cells should have, on average, less than one grain in their nucleus. Cells with more than four grains are deemed positive.

For the continuous label experiment, the percentage of unlabeled cells decreases as the G1 cells progress into the S phase and become labeled. Because the Colcemid block prevents G2 and mitotic cells from entering G1, then the percentage of unlabeled cells decreases only for the duration of G1 after which this percentage remains constant. Plotting the percentage of unlabeled cells against time gives a negative slope which plateaus at time t_p. The time $t = 0$ to t_p is equivalent to the mean G1 duration. The percentage of cells which remains unlabeled

is equivalent to the percentage of noncycling cells within the population.

With the pulse chase experiment, on removal of the HTdr there is initially an increase in the ratio of unlabeled to labeled cells which is then followed by a decrease. The time from removal of the thymidine to the peak ratio is equivalent to the G2 duration, whereas the time from the peak ratio to the time when the ratio equals zero is equivalent to the S phase duration.

Methods for synchronization and partial synchronization of cell cultures can be conveniently divided into two broad groups: those that utilize chemical manipulation of the culture environment and those where separation is based on physical properties of the cells, distinctions which are used in this review. The former group is further subdivided into methods that manipulate the nutrient medium without addition of exogenous cytostasis inducers. These are discussed first followed by discussion of exogenous synchrony inducers. The latter group is arranged according to the cell-cycle phase in which the cells are arrested.

CHEMICAL METHODS

Medium Manipulation

Nutrient Limitation. Synchronization is seen most simply during the late exponential phase and at the peak of a batch culture, and nutrient limitation is the most widely used method for achieving cell-cycle synchronization (9). Nutrient limitation slows and eventually halts cell cycling and division, and the cells arrest in G1. In these nutrient-limited batches, G1 populations of 80% or more can be achieved compared to exponential cultures where populations are typically 50, 40 and 10% G1, S, and G2, respectively (10). Studies of this type with hybridoma and myloma cells suggest that production of antibodies and recombinant protein respectively is greatest during G1. However, as these cells peak and arrest, apoptosis frequently begins (11), and the cells enter a decline phase. Secondary necrosis of apoptotic cells results in passive release of product. Consequently, it is questionable whether true productivity of G1 or passive release of product from apoptotic cells in a population which are G1 arrested is measured. This latter argument is supported by at least two other observations; first, in a range of hybridoma cell lines with internal product concentrations differing by an order of magnitude, secreted product concentration was similar for both lines during the growth phase. However, in the decline phase of the culture, external product concentrations of the cell line with the greatest intracellular concentration was by far the highest (12). Second, with cells which do not undergo traditional apoptosis quite so rapidly, such as most CHO cell lines (11), product concentration remains constant as cells arrest in G1. Productivity here is taken to be associated with growth and the S phase. Clearly though, this approach to studying cell cycle phase and productivity is of limited use because it is not apparent if the results observed are a consequence of cell cycle effects, nutrient limitation, cytostasis induction, or culture background.

Subculture of Nutrient-Limited Cells. Simultaneous release of nutrient-limited G1 arrested cells by subculturing into fresh medium results in a population of cells which begin to cycle from G1 at approximately the same time. The result is synchronized cells moving through all stages of the cell cycle and remaining so over two or three division cycles after which time sychronicity is lost (13). Such cultures are characterized by apparent stepwise increases in cell numbers at approximately 24-hour intervals during early exponential growth, a phenomenon resulting from mainly synchronous mitoses. By frequent sampling, say every 4 to 6 hours, it is possible to follow progression through the cell cycle. Flow cytometric cell-cycle analysis does indeed show that cells are enriched for cell-cycle phases, that these phases vary sequentially as the cells progress through the cell cycle, and also that any synchrony is lost over the three or less rounds of cell division. This type of study suggested that productivity in hybridoma cells is greatest in G1 (12), whereas the S phase is most productive for CHO cells (13,14). Compared to cells synchronized in G1 by simple nutrient limitation, the effects of nutrient limitation and cytostasis induction are removed as possible contributing factors to the observed results. Because the cells are progressing through all cell-cycle stages, it is also possible to assess the relative effects of each stage, as opposed to just G1. However, it must be remembered that these cells are derived from a culture which has been stressed by nutrient limitation, and it has been shown that inoculum quality has greater influence than cell cycle (13,14). Synchronicity is also not absolute and is relatively short lived. Additionally, the results of this nature look at the bulk properties of each cell-cycle phase. Effects due to physiological variation within a cell-cycle phase, for example, cell size, are not normally discerned with this kind of experiment. This may lead to misleading results because cell size seems to be critically

important, as discussed below in the section on centrifugal elutriation (15).

Chemostatic Culture. Though it does not produce truly synchronous cultures, cell cycle enrichment can be achieved by utilizing continuous cultures in chemostats. Here, cell growth rate is influenced by the limiting nutrient concentration and so controlling nutrient feed rates into the chemostat can manipulate cell-cycle duration. It is G1 phase duration which varies when cell cycle duration is changed whereas S and G2 remain fairly constant (16). This technique then can be used to manipulate the proportions of cell-cycle populations giving partial synchronization. Additionally, the principle of using the percentage of the S phase as a correlation of growth rate can be used as an indirect FC method to obtain rapid measurements of cell growth rates with potential for improved control of bioprocessing vessels (12). High dilution rates give greater growth rates and hence short cell-cycle durations. G1 populations in such cultures are at their lowest, typically around 50%, and S-phase populations are at their highest, up to 45% of the cell population is typical. Conversely, dilution rate reduction decreases growth rate by extending G1 and cell-cycle duration, S phase populations are reduced to around 20%, and G1 cultures increase to 70% (17).

Though they yield a large amount of data, it must be remembered that chemostat cultures are run over relatively long periods of time compared to simple batches and such cultures inherently exert considerable environmental pressure on the cells. Thus, guarantees cannot be made that cells at $t > 0$ are the same as $t = 0$ (14). It must also be remembered that the cell-cycle fractions are only enriched. The maximum percentages of G1 and S-phase cells are not as high as can be achieved by other synchronization methods, and G2 cells cannot be enriched above 10% of the total cells. Additionally, heterogeneity within a cell-cycle phase is not taken into account and also, specific growth rate is altered by altering the dilution rate.

Isoleucine Limitation. A widely used method for arresting cells using nutrient limitation for subsequent synchronization is that developed by Tobey and co-workers (18) which utilizes limitation of a single essential amino acid, namely, isoleucine (iℓe). This technique is widely applied to hamster and mouse cell lines but can also be used with permanent human cell lines though viability of the cells after treatment needs careful monitoring because many cell lines, CHO cell being a notable exception (11), rapidly undergo apoptosis when nutrient-limited. Consequently, it is often necessary to determine the duration of isoleucine depletion for each cell type used to gain good viabilities of treated cells. Frequently, isoleucine-free Ham's F12 is used to deprive the cells of iℓe, although other isoleucine-free media can be used if required. Should media supplements, for example Fetal Calf Serum (FCS), be used, it is vital to ensure that they are not a source of iℓe and so should be dialyzed. Dialysis is best done over 4 days at $4\,^{\circ}$C with GE buffer ($20 \times$ GE = 148 g/L NaCl, 5.7 g/L KCl, 5.8 g/L Na$_2$HPO$_4 \cdot$7H$_2$O made in MilliQ or

RO water). Alternatively, funds permitting, predialyzed serum can be purchased. Deprivation for 30–36 hours is usually enough time to arrest cells and allows for up to 95% of CHO and murine cells to be accumulated at the G1 restriction point. It has been reported that synchrony for human cell lines is less tight, and up to 10% of cells are present in S phase, though better reversibility synchrony and viability can be achieved by supplementing isoleucine-free F12 with deoxycytosine, deoxyguanosine, deoxyadenosine, and thymidine, each at 5 µM (19). Despite excellent G1 synchrony and high viability and reversibility in CHO cells, problems may be encountered by unbalanced growth which is seen as enlargement of the cytoplasm but not the nucleus of the cells and which may have a profound influence on studies of cells synchronized by these means (13).

Exogenous Cytostasis Inducers

Addition of cytostatic agents can arrest cells at different specific cell-cycle stages and followed by release from that block can produce synchronised cultures in G1 (mimosine, HTDCT and DMSO and following mitotic arrest with nitrous oxide, Colcemid, and nocodazole), S phase (thymidine, amethopterin, hydroxyurea, and aphidicolin), or G2 (roscovitine, butyrolactone I, and DNA topoisomerase). To achieve tight S and G2 synchrony, cells are normally subjected to sequential arrest and release at progressively phased stages of the cell cycle.

G1 Arresters

DMSO. Recently, Ponzio et al. (20) reported that low levels of DMSO (1.5%) can effectively arrest B cells in early G1, approximately 12 hours before the S phase. The effect, which is both reversible and essentially nontoxic for 15 hours of treatment, gave more than 90% G1 cells when the cells were washed and subcultured in fresh medium. Arrest was attributed to inhibition of cyclin D2 neosynthesis which decreases cyclin D2/CDK4 complex formation allowing redistribution of p27$^{(KIP1)}$ from cyclin D2/CDK4 to cyclin E/CDK2 complexes. Additionally, simultaneous accumulation of p27$^{(CIP1)}$ entails increasing association with cyclin D3/CDK4 and cyclin E/CDK2. Thus p27$^{(CIP1)}$ p27$^{(CIP1)}$ act together to inhibit cyclin E/CDK2 activity which, together with CDK4 inactivation, produces G1 arrest. Recent, as yet unpublished work in our lab has, however, shown that DMSO is not suitable as a synchronization/cytomodulation agent for all cell types. For example, when CHO cells were arrested by up to 1% DMSO there is not an accumulation in any stage of the cell cycle.

L-Mimosine and HTDCT. L-Mimosine, a plant amino acid, and the Hoechst compound 768159 (HTDCT) can be used to arrest cells in late G1, approximately 15 min to 2 hours before the onset of S phase. Both of these compounds are believed to inhibit post-translational modification of lysine to form the rare amino acid hypusine which affects activity of the protein initiation factor eIF-5A (19). Typical concentrations used to arrest cells, both transformed and nontransformed are between 200 and 400 µM. For different cell types, the concentration which causes reversible arrest of the cells will need to be determined. Arrest is reversed simply by washing the cells and resuspending them in fresh medium.

S-Phase Arresters

In general inhibitors of DNA synthesis will arrest cells in S phase and are commonly inhibitors of deoxyribonucleotide triphosphate synthesis such as thymidine, amethopterin (methotrexate) (21), or hydroxyurea (22), or are direct inhibitors of DNA polymerase such as aphidicolin.

Thymidine. In the early 1960s, excess thymidine was first widely accepted as a method for reliably synchronizing cells in S phase as is still commonly used today. Thymidine is rapidly taken up by the cells and is converted to dTTP which is an allosteric inhibitor of ribonucleotide reductase. Consequently high levels of dTTP result in reduced conversion of all four ribonucleotide diphosphates (NDP) to the corresponding deoxyribonucleotide diphosphates and so halts DNA synthesis. Typically, addition of 2 mM thymidine for 17 hours can be used to reversibly arrest cells resulting in cells in phases other than S progressing through the cell cycle and arresting at the G1/S boundary, whereas cells already in S when the block was imposed remain at this stage. Consequently when the block is reversed, cells can be out of synchrony by the length of S phase, typically around 9 hours (19). A double thymidine block increases the degree of synchrony by arresting the cells for 17 hours with 2 mM thymidine, washing them and allowing growth for 9 hours, and then blocking them again with thymidine for 15 hours (22). The rationale here is that imposition of the first thymidine block allows cells in G2, mitosis, and G1 to progress through to the G1/S boundary. Cells already in S will be arrested immediately. Consequently when the block is removed, there can be up to 9 hours difference between the cells. Removing the cells from the block for a duration longer than the S phase allows all of the cells to pass through S phase so that they are all arrested at the G1/S boundary when they are blocked for the second time.

Amethopterin. Amethopterin is a 4-amino analogue of folic acid and a potent inhibitor of dihydrofolate reductase, an enzyme that catalyzes the reduction of folic acid and dihydrofolic acid to tetrahydrofolic acid, which is directly involved in metabolism of the amino acids glycine and methionine, carbons 2 and 8 of the purine ring, the methyl group of thymidine, and indirectly in choline and histidine synthesis (22). Directed blockade of DNA synthesis occurs because of a reduction in the pool of methylated thymidine, whereas other actions are bypassed by supplementation of hypoxanthine or adenosine and glycine to the medium and addition of thymidine can reverse the effects of amethopterin. The medium is supplemented with glycine and adenosine or hypoxanthine to final concentrations of 100, 200, and 30 µM, respectively, and amethopterin is used at a final concentration of 10 µg/mL. Cells are allowed to accumulate in the S phase and after a period of time approximately equal to the duration of G2, M and G1 phases, that is typically around 16 hours but depends on cell type, approximately 90% of the cells are accumulated in the S phase. Amethopterin blockade is reversed by adding thymidine to a final concentration of 5 µg/mL and will result in cells cycling from S phase showing a burst of mitotic activity after 6 to 10 hours depending on the G2 duration of the cell line in question. As mentioned before for thymidine block, blockade of DNA synthesis results in blocking a large proportion of cells at the start of S phase. As they progress from through G2, M, and G1 phases, the remainder will be blocked where they were in S phase. Consequently there may be up to 6 to 9 hours ahead of cells blocked at the G1/S boundary. This problem can be particularly acute for cells with short doubling times (10–12 hours) where up to 60% of the cells from an asynchronous culture can be in this 6-hour window at any one time (22). Again, this can be overcome by using a double-block protocol where the block is released then reapplied once all of the cells have passed through the S phase. Both of these sequential blocks can be of the same type, for example, double thymidine block or alternatively of different types, for example, thymidine block followed by amethopterin block. Alternatively, preliminary phasing of the cells, for example, subculturing predominantly G1 cells from a nutrient-limited medium can increase the degree and tightness of synchrony.

Hydroxyurea. Hydroxyurea also inhibits DNA synthesis through action on ribonucleotide reductase but instead inhibits purine nucleoside diphosphates. Hydroxyurea 2 mM is usually sufficient to arrest the cells which normally remain viable for at least 16 hours, normally long enough for the majority of the cells to pass through the cell cycle and accumulate at the G1/S phase boundary. Cells are easily released from the block by washing and resuspending them in fresh medium (22).

Although these synchronization methods are relatively straightforward, are easily applied to most cell types, and especially for suspension cultures are easy to scale up, problems may be encountered because of directed inhibition of DNA synthesis while allowing cellular actions not associated with DNA synthesis to continue. A most notable example is unbalanced growth where cells are arrested at the G1/S boundary in terms of DNA synthesis while other cellular aspects may reflect G2 or even G1 events. As such, this may have significant implications for experiments using cells synchronized by the methods described.

G2 Arresters

Traditionally, G2 is seen as the most difficult phase in which to collect cells. This frequently has been done by reversing a double thymidine block and then collecting cells 6 to 8 hours later (22). More recently, Leno et al. (23) reported how the yield of G2 cells in adherent cultures can be improved by adding of nocodazole (0.04 µg/mL) after release from the second thymidine block. Addition of this microtubule antagonist induces arrest of cells at mitosis and so the most rapidly cycling cells, which would normally undergo mitosis and pass into G1, thereby contaminating the G2 cells, are arrested in mitosis and can be easily removed by virtue of the reduced adherence of mitotic cells.

Interference with cell cycle regulators, most notably cyclin-dependent kinases is now seen as an effective way to reversibly arrest cells in G2. Recently, butyrolactone I and isomers of roscovitine and have been used to arrest cells in G2 by selective inhibition of p34cdc2/cyclin B, p33cdk2/cyclin A, p33cdk2/cyclin E, and p33cdk2/p35kinases although they do not significantly affect the activity of other protein kinases such as *erk1* and *erk2* (24). Both of these compounds should be used from concentrated stocks (10 mg/mL in DMSO) at a final concentration in the region of 20 µM dependent on cell line. Antimitotic activity of these compounds results in accumulation of cells at the G2/M boundary.

Topoisomerase II Inhibitors. p34cdc2/cyclin B kinase can be inhibited by topoisomerase II inhibitors (19). Mechanistically, these inhibitors produce covalent complexes between topoisomerase and DNA at the sites of DNA strand breaks. The levels of topoisomerase inhibitors required to achieve efficient and reversible arrest depends on cell type, and care is needed to determine the most effective concentration for arrest while ensuring maximum viability and minimal chromosomal damage, the latter inevitable with topoisomerase II inhibitor exposure. A frequently used topoisomerase inhibitor is the Hoechst drug 33342 which is typically used at a final concentration of 1.3×10^{-5} M for CHO cells or 1.3×10^{-7} M for human fibroblasts. A commonly used topoisomerase II synchronization protocol does so following progressive G1 and S-phase arrests. For example, CHO cells are transferred to a suitable isoleucine-free medium with any supplements required for the cell line in question. If used, FCS should be dialyzed to eliminate this as an isoleucine source. After 36 hours the majority of cells should be arrested in mid-G1. The cells are then resynchronized in S phase by washing and replacing them in a medium containing 10^{-3} M hydroxyurea for 10 hours. When replaced with a medium containing 1.3×10^{-5} M Hoechst 33342, the cells begin to accumulate in G2 after about 8 hours. Such a protocol can give G2 populations in excess of 80–90%, and by careful manipulation of drug concentration and exposure duration, it is possible to generate cells in early mid-or late G2 (19).

Nitrous Oxide. Nitrous oxide can be used to arrest cells in mitosis. Release from this block provides cycling cells which rapidly enter G1 as a highly synchronous population (26). Using a system for automatic delivery of nitrous oxide at around midnight on the day before synchronized cells are required allows synchronous cells to be available for experimentation in the morning (19). However, nitrous oxide is narcotic and can be explosive, and for safety reasons, it is not desirable to use gas straight from a high pressure main cylinder. So an intermediate pressure cylinder is used which is charged from the main cylinder by hand. This intermediate cylinder contains enough gas to charge the pressurized culture chamber vessel to 80 psi. Thus, should a leak develop, dangerous quantities of gas will not be released. Between the intermediate pressure vessel and the culture chamber is a timer and solenoid valve to facilitate automatic charging of the culture chamber.

A nitrous oxide block can be preceded by a thymidine block (2.5 mM) the day before. After 24 hours, the thymidine is removed from the culture medium by washing the cells twice and resuspending them in fresh medium (19). The cells in dishes or T-flasks are then transferred to the prewarmed culture chamber, and the lid is bolted down. One-Twentieth of the chamber's volume of CO_2 is injected into the chamber to maintain the pH. The timer is set to allow sufficient time for the S-phase cells to pass into the late S phase or very early G2. Then the valve opens and allows the culture chamber to charge to with nitrous oxide to 80 psi. After 8 to 10 hours, the chamber can be opened. This must be done slowly over about 15 minutes to avoid cell damage caused by production of microbubbles as the solubility of the gas decreases at lower pressure.

This method provides more than 90% of cells in G1 and can also be used to generate large numbers of synchronized cells. Although specialized equipment is required, it is entirely possible to have this constructed in-house to a user's specific requirements.

PHYSICAL METHODS

Centrifugal Elutriation

With all experiments using chemical methods of synchronization, it can be argued that it is not cell-cycle phase that is being investigated but the effects of synchrony induction. An example of this is unbalanced growth cause by some nutrient limitation or DNA synthesis blockade protocols. Also, bulk properties of cell populations are investigated. For example, all G1 cells are assumed to be of similar size when there is considerable evidence to show

that in reality this is not the case (15). Synchronization of cells by physical methods overcomes these problems and centrifugal elutriation (CE) is probably the best method of physical separation. CE separates cells based on size and hence cell cycle because G2 generally are larger than S which are in turn generally larger than G1. Separated cells can then be cultured further as highly synchronized cell populations with high viabilities (approximately 100%).

CE utilizes centrifugation in an elutriation rotor which provides G1 cell fractions at high purity (95–100%) and highly enriched fractions (up to 80%) of other cell cycle phases (15). Mechanistically, isotonic buffer, normally either PBS or growth medium, is pumped through the elutriation rotor, thereby creating a force on the cells which opposes the sedimentation force created by centrifugation. When the force of fluid flow becomes greater than that of sedimentation, then cells are eluted from the rotor. With incremental flow increases, it is possible to elute fractions of cells, the smallest first, dead cells elute before the first fraction of viable cells. Hence, it is possible to obtain a highly pure fraction of G1 cells and greatly concentrated fractions of other cells with minor G1 contamination, as demonstrated by flow cytometric analysis of DNA content. Typically, G1 fractions can be well over 90%, and S and G2 cells can be between 40 and 80% depending on cell type. Although these latter values may appear low, they are comparable or better than enrichments seen with other S and G2 synchronization methods but the advantage is that all of the cells are viable and have not been subjected to nutrient medium manipulation or cytomodulator addition.

The elutriation system, shown diagramatically in Figure 4, is assembled and, after washing, is sterilized with 70% ethanol and then rinsed with sterile elutriation buffer. It is essential to ensure that the system is free of bubbles and remains so. Typically for CHO, myloma, and hybridoma cells, the rotor speed is adjusted to 366 g at

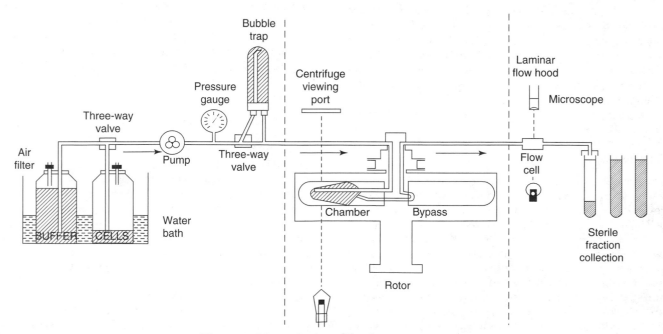

Figure 4. Schematic view of the elutriation system.

the elutriation boundary, and the flow rate is adjusted to less than 11 mL/min. For other cell types, however, these conditions may need to be adjusted for efficient elutriation. A cell suspension is transferred into the loading chamber using a 20 mL syringe, and the diverter valve is opened to load the cells into the chamber of the elutriation rotor. Cells are collected differentially either by incremental increases of buffer flow rate through the rotor or by decreases in rotor speed. For simplicity and practicality, flow rate manipulation appears to be the favored method, and usually 50 mL fractions are collected and 50 mL is diverted to waste before the next increment. The actual increments will be determined by the number of cells in each phase of the cell cycle. Normally, smaller increments are possible at lower flow rates when G1 and S phase cells are eluted, whereas larger increments are desirable to provide fractions containing useful numbers of G2 cells. For CHO, NSO, and hybridoma cells, dead cells are eluted at around 13–14 mL/min, and G1 cells are eluted at around 20 mL/min. The largest single cells, mostly G2, are eluted at around 80 mL/min or so.

Maintaining cells for short culture periods (2 hours) following elutriation, Lloyd et al. (15) revealed that specific productivity of cells commonly used for recombinant protein production is related to cell size and only indirectly to cell cycle (because the mean cell size of G2 cells is greater than that of S-phase cells which is greater than the G1 mean cell size). This size correlation appears to hold for all industrially important cell lines tested to date (CHO, NSO, and hybridoma) and is completely independent of promoter type (SV40 early, CMV, endogenous immunoglobulin) and product (tPA, chimeric monoclonal antibody, interferon gamma, and IgG). This study also revealed the presence of an important subpopulation of cells, less than 1% of the total cells, which have G1 DNA content but which are abnormally large, over twice the size of normal G1 cells and larger than G2 cells. Being the largest cells in the population, they had the highest specific productivity. Currently, these cells are under investigation to enhance productivity by increasing the percentage of these large cells within the population.

Relatively large numbers of cells (2×10^8) can be rapidly processed in an elutriator producing between 10 and 16 synchronized cell fractions representative of all stages of the cell cycle and containing enough cell, cycle, enriched cells of high viability (98–100%) for subsequent experimentation. This technique, however, is limited by its requirement for single cells in suspension. Because many industrially important cell lines have been adapted to suspension culture, this does not represent a problem. Adherent cells can be elutriated only if they have been detached from their support, for example, by trypsinisation, though this additional process may present some form of artifact within the results. Cell aggregation may also be problematic because such aggregates will be retained in the rotor, thus reducing cell yields in each fraction.

Zone Sedimentation

Cells can also be separated by size by zonal sedimentation. Here gravity or centrifugation opposes a force created by a concentration gradient of either Ficoll, serum, or sucrose (22,27). Separation is not as good as for CE, though like CE, this method does give relatively pure G1 cells and large numbers of cells (10^8 to 10^9) can be processed. Additionally, there is no requirement for the specialized equipment needed for CE. Schindler and Schear's method (27) provides for rapid processing by implementing centrifugation as opposed to gravity settling and is the method described now.

A linear sucrose gradient is created by mixing a medium lacking sodium chloride but containing 95 mg/mL sucrose with a medium containing 6.4 mg/mL NaCl and 19 mg/mL sucrose. Thus, the medium remains isotonic throughout the gradient and, barring NaCl, contains all medium constituents at optimal concentrations. Exponentially growing cells, between 10^8 and 10^9, are pelleted from their culture medium and resuspended in a small volume of between 5 and 15 mL of fresh medium. For good sedimentation, it is essential to avoid a sharp interface between the top of the gradient and the cell suspension. This is achieved as follows. The cell suspension is pumped into the bottom of the centrifuge tube (200 mm × 10 to 45 mm) at the same rate as the low density medium. Thus the cells are gradually diluted by the top portion of the gradient. The linear gradient is then produced by gradually mixing the low- and high-density media and again pumping the medium into the bottom of the centrifuge tube which lifts the cell suspension to the top. The tube is then centrifuged for 5 minutes at 80 and 400 g at the top and bottom of the tube, respectively. Turbulence in the gradient should be reduced by ensuring that acceleration and deceleration are gradual. After centrifugation, larger cells are pelleted at the bottom of the test tube, whereas increasingly smaller cells are distributed throughout the lower half of the gradient. G1 cells are uppermost. These slower sedimenting cells are collected, centrifuged, and resuspended in a sucrose-free medium at an appropriate cell density.

Similar to CE, this method is rapid, has high throughput potential and produces cells at very high viability, but, unlike CE, is suitable only for collecting G1 cells. Zonal sedimentation, however, has the advantage that specialized equipment is not necessary. Evidence suggests that better synchrony is achieved for cells grown in suspensions as opposed to those grown in nonagitated or loosely adhered cultures (27).

Mitotic Detachment

Adherent cells undergoing mitosis frequently round up and for many cells become less adherent or even unattached. Mitotic detachment, that is, gentle agitation of adherent cultures in exponential growth to remove these loosely attached cells, has provided highly synchronous (>90%) cultures of mitotic cells (28). Practically, synchronization by mitotic detachment is limited by the low numbers (approximately 2%) of mitotic cells within a population, so pooling of cryopreserved cells from frequent mitotic detachments over several hours can produce larger numbers of mitotic cells for study, but this drastically increases the timescale of what initially appears to be a rapid and straightforward protocol. Inhibitors of spindle formation, such as Colcemid or vinblastine sulphate, that arrest cells in mitosis can increase the percentage of mitotic cells within the population up to about 15% of the total cell number. Synchrony of cells is generally good, typically in the region of 95%, but depends on many factors, including cell type and detachment conditions.

Gaffney (28) describes an effective method for mitotic detachment of A244 (human annion) cells as follows: to exponentially growing cells in 75-mL T-flasks containing 30 mL of medium, Colcemid is added to a final concentration of 0.01 µg/mL. After 4 hours, this medium is replaced with 20 mL of Hank's balanced salts solution (HBSS) diluted 1 to 4 with MilliQ water. Cultures are then gently shaken for 90 seconds, and the hypotonic medium is transferred to a conical centrifuge tube containing 10 mL of the Colcemid-containing medium and 10 mL of HBSS at 1.8 times its normal strength. Cells are pelleted at 600 g for 5 minutes and resuspended in prewarmed medium.

Yield and synchronicity are affected by many factors, including sensitivity of cells to Colcemid and relative adherence of interphase and metaphase cells. Colcemid concentration will need to be optimized on a cell-line-specific basis to find the minimum concentration and exposure time to produce reversible mitotic arrest. In general, lower concentrations are better whereas higher concentrations and exposures longer than about 4 hours result in increased aberrant mitoses. The osmotic strength of the buffer salts is also critical. Decreased osmotic potential increases yield by reducing adherence of the cells but can also decrease synchronicity by reducing adherence of interphase, as well as metaphase cells.

Clearly mitotic detachment is useful only for adherent cell lines. This limits applicability because many bioprocessing cell lines have been selected or adapted to anchorage-independent growth so that they can be grown as suspension cultures within a fermentation vessel. Such anchorage-independent cells can be encouraged to adhere, for example, by serum addition, but then the cells are not grown in conditions comparable to those experienced by cells in the bioprocess medium.

Electronic Separation

Synchronized cultures are not necessarily required for the study of cell-cycle-related events, for example,

correlation of gene expression and cell-cycle phase. Dual fluorescent staining of nucleic acid and products of interest followed by analysis by flow cytometry allow analysis of heterogeneous populations, and then specific, cell-cycle, phase populations can separated and analyzed electronically (6). Typically, cells are washed twice in PBS and then fixed for 20 minutes in paraformaldehyde (1% w/v in PBS; this fixative can be stored at 4 °C for no more than 5 days) at 4 °C. After fixing with PFA, it is important that all solutions contain a blocking agent such as 1% BSA to prevent high-background fluorescence caused by nonspecific binding of labeled probes. The fixed cells are then washed twice in blocking buffer (1% BSA w/v in PBS) and permeabilized with 0.1% (w/v) saponin (Sigma S-2149; from a 2% w/v stock in PBS). To enable longer term storage, PFA-fixed cells are permeabilized and fixed with 70% ethanol (−20 °C), and stored at this temperature for up to 12 weeks. Dual staining will incorporate a nucleic acid specific fluorescent dye, such as propidium iodide, while co-staining with a fluorescent tagged (frequently FITC fluorescein isothiocyanate) antibody to the product of interest. Cells are stained with fluorescent tagged antibody, usually at 1/200 dilution in blocking buffer containing 50 µg/mL RNAse A, though exact dilution depends on the antibody preparation used. Staining will take approximately 45 to 60 minutes, and propidium iodide (PI) should be added to a final concentration of 50 µg/mL for the final 15 minutes.

Data plotted as PI integral against PI integral peak fluorescence gated, as described earlier, will allow debris, cell doublets, and apoptotic and dead cells to be removed from the analysis. Events within this gate can be plotted FITC log or FITC log/Forward scatter ratio (giving approximations of content or concentration of target product) against PI fluorescence, and three analysis gates can then be produced along the y axis which correspond to G1, S, and G2 DNA contents (Fig. 5). Additionally, dual staining using an antibody labeled with a different fluorochrome can yield data regarding cell-cycle-specific expression of two-proteins simultaneously.

These same staining protocols can be used for laser-scanning cytometric analysis, which combines the technologies of flow cytometry and image analysis. Cells are fixed to a slide and stained as for flow cytometric

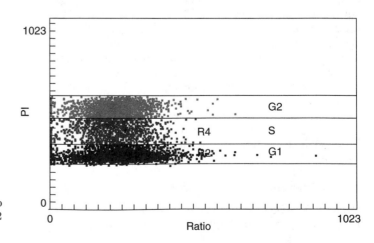

Figure 5. Gating of FITC log/Forward scatter signal vs. PI to determine target protein concentration for G1/G0, S, and G2 cells.

methods described. The slide is then scanned and analyzed digitally. Having cells immobilized on the slide presents a significant advantage over flow-based technologies in that specific events can be revisualized or even restained and reanalysed.

CONCLUSIONS

Although widely accepted as a means of studying cell-cycle, phase-related effects, chemically induced synchronization has a major drawback in that effects of cell-cycle phase and induction of synchronicity are difficult to resolve. Results obtained may merely be an artifact of the synchronization methods used. To this end, methods which do not induce cytostasis, for example, continuous cultures, are likely to yield more reliable data. It must be remembered though that chemostat cultures are usually run over long periods compared to batch cultures and exert considerable selective pressures on the culture which may enrich minor cell populations or mutations. Hence culture history may come into question in this kind of experiment.

Centrifugal elutriation which separates cells based on physical properties seems an ideal alternative to chemically induced synchronization and in our opinion is the method of choice. Here artifactual data resulting from cytostatic induction or culture history are virtually eliminated. The only minor problem of this technology is the requirement for specialized equipment.

Using flow cytometry to analyze a heterogenous population and then to "segregate" cell-cycle populations electronically is an excellent way to study relationships between intracellular proteins and the cell cycle and eliminates the need for synchronization. Additionally, large numbers of cells can be analyzed to provide statistically validated data and multiple stains used.

Synchronization by all of these methods can provide data relating cell cycle and productivity. It is important, however, that cause and effect are established as a reliable relationship. Other parameters, including cell size, also vary with cell cycle, and this effect can be important indeed.

BIBLIOGRAPHY

1. I. Reed, in C. Hutchinson and D.M. Glover, eds., *Cell Cycle Control*, IRL Press, Oxford, U.K., 1995, pp. 40–62.
2. L. Hartwell, in C. Hutchinson and D.M. Glover, eds., *Cell Cycle Control*, IRL Press, Oxford, U.K., 1995, pp. 1–15.
3. J. Pines and T. Hunter, in C. Hutchinson and D.M. Glover, eds., *Cell Cycle Control*, IRL Press, Oxford, U.K., 1995, pp. 144–176.
4. G. Basi and G. Draetta, in C. Hutchinson and D.M. Glover, eds., *Cell Cycle Control*, IRL Press, Oxford, U.K., 1995, pp. 106–143.
5. J.M. Mitchson, in P.C.L. John, ed., *The Cell*, Cambridge University Press, Cambridge, U.K., 1981, pp. 1–15.
6. M. Al-Rubeai, in N. Jenkins, ed., *Animal Cell Biotechnology: Methods in Biotechnology*, Vol. 8, Humana Press, Totawa, N.J., 1999, pp. 145–153.
7. T. Maekawa and J. Tsuchiya, *Exp. Cell Res.* **53**, 55–56 (1968).
8. W. Miller, A. Brulfert, and G.E. Kaufman, *Environ. Exp. Bot.* **18**, 1–8 (1978).
9. S. Cooper, *J. Theor. Biol.* **135**, 393–400 (1988).
10. N.H. Simpson, A.N. Milner, and M. Al-Rubeai, *Biotechnol. Bioeng.* **54**, 1–16 (1997).
11. R.P. Singh, M. Al-Rubeai, C.D. Gregory, and A.N. Emery, *Biotechnol. Bioeng.* **44**, 720–726 (1991).
12. M. Al-Rubeai, A.N. Emery, S. Chalder, and D.C. Jan, *Cytotechnology* **9**, 85–97 (1992).
13. D.R. Lloyd, V. Leelavataramas, A.N. Emery, and M. Al-Rubeai, *Cytotechnology* **30**, 49–57 (1999).
14. V. Leelavatcharamas, Ph.D. Thesis, University of Birmingham, 1997.
15. D.R. Lloyd, P. Holmes, L.P. Jackson, and M. Al-Rubeai, *Bio. Cytotechnology* (submitted).
16. V. Leelavatcharamas, A.N. Emery, and M. Al-Rubeai, *Cytotechnology* **30**, 59–69 (1999).
17. N.H. Simpson, R.P. Singh, A.N. Emery, and M. Al-Rubeai, *Biotechnol. Bioeng.* **64**, 174–186 (1999).
18. R.A. Tobey, *Methods Cell Biol.* **6**, 67–112 (1973).
19. T. Johnson, C.S. Downes, and R.E. Meyn, in P. Fantes and R. Brookes, eds., *The Cell Cycle: A Practical Approach*, IRL Press, Oxford, U.K., 1993, pp. 1–24.
20. N. Ponzio et al., *Oncogene* **17**, 1159–1166 (1998).
21. E. Stubblefield, *Methods Cell Physiol.* **3**, 25–43 (1968).
22. L.P. Adams, in R.H. Burdon and P.H. Van Kippenburg, eds., *Laboratory Techniques in Biochemistry and Molecular Biology*, Vol. 8; Elsevier, Amsterdam, 1990, pp. 211–239.
23. C. Leno, S. Downes, and R.A. Laskey, *Cell* **69**, 151–158 (1992).
24. Y. Furunawa et al., *J. Biol. Chem.* **271**, 28469–28477 (1996).
25. K. Nishio et al., *Anticancer Res.* **16**, 3387–3393 (1996).
26. P.N. Rao, *Science* **160**, 774–775 (1970).
27. R. Schindler and J.C. Schear, *Methods Cell Biol.* **6**, 43–65 (1973).
28. E.V. Gaffney, *Methods Cell Biol.* **9**, 71–84 (1975).

See also CELL AND CELL LINE CHARACTERIZATION; CELL CYCLE EVENTS AND CELL CYCLE-DEPENDENT PROCESSES.

CELL DETACHMENT

OTTO-WILHELM MERTEN
Généthon II
Evry Cedex
France

OUTLINE

Introduction

Use of Enzymes for Cell Detachment

    Trypsin (EC 3.4.4.4.)

    Other Proteolytic Enzymes

    Dispase (EC 3.4.24.4.)

    Collagenase (Clostridiopeptidase, EC 3.4.4.19)

    Accutase

    Reattachement of Detached Cells

INTRODUCTION

Cell detachment is an essential process step for the subcultivation of adherent cells. In contrast to suspension cells, for which the subcultivation can be performed by simple dilution of the cells, anchorage-dependent cells have to be detached and prepared for the subsequent culture. Detachment becomes necessary when all the available substratum is occupied, when the cell concentration exceeds the capacity of the medium, or when an amplification of the culture has to be performed. Although cell detachment can also be understood as cell detachment from other cells (= disposal of aggregates or clumps), this article will emphasize the detachment of anchorage-dependent cells from growth supports for subsequent cultivation.

Although cell detachment is not a reversion of cell attachment (1), cell adhesion will be presented very briefly in the following. Cell adhesion is a dynamic process, leading to an attached and finally spread cell. The rate of cell adhesion depends on the rate of initial contact of cells with the substratum and the subsequent formation of the attachment bonds. Finally an equilibrium is reached, and the cells attain a certain morphology and strength of attachment that depends on the number and organization of the attachment bonds and cytoskeletal structures.

The precise mechanisms of cell attachment will not be discussed here in detail (see, for instance, Refs. 2–5). Cells possess a negative surface charge, which is unevenly distributed and varies with the physiological state of the cells (2). Although cells show a negative charge, they can be cultivated on positively as well as on negatively charged surfaces (polystyrene and microcarriers) (2,6,7). Maroudas (6) pointed out that it is not the charge of the support that governs cell attachment, but the density of the charges on the culture surface. In addition, the surface has to be hydrophilic (2).

In principle, the attachment of cells to surfaces can be due to electrostatic forces when cells and the substratum are oppositely charged (valid for some microcarriers). In most cases, cells and surfaces have the same charge, indicating that the contact will be mediated through ionic interactions (via divalent cations; Ca^{2+}, for instance) or through protein bridges (2,6). This latter interaction between cells and culture surface seems to be the most important one. These proteins, sometimes called substrate adhesion molecules, are essential for adhesion and spreading of the cells. They interact on the cell side with their receptors (focal contact and cell junction molecules), and on the other side the adhesion with the substratum involves one or more extracellular molecules (8). Substrate adhesion molecules include as many as 30 molecules, among them collagen, laminin, fibronectin, and vitronectin (5). It was recognized that these substrate adhesion molecules are partially provided by the serum used in the culture media, affecting cell differentiation and proliferation (e.g., Ref. 2). The other part is synthesized by the cells and deposited as extracellular matrix components on the substratum (9).

As already mentioned, cell detachment is not an inversion of cell adhesion. Cell detachment involves bonds of attachment after they have already been formed and depends on their accessibility and sensitivity to be broken. It is thus independent of the contact interaction. Weiss (10) stated that cell separation takes place within the cell surface rather than between the cell surface and the substratum. In addition, studies on the mechanisms of cell adhesion and cell detachment have revealed that several reagents, like sulfhydryl-binding reagents and cytochalasins, inhibit cell adhesion but do not cause cell detachment. However, other substances, like local anesthetics and trypsin, inhibit cell adhesion and cause cell separation (2). These facts indicate that cell detachment is not the reverse of cell attachment and that cell attachment and cell detachment do not occur at the same site.

This article will present different methods of cell detachment and their principles (enzyme-based cell detachment, nonenzymic cell detachment). Cell detachment and the subsequent cultivation of detached cells will be presented with special emphasis on classical methods as well as on new concepts. This will be followed by a comparison of cell detachment in serum-containing and serum-free media. The article will conclude with a discussion on safety issues.

USE OF ENZYMES FOR CELL DETACHMENT

Trypsin (EC 3.4.4.4.)

In principle, various proteolytic enzymes can be used for the detachment of adherent cells; however, in most cases trypsin is used, mainly due to the fact that after its use it can be easily inactivated by serum. The trypsinization procedure was standardized by Litwin (11) for the detachment and subcultivation of human diploid fibroblasts because of their improvement for viral vaccine production for human use.

Mechanism and Effects on Cells. Trypsin is a serine protease cleaving peptide bonds between lysine or arginine and an unspecific amino acid. Trypsin as well as EDTA treatments lead to cell rounding just as in mitosis (e.g., Refs. 12–15). Although the cells are round after a certain time, points of cell–substrate adhesion become distal tips of long cell processes that are connected to the rounded cell body (16). During this rounding process the position and integrity of the adhesion sites is maintained (Fig. 1) (15). Final detachment occurs either through a release of these processes from the substrate and their integration into the cell body or when the cellular detachments of the processes are severed (12). Following mechanical displacement of trypsinized cells, the substrate surface is covered by remnants of retraction fibers (15) and small padlike structures (12).

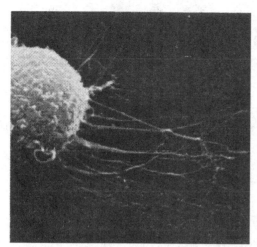

Figure 1. BHK21 C13 fibroblast after 5 min incubation in trypsin. The cell has rounded up and the surface is covered with blebs. Retraction fibers extend radially from the ventrolateral region; they are closely applied to the substrate at their distal ends. 11000x (from Ref. 15, with permission; © The Company of Biologists Ltd.).

These references indicate clearly that trypsin does not detach cells from surfaces but leads to cell rounding (15). This fact suggests that trypsin initially acts on the cell cytoskeleton as well as on the surface components of a membrane–cytoskeletal complex and not on the bonds of attachments (12,16–18). Badley et al. (19) could establish that trypsin leads to a dispersion of stress fibers in advance of shape changes. Fibers higher in the cell and terminating at the cell edge are more sensitive than most basal fibers, and during disintegration, all types of fiber go through an intermediate "beaded" structure. The sequences of events are presented in Table 1. As cells retract, membranes become studded with many blebs and microvilli. Contraction and rounding of the cell results

Table 1. The Sequences of Events During Trypsin and EGTA Detachment of BHK Fibroblasts (earlier events are listed at the top)

Trypsin	EGTA
Upper stress fibers give beaded appearance	Upper and lower stress fibers break up and disappear
Upper stress fibers disappear and lower stress fibers begin to break up to give beaded appearance	—
Lower stress fibers disappear	—
Microtubules and 10-nm filaments draw back; some break up; cell margins draw back	Microtubules and 10-nm filaments draw back, some break up; cell margins draw back
Cell rounds	Cell rounds
Cell detaches	Cell detaches

Source: From Ref. 19, with permission; © The Company of Biologists Ltd.

in numerous filamentous processes stretched between substrate and cell body. These processes are gradually pulled into the rounded cell body [Fig. 2(a,b) (16)], as observed for the human embryo fibroblast cell line HLM18.

The rounded cells are only loosely attached and finally can easily be detached by mechanical means. Using HeLa and L cells, Lamb and Ogden (20) could show that during the rounding up phase cell detachment by trypsin leads to a transient increase in membrane permeability, indicating that the detachment of the "feet" (processes) holding the cells onto the substrate leads to a transient increase in leakiness. Similar facts were described for BHK-21 cells by Whur et al. (15). However, the modification of the membrane permeability during trypsinization can be influenced by the presence of divalent cations. Whereas Ca^{2+} acts mainly on cell to dish attachment, Mg^{2+}-dependent interactions are most important in preserving the impermeability of the cell membrane in nontransformed and SV40-transformed Balb/c 3T3 mouse cells (21). This indicates that Ca^{2+} should be complexed during cell detachment (e.g., by the presence of EGTA [ethyleneglycolbis-(β-amino-ethyl ether)-N,N'-tetra-acetic acid]), whereas the presence of Mg^{2+} ions is desirable.

Vogel (16) observed that trypsinization leads to a release of a large quantity of glycopeptide, hyaluronic acid, and sulfated glycosaminoglycans into the medium [Fig. 2(c)]. In addition, trypsinization leads to an important release (about 39%) of the total sialic acid of the cell (observed by Snow and Allen (22) for BHK-21 cells).

In general, prolonged exposure to trypsin progressively damages cells. In addition to surface damage to cells, trypsin produces internal damage, such as degradation of polyribosomes (23). By using HeLa and CBM17 mouse kidney cells, Hodges et al. (24) documented the variety of cellular alterations that trypsin can produce and have shown that labeled trypsin can be found within the cytoplasm, nucleus, and the nucleolus. However, these damages are not irreversible. Brugmans et al. (25) could show that trypsinized cells carry over trypsin and that a fraction of this remaining trypsin was immunologically indentifiable in the cells one day after cell detachment. In order to reduce potential cell damage by trypsin, McKeehan (26) suggested a reduced temperature during trypsinization. At constant levels of trypsin activity, treatment of cells (normal human and chicken fibroblasts) at temperatures less than 15 °C during all steps from monolayer to subsequent culture markedly improves viability, cloning efficiency, and multiplication potential of single cells (Fig. 3) (27).

General Use. Generally, trypsin is used at a conentration range of 0.01% to 0.5% in PBS; usually the concentration is 0.25%, and the incubation time ranges from 5 to 15 min (28). In the case that trypsin/EDTA is used, the trypsin concentration is the same, whereas EDTA is used at a concentration of 0.02%. This mixture is convenient because it provides the advantages of both components. The pH is critical when harvesting with trypsin, and care must be taken to ensure that harvesting is done between pH 7.4 and 8.0 (29).

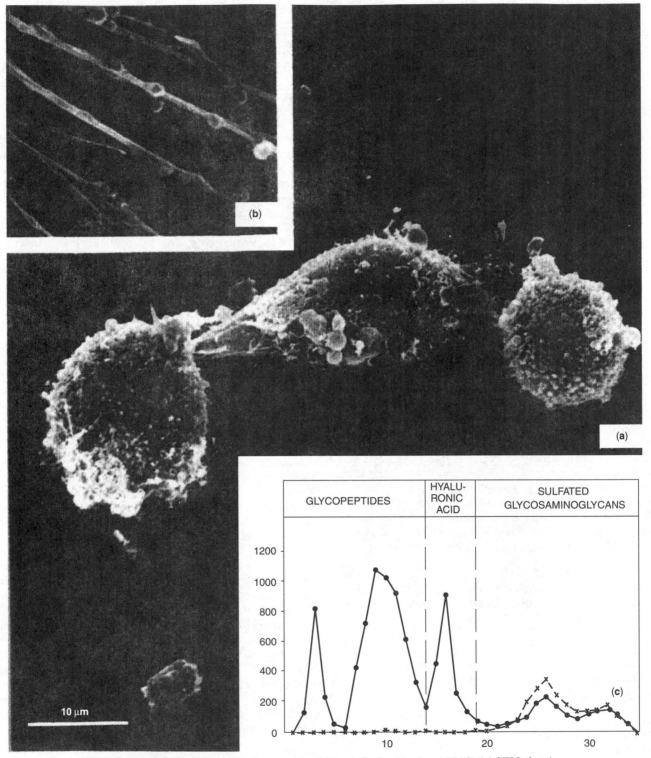

Figure 2. HLM18 cells incubated in trypsin, 0.1 mg/mL, for 30 min at 37 °C. (**a**) SEM showing cellular contraction and appearance of microvilli and blebs on membrane surface; (**b**) SEM of processes formed during early stage of trypsin detachment, showing swellings; (**c**) Abscissa: fraction no.; ordinate: cpm/0.5 mL. ●–●, ^{3}H; x---x, ^{35}S, DEAE–cellulose chromatography of macromolecular material released during incubation of 5×10^6 cells labeled with [^{3}H]-glucosamine and $Na_2{}^{35}SO_4$. Elution in a linear gradient of ammonium acetate 0.2–2 M, 3 mL/fraction, 150 mL total. From Ref. 16, with permission.

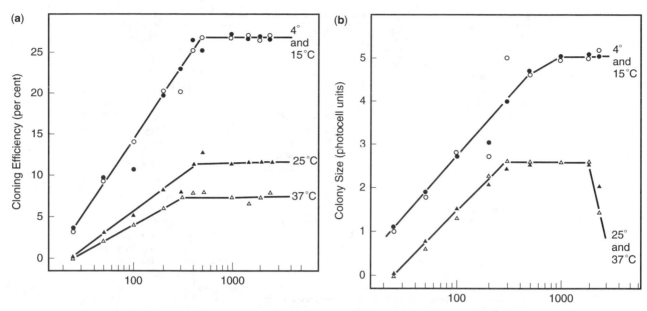

Figure 3. The effect of temperature during harvesting of chicken embryo fibroblast (CEF) cells from monolayer at constant levels of trypsin activity units on sequent clonal growth of single cells. Secondary CEF cells were grown to monolayer in medum MCDB 201 containing 250 μg/mL fetal bovine serum protein prepared as described by McKeehan et al. (27). For trypsin treatment, replicate monolayers of CEF cells were first equilibrated in a water bath to the indicated temperature for 15 min and the medium then discarded. All operations were carried out with solutions and vessels equilibrated with the water bath at the indicated temperature. The cell monolayers were washed twice with 0.2 mL per cm² of solution I [4.0 mM glucose; 3.0 mM KCl; 122 mM NaCl; 1.0 mM Na₂HPO₄; 0.0033 M phenol red; 30 mM HEPES NaOH (pH 7.6)]. The cell sheets were then covered for 30 sec with 0.08 mL per cm² solution I containing the 4 units of trypsin activity per mL. The solution was immediately withdrawn, leaving only a thin film of the solution on the monolayer. The monolayers were then incubated for 8 min at the indicated temperature. After the incubation period, 0.1 mL per cm² of serum free medium was introduced, and the loosened cells were suspended by gently rocking the culture flask. Clumps of cells were broken up by gentle pipetting, and the resulting cell suspension was counted and diluted for use in clonal growth experiments. Clonal growth of 150 CEF cells was carried out as described previously (27). The number of visible clones was scored manually. Cloning efficiency is the number of visible colonies divided by the number of cells inoculated. The average size of colonies was measured photometrically. △–△, 37 °C; ▲ – ▲, 25 °C; ○–○, 15 °C; ●–●, 4 °C. From Ref. 26, with permission.

The damaging effects of trypsin necessitates a short exposure time followed by rapid elimination or inactivation of trypsin. This is generally done by the addition of serum; however, when serum-free media are used, this elimination can be done by the use of trypsin inhibitors, like soybean trypsin inhibitor, pancreatic trypsin inhibitor, or by ovomucoid (30). Rapid dilution of trypsin followed by extensive washing of the cells is also valuable (O.-W. Merten, unpublished results).

Other Proteolytic Enzymes

In principle, other proteolytic enzymes, such as dispase I and II, pronase, and papain, can equally be used for the detachment of adherent cells. The advantage of some of these enzymes is that they are of nonanimal origin (see Table 2) (31), and that they can be used in the presence of serum (e.g., dispase; (32)). Their main disadvantage is that they are not inactivated by serum leading to the necessity

that they have to be eliminated by thorough washing after cell detachment.

Dispase (EC 3.4.24.4.)

Mechanisms and Effects on Cells. Dispase, a neutral protease from *Bacillus polymyxa*, is very powerful. It is activated by Ca²⁺ and several other metal ions, and inhibited by chelating agents like EDTA. It has been classified as an amino-endo peptidase, since it hydrolyzes peptide bonds on the N-terminal sides of nonpolar amino acids (Fig. 4) (33). As described for trypsin and EDTA, dispase leads to a transient increase in the permeability of rounding cells (20). Green et al. (34) showed that dispase cleaves the basement membrane zone of the skin, sharply separating the epidermis from the dermis. Stenn et al. (35) performed studies to define its substrate specificity. They could establish that dispase cleaves fibronectin and type IV collagen, but not laminin, type V collagen,

Table 2. Trypsin and its Potential Replacements not of Animal Origin

Origin	Replacement	Origin	Comment
Porcine pancreas, bovine pancreas	Other means for cell detachment: dispase I and II, papain, pronase, collagenase[a] accutase	*B. polymyxa*, *Papaya latex*, *Streptomyces griseus*, *Chlostridium histolyticum*, from an invertebrate species	After cell detachment by using dispase or papain, the cells have to be washed at least three times to remove the residual activity
	temperature-modulated hydrophilic-hydrophobic polymer surfaces	synthetic (poly(N-iso-propyl acrylamide))	(commercial source: Innovative Cell Technologies) After reduction of the temperature under a certain value, the cells detach because the surface becomes hydrophilic

[a]Collgenase is very useful for cell detachment from collagen- or gelatin-coated growth supports.
Source: From Ref. 31, modified.

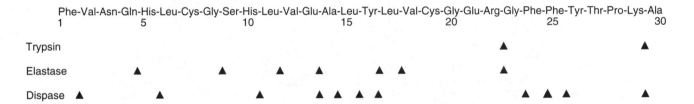

Notes: Although bacterial collagenase and pronase have been used for cell and tissue dispersion, they are not shown here. Bacterial collagenase hydrolyzes peptide sequences unique in collagen molecules, while it does not hydrolyze any peptide bonds of insulin β-chain. Pronase is a mixture of proteases, and hydrolyzes most peptide bonds.

Figure 4. Substrate specificity of some proteases used for cell and tissue dispersion as shown by their points of action on oxidized insulin β-chain. From Ref. 33, with permission.

serum albumin, or transferrin. Type I collagen was only minimally degraded.

It can be efficiently used for detaching cells from their substrata (32). Matsumura et al. (32) showed that the use of dispase leads to a monodisperse suspension mostly in serum-containing medium when used for the detachment of cells of fibroblastic morphology. Epithelial-like cells are detached, but poorly dissociated from each other. The mouse melanoma B16 cell line could not be detached by dispase treatment. Cells are alive and grow in the presence of dispase (36).

The comparison of cell detachment by trypsin and dispase showed that the degree of dispersion was superior for the dispase treated cells than for the trypsinized cells (Table 3). It seems that the incubation conditions are more critical for the trypsin treatment than for dispase treatment (33).

General Use. The concentrations of dispase used were rather different: Whereas Matsumura et al. (32) used 500 U/mL, Cassiman et al. (37) employed a mixture

Table 3. Plating Efficiency of CHO-K1, L6TG, and V79 Cells after Dissociation by Trypsin or by Dispase[a]

Cell line	Protease treatment	Degree of dispersion[b] (%)	Plating efficiency[c] (mean % ±S.D.)
CHO-K1	Trypsin	99.6	52.0 ± 4.0
	Dispase	100.0	45.0 ± 5.1
LGTG	Trypsin	97.3	104.1 ± 7.3
	Dispase	100.0	93.0 ± 4.2
V79	Trypsin	99.0	75.5 ± 6.6
	Dispase	100.0	70.5 ± 10.0

[a]Conditions: Before detachment the cells were rinsed with EDTA (0.02%)-PBS and then the enzyme solutions were added: 0.5 mL of 500 U/mL dispase in growth medium (incubation time: 40 min); 0.5 mL of 0.05% trypsin in EDTA-PBS (incubation time: 15 min). For replating the cells, the detached cell suspension was supplemented with 1 mL of growth medium and centrifuged in order to eliminate residual protease activity. The cells were resuspended in growth medium, counted, and plated (100 cells per 28 cm²).
[b]The degree of dispersion is shown as the percentage of single cells out of total number of cells in suspension.
[c]A value of plating efficiency is the average of values from 6 plates.
Source: From Ref. 33, with permission.

of dispase/EDTA in PBS (4 U/mL/0.02%). The incubation times were 40 and 10 min, respectively. Griffiths (38) reports a normally used concentration range of 0.6–2.4 U/mL.

Collagenase (Clostridiopeptidase, EC 3.4.4.19)

Bacterial collagenases are mixtures of several enzymes. They hydrolyze peptides containing proline, including collagen and gelatin. They are unstable in phosphate buffers and require Ca^{2+} ions therefore, they should not be used with EDTA (39,40). This enzyme can be used for the detachment of cells (e.g., endothelial cells) from extracellular matrix, from collagen, or gelatin (41). Gordon et al. (41) could establish that the rate of the detachment process depended on the composition of the extracellular matrix, the composition of which was influenced by the passage number of the cells. Collagenase detached endothelial cells somewhat less efficient and with slower rate than trypsin (Table 4). Late passage cells (= high cumulative population doubling level) always detached significantly faster than early passage cells. More information can be found in the article by Waymouth (30).

General Use. Collagenase is usually used at a concentration of 0.01–0.15% in PBS (38).

Accutase

Accutase is a mixture of proteolytic and collagenolytic enzymes, extracted from an invertebrate species (Innovative Cell Technologies). Accutase can be used for replacing trypsin for the routine detachment of cells from standard as well as from adhesion coated plasticware. This enzyme mixture is less damaging than trypsin, and in contrast to trypsin, a specific accutase neutralization step is normally not required after dissociation with accutase. It is used following the instructions released by the producer company.

Several groups used *pronase E* for routinely subcultivating animal cells. For instance, human diploid fibroblast cells were more rapidly detached by using 0.05% pronase than using 0.25% trypsin. The durations were seconds at room temperature and 15 min at 37°C, respectively (42). The viability exceeded 90%. As described for EDTA and trypsin, pronase E leads to a transient permeabilization of rounding cells (20). Because there is no antipronase activity in serum, pronase activity has to be eliminated by thorough washing. However, Poste (43) indicated that pronase cannot be removed from the surface of kidney cells by washing and continues to damage the cells in the subculture.

Other enzymes, like *papain* (EC 3.4.22.2.; same detachment efficiency as found for trypsin; O.-W. Merten, C. Fiamma, and C. Rochette, unpublished results) and *elastase* (pancreatopeptidase E, EC 3.4.4.7.; sites of hydrolysis shown in Fig. 4) (30) are also used for cell disaggregation and/or detachment.

Reattachement of Detached Cells

Depending on the protease used for detaching BHK-21 cells, the cells do or do not reattach directly after detachment and dilution. In the case of pronase, protein synthesis is required to restore the adhesive properties towards fibronectin, because membrane proteins have been proteolytically digested. In the case of trypsin, treated cells even start to reattach when protein synthesis is inhibited, signifying that trypsin does not digest the membrane glycoproteins of BHK-21 necessary for cell adhesion (44). Various inhibitions are obtained when cells are treated with proteinase K, chymotrypsin, papain, subtilopeptidase A, and thermolysine.

Human diploid fibroblasts trypsinized in PBS without Ca^{2+}, containing 0.33 M sucrose and traces of Mg^{2+} (10^{-3} M Mg^{2+}) gave rise to better subsequent growth and adsorption than cells from cultures trypsinized in other buffers (11). By subcultivating embryonic lung tissue, Litwin (11) could establish that after 140 days in culture, the following numbers of cell division could be observed: 38 for trypsin in distilled water, 47 for trypsin in Hanks BSS, 51 for trypsin in Eagle's medium, 56 for trypsin in PBS, and 57 for trypsin in 0.33 M sucrose plus 10^{-3} M Mg^{2+}. These data indicate that the composition of the trypsin solution has an important impact on the cell attachment as well as on the total possible number of cell divisions of diploid cell lines. It is evident that the results would differ for other cell lines.

The plating efficiencies for cells detached by trypsin or dispase are comparable, or often somewhat higher for

Table 4. Endothelial Cell Detachment and Reattachment[a]

Incubation time (h)	Agent	Concentration	Cumulative population doubling level	% Cells detached	% Detached cells that reattached
0.5	Trypsin	100 µg/mL	10	85	80
			56	100[b]	70
1.5	Collagenase	0.4%	10	75	80
			56	95[b]	75

[a]Human endothelial cells that detached after treatment with detaching agents were replated and plating efficiency determined. Results of three experiments.
[b]$P < 0.05$.
Source: From Ref. 41, with permission; © John Wiley & Sons, Inc.

the trypsinized cells (Table 3). However, the differences are not significant. Cassiman et al. (37) reported equal growth rates for diploid human fibroblast subcultures after detachment by trypsin or by dispase.

Collagenase and trypsin detached endothelial cells reattached with the same percentage. Whereas about 80% of early passage cells reattached, only 70–75% of the late passage cells reattached (Table 4) (41). For this reattachment, the cells were washed and replated on gelatin-coated plastic dishes in medium supplemented with 20% serum, endothelial cell growth factor, and thrombin.

NONENZYMATIC CELL DETACHMENT AND NEW METHODS

EDTA and *EGTA* are used alone or in conjunction with trypsin for the detachment of anchorage-dependent cell lines. Whereas EDTA is a Ca^{2+} and Mg^{2+} chelating agent, EGTA complexes specifically Ca^{2+}. It is known that omission of Ca^{2+} and Mg^{2+} from the dissociation medium (45) or active removal by chelators (e.g., Ref. 46) leads to a loosening of many cells. Therefore, chelating agents have been very commonly used as a part of disaggregation procedures (30). Both agents lead to cell rounding, as caused by using trypsin. By comparing EGTA and EDTA treatment, it appears that the EGTA treatment is a milder method than EDTA treatment for protecting the cell's permeability barriers and minimizing surface glycoprotein release (21).

The mechanisms of cell detachment by chelating agents (EDTA/EGTA) are different from that by trypsin (Table 1) because they have no proteolytic activity; however, some similarities were observed. EDTA/EGTA exerts its primary effect upon the cytoskeleton. In the presence of EDTA rounding cells become leaky (20); the cells show a much higher permeability of the cell membrane (22) than during trypsinization. Using EGTA, all stress fibers seem to be similarly susceptible, and the beaded stage, observed during trypsinization, is not seen; this probably is a result of the depletion of intracellular Ca^{2+} (19).

By comparing the use of EDTA or EGTA for cell detachment with respect to the use of trypsin, it becomes evident that the surface properties of the detached cells are different from those detached by trypsin, such as agglutinability with lectins (47) and intercellular adhesivity (25,48). EDTA does not release cell-surface components, but causes cell contraction and detachment morphologically similar to that caused by trypsin (16). EGTA detached cells are spherical. In contrast to trypsinization, their glycocalyx appears to remain redistributed over the cell surface (49).

In general, the EDTA dissociation procedure is difficult on normal cells and frequently results in important cell loss. In addition, chelation of divalent ions in confluent cultures leads to a detachment in large sheets, which are difficult to disperse mechanically (16,21). These are the main reasons that chelating agents are most often used in conjunction with trypsin.

For reattachment EDTA and EGTA have to be eliminated to get rid of the chelating activity.

To preserve surface proteins, *sonication* is well adapted to detach cells. Menssen et al. (50) used an ultrasound (water-filled ultrasonic cleaner, 43 kHz) and sonicated culture vessels for 10–50 sec. The detached cells (human melanoma and gastrointestinal carcinoma cells) had the same viability and could be as well replated as trypsin-detached cells [plating efficiency: 86% for the sonicated cells (30 sec), 80% for the trypsinized cells]. The additional advantage is the speed of passaging, which took only 4 min for the ultrasound method, whereas the trypsinization took 16 min, including two centrifugation/washing steps. For detaching BHK-21 cells, Payne et al. (51) used equally low-frequency ultrasonication (50 kHz) and obtained a single cell suspension after 15 sec to 2 min of sonication. These detached cells were comparable in viability and growth characteristics.

For many cell lines, the use of trypsin/EDTA leads to detached cells that are aggregated. In order to release the cells and to obtain monodisperse cell suspension, ultrasonication of 43 kHz for 60–90 sec can be sufficient. By using this approach, Sanford (52) could obtain monodisperse rainbow trout cells.

Recently, *heparin* was proposed as a substance for cell detachment (53). Heparin, a sulfated glycosaminoglycan, is known to interact with various components of the extracellular matrix. The interactions between cells and different collagen types are clearly cell-type and collagen-type specific (54) and may depend upon the presence of determinants in the native collagen triple helix. With respect to Balb/c-3T3 fibroblasts, it could be established that these cells attach readily to substrata of native type I and V collagens, but not to native type III collagen. San Antonio et al. (55) could show that heparin inhibits the attachment and growth of Balb/c-3T3 fibroblasts on collagen substrata due to interaction of heparin — collagen leading to a disruption of the cell collagen attachment. Similar results were shown for CHO cells (56). Heparin inhibits only the adhesion of Balb/c-3T3 fibroblasts to collagens I and V, but not to type III collagen. This inhibition of adhesion is due to the fact that the ectodomain of a heparan sulfate proteoglycan binds to native type I collagen and is displaceable by heparin (57).

Substances with a heparin-like activity are dextran sulfate, dermatan sulfate, and heparan sulfate II. Whereas dextran sulfate was equal to heparin in its cell adhesion inhibitory activity, dermatan sulfate and heparan sulfate also inhibited adhesion, but to a lesser extent (55,56).

Although heparin does not have enzymatic activity, San Antonio et al. (55) could establish that a prolonged incubation of cells in heparin reduced or even inhibited cell growth, probably due to an interaction with the extracellular matrix.

The function of heparin in detaching cells is probably based on the same mechanism, as it inhibits reattachment of animal cells. By using 12 different cell lines, we (53) could show that the detaching efficiencies varied enormously between different cell lines. Whereas recombinant CHO cells (CHO-IC5) were comparably detached in trypsin/EDTA (0.125%/0.01% in PBS), heparin/EDTA (100 IU/mL/0.02% in PBS), or EDTA (0.02% in PBS), cells like MDCK and Vero could practically not be detached by using heparin. The rate of detachment was influenced by the cell density of the cultures. At semiconfluency (about 2.5×10^6 c/25 cm^2) the detachment of CHO-IC5 cells took 1, 15, or 20 min, respectively. However, the rate of detachment of confluent cultures (9×10^6 c/25 cm^2) took 3 min for all detachment conditions. Similar results were observed for BHK21 C13 cells. In this case, the duration for detaching all cells from the polystyrene surfaces was 2–5 min for trypsinization, and 2–10 or 5–15 min for the other detachment conditions, when confluent cultures (10^7 c/25 cm^2) or semiconfluent cultures (3.4×10^6 c/25 cm^2), respectively, were treated. The following cells could easily be detached by using heparin/EDTA (100 IU/mL/0.02%): HeLA R19, ψ-CRIP and derivatives, Sf9, and 143B. For cell line GPen-vAM12, the detachment was possible after optimization of the conditions (J.-L. Salzmann, personal communication).

The subcultivation of cells (CHO-IC5, BHK21 C13) detached by the three detachment agents revealed that the growth was almost not influenced by the detachment conditions, when the cells were washed with culture medium before subcultivation (Fig. 5, bars 1–6). When the detached cells were not washed for eliminating the detachment agent, the cells detached by heparin/EDTA grew better than those detached by trypsin/EDTA or by heparin alone (Fig. 5, bars 7–9).

Lidocaine, a local anesthetic, is relatively nontoxic in contrast to many others (58) and is known to provide reversible inhibition of adherence of cell monolayers. The action of lidocaine and other local anesthetics is probably at the level of the cytoskeleton (59).

Used at a concentration of 0.5%, a successful detachment of human blood monocytes was possible (60). As monocytes are somewhat difficult to obtain and to passage because of their adherence properties, improved non-damaging methods are necessary. A comparison of four different detachment methods: low temperature (4 °C), EDTA (10 mM, 37 °C, 30 min), lidocaine (0.5%, 37 °C, 30 min), and trypsin (0.01%, 37 °C, 30 min), revealed that detachment with lidocaine was associated with the highest

Table 5. Recovery of Blood Monocyes after Detachment by Different Methods; Indicated is the Mean of Four Experiments (Ranges in Parentheses)[a]

4 °C	EDTA	Lidocaine	Trypsin
52	49	67	60
(35–60)	(44–60)	(55–69)	(58–63)

[a]Recovery $= \dfrac{\left(\begin{array}{c}\text{number of monocytes}\\\text{detached from surface}\end{array}\right) \times 100}{\text{number of monocytes in blood sample}}$.

Recovery by lidocaine was significantly higher ($P < 0.05$) than by EDTA and 4 °C. No difference between detachment by lidocaine and trypsin was found.

Source: From Ref. 60, with permission.

recovery, while especially EDTA and cold exposure had disappointing yields (Table 5).

A functionality test (chemotactic responsiveness) revealed that only trypsin had a negative effect on the activity of the cells.

A nonenzymatic detachment method is provided by the use *thermoresponsive polymer surfaces* (actually modified polystyrene surfaces) (Table 2). They are based on the use of poly-N-isopropyl acrylamide (PIPAAm) as substratum for the cultures of animal cells. The characteristics of this polymer is that at low temperature (e.g., under 32 °C) it is fully hydrated and shows extended conformation. Above the lower critical solution temperature it extensively dehydrates and changes to compact chain conformations. So, only by reducing the temperature of the culture do the cells detach very easily when the polymer changes from hydrophobic to hydrophilic characteristics. Because it is nontoxic towards animal cells, it was used as support for cultivating various animal cells (61–63). One disadvantage might be that the cells detach in the form of sheets, probably because the extracellular matrix stays intact owing to the absence of proteases (61,63).

As for nonenzymatic detachment methods, the use of thermoresponsive polymer surfaces provides the

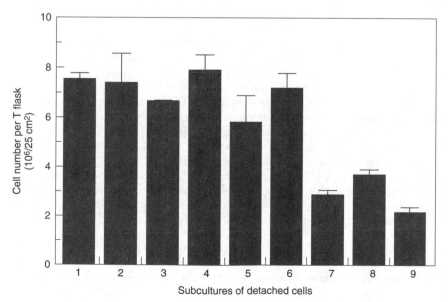

Figure 5. Subcultures of detached CHO-IC5 (bars 1–3) and BHK21 C13 (bars 4–9) cells. Conditions: The detached cells were once washed in culture medium for CHO-IC5 (bars 1–3) and BHK21 C13 (bars 4–6) or were directly subcultivated without washing after detachment (bars 7–9). The cells were plated at 10^6 cells/T flask of 25 cm^2. Time of cell counting: 56, 54.5, and 70.5 h, for cultures 1–3, 4–6, and 7–9, respectively. The detachment solutions were: trypsin/EDTA (0.125%/0.01%) for cultures 1, 4, and 7; heparin/EDTA (100 U/mL/0.02%) for cultures 2, 5, and 8; 0.02% EDTA for cultures 3 and 6; 100 U/mL heparin for culture 9. The detachment solutions were based on Ca^{2+}-and Mg^{2+}-free PBS. From Ref. 53.

advantages that the cells are not damaged by the detachment. This was probably the main reason that such polymers were developed and used for small-scale cultures. Using this approach, Yamada et al. (62) could successfully cultivate and subcultivate bovine hepatocytes, which are normally highly sensitive to enzymatic treatment. The cells attached equally well to normal polystyrenedishes and to PIPAAm grafted dishes, and growth was comparable. However, when the confluent cultures had to be passaged, almost 100% of the cells detached from the PIPAAm dishes were recovered after cooling down the dishes to 4 °C and keeping them at this temperature for 60 min, while only 8% of the cells were detachable from the control dish (treatment with trypsin/EDTA (0.05%/0.02%) at 37 °C for 15 min). The subcultivation and readhesion of cells showed similar differences. Whereas about 73% of the cells detached from the PIPAAm dishes adhered again after 18 h, only 14% of the cells detached by trypsin were able to adhere to the new dish. Similar results were published for detachment of dermal fibroblasts (61), bovine aortic endothelial cells (64), and rat hepatocytes (64).

Another less important nonenzymatic detachment method is the *scraping off* cells from surface with a rubber policeman. However, such mechanical methods can harm the cells to be detached. Anchorage-dependent cells round up during mitosis, although they stay attached to the substratum. This rounding up leads to an important reduction in the strength of cell attachment, signifying that mitotic cells can be detached and separated from the other cells by mechanical shear. This is the basis of the *mitotic selection technique* (65,66). Finally, cell detachment can be provoked by *fluid shear stress* in in vitro cultivation systems. Fluid shear stress in general leads to alterations in growth rate, morphology, metabolism, and genetic expression, and can ultimately lead to cell detachment. Such effects are principally undesirable, but they can happen under nonoptimal culture conditions in reactor cultures (67,68). In addition, research on the effects of shear stress on endothelial cells, which constitute the natural inner lining of blood vessels, is necessary in view of the development and application of vascular prostheses (69,70).

LARGE-SCALE CELL DETACHMENT

The passage of surface adherent cells has the same inherent problems in small-scale as well as in large-scale cultures. These passages are necessary in order to build up volume for the final production stage. In general, the cells have to be released from the flasks, roller bottles or microcarriers using various detachment methods, often with a trypsin/EDTA solution. Because the detachment of small-scale cultures, which includes cultures in flasks and roller bottles, has already been described this paragraph deals with the detachment of cells from microcarrier and fixed-bed cultures mainly used for large-scale cell culture. This detachment is necessary because most of the cells do not detach alone from the carriers and are therefore unable to colonize fresh carriers when the old confluent carriers are put together with new

fresh cell-free carriers [e.g., for MRC-5 cells decribed by Clark and Hirtenstein (71); our own unpublished results obtained in serum-free microcarrier cultures of Vero cells].

For certain cell lines the direct colonization of new carriers is a viable approach and can be the best choice. CHO cells can be directly transferred from one carrier to another. Crespi and Thilly (72) transferred CHO and monkey kidney cells (LLC-MK2) directly from carrier to carrier while stirring in a medium with a low calcium content. Delzer et al. (73) added fresh microcarriers (Cytodex 3) to mouse L-cell cultures grown to confluence and observed movement of cells to the fresh carriers. Kluft et al. (74) increased the volume of a microcarrier culture of human melanoma cells from 3 to 10 and 40 liters by adding fresh medium and microcarriers accordingly. A small fraction of free-floating cells attached to the fresh carriers and colonized them. Xiao et al. (75) used the same effect and scaled up a CHO culture on Cytopore porous microcarriers.

As already indicated for small-scale cell detachment, proteolytic enzymes (trypsin, trypsin/EDTA, and collagenase) are often the first choice for detaching cells from microcarriers. Trypsin, pronase, and dispase can be used for the detachment of cells from all available microcarriers, like dextran-based carriers, as well as collagen-coated carriers (29). In addition, cells can also be detached from collagen-coated carriers by the use of collagenase (76). In order to optimize the subpassaging of MRC-5 and Vero from and to Cytodex 1 microcarriers, Lindner et al. (77) separated the trypsinization from the EDTA-chelating step. In principle, the microcarriers of logarithmic cultures were sedimented. An equal volume of EDTA (0.02%) in PBS (37 °C) was added to the slurry. After a brief wash, the supernatant was discarded and replaced by an equal volume of PBS (37 °C). After sedimentation the PBS was discarded and replaced by an equal volume of 0.2% trypsin in PBS (37 °C). Ten minutes later the cells were detached from the carriers, and the entire mixture was transferred (as an inoculum) to the culture vessel containing freshly prepared microcarriers in fresh culture medium. This method led to the detachment of 95–100% of the cells generally having a viability greater than 95%.

Lindskog et al. (78) compared the use of trypsin and dextranase for detaching cells from Cytodex 1 and 3 microcarriers (uncoated and collagen-coated dextran-based carriers, respectively). Dextranase used alone leads to the release of cell sheets that can be disaggregated when trypsin or EDTA/EGTA are used in the same time as dextranase. Therefore, dextranase was used in conjunction with trypsin. By using MRC-5 and Vero cells, they could establish for all conditions tested (trypsin, collagenase/trypsin, Cytodex 1, Cytodex 3) that there was no difference between the different conditions, and that the cell recovery was about 70%, whereby the viability was approximately 90%. The plating efficiencies were comparable: 71–79% for Vero cells and 73–82% for MRC-5 cells.

Gebb et al. (79) compared detachment methods for cells from Cytodex 3 microcarriers (collagen-coated dextran-based carriers). Cell recovery of Vero cells detached by collagenase and trypsin were comparable, and 75–95%

of the cells were harvested, whereas by using dispase about 60% of the cells were recovered. Furthermore, the viability of the cells was 90–95% for those harvested with trypsin and dispase, whereas the viability of the collagenase harvested cells was lower (54–85%). Similar results were obtained with BS-C-1 cells. The plating efficiencies were best for the cells (Vero, MRC-5) harvested by using collagenase, followed by those detached with trypsin. The use of collagenase for cell detachment could be optimized by prewashing the carriers with an EDTA solution, followed by an incubation with collagenase in the absence of EDTA. This is important because the activity of collagenase depends on the presence of Ca^{2+} ions. Using this protocol the recovery of human fibroblasts can reach values of up to 90%. The use of collagenase, in addition to trypsin and dispase, is recommended by all producers of collagen or gelatin-based or collagen-coated microcarriers.

Manousos et al. (80) compared trypsin with pronase for cell detachment from microcarriers. The detachment of cells from DEAE–Sephadex G-50 microcarriers was efficient when the carriers were incubated in a solution of pronase for 10–15 min, whereas trypsinization did not lead to any significant cell detachment. The pronase–released cells had a viability of greater than 95%.

As described, heparin can be used efficiently for the detachment of animal cells from polystyrene surfaces. In order to evaluate the usefulness of heparin to detach cells from large-scale cultivation systems, cultures of various cells were performed in the Costar CellCube (fixed-bed system based on polystyrene surfaces) and on Cytodex 1 microcarriers. The detachment of cells from the CellCube was evaluated using cell lines used for the production of retroviruses for gene therapy (clones of ψCRIP cells). Confluent cultures of ψCRIP showed an average cell density of 330,000 c/cm^2. When such cultures in a CellCube were detached using trypsin/EDTA (0.125%/0.01%) in PBS (incubation time: 20 min), after one washing step in EDTA about 16–20% of cells could be recovered because these cells started to form enormous aggregates after trypsinization. These aggregates could not be recovered from the CellCube because they were trapped by the spacers between the plates of the culture system. By replacing trypsin/EDTA by heparin/EDTA (100 U/ml/0.02%), all cells were detached within 20 min, and in contrast to trypsinized cells, they stayed as single cells. Such cell suspensions could very easily be recovered from the CellCube and could be used for further applications (e.g., establishment of cell banks for large-scale cell culture; in this case, the cells were frozen, for instance, in Baxter's Cryocyte Freezing Containers). Similar results were obtained with CellCube cultures of clones of the TE FLY cells, which were also used for the production of retrovirus. Concluding, the data obtained from detachment tests in T-flasks could be applied directly to CellCube because the mechanism of the cell attachment was the same in both cases.

The detachment of adherent cells growing on Cytodex 1 microcarriers was less successful. Four different cell lines were tested for evaluating this detachment method. Whereas Vero and CHO-IC5 cells could not be detached from microcarriers when heparin was used, BHK-21 C13

Table 6. Detachment of ψ CRIP M11 Cells from Cytodex 1 by Trypsin and Heparin

Detachment solution[a]	Cells per mL	% of Cells detached by trypsinization
1. Crystal violet[b]	2.15×10^6	
Trypsin/EDTA (0.125%/0.01%)	1.75×10^6	100
Heparin/EDTA (100 U/mL/0.02%)	1.25×10^6	71.4
EDTA (0.02%)	1.43×10^6	82.1
Heparin (100 U/mL)	1.08×10^6	61.8
2. Crystal violet[b]	2.42×10^6	
Trypsin/EDTA (0.125%/0.01%)	1.69×10^6	100
Heparin/EDTA (400 U/mL/0.02%)	1.49×10^6	88.5
Heparin (400 U/mL)	1.38×10^6	81.8

[a]The cells have been washed once with PBS before treatment with detachment solution (based on PBS), the duration for dissociation was 10 to 15 min.
[b]The crystal violet counts were done in order to establish the total number of cells attached to beads. One mL of carrier suspension was treated with 1 mL of a crystal violet solution (0.1%) in citric acid (0.1 M).
Source: From Ref. 53.

and ψCRIP M11 cells could be detached (Table 6) by using heparin or heparin/EDTA; however, the recovery was lower than for trypsinization. Using 100 U/mL heparin with or without 0.02% EDTA detachment was feasible, but the recoveries were 71.4% and 61.8%, respectively, with respect to trypsin (=100%). In addition, the action of detachment was more due to the presence of EDTA (chelation of divalent cations) than to the presence of heparin. At a heparin concentration of 400 U/mL in the presence or absence of EDTA, 88.5% and 81.8%, respectively, of the cells were detached. These data indicate that an increased heparin concentration is important for a more complete cell detachment; however, for the moment, trypsinization is still superior to heparinization of cells from microcarriers.

The basic engineering steps for passaging microcarriers cultures are the following: The microcarriers sediment in the growth vessel or in a separate tank. The disadvantage of using the growth tank is the minimum volume that can be used in this tank. In the case of using a separate tank, its volume normally can be chosen according to the mass of the carriers to be treated. This signifies that the volume of the detachment solution can be minimized. Then the carriers are washed to remove serum, the detachment solution (protease ± EDTA) is added followed by short stirring, and the cells/microcarriers are incubated in this detachment solution for as short a time as necessary to reduce cell damage to a minimum. Then the proteolytic activity is stopped by the addition of, for instance, serum. Now, two different approaches can be undertaken: (1) After detachment, the cells are separated from the carriers and put into contact with new carriers for performing the subsequent culture. At the same time the old carriers are discarded; or (2) after cell detachment, the cells as well as the old carriers are put together with the new carriers and the new culture is started in this way. The advantage of this approach is that no separation process is necessary that potentially can harm the cells and might also lead to a certain loss of cells. The disadvantage is that the colonization of the carriers can be inhomogeneous, because some cells are not detached from the old carriers (81).

The optimal way for passaging cells growing on microcarriers from a smaller to a larger scale is the transfer of the microcarrier culture to a separate detachment vessel, washing of the carriers, addition of the detachment solution, incubation as short as possible in this solution with a brief agitation, followed by the separation of the cells from the beads and transfer of the detached cells to the next scale.

The most difficult step is the separation of the cells from the microcarriers. Although several solutions have been proposed, like filtration through filters with a cutoff of 60–100 μm for retaining microcarriers and letting the cells pass through, there is one special apparatus that allows the detachment of the cells followed by a separation of the cells from the carriers. van Wezel et al. (82) used this apparatus for serial passages of monkey kidney cells (Fig. 6). To reduce the damaging effect of trypsin on these cells, the detachment solution (trypsin/citrate = 0.25%/0.025% M) was added to the cells after having washed them, and the detachment solution was immediately drained off through the packed carriers. After 30 min of incubation at room temperature, the cells were removed from the beads and suspended in tissue culture medium by mixing with a Vibromixer for one or two minutes and filtering through a screen. Subcultures can be inoculated directly with this cell suspension. van Wezel et al. (82) reported a cell recovery of 53–63% for primary monkey kidney cells.

Other methods for cell separation in a laboratory scale were presented and compared by Billing et al. (83). Low-speed centrifugation on Ficoll–Paque gradients, which is based on the differential settling rates between cells and microcarriers, lead to a cell recovery of 65–75%; similar values were obtained for the filtration through an 88-μm pore-size nylon screen. The unit gravity sedimentation led to a recovery of 33–55%. In general, the viability was above 90%. Although these results were obtained from laboratory-scale experiments, the recovery values for the filtration method based on the use of an 88-μm pore-size screen as well as for the gradient centrifugation were slightly superior to those obtained by van Wezel et al. (82) by using the special detachment and separation apparatus. However, the apparatus designed by van Wezel et al. (82) is the only one that has been tested on a large scale and for which the results were published.

CELL DETACHMENT IN SERUM-FREE MEDIA

The use of serum-free or even protein-free culture media is the most appropriate choice to reduce the risk of a potential introduction of adventitious agents into the culture. However, the use of such media has many consequences for the whole technology. Some cells do not grow as adherent cells in the absence of serum but they

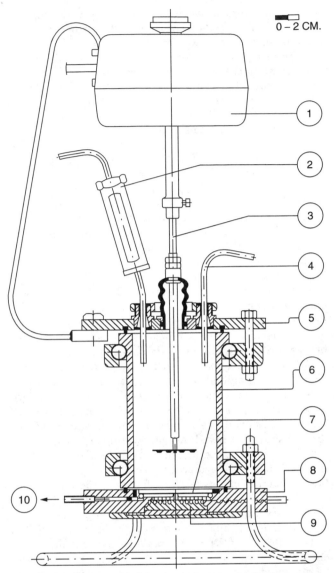

Figure 6. Apparatus for the trypsinization of cells from microcarriers: 1. Vibromixer (model E1, Chemapec Inc.); 2. air filter; 3. Vibromixer shaft with impeller; 5. upper lid with six standardized stainless steel in- or outlet tubes (4); 6. standard Jena glass pipe 80 × 300 mm; 8. bottom lid with stainless steel screen of 60 μm (7) and outlet (9,10). From Ref. 82, with permission.

now grow in suspension cultures as clumps or single cells. For such cells, detachment is no longer an issue. On the other hand, there are cells that still grow as attached cells, when serum- or protein-free media are used. For such cells, detachment is still a necessity for subcultivation.

Although literature data are scarce, it is clear as to how one might proceed to achieve cell detachment for such culture conditions. In principle, the cells are detached from their support as for cultures in serum-containing medium. After detachment, the detaching solution (e.g., trypsin/EDTA (0.25%/0.01%) has to be eliminated, which is generally done by repeated washing and centrifugation of the cells. In order to avoid any adverse effects from residual activity, the cells should be washed at least twice with the medium to be used for the subsequent culture (84,85).

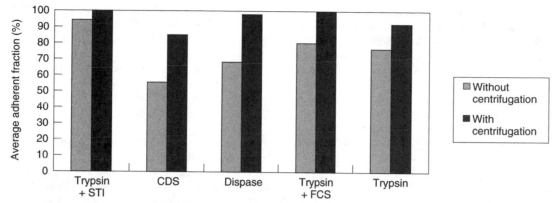

Figure 7. Comparison of different cell-dislodging methods regarding average adherent fraction, i.e., (viable adherent + nonviable adherent)/total cells. These results were obtained after four days of static cell culture in protein-free medium SMIF6. Light bars without centrifugation; dark bars with centrifugation; abbreviations: Tryp., trypsin; STI, soybean trypsin inhibitor; CDS, cell-dissociation solution. From Ref. 86, with permission, © Springer-Verlag.

For the large-scale detachment of cells grown in serum-free medium, the washing step should be done using a continuous centrifuge, such as a Centritech centrifuge (Sorvall), which allows a separation and washing of cells.

Another possibility is the use of inhibitors in order to get rid of residual enzyme activities. For instance, trypsin can be inhibited by trypsin inhibitor from soybean, pancreatic trypsin inhibitor, or ovomucoid (43). However, if such inhibitors are added, one has to be sure that they do not have an adverse effect on cell biology (26) and that they can be eliminated during downstream processing.

Cruz et al. (86) compared different dislodging methods for recombinant BHK21 cells grown in serum-free medium. They compared (in order of the quality for subculturing efficiency) trypsin/EDTA (0.2%/0.2%) followed by an incubation with soybean trypsin inhibitor (1 mg/mL, same volume used as for the trypsin/EDTA solution), trypsin/EDTA followed by inactivation with a ninefold volume of DMEM containing 10% fetal calf serum, trypsin/EDTA followed by an addition of a ninefold volume of serum-free medium, dispase (2.4 U/mL, incubation for 5 min at 37 °C, followed by a subsequent incubation for 10 min at 37 °C after having eliminated the dispase solution), incubation in Ca^{2+} and Mg^{2+}-free phosphate-buffered saline at 37 °C for 5–10 min (Fig. 7). The inactivation of trypsin by soybean trypsin inhibitor was superior to the other methods. All cell dislodging methods followed by a one-step centrifugation were superior with respect to cell viability, total cell concentration, and plating efficiency when compared to dislodging without subsequent centrifugation.

The use of thermoresponsive polymer surfaces for small- and large-scale serum-free cultures would be an important improvement and will considerably facilitate passaging anchorage-dependent cells in the future.

SAFETY ISSUES

The use of animal cell culture products for clinical studies as well as human therapy needs a rigorous control for the absence of adventitous agents. The special field of cell detachment is mainly based on the use of proteolytic enzymes for cell detachment and in some instances on the use of chemicals (like EDTA or EGTA) and other biological substances like heparin. Although adventitous agents can be introduced in different ways into the final product, the raw material used is one potential way of contamination. In order to increase the viral security of a biological, the raw material, which is often of animal origin, should be replaced by raw materials made by chemical synthesis or from plant origin or produced by microbial fermentation. In the latter case, it has to be assured that the fermentation broth does not contain any substances of animal origin, like bactopeptone or tryptopeptone. In such cases, although the substance is of bacterial origin, it is potentially contaminated because animal-derived products were used for its manufacture. With respect to cell detachment, trypsin (see Table 2), elastase, and heparin are of animal origin, whereas all other substances used are of plant or microbial origin or from chemical synthesis. Trypsin and elastase are from pancreas. When these enzymes are used for the industrial cell culture, they have to be tested for the absence of host-specific viruses. For instance, trypsin is mostly of porcine origin, signifying that such trypsin has to be tested for the absence of different porcine viruses; so a test for porcine parvovirus is imperative. If bovine derived trypsin is used, this trypsin has to be tested for the absence of different bovine viruses; so a test for pestiviruses is imperative. The test for other viruses depends mainly on the geographical origin of the product. More details can be found in the article by Eloit (87).

With respect to transmissible spongiform encephalopathies (TSE), bovine pancreas is classified as a category III tissue (= low infectivity) (88), signifying that a replacement of pancreas-derived material by materials of nonanimal origin will improve safety with respect to prion transmission.

Heparin is from porcine intestinal mucosa. This substances is used for medical application (inhibition of blood coagulation) without any untoward incident for several decades. Despite this high safety, a replacement by a nonanimal-derived substance would be desirable. As presented in Table 2 and discussed in the sections on the use of enzymes for cell detachment and nonenzymatic cell detachment and new methods, trypsin, elastase, and heparin can be easily replaced by other proteolytic enzymes of microbial or plant origin or chemical substances. After optimization, such methods show the same cell detachment and plating efficiencies, however, with the added advantage that viral security is improved.

CONCLUSIONS

Cell detachment is a very important step in the small-scale and large-scale cultivation of animal cells. Although trypsinization is largely the method of choice, often this method is not really optimized for the application used or another improved method could be used instead of trypsinization. When large-scale cultures for the production of biologicals are established or serum-free or protein-free media are introduced, the usual cell detachment method has to be reexamined and adapted to the new conditions or replaced by a new more adequate method. In this sense, an emerging technique of great interest will be the use of thermoresponsive polymer surfaces for the large-scale cultivation of animal cells, although for the moment this possibility exists only at the T-flask level. This method and perhaps other new methods will improve and facilitate cell detachment in the future.

BIBLIOGRAPHY

1. F. Grinnell, *Exp. Cell Res.* **97**, 265–274 (1976).
2. F. Grinnell, *Int. Rev. Cytol.* **53**, 65–144 (1978).
3. D. Barnes, in J.P. Mather, ed., *Mammalian Cell Culture: The Use of Serum-Free Hormone-Supplemented Media*, Plenum, New York, 1984, pp. 195–237.
4. L.M. Reid and D.M. Jefferson, in J.P. Mather, ed., *Mammalian Cell Culture: The Use of Serum-Free Hormone-Supplemented Media*, Plenum, New York, 1984, pp. 239–280.
5. D. Barngrover, in W.G. Thilly, ed., *Mammalian Cell Technology*, Butterworth, Sevenoaks, U.K., 1986, pp. 131–149.
6. N.G. Maroudas, *J. Theor. Biol.* **49**, 417–424 (1975).
7. A.L. van Wezel, *Dev. Biol. Stand.* **37**, 143–147 (1977).
8. A. Prokop, in C.S. Ho and D.I.C. Wang, eds., *Animal Cell Bioreactors*, Butterworth-Heinemann, Boston, 1991, pp. 21–58.
9. E.D. Hay, *J. Cell Biol.* **91**, 205s–223s (1981).
10. L. Weiss, *Exp. Cell Res.* **8**, 141–153 (1961).
11. J. Litwin, *Appl. Microbiol.* **21**, 169–174 (1971).
12. J.P. Revel, P. Hoch, and D. Ho, *Exp. Cell Res.* **84**, 207–218 (1974).
13. J.P. Revel, *Symp. Soc. Exp. Biol.* **28**, 447–461 (1974).
14. J.J. Rosen and L.A. Culp, *Exp. Cell Res.* **107**, 139–149 (1977).
15. P. Whur, H. Koppel, C.M. Urquart, and D.C. Williams, *J. Cell Sci.* **24**, 265–273 (1977).
16. K.G. Vogel, *Exp. Cell Res.* **113**, 345–357 (1978).
17. R. Pollack and D. Rifkin, *Cell* **6**, 495–506 (1975).
18. E. Lazarides, *J. Cell Biol.* **68**, 202–219 (1976).
19. R.A. Badley, A. Woods, L. Carruthers, and D.A. Rees, *J. Cell Sci.* **43**, 379–390 (1980).
20. J.F. Lamb and P.H. Odgen, *Q. J. Exp. Physiol.* **72**, 189–199 (1987).
21. L.A. Culp and P.H. Black, *Biochemistry* **11**, 2161–2172 (1972).
22. C. Snow and A. Allen, *Biochem. J.* **119**, 707–714 (1970).
23. H.L. Hosick and R. Strohman, *J. Cell. Physiol.* **77**, 145–156 (1971).
24. G.M. Hodges, D.C. Livingston, and L.M. Franks, *J. Cell Sci.* **12**, 887–902 (1973).
25. M. Brugmans, J.J. Cassiman, F. van Leuven, and H. van den Berghe, *Cell Biol. Int. Rep.* **3**, 257–263 (1979).
26. W.L. McKeehan, *Cell Biol. Int. Rep.* **1**, 335–343 (1977).
27. W.L. McKeehan, W.G. Hamilton, and R.G. Ham, *Proc. Natl. Acad. Sci. U.S.A.* **73**, 2023–2027 (1976).
28. J. Freshney, *Culture of Animal Cells: Manual of Basic Techniques*, Alan R. Liss, New York, 1987.
29. Anonymous, *Microcarrier Cell Culture: Principles and Methods*, Pharmacia Fine Chemicals, 1981.
30. C. Waymouth, *In Vitro* **10**, 97–111 (1974).
31. O.-W. Merten, *Dev. Biol. Stand.* **99**, 167–180 (1998).
32. T. Matsumura et al., *Jpn. J. Exp. Med.* **45**, 383–392 (1975).
33. M. Nagata and T. Matsumura, *Jpn. J. Exp. Med.* **56**, 297–307 (1986).
34. H. Green, O. Kehinde, and J. Thomas, *Proc. Natl. Acad. Sci. U.S.A.* **76**, 5665–5668 (1979).
35. K.S. Stenn et al., *J. Invest. Dermatol.* **93**, 287–290 (1989).
36. T. Matsumura et al., *Jpn. J. Exp. Med.* **45**, 377–382 (1975).
37. J.J. Cassiman, M. Brugmans, and H. van den Berghe, *Cell Biol. Int. Rep.* **5**, 125–132 (1981).
38. J.B. Griffiths, in R.E. Spier and J.B. Griffiths, eds., *Animal Cell Biotechnology*, Vol. 1, Academic Press, London, 1985, pp. 49–83.
39. E. Bidwell and W.E. van Heyningen, *Biochem. J.* **42**, 140–151 (1948).
40. E.Y. Lasfargues and D.H. Moore, *In Vitro* **7**, 21–25 (1971).
41. P.B. Gordon, M.A. Levitt, C.S.P. Jenkins, and V.B. Hatcher, *J. Cell. Physiol.* **121**, 467–475 (1984).
42. D. Weinstein, *Exp. Cell Res.* **43**, 234–236 (1966).
43. G. Poste, *Exp. Cell Res.* **65**, 359–367 (1971).
44. G. Tarone, G. Galetto, M. Prat, and P.M. Comoglio, *J. Cell Biol.* **94**, 179–186 (1982).
45. I. Zeidman, *Cancer Res.* **7**, 386–389 (1947).
46. E. Zwilling, *Science* **120**, 219 (1954).
47. M.M. Burger, in S. Fliescher and L. Parker, eds., *Methods in Enzymology*, Academic Press, New York, 1974, pp. 615–621.
48. J.J. Cassiman and M.R. Bernfield, *Exp. Cell Res.* **91**, 31–35 (1975).
49. S.M. Cox, P.S. Baur, and B. Haenelt, *J. Histochem. Cytochem.* **25**, 1368–1372 (1977).
50. H.D. Menssen, M. Herlyn, U. Rodeck, and H. Koprowski, *J. Immunol. Methods* **104**, 1–6 (1987).
51. N.E. Payne, P. Whur, R.T. Robson, and N.P. Bishun, *Cytobios* **8**, 49–56 (1973).
52. W.G. Sanford, *In Vitro* **10**, 281–283 (1974).
53. O.-W. Merten et al., in M.J.T. Carrondo, B. Griffiths, and J.L.P. Moreira, eds., *Animal Cell Technology: From Vaccines to Genetic Medicine*, Kluwer Academic Publishers, Dordrecht, The Netherlands, 1997, pp. 343–348.

54. H.K. Kleinmann, R.J. Klebe, and G.R. Martin, *J. Cell Biol.* **88**, 473–485 (1981).

55. J.D. San Antonio, A.D. Lander, T.C. Wright, and M.J. Karnovsky, *J. Cell. Physiol.* **150**, 8–16 (1992).

56. R.G. LeBaron et al., *J. Biol. Chem.* **264**, 7950–7956 (1989).

57. J.E. Koda, A. Rapraeger, and M. Bernfield, *J. Biol. Chem.* **260**, 8157–8162 (1985).

58. M. Rabinovitch and M.J. DeStefano, *In Vitro* **11**, 379–381 (1975).

59. G.L. Nicolson, J.R. Smith, and G. Poste, *J. Cell Biol.* **68**, 395–402 (1976).

60. H. Nielsen, *Acta Pathol. Microbiol. Immunol. Scand., Sect. C* **95**, 81–84 (1987).

61. T. Takezawa, Y. Mori, and K. Yoshizato, *Bio/Technology* **8**, 854–856 (1990).

62. N. Yamada et al., *Makromol. Chem., Rapid. Commun.* **11**, 571–576 (1990).

63. G. Rollason, J.E. Davies, and M.V. Sefton, *Biomaterials* **14**, 153–155 (1993).

64. T. Okano, N. Yamada, H. Sakai, and Y. Sakurai, *J. Biomed. Mater. Res.* **27**, 1243–1251 (1993).

65. T. Terasima and L.J. Tolmach, *Exp. Cell Res.* **30**, 344–362 (1963).

66. E. Robbins and P.I. Marcus, *Science* **144**, 1152–1153 (1964).

67. R.E. Spier, C.E. Crouch, and H. Fowler, *Dev. Biol. Stand.* **66**, 255–261 (1987).

68. M.S. Croughan and D.I.C. Wang, in C.S. Ho and D.I.C. Wang, eds., *Animal Cell Bioreactors*, Butterworth-Heinemann, Boston, 1991, pp. 213–249.

69. M.U. Nollert, S.L. Diamond, and L.V. McIntire, *Biotechnol. Bioeng.* **38**, 588–602 (1991).

70. T.G. van Kooten et al., *Med. Eng. Phys.* **16**, 506–512 (1994).

71. J.M. Clark and M.D. Hirtenstein, *Ann. N.Y. Acad. Sci.* **369**, 33–46 (1981).

72. C.L. Crespi and W.G. Thilly, *Biotechnol. Bioeng.* **23**, 983–994 (1981).

73. J. Delzer, H. Hauser, and J. Lehmann, *Dev. Biol. Stand.* **60**, 413–419 (1985).

74. C. Kluft et al., in A. Mizrahi and A.L. van Wezel, eds., *Advances in Biotechnological Processes* Vol. 2, Alan R. Liss, New York, 1983, pp. 97–100.

75. C. Xiao et al., in O.-W. Merten, P. Perrin, and B. Griffiths, eds., *New Developments and New Applications in Animal Cell Technology*, Kluwer Academic Publishers, Dordrecht, The Netherlands, 1998, pp. 389–393.

76. C. Gebb et al., *Dev. Biol. Stand.* **50**, 93–102 (1982).

77. E. Lindner, A.-C. Arvidsson, I. Wergeland, and D. Billig, *Dev. Biol. Stand.* **66**, 299–305 (1987).

78. U. Lindskog, B. Lundgren, D. Billig, and E. Lindner, *Dev. Biol. Stand.* **66**, 307–313 (1987).

79. C. Gebb, B. Lundgren, J. Clark, and U. Lindskog, *Dev. Biol. Stand.* **55**, 57–65 (1984).

80. M. Manousos et al., *In Vitro* **16**, 507–515 (1980).

81. S. Reuveny and R.W. Thoma, *Adv. Appl. Microbiol.* **31**, 139–179 (1986).

82. A.L. van Wezel, C.A.M. van der Velden-de Groot, and J.A.M. van Herwaarden, *Dev. Biol. Stand.* **46**, 151–158 (1980).

83. D. Billig et al., *Dev. Biol. Stand.* **55**, 67–75 (1984).

84. O.-W. Merten et al., in S. Cohen and A. Shafferman, eds., *Novel Strategies in Design and Production of Vaccines*, Plenum, New York, 1996, pp. 141–151.

85. O.-W. Merten, R. Wu, E. Couvé, and R. Crainic, *Cytotechnology* **25**, 35–44 (1997).

86. H.J. Cruz, E.M. Dias, J.L. Moreira, and M.J.T. Carrondo, *Appl. Microbiol. Biotechnol.* **47**, 482–488 (1997).

87. M. Eloit, *Virologie* **1**, 413–422 (1997).

88. Committee for Proprietary Medicinal Products: Ad Hoc Working Party on Biotechnology/Pharmacy and Working Party on Safety Medicines. EEC Regulatory Document. Note for Guidance, *Biologicals* **20**, 155–158 (1992).

See also CELL CYCLE EVENTS AND CELL CYCLE-DEPENDENT PROCESSES; CELL CYCLE SYNCHRONIZATION; CELL DIFFERENTIATION, ANIMAL.

CELL DIFFERENTIATION, ANIMAL

NANCY L. PARENTEAU
Organogenesis, Inc.
Canton, Massachusetts

OUTLINE

Introduction

The Epidermal Keratinocyte: A Paradigm of Cell Lineage, Differentiation, and Regeneration

   Cell Communication via Cytokines Controls Maturation, Renewal, and Repair

   The Heterogeneous Basal Cell Population: A Further Dissection of Cell Lineage

   Working with Cell Populations and Cell Lineage — Proliferation versus Differentiation

   Response to Environment

Organs are Collections of Multiple Cell Types: The Role of Cell–Cell Interaction in a Corneal Model

Hepatocyte Cell Biology and Cell Lineage

   Hepatic Organization

   Enhanced Function through Cell and Matrix Interaction

Conclusion

Acknowledgments

Bibliography

INTRODUCTION

We have learned a great deal about stem cell biology, cell lineage, and cell communication from the hematopoietic system (1). By contrast, our understanding of parenchymal and connective tissue cell lineage is at the relative beginning. This section will focus on growth, differentiation, and lineage of cells in organized tissues, namely, the skin, cornea, and liver. The in vitro behavior of those cell populations and their patterns of adaptation, growth, and differentiation help shed further light on principles of cell biology in vivo. The cell's remarkable ability to adapt in vitro when presented with a new set of circumstances is part of a basic biological drive to survive and maintain, characteristics usually used to describe the whole organism. It is an outcome of evolution and is present in living systems at all levels. For those of us working with complex cell populations or systems outside the body, the plasticity of cell response in vitro is a curse if ignored, a blessing if one knows how properly to interpret it, and is almost always a technical challenge.

THE EPIDERMAL KERATINOCYTE: A PARADIGM OF CELL LINEAGE, DIFFERENTIATION, AND REGENERATION

Keratinocytes are epithelial cells that form stratified, keratinizing epithelium. The epidermal keratinocyte is the primary cell type of the epidermis, the outer layer of skin. These cells stratify and terminally differentiate to form the dead outer cornified layer of skin. Their function is to maintain closure and protect from desiccation and the environment via the barrier established by the cornified layers. The keratinocyte matures as it moves up in the strata of the epidermis (Fig. 1). Keratinocytes build a strong cytoskeletal framework by changing expression of keratin microfilament species from keratins 5 and 14 to keratins 1 and 10. They accumulate cornified envelope proteins such as involucrin and loricrin in preparation for corneocyte formation (2,3). Filaggrin is synthesized and forms keratinohyalin granules that will later aggregate the keratin filaments in the uppermost layers (4,5). Specialized lipid species such as ceramides (6) are synthesized and concentrated into lipid lamellar bodies. The lamellar bodies will be extruded before cell death to form the lipid "mortar" to the corneocyte "bricks" of the stratum corneum (7). These changes are accompanied by a change in morphological appearance that is clearly seen in histological sections. Five strata can be distinguished by their morphological features: the basal, suprabasal, spinous, granular, and cornified layers (Fig. 1). The cell orientation changes from cuboidal or columnar in the basal layer to squamous as maturation proceeds. There is a distinct enlargement of the cells during this process. The cells become increasingly flattened but packed with keratin filaments, and firmly attached to each other via numerous desmosomes. As filaggrin accumulates and forms keratohyalin granules (4), the cells take on a distinct granular appearance followed by an even further flattening and increase in intracellular density. The lamellar lipids are extruded and the cell undergoes terminal differentiation

and cell death. The calcium influx associated with an increased membrane permeability activates membrane-bound transglutaminase K (8,9), which cross-links the cornified envelope proteins at the cell's periphery (10). This forms a tightly packed sack of keratin filaments encased in a strong, heavily cross-linked protein envelope (11). The extruded lipids interact with the proteins of the envelope (7,12) and organize in specific lipid lamellae with a broad–narrow–broad ultrastructural pattern (13). Together, the corneocytes and the lipid lamellae provide a functional barrier.

Cell Communication via Cytokines Controls Maturation, Renewal, and Repair

While the cells of the suprabasal and spinous layers are readying themselves for eventual cornified envelope formation, they also participate in important functions of immune regulation and inflammation through the production of cytokines and their interaction with Langerhans cells (14). In addition to recruiting the necessary immune cells in response to injury, cytokine production by these cells also provides biological feedback both to the epidermal keratinocyte and the dermal fibroblast with respect to the "status of operations."

Cytokines that can act between the keratinocytes and the fibroblasts are summarized in Table 1 (15–18). They can be placed in two general categories, the inflammatory mediators: interleukin-1 (IL-1), tumor necrosis factor-α (TNF-α), and interleukin-6 (IL-6); and the non-inflammatory cytokines: transforming growth factor-α (TGF-α), amphiregulin, heparin-binding epidermal growth factor (HB-EGF), platelet-derived growth factor (PDGF), transforming growth factor-β (TGF-β), keratinocyte growth factor (KGF), neu differentiation factor (NDF), nerve growth factor (NGF), basic fibroblast growth factor (basic FGF), insulin-like growth factor-1 (IGF-1), and interferon α,β (IFN α,β).

Both IL-1 and TNF-α act as primary mediators of inflammation (14). They are the first to communicate that the tissue is compromised in some way. Through their action, the production of other cytokines is stimulated and the dermal fibroblast is activated. For example, barrier disruption initiates a cytokine cascade, resulting in an increase in TNF-α, IL-1α, and IL-1β (19,20), and increased mRNA encoding for TGF-α and TGF-β (15), as well as other immune modulating cytokines (15,19). This manifests itself through increased proliferation (even though TNF-α

The Epidermal Keratinocyte Exhibits a Distinct Cell Lineage Within the Epidermis

Figure 1. The epidermis is formed of keratinocytes in five morphologically distinct layers. The layers can also be distinguished based on biochemical characteristics associated with terminal differentiation.

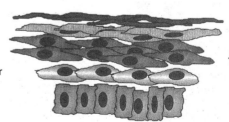

Strata:

Cornified layer
Granular layer
Spinous layer

Suprabasal layer

Basal layer

Some key features:

Cross-linked cell envelope
{ extruded lipid lamellae
{ Keratohyalin, lamellar bodies

{ Transglutaminase Type I
  Envelope proteins
Keratins 1,10

Keratins 5,14

Table 1. Paracrine and Autocrine Factors in Epidermal Keratinocytes and Dermal Fibroblasts

A. Inflammatory mediators	Produced by	Effects
IL-1 (α,β)	Keratinocytes and fibroblasts	Primary inflammatory cytokine, that modulates fibroblast proliferation and ECM regulation by fibroblasts. Effect can be positive or negative. Induces PDGF in fibroblasts. Stimulates keratinocyte migration. Induces KGF in fibroblasts.
TNF-α	Keratinocytes	Primary inflammatory cytokine that inhibits keratinocyte proliferation. Induces ICAM expression in keratinocytes and promotes growth of fibroblasts. Can stimulate or inhibit collagen biosynthesis in fibroblasts. Induces secondary cytokine release in keratinocytes and fibroblasts.
IL-6	Keratinocytes and fibroblasts	Up-regulates tissue inhibitor of metaloproteinase (TIMP) in fibroblasts. Growth support of keratinocytes (co-mitogen).

B. Noninflammatory cytokines	Produced by	Effects
EGF-like molecules (Erb-β ligands) TGF-α Amphiregulin HB-EGF	Keratinocytes	Growth promotion in keratinocytes. Induction of keratins 6 and 16 in keratinocytes. Growth promotion in fibroblasts. Induction of PDGF in fibroblasts. Stimulation of collagen synthesis in fibroblasts.
PDGF	Fibroblasts	Growth promotion in fibroblasts. Chemotactic to fibroblasts. Stimulation of collagenase by fibroblasts.
TGF-β (1,2,3*)	Keratinocytes and fibroblasts	Growth inhibition in keratinocytes. Inhibits or stimulates growth in fibroblasts. Stimulates matrix synthesis by fibroblasts. Induces TGF-β and PDGF in fibroblasts.
KGF (FGF-7)	Fibroblasts	Growth promotion in keratinocytes. Induces TGF-α in keratinocytes.
NDF (heregulin)	Fibroblasts and keratinocytes	Growth promotion of keratinocytes.
NGF	Keratinocytes and Fibroblasts	Prevents apoptosis of keratinocytes. Growth promotion of keratinocytes. Concentrated in hair rudiment mesenchyme during development.
Basic FGF (FGF-2)	Keratinocytes	Growth support (co-mitogen) for keratinocytes.
IGF-1	Keratinocytes	Growth support (co-mitogen) for keratinocytes.
IFN-α, β	Keratinocytes	Growth inhibitor of keratinocytes and fibroblasts. Upregulation of MHC Class II antigens in keratinocytes and fibroblasts. Induction of keratin 17. Inhibition of collagen synthesis by fibroblasts. Induces IL-1.

Source: Ref. 15–18.

can inhibit keratinocyte proliferation (16)), upregulation of keratin-16 expression (associated with hyperproliferative and healing conditions (21), and increased epidermal sterol, fatty acid, and sphingolipid synthesis (19). All this in an attempt to protect the body from further insult by (1) alerting the immune system and (2) restoring barrier function.

The non-inflammatory cytokines (Table 1.B) play a more substantial role in the recruitment of the epidermal keratinocyte and the dermal fibroblast during injury and in the maintenance of homeostasis under normal conditions. Their functions are both stimulatory and inhibitory, and it is the interplay between them that leads to a desired result. They always work in relationship to one another—a fact that is sometimes lost in in vitro experiments that separately probe each factor's actions.

Their action is concentration dependent, ratio dependent, and cell receptor dependent (22–25). Given that caveat, an attempt has been made to highlight some of each factor's purported actions.

TGF-β produced by the keratinocytes acts locally to regulate epidermal turnover, whereas TGF-β produced by the fibroblasts most likely acts on regulation of fibroblast proliferation and matrix biosynthesis. TGF-β is the primary inhibitory factor for the keratinocyte. It is produced by the basal keratinocytes (TGF-β1) and the maturing keratinocytes (TGF-β2) in the suprabasal layers of the epidermis (26). It is a potent inhibitor of keratinocyte proliferation in vitro (27) and appears to be an important regulator controlling proliferation in vivo (28). Squamous cell carcinomas are devoid of TGF-β. In precancerous papillomas, loss of TGF-β1 expression is associated with hyperproliferation of the basal keratinocytes, whereas loss of TGF-β2 results in suprabasal hyperproliferation and is associated with high risk of tumor formation (26). TGF-β can either stimulate or inhibit fibroblast proliferation depending on the concentration. Perhaps a more important effect is that of stimulation of matrix biosynthesis by the fibroblast (29).

The heparin-binding EGF-like family of keratinocyte-produced factors stimulate keratinocyte proliferation via the same receptor (Erb-β1) (14). These molecules are produced by the basal cells. There appears to be a great deal of redundancy in having three self-produced molecules that act in essentially the same way (30). Elimination of TGF-α in knockout mice results in only mild phenotypic change, wavy hair, for example (31). This redundancy may be a result of evolution and illustrates how important proliferation of the amplifying population of the epidermis must be to the organism. Elimination of the Erb-β1 receptor leads to profound defects in epithelial development (32). Neu differentiation factor-beta isoform (NDF, heregulin) has been demonstrated to have proliferative effects on keratinocytes mediated via the Erb-β3 receptor (33).

KGF has also been reported to stimulate keratinocyte growth (34,35) but is distinctly a paracrine factor produced by the dermal fibroblast (34,36). Overexpression of a dominant negative form of KGF in transgenic mice results in epidermal atrophy and impaired wound healing. Interestingly, others have found little impairment of wound healing in KGF knockout mice (37). One pathway of paracrine KGF action is through stimulation of autocrine TGF-α (38). IL-1 has been shown to be a potent inducer of KGF by fibroblasts (39), also linking KGF action to mechanisms of wound healing.

Basic FGF and IGF-1 act as comitogens (40,41). Basic FGF has been found to be more (42) or less (43) effective in stimulation of keratinocyte proliferation depending on the culture conditions and production of intrinsic growth factors. IGF-1 is thought to play a role in dermal–epidermal interactions, particularly in cases such as psoriasis (44), where it has been documented that the fibroblast population can stimulate the psoriatic phenotype in nonpsoriatic epidermis in vitro (45).

Nerve growth factor is produced by keratinocytes (46). It is seen in increasing amounts during growth in vitro (47). NGF produced by fibroblasts has been implicated in the morphogenesis of hair rudiments (48). NGF has also been reported to block apoptosis, acting as a "survival factor" for human keratinocytes in vitro (49).

Even after cursory review of autocrine and paracrine cytokines affecting the epidermis, it is clear that the important factors are the balance and timing of expression of these elements. Recent work using complementary DNA microarray techniques representing approximately 8600 different human genes show the temporal program of gene expression of dermal fibroblasts in response to serum to consist of a wide array of genes associated with the physiology of wound healing. Serum stimulation caused a temporal increase in transcription factors followed by cell-cycle-associated factors, which were followed by wound-healing-associated factors including proteins involved in

matrix remodeling and intercellular signaling. There appeared to be coordinated regulation of clusters of genes, each cluster covering a range of functions (50). Although this system is an artificial situation, it does imply complex linkage in gene expression.

The Heterogeneous Basal Cell Population: A Further Dissection of Cell Lineage

The stem cells of the epidermis reside in the basal layer (51). In addition, there are stem cells residing in the bulge region of the hair follicle (52), which can also give rise to epidermis, particularly in cases of loss of the interfollicular epidermis (i.e., partial thickness skin loss). Much of the evidence for the existence of stem cells has relied upon the fact that stem cells by their nature will be naturally slow cycling in vivo (53,54). Labeling of neonatal (55,56) and regenerating (57) mouse skin and fetal human skin (58) have demonstrated the existence of the epidermal proliferative unit (EPU) (Fig. 2) (59–63). Close examination of cells in vitro grown using the 3T3 feeder cell system of Rheinwald and Green (64) finds a similar kinetic heterogeneity (59), further substantiating the existence of an EPU in humans. This proliferation pattern of stem cell turnover and subsequent amplification of a proliferative population is outlined in Figure 2. In organotypic culture, "columns" of epidermal cells, presumably a result of an EPU pattern of growth, can be distinguished similar to mouse skin (Fig. 3). In tissues where there is a high turnover, such as the epidermis, it makes sense that it would be most efficient to have amplifying populations between the stem cell and the mature cell (53). The amplifying population has a clearly limited proliferative potential and is regulated by the cytokine environment of the epidermis discussed previously. Self-maintenance would be an important aspect to the epidermal stem cell to assure adequate generation of epidermis throughout the lifespan of the individual. Therefore, it may, like the hematopoietic stem cell, be subject to different activation and inactivation signals for cell division, either for purposes of self-renewal or for replenishment of the amplifying population (53).

It stands to reason that much of what we study as keratinocyte growth regulation in vitro is actually a study of the transit amplifying population. Keratinocytes cultured in the 3T3 feeder cell system permit clonal growth from single cells (64). Keratinocytes grown in this way give rise to three different types of colonies (65). Holoclone is the name given to large colonies with the highest proliferative potential, meroclones to those with moderate potential, and paraclones to small abortive colonies. The holoclone was thought to originate from a stem-like cell, based on its high proliferative potential. However, it must be kept in mind that there will be amplifying cells with different proliferative potential based on their point in the amplification pathway [Fig. 2, (2)]. Indeed, some have suggested that because stem cells may require a special niche environment to be maintained, there may be no true stem cells in the cultured population, that is, that all stem cells may be "activated" to become amplifying cells (66). The evidence against this is the fact that epidermal sheet grafts cultured using this method are not only capable

Stem Cells and Heterogeneity of the Basal Keratinocyte

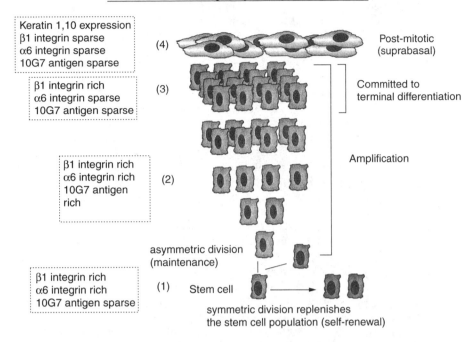

Keratin 1,10 expression
β1 integrin sparse
α6 integrin sparse
10G7 antigen sparse

(4) Post-mitotic
(suprabasal)

β1 integrin rich
α6 integrin sparse
10G7 antigen sparse

(3)

Committed to
terminal differentiation

Amplification

β1 integrin rich
α6 integrin rich
10G7 antigen
rich

(2)

Figure 2. The basal keratinocyte population consists of both slow-cycling stem cells and proliferative transit amplifying cells. The transit amplifying cells have a finite proliferative potential. The stem cell and its progeny form the epidermal proliferative unit (EPU) (54–56, 59). Basal cell populations have been distinguished by basal cell surface markers (60–62) and keratins (63).

asymmetric division
(maintenance)

β1 integrin rich
α6 integrin rich
10G7 antigen sparse

(1)   Stem cell

symmetric division replenishes
the stem cell population (self-renewal)

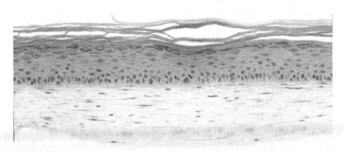

Figure 3. Histological section of epidermal keratinocytes in organotypic culture. Cell proliferation is controlled and terminal differentiation of the suprabasal cells is favored by culture at the air–liquid interface. All strata of the epidermis are once again represented. Note the ordered "stacking" of cell nuclei in the epidermis as a result of the EPU. Hematoxylin and eosin, mag: 62.5x.

of re-establishing epidermis on severely burned patients, but of maintaining it for many years (67). This does not rule out, however, that the vast majority of cells in a holoclone are transit amplifying cells, nor that some clones established from early stage (2) amplifying cells will result in a holoclone morphology.

The study of the keratinocyte stem cell has been hampered by the lack of markers to distinguish them from the transit amplifying and maturing subpopulations. Concentration of integrin expression has been a useful tool in this regard. Using cell–cell associated β1 integrin, Jones and Watt (60) were able to discern two cell populations within the proliferative in vitro population (again cultured in the feeder cell system) using fluorescence-activated cell sorting (FACS): a β1-rich population and a β1-sparse population. The β1-sparse population represented those cells committed to terminal differentiation [refer to (4) in Fig. 2]. Those that were β1-rich bound more readily

to basement membrane components in culture and had a higher proliferative potential as would be expected of subpopulations (1) and (2). The majority of β1-rich, rapid binding cells, however, still contain the amplifying cell subpopulation (2–3). Very recent developments have now made it possible to distinguish subpopulations (1) and (2).

Using an integrin that is cell-matrix associated (i.e., the hemidesmosome-associated α6 integrin), Li et al., (61) were able to distinguish an α6-bright (rich) and α6-dim (sparse) population within the β1 (bright) -rich population by FACS. Subsequent culture of this population indicated that the α6-sparse population within the β1-rich population was already postmitotic in vitro (61). In addition, Kaur et al., (68) had previously developed a monoclonal antibody to a surface antigen, 10G7, present on rapidly dividing normal and tumor cells. The level of 10G7 expression appeared to correlate well with the degree of proliferation. They then hypothesized that stem and noncommitted cells would preferentially express high amounts of α6 integrin, and further hypothesized that if the stem cell was slow cycling, then it should concomitantly express low levels of 10G7 antigen. A fresh dissociate of neonatal foreskin keratinocytes was sorted into subsets using these markers in combination. Subsequent culture of these populations verified that the α6-rich/10G7-sparse population generated the most cell progeny over multiple passages using the 3T3 feeder cell system (61). This subpopulation was estimated to represent approximately 10% of the basal cell layer, and it was estimated that a single α6-rich/10G7-sparse stem cell was capable of generating approximately 5.8×10^{11} cells. It should be kept in mind that none of these studies implies that the stem cell is replenished during the culture period. In fact, evidence has shown that the number of clonogenic cells remains remarkably constant when cultured in vitro, grafted 40 weeks in vivo, and then recultured in vitro (69). Mechanisms of stem cell down-regulation and replenishment, and the factors involved in each, remain to be determined.

Working with Cell Populations and Cell Lineage — Proliferation versus Differentiation

Once dissociated and placed in culture, the order of cell lineage and the spatial relationships present in the tissue are lost; the proportion of each subpopulation may also be drastically altered. This is again easily illustrated using the keratinocyte with its well-defined lineage in vivo and ample markers that can be used in vitro to follow phenotypic changes within the population. When keratinocytes are dissociated from the epidermis and placed in culture, the genetics that drive the differentiation program are present but can become skewed or short-circuited by the keratinocyte's adaptation to the environment. To illustrate, at the time of dissociation, the keratinocyte population is very heterogeneous due to the fact that the suspension contains cells from each of the clearly defined cell layers (Fig. 1).

As we now appreciate, only a relatively small proportion of these cells is capable of proliferation. This fact gives keratinocytes a technical reputation of being difficult to grow. When a feeder cell layer of transformed mouse 3T3 cells is used for coculture, the proliferative cells are able to plate and grow at low density, supported and stimulated by the feeder cell interaction (64,70) (a model used in many of the studies already alluded to in this article). Colonies of growing keratinocytes are generated from the proliferative cell compartment. The growing colonies eventually merge and form an epithelial layer pushing the feeder cells from the dish as they proceed. During this time, differentiation and modest stratification have occurred within the colonies since the feeder cell system permits some stratification owing to a calcium concentration that allows cell junctions to form (71). When the colonies have merged to confluence in the culture vessel, differentiation is further stimulated (72,73) as the mechanisms of keratinocyte behavior dictate an attempt at barrier formation. The high calcium concentration stimulates the enzyme transglutaminase K, which is needed to cross-link the cornified envelope proteins in the cornification process (9). Envelope precursors such as involucrin are increased at confluence, as are most of the differentiation markers (36). Involucrin expression is not dependent on stratification (74), although normally it would only be present in suprabasal keratinocytes. Cornified envelopes can be observed and even a larger proportion of cells are envelope competent (72), meaning that they are expressing transglutaminase K and envelope precursors in functional quantities so that if the cells are exposed to a calcium ionophore transglutaminase is activated and a cornified envelope is formed prematurely. In fact, one of the barriers to complete differentiation of the keratinocyte in vitro is this ability to undergo, for lack of a better term, "premature differentiation." This can occur uncoupled to necessary changes in lipid biosynthesis, formation of lamellar bodies or (pro)filaggrin (75) synthesis, and formation of keratohyalin granules.

It is of interest to note that this same culture system is capable of promoting the keratinocyte phenotype in epithelial cells from nonkeratinizing tissues from the rat (76,77). Therefore, this culture system promotes adaptation to a squamous phenotype — at least in the rat. Much of what we know about the keratinocyte was gained by observing and manipulating their behavior in the feeder cell environment. This system has proven invaluable for clonal analysis and for generation of cultured epidermis from adult tissue. The limitation is that not all processes are equally enabled, and certainly not all strata of the epidermis are clearly represented in the population. Therefore, caution must be used in interpreting cell responses. When grafted in vivo, however, the rudimentary epidermal sheets are able to "take" and rapidly establish differentiated epidermis (78); therefore, the potential for full maturation is present, but limited by the culture environment. While this system may be appropriate for the study of normally squamous epithelium, in vitro study of nonsquamous epithelial cells in this system must be approached with caution, and an appreciation of the underlying cell adaptation toward squamous differentiation that can occur.

Certain parts of a cell's differentiation program can be uncoupled by a limiting culture environment. This was particularly important for our laboratory's work on both defined growth of keratinocytes and formation of a differentiated epithelium in a skin equivalent model (79) — two goals that may seem diametrically opposed. Our first goal was to minimize the heterogeneity in the growing keratinocyte culture by stimulating the regenerative response while limiting the subsequent push towards terminal differentiation. To achieve this, we wished to define a culture system that would rely primarily on paracrine and autocrine mechanisms for generation of the epithelial cells, taking advantage of the regenerative response (Fig. 4). Calcium levels were reduced to levels below that which had been previously shown to promote differentiation in human (80) keratinocytes. This calcium concentration would also not permit stratification. Small explants of tissue were used to start the cultures. We felt that explants maintained a critical cell density and would be a more natural way to promote the regenerative response. As covered earlier, previously scientists had used mouse 3T3 feeder cells to make up for a lack of critical cell density; however, the use of the transformed feeder cells required, until recently (81), the use of serum for optimum feeder performance.

In our efforts to define the medium formulations, we found that serum promoted differentiation with respect to transglutaminase content and envelope competence, particularly when calcium concentrations were permissive (43). Therefore, we wanted a method that would avoid feeder cells and serum. Previous work on a serum-free medium (80) had demonstrated that growth in a serum-free, feeder-free medium was possible if the calcium levels were kept below 1.0 mM. However, the heterogeneity of the primary dissociate required either the support of the feeder cells or supplementation with bovine brain extract to stimulate growth of the proliferative population, which was at or below critical cell density to support autocrine

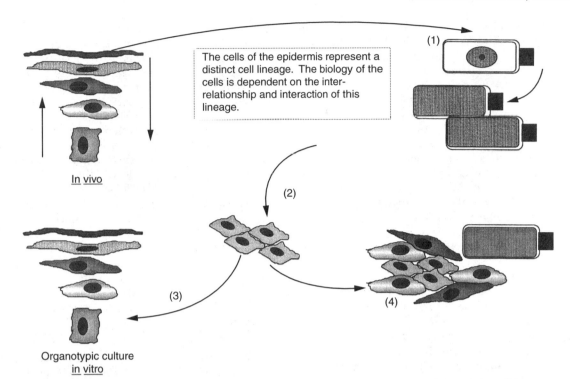

The cells of the epidermis represent a distinct cell lineage. The biology of the cells is dependent on the inter-relationship and interaction of this lineage.

In vivo

Organotypic culture in vitro

Figure 4. The keratinocyte population reflects its heterogeneity in vitro. The interaction of the subpopulations in vitro: (*1*) The culture environment removes the proliferative inhibition on the basal layer. This stimulates the regenerative response in the form of an explant outgrowth within a few days. (*2*) The proliferative response continues for several cell passages as the basal cell compartment attempts to regenerate epidermis in the absence of stratification and inhibition of differentiation signals. Differentiation is not promoted in this environment but is also not entirely blocked. (*3*) Proliferative cells put into an organotypic culture give rise to stratified layers that recapitulate the full cell lineage and reform an organized epidermal tissue. Although their proliferative rate is down-regulated by the presence of the subrabasal strata, they can be stimulated to proliferate once again if given the impetus such as wounding or injury. (*4*) If cells are not placed into organotypic culture, eventually the increase in differentiated cells supplies feedback inhibition to the proliferative cells and limits further growth.

and paracrine response (82). By maintaining the tissue in the form of an explant, a supportive microenvironment could be maintained while stimulating the natural regenerative response during this critical period. Our result was a robust outgrowth of proliferative cells after a few days in a defined medium (43).

The cell population generated was markedly more homogeneous than we had achieved using the undefined systems, and had a strong "basal" cell character as evidence by a lack of differentiation in response to calcium (43). The proliferation response could be maintained in the defined environment for multiple cell generations, although there was a gradual shift in the population as more differentiated cells formed and inhibited the regenerative response from continuing in the remaining proliferative cell population. We were able to favor the proliferative response and inhibit, but not completely block, the differentiative response (e.g., see (4) in Fig. 4). The next goal was to be able to apply this population of keratinocytes, which favored maintenance of basal cell proliferation and enable stratification and differentiation of the cells in an ordered way to reform the multicellular/multiphenotype epidermis (e.g., see (3) in Fig. 4), again largely relying on autocrine, paracrine, and

environmental mechanisms. Epidermal growth factor and fibroblast growth factor were found to have little effect when growth was maximally self-stimulated (43). Even TGF-β, a potent inhibitor of proliferation, was marginally effective in attenuating growth when cultures were maximally stimulated at densities of approximately 60% confluence (83).

Response to Environment

Cytokines are responsible for communicating circumstances within the tissue or cell population. Physical signals trigger changes in cytokines, as illustrated earlier in the case of barrier disruption. With epidermis in vitro, an air interface promotes phenotypic expression, differentiation, and tissue organization of an epidermal keratinocyte culture by stimulating cornification (84). The rate of water loss through the epidermis is thought to act as a stimulus for barrier formation and is most likely communicated through a flux in cytokines. These dynamic interactions are also at play in vitro when we attempt to reconstruct epidermis, as we wished to do in the formation of a skin substitute. Therefore, the goal was to allow normal processes to occur. To understand this we studied the transition epidermal keratinocytes underwent as we took them from a very proliferative, differentiation-inhibiting culture environment (43) and placed them in three-dimensional organotypic culture (85–87).

As the keratinocytes are taken from a rapidly proliferating condition, they are a more homogeneous population. They are placed at a near-confluent density onto a collagen lattice containing dermal fibroblasts (88). This serves as a culture substrate for the epidermal keratinocyte. The conditions are changed to promote epidermal differentiation (86,88). The calcium concentration is increased to 1.88 mM to allow cell junctions to form and cell stratification to occur. Culture at an air interface promotes characteristics of terminal differentiation (84). The epidermis will form by the basal cells dividing, leaving the basal compartment, and differentiating as they rise in the strata to eventually form a corneocyte. It is not formed by mere segregation of the keratinocyte subpopulations. Stratification, differentiation, and maturation are achieved in an active process through the proliferation and transit of cells generated by the basal layer (86). Examination of the morphology of the uppermost squames reveals enlarged, poorly shaped corneocytes, most likely generated by the original late transit amplifying cells that were present in the culture dish (87). With time, the basal population is replenished through proliferation within the basal cell layer. The proliferative rate during this process gradually diminishes to reach a level closer to what is seen in vivo (86,89).

The diminution of proliferative rate is the result of feedback from the stratified layers of epidermis above and the physical feedback of the stratum corneum, which is gradually establishing barrier function (82,87). Proliferation is quelled through normal feedback mechanisms, not due to a loss of proliferative or regenerative capacity. The basal cell labeling index can be stimulated by agents such as sodium ascorbate and cholera toxin, a cyclic AMP elevating agent, but not EGF (90). The regenerative response can also be stimulated by wounding of the tissue. When wounded, IL-1 and PDGF levels rise in a response leading to rapid migration of the keratinocytes followed by re-establishment of the epidermis through proliferation (91; Hardin-Young, personal communication).

Under standard conditions the "mature" skin equivalent has a phenotype reminiscent of a freshly healed wound. Keratin 19, present during rapid proliferation, is down-regulated; however, keratin 16, present in healing or hyperproliferative conditions (21), is still expressed (90) and barrier function is approximately 10–30-fold more permeable than normal skin (92). Basement membrane is not continuous, even though rudimentary structure is present and the components of the basement membrane appear to be synthesized and localized at the dermal–epidermal junction (85; J.D. Zieske, personal communication). Upon grafting, basement membrane assembles (93) and the barrier is equivalent to human values within days (K. Kriwet and N.L. Parenteau, personal communication). In skin, or constructs made of human keratinocytes and de-epidermized dermis, lipid metabolism approaches normal levels within weeks, yet does not fully attain normal lamellar structure upon grafting (94). This may be due to the influence of the murine metabolism to which the skin graft is engrafted.

The in vitro environment does not yet appear to adequately establish all parameters of maturation. Deficient lipid metabolism appears to be a key element. Characterization of a variety of organotypic skin cultures found them to be deficient in free fatty acids (95). Linoleic acid supplementation is important to help direct proper lipid biosynthesis (96), and serum appears to be detrimental to proper terminal differentiation (86), even if delipidized to remove retinoids (90). Reduction of temperature to more physiologic levels (33 °C) is beneficial in helping to maintain a stable culture for longer periods of time (97). This is essential, since even in vivo a freshly healed wound takes approximately three weeks to re-establish full barrier function. Therefore, some of the characteristics we see in vitro are not entirely due to inadequacies in the nutritional base or environment, but are also a manifestation of the physiology, cytokine milieu, and regulation present in the maturing or "repairing" epidermis.

ORGANS ARE COLLECTIONS OF MULTIPLE CELL TYPES: THE ROLE OF CELL–CELL INTERACTION IN A CORNEAL MODEL

The role of communication between different cell types had been appreciated as important ever since embryologists observed the influence of epithelial and mesenchymal interaction. It is obvious from Table 1 that the keratinocyte and fibroblast have ample opportunity for communication through the use of multiple cytokines. Communication can also be indirect through an additional cell type such as an inflammatory cell (14) or through matrix deposition, which can sequester factors or act directly on cell receptors (98). Our experience with a multicellular corneal cell construct (99) presents an intriguing model to study such interactions.

The cornea, the clear outer surface of our eye, consists of three cell types, the corneal epithelial cell, the stromal fibroblast, and the corneal endothelial cell. The corneal extracellular matrix is highly ordered, resulting in a clear appearance. The corneal epithelium is stratified but not cornified. Its barrier properties are established through tight junctions rather than corneum formation. However, when there is substantial loss of tear formation, the corneal epithelium will cornify in response to the dry environment. The location of the regenerative population of the cornea differs from that of the skin in that the proliferative cell compartment is limited to the outermost region of the cornea, the limbal region (100). The stem cells of the cornea also reside in the limbal region (101). As the epithelium is renewed, cells from the limbal region proliferate and move centrally to replenish the central cornea (102). The corneal endothelial layer in vivo is a single layer of specialized cells that establish a barrier on the underside of the stroma and regulate fluid transport into and out of the matrix. This function is essential for maintenance of clarity of the corneal stroma.

Our study of rabbit corneal keratinocytes, rabbit stromal fibroblasts, and mouse endothelial cells in

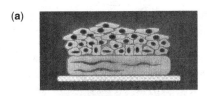

(a)

Monolayer, serum-free cultures derived from the limbal region are placed in 3-dimensional culture. Lacking adequate environmental and cell signals, the culture stratifies but does not undergo differentiation.

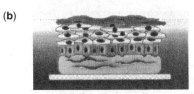

(b)

Providing the environmental signal of a moist interface leads to improved growth and differentiation of the corneal epithelium.

(c)

Addition of corneal endothelial cells dramatically enhances differentiation to a limbal-like phenotpye. Complete basement membrane is formed.

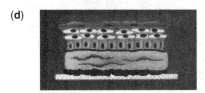

(d)

Corneal endothelium alone is able to trigger full differentiation in the epithelium in the absence of the environmental stimulus of a moist interface.

Figure 5. In an organotypic culture model, rabbit corneal epithelial cells require the presence of the underlying corneal endothelial cells to achieve true differentiation in vitro. Certain characteristics of differentiation can be stimulated by environmental change to a moist interface, although the effect appears to be adaptive rather than specific (summarized from Zieske et al. (99).

organotypic culture has led to surprising observations with respect to cell-cell interaction and effects of the environment (99). Cultures are formed similar to skin equivalents (Fig. 5). The stromal fibroblasts are contained within the collagen lattice, and the epithelial cells are plated on the surface. Cultures have been constructed with or without an endothelial cell layer below the collagen lattice, which consists of mouse endothelial cells transformed with large T antigen to permit proliferation (103).

When the rabbit corneal keratinocytes (CK) are used in conjunction with a collagen matrix containing stromal cells, and grown *submerged* in organotypic culture, the results are poor. The CKs exhibit difficulty spreading over the surface and exhibit uneven stratification. The histological appearance of the stratified corneal epithelium, however, looks similar to that in vivo in some areas. Is this, then, a problem of growth conditions or one of differentiation? Looking at some of the characteristics of these cultures, we find that alpha-enolase, a marker of cell proliferation (100), is expressed throughout the epithelium. Keratin 3, the specialized keratin of the cornea normally expressed in all suprabasal layers (104), is minimally expressed only in the uppermost squamous cells. There is little to no synthesis of the basement membrane components collagen type VII and laminin, which in part explains the cells' difficulty in migration. In keeping with this observation, there is no ultrastructural evidence of basement membrane at the epithelial–stromal junction. The fibroblasts react in

response to the struggling epithelium by migrating out of the stroma and proliferating to form cell mounds under the struggling epithelium. What results, then, is an unstable mixture of cultured cells with little true organization or differentiation [Fig. 5(a)], although at first glance it might look like corneal epithelium.

When the corneal culture is provided a moist interface environment and appropriate environmental stimulus, spreading of the epithelium is improved, and patchy distribution of collagen type VII and laminin is seen [Fig. 5(b)]. Interestingly, matrix molecules also remain within some of the basal cells. The moist interface is sufficient to stimulate synthesis of corneal keratin 3 in the suprabasal layers as in vivo. Alpha-enolase is inhibited from the uppermost layers, although distribution can still be seen suprabasally. Although the culture is improved, there does not appear to be true organization of differentiated characteristics. No basement membrane is formed. It should be noted here that cells may express a particular element without being able to use it properly. Therefore, localization of structural antigens is key—not merely expression—and one should be aware that localization and level of expression may be significantly different in vitro.

The addition of corneal endothelial cells below the stromal cell layer leads to marked improvement in organized growth and spreading of the corneal epithelium [Fig. 5(c)]. Examination of alpha-enolase distribution now shows tight distribution limited to the basal layer, similar to what is seen in the limbal region of the cornea (100). Localization of collagen type VII and laminin show contiguous lines of strong staining at the epithelial–stromal junction. Full structural basement membrane can be seen by electron microscopy.

It had long been recognized that environment was an important component in human skin equivalent maturation; however, much to our surprise, when the moist interface was removed from the three-cell construct of cornea, full differentiation proceeded in the absence of the environmental stimulus [Fig. 5(d)]. This again illustrates the cell's ability to adapt since the cultures in Figure 5(b) had responded to the change to a moist environment, although the changes appear to be adaptive. It was only with the proper interaction of the endothelial cells that specific differentiation occurred. This could be due to a resulting cytokine environment either directly or indirectly mediated by the endothelial cells. The exact mechanism of endothelial cell action remains to be determined. Conditioned medium from the endothelial cells and culture of the corneal model with a physical separation of the endothelium and the stroma leads to a diminution but not complete abolishment of effect (Mason and N. Parenteau, personal communication), indicating there may be more than one element to the mechanism.

HEPATOCYTE CELL BIOLOGY AND CELL LINEAGE

Hepatic Organization

The liver is organized into plates of adjacent hepatocytes bordered by sinusoidal spaces (Fig. 6) (105,106). These plates extend from the portal triad to the central vein in a radial arrangement around the central vein. The bile cannaliculi are formed between the hepatocytes in each

plate. Blood flows from the portal triad to the central vein. Bile flows in the opposite direction to the bile duct. Hepatocyte lineage consists of a stem cell population and three zones of maturing hepatocytes. A few of the differences are highlighted. The hepatocytes mature as they progress from Zones 1 to 3. Zone 3 is the most specialized with a high level of detoxification function. One cannot help but notice the similarities in concept between the zones of hepatocyte lineage and the strata of the epidermis (107). The less mature cells are the most proliferative. Travel, or "streaming," can be observed from Zone 1 to Zone 3 in vivo (108), and the most mature cells exhibit the most specialized functions of the organ.

Enhanced Function through Cell and Matrix Interaction

When hepatocytes are dissociated for cell culture, the architecture and spatial relationships are lost. The hepatocyte has been difficult, if not impossible, to cultivate in vitro until recently (109); therefore, the majority of the work on hepatocytes has focused on enhanced hepatocyte function in vitro with minimal growth. Hepatocytes plated on collagen substrates retain some hepatocyte functions (110). However, hepatocytes in vivo are naturally polarized to two surfaces. If cultured in a double collagen gel with matrix above and below, function of rat hepatocytes is further improved, presumably due to a more enabling environment (110,111). Synthesis of matrix also appears important in enhancing function of rat hepatocytes (111). Therefore, it would appear that factors that enhance the cell's ability to polarize are key.

Work using coculture methods employing mouse 3T3 feeder cells has shown a distinct benefit of coculture in rat hepatocyte function (112). The level of cell–cell contact can be controlled and specifically designed using

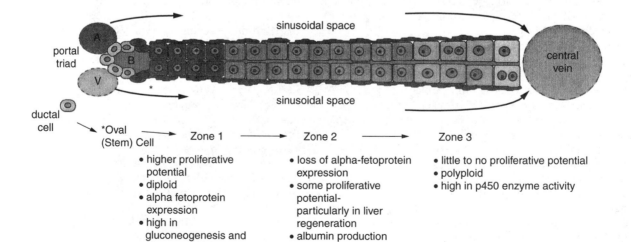

Figure 6. Hepatocyte lineage consists of a stem cell population and three zones of maturing hepatocytes organized in plates radiating from a central vein. A few of the differences are highlighted. The hepatocytes mature as they progress from Zones 1 to 3. Zone 3 is the most specialized, with a high level of detoxification function. Note the similarities in cell lineage biology between the keratinocyte and the hepatocyte: A: hepatic artery; B: bile duct; V: portal vein. Adapted from: Sell, S and Sigal, S.H. (105, 106).

microfabrication techniques. Using this method, Bhatia et al., (112) were able to increase hepatocyte fibroblast interaction in a controlled manner. Keeping relative cell numbers constant, they found that increased interaction resulted in increased metabolic, excretory, and synthetic function in the hepatocytes. The primary signal is most likely tightly associated with the fibroblast surface. These findings are similar to those found for keratinocyte interaction with the 3T3 fibroblasts (70,113).

We now know that keratinocytes do not require the support of the 3T3 fibroblast to proliferate or recapitulate their function; therefore, it should be possible to achieve hepatocyte function without them as well. A recent development in this area has been in the development of a serum-free medium that supports growth of the hepatocyte (109). Clonal growth is achieved with the addition of several growth factors: hepatocyte growth factor (scatter factor) (HGF), EGF, and TGF-α. When proliferating, the cells lose most of their differentiated function and become more ductal in character, although they retain liver-associated transcription factors. When the cells are then cultured with an overlay of EHS gel (Matrigel; Collaborative Research, Bedford, MA) composed primarily of laminin with some type IV collagen, entactin, decorin, TGF-β1, and glycosaminoglycans (114), the cells exhibit a hepatocyte phenotype and secrete albumin. This is also seen in areas of the culture where hepatocytes are in contact with fibroblastic cell contaminants. Hepatocytes cultured between two collagen gels, as described, displayed different phenotypes depending on whether HGF or EGF was present in the serum-free medium. HGF promoted a ductal phenotype, whereas EGF promoted a hepatocyte phenotype (i.e., granularity, albumin secretion, and cord-like arrangement). It appeared that all the cells underwent a similar phenotypic change, ruling out the presence of two distinct populations. From these studies, the authors concluded that the majority of hepatocytes were able to "dedifferentiate" into proliferative cells that were multipotent. Whether the described phenomenon is a real mechanism or the result of adaptation remains to be determined. Examination of cell lineage, additional aspects of differentiation (as we now know, we can be easily fooled by appearance), and attempts at specific cell subpopulations will be needed, particularly in species other than rat, where the genotype is notoriously "unstable" (76,82).

CONCLUSION

The keratinocyte can be an informative paradigm as we forge new ground in other cell systems. We now appreciate that cell lineage affects not only the tissue in vivo, but also the dynamics of the cell population in vitro. Adaptation is a factor that can result in cell survival and growth in the foreign in vitro environment, along with partial expression of phenotype, changes in expression, and different patterns of expression of specific cell components. When conditions foster *true* self-regulation, exogenous agents over and above nutritional components have little effect. If we appreciate and understand this framework, we can learn to work within it to achieve remarkable results

in cell and developmental biology for the goals of tissue engineering.

ACKNOWLEDGMENTS

I would like particularly to acknowledge my colleagues who through the years helped advance our laboratory's understanding of cell differentiation and whose work was highlighted in this article. They are Eric Johnson, Valerie Mason, Patrick Bilbo, Susan Meunier, Marjorie Olsen Hughes, Claire Hastings, and Stefan Prosky; Dr. Cynthia Nolte and Dr. Katrin Kriwet; and collaborators Dr. James Zieske and Dr. Bjorn Olsen.

BIBLIOGRAPHY

1. M.R. Koller and B.O. Palsson, *Biotechnol. Bioeng.* **42**, 909–930 (1993).
2. R.H. Rice and H. Green, *J. Cell Biol.* **76**, 705–711 (1978).
3. T. Mehrel, D. Hohl, and J.A. Rothnagel, *Cell* **61**, 1103–1112 (1990).
4. A.M. Lynley and B.A. Dale, *Biochim. Biophys. Acta* **744**(1), 28–35 (1983).
5. M. Manabe et al., *Differentiation* **48**(1), 43–50 (1991).
6. J. Brod, *Int. J. Dermatol.* **30**, 84–90 (1991).
7. P.W. Wertz et al., *Arch Dermatol.* **123**(10), 1381–1384 (1987).
8. S.M. Thatcher and R.H. Rice, *Cell* **40**, 685–695 (1985).
9. R.H. Rice et al., *Epithelial Cell Biol.* **1**, 128–137 (1992).
10. M. Simon and H. Green, *Cell* **36**, 827–834 (1984).
11. D. Hohl, *Dermatologie* **180**, 201–211 (1990).
12. N.D. Lazo, J.G. Meine, and D.T. Downing, *J. Invest. Dermatol.* **105**(2), 296–300 (1995).
13. D.C. Swartzendruber et al., *J. Invest. Dermatol.* **92**(2), 251–257 (1989).
14. I.R. Williams and T.S. Kupper, *Life Sci.* **58**(18), 1485–1507 (1996).
15. B.J. Nickoloff and Y. Naidu, *J. Am. Acad. Dermatol.* **30**(4), 535–546 (1994).
16. F.W. Symington, *J. Invest. Dermatol.* **92**(6), 798–805 (1989).
17. R.M. Grossman et al., *Proc. Nat. Acad. Sci. U.S.A.* **85**, 6367–6371 (1989).
18. J.D. Chen et al., *J. Invest. Dermatol.* **104**, 729–733 (1995).
19. L.C. Wood et al., *J. Clin. Invest.* **90**(2), 482–487 (1992).
20. J.C. Tsai et al., *Arch. Dermatol. Res.* **286**, 242–248 (1994).
21. J.N. Mansbridge and A.M. Knapp, *J. Invest. Dermatol.* **89**, 253–262 (1987).
22. J. Ansel et al., *J. Invest. Dermatol.* **94**, 101s–107s (1990).
23. E.J. Battegay et al., *J. Immunol.* **154**, 6040–6047 (1995).
24. X.H. Feng, E.H. Filvaroff, and R. Derynck, *J. Biol. Chem.* **270**(41), 24237–23245 (1995).
25. D.A. Lauffenburger et al., *Ann. Biomed. Eng.* **23**(3), 208–215 (1995).
26. A.B. Glick et al., *Proc. Natl. Acad. Sci. U.S.A.* **90**(13), 6076–6080 (1993).
27. X.H. Feng and R. Derynck, *J. Biol. Chem.* **271**(22), 13123–13129 (1996).
28. K. Sellheyer et al., *Proc. Natl. Acad. Sci. U.S.A.* **90**(11), 5237–5241 (1993).

29. S. Karmiol and S.H. Phan, in S.L. Kunkel and D.G. Remick, *Cytokines in Health and Disease*, Dekker, Ann Arbor, Mich., 1992, pp. 271–296.

30. M. Piepkorn, C. Lo, and G. Plowman, *J. Cell. Physiol.* **159**, 114–120 (1994).

31. G.B. Mann et al., *Cell* **73**, 249–261 (1993).

32. P.J. Miettinen et al., *Nature (London)* **376**, 337–341 (1995).

33. M. Marikovsky et al., *Oncogen* **10**(7), 1403–1411 (1995).

34. P.W. Finch et al., *Science* **245**, 752–755 (1989).

35. J.S. Rubin et al., *Cell Biol. Int.* **19**(5), 399–411 (1995).

36. A.L. Rubin, N.L. Parenteau, and R.H. Rice, *J.Cell. Physiol.* **138**, 208–214 (1989).

37. L. Guo, L. Degenstein, and E. Fuchs, *Genes Dev.* **10**(2), 165–175 (1996).

38. A.A. Dlugosz et al., *Proc. Annu. Meet. Am. Assoc. Cancer Res.*, (1994).

39. S. Werner et al., *Science* **266**, 819–822 (1994).

40. H.-J. Ristow and T.O. Messmer, *J. Cell. Physiol.* **137**, 277–284 (1988).

41. D.A. Vardi et al., *J. Cell. Physiol.* **163**(2), 257–265 (1995).

42. E.J. O'Keefe, M.L. Chiu, and R.E. Payne, Jr., *J. Invest. Dermatol.* **90**(5), 767–769 (1988).

43. E.W. Johnson et al., *In Vitro Cell Dev. Biol.* **28A**, 429–435 (1992).

44. B. Nickoloff et al., *Dermatologica* **177**(5), 265–273 (1988).

45. P. Saiag et al., *Science* **230**, 669–672 (1985).

46. E. DiMarco et al., *J. Biol. Chem.* **266**(32), 21718–21722 (1991).

47. C. Pincelli et al., *J. Invest. Dermatol.* **103**(1), 13–18 (1994).

48. M. Akiyama, L.T. Smith, and K.A. Holbrook, *J. Invest. Dermatol.* **106**(3), 391–396 (1996).

49. R. Polakowska, C. Franceschi, and A. Giannetti, *J. Invest. Dermatol.* **109**(6), 757–764 (1997).

50. V.R. Iyer et al., *Science* **283**, 83–87 (1999).

51. R.M. Lavker and T.-T. Sun, *Science* **215**, 1239–1241 (1982).

52. J.-S. Yang, R.M. Lavker, and T.-T. Sun, *J. Invest. Dermatol.* **101**, 652–659 (1993).

53. L.G. Lajtha, *Differentiation* **14**, 23–34 (1979).

54. C.S. Potten and R.J. Morris, *J. Cell Sci.*, Suppl. **10**, 45–62 (1988).

55. J.R. Bickenbach, *J. Dent. Res.* **60C**, 1611–1620 (1981).

56. I.C. MacKenzie, K. Zimmerman, and L. Peterson, *J. Invest. Dermatol.* **76**, 459–461 (1981).

57. S.E. Al-Barwari and C.S. Potten, *Int. J. Radiat. Biol.* **30**(3), 201–216 (1976).

58. J.R. Bickenbach and K.A. Holbrook, *J. Invest. Dermatol.* **88**(1), 42–46 (1987).

59. K.M. Albers, R.W. Setzer, and L.B. Taichman, *Differentiation* **31**, 134–140 (1986).

60. P.H. Jones and F.M. Watt, *Cell* **73**, 713–724 (1993).

61. A. Li, P.J. Simon, and P. Kaur, *Proc. Natl. Acad. Sci. U.S.A.* **95**(7), 3902–3907 (1998).

62. F.M. Watt, *J. Cell Biol.* **98**(1), 16–21 (1984).

63. M. Regnier et al., *J. Invest. Dermatol.* **87**, 472–476 (1986).

64. J.G. Rheinwald and H. Green, *Cell* **6**, 331–343 (1975).

65. Y. Barrandon and H. Green, *Proc. Natl. Acad. Sci. U.S.A.* **84**, 2302–2306 (1987).

66. R.M. Lavker and T.-T. Sun, *J. Invest. Dermatol.* **81**(1 Suppl.); 121s–127s (1983).

67. C. Compton, *Wounds* **5**, 97–111 (1993).

68. P. Kaur et al., *J. Invest. Dermatol.* **109**, 194–199 (1997).

69. T.M. Koladka et al., *Proc. Natl. Acad. Sci. U.S.A.* **95**, 4356–4361 (1998).

70. B.J. Rollins et al., *J. Cell. Physiol.* **139**(3), 455–462 (1989).

71. J.E. Lewis, P.J. Jensen, and M.J. Wheelock, *J. Invest. Dermatol.* **102**(6), 870–877 (1994).

72. P.R. Cline and R.H. Rice, *Cancer. Res.* **43**, 3203–3207 (1983).

73. Y. Poumay and M.R. Pittelkow, *J. Invest. Dermatol.* **104**, 271–276 (1995).

74. F.M. Watt, D.L. Mattey, and D.R. Garrod, *J. Cell Biol.* **99**(6), 2211–2215 (1984).

75. P. Fleckman, B.A. Dale, and K.A. Holbrook, *J. Invest. Dermatol.* **85**(6), 507–512 (1985).

76. M.A. Phillips and R.H. Rice, *J. Cell Biol.* **97**, 686–691 (1983).

77. N.L. Parenteau, A. Pilato, and R.H. Rice, *Differentiation* **33**, 130–141 (1986).

78. S. Banks-Schlegel and H. Green, *Transplantation* **29**(4), 308–313 (1980).

79. E. Bell et al., *Science* **211**, 1042–1054 (1981).

80. S.T. Boyce and R.G. Ham, *J. Tissue Cult. Methods* **9**(2), 83–93 (1985).

81. F. Castro-Munozledo et al., *Biochem. Biophys. Res. Commun.* **236**(1), 167–172 (1997).

82. N. Parenteau et al., *Biotechnol. Bioeng.* **52**, 3–14 (1996).

83. U.S. Pat. Office. (1998), N.L. Parenteau et al.

84. D. Asselineau et al., *Exp. Cell Res.* **159**, 536–539 (1985).

85. N.L. Parenteau et al., *Cytotechnology* **9**, 163–171 (1992).

86. P.R. Bilbo et al., *J. Toxicol.* **12**(2), 183–196 (1993).

87. C.J.M. Nolte et al., *Arch. Dermatol. Res.* **285**, 466–474 (1993).

88. N. Parenteau, in I. Leigh and F. Watt, eds., *Keratinocyte Methods*, Cambridge University Press, Cambridge, U.K., 1994, pp. 45–54.

89. N.A. Wright in L.A. Goldsmith, ed., *Biochemistry and Physiology of the Skin*, Oxford University Press, New York, 1983.

90. P.R. Bilbo et al., *J. Invest. Dermatol.* **96**, p 618a (1991).

91. J.A. Garlick and L.B. Taichman, *J. Invest. Dermatol.* **103**, 554–559 (1994).

92. R. Gay et al., *Toxicol. In Vitro* **6**, 303–315 (1992).

93. C.J.M. Nolte et al., *J. Anat.* **185**, 325–333 (1994).

94. J. Vicanova et al., *Wound Repair Regeneration* (in press).

95. K. Kriwet and N.L. Parenteau, *Cosmet. Toiletries* **111**, 93–102 (1996).

96. M. Ponec, *Toxicol. In Vitro* **5**, 597–606 (1991).

97. S. Gibbs et al., *Arch. Dermatol. Res.* **289**, 585–595 (1997).

98. C. Nathan and M. Sporn, *J. Cell Biol.* **113**(5) (1991).

99. J.D. Zieske et al., *Exp. Cell Res.* **214**, 621–633 (1994).

100. J.D. Zieske, *Eye* **8**, 163–169 (1994).

101. G. Cotsarelis et al., *Cell* **57**, 201–209 (1989).

102. R.M. Lavker et al., *Invest. Ophthalmol. Visual Sci.* **32**(6), 1864–1875 (1991).

103. Y. Muragaki et al., *Eur. J. Biochem.* **207**(3), 895–902 (1992).

104. A. Schermer, S. Galvin, and T.T. Sun, *J. Cell Biol.* **103**(1), 49–62 (1986).

105. S. Sell, *Mod. Pathol.* **7**(1), 105–112 (1994).

106. S.H. Sigal et al., *Am. J. Physiol.* **263** (26), G139–G148 (1992).

107. L.M. Reid, *Curr. Opin. Cell Biol.* **2**, 121–130 (1990).

108. N. Arber, G. Zajicek, and I. Ariel, *Liver* **8**(2), 80–87 (1988).

109. G.D. Block et al., *J. Cell Biol.* **132**(6), 1133–1149 (1996).

110. J.C. Dunn, R.G. Tompkins, and M.L. Yarmush, *J. Cell Biol.* **116**(4), 1043–1053 (1992).

111. F. Berthiaume et al., *FASEB J.* **10**(13), 1471–1484 (1996).

112. S.N. Bhatia, M.L. Yarmush, and M. Toner, *J. Biomed. Mater. Res.* **34**(2), 189–199 (1997).

113. J.G. Rheinwald, *Int. Rev. Cytol, Suppl.* **10**, 25–33 (1979).

114. H.K. Kleinman et al., *Biochemistry* **21**(24), 6188–6193 (1982).

See also CELL AND CELL LINE CHARACTERIZATION; CELLULAR TRANSFORMATION, CHARACTERISTICS; ENRICHMENT AND ISOLATION TECHNIQUES FOR ANIMAL CELL TYPES.

CELL FUSION

JOHN A. WILKINS
University of Manitoba
Winnipeg, Manitoba
Canada

OUTLINE

INTRODUCTION

Cell fusion was first observed in vitro in cultured cells infected with Sendai virus, where it was noted that multicellular syncitia formed as a result of the fusion of several cells. The use of somatic cell hybridization was developed in the early 1960s as a means of transferring genetic elements. More efficient methods of fusion and hybrid selection were subsequently developed, and somatic cell hybridization is now widely used in basic research, biotechnology, and medicine. Recent developments have focused on increasing the range of cell types that can be used and the efficiency of fusion methods. The use of cell hybrids promises to continue to be an important adjunct

in cell biology and biotechnology. The aim of the present article is to present selected aspects of cell fusion and some of the applications of this technology.

Cell hybridization (fusion) is the process by which two or more cells fuse to form a single cell. The combining of gametes during fertilization, in many ways, represents the prototypic example of cell fusion. In contrast, somatic cell hybridization refers to the fusion of somatic cells, as opposed to gametes, for the production of hybrids. The latter will be the topic of the present discussion.

Hybrids resulting from the fusion of the same cell type are called homokaryons, while those of different cell types are designated heterokaryons (Fig. 1). The actual process of cell fusion involves multiple steps that include the close apposition of the membranes of the fusion partners, the local disruption of membrane structure with the subsequent formation of a continuous membrane that bounds the fused cells. Although there have been a number of criteria proposed for assessing cell fusion, it seems that the minimum criteria for designating cell fusion should include the exchange of membrane components and cytoplasmic contents and the continuity of electrical conductivity between the fusion partners (1).

Somatic cell hybridization is observed in a number of physiologically important processes (2,3). As discussed, gamete fusion represents the initial event of fertilization. The generation of mature muscle fibers involves the fusion of several individual myoblast cells to form the multinucleated myocytes that form the mature muscle fiber. The response of a host to certain infectious agents and foreign bodies can involve the encapsulation of the affected site by macrophages. The macrophages may fuse to form syncitia of giant cells, which clinically present as a granuloma. Many enveloped viruses employ direct fusion with the host cell membrane as the mechanism of viral entry (2). Certain viruses induce the formation of syncitia as part of their cytopathogenic effects.

Although the potential use of somatic cell hybridization as an approach to addressing a number of questions relating to the role of cytoplasmic factors on cell growth

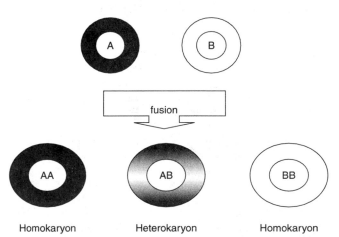

Figure 1. Cell fusion procedures can result in the generation of hybrids containing both partners, heterokaryons, or the same cell types, homokaryons. The former are the desired products and they must be selected from the mixtures.

and differentiation was appreciated in the late 1950s, it was not until the following decade that methods were developed for achieving this. Subsequent developments in the field related to the introduction of new fusion methods, refinements to the procedures, and the development of new strategies for the application of the approach (4–7).

HYBRID FORMATION

The process of cell fusion is dependent upon the close apposition of the outer lamellae of the membranes of the fusion partners, the fusion of these membranes with the formation of pores, and the subsequent exchange of cytosolic contents. Concomitant with these membrane changes is the establishment of a contiguous membrane that permits the ready exchange of membrane constituents through lateral diffusion. There are currently two predominant methods of inducing cell fusion (i.e., chemical and electric field mediated). The methods will be discussed in terms of their relative merits and limitations. The reader is directed to several excellent publications that discuss selected aspects of each of these methods (8–12).

Polyethylene Glycol

Polyethylene glycol (PEG) was introduced as a fusogen in 1976 (7), and at this time it probably represents the most frequently used fusion method. The linear PEG polymer $HO(CH_2CH_2O)_n-CH_2CH_2OH$ can be obtained in a wide range of molecular weights. However, PEGs of 1000–6000 Da are the most frequently used sizes for fusion. This size range offers the most acceptable compromise between fusogenic potential, toxicity, and viscosity. The larger polymers are better fusogens and less toxic to the cells, but the viscosity of the solutions makes them more difficult to handle at the high concentrations required for inducing cell fusion.

The steps in a typical fusion involve bringing the fusion partners into close proximity, usually by centrifugation, although cross-linking agents have also been successfully employed as a adjunct. The cell pellets are then exposed to a concentrated PEG solution for a brief period with subsequent dilution and removal of the PEG.

The conditions of fusion represent an empirical compromise between fusion efficiency and cell survival. Cell pellets are exposed to concentrated PEG solutions, 30 to 50% PEG, for periods ranging from 1 to 10 min. Often the methods involve dropwise addition of PEG to packed cell pellets followed by a brief incubation in the concentrated solution for 30–60 sec. This is followed by a gradual addition serum-free media over several minutes. The cells are gently mixed by shaking or stirring during the procedure as the PEG causes the cells to clump. Finally the cells are washed and plated for growth and selection.

The PEG method is relatively straightforward and involves procedures that are readily adaptable by any laboratory familiar with tissue culture. If desired, large quantities of PEG can be obtained from the manufacturer. However, since PEG preparations even from a single supplier can display marked batch-to-batch variation in their cytotoxicity and fusogenic activities, batches must be preselected by screening. Alternatively, there are several commercial sources of pretested PEG, which provide reliable consistent reagents. However, as with all fusion methods, it is still important to optimize conditions for each cell pair to be fused.

One of the major drawbacks of the method is the lack of fully standardized conditions. There can be "operator" variability even with a so-called standard protocol, but the constancy for a given individual can be very high. Although this may influence the hybrid yields for different individuals, it generally does not prevent the generation of useful hybrids under most circumstances.

Electrofusion

The exposure of cells to a high electric field pulse leads to the transient formation of pores in regions of their membranes. This is thought to arise as a result of electrical-gradient-induced destabilization of the membrane. If cells are brought into close proximity (i.e., touching) during the application of the electric pulse, the membranes of some of the cells will fuse at the region of cell contact, leading to the generation of a hybrid (9).

Operationally cells are placed in an electrofusion chamber and a low-intensity, high-frequency oscillating electric field is applied to the sample. This results in the alignment of cells into a "pearls on a string" like array through the process of dielectrophoresis. A short high-intensity pulse is then applied to generate the hybrids. Most commercial apparatus are equipped to use dielecrophoretic alignment. However, dielectrophoresis is not an obligate component of the procedure, as a range of other methods can be used to bring the cells into close proximity such as centrifugation, cross-linking by antibodies, or micromanipulation of the fusion partners.

This method in principle is very simple. Numerous advantages have been proposed relating to the use of electrofusion: (1) applicability to a wide range of cell types; (2) the absence of exogenous agents (such as PEG); (3) relative nontoxicity; (4) higher efficiency of the procedure; (5) adaptability for the handling of small numbers of cells. The reproducibility of the procedure relates to the ability to control the parameters of fusion such as pulse strength and duration and ionic composition of the media. This is a major advantage over the case using chemical fusion methods.

Perhaps the greatest potential deterrent to using the method is the initial cost for the dedicated equipment. This may make the method non-cost-effective for laboratories that only intend to carry out a limited number of fusions. There can also be difficulties in fusing cells of different sizes since the magnitude of the electric pulse required for electrofusion increases as the cell diameter decreases. The field strength required to effectively permeabilize the smaller cell type may be too great for the larger cell type, resulting in cell death. There is also a tendency for fusion partners of different sizes or membrane properties to preferentially align and to fuse with homologous cell types (i.e., to generate unwanted homokaryons). Difficulties have also been reported in fusing cell types with membranes of very different compositions. This arises from the difference in field strengths required for membrane destabilization. Approaches to circumventing this problem have been published, but they are not necessarily convenient nor is their generality known (13). Since there can be considerable cell type variation in the fusion parameters, the conditions must be optimized for each set of fusion partners.

The use of pulsed electric fields oscillating at radio frequencies have been proposed as a means of circumventing some of the problems described for direct-field-mediated fusions. The method appears to offer several advantages in that it is less sensitive to cell size variation and the yield of fused cells is reportedly better (14).

The amount of manipulation of the cells is low in electrofusion relative to other procedures. This is particularly useful for handling small numbers of cells where the risk of loss during centrifugation steps might be a consideration. Methods using optical tweezers and optical scissors have been described for aligning and fusing cells (15). It thus becomes possible to fuse a single pair of cells. Although such methodologies are not currently within the realms of most laboratories, the results highlight the potential of these procedures.

One of the original limitations of electrofusion was the low cell handling capacities of the fusion chambers. Fusion involving large numbers of cells had to be performed in several batches. This added considerable time, labor, and potential for variability. Recently higher-capacity fusion chambers have been described such that it is feasible to carry out high cell number fusion (16,17).

Comparison of the Methods of Cell Fusion

PEG-based fusion methods were introduced earlier than EF-based ones, and as such more fusions have been done with the former method. However, with the acquisition of experience in EF and the development of simple-to-use, less expensive equipment, this method has been showing increased use in a variety of cells systems.

Several groups, largely proponents of EF, have compared the two methods of fusion (8,9). Collectively their results indicate that EF is at least as efficient in yielding hybrids as PEG with the majority of studies suggesting that EF is considerably better with fusion frequencies ranging from 10–50 times better than with PEG.

In a comparison of PEG- and EF-generated hybridomas, no significant differences in either of the classes of immunoglobulins produced or in the proportions of antigen-specific hybridomas were observed. The major differences appeared to be in the higher frequencies of hybridomas that were generally observed with EF. Some authors have suggested that the appearance and growth rates of hybridomas generated by EF were significantly better than those produced chemically, although these proposed differences were not quantitatively examined.

A comparison of the fusion frequencies using plant protoplasts suggested that there was little difference between PEG and EF (10). However, it was suggested that EF offered several advantages over PEG in terms of reproducibility and convenience.

The decision as to which fusion method is to be used is predicated on a number of considerations. These include the number of fusions to be performed, the cost, the frequency of useful hybrids expected in a given fusion, the number of cells available for a fusion, and the properties of the cells to be fused. It would seem unnecessary to purchase expensive pieces of equipment for very limited numbers of fusions.

Conversely, if only a limited number of cells are available and high fusion rates are critical, then EF appears to be the method of choice. If large numbers of such fusions are to be carried out, EF may be the method of choice because of the ability to stringently control the experimental conditions. An added advantage of EF is the ability to monitor microscopically the early stages of the cell fusion. These considerations coupled with the minimal cell handling required and the elimination of potentially toxic effects PEG further points to this as the preferred method.

It seems fair to suggest that both EF and chemical fusion offer methods that have been shown to work efficiently in a wide range of systems. In the majority of applications, it should be possible to obtain useful hybrids by either approach. It is our experience that the limiting factors in acquiring the desired hybrids are the identification and selection steps, not the numbers of hybrids available for screening. Thus either method should provide acceptable results.

Protoplasts

Plant, bacterial, and yeast cells are surrounded by a stiff wall of carbohydrate and proteoglycans that render the cells rigid and highly resistant to changes in osmotic pressure. However, the cell wall also interferes with the processes required for cell hybridization. In order effectively to fuse cells from these sources, it is necessary to remove the cell wall to expose the cell membrane. The resulting denuded structures are called protoplasts. The term spheroplast has also been used to describe bacteria with cell walls removed.

Plant cells from most tissues are pluripotent, with the capacity to generate fully differentiated plants. These properties have provided the basis for attempting to generate cell hybrids between sexually incompatible plant species (18–20). The fusion of plant cells requires the enzymatic removal of the cell wall with cellulase to expose the cell membrane (Fig. 2). The resulting protoplasts can then be fused by electrical or chemical methods. The subsequent culture and induction of differentiation of these hybrids can lead to reformation of the cell wall and subsequent formation of an entire plant.

Another application of cell fusion to plant biology derives from the fact that aspects of plant herbicide resistance and of male sterility are controlled by genetic material present in the chloroplasts and mitochondrion. In such cases, it may be desirable selectively to transfer this cytoplasmically encoded material onto a desired nuclear genetic background. This can be achieved by treating the recipient cell with iodoacetate to inactivate the chloroplast and mitochondrial DNA while leaving the nuclear material functional. The donor cell is lethally irradiated at doses that do not impact on organelle genetic material but irreversibly damage nuclear DNA. The cells

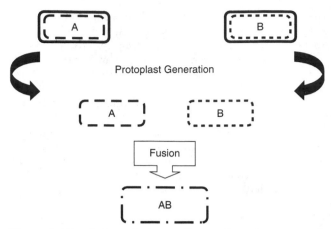

Figure 2. The fusion of plant cells requires the enzymatic removal of the cell walls to generate protoplasts, which are subsequently fused. The hybrids then regenerate the cell wall and develop into totipotent cells.

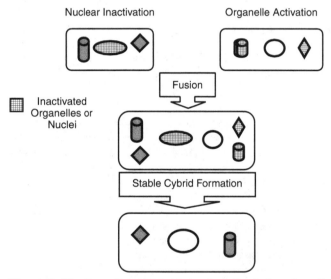

Figure 3. The formation of cybrids involves the fusion of a cell (A) that has had the replicative potential of the chloroplasts and mitochondria inactivated with a cell (B) that has had the nuclear replicative potential inactivated. Protoplasts from these cells are fused to generate a proliferation competent cybrid, which expresses the nuclear activity of cell A and the organelles of B.

are fused to generate a "cybrid," which contains a recipient nucleus and donor organelles as stable genetic elements (Fig. 3) (19).

The generation of plant protoplasts and their fusion presents many technical difficulties that can impact one or both of the fusion methods. The differences in fusion partner size as well as the propensity for protoplasts of one species to align with one another in dielectric fields can often lead to the generation of homokaryons. Although electrofusion has been reported by some to provide a better level of fusion, there is some dispute over this point, and both methods of fusion continue to be used effectively (21).

Selection Markers

Following hybrid generation, there may be fusion between the desired partners or between homologous pairs of either of the fusion partners. The latter, homokaryons, are not desired products. Depending upon the frequencies of heterokaryons produced, the former may overgrow the desired hybrids. If possible, it is advantageous to introduce a selectable marker into one of the fusion partners. The most frequently used system is based on the hypoxanthine, aminopterin, and thymidine (HAT) system (6).

Aminopterin inhibits the main biosynthetic pathway for guanosine, but there is a "salvage" pathway by which hypoxanthine or guanine can be converted to guanosine monophosphate by hypoxanthine guanine phosphoribosyl transferase (HGPRT). Cell lines that lack this enzyme will die in HAT-containing media because they cannot generate guanosine. However, if such a cell line is hybridized with a partner that can contribute the hypoxanthine guanine phosphoribosyl transferase, the hybrids will survive. This method is particularly useful if one of the parents is an immortal line and the other is a normal cell type with limited proliferative potential. The immortal line is rendered HAT sensitive by selecting for hypoxanthine phosphoribosyl transferase negative lines in the presence of toxic analogues such as 8-azaguanine or 6-thioguanine. These mutants grow normally under conditions that do not require the salvage pathway. This system represents the selection basis for majority of hybridoma production in rodent systems.

The selection of plant hybrids was originally dependent upon the introduction of morphological characteristics, which might ultimately be identified in the mature plant. Alternatively, complementation markers such as those observed between chlorophyll deficient mutants were used. Other selectable markers such as hormone requirements or complementation of auxotrophic mutants have also been employed. However, the limitation of this approach is the lack of generality of the markers (i.e., not all cell combinations harbor the necessary complementation groups). The development of methods for the introduction of selectable markers by transformation of plant cells with plasmids carrying antibiotic-resistant genes affords a much greater potential for fusion partner combinations (21).

Following the fusion of two cells, there is the opportunity for the expulsion of genetic material such that not all progeny will necessarily have the same genotype or phenotype. Thus, while hybrids must retain the markers required for their survival in the selection media, they may lose desired phenotypic properties. If this loss of phenotype imparts any growth or survival advantage to such cells, they may overgrow the desired cell types resulting in the loss of the populations of interest.

The preceding considerations make it essential that stable clones with desired phenotype are selected as rapidly as possible following identification. The most common approach is to select individual clones with the desired properties. Clones can be generated by limiting dilution or by plating at less than one cell per culture well. Alternatively, hybrids can be grown in a semisolid medium such agar or methyl cellulose. Individual colonies, presumably (but not necessarily) the progeny of a single cell, can then be physically collected and expanded and analyzed. As with all cell lines, it is advisable to maintain frozen stocks of cells that have undergone limited expansion. It is important to produce reasonable reserves of low-passage frozen stocks of clones of interest. This provides a stock with the desired properties in case of clonal drift or contamination of long-term cultures.

One of the major considerations in determining the likelihood of successfully obtaining hybrids with the desired properties is the method used for identifying/selecting them. In the case of hybridomas or cells that produce soluble mediators, screening for product can be relatively straightforward. However, if one is dependent on the identification of a phenotype that requires maturation of the hybrids, this can be a long and labor-intensive process. Under no circumstances should one undertake hybrid formation without fully establishing and validating the methods to be used for identifying the hybrids of interest. In cases where the screening method is limited or the frequency of hybrids of interest is expected to be low, such as in the case of antigen-specific human hybridomas, it may be advisable to enrich the cells of interest before fusion using a positive selection method such as antigen binding.

Applications

There have been and continue to be a wide range of applications for which cell hybridization has been used. Many of the original applications may now be replaced using gene transfer technology, but this is dependent upon availability of the appropriate constructs. In those cases where multiple interacting genetic elements are involved in the generation of a specified phenotype, cell hybrids still afford one of the best means of beginning to analyze such processes. There may also be cases in which it is technically easier and of greater generality to use cell hybridization to isolate desired properties or products than it is to use molecular biological approaches. The following is meant to offer examples of some of the applications of cell fusion rather than an exhaustive list.

HYBRIDOMAS

The term *hybridoma* refers to any continuously growing cell line, which is a hybrid between a malignant cell and a normal cell. However, the term is often used in a more restricted fashion to describe a somatic cell hybrid between a normal antibody-producing B lymphocyte and a continuously growing myeloma (an antibody-forming tumor cell line).

Monoclonal Antibody Production

The introduction of monoclonal antibody production through the generation of hybridomas is perhaps the best-known application of cell fusion (22). Monoclonal antibodies have become critical tools in medicine, biology, and chemistry. The general approach is to fuse B-cell-containing populations from the lymphoid organs of an animal that has been immunized, with the antigen

of interest, with a myeloma cell line. The resulting hybridomas are then cultured and their supernatants are assayed for reactivity with the antigen of interest. Once a positive culture has been identified, this is cloned to generate hybridomas of a single antibody-producing specificity. These hybridomas are the source of the monoclonal antibody.

The benefits of the use of hybridomas for the generation of antibodies relate to the polyclonal and dynamic nature of an in vivo antibody response. Immunization of an animal results in the activation of B cells with receptors that recognize the immunogen. In the case of most antigens, this means that many B cell clones of different specificities are activated to produce antibodies to the immunogen. This results in a polyclonal antibody response. However, with time or following repeated immunizations the patterns of antibodies to a given antigen can change. This means that samples of immune sera from an individual animal can vary with each bleed. Furthermore the serum contains antibodies to all antigens that the cells have recently been exposed to such that the majority of antibodies are not of the desired specificities. These may be of no consequence to a specified use, or they may interfere with the specific application. Also, the animal host has a finite lifetime, which means that the generation and selection of immune hosts must be an ongoing process.

The advantages of the hybridoma technology for antibody production are severalfold. (1) The cell lines provide a potentially continuous source of antibody of a desired specificity in the absence of other antibodies or serum components. (2) The antibody specificity is a stable property such that it is possible to generate highly reproducible immunological reagents. (3) Large quantities of antibodies can be produced.

It is important to bear in mind that monoclonal antibodies are not necessarily the best immunological reagents for all applications. These antibodies detect a single determinant, epitope, on a molecule. If the epitope is masked or is a site of genetic variation, then the antibody will not react with antigen. The majority of antibodies detect discontinuous (conformational) epitopes that are lost on antigen denaturation; so these antibodies are not useful for applications such as immunoblotting. As with all other antibodies, there is also the potential for monoclonal antibody cross-reactivity with totally unrelated antigens. It is essential to establish that a given monoclonal antibody performs appropriately in a given assay.

The development of methods in molecular biology has afforded the means of "humanizing" monoclonal antibodies for therapeutic applications. This procedure involves shuttling the complementarity-determining regions (CDRs) of an antibody onto a human immunoglobulin framework. This is necessary because most monoclonal antibodies are made in species other than humans. The use of such antibodies in vivo for therapeutic or diagnostic procedures results in the immunization of the recipient with these antibodies. This often renders subsequent use of the antibody in the same individual impossible, as it is rapidly cleared from the circulation by antibodies against it.

A possible solution to this problem is the production of human monoclonal antibodies to antigens of interest.

There are obvious problems of obtaining immune cells from individuals with the desired immunity. Peripheral blood is often used as a source of lymphocytes; this may be followed by in vitro stimulation with antigen to increase the frequency of B cells of the desired specificity before generating hybridomas. Generally, this approach has been less successful than in other species. However, there are several groups that have made significant developments in this field (11).

Chromosomal Assignment and Genetic Mapping

One of the earliest applications of cell hybrids was in the mapping of genes using panels of somatic hybrids (23–26). These panels arise from the fact that most interspecies hybrids tend to lose chromosomes as the cells do not remain tetraploid. The pattern of loss is more or less random such that in a population of hybrids there will be individual cells that contain a limited number of foreign chromosomes from one of the partners. Furthermore, with certain interspecies fusion combinations there appears to be a preferential loss of one of the species chromosomes (e.g., human/rodent hybrids tend to lose the human chromosomes). Since the stained chromosomes are morphologically identifiable, it is possible to generate panels of hybrids that contain known chromosomes from one of the parental cells. Through the selection of clones with stable chromosomal patterns, it is possible to generate panels with known foreign chromosomal patterns. If a panel contains a sufficiently large set of independent clones, it is possible to assign genes coding for specific traits or molecules to a given chromosome by examining the correlation between phenotype and chromosomal expression patterns.

A further refinement in the approach involved the development of "monochromosomal" hybrids that contain a single foreign chromosome. A panel of such hybrids allows for the more rapid and unambiguous assignment of the locations of specific genes. Since the process of chromosome loss is not totally random, in that some chromosomes are more readily lost than others, it may be necessary to generate large panels before a full library can be achieved. As an approach to increasing the likelihood of generating the desired monoclonal hybrids, microcell-mediated transfer of single chromosomes was developed. In this case, microcells are generated by inducing the enucleation of mitotic cells, resulting in the formation of minicells with one or a limited number of chromosomes. These minicells are then fused with a cell line of another species to that of the chromosomal origin selected and screened for their chromosomal content.

The generation of detailed maps of individual chromosomes has been achieved using the irradiation gene transfer (IGFT) technique (25). In this scheme, monochromosomal human/hamster hybrids are lethally irradiated so as to induce chromosomal breakage. These hybrids are then rescued by fusing with a mutant hamster that lacks a resistance gene. The resulting trioma (i.e., hybrid of three cells) is selected based on drug resistance transferred from the original hybrid. However, the line will also carry fragments of the human chromosome, which can be screened for the desired genes.

The use of expressed sequence tag (EST) libraries in conjunction with interspecies hybrids expressing single foreign chromosomes has been used to hybridization map unknown ESTs to specific chromosomes (27). This information may be useful in identifying genetic disease markers or in defining the possible function of a gene based on the position relative to other known gene clusters.

Cell hybrids have also been used extensively to examine aspects of cell differentiation, senescence, and oncogenic pathways (27–31). The identification of regulators of differentiated function regulators as well as oncogenes and tumor suppression sequences has been heavily dependent on the use of this approach. While many of the newer gene transfer methods may supersede the case of hybrids, the latter is particularly well suited for complex multigenic processes where it is initially important to localize the genetic regions of interest before attempting detailed genetic mapping.

Plant Genetics

Cell fusion has been used as a means of circumventing some of the restrictions encountered in conventional plant breeding (20,32). As discussed, the structure of plant cells poses some unique problems for fusion. The limited numbers of selectable markers currently available has led to the development of a number of novel hybrid selection schemes. Frequently genetic complementation schemes in which complementary recessive selectable markers carried by each fusion partner are used. The contributions of the genomes of the partners leads to a functional cell that can be propagated (e.g., light handling mutants, biochemical markers/resistance, antimetabolites) under selection conditions in which homokaryons will not survive. Fluorescent labeling of fusion partners with different dyes followed by subsequent selection of dual labeled hybrids by fluorescence either activated cell sorting or by direct micromanipulation has also been employed. FACS offers the means of screening large numbers of cells and selecting for relatively low-frequency events. This is important as the reported hybrid generation rates are often very low and the selection of hybrids by more conventional markers can be very labor intensive. More recently, methods have been developed for the ready introduction of selectable markers into plant cells.

Embryo Cloning

The ability to generate large numbers of genetically identical animals from superior stocks has been an objective of the livestock industry (33). However, there are several technical constraints that must be overcome before this approach becomes economically viable. The recent realization of livestock cloning using nuclear transplantation via cell fusion indicates that in a non production setting the approach is feasible.

CONCLUSION

Since its introduction, somatic cell hybridization has been widely used as an approach to examining the impact of introduced genetic materials on cellular functions and properties. Apart from the established applications, it appears that the approach will continue to be an important adjunct to the research community with an ever-increasing role in plant and animal biotechnology.

BIBLIOGRAPHY

1. A.E. Sowers, in D.C. Chang, B.M. Chassy, J.A. Saunders, and A.E. Sowers, eds., *Guide to Elecroporation and Electrofusion*, Academic Press, San Diego, Calif., Chapter 8, 1992, pp. 119–138.

2. L.D. Hernandez, L.R. Hoffman, T.G. Wolfsberg, and J.M. White, *Annu. Rev. Cell. Dev. Biol.* **12**, 627–661 (1996).

3. J.M. White, *Science* **258**, 917–924 (1992).

4. G. Barski, S. Sorieul, and F. Cornefert, *C.R. Hebd. Seances Acad. Sci.* **251**, 1825–1827 (1960).

5. S. Sorieul and B. Euphrussi, *Nature (London)* **190**, 653–654 (1961).

6. J.W. Littlefield, *Science* **145**, 709–710 (1964).

7. G. Pontecorvo, *Somatic Cell Genet.* **1**, 397–400 (1976).

8. D.C. Chang, B.M. Chassy, J.A. Saunders, and A.E. Sowers, eds., *Guide to Electroporation and Elecrofusion*, Academic Press, San Diego, Calif., 1992.

9. U. Zimmermann, *Biochim. Biophys. Acta* **694**, 227–277 (1982).

10. N. Duzgunes, ed., *Methods Enzymol.* **221**, Academic Press, San Diego, Calif., 1993.

11. J.W. Goding, *Monoclonal Antibodies: Principles and Practice*, 3 ed., Academic Press, San Diego, Calif., 1996.

12. A. Cassio, *Cell Biology: A Laboratory Handbook*, 2 ed., Vol. 3, 1998, pp. 399–402.

13. L.H. Li, M.L. Hensen, Y.L. Zhao, and S.W. Hui, *Biophys. J.* **71**(1), 479–486 (1996).

14. D.C. Chang, P.Q. Gao, and B.L. Maxwell, *Biochim. Biophys. Acta* 153–160 (1992).

15. M.W. Berns, Y. Tadir, H. Liang, and B. Tromberg, *Methods Cell Biol.* **55**, 71–98 (1998).

16. B. Jones et al., *BioTechniques* **16**(2), 312–321 (1994).

17. R.D. Shillito, J. Paszkowski, and I. Potrykus, *Plant Cell Rep.* **2**, 244–247 (1983).

18. G. Spangenberg and I. Potrykus, in I. Porykus and G. Spangenberg, eds., *Gene Transfer to Plants*, Springer, Berlin, 1995, pp. 58–65.

19. P.T. Lynch, M.R. Davey, and J.B. Power, *Methods Enzymol.* **221** (Part B, 29), 379–393 (1993).

20. G.W. Bates, in D.C. Chang, B.M. Chassy, J.A. Saunders, and A.E. Sowers, eds., *Guide to Electroporation and Electrofusion*, Academic Press, San Diego, Calif., 1992, pp. 249–264.

21. G. Spangenberg, Z.-Y. Wang, and I. Potrykus, in J.E. Celis, ed., *Cell Biology*, Academic Press, San Diego, Calif., 1998, pp. 478–484.

22. G. Kohler and C. Milstein, *Nature (London)* **256**, 495–497 (1975).

23. C. Abbott and S. Povey, eds., *Somatic Cell Hybrids*, IRL Press, Oxford, U.K., 1995.

24. S.J. Goss and H. Harris, *Nature (London)* **225**, 680 (1975).

25. M.A. Walter and P.N. Goodfellow, *Trends Genet.* **9**, 352–356 (1993).

26. O.W. McBride and H.L. Ozer, *Proc. Natl. Acad. Sci.* **70**, 1258–1262 (1973).

27. J.E. Lamerdin et al., *Genome Res.* **5**, 359–367 (1995).

28. I. Yoshida, Y. Nishita, T.K. Mohandas, and N. Takagi, *Exp. Cell Res.* **230**, 208–219 (1993).

29. D.T. Zallen and R.M. Burian, *Genetics* **132**, 1–8 (1992).

30. E.J. Stanbridge, *IARC Sci. Publ.* **92**, 23–31 (1988).

31. R.S. Accolla et al., *Int. J. Cancer Suppl.* **6**, 20–25 (1991).

32. S.L. Van Wert and J.A. Saunders, *Plant Physio.* **99**, 365–367 (1992).

33. S.L. Stice and N.L. First, *Anim. Reprod. Sci.* **33**, 83–98 (1993).

See also CELL PRODUCTS—ANTIBODIES; PLANT PROTOPLASTS; PROTOPLAST FUSION FOR THE GENERATION OF UNIQUE PLANTS.

CELL GROWTH AND PROTEIN EXPRESSION KINETICS

DHINAKAR S. KOMPALA
University of Colorado
Boulder, Colorado

OUTLINE

INTRODUCTION

This article will summarize the experimental observations and mathematical descriptions of mammalian-cell growth kinetics, as well as protein-expression kinetics, in batch, fed-batch, continuous, and perfusion cultures. Focusing primarily on the experimental data from freely suspended single-cell cultures which are being used increasingly in the biotechnology industry because of their simplicity and easy scalability (1), this article will discuss the development and utility of mathematical models of mammalian-cell growth kinetics in operating and controlling bioreactors. Further, it will address the different observed patterns of protein-expression kinetics as they relate to cell-growth kinetics for hybridoma versus recombinant mammalian cells and discuss the different bioreactor operating strategies necessary to maximize the production of desired proteins from any production pattern.

Mathematical models of animal-cell growth kinetics may be classified along the lines of microbial growth models, which are either structured or unstructured and segregated or unsegregated (2). The unstructured cell growth models consider cell mass as a single entity, compositionally uniform and unchanging even at different growth conditions, whereas the structured growth models include some details on the varying composition of intracellular components (such as ribosomes, proteins, metabolites, etc.). The unsegregated cell models consider all cells identical, whereas the segregated cell growth models allow for cellular variations or different cell types (such as high and low producers, cells at different cell-cycle phases, or cells with different target gene copy numbers, etc.). Therefore, the unstructured, unsegregated models are simpler in their kinetic expressions and numeric simulations but can become unrealistic when dealing with more complicated growth phenomena. When the models include more details on the intracellular composition of the average cell (structured growth models) or allow for the different cellular types (segregated growth models), the mathematical descriptions become more realistic but at the same time more difficult for estimating model parameters, independent

model verifications, and numerical computations. Therefore, we will focus in this article primarily on the simplest types of mathematical models for simulating or predicting the different cell-growth and protein-expression kinetics.

BATCH-CULTURE KINETICS

Experimental Data

Typical batch-culture data for cell-growth and product-expression kinetics of a murine hybridoma cell line grown in homogeneous suspension cultures (3) are shown in Figure 1. The salient features in the cell-growth curve include an early exponential cell-growth period, a gradual slowdown in cell growth as the viable cell concentration reaches its maximum value because a key nutrient component is depleted or a toxic metabolite has accumulated, and a subsequent cell-death period. The product (monoclonal antibody) accumulation curve shows more subtle features, which are harder to interpret. One common observation is that the antibody production rate (the slope of the curve marked X) is maximum when the viable cell concentration is at its maximum, and the production rate is noticeably slower when the viable cell concentration is lower. A second common observation made from these data is that the antibody production rate increases again as more cells die.

The first observation on the antibody production curve suggests that the antibody production rate is proportional to the viable cell concentration, which can be expressed

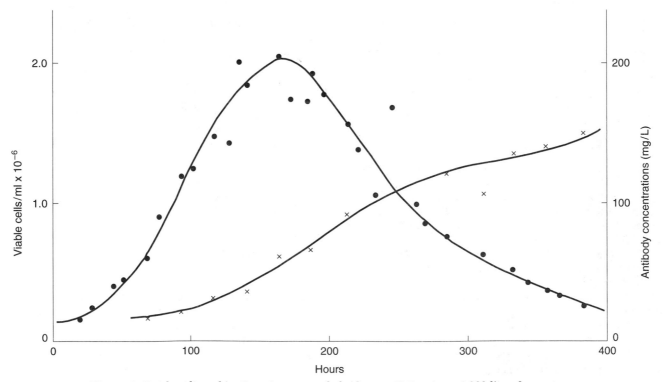

Figure 1. Batch culture kinetics of a mouse hybridoma cell line in a 1,000-liter fermenter, showing viable cell (●) and monoclonal antibody (×) concentrations. (Reproduced from Ref. 3 with permission.)

mathematically as

$$\frac{d[\text{MAb}]V}{dt} = q_{\text{MAb}} X_v V \qquad (1)$$

where [MAb] is the antibody concentration, t is time, V is culture volume, q_{MAb} is the specific (per cell) antibody production rate, and X_v is the viable cell concentration. Assuming constant culture volume (for batch cultures) and a constant specific producton rate (as a first guess), this equation can be simplified to give

$$[\text{MAb}] = q_{\text{MAb}} \int X_v \, dt \qquad (2)$$

This equation has been proposed by different researchers (4,5) who showed that the product titer is roughly proportional to the integral of viable cell concentration over time. This simple relationship immediately suggests some strategies for increasing antibody titers in batch cultures, such as prolonging the viability of the cells, increasing the viable cell concentration, or a combination of the two. These strategies have been implemented successfuly in a fed-batch culture operation by adding medium components to achieve higher viable cell concentrations, as well as to delay the depletion of key nutrient components (6,7).

The second observation on the slight increase in product accumulation during the late culture period, when most cells are dying, has prompted a number of experimental studies to characterize this phenomenon. However, this increase in product accumulation caused by the release of stored products from dying cells is a small fraction of the total product accumulated in the batch or fed batch cultures (5,8).

Unstructured Kinetic Models

Factors that limit the maximum viable cell concentration and the prolonged viability of these cells were investigated by several investigators (6,9,10). It is commonly found that the growth-limiting substrates are glutamine and glucose and that the toxic metabolic by-products are ammonia and to a lesser extent, lactic acid. Based on these observations, extended Monod saturation kinetic expressions for multiple substrates and inhibitors were proposed (11–13) to describe the specific growth rate μ of hybridoma cells as a function of the concentrations of the limiting nutrients and toxic by-products. A typical expression (12) for μ is

$$\mu = \mu_{\max} \left(\frac{G}{K_G + G} \right) \left(\frac{Gn}{K_{Gn} + Gn} \right) \left(\frac{K_A}{K_A + A} \right) \left(\frac{K_L}{K_L + L} \right) \qquad (3)$$

where μ is the specific growth rate of hybridoma cells; $\mu_{\max}$ is the maximum specific growth rate; and G, Gn, A, and L are the concentrations of glucose, glutamine, ammonia, and lactate, respectively. The different K's represent the Monod saturation constants for growth or inhibition by each of these chemicals. A similar multiple saturation kinetic expression has been proposed (11) for the experimentally observed death of these cells in the presence of the two metabolic by-products and in the

absence of growth-limiting nutrient:

$$k_d = k_{d,\max} \left(\frac{A}{K_{Ad} + A} \right) \left(\frac{L}{K_{Ld} + L} \right) \left(\frac{K_{Gn,d}}{K_{Gn,d} + Gn} \right) \qquad (4)$$

where k_d, $k_{d,\max}$ represent the specific death rate and its maximum death rate constant, respectively; and the different K_d's represent the saturation constants for ammonia, lactate, and glutamine, respectively.

The cell-growth and death-rate expressions (eqs. 3,4) can be coupled with material balance equations for glucose, glutamine, ammonia, and lactate through the use of yield coefficients Y's and maintenance coefficient m (14) to describe the dynamic variations of all these chemical and cell concentrations in batch or continuous cultures. Bree et al. (11) simulated the dynamics of cell growth and death, consumption of the limiting nutrients, accumulation of by-products, as well as the product antibody concentration in a glutamine-limited serum-supplemented batch culture. The rate expression for intracellular antibody production was proposed as

$$\frac{d[\text{MAb}]}{dt} = k_P X_v \left(\frac{K_P}{K_P + Gn} \right) \left(\frac{X_T}{K_{XI} + X_T} \right) \qquad (5)$$

where k_p is the maximum specific antibody production-rate constant, K_p is an inhibition constant for glutamine, X_T is the total cell concentration, and K_{XI} is an inhibition constant for total cell number. This proposed expression suggests that the specific monoclonal antibody production-rate constant q_{MAb} used in equations 1 and 2 is not a true constant and has to be modified to reduce the specific production rate at high glutamine concentration and at low total cell concentrations. Many of the parameter values, particularly the yield coefficients, saturation coefficients, and the rate constants, were determined from experimental data, and the remaining six parameters were adjusted to fit the experimental data. Although the model development and consequently the simulations miss the slight reutilization of lactate in the late batch-culture period, it is remarkable that this unstructured kinetic model can describe the experimentally measured consumption of glucose and glutamine relatively closely. Model simulations ignore the early batch data, recognizing the difficulty of fitting the initial lag phase with the Monod saturation kinetic expressions. The complexities of animal-cell culture, such as the complex media requirements, are manifested in the formulation of multiple saturation or inhibition kinetic expressions and difficulties in determining model parameters.

Dalili et al. (15) subsequently developed simpler kinetic expressions for these three rates (cell growth, cell death, and product expression) in glutamine-limited serum-supplemented batch cultures. These model simplifications recognize that glucose is typically in excess and therefore not rate-limiting (as recognized by Bree et al. as well). Further, ammonia and lactate saturation terms are also eliminated in both growth- and death-rate expressions because these toxic metabolites do not build up to growth-limiting or toxic levels in their batch cultures, but the death-rate expression has an additional low constant death rate. A significant difference from the previous model by Bree et al. is proposed in the antibody production-rate expression. Based on their experimental observations of a relatively constant specific antibody production rate for a given initial serum concentration and a reduced production rate as glutamine is depleted, they propose the following rate expression for specific antibody production:

$$q_{\text{MAb}} = m(S_i)\frac{Gn}{K_{\text{MAb}} + Gn} \qquad (6)$$

where $m(S_i)$ is a production-rate constant, which is a function of the initial serum concentration, and K_{MAb} is a Monod saturation constant for antibody production. Most of the simplified model parameters were estimated from a single batch-culture experiment, and a glutamine degradation-rate constant was estimated from available literature. With these rate expressions and associated material balance equations, they simulated the dynamic variations in the cell number, glutamine, and monoclonal antibody concentrations in batch cultures with the same initial serum concentration. The model simulations agree reasonably with the set of batch-culture data in which glutamine is the limiting nutrient.

Following these early modeling attempts, many other unstructured models (16–21) were proposed with many small differences in their rate equations. Portner and Schafer (22) analyzed about ten different unstructured kinetic models for hybridoma cell growth and metabolism published in the literature that vary significantly in each of the rate expressions and compared some model simulations with different sets of experimental data. Though these models were developed on the basis of different sets of experimental data from different hybridoma cell lines growing on different medium formulations and varying serum content, their quantitative comparisons provide some useful general conclusions.

Comparison of Model Simulations with Experimental Data

Despite the use of different rate expressions, many of these models agree well for specific growth rates with experimental data from continuous cultures at different dilution rates. However, significant differences are found between the model simulations and experimental data in the trends of specific death rate, glucose and glutamine consumption rates at low growth rates, and lactate yield from glucose and monoclonal antibody production rate plotted against the specific cell-growth rate.

A comparison of the model prediction with experimental data for a specific death rate indicates the adequacy of each proposed rate expression within a limited range of specific growth rates but shows diverging trends outside that range. These large differences strongly indicate that the proposed expressions for cell death (typically saturation terms with ammonia and lactate and an inhibition term for glutamine) are not quite representative of apoptosis, which is the prevalent mode of gradual cell death due to the buildup of toxic metabolites or depletion of nutrients during the long batch and fed-batch culture times. Necrosis, or sudden traumatic death more common

during excessive sparging or stirring, is not addressed by these proposed rate expressions. Even with these barely adequate specific death-rate expressions, these unstructured kinetic model simulations were applied to develop or optimize fed-batch culture strategies (23,24) to increase hybridoma cell densities, prolong their viabilities, and thereby increase antibody titers.

Nutrient consumption-rate expressions have been typically based on the successful models of microbial growth kinetics. For the simpler microbial growth on a single rate-limiting carbon and energy substrate, the substrate consumption rate in batch cultures is typically modeled as proportional to the cell-growth rate, and the proportionality coefficient is usually called the yield coefficient Y (grams of cell mass dry weight generated per gram of substrate consumed). In continuous cultures, an additional maintenance coefficient m is necessary (14) to account for the lower observed yield coefficients at low growth rates. For mammalian-cell cultures, however, the each of the two major required nutrients glucose and glutamine can provide the carbon and energy requirements to different degrees, depending on the availability of the other substrate, as summarized in Figure 2.

The model simulations with the standard method of using a yield coefficient and a maintenance term in each nutrient consumption equation agree with the chemo-static data on nutrient consumption rates over a limited range of dilution rates (20). Glucose consumption rates exhibit a saturation-type dependence on residual

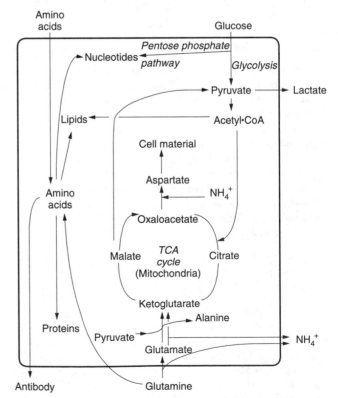

Figure 2. Summary representation of metabolic pathways for the utilization of glucose and glutamine in mammalian cells. (Reprinted from Ref. 19 with permission.)

glucose concentration, even at a constant cell-growth rate μ. Therefore, some modelers proposed saturation-type kinetics for nutrient consumption rates as a function of the residual nutrient concentration (18,21,25). However, these different modifications to nutrient consumption rates describe a set of observed data and do not represent the intrinsic flexibility of intracellular metabolism summarized in Figure 2. The recent observations of multiple steady states through controlled nutrient feeding procedures in fed-batch cultures (26) highlight the feasibility of exploiting metabolic flexibility to minimize the toxic metabolite buildup, and to increase the cellular yield. Such strategies are developed only with a more detailed analysis of intracellular metabolism summarized in Figure 2 and cannot be predicted with any of the simpler unstructured kinetic models discussed before.

Similarly, the two different rate expressions (equations 5,6) proposed before for antibody production kinetics in batch cultures represent simplified descriptions of more complex intracellular mechanisms governing synthesis and secretion of monoclonal antibody. Because these expressions do not directly represent the details of intracellular mechanisms, they are expected to be valid mainly in the cases for which they are developed and tested. Subsequent modelers (16–18,22) have used the classical Leudeking–Piret equation for product-expression kinetics, typically consisting of a constant (nongrowth-associated) q_{MAb} plus in some cases a growth-associated term, which is proportional to the specific growth rate μ.

CONTINUOUS-CULTURE KINETICS

Experimental Data

In batch cultures, all of the culture conditions such as cell number, nutrient concentrations, and product concentrations change throughout the culture period. Consequently, it becomes difficult to evaluate whether the key metabolic parameters, such as the cell specific antibody production rate q_{MAb} and the different yield coefficients remain constant or change systematically with time or nutrient concentrations. This uncertainty contributes significantly to the diversity of rate expressions proposed by different researchers using only batch-culture data. More accurate experimental data are obtained from the steady-state conditions obtained over longer periods of continuous cultures. At a constant nutrient flow rate F or dilution rate $D(=F/V)$, the continuous culture gradually reaches a steady state, wherein all the culture conditions remain unchanged, even as the cells are actively growing, consuming nutrients, and secreting products of interest. At each steady state, the specific cell-growth rate can be estimated easily from the material balances using the measured values of D, X_v and X_T as

$$\mu = D\frac{X_T}{X_v} \qquad (7)$$

Similarly, the specific nutrient consumption rates, as well as the antibody production rates, can be estimated through

material balance equations and measured concentrations of nutrient and antibody, respectively. Using this method with their steady-state data from continuous cultures of hybridoma cells, Miller et al. (12) calculated many metabolic quotients or specific nutrient consumption and product formation rates over a range of cell-growth rates or dilution rates.

An interesting highlight from Miller et al.'s (12) continuous-culture data is that growth rate does not immediately adjust to a change in dilution rate but changes gradually to reach a new steady state. Consequently, the cell number and nutrient concentrations can go through maxima or minima as cells adapt slowly to the newly imposed dilution rate. Similar dynamic phenomena have been observed repeatedly in microbial cultures, and the unstructured Monod models are commonly found incapable of predicting these dynamics accurately.

Inverse-Growth-Associated Production Kinetics. The most surprising result from Miller et al.'s (12) steady-state experimental measurements of hybridoma cell metabolism in continuous cultures is that the specific antibody production rate q_{MAb} does not appear to be related to the specific cell-growth rate μ according to standard models of growth-associated or nongrowth-associated production kinetics. Instead, the specific antibody production rate appears to follow a new inverse-growth-associated pattern, that is, the specific (per cell) antibody production rate is about two- to threefold higher at lower cell-growth rates, as shown in Figure 3.

This inverse-growth-associated production pattern obtained from steady-state data in continuous cultures of murine hybridoma cells has been corroborated by other continuous-culture experimental results (17,27,28) and other growth-limiting batch-culture conditions (29–31) such as high osmolarity and suboptimal pH and addition

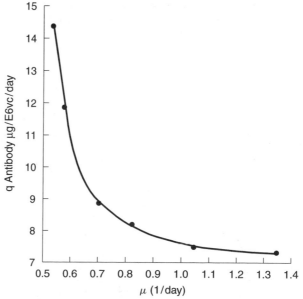

Figure 3. Commonly observed inverse-growth-associated production of monoclonal antibody production by murine hybridoma cells. (Reprinted with permission from Ref. 12.)

of specific growth inhibitors. Although a few experimental studies (20) have observed nongrowth-associated or growth-associated production kinetics, the predominant observation on *monoclonal antibody* production kinetics from *hybridoma* cell cultures has been inverse-growth-associated, as shown in Figure 3.

Growth-Associated Production Kinetics. Among the well-characterized exceptions to the commonly observed inverse-growth-associated production kinetics, the production kinetics of humanized chimeric antibodies in transfected myeloma cells have been found to be growth-associated in steady-state continuous cultures (32), as shown in Figure 4. These growth-associated production kinetics are increasingly observed in the production of recombinant therapeutic proteins, such as γ-interferon and human growth hormone, in batch and continuous cultures of transfected mammalian cells (33,34). An interesting switch in the production pattern has been reported (35) in batch cultures of BHK cells: although the specific production rates of a secreted recombinant antibody and a secreted reported protein (alkaline phosphatase) are growth-associated in suspension cultures, inverse-growth-associated production kinetics are reported for the same cell lines when they are grown attached to microcarriers. However, in the light of the difficulties in getting accurate kinetic results from batch cultures and the added difficulty of estimating the growth rate from attached growth on microcarriers, it is not certain whether this reported switch in the production kinetics on changing the culture environment is reproducible or an experimental artifact.

In summary, it is well established through continuous suspension culture steady-state results that transfected mammalian cells exhibit growth-associated production kinetics for the synthesis of different recombinant proteins, as shown in Figure 4, whereas the production kinetics for the synthesis of monoclonal antibody from hybridoma cell cultures follows an inverse-growth-associated pattern, as shown earlier in Figure 3.

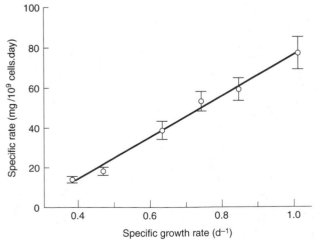

Figure 4. Growth-associated production of chimeric antibody production by transfected myeloma cells. (Reprinted with permission from Ref. 32.)

Structured Kinetic Models

Batt and Kompala (36) developed a structured kinetic modeling framework for simulating the dynamics of cell growth in batch and continous cultures of mammalian cells. Rather than considering the cells as uniform in composition, the structured models consider the dynamic variations of some key intracellular metabolic components, such as DNA, RNA, nucleotides, proteins, amino acids, membranes, and lipids, etc. To simplify their structured model, Batt and Kompala lumped these components into four separate pools and developed different rate expressions in the form of equation 3 for the synthesis of each pool. Then the cell-growth rate is simply calculated as the sum of the rate equations of all intracellular pools. With this framework, the model calculates the dynamics of cell-growth rate even in the unbalanced growth conditions, such as in the initial lag phase of batch cultures and when the dilution rates are shifted up or down in continuous cultures. The structured growth model has been shown capable of simulating the maxima and minima observed in cell number and nutrient concentrations during the shift-up or shift-down experiments. Therefore, this structured model can be useful for simulating simple batch-culture dynamics and steady-state data in continous cultures and also for simulating the dynamics of these cultures due to intermittent additions of nutrients, as well as the fed-batch cultures discussed in the next section.

However, as the single growth-rate expression (equation 3) is now replaced with a rate expression for each of the four intracellular pools, the number of model parameters increases significantly. Although some of these parameter values are determined from the literature on lymphocyte metabolism, many other parameter values were estimated from the same experimental results of Miller et al. (12). For the antibody synthesis rate, this structured kinetic model uses a descriptive rate expression based on the experimental data in Figure 3 in the absence of an exact mechanistic understanding of this pattern. With a more detailed understanding of the mechanisms involved in determining the inverse-growth-associated production kinetics, it is possible to alter that rate expression within the same structured kinetic modeling framework to enhance the utility of model simulations.

DiMasi and Swartz (37) developed a so-called energetically structured model, focusing on energy metabolism of hybridoma cells, represented by four pseudometabolite pools, such as glycolytic intermediates, metabolites derived from glutamine, energy charge (38) of the cell (ratio of ATP, ADP, and AMP), and the redox state of the cell or the ratio of NADH/NAD. Then the specific nutrient uptake rates and waste product synthesis rates are expressed as combinations of saturation and inhibition terms based on these pseudometabolites. The cell growth-rate expression is formulated as a product of multiple exponential saturation terms each based on a single pseudometabolite. Although this approach recognizes the possible variations in the different intracellular metabolites, as in the previous structured kinetic model, the growth-rate expression multiplies different saturation terms, as in the unstructured models represented by equation 3. This energetically structured model satisfactorily describes the steady-state data for different nutrients and metabolites from Miller et al.'s (12) continuous-culture experiments. It is not clear if the simulations of the model are equally successful in simulating the dynamic variations in shift-up or shift-down experiments. As in the previous structure kinetic model, incorporating mechanistic detail into the model comes at the cost of additional model parameters, which need to be estimated through independent experiments to test any predictive capabilities.

Because of the increasing availabilty of detailed mechanistic knowledge, it is becoming possible to develop kinetic models for different metabolic pathways, for example, monoclonal antibody assembly and secretion (39) and regulatory subsystems, for example, intraorganelle pH regulation (40). Because the speed and ease of numerical computations have increased rapidly along with our detailed mechanistic knowledge, it may indeed be possible to develop a more complete structured kinetic model to describe the complex dynamics of animal-cell cultures. Barford and co-workers (41,42) used a modular approach to construct a complex structured kinetic model with three major compartments of cell culture medium, cytoplasm, and mitochondria. After systematically testing the different submodels, it is expected that such a detailed mechanistic model will be useful for optimizing medium formulation (43), developing nutrient feeding strategies (44), and optimizing culture productivity.

Cell-Cycle Models

Suzuki and Ollis (45) proposed a cell-cycle model for antibody synthesis by hybridoma cells to explain the repeated observations of inverse-growth-associated production kinetics of monoclonal antibody in continuous cultures of murine hybridoma cells, as shown in Figure 3. Based on prior literature on cell-cycle-dependent synthesis of antibodies by myeloma or lymphoid cells, they proposed that hybridoma cells secrete monoclonal antibodies primarily in the late G1 and early S phase of the cells. According to the well-established cell-cycle models, cells spend more time in the G1 phase during low growth rates in continuous cultures, as well as in growth-limited batch or fed-batch cultures. Consequently, the higher antibody synthesis in the late G1 phase results in a higher specific (per cell) antibody synthesis rate at lower growth rates. At higher growth rates, the cells spend less time in the productive G1 phase, resulting in a reduced specific antibody production rate. This cell-cycle model for antibody synthesis by hybridoma cells was investigated experimentally by a number of investigators (46–48) with significant support for this central hypothesis.

Following Suzuki and Ollis's successful prediction of the inverse-growth-associated antibody production kinetics by hybridoma cells with the cell-cycle model, Linardos et al. (49) proposed a cell-cycle model for apoptotic cell death, which is observed strongly at low cell-growth rates. They proposed that apoptotic cell death occurs primarily in cells arrested in the G1 phase of the cell cycle and that the death rate is proportional to the fraction of cells arrested in the G1 phase. This cell-cycle model satisfactorily describes the higher cell death rates at low dilution rates in continuous cultures of hybridoma cells.

With these two successful descriptions of antibody productivity and cell death by simpler cell-cycle models, Martens et al. (50) developed a combined cell-cycle and unstructured kinetic model for simulating Miller et al.'s (12) continuous-culture steady-state data. The cell growth-rate expression is similar to equation 3 and the substrate consumption rate equations are based on the yield coefficient and maintenance energy terms described earlier. Although the model simulations follow the trends qualitatively in the steady-state data, the average model parameter values chosen do not fit the experimental data closely enough. Further, the predictions of this unstructured kinetic model for dynamic variations in batch, fed-batch, or continuous culture shift-up or shift-down experiments are expected to be more divergent, as discussed earlier.

The dramatic switch in protein-expression kinetics from the inverse-growth-associated production of antibodies by hybridoma cells (shown in Fig. 3) to a growth-associated production of recombinant proteins by transfected mammalian cells (shown in Fig. 4) has been addressed with a cell-cycle model (51). Gu et al. (51) assumed that although hybridoma cells produce antibodies primarily in the G1 phase, transfected mammalian cells produce their recombinant proteins primarily in the S phase. Gu et al.'s (51) cell-cycle model predictions for an intracellular reporter protein accumulation, shown in Figure 5, have an inverse-growth-associated production pattern for the G1-phase-synthesis assumption and a growth-associated or nongrowth-associated production pattern for the S-phase-synthesis assumption. The S-phase synthesis assumption is supported by flow cytometric investigations on recombinant protein synthesis in transfected Chinese hamster ovary (CHO) cells (52,53). Based on the different cell-cycle phase-specific characteristics of recombinant protein expression from different promoters, Kompala and co-workers (54,55) developed different CHO cell lines transfected with different expression vectors that contain the same reporter gene (lac Z). In

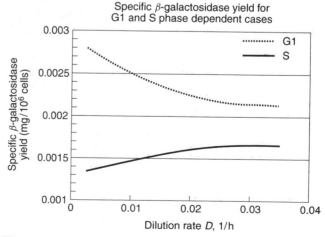

Figure 5. Predictions of the previous two production patterns with different assumptions on the cell-cycle phase-specific expression for intracellular protein accumulation. (Reprinted with permission from Ref. 51.)

continuous cultures of stably transfected CHO cells, the S-phase-specific SV40 promoter-driven expression of β-galactosidase was growth-associated (54), whereas the G1-phase-specific AML promoter-driven expression of the intracellular reporter protein was inverse-growth-associated (55). If these results yielding the dramatic switch between growth-associated and inverse-growth-associated production Kinetics are reproducible with secreted reporter proteins, this phenomenon will have a significant impact on maximizing culture productivities, as discussed in the next section on fed-batch cultures.

Segregated Models

Long-term cultures of some hybridoma cell lines have shown a gradual decrease in antibody productivity due to the appearance of faster growing nonproducer hybridoma cells (56). The loss of culture productivity can be modeled only by explicitly recognizing the growth and production kinetics of these two distinct producer and nonproducer cell lines in a segregated modeling approach (57). Based on a similar approach used successfully to analyze instabilities in recombinant bacterial cultures, new insights have been developed on the rate of conversion from producers to nonproducers and the growth-rate difference between two cell types required to control this culture instability (58).

FED-BATCH AND PERFUSION CULTURES

Optimization of Monoclonal Antibody Production

Successful exploitation of cell-growth and protein-expression kinetics has led to systematic development of fed-batch cultures for maximizing monoclonal antibody productivity from hybridoma cells (59). Maximizing the hybridoma cell concentration and maintaining the viability of these cells by careful feeding of medium concentrates, as suggested by equation 2, represent key ingredients in this optimization process. An equally important ingredient in achieving high product titers has been the *inverse-growth-associated* production kinetics, as shown in Figure 3. During the early batch-culture period, cells grow at a high rate with a low specific production rate. After high cell densities have been achieved, controlled nutrient feeding reduces the cell-growth rate and maintains the cells' viability for as long as possible. During this low growth fed-batch culture period, the specific (per cell) antibody production rate is significantly increased, as shown in Figure 3, resulting in a large production of monoclonal antibodies in fed-batch cultures. Similarly, in high cell-density perfusion cultures, after high cell densities are achieved through the early batch-culture period, the cells are maintained in a lower growth mode once perfusion culture begins. During such a high cell-density perfusion operation, the specific production rate increases (60), resulting again in larger antibody titers than in continuous cultures.

Model Simulations for Fed-Batch Cultures

The preceding development of optimal fed-batch and high cell-density perfusion cultures to maximize culture productivity has taken place without any significant direct input from model simulations. The unstructured models, based on extended Monod Kinetics, are ill suited for dynamic simulations, even though they have been applied for this purpose in fed-batch (23,24), as well as perfusion (61) cultures. Emborg and co-workers (62) simulated the structured kinetic model of Batt and Kompala (36) in batch and fed-batch culture modes and indeed find the fed-batch cultures better suited for maximizing culture productivity, as confirmed by many experimental studies. Ryszczuck and Emborg (63) subsequently compared the fed-batch simulations of the same structured model (36) with an unstructured model to find that the structured model simulations are more responsive to the different fed-batch culture strategies and predict a more significant improvement in antibody production in fed-batch cultures.

Suboptimal Growth-Associated Production Kinetics

Although high cell-density fed-batch and perfusion cultures have been successful in maximizing monoclonal antibody production from hybridoma cells, a necessary requirement for this optimization has been *inverse-growth-associated* production kinetics, as shown in Figure 3. More recently, it has been found that recombinant protein synthesis kinetics from transfected mammalian cells are typically growth-associated, as shown in Figure 4. Some experimental researchers have already recognized the suboptimal nature of fed-batch cultures for these mammalian-cell lines and have used repeated batch cultures (64) to maximize protein synthesis. However, these repeated batch cultures are significantly inferior from the perspective of process optimization. During exponential batch cultures where the growth and specific (per cell) production rates are high, the cell concentrations are typically low. After cell numbers have reached a high enough value and the cell growth rate slows down, the specific productivity drops enormously, if the cells exhibit growth-associated production kinetics. If the successful paradigms of high cell-density fed-batch and perfusion cultures for maximizing the monoclonal antibody synthesis from hybridoma cells are to be applied to recombinant mammalian-cell cultures, then it will be necessary to reengineer them to obtain the same inverse-growth-associated production kinetics. The reengineering of Chinese hamster ovary cells to achieve inverse-growth-associated production kinetics has so far been demonstrated only for an intracellular reporter protein (54,55), and efforts are underway to demonstrate its feasibility for secreted glycoproteins.

CONCLUSIONS

This article highlights the importance of characterizing cell-growth and protein-synthesis kinetics to maximize culture productivity. Unstructured kinetic models, based on extensions of classical Monod kinetics, yield and maintenance coefficients, and Leudeking–Piret production kinetics provide the simplest choice for describing the dynamics of cell growth, nutrient consumption, and product expression. Structured kinetic models address more details of cellular metabolism and consequently can describe more accurately cell-growth and product-expression dynamics in batch, fed-batch, and continuous cultures. With increasing knowledge of animal-cell metabolism, it may soon be more feasible to develop more detailed structured kinetic models to enhance our understanding and optimization of mammalian-cell cultures.

Protein-expression kinetics follow two different patterns of growth-associated and inverse-growth-associated production. Inverse-growth-associated production kinetics more commonly found for the monoclonal antibody synthesis by hybridoma cells have been successfully exploited to maximize productivity in high cell-density fed-batch and perfusion cultures. Growth-associated production kinetics, observed increasingly in the production of therapeutic proteins by recombinant mammalian cells, are more difficult to optimize because repeated batch cultures and continuous cultures are necessary to maintain the high cell growth and specific production rates. It may be possible to employ the successful paradigms of high cell-density fed-batch and perfusion cultures to maximize antibody productivities, if recombinant mammalian cells can be engineered to yield inverse-growth-associated production kinetics.

ACKNOWLEDGMENTS

The author's research on mammalian-cell culture kinetics has been funded by grants BES-9504840 and BES-9817249 from the National Science Foundation. The author also thanks Prof. T. Yoshida of the International Center for Biotechnology, Osaka University, for hosting him while writing this article during a three-month sabbatical visit.

NOMENCLATURE

A	ammonia concentration, g/L
D	dilution rate, day^{-1}
F	nutrient flow rate, mL/day
G	glucose concentration, g/L
Gn	glutamine concentration, g/L
K_d	saturation constant in death rate, g/L
K_{MAb}	Monod saturation constant for antibody production, g/L glut.
K_p	inhibition constant for glutamine, g/L
K_{XI}	inhibition constant for total cell number, cells/mL
$K_d, K_{d,max}$	specific death rate and maximum death-rate constant, day^{-1}
K_p	maximum specific production-rate constant, $\mu g/10^6$ cells/day
L	lactate concentration, g/L
[MAb]	antibody concentration, g/L
m	maintenance coefficient, (g or mM)/day/cells
$m(S_i)$	specific antibody synthesis-rate constant, which is a function of initial serum concentration, $\mu g/10^6$ cells/day
q_{MAb}	specific (per cell) antibody production rate, $\mu g/10^6$ cells/day
V	culture volume, liter
t	time, day
X_T	total cell concentration, cells/mL
X_v	viable cell concentration, cells/mL
Y	yield coefficients, (cells or g or mM)/(g or mM)
μ, μ_{max}	specific growth rate and its maximum value, day^{-1}

BIBLIOGRAPHY

1. J.R. Birch and R. Arathoon, *Bioprocess Technol.* **10**, 251–270 (1990).

2. H.M. Tsuchiya, A.G. Fredrickson, and R. Aris, *Adv. Chem. Eng.* **6**, 125–206 (1966).

3. J.R. Birch, R. Boraston, and L. Wood, *Trends Biotechnol.* **3**, 162–166 (1985).

4. Y.T. Luan, R. Mutharasan, and W.E. Magee, *Biotechnol. Lett.* **9**, 535–538 (1987).

5. J.M. Renard et al., *Biotechnol. Lett.* **10**, 91–96 (1988).

6. S. Reuveny, D. Velez, J.D. Macmillan, and L. Miller, *J. Immunol. Methods* **86**, 53–59 (1986).

7. Y.T. Luan, R. Mutharasan, and W.E. Magee, *Biotechnol. Lett.* **9**, 691–696 (1987).

8. D. Velez, S. Reuveny, L. Miller, and J.D. Macmillan, *J. Immunol. Methods* **86**, 45–52 (1986).

9. M. Butler, *Dev. Biol. Stand.* **60**, 269–280 (1985).

10. K. Low and C. Harbour, *Dev. Biol. Stand.* **60**, 73–79 (1985).

11. M.A. Bree, P. Dhurjati, R.F. Geoghegan, Jr., and B. Robnett, *Biotechnol. Bioeng.* **32**, 1067–1072 (1988).

12. W.M. Miller, C.R. Wilke, and H.W. Blanch, *Biotechnol. Bioeng.* **32**, 947–965 (1988).

13. M.W. Glaken, E. Adema, and A.J. Sinskey, *Biotechnol. Bioeng.* **32**, 491–506 (1988).

14. A.J. Pirt, *Principles of Microbe and Cell Cultivation*, Blackwell Scientific, Cambridge, U.K., 1975.

15. M. Dalili, G.D. Sayles, and D.F. Ollis, *Biotechnol. Bioeng.* **36**, 64–73 (1990).

16. K.K. Frame and W.-S. Hu, *Biotechnol. Bioeng.* **38**, 55–64 (1991).

17. T.I. Linardos, N. Kalogerakis, and L.A. Behie, *Can. J. Chem. Eng.* **69**, 429–438 (1991).

18. M. De Tremblay, M. Perrier, C. Chavarie, and J. Archambault, *Bioprocess Eng.* **7**, 229–234 (1992).

19. J.G. Gaertner and P. Dhurjati, *Biotechnol. Prog.* **9**, 309–316 (1993).

20. M. Harigae, M. Matsumura, and H. Kataoka, *J. Biotechnol.* **34**, 227–235 (1994).

21. R. Portner, A. Bohmann, I. Ludemann, and H. Markl, *J. Biotechnol.* **34**, 237–246 (1994).

22. R. Portner and T. Schafer, *J. Biotechnol.* **49**, 119–135 (1996).

23. M.W. Glaken, C. Huang, and A.J. Sinskey, *J. Biotechnol.* **10**, 39–66 (1989).

24. M. De Tremblay, M. Perrier, C. Chavarie, and J. Archambault, *Bioprocess Eng.* **9**, 13–21 (1993).

25. A.-P. Zeng and W.-D. Deckwer, *Biotechnol. Bioeng.* **47**, 334–346 (1995).

26. W.-S. Hu, W. Zhou, and L.F. Europa, *J. Microbiol. Biotechnol.* **8**, 8–13 (1998).

27. S. Reuveny, D. Velez, L. Miller, and J.D. Macmillan, *J. Immunol. Methods* **86**, 61–69 (1986).

28. R.C. Dean et al., in M. Moo-Young, ed., *Bioreactor Immobilized Enzymes and Cells*, Elsevier Applied Science, Essex, England, 1988, p. 125.

29. S.S. Ozturk and B.O. Palsson, *Biotechnol. Prog.* **7**, 481–494 (1991).

30. S.S. Ozturk and B.O. Palsson, *Biotechnol. Bioeng.* **37**, 989–993 (1991).

31. K. Takahashi et al., *Cytotechnology* **15**, 57–64 (1994).

32. D.K. Robinson and K.W. Memmert, *Biotechnol. Bioeng.* **38**, 972–976 (1991).

33. C.A. Mitchell, J.A. Beall, J.R.E. Wells, and P.P. Gray, *Cytotechnology* **5**, 223–231 (1991).

34. P.M. Hayter et al., *Biotechnol. Bioeng.* **42**, 1077–1085 (1993).

35. A.J. Racher et al., *Appl. Microbiol. Biotechnol.* **40**, 851–856 (1994).

36. B.C. Batt and D.S. Kompala, *Biotechnol. Bioeng.* **34**, 515–531 (1989).

37. D. Dimasi and R.W. Swartz, *Biotechnol. Prog.* **11**, 664–676 (1995).

38. D.E. Atkinson, *Cellular Energy Metabolism and its Regulation*, Academic Press, New York, 1977.

39. T. Bibila and M.C. Flickinger, *Biotechnol. Bioeng.* **37**, 210–226 (1991).

40. P. Wu, N.G. Ray, and M.L. Shuler, *Biotechnol. Prog.* **9**, 374–384 (1993).

41. J.P. Barford, P.J. Phillips, and C. Harbour, *Cytotechnology* **10**, 63–74 (1992).

42. C.S. Saderson, P.J. Phillips, and J.P. Barford, *Cytotechnology* **21**, 149–153 (1996).

43. L. Xie and D.I.C. Wang, *Cytotechnology* **15**, 17–30 (1994).

44. T.A. Bibila et al., *Biotechnol. Prog.* **10**, 87–96 (1994).

45. E. Suzuki and D.F. Ollis, *Biotechnol. Bioeng.* **34**, 1398–1402 (1989).

46. O.T. Ramirez and R. Mutharasan, *Biotechnol. Bioeng.* **36**, 839–848 (1990).

47. M. al-Rubeai and A.N. Emery, *J. Biotechnol.* **16**, 67–85 (1990).

48. R.A. Richieri, L.S. Williams, and P.C. Chau, *Cytotechnology* **5**, 243–254 (1991).

49. T.I. Linardos, N. Kalogerakis, and L.A. Behie, *Biotechnol. Bioeng.* **40**, 359–368 (1992).

50. D.E. Martens et al., *Biotechnol. Bioeng.* **48**, 49–65 (1995).

51. M.B. Gu, P. Todd, and D.S. Kompala, *Ann. N.Y. Acad. Sci., Recombinant DNA Technol. II* **721**, 194–207 (1994).

52. B.D. Mariani, D.L. Slate, and R.T. Schimke, *Proc. Natl. Acad. Sci. U.S.A.* **78**, 4985–4989 (1981).

53. M. Kubbies and H. Stockinger, *Exp. Cell Res.* **188**, 267–271 (1990).

54. G.G. Banik, P. Todd, and D.S. Kompala, *Cytotechnology* **22**, 179–184 (1996).

55. F.W.F. Lee, C.B. Elias, P. Todd, and D.S. Kompala, *Cytotechnology* **28**, 1–8 (1999).

56. K.K. Frame and W.-S. Hu, *Enzyme Microb. Technol.* **13** (9), 690–696 (1991).

57. G.M. Lee, A. Varma, and B.O. Palsson, *Biotechnol. Prog.* **7**, 72–75 (1991).

58. S.J. Kromenaker and F. Srienc, *Biotechnol. Prog.* **10**, 299–307 (1994).

59. T.A. Bibila and D.K. Robinson, *Biotechnol. Prog.* **11**, 1–13 (1995).

60. B.C. Batt, R.H. Davis, and D.S. Kompala, *Biotechnol. Prog.* **6**, 458–464 (1990).

61. F. Pelletier et al., *Cytotechnology* **15**, 291–299 (1994).

62. H.A. Hansen, N.M. Madsen, and C. Emborg, *Bioprocess Eng.* **9**, 205–213 (1993).

63. A. Ryszczuk and C. Emborg, *Bioprocess Eng.* **16**, 185–191 (1997).

64. T. Seewhoster and J. Lehmann, *Biotechnol. Bioeng.* **55**, 793–797 (1997).

See also Flow cytometry of plant cells; Flux analysis of mammalian cell culture: methods and applications; Protein processing, endocytosis and intracellular sorting of growth factors; Protein processing, processing in the endoplasmic reticulum and golgi network.

CELL METABOLISM, ANIMAL

Lena Häggström
Royal Institute of Technology
Stockholm, Sweden

OUTLINE

INTRODUCTION

Mammalian cells with a potential industrial interest for the production of recombinant proteins or monoclonal antibodies are all immortalized, transformed cells, and as such, they harbor mutations affecting the mechanisms that control the progression through the cell cycle. Continuous cell lines have also acquired a number of genetic alterations that cause metabolic changes, phenotypically observed as an increased rate of glycolysis and the production of large amounts of lactate even at completely aerobic conditions; and as an extensive glutamine metabolism accompanied by excretion of inhibitory ammonia/ammonium ions and partially oxidized end products, typically alanine. Glucose and glutamine are the main carbon and energy sources, and glutamine in addition the main nitrogen source for these cells. In culture, large amounts of glucose and glutamine are consumed, and the accumulating waste products negatively affect cell growth, product quality, and productivity. At present no generally accepted opinion explaining the function of the high metabolic rates observed in industrial cell lines, tumor cells, and rapidly dividing normal cells is available. Glucose and glutamine metabolism in tumor cell lines and other rapidly dividing cultured cell lines has been reviewed several times during the past decades (1–8). Other overviews have focused on the metabolism and physiology of industrially important cell lines (9–12). The scope of this article is to review recent progress in understanding the metabolism of cultured mammalian cells with a potential industrial interest such as myeloma, hybridoma, CHO (Chinese hamster ovary), and BHK (baby hamster kidney) cells. This will be done first from a basic point of view, where the involved metabolic pathways and their functions are described, and then from an applied point of view, where the manifestations and consequences in culture are considered. More emphasis has been put on glutamine metabolism than on glucose metabolism, as the latter is likely familiar to most readers.

THE CENTRAL METABOLIC PATHWAYS

Glucose Metabolism: Glycolysis, the Hexose Monophosphate Shunt, and the Pentose Phosphate Pathway

Glucose is metabolized via the well-known cytosolic pathways of glycolysis, the hexose monophosphate shunt (HMS) and the pentose phosphate pathway (PPP) (Fig. 1). Glycolysis is the source of the biosynthetic precursor metabolites glucose 6-phosphate, fructose 6-phosphate, dihydroxyacetone phosphate, and 3-phosphoglycerate

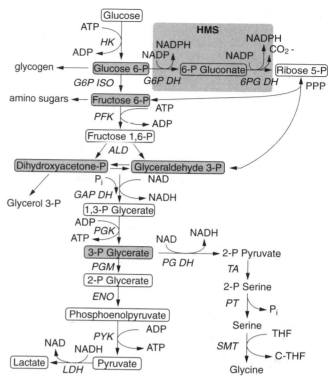

Figure 1. An overview of glycolysis, the hexose monophosphate shunt (HMS) and the pentose phosphate pathway (PPP), including the biosynthetic pathway to serine and glycine. Shaded compounds constitute precursor metabolites. Enzyme abbreviations; Glycolysis: HK = hexokinase, G6P ISO = glucose 6-phosphate isomerase, PFK = phosphofructokinase, ALD = aldolase, GAP DH = glyceraldehyde 3-phosphate dehydrogenase, PGK = phosphoglycerate kinase, PGM = phosphoglycerate mutase, ENO = enolase, PYK = pyruvate kinase, LDH = lactate dehydrogenase, HMS: G6P DH = glucose 6-phosphate dehydrogenase, 6PG DH = 6-phosphogluconate dehydrogenase; serine/glycine biosynthesis: PG DH = phosphoglycerate dehydrogenase, TA = transaminase, PT = phosphatase, SMT = serine-methyl transferase.

required for synthesis of glycogen, amino sugars, glycerol-3-phosphate, and serine/glycine/one-carbon units, respectively. For a fast-growing continuous cell line, an ample supply of glycerol-3-phosphate, the backbone of membrane lipids, the amino acids serine (precursor for ethanolamine synthesis) and glycine (used in purine synthesis), as well as tetrahydrofolate-carried one-carbon units (for purine and pyrimidine nucleotide synthesis), is certainly most important. However, certain CHO cell mutants are reported to be partial glycine auxotrophs (13). The enzyme, serine hydroxymethyl transferase, converting serine to glycine and tetrahydrofolate-bound one-carbon units, is present both in the cytoplasm and in mitochondria of normal cells (14). The mitochondrial isoenzyme is absent in these partial auxotrophs, which are self-supporting in one-carbon units through the cytoplasmic enzyme activity, but that needs medium glycine for protein synthesis. Significant for the glycosylation of recombinant proteins or monoclonal antibodies is the formation of fructose-6P as precursor for UDP-activated amino hexoses. Phosphoenolpyruvate and pyruvate are important precursors for biosynthesis of amino acids in microorganisms, but not in mammalian cells, which have lost the ability to synthesize most amino acids. Pyruvate is, however, the last metabolite of glycolysis and the link to mitochondrial metabolism.

The function of HMS and PPP in higher eukaryotic cells is to provide NADPH for reducing power for biosynthesis and ribose-5P for synthesis of nucleotides and nucleic acids (Fig. 1). This is in sharp contrast to microorganisms, which, in addition, use the HMS and PPP pathways for catabolism of a large number of substrates. As cytosolic NADPH can be formed via other enzymes (malic enzyme and isocitrate dehydrogenase), this function of HMS may be dispensable. For formation of NADPH, the pathway must be fed with glucose-6P, while ribose-5P can be formed either by this route or from carbon entering at fructose-6P and glyceraldehyde-3P. This cyclic mechanism ensures metabolic flexibility and allows surplus intermediates to be fed back to glycolysis.

Glycolysis is also the source of cytosolic ATP: two mol ATP per mol glucose metabolized to pyruvate. Glycolysis, as well as the pathway for serine/glycine synthesis, releases NADH: two mol NADH per mol glucose metabolized to pyruvate, and one mol NADH for each mol serine/glycine formed. To keep glycolysis running, NADH must be reoxidized to NAD. Cytosolic NAD can be regenerated via shuttle mechanisms transporting NADH into mitochondria, or by reduction of pyruvate to lactate. If all cytosolic NADH were transported into mitochondria, no lactate would be formed. The large amounts of lactate produced in mammalian cell cultures show that this is not the case. In fact, lactate formation appears essential for cell growth and continued glucose metabolism, as inhibition of the lactate dehydrogenase reaction (Fig. 1) retards both glucose consumption and proliferation (15). Recent studies with labeled substrates confirm earlier reported results that a large proportion of consumed glucose, 92–96%, enters the glycolytic pathway in hybridoma cells (16–18), while the remaining fraction, 4–8%, enters HMS (16–20).

Shuttle Mechanisms for Transporting NADH into Mitochondria

The mitochondrial membrane is impermeable to NADH, but shuttle systems such as the glycerol–phosphate shuttle, the malate–aspartate shuttle, and the malate/citrate shuttle circumvent this problem by indirectly transporting NADH into mitochondria for further oxidation by the respiratory system (Fig. 2). Although some controversy exists as to the presence and activity of these redox shuttle systems in tumor cell lines, it now appears that, as, for example, Ehrlich ascites tumor cells possess all three systems (21–23). The malate–aspartate shuttle is also functioning in many other tumor cell types, including HeLa cells (24), and the activity of the glycerol–phosphate shuttle is even higher in some transformed cells as compared to the normal counterparts (25,26). The malate–citrate shuttle activity corresponds to 15% of the glucose uptake in Ehrlich ascites cells (23). Therefore, the high rates of lactate production that occur in all continuous cell lines, and the decreased NAD:NADH

ratio (27), are probably not due to a lack of activity of the shuttle systems.

Likely, some other factors restrict the transfer of reducing equivalents from the cytosol to mitochondria. For example, the presence of high levels of lactate dehydrogenase may simply compete for reducing equivalents. It has also been suggested that, as the K_m of cytosolic aspartate transaminase is rather high (5 mM), the malate/aspartate shuttle activity is limited by too low a concentration of cytosolic aspartate (28). Further, owing to glutamine metabolism, malate will accumulate inside mitochondria, thereby rendering malate influx more difficult. Also, the respiratory system may actually be saturated with reducing equivalents, originating from the mitochondrial glutamine metabolism. Saturation of the respiratory system has been the subject of lively discussion as being a cause of acetate and ethanol formation in *Escherichia coli* and *Saccharomyces cerevisiae*, respectively. Finally, respiration may be suppressed by the availability of ADP, as the high flux in glycolysis consumes large amounts of ADP. However, the functioning of the redox shuttle systems and the transfer of NADH into mitochondria has not been studied in detail in industrial continuous cell lines.

The Tricarboxylic Acid Cycle

The function of the tricarboxylic acid (TCA) cycle, located in mitochondria, is to provide precursor metabolites and cofactors for anabolism, as well as metabolic energy (Fig. 3). Citrate, when exported to the cytosol, is the source of the cytosolic acetyl coenzyme A (acetyl CoA) needed for synthesis of cholesterol and fatty acids. Isocitrate can also be exported to the cytosol for NADPH production via cytosolic isocitrate dehydrogenase. α-Ketoglutarate (α-KG) is the precursor for glutamate, and oxaloacetate the precursor for aspartate and asparagine biosynthesis. However, it is doubtful if there is any significant net flux from α-KG to glutamate via glutamate dehydrogenase (GDH) in continuous cell lines during normal cultivation conditions. This matter will be further discussed together with glutamine metabolism. The TCA cycle is also the major source of metabolic energy, in terms of ATP, via NADH fed to the respiratory system after being released from the oxidation of substrates in the cycle. Acetyl CoA, formed from pyruvate via pyruvate dehydrogenase (PDH), from fatty acids via β-oxidation or from breakdown of amino acids, is the only TCA cycle substrate that can be completely oxidized to carbon dioxide and water. All other substrates entering the cycle at, for example, the level of citrate, α-ketoglutarate, or malate must also leave the cycle (otherwise intermediates would accumulate) unless compensating for precursor metabolites withdrawn from the cycle. The pathways involved in converting a four-carbon TCA cycle intermediate to a three-carbon glycolytic intermediate or vice versa will be discussed in conjunction with glutamine metabolism.

The concept of a truncated TCA cycle in tumor mitochondria was introduced by Coleman and Lavietes (29). Briefly, this phenomenon is manifested as a low flux from citrate to α-KG and a much higher flux from α-KG to oxaloacetate. Citrate exits mitochondria for cytosolic

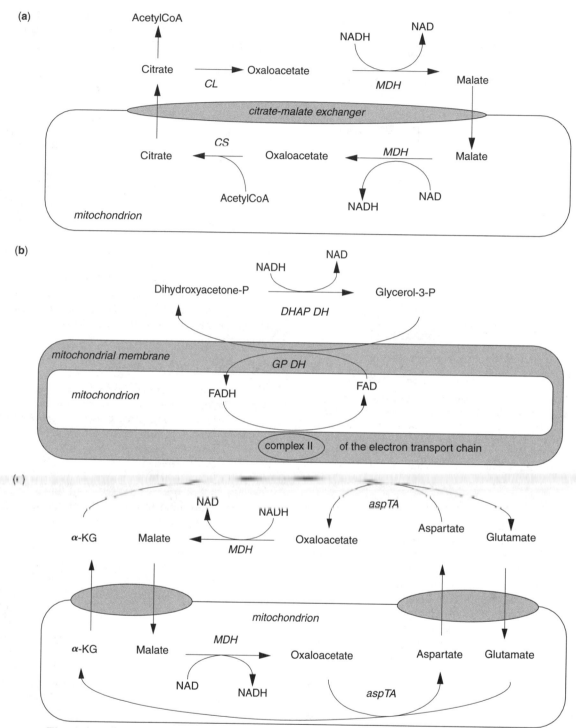

Figure 2. Redox shuttle systems. (**a**) The malate–citrate shuttle. (**b**) The glycerol–phosphate shuttle. (**c**) The malate–aspartate shuttle. Enzymes: CL = citrate lyase, CS = citrate synthase, MDH = malate dehydrogenase, DHAP DH = dihydroxyacetone-phosphate dehydrogenase, GP DH = glycerol phosphate dehydrogenase, aspTA = aspartate transaminase.

formation of acetyl CoA (Fig. 3), required for the extensive, deregulated biosynthesis of cholesterol (and fatty acids) in tumor cell lines (29). Tumor cell membranes and tumor mitochondrial membranes are several-fold richer in cholesterol than normal membranes, a property that in turn influences transport mechanisms—among others, enhancing the export of citrate via the mitochondrial

anion exchanger (30). In addition, a low input of acetyl CoA via pyruvate dehydrogenase contributes to the truncation of the cycle (Fig. 3). No PDH activity has been detected in BHK or CHO cells (20). Only a few percent of the large amounts of glucose metabolized pass this metabolic bottleneck. The fraction of glucose entering the TCA cycle was recently estimated to be in the

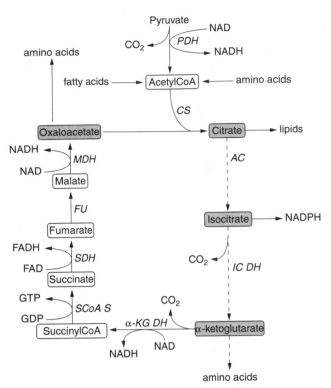

Figure 3. The tricarboxylic acid cycle inclusive of pyruvate dehydrogenase (PDH). Shaded compounds constitute precursor metabolites. Enzymes: CS = citrate synthase, AC = aconitase, IC DH = isocitrate dehydrogenase, α – KG DH = α-ketoglutarate dehydrogenase, SCoA S = succinyl coenzyme A synthetase, SDH = succinate dehydrogenase, FU = fumarase, MDH = malate dehydrogenase.

range of 0.2–0.6% by using labeling techniques (16,17,20), results also confirming earlier reported data. Whether this depends on (1) pyruvate unavailability, pyruvate being withdrawn for lactate formation, (2) a low PDH level, (3) down-regulation of PDH activity via phosphorylation due to high mitochondrial NADH levels generated by the glutamine metabolism, (4) the sensitivity of PDH to superoxide (see further in the following), or (5) a combination of these factors, has not been conclusively shown. In cells with elevated lipogenesis (and a truncated TCA cycle), other substrates such as acetoacetate (31) or catabolized amino acids (18) are used as sources of acetyl CoA to furnish the metabolic machinery with carbon for lipid synthesis. The flux from citrate to α-KG may also be restricted by the sensitivity of isocitrate dehydrogenase to superoxide (O_2^-) (32,33) as tumor cell lines have low levels of protecting superoxide dismutase (34). A truncated TCA cycle exists, for example, in HeLa cells (35), Morris hepatoma cells (30), and in AS-30D hepatoma cells (31). The significance of a truncated TCA cycle would be that acetyl CoA is not able to support respiratory energy generation, but some other substrate must be used. This role is fulfilled by glutamine metabolism, which provides carbon entering the TCA cycle at α-KG. As a consequence, the flux in the left branch of the cycle is increased simultaneously as NADH, and FADH respiratory substrates, are released (Fig. 3).

In contrast to other reports, a substantial flux from glucose into the TCA cycle was found in hybridoma cells in a study of metabolic fluxes using in situ ^{13}C NMR, as shown by the label at [4-^{13}C]-glutamate originating from [1-^{13}C]-glucose (and, consistently, the label in [5-^{13}C]-glutamate and [5-^{13}C]-proline originating from [2-^{13}C]-glucose). Glucose-derived acetyl CoA was reported to enter the TCA cycle at a rate of 0.09 mmol/10^9 cells/h. Less than 10% (0.007 mmol/10^9 cells/h) of this flux was calculated to exit the cycle as citrate and about 30% as isocitrate, resulting in a flux of approximately 0.05 mmol/10^9 cells/h to α-KG, the flux to α-KG from glutamine being of the same range (19,36). Sharfstein and co-workers calculated that the total flux into TCA (0.12 mmol/10^9 cells/h), including amino acids entering at acetyl CoA, was even higher than the uptake rate of glucose (18). These data do not indicate that the TCA cycle should be severely truncated in the hybridoma cells. However, lipogenesis is associated with rapid proliferation and cell division, but as the cells studied grew very slowly during the experimental conditions, it is possible that the data are not representative for cells growing at maximum rates in suspension cultures.

Respiration and ATP Generation

Mitochondrial NADH is oxidized by the respiratory system located in the inner mitochondrial membrane, with the concomitant consumption of oxygen, generation of the transmembrane proton gradient, and ATP formation. These well-known processes will not be treated further here, but the reader is referred to textbooks for details. However, an important aspect is the efficiency of the energy metabolism in industrial cell lines. The P/O ratio, which is the number of ATP molecules formed per oxygen atom used (or NADH molecule oxidized) is a measure of the respiratory efficiency. It is generally agreed that the maximum theoretical P/O ratio is 3, but a great deal of controversy exists regarding actual P/O ratios (37). In HeLa cells, which may have a defect in complex I of the electron transport chain (35), the maximum P/O ratio would be limited to 2. Hybridoma cells, on the other hand, were suggested to be able to increase the P/O ratio from 2 to 3 when oxygen availability was limited (38). This appears plausible, as elevated respiratory efficiency at low dissolved oxygen tension (DOT) levels is a phenomenon known to exist in microorganisms such as *E. coli*. However, not until recently has it been discussed whether the P/O ratio in continuous cell lines may vary depending on the availability of energy-yielding substrates, that is, glucose and glutamine, the critical question being if the P/O ratio decreases at conditions of substrate excess. So-called metabolic uncoupling occurs in *Saccharomyces* yeast cells, where the P/O ratio was calculated to vary by one order of magnitude (from 1.5 to 0.2) in a chemostat culture with energy source limitation or with nitrogen limitation, respectively (39). Likewise, metabolic uncoupling in bacteria at substrate-sufficient conditions is rather regarded as the rule than the exception. Thus it may not be implausible that cultured mammalian cells are able to adjust the P/O ratio accordingly (38).

GLUTAMINE METABOLISM

Function of Glutamine Metabolism

Glutamine metabolism generates energy, provides precursors for biosynthesis, and is associated with the increase in cell volume during progression through the cell cycle. An overview of glutamine metabolism is presented in Figure 4. Apart from being used for protein synthesis, glutamine is the nitrogen donor in pyrimidine, purine, amino sugar, NAD, and asparagine biosynthesis. The biosynthetic reactions are catalyzed by amidotransferases, and as result of the amido-nitrogen removal, glutamate is formed from glutamine. Glutamate is also derived from glutamine by the action of glutaminase. Glutamate is the direct precursor of proline and ornithine, and the principal amino-group donor in the cell. In cells of the immune system (macrophages) glutamine is to be likely the precursor of arginine, and thus indirectly of nitric oxide (40).

Part of the glutamine metabolism is mitochondrial, some of the involved enzymes being exclusively located in mitochondria: phosphate-activated glutaminase (PAG) (41), glutamate dehydrogenase (GDH) (42), and the TCA-cycle enzymes α-ketoglutarate dehydrogenase, succinylCoA synthetase, and succinate dehydrogenase. The mitochondrial metabolism is energy yielding as discussed before, and the TCA-cycle intermediates α-KG, succinyl CoA, succinate, fumarate, malate, and oxaloacetate are formed as a result of glutamine metabolism. Of these, mitochondrial oxaloacetate is presumably the most important precursor, being used for aspartate, which in turn is a precursor for purine and pyrimidines, and asparagine biosynthesis. The reason for the dignity of this function is that the anaplerotic enzyme pyruvate carboxylase (PC), which, in normal cells, substitutes oxaloacetate withdrawn from the TCA cycle to anabolic reactions, is lacking in

the continuous cell lines tested so far. In normal diploid fibroblasts possessing PC, aspartate is synthesized from glucose when glucose is abundant, but from glutamine when glucose is limiting (43). However, no activity of PC could be found in BHK, hybridoma or CHO cells (20), or in Sp2/0-Ag14 myeloma cells (44), and no flux via PC was detected in hybridoma cells, as judged by NMR analysis of the ^{13}C-carbon labeling pattern in intermediates of the TCA cycle resulting from the metabolism of 1-^{13}C-glucose (18,19).

Thus taking up glutamine is a way to deliver, not only glutamine, but also intracellular glutamate and aspartate. Why then could not medium glutamate and aspartate fulfill this role? These amino acids are regarded as "intracellular," and the reason for this may be sought in the regulation of System X_{AG}^-, the amino acid transport system specific for anionic glutamate and aspartate (45,46). While the activity of System X_{AG}^- is low in proliferating cells and extremely low in transformed cells but high in quiescent cells (47,48), the activity of System A, a concentrative amino acid transport system that imports glutamine, increases in response to mitogenic stimuli (49). Thus formation of glutamate and aspartate intracellularly from glutamine metabolism may be the only possibility for proliferating transformed cells to obtain these amino acids. Moreover, the intracellular accumulation of amino acids transported by System A, in particular glutamine and glutamate derived from the glutamine metabolism, is indispensable for the increase in cell volume leading to progression through the cell cycle (49). In fact, the sum of glutamine and glutamate constitutes about half the intracellular amino acid pool (49). It has also been suggested that glutamine exerts its main regulatory effects on cell proliferation by acting as a precursor for adenine and adenosine (50).

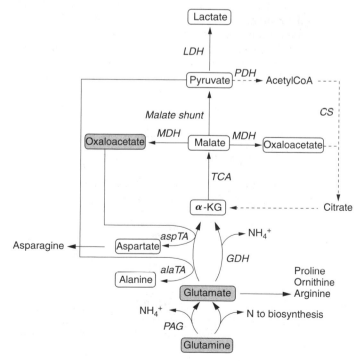

Figure 4. An overview of glutamine metabolism. Shaded compounds constitute precursor metabolites. Enzymes: PAG = phosphate-activated glutaminase, GDH = glutamate dehydrogenase, alaTA = alanine transaminase, aspTA = aspartate transaminase, TCA = enzymes of the TCA cycle; see Figure 3, MDH = malate dehydrogenase, PDH = pyruvate dehydrogenase, CS = citrate synthase, LDH = lactate dehydrogenase.

Mitochondrial Glutamine Transport and Localization of Glutaminase

Mitochondrial glutamine transport mechanisms may influence the first step of glutamine catabolism, catalyzed by the mitochondrial enzyme glutaminase. Glutamine transport into mitochondria has therefore been extensively studied, particularly in kidney cells, where glutamine metabolism has a pH regulating function (via excretion of NH_4^+). It is generally agreed that mitochondrial glutamine transport occurs by a neutral uniport carrier in these cells (51). Also, tumor cell mitochondria possess an efficient uniport transport system, located in the inner mitochondrial membrane (52). However, the mitochondrial neutral amino acid carrier, a transport system with high activity and broad specificity, may be the same as the glutamine uniporter (53). The existence of other transporters, such as an energy-dependent uniport carrier (54) and glutamine/glutamate and glutamine/malate antiporters (55), have also been suggested. The resolution of these anomalies awaits purification, cloning, and sequencing of the transport system(s). Kinetic studies performed over the years converge to the conclusion that mitochondrial glutamine transport is not rate limiting for glutamine catabolism.

The catabolism of glutamine is initiated by deamidation, catalyzed by phosphate-activated glutaminase (PAG, EC 3.5.1.2), a reaction that is thermodynamically favorable, the equilibrium constant being 320 at 25 °C (56). The reaction products are glutamate and a free ammonium ion. PAG in glutaminolytic cells (i.e., cells that catabolize glutamine) is of "kidney-type," the K_m for glutamine being 2–5 mM (3,7), in contrast to "liver-type" glutaminases, which have a much higher K_m for glutamine (around 20 mM). The intracellular glutamine concentration varies in the range of 2–11 mM in hybridoma cells, depending on the extracellular glutamine concentration (57), indicating that the metabolic flux via PAG also is affected by the extracellular glutamine concentration. The activity of glutaminase is correlated with malignancy of tumors and with growth rate, as witnessed in many investigations.

The exact submitochondrial localization of PAG has been a matter of controversy, but it now appears that there exist two pools of glutaminase that are interconvertible, one soluble pool in the matrix and a membrane-bound form (41,58). The heterotetrameric enzyme (59) is membrane bound, but not an integral membrane protein, and is enzymatically dominant over the soluble (inactive) species (41,58). The group of Kvamme suggested that PAG has a *functionally* external localization in the inner mitochondrial membrane (60,61). This conclusion was drawn from the observation that, in isolated brain mitochondria, glutamine-derived glutamate did not readily mix with the mitochondrial matrix pool of glutamate, but was preferentially released into the incubation medium (61). Further, no correlation between mitochondrial glutamine transport and glutamine hydrolysis was found. A possible explanation could be that PAG was localized to a microcompartment in the matrix, connected with the outside by a channeling mechanism involving glutamine and the reaction products. Also, in rat kidney

mitochondria supplied with external glutamine, most of the glutamate formed by PAG was exported from the mitochondria (the remaining glutamate was deaminated intramitochondrially by GDH) (62). These mitochondria contained two separate glutamate pools that did not mix; one pool is related to aspartate aminotransferase and the other to GDH (see further in the following). Thus it appears that, in the investigated cell lines, glutamate produced by mitochondrial PAG mainly is exported to the cytosol. Understanding the site of glutamate formation from glutamine by PAG (mitochondrial intramembrane space and/or mitochondrial matrix) is fundamental for understanding the overall glutamine metabolism, as will be shown in the following. No investigations on these issues seem to have been performed on industrial cell lines.

Pathways of Glutamine Metabolism

The metabolism of glutamine/glutamate involves two major pathways—the transamination (TA) pathway and the glutamate dehydrogenase pathway (Fig. 5). Although the carbon skeleton of glutamate is forwarded into the TCA cycle via the conversion to α-KG in both cases, there is a fundamental difference between the two pathways. The GDH reaction liberates the amino-nitrogen of glutamate as a free ammonium ion. The resulting α-KG can be used freely in the intermediary metabolism, as, for example, for gluconeogenesis, or it can be converted to lactate or completely oxidized in the TCA cycle to carbon dioxide and water (Fig. 6). All these activities require that a four-carbon TCA-cycle intermediate (C4) is converted to a glycolytic three-carbon intermediate (C3), as will be discussed later.

In the TA pathways, the entry of glutamate carbon into the TCA cycle is stoichiometrically coupled to the transamination of oxaloacetate or pyruvate to yield aspartate (the aspTA pathway) and/or alanine (the alaTA pathway) as end products of the glutamine metabolism. Thus the transamination reaction constitutes at the same time the last and the second step of the pathway, creating a closed loop, as illustrated in Figure 7. Whereas alanine formation implies conversion of a C4-TCA-cycle intermediate

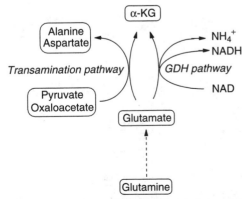

Figure 5. The two main routes of the glutamine metabolism—the transamination pathway and the glutamate dehydrogenase (GDH) pathway.

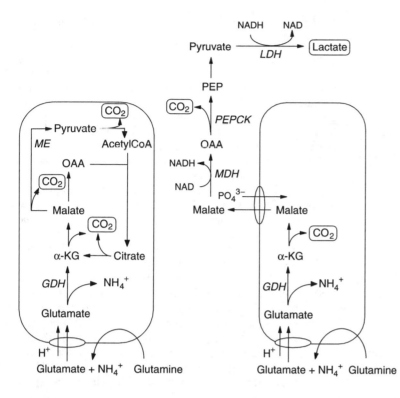

Figure 6. Examples of GDH pathways: Left: Complete combustion of glutamine to carbon dioxide and water by mitochondrial enzymes. Key enzymes: GDH = glutamate dehydrogenase, ME = malic enzyme. OAA = oxaloacetate. Right: Lactate formation from glutamine via cytosolic malate dehydrogenase (MDH), phosphoenolpyruvate carboxykinase (PEPCK), and lactate dehydrogenase (LDH).

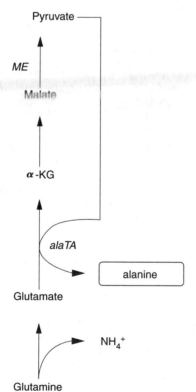

Figure 7. The transamination pathway to alanine. Enzymes: alaTA = alanine transaminase, ME = malic enzyme.

to C3-pyruvate, aspartate formation does not. McKeehan suggested the pathway of incomplete oxidation of glutamine to pyruvate should be named glutaminolysis (63). Other (minor) TA pathways can result in serine/glycine and asparagine formation, the latter being a product of aspartate that has undergone transamidation with glutamine (64,65). In practice, glutamine metabolism is a mixture of different pathways.

The Link between the Four-Carbon Compounds of the TCA Cycle and Three-Carbon Compounds of Glycolysis: The Malate Shunt

The malate shunt is an integral part of glutaminolysis (7,19), its existence being obligatory for the alanine TA pathway. Several possibilities have been discussed in the literature: It was recognized early that $NAD(P)^+$-linked mitochondrial malic enzyme (EC 1.1.1.39) played an important role in the glutamine metabolism of tumor cells, its activity being progression linked in hepatomas (66,67), and it was suggested that this enzyme played the critical role for conversion of malate to pyruvate (7,63,68). This proposal was later rejected because Ehrlich ascites tumor cell mitochondria appeared not to catalyze the conversion of intramitochondrially formed malate, but only of malate entering mitochondria from outside (69). A more recent interpretation of these results involves both the thermodynamics (35) and mitochondrial transporters, as will be discussed later. Other investigations have shown that the role of $NAD(P)^+$-dependent mitochondrial malic enzyme is different in different cell lines: In brain and hepatoma mitochondria it catalyzes the conversion of both extra- and intramitochondrial malate (70,71), while in HeLa cells it does not participate in glutamine metabolism to any great extent (35). The malate shunt was shown to be operative in hybridoma cells, as judged by NMR analysis of the label from [1-^{13}C]-glucose appearing in [2-^{13}C]-lactate (one turn of [1-^{13}C]-glucose label in TCA); and the label from [3-^{13}C]-glutamine appearing in [2-^{13}C]-lactate as well as in [3-^{13}C]-lactate (18,19). The flux via the malate shunt was estimated to be of the same order as the uptake of glutamine (19).

Other possibilities for conversion of C4-TCA-cycle compounds to the C3s of glycolysis include cytosolic (and mitochondrial) strictly $NADP^+$-dependent malic enzyme (EC 1.1.1.40); or phosphoenolpyruvate carboxykinase (PEPCK, cytosolic and/or mitochondrial). Significant activity of cytosolic PEPCK was found in myeloma Sp2/0-Ag14 cells during all growth conditions tested (44,72). However, hybridoma, CHO, and BHK cells lack detectable PEPCK activity (20), and PEPCK was not involved in the glutamine metabolism in rat enterocytes (73).

GDH versus TA Pathways: Origin of NH_4^+ in Cell Cultures

A critical question for the evaluation of energy metabolism is which, or in which proportions, the two main routes of glutamine metabolism are used. Theoretically, the complete combustion of glutamine via the GDH pathway (Fig. 6) generates more energy (maximum 27 ATP/glutamine) than incomplete oxidation via the TA pathways (Fig. 7; 9 ATP/glutamine for both alaTA and aspTA) (10,74). Really, glutaminolytic cells use mainly the TA pathways, as supported by an abundance of evidence. Therefore, most of the ammonium ions in cultures of HeLa, CHO, hybridoma, and Sp2/0-Ag14 myeloma cells are derived from the amido-nitrogen of glutamine, that is, from the PAG reaction, as confirmed by NMR detection of $^{15}NH_4^+$ released from glutamine, labeled with ^{15}N either in the amide or amine position (65,75). The importance of the alaTA pathway is demonstrated by the observation that growth and glutamine oxidation in a hybridoma cell was even inhibited by decreasing the availability of pyruvate for transamination (76).

On the other hand, hybridoma, myeloma, BHK, CHO, and HeLa cells display GDH activity, although no flux via the pathway could be detected (20,44,65,77). As GDH is a universal enzyme, present in most cells throughout the living world, it would be most surprising if it was not needed at all. In fact, during glucose starvation the flux via the GDH reaction was increased in both myeloma and hybridoma cells (75). The uptake of glutamine and the total production of NH_4^+ was significantly increased in both glucose-starved myeloma and fructose-grown hybridoma cells, as compared to glucose-sufficient cells. The increased NH_4^+ production was due to an increased metabolism via PAG (1.6–1.9-fold in the hybridoma, and 2.7-fold in the myeloma cell line) and an even further increased throughput via GDH (4.8–7.9-fold in the hybridoma cells, and 3.1-fold in the myeloma cells). The data indicate that both PAG and GDH are down-regulated when glucose is in excess, but up-regulated in glucose-starved cells. Thus, in times of severe energy limitation, cells use the most energy-efficient pathway. This may be explained by the complex allosteric regulation of both PAG and GDH, the main inhibitor of GDH being α-KG, which accumulates in cells with extensive TA activity (78). The GDH reaction, in the direction toward α-KG, is also inhibited by elevated concentrations of NH_4^+ for thermodynamic reasons.

Aspartate or Alanine as an End Product of the TA Pathway?

The activity of aspartate transaminase in a variety of cell types, including industrial cell lines, is clearly much higher, often about 10-fold, than that of alanine transaminase. This observation has often been interpreted as aspartate being the principal end product of the glutamine metabolism. Although this certainly is the case in a number of cell types such as kidney cells, platelets, lymphocytes, mouse macrophages, Ehrlich ascites tumor cells, and HeLa cells, an overwhelming number of studies show that in industrial cell lines such as hybridoma, myeloma, and CHO cells, alanine is the major end product of the glutamine metabolism. In many such cell lines no extracellular aspartate is produced. How can this be explained? First, the activity of alaTA is obviously not limiting in the sense that it prevents alanine formation, and the relatively high activity of aspTA is not a controlling factor. To distinguish further between factors that lead to the formation of aspartate or alanine as the end products of glutamine metabolism, it is necessary to consider the compartmentalization of the TA reactions (aspTA and alaTA isoenzymes are present in both the cytosol and mitochondria), the overall stoichiometry of the pathways including that of cofactors, the site of glutamine-derived glutamate formation, mitochondrial transport mechanisms, the activity of the malate shunt, and the thermodynamics of the reactions.

Mitochondrial TA Pathways

Supposing transamination is mitochondrial, then gluta-mine-derived extramitochondrial glutamate must enter mitochondria through one of the two glutamate trans-porters, the electrogenic glutamate–aspartate exchanger or the electroneutral glutamate carrier (51,53,79). In both cases glutamate transport is proton coupled, greatly favoring inward transport of glutamate in energized mitochondria. When glutamate enters mitochondria via the glutamate–aspartate exchanger, the end product of the glutamine metabolism is by necessity aspar-tate, otherwise the mitochondrial aspartate pool would be depleted (by export to the cytosol) and the process come to a halt (Fig. 8). Mitochondrial aspTA is local-ized near the mitochondrial periphery (80), indeed bind-ing to the inner mitochondrial membrane (81), thereby increasing the likelihood for transamination of incom-ing glutamate. Moreover, aspTA associates with the α-KG dehydrogenase complex and malate dehydrogenase (MDH), further facilitating the conversion of glutamate to aspartate (82). Furthermore, one functional mitochon-drial glutamate pool is related to the mitochondrial aspTA, with external glutamate channeled directly to this pool by the glutamate–aspartate exchanger (62). Extramitochondrial liberation of glutamine-derived glu-tamate and transport back to glutamate by the glu-tamate–aspartate exchanger readily explains the rapid release of glutamine-derived glutamate into the incu-bation medium, and the much slower accumulation of aspartate, as observed in experiments with isolated mito-chondria from Ehrlich ascites tumor cells (69). In fact, mitochondrial transamination appears to be the major

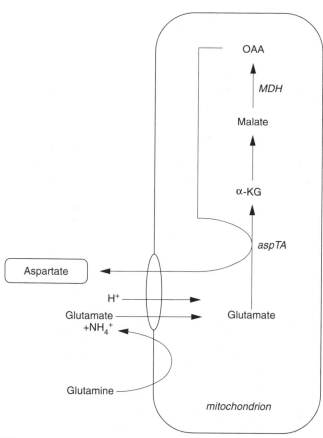

Figure 8. The mitochondrial transamination pathway to aspartate in which the glutamate–aspartate exchanger plays a central role. Enzymes: aspTA = aspartate transaminase, MDH = malate dehydrogenase; OAA = oxaloacetate.

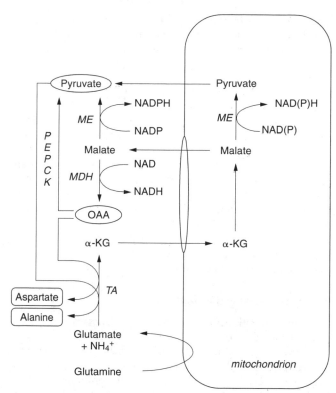

Figure 9. Cytosolic transamination pathways. See text for further explanations. Enzymes: TA = aspartate and alanine transaminases, MDH = malate dehydrogenase, ME = malic enzyme, PEPCK = phosphoenolpyruvate carboxykinase; OAA = oxaloacetate.

pathway of glutamine metabolism in aspartate-producing cells (62,69).

In the case when glutamate enters the mitochondria via the proton symport system or is released intramitochondrially, the subsequent metabolism proceeds either via deamination by GDH or via transamination to alanine. A second mitochondrial glutamate pool located in the mitochondrial matrix and associated with GDH has been identified (62). It has also been clearly shown that when pyruvate is available, mitochondria from various cell types form alanine (35,69,83–85).

Cytosolic TA Pathways

Glutamine-derived glutamate released outside mitochondria could readily be metabolized via transamination in the cytosol. The availability of the substrates pyruvate and oxaloacetate likely determines the proportions formed between alanine and aspartate. Formation of cytosolic glutamine-derived pyruvate or oxaloacetate requires that glutamine carbon be metabolized in the TCA cycle. The transamination product α-KG crosses the mitochondrial membrane in exchange for malate on the α-ketoglutarate/malate carrier (51,53,79), the reactions from α-KG to malate taking place intramitochondrially (Fig. 9). In the cytosol, malate may be converted to pyruvate by the malate shunt enzymes or to oxaloacetate by MDH.

As discussed previously, the major pathway of aspartate formation is mitochondrial TA. An explanation to this preference may well be that the equilibrium of the MDH reaction strongly favors malate formation (35), and that cytosolic NADH, which needs to be reoxidized, is generated. This can neither be done by lactate formation from glucose-derived pyruvate, as NADH is also formed in glycolysis in stoichiometric amounts to pyruvate, nor by transporting the NADH back into the mitochondrion by the malate–aspartate shuttle, as the cofactor was actually released in the cytosol by the reversal of this shuttle. However, co-oxidation of a second molecule of glutamine in pathways that consume cytosolic NADH, such as lactate formation (Fig. 6), would be possible. This route depends on glutamate deamination by GDH, and as discussed, GDH is not very active in rapidly proliferating glucose sufficient cells. Interestingly, increased aspartate (and asparagine) production was observed in myeloma and hybridoma cells (normally producing alanine as the major end product of the glutamine metabolism) during glucose starvation, conditions that increase the GDH activity (75).

By contrast, a pathway involving cytosolic transamination of pyruvate to alanine via a cytosolic or mitochondrial malate shunt is fully feasible, according the scheme in Figure 9. In Sp2/0 myeloma cells, pyruvate could be formed by cytosolic PEPCK, which is present in these cells (72), but as this pathway releases cytosolic NADH, it can, for the same reasons as for cytosolic aspartate transamination, be excluded as a major pathway. However, for hybridomas and other cells lacking PEPCK, malic enzyme is the only possibility. A cytosolic location of the malate shunt activity was recently suggested, implying that transamination to alanine mainly takes place in the cytosol. Mancuso and co-workers noted that very little malate-shunt-derived pyruvate was cycled back to the TCA cycle (19). A further support for this idea is that in some alanine producing cells (enterocytes, myeloma Sp 2/0-Ag14) the alaAT activity is higher in the cytosol than in mitochondria (72,85). However, cytosolic malic enzyme in normal cells is strictly NADP$^+$ dependent, while the NAD(P)$^+$-dependent activity is confined to the mitochondria of tumor cells. As no NADP$^+$-dependent MDH activity (EC 1.1.1.40) was detected in hybridoma, BHK, or CHO cells (20), the malate shunt activity may be located in the mitochondria after all, as originally proposed (7,63). Unfortunately no experiments with isolated mitochondria from industrial cell lines seem to have been performed.

Pathways Involving the Lipid Cycle

Continuous cell lines with elevated lipogenesis (29,86) need to generate cytosolic acetyl CoA, the precursor for cholesterol and fatty acid synthesis. The extent of this activity is not known but is assumed to be significant. Oxaloacetate formed by glutamine metabolism may combine with mitochondrial acetyl CoA to yield citrate, which is exported and cleaved into cytosolic acetyl CoA and oxaloacetate (Fig. 10). The combination of a high

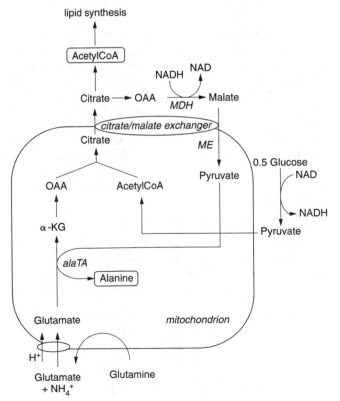

Figure 10. Glutamine metabolism involving the lipid cycle. See text for further explanations. Enzymes: alaTA = alanine transaminase, ME = malic enzyme, MDH = malate dehydrogenase. OAA = oxaloacetate.

cytosolic NADH/NAD ratio, with the equilibrium for the MDH reaction strongly favoring malate formation (35), suggests that cytosolic oxaloacetate most likely is reduced to malate. The further fate of this cytosolic malate depends on the overall glutamine metabolism, the site of glutamate formation from glutamine, and the dominating pathway of glutamate metabolism (aspTA, alaTA, or GDH). In aspartate-producing cells with low GDH activity, malate is transported back into the mitochondria in exchange for the citrate leaving on the malate–citrate exchanger. Incoming malate is converted to pyruvate by mitochondrial malic enzyme and transaminated to alanine simultaneously as aspartate production is decreased (69, 83–85,87) (Fig. 10). The reason why incoming malate does not regenerate mitochondrial oxaloacetate [as it does in normal cells lacking mitochondrial NAD(P)$^+$-dependent malic enzyme activity (69)] depends again on the thermodynamics, pyruvate formation being more favorable (35). As discussed, this implies that some glutamate enters mitochondria via the proton symport, or is formed intramitochondrially, in addition to that entering by the glutamate–aspartate exchanger. In alanine-producing cells with low GDH activity malate may be converted to pyruvate in the cytosol and subsequently transaminated to alanine (18). However, in both alanine- and aspartate-producing cells (with low GDH activity) glutamine metabolism cannot supply the acetyl CoA required for lipid synthesis, but some other carbon source is needed. In hybridoma cells, glucose-derived acetyl CoA was shown to be incorporated into lipids (19). Glycolytic NADH, formed in stoichiometric amounts (1:1) to the glucose-derived acetyl CoA, is then regenerated by the reduction of oxaloacetate to malate (Fig. 10). In cells lacking PDH activity glucose cannot be the source of acetyl CoA, which consequently must be derived from other sources, as, for example, catabolized amino acids (18). It should also be pointed out that the continuous cell lines lacking PC and with mitochondrial malic enzyme are unable to run the lipid cycle from glucose only, as this enzyme pattern prevents formation of oxaloacetate from glucose even if some PDH activity were present. Therefore, the oxaloacetate must be formed from glutamine metabolism.

Cells that normally produce alanine as the major end product of glutamine metabolism do so in parallel to the lipid cycle. This conclusion can be drawn from the observation that in hybridoma and myeloma cells the flux to alanine was the same whether glucose was present or not (75). During these conditions, the lipid cycle is actually self-sufficient. Glutamine is metabolized via GDH, and the stoichiometry of this pathway is:

$$1 \text{ gln} \longrightarrow 2\text{NH}_4^+ + 3\text{CO}_2 + \text{acetyl CoA}_{\text{cytsolic}}$$

Pyruvate as a Precursor for Lactate and Alanine: Compartmentalization and Stoichiometry of Metabolism

In the foregoing it has been argued that glucose-derived pyruvate is mainly converted to lactate and that glutamine-derived pyruvate is mainly transaminated to alanine both in the cytosol and in mitochondria. However, in a hybridoma culture fed with [1-^{13}C]-glucose the fractional labeling in lactate was equal to the fractional labeling in alanine (19), strongly indicating that alanine–pyruvate–lactate are at, or near, equilibrium via lactate dehydrogenase and alaTA. The couple alanine–glutamate also should be at equilibrium through alaTA, as supported by the observation that the fractional labeling (^{15}N) in alanine and glutamate in HeLa cells was very similar irrespective of the label being derived from L-[2-^{15}N]-glutamine, L-[5-^{15}N]-glutamine, or from ^{15}NH$_4^+$ (65). Taking into account that the aspTA reaction is also at equilibrium (1) in the cytosol (but not in mitochondra), it may be possible to conclude that cytosolic glutamate, aspartate, alanine, pyruvate, and lactate are all near equilibrium through the combined actions of the TAs and LDH. This can be consistent with compartmentalization of the metabolism because cytosolic and mitochondrial alanine are readily exchanged via the neutral amino acid carrier. By this mechanism carbon label would rapidly be redistributed in the cytosolic pool of alanine, pyruvate, and lactate, and nitrogen label in alanine, glutamate, and aspartate.

On the other hand, do the labeling data imply that lactate is formed from glutamine, or that alanine is formed from glucose? To understand this further, the cofactors involved must be considered. As discussed before, lactate formation regenerates NAD with a stoichiometry of 1:1. Each glucose-derived pyruvate is accompanied by formation of NADH, also with 1:1 relation. This means that the pathway from glucose to lactate is redox neutral. Can alanine then be formed from glucose, or lactate from glutamine on a stoichiometric basis involving the cofactors? Alanine can be formed from glucose (via transamination of glucose-derived pyruvate by glutamine-derived glutamate) only if glycolytic NADH (in stoichiometric amounts) can be transported into mitochondria via the shuttle systems, and at the same time the resulting α-KG is completely oxidized to carbon dioxide and water or converted to glutamate again by incorporation of NH$_4^+$ via GDH. In view of the finding that very little malate shunt pyruvate is returned to the TCA cycle (which would be necessary for oxidation of α-KG) and the low activity of GDH, this alternative appears not realistic. However, lactate formation from glutamine is a reality as judged by the lac/glc ratio, which sometimes exceeds 2 in hybridoma and myeloma cells (16,44,64).

Conclusions

In conclusion, there is no unifying concept that can describe the glutamine metabolism in all cell lines, but a diversity of mechanisms exist, operating in different cells and under different growth conditions. During normal growth conditions with excess substrates, the TA pathways dominate, while the GDH pathway is up-regulated at severe glucose limitation. It is not altogether clear what factors determine which of the TA pathways — alanine or aspartate — a particular cell line uses. However, in aspartate-producing cells, glutamate enters mitochondria via the glutamate–aspartate exchanger, and transamination is (mainly) mitochondrial. Transamination to alanine occurs both in mitochondria and in the cytosol. Transamination to alanine is associated with the lipid cycle, but

also occurs independently. Whether aspartate-producing cells have higher activity of the glutamate–aspartate exchanger than alanine-producing cells and/or lack an efficient malate shunt has not been investigated. Clearly, these mechanisms need further characterization in industrially important cell lines.

Glutamine Synthetase

Cell lines such as CHO and BHK cells that possess glutamine synthetase (GS) are able to proliferate in the absence of exogenous glutamine. GS also serves as a selection/amplification marker for transfection, resulting in engineered cells (e.g., NSO myelomas) overexpressing GS. Hybridoma cells are usually not able to proliferate in the absence of glutamine, although a rather significant GS activity was found in some cell lines (16,20). HeLa cells grown in a glutamine-containing medium did not display any GS activity at all, while they did after having been adapted to a medium with a low glutamine concentration (65). These results indicate that GS is downregulated when glutamine is present in the medium. The substrates for GS are glutamate, NH$_4^+$, and ATP. Cells growing without glutamine consume more aspartate and asparagine than cells with glutamine. This is likely due to (1) biosynthesis of aspartate and asparagine being restricted in the absence of exogenous glutamine, and (2) aspartate and asparagine being required to generate intracellular precursors for glutamine biosynthesis. No futile cycling of glutamine occurred in cells possessing both PAG and GS (65).

DYNAMICS OF METABOLISM IN CULTURE

Mechanisms and Kinetics of Glucose and Glutamine Uptake

In batch cultures of most industrial cell lines accumulation of lactate, ammonium ions, and alanine accompany cell growth. The final concentrations of the byproducts depend on the initial concentrations of glucose and glutamine; typical figures are 8–35 mM lactate, 2–3 mM ammonium, and 1–2 mM alanine in media with 5–25 mM glucose and 4 mM glutamine. Spontaneous decomposition of glutamine further contributes to the accumulation of ammonium ions. The typical pattern of specific glucose and glutamine consumption rates (q_{glc}; q_{gln}) follow the same trend as the specific growth rate (μ), being initially high and decreasing throughout the culture (88,89). The substrate consumption rates follow this pattern even if growth is forced to be exponential by intermittent additions of serum or insulin (90,91).

The kinetics of glucose and glutamine uptake in cultures of industrial mammalian cells are determined by a number of factors, such as the nature of the substrate uptake systems and their specific regulators, the substrate concentrations, the proportions between glucose and glutamine, the cellular need for energy and precursors, growth rate, the dissolved oxygen concentration, culture pH, and temperature. The characteristics of glucose and glutamine consumption may also be cell line specific. Glucose uptake is mediated by facilitated diffusion in most mammalian cells. The transporter exists in five isoforms (GLUT1–GLUT5) with different kinetic properties: saturable transporters with K_m from 1–2 to 15–20 mM and unsaturable transporters with linear kinetics. The driving force for glucose uptake is solely the glucose gradient over the cell membrane. Glucose may also be taken up by a high-affinity (K_m below 0.2 mM) saturable Na$^+$-symport process driven by the transmembrane sodium ion gradient. Glucose uptake rates by saturable and unsaturable systems are increased in transformed cell lines due to an increased number of the transport protein molecules, particularly the facilitated diffusion transporter. Transport of glucose across the plasma membrane appears not to be rate limiting (for glycolysis), as the measured (potential) transport rate was significantly higher than the glucose utilization rate (16,20). Glutamine is taken up by all three major amino acid transport systems, Systems A, ASC, and L (92). While Systems A and ASC are Na$^+$ symports driven by the sodium gradient, transport by System L, which is an antiport, is dependent on intracellular accumulation of methionine. System A activity is also increased in transformed cell lines (46). Insulin, present in serum and at even higher levels in serum-free media, up-regulates glucose uptake and increases glucose consumption in hybridoma cells (91,93). The activity of system A is also up-regulated by insulin in a wide range of cell types (46).

It is now well known from a number of investigations that q_{glc} and q_{gln} increase in response to increasing glucose and glutamine concentration. The kinetics of this relation are not easily defined, as actual q values depend on a number of factors and results from different combinations of these factors are virtually unpredictable. For example, the high initial rates of substrate consumption seen in batch cultures are certainly a combination of the actual substrate concentration and the fact that cells are stimulated by serum and/or growth factors, an effect that fades with time. A typical maximum value for q_{glc} with glucose initially at 5 mM is 7 mmol/10^9 cells/day, and for q_{gln} with glutamine at 3.5 mM is 2.4 mmol/10^9 cells/day. A compilation of q_{glc} data from several studies of hybridoma cells shows that q_{glc} increases up to a glucose concentration of at least 16 mM, at which level q_{glc} is around 24 mmol/10^9 cells/day (94). In chemostat cultures a Monod type of relation was found for glutamine at limiting concentrations (95), but q_{gln} increases also above this level (96). In perfusion culture both q_{glc} and q_{gln} increased as a function of the growth rate, although neither glucose nor glutamine was limiting (97). The value for q_{glc} is almost always higher than q_{gln}, sometimes up to 10-fold (98).

Yield Coefficients and Metabolic Ratios

Yield coefficients and metabolic ratios are frequently used to evaluate cell metabolism and to interpret the effects of changes in culture conditions. The cellular yield coefficients, that is, $Y_{N/glc}$ and $Y_{N/gln}$, show the number of new cells (N) formed from a unit amount of glucose or glutamine consumed by the cells. A high number indicates that the substrate is used more efficiently for growth than when Y is low. $Y_{N/glc}$ and $Y_{N/gln}$ vary over a relatively wide range depending on actual substrate concentrations

and the proportions between glucose and glutamine in the culture medium (see further in the following). In batch culture of hybridoma cells with excess glucose and glutamine typical overall numbers are 0.3×10^9 and 0.5×10^9 cells/mmol for $Y_{N/glc}$ and $Y_{N/gln}$, respectively. In a fed batch culture with glucose and glutamine limitation the yield coefficients may increase more than twofold (89,91). In a chemostat culture of myeloma cells with glucose or glutamine limitation $Y_{N/glc}$ and $Y_{N/gln}$ were 1.7×10^9 and 1.4×10^9 cells/mmol, respectively (44).

The quotients NH$_4^+$/gln, ala/gln, and lac/glc (all in mol/mol) witness the metabolic performance of the cell and indicate the formation of NH$_4^+$ and alanine from glutamine and lactate from glucose. The theoretical maximum values are: $Y_{NH_4^+/gln} = 2$, $Y_{ala/gln} = 1$, and $Y_{lac/glc} = 2$. Typical overall molar yields in batch culture are in the range of 1.4–2 for $Y_{lac/glc}$, 0.4–1 for $Y_{NH_4^+/gln}$, and somewhat lower, 0.2–0.6, for $Y_{ala/gln}$. It should be noted that these quotients are calculated from analyses of products and substrates in the culture medium and are therefore only apparent. Glutamine, glucose, and alanine used for synthesis of biomass cannot be distinguished but are included (or excluded as for alanine) in the quotients. This means that when glucose or glutamine is consumed at high rates, a lower fraction goes to biomass than when the consumption rates are lower. Consequently, the lac/glc ratio will increase at high glucose consumption rates, as demonstrated in the work by Hiller and co-workers (97). The glutamine-related ratios are more difficult to interpret, but an increase in the NH$_4^+$/gln ratio approaching one or even exceeding one is a strong indication that the GDH pathway is used for catabolism of glutamine. The ala/gln ratio does not follow the same pattern as the lac/glc ratio but may in fact decrease as q_{gln} increases (75,97), indicating that an upper limit for alanine formation may exist, as illustrated by the following example: As glucose was exchanged by fructose in a hybridoma cell line, the glutamine consumption was increased and the ala/gln ratio decreased, yet the throughput to alanine was the same (75). Therefore, it must be emphasized that the metabolic quotients alone cannot serve as a basis for drawing firm conclusions on the usage of particular metabolic pathways. Nevertheless, changes in metabolic ratios can be used to recognize changes in the metabolism.

Interactions between Glucose and Glutamine Metabolism

The interaction between glucose and glutamine metabolism has attracted considerable interest, not the least from a biotechnological perspective, ever since the first report about the reciprocal regulation of glucose and glutamine metabolism. Understanding this relation provides the means for controlling production processes optimally with respect to energy formation for growth and product formation, and for minimizing byproduct formation. Zielke and co-workers found that a low glucose concentration (25–70 μM) stimulated glutamine utilization, but that oxidation of glutamine was depressed by 83% in the presence of 5.5 mM glucose (as compared to without glucose) in human diploid fibroblasts (99). Conversely, glutamine (2 mM) inhibited [6-^{14}C]-glucose oxidation by

88% while it increased [1-^{14}C]-glucose oxidation to CO_2 by 77% as compared to the case without glutamine; that is, the oxidation via PDH was inhibited while the flux via HMS was stimulated.

The dynamics of glucose and glutamine metabolism in industrial cell lines have been studied since the mid-1980s, one of the most comprehensive studies being that of Miller and co-workers using continuous cultures of hybridoma cells with transient and step changes in glucose and glutamine concentrations (98,100). Hybridoma cells respond to a sudden transient increase in glucose concentration (from 0.1 to 6.6 mM) by an immediate increase in q_{glc} (100–200%), a corresponding increase in lactate production and an inhibition of the specific oxygen consumption rate (q_{o_2}) seemingly without any effect on q_{gln} or the specific ammonium production rate ($q_{NH_4^+}$) (100). However, an abrupt and persistent increase in the glucose concentration shifted the metabolism toward an increased glycolytic flux and a decreased q_{gln} and q_{o_2} that persisted in the new steady state (100). The specific ATP production rate (q_{ATP}) was essentially constant during these metabolic changes, although the origin of ATP shifted to an increased contribution from glycolysis and a decreased contribution from oxidative metabolism. A similar repression of q_{o_2} by glucose addition was observed in another hybridoma (101). The metabolic explanation for the inhibition of respiration by excess glucose involves an increased production of cytosolic ATP due to the enhanced glycolytic flux that depresses oxidative phosphorylation by lowering the availability of ADP for mitochondria. Conversely, at limiting concentrations of glucose, glutamine and oxygen consumption, as well as ammonium formation, are significantly increased in hybridomas and myelomas (44,75,89,102). The metabolism of glutamine shifts from the AlaTA pathway to an increased flux via the GDH route, which increases energy generation from glutamine; glutamine carbon may also be forwarded into the glycolytic pathway (75). As discussed, this is revealed as an increased NH_4^+/gln ratio, as seen in many reports, for example, in chemostat cultures with glucose limitation (44). When glucose is limiting or replaced by fructose, glutamine metabolism supplies all the energy (75).

The response of hybridoma cells to changes in the glutamine concentration seems to depend on whether glutamine is anabolically limiting or not. A glutamine pulse to a steady-state continuous culture at a glutamine concentration of 0.2–0.3 mM in the presence of excess glucose immediately increased q_{gln}, $q_{NH_4^+}$ and q_{ala}, without significantly affecting q_{O_2}, q_{glc} or the lac/glc ratio. On the other hand, starting from a steady-state level of approximately 0 mM glutamine, an increased consumption rate of glucose, glutamine, and oxygen and an increased production rate of ammonium, alanine, and lactate resulted from an increase in the glutamine concentration (98). The stimulation of the utilization of glucose, as compared to the utilization in the absence of glutamine, leading to increased fluxes in the glycolytic and HMS pathways but not in the TCA cycle, has been observed in many other investigations as well. Limiting concentrations of glutamine are also associated with an increased use of other amino acids. Some of these amino acids are used for catabolism, but others for anabolism, in particular, aspartate and asparagine, for which glutamine is both the carbon and nitrogen precursor. In a recent (NMR) study of hybridoma cells, glutamine at 0.08 mM was found to be limiting for some anabolic pathways, which require glutamine nitrogen, but this concentration was not considered energetically limiting (36). The Monod constant for glutamine regarding the specific growth rate is reported to be in the same range (0.089 mM) (103).

Miller and co-workers further showed that q_{ATP} was almost constant over a range of q_{glc}/q_{gln} ratios from 1.4 to 10.9, with q_{gln} varying from 0.27 to 1.31 and q_{glc} from 1.8 to 4.6 mmol/10^9 cells/day (98). Similarly, q_{ATP} was constant in a recombinant myeloma for q_{glc}/q_{gln} ratios between 0.4 and 5.5 (102)—a clear indication of efficient regulatory mechanisms for maintaining ATP homeostasis. The largest fraction of ATP (67–89%) is derived from oxidative metabolism irrespective of the culture conditions (oxygen, glucose, and glutamine limitation, transient situations, all substrates in excess) (97,98). However, it should be pointed out that numbers for q_{ATP} depend on the calculation methods used, including the chosen P/O ratio, with reported numbers varying from 16 to 60 mmole ATP/10^9 cells/day (97,100).

Another important question is how glucose and glutamine metabolism interact at steady state or more slowly changing conditions, as is the case in fed batch and continuous culture systems. A gradual decrease in glutamine concentration was found not to affect the glycolytic flux much in hybridoma cells, as compared to a sudden step-down in glutamine concentration, which induced a large increase in the glycolytic flux (36). Therefore, the actual metabolic situation may well be a reflection of the previous history of the culture, a circumstance that undoubtedly would explain many of the disparities that exists in the literature.

To obtain a more quantitative description of the metabolic interactions, modeling approaches have been undertaken. Zeng and Deckwer developed a kinetic model that could describe experimental results from five independent reports. In brief, the model predicts that q_{glc} can be expressed as the sum of three parts owing to (1) cell growth, (2) glucose excess, and (3) glutamine regulation, and q_{gln} as the sum of only two parts owing to (1) cell growth and (2) glutamine excess (104). Obviously, this does not agree with the preceding verbal description. This is due to the inability of the model to account for metabolic effects such as up-regulation of GDH at low glucose concentrations, the history of the culture, and small, but important, cell-line-specific differences. For example, in steady-state chemostat cultures of the Sp2/0 myeloma, increasing feed concentrations of glutamine did not affect q_{glc}, while in a hybridoma cell line, q_{glc} co-varied with q_{gln}; that is, an increase in q_{gln} increased q_{glc} (105). The latter response would be expected if glutamine were anabolically limiting, suggesting that the level at which this occurs may vary between cell lines. Metabolic flux analysis based on a stoichiometric reaction network is a further tool to increase our understanding of metabolism. A problem associated with this kind of model is that the set of linear equations defined by the network is underdetermined, which requires that certain assumptions must be included in the model in order to solve the system. This may lead to erroneous results that do not fit with experimental results (106).

In conclusion, an increase in the glucose concentration increases the glucose consumption rate and the lactate production rate, and depresses oxidative metabolism, sometimes without inhibiting glutamine uptake. At low glucose levels glutamine utilization is enhanced and the efficiency of glutamine metabolism further increased by up-regulation of the GDH pathway. Glutamine consumption rates increase in response to an increase in the glutamine concentration without affecting glucose metabolism or oxidative metabolism when glutamine is above some critical value. It is now generally agreed that continuous cell lines are metabolically hyperactive—producing more ATP and NADH than they actually need, the strategy seemingly being to maximize ATP and NADH production (106,107). How the cell disposes of surplus glutamine has not been experimentally verified, but energy may be dissipated through futile cycling (36) and/or through a reduced P/O ratio. Other, thus far unknown, sinks for glutamine may also exist, as evidenced by the following example: The Sp 2/0 myeloma cell line increased the glutamine consumption rate by 50% in response to ion stress (10 mM K^+ + 10 mM NH_4^+), although the fluxes via PAG, alaTA, or GDH did not increase during the experimental conditions (108). These results also emphasize that the energy metabolism of cultured mammalian cells cannot be understood from judging only the consumption rates of glucose and glutamine, but must be investigated by more comprehensive methods.

Effects of Oxygen, pH, and Temperature on Energy Metabolism

Oxygen is, from several points of view, an important molecule in animal cell cultures. Apart from the harmful effects of air bubbles, which will not be dealt with here, the level of dissolved oxygen tension (DOT) influences cell metabolism. For example, q_{o_2} is constant above 10% DOT (38), but decreases as hybridoma cells become oxygen limited, around 1% DOT (109). At limiting DOT, glucose consumption and lactate production increase, while oxidative metabolism is restricted (109,110). However, the efficiency of respiration may increase at limiting DOT by means of an increased P/O ratio (38), while prolonged exposure to anaerobic conditions decreases (CHO) cells' potential to use oxygen (111). Typical values of qO_2 for hybridoma and myeloma cells at nonlimiting conditions are 0.2–0.3 mmol/10^9 cells/h (97,98,102,112), and for CHO cells somewhat lower, around 0.15 mmol/10^9 cells/h (111).

The culture pH is primarily more important for growth and production characteristics than for the metabolism per se. The exception to this rule is the dramatic restriction of glucose consumption that occurs at pH 6.8 (113), depending on the inhibition of PFK at this particular pH (114). As a result, glutamine consumption increases and the GDH pathway is up-regulated (108).

The culture temperature is normally kept at 37 °C, but lowering the temperature (to 32–34 °C) will decrease substrate consumption rates, and thereby also the production of metabolic byproducts—effects that can be utilized in large-scale processes, especially if product formation is non-growth associated.

Enzyme Levels, Activities, and Effectors

To find a rationale for observed metabolic fluxes, activities of key enzymes in the central metabolic pathways have been measured in crude extracts of industrial cell lines (16,20,44,72). Variations in enzyme activity occur at different growth conditions, with many enzymes' activities being highest during the exponential growth phase of a batch culture (16). Cells grown in serum-free media have higher activities of the glycolytic enzymes than when grown in serum-containing medium (20). Glutamine limitation, oxygen limitation, and glucose limitation in chemostat cultures brought about a 1.3- to 2.9-fold change in the activity of the enzymes of glycolysis, the HMS, the TCA cycle, and glutamine metabolism in myeloma Sp 2/0 cells. Nevertheless, enzyme activities were well above the calculated maximum fluxes, for the glycolytic enzymes by one order of magnitude (44). The general conclusion of these investigations is that metabolic fluxes in glycolysis, HMS, PPP, the TCA cycle, and glutamine metabolism of the actual cell lines are not controlled at the enzyme level. In fact, even in normal tissue, such as the working rat heart, enzyme activities 10–100-fold higher than the glycolytic flux were measured in vitro (115). It should also be pointed out that measured maximum enzyme activities cannot be used for drawing conclusions on flux-limiting steps because in vitro activities may not represent the actual in vivo activity.

High membrane transport rates, high enzyme activities, and "deregulated" enzymes all contribute to observed metabolic fluxes. It would be tempting to build a chain of evidence based on the properties of the individual enzymes and their effectors for explaining the impact of key steps on the flux. However, since this so-called reductionist approach may lead to erroneous conclusion because it is impossible to evaluate the results of numerous combinations of effector and substrate concentrations, I will refrain from doing this. Understanding the control of metabolism requires a more quantitative approach, as described by the concept of "metabolic control analysis" (116) or similar methods. However, this should not restrict us from collecting data on properties of individual enzymes and reactions, and some important features of immortalized cells should be mentioned.

Hybridoma and myeloma cells possess mitochondrial hexokinase (72,117). Mitochondrial hexokinase, binding to porin in the mitochondrial outer membrane, has been shown to be responsible for the high aerobic glycolytic rate in Ehrlich ascites tumor cells (118), and is suggested to play an important role in proliferating cells (119). This enzyme, which is not so sensitive to feedback inhibition by glucose as the soluble hexokinase, uses ATP derived from mitochondrial energy metabolism for phosphorylation of glucose. It is suggested that the use of mitochondrial-lerived ATP also stimulates the TCA cycle by supplying

ADP to mitochondria. Maybe this could be an explanation as to why increased glutamine consumption sometimes seems to stimulate glucose utilization.

The role of phosphofructokinase (PFK) as a regulatory enzyme of glycolysis is a classical topic. For animal cells in culture an interesting feature is the stimulation of PFK (120), the enhancement of glucose-induced inhibition of respiration and increased glycolytic flux (121) by NH_4^+, a factor almost omnipresent in animal cell cultures. On the other hand, glutamine (and asparagine) caused a decrease in glycolytic flux and PFK activity in ascites tumor cells (122). Accumulation of negative effectors (phosphoenolpyruvate, ATP, and citrate) and a decreased level of positive effectors (AMP and fructose-2,6-P) were thought to be responsible for the observed effects. Further, the effects were prevented by a transaminase inhibitor, indicating that transamination — and indirectly oxidation of glutamine in the TCA cycle — is necessary for the observed effects. However, as pointed out, a verbal interpretation of these results may not be feasible.

The regulation of phosphate activated glutaminase (PAG), a key enzyme in the metabolism of glutamine, is complex, with a large number of in vitro effectors, but the most important in vivo effector, except for phosphate, is not known (3). As phosphate is necessary for activity (20–30 mM causes half-maximal activation), it is possible that a high glycolytic flux indirectly controls PAG activity by sequestering phosphate in glycolytic intermediates (123). This conclusion is further supported by the results of an NMR study conducted by Martinelle and co-workers, showing that the metabolism via PAG was significantly enhanced in glucose-deprived hybridoma and myeloma cells (75). Further, NH_4^+ inhibits kidney-type PAG (in contrast to liver-type PAG, which is stimulated by NH_4^+) (54). Thus it may appear that PAG activity is restricted during conditions of high glucose and ammonium ion concentration.

Effects of Ammonia and Ammonium Ions

Ammonia and ammonium ions have been recognized for some time as negative effectors of cell growth, metabolism, and production, and present a complex problem with both specific and nonspecific effects on the cells. Lactate, the major metabolic byproduct, is considerably less inhibitory at the concentrations reached in cell cultures. Knowledge about the mechanisms of inhibition may provide means for developing control strategies in animal cell processes.

Several hypotheses have been put forward to explain the negative effects of ammonia and ammonium ions such as an increased pH of acidic organelles, decreased cytoplasmic pH, and disturbance of electrochemical and proton gradients. Other effects are associated with anomalous glycosylation (124–126), inhibition of endo- and exocytosis (127), and the thermodynamic inhibition of the GDH reaction. UDP-activated aminohexoses, formed as a result of elevated ammonium concentrations, may also be involved in ammonium-mediated cell growth inhibition (128). It has been suggested that the effects of ammonia are more severe than those of ammonium. However, it is not a matter of ammonia or ammonium, but rather the combination of the two.

Most of the negative effects of ammonium and ammonia can be ascribed to intracellular pH changes. To understand this, the processes involved in the transport of ammonia and ammonium ions across cell membranes must be considered. NH_3, a small uncharged molecule, diffuses rapidly across cell membranes. As it does so, protons are left behind and other protons are picked up in the entered compartment (Fig. 11). Thus NH_3 diffusion causes acidification of the anterior compartment and alkalization of the compartment into which it diffuses, whether this is the cytosol, an organelle, or the surrounding medium. The driving force for diffusion is the concentration gradient over the membrane. New equilibria between NH_3 and NH_4^+ are instantaneously established. The concentration of NH_3 is determined by the pH and pK_a according to $pH = pK_a + \log_{10}([NH_3]/[NH_4^+])$; pK_a being 8.95 at 35 °C (129). NH_3 concentrations will be the same in all cellular compartments as in the surrounding medium, but the total concentration of $NH_3 + NH_4^+$ will vary depending on the local pH in each organelle. Given an external pH of 7.2 and a total extracellular concentration of 5 mM, it can be calculated that the concentration of NH_4^+ in an acidic compartment with pH 5.5 would be 246 mM, an extremely high concentration, with unknown effects on the local environment.

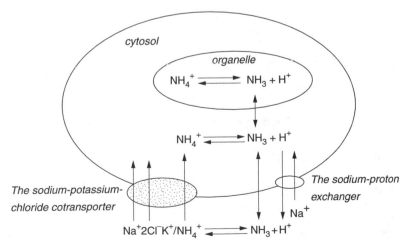

Figure 11. Model for transport of NH_3 and NH_4^+ across cell membranes.

The ammonium ion, a charged species, diffuses much more slowly than ammonia across cell membranes, the difference being five orders of magnitude (130). On the other hand, NH_4^+ can be transported into animal cells by ion pumps in the plasma membrane. Martinelle and co-workers showed that the $K^+/Na^+/2Cl^-$-cotransporter-mediated saturable transport of NH_4^+ in both hybridoma and myeloma cells (Fig. 11) (131,132). The ammonium ions are transported on the binding site for K^+ in competition with K^+ (130,132). Inward NH_4^+ transport causes elevated cytoplasmic levels of NH_3, which immediately diffuses out of the cell, resulting in a futile cycle with cytoplasmic acidification and extracellular alkalization (Fig. 11). Cytoplasmic acidification as a result of ammonium addition to hybridoma cells had previously been shown to occur (133). It could also be confirmed that the plasma membrane Na^+-H^+ exchanger actively extruded protons during exposure to NH_4^+ (131,132). As it now seems that ammonium formed by glutamine metabolism is released in the cytosol, diffusion of NH_3 into other cellular compartments and out into the medium will further contribute to cytosolic acidification. The consequence of the movements of NH_4^+ and NH_3 across cell membranes is an increased demand for maintenance energy caused by the need to keep up ion gradients and to maintain pH homeostasis (cellular compartments are equipped with proton pumps for this purpose). This conclusion is supported by the fact that elevated ammonium levels increase the cells' consumption of glucose and glutamine (12,108,133,134). During culture, with moderate production of ammonium, intracellular pH changes occur so slowly that the cellular machinery can keep up with the changes. However, if ammonium ions are added to a culture, the cell is exposed to a sudden pH shock that may even lead to apoptosis caused by inward transport of ammonium ions by the $K^+/Na^+/2Cl^-$-cotransporter (135). Ammonium-ion-induced apoptosis was extremely rapid, with 40% of the cells becoming apoptotic within 200 min. The fact that initial addition of ammonium to batch cultures causes growth inhibition (even at 5 mM) while slower addition (by feeding) to the same final concentration has no obvious effect (134), and the observation that cells are able to adapt to higher concentrations of ammonium (than those inhibitory in batch) in continuous or perfusion cultures, can all be explained by the different time scales over which cells have to adjust the intracellular pH.

Furthermore, added ammonium stimulates alanine production and decreases ammonium formation from glutamine, as observed in many independent investigations. This has previously been interpreted as an inhibition of the GDH reaction (for thermodynamic reasons). However, as it now is generally agreed that GDH is not active during most cultivation conditions and that kidney-type PAG is negatively affected by ammonium ions (54), this explanation may not be correct. No experimental evidence is available that provides alternative interpretations, but it cannot be excluded that the glutamine amide nitrogen is transferred to other nitrogen sinks as, for example, asparagine and intracellularly accumulated hexoseamines, via amidotransferases not liberating free ammonium.

Controlling Glucose and Glutamine Metabolism

The high glucose and glutamine concentrations normally used in batch cultures enhance substrate uptake rates and metabolic fluxes, thereby leading to overflow metabolism and accumulation of the inhibitory byproducts lactate and ammonia/ammonium ions. Although the levels of waste products in a batch culture are not critical, precautions must be taken if higher cell densities than those reached in a batch culture would be achieved. At some level (about 40 mM) even lactate will become limiting for cell growth and production. It is now well known from several studies that lowering the medium concentrations of glucose and glutamine will decrease byproduct formation (88,89,91,96). The fact that glucose and glutamine are consumed in excess of the cell's anabolic and energy requirements at high substrate concentrations makes it possible to restrict the substrate uptake rates without negatively affecting growth or product formation. Ljunggren and Häggström (89,91) showed that by controlling the concentrations of glucose and glutamine in fed batch or continuous (chemostat, perfusion) cultures, the metabolism of animal cells can be controlled in much the same way as is done in microbial processes. Glutamine limitation reduces not only ammonium formation in myeloma and hybridoma cultures, but also excess formation of amino acids (64,96,136). Simultaneous glucose and glutamine limitation restricts both q_{glc} and q_{gln} and thereby also ammonium and lactate formation (89,91,137,138). In addition, product formation is enhanced by these conditions. Sharfstein and co-workers found a significant increase in the antibody productivity, but no effect on glucose metabolism, by decreasing the glutamine feed from 4 mM to 1.7 mM, indicating that antibody production was not energy limited (18). Similarly, in a glucose- and glutamine-limited fed batch culture (91) or in continuous culture (137) antibody production was not negatively affected. Using this strategy, in combination with amino acid feeding, much higher cell densities and antibody levels were obtained than would have been possible without restricting accumulation of waste products (139,140).

BIBLIOGRAPHY

1. M.S.M. Ardawi and E.A. Newsholme, *Essays Biochem.* **21**, 1–44 (1985).

2. L.G. Baggetto, *Biochimie* **74**, 959–974 (1992).

3. N.P. Curthoys and M. Watford, *Annu. Rev. Nutr.* **15**, 133–159 (1995).

4. Z. Kovacevic and J.D. McGivan, *Physiol. Rev.* **63**, 547–605 (1983).

5. M.A. Medina and I. Núñez de Castro, *Int. J. Biochem.* **22**, 681–683 (1990).

6. T. Matsuno, *Int. J. Biochem.* **19**, 303–307 (1987).

7. W.L. McKeehan, in M.J. Morgan, ed., *Carbohydrate Metabolism in Cultured Cells*, Plenum, New York, 1986, pp. 111–150.

8. M.J. Morgan, *Carbohydrate Metabolism in Cultured Cells*, Plenum, New York, 1986.

9. A. Fiechter and F.K. Gmünder, in A. Fiechter, ed., *Advances. Biochemical. Engineering.* 39, Springer–Verlag, Berlin, 1989, pp. 1–28.

10. M.W. Glacken, *Bio/Technology* **6**, 1041–1050 (1988).

11. M. Newland, P.F. Greenfield, and S. Reid, *Cytotechnology* **3**, 215–229 (1990).

12. M. Schneider, I.W. Marison, and U. von Stockar, *J. Biotechnol.* **46**, 161–185 (1996).

13. L.A. Chasin, A. Feldman, M. Konstam, and G. Urlaub, *Proc. Natl. Acad. Sci. U.S.A.* **71**, 718–722 (1974).

14. D.R. Appling, *FASEB J.* **5**, 2645–2651 (1991).

15. A. Sanfeliu, C. Paredes, J. Joan Cairó, and F. Godià, *Enzyme Microb. Technol.* **21**, 421–428 (1997).

16. L. Fitzpatrick, H.A. Jenkins, and M. Butler, *Appl. Biochem. Biotechnol.* **43**, 93–116 (1993).

17. D. Petch and M.A. Butler, *J. Cell. Physiol.* **161**, 71–76 (1994).

18. S.T. Sharfstein et al., *Biotechnol. Bioeng.* **43**, 1059–1074 (1994).

19. A. Mancuso et al., *Biotechnol. Bioeng.* **44**, 563–585 (1994).

20. J. Neermann and R. Wagner, *J. Cell. Physiol.* **166**, 152–169 (1996).

21. M.L. Eboli and T. Galeotti, *Biochim. Biophys. Acta* **638**, 75–79 (1981).

22. A.R. Grivell, E.I. Korpelainen, C.J. Williams, and M.N. Berry, *Biochem. J.* **310**, 665–671 (1995).

23. J. Perez-Rodriguez et al., *Biochimie* **69**, 469–474 (1987).

24. M.T. Bastos, M.B. Oliviera, A.P. Campello, and M.L. Kluppel, *Cell. Biochem. Funct.* **8**, 199–203 (1990).

25. M.J. Bissell, W.A. Rambeck, R.C. White, and J.A. Bassham, *Science* **191**, 856–858 (1976).

26. M.J. MacDonald, T.F. Warner, and R.J. Mertz, *Cancer Res.* **50**, 7203–7205 (1990).

27. J.P. Schwartz, J.V. Passonneau, G.S. Johnson, and I. Pastan, *J. Biol. Chem.* **249**, 4138–4143 (1974).

28. Z. Kovacevic, *Eur. J. Biochem.* **25**, 372–378 (1972).

29. P.S. Coleman and B.B. Lavietes, *CRC Crit. Rev. Biochem.* **11**, 341–393 (1981).

30. H.A. Parlo and P.S. Coleman, *J. Biol. Chem.* **259**, 9997–10003 (1984).

31. D.A. Briscoe, G. Fiscum, A.L. Holleran, and J.K. Kelleher, *Mol. Cell. Biochem.* **136**(2), 131–137 (1994).

32. P.J. Hornsby and G.N. Gill, *J. Cell. Physiol.* **109**, 111–120 (1981).

33. P.J. Hornsby, *J. Cell. Physiol.* **112**, 207–216 (1982).

34. L.W. Oberley, T.D. Oberley, and G.R. Buettner, *Med. Hypotheses* **6**, 249–268 (1980).

35. T.J. Piva and E. McEvoy-Bowe, *J. Cell. Biochem.* **68**, 213–225 (1998).

36. A. Mancuso et al., *Biotechnol. Bioeng.* **57**, 172–186 (1998).

37. F.M. Harold, *The Vital Force: A Study of Bioenergetics*, Freeman, New York, 1986.

38. W.M. Miller, C.R. Wilke, and H.W. Blanch, *J. Cell. Physiol.* **132**, 524–530 (1987).

39. C. Larsson, U. von Stockar, I. Marison, and L. Gustafsson, *Thermochim. Acta* **251**, 99–110 (1995).

40. C. Murphy and P. Newsholme, *Biochem. Soc. Trans.* **25**, 404S (1997).

41. J.C. Aledo et al., *Biochim. Biophys. Acta* **1323**, 173–184 (1997).

42. V. Vasta, E. Meacci, M. Farnararo, and P. Bruni, *Biochim. Biophys. Acta* **1243**, 43–48 (1995).

43. H.R. Zielke, C.M. Submilla, and P.T. Ozand, *J. Cell. Physiol.* **107**, 251–254 (1981).

44. N. Vriezen and J.P. van Dijken, *Biotechnol. Bioeng.* **59**, 28–39 (1998).

45. G.G. Guidotti and G.C. Gazzola in M.S. Kilberg and D. Häussinger eds., *Mammalian Amino Acid Transport: Mechanisms and Control*, Plenum, New York, 1992, pp. 3–29.

46. J.D. McGivan and M. Pastor-Anglada, *Biochem. J.* **299**, 321–334 (1994).

47. O. Bussolati et al., *Biochim. Biophys. Acta* **1151**, 153–160 (1993).

48. N. Longo et al., *Ann. N. Y. Acad. Sci.* **551**, 374–377 (1998).

49. O. Bussolati et al., *FASEB J.* **10**, 920–926 (1996).

50. W. Engström and A. Zetterberg, *J. Cell. Physiol.* **120**, 233–241 (1984).

51. A.C. Schoolwerth and K.F. LaNoue *Annu. Rev. Physiol.* **47**, 143–171 (1985).

52. M. Molina et al., *Biochem. J.* **308**, 629–633 (1995).

53. J.D. McGivan, in M.S. Kilberg and D. Häussinger, eds., *Mammalian Amino Acid Transport. Mechanisms and Control*, Plenum, New York, 1992, pp. 101–112.

54. E. Kvamme, B. Roberg, and I.A. Torgner, in D.J. O'Donovan, et al., eds., *Nutritional and Acid-Base Aspects of Amino Acid Metabolism*, Karger, Basel, 1997, pp. 69–78.

55. A. Atlante, S. Passarella, G.M. Minervini, and E. Quagliariello, *Arch. Biochem. Biophys.* **315**, 369–381 (1994).

56. T. Benzinger, C. Kitzinger, R. Hems, and K. Burton *Biochem. J.* **71**, 400–407 (1959).

57. G. Schmid and T. Keller, *Cytotechnology* **9**, 217–229 (1992).

58. B. Roberg, I.A. Torgner, and E. Kvamme, in D. O'Donovan, et al., eds., *Nutritional and Acid-Base Aspects of Amino Acid Metabolism*, Karger, Basel, 1997, pp. 11–18.

59. S.Y. Perera, T.C. Chen, and N.P. Curthoys *J. Biol. Chem.* **265**, 17764–17770 (1990).

60. E. Kvamme, I.A. Torgner, and B. Roberg, *J. Biol. Chem.* **266**, 13185–13192 (1991).

61. B. Roberg, I.A. Torgner, and E. Kvamme, *Neurochem. Int.* **27**, 367–376 (1995).

62. A.C. Schoolwerth and K.F. LaNoue, *J. Biol. Chem.* **255**, 3403–3411 (1980).

63. W.L. McKeehan, *Cell Biol. Int. Rep.* **6**, 635–650 (1982).

64. J. Ljunggren and L. Häggström, *Cytotechnology* **8**, 45–56 (1992).

65. C. Street, A.-M. Delort, P.S.H. Braddock, and K.M. Brindle, *Biochem. J.* **291**, 485–492 (1993).

66. W.E. Knox, M.L. Horowitz, and G.H. Friedell, *Cancer Res.* **29**, 669–680 (1969).

67. L.A. Sauer, R.T. Dauchy, W.O. Nagel, and H.P. Morris, *J. Biol. Chem.* **255**, 3844–3848 (1980).

68. L.A. Sauer, R.T. Dauchy, and W.O. Nagel, *Biochem. J.* **184**, 185–188 (1979).

69. R.W. Moreadith and A.L. Lehninger, *J. Biol. Chem.* **259**, 6215–6221 (1984).

70. I.J. Bakken, U. Sonnewald, J.B. Clark, and T.E. Bates, *NeuroReport* **8**(7) 1567–1570 (1997).

71. D.J. Dietzen and E.J. Davis, *Arch. Biochem. Biophys.* **305**, 91–102 (1993).

72. N. Vriezen and J.P. van Dijken, *J. Biotechnol.* **61**, 43–56 (1998).

73. M. Watford, *Biochim. Biophys. Acta* **1200**, 73–78 (1994).

74. L. Häggström, in R.E. Spier, J.B. Griffiths, and B. Meigner eds., *Production of Biologicals from Animal Cells in Culture*, Butterworth-Heinemann, Oxford, 1991, pp. 79–81.

75. K. Martinelle et al., *Biotechnol. Bioeng.* **60**, 508–517 (1998).

76. K. Murray and A.J. Dickson, *Metab., clin. Exp.* **46**, 268–272 (1997).

77. H.A. Jenkins, M. Butler, and A.J. Dickson, *J. Biotechnol.* **23**, 167–182 (1992).

78. A.C. Schoolwerth, W.J. Hoover, C.H. Daniel, and K.F. LaNoue, *Int. J. Biochem.* **12**, 145–149 (1980).

79. K.F. LaNoue and A.C. Schoolwerth, *Annu. Rev. Biochem.* **48**, 871–922 (1979).

80. J. Duszynski, G. Mueller, and K.F. LaNoue, *J. Biol. Chem.* **253**, 6149–6157 (1978).

81. J.K. Teller, L.A. Fahien, and E. Valdivia, *J. Biol. Chem.* **265**, 19486–19494 (1990).

82. L.A. Fahien et al., *J. Biol. Chem.* **263**, 10687–10697 (1988).

83. D.F. Evered and B. Masola, *Biochem. J.* **218**, 449–458 (1984).

84. Z. Kovacevic, O. Brkljac, and K. Bajin, *Biochem. J.* **273**, 271–275 (1991).

85. B. Masola, T.J. Peters, and D.F. Evered, *Biochim. Biophys. Acta* **843**, 137–143 (1985).

86. S.R. Cerda, J. Wilkinson, and S.A. Broitman, *Lipids* **30**, 1083–1092 (1995).

87. Z. Kovacevic, *Biochem. J.* **125**, 757–763 (1971).

88. W.S. Hu, T.C. Dodge, K.K. Frame, and V.B. Himes, *Dev. Biol. Stand.* **66**, 279–290 (1987).

89. J. Ljunggren and L. Häggström, *Biotechnol. Bioeng.* **44**, 808–818 (1994).

90. M. Doverskog, J. Ljunggren, L. Öhman, and L. Häggström, *J. Biotechnol.* **59**, 103–115 (1997).

91. J. Ljunggren and L. Häggström, *J. Biotechnol.* **42**, 163–175 (1995).

92. V. Dall'Asta et al., *Biochim. Biophys. Acta* **1052**, 106–112 (1990).

93. W. Zhou and W.S. Hu, *Biotechnol. Bioeng.* **47**, 181–185 (1995).

94. R. Pörtner, A. Bohmann I. Lüdemann, and H. Märkl, *J. Biotechnol.* **34**, 237–246 (1994).

95. Y.-K. Lee, A.-P. Teoh, and P.-K. Yap, *Enzyme Microb. Technol.* **21**, 429–435 (1997).

96. M.W. Glacken, R.J. Fleishaker, and A.J. Sinskey, *Biotechnol. Bioeng.* **28**, 1376–1389 (1986).

97. G.W. Hiller, D.S. Clark, and H.W. Blanch, *Biotechnol. Bioeng.* **42**, 185–195 (1993).

98. W.M. Miller, C.R. Wilke, and H.W. Blanch, *Biotechnol. Bioeng.* **33**, 487–499 (1989).

99. H.R. Zielke et al., *J. Cell. Physiol.* **95**, 41–48 (1978).

100. W.M. Miller, C.R. Wilke, and H.W. Blanch, *Biotechnol. Bioeng.* **33**, 477–486 (1989).

101. K.K. Frame and W.-S. Hu, *Biotechnol. Lett.* **7**, 147–152 (1985).

102. J.J. Meijer and J.P. van Dijken, *J. Cell. Physiol.* **162**, 191–198 (1995).

103. Y.-H. Jeong and S.S. Wang, *Enzyme Microb Technol.* **17**, 47–55 (1995).

104. A.-P. Zeng and W.-D. Deckwer, *Biotechnol. Bioeng.* **47**, 334–346 (1995).

105. N. Vriezen, B. Romein, K.C.A.M. Luyben, and J.P. van Dijken, *Biotechnol. Bioeng.* **54**, 272–286 (1997).

106. H.P.J. Bonarius, B. Timmerarends, and C.D. de Gooijer, *Biotechnol. Bioeng.* **58**, 258–262 (1998).

107. J.M. Savinell and B.Ø. Palsson, *J. Theor. Biol.* **154**, 421–473 (1992).

108. K. Martinelle, Ph.D. Thesis, Royal Institute of Technology, Stockholm, 1997.

109. S.S. Ozturk and B.Ø. Palsson, *Biotechnol. Prog.* **6**, 437–446 (1990).

110. W.M. Miller, C.R. Wilke, and H.W. Blanch, *Bioprocess. Eng.* **3**, 103–111 (1988).

111. A.A. Lin and W.M. Miller, *Biotechnol. Bioeng.* **40**, 505–516 (1992).

112. D. Wohlpart, D. Kirwan, and J. Gainer, *Biotechnol. Bioeng.* **36**, 630–635 (1990).

113. W.M. Miller, H.W. Blanch, and C.R. Wilke, *Biotechnol. Bioeng.* **32**, 947–965 (1988).

114. B. Trivedi and W.H. Danforth, *J. Biol. Chem.* **241**, 4110–4114 (1966).

115. Y. Kashiwaya et al., *J. Biol. Chem.* **269**, 25502–25514 (1994).

116. D. Fell, *Understanding the Control of Metabolism* in K. Snell, ed., *Frontiers in Metabolism*, Vol 2. Portland Press, London, 1997.

117. M.N. Widjojoatmodjo, A. Mancuso, and H.W. Blanch, *Biotechnol. Lett.* **12**, 551–556 (1990).

118. E. Bustamante, H.P. Morris, and P.L. Pedersen, *J. Biol. Chem.* **256**, 8699–8704 (1981).

119. S.G. Golshani-Hebroni and S.P. Bessman, *J. Bioeneg. Biomembr.* **29**, 331–338 (1997).

120. P.A. Lazo and A. Sols, *Biochem. Soc. Trans.* **8**, 579 (1980).

121. J.S. Olavarria, E. Chico, G. Gimenez-Gallego, and I. Nunez de Castro, *Biochimie* **63**, 469–475 (1981).

122. F. González-Mateos et al., *J. Biol. Chem.* **268**, 7809–7817 (1993).

123. R.M. Sri-Pathmanathan, P. Braddock, and K.M. Brindle, *Biochim. Biophys. Acta* **1051**, 131–137 (1990).

124. D.C. Andersen and C.F. Goochee, *Biotechnol. Bioeng.* **47**, 96–105 (1995).

125. M.C. Borys, D.I.H. Linzer, and E.T. Papoutsakis, *Biotechnol. Bioeng.* **43**, 505–514 (1994).

126. B. Thorens and P. Vassalli, *Nature (London)* **321**, 618–620 (1986).

127. P.A. Docherty and M.D. Snider, *J. Cell. Physiol.* **146**, 34–42 (1991).

128. T. Ryll, U. Valley, and R. Wagner, *Biotechnol. Bioeng.* **44**, 184–193 (1994).

129. K. Martinelle and L. Häggström, *Biotechnol. Tech.* **11**, 549–551 (1997).

130. K. Martinelle and L. Häggström, *J. Biotechnol.* **30**, 339–350 (1993).

131. K. Martinelle, A. Westlund, and L. Häggström, *Cytotechnology* **22**, 251–254 (1996).

132. K. Martinelle, A. Westlund, and L. Häggström, *Biotechnol. Lett.* **20**, 81–86 (1998).

133. S.S. Ozturk, M.R. Riley, and B.O. Palsson, *Biotechnol. Bioeng.* **39**, 418–431 (1992).

134. M. Newland, M.N. Kamal, P.F. Greenfield, and L.K. Nielsen, *Biotechnol. Bioeng.* **43**, 434–438 (1994).

135. A. Westlund and L. Häggström, *Biotechnol. Lett.* **20**, 87–90 (1998).

136. J. Ljunggren and L. Häggström, *Biotechnol. Lett.* **12**, 705–710 (1990).

137. H. Kurokawa et al., *J. Ferment. Bioeng.* **76**, 128–133 (1993).

138. H. Kurokawa, Y.S. Park, S. Iijima, and T. Kobayashi, *Biotechnol. Bioeng.* **44**, 95–103 (1994).
139. L. Xie and D.I.C. Wang, *Biotechnol. Bioeng.* **43**, 1175–1189 (1994).
140. L. Xie and D.I.C. Wang, *Biotechnol. Bioeng.* **51**, 725–729 (1996).

See also ANIMAL CELL CULTURE MEDIA; CELL GROWTH AND PROTEIN EXPRESSION KINETICS; FLUX ANALYSIS OF MAMMALIAN CELL CULTURE: METHODS AND APPLICATIONS; MEMBRANE STRUCTURE AND THE TRANSPORT OF SMALL MOLECULES AND IONS.

CELL PRODUCTS—ANTIBODIES

JOHN R. BIRCH
Lonza Biologics plc
Berkshire, United Kingdom

OUTLINE

INTRODUCTION

The applications of monoclonal antibodies are now widespread, as research tools, as in vitro diagnostic reagents (1), and increasingly as in vivo diagnostic and therapeutic reagents. The level of interest in antibodies as biopharmaceuticals is reflected in the fact that of 350 biological medicines in development in 1998 in the U.S.A., 74 were monoclonal antibodies (2). Antibodies are being evaluated in a wide range of disease indications, for example, cancer, transplantation, and infectious disease (3). The scope of antibody use has increased enormously with the development of recombinant technology to improve the therapeutic potential of these proteins. Now it is possible, for example, to "humanize" antibodies, to produce useful antibody fragments, and to couple antibodies to other drugs or isotopes for targeting purposes. An exciting development in the last year or two (1997/8) has been the number of monoclonals gaining regulatory approval. Recent examples include ReoPro (for use in angioplasty), Rituxan (for treating non-Hodgkins lymphoma), Herceptin (for treating of breast cancer), and Zenapax and Simulect (for preventing transplant rejection) (4). Clearly these wide ranging applications present different manufacturing challenges, particularly in terms of quantities required which may vary from grams to tens or hundreds of grams per year for diagnostic applications to tens or even, potentially, hundreds of kilograms for some therapeutic applications (5).

CELL TYPES AND EXPRESSION SYSTEMS USED TO PRODUCE ANTIBODIES

Rodent Antibodies

The most common cell type used for monoclonal antibody production is the mouse hybridoma. Mouse antibodies are widely used in research and diagnosis. They are also used in some therapeutic applications (for example, the OKT3 antibody which is used to prevent transplant rejection). A number of myeloma cell types have been used as fusion partners to create hybridomas, most of them descended from the parental MOPC21 cell line (Fig. 1). The original MOPC 21 cell line already produces an immunoglobulin. Clearly there is a disadvantage in using such a cell line as a fusion partner because the resultant hybridoma produces a mixture of the desired antibody and an irrelevant antibody. In addition, hybrid molecules that contain chains from each of the antibodies is produced. For this reason, variants of the parent myeloma were created which no longer produced or secreted their own antibody heavy or light chains. These myelomas (NSO, Sp2/O, P3-X63Ag8.653) are most commonly used to create hybridomas or as host cells for recombinant antibody production.

Rat hybridomas have also been used to produce antibodies (6), albeit less frequently than mouse cell lines. As with the mouse system, rat myelomas have been created for use as fusion partners that no longer express the immunoglubulin present in the parent tumor cell line (6). The YO myeloma is an example of a rat cell line that has been used to create hybridomas.

Human Antibodies

The immunogenicity of rodent antibodies in humans, limits their use for some therapeutic applications particularly when multiple administrations are required. For this reason, much attention has focused on the development of human monoclonal antibodies. Several approaches have been adopted using human hybridomas, human–mouse

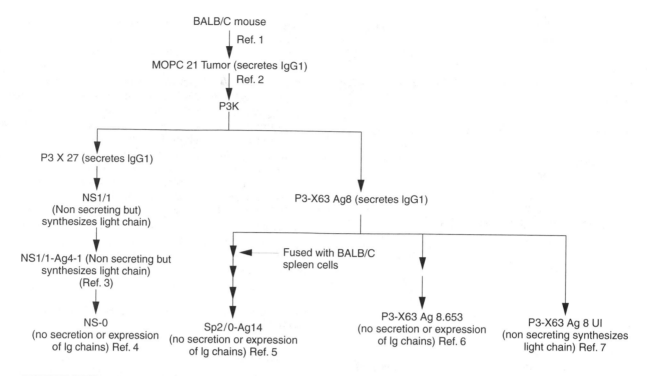

BALB/C mouse
↓ Ref. 1
MOPC 21 Tumor (secretes IgG1)
↓ Ref. 2
P3K

P3 X 27 (secretes IgG1)

NS1/1
(Non secreting but)
synthesizes light chain)

NS1/1-Ag4-1 (Non secreting but
synthesizes light chain)
(Ref. 3)

NS-0
(no secretion or expression
of Ig chains) Ref. 4

Fused with BALB/C
spleen cells

Sp2/0-Ag14
(no secretion or expression
of Ig chains) Ref. 5

P3-X63 Ag8 (secretes IgG1)

P3-X63 Ag 8.653
(no secretion or expression
of Ig chains) Ref. 6

P3-X63 Ag 8 UI
(non secreting synthesizes
light chain) Ref. 7

REFERENCES
1. M. Potter, Physiol. Rev. **52**, 631-719 (1972).
2. K. Horibata And A.W. Harris, Exp. Cell Res. **60**, 61-77 (1970).
3. G. Kohler and C. Milstein, Nature (London), **256**, 495-497 (1975).
4. G. Galfre and C. Milstein, Methods in Enzymol. **73**, 3-75 (1981).
5. M. Shulman, C.D. Wilde, and G. Kohler, Nature (London), **276**, 269-270 (1978).
6. J.F. Kearney, A.D. Radbruch, B. Liesegang and K. Rajesky, J. Immunol. **123**, 1548-1550 (1979).
7. D.E. Yelton, B.A. Diamond, S.P. Kwan and M.D. Scharff, Curr. Top. Microbiol. Immunol. **81**, 1-7 (1978).

Figure 1. Derivation of mouse cell lines commonly used for antibody production.

heterohybridomas, and human lymphoid cells immortalized with Epstein–Barr virus to make "natural" human antibodies. There are a few examples of processes which have been developed using such cell lines. Davis et al. (7), for example, described the production (using hollow-fiber reactors) of rhesus anti-D antibody by EBV-transformed human lymphoblastoid cells. This product is being developed to prevent hemolytic disease of the newborn. Birch and Boraston (8) described the production of an IgG antibody from an EBV-transformed human line grown in an airlift reactor that gives a titer of 90 mg/L. However, human cell lines and heterohybridomas often have technical limitations that restrict their use. These limitations include low productivity and instability. There is also limited potential to obtain human lymphocytes that express desired antibodies, although some progress has been made with *in vitro* immunization techniques. The potential and limitations of these techniques for human antibody production have been reviewed by Boyd and James (9).

The ability to produce human antibodies has improved dramatically with the development of new technologies. In recent years, the use of recombinant DNA technology to humanize rodent antibodies has become the preferred route to many therapeutic antibodies (10,11). A large number of humanized antibodies, mostly IgGs, have been created now. Some have become commercial products,

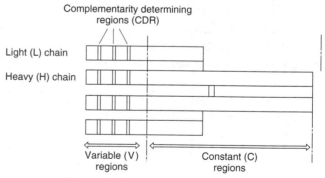

Figure 2. Structure of an IgG molecule.

and many more are being tested in the clinic. The structure of an IgG molecule is shown in Figure 2. Initially, the immunogenicity of rodent antibodies was reduced by creating "chimeric" antibodies in which the constant regions of the rodent antibody were replaced by human equivalents. Subsequently, methods were developed for "humanizing" antibodies more completely by grafting into a human antibody just those small regions of the rodent antibody (the "complementarity-determining" regions (CDR)) that are required for antigen binding. Both chimeric and humanized antibodies are used as therapeutics. The recombinant approach allows great flexibility in protein design and also enables the use of efficient gene expression systems for creating highly productive cell lines. Apart from humanization, genetic engineering can be used to modify antibodies in other ways to improve their therapeutic utility. For example, it is possible to make antibody fragments which retain their antigen-binding characteristics but have altered pharmacological properties, such as modified clearance rate or increased tumor penetration.

Human antibodies can also be created by using phage display techniques. Combinatorial libraries of human antibody genes are expressed in bacteriophages. Then the phages are screened for expression of antibodies with the desired specificity, and the relevant genes can be expressed in a suitable production host cell.

Transgenic technology has also been applied to the production of human antibodies. Transgenic mice have been created that have human antibody genes and that produce human antibodies when immunized. Then B cells from the mice can be immortalized by creating hybridomas.

Recombinant Technology

Many factors influence the choice of cell line and genetic expression systems for recombinant antibody production, including productivity, cell-line stability, suitability for large scale culture, absence of undesirable adventitious agents, and the ability of cells to provide appropriate glycosylation. Whole antibodies and antibody fragments (Fv and Fab—see Refs. 10 and 12) may be produced in animal cells, although in the case of fragments there may be economic advantages in using microbial expression systems. At present the complexity of whole antibodies prevents their production in microbial systems. The most commonly used animal cell lines for recombinant antibody expression are Chinese hamster ovary (CHO) and mouse myeloma NSO or SP2/O cells (for a review of expression systems, see Refs. 10 and 13). All of these cell types have been used in large-scale processes. Rat YO cells have also been used to express antibodies (14). Production in the milk of transgenic animals and in plants is also being evaluated, particularly where extremely large quantities of antibody are required.

Before creating permanent cell lines, it is common practice to use a transient expression system to allow rapid evaluation of different gene constructs and to generate small (microgram) quantities of antibody for preliminary evaluation. Transient systems for antibodies have been described using COS (monkey kidney) cells (15) and human 293 cells (12). Vectors used to transfect immunoglobulin genes into cells are constructed with marker genes to allow the selection of cells that have retained the vector in the host-cell genome. These selectable markers include dihydrofolate reductase (DHFR) (16,17), glutamine synthetase (GS) (18), and antibiotic-resistance genes such as NEO. In addition, the vectors contain promoter sequences to drive product gene expression. With the CHO cell, it is common to use a variant of the cell line that is deficient in dihydrofolate reductase. DHFR co-expressed with the product gene

of interest can be used as a selectable marker and amplified (along with the product gene) by exposure to increasing concentrations of methotrexate (MTX, an inhibitor of DHFR). Isolation of gene-amplified variants using MTX selection is usually necessary to achieve high productivity in CHO cells. Page and Sydenham (16) report an antibody yield of 200 mg/L in batch culture of CHO cells using an amplified dihydrofolate reductase (DHFR) expression system with the human beta actin promoter driving product expression. DHFR gene expression systems with MTX amplification have also been used with mouse Sp2/O cell lines (12,17). Yields up to and exceeding 1 g/L have been achieved using mouse NSO cells in combination with the glutamine synthetase (GS) expression system (5,18,19).

This system exploits the fact that NSO cells do not express sufficient glutamine synthetase to survive in a glutamine-free culture medium. Consequently, GS can be conveniently used as a selectable marker because only cells transfected with GS (and hence the product gene that is also contained in the vector) survive in glutamine-free culture conditions. In the GS system, a viral promoter (hCMV-MIE) is used to drive product expression. The GS vector can be amplified by selecting for resistance to increasing concentrations of an inhibitor of GS (methionine sulphoximine, MSX), although in practice high yields can be achieved without amplification. This can save significant amounts of time in a cell-line development program. The GS system can also be used in CHO cells provided that an inhibitor of GS (MSX) is used to inhibit an endogenous enzyme. Yields in excess of 200 mg/L have been described using vector amplification in such a system (15). Yields from recombinant cell lines can be significantly higher than from conventional hybridomas which typically produce tens to hundreds of milligrams per liter in batch culture (20). In some instances, recombinant DNA technology has been used to "rescue" antibody genes from unsuitable cell lines. The original cell line might be unsuitable because of low productivity, cell-line instability, or because of the presence of undesirable adventitious agents. The rescue of immunoglobulin genes from human cell lines into mouse and rat myeloma cell lines has been described by Gillies et al. (17) and Lewis et al. (21).

Precise interpretation of published yields for different cell lines and expression systems is difficult because several factors contribute to yield: the expression system, the site of vector integration in the host genome, the secretory capabilities of the chosen host cell, and the capability of the chosen production process. Table 1 (taken from Ref. 10) gives examples of published yields for a range of recombinant antibodies and antibody fragments all containing V-region sequences from the murine B72.3 antibody and expressed in a variety of systems. Usually, integration into the host genome is assumed to be random, and one relies on a combination of an efficient selection system and appropriate screening to identify rare, highly productive clones. When an efficient integration site is found, it is possible, by using an appropriate promoter, to achieve efficient expression from a very small number of gene copies and sometimes even from a single copy. This has several advantages; it can

Table 1. Expression of Recombinant B72.3 Antibodies[a]

Molecule	Promoter	Host	Selection	Yield (mg/L)
IgG1	Ig	SP2/O	gpt + neo	60
	Ig	SP2/O	gpt + neo	20
	Ig	SP2/O	DHFR[b]	150
IgG4	CMV-MIE	CHO	gpt + neo	100
	CMV-MIE	CHO	GS[b]	200
	CMV-MIE	NSO	GS[b]	560
Fab	CMV-MIE	CHO	gpt + neo	120
Fv	CMV-MIE	CHO	gpt	2

Source Note:
[a]From 10.
[b]Selection for gene amplification.

save time (by avoiding multiple rounds of amplification), improve the likelihood of obtaining a stable cell line, and reduce cell growth problems that may arise in highly amplified lines especially in the presence of high concentrations of selective drug (13). The efficiency of identifying clones where integration has occurred in a good position can be enhanced by increasing the stringency of the initial selection. For example, in the GS system the GS gene is presumably weakly expressed because of the use of the relatively weak SV40 promoter (18). When transfected cells are grown in a glutamine-free medium in the presence of an appropriate concentration of the GS inhibitor methionine sulphoximine (MSX), then transfectants are selected in which the vector has integrated into transcriptionally highly active sites because only these cells produce enough GS for survival. Another approach (22) has been to use an altered Kozak sequence upstream of the marker gene to impair translation of the marker gene product. The consequence is that only those cells that overexpress the dominant selectable marker gene survive. It has been reported that the efficiency of finding highly producing clones can be further increased by using a specific targeting sequence in the vector which directs the vector to integrate in a chosen site by homologous recombination. Such a site has been identified within the IgG2a locus in a GS NSO cell line, and GS vectors incorporating this region have been used successfully to produce highly yielding cell lines (23,24).

Insect Cells

Insect cells have also been used to a limited extent to express recombinant antibodies (25). Baculovirus expression vectors allow rapid expression of product. However, a disadvantage of insect cells for therapeutic products is that glycosylation is likely to be different from that achieved with mammalian cells. It also seems that IgG proteins tend to accumulate as insoluble complexes in insect cells, although a recent report suggests that this problem might be overcome by the over-expression of a cytosolic chaperone (26).

Screening

Efficiency in finding useful clones might also be improved by using high-throughput screening techniques. Such a

method that combines flow cytometry with a "secretion capture report web" has been described for hybridoma screening (27). In this system individual cells are encapsulated within an agarose drop. Antibody secreted by the cell is captured by a capture antibody immobilized in the agarose and can be detected by a fluorescent reporter. The binding of the reporter can be quantified and desired clones identified and isolated by flow cytometry.

Productivity is not the only criterion used to screen cell clones for their utility. Account also has to be taken of cell-line stability (see section below) and of the natural clonal variation seen with respect to growth characteristics. It is important to have scaled-down systems that give some indication of the probable performance of the cell line in the eventual production process. Hence, for example, if the cells are going to be grown in suspension culture, this feature should be designed into the screen (28).

Cell-Line Stability

The stability of cell lines in terms of productivity and antibody characteristics is an important issue which must be monitored closely during the creation of cell lines and during process development. Stability is particularly important for large-scale processes where cells may be required to grow through many generations, particularly in continuous culture production systems. Instability of productivity can presumably have many causes and can be influenced by several factors, notably the presence of selective drugs (29), the degree of gene amplification, the medium composition (30,31), the serum level (30), and the state of the inoculum (30). Attention also has to be given to stability of the product over time in culture and the appearance of variants. Harris et al. (32) reported the appearance of a Tyr to Gln sequence variant in a recombinant antibody produced in CHO cells. The proportion of variant in the population decreased with increased culture passage.

Quality Control of Cell Lines

When cell lines are being used to make a therapeutic product, their creation and subsequent testing needs to be rigorously controlled and recorded for subsequent review to ensure the safety and efficacy of the product. The details of these requirements are covered more thoroughly in guidelines and "points to consider" produced by regulatory agencies (33,34). Biosafety considerations are reviewed in detail by Lees and Darling (35). The following are examples of areas which require attention during a product development program.

Origin and History of Cell Line. In the case of hybridomas, the history of the fusion partner and the source of the immune parental cells should be described and documented. Immunization procedures also need to be documented, as well as details of the methods used to create the cell line and to ensure its clonality.

Similarly, for recombinant cell lines, the history of the host cell line needs to be recorded to allow, for example, assessment of any potential viral contaminants. Details of the cloning and characterization of the antibody genes used to create the recombinant cell line, as well as vector construction, cell-line transfection, and subsequent cell cloning, need to be recorded.

Cell Bank System. Manufacture of therapeutic proteins is based on the use of well-defined and highly characterized cell banks, usually a master cell bank (MCB) and a manufacturer's working cell bank (MWCB) derived from the MCB.

Control of Virological and Microbiological Contamination. Cell banks (and cells taken at the end of production) are rigorously tested for viral, bacterial, fungal, and mycoplasmal contaminants. Many of the rodent cell lines commonly used to make antibodies produce endogenous rodent retroviruses or retroviruslike particles. Steps are built into production processes to inactivate and to remove these viruses. During manufacture, it is also essential to monitor for adventitious contaminants. Rigorous testing is carried out during the process and on relevant raw materials. In addition, robust steps are built into the process to remove potential adventitious viruses.

Stability of Cell and Product Characteristics During Propagation from Cell Banks. As part of the cell bank characterization program, it is usual to measure the stability of the cell line with respect to productivity, growth characteristics, and product characteristics. This is done under conditions that mimic, as far as possible, the manufacturing process and is used to validate the bank for the number of cell generations required to meet manufacturing requirements. The number of generations required in a given process will depend particularly on whether a batch or continuous process is used.

Genetic Characterization and Genetic Stability (for Recombinant Cell Lines). Characterization of cell banks for recombinant cell lines also includes analysis of the inserted genetic material to ensure fidelity of the DNA sequence for the product and to ensure stability of the genetic insert throughout the manufacturing process. Tests are carried out to determine copy number, to monitor the physical state of the vector, and to confirm the correct product-coding sequence.

REACTOR SYSTEMS USED FOR ANTIBODY PRODUCTION

Many antibodies are still produced by growing hybridoma cells as ascitic tumors in mice or rats. Antibody typically accumulates to concentrations of 10 g/L in the ascitic fluid. For the therapeutic antibodies in particular, most manufacturers have now moved to in vitro cell culture methods which reduce the risk of introducing adventitious agents and irrelevant antibodies from the animal host. Ethical pressures to reduce the unnecessary use of animals in research and manufacture are leading to a continuing shift to in vitro methods even for nontherapeutic applications. For small- to medium-scale production of antibodies for research and for some diagnostic applications, a large number of culture systems are used ranging from small flasks to spinner vessels

and proprietary high-density reactors using hollow fiber systems (36) and other technologies (see, for example, Ref. 37). For large-scale (tens of grams upward), systems with potential for scale-up are used. Antibody doses for some therapeutic applications may be quite high (from 0.5 to more than 5 mg/kg), and this can translate into a requirement for tens to hundreds of kilograms per year (5). In general, suspension culture either in batch or perfused mode has been favored for large-scale production. The relative homogeneity of such systems allows for more precise control of the culture environment compared with nonhomogeneous high-cell-density systems and leads to more straightforward scale-up and process optimization.

Batch and Fed-Batch Suspension Culture

A variety of culture systems have been proposed for antibody production (Table 2), (38–53), although the most common is batch or fed-batch suspension culture. This type of culture has the advantage of relative simplicity and ease of scale-up, leading to relatively rapid process development. The operation of a batch process also leads to a straightforward definition of a product "lot." Suspension culture reactors in excess of 10,000 liters of working volume are now used industrially to produce therapeutic proteins. Antibodies have been produced in both stirred and airlift reactors. Backer et al. (38) describe the production of monoclonal antibodies from hybridomas in stirred, air-sparged reactors at the 1,300-liter scale. Basic design features of the vessels are given, as are operating conditions. Ray et al. (39) described the use of a 2,000-liter stirred tank reactor to produce a humanized antibody from a NSO myeloma cell with the glutamine synthetase expression system. Keen and Rapson (40) describe the production of the humanized antibody, CAMPATH 1H, from CHO cells in 8,000 liter stirred reactors. The use of airlift reactors up to a 2,000 liter scale to produce antibodies from rodent hybridomas, human cell lines, and from genetically engineered mouse myeloma cells is described (54,55). A 2,000 liter airlift fermenter is shown in Figure 3. Productivity of cultures can be greatly increased by systematic modification of the culture medium and physical operating parameters and by operating a fed-batch strategy. In fed-batch cultures,

Table 2. Reactor Systems for Monoclonal Antibody Production

Reactor	Reference
1. Stirred reactor	
a. Batch and fed-batch	38–41
b. Perfusion	42,43
2. Airlift reactors	44
3. Hollow-fiber reactors	7,36,45–47
4. Perfusion, fluidized bed of macroporous matrix	48
5. Perfused ceramic matrix	49
6. Perfused cultures of cells in alginate microcapsules and beads	50–52
7. Perfused fixed beds (porous siran carriers)	53

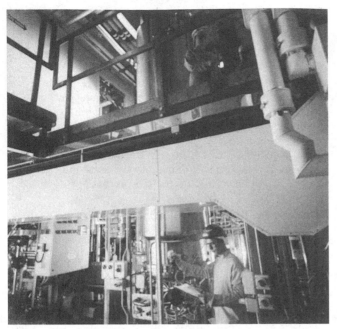

Figure 3. A 2000-liter airlift fermenter used for antibody production (*courtesy of Lonza Biologics plc*).

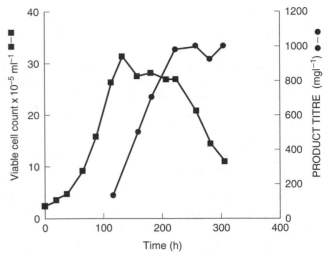

Figure 4. Production of a recombinant antibody from a GS-NSO cell line in a serum-free fermentation (from 44, *courtesy of Lonza Biologics plc*).

small volumes (relative to the total culture volume) of concentrated nutrients are fed into the fermenter to maintain nutrient sufficiency. The feed may be designed to increase cell numbers and/or extend the duration of the culture. Productivity of cultures is determined by the number of cells, the duration of the culture, and the specific production rate. Production kinetics vary from one cell line to another and will determine the types of production systems which can be used. In some instances (56), antibody synthesis seems to be largely growth-related. However, it is more commonly observed that antibody production is partially non-growth-related and continues into the decline phase of batch culture, and indeed this phase may account for a significant proportion of total antibody (44,56,57; and see Fig. 4). This observation has raised a question as to the origin of the antibody in the culture supernatant. Is it produced by secretion or by release of intracellular antibody on cell lysis? Emery et al. (58) showed that product accumulates within the cell at least for some antibodies. However, measurements of intracellular antibody (59) and pulse-chase experiments (60) suggest that most of the antibody in the culture supernatant arises by secretion from viable cells. It is often useful to measure the integrated number of viable cells over the course of the fermentation to establish the number of cell culture hours (CCH). The integral of viable cell count with respect to time is also known as the viability index (VI (t)) (61). For many cell lines, antibody production is linearly related to CCH (57,61). This linearity of response, independent of growth phase, also indicates that antibody is probably produced in most cases by secretion rather than by release of accumulated protein during cell lysis. Feeding can be used to increase CCH either by increasing the maximum cell population density or by increasing the duration of the culture. Several

papers describe the development of feeding strategies for monoclonal antibody production from recombinant mouse NSO cell lines (5,19,62). Bibila and Robinson (5) showed that supplementation with complete 10X medium concentrates provided a rapid route to yield improvement (up to seven fold compared with batch culture) but the highest productivity (1.8 g/L for an amplified NSO cell line) was obtained when iterative nutrient depletion analysis was carried out to design multicomponent feeds to maintain nutrient sufficiency. Xie and Wang (63) describe the use of a stoichiometric model for hybridoma cell growth to design a feed to maximize productivity and minimize accumulation of toxic catabolites. Antibody titers of 2.4 g/L were achieved. An interesting development of fed-batch technology is the use of on-line measurements to control the feed. In one example (64), on-line measurement of oxygen uptake rate was used to provide on-line control of nutrient feeding. This approach prevented excessive accumulation of nutrients and their catabolites and led to cell densities of 1.36×10^7 cell/ml in the fed culture compared with 2×10^6 cell/ml in simple batch culture. In another example (65), on-line HPLC measurement of glucose and glutamine was used to control the feed rate of these nutrients to a hybridoma culture thus maintaining low concentrations of the nutrients and of the catabolites, ammonia and lactate. This led to a twofold increase in cell numbers and antibody concentration (to 172 mg/L).

Perfusion Reactors

Stirred reactors may also be operated in the perfusion mode by using a variety of internal or external separation devices such as spin filters, external tangential flow filtration units, and acoustic resonance sedimentation (42,66 67). Perfusion systems are being operated at large scale for antibody production. Deo et al. (42) describe the operation of large-scale (up to 500 liters), spin filter cultures for antibody production from a hybridoma and from recombinant myelomas. They outline the operating parameters for such a system, including measures to prevent fouling of the spin filter which can be a problem with this type of device. They ran the reactor for 15 to 35 days and generated kilogram quantities of antibody. The authors estimate that the perfusion system has approximately ten times the volumetric productivity of a batch or fed-batch system, and leads to a significantly reduced requirement for plant capacity to produce large volumes of antibody. Van der Velden-de Groot et al. (68), describe the evaluation of spin-filter technology for 13 different hybridoma cell lines.

Hollow Fiber Reactors

Commercially available, hollow-fiber reactor systems are widely used for small- to medium-scale production of antibodies. Davis et al. (7,47) describe the use of such reactors to produce antibody from cell lines making human antibodies. Hollow-fiber systems have the advantage of producing antibodies in high concentrations within the reactor but have disadvantages with respect to gradients in the reactor and the inability to directly monitor and control the cells and their environment. This is especially an issue in optimizing and scaling-up a process.

Jobses et al. (46) describe the development and optimization of a relatively large-scale, hollow-fiber system containing 10 one-liter modules. The system has a capacity to produce 2.5 to 6.5 kg of antibody per year depending on the cell line used.

CULTURE MEDIA AND CULTURE CONDITIONS

Culture Media

The issues driving the design of culture media for the production of monoclonal antibodies are much the same as for other cell culture systems, namely, a requirement for appropriate definition of the medium and a continuing need to increase productivity.

The use of serum-free formulations has long been recognized as desirable. Examples of such media that have been used with antibody-producing cells are shown in Table 3 (69–79). In addition to these examples, most media suppliers can also provide proprietary serum-free and protein-free formulations for a wide range of cell types. Now it is normal to use serum-free media in large-scale manufacture. Birch et al. (44), for example, describe the production of recombinant monoclonal antibodies in large-scale, serum-free, airlift culture. It has been common to replace serum with purified serum-derived proteins such as insulin, albumin and transferrin. Some of these proteins act as carriers of low molecular weight nutrients such as iron or lipids, and by careful design of the medium it has proved possible in a number of cases to develop completely protein-free formulations (examples are given in Table 3). The use of serum-free or protein-free media has several advantages in addition to reducing costs. Serum is variable in composition and therefore can lead to process variability. Avoidance of serum proteins reduces the contaminant load (including irrelevant immunoglobulins derived from serum) to be handled in purification and also reduces the risk of inadvertently introducing animal-derived adventitious agents. The use of low-protein or protein-free media for industrial-scale use is developing rapidly. Keen and Rapson (40), for example, describe the production of a monoclonal antibody from CHO cells in an 8,000-liter reactor using a medium which contained human recombinant insulin but was otherwise protein-free. Interestingly, these authors found that cell growth and productivity were enhanced in the protein-free medium compared with medium containing serum. As noted previously, proteins (notably albumin) may be important as carriers of lipids such as cholesterol which may be required by some mouse myelomas and hybridomas. Addition of insoluble nutrients such as cholesterol is not completely straightforward in a

Table 3. Serum-Free Media for Antibody Production

Medium	Cell Line	Culture System	Protein Supplement	Reference
WCM5	CHO producing Rec. MAb	Stirred bioreactor to 8,000-L scale	Rec. human insulin	40
SFFD-ITES	Hybridomas (mouse)	Dish culture and suspension culture	Insulin; transferrin	69
—	Hybridomas (mouse and rat/mouse)	Static flasks	None	70
KSLM	NSI mouse myeloma, hybridomas	Plate culture	Insulin; transferrin; bovine serum albumin; human LDL	71
PHFM	Mouse and human hybridomas	Hollow-fiber reactor; ceramic cartridge reactor	None	72
PFH	Mouse hybridoma and myelomas	Static plates and dishes; stirred suspension culture	None	73
—	Human hybridomas	Static flasks	None	74
W38	Rat myeloma and hybridoma	Shake flasks	None	75
	Cholesterol independent NSO mouse myeloma	Stirred reactor	None	76
ABC	Murine hybridomas	Static flasks	None	77
BDM3	Mouse hybridoma	Semicontinuous spinner culture	None	78
CDSS	Mouse hybridomas	Shaker and spinner flasks; stirred bioreactor	None	79

protein-free medium. It does, however, seem possible in several cases to adapt cells or isolate variant clones which no longer require cholesterol. This has been described for mouse NS1 cells (80) and for mouse NSO cells (76,81).

In addition to developing serum-free media, significant progress has been made in improving the productivity of media. It is apparent that the nutrient concentrations in traditional media such as DMEM and RPM1 1640 may limit the maximum cell population density. Jo et al. (82) increased the concentration of whole groups of nutrients in RPM1 1640 (five times for amino acids, vitamins, and glutathione and 2.5 times for glucose), while reducing sodium chloride to maintain osmolality. This resulted in the growth of a hybridoma to nearly 10^7 cells per ml in suspension culture and an increase of 5 to 8 times in antibody productivity (to 450 mg/L). Further development of this strategy (83,84) led to antibody titers in excess of 1 g/L. Brown et al. (19) describe how iterative improvements to a culture medium for a mouse myeloma cell line making a recombinant antibody improved productivity by a factor of 10 to approximately 600 mg/L.

In some instances, it may be possible to use metabolic engineering approaches to advantageously alter the nutritional requirements of a cell line. Glutamine synthetase is used as a selectable marker for isolating cell lines transfected with genes that express recombinant antibodies (18). The enzyme confers glutamine independence on the mouse myeloma cell line allowing growth in a glutamine-free medium. The use of such a medium also has process advantages because glutamine is an unstable nutrient and generates ammonia which can accumulate to toxic levels in culture. Birch et al. (81) transfected the glutamine synthetase gene into a murine hybridoma cell line and demonstrated both reduced accumulation of ammonia and increased antibody titer in glutamine-free culture conditions. In the future we are likely to see increased use of metabolic engineering to improve the characteristics of cells in culture. One approach that is already showing promise is the use of genes such as bcl-2 to inhibit the apoptotic response of cells and so prolong their viability in culture (85). There are reports that this may lead to enhanced antibody production (86). An alternative approach to metabolic engineering is to use pharmacological intervention to modify metabolism. It has been reported (87) that, under conditions where glutamine limited maximum population density, dicholoroacetate (an activator of pyruvate dehydrogenase), decreased the specific glutamine utilization rate of a hybridoma that lead to an increase in maximum cell population density and an increase (55%) in antibody yield.

In addition to optimising conditions for cell growth, it is also important to consider those factors that may influence the antibody synthesis rate. It is known that a number of chemical and physicochemical factors can stimulate antibody production. There are several reports that adjusting the osmolality of the culture medium can enhance antibody production. Oyaas et al. (88) reported that mouse hybridomas cultured under hyperosmotic conditions, induced by addition of sucrose or sodium chloride, demonstrated rates of antibody production approximately twofold the rates

observed at physiological osmolality. However, this was not reflected in higher volumetric titers because cell numbers were reduced. When glycine betaine was added to the hyperosmotic medium as an osmoprotective agent, the antibody production rate was increased 2.6-fold and the maximum volumetric titer was increased twofold. Oh et al. (89) also found that antibody titers were significantly increased when hybridomas were grown in hyperosmotic media (350 mOsm) following a period of adaptation to the medium. Productivity was further enhanced by the addition of sodium butyrate (0.1 mM). It has been noted, however, that the response to hyperosmotic conditions may be cell-line-specific (90).

Reactor Operating Conditions

In optimizing antibody-producing cultures, attention should be given to other environmental factors, especially pH (91,92), dissolved oxygen concentration (92,93), and temperature (94), particularly because these factors may influence growth and productivity differently. The effects of these factors will also vary depending on the cell type and culture medium used for antibody production. Depending on the cell line, antibody production can be particularly sensitive to small changes in pH. Wayte et al. (91) examined the effect of pH on the growth and productivity of two hybridomas and a recombinant NSO cell line. For the hydridomas, extremely small changes in pH, as small as 0.1 unit, were sufficient to affect growth and productivity dramatically, albeit differently, in the two cell lines. The specific production rate for one hybridoma doubled when the pH was lowered from 7.2 to 7.1 whereas the growth rate and maximum population density were reduced. For the second hybridoma, the same reduction in pH had little effect on antibody production rate, growth rate, or maximum population density, but the culture had a longer duration leading to a substantially higher antibody titer. The NSO cell line was less sensitive to pH changes although productivity was increased when the pH was reduced from 7.4 to 7.1. These results emphasize the care that must be taken in ensuring precise measurement and control of the pH in bioreactors.

In general, cells grow over a wide range of dissolved oxygen (DO) concentrations, and it is common to control DO in the midrange of saturation with air. Boraston et al. (93) did not find a significant effect on the growth of a hybridoma over the range of 8 to 100% saturation with air. Similarly Ozturk and Palsson (92) did not find an effect on hybridoma-cell growth rates over the range of 20 to 80% air saturation, although specific cell death rates were lowest over the range of 20 to 50%. The highest antibody concentrations were found in the range of 20 to 40%. The response to dissolved oxygen may vary with cell type, however; there is a report (95) of an antibody-secreting human lymphoblastoid cell line that grows optimally at 100% saturation. There are also some examples of a different optimum for antibody synthesis compared with growth. In an experiment using continuous culture of the Sp2/0 cell line (96), it was found that the optimum DO level for antibody production was 50% air saturation whereas a much lower level was optimal for growth. In another report (97), it was shown that a DO level of 60%

air saturation for a particular hybridoma was optimum for growth but antibody titer was increased by 35% when the DO was reduced to 25% saturation.

Cells are generally cultured at a temperature close to 37 °C although, as noted for other parameters, this may not always be optimal for antibody synthesis. It has been found that the specific antibody production rate for a hybridoma (97) increased by almost 100% when the temperature was raised from 37 to 39 °C, although the growth rate was substantially reduced at the higher temperature. Bloemkolk et al. (94), who studied a trioma, found that the specific antibody production rate was not greatly affected over the range of 34 to 38 °C. No growth was observed at 39 °C, and the optimum temperature (37 °C) for growth and antibody production was similar. Reuveny and Lazar (56) note that antibody production for a hybridoma was drastically reduced if the temperature was lowered from 37 to 34 °C.

One parameter which remains little studied is hydrostatic pressure and the effect on cells of varying pressures within large-scale reactors. Takagi et al. (98) studied the effect of hydrostatic pressure on the growth and metabolism of a hybridoma over the range of 0.1 to 0.9 MPa. There was little effect on growth over this range, but the specific production rate of monoclonal antibody increased from 4.5 to 5.6×10^{-13} g/cell/h.

ANTIBODY INTEGRITY AND HETEROGENEITY

Antibody Integrity

It is normal to find microheterogeneity in a monoclonal antibody population, as it is with other proteins. This heterogeneity can result from a number of factors, for example, variation in glycosylation, deamidation of glutamine or asparagine residues, and proteolytic cleavage. Many analytical methods are available to characterize the structure and integrity of antibodies, and these are used to monitor the product during process development and to ensure consistency of the manufactured product (44). It is important to be aware of the impact that process design can have on the characteristics of the product. This is discussed in some detail later with respect to glycosylation. The process may affect other aspects of antibody structure. For example, attention has been drawn to the presence of proteases in cell culture processes. These may be produced both by secretion during proliferation and as a result of cell lysis. Ackermann et al. (99) studied the effect of culture supernatants from an insect-cell line and from four commonly used mammalian-cell lines (CHO, BHK, C127, and P3-X63-Ag8.653) on the integrity of a human monoclonal antibody. Only the insect cell supernatant led to any measurable modification of the antibody despite the fact that this supernatant had the lowest total proteolytic activity. However, the nature of the modification was not determined. Schlaeger et al. (100) identified two cathepsinlike proteases in the culture supernatants of a hybridoma and established that this activity could digest an IgG$_1$ antibody to give F(ab')$_2$ fragments at pH values below 4.5. The presence of

serum reduced the proteolytic activity, and there was no activity at pH7. Subsequent studies (101) also showed the presence of proteases (including serine protease) in the supernatant of a heterohybridoma. Although proteolysis has not been reported as a significant problem in low-density cell culture, it has been commented that the issue could be more significant in high-density cell culture systems (99,101).

Antibody Glycosylation

Antibodies are glycoproteins, and although the carbohydrate component of the molecule may represent a small proportion of the total mass (2 to 3% for an IgG), it nevertheless contributes significantly to the physicochemical and biological properties of the immunoglobulin. For therapeutic applications, therefore it is important to characterize the carbohydrate, as well as the protein moieties of the product, and to include such studies in monitoring product consistency, for example, during process development. IgG molecules have an N-linked glycan attached to the Asn residue 297 in the Fc region of the heavy chain. Occasionally additional N-linked sites are found in the hypervariable region of IgG molecules. Other immunoglobulin classes (IgM, IgA, IgD, IgE) demonstrate additional variety in type and site of glycosylation (102). It has been shown that the glycan component in the Fc region plays an important role in several effector functions of antibodies such as binding and activation of complement and induction of antibody-dependent cellular cytotoxicity (ADCC) (see Ref. 102 for a review of this topic). Glycosylation may also influence antibody half life and clearance of antibody–antigen complexes in vivo (102, 103). In those unusual instances where an IgG exhibits glycosylation in the variable region, the carbohydrate may strongly influence antigen binding (104). The detailed architecture of the glycans found on a particular IgG will show heterogeneity even for a given cell type. Further variation may be imposed by the cell type used to express the antibody and by culture conditions. Yu Ip et al. (105) analyzed the N-glycans of a humanized murine IgG produced by a mouse NSO cell line and identified 13 different structures. There is still a great deal to be learned about the effect of detailed carbohydrate structure on antibody effector functions, but sufficient examples in the literature suggest that this is important. For example, Lifely et al. (14) compared the properties of the humanized IgG antibody Campath 1 produced in rat YO cells, hamster (CHO) cells, and mouse NSO cells. The antibody from rat cells had relatively high levels of dissecting GlcNAc and also had increased activity in ADCC assays. Other groups have reported that the galactose content of outer arm sugars influences biological activity (106). The practical relevance of these differences, of course, depends on the biological properties required for the proposed use of a particular antibody. In some instances, particular glycan structures may be undesirable because they are immunogenic. For example, mouse cell lines such as hybridomas (in contrast to CHO and human cells) can make oligosaccharides that terminate in Gal (alpha 1,3)-Gal (beta 1,4) particularly when grown in a static culture (107). This structure is

Table 4. Carbohydrate Profiles—Comparison of NSO and CHO Cell-Line-Derived IgGs[a]

Terminal Sugars	% Di 1,3 gal-gal[b]	% Mono 1,3 gal-gal[c]	% Di gal[d]	% Mono gal[e]	% Agal[f]	% Other structures	% Sialylated[g]
Antibody Cell line Source							
NSO	1.8	1.6	10.2	28.3	53.5	4.4	1.6
NSO	2.7	2.8	14.2	31.8	44.4	4.0	2.2
CHO			10.0	35.5	51.2	2.7	4.0
CHO			14.7	51.7	28.9	4.0	2.7

Notes:

[a]Results determined by FACE analysis of neutral glycans. Similar results were obtained by GPC. Other structures represent smaller truncated structures of the biantennary form but not high mannose. All structures were biantennary and fully core fucosylated. Cells were grown in a serum-free suspension culture.

[b]
```
G – G – GN – M \           F
                \          |
                 M – GN – GN.
                /
G – G – GN – M /
```

[c]
```
    ┌ G – GN – M \           F
    │             \          |
G  │              M – GN – GN.
    │             /
    └ G – GN – M /
```

[d]
```
G – GN – M \           F
            \          |
             M – GN – GN.
            /
G – GN – M /
```

[e]
```
    ┌ GN – M \           F
    │         \          |
G  │          M – GN – GN.
    │         /
    └ GN – M /
```

[f]
```
GN – M \           F
        \          |
         M – GN – GN.
        /
GN – M /
```

Key: G = galactose; GN = *N*-acetylglucosamine; M = mannose; F = fucose
Data presented by Flatman and King (108).
[g]% Sialylation determined by anion-exchange HPLC.

recognized by antibodies in human serum. However, in practice mouse cell lines such as NSO, used to make recombinant antibodies, produce this structure at low or undetectable levels in several suspension culture systems (14,85 and Table 4) (108). Antibodies made in NSO and other mouse cells are used clinically. A study with a humanized antibody produced in NSO showed that the antibody was well tolerated in humans without a major immune response to the antibody (109).

As noted before, antibody glycosylation may be influenced by a number of factors including cell type and culture conditions, and these factors need to be taken into account when choosing a process and when changing from one process to another (for example, for larger scale production). Patel et al. (110) studied a monoclonal IgG1 produced from a hybridoma grown in an ascitic culture and by *in vitro* culture with and without serum. Clear differences in carbohydrate structure were found, notably in sialic acid content (lowest in ascites-derived material). Maiorella et al. (111) compared the biological properties and glycosylation of an IgM molecule produced by a human/human/mouse trioma in ascites, a hollow-fiber perfusion culture, and a serum-free, airlift, suspension culture. Compared with material produced *in vitro*,

antibody produced in ascites exhibited reduced sialylation and greater glycoform heterogeneity. Other properties of the antibody were also affected; the ascites-derived product had reduced conformational stability, reduced specific binding activity, and an increased residence time *in vivo* compared with airlift-derived material. The pharmacokinetic properties of material made in hollow-fiber reactors were intermediate between ascites- and airlift-derived material. It was noted that it is difficult to generalize from these observations because two other IgM antibodies did not differ in in vivo residence time when produced in vitro or in ascites. A comparison of antibody glycosylation for human IgGs made by EBV-transformed human lymphoblastoid cell lines in both a low-density static culture and in a high-density, hollow-fiber culture significantly differed (106). Compared with the static culture, the antibodies produced in the hollow-fiber culture had a much higher proportion of oligosaccharides with a reduced galactose content.

A great deal is still to be known about the factors that control glycosylation in cell culture. However, there is a growing literature on specific factors that influence glycosylation. Some of this literature relates to proteins other than antibodies but may nevertheless be relevant (see Ref. 112 for a general review), and some is specific to antibodies. Robinson et al. (113) characterized a recombinant IgG1 produced by a mouse NSO cell line during a serum-free, fed-batch process. They observed that the antigen-binding characteristics of the antibody remained constant throughout the process but that carbohydrate composition changed. In particular, antibody produced later in the culture contained more truncated *N*-acetylglucosamine and high-mannose glycans. It was reported that the proportion of high-mannose forms could be reduced by modifying the feeding strategy (114). The appearance of under glycosylated species was apparently not caused by cell lysis or synthesis from non viable cells. Although there is little evidence that carbohydrate structures on antibodies are modified by glycosidase activities from cell lysates, it should be borne in mind that such enzymes may be present. Gramer and Goochee (115) measured the glycosidase activities of a number of cell types including a hybridoma, CHO cells, and mouse NSO cells. There are also reports that specific environmental factors influence antibody glycosylation. Tachibana et al. (116) showed that glucose availability affected the glycosylation of a monoclonal antibody produced by a human hybridoma.

Antibody glycosylation will be influenced by the type of cell in which it is produced. The glycosylation capabilities of various types of cell have been reviewed by Jenkins et al. (117). Glycosylation may even vary from one subclone to another for a given cell line. Bergwerff et al. (118) demonstrated differences in glycosylation for two chimeric antibodies with identical Fc regions and both produced in mouse SP2/O cells. They believe that the differences most probably result from clonal variation in the SP2/0 cell line used for transfection. Clonal variation with respect to glycosylation may also affect productivity. Cole et al. (119) studied two hybridomas derived from the same parental cell line. One cell line had a threefold higher specific antibody production rate. This hybridoma also had significantly higher levels of three glycosyltransferase enzymes (galactosyl-, sialyl-, and fucosyltransferase). Looking to the future we expect to see increasing use of metabolic engineering of cells to advantageously modify glycosylation of proteins (120) and the use of protein engineering of antibodies to produce more homogeneously glycosylated molecules (121). Umana et al. (122) describe the engineering of the glycoforms of an IgG produced by CHO cells by optimizing the expression of a Gn TIII transferase enzyme. This allowed optimization of the antibody-dependent cellular cytotoxic activity of the antibody.

CONCLUSIONS

The number of applications of monoclonal antibodies has grown rapidly since their invention more than twenty years ago. The most exciting current trend is the increasing use of antibodies in therapeutic applications. This trend is leading to the development of improved technology to manufacture large quantities of antibody and to the use of recombinant technology to tailor antibodies to particular applications.

BIBLIOGRAPHY

1. L. Wang, H. Ben-Bassat, and M. Inbar, in A. Mizrahi, ed., *Monoclonal Antibodies: Production and Application*, Alan R. Liss, New York, 1989, pp. 213–304.
2. *Survey of Biotechnology Medicines in Development*, Pharmaceutical Research and Manufacturers of America (PhRMA), Washington, D.C., 1998.
3. D.A. Scheinberg and P.B. Chapman, in J.R. Birch and E.S. Lennox, eds., *Monoclonal Antibodies: Principles and Applications*, Wiley-Liss, New York, 1995, pp. 45–105.
4. P. Holliger and H. Hoogenboom, *Nat. Biotechnol.* **16**, 1015–1016 (1998).
5. T.A. Bibila and D.K. Robinson, *Biotechnol. Prog.* **11**, 1–13 (1995).
6. H. Bazin, ed., *Rat Hybridomas and Rat Monoclonal Antibodies*, CRC Press, Boca Raton, Fl., 1982.
7. J.M. Davis et al., in E.C. Beuvery, J.B. Griffiths, and W.P. Zeijlemaker, eds., *Animal Cell Technology; Developments Towards the 21st Century*, Kluwer, Academic Publishers, Dordrecht, The Netherlands, 1995, pp. 149–153.
8. J.R. Birch and R. Boraston, in P. Rouger and C. Salmon, eds., *Monoclonal Antibodies against Human Red Blood Cell and Related Antigens*, Arnette, Paris, 1987, pp. 55–68.
9. J.E. Boyd and K. James, in A. Mizrahi, ed., *Monoclonal Antibodies: Production and Application*, Alan R. Liss, New York, 1989, pp. 1–43.
10. S.E. Ward and C.R. Bebbington, in J.R. Birch and E.S. Lennox, eds., *Monoclonal Antibodies: Principles and Applications*, Wiley-Liss, New York, 1995, pp. 137–145.
11. M. Schaffner, B. Kaluza, and U.H. Weidle, in H. Hauser and R. Wagner, eds., *Mammalian Cell Technology in Protein Production*, de Gruyter, Berlin, 1997, pp. 159–169.
12. H. Dorai et al., *Bio/Technology* **12**, 890–897 (1994).
13. A.J. Racher et al., in A. Mountain, U. Ney, and D. Schomburg, eds., *Biotechnology*, Vol. 5a, Wiley-VCH, New York, 1998, pp. 245–273.
14. M.R. Lifely et al., *Glycobiology* **5**, 813–822 (1995).
15. C.R. Bebbington, *Methods: Companion Methods Enzymol* **2**, 136–145 (1991).
16. M.J. Page and M.A. Sydenham, *Bio Technology* **9**, 64–68 (1991).
17. S.D. Gillies et al., *Bio Technology* **7**, 799–804 (1989).
18. C.R. Bebbington et al., *Bio Technology* **10**, 169–175 (1992).
19. M.E. Brown, G. Renner, R.P. Field, and T. Hassell, *Cytotechnology* **9**, 231–236 (1992).
20. J.R. Birch et al., in B.K. Lydersen, ed., *Large Scale Cell Culture Technology*, Hanser, Gardiner, N.Y. 1987, pp. 1–20.
21. A.P. Lewis, N. Parry, T.C. Peakman, and J. Scott-Crowe, *Hum. Antibody. Hybridomas* **3**, 146–152 (1992).
22. Int Pat. WO 94/11523 (1994), M.E. Reff.
23. W. Zhou et al., *Cytotechnology* **22**, 239–250 (1996).
24. D.K. Robinson et al., in H. Wang and T. Imanaka, eds., *Antibody Engineering*, ACS, Worthington, 1995, pp. 1–14.
25. J. Zu Putlitz et al., *Bio Technology* **8**, 651–654 (1990).
26. E. Ailor and M.J. Betenbaugh, *Biotechnol. Bioeng.* **58**, 196–203 (1998).
27. J.S. Kenney, F. Gray, M.H. Ancel, and J.F. Dunne, *Bio Technology* **13**, 787–790 (1995).
28. H.N. Brand et al., in R.E. Spier, J.B. Griffiths, and W. Berthold, eds., *Animal Cell Technology: Products of*

Today, Prospects for Tomorrow, Butterworth-Heinemann, London, 1994, pp. 55–60.

29. T. Hassell et al., in R.E. Spier, J.B. Griffiths, and C. MacDonald, eds., *Animal Cell Technology: Developments, Processes and Products*, Butterworth-Heinemann, London, 1992, pp. 42–47.

30. A.S. Chuck and B.O. Palsson, *Biotechnol. Bioeng.* **39**, 354–360 (1992).

31. N. Kessler, S. Bertrand, and M. Aymard, *In Vitro Cell Dev. Biol.* **29A**, 203–207 (1993).

32. R.J. Harris et al., *Bio Technology* **11**, 1293–1297 (1993).

33. Food and Drug Administration, Center for Biologics Evaluation and Research, *Points to Consider in the Manufacture and Testing of Monoclonal Antibody Products for Human Use*, FDA, Washington, D.C., 1997.

34. European Commission, *Note for Guidance, Production and Quality Control of Monoclonal Antibodies*, European Commission, 1994.

35. G. Lees and A. Darling, in J.R. Birch and E.S. Lennox, eds., *Monoclonal Antibodies: Principles and Applications*, Wiley-Liss, New York, 1995, pp. 267–298.

36. M.D. Hirschel and M.L. Gruenberg, in B.K. Lydersen, ed., *Large Scale Cell Culture Technology*, Hanser, Gardiner, N.Y., 1987, pp. 113–144.

37. M.L. Nolli et al., in O.W. Merten, P. Perrin, and B. Griffiths, eds., *New Developments and New Applications in Animal Cell Technology*, Kluwer, Academic Publishers, Dordrecht, The Netherlands, 1998, pp. 409–415.

38. M.P. Backer et al., *Biotechnol. Bioeng.* **32**, 993–1000 (1988).

39. N.G. Ray et al., in M.J.T. Carrondo, B. Griffiths, and J.L.P. Moreira, eds., *Animal Cell Technology*, Kluwer Academic Publishers, Dordrecht, The Netherlands, 1997, pp. 235–241.

40. M.J. Keen and N.T. Rapson, *Cytotechnology* **17**, 153–163 (1995).

41. J.D. Macmillan, D. Velez, L. Miller, and S. Reuveny, in B.K. Lydersen, ed., *Large Scale Cell Culture Technology*, Hanser, Gardiner, N.Y., 1987, pp. 21–58.

42. Y.M. Deo, M.D. Mahadevan, and R. Fuchs, *Biotechnol Prog.* **12**, 57–64 (1996).

43. J. Shevitz, S. Reuveny, T.L. LaPorte, and G.H. Cho, in A. Mizrahi, ed., *Monoclonal Antibodies: Production and Application*, Alan R. Liss, New York, 1989, pp. 81–106.

44. J.R. Birch, J. Bonnerjea, S. Flatman, and S. Vranch, in J.R. Birch and E.S. Lennox, eds., *Monoclonal Antibodies: Principles and Applications*, Wiley-Liss, New York, 1995, pp. 231–265.

45. G.L. Altshuler, D.M. Dziewulski, J.A. Sowek, and G. Belfort, *Biotechnol. Bioeng.* **28**, 646–658 (1986).

46. I. Jobses et al., in R.E. Spier, J.B. Griffiths, and C. MacDonald, eds., *Animal Cell Technology: Developments, Processes and Products*, Butterworth-Heinemann, London 1992, pp. 517–522.

47. J.M. Davis et al., in R.E. Spier, J.B. Griffiths, and C. MacDonald, eds., *Animal Cell Technology: Developments, Processes and Products*, Butterworth-Heinemann, London, 1992, pp. 130–133.

48. P.W. Runstadler et al., in A.S. Lubiniecki, ed., *Large Scale Mammalian Cell Culture Technology*, Dekker, New York, 1990, pp. 363–391.

49. B.K. Lydersen, in B.K. Lydersen, ed., *Large Scale Cell Culture Technology*, Hanser, Gardiner, N.Y., 1987, pp. 169–192.

50. R. Rupp, K. Kilbride, and M. Oka, in B.K. Lydersen, ed., *Large Scale Cell Culture Technology*, Hanser, Gardiner, N.Y., 1987, pp. 81–94.

51. E.G. Posillico, *Bio Technology* **4**, 114–117 (1986).

52. P.C. Familletti and J.E. Fredericks, *Bio Technology* **6**, 41–44 (1988).

53. D. Looby and B. Griffiths, *Trends Biotechnol.* **8**, 204–209 (1990).

54. K.J. Lambert, R. Boraston, P.W. Thompson, and J.R. Birch, *Dev. Ind. Microbiol.* **27**, 101–106 (1987).

55. M. Rhodes and J. Birch, *Bio Technology* **6**, 518–523 (1988).

56. S. Reuveny and A. Lazar, in A. Mizrahi, ed., *Monoclonal Antibodies: Production and Application* Alan R. Liss, New York, 1989, pp. 45–80.

57. J.M. Renard et al., *Biotechnol. Lett.* **10**, 91–96 (1988).

58. N.A. Emery, M. Lavery, B. Williams, and A. Handa, in C. Webb and F. Mavituna, eds., *Plant and Animal Cells, Process Possibilities*, Ellis Horwood, London, 1987.

59. J.R. Birch et al., in C. Webb and F. Mavituna, eds., *Plant and Animal Cells, Process Possibilities*, Ellis Horwood, London, 1987, pp. 137–146.

60. A.G. Walker, W. Davison, and C.A. Lambe, *Proc. 4th Eur. Congr. Biotechnol.* 1987, Vol. 3, p. 587.

61. Y.T. Luan, R. Mutharasan, and W.E. Magee, *Biotechnol. Lett.* **9**, 535–538 (1987).

62. T.A. Bibila et al., *Biotechnol. Prog.* **10**, 87–96 (1994).

63. L. Xie and D.I.C. Wang, *Biotechnol. Bioeng.* **51**, 725–729 (1996).

64. W. Zhou, J. Rehm, and W.-S. Hu, *Biotechnol. Bioeng.* **46**, 579–587 (1995).

65. H. Kurokawa, Y.S. Park, S. Iijima, and T. Konayashi, *Biotechnol. Bioeng.* **44**, 95–103 (1994).

66. D. de la Broise, M. Noiseux, and R. Lemieux, *Biotechnol. Bioeng.* **38**, 781–787 (1991).

67. P.W.S. Pui et al., *Biotechnol. Prog.* **11**, 146–152 (1995).

68. T. van der Velden-de Groot, W. Witterland, E.C. Beavery, and T.L. van Wezel, in R.E. Spier and J.B. Griffiths, eds., *Modern Approaches to Animal Cell Technology*, Butterworth, London, 1987, pp. 513–518.

69. H. Murakami, in *Methods for Serum-Free Culture of Neuronal and Lymphoid Cells*, Alan R. Liss, New York, 1984, pp. 197–205.

70. W.L. Cleveland, I. Wood, and B.F. Erlanger, *J. Immunol. Methods* **56**, 221–234 (1983).

71. T. Kawamoto, J.D. Sato, D.B. McClure, and G.H. Sato, *Methods Enzymol.* **121**, 266–277 (1986).

72. K. Inouye, K. Toyoda, M. Konda, and K. Maki, in Murakami, et al., eds., *Animal Cell Technology: Basic and Applied Aspects*, Kluwer, Academic Publishers, Dordrecht, The Netherlands, 1992, pp. 353–359.

73. J. Kovar and F. Franek, *Biotechnol. Lett.* **9**, 259–264 (1987).

74. S.P.C. Cole, E.H. Vreeken, S.E.L. Mirski, and B.J. Campling, *J. Immunol. Methods* **97**, 29–35 (1987).

75. M.J. Keen, *Cytotechnology* **17**, 193–202 (1995).

76. M.J. Keen and T.W. Steward, *Cytotechnology* **17**, 203–211 (1995).

77. F.J. Darfler, *In Vitro Cell Dev. Biol.* **26**, 769–778 (1990).

78. Y.-J. Schneider, *J. Immunol. Methods* **116**, 65–77 (1989).

79. Y.M. Qi, P.F. Greenfield, and S. Reid, *Cytotechnology* **21**, 95–109 (1996).

80. T. Kawamoto et al., *Anal. Biochem.* **130**, 445–453 (1983).

81. J.R. Birch et al., *Cytotechnology* **15**, 11–16 (1994).

82. E.-C. Jo, H.-J. Park, J.-M. Park, and K.-H. Kim, *Biotechnol. Bioeng.* **36**, 717–722 (1990).

83. E.-C. Jo, D.-II. Kim, and H.M. Moon, *Biotechnol. Bioeng.* **42**, 1218–1228 (1993).

84. E.-C. Jo, H.-J. Park, D.-II. Kim, and H.M. Moon, *Biotechnol. Bioeng.* **42**, 1229–1237 (1993).

85. H. Bierau, A. Perani, M. Al-Rubeai, and A.N. Emery, *J. Biotechnol.* **62**, 195–207 (1998).

86. Y. Itoh, H. Ueda, and E. Suzuki, *Biotechnol. Bioeng.* **48**, 118–122 (1995).

87. K. Murray, K. Gull, and A.J. Dickson, *Biotechnol. Bioeng.* **49**, 377–382 (1996).

88. K. Oyaas, T.E. Ellingsen, N. Dyrset, and D.W. Levine, *Biotechnol. Bioeng.* **44**, 991–998 (1994).

89. S.K.W. Oh et al., *Biotechnol. Bioeng.* **24**, 601–610 (1993).

90. G.M. Lee and S.Y. Park, *Biotechnol. Lett.* **17**, 145–150 (1995).

91. J. Wayte et al., *Genet. Eng. Biotechnol.* **17**, 125–131 (1997).

92. S.S. Ozturk and B.O. Palsson, *Biotechnol. Prog.* **7**, 481–494 (1991).

93. R. Boraston, P.W. Thompson, S. Garland, and J.R. Birch, *Dev. Biol. Stand.* **55**, 103–111 (1984).

94. J.-W. Bloemkolk, M.R. Gray, F. Merchant, and T.R. Mosmann, *Biotechnol. Bioeng.* **40**, 427–431 (1992).

95. A. Mizrahi, G.V. Vosseller, Y. Yagi, and G.E. Moore, *Proc. Soc. Exp. Biol. Med.* **139**, 118–122 (1972).

96. W.M. Miller, C.R. Wilke, and H.W. Blanch, *J. Cell. Physiol.* **132**, 524–530 (1987).

97. N. Barnabe and M. Butler, *Biotechnol. Bioeng.* **44**, 1235–1245 (1994).

98. M. Takagi, K. I. Ohara, and T. Yoshida, *J. Ferment. Bioeng.* **80**, 619–621 (1995).

99. M. Ackermann, U. Marx, and V. Jager, *Biotechnol. Bioeng.* **45**, 97–106 (1995).

100. E.J. Schlaeger, B. Eggimann, and A. Gast, *Dev. Biol. Stand.* **66**, 403–408 (1987).

101. W. Lind, M. Lietz, V. Jager, and R. Wagner, in R. Sasaki and K. Ikura, eds., *Animal Cell Culture and Production of Biologicals*, Kluwer, Academic Publishers, Dordrecht, The Netherlands, 1991, pp. 319–327.

102. R. Jefferis and J. Lund, *Antibody Eng. Chem. Immunol.* **65**, 111–128 (1997).

103. M. Nose and H. Wigzell, *Proc. Natl. Acad. Sci. U.S.A.* **80**, 6632–6636 (1983).

104. H. Murakami and H. Tachibana, in R.E. Spier, J.B. Griffiths, and W. Berthold, eds., *Animal Cell Technology, Products of Today, Prospects for Tomorrow*, Butterworth-Heinemann, London, 1994, pp. 670–675.

105. C.C. Yu Ip et al., *Arch. Biochem. Biophys.* **308**, 387–399 (1994).

106. B.M. Kumpel et al., *Hum. Antibod. Hybridomas* **5**, 143–151 (1994).

107. J. Lund et al., *Hum. Antibod. Hybridomas* **4**, 20–25 (1993).

108. S. Flatman and R. King, *Cell Culture Engineering*, San Diego, Calif., 1996.

109. S. Stephens et al., *Immunology* **85**, 668–674 (1995).

110. T.P. Patel, R.B. Parekh, B.J. Moellering, and C.P. Prior, *Biochem J.* **285**, 839–845 (1992).

111. B.L. Maiorella et al., *Bio Technology* **11**, 387–392 (1993).

112. C.F. Goochee and T. Monica, *Bio Technology* **8**, 421–427 (1990).

113. D.K. Robinson et al., *Biotechnol. Bioeng.* **44**, 727–735 (1994).

114. D.K. Robinson et al., in R.E. Spier, J.B. Griffiths, and W. Berthold, eds., *Animal Cell Technology, Products of Today, Prospects for Tomorrow*, Butterworth-Heinemann, London, 1994, pp. 763–767.

115. M.J. Gramer and C.F. Goochee, *Biotechnol. Bioeng.* **43**, 423–428 (1994).

116. H. Tachibana et al., *Cytotechnology* **16**, 151–157 (1994).

117. N. Jenkins, R.B. Parekh, and D.C. James, *Nat. Biotechnol.* **14**, 975–981 (1996).

118. A.A. Bergwerff et al., *Glycoconjugate J.* **12**, 318–330 (1995).

119. C.R. Cole, C.A. Smith, and M. Butler, *Biotechnol. Lett.* **15**, 553–558 (1993).

120. N. Jenkins and E.M.A. Curling, *Enzyme Microb. Technol.* **16**, 354–364 (1994).

121. J. Lund et al., *J. Immunol.* **157**, 4963–4969 (1996).

122. P. Umana et al., *Nat. Biotechnol.* **17**, 176–186 (1999).

See also ANIMAL CELL PRODUCTS, OVERVIEW; CELL FUSION; CELL PRODUCTS—IMMUNOREGULATORS; CELL STABILITY, ANIMAL; ENRICHMENT AND ISOLATION TECHNIQUES FOR ANIMAL CELL TYPES; PLANT CELL CULTURES, SECONDARY PRODUCT ACCUMULATION.

CELL PRODUCTS—IMMUNOREGULATORS

TIM CLAYTON
Glaxo Wellcome
Beckenham, Kent
United Kingdom

OUTLINE

Introduction

The History of Immunomodulators

What is an Immunoregulator?

Why Modulate the Immune System?

How can the Immune System be Modulated?

Immune Regulators as Products

Production of Immune Regulators

Product Quality and Safety Issues

    Product Quality and Safety

    Operator Safety

    Control of Material

Production Methods

    Isolate Material from Natural Sources such as Blood

    Use Natural Producer Cell Lines

    Produce Recombinant Cell Lines (Animal Cell or Microbial)

    Use Gene Therapy Vectors to Introduce Genes into Target Cells

INTRODUCTION

Immunoregulators are important for the treatment of a number of disorders, including autoimmune diseases and cancer. They are produced by a variety of cell culture systems and act in several ways, which may include direct stimulation of an antiviral response such as can be seen in interferon secretion or a down regulation of the immune response by blocking receptor molecules using products such as monoclonal antibodies. Wellferon, a human alpha-interferon product, was the first therapeutic protein to be made on a large scale from continuous animal cell cultures, and since then a range of other products has been developed. As our understanding of the immune system has developed the quantity and quality of knowledge about the control of the immune system has increased dramatically. Our ability to detect and follow the effects of minute quantities of material within cells has led to a greater understanding of the fate of foreign materials within cells, and the role of naturally occurring immunoregulators in causing and curing disease is becoming clearer. Our present abilities to regulate the immune system are at an early stage of development, and the approaches that have been used to date tend to be systemic and crude compared to the fine regulation of the immune response that we would like to see in the future. Our increased understanding of these complex relationships is now being used to develop new therapeutic products such as DNA vaccines. In addition, work is progressing on new strategies for targeting cytokines to specific cell types using gene technologies. The aim of this article is to describe the production of immune regulators in animal cell culture with reference to specific examples where appropriate. The methods used for cell culture, process development, and product manufacture are discussed.

THE HISTORY OF IMMUNOMODULATORS

A huge amount of excitement was created by the discovery of interferons in response to viral infection Isaacs (1) and the potential to produce a wonder drug to prevent or cure viral diseases. Interferons and other interleukins are immensely potent molecules, and natural sources of interferon contain minute quantities of material. A cure for the common cold, influenza, and a host of other viral diseases appeared to be on the horizon. Therefore, the initial work on interferon concentrated on isolating the material and finding a role for it. There was huge interest in isolating and producing significant quantities of material for research and commercial exploitation. Different companies followed diverse paths to commercialize the production and use of interferon. The Wellcome Foundation concentrated on producing interferon from animal cells, while Roche and Schering Plough developed recombinant interferon in *E. coli* expression systems. Unfortunately timing is important, and to be effective in preventing infection interferon would need to be taken before the infection occurred. Therefore, the emphasis moved from preventing the incidence of disease to fighting long-term disorders such as cancers and persistent viral infections such as hepatitis B. In the early 1980s the Wellcome Foundation made an enormous step forward in the field of biotechnology. Building upon an extensive body of expertise in vaccine development and production, they developed a process for producing human alpha-interferon using a continuous human cell line (2). The culture of the cells was, and is still, performed in 8000–10,000-L cell culture vessels. The human cell line in question is Namalwa, and the cells are induced to produce commercial levels of human alpha-interferon using a multistage process.

The use of a nonengineered cell line to produce a biopharmaceutical may seem strange, but the advantage of this approach is that the cell is capable of producing the whole range of interferon subtypes, not just a single subtype. In addition, the production of human interferon from a human cell line can reduce the change of producing anti-interferon antibodies. The range of immune regulators available since then has increased to include the interleukins and monoclonal antibodies.

Some cytokines such as interleukin 2 (IL2) were also identified as potential disease-fighting drugs while the chronic overexpression of other interleukins like IL5 are implicated in the inflammatory response in asthma. These materials work by regulating the cell's response to immunological challenges and can be used in fighting persistent viral diseases and cancers. Our understanding of the immune system is advancing, and now a considerable body of work is being produced on the use of cytokines in gene therapy.

WHAT IS AN IMMUNOREGULATOR?

An immunoregulator is anything that modifies the response of the immune system and includes cytokines, antibodies, gene therapy vectors, agonists, and leptin. In this article we will discuss molecules such as interferon, antibodies, and DNA.

WHY MODULATE THE IMMUNE SYSTEM?

There are several reasons to modulate the immune system. The immune system may be faulty and need to be up or down regulated to produce a normal immune response. Alternatively the normal immune response may need to be increased to deal with persistent diseases. Down regulation of the normal immune response is essential to the success of many transplant operations and may be essential to the success of some gene therapy vectors (3,4).

Many common disorders are caused by an inability of the immune system to work effectively. This may be as a result of infection such as seen in Human Immunodeficiency virus HIV cases. The destruction of white blood cells and collapse of the immune response caused by Acquired Immune Deficiency Syndrome AIDS enables a large number of opportunistic infections to occur. Latent viral infectious such as herpes simplex virus (HSV) and chickenpox virus can cause great problems and rare tumors such as Kaposi's sarcoma become problems.

An active immune system is no guarantee of safety, and huge health and economic consequences are suffered as a result of autoimmune diseases. Autoimmune diseases such as rheumatoid arthritis cause untold suffering and progressive disability. Disorders such as asthma now affect a significant proportion of the Western population and can cause disability and death.

In most cases the body can recognize abnormal and infected cells, but when the surveillance fails, the body may fail to recognize some cells as normal and raise an immune response. Immune regulation is used to enhance the immune response to viruses and tumors (5–7) and to reduce the effect of autoimmune diseases such as rheumatoid arthritis.

The usefulness of gene therapy vectors may be increased by the use of immunomodulators to block the immune response to the vector during treatment (3,4,8). The dual effect of this treatment is to reduce the production of antibodies which will limit the usefulness of the vector for repeat dosing of the virus Why use cytokines when the body already produces them? In some tumours the ability to activate the immune system is missing because the tumour cells lack part of their signalling system. Some persistent viruses such as hepatitis B are good at hiding themselves.

HOW CAN THE IMMUNE SYSTEM BE MODULATED?

Modulation of the immune system should either enhance the ability of the patient to respond to foreign material or to down regulate the immune response. For example, the treatment of an autoimmune disease can be achieved by destroying or blocking the action of a proportion of the immune system that causes the disease. In theory this can be achieved by targeting antibodies or other blocking agents to receptors such as CD 4, which are present on the surface of the target cells. Cell surface antigens are presented on more than one cell type, and this approach to therapy is not as specific as was originally hoped. The opposite approach is to up regulate the immune response by adding a biologically active molecule that will act on the immune system. The objective of up regulation is to enhance the body's own response to disease. The use of interferon and interleukins for the treatment of chronic viral infections and cancers is based on the stimulation of the immune system to recognize virus-infected or tumor cells as foreign. The side effects of systemic treatments of cytokines to achieve immune regulation can be unpleasant, and the future for immune system regulation may well lie in gene therapy. This approach can achieve a local effect, and the active molecule is produced inside the target cell. This can be done by using viruses or plasmid DNA to infect tumor cells. Tissue-specific promoters can be used to limit the action of the vector and prevent inappropriate expression of the gene therapy product in nondiseased cells.

IMMUNE REGULATORS AS PRODUCTS

Individual proteins can be characterized and sold as well-characterized proteins. This means that the degree of control on the production method and change of production sites is strict but not as strict as for an undefined product such as a viral gene therapy vector, where precise definition of the final product can be impossible to achieve. Different production methods may be used for different products because of the differing potencies and hence quantities of the products. The quantities of interferon that are needed for a therapeutic effect are small but produced in relatively low concentrations and must be recovered from a dilute solution. Antibodies, on the other hand, are required in milligram quantities, and production methods must be capable of producing kilograms of material per year. Proteins do not easily pass into the body because they cannot pass through the skin, and a major limitation of these protein products is that they need to be injected into patients or used to treat blood or other cells outside the body. Systemic delivery is a cause of some concern because the effects of cytokines can make patients feel extremely ill. There is now a huge amount of work being done on the use of gene therapy vectors to achieve cytokine production in tumor cells in the hope that the immune response to these cells will be boosted.

The use of immunoregulators in treating long-term disorders such as asthma and arthritis requires repeated regular dosing of the immunoregulator. It is essential that the patient does not develop antibodies to the product and that the side effects of the treatment are preferable to the effects of the disease. Treatment of persistent viral diseases and cancer treatments are all likely to require repeat dosing, and therefore the same concerns have to be addressed, but the exposure of the patient to product will be for a defined time, and a greater degree of unpleasant side effects may be acceptable for these immediately life-threatening diseases.

PRODUCTION OF IMMUNE REGULATORS

Small volumes of proteins can be prepared by a variety of methods. As the scale of production increases, the viable options for production are reduced. The assumption in this article is that the reason for making immune regulators is to put a clinical program in place. To achieve the final objective of a working production process a development program has to be undertaken that will ensure that a cell bank is laid down, stored, and controlled to protect the essential raw material on which the rest of the work is based. A process will be defined and validated prior to the sale of the product, and this process will be optimized for the cell line used to make the immunoregulator. It is critically important that the cell culture medium is appropriate for use in commercial protein production and that enough of the medium can be supplied to maintain production. The commonly available serum-containing cell culture media that have provided the basis of culture methods in the past are now of limited usefulness. Products that use serum as a component of the growth medium can be, and are being, licensed, but there are a number of issues that need to be addressed:

- The security of the cell line and product quality can both be compromised by the presence of serum. Products containing large concentrations of serum can be difficult to define, and the isolation of the product from a protein containing harvest is harder than isolating material from protein-free medium.
- Sourcing animal-derived products that are considered safe now (and will be considered safe in 10 years time) is a major issue. Serum must be supplied from a limited number of safe sources, and the demand for material may well outstrip supply as the number of products requiring serum increase. Some products can utilize adult bovine serum, but many processes require fetal calf serum because of its lower antibody load. Fetal sera are seasonal, and the quantity of material available is far less than could be gathered from adult animals.
- Because the glycosylation of the protein that is being produced can vary with culture conditions, it is essential to investigate fully the effects of culture medium and cell culture conditions on the final product (9–11).

PRODUCT QUALITY AND SAFETY ISSUES

The production of pharmaceutical products is controlled by a number of regulatory agencies, and the standards that must be met during production are determined by these agencies. When a product is developed, the process and documentation must be designed to answer any concerns that the regulators may have about the product and the controls that are in place to ensure that the process is under control. It is essential to remember that the requirements of different regulators vary and that development and production processes must take these differences into account. Quality must be built into the system. This extends as far back as the maintenance of good records of cell line derivation, audit trails for components of the system, appropriate validation records, and programmed maintaintence plans. Quality and safety are linked because they all require monitoring control of the production process. Organizations such as the Medicines Control Agency (UK) MCA and the Food and Drug (USA) Administration FDA wish to ensure that products are safe and efficacious in clinical use and that they are made in a controlled and reproducible manner. This objective is achieved by ensuring that manufacturers make material to cGMP and comply with various guidelines and regulations that govern particular product groups. The safety of process operators and the environment must also be given high priority when designing a process.

Product Quality and Safety

The aim of making a therapeutic product is to produce something that is safe and active. Because the mechanisms that are used to make biological products are complex and incompletely understood, the quality of product is less easy to control than that of a small molecule. The same protein can be produced in several different glycoforms and may be degraded in culture or during production. Glycosylation is not only important for its effect on the clearance of proteins from the blood, but has also been demonstrated to have direct effects on biological activity of immunomodualting antibodies such as Campath 1H (10).

Because of the lack of control of the biosynthetic pathways and relative complexity of protein products, they are only now coming to be considered as well-characterized products. As a result of the complexity of protein production systems, biological systems used to produce proteins are subject to many controls (12). To ensure product safety and quality a matrix of systems is applied to control the materials, the methods used, the environment and the flow of material through the process, and the production area.

One of the main concerns is that the raw materials used to make product are free from any materials that can cause disease. These components include viruses, proteins, DNA, and transmissible spongiform encephalopathies (TSEs) such as Bovine Spongiform Encaphalopathy (BSE) and a comprehensive strategy that involves sourcing and testing of raw materials is central to controlling this risk. Well-documented audit trails and sensitive tests are key components of this strategy, in the case of TSEs, where rapid detection has not been possible, materials are sourced from areas where no recorded incidences of the disease are known. Processes are controlled and actions recorded using documentation systems. Equipment used in the production of clinical material must meet exacting standards and be validated prior to use. Standard Operating Procedures (SOPs), Process Operation Instructions (POIs), and change control mechanisms are operated to ensure that any changes to process steps or equipment and facilities are assessed and authorized and recorded. Regular calibration and maintenance must also be carried out. In parallel with these activities the supply of raw materials is controlled and materials have to reach agreed specifications prior to use in the process. Raw materials include cell lines and cultures media used in the process.

By analyzing process flows the likelihood of contamination of the culture and downstream processes can be assessed and the relevant controls introduced. Using the appropriate tests at the appropriate stage in the process can monitor the success of these controls.

The testing that may be carried out on the starting materials used in the process are indicated below. Not all materials will be sterile when received, and there are acceptable levels of bacterial contamination prior to sterilization, but the presence of some pathogenic species might preclude the use of the material in a pharmaceutical process. The details of the tests that are performed will be dictated by the product and current regulatory guidelines.

Cell lines should be tested for clonality, stability, identity, virus contamination by relevant viruses including bovine viral diarrhoea (BVD) (13,14) and porcine parvovirus (especially bovine or porcine viruses if serum and trypsin are used), bacterial and mycoplasma contamination. Serum and other biological components should be tested for the ability to grow cells, if necessary, tested for product identity where possible, virus contamination, bioburden, the presence and identity of any mycoplasma or bacterial contaminants (14). The levels of endotoxin in raw materials should be specified and closely monitored.

Operator Safety

Operator safety is a major issue in the design of biological processes. Many of the products such as interferon that are highly active *in vivo* are potentially harmful to operators. Fortunately the main route of administration is by injection because the large proteins cannot easily pass through the skin, and accidental ingestion or absorption by the operator is not a major issue. There are three aspects of the cell culture material that can affect operator safety:

The Cell Line and Contaminants. The cell lines should be selected to be as safe as possible. In the case of animal cell culture the cell line will be considered to be safe, but it is well known that cells of rodent origin will contain endogenous retrovirus (15,16). Other cell lines have become immortalized by being exposed to viruses, and a risk assessment needs to be carried out to confirm that the risk to operators is acceptable. In most cases the risk to operators is minimal and the cell lines can be considered safe. Consideration of biological hazards can be

emotive because of the perception of biotechnology, and it can help to put the risk of growing all in culture in terms of everyday activities such as keeping mice as pets.

The Product. Antibodies are required in relatively high doses, and their activity is dependent on the binding of the antibody to a large number of receptor sites. The chances of accidentally being exposed to damaging doses of antibody during culture are remote. Cytokines are highly potent, and the effect of accidental exposure to the product could be much greater. Like most biologicals, antibodies and cytokines are delivered directly (normally by injection), and therefore it is unlikely that accidental exposure to the product will have any effect related to the biological activity of the product because of the barriers to the transport of the product into the body. However, operators may develop an allergy to the product as a result of inhalation or skin contact.

The Culture Medium. Historically the cell culture medium and the stabilizing agent used in the formulation of product have been the greatest potential sources of infection and allergy. The extensive control and testing carried out on raw materials should have considerably reduced the risk from this source.

Control of Material

The origin and treatment of raw materials is a major component of any strategy for control of product quality. Detailed raw materials specifications must be written, and the materials delivered to the site should not be released until testing is complete and a satisfactory result to obtained. Suppliers should be audited to check the methods of production and storage of material. It may also need to be specified that any material that has been rejected by another manufacturer will not be accepted. Animal-derived materials pose the greatest risk to the product, and therefore they should be tightly controlled. Audit trails are required to confirm the origin and treatment of animal-derived material. Any contamination of the process with TSE-containing material will cause loss of the material to that stage and require extensive TSE decontamination of equipment using heat (134 °C) and sodium hydroxide solutions. Cell banks are particularly vulnerable, and the whole product could be lost if there is a risk of contamination with TSE agents. A large number of cell culture processes depend on animal-derived products as part of the growth medium, and it may be difficult to develop serum or protein-free media for some purposes.

The Wellferon process was developed before serum-free media were available and uses adult bovine serum in the growth medium. Because of this Glaxo Wellcome uses a herd of cattle in New Zealand, and the monitoring of the serum supply chain is given a high priority. Sendai virus, which is used to induce interferon production by Namalwa cells, is produced in eggs, and the egg source must also be carefully controlled and monitored.

Controlling the inputs into the process by using a protein-free medium may significantly reduce process risk by removing the source of contamination in the first place. Serum-free media may not contain serum, but they may contain proteins from a number of sources, some of which may be as high a risk as serum. The control of this material should be checked if the commercially available medium is to be used in producing a material for clinical purposes. The removal of protein from the culture medium also facilitates the downstream processing of the product.

PRODUCTION METHODS

Immune modulators can be produced by a variety of methods, including the isolation of interferon from blood. Most options for producing immunoregulators involve the use of cell culture equipment. The development and operation of these methods has enabled huge quantities of antibodies and interferon to be produced over the years.

Isolate Material from Natural Sources such as Blood

This is a particularly useful method in the early stages of investigation, and in the early years of interferon investigation the production of workable quantities of material was a long and laborious process that could never produce the quantities of material needed for therapeutic purposes. In addition, the extraction of drug compounds from sources that may also contain transmissible diseases such as AIDS, hepatitis, and TSEs will never be the method of choice.

Use Natural Producer Cell Lines

Fermentation methods are the preferred option for production of immune regulators. The product is made in a controlled environment, and scaleup of production can be achieved by using more fermenters or larger fermenters or by increasing production rates. All the inputs to the process can be controlled and defined and the quality of the process monitored regularly. The process can be developed and improved prior to the product reaching the clinic. Biotechnologists prefer to use genetically engineered cell lines, which give a consistent production of material from a single gene. However, in the early days of large-scale interferon production the potential for enhancing the production of human interferon from a human cell line to levels suitable for production was pursued. The advantage of this approach (that interferon which was similar in subtype composition to naturally produced human interferon) and that the production of 'human protein' was considered to provide a significant advantage to the product profile to make the animal cell process viable.

It is unusual to find an immortal cell line that makes product and even rarer to find one that makes the product in large enough quantities to warrant the development of a production process around such a cell line. Some cell lines such as Namalwa cells produce larger quantities of interferon than others. The total amount of product may be too small for commercial exploitation unless the compound is highly active or the cells can be induced to overproduce by modifying the growth conditions or using inducers and a more common approach is to produce recombinant cell lines.

Produce Recombinant Cell Lines (Animal Cell or Microbial)

The most common strategy is to produce genetically modified cell lines that will make the product of choice.

Animal cell cultures have the ability to carry out post-transcriptional modification of proteins, but some products such as cytokines can have severe effects on cell growth. By producing them in a system outside the normal cell line, it may be able to minimize the toxic effects of the cytokines and produce much higher concentrations than would be possible in a natural producer. In microbial cells the use of inducible production systems is well established. These systems will allow rapid growth of cells that would be killed by the toxic effects of constitutive protein production. The production of material in genetically modified mammalian cells is normally from cell cultures that are producing at a constant level. The danger of using GMOs to produce proteins is the ability of the body to recognize foreign proteins and elicit an immune response the therapeutic protein. This limits the suitability of a product for long-term therapy but may also cause cross-reactions with the patients' natural immunoregulators.

Monoclonal antibodies can be used to regulate the immune system by selectively targeting cell surface receptors and interfering with cell signalling. They can also be used to destroy a proportion of a target cell subpopulation and down regulate immune activity in this way. The antibodies can be modified to make them more human like and engineered into a producer cell line. The antibodies do not seem to have a feedback effect on cell growth that would limit the usefulness of fed batch processes, and the antibodies can be produced in large quantities. Compared to cytokines the antibodies are required in large quantities (mg/dose), and the advantages of high productivities are offset by a larger demand. The quality targets for larger quantities of material may be harder to meet because of the larger contaminating load of protein and nucleic acid derived from cell components. In particular DNA and cellular proteins have to be eliminated from the final product.

Use Gene Therapy Vectors to Introduce Genes into Target Cells

Nontargeted use of immune modulators such as interferons can make patients extremely unwell. One approach that is being actively investigated is to replace the shotgun approach and transfer the immune modulator gene, including receptor molecules such as B7 into target cells using gene therapy vectors (3,6,9). These vectors may be based around viruses or plasmid DNA. This approach should be able to achieve prolonged expression of high local concentrations of active product in the areas it is needed and allow the protein to work in the way it is supposed to work. It is hoped that gene therapy can be made to work selectively by using a combination of tissue-specific expression systems and targeting cell surface receptors. The production of a human protein by a human cell should also reduce the changes of developing an immune reaction to the protein (however, the delivery system may produce an immune response, and this will limit repeat dosing with some vectors).

PROCESS DESIGN

There are not any secrets to successful process design, if you keep a process simple and remember the objective is to make a product that has to be put into people. Factors affecting process design such as cell type and product titers are determined early in the development process. Wide-ranging and effective collaboration needs to operate at all stages between the research, development, and production functions for process design to be effective. Minimize the risk of contamination and process failures by using the minimum number of process steps. Try and ensure that the process steps used are robust and reliable.

Process design covers the design of the process flow and the operating philosophy used for the process. In general, the production processes are designed to provide a well-defined flow of material with fixed testing points that can be used to provide information about the progress of the operation. The number of stages and the number of manipulations involved in a production process should be reduced to a minimum. Each stage in the process should be designed to minimize process risk. Containment is an issue that needs to be considered from the perspective of operator and environmental safety and process containment.

The major process risks vary in type and impact at different stages in a process. Processes that involve many manual operations and transfers of sterile material using pipettes are more likely to cause contamination than large-scale operations where transfers are carried out through steam-sterilizable transfer lines. A contamination caused during cell revival and scaleup is likely to be a result of a breakdown in sterile operating techniques, while sterility problems at the larger scale may reflect an engineering problem note introduction of 5ntom indin vià a process fluid. The risk of failure is reduced by defining the production process and by introducing controls into the process. Standard operating procedures are written and operating staff are trained in their use to ensure that the process is operated consistently. Validation of the equipment and process ensures that if the equipment and process are operated to the SOPs the process will work reliably. Specific risks such as the accidental contamination of the cultures by filter breakdown are reduced by integrity-testing equipment immediately prior to use and replacing filters and resterilizing associated equipment if they fail the test.

The objective of the development work that is carried out on a process is to provide a unified operation that can operate and continue to provide product of a uniform quality.

There are many approaches to process design, and the choice of highly automated against manual operations is not always straightforward. For instance, a fermenter sterilization cycle requires the opening and closing of several valves, which can be done automatically or manually, and the relative process risk of using each option needs to be assessed. Automated equipment will always carry a validation cost and computer systems validation issues need to be resolved prior to ordering equipment. Equipment manufacturers are always proud of their advanced instrumentation and control packages that will deliver an excellent result but may not be the best option if a process needs to be set up quickly, because there will be a greater validation requirement for the more sophisticated equipment. There are pieces of equipment where an equally good result is achieved with a fraction of the control equipment. The art of designing a good process is to put the right degree of complexity at the appropriate stage in the process.

Process plants may need to be operated on a campaign basis, and several different products may be run through a plant in a year. This means that initial plant designs should be flexible enough to allow the use of the facility to make several different products. Plant and process flexibility can be achieved by a number of means, one of the most useful being the simple expedient of ensuring that culture vessels are capable of being operated at more than one volume. This reduces the number of vessels needed to achieve scaleup of cultures, allows extra capacity to perform fed batch processes, and allows a range of production volumes to be harvested in a small facility. For instance, the 8000-L cell culture vessels at GlaxoWellcome are used for production of animal cell products in ranges of 2000–8000 L (19,20). It is essential that downstream processing facilities are sized to deal with the quantity of biomass and protein produced by the culture. Development process equipment and buildings should be capable of being used for more than one product because the product that is originally developed will be moved from the development program at some time. This will be because the product has progressed to the market or has been dropped from development. A number of development campaigns will need to be conducted in the same facility, and the use may well need to be scheduled to allow the running of several campaigns per year.

Development and Operation

The development and operation of a production process based on a cell line should follow a well-defined path.

This can be broken into three activities, which will be performed in parallel or overlap and are mutually dependent.

- Cell line development
- Process development
- Production process

Each activity has its own priorities and associated testing and control requirements.

Cell Line Development. Cells need to be from a well-documented source and be tested to assure us that the cells are free of bacteria, mycoplasma, and adventitious virus. If they are grown in serum-containing media, a record of the serum and its source and testing is essential. Initially the cells are cloned and selected to identify suitable cell lines for further development.

Many production systems are based around the methotrexate selection and amplification system that uses chinese hamster ovary (CHO) cells, and the NSO-GS, (glutamine synthetase) system that can be licensed from Celltech. (Slough, UK) Other production systems are based around the growth of murine hybridoma cells and Namalwa (a cell line which was immortalised as a result of a natural viral infection) are used as a nonengineered cell line for interferon production. The importance of cell lines to biotechnology processes cannot be overstated. Lose a cell line and a product can be lost. Cell lines and genes are the basis of a biotechnological process. If the producer cell line is not suitable, the rest of the process will fall down. Therefore, much work is done to select and test the cell line that will be used for production of therapeutic product. The two major concerns about cell banks are whether the cells in the bank can be used in a viable production process and whether they will comply with regulatory requirements.

The development and design of production processes depends on having a reliable, consistent starting material. The cell line used should be stable or change in predictable way so that cell culture and down stream processing (DSP) can be sized correctly. Engineered cell lines should be selected for titer, stability, and ability to grow in the culture system. A cell bank that gives continually varying qualities or quantities of material will be extremely difficult to use in a cGMP process.

Regulators are concerned with the control of pharmaceutical products and ensuring that the material is safe and efficacious and so be produced in a consistent manner. The regulatory bodies lay down the minimum testing requirements, and it is essential that when the tests are carried out the product be of an acceptable standard.

Animal cells produce immunoregulators that are soluble, may be glycosylated, and will be exported into the medium, and this gives a structure to the proteins that can make them more natural and less likely to be identified as foreign by the immune system. Namalwa cells are an immortalized human cell line, and they produce natural human interferon. There is not any genetic engineering involved in the process.

Antibodies used for therapeutic purposes tend to be of murine origin and may then be extensively modified to humanize them prior to use in a therapeutic setting.

Cell Line Production and Selection. The production of cell lines and their selection for factory operations is a keystone of any biotechnology process. This is because a consistent process requires a well-characterized cell line that produces an economically viable quantity of product. A number of candidate cell lines will have to be screened for desirable characteristics and the selection narrowed down to the cell line that will be used to make therapeutic material. A consistent quality in terms of ratios of subtypes and glycoforms is essential to producing a therapeutic material, and cloning and selection are an integral part of that process. The cell line is chosen on the basis of a number of growth and production criteria.

Cloning

Cloning and selection of producer cells take time, and development work must continue while the cloning is being performed. When cells are received, a number of vials of cells are laid down to make an initial cell bank and used as a working stock. Some of these cells will never be used to construct cell banks and are used to carry out initial development work, while other cells from the same stock are kept in tightly controlled conditions and used in the cloning exercise. The initial cell bank can be used to make early phase material for testing, while the other activities are ongoing. The cloning and selection exercises will take approximately 3 months per round of cloning, and stability work will need to be performed on the best clones.

Genetically engineered cell lines that integrate the product gene into their own genetic material will show differences in both growth characteristics and productivity. Cloning will enable the best-growing high producers to be selected and expanded. Each cell line derived from the cloning exercise should be derived from a single cell. By using cells derived from a single cell, the variability that a mixed cell population would introduce to a process is avoided and any changes in cell growth rate or productivity will be attributable to cell line stability rather than competition between existing subtypes of cells.

Any continuous cell line may be cloned, and the process used involves selecting cells by plating them at on appropriate dilution to give on overall cell/well into culture vessels (typically 96 well plates) and selecting colonies that grow from single cells. A range of manual methods are used to ensure that the colonies are derived from single cells [serial dilution (21), picking individual cells (22)]. The use of automated methods (22) such as cell sorting to increase the chance of selecting single clones of high producers has been advocated. The success of cloning is totally dependent on the ability of single cells to form colonies in the cloning environment (the plating efficiency). The plating efficiency can be improved considerably by maximizing the growth conditions before and during cloning. A high plating efficiency provides the opportunity to select from a wide range of cell clones because there will be a greater percentage of recoverable cells in these cultures. Undertaking a round of cloning and selection requires a major investment in time and facilities. Animal cell cloning and stability studies will take several months, and it is essential to have the maximum chance of selecting a clone at each round of cloning. If the cell lines are not clonal, the cloning and selection loop will need to be repeated. This is different from bacterial systems, where a selection and stability and excercise can be performed in two weeks. Several clones that appear to be promising candidates may be eliminated by stability studies (23). Although the principle of clonal selection are the same for all cell lines, the timing varies enormously. As a rule of thumb, a process that takes an hour with *E. coli* will take a day in animal cell culture. A 150-generation stability study would take 6 days and 6 hours for *E. coli* and 150 days for an animal cell culture. The implications for work planning are significant because the development of a process cannot wait for the sequential completion of all the component parts of the development program, and the limiting step is the cell line cloning.

A typical process would involve receiving cells that are genetically modified to produce an immunoregulator from a research group and selecting promising lead cell lines by cloning. There are different methods for cloning cells, and these methods can be divided into two basic types, dilution cloning and selection of individual cells. We have spent a considerable amount of time and effort in the optimization of cloning to give us the maximum recovery of material from single cells. All the methods used depend on the ability of single cells to grow in a small volume of culture medium but in the absence of other cells. The use of feeder cell lines is not acceptable, and the rate of cell recovery during cloning can be low. It is possible to raise the rate of recovery by manipulating the growth conditions used for the cell seed to ensure that the cells are in the the most propitious condition for growth as single cells after the cloning. Maximizing the medium and culture conditions in the cell culture phase directly after aliquoting single cells into a well will also enhance cell recovery. Cell recovery can be checked against the expected recovery based on the number of cells being used in the process.

Limited dilution cloning is a method of cloning that works by diluting cells in culture medium and transferring cells into 96-well plates. The wells can be inspected at regular intervals to check that cells are growing and that colonies, which have grown from single cells, are selected. The chance of achieving a single cell per well can be calculated statistically using the Poisson distribution, and a fairly low number of cells/well are required to give a high confidence of clonality (21). Some cells clump, and the production of single cell isolates will need to be checked visually and the well containing single cells recorded.

Cell sorting by manual means or flow cytometry can give a single cell per well and significantly decreases the number of plates that need to be used to find a clone.

When a large number of clones are isolated, the clones must be differentiated from each other to select the most desirable clone. Selection of the clone is usually on the basis of some simple testing for titer, growth, and maximum cell density. Some other factors, which may be important at later stages of development, such as acceptable pH ranges, will not be identified at this stage.

SELECTION CRITERIA FOR CELL LINES

Stability

Cell line stability is measured in terms of the ability of the cell line to behave in a stable and consistent manner. From a process engineering standpoint the design and operation of plant and equipment is aided by having a constant growth rate and predictable yield of product. Not all cells make product at a constant rate (24–26), and some have a complex induction sequence that can make productivity difficult to measure (2). Once the cells are in a production phase, a high level of production, which gives a consistent quality of product, is desirable. If the production rate is not constant and the cell growth rate varies, this will be acceptable as long as the changes happen in a predictable manner and the quality of the product is consistent. From a process viewpoint the useful life of the cell is determined by the titer produced by the process. The cell must continue to produce at levels above the minimum for a period that will allow for the production of cell banks, scaleup, and production of material.

Productivity

The methods used for testing cell productivity will depend on the type of product made by the cell. ELISA or nephelometry can detect the production kinetics and maintenance of productivity of therapeutic proteins such as antibodies, which are continually secreted by the producer cell line. This allows an estimation of productivity of such cell lines relatively easy. Products with a complex production route such as Wellferon will need the producer cell line to be grown to specified cell densities and induced. The results at any early stage of selection are not necessarily an indication of either the final titer that can be obtained in cell culture or the relative performance of different cell lines relative to each other. (At this stage) big differences are saught. Initial screening using a single test will not be expected to give information about glycoforms and biological activity, and a more detailed examination of the product would be performed at a later date after the options had been narrowed down. The material that is purified and injected into the patient needs to be consistent, but this is much further down the line than the cloning exercise.

Growth Kinetics and Subculture Regime

Cells should be chosen on the basis of cell growth rates that are high enough to enable scaleup to production volumes within a target time scale. Growth should be stable, and the occurrence of useful/useless cycles should be capable of being eliminated by controlling the cell growth regime. In addition, cells that need a high seeding density and grow to low final densities and hence need frequent subculturing are not favored. A cell line that can achieve 2–4 doublings in a 6-day period is typical of the type of cell line that would be chosen.

Oxygen Utilization

Most of animal cell lines used to make biopharmaceuticals have similar oxygen utilization rates. A cell line with significantly higher utilization rates would require the use of oxygen-enriched air and may be difficult to use in fed batch. In addition, a higher than expected oxygen utilization rate indicates that the cell metabolism is different from other cell lines produced from the parental cell line. This may reflect significant changes in the cell metabolism, which are related to the site of integration of the therapeutic gene.

Nutrient Utilization Patterns

The production of high concentration of toxic metabolites or the use of a nutrient that would bypass selection pressures will prevent a cell from being used in the production process.

CULTURE MEDIA

The culture medium is one of the cornerstones of a biotechnology process. It is also a component over which it is difficult to maintain control because the materials used to make the medium are often sourced by a third party from a limited number of sources. An ideal solution is to source and produce medium components in house, but this can be a very expensive option. A workable alternative is to develop partnerships with the commercial medium supplier of your choice and ensure that the medium is produced to a mutually agreed standard and that ingredients used in the production of media are obtained from appropriate sources.

The role of cell culture medium is to feed cells and maintain a suitable osmotic environment and help to maintain pH. A further role may be to provide an external source of nutrients that do not appear to be utilized in any gross way but are essential to cultured cell growth because they provide an environment that is similar to the cells natural environment. Most media are developed for use in a noncontrolled environment, and the same media are used in tissue culture flats, roller bottles, shake flasks, and fermenters, and commercially available media are usually capable of supporting the growth of a range of cells. Standard catalogue media can be used as part of a commercial production process, which removes the need to develop specific formulations for each process. However, the media may not be optimal for growth or product formation, and development of the culture medium may increase the productivity and reproducibility of the process. Medium can be a major source of contaminants that have to be removed downstream, and so the simpler the medium is, the easier the burden on downstream processing.

Formulating appropriate media is an essential part of process development and can affect the success of a product by increasing yield but also by helping to control raw materials costs and risks to the process and product caused by medium components. Decisions made when licensing a product today will stay with a biotechnology product for many years, and decisions on the composition and risk associated with the medium will have to be supported in an ever-changing regulatory environment.

Sterilization

Cell culture medium contains labile components and must normally be sterilized by filtration. Therefore the chance of transferring virus or other contaminating materials into the culture is much more significant than in a process where the medium is heat sterilised *in situ*. It is important to keep the number of components that can place the process at risk to a minimum. Figure 4 shows some of the medium components and environmental factors that could compromise the product.

Serum

Many culture media contain serum because it provides a range of nutrients and growth factors that include hormones such as insulin, transferrin for iron transport, attachment factors such as fibronectin, and serum albumin, which acts as a carrier for any compounds including lipids. Serum also helps to protect cells from the damaging effects of bubble burst during suspension culture. This complex cocktail of ingredients has to be removed from the cell product during purification. Some cell lines can survive and prosper in the absence of

serum, but others need growth factors that are normally supplied by the serum. Serum independence can be aided by engineering the gene for producing the essential growth factor into the producer cell line. The addition of Pluronic F68 has been demonstrated to protect cells from bubble burst damage (27,28).

Medium Formulations

As the medium must support cell growth and product formation and the two functions may have different nutrient requirements, a production process may require different medium formulations. The production formulation may also be modified to aid in the downstream processing of the product by removing components that are needed for prolonged active growth but can be left out for the dead end production phase.

pH Buffering

Basic growth and production media will have a complex range of components that help to buffer the pH of the media and feed the cells in the culture. Most mammalian cell formulations are designed to be used in small-scale culture vessels such as tissue culture flasks, roller bottles, and shake flasks. They need to compensate for the change in pH caused by cellular metabolism. Lactic acid and carbon dioxide are both produced as a result of glucose metabolism, and pH buffering systems have been developed to maintain the pH of media at acceptable ranges in these systems. It should be remembered that carbon dioxide can have a significant effect on cell cultures (29,30). The use of an overlay of carbon dioxide containing air (3–10%) is also used to prevent the pH of medium from dropping too far during the early stages of the culture. Buffers such as (N-[2-Hydroxyethyl]piperazine-N'[2-ethanesulfonic acid]) (HEPES), and (3-[N-morpholino]propanesulfonic acid) (MOPS) are used in some cell culture formulations but should be avoided where possible in formulations intended for large-scale use, as they are expensive.

Some medium components may be used out of habit and because they make small-scale culture easier to monitor. It is common practice to add phenol red to culture media because it acts as a pH indicator. A red culture is in a suitable pH range; a yellow culture is too acidic; and a purplered culture is too alkaline. Large-scale culture systems control pH by using pH probes to monitor the pH of the medium, and the role of phenol red in monitoring the pH can be useful as a check but is functionally redundant. The control system uses the addition of weak acids and bases such as CO_2 and sodium carbonate to control the pH and to compensate for the effect of sparging of the cultures with air, which will strip CO_2 from the medium and drive the pH of the medium down.

Osmolality and Feed Media

The sensitivity of animal cells to osmotic shock dictates the osmolality of the culture medium and limits the amounts of nutrient that can be added to any cell culture medium (29–31). Medium development allows the identification of essential nutrients that are utilized in the culture and need to be replaced; hence the production of minimal feedstock for fed batch processes. Feeding nutrients can increase cell densities and increase the concentration of product that is harvested in a batch. Modifications can be made to media to facilitate fed batch processes. These modifications involve making concentrates of essential nutrients in a format that keeps the components soluble and allows them to be autoclaved or filter sterilized. It is also convenient if the concentrates are stable for some weeks after production so that the batches can be made and tested prior to use. When fed batch processes are being developed, it is of great importance to ensure that the osmolality of the medium is balanced and the osmotic pressure at the end of the process is not so high that it kills the cells.

Intellectual Property

The use of some approaches such as the use of cyclodextrins (32) as lipid carriers to make defined culture media may be restricted by patent coverage, and the development of processes destined for commercial use may be affected by this limitation. Proprietary media are frequently protected as a trade secret by not disclosing the composition of the media.

Supply and use of Media

Dried powdered medium and frozen concentrates are used for large-scale cell culture; they are normally nonsterile and are supplied with a low bioburden and endotoxin content that should be specified and confirmed by the manufacturer. Any materials that need to be added directly to the vessel without further filtration must be presterilized and in suitable containers and then connected to the vessel using a sterile addition rig. The bulk medium components are diluted in water and mixed to ensure homogeneity and then filter sterilized into the culture vessels. This is a point at which there is potential for virus and mycoplasma contaminants to enter the culture. Environmental control and product testing can reduce the risk of contamination, but a small number of mycoplasma or viruses may not be detected in preculture assays. The medium will be produced and used within 4 hours, and the chances of any detection system picking up contamination on time are limited. In addition, the viral contaminants will not increase in number until they come into contact with an animal cell. Membrane filters are used to reduce the risk of bacterial and mycoplasma contamination and will also reduce the viral load.

Properties of a Cell Culture Medium

An ideal culture medium should have properties as discussed in the following.

Defined. A defined medium formulation is desirable because is should allow rational medium design and mass balances to be performed. This in turn should allow optimization and design of media for specific purposes.

Free of Animal-Derived Products (Including Amino Acids). It is possible to design animal cell culture media that are free of animal-derived products. We use these

successfully for the production of monoclonal antibodies in batch and fed batch processes up to the 8000-L scale.

The use of serum-free and protein-free media is an immense advantage for process operation for the following reasons:

- All the issues surrounding the supply of serum (virus contamination, TSE contamination, predicting requirements, audit trails, variability of growth, total quantity of serum available) are no longer of concern.
- Downstream processing is easier. The background protein levels in the harvest are much lower than in a serum-containing harvest.
- Copurification of transferrin or antibodies with the product is not an issue if serum and additional protein are not present.
- Growth and productivity should be more consistent than in a serum-containing system.
- Protein-free medium formulations are less likely to cause foaming problems than protein-containing formulations. This is particularly important in animal cell products where the use of antifoam is less common than in bacterial culture.

Available in Adequate Quantities and the Medium Economically Viable. It is essential that medium components are available in sufficient quantities to meet demand and that these components are all of high enough quality to be used in a biotech process. This issue is providing enormous challenges to companies that have developed serum-based fermentation processes and depend on serum from a limited range of suppliers. Where possible at least two sources of supply and manufacture should be identified. The logistics and economics of medium production and supply will help to determine the commercial viability of a process.

CULTURE MEDIA AND CELL LINE DEVELOPMENT

The production medium should be used in cell line development and any requirements for modification identified at an early stage. We have found that even though we have generic processes that will enable us to grow cells and make product, the specific requirements for cell growth and product formation may vary from clone to clone. New variations will be needed to allow growth of cells or to remove components that are not essential for some clones. This may reflect the effect of integration of the therapeutic construct at different sites in the host cell genome.

Handling of Cell Lines

There are two concerns when handling cell lines: operator safety and cell line integrity. Most cell culture lines are innocuous, but care must be exercised when selecting immortalized human lines, as there may be the potential to produce infective virus from a cell line immortalized by EBV. Human cell lines may also be excellent hosts for human viruses; the risk to human health caused by exposure to cell cultures that are accidentally infected with human pathogens is always a factor that needs to be considered when designing processes and conducting risk assessments. The risk of infection with hardy viruses such as parvovirus is real, but there are few reported occurrences of environmental contamination, and the infecting virus would need to be able to infect the cells in culture in order to do any damage to the cell culture. There are two possible reasons for this lack of evidence: The contaminations may be rare, or the infections have not been recognized. Environmental control, pest control, strict control of the health of operators, and restrictions on the movement of staff between facilities help to maintain culture security. Quarantine areas are essential for the successful handling and segregation of cell lines. Whenever a cell line is received, it should be kept in a segregated area until basic sterility and purity testing has been carried out. The degree of control exerted on a cell line will depend on the final destination of the cell. Material that is used for experimentation should go into a dead end system to ensure that the cell material will not become confused with cell bank material. The environment and raw materials used to make and store a cell bank for use in production will be tightly controlled and run to cGMP. Clear labeling of tissue cultures and the segregation of cell lines by providing separate facilities for each cell line or separating cultures temporally should avoid any cross-contamination problems.

PROCESS DEVELOPMENT

The aim of process development is to enable a laboratory-scale operation to be converted to a production-scale process. In a pharmaceutical process development environment, the major issues of producing toxicology and early-phase development material need to be addressed as part of the process development program. Negative results from any of the trials will be enough to stop further development of that particular product. It is vital to reach these decision points early and avoid diverting resources from promising products to a product that is not going to progress. There is a considerable advantage in developing each process to have the same overall shape so that resource is not wasted and the lessons learned by early-stage development programs continue to have value to other programs.

By adopting a generic approach to development the core resources of a group can be used effectively. A decision to use cell lines in suspension culture will rule out the need to develop technology for microcarrier, hollow-fiber, and other nonstandard technologies. A decision to use cell lines that will grow in serum-free medium will remove the burden of serum testing and sourcing. However, generic approaches must not be developed at the expense of flexibility or appropriates, and new products and cell lines may require a fresh approach.

Regulatory requirements for cell bank production and testing and the production of clinical trial material provide a natural framework for a generic approach to development. The process for developing and testing animal cell products and bacterial products is similar in outline, and the same concerns are apparent in each type of product.

Generic Process Development

Where possible we use a generic approach to production and development. The use of a limited number of host cells for genetic engineering makes the development of a generic process easy and effective. Every new clone and product builds up a matrix of information about the variations that will be seen between different engineered versions of the same cell line. A cell culture medium developed for one clone has a high chance of working effectively for another clone with only minor adjustments to the formulation. The cell lines will be produced using a well-understood cell line for which optimal cloning conditions have already been defined. This aids in the cloning and selection procedure because the number of surprises that the cell line can produce has been limited as much as possible. Whie the cell line is under development initial assessments of growth and productivity is carried out using the existing cell lines. Material from this work is passed on to the groups developing the assay and downstream processes. High turnover assays that give reliable results are essential to developing and operating a process. A product will frequently arrive with an assay that has been used in a research environment. This will need to be developed to enable its use in automated equipment and converted to a form that can be run as a validated assay in the production campaigns. Additional methods for activity and structure will also need to be developed.

The new product will be tested against a scaled-down version of the generic DSP, and the changes in the process can then be tested. Different proteins will have different sizes, charges, and solubilities that will influence DSP. The primary separation stage may include a product-specific ligand, and the medium must be tested to ensure that it does not interfere with this or subsequent stages. It is not unusual to identify changes in buffer composition that need to be made at this stage. The presence of polymers and lipids can have significant effects on ultrafiltration steps. As the cell line development progresses, the cell line used in the process development work will be changed, and the effects of these changes on the DSP will need to be checked.

The initial process development centers on selecting a high-yielding medium and growth regime for the cells and investigating production kinetics. This work can be performed in shake flasks, the use of fermenters will initially be limited to checking that the cells would grow in the fermentation environment, establishing baseline growth kinetics, and providing material for other groups. Once the baseline data are established, some initial work on the effects of control variables should be performed.

The initial work should provide estimates of the following data:

- Subculture regime
- Seeding density
- Final cell density
- Glucose utilization
- Lactate production
- Glutamine utilization
- Glutamate utilization

- Product titer
- Cell doubling time
- Stability of productivity and growth
- Production kinetics
- Baseline fermenter growth data

The next stage involves gathering more information about the control variables needed to optimize the growth and productivity of the cell line. There may also be a requirement to make material for toxicology and early-phase clinical trials, and the understanding of the process must be sufficient to allow the production to be achieved. A good understanding of the effects of control variables and the identification of critical parameters can be achieved using a set of matched fermenters in statistically designed experiments. The final stage will involve the validation of the process and using the working cell bank (WCB) derived cells to make material for phase III and beyond.

Timing. The time scales for large-scale mammalian cell culture are long in comparison to microbial cultures. It takes approximately 50 days to reach full production scale from cell revival. The implication of this is that stability studies that may only look at 150 population doublings may take 100–200 days to perform. Several stages of the development process will have to be performed in parallel to progress an animal-cell-based process in a realistic time. The cell line used for early-phase development and initial trials may not be the line that is used to produce phase III material. The ability to define products as well-characterized proteins has given much greater flexibility to the development process.

Cell Line Qualification. Once a cell bank has been made, the cells will be revived and tested to ensure that they are still exhibiting the properties they showed prior to making the cell bank and are still capable of stable growth and production.

Culture Methods. Appropriate cell culture methods should be chosen for the production of immunomodulators. The methods used can be divided into hardware options and process options, and these in turn are influenced by the cell type and productivity. Suspension culture is the most common method because it is easily scaled up and is simple to operate. The use of suspension cells is described here. Techniques for growing attachment-dependent cell lines can be found in the article on cell products—viral gene therapy vectors.

Culture methods vary depending on the scale and degree of control expected from a process. There are a huge number of variations on the small number of basic of methods available for growing cells and making proteins. Many strange and novel devices have been suggested as the ideal vehicles for animal cell culture, but the options for viable production processes are small and limited by scale issues.

It is unlikely that the choice of culture methods used to make a product will be unconstrained. The existence of plant and development equipment will drive different groups towards developing processes that fit their existing process shapes. The scale of production and the relative costs of different approaches to production are also relevant, and the larger the production scale, the freer the options for production. The choice of culture method for cell culture products has been dominated by concerns about cell damage caused by culture equipment. Various devices were advocated as answers to shear sensitivity. These included low-shear agitation systems, bubble-free gas transfer devices, segregated systems such as hollow fibers, fixed-bed systems, and fluidized beds. It is well known that shear sensitivity does exist and that animal cells are rather less tolerant of high shear than microbial cells, but it is irrelevant in our normal culture systems. We have determined that the oxygen requirements for animal cell cultures are relatively low and that most of the products that we are interested in are capable of being made by suspension cell cultures up to our maximum vessel capacity (8000 L).

Large volumes of culture are required to make commercial volumes of most immune regulators. This has significant implications for the whole process development area.

Factors in Choosing Culture Systems. A range of factors must be considered when choosing a cell culture system and include the following:

- Cell type and product yield
- Projected requirements
- Appropriateness
- Complexity
- Cost
- Capacity
- Scalability
- Consistency
- Operability
- Product quality
- Monitoring
- Control
- Support

Small-scale culture can be performed in a number of ways. Tissue culture, spinner culture, and shake flasks can all be used for small-scale culture of these cells. These processes may be sufficient for production of small quantities of material. Cells can be revived directly into shake flask or spinner culture as the first part of a scaleup process. Commercially available bioreactors can be used for the scaleup, and there are a number of types available. Some of the commercially available fermenter designs are discussed below. Purchasing and specification of a culture system can be a complex task, and there are some issues that are high priority when selecting equipment. Compliance with computer validation guidelines is essential for equipment with software-based control and monitoring systems. A cGMP-compliant system will always be more expensive than a noncompliant system because of the documentation that has to be provided with the controller. Mechanically the culture vessel needs to comply with appropriate engineering standards and must also be cleanable, be sterilizable, and allow sterile connections to be made to the vessel. The method used depends on scale, and at larger scale, the preference is to make connections through hard piping or by using steamable connectors.

Stirred Tank. The simplest way to control the temperature, pH, and DOT of a culture is to use a stirred tank. Animal cell cultures are relatively easy to control. An agitation system mixes the culture medium and keeps the cells in suspension, a gas line is used to blow air into the culture medium on demand, and the agitator can help the mass transfer by breaking up gas bubbles. Sodium carbonate solution and carbon dioxide gas are used as pH control agents in animal cell cultures and can be added to the vessel on demand. Compared to bacterial fermenters, the mixing times in these vessels is very slow. The worst-case mixing time for the 8000-L vessels is approximately 2 min. We have demonstrated that the distribution of pH control agents and any other additives to the surface of the culture liquid will be dispersed slowly, but because of the slow growth of the cells, their relatively low density and the use of weak acids and bases for pH control enable tight environmental control to be achieved in these vessels.

The control of the system is as simple or as complicated as the control equipment used with the vessel. At GlaxoWellcome stirred tank reactors are the preferred method for the production of commercial quantities of proteins. The reasons for this preference are outlined below:

- They work reliably.
- The vessels are relatively short and the maximum aspect ratio during operation is 1.3 : 1.
- We can scale from 1 to 8000 L.
- We can use an 8000-L vessel at culture volumes of 2000 to 8000 L.
- Namalwa, CHO, and NS0 cells all grow well in these systems.
- Processes such as fed batch that result in changes in liquid level do not cause any difficulties to the vessel operation.
- Magnetically coupled agitation systems are used at all scales, and the risk of contamination through mechanical seals on the agitation system is eliminated.
- Factors influencing oxygen transfer and mixing are well understood, and we have confidence that the system is robust enough to continue providing high cell growth for many years.

Recent improvements in fermenter design have improved oxygen transfer sufficiently for us not to need oxygen-enriched air to feed our fed batch cultures (33). A major factor in achieving this improvement was the design of an improved agitation system that still retained the advantages of magnetic couplings but enabled us to use a larger-diameter high-shear impellers than was previously possible.

Airlift Fermenter. The principle of the airlift fermenter is elegant and attractive. A column of liquid is made lighter than the surrounding liquid (by mixing it with air bubbles), and the difference in density combined with the transfer of momentum from the inlet gas drives the circulation of liquid around the fermenter (34,35).

Airlift fermenters were initially developed for use with animal cells when it was believed that the use of stirred tanks for culturing animal cells could be difficult because the shear fields around the vessel impellers would damage the cells in culture. The design of the fermenter vessel is mechanically simpler than the stirred tank, but to ensure consistent agitation there needs to be a constant flow of air throughout the system. This needs to be balanced to allow a fixed dissolved oxygen tension to be achieved and prevent stripping of carbon dioxide in the early stages of the fermentation. Airlift fermenters work at a fixed volume, and even small changes in volume can reduce the efficiency of the liquid circulation in the vessel. Because the flow of liquid around the fermenter is driven by density differences and will be affected by the friction between the liquid and internal components of the vessel, the minimum operating volume for an airlift vessel is approximately 5 L. Mixing of reagents in an airlift fermenter is a complex process with four different hydrodynamic zones existing in these fermenters. A turbulent, well-mixed, disengagement zone at the top of the vessel allows the air to separate from the culture medium. The downcomer should be mostly gas free to give the maximum differential density between the two sides of the circulation system. The bottom zone is the area where gas and liquid mix and the flow changes direction. This is a well-mixed area. A mixture of gas and liquid flows up the draught tube, and there is little mixing in this volume. To prevent material being held in any part of the airlift system the fermenter should be designed to allow a constant volumetric flow rate through all parts of the system. Measuring instruments must be carefully sited and the proportional, integral, derivative control (PID) loops tuned to allow for the pulsing effect as slugs of pH control agent circulate around the system. Lonza uses airlift fermenters for the culture of cells and production of some commercial products.

Hollow Fiber. Hollow fibers are widely used for the small-scale production of materials from animal cells, including cytokines or antibodies (36). When used with a stable cell line the hollow-fiber system is excellent for the production of small amounts of antibody over long periods. The systems are not scaleable, and the consumable costs increase rapidly if the systems are used for short runs, as new cartridges need to be used for every run. The cell concentration in hollow-fiber systems can become high ($>10^8$ cell, 1 mL), and the environment that the cells are exposed to will vary depending on the position of the cell in the concentrated cell mass. This means that the system is not well controlled and the culture is not homogeneous.

Other Immobilization Methods. Cells can be immobilized in a range of systems, which will allow them to be maintained in a state of growth and production. The cells do not need to be adherent as physical entrapment can be used. Although these approaches are novel, it is difficult to see what advantage they may offer in scaleability, product handling, or process operation for cell lines that constitutively produce high concentrations of immunomodulators. The ability to achieve total medium exchange by draining the system and refilling with fresh media may be useful in situations where a serum-containing growth medium needs to be replaced with a serum-free production medium and when high cell densities are required to make a usable amount of product.

Process Operations. The choice of method used for cell products is frequently batch or fed batch because of the relative simplicity of operation and the ability to run these processes in a standard design of bioreactor. A number of approaches can be used to modify the operation of the cultures to enhance productivity and optimize the process.

MODIFY PHYSICAL CONDITIONS

The most common approach to increasing productivity is to divide the process into more than one stage and to maximise the different stages for either cell growth or product formation. Changing the pH temperature and dissolved oxygen tension have all been shown to affect growth and productivity. The more complex the process, the more options exist for modifying the conditions, as can be seen with the Wellferon process, where a number of factors affect Wellferon production and the effect of temperature on the production phase is significant, with greatly reduced temperatures (28 °C for example) producing an improvement in titer (37).

Medium Development. Improved cell culture media, which support the growth of higher cell concentrations and consequently allow the batch to run longer, can be used to increase the product titer obtained from a culture.

Production Enhancers. The use of sodium butyrate has been shown to increase protein production from some cell lines and is an essential part of the Wellferon process where it is added at high concentrations to restrict cell growth and enhance interferon production (2,38). Sodium butyrate action is cell line specific and affects the methylation of DNA, and the use of low levels of sodium butyrate (0.075 mM) have been shown to act on the production of antibody from NS0 cells by improving the stability of antibody production and recovering antibody production from cell lines that had lost their ability to make antibody (39). At the molecular level the effects of sodium butyrate on antibody production can be seen in an increase of mRNA within the cultured cells.

In some cases the production of interferon can be increased by manipulating the time and level of exposure to the Sendai virus (40).

Fed Batch. Fed batch is a classic method for increasing product concentration and has been demonstrated to work well. Operation of a fed batch process requires a concentrated medium formulation that contains all the essential medium components required for continued cell growth and protein production. The factors that determine the success of an animal cell based fed batch process are:

1. *Production of a usable medium concentrate.* In animal cell processes the cells in a fed batch culture require a complex mix of feeds materials, unlike microbial fed batch processes, where a single carbon or nitrogen source may be added to enhance biomass production or product formation. Many cell culture medium components are sparingly soluble, and the formulation of feed concentrates is limited by their solubilities.

2. *Osmolality.* Osmolality is essential in the design of such feeds, and nonessential components that increase osmolality should be removed to limit the increase in osmolality that occurs as a result of feeding and will eventually cause cell death.

3. *Mixing.* Care must be taken to ensure that vessels are adequately mixed to prevent the differences in pH and osmolality between feed medium and bulk culture from damaging the cells.

4. *Medium sterilization.* The simplest method of using a feed medium is to develop a medium formulation that is stable and can be filter sterilized several weeks or months prior to use. Addition to the culture vessel should then be made through a second filtration train to protect the integrity of the culture. The medium used as a feed must be capable of being sterilized, and filter sterilization is the favored method because of the instability of some cell culture medium components. Some components may be capable of being heat sterilized, and others may be filter sterilized and some may be supplied presterilized. Precipitation of components in the medium will make filter sterilization impossible even though the precipitate may redissolve and allow good cell growth.

However, the process may not work well with products that are naturally produced by the cell and are controlled by feedback loops. Fed batch and can work in two ways:

- The concentration of product is increased because the cell concentration per unit volume is increased and the high cell concentration results in higher titers of product. More cells make more product for longer than in conventional batch process.

- Cells are grown to the production scale as quickly as possible and then the culture is fed with materials. Such as glucose and amino acids. The materials are fed as concentrates and may be fed semi continuously as bolus feeds or fed continuously to control the cell growth rate. This will be used to increase productivity.

Changing culture methods and culture components may affect the structure and glycosylation of the product. The effects of changes need to be tested and assessed to ensure the product quality is maintained. In the case of a product like Wellferon the subtype composition of the interferon is tested for the presence of all the subtypes and to ensure that extra molecules are not produced. In addition, the individual subtypes are compared and the proportion of product that is present as each subtype is tested for compliance with the standard composition. The quality and quantity of antibodies is easier to test, as the antibody is produced from a single gene and the number of different glycoforms is limited. The protein still needs to conform to a reference standard that will be used to monitor changes in the product over time. Simple fed batch can be operated as in two ways: Grow cell up to the highest concentration using the addition of medium concentrates as discrete bolus feeds or on a constantly increasing rate.

Batch and fed batch are not always suitable for making product, and in these cases perfusion methods may be used. Perfusion is the constant addition and removal of material from a culture system and can be performed in a number of ways depending on the cell culture system. Perfusion works most effectively with an immobilized cell system that will not show extensive cell death at high cell densities but can also be used with suspension cells. The requirements for a perfusion system are more complex than for the simple batch and fed batch processes, and medium needs to be added to and removed from the culture medium held in the vessel on a continual basis once the cells have grown to a level at which the perfusion process is started. In addition, there has to be a method of retaining most of the cells in the system while allowing the outflow of spart medium. A consequence of using perfusion is that the culture vessel may be quite small but the medium hold and harvest vessels may be many times larger than the culture vessel. The use of membrane filters has allowed high-molecular-weight medium components and products of large molecular weight to be retained within the culture system while removing spent medium. Perfusion systems can also be used to make labile molecules because the constant flow of medium out of a perfusion reactor to a chilled harvest vessel can enhance the recovery of active material. In other cases the growth of cells may be slow and the maintenance of cell viability and productivity is most effectively achieved using perfusion. The advantages of perfusion are that it allows a constant removal of product from the culture vessel and can also feed fresh growth media into the system. This will result in high cell densities in the culture vessel while removing toxic metabolites and potentially inhibitory products. Typical separation devices used in perfusion systems would be spin filters and microfiltration devices (41). Apart from the complexity of some perfusion devices, defining a batch is the only other issue which needs to be determined. This promoter is normally set on are basis of the scale of the unit operations in the downstream processing.

Process Control and Optimization

When we develop a process, we look at two types of variables, control variables and response variables. A control variable is a value such as a set point or inoculum density that can be controlled, and the response variable is an output from the system that can be measured and will change as a result of changing a control variable.

Animal cell cultures do not grow quickly and only achieve low biomasses in culture; so they change their environment slowly. Therefore, the control agents

and techniques used to maintain control variables are relatively gentle. We only need weak acids and bases to control pH; the temperature control of the system is achieved using very little heating or cooling, and the agitation and sparging conditions are relatively gentle.

The response of the culture to different control variables may not be symmetrical. Temperature is one example; high temperatures may kill cells, while a low temperature may simply reduce the cell growth rate. The use of statistically designed experiments can highlight which control variables are important and also identify the optimum levels for each factor. Typical values for a fast growth rate and high viability for animal cell cultures would be:

- Temperature: 37 °C
- pH 7.2
- Dissolved oxygen tension (DOT): 10–30%
- Viability >90%

The effect of changes in control variables will be judged in different ways depending on the part of the process that is being investigated. Optimization of growth conditions will be judged on cell growth and product titer for cell lines that continually produce proteins but on cell growth and titer at induction for inducible systems. Many cell lines have a wide optimum for DOT. The role of the fermenter is to enable the cells to do whatever the process requires (grow and produce material for harvest). Of all the culture parameters that have been investigated, the supply of oxygen to the culture is the most critical.

Oxygen Control. Oxygen is sparingly soluble in water (~6 ppm), and its solubility decreases with increasing temperature. If a culture vessel does not have enough oxygen transfer capacity, the cells grown in the vessel will become anoxic and die. The transfer of oxygen to the cells is achieved by allowing oxygen to diffuse from a gaseous phase into the liquid phase. The rate at which the gas is transferred to the cell culture medium is influenced by the following factors:

- The driving force—the difference in concentration between the gas and liquid phases, which is a function several factors and strongly affected by the metabolic rate of the cells and cell density
- The partial pressure of oxygen in the supply stream
- The surface area available for mass transfer
- The gas–liquid contact time
- Liquid side resistance to gas transfer

The methods used to achieve gas transfer attack these issues from different angles:

- Stirred tank reactors with high shear impellers can reduce bubble size and increase the interfacial surface area by shearing larger bubbles to create smaller stable bubbles.
- In smaller vessels the tip speed of the impeller may never be high enough to cause bubble breakage,

and increased gas flow rates are needed to produce increases in oxygen transfer rate.

- Bubble columns and airlift reactors rely on contact time and hydrostatic head to enhance oxygen transfer. Any system can use oxygen-enriched air or top pressure to enhance oxygen transfer, but great care must be taken not to interfere with feed and pH control inlets, especially if flexible tubing is used.

- Producing a small bubble using microsparging or reducing the orifice size in a ring sparger will not necessarily reduce bubble size because the surface tension of the medium and the gas flow rate have a strong effect on bubble size. In many media small bubbles will coalesce to form large bubbles, and channeling tends to occur in sintered spargers.

Mechanical methods of gas dispersion are a safe and well-tried way of increasing oxygen transfer, and a combination of increasing agitator speed and higher gas flow rates is used for microbial systems. The limitation on oxygen transfer in animal cell culture is not the inability to provide enough oxygen—the technology is well established—but the ability to provide the oxygen without damaging cells. Most small-scale cell cultures that are performed in fermenters use a short, wide culture vessel to allow for some gas transfer from the headspace, and the agitation system usually includes some sort of low-shear impeller. Animal cells lack a rigid cell wall and are less resistant to shear stress than bacterial or fungal cells, and because of the concern about cell damage, there is a tendency to be conservative when designing cell culture vessels. There have been a large number of devices that are designed to reduce the shear rate that animal cells are exposed to during culture. There has also been considerable debate about bubble burst and the subsequent cell disruption that this can cause (42). These concerns are real, but cells are regularly grown in large culture systems without significant detectable damage being caused to the cells. There are indications that the cells used in suspension culture may be less sensitive to agitation and bubble burst effects because the development process either selects shear-resistant cells or cells that adapt to growth conditions by strengthening their membranes. It must not be forgotten that cells in their natural environment are constantly undergoing deformation in one of several directions, and local shear forces in the cardiovascular system are significant. The information that we have is limited to a fairly low number of cell lines, including Namalwa, CHO, and NSO, and cell type may play a significant role is shear sensitivity. Animal cell culture can be performed in a variety of vessels and configurations, but the simplest method is to use a stirred tank reactor in either batch or fed batch mode. There has been a lot of concern in the past about the importance of shear stress in the culture of animal cells. However, the oxygen demand of animal cell cultures is much less than that seen in microbial cultures; therefore, the power input needed to achieve dissolved oxygen control is relatively low, with typical values of 11 to 30 W/m^3 for 8000-L cultures

compared to 100 times that value for bacterial cultures. When designing a process it would be safe to assume that there is not a cell culture that is likely to use more than 12 mg/o$_2$ 10^{11} cells/min (43) which gives a range of target K_{la} values ranging from about 1.5 to 15/h. We have found that for large-scale cultures normal cell densities requiring K_{la}s of 3–4/h can be supplied by a simple Rushton turbine and point sparger combination (33). In fed batch culture the use of oxygen-enriched air is sometimes necessary. We have never resorted to using exotic sparger or impeller designs to prevent shear damage. Although there are some scale-related differences, we have confidence that a process that will work in a 1-L commercial vessel will also work in an 8000-L vessel with a minimum of development effort even though the oxygen transfer is dominated by gas flow rate at the 1-L scale and stirrer speed and gas flow rate at the larger scale. This vessel design works equally well with serum-containing and serum- and protein-free medium.

Contamination was a major concern when the vessels were first conceived and designed in the late 1960s. Therefore, the agitation system was designed using magnetic couplings, which eliminate the need for mechanical seals around the impeller shafts. The torque transmission of the system is limited, which places upper limits on the agitation speed and impeller diameter of the existing system but does not present a major problem with our batch process.

The limits of the system are governed by the cell concentration in the fermenter and the oxygen utilization rate of the cells. Namalwa cells can easily achieve cell densities of 5×10^6 cell/mL in normal culture. Genetically engineered cell lines will achieve about 4×10^6 cells/mL in a normal culture. The potential of the culture system to provide oxygen is much higher than that (33).

PRODUCT CONCENTRATION AND PRODUCT RECOVERY

As a general principle product recovery is easier when high concentrations of material are present in the harvest. Mammalian cell cultures do not produce high product concentrations, but there are many advantages to using the mammalian approach. Glycosylated, folded, processed protein can be produced and exported into the cell culture medium and in many cases will be stable until the harvest. With animal cell culture most of the product is secreted into the culture medium and the primary recovery stage can involve physical separation of the dilute product form the cells that produce it. Removal of intact cells by centrifugation or filtration removes a large proportion of protein RNA and DNA. The production of high concentrations of product needs to be balanced against the potential difficulties provided by increased cell lysis.

PROCESS MONITORING

Production processes may run for several years, and over that time the biological, physical, or control elements of the system may change. This is a particular issue with products that are made on a campaign basis. For example, the source of medium components may change or components such as pH probes may be sourced from a different manufacturer. Long-term data monitoring needs to be carried out to check that the production system is working and to highlight any changes in the performance of the processes. This may involve plotting data and manually scanning it to look for trends or using techniques such as control charting to monitor the progress of a whole production system.

Control charting is suited to processes that run for long periods. Short campaigns are less suited to this approach, as 20 to 30 harvests may be needed to give a good baseline.

Safety Issues

Safety of biotechnology products is assured by the use of well-characterized, monitored processes and confirmed by the extensive testing of materials at several stages of the production process. TSE contamination, virus contamination, residual DNA, endotoxin levels, and sterility of the product are all concerns.

Production of Seed Cultures for the Production Vessels

There are two approaches to biotechnology production: one vial one batch and one vial many batches. Both approaches are valid, but the one-vial one-batch approach requires a series of scaleup trains to ensure a regular provision of cell to the production vessels. The approach that is best suited to long-term high-throughput cultures is the one-revival, many-batches approach. It makes more efficient use of resources and allows a smaller scaleup train to be operated. One advantage of the one-vial many-batches approach is the reduction in the number of small-scale operations that are performed. Activities such as cell revival and routine tissue culture are more likely to result in contamination than large-scale operations, where all the system can be steam sterilized and engineering controls maintain aseptic conditions.

Scheduling the revivals needed to run 7–10 batches a week is a complicated process, and the segregation of a number of cell revivals during scaleup requires a significant resource commitment. Frequent revivals soon deplete cell banks.

One revival one batch does not have to involve any more complexity than the one-vial many-batch approach: Use bigger vessels and perform fewer production runs. Rather than harvest 10 vessels a week, a single 100,000-L vessel could provide exactly the same harvest volume. There are a number of reasons that large vessels (>10,000 L) are only rarely for animal cell culture. The most important ones are:

- The capital implications of losing large batches and the lack of experience of running large-scale animal cell culture processes.
- Downstream processing often involves affinity processes, and the gels used in these processes can be expensive to make.

- If the harvest from a large vessel had to be processed in one batch or within a short time scale, there would be significant cost implications for the downstream processing stages of the process.

OUTER LIMITS VALIDATION

Once a product has been defined and the process is nearly finalized the outer limits of the process need to be validated and tested to determine the stability of the process, the effects of excursions from the control variables on product quality and quantity. The WCB cell stock needs to be used for this work. The process involves the changing of control variables (e.g., pH, dissolved oxygen tension, temperature, and stirrer speed) and measuring the effects on response variables (cell concentration, viability, growth rate, glucose utilization, product quantity, and quality). There are a large number of combinations of factors to be considered, and a statistical approach to determining the important interactions will reduce the time and effort needed to define the outer limits and identify the major factors.

PROCESS VALIDATION AND CONSISTENCY BATCHES

The final stage in the development process is the checking of the full-scale process and confirmation that all stages of the process can run consistently. This would typically consist of demonstrating that the cell revivals, culture, and purification sections could all run consistently for three operations (subcultures, production runs, etc.). Acceptance criteria that are based on the development campaigns and outer limits validation are used to confirm that the process is running as expected.

PRODUCT STABILITY

Product stability has implications throughout the production process. Stability is essential for high-titer production and survival of the product through the purification process. The lag time between production and final product release can be several months, and the stability of product must be assured.

SUMMARY

The technology exists to allow the large-scale production of immune regulators. The majority of immune regulators can be made using animal cell culture to provide a pure product with well-defined characteristics. The culture process and the cell line chosen to make product determine the quality of the product. Processes can be scaled up and run at large volume using simple suspension culture processes. The control of cell lines and cloning and selection of the lines is central to process development and making a product. The development of serum-free media helps to control the risk to the process from external sources and makes the process more reproducible. The future of immune regulation may lay in the *in situ* production of immune regulators by gene therapy. This will allow the targeting of the proteins to cell masses and cell types where they are needed and hopefully reduce the side effects associated with the effects of systemic treatment.

BIBLIOGRAPHY

1. A. Isaacs and J. Lindenmann, *Proc. roy. Soc. B.* **147**, 258.
2. Eur. Pat. 0097353 1988, 10–25, M.D. Johnston.
3. W.K. O'Neal et al., *Hum. Gene Ther.* **9**, 1587–1598 (1998).
4. G.H. Guibinga et al., *Virol.* **72**, 4601–4609 (1998).
5. Z. Abdel-Wahab et al., *Cancer* **80**, 401–412 (1997).
6. L. Chen et al., *Cell* **71**, 1093–1102 (1992).
7. T. Lemig et al., *Hum. Gene Ther.* **7**, 1233–1239 (1996).
8. M.J. Parr et al., *Neurovirol.* **4**, 194–203 (1998).
9. P. Mulders et al., *J. Immunother.* **21**, 170–180 (1998).
10. P.N. Boyd, A.C. Lines, and A.K. Patel, *Mol. Immunol.* **32**, 1311–1318 (1997).
11. J.P. Kunkel, D.C.H. Jan, J.C. Jamieson, and M. Butler, *J. Biotechnol.* **62**, 55–71 (1998).
12. J. Mordenti, J.A. Cavegnaro, and J.D. Green, *Pharm. Res.* **13**, 1427–1437 (1996).
13. R.L. Levings and S.J. Wessman, *Dev. Biol. Stand.* **75**, 177–181 (1990).
14. S.J. Wessaman and R.L. Levings, *Cytotechnology* **28**, 43–48 (1998).
15. A. Hawerkamp et al., *Cytotechnology* **28**, 19–29 (1998).
16. S.J. Froud et al., in M.J.T. Carrando, J.P. Griffiths, and J.L.P. Moreira, eds., *Animal Cell Technology: From Vaccines to Genetic Medicine*, Kluwer Academic Publishers, Dordrecht, The Netherlands, 1997, pp. 681–686.
17. J.C. Pettricciani, *Cytotechnology* **28**, 49–52 (1998).
18. S.E. Kleiman, I. Pastan, J.M. Puck, and M.M. Gottesman, *Gene Ther.* **5**, 671–676 (1998).
19. A.W. Nienow et al., *Cytotechnology* **22**, 87–94 (1996).
20. C. Langheinrich et al., *Trans I. Chem. E.* **76** (Part C), 107–116 (1998).
21. R. Leitzke and K. Unsicker, *J. Immunol Methods* **76**, 223–228 (1995).
22. K. Wawrzer and D. Suilholmer, *J. Immunol. Methods* **179**, 71–76 (1995).
23. J. Raper, Y. Douglas, N. Gordon-Walker, and C.A. Caulcott, in R.E. Spier, J.P. Griffiths, and C MacDonald, eds., *Animal Cell Technology: Developments, Processes and Products*, Butterwoth-Heinemann, Oxford, U.K., 1992, pp. 51–53.
24. W.M. Miller, H.W. Blanch, and C.R. Wilke *Biotechnol. Bioeng.* **32**, 947–965 (1988).
25. O.T. Ramirez and R. Muthurasan, *Biotechnol. Bioeng.* **36**, 839–848 (1990).
26. D.K. Robinson and K.W. Memmert, *Biotechnol. Bioeng.* **38**, 972–976 (1991).
27. A. Handa, Ph.D. Thesis, University of Birmingham, U.K., 1986.
28. N. Kioukia, A.W. Nienow, A.N. Emery, and M. Al Rubeai, *Trans I. Chem. E.* **70** (Part C), 143–148 (1992).
29. D.R. Gray et al., *Cytotechnology* **22**, 65–78 (1996).
30. R. Kimura and W.M. Miller, *Biotechnol. Bioeng.* **52**, 152–160 (1996).
31. S.S. Ozturk and B.O. Palsson, *Biotechnol. Bioeng.* **37**, 989–993 (1991).
32. Japanese Patent 81600/81 1981 05 28, Tamani et al.
33. T.M. Clayton, I. Jenkins, and P.J. Steward, *16th ESACT Meet.*, Lugano, 1999.
34. R.T. Hatch, Ph.D. Thesis, MIT, Cambridge, Mass 1973.
35. M. Moresi, *Biotechnol. Bioeng.* **23**, 2537–2560 (1981).
36. L.R. Jackson, L.J. Trudel, J.G. Fox, and N.S. Lipman *J. Immunol. Methods* **189**, 217–231 (1995).
37. S.C. Musgrave et al., R.E. Spier, J.P. Griffiths, and C MacDonald, eds., *Animal Cell Technology: Developments, Processes and Products*, Butterwoth-Heinemann, Oxford, U.K., 1992, pp. 189–191.
38. S.C. Musgrave et al., in R.E. Spier, J.P. Griffiths, and C MacDonald, eds., *Animal Cell Technology: Developments, Processes and Products*, Butterworth-Heinemann, Oxford, U.K., 1992, pp. 89–91.
39. S. Islam, Presented at *ESACT U.K.*, Leeds 1999.
40. S.C. Musgrave et al., in R.E. Spier, J.P. Griffiths, and C. MacDonald, eds., *Animal Cell Technology: Developments, Processes and Products*, Butterwoth-Heinemann, Oxford, U.K., 1992, pp. 92–94.
41. S.M. Woodside, B.D. Bowen, and J.M. Piret, *Cytotechnology* **28**, 163–175 (1998).
42. A. Handa-Corrigan, A.N. Emery, and R.E. Spier, *Enz. micro. Technol.* **11**, 230–235 (1989).
43. R.E. Spier and J.B. Griffiths, *Dev. Biol. Stand.* **55**, 81–92 (1982).

See also ANIMAL CELL PRODUCTS, OVERVIEW; CELL PRODUCTS — ANTIBODIES; CELL STABILITY, ANIMAL; ENRICHMENT AND ISOLATION TECHNIQUES FOR ANIMAL CELL TYPES; PLANT CELL CULTURES, SECONDARY PRODUCT ACCUMULATION; TRANSFORMATION OF PLANTS.

CELL PRODUCTS — VIRAL GENE THERAPY VECTORS

TIM CLAYTON
Glaxo Wellcome
Beckenham, Kent
United Kingdom

OUTLINE

INTRODUCTION

Gene therapy is the use of a vector to transfer a gene or group of genes to a cell where the nucleic acids will have a beneficial effect on the patient either by being expressed themselves or by affecting the expression of other genes. Gene therapy is a rapidly expanding field that will significantly improve the treatment of many disorders by allowing the insertion of therapeutic genes directly into cells. The genes may act in a variety of ways, and gene therapy has the potential for treating tumors, infectious diseases, and inherited disease. The aim of this article is to:

- Give some background on gene therapy to place vector production techniques in context
- Outline general strategies for making vectors with specific examples of equipment and techniques where appropriate

WHAT IS A GENE THERAPY VECTOR?

A gene therapy vector can be:

- Naked nucleic acid
- Nucleic acid combined with other agents such as lipids and proteins
- A modified virus containing therapeutic genes

Gene therapy vectors are varied, and a great deal of effort is invested in designing the right combination of components. Regardless of the different starting points, the objective of gene therapies (the expression and action of therapeutic genes in the patient) There are many common features of these therapies. These common features are outlined below.

Vector Core

The core of a gene therapy vector is nucleic acid (DNA or RNA). This is the active component of the vector and may be the only component in the case of plasmid DNA. The construct has to have sufficient information to allow the expression of the gene in a human cell. This will include a promoter sequence (e.g., CMV, CEA). Strong, nonspecific promoters may be used in vectors where there is no concern about target tissue specificity. Where gene expression should only occur in a particular tissue or when tightly controlled gene expression is required, a tissue specific promoter may be used to limit the site of action of the gene or a switchable promoter may be used to allow conditional operation of the gene. A nucleic acid sequence that allows the target cell to produce an active product is also required. In many cases a polyadenylation sequence is also required to help gene expression. Selection markers are used to help select transformed *E. coli* in which initial molecular biology is performed. When vectors are produced in mammalian cells, a second selection system may be used to maintain the production of vector.

For safety reasons the vector should not be capable of replication outside the producer cell system (unless the vector is conditionally replication competent and can only reproduce in specific target cells). Vectors are crippled by removal of a gene sequence(s) essential for independent replication, which is then supplied in *trans* by a producer cell line. The greater the number of genes removed, the lower the chance of producing a replication-competent entity. Expression of viral genes in the host cell can lead to cell damage, and toxic viral proteins may need to be produced transiently in the production phase using an induction technique.

The selection for producer cell lines is linked to their ability to produce the complementary genes necessary for viral replication. In at least one case the complementary viral sequence Adenoviral E1 genes also supplies the selection pressure by allowing the host cell to replicate (1).

Vector Packaging

In its most basic form the vector core is a naked plasmid that can be injected directly into a patient. Viral gene therapy material is packaged during production in the cell. At this point it takes on the characteristics of the parental virus by exhibiting the receptors and antigens that will give it a tropism for the targets of the parental virus. Liposomes can be used to coat naked DNA to enhance absorption and expression of the DNA. There is enormous potential for the modification of nucleic acid packages to change the specificity of the package to different cell types. In addition, it may be possible to bypass immune responses to viral vectors by using modified surface antigens. There are three basic approaches to vector production and delivery, which are production of virus vectors, DNA plasmids, and facilitated delivery of DNA plasmid (liposomes, etc.) (2). The majority of clinical trials are being carried out with viral vectors, and this article will deal with the production of viral vectors.

WHY USE GENE THERAPY?

There are a number of disorders that cannot be treated successfully by traditional pharmaceutical products. The area where gene therapy was first identified as being useful was the inherited disease area where the potential to replace defective gene function and repair a disorder at the molecular level rather than treat the symptoms of the disease was immensely attractive.

Gene therapy is an exciting area that is full of potential and utilizes our rapidly advancing knowledge of biology, the potential uses of gene therapy are varied. In some disorders cell function is abnormal because of a mutated gene, and gene therapy can be used to replace the functional protein by transferring the ability to make functional protein into target tissue. Many gene therapies are targeted to specific cells and tissues either by delivery method, vector choice, vector design (3), or tissue-specific expression (4); in this way the defective gene function can be replaced in appropriate tissue. Treatment of disorders such as cancer can be achieved because highly active or toxic proteins can be produced in situ by the patient's own cells, and the effects of gene action can be highly localized. Products that can be produced in situ include ribosymes and proteins. Gene therapy is therefore an attractive approach for the introduction of molecules like cytokines, which are very active and may be toxic to producer cell lines (5). As with many new biotechnologies, the potential of these approaches is huge.

The challenge of the next few years is to try to meet some of the expectations that have been raised by this technology. Production of useable quantities of pure, potent vector is only part of the battle. Gene therapy agents are only useful if they can be delivered to the correct target in sufficient quantities to cause a therapeutic effect, and in the case of tumors this may only need to be a small proportion of the target tissue (6). For disorders such as cystic fibrosis that are caused by defective genes, therapies will only succeed if the vectors can integrate and give very long-term expression or can be used for repeated dosing over a period of years. It is also essential that the disorder is identified at an early stage to prevent cumulative damage caused by lack of treatment in the first few months of life.

GENE THERAPY TARGETS

Gene therapy products are being tested for their ability to treat diseases in the following areas:

1. *Metabolic disorders.* Metabolic disorders may be treated by replacement of an essential protein necessary for the normal operation of the body. These proteins include Cystic fibrosis transmembrane regulator (CFTR) and factor VIII. Noncirculating proteins such as CFTR must be expressed in a group of target cells to be effective. The target tissue may not be the normal site of action of the gene; for instance, the expression of factor VIII in skeletal muscle may be the best method of achieving a long-term stable expression and concentration of the product.

2. *Cancer treatment.* Several approaches have been suggested for cancer treatment. Approaches that do not require the infection of all the tumor cells are likely to be effective. Therapies that have a dual effect, for instance, cell death and a raised immune response to the remaining cells, have the best chance of success and include the following:

   - Replacement of components of the immune system to regulate the response to tumor cells (7)
   - Suicide genes/enzyme prodrug treatment (8,9)

3. *Vascular disease* (4). Biological bypass therapies are being investigated to bypass blocked arteries in the limbs and heart using genes that allow the expression of vascular endothelial growth factor (VEGF).

4. *Arthritis* (10). Local modification of inflammatory responses. Gene therapy cannot be used to modify the germ cells of patients because of the moral or ethical issues surrounding the concept of genetically "improving" humans and, in many countries, specific legislation that forbids such developments. Many gene therapy vectors can only be maintained and expressed transiently, but the distribution of the vectors within the body has still to be established and the rate of clearance of vector checked to ensure that the vector has been completely eliminated form the patient. There is also a concern that random integration of DNA into human chromosomes may cause tumors.

Viral Gene Therapy Vectors

Viral vectors are a popular method of targeting gene therapies because of the inherent properties of viruses. Evolution has equipped viruses to invade and exploit the biochemistry of cells. They have evolved a range of strategies to allow survival. These range from rapid growth and production of huge numbers of virus, resulting in the destruction of the host cell, to long-term maintenance of the viral genome in the host cell by integration of the genome into the host DNA, as is seen in the retroviruses. Viruses also vary in size, hardiness, and carrying capacity for foreign nucleic acid. It is this range that makes viruses so useful. In addition, the natural tropism of viruses for particular tissues can be a great advantage in directing therapies to target tissues. However, cells respond to viral invasion by signalling to the immune system, and some viruses are more easily recognized by the immune system than others. The residual medium components and cell debris that are also present in viral harvests can be immunogens in their own right.

Successful use of viruses for gene therapy will require strategies to overcome some significant problems. The most significant of these are the immune response (11,12), replication-competent virus (RCV) (13), access to the correct tissue, and the need to grow the virus in animal cell culture. By selecting the correct production system and modifying vectors, many of these issues will be overcome. The ideal gene therapy vector should be stable, easy to make in large quantities, easy to purify, specific to target cells, nonimmunogenic, and nontoxic, should allow the appropriate expression of the therapeutic gene, and should not produce RCV. Other factors governing the acceptability of a vector are target disease, target cells, length of expression, mode of action, stability, immune response, mode of administration, ease of downstream processing, required purity, integration of the genome into target cells, and long-term patient risks.

Commonly Used Viral Gene Therapy Vectors. The two vectors that account for more research than any others are retrovirus and adenovirus. The work has been performed by using variations on the themes of moloney murine leukaemia virus (MoMULV) and Adenovirus type 5. These two viruses illustrate the differences between the production and use of viral vector:

- Retrovirus is a fragile virus that is produced continuously in producer cell lines and is difficult to produce in large quantities. Retrovirus is valued for the potential for integration and expression and relatively low immunogenicity.
- Adenovirus is produced by seeding cells with virus and harvesting large quantities of virus after a relatively short time. The chances of an immune response to adenoviruses is larger than would occur with retroviruses.

There is potential for any virus to be used as a gene therapy vector. The limitations are the size of gene that can be inserted into the viral package and the degree of safety that can be built into the system. A further issue is public perception of the vector. Several groups are interested in using HIV as a basis for gene therapy vectors because of the ability of the virus to infect and integrate into nondividing cells. The consequences of reversion to wild type or the production of a novel virus are not issues that can be dismissed and are likely to continually reemerge.

Replication-Competent Virus. The ability of viral vectors to recombine with other viruses or segments of viral genome that may be integrated in the producer cell line or in the patient is a major concern. The issues are:

- Production of novel viruses with wider host specificity
- Production of wild-type virus
- Transfer of the gene therapy gene to the general population
- Tumor production
- Loss of product efficacy

The production of RCV will be minimized if the opportunity for recombination between the vector and other components necessary to make a wild-type virus are reduced by limiting the overlap between complementary genetic segments.

The risk of transferring the therapeutic gene to an RCV is reduced by using a virus that cannot carry more than its normal size genome. Therefore, when an adenoviral E1 segment is reattached to the rest of the viral genome, the therapeutic gene that is inserted in the E1 position cannot be packaged and used in an RCV.

PRODUCT QUALITY AND SAFETY

The pharmaceutical industry and the biotechnology industry are both heavily regulated, and a number of regulatory bodies can significantly affect the success of a gene therapy product. Safety of the patient, environment, and process workers must always be considered in the development and operation of a process. There are also many patents that may limit the options available for constructing a new vector. The following points should be kept in mind:

- It is not worth developing a sophisticated gene therapy vector and production system if it cannot be used because it will not be approved by the regulatory authorities.
- It is also not worth developing a system based around the intellectual property of another organization unless the owners agree that the property can be used.

Other agencies that may have an impact on the success of a product are agriculture departments, agencies concerned with the control of genetically modified organisms, and agencies concerned with general safety.

Drug Regulators

Drug regulators have a duty to ensure that a product that is licensed for use in their country is fit for use, and this quite clearly sets the agenda for safety and quality testing during drug development and production. Different countries may require different information about products, and the process for performing clinical trials and marketing product varies from country to country. The international committee on harmonization (ICH) has produced guidelines that should ensure that the procedures followed in developing processes and making therapeutic product are acceptable to regulators in most countries. When developing a gene therapy product keep the following in mind:

- The primary concern with a pharmaceutical product is safety, quality, and efficacy. Establish proof of principle at an early stage and confirm the mechanism of action of the vector. It is essential to consider the whole package with a gene therapy vector, and delivery systems are critical to the success of the product. "Would you be happy for this product to be used on yourself or your nearest and dearest relative?" If not, why would an agency responsible for protecting the population of a whole country be willing to allow you to use the product in their jurisdiction?

- Prepare for the regulatory authority that will deal with trials and final product release.
- Do not assume that all regulatory bodies have the same requirements or attitudes towards biotechnology products.
- Be aware of the requirements for each market.
- Information gathered at an early stage of the development cycle will be essential to final licensure and product release.

As new gene therapy products are introduced, they tend to be dealt with on a case-by-case basis, and it is advisable to consult with the regulators at an early stage to ensure that they are fully aware of what you plan to do. There are published guidelines that are not only comprehensive but also very easy to read and informative. Further information on testing and quality issues is frequently available from contract testing houses such as Covance (14) and Q1 Biotech (15). Your process starts with the cell line and virus developed at what may be an early stage in development. It is essential to keep an audit trail (data recorded in retrievable files) for the production of the producer cell line and records of medium and testing carried out at this early stage. The main concern is that the cell line is free of adventitious agents. Therefore, the serum used in cell growth should be from an approved source and tested for the presence of extraneous agents that could contaminate the cell line and hence the final product. Exhaustive records of cell line history, medium sources, and testing results are essential.

QUALITY

A concern with quality is essential to ensure meaningful results can be obtained with any gene therapy vector production system, and this concern starts in research. The philosophy with any pharmaceutical product is that quality starts with the raw materials, the operational facility, and appropriates trained personnel. This approach is particularly vital when dealing with biotech products such as viral vectors, where a low-level contamination in therapeutic product could have an effect out of all proportion to its quantity. Therefore, appropriate levels of testing must be carried out on raw materials and product.

A production process can be considered to have three components: raw materials, process equipment and operating procedures. The other activities in a process revolve around this core.

Raw Materials Supply and Testing

In this case the raw materials for a process include medium components and the biological components. The comments in this section assume that the materials will at some time be used for clinical purposes but the comments are equally valid for a non clinical purpose. Cell lines should produce what they are supposed to, not a contaminating virus.

Materials which will be used in a good manufacturing practice (GMP) environment (including cell lines should be supplied with a clear, well documented audit trail detailing testing, medium batches and origin of raw materials. Potential sources of contamination should be identified and the suppliers audited to ensure that their production methods and record system meet a standard that will satisfy the licencing authorities. To confirm the quality of the raw material appropriate testing should be carried out to check that the material matches the specification in the raw materials specification (RMS).

Medium Components

Viral gene therapy processes should be free of extraneous virus. This includes endogenous retroviruses and RCV, which may already be present in the culture or virus seed stock. Cell and virus stocks are tested for the presence of these contaminants, and they must be confirmed to be free of contaminants before they are used for production. Cell culture medium components are a major potential source of viral contamination, as many of them must be filter sterilized rather than autoclaved.

By their nature gene therapy products are not amenable to downstream processing steps that will remove or inactivate viruses or transmissible spongiform encephalopathies (TSE). To build quality into the process an active virus avoidance policy should be pursued. This will have three components:

Testing strategies
Virus avoidance strategies
Virus barrier/inactivation strategies

Potential sources of virus contamination include medium components of animal origin, contamination in storage or during transit, and operator contamination. Testing for contaminants must therefore take place at an early stage in the development process. A strategy of eliminating the source of contamination is always preferable to testing. A test can determine whether something (what you are looking for) is detectable or not. There are two dangers: Very sensitive tests can detect small quantities of noninfective material; less sensitive tests may miss a low level of infective material. Because they can replicate during culture, levels of virus below the levels of detection can be amplified during cell culture. In addition, the concern with TSEs limits the sources of any animal-derived products to approved suppliers.

Wherever possible, the use of animal-derived products should be avoided to reduce the risk of viral and TSE infection of the culture. When animal products must be used, the products should be sourced from suppliers and countries that are acceptable to the regulatory authorities. Audit trails for materials and validation data for virus removal or destruction steps such as filtration and irradiation are also essential.

Most gene therapy vectors will be injected into patients, and the use of protein-containing material in the medium should be as low as possible to reduce the risk of immune response to foreign proteins. Retoviruses and other encapsulated viruses take a proportion of the host cell membrane with them, and any material that has been absorbed into the cell membrane from the culture medium will remain in the product.

Serum-Requiring Processes

There are situations where a process requires serum: There might not be enough time to find alternatives, or the cell line may refuse to grow without serum. When faced with this situation using a retroviral producer line, we used serum at normal concentrations in the growth phase but then identified the lowest serum concentration that could be used in the production medium. We found that if the cells were initially attached and grown in a medium containing 5% foetal calf serum (FCS) and the medium was then exchanged for our proprietary serum-free medium supplemented with 0.5% serum, we could obtain consistent growth and production of virus in the Costar Cell Cube (16). The very low protein content of the medium also helped the downstream processing of the virus.

Cells and Virus

Where a choice exists the best policy for production of cell lines and virus is to set up and maintain a cGMP stock from which vectors destined to make clinical material can be derived. This approach should be easy with cell lines that are used to produce virus by infection, and some companies supply well-defined cell stock from their cGMP cell banks as part of the licensing agreement. These cells can then be transfected with naked DNA to produce the first virus stock for cGMP use. Cell lines that can be grown in serum-free and preferably protein-free medium should always be selected if there is a choice in the matter. Cell lines that reduce the chance of producing replication-competent virus should be used. Virus production systems contain an enormous number of cells. A 1-liter culture containing 10^6 cells/mL will contain 10^9 cells. If there is a 1 in 10^8 chance of RCV production, then it is likely that 10 cells could produce RCV and may cause the loss of a batch. The testing regimes used will not give results on RCV quickly enough to cause an early abandonment of the run or purification process. Assuming the amount of fermentation harvest required to produce a clinical dose is in the range of 100 mL to 1 L, it becomes obvious that the production of RCV is a major issue. Production of virus for clinical purposes may require the production of hundreds or thousands of liters of material from cell culture. There would not be any point in developing such a process if every batch was ruined by RCV production. As the volumes and consequent cell densities increase, the likelihood of RCV production increase and the process risk increases accordingly.

INTELLECTUAL PROPERTY

A major driver in the gene therapy field is intellectual property. Many companies are based on patents that allow then to use subsets of therapies for a limited number of purposes. This has led to the investigation of a number of vectors that may not otherwise be considered. To make a successful gene therapy vector, it may be necessary to licence technologies from several different companies. It is therefore essential that the final system is one that will allow commercial production and therapeutic success.

It is important to review the literature to check the patent status of therapeutic genes and other components of expression and delivery systems.

Therapeutic Areas

The major barriers to successful gene therapy are access, immune responses to both vector-expressed protein and infected cell, and RCV. Production methods, testing regimes, vector choice, and quantities of vector required are all dependent on the target indications for the product. The nature of the product limits the diseases that can be treated.

Two major constraints on the application of gene therapy to diseases are:

- Delivery and targeting
- Production of pure, active vector

These factors ensure that the initial uses of gene therapy will be for treatments that involve small areas that can be physically targeted by vector or the target cells must produce soluble proteins that can act remotely from the site of production.

Vector Delivery. Why don't we design viral vectors that are injected into the bloodstream, go to every cell in the body, and then are switched on in the target cells? This is because not all cells have equal access to blood vessels and virus has to reach the target cells by physical means. If the perfect vector existed, the laws of physics would limit the access of the vector to the target cells because diffusion and the size of the virus limit access to some cells. Systemic delivery of viral vector to every cell in the body is unlikely with the present state of technology unless germ cells or very early stage embryos are transformed with vector. The access of virus to cells in the body is also limited by the sheer number of cells in an average human body ($\sim 10^{13}$), and the physical difficulty of getting to the cells from the bloodstream. Therapies aimed to access all the cells in the patient would involve large quantities of virus and sophisticated delivery system.

Localized Delivery of Vector. Many diseases are amenable to localized treatment that does not rely on total transformation of all affected cells. Vector can be administered locally by injection or perfusion of organs. The approach used to deliver the vector will be determined by the target size and distribution and the concentration of virus in the dose. Local delivery works because of the ability of biological systems to respond to low levels of product and because there may be responses associated with the direct effects of gene therapy. In addition, there are single organs that are the major cause of death and disability in systemic diseases. The correction of a genetic fault in this organ may produce a benefit out of all proportion to the number of cells infected with vector.

- In cancer treatment the strategies used rely on local killing effects supported by a more general immune response.

- In cystic fibrosis many cells are affected, but the lungs present a well-defined target, and there are major benefits to be gained from transfecting cells with the CFTR gene.
- In many cases levels of protein production may be lower than the natural levels found in most of the population and still provide a significant benefit. Conventional treatment of haemophilia cannot provide a constant level of protection from damage as factor VIII is given after bleeding has started. Gene therapy could be used to transfer the ability to produce factor VIII to skeletal muscle cells. Therefore, local production of factor VIII in muscle cells could provide a significant systemic protection from damage and a huge improvement in quality of life.

Pure, Active Vector. It is wise to choose viruses that will not be unduly dangerous if they produce RCV in the patient. In patients who are immune suppressed such as cancer patients and arthritis patients the production of a normally harmless virus in an inappropriate part of the body could be extremely dangerous. With adenoviral vectors, for instance, the vectors are produced from a known human pathogen.

Many gene therapy treatments will require repeat dosing, and it is essential to avoid actions that will cause an immune response to the vector or the gene product made by the vector. The vector should be as pure as possible with as little inactive virus present as possible to avoid interference from inactive virus and viral subunits because the viral attachment process is dependent on the interaction between virus and cell surface proteins. It is possible that receptor sites can become saturated with noninfective material and prevent infection.

The presence of noninfective virus and host cell components has a much lower effect on production of an immune response than the infection of cells with adenovirus (17,18). The production of antibodies to a vector may also be determined by the site of administration with sites like the central nervous system (CNS) being less likely to produce an immune response than other parts of the body (19).

PROCESS DEVELOPMENT STRATEGY

Introduction

The quantity of virus required to perform initial animal and clinical studies must be realistically assessed. Vector production lines can then be developed and tested to determine if sufficient material can be made and the scale of development and production operations can be estimated. A retroviral producer cell line will produce between 1 and 10 infective viruses per cell on a two-day harvest cycle (based on a 400 stack cell cube equivalent in area to 400×850 cm^2 roller bottles, containing 2×10^5 cells/cm^2 and a titer of 1 to 10×10^6 infective units per ml in a harvest). Perfusion techniques should enable an increase in specific productivity by reducing the number of viruses inactivated during the culture process. Vector has

to be recovered from the culture medium as it is secreted by the cells. This means that the purification process needs to be capable of recovering virus from a dilute suspension. Adenoviruses can produce virus titers in the region of 1×10^4 viruses per cell. However, the recovery of these viruses is more difficult because they are typically recovered from a cell lysate, and there is considerable contamination with host cell protein and nucleic acids.

Interaction and Integration

When developing a production system it is best to use an integrated approach with an active dialogue between all areas of the organization. Process development starts in research and progresses to production. Research decisions profoundly influence the process through the choice of vector, producer cell line, and the history of these process components. Therefore, the choice of vector and producer cell line should be carefully considered prior to starting a project. Figure 1 shows that there are a large number of factors that should be considered when choosing a production cell line and virus.

Timing and Planning Issues

Once patents have been issued, the clock starts on pharmaceutical development projects. This is not to say that the development work has only just begun but that the time in which the company can recover its investment and make a profit is limited by its patent protection. Developing a manufacturing process and testing a product to ensure that it is safe for use in humans is an expensive and time-consuming process and must be dealt with in an efficient and timely manner. Evidence for the efficacy of the product should be gathered quickly to ensure that if the product is not viable, the project can be stopped as soon as possible. Assuming that the product works, the production process should be developed as soon as possible. The process must work, be economically viable, and be acceptable to the regulators in the target countries. Guidance on the range of requirements for different products can be found in guidance documents issued by the regulators. Because the products being produced are a cocktail of viable virus, nonviable virus, and in some cases virus-producing cells, a high degree of assurance and quality needs to be built into the process during the development phase. There are many reasons to develop the product as quickly and efficiently as possible:

- Improve patient care
- Does it work?
- Maximize market share
- Recover costs

In some ways biotechnology products are unique. We know they work and have biological activity in cell-based systems, but there are a limited number of animal systems that can be used for testing the product. Elegant molecular biology will not help if the therapeutic agent cannot be delivered to the correct site in the correct quantities. Until dosing regimes are determined in phase II clinical trials the final scale of production cannot be finalized.

For a GMP-compliant product and process the development of a commercially viable process requires a multiphase development plan to be implemented. Process development for a gene therapy vector can be summarized as follows:

1. Determine the type of vector.
2. Identify the vector producer cell combination.
3. Assay development.
4. Modify an existing generic system if it is already available in house.
5. Develop a new generic system using whatever materials are available and refine as the final vector becomes available.
6. Carry out cell and virus banking in parallel, including selection and stability studies. Use existing in-house systems for production and optimization and check that the medium is the most effective, the system contains the correct construct and highest titer, and retains activity.
7. When the process is developed and final constructs are available, optimize and determine the outer limits of the process and perform process validation at production scale.
8. Determine scale and titer needed for work in toxicology, clinical, etc.
9. Develop DSP and assays.
10. Continually review regulations and the relationship of the project to other projects.

VECTOR PRODUCTION

Aims of Vector Production

The aim of vector production is to make sufficiently large quantities of vector of a quality that enables it to be used as a therapeutic product. The system should follow these basic principles:

1. The GMP audit trail starts early in the process, and quality is as much of an issue in the research phase of producing the vector as it is in the final production process.
2. Robust, reproducible and scaleable.
3. Where possible dependence on proprietary equipment for critical process steps should be avoided.
4. Designed for quality.
5. Clean-every step possible should be taken to ensure that the starting material introduces as few non essential components into the process as possible.
6. Segregation-Cell lines and virus should be kept in strictly defined areas to prevent cross contamination of stock (Fig. 2)

Vector Production Methods

Detailed methodologies of production processes vary enormously, but the basics are simple and are determined by the cell type and virus. The basic requirements are a cell line, a virus vector, and a culture medium with which to develop the process. Any virus has the potential to be used as a viral vector, but the basic process of vector production can follow one of two routes (Fig. 3).

- Some cells are engineered to produce viral vectors. Lines that constitutively produce vector can be used in batch, semicontinuous, and continuous production systems.
- Other vectors depend on a complementing cell line to allow vector production. The complementing cell lines are used to allow production of virus, which is replication incompetent and not produced continuously in cell lines. This means that two master and production banks need to be made and characterized. Cell banks need only be laid down as small volumes and titers, but viral banks need to be provided in sufficient quantity to achieve the desired multiplicity of infection (MOI) without many rounds of replication. At this stage the virus banks must be free of RCV if the final product is to be free of RCV.

Assuming that an adenovirus can produce 10,000 new infective particles per cell, the virus needed to seed a 100-L vessel could be prepared from 0.1 L of culture. This does not introduce any insurmountable obstacles, but the strategy for production of virus seed must be addressed early in the development phase and all issues must be resolved prior to registration of a product. Low-titer virus-producing systems may need to be infected with a low MOI. The virus can then go through at least one round of replication in the vessel prior to harvest.

Production and Storage of Virus Seed

Producing a virus stock is not a trivial exercise. The simplest form of virus seed is a frozen suspension of lysed producer cells containing virus. Although this method is crude and does not produce a pure virus suspension, it is an efficient method for seeding culture vessels. Virus seed must be prepared from a large cell culture, and sufficient seed must be prepared for the process to be operated at the final scale and MOI; any additional purification stages will reduce the quantity of virus that is available for seeding the vessel. The virus seed must be of sufficiently high quality for it to be used in a production process, and testing procedures must be factored into time lines, as some of the tests will take a long time. Stability of virus in storage is an important factor. Retrovirus has shown itself to be temperature sensitive (20–25) and badly affected by freezing thaw cycles; it is more stable at 4 and $-70\,^{\circ}$C than at $-20\,^{\circ}$C, and in our hands and factors that affect the speed of freezing and thawing the virus stock were shown to have severe effects on viral titers. The following factors need to be investigated to verify that the virus seed stock and the storage conditions are suitable for production use:

- Freezing regimes
- Testing of the virus seed for contamination
- Presence of replication-competent virus
- Storage optimization and validation of storage facilities

- Stability of virus during repeated rounds of propagation
- Stability of virus during storage

If a virus seed stock is used during the production of a viral gene therapy vector, it is essential that the virus stock can be produced and stored in a manner that enables the virus to be revived and used to seed the production culture. The effect of storage volume and revival techniques on virus recovery will affect the way in which the virus can be used. If storage at high densities or larger volumes causes a loss of infectivity, the scale of operation will be affected, and the process must be designed to allow for these differences.

Production and Storage of Cell Banks

The first stage in setting up a cell bank is the cloning of the producer cell line. There are three reasons for using a clonal cell line:

1. Cloning is carried out because when mixed populations of cells are used in culture, the growth conditions tend to allow one of the clones to predominate. This is frequently a clone that is growing faster than the rest of the cell lines because it has rid itself of the burden of making product. This is particularly important when the cell line is a constitutive virus producer. We do not want nonproducing cells in our stock, and therefore the selection of cells is performed by cloning.
2. There is a substantial amount of data to demonstrate that the production of clones from a single transfection experiment can result in cells with a wide range of productivities and growth characteristics (26,27). This is an expected consequence of production methods that rely on integration of DNA at random points in the chromosome of the host cell.
3. The provision of stored cell lines and virus that are derived from a single cell or virus and that are then tested to ensure that there is a single population present enables us to be confident that results obtained with cell lines and virus will be valid and reproducible.

It is important to keep a small stock of all the clones that are tested until the final qualification of the master cell bank (MCB) has been performed and the results are known to be satisfactory. Techniques such as plaque purification, dilution cloning, and the use of cloning rings can give high confidence of clonality if performed at the correct dilution.

Cell banking is performed at an early stage in the development process to ensure that stocks of material are maintained and to secure cGMP cell banks to use in the production process. The process for banking constitutive producers or noninfected producers is essentially identical. The cell line is cloned and tested for stability and a cell bank laid down. The method for laying down a cell bank will vary depending on the cell growth requirements and the number of revivals that will be made from the cell bank.

A tissue-culture flat or roller bottle process will probably be appropriate for scaling up and banking attachment-dependent cell lines. To achieve an adequate number of cells for a suspension cell bank a 5–10-L working volume fermenter may be needed. A cell bank must have sufficient cells per vial to make revival reliable and robust. A common mistake is to produce cell banks with low cell densities and small volumes of material. A minimum of 1×10^7 cell/mL is ideal for most circumstances. It allows a 50–100-mL shake flask or spinner culture to be seeded directly from a 1-mL stock. Alternatively two 175-cm^2 tissue culture flasks could be seeded from 1 mL of stock.

It is essential that the MCB is extensively tested for contaminants. Unlike other cell culture products, virus contaminants cannot be removed at later stages, and the cell banks must be tested for endogenous viruses. It is essential to revive cells from cell banks to establish the freezing process has not killed all the cells and to establish a baseline value against which the maintenance of viability during storage can be judged. The continued viability of the cell bank should be checked by reviving vials on a regular basis (say, once every two years) or after any uncontrolled event has happened to the cell storage facility. A controlled revival protocol must be in place for this exercise to give valid results, because factors such as speed of thawing, spinning to remove medium, and temperature of diluents can all have an effect on cell revival. Revival should be assessed in terms of the viability of cells immediately after revival and on the growth performance of the revived cells over several passages.

Cell banks should be stored at −130 °C or less, and the storage facility should be validated and monitored with appropriate alarms in place. The critical cell banks should be identified and split between locations to ensure that in case of equipment failure or natural disaster the cell stocks are not lost. The value of cell banks should not be underestimated. Without the cell bank there is not any product!

Subculture Regimes

The culture of cells during the early stages of development will be in roller bottles, shake flasks, or other small-scale vessels. Careful attention to cell counting and the effects of basic culture parameters at this stage will provide a wealth of useful information for the scaleup process. Unless the subculture regime is investigated and optimized the phenomenon of useful/useless growth cycles will be seen. This phenomenon is common to many nonoptimized cell culture processes and reflects the carryover of a growth inhibitory effect from a culture that has previously grown well, and this may indicate that a portion of the cells in the culture have committed to apoptosis (see ●●●). There is a potential for this effect to carry over into the virus infection phase of the process and hence inhibit virus vector production. For meaningful results to be produced it is not adequate to work on a fixed subculture regime of a 1:3 split every 3 days. The actual cell numbers must dictate the process.

VECTOR PRODUCTION METHODS

A range of cell culture techniques are available for the culture of producer cell lines and the production of virus vectors from these cells. The type of cell used will determine the range of approaches that are available. The choice of a suspension or attachment-dependent cell line can have significant implications for scaleup and downstream processing.

Culture Equipment and Control Variables

The function of culture equipment is to grow cells and make product. Different equipment will achieve different degrees of environmental control, and the capital costs of equipment generally reflect the sophistication of the control achieved using the equipment. The factors controlled by the culture equipment are called control variables, and there are a number of variables that can be controlled and monitored during cell culture. The most important control variables are described in the following.

Temperature. If only one parameter is controlled, it should be temperature. Animal cells grow best at temperatures of 36–37 °C. Cells will tolerate short excursions outside the temperature optimum. However, high temperatures are fatal to animal cells, and temperature controllers that allow overshoots of more than a degree Celsius may either kill or damage the cells in culture. The response of cells to temperature excursions is asymmetric, and many cells can survive and grow at reduced temperatures. Some products such as retroviruses are temperature sensitive, and virus production and recovery are enhanced by reducing the culture temperature to approximately 32 °C (20–25).

Dissolved Oxygen Tension. We normally assume that they need a dissolved oxygen tension of between 10 and 80% to stay healthy. Supplying adequate oxygen to cell cultures is achieved in several ways:

1. Ensure that the air space above the culture is large enough to prevent the cultures' becoming anaerobic during culture (tissue culture flasks, roller bottles).
2. Ensure that the culture medium is not too deep. This prevents local oxygen depletion and CO_2 buildup (tissue culture flasks, roller bottles).
3. Use lids that allow diffusion of air, leave lids loose, open vessels every day (tissue culture flasks, roller bottles).
4. Blow air into the vessels (bioreactors).
5. Agitate the vessels (tissue culture flasks on rockers, roller bottles, shake flask, bioreactors).
6. Aerate medium and then circulate through the culture device (flat plate devices, fixed beds, fluidized beds).

The simpler of these methods aim to supply sufficient air to keep the cultures aerobic during the culture period but do not aim to control DOT. The more sophisticated methods will control and monitor the dissolved oxygen tension in the culture.

High cell densities may require the use of oxygen-enriched air or pure oxygen. This is an option that is less safe than increasing the agitation rate or gas flow rate to achieve oxygen transfer.

pH. pH is important for cell growth and virus stability, and therefore the control of pH may be essential for optimal virus quantity. Cell cultures are typically performed at pH values within the range of 6.8 to 7.4. In an ungassed tissue culture flask or roller bottle the cells may see the full range of pH values within the course of one subculture. The factors that cause pH to change in cell cultures are the production of lactic acid and CO_2 as a result of the metabolism of glucose and amino acids.

Other factors such as culture medium osmolality should be followed during the culture because animal cells are more sensitive to osmotic shock than microbial cells, and it is important to keep the osmolality within reasonable ranges when controlling pH. This becomes more of an issue with high cell densities, where feeds may be used to sustain cell growth and concentrates are fed into the medium.

CHOICE OF CULTURE VESSELS

There are many different types of culture device available to the investigator who wishes to culture cells. Every manufacturer claims some advantage of their equipment over the competition. When trying to choose between different devices consider the following:

- What do you want to do?
- Will it do what you want?
- Is the device going to do what you want for as long as you want it to?
- Do you have the staff/skills/infrastructure/finances to use and maintain the equipment?
- Will the process need to be scaleable?
- Can the manufacturer support you and continue supplying essential components?
- Is the equipment robust and reliable?
- If the material produced in the system is used for clinical purposes, can the equipment be used for GMP purposes?

In many cases tissue culture flats, roller bottles, or shake flasks can be used as cheap and convenient devices for small-scale production and initial development work. If in doubt, keep it simple. Scaleup of the processes can proceed in two directions:

- More sophisticated devices may be used to provide process control and allow scaleup of the process.
- Increase the number of unit operations—more of the same (possibly using automation to increase capacity and reproducibility).

Both methods work well, but the associated risks vary depending of the type of approach used. There are a large number of unit operations involved in the routine use of vessels like tissue culture flasks and roller bottles, but the use of a roller bottle process will allow a facility to be flexible and reduce the amount of validation needed to run a process. This increases the risk of contamination, but the use of robots can eliminate this risk and increase productivity and reproducibility. Fermenter systems offer scope for process control and far fewer unit operations per run, and when the equipment is sterilizable in situ, the contamination risk is reduced even further. Fermenters provide the increased control and reduced unit operations at a price; they are expensive to buy and maintain, and an extensive infrastructure is needed to sustain fermenter use.

Culture Strategy

Once the cell culture parameters are defined, the scaleup process should be robust and reproducible. With a retrovirus-producing cell line we could predict that a scaleup process starting on day 0 would give us enough cells to seed a 400-stack cell cube by day 21 (16). Once the cell cube was seeded, we grew the cells for another 3 days and then entered the production phase.

The aim of scaling up for production is to keep the cells growing as actively and consistently as possible so that scaleup is rapid and predictable. This should reduce the time taken to reach production and reduce the chance of RCR production. It is essential to subculture while the cells are in the growth phase and not to waste cells. Once the production vessel is seeded, use those growth conditions that get the cells to the appropriate stage for production. Switch to the production phase and make the material. The approach is the same for any culture system.

The Effect of Cell Lines on Process Design

Producer cell lines can have a huge effect on process development and the commercial feasibility of the production process. It is much easier to scale up a suspension culture than an attachment-dependent culture. Unfortunately cell line selection is limited to a small number of candidates unless a group has the resources to develop and test their own cell lines. Production of a patentable producer cell line with useful characteristics can produce a huge financial return for a small company willing to exploit the line.

Cell lines should be free of endogenous virus and capable of continuous growth in culture when transformed to producer cell lines. The origin of the cells should be well documented, and in many cases human cell lines may be the best source of producer cell lines which that are more likely to produce vector that is not recognized by neutralizing antibodies.

The choice of viral gene sequence used in producing a producer cell line is important. Some gene sequences such as the adenoviral E1 sequence produce protein that reduced the ability of the cells to attach to solid substrates. Other sequences produce toxic proteins, and if they are expressed must be expressed as inducible proteins.

Gene therapy viruses are normally intended to be replication incompetent. Therefore, a producer cell line must include the missing sequences that are needed to allow production of complete vector. The possibility of recombination between the vector sequence and the complementing gene sequence from the producer cell line to produce infective replication-competent virus should be reduced as much as possible. In the case of retroviruses the initial first-generation vectors such as PA 317 (26) supplied the *gag pol* and *env* genes in *trans*. In many cases RCR production was detected in relatively small batches of virus. Subsequent work reduced the chance of RCV production by splitting the *gag pol* and *env* genes. Additionally the producer cell line was changed from mouse to human (27).

Adenoviral production has only recently been freed of the persistent low-level RCA contamination by the production of the PER C6 cell line by Introgene in Holland (1).

The more disabled a virus is and the more complex the producer line, the greater the chance that the productivity of the vector will drop. Other vectors are even more complicated to make. The adeno associated (AAV) virus system has several potential advantages but needs a significant number of adenoviral genes to be expressed. Either the virus is produced using a coinfection with adenovirus or it is produced in genetically modified producer cells that provide all the necessary gene products. It is then necessary to remove helper virus from the product. More recently the production of hybrid virus has been reported (28,29). This is a significant breakthrough because it opens the way to construction of viruses tailored to a function rather than modifying an existing virus.

Virus Choice

The choice of virus determines the rest of the process because the quantity of virus produced, clinical demand, and stability in culture will determine the scale of operation, cell line used, and the type of culture materials needed. The primary reason for choosing a virus is likely to be its therapeutic potential, not the ease of production of the vector. There are many features of different viruses that will determine the purpose that the virus is used for. Highly immunogenic viruses may be useful in one-off treatments but will be of little use for long-term repeated treatments. Viruses that are maintained for longer times such as AAV, which integrates at a specific site in the human genome (30), and retroviruses that integrate in a more random way may be desirable for many indications such as genetic diseases. Other factors that must be considered are the size of the insert that can be carried, the fidelity with which it can be maintained, the handling properties of the vector, and the scale of production required.

Methods of Comparing Production Systems

Everyone compares production systems, which is the most appropriate, what is a realistic figure for viral titer, etc. Following a few basic rules makes comparisons easier.

In attachment-dependent systems the values for virus/harvest virus/cell, virus/cm^2, or virus/mL are often quoted. A change in production method can easily change titer, and methods that allow reductions in volume of medium used may result in a consequent increase in titer per unit volume.

Choice of Production System

Production methodology is determined by a number of factors. The most important factors are the total viral titer needed for a piece of work, cell type, and scale. Some viruses are produced in large numbers and are stable. Others like the retroviruses are difficult to handle and lose titer quickly. Potentially, the scaleup of gene therapy vector production is not an issue of great concern as long as serum-free suspension systems are available. Assuming that a typical dose of vector is 1×10^{10}, a culture producing 10^9 pfu/mL would only require 100 mL of culture, assuming 10% recovery, to provide a clinical dose for local administration. This would involve producing 1×10^5 L of cell culture material to make a million therapeutic doses. Obviously as the scale becomes larger, downstream processing becomes a major issue, especially for the removal of immunogenic material and cellular DNA.

The choice of cell system used for vector production is governed by the need to produce a large amount of replication-incompetent viral vector. Consideration of the growth and scaleup of the systems are often of secondary importance to groups developing these cell lines and vector systems. Finding the combination of continuous cell line that is free of endogenous viruses and can be used to produce useful quantities of viral vectors is a considerable challenge. How important is the titer obtained from a production system? A retrovirus may be present in the culture medium at 10^6 to 10^7 pfu/mL. An adenovirus may be produced at 10^{10} pfu/mL. Consequently 1–10 L of retrovirus will be needed to produce the same starting material as 1 mL of adenoviral culture. The major issue is whether it is easier to purify unstable material from a dilute low-protein background or whether the disruption of concentrated cells and purification of high concentrations of more stable virus from the resulting cell slurry is easy and gives high recoveries. It is also important that the virus production system is consistent.

PERFUSION SYSTEMS

Perfusion and in-line harvest processes are easy to perform with attachment-dependent cell lines because the cells do not require sophisticated equipment to keep them separate from the harvest stream. This means that the titers of product can be raised by the simple expedient of using lower volumes of culture medium for the production phase than were used for the cell growth phase. It is also easy to change the medium composition when transferring to production phase. Attachment-dependent cultures can be performed using a range of technologies from tissue culture flasks to microcarrier cultures. A popular technology is the multistack parallel-plate-type device such as the cell cube or cell factory. These devices allow rapid scaleup of processes while minimizing the capital expenditure required to set up a large fermenter-based microcarrier system. The cost of consumable equipment for use in parallel-plate systems is high and encourages a multiharvest approach. In addition, the setting up and operation of the parallel-plate devices require a number of aseptic connections to be made. We found the use of a manifold containing a set of single-use connectors greatly improved the operability of the cell cube system. Continuous production is most important for the low-titer nonlytic viruses such as retroviruses. One-off processes such as adenovirus production will achieve higher titers in a short time but may be more suited to a small-volume high-cell-density system.

Continuous production is only feasible using a vector production system where the viral vector is produced by a nonlytic process and the producer cell line is producing the vector continually. A suspension cell line in a chemostat would be the classic production method for a continuous system. However, most gene therapy producer cell lines are attachment dependent. Most of these cells once confluent will die and slough off their support. By seeding cultures at low densities, reducing culture temperature, and feeding with fresh medium, some of these cultures can produce material for several weeks. If the feeding is performed by perfusion rather than as batch feeds, the levels of nutrients in the medium can be maintained above predetermined levels, which should ensure that the cells are not nutrient depleted. Continuous harvest of supernatant from an attachment-dependent cell culture can increase the yield of viable vector without increasing the production rate. This will happen if the vector is unstable at culture temperature and can be stabilized at harvest.

With a parallel-plate system optimal results can be achieved by harvesting into a chilled storage area. This technique works well if the residence time of the harvest between the culture vessel and the controlled harvest vessel is short. This can be aided by using a narrow-bore outlet tube to increase linear velocity and reduce residence time in the transfer line. To reduce the residence time of medium through the culture system, reduce the recirculating volume to a minimum. This also increases titer per unit volume. Use the biggest surface area possible for the system. Break the harvest into manageable, well-defined harvests and process them individually. Results for viral titers will always lag behind the harvest times by a matter of days, and the progress of the culture must be followed by looking at cell culture data. It is essential to develop an understanding of the relationship between culture parameters and vector productivity. Oxygen utilization and other indicators such as nutrient utilization must always be examined critically for signs of cell death (16).

The Production of Retroviruses in a Perfusion System

Retroviral producer cell lines are ideal candidates for continual production systems. Producer cell lines for retoviruses continually produce vector and release the vector into the culture medium. Vector is normally produced at low concentrations, and as has already been described, the vector is unstable in culture medium at 37 or 32 °C. A limited number of cell lines are available for virus production, and most of them grow maximally in attachment culture. Small volumes of material can be produced in tissue culture flasks, but larger quantities are made in a controlled fermentation device of some sort. The choice of culture systems that can be used for these vectors is discussed below.

Flat-Plate Devices (Cell Factory, Cell Cube)

There are a variety of flat-plate culture devices available used for the scaleup of attachment-dependent cells. My own experience is mainly with cell cube systems, but that does not mean that the cell cube is better than other similar systems. All these devices need the following:

1. A medium conditioning vessel (fermenter, specialist device) to control pH, dissolved oxygen tension, and reservoir temperature. If the reservoir does not have independent temperature control, the whole device must be kept in an incubator.
2. The following services: compressed air (dry, oil free), oxygen-enriched air or oxygen, carbon dioxide, nitrogen, power, and cooling water.
3. A circulation pump that can pump medium through the system quickly enough to prevent dissolved oxygen tension or nutrient concentrations dropping below critical levels.
4. An incubator to keep the culture medium in the device at the set temperature.
5. Perfusion will require either two pumps or a single pump with two heads. The pumping should run at different rates either by using different internal diameter tubing on the inlet and outlet of the pumps or by running the pumps at different speeds.

The cell cube is a parallel-plate device that can be used to perform a rapid scaleup of attachment-dependent cell cultures. This device and similar flat-plate devices are immensely useful for scaling up small-scale cultures with a minimum of expertise.

Flat-plate devices require relatively little expertise to set up and run small units but at the largest scale provides a physical challenge for operators and is not easily scaleable.

The operation of the cell cube is simple, and the cube can be operated in a hot room or incubator. Cells are trypsinized from the seed vessels and attached to one side of the plates in the cube. Once the cells are attached, the process is repeated by removing the culture medium and turning the cube through 180 degrees. A second batch of freshly trypsinized cells is then seeded onto the second half of the cube. The attachment period can be determined by performing experiments with tissue culture flats and determining the time at which the number of cells attached to the flasks is stabilizes. It must also be remembered that the attachment of cells will take place in a static culture without opportunity for gas transfer. If a high cell density of a culture that uses oxygen at a high rate is used, it may

be necessary to reduce the attachment phase to prevent oxygen exhaustion.

The degree of confluency will depend on the viral vector being used and the process shape. In most cases the greatest benefit can be obtained by seeding the cell cube at a low density and rebulking the cell numbers in this device prior to the production phase. It is important to start the circulation pump at the maximum rate you intend to use to prevent the peeling off of cell sheets that will occur at later stages if the pump speed is gradually increased to maintain the dissolved oxygen tension in the cube.

After a suitable attachment period the medium is removed and fresh medium is circulated from a controlled reservoir using a peristaltic pump. The input and outlet condition of the culture medium can be monitored and used to control the input to the cube. Monitoring the DOT on the outlet and inlet is very important and we always targeted a DOT of >10% on the outlet side of the system. In theory the oxygen tension at the outlet can be controlled by changing the liquid flow rate through the cube or by increasing the oxygen tension in the reservoir. Because of the requirement to run at the maximum flow rate, the control is achieved by increasing the DOT in the reservoir.

The major advantage of such a system is the small number of operations that need to be performed as compared to using an equivalent number of roller bottles. This is an improvement over traditional roller bottle processes. However, because the control of the medium is carried out remotely from the cells, the medium composition and hence the environment in the cube changes as the medium flows through the system. The position of the cells within the cube will affect the growth environment and hence performance of the cells, and defining the best condition to make virus in these systems is not as simple as optimizing a well-mixed culture. It is not easy to sample this device, and determining the optimum conditions for the cells is difficult.

We found the most efficient way to use this device in a GMP area was to change the culture medium and harvest every two days by completely draining the system. The contents of the system were transferred to a sterile plastic bag, filtered to remove cell debris, and then processed. The only parts of the fermentation system that were in contact with more than one fermentation were the stainless steel triclamp connectors used to connect the filter housings to the harvest bag, the exit dissolved oxygen probe holder, and the cell cube aerator vessel.

Microcarriers and Macrocarriers

More control and monitoring can be achieved using microcarrier cultures because the producer cells are grown in a well-mixed suspension culture. There are also a wide range of options for the operation of microcarrier cultures that can be run in batch, fed batch, drain and fill, and perfusion modes. This is because the cells in the system are attached to larger beads, which can be settled and separated easily from the culture medium. There are disadvantages to microcarrier cultures; the major two disadvantages are the sloughing off of cells that become confluent (which is common to all attachment-dependent culture systems) and the difficulty of scalingup the microcarrier processes to large volumes. In many cases the microcarrier stage of a production process is seeded using cells from roller bottles. Using Pharmacia microcarriers (29) as an example, estimates of the sort of activities that are needed to perform microcarrier culture can be achieved.

One liter of a cytodex 2 microcarrier suspension contains approximately 2.9×10^7 beads. To seed these beads at 5 cells per bead will require 1.45×10^8 cells. Actively growing cells from subconfluent roller bottles are in the best condition to use as a seed, and it would require approximately one $1700\text{-}cm^2$ roller bottle to seed a liter of culture.

Microcarriers are small beads that vary in diameter from approximately $100\,\mu m$ upwards. They can be composed of a range of material, with the most commonly used material being dextran, which is supplied as a nonsterile dry powder. Most of the microcarrier materials shrink when dry, and a swelling procedure is performed to hydrate the beads (this must be done in a vessel large enough to accommodate the swollen volume of the beads; 1 g of cytodex 2 will swell to 15 mL). The swollen beads are also sticky, and it is usually necessary to silanize the glass containers that are used with microcarriers. It is normally quicker to carry out the initial swelling and rinsie in a nonsterile environment. PBS is usually used for swelling, rinsing fines from the beads and heat sterilizing the beads. Once the microcarriers are rehydrated, they should be sterilized as soon as possible to prevent the growth of bacteria and endotoxin production. Microcarriers may be used up to a concentration of 5 g/L dry weight for dextran and higher for glass or polystyrene beads. A large fraction of a rehydrated bead is PBS after swelling and sterilization. Medium can either be used to rinse the beads in the culture vessel or the feed medium can be adjusted to compensate for the volume of PBS retained in the vessel.

The major issue in developing microcarrier systems is determining suitable cell seeding regimes. It is critical to get the attachment phase right. In some cases the cells can be added to an agitated aerated culture system and will attach to the microcarriers without any difficulty. However, if the cells do not attach well in the simple system, it may be necessary to develop an initial attachment phase. Factors that may increase the attachment of cells to the microcarriers are:

1. Reduced culture volumes during the attachment phase
2. Indirect aeration of the culture medium
3. Agitation speed reduction

There is a risk that the combination of a high concentration of cells and microcarriers may lead to clumping caused by the attachment of cells to more than one bead at a time. It is also important to remember that the cells will never populate all the microcarriers. Suppliers of microcarriers give information on the use of their equipment and the appropriate cell types and cell densities to use with their product. This information is extremely valuable and helps in identifying the most likely candidate for a growth matrix, but in our hands cloned cell lines vary enormously from one another in their behavior and the manufacturer's data should only be considered as an indicator for factors such as seeding density.

The choice of the correct microcarrier for any process can be made on the basis of materials of construction, mechanical properties, and the results of small-scale culture methods. The size, charge, and densities of microcarriers will vary depending on the material that the microcarriers are made from. Careful screening of microcarriers from different suppliers will identify the best carrier for each cell line. Carriers that achieve high cell densities because they are porous may be appropriate for cell growth, but virus infection and release may be severely limited. In tests performed with a PA317-derived cell line we found that the cell cultures carried out using microcarriers and macrocarriers that gave high growth as determined by glucose and glutamine estimation did not necessarily produce the most virus.

Microcarrier cultures are appropriate for small- to intermediate-scale processes and allow the exchange of medium and multiple harvests without the use of ultrafiltration equipment. Microcarrier systems are homogeneous, and a sample can be taken at any time to monitor the state of the cells.

Culture vessel may need to be modified slightly for larger-scale microcarrier culture. At the simplest level the microcarriers process can be run at least 100 L operating volume by using a settling process for rinsing of the microcarriers and medium exchange. A dip tube is arranged so that it sits just above the level of the settled microcarriers and liquid is pumped out through the tube. A second method uses a mesh to retain the microcarriers while the liquid is removed from the base of the vessel. The flow of liquid out of the vessel must be controlled with both solid and swollen beads. Dextran and gelatin beads will mash and block the outlet mesh if the flow rate is too high. Glass and plastic beads aggregate around the mesh, and cells can be stripped off the surface of the microcarriers by very high local liquid flow rates. It is also possible for microcarriers to fill and block sparge tubes. Spin filters can be used for the aeration of microcarrier cultures and the removal of medium and product in perfusion mode. Many systems use a mesh cage attached to the agitation shaft, but this does not allow the spinfilter to operate independently of the agitator. To achieve control of the spin filtration process an independently controlled filter needs to be used. These can be fitted into the fermenter vessel or attached via an external loop.

Hollow-Fiber System

Hollow-fiber systems have been used for the production of small amounts of antibody (grams/year) and enable long runs to be set up and operated using stable cell lines. The high densities of packed cells that can be achieved in hollow-fiber culture are difficult to quantify and will be exposed to different levels of nutrient. Access of virus to cells and release of virus from cells is difficult. They have been used for the production of gene therapy vectors but are not scaleable, and the consumable costs for running the process are high.

SUSPENSION CULTURE

A number of virus-producing cell lines is capable of growth in suspension culture. Suspension is an excellent means of culturing cells and offers huge advantages over other methods. There are not any trypsinization issues, cells are distributed evenly throughout the vessel, and the control variables can be easily controlled. The capital investment in vessels and facilities can be high, but the advantages include a great reduction in the use of expensive consumable such as tissue culture plastic, parallel-plate devices, hollow fibers, and microcarriers. This approach also reduces the production of solid waste and decreases the dependence of the process on the continues operation and existence of product manufacturers. Scaleup and operation of suspension systems are simple, and the number of unit operations that cause process risk are small. Suspension culture has an excellent track record, animal cell culture is routinely performed at the multi-thousand-liter scale, and cell densities can be increased by the use of perfusion and fed batch processes. The use of serum-free medium makes downstream processing easier.

Simple stirred tank reactors are well proven as production vessels for animal cell culture and can vary in complexity from a magnetic coated magnetic flea in the bottom of a sterile bottle to a fully monitored vessel attached to a sophisticated control system. All the major fermentor manufacturers supply vessels for animal cell culture. The major differences between the animal cell culture vessels and microbial vessels are outlined below.

- *Agitation rate*: Operates at lower agitation rates (typically 40 to 200 rpm).
- *Dissolved oxygen control*: Use gas flow rate rather than controlling stirrer speed.
- *pH control*: Use weak acid/base combinations (usually CO_2/sodium carbonate).
- Magnetic couplings are preferred on the agitation systems used for animal cell cultures.
- *Gas flow rates*: Much lower than microbial systems (up to 0.2 vvm).

The major process issues with suspension culture are related to the subculture methods used in these systems. It is normal to transfer the seed culture as a cell suspension in the growth medium used for the growth of the seed cells. If there is a requirement for complete medium exchange, a specific removal stage needs to be designed into the process. There may also be a case for the partial concentration of cells to make virus infection more efficient. This will require the design of a specific process step based around a microfiltration.

COMPARISON OF CULTURE METHODS FOR VECTOR PRODUCTION

It is important to be able to compare the different methods of making viral vectors (Table 8) in a meaningful way. The standardization of assay methods between groups, both within and outside, and organization is essential to enable meaningful comparisons of productivities and final titers. The implications of this are much broader than the comparison of titers in a research context. Agreed standard methods need to be used to ensure that the material used in clinical trials can be compared on a level footing with other vectors and that the reported dosages are comparable (32).

Monitoring Productivity

Ideally there would be a method for measuring virus online that would give a measure of the total number of particles and also give an estimate of viable titers. Most assays depend on either an infectivity measure (plaque formation, colony formation, protein expression) or some measure of viral components (antigen, PCR, UV absorption). The most commonly quoted values are for infectivity. In a clinical setting this is the most important measurement of the product. Titer can vary considerably depending on the experimental protocol used. The issues of paramount importance are the reproducibility of an assay, its transferability, and its robustness. Any assay must be fit for the purpose, and the application of robust HPLC assays to give total virus particle measurements is a growing trend that should help to give rapid feedback on process performance (33). Methods suitable for use in monitoring production of virus in culture medium must be robust and unaffected by medium components. The concentration of virus may increase as different stages in the purification process are completed, and at this stage different methods may be appropriate.

A major concern with titer estimates is that different investigators using different methods may see large variations in the titer obtained from the same assays. In our hands the addition of polybrene to retroviral samples increases the titer by one factor of 10. The choice of target cells for the assay system will also have an influence. The concern with assays is related to the efficacy of the products in a clinical trial. Measuring the quantity of nonviable material in a dose may also indicator the likelihood of reducing the chance of producing an immune response. It is unrealistic to expect that a fully developed assay is in place for the start of a project, and one consequence of this is that the initial development work may have to depend on a suboptimal assay method and relatively little titer data. Changing methods may make direct comparison of data from early stage development meaningless, and looking at data as a percentage of the highest result or control result in a run may be the most appropriate method of comparing runs.

Sample handling and storage is one of the most critical areas of the assay work. The storage of material at an inappropriate temperature, failure to clarify sample prior to storage, repeated freeze thawing, and the use of inappropriate volumes of sample can all invalidate the assay results. These issues must be addressed early on. Historically many viruses have been considered to produce a large number of inactive particles per infective unit. This is puzzling. Recent work on virus assay using adenovirus has shown that the ratio of noninfective to infective particles is closer to 3:1 than 99:1. The use of assays such as HPLC for the determination of total virus particles gives a useful check against which viable counts are compared. For these reasons the physical effects of medium depth on the assay plate should be noted and the volume reduced as far as possible.

Detection of replication-competent viruses is also a major task that can require the sacrifice of large amounts of material, and in this case the more sensitive the assay, the greater the chance of having a workable process. Further complications in assay can arise with cells used to assay retrovirus where the virus may only be taken up and expressed by actively dividing cells. The test plates used are frequently not confluent when the virus is added. In addition, the virus is temperature labile and the window of opportunity for the cells to see the virus is small. Multiple infection of target cells will be missed. Noninfective material can bind to virus binding sites on the cells, and there is a possibility that interactions between producer cell components and virus can interfere with the virus directly (34).

Only by taking all these factors into account and developing, and preferably validating, an assay can the results obtained by gene therapy groups be compared and used to draw meaningful conclusions. It is vitally important for groups within the same organization, or different companies who are collaborating, to cross validate assay results to ensure that the same measures giving the some or similar answers to all quoted test samples, are used throughout work at different sites.

The most reliable measure of virus is still an infectivity assay. However, the main concern of process operators and investigators is the monitoring of viral levels and action on the levels at stages during production. These measures are useful for the control of DSP.

THE EFFECT OF CULTURE ON DOWNSTREAM PROCESSING

It is easy to argue that the development of high-efficiency purification systems is unnecessary for gene therapy vectors because high titers of material can be produced from relatively small numbers of cells. Personal experience notes that the estimates of the quantity of material required for therapeutic purposes can vary by factors of 10 to 100 and an assumption that laboratory-scale processes are either scaleable or adequate for commercial production can lead to severe setbacks in the progress to market.

Downstream processing of a gene therapy vector may be limited. In some clinical trials virus-producing cells have been introduced to the site of tumors rather than purifying the virus itself. Retroviruses are fragile compared to many viruses and proteins, and gentle purification methods have to be chosen to purify these viruses (20–25). Therefore,

the use of pH gradients, or high-molality salt solutions to elute columns, is not feasible.

Purification may be limited to a simple concentration and diafiltration step. The consequence of this is that any large molecules present in the culture medium (medium components, cell debris, and inactive virus) will be co-concentrated. Some viruses such as retroviruses and herpesviruses are encapsulated in a lipoprotein membrane. Measuring the degree of purity of these viral vector preparations will be difficult because of the presence of cellular material in the membrane. Some contaminating materials may be removed by specific processes (adding DNAase, for instance).

The simpler the starting material, the easier the purification, and the development of improved adenoviral vector producing cell lines is a positive move. Cells growing in serum-free medium and producing RCV-free product give a chance of producing clinical-grade material. Recent advances in adenoviral purification methods are concentrating on addressing a variety of issues. The use of anion exchange chromatography to purify adenoviruses is a large step forward.

Laboratory methods such as cesium chloride gradient centrifugation are scarcely adequate for a production process. The release of virus from cells by freeze–thaw cycling is also not necessarily the best large-scale method for release. Separating the virus from a lysed cell suspension is a different task from that of purifying virus from a culture supernatant and requires either a specific capture step or a series of less specific purification stages to remove noninfective material.

The degree of purification will determine the losses expected across the process. An optimistic assumption that any purification step will involve 5% losses (as a result of sample removal for testing, a small loss on the process) shows that the harvest titer can soon be reduced by 50%.

More alarmingly, virologists are quite accustomed to talking about log reductions (90%) in titer for various purification stages. An apparently exceptionally high titer can disappear quickly unless efforts on culture development and DSP are integrated.

MAXIMIZATION

All processes need to be maximized, and the producer of a gene therapy product has to demonstrate that the process is both robust and under control. This ensures a consistent supply of high-quality product. The level of optimization will depend on the process type and equipment. The process may depend on a growth phase followed by a production phase and the optima and equipment used for these processes may be very different.

With our cell cube retrovirus production process poorly controlled tissue culture vessels were used for the initial scaleup; then the cells were transferred to the cell cube for the final part of the growth phase and the temperature was dropped and medium changed for the production phase. By controlling the cell density at subculture, we avoided lags in the cell growth rate. The drivers for this process were regulatory, ease of processing, speed to first harvest, and titer of virus and the development phase was constrained by a requirement to provide materials quickly.

As the project reaches the phase three stage, outer-limit validation needs to be performed using systems that will mimic the final process. At earlier stages a lot of valuable information can be gathered using statistically designed experiments to look for the major interactions in a process. Where the process is complicated and involves infection and replication of virus in cells the potential number of interactions is increased. The final stage is process validation at full scale, and one would expect to run at least three consistency batches to prove that the process was under control.

GENERIC PROCESSES

The development of a new product can be speeded up by developing a generic approach to the technology. The main advantage to generic processes is apparent when a single cell line can be defined as a host for all the vectors that are being used within a group. The basic culture process and downstream process can be defined and expertise developed in the area. If the cell type used varies, the generic development can be performed on a different culture system but will obviously be less useful.

SUMMARY

The future for the production of gene therapy vectors is in the hands of the molecular and cellular biologists who make and develop cell lines. The basic technology needed to grow viral vectors is in place and has been used for many years using other cell lines and products. The major issues limiting large-scale production of vectors are:

- *Cell line availability.* The number of suitable cell lines is very small, and the developers of such cell lines quite properly wish to protect their investment with patents and restrictive licencing agreements.

- *Serum dependency.* It is not acceptable to use serum-dependent cell lines.

- *Attachment dependency.* Attachment dependency adds a large number of manipulations to the vector production process.

- *Replication-competent virus* (RCV). Large-scale unit operations can only be a realistic option if the chance of making replication-competent virus is vanishingly small.

- *Targeting of vectors.* The use of vectors in the therapeutic area will depend on developing vectors that target the correct cells.

- *Production of the right vector.* Vector development follows trends and fashions. Retoviruses were in favor a few years ago, and now adenoviruses are in favor. In the longer term the coupling of virus types to disease treatment will result in the need to make a wide range of vector types.

- *Purification.* High-quality purification and delivery systems are essential to the success of viral gene therapy.

BIBLIOGRAPHY

1. F.J. Falloux et al., *Hum. Gene Ther.* **9**, 1909–1917 (1998).
2. P.D. Robbins and S.C. Ghivizzani, *Pharmacol. Ther.* **80**, 35–47 (1998).
3. R. Lodge et al., *Gene Ther.* **5**, 655–664 (1998).
4. F. Griscelli et al., *C.R. Seances Acad. Sci., Ser. 3* **320**, 103–112 (1997).
5. N.A. Wivel and J.M. Wilson, *Hematol. Oncol. Clin. North Am.* **12**, 483–501 (1998).
6. B.E. Huber et al., *Proc. Natl. Acad. Sci. U.S.A.* **91**, 8302–8306 (1994).
7. M.T. Lotze et al., *Ann. N.Y. Acad. Sci.* **795**, 440–459 (1996).
8. B.E. Huber et al., *Cancer Res.* **53**, 4619–4626 (1993).
9. N.K. Green et al., *Cancer Gene Ther.* **4**, 229–238 (1997).
10. C.H. Evans et al., *Hum. Gene Ther.* **7**, 1261–1280 (1996).
11. S. Ghazizadeh, J.M. Carrol, and L.B. Taichman, *J. Virol* **71**, 9163–9169 (1997).
12. S. Worgall, G. Wolff, E. Flack-Pedersen, and R.G. Crystal, *Hum. Gene Ther.* **8**, 37–44 (1997).
13. C.A. Wilson, M.S. Reitz, H. Okayama, and M.V. Eiden, *Hum. Gene Ther.* **8**, 869–874 (1997).
14. C.N. Martin, *Evolutions* **2**, 1–7 1997.
15. D. Onions and G. Lees, *Q-One Tech. Bull.*, No. 17, 1994.
16. N.J. Moy et al., *16th ESACT Meet.*, Lugano, 1999 (in press).
17. J.F. Engelhardt, L. Litsky, and J.M. Wilson, *Hum. Gene Ther.* **5**, 1217–1229 (1994).
18. W.K. O'Neal et al., *Hum. Gene Ther.* **9**, 1587–1598 (1998).
19. M.J. Parr et al., *Neurovirol.* **1**, 194–203 (1998).
20. PCT/US95/00069 1995, H. Kotani, P. Newton, and S. Zhang.
21. R.M. Bieganski, A. Fowler, J.R. Morgan, and M. Toner, *Biotechnol Prog.* **14**, 615–620 (1998).
22. R.W. Paul et al., *Hum. Gene Ther.* **4**, 609–615 (1993).
23. J.R. Morgan et al., *J. Virol.* **69**, 6994–7000 (1995).
24. S.T. Andreadis, D. Brott, A.O. Fuller, and B.O. Pallson, *J. Virol.* **71**, 7541–7548 (1997).
25. S.-G. Lee, S. Kim, P.D. Robbins, and B.-G. Kim, *Appl. Microbiol. Biotechnol.* **45**, 477–483 (1996).
26. A.D. Miller and C. Buttermore, *Mol. Cell. Biol.* **6**, 2895–2902 (1986).
27. F.-L. Cosset et al., *J. Virol.* **69**, 7430–7436 (1995).
28. X. Lin, *Gene Ther.* **9**, 1251–1258 (1998).
29. A. Recchia et al., *Proc. Natl. Acad. Sci. U.S.A.* **96**, 2615–2620 (1999).
30. R.J. Samuelski, X. Zhu, X. Xiao, et al., *EMBO J.* **10**, 3941–3950.
31. *Microcarrier Cell Culture, Principles and Methods*, Pharmacia, Uppsala, Sweden.
32. C. Nyberg-Hoffman, P. Shabram, W. Li D. Giroux, and E. Aguilar-Cordova, *Nat. Med.* **3**, 808–811 (1997).
33. P.W. Shabram et al., *Hum. Gene Ther.* **8**, 453–465 (1997).
34. C.A. Wilson, T.H. Ng, and A.E. Miller, *Hum. Gene Ther.* **8**, 869–874 (1997).

See also ANIMAL CELL PRODUCTS, OVERVIEW; VIRUS REMOVAL FROM PLANTS.

CELL STABILITY, ANIMAL

Martin S. Sinacore
Timothy S. Charlebois
Denis Drapeau
Mark Leonard
Scott Harrison
S. Robert Adamson
Genetics Institute
Andover, Massachusetts

OUTLINE

The genotypic and phenotypic stability of animal cells in culture are of great interest to cell and molecular biologists conducting basic biological investigations and to those in the biopharmaceutical sector engaged in the applied aspects of cell-culture-based process development. The genotypic and phenotypic status of animal cells is under constant assault by both intrinsic and extrinsic factors that conspire to invoke change in the nature of animal cell in culture. Since change is one of few certainties in nature, it is the context in which changes manifest that define their impact. The cell transformation process profoundly changes the genotype and phenotype of animal cells. In the context of a living organism, this can have significant negative effects while, in another context, it can enable cells to grow perpetually *in vitro* to the benefit of basic biological investigations. The focus of this chapter will be to review and discuss the mechanisms by which the genotypes and phenotypes of transformed animal cells change and, once established, the means by which genotypes and phenotypes can be stably maintained in continuous culture.

The starting place for the discussion of animal cell stability is to review factors that affect genomic stability of endogenous genes at the molecular level. This discussion is centered on the incompletely understood mechanisms and manifestations of genetic and phenotypic instability associated with oncogenic transformation of cells in culture. This is by no means a comprehensive review of the field of cell transformation but rather is focused on several key aspects of the transformation process that have been shown to directly impact genetic and phenotypic stability of animal cells during continuous culture.

Cell transformation is a profound alteration in the genotypic and phenotypic attributes of animal cells and represents a critical turning point in the molecular decision making process that signals continued proliferation of cells, cellular senescence, or activation of apoptotic pathways. The balance between the activation of proto-oncogenes and the inactivation of tumor suppressor genes is at the center of the transformation process. Just as important are processes by which the genetic integrity of animal cells is protected and preserved through the action of DNA repair systems activated by errors generated during DNA replication or in response to DNA damage inflicted by radiation or chemical assaults to cells.

With some understanding of the molecular processes that preserve the stability of endogenous genes, we then move on to consider the factors that affect the stability of heterologous genes in recombinant cells and factors that influence maintenance of stable phenotypes during continuous culture. These subjects are of particular interest to scientists focused on the development of biological systems that rely upon the stable expression of heterologous genes in continuously cultured animal cells. Frequently in these systems, the level of foreign protein expression is maximized through gene amplification strategies that create additional challenges with respect to genetic stability that arises from the dispersion of gene integrants throughout the cell genome. Flexibility or adaptability of animal cell phenotypes within the context of cell biology and biopharmaceutical process development offers the opportunity to shape animal cells to fit a particular experimental system or cell-culture process paradigm. However, once the appropriate genotypes and phenotypes are established through cell line adaptation strategies or by direct genetic manipulation, further change (instability) of those attributes is not welcome. Herein lies the paradox of animal cell technologists—practitioners depend and rely on the pliability of cell genotypes and phenotypes in developing for instance, large-scale cell culture processes to produce recombinant proteins, yet once established, strive to make those attributes stable to provide assurance of process and recombinant product consistency.

For the cell and molecular biologists set upon the task of developing cell-culture-based manufacturing processes, the possible linkage between genetic (and phenotypic) stability and recombinant product quality has historically been the subject of discussion and investigation. The consequence of genotypic and phenotypic instability on recombinant protein quality is discussed by way of specific examples derived from experience in cell-line and cell-culture process development at Genetics Institute. These case studies provide a basis for arguments against a linkage between cell stability and recombinant protein quality and exemplify the nature of the challenges that are faced in developing robust cell-culture-based processes used in the manufacture of biotherapeutic proteins. Finally, approaches taken at Genetics Institute to characterize genotypes of recombinant cell lines for evaluating the genetic stability of recombinant cell lines and cell culture processes are discussed.

FACTORS THAT AFFECT GENETIC STABILITY OF ENDOGENOUS GENES

Oncogenic Transformation, Aneuploidy, and Genetic Stability

Genetic instability, manifested as various chromosomal aberrations such as aneuploidy, gene amplification, translocation, deletions and point mutations, is frequently observed during oncogenic transformation of cells. The cause and effect relationship between genetic instability and cell transformation has been a subject of much discussion, particularly with respect to the role of aneuploidy in these processes. Aneuploidy (an abnormal complement of chromosomes) is a massive genetic abnormality observed in nearly all of the many thousands of solid human cancers examined. This fundamental observation has led to the "aneuploidy-genetic instability" hypothesis which holds that the degree of aneuploidy is proportional to the degree of genetic instability associated with oncogenic transformation.

The aneuploidy-genetic instability hypothesis gained support over the years through the tight association between aneuploidy, genetic instability, and carcinogenesis. Recently, investigations using clones of chemically and spontaneously transformed Chinese hamster embryo cells indicated that clones were 50–100% aneuploid. Furthermore, colonies of transformed cells were highly heterogeneous with respect to chromosomal number, suggesting that aneuploidy can perpetually destabilize the karyotypes of these cells (1). Likewise, it was found that karyotypic instability was proportional to the degree of aneuploidy in cultures of continuously passaged clonal Chinese hamster embryo transformants (2). Furthermore, the quasidiploid karyotype of wild-type and a variety of mutant Chinese hamster ovary (CHO) cells were shown to be relatively stable after being cultured for extended periods of time. In contrast, quasitetraploid cells, derived spontaneously or by Sendai virus-induced fusion, were more prone to losses and gains in whole chromosomes, rearrangements of chromosomes, and the appearance of new and distinctive chromosomes (3). A model has emerged from these observations in which the loss of genes (or sets of genes in chromosomes) involved in mitosis perpetuates unequal segregation of chromosomes in subsequent cell divisions. Clearly, under such a model, stable maintenance of a genotype that contains an abnormal number of chromosomes would be improbable.

The notion that aneuploidy is a cause of genetic instability was challenged by data suggesting that an abnormal number of chromosomes in a cell, in and of itself, could not account for the high level of chromosomal instability found in aneuploid colorectal cancer cells (4). The authors proposed that dominant genetic defects in chromosomal segregation result in chromosomal instability, which drives aneuploidy. The basis for this assertion came from experiments in which the genetic stability of cells was not disturbed by the introduction of a single extra chromosome or by gross aneuploidy created by cell fusion techniques. Similarly, using gene amplification as an index for genetic instability, results from another study indicated that the propensity of methotrexate-resistant CHO cells to amplify endogenous dihydrofolate reductase genes was comparable in quasidiploid and quasitetraploid subclones (5). However, it was noted that genetic instability of aneuploid cells created spontaneously or by cell fusions was more pronounced when the resultant chromosomal number was nonmodal (4). Although a cause and effect relationship between aneuploidy and genetic instability has not been firmly established, it seems clear that once an aneuploid karyotype has been established in transformed animal cells, maintaining a stable karyotype in successive culture passages becomes more difficult.

Molecular Regulation of Genetic Stability

Tumor Suppressor Genes and Proto-Oncogenes. A hypothesis opposing the aneuploid-genetic instability hypothesis, termed the "somatic gene mutation" hypothesis, holds that activation of proto-oncogenes and inactivation of tumor suppressor genes trigger transformation and genetic instability. Under the somatic gene mutation hypothesis, animal cell transformation is the result of a multistage process starting with the activation of cellular proto-oncogenes and inactivation of tumor suppressor genes. These and other events arise from (or lead to) an enhanced state of genetic instability in which the genotypic (and phenotypic) diversity among clones increases. Most notably, this diversification manifests itself in changes in cell morphology and karyology among clones in a transformed cell population. The molecular mechanisms that underlie the genotypic and phenotypic instabilities associated with cell transformation are incompletely understood. However, some key molecules have been identified and their roles in regulating and maintaining genetic stability have become clearer. The following discussions focus on these key molecular factors and the mechanisms by which they influence the stability of cell genomes.

The p53 Tumor Suppressor Gene. Genetic alterations that involve inactivation of tumor suppressor genes are among the most thoroughly documented changes associated with animal cell transformation. In particular, mutation in the p53 tumor suppressor gene is frequently observed in human cancers.

The p53 tumor suppressor gene has been associated with critical cellular events involved in cell immortalization, growth arrest at the G1/S boundary in the cell cycle (i.e., cellular senescence), and activation of cell death by apoptosis (6,7). In these capacities, p53 has been considered a central factor of a genomic protection system in which, it has been hypothesized, accumulation of excess somatic mutations is prevented by growth arrest (presumably to facilitate DNA repair) or by the prevention of the further propagation of such DNA defects via apoptotic cell death. Because of the pivotal role of p53 in protecting cells and tissues from the generation and proliferation of genetic defects, it has been called the "guardian of the genome" (8).

Consistent with its role in maintaining genomic stability, the loss of p53 tumor suppressor functions results in genetic instability manifested by changes in chromosomal ploidy (9–14) and gene amplifications and deletions (12,13,15,16). Furthermore, in contrast to normal mouse fibroblast cells, mouse fibroblast cells that lack p53 protein display abnormal centrosomal amplification that causes unequal segregation of chromosomes (17).

The role of the p53 tumor suppressor function in genetic stability may have its basis in its proposed role in DNA repair. The human p53 protein contains both specific and non-specific DNA-binding domains that are hot spots for mutations that inactivate p53 function (6). Wild-type p53 protein can bind to damaged DNA and it has been shown that it plays a role in nucleotide excision repair in response to ionizing radiation (18–22) and bridge-breakage fusion of DNA (23), which has been implicated in gene amplification processes.

The Retinoblastoma Tumor Suppressor Gene. The possible role of tumor suppressor genes in genomic stability was also exemplified in studies involving the retinoblastoma tumor suppressor gene (pRb). Like p53, pRb is central in regulating cell-cycle progression and can act alone or in concert with p53 to arrest cell growth in response to DNA damage induced by gamma or UV irradiation or treatment with chemotherapeutic drugs (24,25). Linkage between pRb - and p53-mediated arrest pathways appears to come about via inhibition of pRb phosphorylation by cyclin-dependent kinases. Hypophosphorylated pRb promotes a growth-suppressive state via E2F transcription-factor-mediated repression of genes required for cell cycle progression (26). Similar to the model proposed for p53, pRb-mediated growth arrest and activation of DNA repair systems prevent replication of damaged DNA.

Proto-Oncogenes and Genomic Instability. The proto-oncogene *c-myc* is a key regulator of both cell growth and apoptosis and has been associated with cell immortalization and transformation (27). Gene amplification, a marker for genetic instability in mammalian cells, is often associated with *c-myc*-mediated transformation. Targets of *c-myc*-induced gene amplification include dihydrofolate reductase (28,29). Amplification of the dihydrofolate reductase gene may be associated with ongoing gene rearrangements and further genomic instabilities that lead to neoplastic transformation (30). Furthermore, it has been shown that *c-myc* has a direct role in activating telomerase by inducing the catalytic subunit, telomerase reverse transcriptase (31). As discussed in a section later in this chapter, it has been shown that telomerase is critical in maintaining genomic stability.

Likewise, expression of the human H-ras oncogene in mammalian cells results in cell transformation and DHFR gene amplification within a single cell cycle (32–34). H-ras tranformation is also associated with loss of contact inhibition of growth and the ability to form colonies in soft agar. Furthermore coexpression of the Ras-related tumor suppressor gene, *Rap1*, suppressed H-ras-induced amplification of the DHFR gene (35). Coexpression of the *Rap1* gene also caused H-ras-transformed cells to revert to a flat morphology and reestablished contact inhibition of growth and loss of the capacity to grow in soft agar. These studies demonstrated the profound influence of proto-oncogenes on cell phenotypes and genotypes.

Phenotypic Change Due to DNA Methylation and Gene Silencing. Enzymatic methylation of the C-5 position of cytosine residues in DNA motifs termed CpG islands may lead to inactivation of genes in a heritable fashion (36). Hypermethylation of promoter sequences may lead to silencing of genes by inhibiting DNA transcription as a result of interference with transcription factor binding. Several tumor-suppressor genes contain CpG islands in their promoters and show evidence of methylation-based gene silencing. In addition, the majority of colorectal carcinomas with microsatellite DNA instability have hypermethylation of the promoter CpG island of the DNA mismatch repair gene MLH1 (37). Hypermethylation-induced attenuation of the DNA mismatch-repair system may, in turn, render the genome susceptible to additional genetic hits that lead to further genomic instability and oncogenic transformation. However, whether hypermethylation of DNA sequences is a cause or consequence of gene inactivation is highly disputed.

Poly(ADP-Ribose) Polymerase. Interruptions of DNA strands brought about by DNA-damaging agents (e.g., γ irradiation, alkylating reagents) elicit a rapid response that involves the mobilization of several DNA repair systems. Components of these DNA repair systems have been implicated in DNA gap repair and religation, DNA integration and recombination efficiency, and genomic stability. One such rapid response to DNA damage involves attachment of poly(ADP-ribose) polymerase (PARP) to DNA strand breaks (38,39). PARP is found in abundance tightly associated with chromatin in the nuclear compartment of eukaryotic cells. The PARP protein contains three functional domains: a DNA-binding domain, an automodification domain, and a catalytic domain. The DNA-binding domain rapidly binds to DNA strand breaks which in turn stimulates the catalytic domain. Using NAD as substrate, the catalytic domain carries out extensive ADP-ribosylation of PARP in the automodification site, as well as in a variety of nuclear target proteins, including topoisomerase and histones. PARP and other ADP-ribosylated proteins are recycled to their active state by the action of poly(ADP-ribose) glycohydrolase.

ADP-ribosylated PARP has a reduced affinity for DNA because of electrostatic repulsion, and the complex falls off the DNA strand and allows other DNA repair enzymes to dock and carry out DNA repair. Therefore, PARP functions as a "molecular nick sensor" in eukaryotes and, by way of direct interaction with nicked DNA, protects the site from the action of nucleases and other DNA-binding proteins. Furthermore PARP may be acting to home in other proteins at the damaged site, such as XRCC1, DNA ligase III, DNA polymerase β, p53 and DNA-dependent protein kinase and, in so doing, primes damaged DNA for repair.

The temporary protection of nicked DNA by PARP and the rapidity of the response have been considered indicative of the critical role of PARP in maintaining DNA integrity. Several lines of investigation have implicated its role in genomic stability (40–42) gene amplification (43), efficient retroviral infection (44), DNA transfection (45,46), and DNA base excision repair (47). Compelling evidence of the critical role of PARP in maintaining genomic stability comes from work done with PARP (−/−) knockouts. Mice that lack the gene that encodes PARP display no phenotypic abnormalities but are sensitive to γ irradiation and treatment with genotoxic agents like N-methyl-N-nitrosourea (48–50). In addition, primary cultures of PARP (−/−) cells are more prone to sister chromatid exchange that indicates a higher rate of rejoining of chromosomal breaks (41,43,48,49). Cells from PARP (−/−) animals also tend to display an increased number of micronuclei (a marker for genomic stability) after γ irradiation or treatment with mitomycin C (41,48) and exhibit a larger number of chromatid breaks and chromatid exchanges (49).

The potential interplay between PARP and other DNA-binding proteins like p53 and DNA-dependent protein kinase (discussed in the section following) has also been a subject of investigation. Agarwall et al. (49) observed a twofold lower basal level of p53 in PARP (−/−) cells and diminished induction of p53 in response to DNA damage that combine to reduce the induced level of p53 protein by four- to fivefold compared to normal cells. However, a decrease in the induction of p53 activity in PARP-deficient cells was not noted in these cells. In contrast, splenocytes from PARP knock-out mice exhibit normal or even elevated p53 (50). These results indicated that p53 protein induction may be regulated by a combination of PARP-dependent and-independent pathways. Finally, using PARP-specific antibody, it was shown that PARP and p53 coimmunoprecipitate in HeLa cell extracts after induction of apoptosis (51). This suggests that PARP and p53 may directly interact under these conditions.

DNA-Dependent Protein Kinase. It is thought that DNA-dependent protein kinase (DNA-PK) is critical in T-cell receptor V(D)J recombination in lymphoid cells and DNA double-strand break repair (52). The recombination process involves site-specific cleavage by recombination activating gene (RAG)1 and RAG2, that generates double-strand breaks that are specifically joined and repaired. DNA-PK binds to DNA double-strand breaks leading to activation of the protein kinase catalytic subunit. Phosphorylation of DNA-binding proteins such as transcription factors Sp1, c-Jun, and p53 takes place along with autophosphorylation of DNA-PK subunits. Localized phosphorylation of transcription factors may arrest transcription in the vicinity of the double-strand break that allows docking and assembly of DNA repair machinery.

Cells derived from animals that bear the spontaneous severe combined immunodeficiency (SCID) mutation lack DNA-PK activity and are defective in V(D)J recombination and DNA repair (53–55). The role of DNA-PK in DNA recombination and repair has been definitively confirmed in murine DNA-PK null knockouts (56,57) and in a variety of murine and CHO cell mutants that lack DNA-PK activity (58–62).

Both DNA-PK and PARP are nuclear proteins that bind to DNA at strand breaks and may, in fact, compete for binding to DNA strand breaks induced by genotoxic agents or by normal recombination. These similarities have led some to suggest that PARP and DNA-PK can substitute for one another under different physiological circumstances (38). Because PARP mutant cells have elevated levels of spontaneous sister chromatid exchange, indicative of a higher rate of rejoining of chromosomal breaks, some have hypothesized that a PARP deficiency might permit processing and ligation of V(D)J recombination intermediates. Indeed, Morrison et al. (63) tested this hypothesis directly by creating SCID/PARP (−/−) double knock-out mice. These mice have a reduced survival rate, and double knock-out survivors appear to have regained, to a limited extent, the ability to carry out V(D)J recombination. These results suggest that PARP deficiency partially rescues animals from the SCID phenotype and may play a direct role in the V(D)J recombination process. In this context, PARP appears to be acting as an "antirecombinogenic molecule" suggesting that PARP and DNA-dependent protein kinase act cooperatively to minimize genomic damage caused by DNA strand breaks. These observations confirm the critical roles of both PARP and DNA-PK in maintaining genomic stability.

Telomeres and Telomerase. Cultures of primary human and other mammalian cells undergo a limited number of cell divisions, cease to divide thereafter, and often go on to die. This phenomenon, termed replicative senescence, has been a model for cellular aging (64). As cells age, the specialized nucleoprotein structures located at the ends of chromosome, called telomeres, become progressively shorter. Telomeres protect chromosomes from DNA degradation, end-to-end fusions, rearrangements, and chromosomal loss.

Telomeres are composed of tandem $(TTAGGG)_n$ nucleotide repeats bound to a unique family of telomere-binding proteins, including telomerase, which maintains telomeric length by synthesizing telomeric repeats (65,66). Telomerase is an enzyme that consists of an RNA-protein complex containing a reverse transcriptase component and an RNA template. The small proportion of cells that become immortalized and emerge from a growth crisis associated with cellular senescence were correlated with the presence of telomerase activity whereas precrisis cells were not (67,68). Therefore, continued maintenance of telomeres by the action of telomerase is considered a critical event responsible for preventing cellular senescence and for continued growth of transformed cells. Indeed, direct demonstration of the role of telomerase activity in overcoming senescence was obtained by overexpressing telomerase in cells (68–70) and the observation that immortal HeLa cells can be forced to crisis when telomerase activity is inhibited by expression of the antisense transcript to the RNA component of telomerase (71).

A critical role of telomerase function in maintaining genomic stability was observed in mice in which the gene that encodes the telomerase RNA component was knocked out (72). As these telomerase-deficient mice aged, an increased frequency of chromosomal fusions and a twofold increase in aneuploidy were observed. A model emerged from these studies in which chromosomal aberrations associated with crisis may be brought about by shortening telomeric structures into subtelomeric regions that lead to the formation of dicentric chromosomes via end-to-end fusion (73). In addition, the observed genomic instability was coincident with an increased frequency of spontaneous cancer in aged animals. All of these observations were correlated with shortening of telomeric length.

Activation of telomerase and cell immortalization has been associated with inactivation of p53 and pRb (73,74,75), and the telomerase gene is a direct target for activation by c-myc (76). The potential interplay between p53 and telomerase functions was also suggested by the observation that inhibition of telomerase activity triggered cell death (77).

GENOTYPIC AND PHENOTYPIC STABILITY OF RECOMBINANT CELL LINES

Stable Integration of Heterologous Genes in Animal Cell Genomes

Animal cells have been used as hosts for expressing a wide variety of heterologous proteins. Introduction of foreign DNA into animal cells by using any one of a number of transfection techniques (e.g., electroporation, lipofection and calcium phosphate coprecipitation) leads to transient expression of genes for up to several days. Stable expression of heterologous genes, however, is associated with integration into the host-cell genome. The mechanism by which foreign DNA integrates into the genome of host cells is not fully understood. Presumably, the integration process involves the normal DNA excision-repair systems, DNA ligase, and possibly eukaryotic analogs of prokaryotic integrase activities (78, 79). Consistent with this model are observations in which perturbations of the eukaryotic DNA replicative and repair systems have been shown to affect the frequency of cell transformation, indicating that integration of foreign DNA (via heterologous recombination) may arise from normal DNA replication and/or double-strand break repair mechanisms (80), including PARP (81).

Isolation of transfected cells that have stably integrated and express heterologous genes can be achieved by coexpressing selectable (and often amplifiable) marker genes that confer resistance to cytotoxic agents on recipient cells. The use of the selectable and amplifiable marker genes dihydrofolate reductase (DHFR) and glutamine synthetase (GS) has been extensively studied (82). In the former, coexpression of DHFR and another gene of interest renders recipient cells resistant to the cytotoxic folate analog, methotrexate (MTX). Further selection of drug-resistant transfected cell populations with increased concentrations of MTX results in coamplifying DHFR and the gene of interest. Drug-resistant cells cultured in the absence of the selective agent are vulnerable to the loss of amplified genes, whereas it has been shown that extended cultivation of cells under selective pressure increases the stability of amplified genes (83,84).

Once integrated, the genomic localization of amplified genes, it has been shown, influences their stability. Unstable amplified genes tend to be found in acentric double minute chromosomes that do not segregate properly during mitosis and are rapidly lost with each cell division. In contrast, stably amplified genes are usually found integrated into the chromosome associated with homogeneously staining regions that replicate in the early part of the S phase (85–87). Furthermore, amplified plasmid DNA was found associated at or near the ends of chromosomes or on unstable dicentric chromosomes. These observations suggested that integration and/or amplification of DNA may disrupt telomeres and facilitate formation of unstable dicentric chromosomes. Instability of amplified genes in the absence of selective pressure may also be related to the tendency of integrated genes to be arranged in a "head-to-tail" mode, that increases the frequency of homologous recombination between the integrated plasmids (87).

The distribution of amplified genes in a recombinant CHO cell line was elucidated using flourescent in situ hybridization (88,89). In highly amplified cells cultured under selective pressure, integrated amplification units were widely dispersed throughout the genome. Upon removing drug selection, subpopulations of recombinant CHO cells displayed genotypic instability marked by the loss of amplified genes in multiple and heterogeneous integrations dispersed throughout the genome. Continued cultivation of unstable cell lines under nonselective conditions resulted in the appearance of a dominant integration, termed the "master integration," which remained stable for up to 100 days in the absence of selective pressure.

Other examples of genomic positional effects come from work done to optimize expression of heterologous genes in cultured cells and transgenic animals. In one example of positional effects, Wurm et al. (90) showed that incorporation of retroviral sequences, endogenous to CHO cells, within expression plasmids significantly improved the efficiency with which high-expressing clones were obtained. The authors speculated that homologous recombination of the retroviral sequences in the expression plasmid with endogenous sequences in the CHO genome targeted the integration of foreign DNA into transcriptionally active regions of the CHO genome. In another example, the targeted insertion of a hypoxanthine phosphoribosyl transferase (HPRT) marker gene at different loci in embryonic stem cells resulted in heterogeneity in expressing stability phenotypes (91). Interestingly, loci displaying expression instability were associated with hypermethylation of the promoter which could be corrected by inserting a different promoter that contains a CpG island.

Phenotypic Heterogeneity and Population Dynamics

Recombinant CHO cells, in which amplified coexpression of DHFR and a heterologous gene of interest has been

achieved, generally display good correlation between gene copy number, RNA transcription levels, and heterologous protein expression (92–94). Although this relationship is generally true, there are examples in which divergence in these parameters has been noted. Gene silencing by hypermethylation may play a role in creating phenotypic divergences within cell populations. However, phenotypic divergences can come about through alternate mechanisms as well. For example, it has been shown that expression levels of recombinant hepatitis B surface antigen in CHO cells became uncoupled from gene copy number because of decreased secretion efficiencies and increased intracellular degradation of recombinant protein (92). In another study, population analysis of recombinant CHO cells after extended culture in the presence of methotrexate indicated an increase in heterogeneity among subclones with respect to their cellular productivity phenotypes. In contrast, the recombinant CHO cell population became more homogeneous with respect to expression of DHFR (95). These results indicate that coexpression of DHFR and heterologous proteins can diverge post-transcriptionally at the level of protein translation, post-translational modifications, or protein secretion.

Instability of amplified genes in recombinant CHO cell populations may result from phenotypic changes caused by the amplification process itself or by the large increase in heterologous gene copy number in the host-cell genome. Results from several investigations have suggested that increased gene copy number may impose a metabolic burden on recombinant cells that manifests in a decrease in cellular growth rates (92,96). The model emerging from these observations would be one in which the population is overcome by faster growing, low-producing subpopulations. This model also has precedent in populations of hybridoma cells in which increased monoclonal antibody secretion rates can be associated with decreased growth rates (97,98) and expression instability associated with the appearance of fast-growing, nonproducing subpopulations (99,100). However, other investigations point to a mechanism of population instability in which no increase in nonproducers was noted, but rather a general decline in heterologous protein secretion rates was observed (95,101). Furthermore, the growth rates of clones isolated from a population of recombinant CHO cells went unchanged despite a decrease in cellular productivity after extended culture in the absence of selective pressure (94).

In general, a positive correlation between cell growth rates and cellular productivity of heterologous proteins has been observed in triphasic batch cultures that experience an initial lag phase followed by a logarithmic growth phase and a plateau phase in which cell growth ceases (102–105). However, it has been shown that an increase in cellular productivity was obtained in growth-arrested recombinant CHO cells (106–108) and hybridoma cells (97). The promoter/enhancer systems used to drive expression of heterologous genes (109) and possibly the chromosomal location/environment into which the genes are integrated, are likely to play roles in the degree to which cellular growth rates and cellular productivity phenotypes are coupled.

Genetic and Epigenetic Manipulation of Cell Phenotypes

Cell phenotypes can, and have been, altered, by genetic modifications in which, for example, cells can be made resistant to cytotoxic drugs by introducing the appropriate drug-resistance marker genes. Furthermore, fundamental phenotypic attributes of animal cells such as the proliferative requirement for attachment strata (anchorage-dependency) and dependence on serum or exogenous growth factors, can be genetically manipulated. Renner et al. (110) showed that the anchorage-dependence of Chinese hamster ovary cells was eliminated by expression of cyclin E. Similarly, a loss in the anchorage-dependency of fibroblasts was associated with overexpression of cyclin A (111). In another series of experiments, the robust growth of hybridoma and CHO cells under a variety of culture stresses was enhanced by overexpression of anti-apoptotic genes such as bcl-2 (112).

Cellular phenotypes can also be shaped in response to environmental conditions imposed upon them during cell culture. Examples of this phenomenon come from successful laboratory attempts to alter a variety of phenotypes through strategies that involve cell line adaptation. Much of the impetus for intentionally reshaping cellular phenotypes comes from the desire to eliminate the serum- and anchorage-dependency of animal cells used in producing protein biopharmaceuticals (113).

Without the use of adaptation procedures, the growth of CHO and BHK cells tends to be both anchorage-dependent and serum-dependent. In addition, once a serum-free suspension-adapted lineage is established, the volumetric productivity of the cell culture process can be optimized by adaptation strategies that lead to significant improvements in the final cell densities of cultures (113). Although the mechanism by which animal cells adapt to high cell-density conditions is unclear, one means is by developing tolerance to growth-inhibiting substances released by the cells, such as lactic acid and ammonia (114–118).

The adaptation to serum-free suspension or high cell-density conditions is generally performed using cell lineages that have already been engineered to express the desired protein biotherapeutic through recombinant technologies or methods based on cell fusion (102,119–124). However, reports have been made of CHO (125–127), murine myeloma (128,129), and Namalwa (130) cell lineages that were adapted to growth in serum-free suspension culture before serving as host cells. This approach greatly reduces or eliminates the need for further adaptation once expression of the desired protein is achieved in these preadapted mammalian-cell host lineages.

The transition to high cell-density, serum-free suspension culture can lead to changes in the growth performance phenotypes of cells and the structural characteristics of secreted proteins. Some decline in growth performance (i.e., decreased growth rate or cell viability) following serum withdrawal and/or removal of cell attachment substrata was reported (126,131). This response to changes in culture conditions is potentially due to a perturbation of events associated with cell-cycle progression and entry into the S phase (132). In the most severe case, recovery from such perturbations may be prevented by activating

apoptotic pathways (133,134). Furthermore, significant differences in the carbohydrate structures of secreted proteins were documented (135–140). Finally, cell populations that undergo a "growth crisis" (i.e., a significant decline in growth rate and cell viability) during the adaptation process may be vulnerable to outgrowth of nonrepresentative or undesirable subpopulations of cells that may not have optimum performance characteristics (126,141).

Cell growth and expression phenotypes can also be significantly influenced by cell-culture parameters such as temperature, medium pH, and dissolved oxygen levels. For instance, lowering the temperature of cell cultures can decrease cell growth rates, increase protein expression rates, and alter nutrient metabolism (106,142). These results suggest that culture temperature is a variable with the potential for improving the performance of cell-culture processes, such as fed-batch and perfusion-based processes, in which static growth and prolongation of relatively high protein-expression rates are desired outcomes.

In addition to the effects of temperature, cell phenotypes can be reshaped by the action of agents that alter gene expression patterns. The most widely studied of these agents is sodium butyrate. Butyrate inhibits cell growth and alters gene expression of butyrate-responsive elements (positively or negatively) by increasing acetylation of nucleosomal histone proteins via inhibition of histone deacetylase (143). Sodium butyrate administration alone or in combination with a temperature shift has been shown to improve the expression of recombinant proteins (143,144).

CONSEQUENCES OF GENETIC INSTABILITY ON RECOMBINANT PROTEIN QUALITY

The genetic stability of recombinant cell lines and its impact on the quality and consistency of protein biopharmaceuticals continues to be a topic of great interest to process scientists and regulatory agencies. Many different technologies have been used in constructing expression vectors and expression systems, as well as manufacturing processes. The genetic stability of recombinant cells, defined within the limits of a particular cell-culture manufacturing process, has historically been a surrogate marker for biopharmaceutical protein consistency. However, the link between genetic stability and recombinant protein quality and consistency has not been absolutely established or may be of reduced importance in view of improved analytical methods for protein product characterization.

Two key phenotypic markers thought to be relevant to protein synthesis have been emphasized when assessing the stability of recombinant cell lines and cell-culture processes: specific growth rate (hours^{-1}) and cellular productivity (units of product produced per cell per unit time). To date, where possible, cell lines have generally been selected and manufacturing processes designed so that the product was made during a period when growth rate and cellular productivity are essentially constant. This approach can be restrictive at times, adds significant time and resources to the development process (e.g., when a number of candidate cell lines need to be evaluated to select a stable production cell line), or disallows flexibility

in a manufacturing process (e.g., when a process has to be limited in duration to ensure that the product is made during a period of stable growth rate and cellular productivity).

The challenges of establishing the stable performance of cell-culture processes used in manufacturing recombinant biotherapeutic proteins are complex because of the variations inherent in biological systems and the metrics used to monitor their performance. An example of this can be seen when one examines cellular productivity data from a recombinant CHO (rCHO)-cell-based process developed and used by Genetics Institute to manufacture recombinant human antihemophiliac factor (rAHF, Recombinate). rAHF is a highly complex glycoprotein used therapeutically in treating hemophilia A (145). Efficient expression of rAHF by CHO cells is greatly facilitated by coexpressing von Willebrand factor (vWF) which binds to and stabilizes the rAHF molecule. In constructing the rCHO cell line used in the commercial production of rAHF, amplified coexpression of rAHF and vWF was driven by separate selectable/amplifiable markers: DHFR and adenosine deaminase, respectively (82). Therefore, complete evaluation of the expression phenotype of rAHF-producing rCHO cells required evaluating both rAHF and vWF cellular productivities. Based on this analysis, a rCHO cell line was identified and used to develop a rAHF commercial manufacturing process.

A conventional two-tiered cell banking system was used in which a Master Cell Bank (MCB) and Working Cell Bank (WCB) were prepared. The batch-refeed cell-culture process developed and used in the largest-scale manufacturing of rAHF starts when cells are inoculated into a production bioreactor (2500 L) after expansion from the frozen WCB. After inoculation, WCB cells are grown for three days (which constitutes a single batch), by which time they have reached maximum cell density (just before the onset of the stationary phase). At this point, approximately 80% of the cell suspension is removed from the bioreactor for cell separation and purification and is replaced with an equivalent volume of fresh medium. The remaining 20% of the cell suspension acts as the inoculum for the next cycle. This process continues up to a maximum number of batches which is established for an individual cell line based on cell-culture performance characteristics and product characterization. Throughout this process, cells are cultured in the absence of selective pressure.

The perspective that genetic stability served as a surrogate marker for recombinant protein quality and consistency led to establishing operational limits for the rAHF cell-culture process within which rAHF expression was relatively stable. To establish process limits, the rAHF cell-culture process was operated at full scale for an extended period in which the number of cumulative population doublings (CPD) allowed for the WCB cell population was extended to more than 120. The rAHF and vWF cellular productivity data were acquired and analyzed for trends. Figure 1a shows the rAHF and vWF cellular productivity data obtained during the operation of the extended process. Upon extended culture of the WCB cell line to greater than 70 CPD, both rAHF and vWF productivity gradually declined. Southern blot and

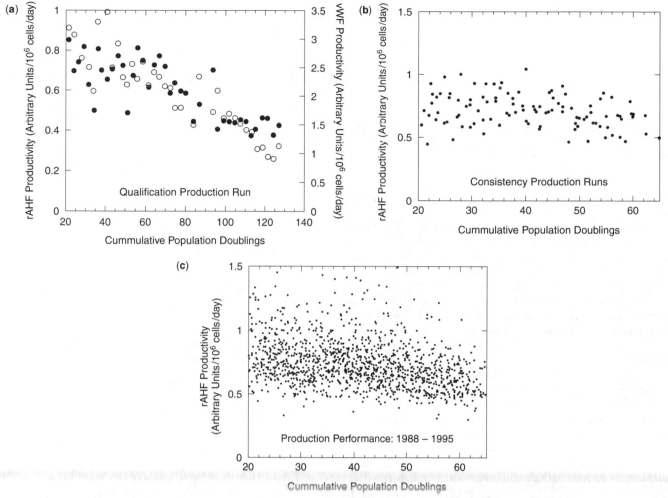

Figure 1. Performance of the rAHF cell culture manufacturing process. The rAHF cell-culture process was operated, as described in the text, for an extended period of time or for periods restricted to 65 total CPD. Expression of rAHF (closed circles) and vWF protein (panel A, open circles) in conditioned medium was determined by using an activity assay and by ELISA, respectively, and the cellular productivities were calculated.

Northern blot analysis of in-process samples indicated that qualitative changes in the genes or RNA transcripts that encode either protein could not be detected (data not shown). The results also indicated that a significant change in rAHF gene copy number or RNA transcript level was not observed over the course of the extended cell culture (approximately 127 CPD from the WCB). However, a reduction in vWF gene copy number and transcript level was observed after 110 CPD of continuous culture (data not shown). Consistent with the physiological role of vWF in stabilizing rAHF protein and because a detectable change in rAHF gene copy number or transcript level was not found, we interpreted these results to mean that the decline in rAHF productivity was driven by vWF expression instability after 70 CPD of continuous culture. With this information in hand, the duration of the cell-culture process was restricted to 65 CPD to ensure consistent process performance.

To validate the consistency of the manufacturing process operationally restricted to 65 CPD, two additional production runs were performed to establish a database for process consistency. The results obtained from the process consistency runs indicated that the cellular productivity phenotype of WCB cells was relatively stable when operationally restricted to 65 CPD (Fig. 1b). However, when cellular productivity data obtained from commercial production runs carried out from 1989 through 1995 are plotted, a slight negative trend (representing a decline of approximately 30%) was observed in the data (Fig. 1c). Therefore, the phenotypic stability of the production cell line and cell-culture process may be difficult to evaluate fully in the absence of a suitably large performance database. Thorough biochemical characterization of the rAHF produced during this period indicated that a detectable qualitative change in the product did not occur, challenging the notion that cellular productivity is an appropriate surrogate marker for recombinant protein quality.

To further investigate whether a correlation exists between changes in the cellular productivity (and growth rate) phenotypes of rCHO cells and recombinant product quality, a series of studies was performed at both bench

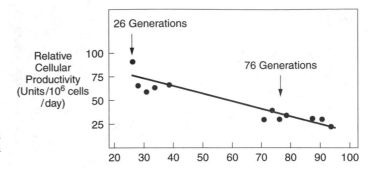

Figure 2. Relative cellular productivity of a rCHO cell line in a 2500-L batch-refeed process. Reprinted from Ref. 146 with permission.

and production scale, and the effect of variations in both phenotypic markers on product characteristics, with emphasis on N-linked glycan, were examined. The focus on protein glycosylation was based on an expectation that this complex, energy-requiring post-translational modification might well be influenced by changes in global phenotypic markers such as growth rate and cellular productivity. In all cases, a host/expression vector system was employed which uses CHO cells genetically engineered to contain multiple genomic copies of the product gene by using DHFR as the selectable/amplifiable marker.

In one such study, a rCHO cell line that produced a thrombolytic protein was cultured using a generic serum-free, suspension-culture-based, batch-refeed cell-culture process. Figure 2 shows the cellular productivity of a rCHO cell line as a function of generations from the WCB in a 2500-L batch-refeed process. Cellular productivity dropped by approximately 50% from the beginning to the end of the process. It was of interest to establish whether the progressive drop in cellular productivity caused a progressive change in the product profile. Accordingly, product derived from sequential batches was characterized by using a number of biochemical and biophysical analyses, including peptide mapping and specific activity (potency) determination. A significant change in product characteristics was not evident from batch to batch. This qualitative similarity is illustrated in Figure 3, which shows that high pH anion-exchange chromatographic (HPAEC) carbohydrate fingerprints (147,148) of complex type, N-linked glycan released from product produced from batches early (26 generations) and late in the process (76 generations) are indistinguishable. This case again illustrates that a substantial change in cellular productivity can take place within a population of continuously cultured cells without causing detectable changes in product characteristics.

Another study involved a large-scale production process employing a rCHO cell line that produced a glycosylated lymphokine. Figure 4 shows a profile of cell growth during a serum-free, suspension-culture-based, batch-refeed process similar to that described before. Population doubling time (inversely proportional to growth rate), calculated as an average over each three-day growth period, increased from an initial value of approximately 22 h in the first batch to approximately 32 h in batch 3 (B3) and then declined in batches 4, 5, and 6. The improved growth rate resulted in significantly higher cell densities in batches 5

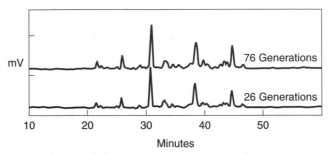

Figure 3. HPAEC carbohydrate fingerprint analysis of N-linked glycans. HPAEC carbohydrate fingerprints analysis was performed on complex type, N-linked glycans released from product produced from batches at the beginning (26 generations) and toward the end (76 generations) of a 2500-L batch-refeed process. Reprinted from Ref. 146 with permission.

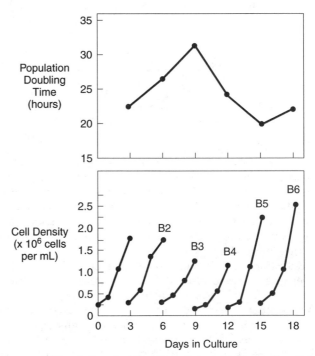

Figure 4. Population doubling time (t_d) and cell density of six batches of a large-scale rCHO production process. Reprinted from Ref. 146 with permission.

and 6. Subsequent product characterization revealed that the observed fluctuations in growth rate and maximum cell density did not cause any detectable change in the product (characterization included peptide mapping and specific activity determination, among others). Consistent with this conclusion, HPAEC-carbohydrate fingerprints of complex type, N-linked glycan from batches 2 to 6 were indistinguishable (Fig. 5). This study illustrates that substantial changes in growth rate can take place under the production conditions described without causing significant changes in product characteristics.

The final study involved a large-scale production process employing a rCHO cell line that produced a glycosylated cytokine in a serum-free, suspension-culture-based, batch-refeed system similar to that described before. Figure 6 shows the growth rate and cellular productivity (in arbitrary units) of the production cell line as a function of generations from the WCB. The growth rate increased as a function of generations from the WCB from approximately $0.025 \ h^{-1}$ (28 hour doubling time) at around 20 generations to approximately $0.033 \ h^{-1}$ (21 hour doubling time) at around 70 generations. In the same time frame, the cellular productivity declined by approximately 40%. It was of interest to establish whether this coordinate change in growth rate and cellular productivity caused any change in the product profile.

Subsequent product characterization, which included peptide mapping and specific activity determination, showed that the observed concurrent increase in growth rate and decline in cellular productivity had no effect on the product profile from batch to batch. HPAEC-carbohydrate fingerprint analysis was carried out in this case on the high-mannose glycan released from product that was purified at five different points in the production process. Figure 7 shows that glycan

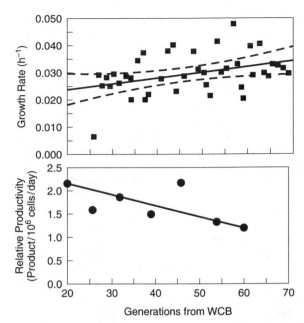

Figure 6. Characteristics of a large-scale rCHO production process in which changes in growth rate and cellular productivity were observed. Growth rate and relative productivity are indicated as a function of increasing generations. Reprinted from Ref. 146 with permission.

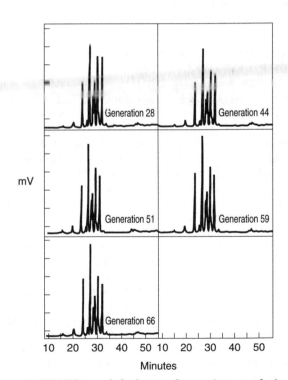

Figure 7. HPAEC carbohydrate fingerprint analysis of high-mannose glycan. HPAEC carbohydrate fingerprint analysis was performed on samples released from product produced at five different points in a large-scale rCHO production process. Reprinted from Ref. 146 with permission.

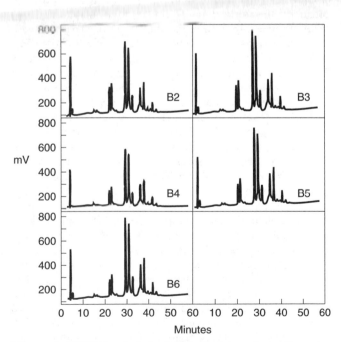

Figure 5. HPAEC carbohydrate fingerprint analysis of complex type, N-linked glycans. HPAEC carbohydrate fingerprint analysis was performed on samples obtained from five consecutive batches of a large-scale rCHO production process in which significant batch-to-batch changes in growth rate were observed. Reprinted from Ref. 146 with permission.

fingerprints were comparable from generation 28 to 66. This case study illustrates that significant changes can take place simultaneously in growth rate and cellular productivity without causing significant changes in product characteristics. The results from these studies clearly demonstrate that significant changes in growth rate and cellular productivity, under the conditions described, do not necessarily predict changes in product quality. Similar observations (with respect to cellular productivity) were made by others working with rCHO production systems (149).

These studies, at both small scale and manufacturing scale, illustrate a key principle pertaining to the development, implementation, and long-term success of a cell-culture-based biopharmaceutical manufacturing process. Manufacturers should not assume that absolute consistency of process performance is necessary to ensure that qualitatively consistent product is produced. In particular, for biopharmaceutical products whose active substance can be characterized in great detail using modern analytical methods, it is possible to determine with a high degree of confidence that variations in process parameters such as growth rate and cellular productivity do not lead to significant variations in the quality of the active substance produced. Therefore the emerging paradigm for "well-characterized" (150) or specified biotechnology products (151) is one that makes it possible for a manufacturer to rely on analytical characterization of an active substance to determine which measures of process performance are important for ensuring consistent product quality.

GENOTYPIC CHARACTERIZATION AND VALIDATION OF GENETIC STABILITY

Genotypic instability has the potential to impact the quality of heterologous proteins expressed by a recombinant cell line. Obviously, changes in the DNA sequence via point mutations (which can occur spontaneously at a frequency of $\geq 10^{-6}$), deletions, or gene rearrangements can alter the structure of the gene product. Accordingly, extensive genotypic characterization of recombinant cell lines needs to be carried out as part of a package that provides assurance of recombinant protein quality. A general practice for characterizing the expression construct is outlined here. Emphasis is placed on characterizing the MCB cells and the transcripts derived from them, so that a comprehensive understanding of the structure of integrated and amplified genes is established at the starting point of the manufacturing process. Once the manufacturing process has been established, end-of-production (EOP) cells taken from a representative run are analyzed in a manner similar to the MCB, to ensure that significant changes in the product genes or closely flanking sequences do not occur during the production process.

Gross Structural Evaluation of Genes and Transcripts

The evaluation of the expression construct includes an analysis of the integrity of the RNA transcripts that encode the protein(s) representing the active substance. This analysis employs standard molecular techniques (such as Northern blot hybridization) to establish that the predicted transcript(s) is produced by the production cell line and that significant qualitative changes in transcript structure do not occur over the course of production.

Similarly, the copy number of the expression construct is evaluated by using standard techniques. This analysis is undertaken not because the exact number of integrated and amplified copies is considered critical (it requires a very substantial effort to credibly obtain even a reasonably precise estimate of gene copy number). Rather, this estimate is useful as part of the general description of the way the cell line was established and how suitable expression levels were achieved. What is much more relevant to establish is that the correct coding sequence of the product has been incorporated into the host-cell genome and is maintained during culture to the end of production. Methods which can be employed to provide this assurance are illustrated here with several examples.

The gross structural integrity of integrated and amplified genes is evaluated by digesting genomic DNA at restriction sites that immediately flank the coding regions, followed by Southern blot analysis to establish the presence of the appropriately sized fragment in genomic DNA (usually compared directly to expression plasmid digested with the same restriction enzymes). Detection of fragments other than those predicted indicate the presence of rearranged or aberrant copies of these genes. By loading various amounts of control (expression plasmid) DNA in adjacent lanes, it is possible to estimate the sensitivity of this method for detecting variant sequences. In our experience, if aberrant fragments are electrophoretically distinguishable, they are detectable at a level of 1–5% of the intact genes.

The second level of characterization is directed toward a more detailed understanding of the initial expression plasmid integrants that are represented in the amplified DNA of the cell line. In particular, we seek to determine if one or more plasmid molecule is integrated into host CHO cell DNA (often in tandem at a single chromosomal site). This question is approached by using genomic Southern blot analysis, taking advantage of the known restriction map of the expression plasmid to permit identifying restriction fragments in which one site is contributed by host-cell genomic DNA flanking the expression plasmid. Such chimeric fragments are generally distinct in size for each of the original integrations. Thus, the number of distinct fragments identified can be used to determine the number of unique sites of integration and potentially, the number of plasmid molecules that were integrated upon transfection and before amplification. This information is valuable in planning a rational DNA cloning and sequencing strategy (discussed in more detail following).

Our approach to integration unit characterization is illustrated by considering the example of a rCHO cell line that produces recombinant human interleukin-12 (IL-12), a novel immunomodulator (152). The IL-12 molecule is a heterodimer of molecular weight 75 kDa, consisting of 35-kDa (p35) and 40-kDa (p40) subunits. IL-12-producing rCHO cells were generated by introducing p35 and p40 cDNAs into independent DHFR expression plasmids and subsequently amplifying DHFR/IL-12 genes with increasing levels of methotrexate. The approach by which we characterized the integration units of both p35 and p40 genes is illustrated in Figure 8, which depicts the integrated linearized expression plasmids for both genes flanked by host-cell genomic DNA.

DNA derived from the MCB was digested with restriction enzymes for which recognition sites are present in the expression plasmid on only one side of the gene of interest. This is illustrated for the p40 gene in Figure 8 (right-hand panel), where a single *Bgl* II site is present within the p40 expression plasmid, just 5′ of the p40 cDNA (indicated by the dotted arrow below the expression plasmid). The next downstream *Bgl* II recognition site cannot come from this same expression plasmid, but instead must derive from DNA that flanks the integrated plasmid. Therefore, visualization of a single p40-positive *Bgl* II fragment on genomic Southern analysis argues that the 3′ DNA segment that flanks all copies of the integrated p40 plasmid is the same (i.e., contains a *Bgl* II site at the same position relative to the 3′ end of the integrated plasmid). This suggests that a single p40 expression plasmid integrated into the host-cell genome (with a *Bgl* II site just downstream) and that the configuration of the gene and local flanking DNA was maintained upon gene amplification. In contrast, the presence of two fragments in the p35 (*Sph* I-digested) Southern blot (Fig. 8, left-hand panel) argues that at least two distinct copies of the p35 plasmid (with *Sph* I sites at unique distances from the 3′ ends of the integrated plasmids) were integrated and amplified in the cell line. In each case, the results shown are representative of a number of experiments using different restriction enzymes which, taken together, allow conclusions to be drawn regarding the number of initial plasmid integrations that are represented in the amplified DNA of the cell line.

The insertion sites described before should not be confused with chromosomal sites that contain multiple copies of amplified genes, observed on cytogenetic analysis and designated as homogeneously staining regions (84–86), or smaller chromosomally amplified sequences detected by fluorescent *in situ* hybridization (88). The insertion sites described before are likely to reflect integrations which took place on transfection of the cells. As discussed previously, subsequent gene amplification and extended

serial propagation of the cell line under methotrexate selection is normally accompanied by translocation of the original plasmid inserts and flanking DNA to different sites in the genome. The average size of such amplicons (including flanking host-cell DNA) has been observed to be on the order of 1 megabase, significantly larger than the fragments evaluated by the Southern blot method. Nevertheless, the value of establishing the number of original sites of integration for a particular gene is that this information directs our strategy in the next step in the genotypic characterization process, gene cloning, and DNA sequence analysis.

Sequence Analysis of Genes and Transcripts Encoding the Product

In the IL-12 p40 gene, cloning and DNA sequence were analyed on genomic DNA prepared from the MCB. Subsequent DNA sequence analysis of the cloned p40 gene was used to establish that the sequence completely agreed with that present in the expression plasmid. This indicated that the initial plasmid integrant represented in all the amplified copies of the p40 gene contained a completely intact coding region at the DNA sequence level. However, because we detected two independent plasmid integrants represented in the amplified p35 genes, it was necessary to carry out an additional investigation to demonstrate that both copies integrated with complete fidelity. Accordingly, the two Sph I fragments shown in Figure 8 (p35 plasmid) were isolated, cloned, and sequenced. DNA sequences from both fragments completely agreed with the p35 coding sequence in the original expression plasmid. Thus the DNA sequencing approach described allowed us to conclude that both p40 and p35 expression plasmids were integrated into the CHO genome with complete fidelity. However, this DNA sequencing strategy does not allow us to ascertain the fidelity of all, or even the majority of multicopy amplified genes or the RNA transcripts derived from these originally integrated plasmid molecules. For this reason, a complementary approach needs to be applied to address the bulk population of transcripts present in the cell line.

Recent regulatory guidance (153) has called for DNA sequencing approaches such as DNA sequence analysis of bulk cDNA derived from reverse transcriptase polymerase chain reaction (RT-PCR) treatment of total RNA. Althought it does not set a required sensitivity limit for such methods, the guidance document requested that the sensitivity of methods employed to detect variant sequences be estimated. Accordingly, DNA sequencing was performed on cDNA clones derived from product genes expressed in a rCHO MCB and end-of-production (EOP) cells (see also Genetic Stability discussion following). Emphasis in these experiments was placed on establishing limits of detection of aberrant sequences.

Total RNA was isolated from the MCB or EOP cells, reverse transcribed into cDNA, and the entire coding region of the recombinant protein was amplified by high-fidelity PCR. The expression plasmid used in constructing the MCB lineage was subjected to the same PCR procedure as a control. Pairs of PCR primers were designed to permit amplification of the coding region

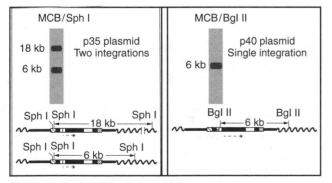

Figure 8. Evaluation of plasmid integration units in a recombinant human IL-12 cell line. The position and orientation of the p35 and p45 cDNA inserts are indicated by dotted arrows. Genomic Southern blots are shown from a representative digest for both p35 and p40. Reprinted from Ref. 146 with permission.

Table 1. Evaluation of the Ability to Detect DNA Sequence Variants

Sample	Mutation 1[a]	Mutation 2[a]	Mutation 3[a]	Mutation 4[a]
Mix 1: actual %[b]	0	0	0	0
% by seq.[c]	0	0	0	0
Mix 2: actual %	30	30	30	30
% by seq.	20–25	30–35	30–35	30–35
Mix 3: actual %	30	0	0	0
% by Seq.	25–30	0	0	0
Mix 4: actual %	0	30	30	30
% by seq.	0	30–40	30–40	30–40
Mix 5: actual %	20	40	40	40
% by seq.	20 or less	40–50	40–50	40–50
Mix 6: actual %	30	20	20	20
% by seq.	25–30	20 or less	20 or less	20 or less
Mix 7: actual %	15	15	15	15
% by seq.	NQ[d]	NQ	NQ	NQ

Notes: [a]Mutation 1: Arg (AGG) to Lys (AAG) at amino acid position −1; mutation 2: Glu (GAA) to Asp (GAT) at amino acid position 33; mutation 3: Glu (GAA) to Asp (GAT) at amino acid position 36; mutation 4: Glu (GAA) to Asp (GAT) at amino acid position 40.
[b]Known starting proportion, % of wild type (w/t) at indicated position.
[c]Proportion of mutant estimated by blinded DNA sequencing (both strands), % of w/t at indicated position.
[d]Mutation was detected but not quantifiable.

in four overlapping fragments. Only fragments of the predicted size were observed upon agarose gel analysis of the RT-PCR products, and these were indistinguishable from the expression plasmid control PCR fragments, further confirming the integrity of the coding region of expressed genes in MCB cells.

The products of each PCR reaction were cloned and sequenced. Similarly, PCR-products from expression plasmids that encode defined structure-function mutants of the product gene were cloned. The cloned mutant PCR-products were used to evaluate the ability of the DNA sequencing methodology to detect sequence variants by spiking the wild-type DNA preparations with known quantities of the mutant DNA preparations. The DNA sequencing of these samples indicated that mutations could be quantitatively detected to a level of about 20% (Table 1).

Population sequencing provides a convenient and sensitive method for evaluating the protein coding sequence in a large and statistically significant number of independently derived cDNA clones. The results confirm that the predominant sequence at each position is correct, but the method cannot demonstrate that the actual sequence of any one clone is correct in its entirety. For this reason, these data are complemented by sequencing individual product genes cloned from MCB genomic DNA, as described before. In conjunction with Northern and Southern blot data (not shown), the sequencing results confirm that expression plasmids with the expected product coding sequence are stably maintained in the genome of these cells during inoculum buildup and full-scale production.

Genetic Stability

The purpose of analyzing the expression construct is to establish that the correct coding sequence for the product

has been incorporated into the host cell and is maintained over the course of production. Our approach, which is consistent with current regulatory guidance (150), is to comprehensively analyze the expression construct at least once at the end of production. The analysis of EOP cells covers all of the major characteristics of the expression construct that are also analyzed in MCB cells: gross structural integrity of genes and transcripts that encode the product, gene organization and copy number, and DNA sequencing. For this analysis, EOP cells at the "limit of *in vitro* age used for production" are characterized, so that the data can be used to establish that no significant change occurs in the product-encoding genes during production. Over the course of several years of commercial production at Genetics Institute, analysis of EOP cells has not revealed any detectable genetic instability in rCHO cells.

BIBLIOGRAPHY

1. R. Li et al., *Proc. Natl. Acad. Sci. U.S.A.* **94**, 14506–14511 (1997).

2. P. Duesberg, C. Rausch Rasnick, and R. Hehlmann, *Proc. Natl. Acad. Sci. U.S.A.* **95**, 13692–13697 (1998).

3. R.G. Worton, C.C. Ho, and C. Duff, *Somatic Cell Genet.* **3**, 27–45 (1977).

4. C. Lengauer, K.W. Kinzler, and B. Vogelstein, *Nature (London)* **386**, 623–627 (1997).

5. M.W. Hashimoto et al., *Carcinogenesis* **17**, 389–394 (1996).

6. L.J. Ko and C. Prives, *Genes Dev.* **10**, 1054–1072 (1996).

7. K.W. Kinzler and B. Vogelstein, *Nature (London)* **379**, 19–20 (1996).

8. R.R. Reddel, *Ann. N.Y. Acad. Sci.* **854**, 8–19 (1998).

9. S.M. Cross et al., *Science* **267**, 1353–1356 (1995).

10. C.A. Reznikoff et al., *Genes Dev.* **8**, 2227–2240 (1994).

11. J.E. Hundley et al., *Mol. Cell. Biol.* **17**, 723–731 (1997).

12. L.A. Donehower et al., *Genes Dev.* **9**, 882–895 (1995).

13. K. Fukasawa, F. Wiener, G.F. Vande Woude, and S. Mai, *Oncogene* **15**, 1295–1302 (1997).

14. M. Harvey et al., *Oncogene* **8**, 2457–2467 (1993).

15. L.R. Livingstone et al., *Cell* **70**, 923–935 (1992).

16. Y. Yin et al., *Cell* **70**, 937–948 (1992).

17. K. Fukasawa et al., *Science* **271**, 1744–1747 (1996).

18. X.W. Wang et al., *Nat. Genet.* **10**, 188–195 (1995).

19. J.M. Ford and P.C. Hanawalt, *Proc. Natl. Acad. Sci. U.S.A.* **92**, 8876–8880 (1995).

20. M.B. Kastan et al., *Cancer Res.* **51**, 6304–6311 (1991).

21. M.L. Smith et al., *Oncogene* **10**, 1053–1059 (1995).

22. P.A. Havre et al., *Cancer Res.* **55**, 4420–4424 (1995).

23. Y. Ishizaka, M.V. Chernov, C.M. Burns, and G.R. Stark, *Proc. Natl. Acad. Sci. U.S.A.* **92**, 3224–3228 (1995).

24. E.A. Harrington, J.L. Bruse, E. Harlow, and N. Dyson, *Proc. Natl. Acad. Sci. U.S.A.* **95**, 11945–11950 (1998).

25. M.L. Smith, Q. Zhan, I. Bae, and A.J. Fornace, *Exp. Cell Res.* **215**, 386–389 (1994).

26. D. Cobrink, *Curr. Top. Microbiol. Immunol.* **208**, 31–61 (1996).

27. C. Taylor, A. Jalava, and S. Mai, *Curr. Opin. Microbiol. Immunol.* **224**, 210–207 (1977).

28. C. Taylor and S. Mai, *Cancer Detect. Prev.* **22**, 350–356 (1998).

29. S. Mai, J. Hanley-Hyde, and M. Fluri, *Oncogene* **12**, 277–288 (1996).

30. S. Mai, M. Fluri, D. Siwarski, and K. Huppi, *Chromosome Res.* **4**, 365–371 (1996).

31. K.-J. Wu et al., *Nat. Genet.* **21**, 220–224 (1999).

32. M.A. Wani, X. Xu, and P.J. Stambrook, *Cancer Res.* **54**, 2504–2508 (1994).

33. N.H. Denko, A.J. Giaccia, J.R. Stringer, and P.J. Stambrook, *Proc. Natl. Acad. Sci. U.S.A.* **91**, 5124–5128 (1994).

34. N. Denko, J. Stringer, M. Wani, and P.J. Stambrook, *Somatic Cell Mol. Genet.* **21**, 241–253 (1995).

35. M.A. Wani, N.C. Denko, and P.J. Stambrook, *Somatic Cell Mol. Genet.* **23**, 123–133 (1997).

36. P.A. Jones and P.W. Laird, *Nat. Genet.* **21**, 163–167 (1999).

37. C.C. Gurin, M.G. Federici, L. Kang, and J. Boyd, *Cancer Res.* **59**, 462–466 (1999).

38. T. Lindahl, Satoh, G.G. Pirier, and A. Klungland, *Trends Biochem. Sci.* **20**, 405–411 (1995).

39. Y.L. Rhun, J.B. Kirkland, and G.M. Shah, *Biochem. Biophys. Res. Commun.* **245**, 1–10 (1998).

40. V. Schreiber et al., *Proc. Natl. Acad. Science U.S.A.* **92**, 4753–4757 (1995).

41. I. D'Agano et al., *Environ, Mol. Mutagen.* **32**, 56–63 (1998).

42. R. Ding and M. Smulson, *Cancer Res.* **54**, 4627–4634 (1994).

43. J.-H. Kupper, M. Muller, and A. Burkle, *Cancer Res.* **56**, 2715–2717 (1996).

44. J.A. Gaken et al., *J. Virol.* **70**, 3992–4000 (1996).

45. B.C. Waldman, J.R. O'Quinn, and A.S. Waldman, *Biochem. Biophys. Acta* **1308**, 241–250 (1996).

46. F. Farzaneh et al., *Nucleic Acids Res.* **16**, 11319–11326 (1988).

47. C. Trucco, F.J. Oliver, G. de Murcia, and J. Menissier-de Murcia, *Nucleic Acids Res.* **26**, 2644–2649 (1998).

48. Z.-Q. Wang et al., *Genes Dev.* **11**, 2347–2358 (1997).

49. M.L. Agarwall et al., *Oncogene* **15**, 1035–1041 (1997).

50. J.M. de Murcia et al., *Proc. Natl. Acad. Sci. U.S.A.* **94**, 7303–7307 (1997).

51. S.R. Kumari, H. Mendoza-Alvarez, and R. Alvarez-Gonzalez, *Cancer Res.* **58**, 5075–5078 (1998).

52. P.A. Jeggo, G.E. Taccioli, and S.P. Jackson, *BioEssays* **17**, 949–957 (1995).

53. T. Blunt et al., *Cell* **80**, 813–823 (1995).

54. N. Nicolas et al., *J. Exp. Med.* **188**, 627–634 (1998).

55. E.K. Shin, L.E. Perryman, and K. Meek, *J. Immnol.* **158**, 3565–3569 (1997).

56. G. Taccioli et al., *Immunity* **9**, 355–366 (1998).

57. C. Jhappan et al., *Nat. Genet.* **17**, 483–486 (1977).

58. S.R. Peterson et al., *Proc. Natl. Acad. Sci. U.S.A.* **92**, 3171–3174 (1995).

59. R. Fukumura et al., *J. Biol. Chem.* **273**, 13058–13064 (1998).

60. S.R. Peterson et al., *J. Biol. Chem.* **272**, 10227–10231 (1997).

61. A. Errami et al., *Nucleic Acids Res.* **26**, 3146–3153 (1998).

62. G.E. Taccioli et al., *J. Biol. Chem.* **269**, 7439–7442 (1994).

63. C. Morrison et al., *Nat. Genet.* **17**, 479–482 (1997).

64. L. Halflick, *Exp. Gerontol.* **33**, 639–653 (1998).

65. J.W. Shay and E.W. Woodring, *Curr. Opin. Oncol.* **8**, 66–71 (1996).

66. D.E. Shippen, *Curr. Opin. Genet. Dev.* **3**, 759–763 (1993).

67. T.L. Halvorsen, G. Leibowitz, and F. Levine, *Mol. Cell. Biol.* **19**, 1864–1870 (1999).

68. C.M. Counter et al., *EMBO J.* **11**, 1921–1929 (1992).

69. A.G.M. Bodnar et al., *Science* **277**, 349–352 (1998).

70. H. Vaziri and S. Benchimol, *Curr. Biol.* **8**, 279–282 (1998).

71. J.W. Feng et al., *Science* **269**, 1000–1301 (1996).

72. K.L. Rudolph et al., *Cell* **96**, 701–712 (1999).

73. B. Van Steensel, A. Smororzewaska, and T. de Lange, *Cell* **92**, 401–413 (1998).

74. H. Vaziri et al., *Mol. Cell. Biol.* **19**, 2372–2379 (1999).

75. J.D. Coursen et al., *Exp. Cell Res.* **235**, 245–253 (1997).

76. R.A. Greenberg et al., *Oncogene* **18**, 1219–1226 (1999).

77. J. Karlseder et al., *Science* **283**, 1321–1325 (1999).

78. F.M. Wurm and Petropoulos, *Biologicals* **22**, 95–102 (1994).

79. Y. Voziyanov, S. Pathania, and M. Jayaran, *Nucleic Acids Res.* **27**, 930–941 (1999).

80. R.T. Schimke, S.W. Sherwood, A. Hill, and R.N. Johnston, *Proc. Natl. Acad. Sci. U.S.A.* **83**, 2157–2161 (1986).

81. A.J. Strain, *Dev. Biol. Stand.* **68**, 27–32 (1986).

82. R.J. Kaufman, *Methods Enzymol.* **185**, 537–566 (1990).

83. R.J. Kaufman and R.T. Schimke, *Mol. Cell. Biol.* **1**, 1069–1076 (1981).

84. R.J. Kaufman, P.C. Brown, and R.T. Schimke, 1979. *Proc. Natl. Acad. Sci. U.S.A.* **76**, 5669–5673 (1979).

85. J.H. Nunberg et al., *Proc. Natl. Acad. Sci. U.S.A.* **75**, 5553–5556 (1978).

86. R.J. Kaufman, P.A. Sharp, and S.A. Latt, *Mol. Cell. Biol.* **3**, 699–711 (1983).

87. U.H. Weidle, P. Buckel, and J. Weinberg, *Gene* **66**, 193–203 (1988).

88. M.G. Pallavicini et al., *Mol. Cell. Biol.* **10**, 401–404 (1990).

89. F.M. Wurm et al., in R.A. Spier, J.B. Griffiths, and B. Meignier, eds., *Production of Biologicals from Animal Cells in Culture*, 1991, pp. 316–323.

90. F.M. Wurm et al., *Ann. N.Y. Acad. Sci.* **782**, 70–78 (1996).

91. D.W. Melton, A.-M. Ketchen, and J. Selfridge, *Nucleic Acids Res.* **25**, 3939–3943 (1997).

92. G.J. Pendse, S. Karkare, and J.E. Bailey, *Biotechnol. Bioeng.* **40**, 119–129 (1992).

93. S.J. Kim et al., *Biotechnol. Bioeng.* **58**, 73–84 (1998).

94. N.S. Kim, S.J. Kim, and G.M. Lee, *Biotechnol. Bioeng.* **60**, 679–688 (1998).

95. M.S. Sinacore et al., in E.C. Beuvery, Griffiths, and Zeijlemaker, eds., *Animal Cell Technology*, 1995, pp. 63–67.

96. M.B. Gu, J.A. Kern, P. Todd, and D.S. Kompala, *Cytotechnology* **9**, 237–254 (1992).

97. P.M. Hayter, N.F. Kirkby, and R.E. Spier, *Enzyme Microb. Technol.* **14**, 454–461 (1992).

98. E. Suzuki and D.F. Ollis, *Biotechnol. Bioeng.* **34**, 1398–1402 (1989).

99. A.S. Chuck and B.O. Palsson, *Biotechnol. Bioeng.* **39**, 354–360 (1992).

100. J.M. Coco-Martin et al., *Hybridoma* **11**, 653–665 (1992).

101. S.W. Bae, H.J. Hong, and G.M. Lee, *Biotechnol. Bioeng.* **47**, 243–251 (1995).

102. P.M. Hayter et al., *Appl. Microbiol. Biotechnol.* **34**, 559–564 (1991).

103. M.I. Cocket, C.G. Bebbington, and G.T. Yarranton, *Bio/ Technology* **8**, 662–667 (1990).

104. D.K. Robinson and K.W. Memmert, *Biotechnol. Bioeng.* **38**, 972–976 (1991).

105. A.L. Smiley, W.-S. Hu, and D.I.C. Wang, *Biotechnol. Bioeng.* **33**, 1182–1190 (1989).

106. K. Furukawa and K. Ohsuye, *Cytotechnology* **26**, 153–164 (1998).

107. X. Mazur, M. Fussenegger, W.A. Renner, and J.E. Bailey, *Biotechnol. Prog.* **14**, 705–713 (1998).

108. M. Fussenegger et al., *Nat. Biotechnol.* **16**, 468–472 (1998).

109. R.G. Werner, W. Noe, K. Kopp, and M. Schluter, *Arzneim.-Forsch.* **48**, 870–880 (1998).

110. W.A. Renner et al., *Biotechnol. Bioeng.* **47**, 476–482 (1995).

111. T.M. Guadagno, M. Ohtsubo, K.M. Roberts, and R.K. Assosian, *Science* **262**, 1572–1575 (1993).

112. R.P. Singh and M. Al-Rubeai, *Adv. Biochem. Eng. Biotechnol.* **62**, 167–184 (1998).

113. M.S. Sinacore, D. Drapeau, and S.R. Adamson, *Methods Biotechnol.* **8**, 11–22 (1999).

114. W.M. Miller, C.R. Wilke, and H.W. Blanch, *Bioprocess Eng.* **3**, 113–122 (1988).

115. M. Matsumura, M. Shimoda, T. Arii, and H. Kataoka, *Cytotechnology* **7**, 103–112 (1991).

116. B. Schumpp and E.-J. Schlaeger, *Cytotechnology* **8**, 39–44 (1992).

117. U.S. Pat. 5,156,964 (1992). D. Inlow, B. Maiorella, and A.E. Shauger.

118. U.S. Statutory Invention Registration H1532 (1996). S.R. Adamson, D. Drapeau, Y.-T. Luan, and D.A. Miller.

119. M. Murata, Y. Eto, and H. Shibai, *J. Ferment. Technol.* **66**, 501–507 (1988).

120. D.T. Berg, D.B. McClure, and B.W. Grinnell, *Bio Techniques* **14**, 972–978 (1993).

121. D. Broad, R. Boraston, and M. Rhodes, *Cytotechnology* **5**, 47–55 (1991).

122. J.P. Mather, *Methods Enzymol.* **185**, 567–577 (1990).

123. M.J. Keen and T.R. Nicholas, *Cytotechnology* **17**, 153–163 (1995).

124. P. Perrin et al., *Vaccine* **13**, 1244–1250 (1995).

125. M.S. Sinacore et al., *Biotechnol. Bioeng.* **52**, 518–528 (1996).

126. M. Zang et al., *Bio/Technology* **13**, 389–392 (1995).

127. C. Gandor, C. Leist, A. Feichter, and F. Asselbergs, *FEBS Lett.* **377**, 290–294 (1995).

128. T. Kawamoto et al., *Anal. Biochem.* **130**, 445–453 (1983).

129. J. Kovar and F. Franek, *Immunol. Lett.* **7**, 339–345 (1984).

130. H. Miyaji et al., *Cytotechnology* **3**, 133–140 (1990).

131. J.B. Griffiths and A.J. Racher, *Cytotechnology* **15**, 3–9 (1994).

132. R.K. Assoian, *J. Cell Biol.* **136**, 1–4 (1977).

133. E. Rouslahti and J.C. Reed, *Cell* **77**, 477–478 (1994).

134. B.D. Jeso et al., *Biochem Biophys. Res. Commun.* **214**, 819–824 (1995).

135. E.M. Curling et al., *Biochem. J.* **272**, 333–337 (1990).

136. E. Watson et al., *Biotechnol. Prog.* **10**, 39–44 (1994).

137. W. Chotigeat, Y. Watanapokasin, S. Mahler, and P.P. Gray, *Cytotechnology* **15**, 217–221 (1994).

138. M.R. Lifely et al., *Glycobiology* **5**, 813–822 (1995).

139. N. Jenkins, R.B. Parekh, and D.C. James, *Nat. Biotechnol.* **14**, 975–981 (1996).

140. M. Gawlitzek et al., *J. Biotechnol.* **42**, 117–131 (1995).

141. S.S. Ozturk and B.O. Palsson, *Biotechnol. Bioeng.* **37**, 35–46 (1991).

142. S. Chuppa et al., *Biotechnol. Bioeng.* **55**, 328–338 (1997).

143. J. Kruh, *Mol. Cell. Biochem.* **42**, 65–82 (1982).

144. U.S. Pat. 5,705,364 (1998), T. Etcheverry and T. Rhyll.

145. R.J. Kaufman, *Trends Biotechnol.* **9**, 353–359 (1991).

146. S.R. Adamson and T.S. Charlebois, *Dev. Biol. Stand.* **83**, 31–44 (1994).

147. R.R. Townsend et al., *Anal. Biochem.* **182**, 1–8 (1989).

148. D.A. Cumming, *Glycobiology* **1**, 115–130 (1991).

149. A.S. Lubiniecki et al., *Dev. Biol. Stand.* **76**, 105–115 (1992).

150. A.S. Lubiniecki, *Dev. Biol. Stand.* **96**, 173–175 (1998).

151. *Guidance for Industry: Changes to an Approved Application for Specified Biotechnology and Specified Synthetic Biological Products*, U.S. Food and Drug Administration, Washington, D.C., 1997.

152. S.F. Wolf et al., *J. Immunolo.* **146**, 3074–3081 (1991).

153. *Final Guideline on Quality of Biotechnological Products: Analysis of the Expression Construct in Cells Used for Production of r-DNA Derived Protein Products*, Int. Conf. Harmonization (endorsed), U.S. Food and Drug Administration, Washington, D.C., 1995.

See also CELL PRODUCTS—ANTIBODIES; CELL PRODUCTS—IMMUNOREGULATORS; SOMACLONAL VARIATION; TRANSFORMATION OF PLANTS.

CELL STRUCTURE AND MOTION, CYTOSKELETON AND CELL MOVEMENT

HELEN M. BUETTNER
Rutgers University
Piscataway, New Jersey

DAVID J. ODDE
Michigan Technological University
Houghton
Michigan

AQUANETTE M. BURT
Rutgers University
Piscataway
New Jersey

OUTLINE

INTRODUCTION

The cytoskeleton is a complex, three-dimensional network of protein filaments distributed throughout the cytoplasm of the cell. It is a dynamic structure that not only provides shape to the cell but also undergoes a remodeling process to achieve the changes in cell shape that accompany cell growth and motility. As an intracellular framework, the cytoskeleton serves to anchor cell components in place and also facilitates their transport from one location to another within the cell at various times. Cytoskeletal behavior is a fundamental aspect of cellular systems and represents a major point of control in manipulating cells and tissues.

This article first describes general aspects of the cytoskeleton, including the three filament systems that serve as the major structural elements of the cytoskeleton, their spatial organization within the cell, and some of the characteristic dynamic behavior that they contribute to the cytoskeleton. The remainder of the article then addresses each of the cytoskeletal filament systems in greater detail, focusing on basic aspects of relevance to cell shape and movement.

THE CYTOSKELETON

Major Structural Components

The cytoskeleton is composed of filamentous proteins connected and coordinated by additional protein molecules. The protein filaments are viewed as the primary components of the cytoskeleton with the other proteins playing an accessory role. There are three main types of filaments in the cytoskeleton: microtubules, actin filaments (also called microfilaments), and intermediate filaments. Although these three populations of filaments are integrated both physically and functionally within the cell, their diverse characteristics enable them to serve different functions within the cytoskeleton, and each is associated with a different aspect of cell structure or movement (1). Microtubules form a structural framework that helps to define the shape of the cell and to organize intracellular organelles, which attach to and often move along individual microtubules. Actin filaments can vary considerably in length depending on the cell type and even the site within the cell (2); they associate into a variety of subcellular structures, including highly aligned bundles and random meshworks of varying density, and play a major role in the motility characteristics of cells (3). Intermediate filaments are viewed as mechanical elements that contribute to the structural integrity of the cell. All three filament systems are dynamic and undergo remodeling behavior that helps to guide cell movement and changes in cell shape.

Organization

The cytoskeleton can be organized conceptually into three spatial domains. (1) the cell interior, (2) the cortex, and (3) the plasma membrane. These domains differ with respect to both occurrence and organization of the three long filament systems of the cytoskeleton (4). Microtubules and intermediate filaments predominate in the interior of the cell. In many cells, both types of filaments project radially from near the center of the cell to the cortex. Microtubules serve to anchor various membrane-bound organelles and intracellular structures, including the endoplasmic reticulum, mitochondria, and Golgi complex, in place within the inner cytoplasm. The major role of intermediate filaments in the interior of the cell appears to be to impart mechanical strength to the cell. Actin filaments are observed in the inner region of cells that are attached to extracellular surfaces. They are organized into bundled structures called stress fibers and project to sites where the plasma membrane is attached to other cells or structures. These sites, called focal contacts, and stress fibers tend to disassemble as cells detach from a surface and begin to move. They can be visualized using microscopy and immunofluorescence techniques and provide a useful indication of the effects of culture surfaces on cell attachment and motility through their effects on the actin cytoskeleton.

The cortex of the cell lies immediately under the plasma membrane. Actin filaments occur at higher density than in the cell interior and in a cross-linked network; other microfilament structures such as bundles may be present as well. Intermediate filaments are observed in the cortex;

microtubules are not a typical constituent of the cortex in animal cells but may interact with it.

The cytoskeleton at the plasma membrane forms specialized structures that help the cell interact with the extracellular environment. Intermediate filaments are a major component of cytoskeletal structures that form where cells are anchored to other cells or surfaces, presumably reinforcing the junction and participating in the transmission of forces. Microfilament bundles form the core of fine projections at the cell surface, including microvilli and filopodia, that help to increase the contact between a cell and its environment.

Dynamic Behavior

Dynamic changes occur in the cytoskeleton during development, cell division, cell motility, and disease. Changes during development involve the growth and rearrangement of cytoskeletal elements, as well as changes in composition. Cell division is characterized by a series of events dominated by changes in the microtubule system. The actin cytoskeleton also undergoes significant reorganization, while the participation of intermediate filaments varies for different cell types (4). Cell motility encompasses a wide range of behaviors that involve the cytoskeletal filament systems to different degrees. Muscle contraction has been studied and can be explained largely in terms of force generation by actin–myosin dynamics. Both actin force generation and assembly dynamics play an important role in nonmuscle cell motility. Microtubule dynamics can also be important. An example is neuronal growth cone motility, in which microtubule dynamic instability appears to provide a mechanism for the migrating growth cone to either stabilize or quickly alter its course, depending on feedback from the external environment (5,6). The interaction between microtubules and actin is likely to be important in this and other examples of cell locomotion, although few specific details are known. Another form of cell motility involves the transport of organelles along microtubules within the cell.

A number of diseases and disorders result from genetic abnormalities in the cytoskeleton, typically a defect in one of the filament systems or its accessory proteins. These disorders, or similar ones that can be produced in the laboratory by genetic manipulation of specific cytoskeletal components, provide a useful means of studying the cytoskeleton (7,8). A complex example occurs in the case of cancer, in which multiple cytoskeletal changes are observed. Key among these is an increase in uniformity of the actin cytoskeleton, changes in intermediate filaments, and loss of cell–cell contacts in which these filaments participate. Studies of the cytoskeleton provide a means of investigating this and

other diseases and may ultimately help in understanding and treating them.

MICROTUBULES

Structure

Microtubules are self-assembled protein filaments in eukaryotic cells that serve to organize the cytoplasm and mediate molecular motor-based transport. They can be of variable diameter, but are typically of about 17 nm inner diameter and 25 nm outer diameter (9). Microtubules can extend over the entire length of the cell, in some cases exceeding 100 μm in length (10). The tubes are quite rigid with a Young's modulus of $\sim 10^9$ Pa, a value comparable to plexiglas (11). Paradoxically, microtubules are often curved in living cells, occasionally to a radius of curvature of <500 nm, implying that relatively strong forces are exerted on microtubules (12).

The protein that constitutes microtubules is an $\alpha\beta$ heterodimer called *tubulin*. Dimeric tubulin has a molecular weight of 100 kD, and its structure has recently been determined at 3.7-angstrom resolution (13). Tubulin is highly conserved across species and has a homolog in bacteria called FtsZ, a protein that mediates contractile ring formation (14), suggesting that it is an ancient protein little changed, by evolution. Tubulin dimers stack head-to-tail and side-to-side to form microtubules as depicted in Figure 1. The head-to-tail stacks are called *protofilaments,* of which there are typically 13 that together form a tube. The lateral contacts between protofilaments are slightly pitched so that there is a helical pitch to the lattice that is typically left-handed with a 1.5-dimer pitch per revolution (15). The nonintegral pitch gives rise to a discontinuous seam in the microtubule lattice where α-subunits are adjacent to β-subunits, instead of the usual $\alpha-\alpha$ and $\beta-\beta$ lateral associations. Of particular importance is the asymmetry of the lattice, which gives rise to polarity, with one end of the microtubule designated as "plus" (functionally fast growing) and the other "minus" (functionally slow growing). Intracellular protein motors are able to recognize the polarity and move specifically toward either one end or the other. Cells exploit this recognition by having their microtubule minus ends generally point toward the cell center and their microtubule plus ends generally point toward the cell periphery.

Function

The force generation associated with motor protein/microtubule interaction gives rise to many of the functional characteristics of microtubules. There are two broad classes into which microtubule motors fall: *kinesins*

Figure 1. Microtubule structure. Microtubules are composed of $\alpha\beta$-tubulin heterodimers in a crystalline lattice structure. Each solid/open pair of balls in the diagram represents a single dimer. Microtubules have a 25 nm outer diameter, a 17 nm inner diameter, and can be many micrometers long. Growth occurs faster at the plus end than the minus end.

and *dyneins*. Both kinesins and dyneins use adenosine triphosphate (ATP) hydrolysis to power their persistent movement along the exterior of the microtubule lattice. Detailed studies of kinesin reveal that it advances in discrete steps of 8 nm and hydrolyzes one ATP per step (16,17). An important difference between kinesins and dyneins is that kinesins (generally) move toward plus ends while dyneins move toward minus ends. This functional distinction allows cells selectively to transport motor-associated material toward either the plus or the minus ends of microtubules, depending on whether the motor is plus end-directed or minus end-directed.

The functional implications of microtubule motor-based transport are myriad. The kinesinbased plus end-directed transport allows neurons effectively to transport material from the cell body, where the biosynthetic machinery is located, to the ends of long cellular extensions called *axons*. Because of the uniform plus end-distal polarity of microtubules in axons, membrane-bound vesicles can be transported over meter distances. Microtubules also work with minus end motors to facilitate transport processes. Two primary examples are the beating of flagella and cilia, which are the result of dynein motor activity inducing microtubule bending. The intracellular localization of the endoplasmic reticulum (ER) and the Golgi apparatus are determined by their coupling to microtubules via motor proteins (18). The ER associates with plus end-directed motors, which induce ER extension toward the cell periphery, while the Golgi associates with minus end-directed motors, which induce Golgi retraction toward the cell center. Perhaps the most critically important function of microtubules and their motors is to mediate the proper segregation of chromosomes during mitosis. Here both plus and minus end-directed motors coordinate to achieve the highly accurate and rapid partitioning of the replicated genome (19).

Organization and Dynamics

Nucleation. Microtubules lengthen as a result of $\alpha\beta$-tubulin self-assembly. For the most part self-assembly is initiated by a structure located near the nucleus called the *centrosome*, which is about 1 μm in diameter. The characteristic organization of microtubules, where the plus end is directed toward the cell periphery, is believed to be primarily the result of centrosomal microtubule nucleation, which is polarity specific. While minus ends are embedded in the centrosome, plus ends extend away from the centrosome toward the cell periphery. This organization of polarity dictates much of the overall microtubule organization in the cell. However, recent results have shown that microtubules can form the typical plus-end-distal organization in the absence of a centrosome (20) and can form by spontaneous nucleation in the cytoplasm (21). In addition, new microtubules can be generated by either releasing microtubules from the centrosome or by breaking microtubules in the periphery (12,22,23).

Self-Assembly by "Dynamic Instability". One of the most unusual features of microtubules is their self-assembly. Most self-assembled structures undergo stochastic subunit addition and loss events, with growth supported by addition being favored over loss. Microtubules, on the other hand, undergo persistent growth, but then abruptly switch to a persistent shortening phase, a phenomenon called *catastrophe*. Shortening microtubules can then abruptly switch back to the growth phase, a phenomenon called *rescue*. Individual microtubules switch stochastically between growth and shortening phases, even at steady state. This switching, called *dynamic instability*, allows for more rapid turnover of the microtubule mass *in vivo* than would be possible by simple equilibrium dimer addition and loss (24,25). In addition, dynamic instability is apparently necessary for nerve growth, because if it is suppressed, then neurons stop growing, even though microtubules are still present in the neurons (26).

Like actin filaments, microtubule assembly requires a nucleotide triphosphate, which for microtubules is guanosine triphosphate (GTP). Once a tubulin dimer is incorporated into the microtubule lattice, one of its two associated GTPs is hydrolyzed to guanosine diphosphate (GDP) and inorganic phosphate (27). The hydrolysis provides free energy to drive microtubule assembly and in addition may give rise to dynamic instability. It is widely believed that the presence of GTP tubulins at the microtubule tip confers stability on the lattice and that when this GTP "cap" is lost through hydrolysis and/or stochastic dimer dissociation, it exposes the labile core of GDP tubulins in the microtubule (28,29). In this model the exposed GDP-tubulin lattice will persistently disassemble until recapped by stochastic association of GTP tubulins (30,31).

Nonideal Assembly Behavior. Although microtubules often exhibit dynamic instability both in purified systems and in living cells, their behavior often deviates from the idealized random switching between phases of constant growth and shortening that characterizes dynamic instability. One way in which they deviate is that their switching is not random, but rather becomes more likely as they persist in their current state (32–34). This history-dependent switching can significantly alter the steady-state length distribution of microtubules and put constraints on the type of exploratory dynamics that microtubules can exhibit (5). For example, history-dependent switching will tend to reduce the relative frequency of brief and long growth phases relative to intermediate duration growth phases, potentially making microtubules more efficient in locating intracellular targets such as chromosomes during mitosis. In some cases it is difficult to identify dynamic instability behavior altogether, and in these cases the assembly is better characterized by a random walk (21,35).

Microtubule Breaking. Microtubules must be able to undergo massive reorganization in a relatively brief time. For example, when animal cells divide, they transit from having an array of long, stable microtubules to having short, dynamic microtubules over a time period of ~10 min, which requires a rate of reorganization faster than could likely be achieved by dynamic instability alone (36). One potential mechanism by which the microtubule array could be broken down more rapidly is by the action of a microtubule severing enzyme, such as *katanin*, a 140-kD heterodimeric ATPase (37). The severing action of katanin could also facilitate the maintenance of the plus ends pointing toward the cell periphery. When microtubules become highly curved, their plus ends tend to point back toward the center of the cell, which in turn disrupts the polarity information "encoded" in the normal microtubule organization. Severing these highly curved microtubules can serve selectively to eliminate these disoriented microtubules and maintain polarity (12).

Microtubule-Associated Proteins. Microtubules interact with many molecules in the cell, a number of which have been discussed above: kinesins, dyneins, and katanin. These proteins are part of a group of proteins that associate with microtubules and are called, appropriately, *microtubule-associated proteins* (MAPs). In addition to the motor activities of kinesin and dynein and the severing activity of katanin, MAPs can exhibit other activities. Some MAPs modulate microtubule assembly dynamics. For example, microtubule-associated protein 2 (MAP2), a protein that localizes to neuronal dendrites (information receiving processes), and tau, a protein that localizes to neuronal axons (information transmitting processes), both serve to promote microtubule assembly by suppressing catastrophe, promoting rescue, and increasing growth rates (38). MAP4 is expressed more ubiquitously among various cell types, and serves to link cyclins, cell cycle regulating proteins, to microtubules (39). Another MAP, XKCM1, tends to destabilize microtubules by promoting catastrophe (40), while yet another, XMAP215, increases both the microtubule growth and shortening rates and suppresses rescue (41). Undoubtedly more MAPs will be identified in the coming years and their functional significance clarified.

Importance in Biotechnology

Understanding the fundamental structure, function, organization, and dynamics of microtubules will be useful in a number of areas of technical interest. First, the requirement of microtubule assembly and dynamics for nerve growth can serve as a potential means of controlling and directing nerve growth and regeneration. Specifically, the application of drugs, such as vinblastine, that suppress microtubule dynamics may serve as a useful means of stopping undesired nerve growth in a tissue-engineered living neural network. Second, the requirement of microtubule assembly, disassembly, and reorganization during cell division serves as a potential target for cancer treatment. In this light it is significant that some of the more effective anticancer drugs, such as vinblastine and taxol, are also drugs that disrupt the normal chemical and mechanical properties of microtubules (42–44). Finally, microtubules may be mediators of mechanochemical signal transduction (12,45). Understanding how this transduction occurs may give us a more rational approach to controlling mechanochemical responses of engineered tissue to fluid shear stresses and material mechanical properties.

Figure 2. Actin filament structure. Actin filaments consist of two linear chains of G actin wound around each other into a double helix. They have a 7 to 9 nm diameter and range from 10 to 1000 subunits long.

ACTIN FILAMENTS

Structure

Actin filaments, or microfilaments, are linear polymers of actin molecules that assemble into a double helix with a 71-nm turn length and a 7 to 9 nm diameter (Fig. 2). The basic structural unit, the actin molecule, is a globular protein referred to as G actin. Subunits of G actin assemble into linear chains, and pairs of these chains associate to form the two strands of the double helix. Actin filament length ranges in order of magnitude from 10 to 1000 molecules, depending on the cell type and cellular structure in which the filaments occur (4).

The actin molecule exists in different forms that vary slightly in their amino acid sequences and properties (46). However, most characteristics of actin are highly conserved between the different forms. Actins are typically 374–376 amino acids long, with a molecular weight of about 42,000, and they contain binding sites for a number of other molecules and ions that participate in actin assembly and dynamics.

Organization

Different types of actin structures can exist within the same cell (3,47), and their relative balance influences the shape of the cell. The organization of actin within the nerve growth cone illustrates the coordination of several different types of structures. Microfilaments lie underneath the plasma membrane of the growth cone, forming a cross-linked array of polymer located primarily at the periphery of the growth cone, where protrusive structures such as lamellipodia and filopodia form. Lamellipodia are broad, veil-like extensions of the growth cone edge, while filopodia are thin spiky protrusions. Both structures continually remodel by extending and retracting; filopodial dynamics tend to be faster than lamellipodial dynamics. The organization of actin filaments in lamellipodia and filopodia differs. Lamellipodia contain two populations of actin filaments: bundles 40–100 nm in width that radiate from the leading edge, and a second population of shorter, branching filaments that fill the volume between the dorsal and ventral membrane surfaces of the growth cone (48; Fig. 3). The filaments form a relatively loose network linked together through cross-linking proteins. The orientation of the longer filaments is primarily with the faster-growing plus ends facing the periphery while the minus ends point toward the central region of the growth cone. This is thought to be of significance in the outward, or forward, movement of the growth cone. There are, however, filaments in the opposite orientation or which run parallel to the leading edge. The shorter filaments are more randomly oriented and thought to contribute

Dorsal surface

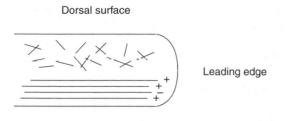

Leading edge

Ventral surface

Figure 3. Organization of actin filaments in the nerve growth cone lamellipodium. The longer filaments are oriented with their plus end mostly facing the membrane and are thought to play the major role in lamellipodial advance. The ventral surface of the lamellipodium contacts the extracellular surface on which the growth cone is migrating.

less than the longer filaments to lamellipodial extension. Parallel bundles of closely spaced microfilaments underly the filopodia, with the vast majority of filaments oriented such that they have the same polarity, the plus ends pointing away from the central region.

Assembly Dynamics

Actin filament assembly is a reversible polymerization reaction that requires ATP. The ATP hydrolyzes shortly following polymerization, an event that appears to promote depolymerization (49). As with microtubules, nucleoside triphosphate hydrolysis is believed to provide a necessary energy source for maintaining the polymerization–depolymerization cycle integral to the dynamic reorganization of actin cytoskeletal components.

Filaments of actin have a defined polarity with respect to polymerization rates, with the faster growing plus end polymerizing at about ten times the rate of the slower-growing minus end. This polarity can be visualized by reacting microfilaments with myosin segments. The orientation of the myosin segments along the microfilament produces a characteristic "arrowhead" pattern when the filaments are viewed by electron microscopy. The heads of the arrows point toward the minus, or pointed, end of the filament while the tails point toward the plus, or barbed, end.

The difference in polymerization rates is due in part to a difference in conformation at the two ends. In addition, because hydrolysis occurs slightly after subunit addition to the filament, the newest subunits in the filament are bound to ATP while older subunits are bound to ADP, increasing the polymerization rate at the plus end relative to that at the minus end. This reduces the critical concentration of monomer required for growth at the plus end relative to that at the minus end, so that it becomes possible to find windows of monomer concentration in which the plus end polymerizes while the minus end depolymerizes (50). These conditions give rise to a dynamic behavior called treadmilling, in which the simultaneous addition of subunits at the plus end and loss at the minus end gradually moves subunits through the filament from front to back without changing the filament length. Treadmilling may play an important role in lamellipodial protrusion by motile cells (50–52).

The structure and assembly dynamics of the actin network in the cytoskeleton can be modified by a number of proteins that occur naturally within the cell and bind to actin. Some of these actin-binding proteins (ABPs) alter the actin monomer–polymer equilibrium. Others provide structural links between actin filaments and other cytoskeletal components or the plasma membranes (53–55). ABPs affect the monomer–polymer equilibrium in three possible ways: (1) by binding to monomer and preventing its assembly into filaments, (2) by capping assembled filaments and preventing further growth, and (3) by severing assembled filaments. Structural links between actin filaments can serve either to cross-link filaments into a network, or to form parallel arrays of actin filaments such as those forming the core of filopodia, microvilli, and other similar cellular extensions.

Actin assembly dynamics can also be altered by drugs that bind to actin (49). Two such drugs have been used extensively in studying actin dynamics within the cell. Both are fungal products that affect the actin monomer–polymer equilibrium. Cytochalasins are destabilizing compounds that bind to the plus end of microfilaments, capping it against further polymerization and promoting disassembly. Phalloidins, on the other hand, bind along the length of microfilaments and stabilize them against disassembly. Both drugs exist in forms that are quite specific for actin and have consequently been very useful in studying cell motility. Cytochalasin can be used to reduce or eliminate actin filaments from locomoting cells, sometimes in a dose dependent manner (56), and thus is used to examine the effects of the actin cytoskeleton on the process. Phalloidins have been most useful for staining actin filaments to examine their distribution in the cell.

Actin-Based Motility

In addition to ABPs, which modify the structure and assembly dynamics of the actin cytoskeleton, motor proteins exist that move actin filaments within the cell (57). The motor proteins associated with actin belong to the myosin family, a set of molecules that bind ATP and undergo an ordered series of three conformational changes corresponding to the following cycle of events: (1) ATP binding, (2) ATP hydrolysis to ADP, and (3) ADP release. During this sequence, the myosin head repeatedly attaches to, pulls, and releases from the same actin filament, resulting in an irreversible translocation of the actin filament with respect to the myosin molecule. This myosin-powered mechanism of actin movement is referred to as the sliding filament model (1).

The sliding filament model is believed to contribute to actin-based motility in a wide variety of cell types, including skeletal and cardiac muscle, smooth muscle, and a number of motile nonmuscle cells. A second way in which actin contributes to cell movement is through the assembly and disassembly of actin filaments. A complete description of various models of actin-based motility, which include muscle contraction, cell division, cell development, and cell migration, is beyond the scope of this article. Thus we will focus on the example of cell migration, which incorporates both microfilament sliding and growth.

A general model of cell migration consists of three basic processes: protrusion, attachment, and traction (58,59). The underlying cellular mechanisms responsible for these components are thought to be the same or similar for various types of cells (59). Protrusion of lamellipodia and filopodia at the leading front is believed to occur by actin polymerization, with myosin or myosinlike motors also playing a role. The actin cytoskeleton of the cell is connected to the substrate via transmembrane receptors. Inside the cell these receptors are attached to linker proteins such as talin, α-actinin, and vinculin, which in turn are attached to the actin cytoskeleton. The protein–protein interactions are continually made and broken as the cell advances. The rear of the cell moves forward during traction.

A simple model of cell migration based on actin polymerization has been proposed for fish epidermal keratocytes (60). The keratocyte was modeled as a semicircular or ellipsoidal cell that extended in the front and retracted at the rear, both occuring orthogonally to the cell edge. Actin movement was observed using fluorescently labeled actin at the leading edge of cells *in vitro*. Fluorescent actin from the front of the cell moved in a circumferential manner to the rear, suggesting an association between actin motion and cellular movement through similarity.

A more complex model is needed to describe neuronal growth cone migration (61; Fig. 4). The exact mechanism of growth cone lamellipodial and filopodial movement due to actin is not known, but it is believed to be determined by three processes: (*1*) actin polymerization at the membrane, (*2*) depolymerization in the growth cone center, and (*3*) retrograde flow of actin (62). Actin polymerization at the leading front of the growth cone occurs continually during growth and extension of the neurite and is balanced by polymerization and depolymerization at the rear. A steady movement of actin from the front to more distal regions is also commonly observed. This constant flow of polymer is modulated by protein receptor links with the substrate. Lin and co-workers have shown that for bag cell neurons, myosin motors move actin filaments away from the leading front membrane and hence are responsible for retrograde flow, while receptors on the growth cone bind actin filaments to the membrane (63). It is the longer, bundled filaments of the lamellar region that are involved in these membrane associations (48). Because the actin filaments are bound, retrograde flow is diminished, thus creating a net advance of actin toward the membrane of the growth cone. This net advance of actin leads to protrusions in the membrane and thus the formation of both lamellipodia and filopodia.

INTERMEDIATE FILAMENTS

Structure

Intermediate filaments, with a diameter of 8 to 12 nm, are thicker than actin filaments and thinner than microtubules. Like actin filaments and microtubules, intermediate filaments are linear protein polymers that contribute to the structure of the cell. However, there are numerous contrasts between intermediate filaments and the other two major filament systems of the cytoskeleton: Intermediate filaments lack polarity and exhibit a much more heterogeneous subunit composition, as well as greater mechanical strength, flexibility, and stability than either microfilaments of microtubules.

Nearly 50 different proteins have been identified as subunits of intermediate filaments. These proteins can be grouped into six major types based on similarities in their amino acid sequences (see Table 1; 1,4,64). As a rough rule of thumb, proteins of a given type exhibit greater than 50% sequence homology and those of different types less than 50%. However, the similarity within each major type is often much greater than 50%, and the similarity between different types is often much less than 50% (65). These six protein types assemble into five different classes of filaments. Keratin protein types I and II do not exist independently of one another in assembled filaments,

Table 1. Intermediate Filament Protein Types

Protein	Type	MW (RD)	Cellular occurrence
Keratin	I and II	40–70	Epithelia
Vimentinlike proteins			
Vimentin	III	57	Mesenchymal cells
Desmin	III	55	Muscle
Glial fibrillary acid protein (GFAP)	III	50	Glial cells
Peripherin	III	57	Peripheral neurons
Neurofilament	IV	60–130	Neurons
Lamin	V	60–70	All eukaryotic cells
Nestin	VI	240	Neuronal stem cells

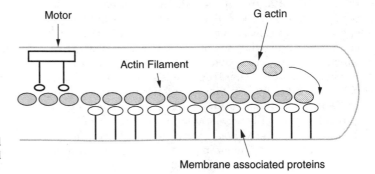

Figure 4. Key features in model of growth cone lamellipodial advance, based on actin dynamics in neuronal growth cones and other cell types.

but rather always combine in equal numbers within a filament.

Different intermediate filament proteins vary in their amino acid sequence but have a common secondary structure consisting of a rodlike central region that terminates in globular domains at either end. The central domain is approximately 300 amino acid residues (50 nm) long, with some variation according to type, and exhibits an α-helical structure. The α-helical structure is discontinuous, occurring within four separate segments connected by nonhelical linker segments that increase the flexibility of the central domain. The globular head (amino terminal) and tail (carboxy terminal) exhibit much more variability, ranging in approximate length from 10 to 1500 amino acid residues (66).

The coil motif is repeated at several structural levels to arrive at the final filament structure. First, the central rod domains of two protein molecules align in a parallel fashion and wind around one another to form a coiled-coil dimer. Two dimers then align head-to-tail, and may coil around one another, to form a tetramer (4). Protofilaments may be formed by linking tetramers end-to-end, with these protofilaments packing together into a helical structure to yield the completed filament (65). The final filament consists of approximately 8 tetramers, or 32 protein molecules in cross-section (Fig. 5). The resulting nonpolar, "twisted rope" filament structure is stronger, more flexible, and more stable than are the polar chains of globular proteins comprising actin and microtubule filaments.

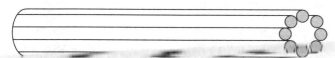

Figure 5. Intermediate filament structure. Intermediate filaments are typically composed of 8 nonpolar tetramers of intermediate filament protein polypeptides. The diagram here does not show the helical structure believed to be formed when the tetramers are assembled into the final filament form.

Assembly Dynamics

Assembly of intermediate filaments also appears to be less complex than actin or microtubule assembly, without a requirement for an energy source or for auxiliary proteins. Intermolecular associations between the protein molecules, especially near the ends of the central rod domain, are believed to be important. Many details about the assembly process remain to be elucidated.

Intermediate filaments are quite stable in vitro, requiring nonphysiological ionic conditions to promote disassembly. In addition, the in vivo monomer pool appears to be small. These observations supported the view, until recently, that intermediate filaments contributed little to the dynamic nature of the cytoskeleton. However, the kind of extensive reorganization of the cytoskeleton known to occur during the cell cycle, differentiation, or cell repair argued for a more active system of filaments. A number of studies now suggest that a dynamic equilibrium between soluble pools of subunits and assembled intermediate filaments is important in maintaining the intermediate filament network (67). It is likely that phosphorylation plays a significant role in regulating intermediate filament assembly. Phorphorylation, which promotes intermediate filament disassembly, accompanies filament disassembly during mitosis and provides a possible mechanism for countering the inherent stability of intermediate filaments noted in vitro (68).

Intermediate Filament Associated Proteins

As with microtubules and actin filaments, intermediate filaments are linked to the cytoskeletal network through a set of associated proteins, in this case, intermediate filament associated proteins (IFAPs). IFAPs serve several functions. They may cap intermediate filaments to prevent further growth, cross-link intermediate filaments, or link them to other cytoskeletal structures. A number of IFAPs that have been identified are listed in Table 2 (1,4,69,70). Some specific roles of IFAPs have been identified at least tentatively. For example, filaggrin bundles keratin

Table 2. Intermediate Filament Associated Proteins

IFAP	MW (RD)	IF associations	Other cytoskeletal associations	Cellular associations
Filaggrin	26–64	Keratin		Epidermal
Plakin family				
Desmoplakin	210–285	Keratin, Vimentin		Epithelial
Plectin	300	Vimentin Keratin Neurofilaments	Actin, microtubules	Mesodermally derived cells
BPAG	230	Keratin Neurofilaments	Actin	Epidermal Neuronal
Envoplakin	210	Keratin		Keratinocytes
Epinemin	44.5	Vimentin		Muscle
Paranemin	280	Desmin Vimentin		Myocytes
Synemin	230	Desmin Vimentin		Muscle Erythrocytes

filaments in epidermal cells (65); plakins localize to intermediate filament attachment sites at the plasma membrane (71); paranemin and synemin appear to serve as cross-linkers of intermediate filaments. Future work will be required to better understand the ways in which IFAPs contribute to cytoskeletal behavior.

Function

Intermediate filaments have been assigned several functions within the cell. The most basic of these is to provide mechanical integrity to the cell. This applies to both the nuclear lamins, which form a meshwork surrounding the nucleus, and the cytoplasmic intermediate filaments. More specialized roles may include the determination of axonal caliber by neurofilaments, support of DNA replication by nuclear lamins, and provision of attachment sites for cytoplasmic molecules and structures (65,72). Intermediate filaments appear to be nonessential elements of the cytoskeleton in some cases, though their deletion in other cases leads to cell and tissue abnormalities. Much of what we know about intermediate filament function has been developed through studying diseases caused by intermediate filament defects (8). The most commonly observed abnormalities are skin and hair disorders caused by keratin defects (66). In addition, neural dysfunctions have been linked to mutations or altered expression patterns in neurofilaments (66,73,74). Various other links between intermediate filament defects and disease have been reported anecdotally. Intermediate filaments are the least understood of the major filament systems of the cytoskeleton, but current research suggests an interesting story that is in the process of unfolding.

BIBLIOGRAPHY

1. B. Alberts et al., *Molecular Biology of the Cell*, Garland, New York, 1994.
2. V.M. Fowler, *Curr. Opin. Cell Biol.* **8**, 86–96 (1996).
3. J.V. Small, K. Rottner, and I. Kaverina, *Curr. Opin. Cell Biol.* **11**, 54–60 (1999).
4. S.L. Wolfe, *Introduction to Cell and Molecular Biology*, Wadsworth, New York, 1995.
5. D.J. Odde and H.M. Buettner, *Biophys. J.* **75**, 1189–1196 (1998).
6. T. Mitchison and M. Kirschner, *Neuron* **1**, 761–772 (1998).
7. P.A. Jamney and C. Chaponnier, *Curr. Opin. Cell Biol.* **7**, 111–117 (1995).
8. W.H.I. McLean and E.B. Lane, *Curr. Opin. Cell Biol.* **7**, 118–125 (1995).
9. L.A. Amos, in K. Roberts and J.S. Hyams, eds., *Microtubules*, Academic Press, London, 1979, pp. 1–64.
10. P.C. Letourneau, *J. Neurosci.* **2**, 806–814 (1993).
11. F. Gittes, B. Mickey, J. Nettleton, and J. Howard, *J. Cell Biol.* **120**, 923–934 (1993).
12. D.J. Odde, A. Briggs, and A. DeMarco, *Ann. Biomed. Eng.* **25**, S-35 (1997).
13. E. Nogales, S.G. Wolf, and K.H. Downing, *Nature (London)* **391**, 199–203 (1998).
14. C.A. Hale and P.A.J. de Boer, *Cell* **88**, 175–185 (1997).
15. R.H. Wade and A.A. Hyman, *Curr. Opin. Cell Biol.* **9**, 12–17 (1997).
16. M.J. Schnitzer and S.M. Block, *Nature (London)* **388**, 386–390 (1997).
17. K. Svoboda, C.F. Schmidt, B.J. Schnapp, and S.M. Block, *Nature (London)* **365**, 721–727 (1993).
18. N.B. Cole and J. Lippincott-Schwartz, *Curr. Opin. Cell Biol.* **7**, 55–64 (1995).
19. S. Inoue and E.D. Salmon, *Mol. Biol. Cell* **6**, 1619–1640 (1995).
20. V.I. Rodionov and G.G. Borisy, *Nature (London)* **386**, 170–173 (1997).
21. I.A. Vorobjev, T.M. Svitkina, and G.G. Borisy, *J. Cell Sci.* **110**, 2635–2645 (1997).
22. T.J. Keating et al., *Proc. Natl. Acad. Sci. U.S.A.* **94**, 5078–5083 (1997).
23. C.M. Waterman-Storer and E.D. Salmon, *J. Cell Biol.* **139**, 417–434 (1997).
24. T.L. Hill, *Linear Aggregation Theory in Cell Biology*, Springer-Verlag, New York, 1987.
25. T.J. Mitchison and M.W. Kirschner, *Nature (London)* **312**, 237–242 (1984).
26. E. Tanaka, T. Ho, and M.W. Kirschner, *J. Cell Biol.* **128**, 139–155 (1995).
27. M. Carlier, D. Didry, C. Simon, and D. Pantaloni, *Biochemistry* **28**, 1783–1791 (1989).
28. M. Caplow and J. Shanks, *Biochemistry* **34**, 15732–15741 (1995).
29. D.N. Drechsel and M.W. Kirschner, *Curr. Biol.* **4**, 1053–1061 (1994).
30. S.R. Martin, M.J. Schilstra, and P.M. Bayley, *Biophys. J.* **65**, 578–596 (1993).
31. R.A. Walker et al., *J. Cell Biol.* **107**, 1437–1448 (1988).
32. M. Dogterom, M.-A. Felix, C.C. Guet, and S. Leibler, *J. Cell Biol.* **133**, 125–140 (1996).
33. B. Howell, D.J. Odde, and L. Cassimeris, *Cell Motil. Cytoskel.* **38**, 201–214 (1997).
34. D.J. Odde, L. Cassimeris, and H.M. Buettner, *Biophys. J.* **69**, 796–802 (1995).
35. D.J. Odde, E.M. Tanaka, S.S. Hawkins, and H.M. Buettner, *Biotechnol. Bioeng.* **50**, 452–461 (1996).
36. N.R. Gliksman, R.V. Skibbens, and E.D. Salmon, *Mol. Biol. Cell* **4**, 1035–1050 (1993).
37. F.J. McNally and R.D. Vale, *Cell* **75**, 419–429 (1993).
38. N.K. Pryer et al., *J. Cell Sci.* **103**, 965–976 (1992).
39. K. Ookata et al., *J. Cell Biol.* **128**, 849–862 (1995).
40. C.E. Walczak, T.J. Mitchison, and A. Desai, *Cell* **84**, 37–47 (1996).
41. R.J. Vasquez, D.L. Gard, and L. Cassimeris, *J. Cell Biol.* **127**, 985–993 (1994).
42. S.B. Horwitz, *Annl. Oncol.* **5**, S3–S6 (1994).
43. P.T. Tran et al., *Mol. Biol. Cell* **6**, 260a (1995).
44. L. Wilson and M.A. Jordan, (1994). in J.S. Hyams and C.W. Lloyd, eds., *Microtubules*, Wiley-Liss, New York; pp. 59–83.
45. A.J. Putnam et al., *J. Cell Sci.* **111**, 3379–3387 (1998).
46. I.M. Herman, *Curr. Opin. Cell Biol.* **5**, 48–55 (1993).
47. A. Spiros and L. Edelstein-Keshet, *Bull. Math. Biol.* **60**, 275–305 (1998).

48. A.K. Lewis and P.C. Bridgman, *J. Cell Biol.* **119**, 1219–1243 (1992).

49. T.D. Pollard, *Curr. Opin. Cell Biol.* **2**, 33–40 (1990).

50. M.F. Carlier, *Curr. Opin. Cell Biol.* **10**, 45–51 (1998).

51. Y. Wang, *J. Cell Biol.* **101**, 597–602 (1985).

52. J.V. Small, *Trends Cell Biol.* **5**, 52–55 (1995).

53. K.R. Ayscough, *Curr. Opin. Cell Biol.* **10**, 102–111 (1998).

54. J.J. Otto, *Curr. Opin. Cell Biol.* **6**, 105–109 (1994).

55. J.H. Hartwig and D.J. Kwiatkowski, *Curr. Opin. Cell Biol.* **3**, 87–97 (1991).

56. P.C. Letourneau, T.A. Shattuck, and A.H. Ressler, *Cell Motil. Cytoskel.* **8**, 193–209 (1987).

57. L.P. Cramer, T.J. Mitchison, and J.A. Theriot, *Curr. Opin. Cell Biol.* **6**, 82–86 (1994).

58. T. Oliver, J. Lee, and K. Jacobson, *Semin. Cell Bio.* **5**, 139–147 (1994).

59. J. Lee, A. Ishihara, and K. Jacobson, *Trends Cell Bio.* **3**, 366–370 (1993).

60. J. Lee, A. Ishihara, J.A. Theriot, and K. Jacobson, *Nature (London)* **362**, 167–171 (1993).

61. A. Burt and H.M. Buettner, *AIChE Annu. Meet.*, 1997.

62. S. Okabe and N. Hirokawa, *J. Neurosci.* **11**, 1918–1929 (1991).

63. C.H. Lin, E.M. Espreatico, M.S. Mooseker, and P. Forscher, *Neuron* **16**, 769–782 (1996).

64. K. Albers and E. Fuchs, *Int. Rev. Cytol.* **134**, 243–279 (1992).

65. E. Fuchs and K. Weber, *Annu. Rev. Biochem.* **63**, 345–382 (1994).

66. E. Fuchs, *J. Cell Biol.* **125**, 511–516 (1994).

67. J.E. Eriksson, P. Opal, and R.D. Goldman, *Curr. Opin. Cell Biol.* **4**, 99–104 (1992).

68. S. Heins and U. Aebi, *Curr. Opin. Cell Biol.* **6**, 25–33 (1994).

69. R. Foisner and G. Wiche, *Curr. Opin. Cell Biol.* **3**, 75–81 (1991).

70. M.K. Houseweart and D.W. Cleveland, *Curr. Opin. Cell Biol.* **10**, 93–101 (1998).

71. C. Ruhrberg and F.M. Watt, *Curr. Opin. Genet. Dev.* **7**, 392–397 (1997).

72. M. Stewart, *Curr. Opin. Cell Biol.* **5**, 3–11 (1993).

73. M.K. Lee and D.W. Cleveland, *Curr. Opin. Cell Biol.* **6**, 34–40 (1994).

74. S. Singh and P.D. Gupta, *Biol. Cell.* **82**, 1–10 (1994).

See also ANATOMY OF PLANT CELLS; CELL STRUCTURE AND MOTION, EXTRACELLULAR MATRIX AND CELL ADHESION; CELL-SURFACE RECEPTORS: STRUCTURE, ACTIVATION, AND SIGNALING; CRYOPRESERVATION OF PLANT CELLS, TISSUES AND ORGANS.

CELL STRUCTURE AND MOTION, EXTRACELLULAR MATRIX AND CELL ADHESION

W. MARK SALTZMAN
MELISSA J. MAHONEY
Cornell University
Ithaca, New York

OUTLINE

INTRODUCTION

Cells are membrane-bounded aqueous compartments that interact cooperatively with one another and their environment. Cells in tissues interact and communicate with other cells and the surrounding matrix through specialized sites of cell–cell or cell–matrix contact. The molecular factors that mediate these interactions have been the focus of considerable study; therefore, many of the mechanisms of contact formation and the functional consequences of contact are understood. In this article we summarize these findings by reviewing the physical forces governing cell adhesion, the structure and function of specialized cell–cell and cell–matrix contacts, and the molecules involved in cell adhesion. Many biotechnological processes involve culture of tissue-derived cells that have been dispersed and suspended; therefore, understanding (and sometimes replacing) the normal structure and function of cell contacts is essential.

PHYSICAL FORCES DURING CELL BINDING AND AGGREGATION

Organs, the principal functional unit of higher organisms, are composed of tissues, which consist of collections of cells distributed throughout a three-dimensional assembly of biopolymers called the extracellular matrix. The external surface of the cell carries a carbohydrate-rich coat called the glycocalyx (Fig. 1) (1); ionizable groups within the glycocalyx, such as sialic acid (*N*-acetyl neuraminate), contribute a net negative charge to the cell surface. Many of the carbohydrates that form the glycocalyx are bound to membrane-associated proteins. Each of these components — phospholipid bilayer, carbohydrate-rich coat,

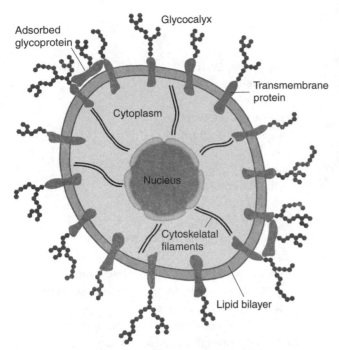

Figure 1. Schematic of a simplified animal cell [adapted from Panel I-1 and Figure 6–40 of Ref. 1]. The cell is surrounded by a glycocalyx that consists of carbohydrates on transmembrane and absorbed proteins. Transmembrane proteins can interact with elements of the filamentous cytoskeleton, providing a mechanism for transmitting extracellular information to the nucleus.

be overwhelmed by the mechanical forces present in most biological systems.

The most stable and versatile mechanism for cell adhesion involves the specific association of cell surface glycoproteins, called receptors, and complementary molecules in the extracellular space, called ligands (3). Ligands may exist freely in the extracellular space, they may be associated with the extracellular matrix, or they may be attached to the surface of another cell (Fig. 2). Cell–cell adhesion can occur by homophilic binding of identical receptors on different cells, by heterophilic binding of a receptor to a ligand expressed on a different cell's surface, or by association of two receptors with an intermediate linker. Cell–matrix adhesion usually occurs by heterophilic binding of a receptor to a ligand attached to an insoluble element of the extracellular matrix.

Receptors bind specifically to ligands by mechanisms that are similar to antibody–antigen or enzyme–substrate binding, but the affinity of a single cell adhesion receptor and ligand combination is usually much lower than the affinity associated with receptors for hormones or enzyme–substrate pairs (Table 1) (4–10). A typical integrin receptor for extracellular matrix, for example, has an affinity of $\sim 10^{-6}$ M, whereas the receptor for epidermal growth factor has an affinity of $\sim 7 \times 10^{-10}$ M (when affinity is represented by the dissociation constant, K_D, higher affinity corresponds to a lower number). The total strength of specific cell adhesion depends on the

membrane-associated protein — has distinct physicochemical characteristics and is abundant. Plasma membranes contain $\sim 50\%$ protein, $\sim 45\%$ lipid, and $\sim 5\%$ carbohydrate by weight. Therefore, each component influences cell interactions with the external environment in important ways.

Cells near a surface experience electrostatic, steric stabilization, and van der Waals forces (2). Electrostatic forces may be either attractive or repulsive, depending on the charge associated with the surface. When the opposing surface is on another cell, the electrostatic forces tend to be repulsive because the glycocalyx of each cell carries a net negative charge. Steric stabilization forces result from the osmotic imbalance developed in the gap between cell and surface. To close the gap, water must be forced out of this region; therefore, cell-surface macromolecules in the shrinking gap become concentrated, resulting in an osmotic imbalance that pulls water into the gap and generates a repulsive force. This repulsive force can become large as the gap region becomes small. Cell-surface molecules also become compressed during close approach, contributing further to the repulsive steric stabilization force. Charge interactions between polarizable but uncharged molecules produce van der Waals forces, which are attractive. These forces are important at separation distances on the order of 100 to 300 Å, the separation distance that is usually associated with cell–cell and cell–surface adhesion. But van der Waals forces are not sufficiently strong to allow stable multicellular assemblies to form; they are weak enough to

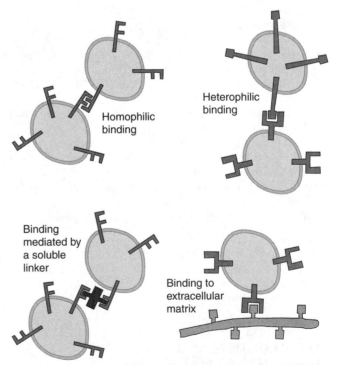

Figure 2. Cell–cell and cell–matrix binding [adapted from Figure 14–64 of Ref. 1]. Cell–cell binding can occur by the association of identical receptors (homophilic binding), by the association of complementary receptors (heterophilic binding), or through the cooperation of a multifunctional molecule dissolved in the extracellular space. Cell–matrix binding involves the association of a cell surface molecule with a complementary ligand on the matrix.

Table 1. Binding Affinity for Cell-Adhesion Receptors

Ligand	Receptor	K_D (nM)	Reference
Integrins			
Fibrinogen	$\alpha IIb\beta 3$	>7,000,000	(4)
RGDS	Fibroblast cell surface	600,000	(5)
cHarGD	$\alpha IIb\beta 3$	10	(6)
Cadherin			
Cadherin	Cadherin	N.A.	
β-Catenin	α-Catenin	100	(7)
Ig-family			
Phosphacan	N-CAM/Ng-CAM	0.1	(8)
Neurocan	N-CAM/Ng-CAM	1	(8)
Selectins			
GlyCAM-1	L-Selectin (CD62L)	108,000	(9)
sLe^x	E-Selectin (CD62E)	720,000	Reported in (10)
sLe^x	P-Selectin (CD62P)	7,800,000	Reported in (10)

Notes: K_D is the dissociation constant. Phosphacan and neurocan are proteoglycans.

Abbreviations: N.A., not available; R, arginine; G, glycine; D, aspartic acid; S, serine; Har, homoarginine.

strength and number of individual receptor–ligand bonds formed. Because cell-surface glycoprotein is abundant and accounts for ~50% of the membrane mass, a single cell can display distinct receptors that recognize thousands of different ligands; in addition, there may be thousands of copies of the receptor for any particular ligand. For example, each fibroblast has ~500,000 receptors for the extracellular matrix protein fibronectin. The combined effect of many receptor–ligand interactions produces a strong adhesive force, sufficient to permit stable cell–cell and cell–matrix adhesion.

CELL JUNCTIONS

Binding of cell-surface receptors to complementary ligands is a principal mechanism for initial association of cells with other cells and with matrices. This mechanism of biological recognition is particularly important during development of the embryo; cells proliferate, migrate, and use specific binding signals to localize in the developing organism. In the adult organism, stability of the cell–cell or cell–matrix assembly is enhanced by the formation of specialized contact regions called cell junctions. Through these junctions, cells in tissue are linked to one another and to their surrounding extracellular matrix.

Three types of junctions are found in tissues: tight cell junctions, anchoring cell junctions and communicating cell junctions. This overview focuses on junctions that function to create mechanically stable tissues from individual cells: tight cell junctions and anchoring cell junctions. Communicating cell junctions, such as gap junctions, provide a mechanism for regulated exchange of molecules between adjacent cells. In gap junctions in the liver, aqueous channels are created by a coordinated assembly of 12 transmembrane proteins called connexins: six connexin molecules are contributed by each cell. The channel

permits the passage of small molecules (molecular weight <1,200) from the cytoplasm of one cell to the other (11).

Tight Junctions

Epithelial cells are sealed together via tight junctions into a continuous sheet that forms the barrier surface of all mucosal tissues, such as the intestinal, reproductive, and respiratory tracts. The tight junction is composed primarily of interconnected transmembrane proteins, contributed equally by both cell partners, which proceed continuously across the space between adjacent cells (Fig. 3) In this arrangement, tight junctions act as barriers to the passage of small molecules in the extracellular space. Because the permeability of tight junctions decreases logrithmically with protein density in the junction, tissues can have junctional complexes of different structure that permit molecules of a specific size or charge to diffuse; some tight junctions are so dense that they are essentially impermeable. Within the plasma membrane, tight junctions also act as physical barriers which confine transmembrane proteins involved in carrier-mediated transport to specific regions of the cell (e.g., the apical and basolateral surfaces; Fig. 3). In this way, tight junctions facilitate the transfer of water-soluble molecules, such as proteins or glucose, from the lumen of the gut into the blood and even permit transport against a concentration gradient.

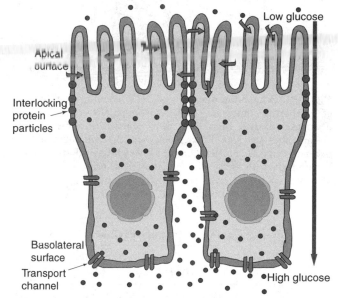

Figure 3. Tight junctions form a barrier to the movement of molecules in the space between adjacent cells. The junction is formed by arrays of interlocking protein particles that are contributed by both of the neighboring cells. The junction prevents the diffusion of water-soluble molecules in the extracellular space, as well as membrane proteins. The presence of tight junctions creates segregated domains in each cell, and glucose active transport proteins (arrows) are confined in the apical domain and glucose-facilitated transport channels confined in the basolateral domain. Tight junctions enable transport of glucose up a concentration gradient by segregating functional transport proteins in the membrane and prohibiting the back diffusion of glucose through the gaps between cells.

Anchoring Junctions

Cells are mechanically attached to another cell or an extracellular matrix protein over discrete regions called anchoring junctions. Anchoring junctions serve as mechanical links between cells, as well as points of intersection of cell scaffolding proteins (i.e., cytoskeletal filaments) and the plasma membrane. Each junction contains intracellular attachment proteins, which form the physical connection between the cytoskeleton and the membrane, and transmembrane linker proteins, which tether the external face of the membrane to a complementary protein on an adjacent cell or matrix (Fig. 4). Cellular attachment occurs when the extracellular domain of the transmembrane linker protein associates with an extracellular matrix molecule or with the extracellular domain of a transmembrane linker protein on another cell. Adheren junctions, desmosomes, and hemidesmosomes are distinct types of anchoring junctions that differ in function, as well as in composition of the junctional complex (Fig. 4 and Table 2).

Adherens Junctions. Adherens junctions connect actin filaments, which are intracellular contractile cytoskeletal elements, in one cell either to an extracellular matrix molecule or to the extracellular domain of a transmembrane linker protein on another cell. In cell–cell adhesion, the transmembrane linker proteins are cadherins. In cell–matrix adhesion, the transmembrane linker proteins are members of a family of cell-surface matrix receptors called integrins.

Adherens junctions occur in several forms in mammalian tissue. Epithelial sheets have continuous beltlike junctions, or adhesion belts, which occur just below the tight junction between adjacent cells. Adhesion belts, formed from contractile actin filaments, generate folding movements in sheets of cells during tissue morphogenesis. In other cells, focal contacts, which are punctate attachment sites, occur on the surface of cells that are associated with extracellular matrix. Focal contacts form in specialized regions of the plasma membrane that coincide with the termination sites of actin filament bundles. The contacts function as mechanical anchors, but they can also translate signals from the extracellular matrix to the cytoskeleton. This signaling capability is an essential element in regulating cell functions, such as survival, growth, morphology, movement, and differentiation.

Desmosomes. Desmosomes provide an indirect link from the intermediate filaments of one cell to those of another. Intermediate filaments are assemblies of fibrous protein (e.g., vimentin, keratin, or desmin) that form a ropelike intracellular network; these filaments are responsible for much of the structural framework of the cell. The extracellular domain of cadherin receptors form the link from one cell to another. The physiological importance of desmosomes is observed in patients with pemphigus, a serious skin disease that involves

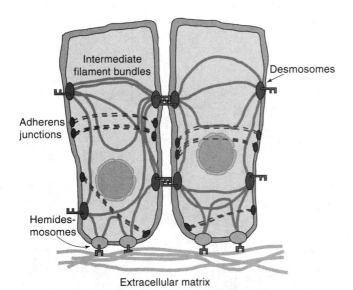

Figure 4. Schematic diagram of cells containing a variety of anchoring junctions. Adherens junctions connect actin filaments (dashed line) to form a contractile adhesion belt. Desmosomes provide a linkage of intermediate filament bundles between cells. Hemidesmosomes connect intermediate filament termination with extracellular matrix. Anchoring junctions are also characterized by the protein receptor, represented here as cadherins (F-shaped) and integrins (Y-shaped).

Table 2. Summary of Characteristics of Anchoring Junctions

Junction type	Transmembrane protein	Extracellular ligand	Intracellular linkage (accessory proteins)	
Adherens	Cadherin	Cadherin	Actin filaments (catenins)	Figure 12
Desmosomes	Cadherin	Cadherin	Intermediate filaments (desmoplakin, plakoglobin)	Figure 4
Hemidesmosomes	$\alpha_6\beta_4$ Integrin	ECM protein	Intermediate filaments	Figure 4
Focal contact	Integrin	ECM protein	Actin filaments (α-actinin, talin, vinculin)	Figure 12

Source: Adapted from Ref. 1.

a destruction of desmosomes between epithelial cells. Patients develop scattered bullae, or blisters, that rupture easily leaving denuded areas of skin and creating, in the worst cases, risk of severe infection and life-threatening fluid loss.

Hemidesmosomes. Hemidesmosomes connect the basal surface of epithelial cells and intermediate filaments in the cytoplasm to an underlying thin sheet of extracellular matrix called the basal lamina. Therefore, basal lamina, separates the epithelium from connective tissue. The transmembrane linker proteins that mediate this type of adhesion are integrins.

CELL ADHESION RECEPTORS

When the net interaction potential between a cell and a surface is attractive, the two can approach one another and become stably associated. This initial association usually involves the formation of specific receptor-ligand bonds (Fig. 2). Further maturation of cell–cell or cell–matrix contact involves receptor–ligand bonds that accumulate in specialized junctional complexes, as shown in Figure 4. Over the past several decades, a variety of receptor–ligand systems for cell–cell and cell–matrix binding have been identified and characterized (Fig. 5).

Integrins

The integrin family of cell-surface molecules is involved in cell adhesion and cell motility. Integrin-mediated cell adhesion is usually associated with cell–matrix interactions, but certain integrins (particularly those on white blood cells) also play an essential role in cell–cell adhesion. All integrins are heterodimeric membrane proteins that consist of noncovalently associated α- and β-subunits; different homologs of the subunits are identified by a numerical suffix $(\alpha_1, \alpha_2, \alpha_3, \ldots, \beta_1, \beta_2, \ldots)$. Each member of the integrin family can be characterized by the combination of subunits involved: for example, $\alpha_1\beta_1$ binds to collagen and laminin, $\alpha_5\beta_1$ binds to fibronectin, and the β_2-integrins bind to ligands on cell surfaces. Integrin binding is Ca^{2+}-dependent.

In cell-matrix binding, the extracellular domain of the integrin receptor binds to an extracellular matrix protein, and the cytoplasmic integrin domain binds to the protein cytoskeleton (Fig. 4). Therefore, integrin receptors provide a critical mechanical connection between the extracellular and intracellular environment. Certain aspects of integrin binding are conserved among a variety of extracellular matrix (ECM) proteins. For example, several matrix proteins (fibronectin, collagen, vitronectin, thrombospondin, tenascin, laminin, and entactin) contain the three amino acid sequence Arg-Gly-Asp (RGD), which is critical for cell binding. Integrin-ECM binding can be reduced and sometimes eliminated by addition of dissolved RGD peptides, which compete for the integrin receptor site and, hence, encourage dissociation of the integrin-matrix bond. In addition, cells will adhere to immobilized peptides containing the RGD sequence. RGD is not the only sequence involved in integrin interaction with matrix proteins; collagens contain many RGD sequences, but soluble RGD does not alter cell binding to collagen. In addition, different integrins bind

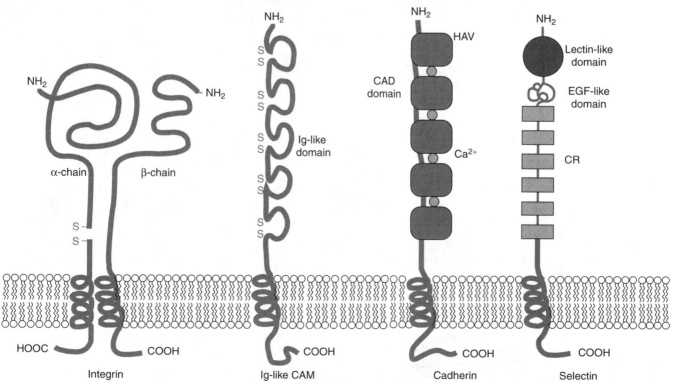

Figure 5. Schematic structure of protein cell receptors involved in adhesion.

selectively to different proteins at RGD-containing sites, suggesting that binding is also influenced by polypeptide domains adjacent to the RGD domain, but specific to each protein.

Cell-adhesion domains are often associated with the termination points for filaments of the cytoskeleton (Fig. 4). In the case of integrin-mediated cell adhesion, the association point is an integrated assembly of cytoplasmic proteins, such as vinculin and talin (Fig. 6) (12,13). The composition of the analogous protein assembly in hemidesmosomes, which connects the integrin-ECM complex to intermediate filaments, is unknown. However, $\alpha_6\beta_4$, the β-subunit of the integrin involved in hemidesmosomes, has an unusually long cytoplasmic tail, which may interact directly with intermediate filament proteins or facilitate assembly of accessory proteins. For more details on the role of the cytoskeleton in the maintenance of cell shape and motility.

Cadherins

Cadherin receptors are ~700 residue transmembrane proteins that mediate cell–cell adhesion through homophilic binding. Binding of cadherins is Ca^{2+}-dependent and ubiquitous. Cadherins can be identified in almost all vertebrate cells and account for the Ca^{2+} dependence of solid tissue form. In the absence of extracellular Ca^{2+}, cadherins are subject to rapid proteolysis and, therefore, loss of function. The best characterized cadherins are E-cadherin (found predominantly on epithelial cells), P-cadherin (found on cells of the placenta and skin) and N-cadherin (found on nerve, lens, and heart cells). These molecules have common structural features, including five extracellular repeats (CAD domains), Ca^{2+} binding regions, a membrane-spanning region, and a cytoplasmic domain (Fig. 5). Each of the CAD domains has structural similarity, although not sequence homology, with the immunoglobulin folds. Each CAD domain contains an HAV (His-Ala-Val) motif that is important for binding; the specificity of each cadherin is determined by the residues flanking this HAV region.

The affinity of cadherin–cadherin binding is not known, but available evidence suggests that the affinity of individual bonds is low. In adherens junctions, the cytoplasmic tail of cadherins interact with the actin cytoskeleton through the proteins α-catenin and β-catenin, which form a coordinated assembly. The interaction of the cytoplasmic catenin complex with cadherin enhances cadherin–cadherin affinity. Other proteins, such as desmoplakin and plakoglobulin, are associated with cadherin-intermediate filament contact in desmosomes (14).

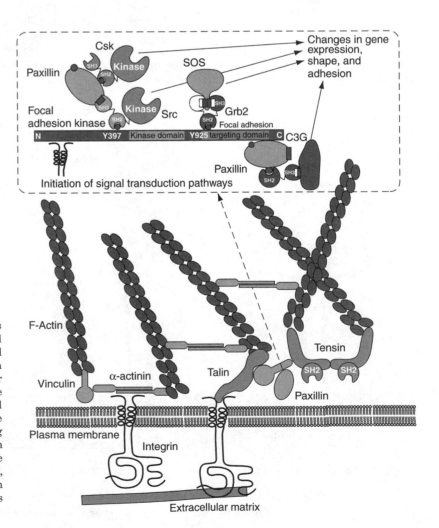

Figure 6. Association of cytoplasmic proteins (tensin, talin, vinculin, α-actinin, and F-actin) and extracellular proteins with the cytoplasmic and extracellular domains of a transmembrane integrin receptor. The inset shows a hypothetical model for signal transduction based on focal adhesion kinase (FAK). The model includes tyrosine kinases (Src and Csk), adapter proteins (Grb2 and Crk), and guanine nucleotide exchange factors (SOS and C3G). Binding occurs through SH2 (Src homology 2) domains, which bind to proteins that contain a phosphotyrosine (small circles), and SH2 (Src homology 3) domains, which bind to proteins that contain a proline-rich peptide motif (smal rectangles). The diagram was adapted from a variety of sources (2,5,17,18).

Ig-like Receptors

Members of the immunoglobulinlike (Ig-like) family of cell-surface receptors mediate cell–cell adhesion via a Ca^{2+}-independent process. The most well-studied of these receptors are cell-adhesion molecules in the nervous system (e.g., neural cell-adhesion molecule, N-CAM, and L1). All members of this family have structural similarities to immunoglobulins (Fig. 5); in addition, N-CAM has structural similarities to extracellular matrix molecules, such as fibronectin. N-CAM functions in cell adhesion and neurite outgrowth predominantly by homophilic binding.

Both cadherins and Ig-like receptors are present on the surfaces of cells during development and probably provide signals for cell assembly into organized structures. However, cell adhesion mediated by cadherins is much stronger than that mediated by Ig-like proteins; overexpression of cadherins, but not Ig-like receptors, leads to abnormalities in embryonic development. These observations suggest that the Ig-like proteins provide fine control over cellular attachments, which are dominated by cadherins.

Selectins

The selectins contain a carbohydrate-binding lectin domain (Fig. 5) and are transiently expressed during the inflammatory response. These cell-adhesion receptors are usually expressed on the surface of endothelial cells, where the lectin domain recognizes specific oligosaccharides expressed on the surface of neutrophils. Neutrophils bind to endothelial cells at the site of inflammation and roll along the blood vessel surface until other adhesion mechanisms initiate vessel wall transmigration into and participation in the local inflammatory response.

EXTRACELLULAR MATRIX

The extracellular space contains a three-dimensional array of protein fibers and filaments (Fig. 7) (15), which are embedded in a hydrated gel of glycosaminoglycans (GAGs). Extracellular proteins and polysaccharides are secreted locally by cells and assemble to form a scaffold that supports cell attachment, spreading, proliferation, migration, and differentiation. Cells influence the chemistry of the matrix by secreting protein and polysaccharide elements, but they also modify the physical characteristics of the matrix by releasing modifying enzymes or by applying physical forces. This section reviews the properties of the molecular constituents of the extracellular matrix.

Glycosaminoglycans and Proteoglycans

GAGs are high molecular weight polysaccharides, which are usually highly sulfated and, therefore, negatively charged. Each GAG is a poly(disaccharide), with sugar residues A and B (Table 3), repeated in a regular pattern, -(-A-B-)$_n$- (Fig. 8). Most GAGs also contain other sugar components, such as D-xylose, which create additional complexity in the linear chemical structure. GAGs are unbranched, relatively inflexible, and highly soluble in water; they adopt random coil conformations that occupy large volumes in aqueous media. Because individual

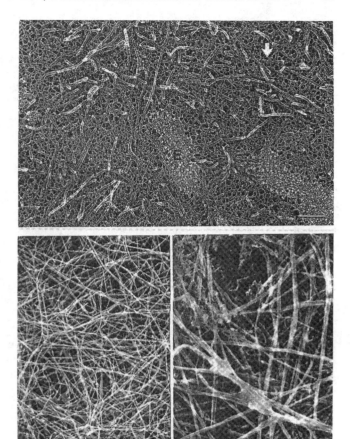

Figure 7. (Top) Electron micrograph of the extracellular matrix of ear cartilage showing collagen fibers and elastin (E) fibers. Arrowhead indicates point of attachment to collagen filaments in the plane below the section. Bar = 0.2 μm. From Ref. 16. (Left) Scanning electron micrograph of a reconstituted collagen gel (0.4 mg/mL collagen) formed from type I collagen extracted from rat tail tendons. Bar = 1 μm. (Right) Identical to image on left, but with 40 μg/mL of heparin added during reconstitution. Notice that the presence of heparin causes assembly of the collagen fibers into larger structures. Bar = 1 μm.

chains are extended and interact with each other to form additional structure, GAGs form hydrated gels at low GAG concentration. Counterions in the aqueous phase that surrounds the GAG chain, principally Na^+, create osmotic forces that pull water into the gel. Because of this property, tissues that are high in GAG content, such as cartilage, have high resistance to compressive stress.

Hyaluronic acid is an abundant polysaccharide that is unusual among the GAGs because it is unsulfated, not covalently attached to protein, and free of other sugar groups. Hyaluronic acid is present during embryogenesis and is frequently associated with tissues undergoing repair. In both of these situations, the presence of hyaluronic acid facilitates cell migration through the extracellular matrix, perhaps by modulating the level of hydration in the tissue.

Except for hyaluronic acid, GAGs are covalently attached to a protein; this macromolecular complex is called a proteoglycan (Table 4). Proteoglycans can have extremely high sugar contents, up to 95% by weight.

Table 3. Characteristics of Glycosaminoglycans (GAGs)

GAG	M.K. (kDa)	Sugar A	Sugar B	Sulfates	Links to protein	Other sugars	Tissue distribution
Hyaluronic acid	4 to 8,000	D-Glucuronic acid	N-Acetyl-D-glucosamine	−	−	−	Connective tissues, skin, vitreous body, cartilage, synovial fluid
Chondroitin sulfate	5 to 50	D-Glucuronic acid	N-Acetyl-D-galactosamine	+	+	+	Cartilage, cornea, bone, skin, arteries
Dermatan sulfate	15 to 40	D-Glucuronic acid or L-iduronic acid	N-acetyl-D-galactosamine	+	+	+	Skin, blood vessels, heart
Heparan sulfate	5 to 12	D-Glucuronic acid or L-iduronic acid	N-acetyl-D-glucosamine	+	+	+	Lung, arteries, cell surfaces, basal laminae
Heparin	6 to 25	D-Glucuronic acid or L-iduronic acid	N-acetyl-D-glucosamine	+	+	+	Lung, liver, skin, mast cells
Keratan sulfate	4 to 19	D-Galactose	N-acetyl-D-glucosamine	+	+	+	Cartilage, cornea, intervertebral disc

Source: Adapted from Ref. 1.

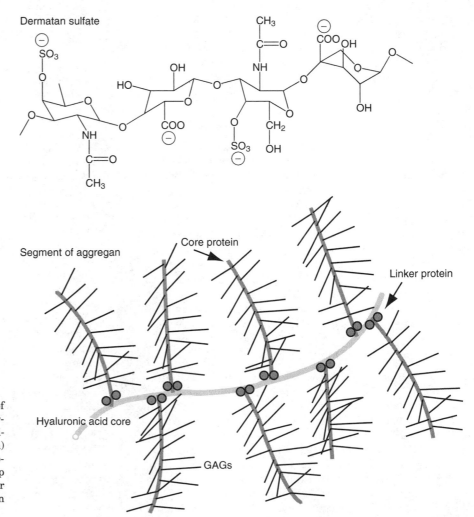

Figure 8. (Top) Chemical structure of repeated disaccharide of the GAG dermatan sulfate (protons are not indicated in this schematic diagram). (Bottom) Schematic diagram of the aggrecan proteoglycan complex showing the relationship between the core (thick lines) and linker proteins (spheres), GAGs (straight thin lines), and hyaluronic acid core.

Table 4. Characteristics of Proteoglycans

Proteoglycan	MW of core protein (kDa)	Type of GAG	Number of GAGs	Tissue distribution
Aggregan	210	Chondroitin sulfate Keratan sulfate	~130	Cartilage
Decorin	40	Chondroitin sulfate Dermatan sulfate	1	Connective tissues
Syndecan-1	32	Chondroitin sulfate Heparan sulfate	1–3	Fibroblast and epithelial cell surfaces

Source: Adapted from Ref. 1.

The diversity of the core protein structure is compounded by the potential for linkage to combinations of GAGs of different chain length, making proteoglycans a highly heterogeneous group of molecules. Proteoglycans often have an additional level of structural organization in tissues: for example, the individual proteoglycans in aggrecan are arranged systematically around a hyaluronic acid core (Fig. 8). This structural versatility translates into a variety of functions in tissues: proteoglycans serve as reservoirs of biological activity by binding growth factors; as size- and charge-selective filters in the glomerulus; and as mediators of adhesion on cell membranes.

Proteins

Proteins in the extracellular space are either of the structural type (e.g., collagen, elastin) or the adhesive type (e.g., laminin and fibronectin).

Collagen. Collagen, the most abundant protein in the extracellular space, is secreted by chondrocytes, fibroblasts, and other cell types. Several chemically distinct forms of collagen have been identified, each of which contains the same basic macromolecular unit: an alpha helical chain formed by the interaction of three polypeptides (Fig. 9). These polypeptide chains are 1000 amino acids long and are specific to each type of collagen. The most common forms of collagen within the extracellular space are collagens type I, II, III, and IV. Following secretion into the ECM, molecules of collagen types I, II, and III organize into larger fibrils 10 to 300 nm in diameter (Fig. 9). These fibrils are stabilized by cross-links which connect lysine residues within or between adjacent collagen molecules. In some tissues, these fibrils become further organized, and form larger collagen fibers several micrometers in diameter. Fibrillar collagens interact with cells through integrin receptors on cell surfaces. Cell differentiation and migration during development are influenced by fibrillar collagens.

In contrast to the fibrillar collagens, collagen type IV forms a meshlike lattice that constitutes a major part of the mature basal lamina, the thin mat separating epithelial sheets from other tissues. Collagen type IV interacts with cells indirectly by binding to laminin, another major component of the basal lamina.

Fibronectin. Fibronectin is a dimeric glycoprotein composed of similar subunits, each of which contains 2500 amino acids (Fig. 10). The two similar polypeptide chains are linked by disulfide bonds at the carboxyl termini and folded into a number of globular domains. The biological activities of certain polypeptide domains within the fibronectin macromolecule have been defined by observing the properties of fibronectin fragments. Fibronectin binds to collagen and heparin; this binding contributes to the organization of the extracellular matrix. Integrin receptors on cell surfaces bind to a fibronectin domain that contains the tripeptide sequence RGD. It appears that complete cellular adhesion also requires the participation of another region on fibronectin, the "synergy" region on the amino terminus side of the RGD-containing region. A second cell-binding region on fibronectin, the IIICS region, contains sequences that permit the adhesion of specific cell types. Usually fibroblasts do not adhere to the IIICS region but other cells (e.g., neural cells and lymphocytes) do. Cellular interactions with fibronectin affect cell morphology, migration, and differentiation.

Laminin. Laminin is a large cross-shaped protein composed of three polypeptide subunits: A, B1, and B2. In the long arm of laminin, the three subunits form a coiled-coil α-helical domain (Fig. 10). Homologs of the laminin subunits have been identified including merosin (containing the A-chain variant, Am), K-laminin (Ak), and homologs of the B1- and B2-chains. The existence of homologous forms of laminin, which have different distributions within the adult and embryo, suggests that these laminin isoforms may also have different functions.

Laminin contains binding sites for cell attachment and binding sites that promote neurite outgrowth. Regions that promote cell attachment, heparin binding, and neurite outgrowth have been identified and involve the specific peptide sequences RGD, YIGSR, IKVAV on the A-chain. Several integrin receptors ($\alpha_6\beta_1$, $\alpha_3\beta_1$) have been implicated in cell interactions with laminin, but the IKVAV region appears to interact with non-integrin receptors.

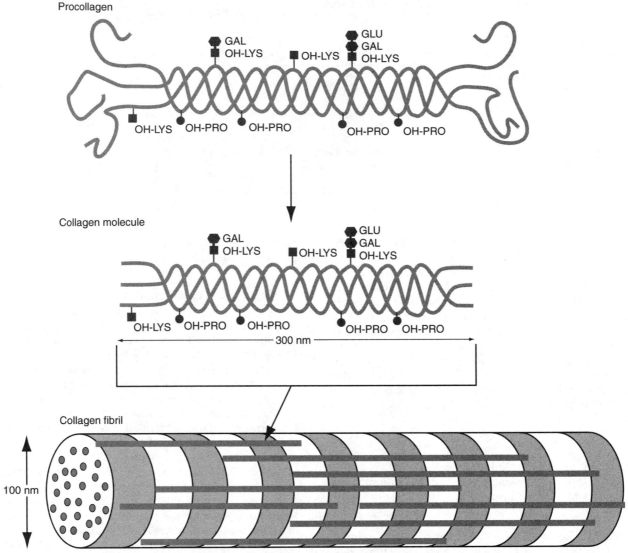

Figure 9. Schematic structure of collagen fibers. Procollagen molecules are produced by cells and secreted. The collagen triple helix is produced by enzymatic cleavage in the extracellular environment. Collagen molecules assemble into fibrils and fibers by an orderly arrangement that is stabilized by cross-linking between individual molecules.

Elastin. Elastin is a hydrophobic, nonglycosylated protein composed of 830 amino acids. After secretion into the ECM, extended elastin molecules form cross-linked fibers and sheets which can stretch and relax upon deformation (Fig. 7). Elastin is abundant in tissues that undergo repeated stretching, such as blood vessels.

Tenascin. Tenascin, a multiunit glycoprotein of 1.9×10^6 daltons, has both cell-adhesive and antiadhesive properties. It is present in embryonic tissues, but is confined primarily to the nervous system in adults, and may be involved in modulating cell migration during embryogenesis. Tenascin is composed of six polypeptide chains, some containing the RGD sequence, which are disulfide-linked to produce a windmill-shaped complex.

Vitronectin. The intact form of vitronectin, a 75,000 dalton monomer, is proteolytically converted into 65,000

and 10,000 dalton fragments. These molecules are present in the blood (0.2 to 0.4 g/L) and in certain tissues, usually associated with fibronectin. Vitronectin contains the RGD tripeptide sequence and can promote the attachment of many cell types, presumably through integrin receptors other than the common fibronectin binding integrin.

Thrombospondin. Thrombospondin is a trimer that consists of three identical 140,000 dalton subunits and is presumed important in control of cell growth. The cell-adhesion activity of thrombospondin is associated with a heparin-binding domain and an RGD sequence. Certain regions can also interact with collagen, laminin, and fibronectin.

Entactin. Entactin (nidogen) is a protein of 150,000 daltons that is found in basement membranes, where

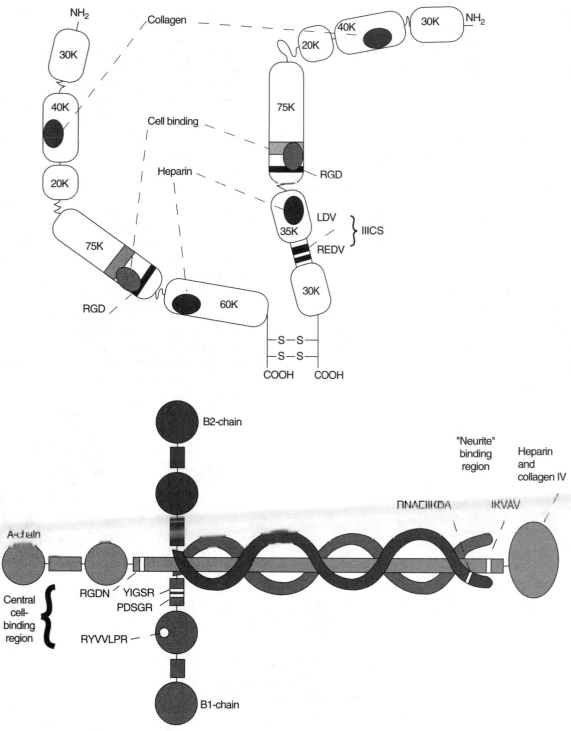

Figure 10. Schematic structure of the major adhesive proteins, fibronectin (top) and laminin (bottom).

it is strongly associated with laminin. Almost all laminin preparations contain entactin, which is associated stoichiometrically with the laminin central cell-binding region. Entactin contains the RGD sequence and a region that permits self-association.

The composition and structure of the extracellular matrix in a particular tissue is related to its function. In general, the extracellular matrix of connective tissue consists of collagen and/or elastin fibers coursing through a GAG-rich ground substance, in which mesenchymal cells, such as fibroblasts, migrate. Protein fibers give the tissue tensile strength (collagen) and elasticity (elastin), and the GAG gel resists compression; mesenchymal cells secrete protein and carbohydrates, facilitate organization of the matrix structure, and infiltrate into damaged tissue to initiate wound healing. Epithelial tissues are typically organized around basal laminae; these thin gel layers are rich in collagen IV and laminin. When present on one face of a cell monolayer, a basal lamina can induce cell polarity. For example, columnar epithelial cells in the intestine rest on a basal lamina that induces differentiation of the cell surface into basal and apical domains. The development and maintenance of polarity is essential for intestinal tissue function, as described more completely following.

Extracellular matrix interactions with cells are critically important in the developing organism. The extracellular matrix composition varies with time and location during embryogenesis (Fig. 11) (16). Changes in composition are produced by protein secretion from embryonic cells. Cell-derived matrix cues provide guides for cell migration and assembly into specialized tissues. When these cues are missing, for example, when fibronectin-rich segments of tissue are removed or transplanted during development, mesodermal cells cannot migrate properly during the early stages of gastrulation. Similar experiments (in the case of fibronectin, for example, using microinjection of antifibronectin antibodies, adhesion-blocking RGD-peptides, or embryos with the fibronectin gene eliminated) have defined the critical importance of matrix composition in cell migration, cell differentiation, and normal progression through almost every stage of animal development.

CELL ADHESION AND INTRACELLULAR SIGNALLING

The previous sections catalogued properties of cell-adhesion receptors and extracellular matrix proteins and presented a simplified description of cell interactions with their surrounding environment. Nonspecific forces bring cells near a surface; receptor–ligand interactions add strength and specificity to binding; stabilization is often produced by the development of specialized junctional complexes (Fig. 12). Specific cell binding is accomplished with transmembrane receptors: cell–cell binding is typically mediated by cadherins, and cell–matrix binding is mediated by integrins.

Receptor extracellular domains are continuous with cytoplasmic domains, which are often linked to the filamentous cytoskeleton. In this way, binding at the cell surface is united with intracellular signal transduction and metabolic pathways. For example, focal contacts have an important function as anchors for cells (Fig. 12), but focal contacts also relay signals from the extracellular matrix into the cytoplasm. The focal contact region is rich in a variety of proteins that have enzymatic activity, such as focal adhesion kinase (FAK) and Src, the tyrosine kinase that is encoded by the *src* gene (Fig. 6). These kinases phophorylate other proteins, including components of the

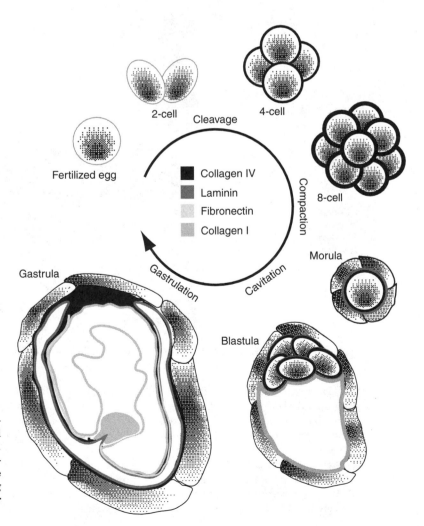

Figure 11. Highly schematic diagram of the distribution of collagen (types I and IV), fibronectin, and laminin during the early stages of mouse development. Collagen IV is the first protein to appear during development; fibronectin on the inner blastocoel surface is critical for migration of mesodermal cells during gastrulation; collagen I does not appear until after implantation (adapted from 1. Ref. 16).

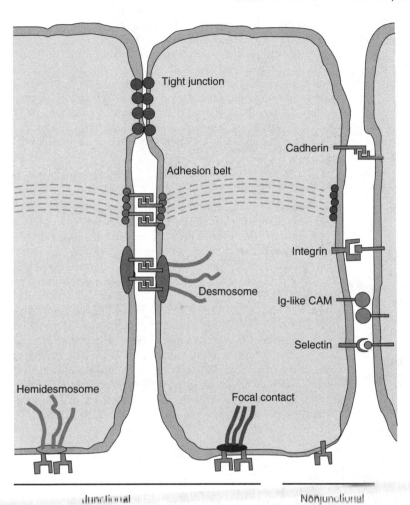

Figure 15. Summary of cell junctions and nonjunctional adhesion.

cytoskeleton; phosphorylation modulates binding activity, providing sites for association of additional proteins and localizing powerful enzymatic activities to the adhesion site.

Considerable experimental evidence supports the following model for integrin-mediated signal transduction (12) (Fig. 6 inset): integrin–ECM binding promotes FAK autophosphorylation of the Tyr residue at position 397, which stimulates Src binding. Once bound, Src catalyzes additional phosphorylation (at the Tyr residue 925), creating a binding site for the Src homology 2 (SH2) domain of Grb2, which presents proline-rich polypeptide motifs that facilitate SOS binding (through the Src homology 3(SH3) domains). This regulated cascade of kinase activity produces an assembly of activated proteins at the focal adhesion site (Fig. 6); these proteins mediate cell shape and cell migration by modulating local assembly of the cytoskeleton, as well as the binding affinity of the integrin extracellular domains. In addition, important intracellular signal transduction proteins, such as Ras and Rho, are activated. Ras and Rho are members of the GTPase superfamily: Ras activates proteins that control cell growth and differentiation, whereas Rho influences actin stress fiber assembly. Although many elements of the overall process are poorly understood now, it is clear that integrin binding has consequences beyond adhesion:

binding of integrins to ECM can influence cell processes as vital and complex as growth and movement. Similar mechanisms are also present in cadherin-mediated adhesion complexes.

BIOTECHNOLOGICAL SIGNIFICANCE OF CELLULAR ADHESION

Adhesive interactions among cells and extracellular matrix molecules are intimately associated with the molecular machinery that controls cell viability, movement, differentiation, and growth. Because cell–cell or cell–matrix contact can influence the function of cultured cells, cell adhesion is of considerable consequence in biotechnology. In this final section, we provide an overview of several approaches for exploiting our understanding of the molecular mechanisms of cell binding and adhesion.

Cell Adhesion to Surfaces

The growth and function of most tissue-derived cells requires adhesion, attachment, and spreading on a solid substrate. The nature of the substrate influences cellular function (17). Cell attachment, migration, and growth on polymer surfaces is mediated by proteins that are precoated onto the surface, adsorbed from the culture medium, or secreted by the growing cells. Because it is difficult to control adsorption or secretion during cell culture, the polymer surfaces are often pretreated with purified protein solutions. In this way, investigators hope that subsequent cell behavior on the surface will mimic some aspects of cell behavior *in vivo*.

Surfaces can be made attractive to cells by modifying their physical properties; electrostatic forces are frequently used to encourage cell association with a surface. For example, after a surface is coated with poly-L-lysine, it has a positive charge resulting from the primary amine groups on lysine; many cells will adhere nonspecifically to such a surface because the cell surface is negatively charged. Commercially available tissue culture plasticware is often subjected to surface treatments that create a favorable surface charge (18).

Specific receptor–ligand binding interactions are also used to encourage cell adhesion to synthetic surfaces (19). For example, the cell-binding domain of fibronectin contains the tripeptide RGD. Cells attach to surfaces containing adsorbed oligopeptides with the RGD sequence or covalently coupled synthetic peptides. In addition, cell adhesion can be reduced by adding soluble, synthetic peptides containing the RGD sequence to a culture medium. The addition of cell-binding peptides to a polymer can induce cell adhesion to otherwise nonadhesive or weakly adhesive surfaces. Cell spreading and focal contact formation are also modulated by the addition of peptide. Because cells contain cell-adhesion receptors that recognize only certain ECM molecules, using an appropriate cell-binding sequence can lead to cell-selective surfaces as well, where the population of the cells that adheres to the polymer is determined by the peptide.

Cell–Cell Adhesion and Aggregation

Cell aggregates are important tools in the study of tissue development and permit correlation of cell–cell interactions with cell differentiation, viability and migration, as well as subsequent tissue formation. The aggregate morphology permits reestablishment of cell–cell contacts normally present in tissues; therefore, cell function and survival are frequently enhanced in aggregate culture. Because of this, cell aggregates may be useful in biotechnology to enhance the function of cell-based hybrid artificial organs or reconstituted tissue transplants. To employ cell aggregates in this manner, techniques for controlling the formation of aggregates from specific cells of interest must be developed.

Aggregates are usually formed by incubating cells in suspension, using gentle rotational stirring to disperse the cells. Although this simple method is suitable for aggregating many cells, serum or serum proteins must be added to promote cell aggregation in many cases, making it difficult to characterize the aggregation process and to control the size and composition of the aggregate. Specialized techniques can be used to produce aggregates in certain cases, principally by controlling cell detachment from a solid substratum. For example, stationary culture of hepatocytes above a nonadherent surface or attached to a temperature-sensitive polymer substrate have been used to form aggregates.

As previously mentioned, nonadhesive substrata can be made adhesive by adsorption or chemical conjugation of synthetic peptides containing RGD and YIGSR. These hybrid surfaces can serve as cell-selective templates for cell culture and may someday lead to novel strategies for tissue regeneration. These same synthetic cell-binding peptides can be engineered to enhance the formation of cell aggregates in suspension by coupling the peptides to water-soluble, inert polymers, such as poly(ethylene glycol).

Cell-Virus Adhesion and Gene Transfer

Viruses are small, obligate cellular parasites. For many classes of viruses, infection is initiated when a surface protein on the virus binds to a normal cell protein, which serves as a viral protein receptor, on the host-cell plasma membrane (Table 5). Then, the virus uses the cell's normal receptor-mediated endocytosis mechanisms to enter the cell. A possible result of entry of the viral genome into a cell is the production of large quantities of viral progeny, which may infect more cells. Therefore, interventions that enhance viral entry can enhance viral productivity, and interventions that inhibit viral entry can block infection. In some cases, soluble proteins that mimic either the viral ligand or the cell receptor have been tested as agents to block propagation of infection. This approach can work: soluble CD4 proteins inhibit HIV

Table 5. Cell-Surface Receptors for Viruses

Virus	Host	Receptor
Human immunodeficiency virus	T lymphocytes	CD4 protein
Influenza virus Reovirus Rotavirus		Sialic acid containing glycoproteins or glycolipids
Herpes simplex virus	Epithelial cells	Heparan sulfate proteoglycan (?) Fibroblast growth factor receptor protein
Epstein–Barr virus	B lymphocytes	C3b receptor protein
Rabies virus		Acetylcholine receptor (?)
Human rhinovirus (major group)		Intercellular adhesion molecule-1 (ICAM-1)
Human rhinovirus (minor group)		Low-density lipo-protein receptor protein
Reovirus	Erythrocyte	Glycophorin A
Vaccinia virus		EGF receptor (?)
Human parvovirus	Erythrocyte	B19P antigen
Influenza virus	Erythrocyte	MN antigen
Measles virus		CD46 protein (?)
Rotavirus		Three-sugar ganglioside

Notes: In some cases, evidence for the identity of the receptor is not conclusive (indicated by ?). Information is compiled from a variety of sources including B.N. Fields and D.M. Knipe, eds., *Fundamental Virology*, 2nd ed., Raven Press, New York, 1991.

infection of cells in culture and in animals (20), although the concentrations of soluble protein required for blockade are quite high, making this a difficult approach for disease prevention.

Finding methods for enhancing the efficiency of gene expression in cultured cells is of considerable interest. This is often accomplished by selecting host-virus systems which already possess natural receptor–ligand pairs that accommodate viral entry. Bacteriophages are used as vectors to incorporate foreign DNA into microorganisms, and retroviruses are used to pass genes into human cells. Certain virus-host systems have been optimized to produce heterologous proteins. For example, cells from the fall armyworm (*Spodoptera frugiperda*) and a baculovirus (*Autographa californica* nuclear polyhedrosis virus) are widely used now in biotechnology. This system has high protein expression levels and is safer than mammalian-retrovirus systems. Synthetic systems are also being used with considerable success. For example, DNA plasmids will enter cells, particularly if the plasmid is complexed with cationic lipids (21). The lipids impart a positive charge on the complex, which facilitates fusion with the negatively charged cell surface.

BIBLIOGRAPHY

1. B. Alberts et al., *Molecular Biology of the Cell*, 3rd ed., Garland Publishing, New York, 1994.

2. P. Bongrand and G.I. Bell, *Cell Surface Dynamics: Concepts and Models*, A.S. Perelson, C. DeLisi, and F.W. Wiegel, eds., Dekker, New York, 1984, pp. 459–493.

3. D.A. Lauffenburger and J.J. Linderman, *Receptors: Models for Binding, Trafficking, and Signaling*, Oxford University Press, New York, 1993.

4. S.P. Palecek et al., *Nature (London)* **385**, 537–540 (1997).

5. M.D. Pierschbacher and E. Ruoslahti, *Nature (London)* **309**, 30–33 (1984).

6. K. Suehiro, J.W. Smith, and E.F. Plow, *J. Biol. Chem.* **271**(17), 10365–10371 (1996).

7. E.R. Koslov et al., *J. Biol. Chem.* **272**(43), 27301–27306 (1997).

8. P. Milev et al., *J. Cell Biol.* **127**(6), 1703–1715 (1994).

9. D.R. Friedlander et al., *J. Cell Biol.* **125**(3), 669–680 (1994).

10. M.W. Nicholson et al., *J. Biol. Chem.* **273**(2), 763–770 (1998).

11. N.M. Kumar and N.B. Gilula, *Cell (Cambridge, Mass.)* **84**, 381–388 (1996).

12. E.A. Clark and J.S. Brugge, *Science* **268**, 233–239 (1995).

13. C.M. Longhurst, and L.K. Jennings, *Cell. Mol. Life Sci.* **54**, 514–526 (1998).

14. N.A. Chitaev, A.Z. Averbakh, R.B. Troyanovsky, and S.M. Troyanovsky, *J. Cell Sci.* **11**(14), 1941–1949 (1998).

15. R.P. Mecham and J. Heuser, *Connect. Tissue Res.* **24**, 83–93 (1990).

16. E.D. Adamson, J.B. Weiss and M.I.V. Jayson, eds., *Collagen in Health and Disease*, Churchill-Livingstone, Edinburgh, 1982, pp. 219.

17. W.M. Saltzman, in R. Lanza, W.L. Chick, and R. Langer, eds., *Textbook of Tissue Engineering*, R.G. Landes, New York, 1996, pp. 227–248.

18. R.E. Spier, *Adv. Biochem. Eng.* **14**, 120–162 (1980).

19. J.A. Hubbell, S.P. Massia, N.P. Desai, and P.D. Drumheller, *Bio/Technology* **9**, 568–572 (1991).

20. D.J. Capon et al., *Nature (London)* **337**, 525–531 (1989).

21. R.J. Lee and L. Huang, *Crit. Rev. Ther. Drug Carrier Syst.* **14**(2), 173–206 (1997).

See also CELL STRUCTURE AND MOTION, CYTOSKELETON AND CELL MOVEMENT; CELL-SURFACE RECEPTORS: STRUCTURE, ACTIVATION, AND SIGNALING; TRANSCRIPTION, TRANSLATION AND THE CONTROL OF GENE EXPRESSION.

CELL-SURFACE RECEPTORS: STRUCTURE, ACTIVATION, AND SIGNALING*

LAURA A. RUDOLPH-OWEN
GRAHAM CARPENTER
Vanderbilt University School of Medicine
Nashville, Tennessee

OUTLINE

Introduction

Structure of Cell-Surface Receptors

    Growth Factor Tyrosine Kinase Receptors

    Cytokine Receptors

    Heptahelical G Protein Coupled Receptors

Receptor Activation

    Activation of Tyrosine Kinase and Cytokine Receptors

    Activation of Heptahelical Receptors

Second Messengers and Effectors

Protein:Protein Interacting Domains

    Ras/MAP Kinase Cascade

    STAT Pathway

    Phosphatidylinositide Signaling

    Heterotrimeric G Protein Signaling

Summary and Prospectus

Bibliography

INTRODUCTION

Cells have evolved that organize biological responses to numerous but specific environmental stimuli. This allows cells within multicellular organisms to communicate and coordinate physiological functions. The recognition and conveyance of signals at the cell surface are achieved by a variety of plasma membrane receptors. Almost all cells have developed several related but distinct receptor mechanisms to identify these extracellular signals and

*Supported by NIH grants T32AR07491, R01CA24071 and R01CA75195.

transduce the information to points of signal reception within the cell, such as the nucleus. In the majority of cases, extracellular molecules, or ligands, associate with specific receptors located on the cell surface. The formation of ligand:receptor complexes stimulates intracellular signaling pathways, including the activation of cytosolic proteins and the release of metabolic second messengers, thereby provoking intracellular responses such as the induction of gene expression and changes in cell shape. Lipid-soluble signals, such as steroid hormones, pass through the plasma membrane and are recognized by specific receptors within intracellular compartments. This article will focus on several types of cell-surface receptors for protein ligands, while steroid hormone receptors are described elsewhere in this volume.

STRUCTURE OF CELL-SURFACE RECEPTORS

The organization of nearly all cell-surface receptors is similar in overall structure and function. All are transmembrane proteins having an ectodomain that functions as a ligand recognition site and a cytoplasmic domain that mediates the initiation of signal transduction pathways. However, evolution has devised variations of this general model to facilitate the requirements of individual ligands and cells. Based on relative structural topology, transmembrane receptors can be described as one of three types: (1) growth factor tyrosine kinase receptors, (2) cytokine receptors, or (3) heptahelical G-protein coupled receptors. The structural organization of these receptor groups is discussed individually in the following sections.

Growth Factor Tyrosine Kinase Receptors

In general, growth factor tyrosine kinase receptors have a relatively simple organization consisting of a single hydrophobic transmembrane domain that separates an extracellular ligand-binding domain and a cytoplasmic tyrosine kinase catalytic domain. However, there are variations on this basic organization, a few of which are shown in Figure 1(a). The examples depicted are the epidermal growth factor (EGF) receptor, the insulin/insulin-like growth factor (IGF) receptor, and the glial-derived neurotrophic factor (GNDF) receptor. These three examples are presented to illustrate the manner in which nature has produced structurally varied tyrosine kinase receptors to recognize different extracellular signals, but yet transduce this primary signal by the common mechanism of tyrosine kinase activation.

As shown in Figure 1(a), the EGF receptor has the most straightforward organization, consisting of a single polypeptide chain that contains a ligand-binding ectodomain, a transmembrane domain, and a tyrosine kinase cytoplasmic domain (1). The structure of the insulin/IGF receptor is functionally similar, but represents a dimeric receptor, consisting of two α and two β chains linked by disulfide bonds. Each β subunit has a transmembrane domain and a tyrosine kinase domain that is activated following ligand binding to the α subunits (2). A single precursor protein encodes the α and β

subunits, which are formed by post-translational cleavage. The GNDF receptor consists of two proteins, encoded by separate genes, that together form a functional receptor. One receptor component, GDNFR-α, is an extracellular protein tethered to the outer surface of the plasma membrane by a glycosyl-phosphatidylinositol (GPI) anchor and is responsible for primary ligand recognition. GNDFR-α lacks transmembrane and cytoplasmic sequences and is not, therefore, expected to transduce intracellular signals on its own. Signal transduction is mediated by a second component, Ret, a single transmembrane protein containing an ectodomain and a cytoplasmic tyrosine kinase domain. Ligand binding to the GNDFR-α subunit provokes association with Ret and transduction of intracellular signals by activation of the Ret tyrosine kinase (3).

The ectodomains of receptor tyrosine kinases are often characterized by several copies of protein motifs that enhance protein stability and may also be important in the process of protein:protein recognition for ligand binding (1). Examples are the two cysteine-rich regions in the EGF receptor or five immunoglobulin-like domains in the platelet-derived growth factor (PDGF) receptor. Without exception, these receptors are complex proteins as the ectodomains are highly glycosylated. For example, the EGF receptor ectodomain is N-glycosylated at twelve sites, which contributes significantly to receptor molecular mass, particularly as observed on an SDS gel. However, the role of carbohydrate in the process of ligand binding or other receptor functions is not clear.

For the most part, the transmembrane domains have an essential but passive role in receptor function and do not contribute significantly to the signaling specificity of tyrosine kinase receptors. Numerous experiments with chimeric receptors (i.e., fusion proteins containing the extracellular domain of one receptor and the cytoplasmic domain from another receptor) show that heterologous ligand binding will activate the cytoplasmic tyrosine kinase regardless of the origin of the transmembrane domain (4). This does not, however, imply that the transmembrane domains are biologically inert. The basal tyrosine kinase activity of ErbB-2, a member of the EGF receptor family, is enhanced when a single valine residue within the transmembrane domain is changed to glutamic acid. This amino acid substitution leads to constitutive dimerization of ErbB-2 and activation of its tyrosine kinase, resulting in an oncogenic form of the receptor (5).

The cytoplasmic region of growth factor receptors and, in particular, the tyrosine kinase domain is essential for transducing the first messenger information of ligand binding into intracellular biochemical information, termed second messengers. Within the kinase domain is a highly conserved consensus sequence, GlyXGlyXXGlyX$_{(15-29)}$ Lys, found in all protein kinases (6). This sequence motif functions as part of the binding site for ATP, the phosphate donor for the formation of phosphotyrosine on substrate proteins. Several studies show that replacement of the lysine residue within this sequence abolishes or greatly attenuates receptor kinase activity in vivo and in vitro (4).

The carboxy terminal of the kinase domain is a region of 200–300 residues containing several tyrosine residues

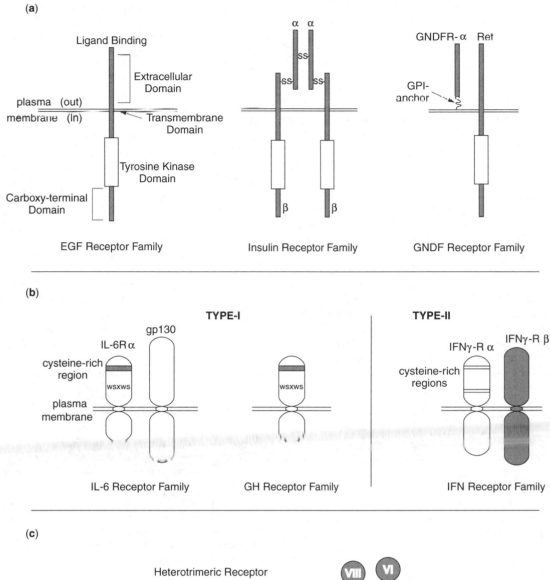

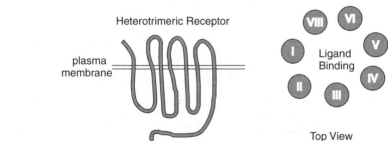

Figure 1. Structure of cell-surface receptors. Diagrammatic representation of the basic structural topology of (**a**) growth factor tyrosine kinase receptors, (**b**) cytokine receptors, and (**c**) heptahelical G protein-coupled receptors.

that function as sites for receptor autophosphorylation. Autophosphorylation occurs not only at multiple tyrosine residues in the carboxy terminal region, but in some instances also occurs at tyrosine residues within the kinase domain itself and the juxtamembrane region. In the EGF receptor, all five autophosphorylation sites are located within the carboxy-terminal domain, while the insulin receptor has eight autophosphorylation sites found throughout the cytoplasmic domain of each β chain (4). The

autophosphorylation sites provide important recognition information for downstream signal transduction molecules and in some cases have a role in tyrosine kinase activation (see the following).

Cytokine Receptors

Members of the superfamily of cytokine receptors are usually classified as Type I or Type II based on structural distinctions (7). Type I receptors, also referred to as hematopoietin receptors, include receptors for most of the interleukins (ILs), such as IL-6, as well as proteins that mainly function outside the immune system, such as growth hormone [GH; Fig. 1(b)]. An example of a Type II cytokine receptor is the receptor for interferon gamma [IFN-γ; Figure 1(b)].

All Type I cytokine receptors have an extracellular domain with a WSXWS (TrpSer $\times$ TrpSer) sequence motif close to the plasma membrane and a conserved pattern of cysteine residues near the amino terminus. Also, the extracellular domains contain two fibronectin type III modules connected by a hinge region, which contains the WSXWS motif and functions as a ligand interaction site. Mutagenesis studies show that mutations within the hinge region, including residues in the WSXWS motif, attenuate the ligand binding capacity of the receptor (8). Unlike Type I receptors, Type II cytokine receptors do not contain the conserved WSXWS motif in their extracellular domain, but do possess conserved cysteine residues within this region [Fig. 1(b)].

Ligands that interact with cytokine receptors produce a wide range of biological effects on various tissues and cells, often in a seemingly redundant manner. For example, cytokines such as IL-6, IL-2, IL-4, IL-5, and IFN-γ all initiate antibody production in B cells. The functional pleiotropy and redundancy of cytokines are explained, in part, because a common receptor subunit is shared by several different ligands. Therefore, cytokine receptors usually consist of two polypeptide chains, one that provides a ligand-specific binding site and a second that mediates signal transduction (7, and references therein). For example, IL-6 is first recognized by IL-6Rα, and subsequently associates with gp130 to form a signaling competent complex of IL-6:IL-6Rα:gp130 [Fig. 1(b)]. Other cytokines whose receptors associate with gp130 include IL-11, leukemia inhibitory factor, oncostatin, cardiotropin, and ciliary neurotropic factor. Similarly, the IFN-γ receptor also consists of α and β chains, where association of two receptors is provoked by ligand binding to the IFN-γRα molecules. The growth hormone, prolactin (PRL), and erythropoietin (Epo) receptors, however, do not share a common receptor subunit with other cytokine receptors, but instead homodimerize [Fig. 1(b)].

Unlike growth factor tyrosine kinase receptors, the cytoplasmic domains of cytokine receptors do not contain protein kinase catalytic motifs. To transmit downstream signals, therefore, cytokine receptors require association with an additional protein(s) to constitute high-affinity, signal-transduction-competent receptors. This auxiliary protein, usually a protein tyrosine kinase, is recruited to the plasma membrane from the cytosol to mediate intracellular signaling (9, and references therein). This topic will be discussed in detail later in this article.

Heptahelical G Protein Coupled Receptors

Heptahelical G protein coupled receptors present a unique membrane topology compared to receptors previously discussed. As the name implies, each of these receptors is composed of a single polypeptide chain with seven transmembrane domains. Site-directed mutagenesis has shown that amino acid residues within some of these transmembrane domains are necessary for ligand binding, particularly for small ligands. For example, the binding site for neurotransmitters, such as acetylcholine and noradrenaline, is located in the upper two-thirds of the transmembrane domains, on the extracellular side of the membrane [Fig. 1(c)]. In these cases, residues in all helices, except helix I, contribute to ligand binding, with strong involvement of residues in helices III, V, VI, VII and smaller influences from residues in helices II and IV (10). There is less information on the binding site for large ligands, such as glycoproteins and neuropeptides. However, it is known that glycoprotein hormone receptors, for example, the chorionic gonadotropin receptor, have an extended extracellular region at the amino terminus. This ectodomain can, by itself, bind ligand with high affinity, suggesting that in these cases residues in the transmembrane domain are not necessary to form the ligand binding site (11).

Ligand occupancy induces conformational changes in the cytoplasmic domain of heptahelical receptors that bring about association with cytosolic heterotrimeric G proteins. These G proteins consist of an α subunit that binds the guanine nucleotides GDP and GTP, plus a β and a γ subunit. The β and γ subunits are tightly associated and are usually referred to as a $\beta\gamma$ complex.

RECEPTOR ACTIVATION

While the ligand recognition and binding strategies of cell-surface receptors are apparent, the means by which the intracellular domains are activated is less clear. The complicating feature is the transmembrane nature of the receptors, which restricts, topologically and energetically, the transmission of ligand-induced conformational changes from the ectodomain to the cytoplasmic domain of the molecule. Activation of receptor cytoplasmic domain functions is thought to be achieved by a mixture of mechanisms, including dimerization, kinase activation, autophosphorylation, and conformational changes. Growth factor tyrosine kinase receptors and cytokine receptors share receptor dimerization, tyrosine kinase activation, and autophosphorylation as initiating events. In the case of heptahelical receptors, ligand binding is postulated to stimulate conformational changes in the cytoplasmic domains facilitating interaction with heterotrimeric G proteins. This portion of the review will focus on mechanisms that activate cell-surface receptors following ligand binding.

Activation of Tyrosine Kinase and Cytokine Receptors

During or following ligand binding, receptor dimerization is considered a general initiating mechanism for the activation of most growth factor and cytokine receptors. The immediate consequence of receptor dimerization is tyrosine kinase activation (12), involving receptors with an intrinsic tyrosine kinase (for example, EGF and insulin receptors) as well as receptors in which the tyrosine kinase is extrinsic or physically separate from the ligand binding molecule. Certain cytokine family receptors (Epo, interferons, or GH, for example) associate with tyrosine kinases to form receptor:kinase complexes. In these cases, cytoplasmic tyrosine kinase molecules, termed Janus kinases (JAKs), are either noncovalently preassociated with the cytoplasmic region or become associated following ligand binding and dimerization (13). The number of different JAKs is small relative to the number of different cytokine binding molecules. Hence, different cytokine receptors may activate the same JAKs.

There are at least three means by which ligand binding produces receptor dimerization. First, ligands may stimulate dimerization as a consequence of their dimeric nature. For example, growth factors such as PDGF, colony stimulating factor-1, and stem cell factor, exist as covalent dimers with each monomer containing a receptor-binding site. One ligand molecule, therefore, forms a receptor dimer by simultaneously interacting with two receptors. Other ligands, such as EGF, are monomeric, and receptor dimerization is thought to be mediated by a ligand-dependent conformational change in the receptor extracellular domain. This promotes the association of two occupied receptors to stabilize a dimeric configuration (12). A third dimerization mechanism is best illustrated by growth hormone, which has two asymmetric receptor binding sites within the monomeric ligand and thereby functions similar to a dimeric ligand with regard to receptor interaction and dimerization (14). An exception to ligand-induced receptor dimerization exists for the insulin/IGF receptors. Since these receptors pre-exist as covalent dimers, ligand binding activates the receptors by the modulation of dimer conformation in a manner that remains unclear.

Dimerization is thought to facilitate tyrosine kinase activation by the apposition or juxtapositioning of two kinase domains that are then able to engage in cross- or trans-phosphorylation. Hence, two closely positioned kinase molecules phosphorylate each other in an intermolecular process, which is nevertheless referred to as autophosphorylation. High-resolution three-dimensional structures of tyrosine kinase domains, produced by x-ray diffraction studies, reveal that many of these enzymes have an "activation loop" with a tyrosine side chain in close proximity to the kinase activation site (4). This suggests a mechanism to explain how autophosphorylation may activate a tyrosine kinase; that is phosphorylation of the tyrosine side chain in this activation loop alters its relationship to the active site such that substrate access at the active site is enhanced. This model has been best developed from structural studies of the insulin receptor tyrosine kinase domain. Also, in some cases kinetic studies have indicated that autophosphorylation, which is actually trans-phosphorylation, increases tyrosine kinase catalytic activity.

The fact that certain growth factors are bivalent in terms of receptor binding permits certain predictions about how ligand concentration relates to biological activity. Growth hormone has two different sites (site 1 and site 2), each of which recognizes the same binding epitope on a receptor monomer. In this case, recognition of receptors by growth hormone is of higher affinity at site 1 than site 2. Dimers are formed by ligand site 1 recognition of one receptor and subsequently site 2 interaction with a second receptor molecule (15).

The preceding binding mechanism means that the concentration curve of growth hormone for a biological response should be bell-shaped such that at high ligand concentrations the response is diminished. In fact, this is seen both in experimental and clinical situations. At high hormone concentrations all receptors are occupied by growth hormone through site 1, and there are, therefore, no free receptors to interact with site 2 on the ligand to form a receptor dimer. An important point to understand is that sites 1 and 2 on the ligand bind to the same receptor epitopes. This information has been used to create antagonists of growth hormone by mutagenesis of site 2 such that it is incapable of receptor recognition. This ligand mutant, therefore, binds one receptor molecule through site 1 but is unable to form receptor dimers (14,15). At high concentrations, this altered growth hormone saturates receptor molecules and prevents low levels of endogenous growth hormone from forming dimers.

Another example of the biological significance of dimerization is exemplified by mutations in the Ret component of the GNDF receptor [Fig. 1(a)]. The extracellular domain of Ret has four cysteine residues that normally form two intramolecular disulfide bonds. A point mutation that changes one cysteine to another amino acid leaves an unpaired cysteine residue in the Ret ectodomain. As a consequence, an intermolecular disulfide bond is formed between unpaired cysteine residues on two separate receptor molecules (16). This results in the formation of a covalent Ret dimer and constitutive activation of the Ret tyrosine kinase domains. In humans, this type of mutation produces an oncogene that gives rise to the malignancy Multiple Endocrine Neoplasia 2A (MEN2A).

In certain cases a receptor may heterodimerize with another receptor kinase as well as homodimerize. For example, within the EGF receptor family of ErbB receptor tyrosine kinases heterodimeric complexes between ErbB-2 and the EGF receptor (ErbB-1) are induced by EGF concurrently with the formation of EGF receptor homodimers (12). The proportion of EGF-dependent homo- and heterodimers will depend mostly on the relative concentrations of the two receptor species. Since no ligand has been identified that directly binds to ErbB-2, this molecule is generally thought of as a co-receptor. Heregulin, a growth factor that binds directly to ErbB-3 and ErbB-4 in this receptor family, also induces heterodimeric complexes of ErbB-2:ErbB-3, ErbB-2:ErbB-4, and ErbB-3:ErbB-4 as well as homodimers of ErbB-4. Receptor heterodimerization expands the possibilities of receptor–ligand interactions and extends the potential diversity of downstream signals.

While autophosphorylation may occur in heterodimers as well as homodimers, both receptors in the dimer must be active to induce a significant level of kinase activation and biological responses. Cells expressing both kinase-inactive EGF receptor mutants and wild-type EGF receptors form dimers of wild-type receptors, dimers of mutant receptors, and dimers containing one wild-type and one mutant receptor. However, only the wild-type homodimer, and not a heterodimer composed of a wild-type and a mutant receptor, displays autophosphorylation and biological activity, indicating that both receptors in the dimer must be functional for activation to occur (17).

Kinase-inactive receptor mutants can also function as dominant-negative molecules when overexpressed relative to the wild-type receptor. Under these circumstances, the only dimers formed are composed of either two mutant receptors or one mutant and one wild-type receptor. Due to overexpression of the mutant receptor, it becomes statistically unlikely that two wild-type receptors will dimerize. These dominant-negative mutants represent an active experimental approach to interrupt endogenous ligand-induced receptor activation and to prevent biological responses to the ligand. Kinase-inactive mutants of the EGF receptor, for example, have been produced by point mutations in the kinase active site or by deletion of the entire cytoplasmic domain (4). The latter mutant is additionally instructive because it dimerizes with itself and wild-type receptors, indicating that only the ectodomain and transmembrane domain are necessary for the dimerization process. It seems likely that the transmembrane domain has a passive role in this process and that dimerization is driven by conformational changes in the ectodomain following ligand binding.

An important family of growth factor receptor kinases is the one that mediates the activity of the transforming growth factor beta (TGF-β) family of ligands. These growth factors are especially important in processes such as cell growth inhibition and embryonic development. The two TGF-β receptors (Type I and Type II) are both single transmembrane proteins with a ligand-binding ectodomain and an intracellular domain that encodes a serine/threonine kinase, instead of a tyrosine kinase. Ligand binding promotes the heterodimerization of Type I and Type II molecules. The Type I receptor is a constitutively active kinase, but only phosphorylates the Type II receptor in the context of a ligand-induced dimer (18). Transphosphorylation of the Type II receptor activates its kinase activity and leads to signal transduction and biological responses.

Although the intracellular domains of cytokine receptors lack intrinsic kinase activity, the overall mechanism of activation appears to be similar to that of tyrosine kinase receptors. Ligand binding induces dimerization of cytokine receptors that initiates the association of cytoplasmic tyrosine kinase JAK molecules with the receptor intracellular domains (9). Interestingly, dimerization of cytokine receptor components is required, but may not be sufficient, for the initiation of signaling processes in all cases. For example, as previously outlined, the ligand-bound IL-6 receptor heterodimerizes with gp130 [Fig. 1(b)]. Studies also indicate that ligand-induced dimerization of cytokine receptors, such as IL-6R:gp130, recruits JAK tyrosine kinases from the cytoplasm to the receptor complex. Examination of sequences in the cytoplasmic portion of cytokine receptors has revealed a conserved proline-rich sequence of eight residues that may mediate JAK association (19). Mutations within these proline residues in the cytoplasmic region of gp130 do not affect association with IL-6R, but do attenuate JAK binding and intracellular signaling. Hence, both dimerization and the separate process of JAK

association are essential for receptor activation. Although it is known that the intracellular region of cytokine receptors is critical for JAK association, it is not yet known if this proline-rich motif is the only critical JAK interacting site within the cytoplasmic domain.

There are four known members of the JAK family: JAK1, JAK2, JAK3, and Tyk2. The binding of JAKs to the intracellular domains of cytokine receptor complexes places these molecules in close proximity, thus facilitating transphosphorylation and activation of tyrosine kinase activity, similar to that for growth factor tyrosine kinase receptors. The activated JAKs subsequently tyrosine phosphorylate specific sites in the cytoplasmic domain of the cytokine receptors, creating docking sites for other proteins to interact with the receptors.

Activation of Heptahelical Receptors

Heptahelical receptors are probably more ancient than the growth factor receptor/cytokine receptors previously described. Yeast, for example, display heptahelical receptors for mating factors, but do not have tyrosine kinase-coupled receptors. While the exact activation mechanism is unclear, the cytoplasmic domains of heptahelical receptors, following ligand binding, adopt an altered conformation, which results in enhanced interaction with signal-transducing heterotrimeric G proteins. As shown in Fig. 1(c), the intracellular domain of a heptahelical receptor contains three loops and a free carboxy-terminal polypeptide. Ligand-dependent activation of G proteins requires their association with these receptor cytoplasmic regions, the third cytoplasmic loop usually being the most critical (10).

To transduce a signal from the heterotrimeric G proteins to a downstream molecule, or effector, G protein subunits cycle between a GTP-bound and a GDP-bound state. G proteins contain α, β, and γ subunits with the α subunit binding GDP or GTP (20). When GDP is bound, the α subunit is associated with the $\beta\gamma$ subunits to form an inactive heterotrimer that can only weakly associate with heptahelical receptors. Following ligand binding and conformation changes in the receptor cytoplasmic domain, the α subunit is altered such that its conformation has a low affinity for GDP. Because the concentration of GTP in cells is much higher than that of GDP, the GDP is replaced with GTP. Binding of GTP to the α subunit produces dissociation of the α subunit from the receptor and from the $\beta\gamma$ subunits. The β and γ subunits, however, remain tightly associated. Hence, this "activation" step produces two new components—the GTP-bound α subunit and the free $\beta\gamma$ complex. While the former is a well-recognized activator of downstream signaling components, it has more recently become appreciated that $\beta\gamma$ complexes also have a signal-transducing role (21).

The activated state of a G protein lasts until GTP on the α subunit is hydrolyzed to GDP by interacting with GTPase-activating proteins or GAPs. A large family of GAPs specific for $G\alpha$ proteins has recently been discovered and termed regulators of G protein signaling, or RGS proteins. RGS proteins were first identified as molecules that negatively regulate G protein signaling due to their function in accelerating GTP hydrolysis (22). GAPs, such as the RGS proteins, can be viewed as enzymes that bind substrate and facilitate its conversion to a specific product. X-ray crystallography and mutational studies have shown that RGS proteins act as GAPs by binding to the Gα protein and stabilizing the molecule in a lower-energy conformation, resulting in an acceleration of GTP hydrolysis. Once GTP is cleaved to GDP, the α and βγ subunits reassociate, as an inactive ternary complex (20). Although the βγ subunit complex does not bind GTP, its active lifetime depends on the concentration of GDP-bound α subunits, and therefore the rate of GTP hydrolysis by RGS: α subunit complexes.

SECOND MESSENGERS AND EFFECTORS

Activation of cell-surface receptors by ligand binding, receptor dimerization, autophosphorylation, and conformational change initiates the transmission of intracellular signals to allow for the flow of information from the cell surface to points of signal reception, such as the nucleus and cytoskeleton. This cellular process involves the biochemical modulation of an array of cellular proteins by a variety of mechanisms, including protein:protein association, guanine nucleotide binding, post-translational modification, and topological relocalization within the cell. In particular, protein:protein interactions are an important feature of receptor proximal signaling.

The following sections outline mechanisms of signal transduction in selected well-characterized pathways that illustrate the molecular mechanisms described previously. Downstream pathways activated by growth factor and cytokine receptors will be described first, followed by the signaling effectors stimulated by heptahelical G protein-coupled receptors. Although a condensed view of each separate pathway will be presented, current data show that communication or "cross-talk" does occur between these pathways.

PROTEIN:PROTEIN INTERACTING DOMAINS

Before describing the mechanics of individual pathways, it is worthwhile to outline the biochemical means by which specific protein:protein interactions are promoted in these pathways. Activated growth factor receptors contain cytoplasmic domains that are tyrosine phosphorylated (see preceding discussions). These phosphorylation sites function as high-affinity binding sites for intracellular proteins that contain specialized domains that have evolved specifically to recognize protein sequences containing a phosphotyrosine residue. These domains of approximately 100 residues are termed src homology 2 (SH2) domains or, in a few cases, phosphotyrosine binding (PTB) domains, which are structurally distinct from SH2 domains but serve the same function (23).

Specificity in the association of different molecules with SH2 and/or PTB domains and tyrosine phosphorylated receptors is derived from two complementary structural considerations. First, these domains recognize not just any phosphotyrosine, but also other residues surrounding the phosphotyrosine. Second, each SH2/PTB domain contains both conserved and unique sequence features that allow, respectively, phosphotyrosine binding and specific recognition of adjacent residues. The subtle differences in sequence information surrounding each phosphotyrosine residue plus the unique structural subtleties of each SH2/PTB domain provide the essential specificity characteristic of a signaling system.

Frequently, phosphotyrosine-binding molecules have an additional determinant of protein:protein interaction termed an SH3 domain that mediates recognition and physical association with specific proline-rich sequences in other proteins (23). Also, some signaling proteins contain a specialized domain termed a pleckstrin homology (PH) domain that binds phospholipids, particularly phosphoinositide polyphosphates. PH domain-containing proteins bind these phospholipids with moderate affinity and are thought to function as signal-dependent membrane adaptors. The individual pathways to be discussed illustrate how these protein:protein association mechanisms provide critical specificity to signal transduction mechanisms.

Ras/MAP Kinase Cascade

The most understood signal transduction pathway activated by receptor tyrosine kinases is the Ras/MAP kinase cascade (Fig. 2). This pathway relays, by direct interactions within a series of proteins, transmission of a signal from the plasma membrane to the nucleus, leading to the initiation of gene transcription.

The initial phase of this pathway involves mechanisms that translate receptor activation into the activation of Ras, a guanine nucleotide binding protein. The first intracellular event in this process is the SH2/PTB domain-mediated association of adaptor proteins, which have no catalytic function, with activated phosphotyrosine containing receptors. Adaptor proteins utilize multiple

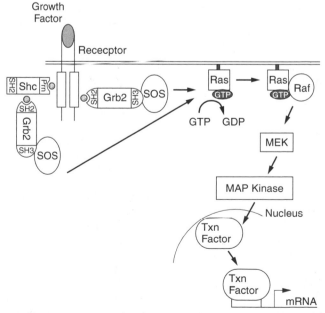

Figure 2. Ras/MAP kinase signaling cascade. ⊙ represents phosphotyrosine.

protein association domains to associate with activated kinase receptors and simultaneously with other proteins. For example, Grb2, an important adaptor in Ras activation, is composed of one SH2 domain that interacts with receptors and two SH3 domains that recognize proline-rich sequences in SOS, a Ras GTP/GDP exchange, and activation factor (24). A second mechanism to modulate SOS involves the adaptor protein Shc, which contains an SH2 domain at its carboxy terminus, a PTB domain at the amino terminus, and a major site of tyrosine phosphorylation in its central region. Shc associates with activated growth factor receptors through its PTB domain and becomes tyrosine phosphorylated (25). Subsequently, the Grb2:SOS complex associates with tyrosine-phosphorylated Shc, by means of the Grb2 SH2 domain. In sum, Grb2 mediates the formation of two ternary complexes: an activated receptor:Grb2:SOS complex and a phosphorylated Shc:Grb2:SOS complex. The latter could actually be a quaternary complex, since Shc through its PTB domain is known to associate with activated receptors.

The functionally important component of these complexes is SOS (an acronym for Son of Sevenless), a Ras guanine nucleotide exchange factor originally identified in *Drosophila*. SOS stimulates the replacement of GDP by GTP on Ras, converting Ras to its active GTP-bound state (26). Similar to trimeric G proteins previously discussed, monomeric G proteins, such as Ras, possess an intrinsic GTPase activity that converts the active GTP-bound molecule to an inactive GDP-bound state. Ras, a major component of the Ras/MAP kinase cascade, is constitutively localized at the cytoplasmic face of the plasma membrane. In contrast, SOS exists as a cytoplasmic protein. Therefore, activation of Ras can be facilitated by the relocalization of SOS to the cytoplasmic face of the plasma membrane. This is accomplished through Grb2:receptor and/or Grb2:Shc complexes with SOS.

The importance of Ras activation in mitogenic signaling is illustrated by the fact that approximately 30% of human cancers contain a mutated constitutively active form of Ras, which is capable of acting as an oncogene in experimental and, perhaps, clinical situations. GTP-Ras inactivation is facilitated by association with a molecule termed RasGAP which accelerates the GTPase activity of Ras, producing GDP-Ras. Activating Ras mutations, of the type frequently found in tumors, most often decrease the capacity of GTP-Ras to interact with RasGAP, which decreases conversion to the inactive form (27). This illustrates the importance not only of activating mechanisms, but also inactivating steps in signal transduction pathways.

In the Ras/MAP kinase signaling pathway, GTP-Ras acts as a switch to initiate the activation of a series of serine/threonine protein kinases. While several molecules have been identified that associate preferentially with GTP-Ras, the most understood and significant is the serine/theorine kinase known as Raf. Experimental data have shown that Raf interacts with GTP-Ras but not GDP-Ras (28). While this interaction leads to Raf activation, however, the mechanism by which Raf is actually activated remains unclear. Since Ras is membrane localized and Raf is cytosolic, one consequence of their interaction is the relocalization of Raf to the plasma membrane. Therefore, to activate Raf, a key event may involve the GTP-Ras-dependent recruitment of Raf to the plasma membrane. Experimentally it is possible to mutate Raf so that it is constitutively localized at the plasma membrane. When this form of Raf is expressed in cells, it is constitutively activated and independent of Ras function, suggesting that Ras does serve to recruit Raf to the plasma membrane, with subsequent events at the membrane actually activating Raf (28). It is worthwhile noting that an activated form of Raf has been identified as an oncogene in retroviral-induced animal tumors.

Raf activation leads to the stimulation of a second protein kinase termed MEK (MAP/ERK kinase). Raf directly phosphorylates MEK on two serine residues, and both phosphorylations are required for full activation of MEK (29). MEK is an unusual kinase since, once activated, it may phosphorylate downstream molecules on both serine/threonine and tyrosine residues. Hence, it is a rare example of a dual- or mixed-specificity protein kinase.

The only known downstream target for activated MEK is mitogen-activated protein (MAP) kinase (also known as ERK), a serine/threonine kinase (Fig. 2). MAP kinase must be phosphorylated on both a threonine and a tyrosine residue to be activated, and once activated, dephosphorylation of either the phosphothreonine or phosphotyrosine residue will inactivate the enzyme (30). MAP kinase has several intracellular substrates, the most significant of which are transcription factors involved in growth factor-dependent gene expression. Key to this activity is the fact that activated MAP kinase is translocated into the nucleus, where it phosphorylates specific transcription factors, altering their DNA binding capacity and thus contributing to the expression of genes responsible for cellular proliferation.

STAT Pathway

In contrast to the multistep MAP kinase pathway, the STAT pathway employs a more direct mechanism to change gene expression. STAT is an acronym for signal transducers and activators of transcription, and there are currently seven STAT proteins identified in mammals.

STATs are activated by tyrosine kinases through a mechanism that relies on SH2 domains and tyrosine phosphorylation. After ligand binding, cytokine:receptor complexes are phosphorylated by associated JAKs, producing an association site(s) for the SH2 domain in a STAT molecule(s) (31). Similarly, activated growth factor receptors may associate with STATs by an SH-2-dependent recognition of receptor autophosphorylation sites, but without the involvement of JAKs (Fig. 3). Specificity in STAT activation is achieved by the specificity of STAT:receptor interactions, which are dependent on STAT SH2 domain recognition of a particular receptor phosphotyrosine-containing sequence. For example, the IFN-γ receptor, which normally activates STAT1, will activate STAT2 if the SH2 domain of STAT2 is exchanged for that of STAT1 (19).

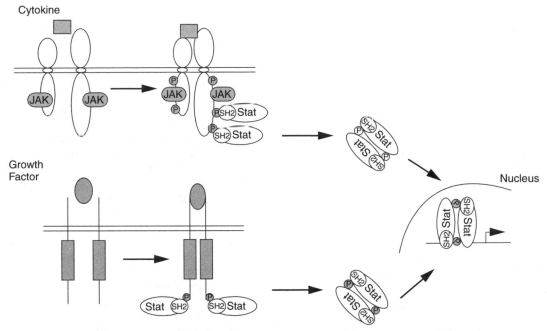

Figure 3. STAT signal transduction cascade. Schematic outline of the STAT signal transduction cascade induced by both activated cytokine and growth factor receptors. ⊖ represents phosphotyrosine.

Following their association with activated receptors, STATs are tyrosine phosphorylated, dissociate from the receptors, and form stable cytoplasmic dimers. These dimers may be either homo- or heterodimeric, adding to the diversity of STAT signaling. Dimerization of STATs reutilizes the SH2 function, such that in the dimer each SH2 domain recognizes the phosphotyrosine residue on the other molecule. Hence, within each dimer two SH2–phosphotyrosine associations are formed, producing added stability to the complex. Unphosphorylated STATs have a high affinity for receptor docking sites, but when tyrosine phosphorylated they bind with a much higher affinity to another STAT molecule (31). After tyrosine phosphorylation and dimerization, STAT complexes rapidly translocate to the nucleus and bind to target DNA sequences in the promoter regions of specific genes. The mechanism of translocation, however, is unclear.

Comparing the MAP kinase and STAT pathways, it is clear that cells have evolved two topologically distinct mechanisms to transduce information from the cell surface to the nucleus. In the former instance, it is the translocation of a protein kinase into the nucleus that promotes gene expression, while STATs, which are transcription factors, are held in the cytoplasm until phosphorylated and then translocated to the nucleus to affect mRNA transcription of specific genes.

Phosphatidylinositide Signaling

The signal transduction pathways outlined thus far describe the transmission of intracellular signals by mechanisms that depend entirely on protein:protein interactions leading to covalent modifications or altered nucleotide binding. However, activated receptors also stimulate the production of small molecule second messengers by activating phospholipases and lipid kinases that, respectively, hydrolyze or phosphorylate specific phospholipids. Phosphatidylinositol 4,5-bisphosphate (PIP$_2$), a minor phospholipid constituent of the plasma membrane, is the immediate precursor for the generation of three metabolite second messengers. The two signaling pathways by which PIP$_2$ gives rise to these messengers are discussed in the following and depicted in Figure 4.

Activated cell-surface receptors stimulate the activity of a phosphoinositide-specific phospholipase C (PLC) to catalyze the hydrolysis of PIP$_2$. Two PLC isoforms have been extensively studied and found to be regulated by separate receptor types. PLC-γ isoforms are activated by receptor and nonreceptor tyrosine kinases, while PLC-β isoforms are activated by heptahelical heterotrimeric G protein-coupled receptors. Both PLC-β and PLC-γ have conserved X and Y domains that are present in all PLC isoforms and that together form the catalytic site for PIP$_2$ hydrolysis. Mutations in either the X or Y domain yield a drastic decrease in enzyme activity, demonstrating that each of these domains is essential for enzyme activity (32). PLC-β will be discussed in a later section.

PLC-γ, but not PLC-β, contains two SH2 domains and one SH3 domain, which are not essential for enzyme activity. Association between PLC-γ and tyrosine kinase receptors, such as the EGF receptor, is mediated by the SH2 domains of PLC-γ and phosphorylation sites at the carboxy terminus of the activated receptor. Following receptor association, PLC-γ is tyrosine phosphorylated to produce an activated form of the enzyme, which then catalyzes cycles of PIP$_2$ hydrolysis. The association step is essential for subsequent tyrosine phosphorylation, and both association and phosphorylation may have functional

(a)

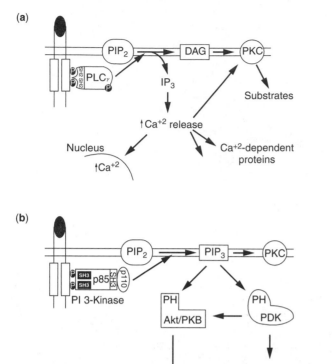

Figure 4. Phophatidylinositol signaling. Outline of phosphatidylinositol signaling mediated by (**a**) PIP$_2$ hydrolysis and (**b**) PIP$_2$ phosphorylation. ⊙ represents phosphotyrosine.

roles in maximizing the activity of PLC-γ. The role of the SH3 domain in PLC-γ function is not known.

The hydrolysis of plasma membrane PIP$_2$ by PLC-γ generates the lipophilic molecule diacylglycerol (DAG) and the hydrophilic molecule inositol 1,4,5-trisphosphate (IP$_3$) as metabolite second messengers [Fig. 4(a)]. DAG activates the serine/threonine-specific protein kinase C (PKC), while IP$_3$ interacts with specific receptors on the surface of the lumen of the endoplasmic reticulum to promote the release of calcium, stored within the endoplasmic reticulum, into the cytoplasm as free Ca^{2+}. This rise in intracellular free-calcium concentration, on the order of 10–50-fold, promotes the activation of a number of Ca^{2+}-dependent molecules, including protein kinases and phosphatases. Also, high levels of cytoplasmic free Ca^{2+} passively raise the levels of Ca^{2+} in the nucleus, where the expression of several genes is reported to require this ion. DAG activates PKC isoforms by increasing the enzyme's affinity for Ca^{2+} and phospholipids, thus reducing the concentration of Ca^{2+} needed to stimulate the enzyme (33). Activated PKC in turn phosphorylates and modulates the activity and/or function of several intracellular proteins. Hence, the hydrolysis of PIP$_2$ ultimately leads to the modulation of pleiotropic targets and does not constitute a specialized linear pathway such as those previously discussed. Both IP$_3$ and DAG are metabolically labile and once formed can be rapidly transformed into biologically inactive products.

In addition to hydrolysis by PLC-γ, PIP$_2$ is also a precursor for the formation of phosphatidylinositol 3,4,5-trisphosphate (PIP$_3$), which is produced by a lipid kinase known as phosphatidylinositol (PI) 3-kinase [Fig. 4(b)]. This kinase catalyzes the addition of a phosphate group to the 3′-hydroxyl group of the inositol ring of PIP$_2$ (34). PI 3-kinase consists of two subunits; an 85-kDa adaptor protein (p85), which is a regulatory subunit, and a catalytic 110-kDa subunit (p110). The p110 subunit is inactive in the absence of the p85 subunit, which contains several domains known to be involved in protein–protein interactions, including an SH3 domain and two SH2 domains. The p110 catalytic subunit, which contains a proline-rich region, associates with the SH3 domain of the p85 subunit. Association of the p85 SH2 domains with activated cell-surface receptors translocates the cytosolic PI 3-kinase to the cytoplasmic face of the plasma membrane and therefore close to its lipid substrate. The association of p85 with phosphorylated receptor also induces a conformational change that contributes to activation of the p110 catalytic subunit (35).

The formation of PIP$_3$ stimulates the activation of several downstream serine/threonine kinases, including certain isoforms of the previously mentioned PKC [Fig. 4(b)]. Unrelated to this effect on PKC, PIP$_3$ activates a series of protein kinases beginning with 3-phosphoinositide-dependent protein kinase (PDK), which has a PH domain that enables interaction with PIP$_3$ (36). Activated PDK then phosphorylates and activates two other serine/threonine kinases. One is the proto-oncogene Akt, also known as protein kinase B (PKB), which also contains a PH domain. Activation of Akt most likely involves specific phosphorylation by PDK as well as PH-mediated PIP$_3$ binding (37). Recently, Akt has been shown to phosphorylate BAD, a protein involved in apoptosis, the process of cell death (38,39). Phosphorylation of BAD induces its dissociation from Bcl-X$_L$, a cell survival factor, allowing Bcl-X$_L$ to resume its normal function of suppressing cell death. A second target of PDK is the protein kinase p70^{s6k}, which is responsible for phosphorylating the S6 protein component of the 40S ribosomal subunit (34). However, the manner in which phosphorylation of this ribosomal protein may influence ribosome function is not clear. In contrast to PIP$_2$, PIP$_3$ is not hydrolyzed by phospholipases. Inactivation of PIP$_3$ is accomplished by lipid phosphatases, which dephosphorylate this second messenger.

Heterotrimeric G Protein Signaling

The signaling pathways outlined previously emphasize protein–protein interactions to elicit long-term biological responses, occurring in minutes to hours, related to the control of cell growth. In contrast, the responses regulated by heterotrimeric G protein-coupled heptahelical receptors are rapid and transient responses, such as neurotransmitter release and muscle contraction, which take place in seconds or less. To transmit rapid signals, G protein-coupled receptors activate rapidly diffusible metabolic second messengers.

Signaling via G proteins is transient as well as rapid since the GTPase activity of Gα subunits converts the active GTP-bound protein to its inactive GDP-bound state, facilitating reassociation of Gα and $\beta\gamma$ subunits (as discussed in the section on activation of heptahelical receptors). The association of various G proteins with different heptahelical receptors is determined by the α subunit. For example, the Gα_s subunit normally interacts with β-adrenergic receptors, while the G$\alpha_{i/o}$ subunit interacts with muscarinic cholinergic 1 (m1) receptors and the Gα_q subunit with m2 receptors (11). G protein subunits modulate the functions of effector molecules, including adenylyl cyclase, phospholipase, and ion channels, which together control the intracellular concentration of second messenger molecules (Fig. 5). Effectors can be regulated by a GTP-Gα subunit, the G$\beta\gamma$ complex, or both. Depending on the effector, the G-protein subunits may stimulate and/or inhibit effector function. Also, different activated α subunits may compete with each other to modulate the some effector function. Examples of individual effector molecules regulated by G proteins will be discussed in the following.

Several different mammalian G protein α subunits have been described and can be divided into four subfamilies based on sequence homologies and functional association with different downstream effector proteins. The Gα_s subfamily stimulates the activity of adenylyl cyclase and opens Ca^{2+} channels, while the G$\alpha_{i/o}$ subfamily increases cGMP phosphodiesterase activity, inhibits adenylyl cyclase activity, and closes Ca^{2+} channels. The Gα_q subfamily stimulates PLC-β activity, while the Gα_{12} subfamily does not have a known target effector (40). In addition to the Gα subfamilies, several isoforms of the β and γ subunits have also been identified. However, the function of these different isoforms is only beginning to be elucidated. Therefore, β and γ subunits will be referred to in a generic manner.

Activation of heterotrimeric G proteins, not unlike the Ras G protein previously discussed, requires localization at the cytoplasmic face of the plasma membrane. Similar to many other proteins involved in receptor proximal steps of

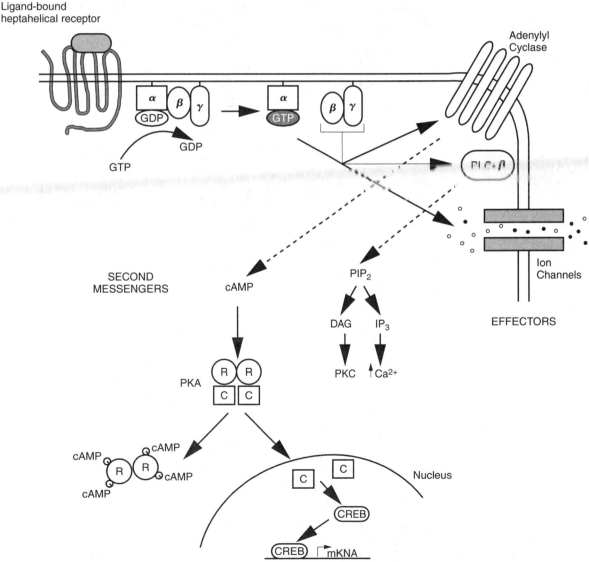

Figure 5. Heterotrimeric G protein signaling. Diagrammatic summary of heterotrimeric G protein signaling, illustrating downstream effectors and second messengers.

signaling pathways, heterotrimeric G proteins are post-translationally modified to facilitate direct interaction with receptors and effectors at the plasma membrane. The carboxy terminus of most Gγ subunits is modified by prenylation, an unsaturated fatty acid addition, that directs membrane localization. This modification is not required for interaction between the β and γ subunits, but is necessary for association of $\beta\gamma$ complexes with the plasma membrane and for high-affinity interactions with α subunits. Gα subunits, such as α_o and α_i, are modified by the addition of myristate, a saturated fatty acid, while Gα_s and Gα_q are modified by the addition of palmitate (41). No lipid modifications have been identified on β subunits. Therefore, both potential signaling components, α subunits and $\beta\gamma$ complexes, are topologically positioned at the plasma membrane to interact rapidly with receptors and effectors.

Adenylyl Cyclase. Adenylyl cyclase is a transmembrane enzyme that catalyzes the conversion of ATP to the second messenger cyclic adenosine monophosphate (cAMP) and is one of the most understood examples of an effector molecule regulated by G protein subunits. At least eight mammalian adenylyl cyclase isoforms have been cloned, and all isoforms are stimulated by the GTP-bound Gα_s. However, GTP-activated G$\alpha_{i/o}$ subunits inhibit certain adenylyl cyclase isoforms. The effects of $\beta\gamma$ subunits on adenylyl cyclases differ slightly with each isoform, though inhibition of the enzyme is the most predominant effect.

Stimulation of adenylyl cyclase catalyzes the formation of cAMP and rapidly activates cAMP-dependent protein kinase (PKA). The inactive holoenzyme of PKA is a heterotetramer containing two regulatory (R) and two catalytic (C) subunits. Activation of PKA involves cAMP binding with high affinity to each of the R subunits, which promotes dissociation of the R and C subunits. When dissociated from the R subunit, the C subunit is active and able to phosphorylate cytoplasmic substrates (Fig. 5). In addition, the free C subunit is able to translocate to the nucleus and phosphorylate proteins important for the regulation of gene transcription, such as the cAMP response element binding protein known as CREB. Therefore, the R subunit not only serves as an inhibitor of catalytic activity, but also serves as a cytoplasmic anchor for the C subunit.

Phospholipase C. As described, receptor tyrosine kinases provoke formation of the second messengers IP$_3$ and DAG through phosphorylation-dependent activation of PLC-γ isoforms. Heptahelical receptors also stimulate the formation of IP$_3$ and DAG, but do so by communicating through heterotrimeric G proteins with PLC-β isoforms. PLC-β can be independently activated either by Gα or by $\beta\gamma$ subunits. Deletion mutagenesis results suggest that the GTP-bound Gα_i subunit binds to the carboxyl-terminal region of PLC-β, while the $\beta\gamma$ complex interacts with the amino-terminal region. Activation of PLC-β by G protein subunits, like stimulation of PLC-γ by receptor kinases, leads to hydrolysis of PIP$_2$, and the generation of DAG and IP$_3$ with the respective activation of PKC and mobilization of intracellular free Ca^{2+}.

Ion Channels. Plasma membrane Ca^{2+} and K$^+$ channels are also known effectors whose functions are modulated by G protein subunits. Different G protein subunits communicate with ion channels in different, even opposite, manners. For example, high-voltage-activated and slowly inactivating Ca^{2+} channels can be inhibited by heptahelical receptors acting through Gα_o subtypes, whereas high-voltage-activated and persistent-current Ca^{2+} channels are stimulated by receptors acting via Gα_s subtypes. Using patch clamp techniques, it has been shown that the frequency with which the K$^+$ channels are opened is greatly enhanced by either purified Gα or G$\beta\gamma$. Gα activates K$^+$ channels at a much lower concentration than G$\beta\gamma$ ($\sim$10 pM compared to $\sim$10 nM). However, G$\beta\gamma$ activates 95% of patches, whereas Gα only activates about 30% of patches (21). These data suggest that the G protein subunits regulate ion channels in different manners, thereby resulting in divergent biological responses. However, it is currently unclear whether G proteins directly or indirectly interact with plasma membrane ion channels.

SUMMARY AND PROSPECTUS

This article has summarized the mechanisms by which extracellular signals are biochemically transduced within cells through different types of cell-surface receptors including growth factor, cytokine, and heptahelical G protein-coupled receptors. In addition, the structural aspects of several major receptor types, as well as prominent signaling pathways associated with these receptor families, have been described. While a great deal is known in these areas, significant issues remain to be clarified within this research field. One major problem is the issue of specificity. For example, EGF and nerve growth factor produce mutually exclusive responses (cell proliferation and differentiation, respectively) on the same target cells. However, the post-receptor signaling elements seem to be the same. This raises the question of how activation of the same signaling molecules is interpreted at points of signal reception within the cell so as to generate different responses.

Another critical issue of signal transduction pathways that has not been fully addressed in this article are the mechanisms that inactive or reverse individual steps in the signaling process to down-regulate these pathways. For example, what regulates the specificity of protein phosphatases and how are ligand:receptor complexes desensitized by endocytosis? In addition, the complexity of signaling cascades that connect cell-surface receptors to the nucleus is just beginning to be appreciated. Available data demonstrate that communication between signaling pathways exists, that is, stimulation of the MAP kinase pathway by the activation of G protein-coupled receptors. However, present and future research needs to define the functional and biological responses mediated by these interrelated pathways.

BIBLIOGRAPHY

1. C.-H. Heldin, in P.J. Parker and T. Pawson, eds., *Cell Signalling*, Cold Spring Harbor Lab. Press, Cold Spring Harbor, N.Y., 1996, pp. 7–24.

2. R. Baserga et al., *Biochim. Biophys. Acta* **1332**, F105–F126 (1997).

3. K.S. Kolibaba and B.J. Druker, *Biochim. Biophys. Acta* **1333**, F217–F248 (1997).

4. A. Ullrich and J. Schlessinger, *Cell (Cambridge, Mass.)* **61**, 230–212 (1990).

5. D.F. Stern, M.P. Kamps, and H. Lao, *Mol. Cell. Biol.* **8**, 3969–3973 (1988).

6. S.K. Hanks, A.M. Quinn, and T. Hunter, *Science* **241**, 42–52 (1988).

7. N. Sato and A. Miyajima, *Curr. Biol.* **6**, 174–179 (1994).

8. T. Kishimoto, T. Taga, and S. Akira, *Cell (Cambridge, Mass.)* **76**, 253–262 (1994).

9. S.S. Watowich et al., *Annu. Rev. Cell Dev. Biol.* **12**, 91–128 (1996).

10. J.M. Baldwin, *Curr. Opin. Cell Biol.* **6**, 180–190 (1994).

11. A.M. Spiegel, in A.M. Spiegel, T.L.Z. Jones, W.F. Simmonds, and L.S. Weinstein, eds., *G Proteins*, R.G. Landes, Austin, Tex., 1994, pp. 6–17.

12. C.-H. Heldin, *Cell (Cambridge, Mass.)* **80**, 213–223 (1995).

13. J.N. Ihle, *Nature (London)* **377**, 591–594 (1995).

14. A.A. Kossiakoff et al., *Protein Sci.* **3**, 1697–1705 (1994).

15. J.A. Wells and A.M. de Vos, *Annu. Rev. Biochem.* **65**, 609–634 (1996).

16. B. Pasini, I. Ceccherini, and G. Romeo, *Trends Genet.* **12**, 138–144 (1996).

17. L. Ihle, *J. Lab. Clin. Med.* **120**, 688–692 (1992).

18. R. Derynck and X.-H. Feng, *Biochim. Biophys. Acta* **1333**, F105–F150 (1997).

19. S. Pellegrini and I. Dusanter-Fourt, *Eur. J. Biochem.* **248**, 615–633 (1997).

20. E.J. Neer, *Cell (Cambridge, Mass.)* **80**, 249–257 (1995).

21. D.E. Clapham and E.J. Neer, *Nature (London)* **365**, 403–406 (1993).

22. D.M. Berman and A.G. Gilman, *J. Biol. Chem.* **273**, 1269–1272 (1998).

23. G.B. Cohen, R. Ren, and D. Baltimore, *Cell (Cambridge, Mass.)* **80**, 237–248 (1995).

24. P. van der Geer and T. Pawson, *Trends Biochem. Sci.* 277–280 (1995).

25. L. Bonfini et al., *Trends Biochem. Sci.* 257–261 (1996).

26. J. Downward, in P.J. Parker and T. Pawson, eds., *Cell Signalling*, Cold Spring Harbor Lab. Press, Cold Spring Harbor, N.Y., 1996, pp. 87–100.

27. W.J. Fantl, D.E. Johnson, and L.T. Williams, *Annu. Rev. Biochem.* **62**, 453–481 (1993).

28. R. Marais and C.J. Marshall, in P.J. Parker and T. Pawson, eds., *Cell Signalling*, Cold Spring Harbor Lab. Press, Cold Spring Harbor, N.Y., 1996, pp. 101–125.

29. J.M. Kyriakis and J. Avruch, in J.R. Woodgett, ed., *Protein Kinases*, Oxford University Press, Oxford, 1994, pp. 85–148.

30. J.M. Tavare, in M.J. Clemens, ed., *Protein Phosphorylation in Cell Growth Regulation*, Harwood Academic Publishers, Chur, Switzerland, 1996, pp. 93–110.

31. J.E. Darnell, *Science* **277**, 1630–1365 (1997).

32. M. Katan, in P.J. Parker and T. Pawson, eds., *Cell Signalling*, Cold Spring Harbor Lab. Press, Cold Spring Harbor, N.Y., 1996, pp. 199–211.

33. G.C. Blobe, S. Stribling, L.M. Obeid, and L.A. Hannun, in p. J. Parker and T. Pawson, eds., *Cell Signalling*, Cold Spring Harbor Lab. Press, Cold Spring Harbor, N.Y., 1996, pp. 213–248.

34. B. Vanhaesebroeck, S.J. Leevers, G. Panayotou, and M.D. Waterfield, *Trends Biochem. Sci.* **22**, 267–272 (1997).

35. R. Kapeller and L.C. Cantley, *BioEssays* **16**, 565–576 (1994).

36. M.A. Lemmon, M. Falasca, K.M. Ferguson, and J. Schlessinger, *Trends Cell Biol.* **7**, 237–242 (1997).

37. T.J. Franke, D.R. Kaplan, and L.C. Cantley, *Cell (Cambridge, Mass.)* **88**, 435–437 (1997).

38. L. del Peso et al., *Science* **278**, 687–689 (1997).

39. H. Dudek et al., *Cell (Cambridge, Mass.)* **91**, 231–241 (1997).

40. S. Offermanns and M.I. Simon, in P.J. Parker and T. Pawson, eds., *Cell Signalling*, Cold Spring Harbor Lab. Press, Plainville, N.Y.) 1996, pp. 177–198.

41. J. Iniguez-Lluhi, C. Kleuss, and A.G. Gilman, *Trends Cell Biol.* **3**, 230–236 (1993).

See also CELL PRODUCTS—ANTIBODIES; CELL PRODUCTS—IMMUNOREGULATORS; CELL STRUCTURE AND MOTION, EXTRACELLULAR MATRIX AND CELL ADHESION; CELLULAR TRANSFORMATION, CHARACTERISTICS; MEMBRANE STRUCTURE AND THE TRANSPORT OF SMALL MOLECULES AND IONS; RECEPTORS AND CELL SIGNALING, INTRACELLULAR RECEPTORS—STEROID HORMONES AND NO; TRANSFORMATION OF PLANTS.

CELLULAR TRANSFORMATION, CHARACTERISTICS

SUZANNE D. CONZEN
University of Chicago
Chicago, Illinois

OUTLINE

INTRODUCTION

Neoplastic transformation of animal cells from a benign to a malignant state involves the perturbation of multiple genetic growth controls. Cells maintained in tissue culture can undergo neoplastic transformation in a variety of ways. Transformation can result from an infection with an oncogenic RNA retrovirus or DNA virus, from transfection (chemically mediated uptake) of DNA encoding oncogenes, or from the inactivation of tumor suppressor genes or their protein products. Chemical carcinogens can also contribute to malignant transformation through their ability to mutate cancer-related genes (1).

Although the mechanisms that contribute to the transformed phenotype are many, two major types of pathways have been identified: (1) abnormal expression of growth promoting oncogenes and (2) inactivation of growth inhibitory genes, termed tumor suppressor genes. The ultimate result of the combined expression of oncogenes and/or the loss of tumor suppressor gene expression or protein function is that transformed cells exhibit a variety of abnormal genetic, physical, and physiological characteristics. First, transformed cells exhibit abnormal chromosomal characteristics including changes in absolute chromosome number (aneuploidy) and loss and/or gain of chromosomal material from individual chromosomes. Chromosome number and morphology can be assessed in a karyotype analysis (Experiment 1.1). The total DNA content of a population of transformed cells can be measured by flow cytometry, a method that also provides information about the percentages of cells in the various phases of the cell cycle by virtue of their DNA content (Experiment 2.1). A second quantifiable abnormality of the chromosomes in transformed cells is that they often exhibit a change in the normal rate of shortening of telomeres, which are repetitive sequences at the ends of chromosomes. Differences in telomeric length and the activity of telomerase, an enzyme that maintains telomere length, can be compared between normal and transformed cells (Experiment 3.1).

In addition to altered chromosome characteristics, transformed cells cultured in vitro exhibit a number of characteristic growth changes. Most of the changes described in this article pertain to adherent fibroblasts and epithelial cells. These cells are those most commonly assayed in studies of malignant transformation since they have the most dramatic changes in physical characteristics following transformation. First, adherent cells grown in tissue culture conditions exhibit a loss of contact inhibition, meaning that they continue to proliferate despite cell-to-cell contact. The overgrowth of transformed cells results in the formation of dense heaps or foci of cells that grow on top of a confluent monolayer of underlying cells (2). A focus assay (Experiment 4.1) can be used to count the number of colonies of transformed cells, and the relative foci number can be used to compare the degree of transformation between two experimental conditions. Second, transformed cells, in contrast to the parental cell line from which they were derived, frequently do not require a solid (plastic or glass) matrix to grow on because they can exhibit "anchorage-independent" growth and thus can proliferate even when plated in a semi solid medium (3). This phenotype can be assayed by growing cells in either low-concentration methocellulose or agarose (Experiment 4.2). Third, unlike "primary" cells that are harvested directly from an animal and exhibit a limited number of cell divisions in culture before undergoing senescence (4), transformed cells have the potential to continue dividing indefinitely. However, not all immortalized cells are transformed, and many cell lines exist that are capable of indefinite division but do not have transformed characteristics. Under adverse environmental conditions (for example, growth factor deprivation or ultraviolet radiation), transformed cells can stop their continuous cell cycling and undergo programmed cell death or apoptosis (5). In fact, many oncogene products are able to contribute to either a transformed phenotype or to an apoptotic one depending on both the intracellular environment (for example, whether or not complementing transforming oncogenes are expressed in the same cell) and the extracellular conditions (for example, the availability of growth factors). Finally, transformed cells are able to form solid tumors in vivo in either syngeneic animals that cannot mount an immunologic rejection of the tumor or in congenitally immunocompromised animals (for example, nude mice that are athymic) (6).

Biochemical changes common to malignant tumors also frequently accompany cellular transformation. Increased aerobic glycolysis by tumors was first observed by Warburg over half a century ago (7). The transport of glucose into cells, the glycolytic activity of transformed cells, and the ultimate production of lactic acid can all be measured in tumor cells (8). While the details of the molecular pathways linking many of the known oncogenic events common to tumor cells with the physiological and biochemical changes in metabolic pathways are not known, biochemical characteristics of malignant cells provide a framework within which to analyze transformed cells. Characterization of the cellular and biochemical changes accompanying neoplastic transformation has led to the development of several assays that measure these hallmarks of transformation in a quantitative and reproducible way. In this article, the major mechanisms contributing to cellular transformation will be reviewed. The experimental techniques that may be used to quantify the changes associated with the transformed phenotype will then be described.

MECHANISMS OF CELLULAR TRANSFORMATION

Retroviruses

Retroviruses are small RNA viruses that encode their genetic information in RNA; following host cell infection, the viral RNA serves as a template for DNA synthesis and subsequently virally encoded proteins are produced. Retroviruses are a common cause of cancer in some animal species and can contribute to the transformation of mammalian cells in tissue culture (9). Following infection of the target cell, reverse transcriptase, a retroviral enzyme packaged in the viral capsid, is released and catalyzes the production of the first DNA copy of a viral RNA strand. Following reverse transcription of viral RNA into DNA, a virus-encoded integrase catalyzes the integration of the linear double-stranded retroviral DNA into the animal cell genome. This newly introduced genetic material contributes to carcinogenesis in one of two ways: Either the viral DNA causes the transcription and expression of a host cell gene that is potentially transforming (10), or the viral gene itself may encode an oncogenic protein (11). Oncogenes identified through their presence in transforming retroviruses include growth factor receptors (e.g., the EGF receptor), protein kinases (e.g., raf) and GTP-binding proteins (e.g., K-ras).

Engineered Expression of Oncogenes

Using recombinant DNA technology, retroviruses can also be engineered to express a desired transforming oncogene by insertion of a designated cDNA (encoding a gene of interest) into a retroviral-based DNA vector. Retroviral vectors and complementary "packaging" cell lines have been developed to facilitate the efficient introduction of foreign DNA into tissue culture cell lines. The retroviral vector is transiently transfected into a retroviral "packaging" cell line. Several extremely high-efficiency viral packaging cell lines have been developed that can be used to make high-titer retroviral stocks easily and rapidly (12). High-titer retroviruses, released into the media in which the packaging cell lines are propagated, can then be used to infect mammalian cells in order to make stable pools or clones of cells expressing a specific gene of interest. If the retroviral vector is engineered to express an antibiotic resistance gene in addition to the oncogenic gene of interest, transformed cells can be selected for by subjecting cells to antibiotic selection. Only cells surviving the exposure to antibiotic carry the gene of interest. Alternatively, the retroviral vector can be engineered to co-express green fluorescent protein (GFP), and green cells can be selected by fluorescence-activated cell sorting (FACS analysis) in order to isolate subpopulations of cells expressing the desired retroviral genes. Through this approach, complementary combinations of oncogenes can be expressed (for example, v-Myc and Ras) efficiently in cells and the characteristic changes of transformation evaluated.

Foreign DNA encoding transforming oncogenes can also be introduced into mammalian cells by chemically mediated transfection. Calcium phosphate–DNA precipitates and commercially available liposome-mediated transfection methods have been developed to facilitate introduction of recombinant DNA plasmids into mammalian cells. Cells expressing the ectopic DNA plasmids can then be selected by either antibiotic exposure or coexpression of a fluorescent marker. In this way, populations of cells expressing different oncogenes can be studied for transformation phenotypes using the assays described in this article.

Small DNA Tumor Viruses

Small DNA tumor viruses are also capable of transforming mammalian cells. The oncogenic adenoviruses, the Simian Virus 40 (SV40), and the human papilloma viruses (HPV) are examples of small DNA tumor viruses capable of transforming rodent embryo fibroblasts and selected other mammalian cells. These tumor viruses infect cells and subsequently use the host cell's machinery to produce viral proteins capable of eliciting cellular transformation (13). Expression of transforming proteins can alter the proliferative pathways of their host cells in multiple ways. The most common mechanism used by small DNA tumor virus is the production, soon after infection, of transforming proteins that bind to endogenous cellular proteins critical for growth control and either inactivate them through sequestration or target them for degradation. Many of these growth inhibitory proteins have been identified genetically as tumor suppressor proteins, meaning their loss (or inactivation by viral proteins) is associated with the development of cancer. For example, following infection, the adenovirus early genome encodes two major viral proteins products, E1A and E1B (early region 1A and 1B proteins). E1A is able to bind to and inactivate the p300/Creb binding protein (CBP) -associated factor, a transcriptional coactivator that connects the basal transcriptional machinery to various DNA-binding transcriptional factors. E1A can also bind to and inactivate the retinoblastoma (Rb) tumor suppressor protein, an important cell cycle regulated protein (14). Rb is encoded by a tumor suppressor gene that undergoes homozygous deletion in human retinoblastoma tumors and is altered in approximately 40% of other human tumors (15). Binding of E1A to the Rb protein in turn prevents Rb from associating with the E2F family of transcription factors (16). By preventing Rb from inactivating the E2F transcription factors, E1A (and, as will be seen, other Rb binding viral transforming proteins) is able to promote inappropriate cell proliferation.

The adenovirus uses yet an additional strategy to promote cell transformation of infected cells. The E1B 55K protein product of the adenovirus E1B gene can bind to and functionally inactivate the important tumor suppressor protein, p53. Mutation or loss of the p53 gene is one of the most commonly occurring events in human tumors (17). The p53 protein is thought to be a critical regulator of the transition from the G1 stage of the cell cycle to S phase; disruption of p53 function can lead to inappropriate progression through the cell cycle (18). Thus sequestration and inactivation of p53 by the adenovirus E1B 55K protein further deregulates the normal checkpoint growth controls of the host cell cycle, contributing to uncontrolled proliferation and to transformation.

SV40 is another small DNA tumor virus capable of transforming rodent embryo fibroblasts in cell culture (19). SV40, originally discovered as a cytopathic contaminant in African green monkey epithelial cells that were being used to passage attenuated polio virus, encodes several viral proteins. The first twenty hours following SV40 infection is dedicated to production of three alternatively spliced mRNA products encoding large, small, and tiny tumor (T) antigens (20). In cell culture, only SV40 large T antigen is required for transformation, although small t antigen increases the efficiency of transformation (21). The role of tiny t remains undefined. Analogous to the adenoviral E1A and E1B protein functions, the single SV40 large T antigen also binds the Rb family of tumor suppressor proteins (pRb, p107, and p130), as well as p53 and p300 (22–24).

The human papilloma virus strains 16 and 18 are known etiological agents in human cervical cancer. HPV encodes two known transforming proteins designated E6 and E7; these proteins are functionally equivalent to the SV40 large T antigen and the adenovirus E1A and E1B proteins. The E7 gene product binds to and inactivates the Rb family members while the HPV E6 protein targets p53 for degradation, contributing to the transformation of epithelial cells. The DNA of high-risk HPV strains 16 and 18 is frequently (if not always) integrated into the genome of cancer cells; it is normally episomal in premalignant lesions (25).

Thus several small DNA tumor viruses use common mechanisms to inactivate growth regulatory proteins and promote transformation. The individual transforming proteins of DNA tumor viruses can also be introduced by chemical transfection of cells or by retroviral or viral infection. These protein products, once expressed in rodent embryonic fibroblasts and some mammalian epithelial cells, have the ability to transform them by the criteria outlined in Table 1. Recombinant DNA technology has allowed the production of deletion and point mutations of these virally encoded genes; the resulting structure–function analyses have repeatedly revealed the crucial role of tumor suppressor

inactivation in DNA tumor virus-mediated transformation. Cell culture assays described in this article have played a valuable role in characterizing the transformed phenotype following expression of DNA tumor virus transforming proteins. Furthermore, organ-specific expression of DNA tumor virus transforming proteins in transgenic animals has revealed fascinating differences in the consequences of inactivation of tumor suppressor proteins (26). Analysis of tumor development in transgenic mice has allowed investigators to study the multistep process of transformation in a controlled in vivo setting (27,28).

CHARACTERISTICS OF TRANSFORMED CELLS

Chromosome Characteristics

Karyotypes. Every species has a characteristic chromosome number, size, and shape which comprise the normal karyotype for that species. However, transformed cells commonly exhibit karyotypic abnormalities that result from a gross change in chromosome content such as the loss, addition or rearrangement of chromosomal material (Fig. 1). The first consistent chromosomal abnormality was discovered nearly 40 years ago when Nowell and Hungerford detected a small karyotypic marker, the Philadelphia chromosome, in patients with chronic myeloid leukemia (29). Since then, more than 22,000 neoplasms with corresponding karyotypes have been catalogued (30). Although this is a large number of karyotypes, the majority of chromosomal information is on hematologic abnormalities, and more common tumors such as uterine, cervix, breast, lung, and other epithelial cancers are less represented. Many cell lines commonly

Table 1. Characteristics and Assays of Cellular Transformation

Transformation characteristic	Experimental assay
Genetic abnormalities	
Chromosome abnormality	Karyotype analysis (Experiments 1.1–1.4)
Aneuploidy	DNA content by flow cytometry (Experiment 2.1)
Telomerase activity	TRAP assay (Experiment 3.1)
Growth characteristics	
Loss of contact inhibition	Focus assay (Experiment 4.1)
Anchorage-independent growth	Soft agarose assay (Experiment 4.2)
Reduced growth factor requirements	Doubling time/saturation density (Experiment 4.3)
In vivo tumor formation	Nude mouse assay (Experiment 4.4)
Biochemical changes	
Increased glucose transport	Measurement of glucose transport (Experiment 5.1)
Increased lactate production	Assay for lactate production (Experiment 5.2)

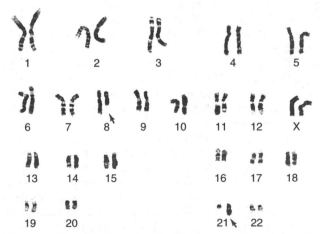

Figure 1. Karyotypic analysis with trypsin–Giemsa banding technique. A characteristic human karyotype of metaphase tumor cells from a patient with acute myeloid leukemia. Karyotypic analysis and trypsin–Giemsa banding reveals a balanced translocation between chromosomes 8 and 21. The t(8;21) interrupts two genes, AML1 on chromosome 21, and ETO on chromosome 8, joining them to form a new gene product. This new product is likely to deregulate the normal growth and differentiation of hematopoietic precursor cells, contributing to malignancy. (Courtesy of Dr. Michelle Le Beau, University of Chicago.)

used in in vitro tissue culture experimentation have also been karyotyped. Some of these cell lines as well as their karyotypic information are catalogued and are available from a centralized tissue culture bank maintained by the American Type and Tissue Culture bank (ATCC) in Maryland and are available for purchase. The ATCC is an excellent and reliable source of transformed cell lines and provides technical assistance on growth conditions.

Chromosomes are characterized according to their size (for example, in humans, #1 is the longest, #22 the shortest), the location of the centromere (separating the chromosome into two arms, a short and long arm), and the banding pattern on each arm (Fig. 1). Comprehensive cytogenetic studies in experimental tumors, including more than 200 primary sarcomas induced by the Rous sarcoma virus in various species of animals, supported the conclusion that non-random as well as random cytogenetic changes occur during the process of transformation (31). However, a whole new level of precision in chromosome description was achieved when the technique of chromosome banding (32) was developed. This allowed for the identification and cataloguing of specific regions of individual chromosomes that undergo gross mutations in transformed cells. Several chemicals have been used to stain bands including the fluorescent quinicrine dihydrochloride (Q-banding) and the Giemsa colorimetric dyes (G-banding) (see Experiment 1.1).

Structural abnormalities of chromosomes in transformed cells include a loss of chromosomal material, a gain of chromosomal material, or relocation without gain or loss of genetic material (33). Transformed cell lines, pools of individually transformed clonal cell populations, and in vivo–derived tumors do not represent a homogeneous population, since subclones may develop with serial passage in tissue culture or with subsequent population doublings in vivo. Thus the chromosomal number may vary among individual cells in a transformed population. The modal number refers to the most common chromosome number in the cell population; this number may be diploid (normal number of chromosomes) or near diploid. It is hypodiploid if it is less than the standard number of chromosomes for a given species and hyperdiploid when it is more than normal. The absolute number of chromosomes in individual cells can be counted and described in the karyotype analysis. The amount of DNA in a population of cells can be obtained using flow cytometry analysis. Unlike a karyotypic analysis, flow cytometry gives no information on actual chromosomal structure, but it quantifies the percentage of cells in the various cell cycle subsets (34).

In addition to the relatively new banding techniques, a molecular biology–based technique termed fluorescent in situ hybridization (FISH) has been developed that allows hybridization with a specific chromosomal sequence (35). This technique can be used on both dividing and interphase cells and can give specific information about the number of copies of specific genes as well as their locations. It is, therefore, an adjunct to the initial chromosomal evaluation by banding.

Chromosome Analysis (36,37)

Experiment 1.1: Karyotype Analysis. Purpose: To identify chromosome number and morphology

Materials

- Colcemid solution (5 ug/ml)
- 0.075 M KCl
- Trypsin–EDTA
- 3 : 1 Methanol–acetone
- Giemsa solution diluted 1 : 50
- Mounting solution (see the following)

Methods

Harvesting of cells.

1. The day before the experiment seed cells at 10^6 cells/100 mm dish.
2. Add 0.1 ml of colcemid solution (5 μg/ml in dH$_2$O) to each 100 mm dish. Incubate at 37 °C for 90 min to 2 h.
3. Prepare hypotonic solution of 0.075 M KCl at 37 °C.
4. After incubation, remove medium to a 15-ml conical centrifuge tube. Trypsinize cells from the dish and combine with medium from that dish.
5. Spin down cells at 1000 rpm for 5 min. Aspirate supernatant, leaving a small residual.
6. Gently resuspend the pellet by tapping the tube, and slowly pipette the hypotonic KCl solution into the tube to a total volume of 8 ml.
7. Incubate in a 37 °C waterbath for 15–20 min.
8. Spin down cells at 1000 rpm for 5 min. Aspirate supernatant, as previously.
9. Gently resuspend the pellet by tapping the tube. Slowly, with agitation, add *freshly* prepared fixative solution (3 : 1, methanol:acetone) to a total volume of 4 ml. Allow to stand at room temperature for 15 min or longer.
10. Spin down cells at 1000 rpm for 5 min. Aspirate supernatant.
11. Resuspend pellet in fresh methanol:acetate to a total volume of 3 ml. Allow to stand for 5 min.
12. Spin down cells at 1000 rpm for 5 min. Aspirate supernatant.
13. Add approximately 0.5 ml of fresh fixative, more or less depending upon the size of the cell pellet. This will be used for the slide preparation. The solution can be stored at 4 °C for some time.

Preparation of Slides.

1. Dip a 3″ × 1″ microscope slide into dH$_2$O. Remove and shake off excess dH$_2$O.
2. Holding the slide at approximately 30 °C from the horizontal, drop 3 drops of the cell suspension from about 6 in. The liquid should run lengthwise across the slide surface.
3. Blow on the slide several times. Stand the slides upright and allow to air dry.

4. Examine under phase-contrast microscope for the presence of metaphase chromosomes. Additional staining (see the following) is usually needed.

Experiment 1.2: Giemsa Staining. Purpose: To induce G-bands on unbanded chromosomes for more detailed analysis.

Materials

- Giemsa stain solution (1 : 50 in dH$_2$O)
- Mounting solution
- Add 500 µl of 20 mg/ml *p*-phenylenediamine (1,4-phenylenediamine) in 10X PBS to 8.5 ml of glycerol.
- Stir until *p*-phenylenediamine has dissolved.
- pH to 8.0 with 0.2 M sodium bicarbonate (1.68 g/ml). Check pH with pH paper.
- Store wrapped in aluminum foil at $-20\,^{\circ}$C

Procedure

1. Place slides in Coplin jar.
2. Stain with Giemsa, diluted 1 : 50 with dH$_2$O, for 5 min.
3. Wash in several changes of dH$_2$O.
4. Add several drops of mounting solution and affix a cover slip.
5. Score for chromosome number using light microscopy.

Experiment 1.3: Quinicrine Banding Technique (38). Purpose: Fluorescent banding detail of chromosomes.

Materials

- Quinicrine–HCl 20 mg/ml in 1X PBS
- Mac Ilvane's citrate–phosphate buffer, pH 5.4, prepared from 84 mls of disodium phosphate solution (200 mM) and 66 mls of citric acid solution (100 mM)

Methods

1. Soak slides in methanol for 5 min. Remove and allow to dry.
2. Stain with quinicrine–HCl 1 (Sigma) at 20 mg/ml for 5 min.
3. Wash with dH$_2$O.
4. Soak in Mac Ilvaine's citrate–phosphate buffer for 20 sec (longer may destain the slides).
5. Wash with dH$_2$O.
6. Mount in a 1 : 5 dilution of Mac Ilvaine's buffer in dH$_2$O. Full-strength buffer will destain the slide, while a higher dilution causes the chromosome to swell.
7. To destain, wash slides by dipping into two Coplin jars containing Mac Ilvane's buffer, pH 5.4.
8. Place several drops of buffer on the slide and mount with a coverslip. Remove excess buffer by holding slide with edge touching absorbant paper.
9. Observe under fluorescent microscope.

10. Remove coverslip under gently running distilled and deionized H$_2$O. Slides can be stored in the dark for up to 2 weeks in 95% ethanol.

Experiment 1.4: Trypsin–Giesma Banding Technique (39). Purpose: More detailed banding technique.

Materials

- Hank's balanced salt solution (HBSS)
- Trypsin (0.05%)/EDTA (0.02%), pH 7.0 diluted in HBSS
- Prepare Gurr buffer, by dissolving one Gurr buffer tablet in 1 L dH$_2$O, pH = 6.8

Protocol

1. Rinse slides in Hank's BSS for 1–1/2 min.
2. Incubate slides in trypsin–EDTA solution at room temperature. Time varies from 2 to 5 min; so experiment with test slides for optimal results.
3. Stop trypsinization in medium +10% serum for 15 sec.
4. Rinse in Hank's BSS for 1 min.
5. Stain in Giemsa solution for 3–1/2 to 5 min. Time in stain varies accordingly with time in trypsin.
6. Rinse in Gurr buffer for 30 sec.
7. Dry slides. Mount with mounting solution.
8. Observe and record chromosomes under light microscope.
9. Results are enhanced by storing prepared slides for 2 days at room temperature prior to trypsin treatment.

Flow Cytometric Analysis of Cell Populations to Determine DNA Content

A flow cytometer is an extremely valuable machine that can be used to measure a variety of phenotypic characteristics of cells, including total DNA content (40). Because transformed cells are often aneuploid, the degree of aneuploidy can be determined with flow cytometry (Fig. 2). Propidium iodide, a DNA and RNA avid fluorescent dye, can be used to stain cellular DNA following trypsinization of adherent cells. The flow cytometer aspirates a thin stream of stained cells and allows only one cell at a time to pass in front of a laser. The scattered light gives information on the size of the cell and the intensity of the staining. A histogram of fluorescent intensity versus cell numbers is then plotted. Thus cells in G$_0$ and G$_1$ have a normal (diploid DNA) content, cells in S phase have between a 2*n* and 4*n* complement of DNA, and cells in G$_2$ have a 4*n* DNA content prior to cell division. Thus the percentage of cells in each phase of the cell cycle can be plotted (Fig. 2).

Flow cytometric analysis also allows the detection of cells that are aneuploid, since cells with more than 4*n* DNA content can be plotted as well. In addition, the percentage of the aneuploid population that is in S phase can also be determined. Flow cytometers can also be used with an attached cell sorter (fluorescence-activated cell

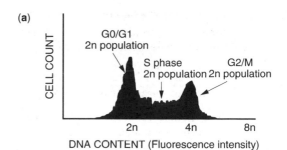

(a)

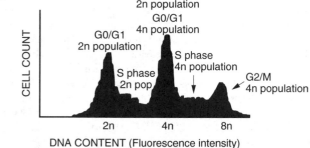

(b)

Figure 2. Histogram of normal and aneuploid cells by flow cytometry. Untransformed cycling cells have between a $2n$ and $4n$ complement of chromosomes, depending on which phase of the cell cycle the cells are in. The content of DNA can be estimated by flow cytometric measurement of the fluorescence intensity emitted by individual cells stained with propidium iodide. (**a**) Characteristic histogram of cells with a normal $2n$ DNA content as measured by flow cytometry. The percentage of cells in each phase of the cell cycle can be calculated. (**b**) DNA histogram of two populations of cells, a diploid cell population ($2n$) and an aneuploid (tetraploid or $4n$) population. Transformed cell lines can contain heterogeneous subpopulations of aneuploid cells.

sorting, FACS), in which specific cell subpopulations can be selected (sorted) for further study.

Experiment 2.1: Flow Cytometric Analysis of DNA Content. Purpose: To measure the DNA content of a population of cells in order to determine the distribution of cells in the cell cycle and the ploidy of the population.

Materials
- Trypsin–EDTA
- Cell culture media with trypsin inhibitors or 10% serum added
- PBS 1X
- 100% Ethanol

Methods
1. Harvest cells by trypsinization and disperse to the single cell state by dilution with a trypsin inhibitor or serum-containing media.
2. Spin down cells in a 15-ml tube at 800 RPM in a clinical centrifuge.
3. Rinse cells once in media with a trypsin inhibitor or serum in order to inactivate trypsin.
4. Rinse cells in PBS × 2 and resuspend in 300 μl PBS.

5. Add 700 μl cold ethanol dropwise while vortexing gently.
6. Fixed cells can be stored at −20 °C until analysis.
7. Pellet cells at low speed in a microcentrifuge, resuspend in 1 ml PBS, and pellet again.
8. Resuspend cells in 1 ml PBS containing 10 μg/ml propidium iodide and 100 μg/ml DNAse-free RNAse A.
9. Incubate at 37 °C for 30 min.
10. Resuspend cells well by gentle micropipetting with a 1000 μl pipette tip.
11. Determine DNA content and cell cycle profiles using a flow cytometer.
12. Analyze DNA content using a DNA content program (for example, Lysis II, Beckton-Dickinson).

Telomeres

Telomeres are the physical ends of chromosomes. They consist of DNA sequences that are highly conserved and differ greatly from the ends of nonchromosomal DNAs (41). The repeated TTAGGG DNA sequences are synthesized by an RNA-dependent DNA polymerase called telomerase (42). Due to the gaps in the primers that are used to begin DNA polymerization prior to each cell division, telomere length diminishes over the course of a somatic cell's lifespan, becoming progressively shorter (43). Telomeric length in cultured mammalian cells can be measured by digesting genomic DNAs at points outside, but not within, the telomeric TTAGGG repeats present in the telomeres. The resulting fragments are separated by electrophoresis and hybridized to a ^{32}P labeled TTAGGG probe that detects terminal restriction fragments and includes the terminal repeats and the subtelomeric regions. These fragments appear as a smear of different-sized fragments because of the interchromosomal variations. Analysis of these chromosome terminal restriction fragments (TRFs) provides the composite lengths of all telomeres in a cell population. The general size predominance of the fragments varies from immortal cells (which show no loss of telomeres with subsequent cell divisions) in comparison to the successive shortening of telomeres in dividing normal cells (44).

Comparing the composite length of telomeres in cell populations by gel electrophoresis patterns is a somewhat difficult assay which has largely given way to an assay designed to measure differences in telomerase activity between cell populations. Telomerase activity is usually repressed in somatic tissues, although self-renewing tissues are an exception (45). In contrast to primary cells, both immortalized cells and tumor cells often exhibit enhanced telomerase activity and a consequent relative lack of telomere shortening with successive cell divisions compared with normal senescing cells (44). In the telomeric repeat amplification protocol (TRAP) assay, cells are lysed and the detergent extracts evaluated for telomerase activity. Telomerase in the extracts synthesizes telomeric repeats onto a nontelomeric oligonucleotide. Such telomerase products are specifically amplified by a thermal cycling reaction incorporating a downstream primer and an upstream primer (see Experiment 3.1) and using $\alpha - {}^{32}PdCTP$- and $\alpha - {}^{32}PdGTP$-labeled nucleotides. Telomerase activity is measured as positive or negative based on the presence or absence of radioactive amplified product. Telomerase activity has been shown to be reproducibly elevated in tumor cells, where immortalized cells are likely to be present to maintain the tumor's growth.

Experiment 3.1: Measurement of Telomerase Activity (46). Purpose: To measure comparative telomerase activity in the lysates of distinct cell populations.

Materials: Buffer #1
- 10 mM HEPES–KOH (pH 7.5)
- 1.5 mM $MgCl_2$
- 10 mM KCl
- 1 mM EGTA
- 100 mM Dithiothreitol

Methods
1. Wash cells of interest once in PBS.
2. Pellet at 10,000 g for 1 min at 4 °C.
3. Resuspend in Buffer #1.
4. Pellet.
5. Resuspend 1×10^6 cells in 20 µl of ice-cold lysis buffer.
6. Incubate the suspension for 30 min on ice and then centrifuge for 30 min in a microultracentrifuge at 100,000 g at 4 °C.
7. Remove the supernatant, quick freeze on dry ice, and store at −70 °C.
8. Place 0.1 µg of CX primer [5′-(CCCTTA)₃CCCTAA-3′] in the bottom of a PCR tube and seal with molten wax. Allow to solidify.
9. Place at 4 °C.
10. Place 50 µl of TRAP reaction at the top of the wax. Trap reaction:

    - 20 mM Tris–HCl (pH 8.3)
    - 1.5 mM $MgCl_2$
    - 63 mM KCl
    - 0.005% Tween-20
    - 1 mM EGTA
    - 50 µM Deoxynucleotide triphosphates
    - 0.1 µg of TS oligonucleotide
    - 1 µg of T4g 32 protein (Boerringer–Mannheim)
    - Bovine serum albumin at final concentration of 0.1 mg/ml
    - 2U of Taq polymerase
    - 2 µl of CHAPS cell extract
    - 0.4 µl of $\alpha - {}^{32}P$ dCTP

11. Incubate for 10 min at 23 °C to allow extension of oligonucleotide TS by telomerase.
12. Transfer to thermal cycle for 27 rounds at 94 °C for 30 sec, 50 °C for 30 sec and 72 for 1.5 min.

13. The CX primer is liberated when the barrier melts.
14. Analyze reaction by electrophoresis in 0.5X Tris borate EDTA on 15% polyacrylamide nondenaturing gels.

Physical Characteristics of Transformed Cells

Loss of Contact Inhibition. Transformed cells exhibit a number of physical properties that distinguish them from their normal counterparts. As mentioned previously, transformed cells lose the normal density dependent inhibition of cell growth; in other words, transformed cells continue to divide although they have reached a confluent monolayer. This results in the formation of heaps of cells that can be seen as dense colonies or foci. In addition, some human tumor cells can form colonies on confluent layers of normal cells (47). The counting of visible colonies allow one to quantitate the degree of transformation by the number of dense colonies formed per plate, starting with a constant number of cells for each dish (sometimes referred to as the plating efficiency) (48). Thus, loss of contact inhibition can be measured in an assay called a colony forming (focus) assay which is simply a measurement of the frequency or degree of loss of contact inhibition of a particular group of cells (Experiment 4.1).

Experiment 4.1: Focus Assay (Cell Overgrowth Assay). Purpose: To quantify the number of contact-uninhibited colonies arising in a population of transformed cells.

1. Trypsinize a flask of cells and plate 5×10^5 cells on each of three 10-cm dishes in media supplemented with calf serum. Smaller dishes (e.g., 6 cm) may also be used.
2. Maintain the cultures in the incubator for 4 weeks, changing the media every 3–4 days. To change the medium aspirate off the medium, being careful not to disturb the cell layer with the tip of the pipette or the stream of new media. Add 10 ml of fresh medium gently with a wide-bore pipette, holding the tip of the pipette against the inner surface of the side of the dish and slowly let the medium run down the side of the dish.
3. Examine the dishes with the naked eye and under low-power magnification for piles of cells at 2, 3, and 4 weeks, respectively. Foci are areas where transformed cells pile up on top of each other rather than growth arresting at confluence. Thus foci are generally seen as dense, heavily stained areas built up upon the monolayer.
4. Alternatively the dishes can be stained with 1% Giemsa stain (see the following) in ethanol and counted or photographed.

Staining Dishes with Giemsa Stain
1. Aspirate off the medium, rinse cells twice with PBS, and add 1–2 ml of Giemsa stain (1% weight/volume in 100% ethanol).
2. Rotate to cover the dish surface.

3. Allow stain to sit for about 30 sec and then pour off.

4. Wash in a sink under a gentle stream of water.

5. Set tilted plates on paper towels to air-dry.

Anchorage-Independent Growth in Semisolid Media. The loss of the requirement for attachment to the tissue culture substratum is a frequently noted in vitro characteristic of malignant transformation. This fundamental change of cellular growth requirements, termed *anchorage-independent growth*, applies to transformed epithelial cells and fibroblasts only, since hematopoietic cells do not normally need a substratum to grow on. For many years, it has been known that transformed cells lose cell surface expression of fibronectin (49,50). Subsequently, a large family of cell adhesion molecules (CAMs) has been identified that exhibit deregulated expression in transformed cell types (51). More recently, anchorage-dependent growth has been demonstrated to be mediated by activation of integrin-dependent cell signaling pathways (52). Thus anchorage-independent growth appears to result from cellular mutations affecting the physical relationship between a transformed cell and its extracellular environment. Anchorage-independent growth can be assessed in the laboratory by growing cells in a semisolid medium combining tissue culture media with soft agarose at an elevated temperature and then allowing the medium to solidify at a lower temperature. The soft agarose conditions support the growth of colonies of dividing transformed cells while untransformed cells are unable to proliferate (see Fig. 3). Although anchorage-independent growth was once thought to correlate directly with the ability of transformed cells to grow in nude mice (53), the correlation is not perfect. For example, reintroduction of the wild-type Von Hippel Landau (VHL) tumor suppressor protein into a renal cell carcinoma cell line lacking a wild-type VHL protein has no demonstrable effect on anchorage-independent growth but inhibits these cells' ability to form tumors in nude mice (54). Second, some untransformed adherent cells can form colonies in suspension (55). Third, not all human tumors can form anchorage-independent colonies despite their clearly malignant phenotype in vivo (56). Differences in the nature of the semisolid support may also be important in determining the ability of cells to grow in suspension (for example, agar versus agarose) (57). Despite these limitations, soft agarose colony formation remains a useful assay for assaying in vitro transformation by selected oncogenes and tumor viruses.

Experiment 4.2: Soft Agarose Assay. Purpose: To measure the ability of transformed cells to form anchorage-independent colonies of proliferating cells.

Materials

• Mix a 1.4% NuSieve Agarose (FMC Biochemicals) solution in dH$_2$O. On the day of the experiment, autoclave to sterilize. Remove from autoclave, swirl gently and place agarose in 50 °C water bath.

• Prepare 2X DMEM with 2X additives (antibiotics and serum should be double concentrated).

(a)

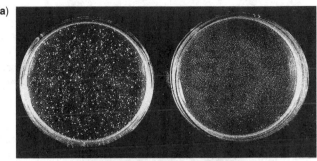

(b)

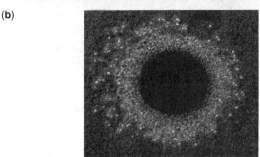

Figure 3. Anchorage-independent growth assay. A soft agarose assay can be used to quantify anchorage-independent growth by transformed cells. Rat1a cells (a rat embryonic fibroblast cell line) can be transformed by overexpression of c-Myc alone. (**a**) A soft agarose containing tissue culture plate seeded with c-Myc-overexpressing cells and allowed to incubate for 3 weeks. On the right, an identical plate was seeded with the parent Rat1a cells and shows no evidence of colony formation. (**b**) Magnification of a single transformed colony shows the symmetrically spherical growth pattern common to transformed cells grown in semisolid media.

Protocol

1. Mix 1.4% autoclaved Sea-Kem agarose with an equal volume of prewarmed 37 °C 2X DMEM/FCS to make 0.7% agarose/1X DMEM solution. Use immediately.

2. Place 5 ml of this mixture in each 60-mm plate. Work quickly so that mixture does not harden.

3. Allow to solidify at 4 °C.

4. Trypsinize enough cells to plate 1×10^4 cells/plate in a single cell suspension.

5. Count cells and resuspend single cells in desired media to allow 1×10^4 cells/ml. Keep warm at 37 °C.

6. Combine resuspended cells with an equivalent amount (2 ml/plate) of 0.7% agar/DMEM to make a 0.35% cell/agarose solution and quickly aliquot 4 ml of this mixture per dish into dishes filled with 0.7% agarose base. This will give 2×10^4 cells/plate.

7. Incubate at 37 °C. Do not move dishes for one day to allow top layer of agarose to semisolidify.

8. Feed each plate gently with 2 ml of $1 \times$ DMEM/FCS twice weekly to prevent plates from drying out.

9. After 3 weeks of growth, remove plates from incubator and place on overhead projector to project colonies onto screen. Colonies will appear as dark shadows on the screen.

10. Count colonies on screen using predetermined size parameters to classify colony size (for example, small, <2 mm; medium, 2–5 mm; and large, >5 mm).

11. Determine average number of anchorage-independent colonies per plate and compare experimental groups.

12. Individual colonies can be plucked with a sterile Pasteur pipette and released into media without agar. Cells will eventually adhere to plastic dish and can be propagated as clonal cell lines.

Growth Factor Requirements

Transformed cells have a reduced requirement for growth-factor-dependent proliferation compared with untransformed cells (58). This is likely due to the fact that transformed cells have constitutive activation of cell signaling pathways from growth factor receptors, either due to mutations in the pathway components or the expression of transforming proteins that activate these components. Thus lower concentrations of growth factors are needed to retain cell viability and proliferation in transformed cells compared with normal cells (59). This reduced requirement for growth factors can be quantified in an experiment in which varying numbers of cells are grown at limiting growth factor concentrations, and then cell numbers are counted periodically to measure proliferation. Cell lines can be grown in defined medium supplemented with increasing concentrations of growth factors. The cell densities and doubling times can then be measured for a defined period of time. This procedure is described in Experiment 4.3.

Experiment 4.3: Determination of Doubling Time and Saturation Density Using Adherent Cell Cultures.

Purpose: To measure growth characteristics of transformed cells.

1. Trypsinize and count cells from a subconfluent culture. Bring cells to a concentration of 10^4 cells/ml in tissue culture medium.

2. Place 2 ml (2×10^4 cells) of the preceding cell suspension into each of approximately 10–16 small (35-mm) tissue culture dishes.

3. Each day, trypsinize (see procedure to follow) and count the cells in duplicate dishes, using either a Coulter Counter or hemacytometer. Record the counts and the volume of the cell suspension counted (this will be 1 ml if the trypsinization procedure to follow is followed). Calculate the total cell number in the dish as follows: cell number (per dish) = cell count (cells/ml) × volume (ml) of cell suspension.

4. On the third day of the experiment, feed the remaining dishes by removing the medium and placing fresh medium onto the cell monolayer.

5. Terminate the experiment when daily cell counts begin to decrease.

6. Plot the data on semilog graph paper with time (days) on the x axis and total cell number on the y axis (the y axis is on a log scale).

7. Calculate doubling time from the linear portion of the curve as follows:

$$\text{doubling time} = \frac{(0.301)t}{\log_{10} N_2 - \log_{10} N_1}$$

where N_1 is the cell number at the beginning of the linear portion of the curve; N_2 the cell number at the end of the linear portion of the curve; and t the time interval between the two observations (N_1 and N_2).

8. Calculate saturation density from the peak of the curve, as follows: saturation density = (number of cells at the peak of curve)/8.04 cm² where: 8.04 cm² is the area of a 35-mm tissue culture dish. Note that this area may differ slightly with different brands of dishes.

To Trypsinize Cells for Determination of Cell Count by Coulter Counter

• Wash cell monolayer twice with 0.5 ml trypsin/EDTA.

• Add 0.5 ml trypsin/EDTA and allow cells to detach (check under microscope). Detachment can be facilitated by placing cells at 37 °C for a few minutes, by tapping the plate, or by flushing the monolayer by pipetting up and down.

• Stop the action of the trypsin by adding 0.5 ml of serum-containing medium (final volume = 1.0 ml).

• Triturate (i.e., pipette up and down) carefully but well using a 1-ml pipette. Isoton from the counting vial (step f) can be added to facilitate triturating.

• Place a drop of the cell suspension on a slide, and check under a microscope to make sure that a single cell suspension has been achieved.

• Transfer the entire cell suspension to counting vial containing 9.0 ml Isoton (1 : 10 dilution). Count using a Coulter Counter.

Formation of Tumors in Nude Mice

The growth of transformed cells in immunologically compromised mice allows an in vivo assay for cell transformation and invasiveness into surrounding stroma and is an important aspect of cellular transformation (60). The genetically athymic "nude" mouse (61) is commonly used to assay transformation in vivo (Fig. 4). Formation of tumors in nude mice does not always correlate perfectly with other in vitro assays of transformation (54), but it still represents a valuable assay for comparing the effect of various treatments or ectopically expressed genes on different cell lines.

Experiment 4.4: Tumor Formation in Nude Mice.

Purpose: To compare the ability of cells to grow (malignant potential) as independent tumors in immunocompromised mice.

1. Grow subconfluent cultures in optimal conditions to achieve exponential cell growth.

2. Trypsinize cells and keep on ice until injection.

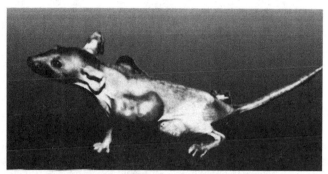

Figure 4. Transformed cell line growth in nude mice. A nude mouse is seen 4 weeks following subcutaneous injection with 1×10^6 PC-3 (a human prostate carcinoma) cells in each flank. Large subcutaneous tumors appear as shown. (Courtesy of Dr. Shutsung Liao, University of Chicago.)

3. Dilute in PBS or media without FCS to a concentration of 3×10^6/ml. Inject female nude mice (4–6 wk old) subcutaneously with 0.2 ml of the cell suspension. Control mice should receive PBS or media alone.

4. After 4 weeks, sacrifice mice and measure size of subcutaneous tumor masses in three dimensions.

5. Alternatively, tumors can be excised and weighed for comparison.

Glucose Uptake, Transport, and Metabolism in Transformed Cells. Increased uptake and consumption of glucose is known as a characteristic feature of transformed cells (62). The increased uptake of glucose may be related to quantitative as well as qualitative changes of four glucose transporter proteins (Glut 1–4) (63). It is still unknown exactly how the altered expression of glucose transporters (known as Glut proteins) is linked to the process of transformation (64). The level of glucose uptake of transformed cells can be assayed by measuring the uptake of a radiolabeled glucose analogue 2-deoxy-D-[^{1,3}H] glucose in the media of actively growing transformed cells. The radioactivity incorporated can be measured by lysing the cells and measuring the tritiated glucose by liquid scintillation spectroscopy (see Experiment 5.1).

Glucose and glutamine serve as major energy sources for most in vitro cultivated cell lines (65). Increased channeling of glucose from glycolysis into the pentose phosphate pathway has been reported for transformed cells. In addition, transformed cells are able to produce lactic acid aerobically, while normal cells produce most lactic acid anaerobically (66). The ability of transformed cells to undergo aerobic glycolysis was recognized almost seven decades ago by Warburg, although the molecular mechanism has remained elusive. One clue to the mechanism of aerobic glycolysis is that many tumor cells overproduce the lactate dehydrogenase A gene and produce concomitant high levels of this protein. This may be the result of relative hypoxia, since LDH is a hypoxia-induced gene (67). Lactic acid production can be assayed for various cell lines and compared. Transformed cells often produce significantly more lactate than normal cells.

Experiment 5.1: Measurement of Glucose Transport by Cells. Purpose: To measure the rate of glucose uptake in transformed versus normal cells.

Materials

- Serum-free media
- PBS with 0.2% bovine serum albumin (BSA)
- 2–^{3}H deoxy-D-glucose (2-DOG) (Amersham)
- 1 μM porcine insulin (Hoechst) if stimulation experiments done

Protocol

1. Trypsinize stock plates to be tested for glucose uptake.
2. Plate 10^5 cells in 24-well dishes for 24 h.
3. Preincubate with serum-free medium for 2 h and wash with PBS (0.2% bovine serum albumin).
4. Add 1 μCi -^{3}H deoxy-D-glucose in PBS +1% BSA.
5. Allow 2-DOG uptake for 10, 20, and 30 min.
6. Wash cells with PBS at various time points to remove DOG.
7. Solubilize the cells with 0.2% SDS in PBS at 10, 20, 30 min.
8. Count radioactivity of an aliquot of lysate in a β-scintillation counter.
9. Determine protein content per sample using a BCA protein assay (Pierce).
10. Determine DOG transport by calculating mMoles of DOG per μg of protein assayed. Plot DOG transport versus time.
11. To assess the integrity of insulin responsive uptake in transformed cells, stimulate with 1 μM insulin × 30 min and repeat preceding procedure.

Experiment 5.2: Assay for Lactate Production. Purpose: To measure the amount of lactate produced by transformed cells.

1. Wash cells to remove serum and incubate overnight in serum-free medium.
2. Replace with sodium phosphate containing buffer supplemented with 20 mM glucose.
3. After 1 h incubation measure lactate levels by quantifying lactate-dependent reduction of nicotine adenine dinucleotide in the presence of lactate dehydrogenase and quantifying by measurement of OD at 339 nm.
6. Trypsinize monolayer, count cells, and normalize lactate determinations to cell number (mmol of lactate/10^5 cells/h).

ACKNOWLEDGMENTS

I am indebted to Ruth Craig for generous help in identifying protocols. I also thank Brad Arrick for critical review of the manuscript, Rafael Espinosa for expert help with illustrations, and Michelle Daoust for skilled secretarial assistance.

BIBLIOGRAPHY

1. C.C. Harris, *Cancer Res.* **51**, 5023s–5044s (1991).
2. G. Todaro and H. Green, *J. Cell Biol.* **17**, 299–313 (1963).
3. I. Macpherson and L. Montaigner, *Virology* **23**, 291–294 (1964).
4. L. Hayflick and P.S. Moorhead, *Exp. Cell Res.* **25**, 585–621 (1961).
5. G. Evan and T. Littlewood, *Science* **28**, 1317–1322 (1998).
6. J. Rygaard and C.O. Povlsen, *Acta Pathol. Microbiol. Scand.* **77**, 758–760 (1969).
7. O. Warburg, *The Metabolism of Tumors*, Constable, London, 1930.
8. J. Nerrman and R. Wagner, *J. Cell. Physiol.* **166**, 152–169 (1996).
9. H.M. Temin and H. Rubin, *Virology* **6**, 669–688 (1958).
10. J.D. Rosenblatt, S. Miles, J.C. Gasson, and D. Prager, *Curr. Top. Microbiol. Immunol.* **193**, 25–49 (1995).
11. P.N. Tsichlis and P.A. Lazo, *Curr. Top. Microbiol. Immunol.* **171**, 95–171 (1991).
12. W.S. Pear, G.P. Nolan, M.L. Scott, and D. Baltimore, *Proc. Natl. Acad. Sci. U.S.A.* **90**, 8392–8396 (1993).
13. A.J. Levine, *Viruses*, Scientific American Library, New York, 1990.
14. M. Hatakeyama and R.A. Weinberg, *Prog. Cell. Cycle Res.* **1**, 9–19 (1995).
15. S.H. Friend et al., *Nature (London)* **323**, 643–646 (1986).
16. S. Chellappan et al., *Proc. Natl. Acad. Sci. U.S.A.* **15**, 4549–4553 (1992).
17. J.M. Nigro et al., *Nature (London)* **7**, 705–708 (1989).
18. D.P. Lane, *Nature (London)* **358**, 15–16 (1992).
19. S.D. Conzen and C.N. Cole, *Semin. Virol.* **5**, 349–356 (1994).
20. C.N. Cole, in B.N. Fields, D.M. Knipes, and P.M. Howley, eds., *Fields Virology*, 3rd ed., Lippincott-Raven, New York, 1995, pp. 1997–2026.
21. K. Rundell, S. Gaillard, and A. Porras, *Dev. Biol. Stand.* **94**, 289–295 (1998).
22. R. Eckner et al., *Mol. Cell. Biol.* **16**, 3454–3464 (1996).
23. J.A. DeCaprio et al., *Cell* **54**, 275–283 (1988).
24. J. Zalvide and J.A. DeCaprio, *Mol. Cell. Biol.* **15**, 5800–5810 (1995).
25. D.C. Swan, S.D. Vernon, and J.P. Icenogle, *Arch. Virol.* **138**, 105–115 (1994).
26. T.A. Van Dyke, *Semin. Cancer Biol.* **5**, 47–60 (1994).
27. P. Naik, J. Karrim, and D. Hanahan, *Genes Dev.* **1**, 2105–2116 (1996).
28. P.A. Furth, *Dev. Biol. Stand.* **94**, 281–287 (1998).
29. P.C. Nowell and D.A. Hungerford, *Science* **132**, 1497 (1960).
30. F. Mitelman, *Catalogue of Chromosome Aberrations in cancer*, 5th ed., Wiley-Liss, New York, 1994.
31. F. Mitelman and J. Mark, *Hereditas* **65**, 227–235 (1970).
32. T. Caspersson, G. Gahrton, J. Lindsten, and L. Zech, *Exp. Cell Res.* **63**, 238–240 (1970).
33. S. Heim and F. Mittelman, *Nonrandom Chromosome abnormalities in Cancer-an overview in Cancer Cytogenetic — An Overview*, 2nd ed., Wiley-Liss, New York, 1995 pp. 19–32.
34. J.S. Ross, *Oncology* **10**, 867–882 (1996).
35. T. Cremer et al., *Hum. Genet.* **80**, 235–246 (1988).
36. R.S. Verma and A. Babu, *Human Chromosomes. Manual of Basic Techniques*, Pergamon, New York, 1989.
37. R. Crowe, H. Ozer, and D. Rifkin, *Experiments with Normal and Transformed Cells*, Cold Spring Harbor Lab., Cold Spring Harbor, N.Y., 1978.
38. R. Czaker, *Humangenetik* **19**, 135–144 (1973).
39. M. Seabright, *Lancet* **1**, 1249–1250 (1973).
40. L. Rodgers, *Flow Cytometry in Cells: A Laboratory Manual*, Vol. 1, Cold Spring Harbor Lab., Cold Spring Harbor, N.Y., 1997, pp. 1–12.
41. E.H. Blackburn, *Nature (London)* **350**, 569–573 (1991).
42. C.W. Greider and E.H. Blackburn, *Nature (London)* **337**, 331–337 (1989).
43. R.C. Allsopp et al., *Exp. Cell Res.* **220**, 194–200 (1995).
44. O. Yamada, *Int. J. Hematol.* **64**, 87–99 (1996).
45. S. Kyo, M. Takakura, T. Kohama, and M. Inoue, *Cancer Res.* **15**, 610–614 (1997).
46. N.W. Kim et al., *Science* **23**, 2011–2015 (1994).
47. S.A. Aaronson, G.J. Todaro, and A.G. Freedman, *Exp. Cell Res.* **51**, 1–5 (1970).
48. G.J. Todaro and H. Green, *Virology* **23**, 117–119 (1970).
49. A. Vaheri, and D.F. Mosher, *Biochim. Biophys. Acta* **18**, 1–25 (1978).
50. S.L. Dalton, E.E. Marcantonio, and R.K. Assoian, *J. Biol. Chem.* **25**, 8186–8191 (1992).
51. T.J. Baldwin, M.S. Fazeli, P. Doherty, and F.S. Walsh, *Cell. Biochem.* **15**, 502–513 (1996).
52. M.A. Schwartz, D. Toksoz, and R. Khosravi-Far, *EMBO J.* **2**, 6525–6530 (1996).
53. V.H. Freedman and S. Shin, *Cell* **3**, 355–359 (1974).
54. O. Iliopoulos, A. Kibel, S. Gray, and W.G. Kaelin, Jr., *Nat. Med.* **1**, 822–622 (1995).
55. W.E. Laug, Z.A. Tokes, W.F. Benedict, and N. Sorgente, *J. Cell Biol.* **84**, 281–293 (1980).
56. A.W. Hamburger and S.E. Salmon, *Science* **197**, 461–463 (1977).
57. A.I. Neugut and I.B. Weinstein, *In Vitro* **15**, 351–355 (1979).
58. H.M. Temin, *J. Natl. Cancer Inst. (U.S.)* **37**, 167–175 (1966).
59. D. Greenwood, A. Srinivasan, S. McGoogan, and J.M. Pipas, in R. Baserga, ed., *Cell Growth and Division: A Practical Approach*, Oxford University Press, Oxford, U.K., 1989, pp. 37–59.
60. M.M. Tomayko and C.P. Reynolds, *Cancer Chemother. Pharmacol.* **24**, 148–154 (1989).
61. B.C. Giovanella, J.S. Stehlin, and L.J. Williams, *J. Natl. Cancer Inst. (U.S.)* **52**, 921–930 (1974).
62. G. Weber, *Cancer Res.* **43**, 3466–3492 (1983).
63. C. Binder et al., *Anticancer Res.* **17**, 4299–4304 (1997).
64. D.J. Templeton and R.A. Weinberg, In A.I. Holleb, D.J. Fink, and G.P. Murphy, eds., *American Cancer Society Textbook of Clinical Oncology*, American Cancer Society, Atlanta, Ga., 1991, pp. 678–690.
65. K.W. Lanks and P.W. Li, *J. Cell. Physiol.* **135**, 151–155 (1988).
66. R.J. Resnick, R. Feldman, J. Willard, and E. Racker, *Cancer Res.* **46**, 1800–1804 (1996).
67. C.V. Dang et al., *Bioenerg. Biomembr.* **29**, 345–354 (1997).

See also CELL AND CELL LINE CHARACTERIZATION; CHARACTERIZATION OF CELLS, MICROSCOPIC; ENRICHMENT AND ISOLATION TECHNIQUES FOR ANIMAL CELL TYPES; GENETIC ENGINEERING: ANIMAL CELL TECHNOLOGY; TRANSFORMATION OF PLANTS.

cGMP COMPLIANCE FOR PRODUCTION ROOMS IN BIOTECHNOLOGY: A CASE STUDY

Claude Artois
Jean Didelez
Patrick Florent
Guy Godeau
SmithKline Beecham Biologicals
Rixensart, Belgium

OUTLINE

Viral vaccine development began approximately fifty years ago with virus propagation in stationary cell culture. It was the ability to grow human virus outside a living host that led to the development of viral vaccines. This article describes design considerations for an inactivated viral vaccine bulk manufacturing facility. Many of the design features described in the text may also be applicable to manufacturing other biological products. The drawings are published by the courtesy of Dr. C. Vandecasserie, SmithKline Beecham Biologicals, Rixensart, Belgium. The authors thank Mrs. Doris L. Conrad and Dr. Michael Korcyznski for their constructive comments. They were beneficial in preparing of this article.

The design of the manufacturing areas and the HVAC system is based on the concepts and technical data described in the section "General Points for Facility Design" and the section "Design of HVAC System parameters." The facility design could fit the requirements of various production processes involving cell culture, viral replication and viral inactivation, for instance, hepatitis A or noninfectious polio vaccines. The facility can accept a multistrain manufacturing strategy, as described later. The facility contains all of the equipment and support services necessary for cell culture, viral culture, purification, and inactivation of the vaccine. In the design of an industrial-scale facility for cell culture, viral replication, purification, and inactivation operations, engineering, manufacturing, maintenance, validation, quality control, and quality assurance disciplines are joined to establish a satisfactory production and control system. Due to the different perspectives and priorities of vendors, contractors, design engineers and users, some of the operational, safety, and quality requirements may be overlooked.

In addition, both facility and equipment designs should meet operator expectations and GMP requirements. The system features must aid the operator in achieving training, operational reproducibility, safety, product quality, and maintenance requirements.

This chapter attempts to provide the reader with a global view of design considerations. Emphasis is put on HVAC parameters and operational flows which satisfy cGMPs (1,2), regulatory (3,4), and biosafety requirements (5). The design considerations are illustrated by using an example of an inactivated viral bulk manufacturing facility that combines three major biotechnological challenges: cell culture, aseptic operations, and biosafety level 3 associated with handling pathogenic viruses.

GENERAL POINTS FOR FACILITY DESIGN

Points to Consider

Primary manufacturing of biotechnological products generally involves operations starting from cell culture, including biological expression of the product, extraction, and purification. A recombinant cell culture or viral replication system usually includes one or more viral clearance steps to ensure that product safety requirements are met.

In addition, a final sterile filtration step is usually included in the manufacture of the biological product to obtain a sterile product. The sterile filtration step may or may not be part of the primary manufacturing operations depending on the type of product, its final use, and the organization of the whole manufacturing process.

To comply with GMPs, each manufacturing area is designed keeping in mind the following biopharmaceutical facility features:

- separate access for personnel and raw materials
- separate exit for bulk product
- dedicated HVAC units
- dedicated water systems
- need for containment, decontamination, and waste handling
- dedicated multiuse or multistrain manufacturing strategy

Cell cultures are lengthy and highly susceptible to contamination. Viral replication offers an additional challenge, especially when a live pathogenic virus is handled. Segregation of areas, dedicated HVAC units, material decontamination, and effluent treatment systems represent additional engineering requirements that must comply with the various applicable codes.

The codes for Good Manufacturing Practices applicable to the manufacture of biological products are specific:

- CFR 21 Parts 600 "Biological Products" (United States).
- Annexe 2 of European Union, code of Good Manufacturing Practice for Medicinal Products:
  Manufacture of Biological Medicinal Products for Human Use.

If a multiuse or multiproduct manufacturing strategy is developed, then the design features will incorporate the principles described in the codes mentioned. The section "Dedicated, Multiuse or Multistrain Manufacturing Strategy ?" describes how the manufacturing strategy impacts the design of the facility.

Sterile filtration involves all requirements associated with drug aseptic operations. Guidelines and regulatory requirements for Good Manufacturing Practices applicable to the manufacture of sterile drug products are also specific (see references at end of chapter). Facility design philosophy should attain primary containment in the production vessels and transfer piping by using closed systems. This maintains containment and protects the environment, personnel, and product. The piping slopes, welds, joints and seal materials used throughout the system are designed to eliminate leakage and maintain overpressure to assure sterility of the system. Sanitary valves need to be welded.

Secondary containment is achieved by maintaining the rooms, where live pathogenic viruses are handled, under negative pressure. Refer to the section "Design Features and Safety Precautions for Personnel Working in BL-2 and BL-3 Areas."

Sterility of cell culture or aseptic operations performed in open systems should be accomplished by using laminar air flow (Class 100) by personnel appropriately dressed and trained in aseptic operations and techniques.

The authors of this article propose a room class description (Table 1) that complies with EU GMP Annex 1 "Manufacture of Sterile Medicinal Products (January 1997)" and U.S. FDA Guidelines for Drug Aseptic Processing and U.S. Federal Standard 209E (6).

Figure 1 illustrates the personal, materials, and product flow concepts used to design the inactivated viral vaccine bulk manufacturing facility. Figure 2 provides a view of the layout of this facility.

Table 1. A Guide to Classify a Series of Processing Rooms Commonly Found in a Biopharmaceutical, Cell-Culture, Production Facility[a]

EU Class	US Class	Production Area Classification				Microbial Requirements	
		Static	Dynamic	Type of Operation		per (m^3)	per ($10\,ft^3$)
A	100	100	100	Aseptic operation during which sterile product is exposed to the environment		<1	<0.3
B	10 000	100	10 000	Aseptic area surrounding a class 100 A. Room pressure is positive versus adjacent rooms of lower cleanliness level		<10	<3
B	10 000	1000[b]	10 000	Aseptic area surrounding a class 100 A. Room absolute pressure is negative due to the handling of pathogenic microorganisms. The room is used for aseptic bulk cell culture and for viral replication.		<18	<5
C	100 000	10 000	100 000	Fermentation/Extraction, nonsterile media preparation, purification of antigens, capping of filled containers, storage of clean and sterilized Materials		<85	<25
D	100 000	100 000	100 000	Materials wrapping room equipped with a laminar flow hood Any room used for sterilization of equipment and materials		<85	<25
Unclassified	Unclassified	100 000	NA[c]	Materials Washing Room, nonsterile corridor, office, in-process control laboratory, visual inspection room, storage of materials, Product, and equipment		NA[c]	NA[c]

Notes: [a]Areas are classified under static and dynamic conditions. Static condition: the room classification is determined at rest with all equipment in operation but no personnel inside the area. Dynamic condition: the room classification is determined while equipment is in operation and personnel perform production operations.

[b]The authors propose a class 1000 for a room kept under negative pressure. Acceptance of this concept will depend on the risk analysis that the designer of a new biotechnology facility performs during the conceptual study phase.

[c]Not applicable.

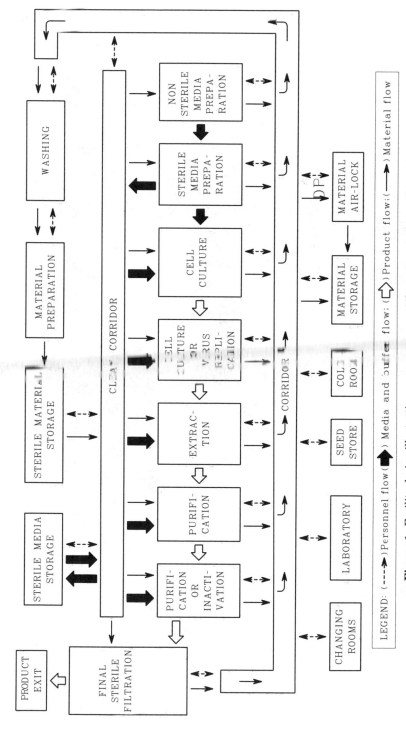

Figure 1. Facility design illustrating personnel, materials, and product flow Concepts.

LEGEND: (- - - ➤) Personnel flow (➤) Media and buffer flow; (⇧) Product flow; (➤) Material flow

521

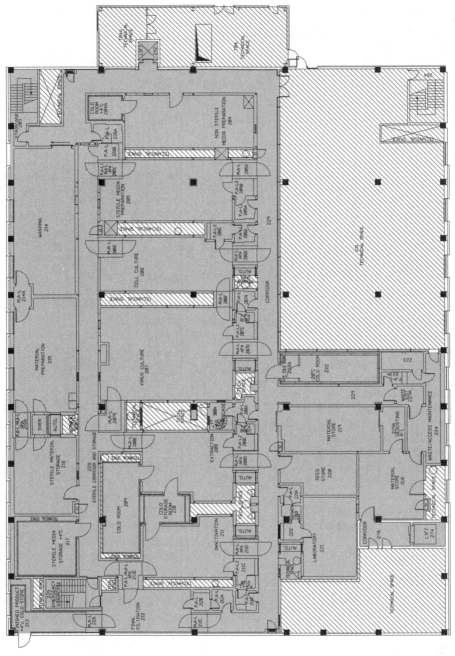

Figure 2. General layout.

INACTIVATED VACCINE MANUFACTURING AREA

TECHNICAL SPACE

Room Class Descriptions. In a pharmaceutical environment, room air pressures and directional air flows are designed and balanced to create a positive pressure cascade from more controlled or critical areas to less controlled areas (7,8). Rooms are classified according to their level of cleanliness (6–8).

In a biopharmaceutical environment where live pathogenic organisms are manipulated, HVAC systems should be segregated for production areas according to biosafety (5) and cGMP (3,4) requirements. HVAC systems should be segregated to prevent recirculation of air between these designated production areas (9). Airlocks are widely used to meet biosafety requirements and to maintain an aseptic environment. Rooms might be kept at negative pressure with respect to adjacent rooms. Exhaust air is HEPA filtered.

Personnel Flow

Personnel should access the facility through one general entrance. The facility should be designed so that rest rooms, ladies' and gentlemen's changing rooms, and administrative offices are located outside of the Production core. Entrance to the facility should be secured and limited to authorized personnel. This is usually achieved by an electronic lock requiring a coded badge entry. Personnel enter the production core through access to changing rooms. Three changing rooms will meet requirements for female personnel, male personnel, and visitors. Entrance to changing rooms should also include electronic locks requiring coded badge entry.

Before entering the production core, production personnel should remove their city garments and put on production garments. Personnel can then access various process rooms including aseptic production rooms. Personnel should move between areas of a biofacility according to detailed procedure established in compliance with biological regulatory requirements (10,11). Before entering aseptic rooms, production operators enter the gowning room (Personnel AirLock) where they remove their production garments and put on sterile gown, sterile boots, sterile hood, sterile gloves, and a mask or helmet (10).

Figure 2 illustrates the personnel flow in the viral vaccine bulk manufacturing facility. Personnel enter the building on level 1 (ground floor) where three changing rooms are available. Personnel access level 2 using a staircase or an elevator.

Design Features and Safety Precautions for Personnel Working in BL2 and BL3 Areas

Four biosafety levels (BL) are recommended for activities involving infectious microorganisms. The main features of the levels BL2 and BL3 are summarized in Table 2. This information is directly extracted directly from the publications of the U.S. Centers for Disease Control and Prevention (CDC) and the U.S. National Institutes of Health (NIH).

The features that were employed to meet the BL-3 requirements for the live virus manipulation areas of the inactivated virus vaccine facility are described as follows:

1. Primary barriers

   - The access to live virus manipulation areas is limited to personnel involved in production operations and supervision. Each entry into BL-3 area is recorded on a checklist. The use of an electronic controlled access system for each live

Table 2. Summary of Recommended Criteria for Biosafety Levels 2 and 3

Biosafety Level	Agents	Practices	Safety Equipment (Primary Barriers)	Facilities (Secondary Barriers)
2	Associated with human disease hazard comes from autoinoculation, ingestion, mucous membrane exposure	Standard microbiological practices plus • biohazard warning signs. • "sharps" precautions. • biosafety manual defining any needed waste decontamination or medical surveillance policies	Primary barriers involving Class 1 physical containment devices used for all manipulations of agents that cause splashes or aerosols of infectious materials; personnel protection equipment involving protective lab clothing, gloves, respiratory protection as needed	Open bench top sink required plus autoclave available
3	Indigenous or exotic agents with potential for aerosol transmission; disease may have serious or lethal consequences.	BL-2 practice plus • controlled access. • physical • decontamination of all waste • decontamination of lab clothing before laundering • baseline serum	Similar to BL-2	BL-2 plus • Physical separation from access corridors • self-closing double door access • exhausted air not recirculated • negative airflow into laboratory

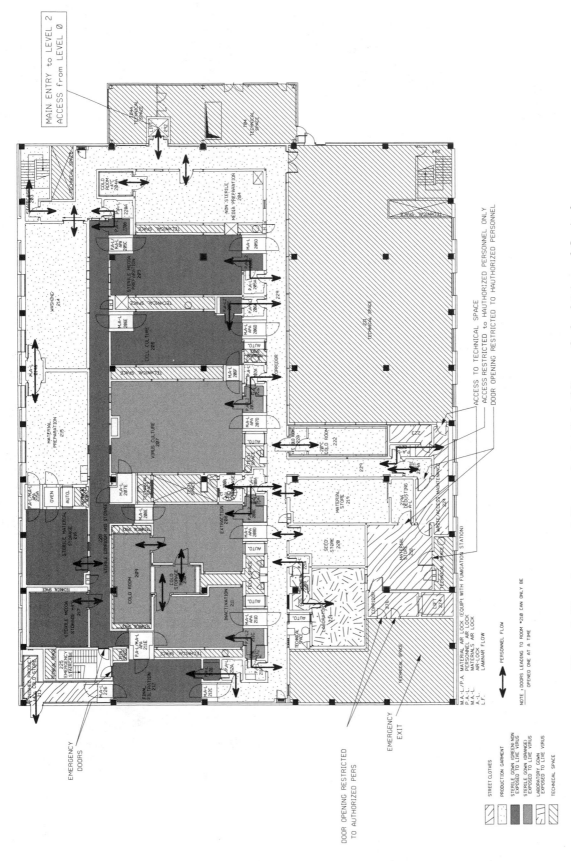

Figure 3. Flow of personnel. Personnel enter the building on the level 1 (ground floor) where three changing rooms are available (Female, male and staff/visitors). The entrances to changing rooms are equipped with electronic locks requiring coded badge entry. Personnel access level 2 using a staircase or an elevator. Before entering aseptic rooms (Sterile Media Preparation, Cell Culture, Virus Culture, Purification, Inactivation, Final Filtration, Sterile Material Storage, and Sterile Media Storage), production operators enter the gowning room (Personnel Air Lock 1) where they remove their blue production garments and put on a color-coded sterile gown, sterile boots, sterile gloves, and a mask.

virus manipulation area was not selected during the detail design phase because it was decided to equip the entrances to the facility and the changing rooms with electronic locks requiring coded badge entry. The facility manager controls access to the areas and restricts access to persons whose presence is required for production purposes.

- Liquid wastes are decontaminated through dedicated effluent decontamination stations.
- Solid materials are decontaminated through double-door autoclaves installed in each live virus manipulation area. The autoclaves are designed to maintain the steam condensates within the autoclave chamber during the decontamination cycle.
- Personnel are gowned with overalls, boots, masks, and hoods which are decontaminated through autoclaves.
- Personnel exposure to live virus is reduced through the use of BL-3 Laminar Air Flow Bench (Cabinet), Closed system (Mobile or fixed tanks), and specific training in the procedures related to the operations.
- Personnel are vaccinated against the live virus manipulated in the facility.
- Floor and works surfaces are sanitized each day using a sodium hypochlorite solution.

2. Secondary barriers (Facility)

- Segregation of air handling units: each live virus manipulation area is equipped with a dedicated air-handling unit.
- Exhausted air is not recirculated. Air supply and return are HEPA filtered.
- Air pressure differentials: the room pressures are negative at 30 Pa or −0.12 inch of water versus atmospheric pressure. Air locks have positive air pressure or are designed as decontamination air locks (see section on "Material Flow").
- Each air locks is equipped with a self-closing double-door access.
- Personnel air locks consist of three physically separate compartments (See section on "Design of HVAC System Parameters")
Personnel air lock, first part: personnel remove production garment. Air pressure is positive (+30 Pa, or +0.12 inch of water).
Personnel air lock, second part: this is equipped with a safety shower. Air pressure is negative (+30 Pa, or +0.12 inch).
Personnel air lock, third part: personnel put overalls under laminar air flow. Air pressure is negative (−15 Pa or −0.6 inch).
- Sealed windows and sealed penetrations.
- HEPA filters are tested every twelve months.

Materials Flow

All raw materials and components should be sampled, tested, and released by the Quality Control Unit prior to entering the production facility. QA/QC released materials should be transferred to Production using a dedicated Material Cardboard Lock. External packaging protection like cardboard and plastic films should be removed in the Air Lock. Cardboard and wood should not be introduced in the Production Core to minimize the risk of microbial, rodent, and insect ingress into the facility. The Production Core should include provisions for raw materials storage at controlled room temperature of 2 to 8 °C or at adequate freezing temperatures. Exit of used raw materials from the Production Core might be achieved using the same Air Lock providing that a time-based segregation procedure is strictly followed by production operators.

Materials that are used in sterile conditions should be prepared and sterilized using validated equipment and procedures. The Code of Federal Regulations 21 CFR 600.11 (12) describes specific requirements for material sterilization procedures used in the manufacture of biological product: "The effectiveness of the sterilization procedure shall be no less than that achieved by an attained temperature of 121.5 °C maintained for 20 minutes by saturated steam." That requirement is higher than the usual rule of 15 minutes at 121.5 °C expected for sterile drug products.

To maintain the sterile status of wrapped materials, the production unit should be designed with a dedicated sterile material storage room. Sterilized equipment and materials should then be taken from the sterile material storage room to the appropriate manufacturing area via a clean corridor and the corresponding Material Air Lock, so-called "Air Lock In."

After use, soiled equipment and material should be appropriately removed from the processing rooms via corresponding Air Locks, so-called "Air Lock Out." Decontamination, if required, should take place in decontamination autoclaves or decontamination material Air Locks. Decontamination Air Locks are used to decontaminate the external surfaces of mobile vessels when they are removed from a live virus manipulation area. After use, the mobile vessel is decontaminated-in-place (DIP) using clean steam. The mobile vessel is then transferred into the decontamination Air Locks in which a fumigation cycle is applied. Fumigation can be carried out using a peracetic acid or hydrogen peroxide generator.

Within the Production Core, materials should follow a unidirectional flow. Figure 4 illustrates the following flow of equipment, materials, and waste in the viral vaccine bulk manufacturing facility.

- Sterilized equipment and materials are taken from sterile material storage room 216 to the appropriate manufacturing area via corridor 228, and the corresponding Material Air Lock.
- The fixed vessels used for non sterile media preparation are cleaned-in-place per the appropriate procedure.
- After use, soiled equipment and material are appropriately removed and decontaminated from the processing rooms via corresponding fumigation Material Air Locks or decontamination autoclave, corridor 229, and are transferred to Washing Room 214.

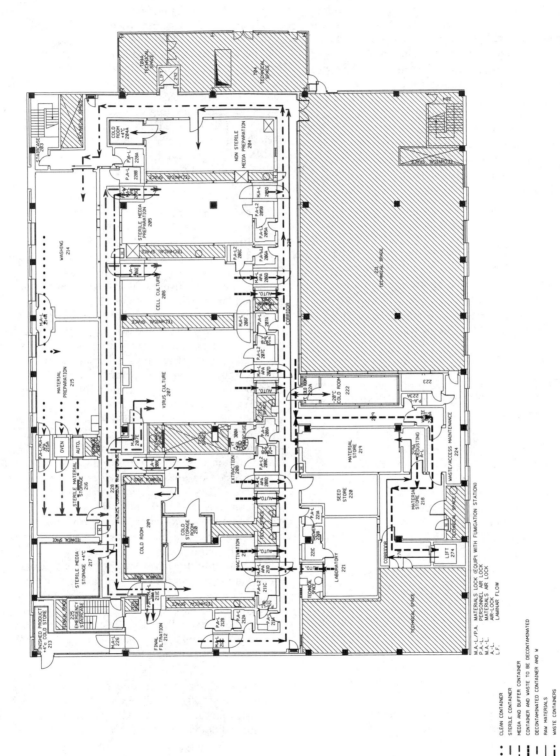

Figure 4. Flow of equipment, materials, and waste.

526

- Materials are washed following appropriate cleaning procedures. Cleaned equipment and materials are then transferred into Materials Preparation Room 215 via Material Air Lock 214A.
- As appropriate, materials and equipment are wrapped, assembled, and sterilized by autoclave, dry heat oven, or in a fumigation air lock (215A) after which they are stored until use in room 216.
- Waste and single-use materials that are decontaminated are discarded via corridor 229, waste exit 229A, Waste/Access Maintenance 224, Material Store 218, corridor 276 and removed from the Production Core using lift 274.

Product Flow

When considering the design of a facility, care should be taken to address the following concerns about product flow:

- dedicated storage equipment for Working Cell Bank and Working Seed.
- Unidirectional product flow from Media and Buffer preparation up to product exit.
- Segregation between areas where live organisms are handled and areas where no live organism is manipulated.
- Dedicated product storage room and product exit.

Dedicated Storage Equipment for Working Cell Bank and Working Seed. Master and Working Cell Banks and/or Seed should be stored in two separate buildings to minimize the risk of loss due to potential disaster. Within each building, care should be taken to store the biological material to prevent any cross-contamination or damage (13). The 21 CFR 600.11 standard states that cold or freezer rooms shall be maintained at temperatures within applicable ranges and shall be free of extraneous material that might affect the safety of the product. An accurate inventory log of the Master and Working Cell Banks and Seed must be maintained at each storage site.

The following practices are frequently observed in industry to prevent cross-contamination and mix-ups:

- Dedicated freezer or nitrogen tank for each type of micro organism strain.
- Dedicated freezer or nitrogen tank for bank and seed lot awaiting QA/QC release (Quarantine).
- Spare freezer ready to use in case of mechanical problems on the main storage units.
- Chemical sanitization of freezer internal chamber after each shutdown and maintenance intervention.
- Limited access to storage units.
- Log book to record entries in and removal from each storage unit.

Media and Buffer Flow. Media and buffers used for cell-culture operations are critical components. Right from the start of their preparation in sampling, weighing, and powder dissolution, strict procedures should be followed to minimize the risk of contamination. Mycoplasma, adventitious viruses, and some bacteria strains may be difficult to detect in cell cultures. Mycoplasma and viruses may not be retained on 0.2-μm sterile filter membranes. Therefore, the design of media and buffers preparation rooms should address the following concerns:

- Limited access to authorized personnel.
- Two-stage media preparation suite: nonsterile media and sterile media preparation areas.
- Specific gowning practices for nonsterile media preparation similar to those followed for aseptic operations.
- After sterile filtration, validated procedures and container/closure integrity systems should be used to protect media and buffers during storage, transfer to, and distribution into the processing rooms.
- Equipment design: media residues accumulate in badly drained or inaccessible locations. Residues dry out and become baked on in decontamination, cleaning, and sterilization procedures. Therefore, design of equipment should include provision for good draining of all equipment parts. Vessels should be made of stainless steel (316L) and should be vented using 0.2-μm air filters. Other materials, single-use irradiated plastic bags, for instance, can be successfully used providing that compatibility between materials and the cell or virus culture has been validated. Single-use sterile plastic bags can be used to store media and buffers. That reduces the need for stainless steel mobile vessels and the need for cleaning capacity associated with CIP stations (clean-in-place). All valves installed on equipment must meet sanitary design requirements. Stationary tanks are cleaned and sterilized using CIP/SIP stations.
- Process design: media and buffers are usually sterilized through 0.2-μm liquid sterile filters. The manufacturer will assess the needs for stationary and mobile tanks. The decision to use installed tanks instead of mobile vessels will be based on the volume of media, buffer, or product transferred between the various processing units. The higher the volume of one specific medium, the higher will be the need for a stationary tank. However, the complexity of CIP/SIP stations for big vessels and multiple transfer piping need to be evaluated against the benefit of using a fixed vessel. Containers of media and buffers that are partially used in areas where live organisms are manipulated should never be transferred back to the central storage room (14).

Product Flow. Within the Production Core, product flow should be unidirectional. Areas where cells are cultured should be segregated from other processing areas. Areas where live viruses are manipulated should also be segregated from other processing rooms. Figure 5 provides a view of the live virus manipulation areas.

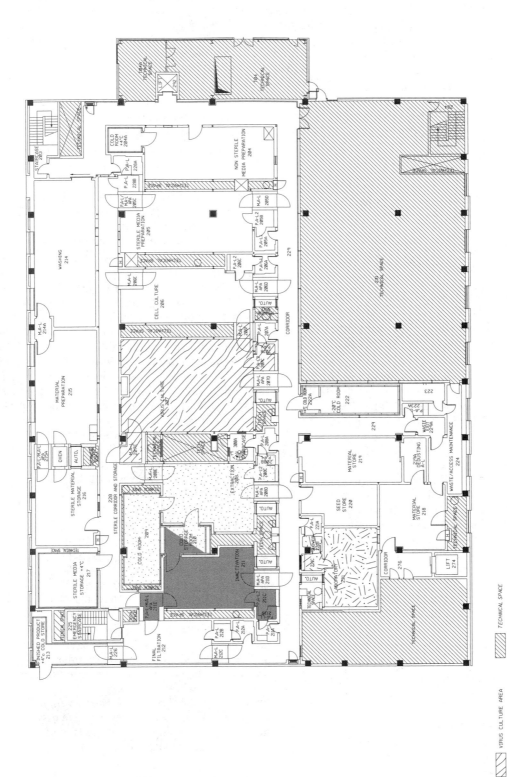

Figure 5. Live virus manipulation areas.

Table 3. Water Systems: Sources and uses of Water

Water Quality	Use	Source	Microbial Limits
Iron-free water	Production of chilling water	Municipal drinking water or treated well water	NA[a]
Purified water	In-feed water for clean steam and water for injection production Rinsing of materials and equipment before washing	Municipal drinking water or treated well water that complies with regulations 40 CFR 141 (National Primary Drinking Water Regulatory)	Less than 10 000 CFU per 100 mL
Water for injection	Cleaning; last rinse water; water for media/buffer preparation.	Purified water	Less than 10 CFU per 100 mL

Note: [a]NA = Not Applicable.

Water Systems

Many of the industry questions about producing water for pharmaceutical purposes were clarified by the publication of the USP Workshop on Microbiology and Pharmaceutical Water in April 1997 (15). The USP publication, Monograph <1231> "Water for Pharmaceutical Purposes," provides guidance for compliance with the current Good Manufacturing Practice (cGMP) regulations for drug products codified under Title 21 of the Code of Federal Regulations (CFR), Food and Drugs, Chapter 1, Parts 210 and 211 (21 CFR 211). Bearing in mind that cell cultures are highly susceptible to contamination, especially viral contamination, Water for Injection should be used at all process stages, including media/buffer preparation. Table 3 provides a summary of water systems commonly found in biotechnology facilities dedicated to cell culture.

Dedicated Multiuse or Multistrain Manufacturing Strategy?

Multiuse Facility. If a multiuse manufacturing strategy is developed, the facility will be designed for successive production of different strains of live organisms. Detailed and validated changeover procedures should then be followed to prepare the next product campaign. Depending on the type of organism handled during the previous campaign, the changeover procedures should address the following concerns:

- Removal of Working Seeds.
- Decontamination of all materials and equipment.
- Room cleaning and sanitization.
- Room fumigation (If environmental safety codes permit) using paraformaldehyde, peracetic acid or other validated fumigation method. Room fumigation using paraformaldehyde involves the sublimation of paraformaldehyde over a period of ±6 hours. HVAC unit is shut down during the fumigation. This phase is followed by a neutralization step involving the sublimation of ammonium carbonate. Cleaning the surfaces before room fumigation is recommended. The efficiency of this method is increased when the temperature and the relative humidity of the air are increased up to 25 ± °C 80% RH, for instance. The reader should note that

fogging with formaldehyde is no longer permitted in the United State America. Other alternatives are available that involve peracetic acid, hydrogen Peroxide, or glutaraldehyde. Whatever the method, the efficiency of fogging against the microorganism should be validated both in laboratory experiments and production condition.

- Air and surface sampling to test for the absence of previously handled organisms.
- Room cleaning and sanitization.
- Equipment cleaning and sterilization or sanitization.
- Introduction of new organism strains.

Multistrain Facility. If a multistrain manufacturing strategy is developed, then the design of the facility should allow producing different strains concurrently in the different manufacturing areas within the limits defined by regulatory requirements (11,14). In addition to detailed and validated changeover procedures, the principles allowing concomitant presence of different strains in different manufacturing areas are

- Separate air handling units (HVAC): each room, together with its respective air locks, should be serviced by at least one HVAC which prevents air recirculation between rooms where different strains are handled (14).
- Dedicated air locks for personnel and material.
- Dedicated system for material decontamination and effluent treatment.
- Dedicated personnel: movement of personnel within the Production Core should be restricted to ensure segregation between the different areas (6,7).
- Validated cleaning procedures, especially for multiuse equipment.

Facility Finishes

Areas in a facility for bio production should be designed with smooth, readily cleanable surfaces (floors, walls, ceilings). Lighting fixtures, windows, doors and control panels should be flush, minimizing the potential for dust buildup.

DESIGN OF HVAC SYSTEM PARAMETERS

Introduction

In any biotechnological production facility, the HVAC system is crucial to good operations. The system must be capable of providing a uniform flow of conditioned air at sufficient velocity and pressure to meet both biosafety and GMP requirements. Requirements for maintaining room cleanliness have become more strict. Therefore, a series of operating parameter must be carefully fixed when designing modern facilities. The authors do not claim that the proposed description is the only way to comply with cGMPs and Biosafety Level requirements. The document is a translation of their experience in design, validation, and registration of biotechnology facilities.

Four major factors determine the operating parameters of the process rooms. The airflow is the first aspect to consider. Sufficient airflow to provide at least 20 to 40 air changes per hour is required to meet Class 100 000 or 10 000 requirements. The specification of 90-feet-per-minute air velocity for class 100 can be achieved only by using a vertical unidirectional flow hood or cleanroom. A vertical unidirectional flow Class 100 clean room requires 20 times more airflow than a Class 10 000 clean room of the same size. Energy costs just to operate the blowers to provide the airflow make up an important part of clean room cost. The authors recommend the installation of laminar air flow hoods for cost and technical reasons associated with BL-2/BL-3 requirements. Secondly, heating and cooling capacities are required to maintain air temperature and relative humidity within a specified range: 18 to 22 °C and 30 to 50 RH% for Class 100 and 10 000, for instance. These values have been found satisfactory for controling microorganism growth with reasonable worker comfort. To maintain the values of temperature and relative humidity within the specified range, the engineers should take into account the variations of external ambient conditions, the number of persons, and the heat and moisture that the process may generate. The internal design of the room is the third major factor that determines the capacity to maintain the cleanliness of a production area. An adequate positioning of the air delivery inlets and the air exhaust outlets versus the positioning of the process equipment and personnel within the area will directly impact the capacity to meet the classification requirements. This capacity is measured by the decontamination time which represents the time required to recover the room class after contamination with a chemical aerosol. The test involves the use of nonviable particle counting devices which are placed at specific locations selected as representative of the clean room area (6). And last, differential pressures and air locks adjacent to manufacturing areas play a major role in protecting the cleanliness of the room and ensuring the containment of the live microorganisms within the area. Air locks are helpful in separating adjacent areas. They should be designed with sufficient airflow velocity and self-closing doors to make a real barrier; the interlocking system will maintain a minimum time between the closing

and the opening of the two doors of the air lock. This system will ensure the efficacy of the so-called, "air shower." Differential pressures between adjacent manufacturing areas will be chosen to compensate for the variability of the air pressure and to ensure a unidirectional airflow.

Basic Description of Production Areas and their Operating Functions

1. Class 100 and 10 000 (EU class A and B) kept at positive air pressure

   - Scope: Filtration of inactivated product, filling, formulation, sterile storage if included in a neutral aseptic production room, cellular culture, and nonpathogenic virus replication if aseptic processing, purification if required by the process, aseptic media preparation. The room is equipped with a Class 100 (EU class A) laminar airflow hood.
   - Access via:
     - Personnel Air Lock three doors: two parts
     - Materials Air Lock In
     - Materials Air Lock Out with fumigation for live organism handling areas.

2. Class 10 000 (EU class B) kept at negative air pressure

   - Scope: Culture of pathogenic virus and inactivation of pathogenic virus if aseptic operations, purification if requested by the process. The room is equipped with a class 100 (EU class A) laminar airflow hood.
   - Access via:
     - Personnel Air Lock four doors: two parts
     - Materials Air Lock In
     - Materials Air Lock Out with fumigation.

3. Class 100 000 (EU class C) kept at positive air pressure

   - Bioreactor culture and purification of non-pathogenic products, capping, clean materials storage, media preparation.
   - Access via Air Lock two doors: one part if access from a nonclassified production room.
   - Direct access if access from a 100 000 (EU class D) production room.
   - The access to media preparation production rooms proceeds via an Air Lock if the facility is designed for cell culture and viral replication.

4. Class 100 000 (EU class C) kept at negative air pressure

   - Scope: Bioreactor culture and purification of pathogenic products. If the production room is classified BL3,

Access via:
- Personnel Air Lock four doors: three parts
- Materials Air Lock In
- Materials Air Lock Out with fumigation.

If the production room is classified BL2,

Access via:
- Air Lock two doors: one part

5. Class 100 000 (EU class D)

- Scope: Materials preparation with laminar flow, washing room, products storage, materials storage, corridor close to the production rooms.

6. Specific consideration

Whatever the classification of a room, it can often be maintained at a low temperature but not exclusively between 2 °C to 8 °C to meet process requirements. In this case, the relative humidity needs to be kept lower than 90% to avoid condensation.

HVAC Characteristics

The characteristics of HVAC systems are described on the basis of dynamic room class and Biosecurity level (U.S. Standards). The following HVAC characteristics are described for each type of production area:

- Pressure.
- Air changes.
- Decontamination time (= time required to recover the room class after contamination by a chemical aerosol).
- Partial recycling of the air.

- Temperature.
- Relative humidity.
- Efficacy of air delivery filter (final filters).
- Efficacy of air exhaust filter (final filters).
- Location of air delivery inlets outlets.
- Location of air exhaust outlets.
- Fumigation.
- HVAC group in standby for the critical cold rooms.

Pressure Changes

The principle is a pressure decrease by steps of 15 Pa ±5 Pa (= 0.06 inch of water ±0.02).

HVAC Technical Data

Class 100 and 10 000 (EU Class A and B) Kept at Positive Air Pressure.

- Layout

Sterile storage

Material Air Lock In		
Class 100 Production Room		
Personnel Air Lock 1	Personnel Air Lock 2	Material Air Lock Out

Corridor

Operating parameters for Class 10 000 (EU Class B) area kept at positive air pressure.

	Production Room	Materials Air Lock In	Materials Air Lock Out	Personnel Air Lock 1	Personnel Air Lock 2
Pressure (Pa)	+45	a	b	+30	e
Min. changes (AC/h)	40	200	200	20	600
Decontamination time (min)	15	3	3	12	1
Partial air recycling	Yes	100%	c	yes	100%
Temperature (°C)	18–22	NA	d	18–22	18–22
Relative humidity (%)	30–50	NA	d	40–60	40–60
Air delivery filter (% DOP retention)	99,995	99,995	99,995	99,995	99,995
Air exhaust filter	99,995	G 85	G 85	G 85	G 85
Location of delivery inlets	Ceiling	Ceiling	Ceiling	Ceiling	Ceiling
Location of exhaust outlets	Bottom of wall	Bottom of wall	Bottom of wall	Ceiling	Bottom of wall
Fumigation	Yes	Yes	Yes	No	Yes

Notes: [a]Between sterile storage and production room by balancing.
[b]Between production room and corridor by balancing.
[c]Delivery and haust separated for the fumigation cycle.
[d]The parameter must be controlled during the fumigation cycle.
[e]Between PAL 1 and production room by balancing.

Class 100 (EU Class B) Kept at Negative Air Pressure.

- Layout

Sterile storage

Materials in

Production Room

Personnel	Personnel	Personnel	Materials
Air Lock 1	Air Lock 2	Air Lock 3	Air Lock

Corridor

Class 100 000 (EU Class C) Area Kept at Positive Air Pressure.

- Layout:

Production Room
Air lock

Operating parameters for Class 100 000 (EU Class C) area kept at positive air pressure:

	Production Room	Air Lock[a]
Pressure (Pa)	+30	[b]
Min. changes (r/h)	20	20
Decontamination time (min)	12	3
Partial air recycling	yes	yes
Temperature (°C)	18–22	18–22
Relative humidity (%)	40–60	NA
Air delivery (% DOP retention)	99,995	95
Air exhaust filter	G 85	G 85
Location of delivery inlets	Ceiling	Ceiling
Location of exhaust	Bottom of wall	Bottom of wall
Fumigation	Occasional	Occasional

Notes: [a]An Air Lock is justified if the corridor is not classified (100 000) or if the room is a production room for media preparation for cell culture and viral production.
[b]Between production room and corridor.

Operating parameters for Class 10 000 (EU Class B) area kept at negative air pressure:

	Production Room	Materials Air Lock In	Materials Air Lock Out	Personnel Air Lock 1[e]	Personnel Air Lock 2[e]	Personnel Air Lock 3[e]
Pressure (Pa)	−30	[a]	[b]	+30	−30	−15
Min. changes (r/h)	40	200	200	20	20	600
Decontamination time (min)	15	3	3	12	12	1
Partial air recycling	No	100%	[c]	Yes	No	No
Temperature (°C)	18–22	NA	[d]	18–22	18–22	18–22
Relative humidity (%)	30–50	NA	[d]	40–60	40–60	40–60
Air delivery filter (% DOP retention)	99,995	99,995	99,995	99,995	99,995	99,995
Air exhaust filter	99,995	G 85	G 85	G 85	99,995	99,995
Location of delivery inlets outlets	Ceiling	Ceiling	Ceiling	Ceiling	Ceiling	Ceiling
Location of exhaust outlets	Bottom of wall	Bottom of wall	Bottom of wall	Ceiling	Ceiling	Bottom of wall
Fumigation	Yes	Yes	Yes	No	Yes	Yes

Notes: [a]Between sterile storage and production room by balancing.
[b]Between production room and corridor by balancing.
[c]Delivery and exhaust separated for the fumigation cycle.
[d]The parameter must be controlled during the fumigation cycle.
[e]The authors have proposed and implemented an original approach with respect to the differential pressures of the Personnel Air Locks. The set points are fixed so that the differential pressures ensure two opposite directional air flows at two different locations: PAL 1 and PAL 3. Such design provides a double-barrier protection system working on two separate air handling units. The following schema illustrates the concept:

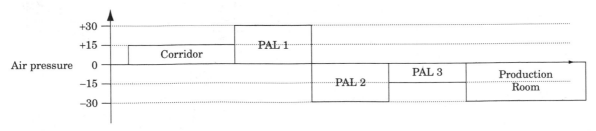

Class 100 000 (EU Class C) Area Kept at Negative Air Pressure (Biosafety Level 3).

- Layout:

Production Room			
Personnel Air Lock 1	Personnel Air Lock 2	Personnel Air Lock 3	Materials Air Lock Out

Class 100 000 — Pathogenic BL2 (EU Class C) Kept at Negative Air Pressure.

- Layout:

Production Room
Air lock

Operating parameters for Class 100 000 (EU Class C) area kept at negative air pressure:

	Production Room	MAL Out	PAL 1	PAL 2	PAL 3
Pressure (Pa)	−30	a	+30	−30	−15
Min. changes (r/h)	20	200	20	20	600
Decontamination time (min)	12	3	12	12	1
Partial air recycling	No	b	Yes	No	No
Temperature (°C)	18–22	c	18–22	18–22	18–22
Relative humidity (%)	40–60	c	40–60	40–60	40–60
Air delivery filter (% DOP)	99,995	99,995	99,995	99,995	99,995
Air exhaust filter	99,995	G 85	G 85	99,995	99,995
Location of delivery inlets	Ceiling	Ceiling	Ceiling	Ceiling	Ceiling
Location of exhaust outlets	Bottom of wall	Bottom of wall	Bottom of wall	Bottom of wall	Bottom of wall
Fumigation	Yes	Yes	No	Yes	Yes

Notes: [a]Between production room and corridor by balancing.
[b]Delivery and haust separated for fumigation cycle.
[c]Control requested during the fumigation cycle.

Operating parameters for Class 100 000 (EU Class C) area kept at negative air pressure:

	Production Room	Air Lock[a]
Pressure (Pa)	−15	−30
Min. changes (r/h)	20	20
Decontamination time (min)	12	3
Partial air recycling	No	No
Temperature (°C)	18–22	18–22
Relative humidity (%)	40–60	NA
Air delivery filter (% DOP)	99,995	99,995
Air exhaust filter	99,995	99,995
Location of delivery inlets	Ceiling	Ceiling
Location of exhaust outlets	Bottom of wall	Ceiling
Fumigation	Yes	Yes

Note: [a]Air lock on independent HVAC group.

Class 100 000 (EU Class D).

	Production Room
Pressure (Pa)	Steps of 15 Pa
Min. changes (r/h)	20
Decontamination time (min)	3
Partial air recycling	Yes
Temperature (°C)	18–22
Relative humidity (%)	40–60
Air delivery filter (% DOP)	95
Air exhaust filter	G 85
Location of delivery inlets	Ceiling
Location of exhaust outlets	Ceiling
Fumigation	Occasional

HVAC Systems of the Viral Vaccine Bulk Manufacturing Facility

The production unit is serviced by twelve HVAC units as depicted in Figure 6. Each system has fresh air intake and all rooms classified as 10,000 or better are equipped with low wall air returns. Exhaust air is HEPA filtered.

- In production areas exposed to virus (Virus Culture area, Purification area, Inactivation area and Laboratory 221 previously described:
  ⇒ HVAC systems are segregated, and there is no recirculation of air between these designated production areas.
  ⇒ These areas are kept at negative pressure with respect to adjacent areas.
- In production areas with no live virus,
  ⇒ Room pressure and directional air flows are designed and balanced to create a positive pressure cascade from more controlled, or critical areas, to less controlled areas; see Figure 7 for pressure differentials and airflows.

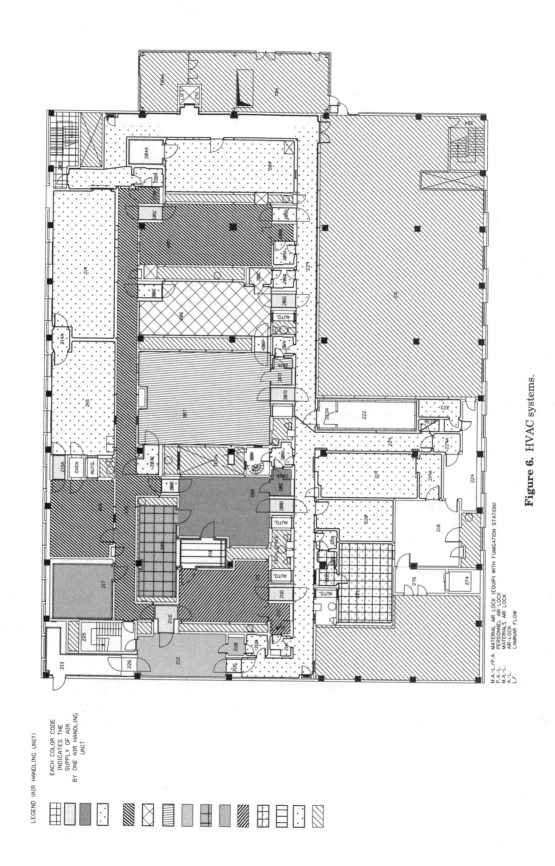

Figure 6. HVAC systems.

M.A.-L./P.A. MATERIAL AIR LOCK (EQUIPT WITH FUMIGATION STATION)
P.A.-L.   PERSONNEL AIR LOCK
M.A.-L.   MATERIALS AIR LOCK
A.-L.   AIR-LOCK
L.F.   LAMINAR FLOW

LEGEND (AIR HANDLING UNIT)

EACH COLOR CODE
INDICATES THE
SUPPLY OF AIR
BY ONE AIR HANDLING
UNIT

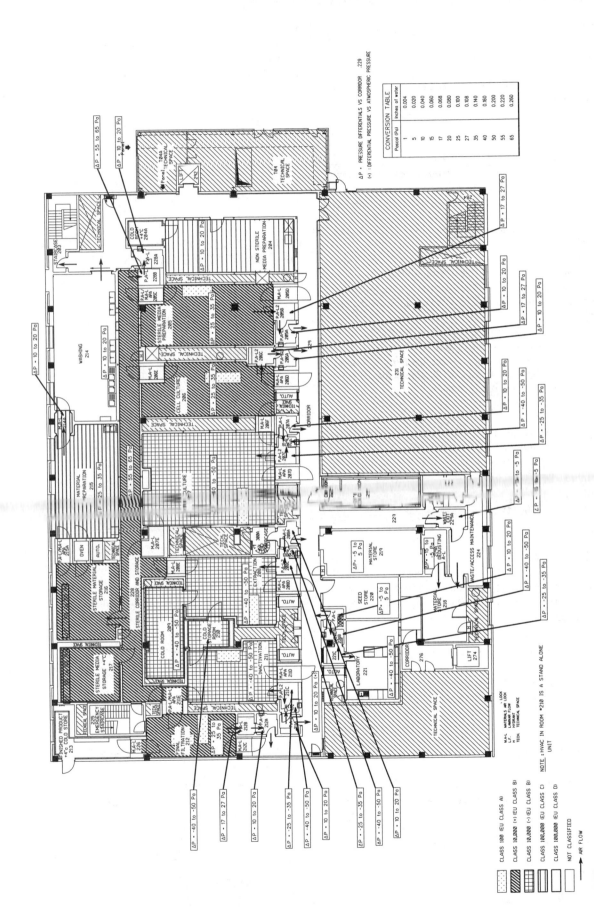

Figure 7. Room classification (dynamic conditions), pressure differentials, and Airflows.

535

BIBLIOGRAPHY

1. Code of Federal Regulations 21 Parts 210 and 211 (United States of America): *Good Manufacturing Practices for Finished Pharmaceuticals*, U.S. Government Printing Office, Washington, D.C.

2. *Good Manufacturing Practices for Medicinal Products, European Union*.

3. CFR 21 Parts 600.

4. *Annexe 2 of European Union Code of Good Manufacturing Practice for Medicinal Products: Manufacture of Biological Medicinal Products for Human Use*.

5. *Biosafety in Microbiological and Biomedical Laboratories*, Center for Disease Control and Prevention (CDC), U.S. Department of Health and Human Services (NIH), Washington, D.C.

6. *Airborne Particulate Cleanliness Classes for Clean Room Production Rooms*, U.S. Federal Standard 209E.

7. *FDA Guidelines on Sterile Drug Products Produced by Aseptic Processing*, Food and Drug Administration, Washington, D.C., 1987.

8. *Annex 1 of the EU Guide to Good Manufacturing Practice: Manufacture of Sterile Medicinal Products*, 1997.

9. *Annexe 2 of EU Guide to Good Manufacturing Practice: Manufacture of Biological Medicinal Products for Human Use, Section "Premises and Equipment"*.

10. 21 Parts CFR 600.10, Section C.

11. *Annexe 2 of EU Guide to Good Manufacturing Practice: Manufacture of Biological Medicinal Products for Human Use*, Section 5 (Personnel).

12. 21 Parts CFR 600.11, Section (b).

13. 21 CFR 600.11, Section (a).

14. 21 CFR 600.11, Section (c).

15. USP XXIII Monograph <1231>.

ADDITIONAL READING

FDA Biotechnology Inspection Guide, November 1991, Food and Drug Administration, Washington, D.C., 1991.

ISO/FDIS 13408-1, Aseptic processing of health care products.

Part 1. General requirements.

ISO/FDIS 14644-1, Cleanrooms and associated controlled environments.

Part 1. Classification of airborne particles.

ISO/FDIS 14644-2, Cleanrooms and associated controlled environments.

Part 1. Specifications for testing and monitoring to prove continued compliance with ISO 14644-1.

ISO/FDIS 14644-4, Cleanrooms and associated controlled environments.

Part 4. Design, construction, and start-up.

ISO/FDIS 14698-1, Cleanrooms and associated controlled environments.

Biocontamination control. Part 1. General principles.

ISO/FDIS 14698-2, Cleanrooms and associated controlled environments.

Biocontamination control. Part 2. Evaluation and interpretation of biocontamination data.

ISO/FDIS 14698-3, Cleanrooms and associated controlled environments.

Biocontamination control. Part 3. Measurement of the efficiency of processes of cleaning and/or disenfection of inert surfaces bearing biocontaminated wet soiling or biofilms.

USP XXVIII, Supplement 8, Section <1116>, Microbial Evaluation of Clean Rooms and other Controlled Environment.

See also CONTAMINATION DETECTION IN ANIMAL CELL CULTURE; ETHICAL ISSUES IN ANIMAL AND PLANT CELL TECHNOLOGY; ICH GCP GUIDELINES: PREPARATION, CONDUCT AND REPORTING OF CLINICAL TRIALS; STERILIZATION AND DECONTAMINATION; VIRAL INACTIVATION, EMERGING TECHNOLOGIES FOR HUMAN BLOOD PRODUCTS.

CHARACTERIZATION AND DETERMINATION OF CELL DEATH BY APOPTOSIS

JASON E. REYNOLDS
JOHN A. MCBAIN
PING ZHOU
ALAN EASTMAN
RUTH W. CRAIG
Dartmouth Medical School
Hanover, New Hampshire

OUTLINE

INTRODUCTION

Cell death can occur by two morphologically dissimilar pathways, necrosis and apoptosis. Necrotic cell death is characterized by cell swelling and loss of membrane integrity. These occur early and are followed by the release of lysosomal enzymes and disintegration of the nucleus. In many cases, necrotic cell death is induced by exogenous insults or poisons. It can occur as a response to wounding or tissue injury, hypoxia or ischemia, hyperthermia, or acute exposure to toxic chemicals, such as carbon tetrachloride (1). Necrosis generally appears in tissues within the whole animal, as a synchronous process involving multiple contiguous cells. In addition, necrosis produces an inflammatory response, presumably due to leakage of cytoplasmic proteins. Because of these latter two features, necrosis is readily apparent, often striking, upon histologic examination.

About a quarter of a century ago, Kerr et al. outlined the characteristics of a type of cell death, termed apoptosis, that differed morphologically from necrosis (2). In contrast to the swelling and membrane lysis seen in necrotic cells, apoptotic cells exhibit cell shrinkage and nuclear condensation. This can occur without loss of membrane integrity or evidence of lysosomal damage. Ultimately, the nucleus may become blebbed and break into fragments, and these nuclear fragments may break off from the cell to form membrane-bound "apoptotic bodies." In animal tissues in vivo, apoptotic cells and the apoptotic bodies they shed are engulfed by macrophages or neighboring cells. Because the plasma membrane of the apoptosing cell is not lysed during this process, cytoplasmic contents are not released and apoptosis is not accompanied by an inflammatory response. In cells in tissue culture, however, apoptosis may be followed by eventual membrane lysis or "secondary necrosis." Because apoptosis occurs rapidly and because the products are rapidly engulfed without producing an immune response, apoptosis may not be striking upon histologic examination. In fact, apoptosis in vivo is frequently observed in scattered individual cells, and thus even tissues that have a relatively high apoptotic rate may exhibit only a minority of apoptosing cells at any time.

Apoptotic cell death is involved in a host of physiological as well as pathological processes (2,3). Physiological apoptosis is an important determinant of tissue homeostasis. An example of this is seen in the colonic crypts, where cell division and death are tightly regulated to maintain normal architecture. Specifically, cells in the base of the crypts can undergo division, cells progressing outward toward the tips of the villi are undergoing differentiation, and cells at the surface eventually undergo apoptotic cell death and are sloughed into the lumen (4). Therefore, alterations that affect apoptosis can disrupt homeostasis and have pathological effects. On one hand, alterations that interfere with the ability of cells to undergo apoptosis can lead to disorders involving cell accumulation (e.g., hyperplasia or neoplasia) (5). In fact, it has been postulated that the relative number of apoptotic cells in tumors is a prognostic indicator in breast cancer (6). On the other hand, alterations that increase the propensity of cells to undergo apoptosis can lead to disorders of cell loss, such as the loss of $CD4^+$ cells in patients who have AIDS (7). Apoptosis can be induced by pharmacological as well as physiological stimuli; for example, numerous cancer chemotherapeutic agents induce apoptosis (8). Accordingly, it is suggested that suppression of apoptosis plays a role in the multidrug resistance phenotype in many human malignancies (5). The role of apoptosis in disease pathogenesis, as well as treatment, suggests that studies aimed at elucidating the mechanisms involved in this process may lead to the identification of new prognostic indicators in cancer, as well as novel therapeutic drug targets. Overall, the fact that a variety of diverse stimuli, acting on multiple targets, induce a similar sequence of events that culminates in cell death suggests the existence of a final common pathway that regulates cell viability. Because this pathway is involved in many critical physiological, pathogenic, and pharmacological phenomena, the ability to recognize and quantify apoptotic cells is becoming increasingly important.

A wide variety of assays are available to detect apoptotic cells. Some of these rely on identifying the characteristic morphological changes, such as chromatin condensation and nuclear fragmentation. Others were developed following the identification of biochemical events that occur during apoptosis, such as the appearance of fragmented DNA (a "DNA ladder") or the cleavage of specific proteins that occurs during the effector phase of apoptosis (e.g., the cleavage of caspase substrates). Because apoptosis was originally defined on the basis of morphology, the morphological assays are considered the "gold standard" for detecting apoptosis. However, there are limitations to such assays. First, they depend assessment by visual inspection. Biochemical assays are less subjective but are usually performed on broken cell preparations and thus cannot be directly equated to a fraction of the cell population that is undergoing apoptosis. Assays that avoid some of these problems, such as flow cytometric assays that yield objective data and can be quantitated on a per cell basis, are being developed. In the following sections, the assays for apoptosis have been grouped into three broad categories, morphological assays, biochemical assays, and flow cytometric assays. The various assays vary tremendously in their technical difficulty. Some are simple, and others require sophisticated technical expertise, as well as specialized equipment. The assays also vary in their ability to provide qualitative versus quantitative information, as well as in their sensitivity to detect apoptotic cells. Although objective data are obtained from many of the assays, in some cases a subjective judgment enters into determining whether cells are apoptotic. Finally, some of the assays are highly specific for detecting apoptosis, whereas others may also detect necrotic cells. These assays have been widely tested in a variety of systems, where it has been observed that certain assays are more readily applicable to specific apoptotic cell systems and others are better suited to other systems. In most cases, it is advisable to use multiple assays for each system because the complementary strengths of several assays can compensate for their individual shortcomings.

MORPHOLOGICAL ASSAYS OF APOPTOSIS

Staining for Nuclear Morphology using Fluorescent DNA-Binding Dyes

Apoptosis was initially defined by morphological criteria. Therefore, morphological assays, such as those described below, incontrovertibly test whether cell death is proceeding through an apoptotic process. Fluorescent DNA-binding dyes, such as acridine orange, 4′,6-diamidino-2-phenylindole (DAPI), Hoechst 33342, and ethidium bromide stain the cell nucleus and allow visual assessment of nuclear morphology. The first three of these, acridine orange, DAPI, and Hoechst 33342, are membrane-permeable dyes and can freely enter the cell (i.e., in the absence of loss of membrane integrity and the absence of cell permeabilization). Therefore, these dyes stain the nuclei of all cells, including apoptosing cells that have not lost plasma membrane integrity. Ethidium bromide does not cross intact cell membranes. Therefore in the absence of cell permeabilization, ethidium bromide does not stain viable cells or cells in the early stages of apoptosis but does stain late apoptotic or necrotic cells (i.e., cells that have lost membrane integrity). In this respect, ethidium bromide staining of cells is an indicator of membrane permeability, much like trypan blue. These dyes are used in the various methods described later. The first of these methods employs two dyes simultaneously, acridine orange and ethidium bromide, whereas the other methods employ a single dye. Detailed protocols for these methods are in Appendix I.

The acridine orange/ethidium bromide method is carried out by adding these two dyes directly to an unfixed suspension of cells. The principle that underlies the use of these two dyes is as follows (9). Acridine orange is taken up by all cells, whereas ethidium bromide can only be taken up by cells that have lost plasma membrane integrity. Thus, cells that retain membrane integrity take up only acridine orange, which causes green nuclear staining. Cells that have lost membrane integrity additionally take up ethidium bromide which causes the nuclei to stain orange/red instead of green. It is the color of the nuclear staining (green or orange/red) that is important in this assay. Cytoplasmic staining can occur because the dyes bind to RNA, as well as DNA. However, cytoplasmic staining is not taken into account in the scoring. Based on this principle, scoring is carried out by viewing the stained cells using an ultraviolet fluorescence microscope. Cells are categorized as viable cells (Category I), cells undergoing apoptosis (Category II), and dead cells (Category III), as follows. Cells that exhibit orange/red nuclear staining have lost membrane integrity and are scored as Category III. Cells that exhibit green nuclear staining have not lost membrane integrity. Among these, one can distinguish viable nonapoptotic cells (Category I) from apoptotic cells (Category II) based on the fact that the former exhibit a normal, lacy pattern of nuclear staining whereas the latter exhibit dense/condensed nuclei and may exhibit nuclear fragmentation or blebbing.

As mentioned earlier, the later stages of apoptosis in tissue culture frequently involve plasma membrane lysis or "secondary necrosis" (10,11). One drawback of the acridine orange/ethidium bromide method is that cells that have undergone initial apoptosis followed by "secondary necrosis" cannot be distinguished from cells undergoing primary necrosis. A time study using this method may be informative, because primary apoptosis followed by secondary necrosis may appear in this sequence (12).

Additional methods for assessing nuclear morphology involve using other fluorescent DNA-binding dyes, such as DAPI or Hoechst 33342. With either of these dyes, the condensing/fragmenting nuclei in cells undergoing apoptosis are unmistakable (Fig. 1). Because these dyes enter all cells, cells that have lost membrane integrity (e.g., during secondary necrosis) cannot be distinguished from those that have not.

An advantage of these staining methods is that they allow scoring individual cells for apoptosis on the basis of the original definition of this process. These methods can be used quantitatively, although a disadvantage is that quantitation is subjective to some extent. One reason for this is that cells in the earliest stages of apoptosis may not exhibit fully condensed nuclei and the observer must judge where to draw the line in scoring such cells. A related point is that different cell types or cells exposed to different apoptosis-inducing agents may differ morphologically. For example, apoptotic cells can show simply condensation of DNA to the periphery of the nucleus, highly condensed nuclei, or different degrees of nuclear fragmentation (see, for example, Ref. 10). These differences may be due to variations in the extent of DNA digestion or to secondary effects of the inducing agent, such as inhibition of lamin degradation required for nuclear fragmentation. Another disadvantage of these staining methods, particularly later in the cell death process, is that secondary necrosis resulting from apoptosis is not objectively distinguishable from primary necrosis.

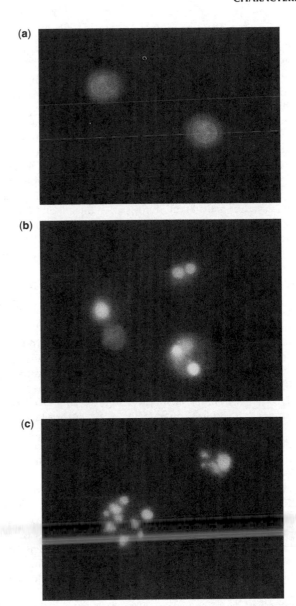

Figure 1. Analysis of apoptotic nuclear morphology using a fluorescent DNA-binding dye. Viable murine FDC-P1 hematopoietic progenitor cells (**a**) are compared to FDC-P1 cells induced to undergo apoptosis by incubation with 1.5 µM A23187 (**b**) or 20 µg/mL etoposide (**c**). All cells were stained with DAPI using the protocol described in Appendix I. Cells in (b) and (c) exhibit the dense, highly fluorescent, fragmenting nuclei that typify apoptotic cells.

BIOCHEMICAL MARKERS OF APOPTOSIS

DNA Fragmentation

The first biochemical hallmark of apoptosis discovered was the "DNA ladder", which is seen upon agarose gel electrophoresis of DNA from apoptotic cells. The oligonucleosomal-sized DNA fragments observed result from apoptosis-induced activation of endogenous endonuclease(s), which cleave genomic DNA in the linker regions between nucleosomes (13).

Early methods for assessing genomic DNA laddering used intricate procedures to purify fragmented DNA from intact DNA. However, using a modification of a plasmid DNA isolation technique, now it is possible to carry out the assay without special DNA purification techniques (14). The modification has been described in Ref. 8 and is detailed in Appendix II. Briefly, 10^6 cells are added directly to the wells of a 2% agarose gel where they are lysed and digested with ribonuclease A (RNase A) and proteinase K. The proteinase K and sodium dodecyl sulfate (SDS) are within the gel matrix, and the RNase A is in the sample loading buffer. Upon electrophoresis, high molecular weight DNA (> 20 kb) remains at or near the top of the gel. Shorter, endonuclease-digested fragments are resolved in the gel. The presence of a series of fragments with sizes representing multiples of 180 base pairs (one interoligonucleosome length) is consistent with the endonuclease digestion that typifies apoptosis (Fig. 2). DNA broken at random intervals, such as in necrosis, yields a continuous smear running the length of the gel. Large format gels (20 × 34 cm) yield optimal resolution of the apoptosis-associated DNA fragments. However, smaller format gels can also be used to obtain results in a shorter time (e.g., 3 to 5 hours). In either case, this DNA fragmentation assay qualitatively assesses apoptosis and can be used as a screen for apoptotic cells.

Although the above method provides a simple means of screening for apoptosis, the assay has several drawbacks or limitations. Specifically, the absence of DNA fragmentation is not diagnostic for the absence of apoptosis because some cells that have distinct apoptotic morphology do not exhibit internucleosomal DNA fragmentation. In some cases, short internucleosomal DNA fragments appear late, preceded by the appearance of large molecular weight fragments detectable using pulse-field

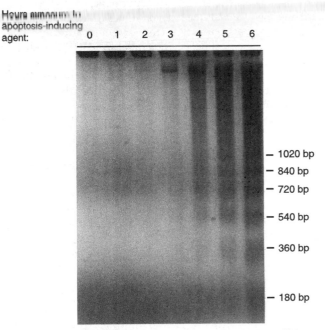

Figure 2. Analysis of apoptotic DNA fragmentation. Chinese hamster ovary (CHO) cells were incubated with the apoptosis-inducing agent staurosporine (800 nM) for up to 6 hours and then analyzed for DNA fragmentation using the protocol in Appendix II. DNA was visualized by ethidium bromide staining. Oligonucleosomal length DNA fragments are indicated on the right of the gel.

gel electrophoresis (15). Therefore, in the absence of a DNA ladder, it is useful to screen for these larger fragments or other markers of apoptosis. Another limitation of the DNA fragmentation assay is that it is qualitative rather than quantitative. In addition, the limit of detection has not been well defined. Thus, as compared to the other methods discussed later, relatively large numbers of cells are required. However, in one system, DNA fragmentation was detected with as few as 1.5×10^5 apoptotic cells, representing 15% of the population (16). Overall, if cell number is not a limiting factor, the rapidity and ease of the DNA fragmentation assay make it a good choice as an apoptotic marker for initial or screening experiments.

Proteolytic Cleavage of Apoptosis-Related Proteins

A rapidly advancing area in apoptosis research is the cascade of proteases that are the "effectors" of the apoptotic program. The first protease identified in this regard was CED3, a protease required for apoptosis in the nematode *Caenorhabditis elegans* (17). It was realized that CED3 is homologous to the human interleukin 1β converting enzyme (ICE), which led to the characterization of a large family or proteases now known as cysteine-aspartate proteases or caspases. A variety of caspases are activated during apoptosis (18), resulting in the cleavage of a variety of specific intracellular substrate proteins. The first protein found to be specifically cleaved during apoptotic cell death was the nuclear protein, poly-ADP-ribose polymerase (PARP) (19,20). During apoptosis, caspase 3 cleaves PARP from the full length of 116 to 85 kDa. Caspase 3 itself is activated by proteolytic cleavage by another member of the caspase family. Studies on the timing of protease activation, as it relates to other events in apoptosis, have consistently shown that the cleavage of PARP and caspase 3 coincides with the appearance of DNA fragmentation and other markers of apoptosis (21). Cleavage of PARP consistently correlates with other markers of apoptosis, such as intracellular acidification (21). It has been shown that PARP degradation occurs during necrotic cell death. However, the size of the fragments differs from those observed during apoptosis (22). Thus, generation of the 85 kDa PARP cleavage fragment is essentially diagnostic for apoptosis.

The protocol for assaying the proteolytic cleavage of PARP is detailed in Appendix III. It involves Western blotting using an antibody that detects both full length PARP (116 kDa) and its cleavage product (85 kDa). Non-apoptotic cells contain only full length PARP, whereas apoptotic cells also contain the cleaved product. Increased proteolytic cleavage is observed with increasing concentrations of apoptosis-inducing agents and increasing length of exposure.

An advantage of the PARP cleavage assay is that, unlike the DNA fragmentation assay, quantitation is possible. Thus, the proportions of the cleaved versus the intact protein can be estimated. Unfortunately, this information cannot be used to calculate the percentage of apoptotic cells because low levels of cleavage in a substantial fraction of cells cannot be distinguished from complete cleavage in a minor fraction of cells. The limits of detection of the two assays may be in the same approximate range, although little data is yet available.

Percoll Gradient Fractionation of Apoptotic Cells

As discussed before, cell shrinkage is one of the earliest detectable changes in apoptotic cells. This occurs before a loss in membrane integrity and probably involves the selective loss of salt and cellular water. Cell shrinkage results in an increase in buoyant density, that is, an increase in cell mass per volume (relative to the same volume of water). Based on the increase in buoyant density, apoptotic cells can be separated from viable cells by centrifugation through an appropriate density gradient. Apoptotic and viable cells can also be separated from necrotic cells, which swell and therefore decrease in density. Separation of dying cells based on density yields better results than alternative methods involving fractionation by size alone because fractionation by size fails to distinguish between early volume reduction and subsequent cell fragmentation. The latter process results in small blebs or apoptotic bodies, as well as debris, which sediment at very low density in density gradients.

Density ranges of interest in terms of cell death include values from 1.00 to 1.15 g/cc. Adenocarcinoma cells normally maintain densities of approximately 1.06 g/cc, whereas lymphocytes have densities of approximately 1.075 g/cc (23) and erythroleukemia cells have densities of 1.09 g/cc. This density is in the range of that of Percoll (Pharmacia), a colloidal suspension of silica that has been treated to reduce association with cells. Percoll has an osmolality and viscosity appropriate for cells and has the additional benefit of forming a continuous gradient of appropriate shape upon centrifugation.

Cells to be separated by this method are suspended in a dilution of isotonic Percoll and subjected to centrifugation (protocol in Appendix IV). The optimal Percoll density must be determined empirically. A reasonable starting point is to use Percoll diluted to the approximate density of the cells to be separated. Varying centrifugation times can also be tested to optimize separation. It is preferable to use a single cell suspension, although small aggregates usually do not affect the separation. The cells are maintained under physiological conditions, allowing normal cell volume regulatory processes to operate. Therefore, the cells sediment to the position in the developing density gradient that is equivalent to the buoyant density of the cell (isopycnic sedimentation).

One caveat about this method is that the apoptotic cells must remain in the shrunken state throughout the period of centrifugation for the procedure to work. For colon cancer cells dying from treatment with butyrate, apoptotic cells in the highest density fractions maintain this density with a half-life of one hour. After this time, the cells are recovered from progressively lower density fractions (24,25). Therefore, fractions that have "normal" density can be variably contaminated with cells which had previously shrunken. The presence of such apoptotic cells in the viable cell fraction can be minimized by discarding all nonadherent cells approximately 1 to 2 h before fractionation or performing the fractionation twice. A related caveat is that the time required for centrifugal fractionation can prove limiting (as in examining early apoptosis on a minute scale). In this case, the fractionation can be shortened to 10 to 15 minutes by using preformed Percoll gradients.

This method has proven highly adaptable for answering numerous questions relating to cell death mechanisms. Because these procedures can be performed under sterile, physiological conditions, the cells can be maintained in culture after fractionation. Percoll preparations lack much of the toxicity associated with earlier preparations of colloidal silica (thought to result from binding to cell membranes and internalization). However, certain cell types (monocytes and macrophages) can suffer from inhibition of normal functions due to internalization of the silica. For these cell types, step or continuous gradients of a density media (e.g., Metrizamide) may be more appropriate, where the step gradient can be prepared by gentle layering of multiple 5 to 20 mm steps, starting with a layer that exceeds the density of the highest anticipated cell density. Even with these caveats (and associated methodological modifications), density gradient fractionation is one of the best methods for either microscale (10 to 10,000 cells) or milligram to gram harvests for biochemical analysis.

FLOW CYTOMETRIC ANALYSIS OF CELL DEATH

Apoptosis is a rapid process and, in vivo, apoptotic cells are rapidly engulfed, as mentioned before. Therefore, at any given time, the percentage of apoptotic cells within the population may be small. Thus, in assays of morphology, it may be necessary to score large numbers of cells to accurately estimate the fraction that is undergoing apoptosis. Biochemical assays similarly require a substantial number of cells. Flow cytometric techniques are very useful in this regard in that large numbers of individual cells can be assayed rapidly, allowing apoptosis to be quantitated accurately even in a minute fraction of the population. The high sensitivity of flow cytometry is also particularly useful for defining the onset of apoptosis, for example, in studies designed to understand the mechanisms that trigger this process. Flow cytometric assays have been developed to detect apoptotic cells on the basis of various properties, including DNA fragmentation, altered membrane permeability, decreased intracellular pH, decreased cell size, and altered phospholipid composition of the extracellular membrane.

Detection of Cells with Subdiploid DNA Content

When apoptotic cells are fixed with ethanol and stained with the DNA binding dye propidium iodide (PI), cellular DNA content is reduced compared to that of viable cells. Because PI is fluorescent, this decrease in DNA content can be assayed by flow cytometry. The reduction in DNA content in ethanol-fixed, PI-stained cells depends on ethanol fixation, which causes cell permeabilization and allows oligonucleosomal fragments to leak out of the cell. The PI-stained apoptotic cells appear as a population with a subdiploid DNA content, that is, a population with a DNA content lower than that of viable cells in the G1 phase of the cell cycle.

Assessment of cells with sub-G1 DNA content is carried out by using standard methods for flow cytometric analysis

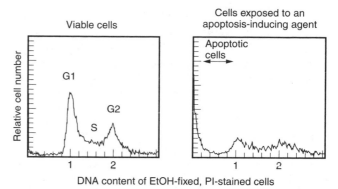

Figure 3. Demonstration of the sub-G1 DNA content of apoptotic cells stained with propidium iodide (PI) after ethanol fixation. CHO cells were incubated in the absence (left panel) or presence (right panel) of the apoptosis-inducing agent staurosporine (800 nM) for 6 hours. DNA content was determined after fixation with ethanol and staining with PI as described in Appendix V. Viable cells (left panel) show the typical cell cycle profile for logarithmic growth. Essentially all cells are in the G1, S, and G2 phases of the cell cycle and very few cells have a sub-G1 DNA content. In contrast, many of the cells exposed to the apoptosis-inducing agent (right panel) show a sub-G1 DNA content.

of cell cycle distribution (protocol in Appendix V). Upon ethanol fixation, most cells can be stored for at least several days before incubation with PI and RNase. The RNase is used to destroy endogenous RNA because PI can bind both DNA and RNA. With this assay, the relative fluorescence due to PI is proportional to the cellular DNA content. Viable exponentially growing cell populations exhibit two characteristic peaks representing G1 and G2 DNA content; cells that have a DNA content intermediate between these two peaks are in the S phase (Fig. 3). Apoptotic cells appear as cells that have sub-G1 DNA content. In some situations, a clear sub-G1 peak is observed, but in other cases as shown in Figure 3, this population may be very heterogeneous in size. It should be recalled that apoptotic cells frequently fragment. Hence this sub-G1 population will include these fragments. Accordingly, the method is not good as a quantitative assay because one viable cell may yield more than one sub-G1 particle.

An advantage of this assay is that it represents a simple, rapid means of screening for dying cells. Thus, one can assay large numbers of samples and large numbers of cells per sample. A drawback is that this assay detects all cells that have sub-G1 DNA content. Thus, necrotic cells cannot be readily distinguished from apoptotic cells because they may also have reduced DNA content. In addition, it has been postulated that this method is most sensitive for detecting cells that die in the G1 phase of the cell cycle because cells that die in the S or G2 phases must lose more DNA than cells in G1 to appear in the sub-G1 population (26). For these reasons, it is advisable to use this method in conjunction with at least one additional measure of apoptosis. The sub-G1 population frequently appears within the same time frame as DNA fragmentation (16,27); thus, this assay can be used for rapid initial screening of a variety of times to select an appropriate window for additional tests.

Detection of Cells that Contain DNA Strand Breaks

The previous assay that employs propidium iodide is somewhat nonspecific in that it is based solely on the loss of DNA from dying cells. A more specific flow cytometric method, the Terminal dUTP Nick End Labeling (TUNEL) assay, is based on the presence of fragmented DNA (i.e., double-stranded DNA breaks) in apoptotic cells (28,29).

The TUNEL assay is based on the following principles: The oligonucleosomal DNA fragments in apoptotic cells contain a 3'-hydroxyl group (and 5'-phosphate) that arise from the cleavage of the phosphoribosyl backbone of the DNA helix. These breaks can be detected on the basis of the ability of the terminal deoxynucleotidyl transferase (TdT) enzyme to add nucleotides to free 3'-hydroxyl groups in genomic DNA. Thus, in the TUNEL assay, cells are incubated with this enzyme along with biotinylated nucleotide triphosphate substrates (protocol in Appendix VI). Incorporation of the biotinylated nucleotides at the 3'-hydroxyl break sites is detected by staining with fluorescein isothiocyanate (FITC)-labeled avidin (note that alternative staining protocols may be used). The number of apoptotic cells is estimated by quantitating the number of cells that exhibit increased fluorescence compared to a viable control cell population.

Advantages of the TUNEL assay are that it is objective and quantitative. In addition, this assay is more sensitive than the PI staining assay for sub-G1 DNA content (29). The increased sensitivity is thought to be due to the fact that the assay detects DNA breaks that occur early in apoptosis, before the loss of a large amount of cellular DNA or the loss of membrane integrity. It should be noted that apoptotic cells do not show decreased DNA content in the TUNEL assay because fixation is with formaldehyde rather than ethanol. Thus, another advantage of this assay is that the cells can be simultaneously assayed for cell cycle distribution (by costaining with PI) to assess the cycle phase of the cells undergoing apoptosis. One of the drawbacks to this assay is that the ability to detect apoptotic cells depends on the activity of an exogenously added enzyme in an in situ reaction. Thus, this assay may be more prone to technical difficulties than the previously discussed DNA content or biochemical assays. It has been reported that strand breaks formed by the immediate action of topoisomerase inhibitors may lead to increased staining in the TUNEL assay, even though the cells have not undergone apoptosis (28). Here, the high sensitivity of the TUNEL assay and its ability to detect early events could lead to false positive results. As with many of these assays, therefore, it is advisable to confirm the results of the TUNEL assay by using alternative assays.

Detection of Apoptotic Cells based on Intracellular pH and Membrane Permeability

Many different types of cells that undergo apoptosis in response to both physiological and pharmacological insults demonstrate the common property of intracellular acidification (16,21,27,30,31). This decrease in pH can be detected by flow cytometry using the pH-sensitive intracellular dye, carboxy-SNARF-1AM. However, the change in pH is small, generally on the order of only 0.5 to 0.8 pH units. Therefore, staining with the pH-sensitive dye alone may not achieve complete separation from the viable population although it demonstrates a shift in the apoptotic cell population.

Accelerated uptake of the fluorescent dye, Hoechst 33342 is another property exhibited by many apoptotic cells (32–35). Therefore, when cells are incubated for short periods (1 to 5 min) with Hoechst 33342, a population of apoptotic cells that exhibits increased Hoechst 33342 fluorescence can be identified (35). When cells are incubated for longer than 5 minutes, Hoechst 33342 uptake becomes maximal; at this time, this flow cytometric assay can no longer be used to distinguish between apoptotic and viable cells (35), although these cells can still be distinguished morphologically, as described above. The mechanism underlying the enhanced uptake of Hoechst 33342 in apoptotic cells is not well understood, but it may relate to alterations in membrane permeability or in DNA conformation (26).

A dual staining method that takes advantage of both of these properties—changes in intracellular pH and in the rate of Hoechst dye accumulation—can more clearly distinguish apoptotic from viable cells in many cases. This method employs both carboxy-SNARF-1AM and Hoechst 33342. The simultaneous use of the two dyes in two-color flow cytometry allows generating of dot plots that display data reflecting pH on one axis and Hoechst 33342 uptake on the other. Because apoptotic cells differ from viable cells in both characteristics, they can be distinguished as a discrete population shifted in both dimensions (Fig. 4).

The development of this assay has been described in detail elsewhere (35), and the protocol is provided in Appendix VII. In brief, cells are first incubated with carboxy-SNARF-1AM for 30 to 60 minutes. This can be added directly to the cell culture medium for cells growing in suspension, whereas adherent cells must be harvested (see Appendix) and placed in suspension. Hoechst 33342 is added to the cell suspension 1 to 10 minutes before flow cytometric analysis. A flow cytometer with dual argon ion lasers is required: pH is assessed by excitation of carboxy-SNARF-1AM at 488 nm, and emission is monitored at 585 and 640 nm. Hoechst fluorescence is assessed by excitation with ultraviolet lines at 351 and 364 nm and emission is monitored at 440 nm.

The principle that underlies the use of the pH sensitive carboxy-SNARF-1AM dye is as follows. When the pH decreases in cells, the fluorescent emission monitored at 640 nm decreases, and the fluorescent emission at 585 nm increases. The ratio of these emission values is calculated because it corrects for variations in dye uptake in different cells. Hence, cells with a decrease in intracellular pH exhibit an increase in the 585 nm/640 nm fluorescent intensity ratio, which appears as a rightward shift in the graph in Figure 4.

Several points must be mentioned about this assay: First, cells cannot be subjected to manipulations that could cause alterations in pH. Thus, until the moment of data collection on the flow cytometer, the cells must be maintained in a 37 °C incubator with an appropriate CO_2 concentration in the atmosphere (e.g., 5% CO_2), using a culture medium containing sodium bicarbonate and

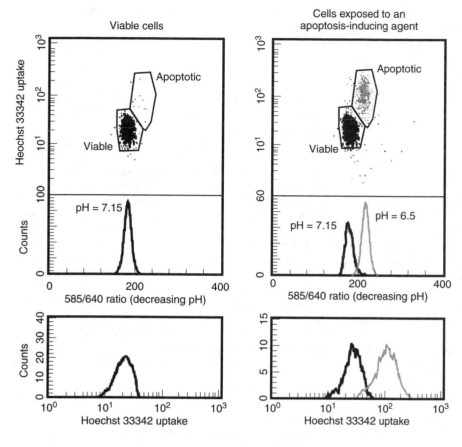

Figure 4. Identification of apoptotic cells by dual staining for increased Hoechst 33342 uptake and decreased intracellular pH. ML-1 cells were incubated in the absence (left panel) or presence (right panel) of 20 μg/mL etoposide for 30 minutes and assayed four hours later for the induction of apoptosis. Intracellular pH and Hoechst 33342 uptake were assayed as described in Appendix VII. The data are displayed as a dot-plot distribution of Hoechst 33342 fluorescence versus carboxy-SNARF-1 585/640 nm emission ratio. The histograms show the separation observed if the same data is analyzed for either carboxy-SNARF-1 or Hoechst 33342 alone. Viable cells (left panel) show only a low level of Hoechst 33342 uptake. Cells undergoing apoptosis (right panel) exhibit an increase in Hoechst 33342 uptake so that they are shifted up. The apoptotic cells also exhibit a rightward shift in the carboxy-SNARF-1 585/640 nm fluorescence ratio, which reflects a decrease in intracellular pH.

serum. In addition, because Hoechst 33342 uptake is a kinetic measure, all samples must be exposed to this dye for an equal amount of time. It is also important to note that the rate of accumulation of Hoechst 33342 varies among cell lines (35). Thus, before attempting the dual staining assay, the kinetics of Hoechst 33342 uptake should be examined as a single parameter. In sum, this assay must be carried out with care and with attention to the specifics of the protocol. It should also be noted that, to estimate the percentage of apoptotic cells, it is not necessary to obtain an accurate value for intracellular pH; in other words, a relative shift manifested by apoptotic compared to viable cells can be used to discriminate between the two cell types. However, if accurate values for intracellular pH are desired for other reasons, a pH curve must be calibrated for each cell line. This standard curve is generated by incubating cells in a high potassium buffer in the presence of 10 μM nigericin at various pH values, as detailed in Ref. 36.

One advantage of this above assay is that it simultaneously analyzes two separate properties that have been associated with apoptosis in numerous cell lines. Thus, it can provide better resolution between apoptotic cells than either method alone (35). The ability of carboxy-SNARF-1AM to resolve apoptotic versus viable cells varies among cell lines (35), and there is always some overlap of these populations on the pH axis (Fig. 4). The dual staining procedure facilitates better discrimination of apoptotic cells when each individual dye alone provides maximum resolution (35). Another point concerns the Hoechst 33342 uptake component of the assay: because this is a kinetic

assay this method cannot be used to sort apoptotic cells, although sorting can be carried out on the basis of the pH by itself.

Detection of Apoptotic Cells based on Changes in Forward Scattering

As discussed previously, apoptotic cells shrink whereas necrotic cells swell and undergo membrane lysis. Because forward light scattering determined by flow cytometry is proportional to cell volume, a decrease in forward light scattering is frequently seen in apoptotic cells (26,35,37). In many cases, however, the forward light scattering histograms of apoptotic and viable cells overlap and make it difficult to discriminate apoptotic cells based solely on this parameter (35). Furthermore, even where discrimination is possible, changes in forward light scattering lag behind changes in cell morphology (26). Finally, not all models of apoptosis change in forward scattering (37). Thus, decreased forward light scattering has limited utility as a measure of apoptosis. However, forward light scattering data are routinely generated in other flow cytometric assays of apoptosis. Thus, forward light scattering can be monitored along with additional markers, as is common when developing novel flow cytometric assays for assessing of apoptosis.

Detection of Apoptotic Cells based on Annexin V Binding

Recently, it was found that apoptotic cell death is associated with a loss of the phospholipid asymmetry of the cell membrane (38) and the appearance of phosphatidylserine on the outer leaflet. Annexin V is a calcium-dependent phospholipid-binding protein that binds preferentially to phosphatidylserine. Thus, apoptotic cells can detected on the basis of increased binding of FITC-conjugated Annexin V (39). The mechanism that underlies the loss of phospholipid asymmetry in apoptotic cells is not known. However, it is thought that the increase in phosphatidylserine on the outer membrane has a physiological function in that it provides a mechanism for macrophages to recognize apoptotic cells allowing for rapid clearance of these cells without lysis. Thus, it is hypothesized that loss of phospholipid asymmetry is important and perhaps characteristic in apoptotic cell death.

The assay involves incubating cells briefly in a solution containing FITC-conjugated Annexin V in a buffer that facilitates its binding (protocol in Appendix VIII). The concentration of $CaCl_2$ in the buffer can be varied to obtain optimal binding for individual cell lines. An example of increased Annexin V binding with increasing exposure to an apoptosis-inducing agent is shown in Figure 5a. Importantly, Annexin V binding is an early marker of apoptosis because it appears in conjunction with the earliest DNA fragmentation.

The Annexin V binding assay has many of the advantages of other flow cytometric assays in that it is simple, rapid, sensitive, and objective. Unfortunately, it is not specific only for apoptosis in that whenever cell membrane integrity is disrupted (even with nonionic detergents), cells may stain with Annexin V (39), probably because sites on the inner membrane become accessible. Thus, even though loss of membrane asymmetry is an early event characteristic of apoptosis, Annexin V alone cannot distinguish cells dying by apoptosis from those dying by necrosis.

One means of assessing whether an apoptotic or a necrotic process is at work is to perform co staining by using Annexin V along with propidium iodide (Fig. 5b). This double staining assay, which involves unfixed cells, allows monitoring changes in phospholipid asymmetry relative to changes in membrane integrity because, in the absence of permeabilization, PI cannot enter intact cells and therefore stains only those that have lost membrane integrity (in contrast to the ethanol fixed/permeabilized cells discussed before, where PI can enter all cells). Upon co-staining of unfixed cells with Annexin V plus PI, apoptotic cells lose phospholipid asymmetry before losing membrane integrity, as shown in Figure 5b. In contrast, necrotic cells, acquire both markers simultaneously. One cautionary note when considering the Annexin V assay is that increased Annexin V binding is not observed in all

(a)   One color staining

Untreated, viable cells

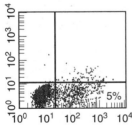

Cells exposed to an apoptosis-inducing agent for varying times

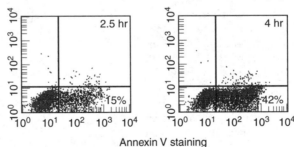

Annexin V staining

(b)   Two color staining

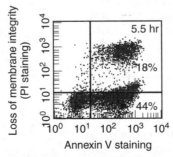

Annexin V staining

Figure 5. Identification of apoptotic cells by increased binding of FITC-conjugated Annexin V. ML-1 human myeloblastic leukemia cells were incubated in the absence (leftmost panel) or presence of the apoptosis-inducing agent etoposide (20 μg/mL) for up to 5.5 hours. (**a**) Apoptotic cells were detected by increased binding of FITC-conjugated Annexin V. Although only a minority of untreated ML-1 cells bound Annexin V, increasing exposure to the apoptosis-inducing agent caused an increasing percentage of cells to stain positively for Annexin V binding. (**b**) Apoptosis was monitored by both Annexin V binding and loss of membrane integrity. Membrane integrity was determined by the ability of the (unfixed) cells to take up and be stained by propidium iodide. Only a portion of the cells with increased Annexin V-FITC fluorescence show loss of membrane integrity because this occurs subsequently. The percentages in the figure indicate the percentage of cells in specific quadrants.

Table 1. Comparison of Various Assays for Cell Death by Apoptosis

Assay	Technically challenging (−1) → Rapid and Simple (+1)	Qualitative (−1) → Quantitative (+1)	Subjective (−1) → Objective (+1)	Nonspecific (−1) → Specific for Apoptosis (+1)
Morphologic Staining	+1	+1	−1	+1
Biochemical Assays				
DNA fragmentation	0	−1	0	+1
PARP cleavage	0	0	+1	+1
Percoll gradient	−1	+1	+1	+1
Flow Cytometric Assays				
EtOH/PI staining	+1	0	+1	0
TUNEL	−1	+1	+1	0
Intracellular pH/ Hoechst uptake	−1	+1	+1	+1
Annexin V binding/PI staining	+1	+1	+1	+1

cells that undergo apoptosis [Zhou and Craig, unpublished observations *and* (37)], and in some models increased Annexin V binding may be a late event compared with other markers of apoptosis, such as decreased intracellular pH (37).

CONCLUSION

This overview of various assays for apoptosis makes it clear that no single method is ideal for every situation. A comparison of the relative strengths and weakness of the various methods is presented in Table 1. The use of more than one method is generally recommended. In this respect, assessing cell death by apoptosis is similar to monitoring cell proliferation or differentiation, which are also most clearly understood by simultaneously using of several complementary assays. It is hoped that this review will be useful, especially to those entering this relatively new field, in deciding which methods are most likely to be appropriate for individual applications.

APPENDIX I

DETERMINATION OF APOPTOTIC MORPHOLOGY

Stock Dyes.

1 mg/mL acridine orange in phosphate buffered saline (PBS)

10 mg/mL ethidium bromide in PBS

1 mg/ml Hoechst 33342 in water

25 µg/mL DAPI in PBS

Cell Preparation.

For ease of assay, cells should be at a density approaching 2.5×10^6 cells/mL. For adherent cells, remove the culture medium (do not discard because apoptotic cells may be floating in the medium). Incubate the monolayer with trypsin for approximately 5 minutes, dislodge the cells, and combine with the cells in the medium removed initially.

Acridine Orange/Ethidium Bromide Assay.

1. Make dye mix by adding 100 µL acridine orange stock and 10 µL ethidium bromide stock to 890 µL PBS, a concentration of 100 µg/mL of each dye in the dye mix.

2. Add 1 µL of dye mix to 25 µL of cell suspension, and incubate for 5 minutes at room temperature. It is important not to leave the cells standing in the dye for extended periods of time, because the ethidium bromide may start to penetrate additional cells.

3. Place 5 to 10 µL of this suspension on a clean glass slide, and cover with a coverslip. Another convenient method of mounting these cells is to make a tripartite chamber by taping two glass slides together using double-sided tape (one strip of tape around all but one of the long sides of the "sandwich" formed by the two slides, plus two short strips of tape across the width of the slide. The latter two strips divide the slide area into three compartments). This arrangement, where cells are placed in the chambers formed by the two slides and the tape, puts less pressure on the cells.

4. Examine the cellular morphology using UV fluorescent microscopy.

5. Count at least 200 cells in several different fields, and score cells with normal versus apoptotic nuclei (Category I versus II), as well as cells that have lost membrane integrity (Category III). Ethidium bromide is taken up only by cells that have lost membrane integrity. The nuclei of these cells stain orange/red.

Hoechst 33342 Staining Assay.

1. Add 1 to 2 µL of the dye stock to 1 mL of cell suspension containing ~2×10^6 cells/mL (1 to 2 µg/mL final concentration), and replace in a 37 °C incubator for 20 to 30 minutes.

2. Place 5 to 10 µL of this suspension on a clean glass slide, and cover it with a coverslip.

3. Examine the cellular morphology using a UV fluorescent microscope.

4. Score cells as to whether they display normal or apoptotic nuclei.

DAPI Staining Assay.

1. Harvest cells by centrifugation and resuspend at approximately 3×10^5 cells/mL in PBS.

2. Add 1 μL of the dye stock to 25 μL of cell suspension (1 μg/mL final concentration), and incubate for 5 minutes.

3. Place 5 to 10 μL of this suspension on a clean glass slide, and cover with a coverslip.

4. Examine the cellular morphology using UV fluorescent microscopy.

5. Score cells as to whether they display normal or apoptotic nuclei.

APPENDIX II

ASSAY OF DNA DIGESTION BY AGAROSE GEL ELECTROPHORESIS

Stock Solutions.

0.5 M EDTA:

    186.1 g $Na_2EDTA \cdot 2H_2O$

    800 mL water

    Add about 20 g NaOH pellets.

    Stir, bring pH to 8.0, and bring final volume to 1 L.

10X Tris-Borate-EDTA (TBE):

    108 g Trizma base (0.89 M Tris-Cl)

    800 mL water

    55 g boric acid (0.89 M boric acid)

    50 mL 0.5 M EDTA (25 mM EDTA)

    Bring pH to 8.0, and bring final volume to 1 L.

20% SDS (w/v):

    100 g SDS

    400 mL water

    Stir until dissolved, using gentle heat if necessary.

    Bring final volume to 500 mL.

10 mg/mL RNase A (1%):

    100 mg RNase A

    10 mL of 10 mM Tris/15 mM NaCl, pH 7.5

    Make solution in a 15 mL tube.

    Place tube in boiling water for 15 minutes to inactivate DNase activity.

    Aliquot 115 μL/ Eppendorf tube, and store at $-20\,°C$.

16 mg/mL proteinase K:

    100 mg proteinase K

    6.25 mL 1X TBE

    When dissolved, aliquot into 210 μL aliquots, and store at $-20\,°C$.

Sample buffer:

    1.0 mL glycerol (10% final concentration)

    0.1 mL 1 M Tris-Cl, pH 8 (10 mM final concentration)

    0.01 g bromophenol blue (0.1% final concentration)

    8.9 mL water

Preparation of Gel.

1. A flat agarose gel is prepared by heating 7 g of agarose until boiling in 350 mL of 1X TBE.

2. Allow agarose to cool, and then pour into a large (20 × 34 cm) horizontal gel support. It is best to have the flat side of the comb toward the top of the gel.

3. Once the 2% gel solidifies, remove the section of the gel above the comb by cutting along the top side of the comb with a scalpel, and discard this piece, while keeping the comb in place.

4. Prepare a 0.9% agarose gel by heating 0.5 g of agarose in 50 mL of 1X TBE and 5 mL of 20% SDS.

5. Allow the agarose to cool to about 45 °C, and then add 200 μL of proteinase K solution. Swirl the solution gently, then use it to replace the section of gel that was removed from the 2% gel prepared in Step 1. Make sure that the comb is in place when pouring this part of the gel. Allow the gel to solidify, and then remove the comb.

Sample Preparation and Gel Conditions.

1. Centrifuge 1×10^6 cells at 1,000 rpm for 5 minutes, and aspirate the medium. If working with adherent cells, collect cells by gently scraping culture dishes with and transferring the medium and cells to a centrifuge tube.

2. Prepare sample loading buffer: two volumes of sample buffer mixed with one volume of 10 mg/mL RNase A.

3. Resuspend cell pellet in 20 μL of loading buffer, and load directly into a well of the agarose gel.

4. Electrophorese in 1X TBE for 16 to 18 hours at 60 volts or 24 hours at 45 volts.

5. Stain the gel with 2 μg/mL ethidium bromide in 200 mL 1X TBE or water for 1 hour. Remove excess ethidium bromide, digested RNA, and SDS by destaining the gel for 48 hours in 2 liters of water. After the first 24 hours, add fresh water. The process may be accelerated by changing the water more frequently.

APPENDIX III

ASSAY OF PARP CLEAVAGE BY WESTERN BLOTTING

Solutions.

Lysis buffer:

    50 mM Tris-HCl, pH 6.8

    2% SDS

    4 M urea

    5% β-mercaptoethanol

Tris-buffered Saline (TBS):

    50 mM Tris-Cl, pH 7.4

    150 mM NaCl

Tris-buffered saline containing Tween-20 TBST:

    50 mM Tris-Cl, pH 7.4

    150 mM NaCl

    0.1% (v/v) Tween-20

Gel loading buffer:

    50 mM Tris-Cl, pH 6.8

    2% SDS

    0.1% bromophenol blue

    10% glycerol

    0.1% β-mercaptoethanol

Preparation of Lysate.

1. If working with adherent cells, gently scrape the culture dish to detach the cells, and transfer to a centrifuge tube. Otherwise go directly to Step 3.
2. Centrifuge cells for 5 minutes at 1,000 rpm, and aspirate the medium.
3. Resuspend cells at 0.5 to 1.0×10^6 cells/mL in PBS, and transfer 1 mL to an Eppendorf tube.
4. Centrifuge for 30 seconds at 15,000 rpm, and then aspirate PBS.
5. Resuspend the cells in 100 μL of lysis buffer.
6. Sonicate for 10 seconds.
7. At this point the samples can be stored at $-20°C$, or Western blot analysis can be done directly.

Western Blot Procedure.

1. Load 40 μL of the lysate prepared above to the wells of a 6% polyacrylamide/SDS minigel.
2. The proteins can be resolved by electrophoresis. Using prestained molecular weight markers, monitor the resolution, and stop electrophoresis at the point when the 60 kDa marker has reached the bottom of the gel.
3. Electrotransfer the proteins to a polyvinylidene fluoride membrane.
*4. Block the membrane for 2 hours at room temperature using TBST containing 5% nonfat dry milk Tris-buffered saline containing Tween-20 and milk (TBSTM).
*5. Probe the membrane for 1 hour at room temperature with a polyclonal antibody (Upstate Biotechnology) to PARP diluted in TBSTM.
6. Remove the antibody, and save for future use.
7. To wash, add TBST to the membrane, incubate for 5 minutes, and then discard the TBST.

* The blocking and primary antibody incubation conditions are specified for using of a PARP polyclonal antibody which can be purchased from Upstate Biotechnology. However, PARP antibodies are available commercially from other sources. Blocking and primary antibody incubation conditions may vary accordingly.

8. Repeat Step 7 twice more.
9. Incubate the membrane for 30 minutes with an appropriate horseradish peroxidase-conjugated secondary antibody diluted in TBST.
10. Discard the secondary antibody/TBST solution.
11. To wash, incubate the membrane with TBS for 5 minutes, and then discard the TBS.
12. Repeat Step 11 twice more.
13. Detect PARP using enhanced chemiluminescence.

APPENDIX IV

PERCOLL GRADIENT FRACTIONATION OF VIABLE AND APOPTOTIC CELLS

Solutions.

Isoperc:

Percoll‖™ (Pharmacia; which has a density of 1.13 g/cc)	86 mL
10× Minimal Essential Medium (MEM; GIBCO)	10 mL
7.5% NaHCO$_3$	1.5 mL
2 M HEPES, pH 7.4	1 mL
10% Pluronic F68 (Sigma; a mild surfactant)	1 mL

Protocol.

1. Detach cells from the growth surface if necessary, and suspend in a tissue culture medium. The number of cells that can be fractionated on one 10 mL gradient ranges from 1×10^4 to 3×10^7. Thus, initially cells should be suspended at a two to four fold higher concentration because they will be diluted in Step 2 below. Some disaggregation of cells may be necessary, but small aggregates of cells will band with single cells of the same density. Note: Apoptotic cells detach as they shrink.
2. Mix cell suspension with isoperc. Use a ratio of between 1 part isoperc/1 part cells and 3 parts isoperc/1 part cells. Optimization of this ratio is critical for optimal separation of viable cells from necrotic cells (1.01 to 1.02 g/cc) and from apoptotic cells (1.08 to 1.10 g/cc).
3. Using 10.5 ml Oak Ridge polycarbonate centrifuge tubes, filled to 1.0 to 1.5 cm of the top, centrifuge at ~5000 g (8000 rpm in type 50 or 70.1 ultracentrifuge rotor) in a 36°C centrifuge. After 60 minutes, deccelerate using low or no brake to avoid disrupting the gradient.
4. For later reference, it is useful to measure the distance of cell-containing layers from the top of the gradient. This is best accomplished using trans- or dark-field illumination.
5. One of two methods can be used to harvest the density fractions:
   a. If the object is simply to collect the cells of the three different densities, one can proceed as follows: Aspirate the top fractions (containing necrotic cells and fragments). Using a 100 μL

pipettor or siliconized Pasteur pipette, collect the band containing viable cells (this will be the major band if viable cells are predominant). Finally, collect and pool the fractions containing apoptotic cells, which begin below the viable cell layer. It is not necessary to collect fractions all the way to the bottom of the gradient because the density of these fractions (>1.1 g/cc) exceeds that of apoptotic cells.

b. If the density of the cells is also to be determined, a 500 μL pipette, calibrated "to contain a known volume" (TC; Lang–Levy or Kirk type, VWR Scientific), can be used. Collect successive 500 μL fractions, starting at the very top of the gradient. Beforehand, tare the dry pipette, then weigh it filled with 500 μL distilled water. Finally reweigh the pipette as each fraction is collected. The weight of the fraction divided by the weight of water is the density in g/cc. Density can also be monitored by using refractive index measurement or colored beads with defined densities if TC calibrated pipettes are not available.

6. For each harvest method, excess Percoll can be eliminated by diluting the fractions (or pools) two to four fold with phosphate buffered saline or other balanced salt solution, and then centrifuging (at 200 g) for 10 minutes. Carefully decant or aspirate the supernatant, and wash again if complete removal of serum protein is desirable.

7. Use a small aliquot of each fraction to assay either cell number or protein content.

APPENDIX V

FLOW CYTOMETRIC DETERMINATION OF DNA CONTENT USING PROPIDIUM IODIDE

Solutions.
10 mg/mL RNase A Stock:
    100 mg RNase A
    10 mL of 10 mM Tris/15 mM NaCl, pH 7.5
    Make solution in a 15 mL tube.
    Place tube in boiling water for 15 minutes to inactivate DNase activity.
    Aliquot 1.15 mL/tube, and store at $-20\,°C$.
Propidium Iodide Stock:
    1 mg/mL propidium iodide in PBS
    Store at $4\,°C$.
DNA staining solution:
    1 mL of propidium iodide stock (0.1 mg/mL final concentration)
    1 mL of 10 mg/ml RNase A Stock (1 mg/mL final concentration)
    8 mL of PBS
Prepare immediately before use.

Sample Preparation and Analysis.

1. If working with adherent cells, remove and save the medium from the tissue culture dish (place in a centrifuge tube). Add trypsin to the adherent cells, and dislodge them from dish (after incubation at $37\,°C$ for 5 minutes). Return the saved medium to the tissue culture dish to stop the trypsin. Transfer the cells in medium/trypsin to the centrifuge tube.

2. Centrifuge cells for 5 minutes at 1,000 rpm, and then aspirate the medium.

3. Resuspend cells in 3 mL of cold PBS, and then add 1 mL of cold 95% ethanol. Mix by gentle inversion of the tubes.

4. Incubate cells at $4\,°C$ overnight to allow all low molecular weight DNA to exit. At this point the cells can be stored for several days up to a week at $4\,°C$.

5. Centrifuge cells as in Step 2.

6. To wash, resuspend in 3 mL of PBS, and centrifuge as in Step 2.

7. Resuspend cells at 0.5 to 1×10^6 cells/mL in DNA staining solution.

8. Incubate at $37\,°C$ for 30 minutes, and protect from light.

9. Place on ice, and analyze by flow cytometry with excitation at 488 nm, and monitor emission at 585 nm.

APPENDIX VI

TUNEL OR TERMINAL DEOXYNUCLEOTIDYL TRANSFERASE (TDT) ASSAY

Solutions.
1% formaldehyde diluted in PBS
Enzyme buffer

Final conc.	Stock	μL of stock / 500 μL final volume
0.1 M Na Cacodylate, pH 7.0	0.5 M	100 μL
0.1 mM dithiothreotol (DTT)	10 mM	5 μL
0.05 mg/ml BoVine Serum Albumin (BSA)	1 mg/mL	25 μL
1.0 mM CoCl$_2$	25 mM	20 μL
*5 units TdT/50μL	25 units/μL	2 μL
*0.5 nmol biotin-dUTP/50 μL	1 nmol/μL	5 μL

*add immediately before use.

Cell Staining Buffer (protect avidin-FITC from light)
    2.5 mL 4X SSC (where 1X SCC consists of 0.15 M NaCl/0.15 M NaCitrate, pH 7.0)
    2.5 μL Triton-X-100 (0.1% final concentration)
    0.125 g nonfat dry milk (5% w/v final concentration)
    Vortex to mix.
    In the dark, add 25 μL 0.25 mg/mL avidin-FITC (2.5 μg/mL final concentration) and mix.

Sample Preparation and Analysis.

1. If working with adherent cells, remove and save the medium from the tissue culture dish (place in a centrifuge tube). Add trypsin to the adherent cells, and dislodge them from dish (after incubation at 37 °C for ~5 minutes). Return the saved medium to the tissue culture dish to stop the trypsin. Transfer the cells in medium/trypsin to the centrifuge tube.

2. Centrifuge cells for 5 minutes at 1,000 rpm, and then aspirate medium.

3. Resuspend cells in cold 1% formaldehyde at 1×10^6 cells/mL. Incubate on ice for 15 minutes to fix the cells.

4. Centrifuge cells as in Step 2.

5. At this point the cells can be stored in 70% ethanol at -20 °C at 1×10^6 cells/mL.

6. If cells are stored, they must be rehydrated before use by centrifuging cells as in Step 2, aspirating the 70% ethanol, and resuspending the cells in PBS. Repeat this wash step one additional time for a total of two washes with PBS.

7. Some cells, such as CHO cells, require a methanol permeabilization step by centrifuging cells as in Step 2. After aspirating the PBS, resuspend the cells at 1×10^6 cells/mL in methanol. Incubate the cells for 5 minutes at room temperature. Then add an equal volume of PBS, and centrifuge as in Step 2. Repeat the rehydration procedure (Step 6) before proceeding to Step 8.

8. Transfer 500 μL of cell suspension to an Eppendorf tube, centrifuge for 30 seconds at 15,000 rpm, and then aspirate PBS.

9. Resuspend pellet in 50 μL enzyme buffer.

10. Incubate at 37 °C for 30 minutes.

11. Wash cells twice with 100 μL PBS.

12. Resuspend pellet in 100 μL staining buffer.

13. Incubate 30 minutes at room temperature.

14. Rinse cells once with 100 μL 0.1% Triton-X-100 (in PBS).

15. If simultaneous monitoring of cell cycle distribution is desired, resuspend in 50 μL propidium iodide stock solution (see previous section), and incubate for 30 minutes at 37 °C in the dark.

16. Flow cytometric analysis is carried out with fluorescent excitation at 488 nm, and fluorescent emission is monitored at 530 nm for FITC and 585nm for propidium iodide.

APPENDIX VII

TWO-COLOR FLOW CYTOMETRIC ANALYSIS OF INTRACELLULAR PH AND HOECHST 33342 UPTAKE

Solutions.
Carboxy-SNARF-1AM:

    50 μg carboxy-SNARF-1AM (Molecular Probes)

    50 μL DMSO

Protect from direct light and use immediately after preparation.

Hoechst 33342 stock:

    1 mg/mL in PBS

    Store at 4 °C and protect from direct light.

Nigericin stock:

    10 mg/mL in ethanol

    Store at 4 °C.

K-buffer:

    17.3 mM 4-(2-hydroxyethyl)-1-piperazine-ethane-sulfonic acid (HEPES)

    17.3 mM 2-(N-morpholino)-ethanesulfonic acid (MES)

    30 mM NaCl

    115 mM KCl

    1 mM $MgCl_2$

    0.1 mM $CaCl_2$

Prepare 250 mL of K-buffer at pH 6.5 and 250 mL at pH 8. These two stock buffers can be mixed at different ratios to produce a range of pH standards between 6.5 and 8.0.

Sample Preparation and Analysis.

1. If working with adherent cells, gently scrape the culture dish to detach the cells and transfer to a centrifuge tube. We have avoided the use of trypsin to reduce the possibility that ion channels are proteolytically cleaved which could alter intracellular pH. An alternative mechanism for detaching cells is to transfer the medium to a centrifuge tube and then add 5 mM EDTA pH 8.0 to the plates to dislodge the cells. Then add the EDTA cell suspension to the centrifuge tube containing medium.

2. Centrifuge cells for 5 minutes at 1,000 rpm, and aspirate medium. Resuspend the cells in fresh medium, centrifuge again for 5 minutes at 1,000 rpm, and aspirate the medium.

3. Resuspend cells at 0.5 to 1.0×10^6 cells/mL in tissue culture medium, and transfer 1 mL to a vial appropriate for flow cytometry.

4. Add 1 μL of the carboxy-SNARF-1AM stock solution to the vial and incubate cells for 1 hour at 37 °C.

5. One to 10 minutes before analysis on a flow cytometer with capacity for two different excitation wavelengths (i.e., FACStar Plus, Becton Dickson), add 1 μL Hoechst 33342 stock to a final concentration of 1 μg/mL.

6. Flow cytometry is carried out with excitation of carboxy-SNARF-1 at 488 nm, and emission is monitored at 585 and 640 nm. Hoechst fluorescence is assessed by excitation with the ultraviolet lines at 351 and 364 nm, and emission is monitored at 440 nm. The protocol can be stopped here if the goal is to assess viable versus apoptotic cells.

If it is additionally desired that intracellular pH be calculated, a standard curve must be generated as follows:

7. Transfer 0.5 to 1.0×10^6 cells to each of five Eppendorf tubes.

8. Centrifuge the tubes for 30 seconds at 15,000 rpm and aspirate the supernatant medium.

9. Resuspend the cells in 1 mL of K buffer titrated to various pH values between 6.5 and 7.8.

10. Add 1 μL of Nigericin (Final concentration of 10 μg/mL) to each vial.

11. Incubate cells for 10 minutes at 37 °C.

12. Analyze for carboxy-SNARF-1AM emission as in Step 6.

13. Plot intracellular pH versus the fluorescent ratio of 585/640 nm to produce a standard curve.

Flow Cytometry Data Analysis. All analysis of flow cytometry data can be done using Cell Quest software (Becton Dickinson).

APPENDIX VIII

FLOW CYTOMETRIC DETERMINATION OF ANNEXIN V BINDING

Solutions.
Annexin V binding buffer:

   10 mM HEPES, pH 7.4 (pH with NaOH)

   140 mM NaCl

   2.5 mM $CaCl_2$

   Filter using a 0.2 μm pore filter.

Annexin V-FITC (can be purchased from many commercial sources, including Beckman-Coulter, Immunotech Division; the recombinant human protein is produced in *E. coli*)

Sample Preparation and Analysis.
1. If working with adherent cells, remove and save the medium from the tissue culture dish (place in a centrifuge tube). Dislodge the adherent cells using 5 mM EDTA, pH 8.0. Return the saved medium to the tissue culture dish. Transfer the cells in the medium/EDTA to the centrifuge tube, and wash to remove the EDTA (because chelation of calcium could interfere with Annexin V binding.) Trypsin can be used, but with great caution, to minimize exposure of the cells to this proteolytic enzyme.

2. Centrifuge cells for 5 minutes at 1,000 rpm, and then aspirate medium.

3. Resuspend cells at 2 to 5×10^5 cells/mL in PBS. Transfer 180 μL to an Eppendorf tube, centrifuge for 30 seconds at 15,000 rpm, then aspirate PBS.

4. Resuspend cells in 180 μL of Annexin V binding buffer.

5. In the dark, add 10 μL of Annexin V-FITC. For simultaneous monitoring of loss of membrane integrity, also add 20 μL of a solution of 20 μg/mL propidium iodide (in PBS). Mix gently and incubate for 10 minutes at room temperature.

6. Centrifuge cells for 30 seconds at 15,000 rpm, aspirate buffer, and resuspend in 1 mL of PBS.

7. Centrifuge cells for 30 seconds at 15,000 rpm, aspirate PBS, and resuspend in 0.5 to 1 mL of binding buffer.

8. Flow cytometric analysis is carried out with fluorescent excitation at 488 nm and monitoring fluorescent emission at 530 nm for FITC and 585 nm for propidium iodide.

BIBLIOGRAPHY

1. R.A. Schwartzman and J.A. Cidlowski, *Endocr. Rev.* **14**, 133–151 (1993).

2. J.F.R. Kerr, A.H. Wyllie, and A.R. Currie, *Br. J. Cancer* **26**, 239–257 (1972).

3. J.J. Cohen, R.C. Duke, V.A. Fadok, and K.S. Sellins, *Annu. Rev. Immunol.* **10**, 267–293 (1992).

4. J. Strater, K. Koretz, A.R. Gunthert, and P. Moller, *Gut* **37**, 819–825 (1995).

5. C.B. Thompson, *Science* **267**, 1456–1462 (1995).

6. S. Bodis et al., *Cancer (Philadelphia)* **77**, 1831–1835 (1996).

7. M.L. Gougeon and L. Montagnier, *Science* **260**, 1269–1270 (1993).

8. M.A. Barry, C.A. Behnke, and A. Eastman, *Biochem. Pharmacol.* **40**, 2353–2362 (1990).

9. R.C. Duke, J.J. Cohen, and J.E. Coligan, et al., eds., *Current Protocols in Immunology*, Vol. 1, Wiley, New York; 1992, pp. 3.17.1–3.17.16.

10. A.H. Wyllie, J.F.R. Kerr, and A.R. Currie, *Int. Rev. Cytol.* **68**, 251–306 (1980).

11. M.J. Arends and A.H. Wyllie, *Int. Rev. Exp. Pathol.* **32**, 223–254 (1991).

12. P. Zhou, L. Qian, K.M. Kozopas, and R.W. Craig, *Blood* **89**, 630–643 (1997).

13. A.H. Wyllie, *Nature (London)* **284**, 555–556 (1980).

14. T. Eckhardt, *Plasmid* **1**, 584–588 (1978).

15. P. Huang, K. Ballal, and W. Plunkett, *Cancer Res.* **57**, 3407–3414 (1997).

16. M.A. Barry, J.E. Reynolds, and A. Eastman, *Cancer Res.* **53**, 2349–2357 (1993).

17. J. Yuan et al., *Cell (Cambridge, Mass.)* **75**, 641–652 (1993).

18. D.W. Nicholson, *Nat. Biotechnol.* **14**, 297–301 (1996).

19. S.H. Kaufmann, *Cancer Res.* **49**, 5870–5878 (1989).

20. Y.A. Lazebnik et al., *Nature (London)* **371**, 346–347 (1994).

21. C.M. Wolf, J.E. Reynolds, S.J. Morana, and A. Eastman, *Exp. Cell Res.* **230**, 22–27 (1997).

22. G.M. Shah, R.G. Shah, and G.G. Poirier, *Biochem. Biophys. Res. Commun.* **229**, 838–844 (1996).

23. A.H. Wyllie and R.G. Morris, *Am. J. Pathol.* **109**, 78–87 (1982).

24. J.A. McBain, A. Eastman, C.S. Nobel, and G.C. Mueller, *Biochem. Pharmacol.* **53**, 1357–1368 (1997).

25. J.A. McBain et al., *Int. J. Cancer* **67**, 715–723 (1996).

26. A.E. Milner, H. Wang, C.D. Gregory, in M. Al-Rubeai and A.N. Emery, eds., *Flow Cytometry Applications in Cell Culture*, Dekker, New York, 1996, pp. 193–209.

27. S. Morana et al., *J. Biol. Chem.* **271**, 18263–18271 (1996).

28. M.A. Hotz, J. Gong, F. Traganos, and Z. Darzynkiewicz, *Cytometry* **15**, 237–244 (1994).

29. Z. Darzynkiewicz et al., *Cytometry* **13**, 795–808 (1992).

30. J. Li and A. Eastman, *J. Biol. Chem.* **270**, 3203–3211 (1995).

31. J.E. Reynolds, J. Li, R.W. Craig, and A. Eastman, *Exp. Cell Res.* **225**, 430–436 (1996).

32. M.G. Ormerod et al., *Cytometry* **14**, 595–602 (1993).

33. C. Dive et al., *Biochim. Biophys. Acta* **1133**, 275–285 (1992).

34. I. Schmid, C.H. Uittenbogaart, and J.V. Giorgi, *Cytometry* **15**, 12–20 (1994).

35. J.E. Reynolds, J. Li, and A. Eastman, *Cytometry* **25**, 349–357 (1996).

36. A. Eastman, *Methods Cell Biol.* **46**, 41–55 (1995).

37. T. Frey, *Cytometry* **28**, 253–263 (1997).

38. V.A. Fadok et al., *J. Immunol.* **148**, 2207–2216 (1992).

39. G. Koopman et al., *Blood* **84**, 1415–1420 (1994).

See also CELL CYCLE EVENTS AND CELL CYCLE-DEPENDENT PROCESSES; CHARACTERIZATION OF CELLS, MICROSCOPIC; FLOW CYTOMETRY OF PLANT CELLS.

CHARACTERIZATION OF CELLS, MICROSCOPIC

ERWIN HUEBNER
University of Manitoba
Winnipeg, Manitoba
Canada

OUTLINE

INTRODUCTION AND PERSPECTIVE

The awe and excitement that cell biologists and biotechnologists experience as they probe the inner secrets of cells with the diverse types of microscopy available today are akin to that felt by mankind seeing dramatic images from distant planets and stars from astronomical telescopes and space exploration.

Microscopy opens vistas not possible with the naked eye and allowed early pioneers like Antoine Leuwenhook, Robert Hooke, and the early giants in cell biology to discover that living organisms are composed of cells thus creating the field of cell biology.

For almost 400 years, microscopes have revealed the microcosm of the cellular world. The revealing of this "inner universe" from the 16th century until today has impacted on virtually all facets of humanity as it probes the essence of life itself. Those of us working in cell biology, biotechnology, and developmental biology have the good fortune of sharing and experiencing the incredible wonder past microscopists, like Hooke, Abbe, Zeiss, Zernicke, Nomarski and the many others who developed and used the new advances of their eras, must have felt.

The spectrum of types of light microscopy and electron microscopy that have become available through the genius and creative efforts of the myriad of scientists since Zacharias Jansen made the first two-lens light microscope in 1595 in Holland is astonishing. The renaissance and revolution in light microscopy we are now experiencing has and continues to generate new technologies and new ways of seeing, measuring structures, and characterizing events in living cells. Images now attainable would have been considered science fiction, even 10–20 years ago. So as we enter the new millennium, we can visualize structures in cells with light microscopy and allied methods that were deemed impossible 50–60 years ago when electron microscopes were being brought to bear to resolve the fine structure of cells. The frontiers have been pushed to where we can now see structures like single microtubules, centrioles, pinocytotic vesicles, live organelles; localize gene sequences; observe biochemical processes *in situ* in living cells, see dynamic changes in ions like calcium; and map intracellular pH, to cite a few. Microscopy continues to be a problem-solving tool that is yielding significant benefits in biomedical research, particularly in cell biology. These benefits are being reaped because rapid advances during the past decade have occurred with spectacular advances in optical systems and components, the inclusion of lasers and scanning devices in optical systems, the marriage between microscopes and electronic imaging devices and detectors, the utilization of computers and image processing, as well as the continual introduction of new fluorescent dye molecules to characterize and highlight cellular components and processes.

In view of the enormity of the field of microscopy, past and present, the many excellent research papers, reviews, and books, as well as informative web sites available, it would be presumptuous to do justice to the topic in as short an article as this. Thus the aim is to provide a selective perspective, including a cursory coverage of the historic aspects, an overview of some basic information on microscopy in general, and to follow this with highlights of the range of light microscopes from bright-field microscopy to the newest multiphoton microscopy, scanning probe and scanning near-field optic microscopes, electronic imaging and image processing; laser tweezers, and electron microscopy, including TEM, SEM, and freeze-fracture.

DEVELOPMENTS AND MILESTONES IN MICROSCOPY

The following highlights some of the milestones in microscopy beginning with invention of a two-lens microscope by Jansen in 1595. The term microscope was coined by Giovani Faber in 1625. Subsequently, cells were observed by Robert Hooke in the mid 1660s, and Malpighi visualized blood capillaries. The Dutch draper Leeuwenhoek built many single lens microscopes in the late 1660s and 1670s with which he observed protozoa, bacteria, sperm, and other cells. The eighteenth century yielded advances in mechanical design of compound microscopes and notably the improvement of lenses particularly by the British scientist Lister in 1829 who developed achromatic lenses by using flint glass. Nicol prisms important for polarizing microscopy were introduced in 1829 by Fox-Talbot as was the first use of reflecting microscopy in the 1820s. A watershed in the improvement of microscopy, key to the advances in cell biology from the 1850s to the turn of the century, was the contributions of Abbe in the 1870s–1880s in providing an understanding of image formation and the importance of the collection angle of light received by the objective. The concept of numerical aperture and its relationship to resolution is expressed by the Abbe formula: $d = \lambda/NA$ where $NA = n \sin \alpha$. n is the refractive index of the medium between the object and the objective, and α is one-half of the collection angle of the objective. Abbe with Zeiss was instrumental in bringing about major improvements in objectives, so by the 1880s objectives of the highest NA (1.4) were reached and structures 0.2 µM apart could be resolved. Subsequently, there has been a plethora of advances in corrected objectives to this day with computer designed optics. A key advance in the efficient use of microscopes was the consideration of uniform illumination and alignment of the optical components. In 1894 Köhler (1) introduced an illumination alignment system still central to microscopy today, namely, Köhler illumination. For in-depth coverage of resolution, optical pathways, and alignment procedures, the reader is referred to Inoué and Spring (2), Keller (3), Lacey (4), Spencer (5), and Bradbury (6), as well as the various web sites noted at the end of this review.

During the twentieth century there was a blossoming of various types of microscopy including a variety of contrasting approaches (dark-field, phase-contrast, interference-contrast, asymmetrical-illumination, etc.), the application of polarizing microscopy, and the introduction of fluorescence microscopy. More recent advances in light microscopy include confocal and multiphoton microscopy, the use of video and image processing, and allied methods. The 1980s saw the introduction of scanning tunneling microscopy (STM) and atomic force microscopy (AFM), as well as scanning near-field optical microscopy (SNOM).

The past few decades are also marked with the imaging of cells using acoustic microscopy, Doppler–shift microscopy, X-ray microscopy, NMR, and others.

The first electron microscope (EM) was built in Germany by Ruska in 1931 (7–10) and rested on the prior findings of de Broglie in 1924, who showed that electrons travel in waves, and those of Busch in 1926 who showed they could be focused with electromagnetic lenses. With the short wavelength attainable in an EM, particularly at higher electron gun accelerating voltages, resolutions of 1–2 Å (10^{-10} m) are possible. Advances in the design and construction of EMs coupled with the advancement of improved fixation, initially due to osmium tetroxide and subsequently glutaraldehyde, made possible the high resolution of cell ultrastructure and descriptions of cell organelles. The 1950s to 1980s are replete with EM studies that characterized hundreds of cell types. Scanning EM (11) made possible three-dimensional viewing of cells, as well as cell interior components when appropriately prepared. The use of ultracryofreezing methods combined with carbon/platinum replica freeze-fracture methods made it possible to visualize the molecular architecture of membranes, cell junctions, and macromolecular arrays such as F-actin filaments and microtubules. Analytical methods have also been incorporated into EMs making possible molecular characterizations using X-ray microanalysis, energy loss spectroscopy, or electron spectroscopic imaging, and other approaches. The combination of immunogold-labeling techniques and the resolving power of EM has impacted significantly on the characterization of cell components, receptor sites, etc. In the 1990s the emphasis has shifted back to light microscopy, in particular with confocal and multiphoton microscopy and the emerging areas of atomic force and scanning near-field optic microscopy. The ability to visualize structures and processes in live cells with fluorescent probes and to utilize digital imaging methods has supplanted electron microscopy in many areas of cell biology. For additional information and perspectives on new developments in microscopy, the reader is referred to Refs. 12–19, Ref. 20(Vol. 3), and the various microscopy web sites.

LIGHT MICROSCOPY

Basic Concepts, Microscope Components and Principles

A superb up-to-date reference source that is comprehensive in coverage of fundamental principles and practical aspects of basic microscopy, as well as specialized types, is by Inoué and Spring (2). This is an invaluable resource in any cell biology lab that uses microscopes. Definitions of the terminology used in microscopy are available in the Royal Microscopical Society Dictionary (21).

The essential components of a light microscope are as follows:

Illumination Source. A light source which usually is built-in and includes a collector lens, a diaphragm, and has focusing and centering capabilities. The specific type of light source will vary depending on the type of microscope and application. Light sources could be tungsten filament lamps (usually 12 V), quartz-halogen lamps (often 12 V 100 watts), high-pressure arc lamps HB0 100 or 50 mercury or XB0 75 Xenon that are operated with a DC power supply, or various high-intensity laser light sources depending upon which wavelengths are needed. Some setups may have more than one light source fitted for multiple applications.

In most microscopes, the light is introduced directly into the optical path via a collector lens, but increasingly fiber optic light guides are being used to bring the light path to the condenser or other optical components. Where a single fiber optic fiber is being used, improved images are attained if the light field is made uniformly homogeneous with minimal loss of luminance by using a "light scrambler" (2,22). A field diaphragm in the illumination pathway is essential for setting up Köhler illumination and centering the condenser.

Condenser. Between the field diaphragm and the condenser lens is a condenser iris diaphragm, and depending upon the specific type of specialized microscopy, there may be other additional components, for example, a polarizing filter in polarizing microscopy, Differential Interference microscopy (DIC), and Hoffman modulation-contrast microscopy, a beam splitter in Nomarski DIC, a center stop in dark-field microscopy, a phase annulus in phase-contrast microscopy, and an off axis aperture in oblique or anaxial illumination, or an excitation filter in fluorescence microscopy.

The quality and numerical aperture of the condenser is important to the overall image quality and resolution. Resolution of the microscope is

$$d = \frac{\lambda}{\text{NA objective} + \text{NA condenser}}$$

where d is the smallest distance between two resolvable points. The best condensers are achromatic-aplanatic condensers. However, Abbe condensers are also common. For condensers with NAs greater than 0.9, optimal results are attained if immersion oil is put between the front lens of the condenser and the specimen. The condenser must also be properly focused as there is a focal length, and the light must be focused at the specimen which will fill the back lens of the objective with even illumination when the iris diaphragm is properly adjusted. High-quality condensers should be free of chromatic and other aberrations. The lens elements should have high transmission properties to minimize light loss.

Stage. Next is the specimen which is held on a stage. Most microscopes have mechanical stages for precise positioning, and more sophisticated systems have X-Y computer-controlled positioning stages. Polarizing microscopes usually have circular stages for specimen orientation. Optical factors important in the specimen itself include the glass slide, coverslip, and mounting medium. Slides are usually glass, but in some special applications may be quartz. An especially important factor is coverslip thickness. The thickness of #0 coverslips is 0.1–0.13 mm, #1 is 0.13–0.17 mm, #1.5 is 0.15–1.20 mm,

#2 is 0.17–0.25 mm, and #3 is 0.25–0.5 mm. Although some objectives have correction collars to adjust for different coverslip thicknesses, most objectives have a fixed correction factor for a 0.17 mm thickness. Usually, #1.5 coverslips should be used for optimal image quality. The refractive index of the medium that contains the cells and the medium between the coverslip and objective lens also are important factors because reducing the refraction of light as it passes from media of differing refractive indexes reduces light loss and background noise due to random scattering. Because the refractive index of glass is 1.52, the use of immersion oil in place of air between the slide and immersion objective lens allows for higher NAs and higher resolution and image quality. In fluorescence microscopy, the use of media without or minimal fluorescence is essential.

Objective Lens. Detailed knowledge of the characteristics, type, and quality of the objective lens is one of the most important aspects in microscopy. Having lenses free of aberrations, with high light-gathering and transmission characteristics and the highest numerical aperture possible are pivotal in attaining high-quality images. Furthermore in various specialized types of microscopy, additional components (e.g., phase plate in phase-contrast microscopy) are positioned in the back focal plane of finite tube length objectives. For in-depth coverage of objectives, refer to Refs. 2,3, and 23. The following only skims the highlights and presents some of the terminology. Unfortunately, markings on objectives may vary among manufacturers.

Until recently, the vast majority of microscopes were fixed tube length of 160 or 170 mm (mechanical tube length from objective nosepiece opening to eyepiece opening). In such finite systems, the objective projects a real image in the microscope. The lens focuses convergent light in the interlens space. Most manufacturers have switched over to infinity-corrected optics (2,3,23). Infinity objectives, marked ∞, are designed to project an image to infinity so that essetially there is parallel light between the objective and eyepiece. The advantage is that one or more optical components such as is needed in polarizing, phase, DIC, or fluorescence microscopy can be inserted without changing the functional tube length. One, two, three or more components can be introduced without affecting microscope performance. So multimode microscopy is easier to design without deleterious effects on image quality. But because the light rays are parallel to make the light convergent and project an image to the eyepiece, a second lens, or so-called tube lens, is needed between the objective and the eyepiece. Finite systems do not have a tube lens. Objectives cannot be interchanged between infinity and finite systems.

Beyond these two fundamental types there are other features important to know for proper selection and use of objectives. Most standard objectives are achromats (sometimes marked Achromat) with chromatic aberration correction for two colors (red and blue) and spherical correction in green. Apochromats are corrected for three spectral colors (blue, green, red). Plan indicates correction for a flat field. Fluorite objectives, often marked FL, fluor, Neofluar, etc., use fluorite glass and are ideal for fluorescence microscopy. Fluorite achromates have better spherical correction than conventional achromates. Etched on the objective could well also be the magnification, coverslip correction (usually 1.7), and numerical aperture (1.4 is highest presently attainable). In terms of final image magnification, a ballpark practical magnification limit should be around 500–1000 times the objective NA, preferably 750 or less. Objectives may also have coverslip thickness correction collars or diaphragms (useful in fluorescence microscopy to adjust image intensity). Some objects are high dry and are to be used in air, and others are designed to have an immersion fluid between the objective and specimen. The immersion medium is most frequently oil (marked oil, oel) but glycerine (gly), water (w), and other immersion objectives are also used. Individual manufacturers often additionally identify these with color-coded rings etched around the objective (see Tables 2–3 in Ref. 2). As the magnification of objectives increases, the working distance (distance between the objective and specimen when at the focal point) decreases as does the depth of focus. Objectives labeled LD are designed to provide longer working distances (useful for microinjection, manipulation, etc.). For polarizing microscopy, objectives must be strain-free and are often marked POL.

As will be dealt with later in the context of the specialized microscopy types, additional components can be incorporated either in the back focal plane of the objective or between the objective and ocular (depending on the type of system). Examples of such elements are Nicol or Wollaston prisms for Nomarski DIC, analyzer (polarizing filter) and compensator for polarizing microscopy and others, phase plate for phase-contrast microscopy, attenuation filter in single sideband edge-enhancement microscopy, barrier or emission filter in fluorescence microscopy, modulator plate in Hoffman modulation microscopy, etc.

Before moving on to the eyepiece or ocular, it is important to stress that using the highest NA objectives possible is essential for critical excellent microscopic imaging. The light-gathering capability, brightness, and resolving power are dramatically improved in higher NA objectives. The micrographs in Figure 1 illustrate that the detail resolved in the diatom test slide is much better with NA 1.4 versus 0.65. This becomes especially important when light intensity is limited (e.g., in polarizing microscopy, certain fluorescence). In addition to the importance of NA in image resolution and spatial frequency, according to Abbe (see Ref. 2), contrast transfer function (CTF) and modulation transfer function (MTF) are also affected. One can determine

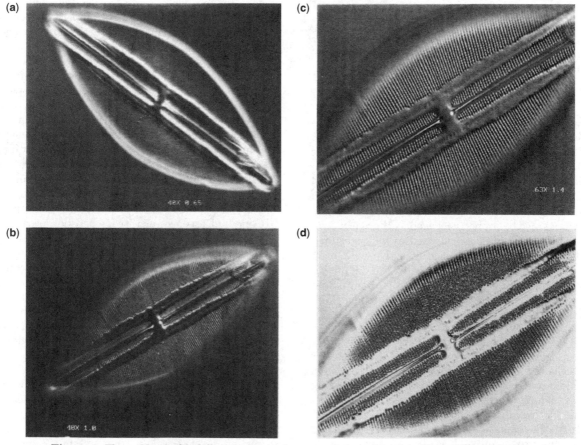

Figure 1. These Nomarski differential interference micrographs of a diatom illustrate the improved resolution with increasing objective NA (**a**) 40 × NA 0.65; (**b**) 40 × 1.0 NA; and (**c**) 63 × 1.4 NA; (**d**) this black and white image of a pseudocolored image of (c) reveals the fine structural detail resolvable.

a specific optical transfer function (OTF) or MTF for any objective under a specific condition of use. The MTF specifies the performance system to image spatial detail. For three-dimensional image reconstruction and deconvolution routines used in image processing (particularly in fluorescence and confocal applications), the determination of the axial intensity distribution above and below the objective's focal point, expressed as a point-spread function, is needed. The axial intensity distribution of a point source above and below focus generates a three-dimensional diffraction pattern. One can calculate a point-spread function from a stack of serial optical sections of a point source (circular aperture or fluorescent bead). Suffice it to say that the quality of the objectives is at the heart and soul of the microscope's performance.

Eyepieces and Imaging. The eyepiece projects and magnifies the image formed by the objective lens. The power and diameter of field of view, as well as information on correction of aberrations and special features, are often indicated. CF indicates chromatic aberration free, W or similar markings indicate wide field, C or K indicates compensation, etched glasses are for high focal point, p or pl is for plan, and Kpl indicates a compensating flat field.

Conventionally observations are made by eye. However, increasingly, images are obtained by using a variety of video cameras, detectors, and photomultipliers with linkage to computer-based image processing and display systems. These will be covered in a subsequent section. Particularly useful are test slides to assess the overall imaging quality and spatial and axial resolution of the overall microscope system. Convenient test slides include diatoms with their precise repeat spatial patterns (e.g., *Amphipleura pellucida* — 0.24 µM) *Pleurosigma angulatum* — 0.62 µM), butterfly scales, specially prepared pattern slides [e.g., Richardson slide (24), MBL-NNF slide (2)].

The preceding are the components that comprise a basic light microscope, including the possible variations in optical components and brief reference to concepts and practical aspects. Microscopes can be of the standard upright configuration or can be inverted with the light source in the upper vertical position, as one finds in inverted microscopes often used for examining cells in culture flasks or chambers.

Survey of the Variety of Light Microscopes and Cell Biology Usage

Visualization of structures in cells and tissues requires magnification because cells range from about 10–50 µM in diameter and are not resolvable by the naked eye. Along with magnification, which is achieved by the two magnifying systems (objective and eyepiece) working in tandem, structure can be discerned only by the presence of contrast within cells. The human eye cannot detect phase differences but can discriminate contrast differences in the range of 2–20% (18). Contrast can be achieved in various ways involving absorption, refraction, diffraction, reflection, light scattering, birefringence, and fluorescence. Central is the interaction between matter and light. In instances where this interaction

results in phase changes, visualization is achieved only if these can be converted to contrast differences. Most living cells lack sufficient inherent contrast, so that bright-field microscopy reveals little without the use of stains to reduce the amplitude of certain wavelengths and render contrast differences preferentially. A number of different strategies have been developed to generate contrast thereby revealing structures in unstained or live cells and tissues (2,18,20,25–28). The following covers the highlights and uses of the array of light microscope types available for cell technology. A variety of useful contributions can be found in the Royal Microscopical Society Handbook Series (27) and Cell Methods Handbooks [28, also Ref. 20(Vol. 3)]. A particularly good web site that covers various types of microscopes is the Molecular Expressions-Microscopy Primer (http://micro.magnet.fsu.edu/primer/webresources.html).

Bright-Field Microscopy. The fundamental components of the bright-field microscope are as indicated earlier. Live cells and tissues are poorly viewed with a bright field because only opaque or naturally occurring pigment granules absorb or refract sufficient light to generate discernible contrast. Although there are a few classical vital dyes (e.g., Janus Green B, Evan's Blue) that reveal selected facets of live cells, the primary use of bright-field microscopy is on fixed cells stained with dyes that bind selectively to cell structures. Figure 2a shows a typical paraffin section stained for bright-field observation. There is a wide array of selective and semiselective dyes available for cytological, histological, and histochemical uses (20(Vol. 3),29). The limitation is that fixed cells are usually used, and the problem of fixation artifacts must be considered.

Contrast visible to the eye in bright-field microscopy is due almost exclusively to the selective absorption of light. Because of differential binding affinities to stains, certain cell components become distinguishable. The classic example is hematoxylin and eosin staining (Fig. 2a) which highlights nuclei in shades of blue, and the cytoplasm pink, depending on the protocol used.

Dark-Field Microscopy. This is one of the oldest and least costly forms of microscopy that renders contrast to living or unstained cells (30–32). It reveals cells (particularly edges and boundaries) in bright contrast against a dark background (Fig. 2b). During the past 20 years, this form of microscopy has received renewed interest because it provides high-contrast viewing of very fine structures below the resolution limit of the microscope (2,3). One can view bacteria, cilia, flagella, single cytoskeletal elements, like F-actin and microtubules, and isolated particles and cell components.

Dark-field microscopes (DF) have special condensers with a center stop which blocks out all of the central rays that create a cone of light. The ray diagrams in Figure 3 show dark-field microscopy compared to bright-field. Without a specimen in place and the condenser and objective properly focused, all of the undeviated light of this cone misses the objective. Once a specimen is introduced into the light path, some of this light

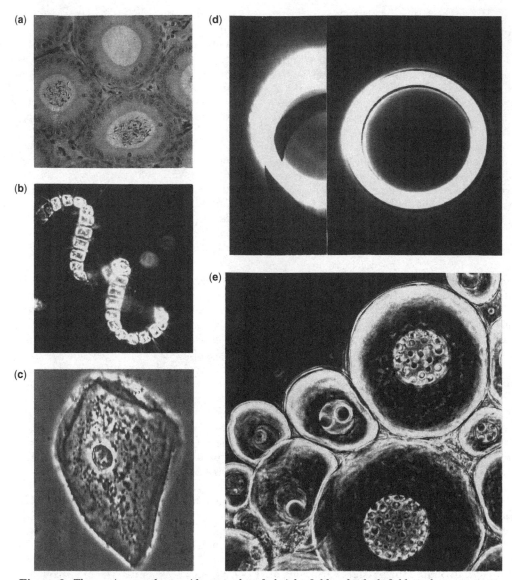

Figure 2. These micrographs provide examples of a bright-field and a dark-field on phase-contrast images. (**a**) an H and E stained epidymus section; (**b**) a live marine algal chain viewed in a dark field; (**c**) a live human buccal cell viewed with phase contrast; the two panels of (**d**) show the phase ring and annulus misaligned and aligned; (**e**) a phase view of live multicellular tissue (goldfish ovary) where nuclei and nucleoli are seen.

is diffracted or refracted (deviated light) and enters the objective. So, only light scattered by the cellular components is seen, and the specimen appears as a bright extremely high-contrast structure against a black background. Dry DF condensers are usually used with objectives of NA <0.75. High NA objectives used for DF need a diaphragm to exclude undeviated light, some of which enters the objective. Not only is DF useful for looking at intact cells but has had renewed impact in visualizing exceedingly small structures. It is valuable in studies of cytoskeletal dynamics *in vitro* and cytoskeletal-based transport. In addition to these applications, DF is also very useful in autoradiography because the silver grains are highlighted as bright spots on a dark background making visualization and quantification clearer.

Phase-Contrast Microscopy. The absorptive light scattering differences between the variety of components in cells (due to refractive index and thickness differences) generate phase shifts. However, these are not detectable by the human eye or detectors. The interaction between light waves not deviated by cell components and deviated light is insufficient to generate amplitude differences large enough to give contrast. The invention of the phase-contrast microscope by Zernicke in 1935 (33) achieved the goal of producing a sufficient phase shift between undeviated light and specimen-deviated light, so that better interference occurred between the two wave fronts thereby providing sufficient amplitude change to give visible enhanced contrast. The interference between the two wave fronts provides a resultant brightness change (2,3,20,25–28,33). Now structures not visible in bright-field microscopy are

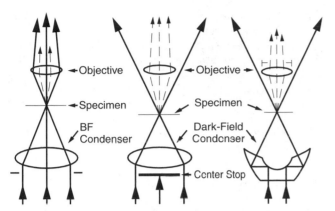

Figure 3. These ray diagrams contrast a bright-field microscope (left) with simple center-stop dark-field and dark-field condenser microscopes. A hollow cone of light is created by either the center stop or a dark-field condenser that has a central reflecting surface. This cone of undiffracted or zero-order light (solid line) bypasses the objective, and only the deviated refracted light (dashed line) enters the objective.

easily seen. This provides an easy to use, cost-effective way of viewing live cells in culture and determining general cell features, nuclear shape, size, cell movements, etc. Figure 2c and 2e show phase-contrast images of both single cells and multicellular tissue. The impact of phase contrast on cell biology was considered so important that Zernicke was awarded the Nobel prize in 1953.

A phase microscope has a special condenser with a series of phase annuli (each matched for its respective objective) (Fig. 4). The phase annulus allows a hollow cone of light (undiffracted zero-order illumination) to enter the objective and fall on a coincident donut-shaped region of a phase plate that is located in the back focal plane of the objective. This circular ring region of the phase plate (often coated by a dielectric) absorbs or attenuates 70–80% of the light and introduces a phase shift of one quarter of a wavelength. Light scattered and altered by

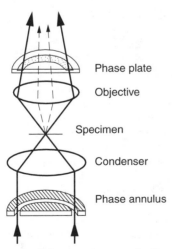

Figure 4. The essential components of the phase-contrast microscope. The condenser annulus produces the cone of undiffracted light that passes through the matched-phase ring or phase plate located at the back focal plane of the objective. The dashed lines represent the diffracted light.

cellular components is deviated and passes through the other transparent regions of the phase plate (Fig. 4). The net result is that the introduction of the phase shift allows sufficient interference between the altered and unaltered light to generate visible amplitude differences, that is, contrast (can be negative or positive contrast). This reveals finer cellular detail (Fig. 2c, 2e). However, an annoying feature is the presence of halos surrounding contrasted structures because some of the diffracted light also passes through the ring area of the phase plate adding to the directly transmitted light. A second limitation is that one cannot do optical sectioning, so out-of-focus details limit the level of detail discernible.

Phase contrast works best on thinner objects (<5 μM thick), and proper images are obtained only if the correct phase annulus is used for each objective phase plate and if the annulus and ring in the phase plate are precisely aligned. This can be easily done with microscopes having a flip-in Bertrand lens (so the back focal plane of the objective can be seen) or by removing an eyepiece and using a phase-contrast telescope lens (Fig. 2d).

The invention of an aperture-scanning phase-contrast system by Ellis in 1988 (described in Ref. 2) uses a small circular fiber optic fiber that scans in a circle at the front focal plane of the condenser. This is conjugate with a phase plate in the back focal plane of the objective where there is a small dot that introduces the 1/4 λ shift instead of the usual ring. The result is improved imaging with less interference of out-of-focus image planes and no high-contrast halos. For detailed coverage of phase-contrast optics, the reader is referred to Ross (34,35), Bennett et al. (36), Inoué and Spring (2), Keller (3), and Sanderson (in Ref. 20(Vol. 3)).

Oblique, Anaxial, or Asymmetrical-Illumination Microscopy. The use of oblique or even lateral illumination to yield contrast differences in unstained cells characterizes various modes of microscopy, as well as specially constructed microscope types.

Oblique or Anaxial (Asymmetrical) Illumination. This is achievable with a bright-field microscope set up for Köhler illumination and then by various means passing only the unaltered (undeviated, zero-order light) through one off-center region of the condenser and objective. Diffracted light passes through other areas of the objective. Oblique illumination can be produced by moving the lamp filament extremely to one side (37), by a variety of asymmetrical stops in front of the condenser or over the field diaphragm (38,39), or simply by offsetting the condenser diaphragm to one side (3) as Abbe did in the 1800s. Asymmetrical illumination due to uneven condenser illumination produces a contrasted differential image that usually has a shadow-cast appearance. This is caused by the components in the cells diffracting the light and creating other orders of diffraction which enhance interference and improve contrast.

Hoffman Modulation-Contrast Microscopy. This type of microscope developed by Hoffman (40) allows more precise control and refined imaging which is particularly useful for single cells. Shaded images with a three-dimensional appearance, high contrast, optical sectioning, and control of contrast result. The essential optical components are (1) a polarizer over the light source (2), a slit aperture with a rectangular polarizer covering about half of it in the condenser and a special amplitude plate called a modulator in the back focal plane of the objective (Fig. 5). The modulator is a trizonal plate with three different transmission regions (3%, 15%, 100%). The condenser slit/pol aperture must be aligned with the modulator. The clear slit is over the 15% zone, and the polarized half of the slit is over the 100% zone. The net result of this off axis illumination is that the small phase-gradient differences within the sample yield deviated or altered light that is converted to intensity variations giving the shadowing and optical sectioning effect.

Single Sideband Edge-Enhancement Microscopy.
This related form of asymmetrical-illumination micrscopy was developed by Ellis (41) and utilizes a half mask (stop) in the condenser and a carrier attenuation filter in the back aperture of the objective. This attenuation or spatial filter is sandwiched between two polarizing filters (see Refs. 2,41, for details). The images obtained are much like differential interference-contrast images and are extremely rich in fine spatial detail. This is an ideal type of microscopy for viewing thin objects and where high spatial detail is required.

Multiple Oblique-Illumination 3-D Microscope.
Three-dimensional images of cells, tissue slices, and live

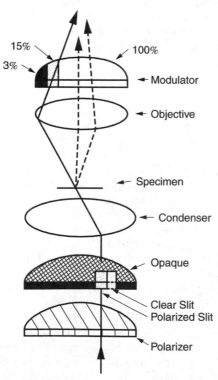

Figure 5. A simplified summary of the Hoffman modulation-contrast microscope (HMCM). Located in the front focal plane of the condenser is an opaque plate that contains an asymmetrically located rectangular slit half of which is covered by a polarizer strip. The back focal plane of the objective has a modulator plate with three zones of transmission. The condenser slit is aligned with the 100% and 15% areas. Diffracted light (dashed lines) passes through the rest of the modulator.

specimens can be produced by using a 3-D microscope that utilizes multiple (4) oblique fiber optic light polarized sources reflected by a pyramidal mirror. The microscope (3-D edge microscope, Edge Science Institute, Santa Monica) utilizes a field and relay lens to image the objective aperture on top of the optical path where it is split and channeled to the two eyepieces that provide a 3-D stereo image. See Greenberg and Boyde (42) for a general review of the optics and applications of this novel use of oblique illumination.

Polarizing Microscopy.
Polarizing microscopy is a powerful qualitative and quantitative tool for examining cells and cell components that have structures with high degrees of molecular order. Structures such as filaments, microtubules, crystal arrays, extracellular matrix fibers, DNA, certain lipids, etc., can be viewed with sufficient contrast generated by polarization microscopy. Ordered structures appear as highly birefringent structures when ideally oriented. Anisotropy, the property of matter which has different values when measured in different directions in the same material, is the basis of polarizing microscopy. Polarizing microscopy has a long history dating from the first polarization microscope built by Smith in 1851 (43). Much of our current use of the polarization microscope in cell biology is heavily influenced by the extensive work of Inoué (2,44) and significant new developments introduced by Oldenbourg and Mei in 1995 (45).

A polarization microscope is basically an ordinary bright-field microscope equipped with a polarizer beneath the condenser and a second polarizing filter called the analyzer above the objective (2,46). Figure 6 shows a schematic of the components of a polarizing microscope. When light is propagated as a wave, it vibrates in two vectors at right angles to each other. When this passes through a Nicol prism (introduced by Nicol in 1860) or an anistropic structure such as a calcite crystal, the two beams (often called the e and o rays) are split. The polarizer preferentially passes light vibrating in a single plane, so that when the analyzer is crossed, little light is transmitted to the eyepiece. Only the alteration of polarized light by ordered structures in the cells placed in the optical path will generate birefringence (see Fig. 8f). The optics should be strain-free, and usually a rotating stage is used for specimen orientation. For analytical work, a compensator is often used between the polarizer and analyzer. This allows one to measure the degree of birefringence. When the polarizer and analyzer are crossed, one has extinction. Extinction is a measure of the amount of stray light that is prevented from passing when the filters are crossed. To visualize anisotropic structures in cells and tissues, a high level of extinction is essential. Because the amount of light resulting from birefringent structures (e.g., microtubule arrays, mitotic spindles, DNA, etc.) is small, one needs to use bright light sources to overcome the significant amount lost by extinction. High NA objectives are used, so that maximum light capture is possible. Unfortunately conventional high NA objectives often reduce extinction. This can be corrected by using rectified optics (2).

With a polarizing microscope, anisotropic structures (e.g., microtubules, etc.) with their ordered molecular

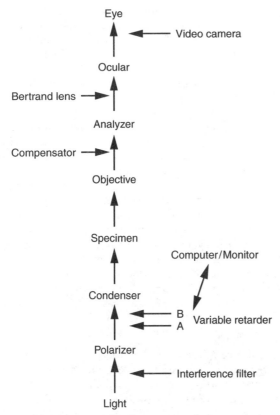

Figure 6. In the center is the sequence of components that comprise a basic polarizing microscope. On the left are optional elements (Bertrand lens, compensator frequently present). On the right are the additional components incorporated by Oldenbourg and Mei in their new Pol-Scope. A key component is the electrically controllable variable retarders (A and B). See text for details.

architecture appear as bright birefringent objects if extinction is high enough to yield a dark background. Birefringence occurs when the refractive index in one axis of the object is different in the orthogonal axis. This birefringence is measured as retardation (birefringence × thickness) and expressed in nM. Values less than 5 nM are hard to measure. However, this has recently been reduced to 0.02 nM by the introduction of a new innovative approach (45). The Pol-Scope is a traditional polarizing microscope augmented by electro-optical devices and imaging algorithms to complete specimen birefringences. In a conventional Pol scope, only those anisotropic structures in a certain orientation are displayed in a single image. The new Pol-Scope uses a precision universal compensator made of two electrically controllable liquid crystal retardation plates (B $\lambda/2$ 0°; A $\lambda/4$ 45°) that overcome the orientation problem. One can measure dynamically and non-destructively in live cells or isolated cell components with a high degree of sensitivity (as low as 0.02 nM) with excellent spatial resolution (0.2 nM). Recently (1997) Shinya Inoué (47) in conjunction with Hamamatsu and Olympus has added a new centrifuge polarizing microscope to the family of polarizing microscopes. This adds to the possibilities by allowing visualization of live cells and their organelles under centrifugation with high fidelity and resolution.

Interference-Contrast Microscopy. Like phase-contrast, interference-contrast microscopy exploits interference differences initially generated by local phase differences in the specimen and converts these to intensity differences. Perhaps the most widely used type of microscopy today to visualize live cells is the differential-interference microscope of which there is the Smith and Nomarski types (see Refs. 3,20(Vol. 3)). Surprisingly, interference microscopy was first introduced in 1893 by Sirks (see Ref. 48). However it was not until after the widespread use of phase-contrast microscopy that interference microscopes were designed and manufactured for general use. As reviewed in Inoué and Spring (2), there are essentially three types of interference microscopes *(1)* the Mach–Zehnder which involves two microscopes for separate reference and modified light paths; *(2)* beam-shearing types where the split beams are displaced laterally as in the Jamin–Lebedeff, Smith, and Nomarski types; and *(3)* the Mirau type where the reference beam is focused to a level differently from the specimen plane. The Jamin–Lebedeff interference microscope initially designed by Piller and built by Zeiss in the 1960s allowed quantitative analysis of live cells making it possible to measure cell volumes, mass of cell components, etc., based on calculations using optical path difference measurements. Such instruments are no longer available. The most widely used modern interference microscopes are the differential-interference microscopes (DIC) (2,18,26,48) developed by Nomarski in 1955 (48–50) and Smith in 1956 (48). In these, the beam splitter at the front focal plane of the condenser is a Wollaston prism that splits the polarized light into two paths displaced laterally relative to each other with shear. These two paths then pass through the specimen and are altered differentially depending on the refractive indexes of the cell components each encounters. At the back focal plane of the objective is another Wollaston prism (which can be adjusted in the Nomarski system) that recombines the two light paths. An analyzer crossed with respect to the polarizer (below the condenser) is next in the optical path. When the two beams are brought together, they interfere (Fig. 7). The phase shifts in each result from refractive index differences and thickness gradients in the cells being viewed. This exploitation of local phase differences results in a shadow-cast effect which reveals very fine cell detail and allows optical sectioning because of the very shallow depth of field. Figure 8c–8d shows DIC views of live cells. For comparison Figure 8 a provides a HMCM image and Figure 8b provides a phase-contrast image. Organelles and other cell structures appear in relief with variable contrast. For general viewing of cells in culture, this is widely used as the preferred method. However, it cannot be used with plastic culture dishes or plastic coverslips because these are anisotropic materials and affect the polarized light necessary in DIC. A recent improvement in DIC has been introduced by Holzwarth et al. (51) which is called polarization-modulated DIC (PM-DIC). This involves using a liquid crystal modulator that can switch the polarization from X to Y rapidly, coupled with a frame grabber and an imaging processing system. The result is a significant improvement in contrast, and with incorporation of background

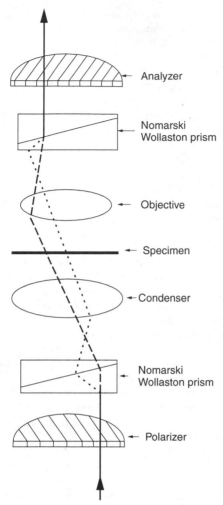

Figure 7. An example of one of the types of interference-contrast microscopes shown here is the Nomarski differential interference-contrast system (DIC). Plane-polarized light is split by Wollaston prisms into two paths separated by shear. These are indicated by the dashed and dotted lines. The final recombination of the two paths is via the second Wollaston prism. The interaction produces a differential-interference image with shading and high contrast.

subtraction, one can resolve very fine spatial details of high contrast.

Reflection-Contrast Microscopy. Reflection-contrast microscopy (RCM; also sometimes called reflection-interference microscopy, RIM) can be used to study live cells, as well as fixed cells. In its simplest form, it involves passing light (usually polarized via an epi-illumination system with a dichroic mirror down through the objective, and reflected light from the specimen passes through the objective, the dichroic mirror, a polarizer at 90° to the first, and to the eyepiece (52). This was first used by Curtis (53) to investigate cell adhesion and improved by Ploem by placing a quarter λ coating on the front of the objective. Improvements in image quality result if special antiflex objectives are used (54). Izzard and Lochner (55) successfully used RCM (also called interference RCM or IRCM) to look at fibroblast cell substrate adhesion and

interactions. Opas and Kalnins (56) also examined cell substrate local adhesion. In RCM, light waves reflected at the sample surface interfere with those reflected from the substrate. The result is an image that highlights contact and adhesion sites in high contrast. Besides visualizing cell–substrate adhesion sites, RCM is very useful for visualizing nonradioactive *in situ* hybridizations where peroxidase or silver/gold labels are used (57).

Reflection microscopy has also involved using a scanning mode via a spinning Nipkow disk containing 24,000 small holes that produces and detects many points of light. These were first developed independently by Minsky and by Petrán in the 1960s (58–62). They are called tandem-scanning reflecting microscopes and are also used for examining cell–substrate adhesion, as well as cytoskeletal structures. The tandem-scanning reflecting-light microscope developed by Petran was the first use of a confocal microscope approach. For more in-depth information on reflection contrast microscopes, see Verschueren (52); Egger and Petrán (58), Curtis (53), Opas and Kalnins (56), Izzard and Lochner (55), Ploem et al. (57), and Bereiter-Hann and Vesely (in Ref. 20(Vol. 3)).

Fluorescence Microscopy. The phenomenon of fluorescence was first discovered in 1833 by Sir David Brewster and named fluorescence by Stokes in 1852. In fluorescence, molecules in specimens absorb light radiation and per Stoke's Law emit light of a higher wavelength. The excitation light of specific wavelengths excites fluorophore molecules such that outershell electrons move to higher orbits. Within 10^{-7} seconds, these drop to lower orbits and lose the energy via emission of light at a higher λ. For a general description, see Refs. 3,20(Vol. 3),63. A major impetus to use fluorescence in cell biology comes from the immunofluorescence studies of Coons in the 1940s who localized antigens in cells using antibodies conjugated to fluorescent dyes like fluorescein isothiocyanate (FITC). With the tremendous development of immunofluorescent approaches, the fluorescence microscope became a routine instrument in many cell biology and biotechnology labs during the past 40–50 years. Initially, instruments were transmitted fluorescence microscopes. However, these are rarely used today and have been supplanted by epifluorescence microscopes. A schematic of a standard epifluorescence microscope is presented in Figure 9.

The essential components of an epifluorescence microscope are a lamp that emits the appropriate range of wavelengths, a filter to select the excitation wavelength, a dichroic mirror to direct that wavelength through the objective to the specimen which will then emit light to be captured by the objective and passed through the dichroic mirror, a barrier or emission filter that blocks wavelengths below the emission λ, and finally imaging by the oculars. Basic information is available from a variety of texts [2,3,20 (Vol. 3),63,64], as well as a number of web sites, particularly the Molecular Dynamics and Molecular Probes sites.

Various types of lamps are usable in fluorescence, depending upon the intensity and the spectral range needed. Tungsten halogen lamps produce a continuous white light and emit from about 350–1000 nM but of relatively low intensity particularly in the 350–400 nM

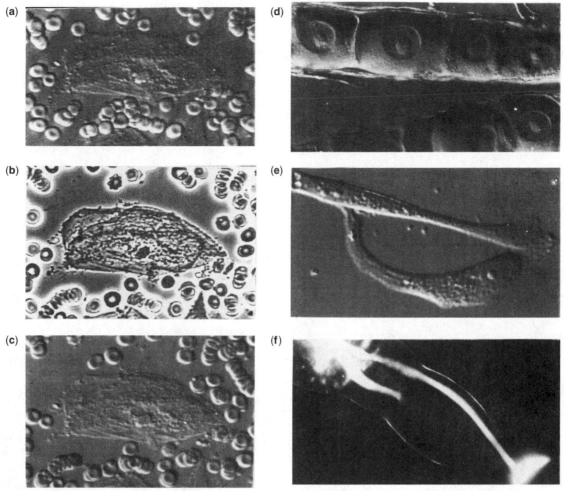

Figure 8. (a–c) HMCM, phase-contrast, and DIC views of a combined preparation of blood and a buccal cell. Note the graded shading in the DIC (c) and HMCM (a) images, and no halos on are prevalent in phase image (b). (d) The optical sectioning ability of DIC of a praying mantis ovary. The oocyte's large nuclei and prominent nucleoli are delineated. (e) A DIC view of HeLa cells. (f) A live insect ovariole (of *Rhodnius proxlixus*) as seen in polarizing microscopy. The brilliant birefringent structures are the intercellular bridges or trophic cords that are packed with aligned microtubules.

range. Such lamps are usable for FITC and a number of newer fluorochromes excitable in the visible spectrum. With the advent of low-light and more sensitive CCD cameras, lower intensity is less of a problem. More widely used lamps are xenon arc lamps (e.g., XBO 75) which have high-intensity output over a fairly continuous spectrum from 250–1000 nM. These are preferred where fluorescence ratioing is needed and multiple λs are needed for different probes. Mercury arc lamps (HB050 and 100) are also widely used. They have high intensity but a more varied spectrum with distinct peaks of intensity at certain wavelengths (e.g., 366, 405, 436, 546, 578). Lasers are being increasingly implemented as excitation sources, each with their specific wavelength(s) rather than a broad spectral array. A heat filter is between the collector lens of the lamp housing and the optical path. Initially, excitation filters used were relatively broad (the UG, BG filter) but now most systems have excitation filters with a reasonably narrowly defined transmission window (this varies among manufacturers but, for, example a 450–490 nM range is common for fluorescein). The fluorescence microscope has an epifluorescence condenser. In some systems, the excitation filters may be located in a computer-controllable filter wheel, and in others the excitation filter along with the dichroic mirror and emission filter is located as a unit (filter cube) in the epifluorescence condenser (Fig. 9). The dichroic mirror will reflect the excitation λ into the objective but will transmit only the higher λ of the emitted light, and the higher λ will pass through to the eyepiece. Ideally, the emission filter would be a narrow band-pass filter and would transmit only the λ characteristic of the fluorochrome used. There is a tremendous range of filter combinations available (see the Omega Optical and Molecular Probes web sites, www.sover.net/~omega and www.probes.com). The inset graph in Figure 9 shows a "generic" example of a narrow band-pass filter set. The terminology can be confusing. Dichroic mirrors may also be called chromatic beam splitters.

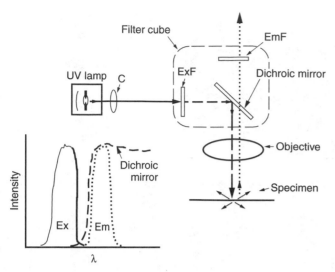

Figure 9. A schematic of a typical epifluorescence microscope with a lamp (usually HBO or XBO) that produces a spectrum of λ s. The λ of choice (depending on the stain used) is selected by the excitation filter (ExF). The selected excitation λ is reflected to the specimen by the dichroic mirror (dashed line). Fluorescence (higher λ) enters the objective lens and passes through the mirror with a final selection of the appropriate emission λ by the emission filter (EmF) or barrier filter. The generalized graph shows the spectral pattern for narrow band-pass Ex and Em filters and the dichroic mirror.

Because the image and its contrast come from emission of light from the specimen it is important that high NA, high transmission objectives are used, so that a sufficiently bright image results. As covered later, the use of low-light video cameras has opened up new vistas for fluorescence microscopy on live cells without the inherent problems of UV damage from the excitation beam or photobleaching because the intensity of the excitation illumination can be reduced with a neutral density filter or other means (13,14,16,20(Vol. 3)).

The application of fluorescence microscopy is exceptionally far-reaching. Although there are relatively few endogenous fluorochromes (when present they often pose problems such as autofluorescence), there is an enormous array of fluorescent dyes now available for many applications relevant to cell technology. In-depth coverage of these is beyond the scope of this review, but there is a large and growing literature [3,20(Vol. 3),63–68].

The following highlights only a few of the applications of the fluorescence microscope.

A number of fluorescent dyes that selectively stain cell organelles (Fig. 10) [20(Vol. 3),68] are available. Mitochondria can be visualized in live cells with cell-permeant dyes like Rhodamine 123, Rhodamine 6G, $DiOC_6$, nonylacridine orange, JC-1 (Fig. 10a,b), and a range of MitoTracker dyes (Fig. 10c) (see Refs. 69–71, and the Molecular Probes web site www.probes.com). Endoplasmic reticulum can be

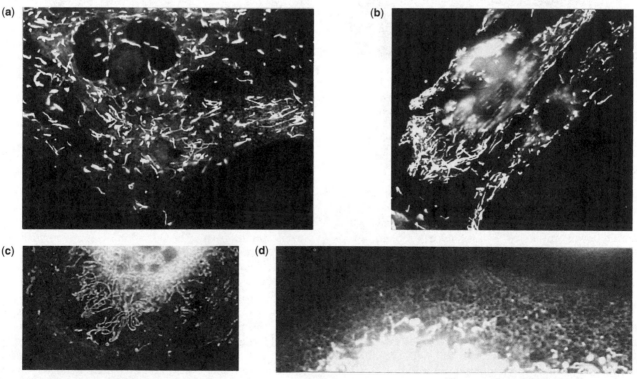

Figure 10. These black and white images of the original color fluorescence images show selectively stained organelles in live cells. (**a**) The mitochondria-specific dye JC-1 highlights mitochondria in drug (dexrozoxane)-treated CHO cells. (**b**) In rat fetal cardiac myocytes. (**c**) Long slender mitochondria in dexrozoxane-treated CHO cells stained with MitoTracker Green FM. (**d**) A similar cell stained with the endoplasmic-reticulum-specific stain, ER-Tracker Blue-White DPX. The ER stains as a lacey network seen here in the thinnest peripheral area of a spread cell (E. Huebner and B. Hasinoff, unpublished data, 1999).

visualized (72,73) with labeled conA, DiOC$_6$, ER-Tracker Blue-White DPX (Fig. 10d); Golgi (74) with labeled C$_5$ and C$_6$ ceramide; DNA in nuclei with DAPI, Hoechst (Fig. 11d) or propidium iodide and so on. Fluorescent probes are used for tracking endocytotic uptake, cell to cell coupling, lysosome and lipid inclusions, and cytoskeletal elements, to name a few (75). There are also probes to quickly determine live and dead cells in culture and to measure levels of various ions, intracellular pH, and biochemical reactions (75–78). Fluorescently labeled probes are also used in *in situ* hybridization methods commonly known as FISH (fluorescence *in situ* hybridization). Immunofluorescence approaches to visualize cell structures, receptor sites, biosynthetic products, etc. is a major area of application [12–16,20(Vol. B),68]. Immunofluorescence has provided a powerful means for visualizing the cytoskeletal organization of cells. Figure 10 provides some

examples of fluorescence microscopy of the microtubule and F-actin cytoskeleton using IF methods and a specific F-actin probe.

The introduction of new fluorescent probes and stains almost daily has resulted in an explosion of possible ways to study almost all facets of cell morphology and molecular functions of cells with fluorescence microscopy. Improved contrast and brightness is attained with newer dyes that are replacing the classical fluorochromes FITC (Fig. 11a) and tetrarhodamine isothiocyanate (RITC). Some of these are Texas Red, the Cy dyes (Fig. 11b), Bodipy, and most recently the Alexa dyes. Table 1 provides some examples of fluorophores. Some dyes require microinjection (e.g., the calcium-sensing cell, Fura 2) or some alternative physical cell loading method to introduce them into the cytosol. Increasingly dye-permeant forms (e.g., Fura 2 AM) have been produced. Although there

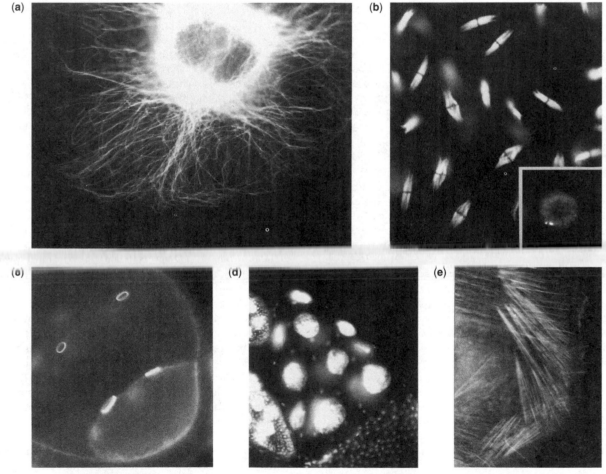

Figure 11. These black and white images of the original color fluorescence images illustrate immunofluorescence and other specific probes. (**a**) Dexrozoxane-treated CHO cells greatly enlarge in size. Their dramatic microtubule arrays are stained for acetylated tubulin using a fluorescein-labeled secondary antibody. (**b** and inset) Immunofluorescence of two stages of spermatogenic cells of the insect *Rhodnius*. The inset shows gamma tubulin labeling of the centrioles, and the main figure shows the tubulin of the spindles. The secondary antibody was labeled with Cy3 (E. Byard and E. Huebner, unpublished data, 1999). (**c**) The F-actin-stained ring canals of the *Drosophila* nurse-cell-oocyte cysts of the ovary. The F-actin-specific probe phalloidin is labeled with rhodamine. (**d**) DNA staining of the large polyploid nurse cell and smaller follicle cell nuclei in a *Drosophila* ovariole cyst. The DNA-specific stain is Hoechst 33342. (**e**) The rhodamine-phalloidin staining of F-actin in the stress fibers of an enlarged flattened dexrozoxane-treated CHO cell.

Table 1. Examples of Fluorophores

Fluorophore	Excitation wavelength	Emission wavelength	Uses
DAPI	358	361	DNA stain
Hoechst 33342	357	461	DNA stain
FITC	495	520	Label
Texas Red[a]	595	615	Label
Cy3[b]	565	590	Label
Bodipy FL[a]	503	513	Label
Alexa 488[a]	495	519	Label
GFP wild-type	396/475	508/503	Expression marker
Lucifer Yellow	428	534	Cell tracer, uptake, gap junctions
MitoTracker[a] Green FM	490	517	Mitochondria stain

[a]Trademarks of Molecular Probes Inc.
[b]Cy3 is trademark of Amersham Intl.

are a number of sources of fluorescent dyes, Molecular Probes has become one of the major suppliers of new probes and continues to introduce an impressive array of new dyes.

Wild-type green fluorescent protein (GFP) was originally isolated in the 1960s from jellyfish *Aequoria victoria* and cloned in the early 1990s. It absorbs at 395 (brighter) and 475 nM and emits at 508 nM. Chalfie showed that GFP can be linked as a reporter to other genes. This provided the catalyst for methods now routinely used to study gene expression in live cells using GFP fluorescence (79–83). Mutant forms of GFP generally called chameleons provide an array of GFPs as endogenous probes that monitor a variety of changes in pH and other factors (80). pH-sensitive GFP mutants called pHluorins are used to investigate endocytosis, for example (84). Recently, another family of fluorescent proteins, the phyofluors, which are much brighter than GFP, has been introduced (85). Fluorescent reagents can be introduced into cells in bound or caged forms and with appropriate wavelength irradiation are released and participate in reactions generating fluorescent molecules, thereby allowing one to study dynamic reactions at appropriate times [20(Vol. 3),86]. Suffice it to say, this is a rapidly advancing field.

A significant drawback in visualizing fluorescently labeled cells with conventional epifluorescence microscopy is the inability to visualize single focal planes. The fluorescence emanating from above and below focus plane levels contributes to the overall image, obscuring fine structural resolution. Image processing can improve visualization, but the introduction of laser scanning confocal microscopy dramatically improved and facilitated fluorescence optical sectioning (see Shotton in Ref. 20(Vol. 3)).

Confocal Microscopy. This form of microscopy is part of the scanning microscopy family. In particular, confocal fluorescence microscopy has become popular as it overcomes the problem of out-of-focus plane fluorescence in fluorescence microscopy. Confocal has its origin in the 1950s and 1960s with its invention by Minsky in 1957 (58–62). Independently, Petran and co-workers in

1968 developed a Nipkow spinning-disc field-scanning or confocal microscope [2,20(Vol. 3),87–91].

A specimen can be scanned with the illumination beam by either stage scanning or the more widely used beam-scanning systems. Of these there are basically two common types, the disk-scanning confocals that use a spinning Nipkow or a modified Nipkow disk. In the Petran type the symmetrically placed pinholes alternatively provide the illumination pinhole and the imaging pinhole which are confocal. For current approaches in aperture scanning confocals, see Refs. 28(Vol. 3),91–93. A newer modification developed by Kino in 1995 (2) uses a beam splitter that makes it possible for each pinhole to act both in illumination and imaging. The incorporation of a second disk with microlenses has further extended the use of the spinning-disk confocal microscope as a significant increase in image brightness was developed (94; also see 2,95). The laser-scanning confocal became the most prevalent confocal in the 1980s and 1990s. Figure 12a provides a simplified schematic of a laser-scanning confocal fluorescence microscope. Basically, a focused laser beam is scanned in raster fashion across the specimen. The laser light is directed through the objective via a dichromatic beam splitter, and fluorescence emission light passes through the beam splitter to a pinhole aperture that is confocal with the point source. Because there is fluorescent excitation along the excitation beam path, there is some out-of-focus fluorescence. So the pinhole aperture before the detector is necessary to collect only fluorescence from the in-focus plane. The illumination-focused spot in the sample and the detector pinhole are confocal with each other. The aperture ensures that only light from the confocal plane hits the detector. The net result is that out-of-focus plane fluorescence is minimized, optical sectioning can be done, and by collecting a Z axis series of images, 3-D reconstruction can be done. The advantages of removing obscuring glare, looking at thicker specimens without cutting sections, and the high degree of resolution has had a major impact in the cell biology literature. There are many excellent reference sources on all facets of confocal microscopy [2,3,20(Vol. 3),87,89,96–100] and also a number of valuable web sites (see end of article).

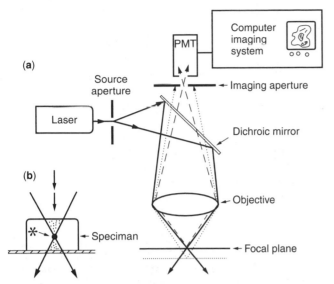

(a)

(b)

Figure 12. (a) A simplified schematic of a fluorescence laser-scanning confocal microscope. For simplicity, the excitation and emission filters routinely present in a fluorescence scope have been omitted (see Fig. 9 for their normal location). Also omitted is the scanning component. The excitation laser light (solid line) stimulates fluorescence in the sample. The fluorescence from the out-of-focus planes (dotted line) does not reach the detector (photomultiplier — PMT) because of the imaging aperture. Only the fluorescence from the confocal plane (dashed line) is deleted. (b) A simplified view of two-photon fluorescence in the specimen where two-femtosecond infrared laser pulses (depicted as two sequential arrows) simultaneously stimulate fluorophores at the focal point. Fluorescence occurs only in the small volume at the focal point (*pointing to solid dot). The stippled area shows, for comparison, the area where fluorophores are stimulated to fluoresce in conventional fluorescence and confocal-fluorescence microscopy.

Figure 13 provides two examples of confocal images of an insect ovariole (*Rhodnius*). In normal fluorescence, detailed resolution is difficult (e.g., Fig. 13c) due to the fluorescence from multiple levels, but in confocal, as seen in an aqueorin-injected nurse-cell chamber (a) and in an immunofluorescence preparation of a plasma membrane calcium ATPase (PMCA) in the follicle (b), clear focus is acheived.

The significant development and impact of confocal has been aided by the refinement of scanning devices, detectors, varieties of lasers, and the improvements in computers necessary to handle the mass of data essential to 3-D imaging (14,16; Shotton in Ref. 20(Vol. 3)).

An alternative approach to laser scanning confocals is the so-called digital confocals. These involve using deconvolution algorithms with a stack of serial optical sections obtained with a fluorescence microscope (see Shaw in Ref. 20(Vol. 3); also see Refs. 101–105). Each focal plane is sharpened by eliminating the blur from above and below the focal planes via such algorithms. Excellent optics, calculation of the point-spread functions, high-quality video cameras, and precise Z axis controllers are needed.

Particularly with single cells which are often fairly thin, the potential to resolve fine subcellular detail is now possible with conventional fluorescence microscopy in conjunction with video and image processing, confocal microscopy disk or laser scanning types, or computer deconvolution based imaging.

Two-Photon and Multiphoton Microscopy. Despite all of the positive features of laser confocal fluorescence microscopy, there are some drawbacks (beyond the cost factor). The intense focused laser beam can cause

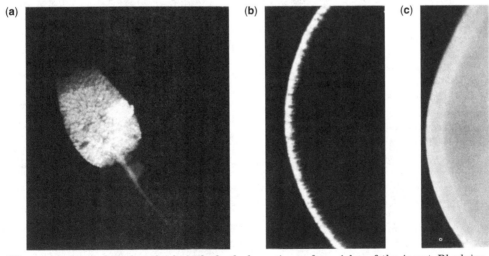

(a)   **(b)**   **(c)**

Figure 13. Examples of confocal single focal plane views of ovarioles of the insect *Rhodnius*. (a) Fluorescence in the nurse chamber and extending intercellular channels (trophic cords) of an ovariole that was injected with fluorescein-labeled aqueorin (Ca²⁺ sensitive photon-emitting protein) (E. Huebner, C. Bjornsson and A. Miller, unpublished data, 1999). (b) and (c) Micrographs of oocytes with surrounding follicular epithelium that has been labeled via immunofluorescence for a plasma membrane calcium ATPase. The confocal image (b) shows that the label is localized in the outer follicle cell surface and (c) a similar preparation with conventional fluorescence microscopy where the out-of-focus plane fluorescence makes definitive localization difficult (Bjornsson and Huebner, unpublished data).

significant photobleaching, and the laser irradiation can cause phototoxicity-generating free radicals and possible cell damage. Fluorescence eminates from all fluorchromes in the excitation pathway, including out-of-focus planes. This puts significant limitations on examining live cells. Two-photon microscopy was invented by Denk, Stickler, and Webb in 1989 (106,107). The suggestion of a two-photon microscope was first proposed in 1978 by Sheppard and Kompfner (108). The principle underlying two-photon and multiphoton absorption is based on the theoretical predictions of Maria Goeppert-Meier in 1931 and used in the field of spectroscopy. Imaging is similar to confocal microscopy except that photobleaching and damage are reduced. The excitation energy is concentrated temporarily and spatially, so that the saturation average power is small. For example, for Rhodamine B this would be 5 mW for a 100-fs pulse of 100 MHz (107). Excitation occurs only at the focal point, so that a pinhole aperture in the imaging path is not necessary because there is no out-of-focus plane fluorescence to remove. Figure 12b compares the area of fluorescence in two-photon versus conventional confocal fluorescence. Any damage to the sample that might occur is also restricted to the focal volume itself.

In two-photon microscopy, infrared wavelength high-energy ultrashort (femtasecond) pulsed lasers are used. These can be Ti: sapphire mode-locked lasers pumped by high-power argon ion gas lasers or diode-pumped 532-nM lasers. The excitation of a fluorescent dye molecule by one photon stimulates it to a partially excited short-lived virtual state from which it can be raised to its excited state if a second or more photons hit virtually simultaneously. Subsequently, thermal decay of the molecule yields fluorescence. The net result is that two or more photons of long wavelength create the same excitement normally achieved by short wavelength stimulation (in confocal), as long as they arrive almost simultaneously [for details, see Refs. 28(Vol. 3),107,109–113]. Thus, unlike confocal and basic fluorescence microscopy, the excitation wavelength is greater than the emission wavelength. The red wavelengths penetrate deeper than UV so that the imaging range is extended, viability is improved, and there is no out-of-focus photo damage. Unlike confocal microscopy, the range of fluorescent probes that can be visualized is extended and excitation wavelength is not a problem because the excitation spectrum is much wider (113). The impact of two- and multiphoton microscopy is only now being felt with exciting work looking at intracellular pH, Ca^{2+} enzymes like glucose 6-phosphatase, ratio measuring of NAD and NADH, serotonin solutions, GFP, mitochondria, etc., in live cells (83,107,110–112,114,115, and web sites).

A major drawback to photon microscopy is the high cost of the femtosecond lasers and a potential problem of focal heating due to the absorption of infrared light. As manufacturers develop new instruments with improved technology and lower costs, multiphoton microscopy will become more accessible to biologists. The potential to study dynamic events in living cells at the light microscope level has been brought to single-molecule sensitivity levels.

A variety of additional types of microscopy have and are being developed that use not only light but other forms of electromagnetic energy for imaging. These include infrared spectromicroscopy, soft X-ray microscopy, laser Doppler microscopy, scanning acoustic microscopy, nuclear magnetic resonance microscopy, and others (116–119). These are beyond the scope of this article.

LASER TWEEZERS AND SCISSORS

Lasers have had an ancillary impact important in microscopy and cell technology in the form of laser tweezers and scissors that in combination with the light microscope provide a new noninvasive tool to measure forces operating in all processes like molecular motor-based transport and membrane characteristics, to isolate individual cells or cell components, and to do ultrafine microsurgery. Structures can be trapped and held by a laser beam-gradient force, and in essence particles are moved by light. A number of reviews and articles cover the principles of laser trapping (120–124). A novel tool uses laser capture to isolate single cells from tissue sections thereby opening new frontiers for molecular analysis (125,126).

SCANNING-PROBE NEAR-FIELD MICROSCOPES

These are the newest microscopy tools impacting cell biology. They have the remarkable potential to resolve structure at the atomic level, thereby revealing structural detail normally seen only with electron microscopy. Involved is scanning of a small probe (1–10 nM) near the surface of structures with images generated as a result of the interaction between the probe and surface. Binnig and Rohrer revolutionized microscopy with the introduction of the scanning-tunneling microscope (STM) in 1981 (127) for which they received the Nobel prize in 1986. Subsequently in 1986, Binnig, Quate, and Gerber (128) invented a related microscope, the atomic force microscope (AFM). Initially these instruments were primarily used in the physical sciences, but within the past five years they are increasingly being used in cell biology. These instruments represent one of the most important breakthroughs in analytical instrumentation technology in decades. Scanning probe microscopes work in a variety of environments (air, liquid, etc.). They magnify and resolve at the atomic level providing high-resolution, three dimensional morphology by "feeling and sensing" surface structures rather than using a photon or electron beam (129–134).

Scanning Tunnelling Microscopes (STM)

This type of scanning probe microscope requires an electronically conductive sample and is not generally applicable in biological systems. The probe tip scans the sample while being kept at a constant tip-to-sample distance of about 8 Å by an electronic feedback circuit. The tunneling current is kept constant, so that the rigid probe moves up and down indicative of the atomic surface contours being scanned. The probe uses an electron tunnelling current between the sample to activate a piezoelectric ceramide that supports the probe, raising and lowering it. Very sensitive measurements of 0.1 Å can be made.

Atomic Force Microscopes (AFM)

These evolved from STM, were introduced in 1986 (128), and are increasingly being used to study cells and cellular components (135–142).

In AFMs, the probe tip is positioned over the sample and then scans the selected area. The tip is mounted on a flexible cantilever arm. As the tip scans the cantilever, it is deflected, and the force of deflection is measured. A laser beam is usually used to detect cantilever deflection. The degree of deflection relates to the three-dimensional atomic structure of the surface scanned. Probe scans are sensitive to forces on the order of 10^{-12} N. Although care must be taken in interpretations by AFM of soft elastic surfaces that are easily deformed, recent applications to live and fixed cells have given dramatic images. Researchers have obtained images of chromosomes, cytoskeletal elements, whole cells, membrane surfaces, and membrane exocytosis (see web sites for images).

The development of three versions of AFMs and the combination of AFMs with other forms of optical microscopy has made the study of soft material (i.e., cells) and dual imaging more amenable (132,133,140,143). The three forms of AFM are (1) contact type, where probe tip friction and adhesion may be factors and bending in the lever is due to van der Waals forces; (2) noncontact type, where the probe hovers 1–10 nM above the surface causing less damage to delicate samples but with lower resulting resolution and (3) the newer tapping mode, which involves an intermittent contact oscillation providing for high resolution while minimizing mechanical distortion (138; also see the Life Sciences, Digital Instruments, and the merged Park Instruments/Topometrics web sites). These web sites provide examples of images.

Suffice it to say that this new form of microscopy is still in its infancy but growing exponentially as the impact on biological systems is realized. The ability to visualize cellular structures at submicron to atomic resolution is remarkable, particularly because live cells in natural media can be viewed and dynamic events recorded. The full range of applications in cell technology is still to be determined.

Scanning Near-Field Optical Microscopes

In 1928 Synge proposed a microscope without lenses that could circumvent the diffraction limit to resolution. This led to development in the 1990s of near-field aperture-scanning microscopes by Betzig and Trautman that gave super-resolution (144). Such microscopes are scanning near-field optical microscopes (SNOM) (145–148). High resolution is achieved by overcoming the far-field diffraction barrier by making an optical aperture much smaller then the λ of the light used and placing it close to the sample. With such a sub-λ aperture extremely close, the light emanating from the aperture illuminates a small area immediately below the aperture (the near zone), and this is scanned over the sample; the imaging is via a photodetecter multiplier system or, as in new

instruments, a high NA objective relays the light to a photodetecter and ultimately, to a computer for image display. The light is carried to the scanning aperture via an optical fiber, and often lasers are used. Resolution depends on the size of the fiber (around 50 nM) and the distance from the sample (within 1 λ). New hybrid instruments that combine SNOM and AFM are becoming available (e.g., Nanomics Instruments) so that both a topographical and an optical image can be obtained. With high NA objectives and appropriate fluorescent dyes, high-resolution fluorescence microscopy is possible with SNOM. Until very recently, SNOM was mainly used on nonbiological material. The potential to use SNOM and continuous-wave two-photon SNOM on cells is in early stages of exploration (147,149–152).

VIDEO MICROSCOPY AND IMAGE PROCESSING

Along with the growing number of improvements and innovations in optical microscopes and the various scanning probe microscopes, the renaissance in microscopy is also due in large measure to the availability and improvements in video cameras, the use of computers, and digital processing. Video microscopy and image processing have become essential routine components of most laboratories in cell biology (2,14–16,20(Vol. 3)). Figure 14 provides a block diagram with some of the components one might find in a typical microscope/computer workstation. Peripheral devices include X-Y stages, Z axis focus, filter wheels, shutters, lamps, liquid crystal filters, cameras, and recording devices, to name a few. The variety of cameras range from tube vidion cameras, CCDs, CMOS, intensified CCDs, cooled CCDs, and digital cameras. Image storage and recording components can also vary considerably.

An invaluable reference source that provides the historic perspective since TV was invented by Zworykin in

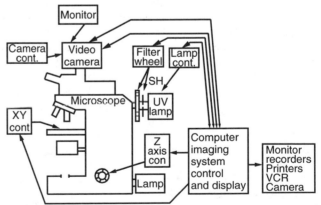

Figure 14. A block diagram of a typical system that integrates a computer and image processing with a microscope and peripheral devices. The system depicted has transmitted and epi illumination, X-Y stage and Z axis focus control, fluorescence with shutter and filter wheel, variable HBO lamp control for intensity, multiport head for various possible video cameras, video camera analog control, analog camera, display for video microscopy, computer image processing, computer control of peripherals, and various output options. The complexity of such systems depends on the applications.

1934, the use of various types of video cameras, microscope interfacing and image processing is Inoué and Spring (2, also see 153). Both theoretical and practical aspects are covered. The work of Inoué and Allen in the early 1980s stimulated the widespread use of video microscopy in cell biology (154–157). In its simplest form, using a TV camera (with manual gain and black level control), high NA microscope optics, and a video monitor gives significant analog video enhancement and reveals excellent cell detail in live cells and allows analysis of fast dynamic events. The subsequent introduction of image processing with the use of A/D conversion of the video signal by frame grabbers in computers expanded the capabilities enormously. Image processing software routines made it possible for the cell biologist to do background subtraction, frame averaging, change contrast and brightness digitally, and a variety of other image-enhancement functions. The range of possibilities is broad. Figure 15 illustrates some of these using three different cell types (buccal cell, fish hepatocyte, and brain pyramidal neuron). Depicted is image enhancement (a), intensity representation (b,c), various manipulations (d–f), and 3-D integration and merging of stacked images. The cellular detail revealed can be astounding, allowing visualization of structures like microtubules, mitochondria, cytoplasmic movements, etc. [20(Vol. 3),157].

The end result is that video-enhanced microscopy is affordable and widely available, and when used on DIC, phase contrast, polarizing, and other forms of light microscopes provides excellent image quality and resolution extending our ability to study cells, live and fixed. Resolution of video systems and printers has improved often achieving photographic film quality. An array of video cameras, including Neuvicon tube cameras, charge-coupled cameras (CCD's), complementary metal oxide semiconductors (CMOS), and digital cameras are used [see 14,16, Shotton and Ladic, in Ref. 20(Vol. 3),158–161].

The development of ultrasensitive low-light video cameras has had a major impact in capitalizing on the use of fluorescence microscopy on living cells. Normal fluorescence microscopy uses levels of UV that cause significant cellular damage, making study of living processes difficult or impossible. Reduction of the level of excitation illumination generates a very weak fluorescent signal not detectable by eye or standard cameras. Low-light cameras such as silicon intensified cameras (SIT), intensified SIT (ISITS), intensified CCD (ICCD), cooled CCDs and cameras capable of integration on chip provide output signals of sufficient brightness that fluorescently labeled cellular structures can be studied in live cells without impairing function [14,20(Vol. 3),158,160,161]. Furthermore components that are not abundant (e.g., a small number of receptors) can become detectable because the sensitivity of detection is very high.

The combined benefits of video-enhanced contrast and video low-light microscopy makes video microscopy applicable to the full range of light microscope types and a broad spectrum of cell biology problems.

The implications of image processing of digitized images with computers adds an additional dimension in the potential to extract information and also for quantitative and morphometric analysis. Possibilities range from simple measurements such as counting objects, length and surface area measurement, and density profiles to intensity measurements in quantifying fluorescence, staining, etc. The microscope and image processing have become indispensable tools in many cell biology laboratories (14,16,20(Vol. 3)).

In-depth details of the terminologies and technical aspects of video microscopy imaging is beyond the scope of this article. The reader is referred to Inoué and Spring (2), Shotton (in 20(Vol. 3)), Ladic (in 20(Vol. 3)), Oshiro (159), CLMIB (158), Lynch (162), Moomaw (161), Lockyer (160), and the web sites of manufacturers. An especially important aspect is the selection of the appropriate camera(s), particularly because new options are becoming available almost daily. There are many manufacturers (e.g., Apogee, Cohu, Cooke, Dage, Dulsa Fairchild, Hamamatsu, Photometrics and Princeton Instruments/Roper Scientific, Kodak, Video Scope, Xybion, RCA, Panasonic, Hitachi, Phillips, and others). Resolution, signal-to-noise ratio improvements, and sensitivity are continually improving, and costs are becoming affordable. The advances in thermoelectric cooling devices or Peltier cooling has allowed for excellent cooled CCD cameras. Digital video cameras are also becoming more commonplace. A broad selection of frame grabbers and imaging software is also available from many manufacturers.

So, on all fronts of video microscopy, this impact on cell biology is strong and growing daily.

ELECTRON MICROSCOPES

The golden age of electron microscopy was ushered in by the world's first electron microscope built by Knoll and Ruska in 1932 in Berlin (7–9). The first micrographs of biological specimens were produced in Belgium by Marton (163). The first North American EMs were built in Canada by Hall in 1936 and Prebus and Hillier 1983, and other microscopes were developed in the late 1930s in England (Martin, Whelpton, and Parnaim) and Japan (Sugata; Hibi) (see 9,10,164).

The excitement that characterized these early days of EM was based on deBroglie's discovery that electrons travel as a wave and Busch's discovery that they could be focused by electromagnetic lenses. One had the potential to exceed the resolving power of the light microscope. The first commercially available electron microscopes became available in the 1938–1940s first from Siemens in Germany and RCA in North America, and from Phillips, AEI, and various Japanese companies. Improvements in electromagnetic lens aberration correction, use of apertures, and vacuum system advances led to the transmission electron microscopes, that provide unprecedented resolution for cell biologists to determine the ultrastructure of cells and tissues. The first scanning electron microscope was developed in the 1930s by von Ardenne in Germany. The predecessor of the SEM, as we know it, was developed by Oatley in Britain (11) and first commercially produced by Cambridge Instruments. Since then there have been innumerable advances in EM with high voltage EMs, analytical

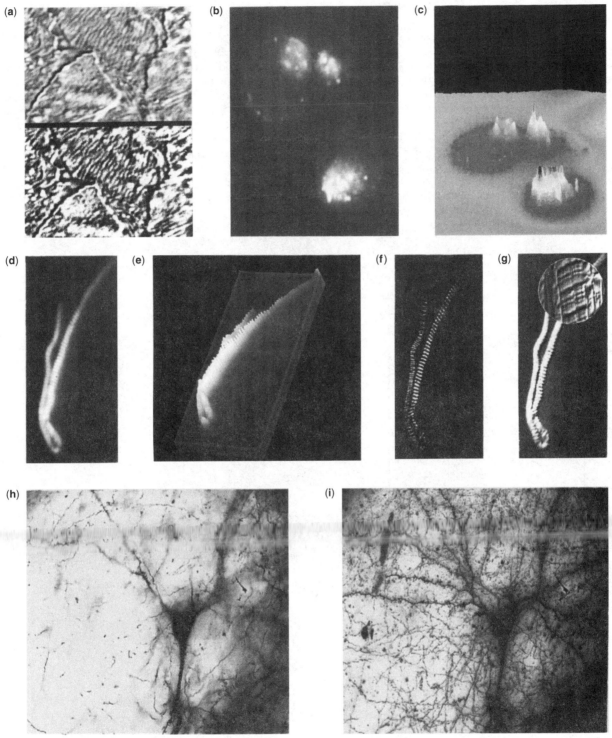

Figure 15. Some examples of image processing. (**a**) Contrast enhancement of the surface membrane feature of live buccal cells. (**b**) Trout hepatocytes that have endocytosed the fluorophore lucifer yellow; the relative brightness of the fluorescent areas is illustrated in this black and white image of a pseudocolored 3-D intensity profile in (**c**). (d–g) Various manipulations of a fluorescence image of an isolated insect striated muscle cell stained with rhodamine-labeled phalloidin (**d**). The F-actin of the sarcomeres fluoresces brightly. (**e**) An intensity profile (brightest areas as peaks). (**f**) Contrast enhancement. (**g**) A DIC-like image produced by subtraction of two images with shear. The inset is at higher magnification to show more clearly the DIC-like view of the sarcomeres (the original nonprocessed image was a fluorescence image). (**h**) and (**i**) 3-D reconstruction of a stalk of focal planes. (**h**) A single focal plane of a Golgi stained pyramidal neuron cell in a thick section of the cerebrum. Only part of the dendritic structure and the axon are seen. (**i**) Merging of a series of focal planes into a single image. On screen this can be viewed in stereo and rotated.

EMs using scanning transmission EM (STEM), imaging based on energy loss spectroscopy, and others. Magnification attainable and routinely used by cell biologists ranges from 1,000 × to 500,000 × with resolving powers as low as to 1 Å or less. Some specialized new EMs in the 300 kV or more range can resolve single atoms. Ease of use has progressed from manual control with manual valves, geiger tubes, etc. to microscopes that are slick, ergonomically designed with microprocessor and digital control, and have sophisticated imaging capabilities. Perusal of the literature and web sites of the major manufacturers of EMs (JEOL, Hitachi, Phillips, Leo, and others) provides some insights into the state of EM technology available today. Many reference texts are available on the theory,

principles, and uses of biological electron microscopy and tissue preparation techniques (165–169).

The application of electron microscopes to cell biological research also hinged on the development of specimen preparation methods and cutting sections thin enough for an electron beam to pass through to yield a transmitted image. The selection of fixatives that preserve cell morphology adequately was essential. Osmium tetroxide initially (used in the 1940s–1950s) and glutaraldehyde (introduced in the 1960s) provide excellent cell ultrastructural preservation and are still in use today. The ability to attain ultrathin sections of TEM rested on finding suitable knives and developing microtomes with ultrafine advance capabilities. The implementation of plastic resins (a wide

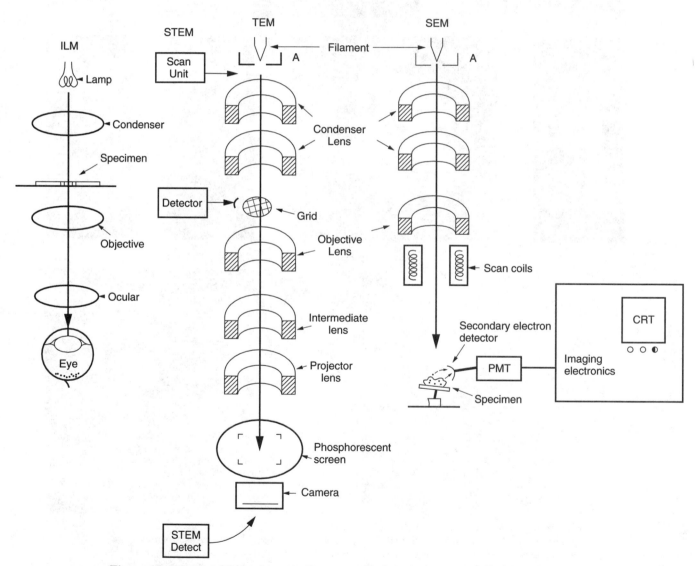

Figure 16. This simplified schematic illustrates the key components of electron microscopes (STEM, TEM, and SEM) compared to an inverted light microscope (ILM). The vacuum system and electronic control components are omitted. Presented is the sequence from the electron gun (filament and anode A) through the various electro magnetic lenses and viewing on a phosphorescent screen (TEM) or cathode ray tube (CRT) in SEM. Adding scanning coils and detectors near the sample (e.g., X-ray detector) or a STEM detector below the phosphorescent screen basically converts a TEM to an STEM and greatly expands its range of capabilities. STEM and other detector-generated images are viewed on a CRT via an electronic control unit (not shown).

range available today), the utilization of glass and diamond knives, and the design of ultramicrotomes have made biological TEM the widely used tool it is today. The selective enhancement of electron density in cellular components is achieved by using heavy metal stains such as uranyl acetate and lead citrate.

The light microscope was overshadowed from the 1940s to 1960s by an enormous explosion of literature based on electron microscopy that provided incredible resolution and led to new understanding of cell organelles and cell processes. It opened new vistas in cell biology. The electron microscope is still a powerful tool with important applications today, but as covered earlier, the advances in light microscopy and scanning probe microscopy have taken more of a lead role. The exciting prospects for a cell biologist is that there is now an arsenal of instruments available to reveal almost any facet of cell structure and function one wishes to visualize, and the TEM is one these tools.

The transmission EM shares a similar plan with a conventional light microscope. It has electromagnetic lenses in place of glass ones and a filament (either tungsten or lanthanum hexaboride) that emits a stream of electrons by thermionic emission upon heating (to 2700 °K) into a vacuum (in the TEM column). In some newer instruments field emission (electrostatic) guns are used. Efficient vacuum is created by various vacuum pumps (e.g., diffusion types, ion getters, turbomolecular pumps). Figure 16 shows a simplified diagram (minus the vacuum system and electronic control systems) of a basic TEM, the components added to make it a scanning transmission EM, STEM, or a standard SEM, and compares them with an inverted light microscope ILM. The literature is replete with conventional electron-dense/electron-lucid images provided in TEM micrographs that reveal fine cellular detail (membranes, mitochondria, cytoskeletal elements, labeled receptors etc.). The advent of peroxidase, colloidal gold, and other electron-dense labeling methods facilitates immunoelectron microscopy. When fixation interferes with immuno methods, ultracryosectioning and labeling methods can be used. With the addition of scanning coils to a TEM, one can position or raster a beam over a sample or selected area, and combining a host of X-ray microanalysis and other detectors, the TEM becomes a STEM (Fig. 16). This extends the EM's capability into the analytical range. Figure 17 provides an example of a conventionally stained TEM micrograph. Energy loss spectroscopy has also been used for electron microscopic imaging (170). Using electron spectroscopic imaging with an EM like the Zeiss CEM 902 makes it possible to image unstained sections based on their composition. Figure 18a shows microtubules imaged at the carbon edge.

Scanning EM also has major uses in cell biology (171). While utilizing electromagnetic lenses (condenser, objective and field lens), the beam is scanned across a sample, and secondary electrons given off by the surface are collected on a detector nearby converted to photons, which are transmitted to a photomultiplier and are displayed on a screen via the SEM electronics (Fig. 16). Usually samples are fixed, dehydrated, dried (critical point drying, freeze-drying, or other means) and have an ultrathin metal coating (gold or gold-palladium) on the surface (by a

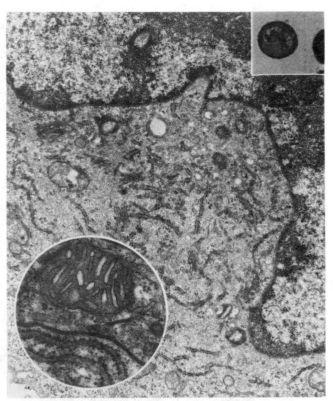

Figure 17. A typical TEM micrograph of a thin section of an ascites tumor cell (inset is LM view) depicting the nucleus and array of organelles and filaments stained with uranyl acetate and lead citrate. The circular insets shows a high magnification view of a mitochondrion and the rough endoplasmic reticulum of a gastric epithelial cell from a jellyfish.

sputter coater). Now, whole cells (Fig. 18c, d), cut surfaces, and isolated organelles can be viewed with resolution at 1 nM or less. Three-dimensional images of high resolution can reveal cell surface specializations, as well as internal cell morphology (e.g., membranous organelles, the cytoskeletal cytoarchitecture, etc.) (172). Figure 18e shows an SEM view of a complex microtubular array isolated from an insect ovariole (Huebner, unpublished data).

Improvements in SEMs and innovative approaches have extended the range of the type of sample that can be viewed. Field-emission SEMs with ultravacuums can give magnifications of one million with high resolution. Another new category of SEM, the environmental SEM (ESEM), has removed the constraint of high vacuum. They operate in the range of 1–20 torr. This opens new avenues to examine unfixed hydrated samples without distortion and thus avoids artifacts due to specimen preparation inherent in conventional SEMs (CSEM) or in low-vacuum conventional SEMs (LV-CSEM). Cells and tissues can be imaged in hydrated form with no fixation under nearly natural conditions at EM resolution. The full impact of the ESEM in cell biology is yet to be realized. A variety of analytical probes can also be incorporated into SEMs making them analytical SEMs. Thus the cell biologist also has a variety of high-resolution instruments available under the SEM umbrella that can produce three-dimensional images of cells and cellular components.

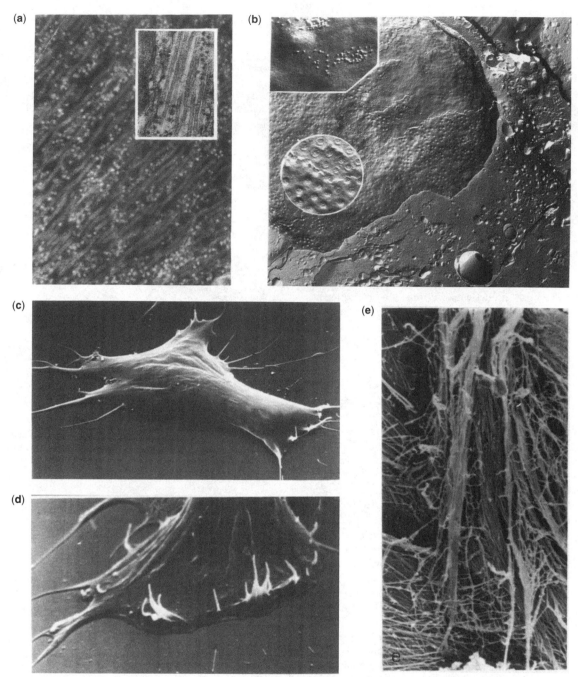

Figure 18. (**a**) A TEM view of an unstained section of insect ovary microtubules and ribosomes in the ovarioles of the insect *Rhodnius* imaged on the basis of their carbon content by electron spectroscopic imaging (U. Heinrich and E. Huebner, unpublished data, 1999). The insets show a comparable stained image in conventional TEM. (**b**) A TEM freeze-fracture micrograph of an insect follicle cell in the insect *Rhodnius*. The fracture plane reveals the nucleus with its many nuclear pores, the inner and outer nuclear envelopes, and vesicles and mitochondria in the cytoplasm. The upper inset shows intramembranous particles of the plasma membrane. (**c–e**) Examples of SEM of whole cell and cell components. (**c, d**) The surface filopodia and ruffles of isolated American lobster hemocytes. (**e**) Isolated microtubular arrays from the ovarioles of the insect *Rhodnius*.

A powerful cryomethod used to provide the molecular architecture of structures like membranes, cell junctions, and cytoskeletal elements, is freeze-fracture. Early workers such as Moor, Steere, and others in the 1950s provided the foundations for freezefracture replica methods (167,173–175). In essence, cells are frozen with ultrarapid cryomethods and placed in a freeze-fracture device which has ultralow temperature control, low vacuum, a cracking or fracturing device, and electrodes for carbon and platinum coating. The frozen sample is cracked or fractured leaving randomly cleaved surfaces (interior of membranes and internal cell components). This is followed by a time invacuo when sublimation of water from the frozen sample surface occurs (etching). This is normally a very short period but can be extended in deep etch procedures. Deep etching reveals internal cytoskeletal elements and membranes below the fractured surface. The exposed surface is then coated with carbon and platinum. The sample with its coating is removed from the freeze-fracture unit, thawed, and the tissue washed away leaving only the thin carbon–platinum coating called a replica, which is a cast of the fractured surface. Astonishingly, such replicas faithfully reveal the finest morphology of submembrane architecture (proteins within the bilayer) and molecular subunits of cytoskeletal elements (F-actin and microtubules), for example. Revealing this level of structural detail necessitates using ultrarapid freezing methods. The best molecular architecture is attainable with freeze-slam methods using a copper block cooled with liquid helium, as in the freeze-slam deep etch methods. More commonly in conventional freeze-fracture, immersion freezing with cryogenic liquids is used.

Freeze-fracture replicas are visualized in transmission EMs. The replicas are thin enough for the beam to pass. Due to the shadowing of platinum preferentially on areas due to their three-dimensional contours, one has a range of electron densities reflective of the differential amounts of platinum. Platinum is coated onto the sample from an angle and carbon is coated from above. The net effect is an image that shows high contrast, high resolution and has a shaded, contoured, three-dimensional appearance. Figure 18b and insets show a freeze-fracture image of an insect follicle cell. Depicted is the nuclear surface with the inner and outer envelopes and nuclear pores. The rectangular inset shows intramembranous particles of the cell membrane.

SUMMARY

It is difficult to do justice to the scope of the field of microscopy today in a few pages, but hopefully this presentation provides the reader with an introduction to the variety of microscopes available today to the cell biologist. The field is growing rapidly, and many exciting prospects undoubtedly await us in the future.

Biology in general and cell biology in particular will be front and center in the science of the next century. It is reassuring to anyone in cell biotechnology, who is exploring the intricacies of cell structure and function to have such an arsenal of microscopical tools as already exists. The long history of microscopes since Janssen, Hooke, and Leeuwenhoek continues with renewed vigor as we explore the microcosm.

WEB SITES

The following provides some of the many web sites with microscopy content. Many of these have links to other sites within the list and outside. Included are various educational, personal, and vendor sites. The list is not all inclusive and is not intended to endorse any particular product or manufacturer. Addresses of web sites may change, companies may merge, new sites may appear and others may close, so the list provides some of the vast array of microscopy resources and information available at this time. Due to the nature of the web, it is not possible to predict the future content and details of any site.

Microscopical Societies

http://www.rms.org.uk   (Royal Microscopical Society)

http://www.MSA.microscopy.com/   (Microscopical Society of America)

http://www.ualberta.ca/~cherbur.msc   (Microscopical Society Canada)

General Microscopy and Variety of Microscopes

http://micro.magnet.fsu.edu/primer/webresources.html (Molecular Expressions — Microscopy Primer — M. Davidson, M. Abramowitz, Olymus & Florida State University excellent interactive tutorials on LM types)

http://www.mih.unibas.ch/Booklet/overview.html (LM, EM, and SPM images)

http://www.pharm.Zrizona.edu/centers/tox_center/ swehsc/exp_path/m-i_onw3.html (D. Cromey, University of Arizona, many links)

http://nsm.fullerton.edu/~skarl/EM/EMLab.html (R. Koch, California State University Fullerton, images, instruction).

http://members.aol.com/BobCat54/index.links.html (many links to world sites, universities, suppliers)

http://www.uq.oz.au/nanoworld/nanohome.html/ (University of Queensland, collection of microscopy resources)

http://www.mwrn.com/ (collection of microscopy web resources, vendors' products)

http://www.ou.edu/research/electron/www-vl/menu. shtml   (University of Oklahoma)

http://www.ou.edu/research/electron/www-vl/educatio. shtml   (University of Oklahoma educational site, many links)

http://www.ou.edu/research/electron/mirror (C. Jeffries, Bristol, United Kingdom, many links)

http://www.microscopy-online.com   (web magazine with resources, vendors, bulletin board)

http://www.microrgc.demon.co.uk   (microscopy and Analysis Publication, info, links to other sites)

http://www.MME-Microscopy.com/education/ (national consortium of microscopy experts)

http://www.yahoo.co.uk/Science/Engineering/ optical_Engineering (list of web sites and many links)

Fluorscence Microscopy, Confocal and Multiphoton Microscopy

http://www.mdyn.com/application_notes/applications.htm# Application Guides http://www.probes.com (Molecular Probes, Oregon, many links to journals and microscopy sites via/sites/

http://www.Fluorescentprobes.com (TEF Labs, Texas)

http://www.molbio.princeton.edu/facilites/index.html (J. Goodhouse, confocal images and biol. EM)

http://www.bocklabs.wisc.edu/imr/home2.htm (D. Wokosin, University of Wisconsin. Institute Microscopy Facility, variety of LM, Em, and Imaging)

http://www.caltech.edu/~pinelab/2photon.html (S. Potter and S. Fraser — excellent two-photon and confocal site)

http://www.cs.ubc.ca/spider/ladic/confocal.html (Lance Ladic — excellent confocal site)

http://www.vaytek.com (digital deconvolution, confocal)

http://www.scanalytics.com (deconvolution, digital confocal)

Scanning Probe, SNOM, and Laser, Tweezers

http://www.cellrobotics.com/cell (Laser tweezers — Cell Robotics Inc.)

http://www.thermomicro.com (SPM — Park Instruments and Topometrix merged)

http://www.di.com (Digital Instruments)

http://www.molec.com (Molecular Imaging Inc.)

http://www.omicron-instruments.com/ (Omicron — Twin SNOM)

Electron Microscopy

A number of the general sites listed above also include EM (e.g., University of Wisconsin IMR, Princeton Molbio, University Oklahoma, U.C. Fullerton etc.). The various EM manufacturers also have informative web sites eg. (Hitachi, JEOL, Phillips, Zeiss, etc.) http://www.mos.org/sln/sem/ (This Museum of Science, Boston, site has good SEM info and links to other sites.)

ACKNOWLEDGMENTS

I am indebted to Everett Anderson for his mentorship and expertise in revealing the wonders of electron microscopy and to Shinja Inoué for exposing me to the world of analytical light microscopy and the power of light microscopy in his MBL course and since. Research funding from the Canadian Natural Sciences and Engineering Research Council and the University of Manitoba has been invaluable. I thank Madeleine Harris for secretarial help in preparing this article.

BIBLIOGRAPHY

1. A. Köhler, Zeitschrift für wissenshaftl. Mikroskopie 10, 433–440 (1893) (reprinted and translated in Royal Microscopical Society Proceedings **28**, 181–185 (1993)).

2. S. Inoué and K.R. Spring, *Video Microscopy: The Fundamentals*, 2nd ed., Plenum, New York, 1997.

3. E. Keller, in D.L. Spector, R.D. Goldman, and L.A. Leinwand, eds., *Cells: A Laboratory Manual*, Vol. 2, Cold Spring Harbor Press, Cold Spring Harbor, N.Y., 1997, Chapter 94, pp. 94.1–94.53.

4. A.J. Lacey, ed., *Light Microscopy in Biology*, IRL Press, Oxford, 1989.

5. M. Spencer, *Fundamentals of Light Microscopy*, Cambridge University Press, London, 1982.

6. S. Bradbury, *An Introduction to the Optical Microscope*, Bios Sci. Publ., Oxford, U.K., 1994.

7. E. Ruska, *The Early Development of Electron Lenses and Electron Microscopy*, Hirzel, Stuttgart, 1980.

8. M. Knoll and E. Ruska, *Physik* **78**, 318–339 (1932).

9. P.W. Hawkes, ed., *The Beginning of Electron Microscopy*, Adv. Electron. Electron Phys., Suppl. 16, Academic Press, New York, 1985.

10. P. Hawkes, *Microsc. Analy.* **28**, 10–13 (1998).

11. C.W. Oatley, *J. Appl. Phys.* **53**, R1 (1982).

12. D.L. Taylor, M.A. Nederlof, F. Lannia, and A.S. Waggoner, *Am. Sci.* **80**, 322–335 (1992).

13. B. Herman and J.J. Lemasters, eds., *Optical Microscopy: Emerging Methods and Applications*, Academic Press, San Diego, Calif., 1993.

14. R. Stevenson, *Am. Lab.* **28**, 23–51 (1996).

15. J. Manni, *Biophotonics Int.* **3**, 44–51 (1996).

16. K. Robinson, *Biophotonics Int.* **4**, 46–52 (1997).

17. W.C. McCrone, *Am. Lab.* **20**, 21–28 (1988).

18. J. Plasek and J. Reischig, *Proc. R. Microsc. Soc.* **33**, 196–205 (1998).

19. T. Wilson, *Microsc. Microanal.* **32**, 7–9 (1998).

20. J.E. Cellis, ed., *Cell Biology: A Laboratory Handbook*, 2nd ed., Vols. 2 and 3, Academic Press, San Diego, Calif., 1998.

21. RMS, *RMS Dictionary of Light Microscopy*, Bios. Sci. Publ., Oxford, U.K., 1989.

22. G.W. Ellis, *J. Cell Biol.* **101**, 83a (1985).

23. M. Brenner, *Am. Lab.* **26**, 14–19 (1994).

24. T. Richardson, *Proc. R. Microsc. Soc.* **33**, 3–8 (1998).

25. M. Abramowitz, in *Basics and Beyond Series*, Vol. 2, 1–31. Olympus Corp., New York, 1994.

26. S. Bradbury and P. Everett, *Contrast Techniques in Light Microscopy*, Bios. Sci. Publ., Oxford, U.K., 1996.

27. S. Bradbury, ed., *RMS Microscopy Handbooks*, Multivol. Ser, Bios Sci. Publ., Oxford, U.K., 1990–1998.

28. D.L. Spector, R.D. Goldman, and L.A. Leinwand, eds., *Cells: A Laboratory Manual*, Vol. 2, Cold Spring Harbor Lab. Press, Cold Spring Harbor, N.Y., 1998.

29. J.K. Presnell and M.P. Schreibman, *Humason's Animal Tissue Techniques*, Johns Hopkins University Press, Baltimore, Md., 1997.

30. F.H. Wenham, *Trans. Microsc. Soc. London* **III**, 83–90 (1852).

31. L.V. Martin, *Microscopy* **36**, 124–138 (1988).

32. W. McCrone and R.D. Moss, *Microscope* **38**, 1–8 (1990).

33. F. Zernicke, *Physica* **9**, 686–698, 974–986 (1942).

34. K.F.A. Ross, *Phase Contrast and Interference Microscopy for Cell Biologists, Edward Arnold*, London, 1967.

35. K.F.A. Ross, *Microscopy* **36**, 97–123 (1998).

36. A.H. Bennett, H. Jupnik, H. Osterberg, and O.W. Richards, *Phase Contrast Microscopy: Principles and Applications*, Wiley, New York, 1951.

37. B. Kachar, *Science* **227**, 766–768 (1985).

38. A. Strange, *Microscope* **37**, 355–376 (1989).

39. A. Strange, *J. Biol. Photogr. Assoc.* **62**, 91–102 (1994).

40. R. Hoffman, *J. Microsc. (Oxford)* **110**, 205–222 (1977).

41. G.W. Ellis, in E. Dirksen, O. Prescott, and C.F. Fox, eds., *Cell Reproduction: In Honor of Daniel Mazia*, Academic Press, New York, 1978, pp. 465–476.

42. G.L. Greenberg and A. Boyde, *Proc. R. Microsc. Soc.* **32**, 87–101 (1997).

43. A.A. Hallimond, *The Polarizing Microscope*, 3rd ed., Vickers Inst. Ltd., York, England, 1970.

44. S. Inoué, in G.L. Clarke, ed., *The Encyclopedia of Microscopy*, Reinhold, New York, 1961, pp. 480–485.

45. R. Oldenbourg and G. Mei, *J. Microsc. (Oxford)* **180**, 140–147 (1995).

46. P.C. Robinson and S. Bradbury, *Qualitative Polarized-Light Microscopy*, Bios. Sci. Publ., Oxford, U.K., 1992.

47. S. Inoué et al., *Am. Soc. Cell Biol.* Poster Abstr. No. H-117 (1997).

48. G.A. Dunn, *Proc. R. Microsc. Soc.* **33**, 189–196 (1998).

49. R.D. Allen, G.B. David, and G. Nomarski, *Z. Wiss. Mikrosk. Mikrosk. Tech.* **69**, 193–221 (1969).

50. J. Padawer, *J. R. Microsc. Soc.* **88**, 305–349 (1968).

51. G. Holzwarth, S.C. Webb, D.J. Kubinski, and N.S. Allen, *J. Microsc. (Oxford)* **188**, 249–254 (1997).

52. H. Verschueren, *J. Cell Sci.* **75**, 279–301 (1985).

53. A.S.G. Curtis, *J. Cell Biol.* **20**, 199–215 (1964).

54. H. Piller, *J. Microsc. (Oxford)* **100**, 35–48 (1974).

55. C.S. Izzard and L.R. Lochner, *J. Cell Sci.* **21**, 129–159 (1976).

56. M. Opas and V.I. Kalnins, *J. Microsc. (Oxford)* **133**, 291–306 (1983).

57. J.S. Ploem, I.C.-T. Velde, F.A. Prins, and J. Bonnet, *Proc. R. Microsc. Soc.* **30**, 185–192 (1995).

58. M.D. Egger and M. Petrán, *Science* **157**, 305–308 (1967).

59. M. Petrán, M. Hadravsky, and A. Boyde, *Scanning* **7**, 97–108 (1985).

60. M. Minsky, *Scanning* **10**, 128–138 (1988).

61. A. Boyde, *Ann. N.Y. Acad. Sci.* **483**, 428–439 (1986).

62. R. Harris, *Bull. —Microsc. Soc. Can.* **26**, 15–17 (1998).

63. B. Herman, *Fluorescence Microscopy*, 2nd ed., Bios Sci. Publ., Oxford, U.K., 1998.

64. F.W.D. Rost, *Fluorescence Microscopy*, Vol. 1, Cambridge University Press, Cambridge, U.K., 1992.

65. W.T. Mason, ed., *Fluorescent and Luminescent Probes for Biological Activity: A Practical Guide to Technology for Quantitative Real-time Analysis*, Academic Press, San Diego, Calif., 1993.

66. D.L. Taylor et al., ed., *Applications of Fluorescence in the Biomedical Sciences*, Alan R. Liss, New York, 1986.

67. Y.-L. Wang, in D.L. Spector, R.D. Godman, and L.A. Leinwand, eds., *Cells: A Laboratory Manual*, Vol. 2, Cold Spring Harbor Lab. Press, Cold Spring Harbor, N.Y., 1998, Chapter 76, pp. 76.1–76.6.

68. Y.-L. Wang and D.L. Taylor, eds., *Methods in Cell Biology*, Parts A and B, Vols. 29 and 30, Academic Press, San Diego, Calif., 1989.

69. L.V. Johnson, M.L. Walsh, and L.B. Chen, *Proc. Natl. Acad. Sci. U.S.A.* **77**, 990–994 (1980).

70. L.B. Chen and S.T. Smiley, in W.T. Mason, ed., *Fluorescent and Luminescent Probes for Biological Activity*, Academic Press, San Diego, Calif., 1993, pp. 124–132.

71. M. Poot, in J.E. Cellis, ed., *Cell Biology: A Laboratory Handbook*, 2nd ed., Vol. 2, Academic Press, San Diego, Calif., 1998, pp. 513–517.

72. M. Terasaki, in J.E. Cellis, ed., *Cell Biology: A Laboratory Handbook*, 2nd ed., Vol. 2, Academic Press, San Diego, Calif., 1998, pp. 501–506.

73. C. Lee and L.B. Chen, *Cell* **54**, 37–46 (1988).

74. R.E. Pagano and D.C. Martin, in J.E. Cellis, ed., *Cell Biology: A Laboratory Handbook*, 2nd ed., Vol. 2, Academic Press, San Diego, Calif., 1998, pp. 507–517.

75. K.A. Giuliano and D.L. Taylor, *Curr. Opin. Cell Biol.* **7**, 4–12 (1995).

76. R.Y. Tsien, *Annu. Rev. Neurosci.* **12**, 227–253 (1989).

77. G. Grynkiewicz, M. Peonie, and R.Y. Tsien, *J. Biol. Chem.* **260**, 3440–3450 (1985).

78. R.B. Silver, *Cell Calcium* **20**, 161–179 (1996).

79. K.F. Sullivan and S.A. Kay, eds., *Methods in Cell Biology*, Vol. 58, Academic Press, San Diego, Calif., 1998.

80. A.B. Cubitt et al., *Trends Biochem. Sci.* **20**, 448–455 (1995).

81. K.A. Giuliano, P.L. Post, K.M. Hahn, and D.L. Taylor, *Annu. Rev. Biophy. Biomol. Struct.* **24**, 405–434 (1995).

82. M. Chalfie et al., *Science* **263**, 802–805 (1994).

83. S. Potter, C.-M. Wang, P.A. Garrity, and S.E. Fraser, *Gene* **173**, 25–31 (1996).

84. G. Miesenböck, D.A. DeAngelis, and J.E. Rothman, *Nature (London)* **394**, 192–195 (1998).

85. J.T. Murphy and J.C. Lagarias, *Curr. Biol.* **7**, 870–876 (1997).

86. S.R. Adam and R.Y. Tsien, *Annu. Rev. Physiol.* **55**, 755–784 (1993).

87. J.B. Pawley, ed., *Handbook of Biological Confocal Microscopy*, 2nd ed., Plenum, New York, 1995.

88. S. Inoué, in J.B. Pawley, ed., *Handbook of Biological Confocal Microscopy*, 2nd ed., Plenum, New York, 1995, pp. 1–17.

89. A. Hall, M. Browne, and V. Howard, *Proc. R. Microsc. Soc.* **26**, 63–70 (1991).

90. B. Matsumoto, ed., *Methods in Cell Biology* Vol. 38, Academic Press, San Diego, Calif., 1993.

91. T. Wilson, *Proc. R. Microsc. Soc.* **33**, 259–264 (1998).

92. R. Juskaitis, T. Wilson, M.A.A. Neil, and M. Kozubec, *Nature (London)* **383**, 804–806 (1996).

93. T. Wilson, M.A.A. Neil, and R. Juškaitis, *J. Microsc. (Oxford)* **191**, 116–118 (1998).

94. A. Ichihara et al., *Bioimages* **4**, 57–62 (1996).

95. S. Inoué and T.D. Inoué, *Biol. Bull. (Woods Hole, Mass.)* **191**, 269–270 (1996).

96. J.G. White and W.B. Amos, *Nature (London)* **328**, 183–184 (1987).

97. J.G. White, W.B. Amos, and M. Fordham, *J. Cell Biol.* **105**, 41–48 (1977).

98. G.J. Brakenhoff, E.A. Van Spronsen, H.T.M. Van Der Voort, and N. Nanninga, in Y.-L. Wang and D.L. Taylor, eds., *Methods in Cell Biology* Part B., Vol. 30, Academic Press, Orlando, Fla., 1989, pp. 379–399.

99. C.J.R. Sheppard and D.M. Shotton, *Confocal Laser Scanning Microscopy*, Bios. Sci. Publ., Oxford, U.K., 1997.

100. P.C. Cheng, T.H. Lin, and W.L. Wu, eds., *Multidimensional Microscopy*, Springer-Verlag, New York, 1994.

101. D.A. Agard and J.W. Sedat, *Nature (London)* **302**, 676–681 (1983).

102. Y. Hiraoka, D.A. Agard, and J.W. Sedat, *J. Cell Biol.* **111**, 2815–2828 (1991).

103. W.A. Carrington et al., *Science* **268**, 1483–1487 (1995).

104. W.A. Carrington, K.E. Fogarty, L. Lifschitz, and F.S. Fay, in J.B. Pawley, ed., *Handbook of Biological Confocal Microscopy*, Plenum, New York, 1995, pp. 151–161.

105. R. Rizzuto, W. Carrington, and R.A. Tuft, *Trends Cell Biol.* **8**, 288–292 (1998).

106. W. Denk, J.H. Strickler, and W.W. Webb, *Science* **248**, 73–76 (1990).

107. W. Denk, *Photonics Spectra* **31**, 125–130 (1997).

108. C.J.R. Sheppard and R. Kompfner, *Appl. Opt.* **17**, 28, 79–83 (1978).

109. M. Gu and C.J.R. Sheppard, *J. Microsc. (Oxford)* **177**, 128–137 (1995).

110. W. Denk, D.W. Piston, and W.W. Webb, in J.B. Pawley, ed., *Handbook of Biological Confocal Microscopy*, 2nd ed., Plenum, New York, 1995, pp. 445–458.

111. M.B. Cannell and C. Soeller, *Proc. R. Microsc. Soc.* **32**, 3–8 (1997).

112. M.K. Robinson, *Biophotonics Int.* **4**, 38–45 (1997).

113. J. Mann, *Biophotonics Int.* **3**, 44–52 (1996).

114. W. Denk, M. Sugimori, and R. Llinas, *Proc. Natl. Acad. Sci. U.S.A.* **18**, 8279–8282 (1995).

115. S. Maiti et al., *Science* **275**, 530–532 (1997).

116. T.G. Rochow and P.A. Tucker, *An introducation to Microscopy by means of Light, Electrons, X-Rays or Acoustics*, 2nd ed., Plenum, New York, 1994.

117. A. Briggs, *An Introduction to Scanning Acoustic Microscopy*, Bios Sci. Publ., Oxford, U.K., 1985.

118. P. Vesely, H. Lüers, M. Riehle, and J. Bereiter-Hahn, *Cell Motil. Cytoskel.* **29**, 231–240 (1994).

119. Y. Wang and C. Jacobsen, *J. Microsc. (Oxford)* **191**, 159–169 (1998).

120. A. Ashkin and J.M. Dziedzic, *Science* **235**, 1517–1520 (1987).

121. S.M. Block, in J.K. Foskett and S. Grinstein, eds., *Noninvasive Techniques in Cell Biology*, Wiley-Liss, New York, 1990, pp. 375–402.

122. K. Svoboda and S.M. Block, *Annu. Rev. Biophys. Biomol. Struct.* **23**, 247–285 (1994).

123. M. Berns, *Sci. Am.* **278**, 62–67 (1998).

124. M.P. Sheetz, ed., *Methods in Cell Biology*, Vol. 55, Academic Press, San Diego, Calif., 1998.

125. N.L. Simone et al., *Trends Genet.* **14**, 272–276 (1998).

126. R.F. Bonner et al., *Science* **278**, 1481–1483 (1997).

127. G. Binnig and H. Röhrer, *Science* **244**, 475–477 (1984).

128. G. Binnig, C.F. Quate, and G. Gerber, *Phys. Rev. Lett* **56**, 930–933 (1986).

129. O. Marti and M.A. Marti, eds., *STM and SFM in Biology*, Academic Press, San Diego, Calif., 1993.

130. R. Coratger, V. Sivel, F. Ajustron, and J. Beauvillain, *Micron* **4**, 371–385 (1994).

131. H.G. Hansana and J.H. Hoh, *Annu. Rev. Biophys. Biomol. Struct.* **23**, 115–139 (1994).

132. J. Yank and Z. Shao, *Micron* **26**, 35–49 (1995).

133. A. Stemmer, *J. Microsc. (Oxford)* **178**, 28–36 (1995).

134. R. Howland and L. Benatar, *A Practical Guide to Scanning Probe Microscopy*, Park Sci. Inst. Pub., Sunnyvale, Calif., 1996.

135. M. Radmacher, R.W. Tillman, M. Fritz, and H.E. Gaub, *Science* **257**, 1900–1905 (1992).

136. D. Ricci and M. Grattarola, *J. Microsc. (Oxford)* **176**, 256–261 (1994).

137. J. Vesenka et al., *BioTechniques* **19**, 240–249 (1995).

138. C.B. Prater, P.G. Maivald, K.J. Kjoller, and M.G. Heaton, *Am. Lab.* **27**, 50–53 (1995).

139. R. Winn Harden, *Biophotonics Int.* **4**, 46–49 (1997).

140. W.R. Bowen, N. Itilal, R.W. Lovitt, and C. Wright, *Microsc. Anal.* **33**, 7–9 (1998).

141. E. Nagao and J.A. Dvorak, *J. Microsc. (Oxford)* **191**, 8–19 (1998).

142. J.A. Dvorak and E. Nagao, *Exp. Cell Res.* 1998, 242, 69–74 (1998).

143. M. Hoffman, *Am. Sci.* **85**, 123–124 (1997).

144. E. Betzig and J. Trautman, *Science* **257**, 189–195 (1992).

145. D. Courjon, *J. Microsc. (Oxford)* **177**, 180–185 (1995).

146. G.J. Collins, *Bull. — Microsc. Soc. Can.* **23**, 12–17 (1995).

147. S.P. Marchese-Ragona and P.G. Haydon, *Bull. — Microsc. Soc. Can.* **24**, 10–15 (1996).

148. R. Harris, *Bull. — Microsc. Soc. Can.* **26**, 18–21 (1998).

149. A.K. Kirsch et al., *Biophysi. J.* **75**, 1513–1521 (1998).

150. P. Moyer, S.P. Marchese-Ragona, and B. Christie, *Am. Lab.*, April, pp. 30–32 (1994).

151. M.H.P. Moers, *J. Microsc. (Oxford)* **182**, 40–45 (1995).

152. P.G. Haydon et al., *J. Microsc. (Oxford)* **182**, 208–216 (1996).

153. G. Sluder and D.E. Wolf, eds., *Methods of Cell Biology*, Vol. 56, Academic Press, San Diego, Calif., 1998.

154. S. Inoué, *J. Cell Biol.* **89**, 346–356 (1981).

155. R.D. Allen, N.S. Allen, and J.L. Travis, *J. Cell Motil.* **1**, 298–302 (1981).

156. R.D. Allen, *Annu. Rev. Biophys. Chem.* **14**, 265–290 (1985).

157. S. Inoué, in L. Taylor and Y.L. Wang, eds., *Methods in Cell Biology*, Part B, Vol. 3, Academic Press, Orlando, Fla., 1989, pp. 85–112.

158. CLMIB, *Am. Lab.* **27**, 25–40 (1995).

159. M. Oshiro, in D.L. Spector, R.D. Goldman, and L.A. Leinwand, eds., *Cells: A Laboratory Manual*, Vol. 2, Cold Spring Harbor Press, Cold Spring Harbor, N. Y., 1997, Chapter 95, pp. 95.1–95.15.

160. E.M. Lockyer, *Photonics Spectra* **31**, 80–90 (1997).

161. B. Moomaw, *Biophotonics* **5**, 48–53 (1998).

162. T.F. Lynch, *Am. Lab.* **26**, 26–32 (1994).

163. L. Marton, *Nature (London)* **133**, 911 (1934).

164. F.W. Doane, G.T. Simon, and J.H.L. Watson, *Canadian Contributions to Microscopy*, Micros. Soc. Can., Toronto, 1993.

165. M.A. Hayat, *Principles and Techniques of Electron Microscopy*, Multi vol., Van Nostrand-Reinhold, New York, 1970–1976.

166. A.M. Glauert, *Practical Methods in Electron Microscopy*, Multi Vol. Ser., North-Holland/Elsevier, New York, 1972–1991.

167. M.J. Dykstra, *Biological Electron Microscopy: Theory, Techniques and Troubleshooting*, Plenum, New York, 1992.

168. J. Bozzola and L. Russell, *Electron Microscopy. Principles and Techniques for Biologists*, 2nd ed., Jones & Bartlett, Boston, 1998.

169. A.B. Maunsbach and B.A. Afzelius, *Biomedical Electron Microscopy*, Academic Press, San Diego, Calif., 1998.

170. R.F. Egerton, *Electron Energy-Loss Spectroscopy in the Electron Microscope*, 2nd ed., Plenum, New York, 1996.

171. M.A. Hayat, *Introduction to Biological Scanning Electron Microscopy*, University Park Press, Baltimore, Md., 1978.

172. G.H. Haggis, *Microsc. Res. Tech.* **22**, 151–159 (1992).

173. R.L. Steere, *J. Biophys. Biochem. Cytol.* **3**, 45–60 (1957).

174. H. Moor, *Z. Zelforsch. Mikrosk. Anat.* **62**, 546–580 (1964).

175. C. Stolinski and A.S. Breathnach, *Freeze-fracture Replication of Biological Tissues: Techniques, Interpretations and Applications*, Academic Press, New York, 1975.

See also ANATOMY OF PLANT CELLS; CELL STRUCTURE AND MOTION, CYTOSKELETON AND CELL MOVEMENT; CELLULAR TRANSFORMATION, CHARACTERISTICS; HISTORY OF ANIMAL CELL TECHNOLOGY.

CONTAMINATION DETECTION AND ELIMINATION IN PLANT CELL CULTURE

ALAN C. CASSELLS
National University of Ireland
Cork, Ireland

OUTLINE

INTRODUCTION

Biotic contamination of plant cell, tissue, and organ cultures can result in the loss of batches of cultures affecting production schedules in micropropagation or in bioreactors, or in the loss of elite genotypes in a genetic manipulation program. The consequential losses can range from financial claims from customers for loss of downstream production to delays in breeding programs. In an academic research program the losses can be equally catastrophic for the individual or group concerned. Biotic contaminants range from plant pathogens to common environmental organisms that enter the cultures in, or on, the initial explant; to laboratory contaminants, that is, cultivable bacteria and fungi including human pathogens, capable of living on the plant culture medium or in the explant tissues; and infestation by microarthropods (1). Expressed contamination usually results in immediate loss of cultures due to overgrowth by the contaminant; however, in some cases contamination is latent and may not be expressed until a change in medium composition, e.g., on transfer of shoots to rooting medium, or until field release (2). The contaminant may not be a pathogen of the crop but may be a source of inoculum for disease in other crops or in crop consumers (1). The strategy for the management of biotic contamination in plant tissue culture advocated here is, first, to establish aseptic cultures and thereafter to maintain clean cultures by good laboratory practice (3). Readers are recommended to consult (2) for a detailed text on plant tissue culture and the website of the American Phytopathological Society (http://www.scisoc.org) for a comprehensive range of reference books and services in plant pathology.

BIOTIC CONTAMINATION MANAGEMENT IN TISSUE CULTURE

The premise here is that all users of plant tissue cultures should be concerned about biotic contamination and the risks it poses to them directly as well as to third parties. Emphasis should be put by the tissue culturist on the initiation of aseptic cultures (3) and subsequently on good laboratory practice to manage laboratory contamination (4). Attention should be paid to the certification standards for the crop. Certification usually sets standards for both genetic quality [absence or low levels of off-types—variability may arise from chimeral breakdown, subliminal pathogen elimination or mutation depending on the tissue culture pathways used (2)] and health status (5). In certification schemes the requirement is freedom from quarantine pathogens, but, depending on the stock grade (see Fig. 1), increasing levels of specified pathogens/pathogen strains may be tolerated with later field generations (5,6).

In the case of vegetatively propagated crops, considerable expenditure can be afforded at the start to select for, or produce, disease-free individuals that form the starting material for mass clonal propagation, that is, the nuclear stock, this and the derived propagation stock, are raised under conditions aimed at maintaining a high health status and are visually inspected, rogued, and indexed, as appropriate. Similarly the mother stock generations are visually inspected, rogued, and indexed during each multiplication cycle. Finally, the progeny for sale are graded according to diseases present and disease levels (as well as other criteria, e.g., percentage of off-types) and certified according to the scheme under which they were produced. In conventional vegetative propagation the phenotype is available throughout the growth cycle in each generation for visual examination; consequently pathogen symptoms, and more important, transient symptoms characteristic of some infections, can be observed.

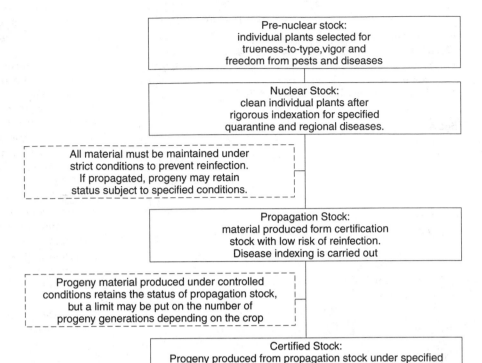

Figure 1. Stages in a plant health certification scheme.

This is not the case in micropropagation, where during repeated cloning cycles there is no opportunity to observe the mature phenotype (Fig. 2). In conventional propagation, if the parental material is clean ("pathogen-free"), then pathogen reentry is likely to be random, and this is addressed by random sampling/indexing of the material prior to certification. In the case of micropropagated stock, if the parental material is pathogen-free, then it is likely to remain pathogen-free, as there is a very low risk of exposure to pathogen inoculum in the tissue culture process. If, however, the initial explant is infected, there is the risk of mass clonal infection at the end of the culture process. It should also be appreciated that plants produced conventionally may be contaminated with environmental bacteria and fungi that are not pathogenic to plants, or at least their effects are subliminal (7). They may be, for example, pathogenic to those in contact with plants derived from tissue cultures (8), or cause plant produce (e.g., cut flowers) to deteriorate. Reflecting inoculum potential and the respective production systems, microplants may be clonally infected as opposed to environmentally grown plants, which may show more random patterns of contamination, reflecting field disease transmission.

In summary, the management of biotic contamination of in vitro culture has common elements with regional disease management in certified stock production and with the broader management of genetic resources in germplasm conservation and global distribution (9), but conventional heterotrophic/mixotrophic plant tissue culture also has unique problems due to the presence of organic substrates in the culture medium that support microbial growth giving rise to laboratory contamination (4). Where micropropagation is used in a crop certification scheme, total freedom from quarantine pathogens is required as for field-produced stock for certification, but while low levels of specified pathogen isolates are tolerated in field-produced crops for certification, in micropropagation, by virtue of the characteristics of the process, the likelihood is absence or total clonal infection. International diseases management in germplasm distribution requires absence from all pathogen strains, not merely specified strains, and this is a shared problem with micropropagation, where trade aspires to be international (9,10). This has major implications for the development of molecular diagnostics (see the following). Micropropagation, specifically heterotrophic or mixotrophic culture, as opposed to sugar-free autotrophic culture, has the unique problem of contamination by cultivable organisms that may infect the culture (2; see laboratory contamination management). Pathogen management in micropropagation is complicated by the problem that even for major crops the causal agents of many diseases are uncharacterized. Disease management where field-grown crops are being grown for certification addresses "local" pathogens, whereas micropropagation may involve the wider problem of "international" pathogens and thus involves potentially more pathogens and more, perhaps less characterized, strains. In potato some 25 viral diseases have been characterized (11), but even in strawberry, many of the pathogens have not been identified (12,13). The situation is even worse for rosaceous fruit trees (14). This problem is very serious for micropropagators, who work with a diverse range of minor crops where pathogens are poorly characterized (akin to the germplasm management problem). The major virus groups (e.g., the potyviruses) have a wide host range but e.g., specific antiserum for the PVY^o strain will not cross react with the PVY^n strain of potato virus Y (a

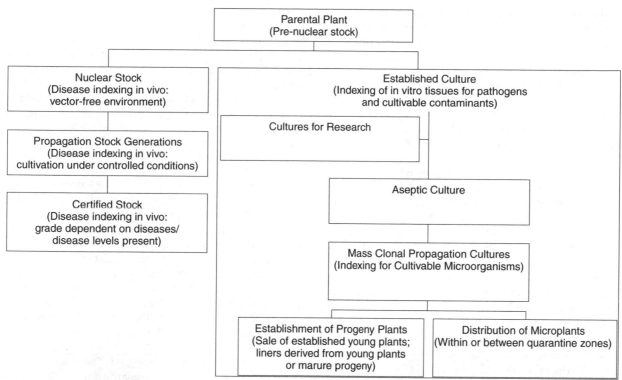

Figure 2. Flow diagram representing the contrast between the production of high-health-status material in vivo (left-hand branch) and in vitro (right-hand branch). In the in vivo system the plant material is available for inspection and indexing at all developmental stages, and indexing is based on validated diagnostic strategies. In the in vitro system, only the juvenile phenotype is available for inspection, the diagnostic protocols have generally not been validated. The in vivo system is open to reinfection with pathogens; the in vitro system is unlikely to be reinfected with pathogens, but pathogens that escape detection may be clonally propagated and cultivable contaminants may become clonally established.

quarantine pathogen in California); so available commercial diagnostics may not detect a major pathogen risk with consequences of introduction of new pathogenic strains into a previously clean area. Pathogens aside, the major concern for all users of tissue culture systems is the universal problem of laboratory contamination management. Many micropropagators cloning crops for which there are no certification guidelines do not carry out pathogen testing and may even ignore cultivable contaminants if they are not perceived as affecting production.

PRINCIPLES OF CERTIFIED STOCK PRODUCTION

Strategies used in, for example, potato (15) and strawberry (16) certified planting material production provide good insights into the issues to be addressed in the management of material free of specified pathogens. In these crops, as with major crops, disease intelligence is high; thus in a given region the identity and risks of specific crop pathogens in specific cultivars is known. Equally important, disease symptoms are characterized and disease diagnostic procedures are standardized. Using European Union strawberry certification as a model, the pathogens that must be tested for are specified and the testing procedures laid down by the European Plant Protection Organization (16). It should be emphasized in the

latter regard that the in vivo tissue to be indexed and the tests to be carried out are specified and that the testing is not "approved" for use on in vitro material, nor do the tests include, as yet, any DNA-based techniques. In general enzyme-linked immunosorbent assay (ELISA) (17) is widely approved, and in some cases it is recommended that this is supported by inoculation of indicator species, where the ELISA result is negative or to confirm the pathovar. Certification schemes have as their objective the production of plants that are true to type, free of specified diseases (e.g., quarantine diseases) and substantially free from other diseases/pests (18). Such schemes involve the following stages (see also Fig. 1):

1. Selection of high-quality individual plants of each cultivar. This is usually based on visual inspection for trueness to type, vigor/yield quality, and absence of symptoms of diseases and pests.

2. The selected individual plants should be isolated in vector-proof glasshouses and maintained in individual pots in sterilized compost. Foliar contact between individuals should be prevented by growing, for example, in vector-proof cages with bottom watering to prevent bacterial spread. Selection for virus-free individuals from among these plants is by testing for "escapes." All plants should be

individually tested for the listed viruses and other pathogens covered by the scheme using prescribed tests. Material imported from outside the quarantine zone should be tested for pathogens occurring in the region of origin. The production of virus-free plants by heat therapy or meristem culture may also be carried out. Virus-free material may be obtained by importation. Virus-free stock is given the designation *nuclear stock*. Plants giving negative results in the test should be transferred to a separate vector-proof house; plants testing positive should be removed immediately and destroyed by burning, etc. Plants from heat therapy should be reindexed and reassessed for trueness to type and vigor. In using meristem culture to eliminate pathogens, the smallest meristem explant compatible with meristem establishment should be used. A similar approach may be used to establish cultures from selected disease-free individuals.

3. Nuclear stock must be maintained under protected conditions where reinfection by pollen, aerial, or soil-borne vectors is prevented and must be routinely reindexed.

4. Nuclear stock, a minimum of five individuals per clone has been recommended, should be multiplied under conditions where reinfection is prevented with disease monitoring to give propagation stock. Micropropagation may be used for this purpose, but usually there is a limited on the number of culture cycles or duration for which in vitro stocks may be maintained (5).

5. Certified pathogen-tested stock production from propagation stock is carried out under official control. If in the field, the candidate crops should be grown in soil substantially free of virus vectors (e.g., nematodes), free of specified soil-borne diseases (e.g., *Phytophthora fragariae* in the case of strawberry) and isolated from commercial fields of the same crop. The crop should be regularly inspected and diseased individuals rogued.

6. For certification of planting material (tubers, runners, etc.) from the pathogen-tested individuals, the material should satisfy the phytosanitary conditions laid down with respect to freedom from specified diseases and be below threshold levels for tolerated diseases.

The key elements in all certification schemes are: (*1*) The stock material is disease indexed, and only pathogen-free material is used to establish the cultures (albeit meristem culture may be used in disease elimination), and (*2*) certification for pathogens (absence, or specified level) and for trueness-to-type (along with screening for secondary factors) is carried out on field progeny.

PLANT DISEASE DETECTION AND DIAGNOSIS

The human eye is capable of detecting and analyzing an array of information in color and in three dimensions; thus symptomatology, based mainly on visual crop inspection, remains the most powerful tool available in plant disease management, and this is reflected in many certification and inspection schemes. Examination of the crop from the planting material to the mature crop is the cornerstone of plant disease management in the field. The trained human eye is used both to select disease escapes and to monitor production for rogues. Crop disease management is based on planting disease-free propagules and keeping disease out until it is no longer economical to do so. Hence the importance of using certified planting material. Where the seed crop is available for inspection throughout its growth cycle, symptoms are not a universally reliable method to assess pathogen risk in all cases; for example, symptoms may be transient as in clear-vein in pelargoniums; masked due to virus complexes as in potato and fruit crops; not expressed in the first growing season as in potato leaf roll virus infection; or suppressed in certain host genotypes (19). To avoid the requirement for expensive labor-intensive multiple field inspections, and to confirm the pathogens present and their levels in planting material, plant pathologists use Koch's postulates for cultivable pathogens and the strategy of using indicator species that give a defined response to a specific pathogen. For pathogens that cannot be mechanically inoculated, vector or graft transmission is used. These methods remain central in plant pathology, being the only methods that can be used to detect uncharacterized pathogens; hence they are used widely in soft fruits crops and woody species and to confirm positive results obtained by other methods that do not confirm microbial viability or pathogenicity (20).

With the application of serological tests to plant pathogens, ELISA has become widely used and approved by certification authorities. ELISA has the advantages of being inexpensive, quick, sensitive, and reliable. It is the standard method for plant virus detection but is not applicable to viroid detection and less reliable for bacteria and fungi, where it is more difficult to raise specific antibodies (21). A special problem in the case of bacteria is that many nonpathogenic isolates occur, which necessitates pathogenicity testing by inoculation of sensitive genotypes (22).

Bacterial detection is widely based on isolation and "culture indexing," and differential media have been developed for keying out the major bacterial groups (22). Biochemical test kits developed for clinical and environmental bacterial identification are also used in plant pathology but are of restricted use, as plant pathogens may not be in the database. Such biochemical test kits are of use in laboratory contamination management, where specific contaminants may be used as indicators of problems such as malfunctional autoclaves (23). Fatty acid profiling is a useful tool in plant pathogen identification and can distinguish pathogens to the pathovar level (24). Fungal plant pathogens are still identified mainly by morphological characters and pathogenicity is tested by inoculation of host lines with known resistance genes (20). Mycelial fungi are generally not a problem in tissue culture if aseptic procedures are observed. Yeasts, however, are common contaminants (25).

The detection of noncultivable latent (symptomless) contaminants in plant tissues in vivo and, indeed, the

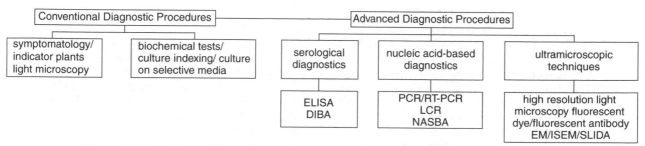

Figure 3. Plant disease diagnostic procedures. Plant disease diagnosis is very dependent on crop intelligence and usually involves known disease confirmation. Detection usually begins with observation of characteristic symptoms and pathogen identification to pathogen type follows, that is, fungus, bacterium, virus, or viroid. The strategy used then is largely dependent on the pathogen type. Fungi are generally identified using light microscopy. Cultivable bacteria are characterized by biochemical tests and growth on selective media. Fastidious bacteria may be detected by in situ tissue staining with fluorescent dyes or fluorescent antibodies; there is a tendency to use nucleic acid probes for all bacteria. Viruses are mainly detected by inoculation of indicator plants and by ELISA, also with the use of electron microscopic and immuno-EM techniques. It is good, if not essential practice, to inoculate indicator plants to confirm ELISA results where test are negative.

diagnosis of disease in plants in general is complicated by low-titer, cyclic changes in titer of the organisms, phloem restriction, or uneven systemic distribution, which carries the risk of false positive tests (19). Recently, the relationship between phytoplasmas and the eubacteria has been elucidated using DNA probes (26). The cost, reliability/reproducibility, and sensitivity of diagnostic tests for plant material are key factors in diagnosis. There is no limit to the sensitivity required in testing for quarantine pathogens; with nonquarantine pathogens, increased sensitivity means that tolerance threshold may be reduced, but the inoculum pressure from a few infectors in the crop may pose no economic threat compared with pressure from imported inoculum. Specificity is equally important, where there is the requirement to identify specific strains (as prescribed in the certification scheme) or for quarantine purposes (all strains) (9). Bearing in mind the characteristics of micropropagation, arguably the requirement is in general for a quarantine approach and quarantine diagnostics since intermediate inspection for symptoms is impossible/impractical and crop disease intelligence may be low (Fig. 3).

DIAGNOSTIC PROCEDURES

Indicator Plants

Indicator plants are used to detect intracellular pathogens and depend on the ability to transmit the suspect organism to the indicator species (20). This classic method has been the subject of several reviews, and successful application depends on the titer and distribution of the organism in its host, which may have a seasonal variation, necessitating appropriate sampling, the presence of inhibitors in the host sap (these may also interfere with molecular diagnostics), the stability of the organism and its transmission characteristics, whether it is mechanically transmissible, vector transmission, or graft transmissible. Direct scion to stock grafting or indirect grafting via dodder, depending on the interplant compatibility or dodder–host compatibility, is widely applicable where the organism is neither sap or

vector transmitted. These techniques are used extensively for strawberry (13) and woody plant (14) pathogens but are labor intensive in that indicator plant populations have to be maintained at the optimum growth stage. For a crop like potato many species are required to separate virus complexes and differentiate different viruses and virus strains (6). In known-disease indexing, selected indicator species are maintained to confirm negative ELISA results and to confirm pathogenicity.

Serological Diagnostics

Serological diagnostics are widely used in plant pathology, with variants of ELISA replacing Latex precipitation and the double diffusion technique on the basis of increased sensitivity and ease of use (11). ELISA has gained wide acceptance because it is inexpensive, with generally satisfactory sensitivity and adequate specificity for most applications. ELISA can be based on monoclonal, mixed monoclonal, or polyclonal antibodies. ELISA gives statistical results, and the accuracy of the results depends on the accurate determination of the positive–negative threshold (27). This is achieved by having a large control sample size to determine the population mean and the positive threshold (usually set at twice the standard deviation, SD, of the mean). Values close to the threshold and random samples should be confirmed by inoculation of sensitive indicator plants. As mentioned, ELISA may be too specific for quarantine purposes but appropriate for certification purposes when supported by inoculation of indicator species. Antibodies are available commercially from a number of suppliers, as are kits and diagnostic services. It is also possible to purchase primary antibodies and to construct kits from enzyme-linked secondary antibodies (double antibody sandwich ELISA, DAS-ELISA). These methods vary in sensitivity (21). ELISA is the benchmark with which DNA diagnostics have to compete. It is free of royalty payments, sap extraction can be semiautomated, and sap can be analyzed without purification; the procedure itself can be automated. Dot immunobinding assays (DIBA) are also used but not as extensively. This method involves spotting infected sap onto nitrocellulose filters and using enzyme-linked antibody as in ELISA to detect the pathogen. The system is suitable for field use, and commercial kits are available. It is reported to be as sensitive as ELISA and has the advantage that it requires minimal laboratory facilities. Tissue sections can be blotted onto membrane also (the technique is called tissue blotting or immuno-tissue printing) in an attempt to localize the pathogen in infected tissues (28). All serological techniques are susceptible to interference from sap components or cross reactivity with host proteins. In these cases loss of sensitivity or high backgrounds may arise. These issues are dealt with extensively in the literature (29).

Nucleic Acid Based Diagnostics

The polymerase chain reaction (PCR) (30) provides the basis for a potentially highly sensitive diagnostic test and it can satisfy the requirement for highly specific and broad-spectrum diagnosis when the primers are based on unique pathovar sequences and conserved pathogen sequences, respectively (31). There are many variants of PCR, such as reverse transcriptase-PCR (RT-PCR) (32) and ligase chain reaction (LCR) (33) for RNA viruses. An attractive aspect of these methods is the possibility of carrying out multiple assays in a single test (multiplex PCR, based on the use of multiple selective primers and using the added dimension over ELISA, of gel separation of products). However, there are many problems to be resolved before PCR or other nucleic acid based diagnostics will be used routinely. First, many of these techniques like PCR are patented, and this adds to the cost. Second, there is the problem that to obtain reproducible results it may be necessary to purify the pathogen nucleic acid. Immunocapture PCR (21), where the pathogen is trapped by antibodies and where inhibitors can be washed out before probing, is a positive advance, but this depends on the availability of appropriate antibodies and raises the problems of low-titer pathogens or uncharacterized pathogens for which antibody is unavailable or not easy to raise (30).

Before PCR, detection by hybridization product was facilitated by the use of radiolabeled probes (31). Routine use of radiolabeling was, however, very restricted by safety factors. This problem was overcome by the introduction of nonradioactive detection methods. But the amplification possible in PCR has rendered this problem obsolete. Nucleic acid sequence based amplification (NASBA; 34) differs from PCR/LCR in that it does not involve a separate reverse transcriptase step and temperature cycles. It is based on RNA and is appropriate for RNA virus detection. This method uses an RNA polymerase and specific primers. The first primer is complementary to the target RNA and contains a promoter for the RNA polymerase. The polymerase binds to the primer–RNA complex giving an RNA–DNA duplex, a second enzyme recognizes the duplex and selectively hydrolyzes the RNA, the second primer binds to the DNA former, and this duplex acts as a template for the amplification of the RNA sequence, which is produced rapidly, along with more copies of the double-stranded DNA. The problems

remaining in the application of nucleic acid diagnostics relate to validation of the results (30). It has been reported that PCR-RAPD had low reproducibility in a European Union interlaboratory evaluation (35). At present, sample preparation and manual gel analysis costs are too high for routine application. Simple methods for nucleic acid preparation, single reaction steps (as for NASBA), and adaptation to automatable microtiter plate systems are among the methods being developed to address these cost components (30).

Ultramicroscopic Techniques

Light microscopy is widely used in mycology and plant bacteriology but has been superseded in the latter to a large extent by the use of biochemical tests kits and selective media. The exception is the gram stain, the first step in bacterial identification, and other preliminary metabolic tests that facilitate bacterial grouping, a prerequisite for the selection of an appropriate test kit (3). The light microscope remains the main tool in fungal identification (20) and continues to be used to detect fastidious bacteria and phytoplasmas based on fluorescent dye or fluorescent antibody stain of thin tissue sections (20). For the latter tests, confocal microscopy (36) offers the great advantage of easier sample preparation and specimen examination, but the capital costs are high. Electron microscopy (EM) also has applications in pathogen detection and diagnosis (37), but again the capital and recurrent costs are high and sample throughput slow. Rapid EM methods such as leaf dips, immunoelectron microscopy, and sphere-linked immunodiagnostic techniques are all useful techniques where available. The leaf dip method, in which sap of test material is expressed onto a formvar grid and stained with a heavy metal solution, is useful for high-titer, rod-shaped particles as a confirmatory test based on particle morphology and dimensions. Immuno EM methods (ISEM; 38) can increase the sensitivity and specificity of detection to that of ELISA and when combined with antibody decoration and can allow resolution of virus strain in virus complexes or strain mixes. The sphere-linked immunodiagnostic assay uses the principles of ELISA on a microscale and is based on binding of viral coat protein onto microspheres stained with gold-conjugate (39). Thin-section electron microscopy, like confocal microscopy, allows localization of contaminants, but as with all microscopic methods the ease of pathogen detection depends on distribution and abundance at the site of occurrence. This is essentially a research tool and is not for routine applications.

Fatty Acid Profiling

Fatty acid extraction and analysis or profiling has useful applications in the management of bacterial contamination of plant tissue cultures, since it can provide information on the identification of both environmental bacteria and bacterial plant pathogens (40). Fatty acid profiles are used to match isolates with records on the database, but also the presence of specific fatty acids in unmatched isolates provides taxonomic information.

Culture Indexing

Endophytic contamination with cultivable organisms is common in perennial plants in nature and is not of general concern to plant propagators. As stressed previously, however, endophytic cultivable contaminants present a serious contamination risk to the plant tissue culturist (1). Consequently, routine bacterial and yeast indexing of cultures arguably must be carried out by those using plant tissue cultures for whatever purpose. For bacterial indexing of cultures a range of media has been recommended from findings that a single medium was not adequate (23). Where contaminant/pathogen identification is required, as for certification or for selection of appropriate antibiotics for chemotherapy, the selective media, fatty acid profiling, or the other tests referred to can be used.

BIOTIC CONTAMINATION INDEXING OF IN VITRO MATERIAL

It is reported in the literature that virus titer and distribution in plants in vivo varies (19), and pathogen testing reflects this; for example, potato tubers are sprouted and the leaf material tested for potato virus, as the titer is too low in the tubers (41). It is also known that hormone applications to plants can influence virus titer (19). Endophytic bacterial contamination may be unevenly distributed, again making indexing unreliable. It is not unreasonable to hypothesize, therefore, that in vitro growth conditions, both media factors and the culture environment, may influence both the titer and distribution of biotic contaminants in the cultures. It is widely observed that cultures may appear contaminant-free until the medium is changed and then contamination is expressed (2). There are also disturbing results that heat therapy in some cases may suppress viruses below the threshold for detection as virus may re-emerge in subsequent tests on progeny plants (28). Indeed, in fruit trees repeated pathogen testing is advocated. While much of the evidence is anecdotal, the risk that in vitro conditions and the use of contaminant suppressive treatments, viz. thermotherapy and chemotherapy (see the following), may reduce pathogens below the detection limits of diagnostic procedures validated on in vivo material should be taken very seriously. It has been recommended here that for pathogen indexing only appropriate tissues from in vivo plant material should be used. In practice this means that pathogen-free, or pathogen-freed, stock plants should be used to initiate cultures.

STRATEGIES TO OBTAIN CLEAN STOCK PLANTS

The classical strategy for initiating a "clean stock program" was to seek disease escapes (i.e., symptomless individuals, 3). This strategy is now generally reinforced by using disease-indexing procedures. The most effective strategy is to obtain pathogen-free plants from certification authorities. If not available, disease escapes should be sought and these sent for official testing or testing to a commercial laboratory. It should be remembered that disease-indexed plants are free of only the specific diseases they were tested for and may contain latent pathogens and endophytic contaminants.

ELIMINATION OF PATHOGENS FROM STOCK PLANTS

Thermotherapy

The use of heat therapy predates the characterization of viruses and has been used for almost 50 years in the elimination of intracellular pathogens, namely, viruses, viroids, and phytoplasmas (42). Despite the limited understanding of the process, it remains, with meristem culture, a standard method. Thermotherapy, except in the case of thermotolerant strains, may inactivate the pathogen, inhibit replication, and/or inhibit movement. Any or all of these effects may serve to restrict the colonization of apical cells and tissue by the pathogen, so that these regions may be used to establish virus-free plants (42). A caution, however, is that the pathogen may merely be suppressed and may be re-expressed as the derived propagule grows. This is a particular problem with woody species, where repeated retesting over several seasons may be required to confirm freedom from disease (14). Plants are often acclimatized to high temperatures before therapy (e.g., are maintained at 28–34 °C before being placed at temperatures over 36 °C). The heat treatment may involve continuous exposure to high temperature for longer than 20 days and up to c. 100 days. Alternatively, alternating high and moderate temperatures may be used or a constant warm temperature c. 30 °C. Natural light has been used in some cases, and in other cases artificial light of varying intensities with varying day lengths has been used. After therapy, shoot tips (5–15 mm), meristems (see the following) have been excised and grown on. Alternatively these explants have been grafted to disease free root stocks. Viroids may tolerate higher temperatures better that other intracellular pathogens and may be difficult to eliminate in thermotherapy (42).

Meristem Culture

The meristem is the preferred explant for the introduction of plants into culture and has the advantage that, in general, it is free of endophytic inter-cellular and intracellular biotic contamination (3). Thus micropropagators may subliminally eliminate pathogens and contaminants. However, elimination of biotic contaminants depends critically on the size of the "meristem" explant taken (43). Ideally this should be the smallest explant capable of establishment; usually this is the apical dome plus the first pair of primordia. Thin-section electron microscopy and DIBA have shown the presence of some viruses in the meristem (e.g. TMV in tomato); however, in these cases virus elimination may occur in tissue culture, perhaps due to the inhibitory effects of media components (43). The risk is that virus suppression can occur and that virus titer may remain low during in vitro culture to re-emerge in the progeny plants as reported in some woody species.

Chemotherapy

Chemotherapy of intracellular pathogens in plants is limited by the lack of specific antibiotic and/or by the difficulty of maintaining an inhibitory concentration of the antibiotics during plant growth (44). The compounds used, antibacterial and antiviral compounds, are generally biostatic rather then biocidal, which increased the problem. Both antibiotics (2) and the broad-spectrum plant antiviral chemical Ribavirin (45) have been used to treat stock plants, with the objective of slowing the movement of endophytic contaminants into the apical region, thus facilitating the excision of contaminant-free tissue explants for establishment of cultures. Ribavirin has also been used in this way in combination with thermotherapy (42). In vitro culture theoretically provides a better application system than in vivo plants for antibiotics as the active concentration can be maintained more easily. Provided that antibiotic selection criteria of: nontoxicity to the plant tissue, low risk of mutation induction in the host, and correct determination of the inhibitory effect and MIC for the antibiotic are determined, then antibacterial chemotherapy in vitro can be successful, provided that it is recognized that the antibiotic may only be bacteriostatic (44). To achieve bacterial elimination new tissue growth only should be excised, and probably passaged through a series of subcultures on antibiotics to obtain bacteria-free cultures. Similarly, Ribavirin is virustatic, and again correct use is essential to obtain satisfactory results (45). In all cases of thermotherapy and chemotherapy or combinations of both, only a percentage of the progeny may be contaminant free, and these require rigorous screening. Classically, heat therapy has been used to eliminate viruses from stock plants. This approach continues to be widely used. Now meristem culture is also widely used and has the advantage, in principle, that all biotic contaminants viz., pathogens and endophytic contaminants may be eliminated.

Elimination of Pathogens Introduced into Culture via the Explant

While, depending on the size of the explant and distribution of the biotic contaminant, most endophytic, inter- and intracellular contaminants, including pathogens, may be eliminated in meristem culture, it is generally useful to encourage bacterial expression by using a bacteriological media component to the plant tissue culture medium. There remains the question of whether successful elimination was achieved, and here the fundamental question of whether in vitro testing is reliable arises. In potato, etc., in vitro disease indexing is not officially approved. The best scenario may be to test the in vitro material; if clean, then grow out a sample of the material and retest under standardized test conditions. If indexing indicates the presence of bacterial contamination, the above protocol should be followed but it is almost impossible to rescue heavily contaminated cultures. If in vitro material is contaminated with viruses/viroids, then antiviral chemotherapy or thermotherapy in vitro may be attempted but taking care to test as indicated.

MAINTENANCE OF CLEAN CULTURES

If the strategy of introducing only clean plants into culture is followed, then good laboratory practice should result in the maintenance of clean cultures (23). Cultures should be monitored visually for the presence of cultivable contaminants. Fungal contaminants indicates poor operative technique, whereas heat-stable bacteria may represent faulty sterilization technique. Micro-arthropod infestation (46), usually associated with maintaining fungal cultures in the vicinity of the micropropagation laboratory/growth rooms or the failure to monitor and discard fungally contaminated cultures, can be recognized by the trail of bacterial and fungal colonies in the cultures. In all cases, cultures should be routinely subjected to bacterial culture indexing to prevent the buildup of endophytic bacterial contamination in the cultures (3).

CONCLUSIONS

It has been argued that while plant tissue culture, and specifically, its application in mass clonal propagation (i.e., micropropagation), shares common problems with vegetative propagation in clean plant production, it also has unique problems that affect all users of plant tissue culture systems. It has been proposed that micropropagators should follow the proven strategies of certified stock producers in indexing the stock plant and establishing pathogen-free cultures because of the risks that follow from the nonexpression of pathogen symptoms in vitro that prevent detection of pathogens during the multiplication cycle. Furthermore, there is uncertainty regarding the use of diagnostics for in vitro material where the influence of the individual laboratory tissue culture protocol may suppress pathogen titer below the level of detection. If pathogen-tested stock is not used to initiate the culture, then a sample batch from production should be grown to the stage at which the conventional testing is done before indexing is carried out. Micropropagators also face similar problems to quarantine germplasm conservationists, where less common or novel pathogen strains and pathovars may occur, and consequently, they require both specific tests to satisfy certification requirements and broad-spectrum tests to satisfy quarantine (or international shipment) requirements. Where meristem culture is used to obtain pathogen-free cultures, samples of the plants should be grown-on for testing as conventional plants (as previously). When thermotherapy or chemotherapy is used, singly or in combination, the progeny plants should be subjected to testing as for conventional material; that is, repeated tests should be carried out. A unique problem in heterotrophic/mixotrophic, that is, conventional tissue culture/micropropagation, is the problem of contamination with cultivable microbial contaminants. This can be controlled usually by excising small shoot-tip explants and by following good laboratory practice with indexing for cultivable contaminants. New nucleic acid–based diagnostic tools offer the potential of more sensitive tests of a wide range of specificity, even to identification of pathovars, and they offer the prospect of multiple pathogen detection in one test. These diagnostics have yet to be confirmed to be economic and reliable and already are under competition from electrochemical biosensors (47).

BIBLIOGRAPHY

1. A.C. Cassells, ed., *Pathogen and Contamination Management in Micropropagation*, Kluwer Academic Publishers, Dordrecht, The Netherlands, 1997.

2. E.F. George, *Plant Propagation by Tissue Culture*, Exegetics, Basingstoke, U.K., 1993.

3. A.C. Cassells, in A. Altman, ed., *Agricultural Biotechnology*, Dekker, New York, 1998, pp. 43–56.

4. C. Leifert and S. Woodward, in A.C. Cassells, ed., *Pathogen and Contamination Management in Micropropagation*, Kluwer Academic Publishers, Dordrecht, The Netherlands, 1997, pp. 237–244.

5. A.C. Cassells, in A.C. Cassells, ed., *Pathogen and Contamination Management in Micropropagation*, Kluwer Academic Publishers, Dordrecht, The Netherlands, 1997, pp. 1–13.

6. J.A. de Bokx and J.P.H. van der Want, eds., *Viruses of Potatoes and Seed-Potato Production*, Pudoc, Wageningen, 1987.

7. A.C. Cassells and V. Tahmatsidou, *Plant Cell Tissue Organ Cult.* **47**, 15–26 (1996).

8. R. Weller, in A.C. Cassells, ed., *Pathogen and Contamination Management in Micropropagation*, Kluwer Academic Publishers, Dordrecht, The Netherlands, 1997, pp. 2435–2458.

9. H.E. Waterworth, in A. Hadidi, R.K. Khetarpal, and H. Koganezawa, eds., *Plant Virus Disease Control*, APS Press, St. Paul, Minn., 1998, pp. 025–331.

10. E.A. Frison and M. Diekmann, in A. Hadidi, R.K. Khetarpal, and H. Koganezawa, eds., *Plant Virus Disease Control*, APS Press, St. Paul, Minn., 1998, pp. 230–236.

11. W.J. Hooker, ed., *Compendium of Potato Diseases*, APS Press, St. Paul, Minn., 1981.

12. J.L. Maas, ed., *Compendium of Strawberry Diseases*, APS Press, St. Paul, Minn., 1984.

13. S. Spiegel, in A. Hadidi, R.K. Khetarpal, and H. Koganezawa, eds., *Plant Virus Disease Control*, APS Press, St. Paul, Minn., 1998, pp. 320–324.

14. M. Barba, in A. Hadidi, R.K. Khetarpal, and H. Koganezawa, eds., *Plant Virus Disease Control*, APS Press, St. Paul, Minn., 1998, pp. 288–293.

15. T.D. Hall, *BCPC Monog.* **54**, 77–82 (1993).

16. Anonymous, *EPPO Bull.* **24**, 875–889 (1994).

17. M.F. Clark and A.N. Adams, *J. Gen. Virol.* **34**, 475–483 (1977).

18. G. Krczal, in A. Hadidi, R.K. Khetarpal, and H. Koganezawa, eds., *Plant Virus Disease Control*, APS Press, St. Paul, Minn., 1998, pp. 277–287.

19. R.E.F. Matthews, *Plant Virology*, 3rd ed., Academic Press, San Diego, Calif., 1991.

20. R.T.V. Fox, *Principles of Diagnostic Techniques in Plant Pathology*, CAB International, Wallingford, 1993.

21. R.R. Martin, in A. Hadidi, R.K. Khetarpal, and H. Koganezawa, eds., *Plant Virus Disease Control*, APS Press, St. Paul, Minn., 1998, pp. 381–391.

22. R.A. Lelliot and D.E. Stead, *Methods for the Diagnosis of Bacterial Diseases of Plants*, Blackwell, Oxford, 1987.

23. C. Leifert and W.M. Waites, in P.J. Lumsden, J.R. Nicholas, and W.J. Davies, eds., *Physiology, Growth and Development of Plants in Culture*, Kluwer Academic Publishers, Dordrecht, The Netherlands, 1994, pp. 363–378.

24. D.E. Stead, J. Hennessy, and J. Wilson, in A.C. Cassells, ed., *Pathogen and Contamination Management in Micropropagation*, Kluwer Academic Publishers, Dordrecht, The Netherlands, 1997, pp. 61–74.

25. S. Danby et al., in P.J. Lumsden, J.R. Nicholas, and W.J. Davies, eds., *Physiology, Growth and Development of Plants in Culture*, Kluwer Academic Publishers, Dordrecht, The Netherlands, 1994, pp. 397–403.

26. J.M. Bove and M. Garnier, in A.C. Cassells, ed., *Pathogen and Contamination Management in Micropropagation*, Kluwer Academic Publishers, Dordrecht, The Netherlands, 1997, pp. 45–60.

27. C.L. Sutula, J.M. Gillett, S.M. Morrissey, and D.C. Ramseu, *Plant Dis.* **70**, 722–726 (1986).

28. E. Knapp et al., in A.C. Cassells, ed., *Pathogen and Contamination Management in Micropropagation*, Kluwer Academic Publishers, Dordrecht, The Netherlands, 1997, pp. 23–30.

29. R. Hampton, E. Ball, and S. DeBoer, eds., *Serological Methods for the Detection and Identification of Viral and Bacterial Plant Pathogens*, APS Press, St. Paul, Minn., 1990.

30. T. Candresse, R.W. Hammond, and A. Hadidi, in A. Hadidi, R.K. Khetarpal, and H. Koganezawa, eds., *Plant Virus Disease Control*, APS Press, St. Paul, Minn., 1998, pp. 399–416.

31. K. Weising, H. Nybom, K. Wolff, and W. Meyer, *DNA Fingerprinting in Plants and Fungi*, CRC Press, Boca Raton, Fla., 1995.

32. A. Hadidi, M.U. Montasser, and L. Levy, *Plant Dis.* **77**, 595–601 (1993).

33. K.J. O'Donnell, E. Canning, and L.G.A. Young, *BCPC Symp. Proc.* **65**, 187–191 (1996).

34. G. Leone, H.B. van Schindel, B. van Gemen, and C.D. Schoen, *J. Virol. Methods* **66**, 19–27 (1997).

35. C.J. Jones et al., in A. Karp, P.G. Isaac, and D.S. Ingram, eds., *Molecular Tools for Screening Diversity*, Chapman & Hall, London, 1998, pp. 176–179.

36. D.M. Shotton, *J. Cell Sci.* **94**, 175–206 (1989).

37. V. Hari and P. Das, in A. Hadidi, R.K. Khetarpal, and H. Koganezawa, eds., *Plant Virus Disease Control*, APS Press, St. Paul, Minn., 1998, pp. 417–427.

38. K.S. Derrick, *Virology* **56**, 652–653 (1973).

39. V. Hari, D.A. Baunoch, and P. Das, *Bio Techniques* **9**, 342–350 (1990).

40. D.E. Stead, in J.M. Duncan and L. Torrance, eds., *Techniques for the Rapid Detection of Plant Pathogens*, Blackwell, Oxford, 1992, pp. 76–114.

41. S.A. Slack and R.P. Singh, in A. Hadidi, R.K. Khetarpal, and H. Koganezawa, eds., *Plant Virus Disease Control*, APS Press, St. Paul, Minn., 1998, pp. 249–260.

42. G.I. Mink, R. Wample, and W.E. Howell, in A. Hadidi, R.K. Khetarpal, and H. Koganezawa, eds., *Plant Virus Disease Control*, APS Press, St. Paul, Minn., 1998, pp. 332–345.

43. G. Faccioli and F. Marani, in A. Hadidi, R.K. Khetarpal, and H. Koganezawa, eds., *Plant Virus Disease Control*, APS Press, St. Paul, Minn., 1998, pp. 346–380.

44. C. Barrett and A.C. Cassells, *Plant Cell, Tissue Organ Cult.* **36**, 169–175 (1994).

45. A.C. Cassells, in G. Reuther, ed., *Physiology and Control of Plant Growth In Vitro*, CEC, Luxembourg, 1998, pp. 144–148.

46. J. Pype, K. Everaert, and P. Debergh, in A.C. Cassells, ed., *Pathogen and Contamination Management in Micropropagation*, Kluwer Academic Publishers, Dordrecht, The Netherlands, 1997, pp. 259–267.

47. J.-M. Kauffmann and M. Pravda, *Bioforum Int.* **2**, 18–25 (1998).

See also ASEPTIC TECHNIQUES IN CELL CULTURE; CONTAMINATION DETECTION IN ANIMAL CELL CULTURE; STERILIZATION AND DECONTAMINATION; VIRUS REMOVAL FROM PLANTS.

CONTAMINATION DETECTION IN ANIMAL CELL CULTURE

CAROL MCLEAN
Protein Fractionation Centre
Scottish National Blood Transfusion Service
Edinburg, Scotland

COLIN HARBOUR
The University of Sydney
NSW, Australia

OUTLINE

INTRODUCTION

This article describes the detection of contamination in cell culture by a range of infectious agents, including bacteria, fungi, mycoplasmas, viruses, and prions. The presence of any form of microbial contamination renders the results of scientific studies with that cell culture invalid. In manufacturing processes designed to produce therapeutic biologics from cell culture systems, it is imperative that all possible steps be taken to ensure that the product is free from any infectious agents. This article provides a historical perspective to demonstrate how contamination and its detection have affected the development of cell culture as a technology and then describes the current regulatory requirements expected of cell-culture-derived biologics. The regulatory framework is described, in part, to draw attention to the current gulf in cell culture practices with regard to contamination detection required by manufacturers compared to that performed by cell culture workers in institutions devoted to research such as universities. The issues of good manufacturing practice (GMP) and good laboratory practice (GLP), which largely determine the procedures adopted by the former group, are usually ignored by the latter. This is an unacceptable situation and it is our hope that this article will indicate what can happen if the detection of contamination is not performed as an essential part of cell culture practice. In addition, the article describes appropriate strategies for detecting contamination and thus avoiding the generation of erroneous results in cell culture research and providing the safest possible biologics.

HISTORICAL PERSPECTIVES

Following the cultivation of nerve cell tissue using the hanging drop technique by Harrison in 1907, the historical development of cell and tissue culture was inhibited by the problem of microbial contamination. Microbes grow rapidly in the enriched media required for cell and tissue growth leading almost inevitably to the destruction of such cultures. Although some practitioners were successful, such as Alexis Carrel who had shown in 1923 that it was possible to culture cells *in vitro* for long periods by employing rigorous aseptic techniques, the problem of microbial contamination was never controlled sufficiently until the introduction of antibiotics in the 1940s. Since then, the introduction of work stations such as laminar air flow cabinets for cell culture manipulations in clean, filtered air, has also contributed enormously to reducing the levels of microbial contamination. Nevertheless, contamination does occur from time to time, and regular screening for the presence of bacteria, fungi, and mycoplasmas should be mandatory practice in all cell culture facilities. This is even more important today when the routine use of antibiotics for cell culture is discouraged because it may make low level contamination difficult to detect, lead to the emergence of resistant micro organisms, and in the production situation, could result in side effects in patients, who receive cell-culture-derived biologics contaminated with trace amounts of antibiotics. The procedures for detecting these microbes are relatively straightforward, and the appropriate protocols are provided in this article. By contrast, the detection of viral contamination is more complex, and thus this article devotes more attention to this problem.

As stated earlier, the introduction of antibiotics in the 1940s led to rapid advances in cell culture, such as the work of Enders and co-workers who published the first report of growing polio virus in nonneural cells in culture (1). This seminal work, recognized with the award of a Nobel prize, showed that it was feasible to produce viruses and hence viral vaccines in cell culture systems and thus provided the stimulus for the development of cell culture as a technology. Rapid advances were then made resulting in the production of the Salk polio vaccine derived from monkey kidney cells, which was first licensed in 1954, followed by the Sabin polio vaccine (a live vaccine) which was first licensed in 1955 (2,3). Unfortunately, it was recognized then that these early polio vaccines were contaminated with a virus, simian virus 40 (SV40), present in the monkey kidney cells. This finding had a salutary effect on the cell culture "industry" and demonstrated clearly that any material of biological origin can be contaminated with an infectious agent (4,5). Since that discovery, various strategies have been introduced to address the issue of viral contamination, including a wide range of different types of viral assays which are described in this article. The problem of viral contamination was also a major driving force in developing more sophisticated cell culture techniques, including the introduction of normal diploid cell lines for producing viral vaccines (6,7). These cells provide the opportunity to establish frozen cell banks and thus the time to perform comprehensive investigations designed to detect microbial contamination. Thus, as a result of these screening procedures, normal diploid cell lines are inherently safer than primary cells in their potential to transmit infectious diseases, but because absolute safety can never be guaranteed, all cell lines must be regularly, monitored. Further advances in cell culture technology and molecular biology have led to the introduction and acceptance of transformed or continuous cell lines for producing a range of biologics for therapeutic use such as interferon (8), monoclonal antibodies, and various recombinant proteins, for example, tissue plasminogen activator (9). Many of these cells are known to express endogenous retroviruses (10–12) which have oncogenic potential, and this issue has been addressed by the introduction of a comprehensive range of screening assays which are described in detail in this article.

After summarizing the historical development of cell culture so as to emphasize the move away from the use of primary cells to improve safety regarding transmitting infectious agents, it is somewhat ironic to note that the most recent developments in this field, broadly categorized as cell and/or gene therapy and tissue transplantation involve, in many cases, a return to the use of primary cells with all, of their inherent safety risks (13,14). These primary cells, which may be autologous, allogenic, or xenogenic cells chemically stimulated *in vitro* in some way or even genetically modified, must be administered to a patient as rapidly as possible after treatment because the cells have a limited life span and the patient is more than likely gravely ill. Therefore, it is a necessary to perform all quality control screening procedures as rapidly as possible. For example, by the end of the 14-day period required for a routine sterility test, the patient may have died or become so sick that the treatment would no longer be effective. The sample sizes required for the recommended screening procedures may constitute an unacceptably large proportion of single-patient treatments, and reducing the sample size so as to preserve the treatment dose will obviously lead to a reduced likelihood of detecting infectious agents. These are important issues created by new developments in cell culture technology, and the ways by which they are being addressed are discussed later, as is the prion problem (15). At this time, the latter has become a major challenge for all those involved in manufacturing biologics and is discussed in more detail later in this article. The prion, or proteinaceous infectious particle, appears to have crossed the species barrier from cows suffering from bovine spongiform encephalopathies (BSE) and resulted in a new variant of Creutzfeldt–Jakob disease (vCJD) in humans. The prion is the latest discovery of an infectious agent that can infect humans and underlines the continuing need and importance of the screening procedures described in this article (16). In the next section the detection of contamination is described in the context of the current regulatory framework which has developed to ensure that cell-culture-derived biologics are as safe as possible.

REGULATORY ISSUES

Biopharmaceuticals are manufactured from three main sources comprising human tissues or plasma, animal tissues or plasma, and mammalian cell lines. Production of biologics in mammalian cell lines offers the advantage over the other systems that the products can be readily manufactured in large amounts and, compared to biologics prepared from human tissues or plasma, are less likely to be contaminated with human pathogens. Mammalian cell culture systems, however, carry the risk that the cell line or raw materials used in cell cultivation can introduce a viral contaminant of animal origin capable of zoonoses or other microbial contaminants pathogenic to humans such as *Mycoplasmas*. For example, some cell lines harbor viral sequences integrated into the genomic DNA that are passed vertically from generation to generation such as endogenous retrovirus sequences which are found in cell lines of many species. Integrated viral sequences alternatively originate from virus that was deliberately introduced during establishment of the cell line such as Epstein–Barr virus (EBV) used to immortalize human B cell lines for establishing human hybridomas and human–murine heterohybridoma. Raw materials or ingredients of animal origin used during cell line cultivation or product purification that may introduce adventitious agents include reagents, such as bovine serum used in cultivation of cell lines, and monoclonal antibodies produced in ascites or cell culture used for affinity purification of the product. Manipulation of cell lines, product intermediates, and materials by human operators during manufacturing provides a further source of adventitious agent contamination, for example, via aerosols or by direct contact.

"Notes for guidance" and "points for consideration" have been issued by various national and international agencies, including the CBER (Center for Biologics Evoluation Research) a branch of the Food & Drug Administration of the USA in the US, the Committee for Proprietary Medicinal Products (CPMP) in the European Commission, and Japanese regulatory authorities, that outline the testing strategy recommendations and requirements for safety evaluation of biological products made in cell line systems (17–21). The strategies recommended vary slightly from one agency to another; therefore integration has been done by the International Committee for Harmonisation (ICH) (18). These guidelines are initially published in draft form and issued to interested parties for comment before being issued as a final version. The final versions of these regulatory documents are not issued simultaneously; therefore when testing cell lines used to produce biologics, the most recent final versions of the relevant regulatory documents, as well as the most recent version of the ICH guidelines, should be consulted.

The regulatory documents mentioned recommend the following strategies as evidence of product safety with respect to viral contamination:

1. Testing cell lines and raw materials for viruses that may be infectious for humans.

2. Testing the product/intermediates at appropriate stages during the production process for viral contaminants.

3. Validation of the production process for its capacity to remove and/or inactivate viral contaminants.

The approach adopted by manufacturers will depend on several factors such as the nature of the particular product, the results obtained during testing (points 1 and 2 above), and will generally be a combination of the three. Some biopharmaceutical products undergo little or no downstream processing and therefore rely on virological testing and characterization of the producer cell line and the final product, for example, replication-incompetent gene therapy vectors (22,23).

When physical and/or chemical purification steps are involved in the manufacture of the final product, validation of the ability of the process to remove or inactivate viral contaminants is performed by process validation studies (also called virus clearance studies) (point 3 above) (18,24). These contaminants may be known to exist in the manufacturing process, such as murine endogenous retroviruses in cell lines of murine myeloma lineage used in the manufacture of therapeutic monoclonal antibodies (19,20), or they may be contaminants for which there is a risk in the manufacturing process but for which testing for low levels of such contamination in process samples and raw materials is limited by the sample size.

Process validation studies are done by first establishing a representative downscale version of a specific step in the manufacturer's process that has been identified as known or likely to be effective in removing or inactivating viral contaminants. The downscale version must represent as faithfully as possible the conditions of the manufacturing process with respect to physical and chemical aspects such as flow rate, pH, protein concentration, etc. The downscale process is then spiked with a range of viruses with varying degrees of resistance to physical and chemical methods of inactivation to evaluate the ability of the process to remove and/or inactivate known and/or unknown viral contaminants. The viruses used in process validation studies can be "relevant" or "model" for example, human immunodeficiency virus (HIV) (relevant) and bovine viral diarrhoea virus (BVDV) (a model for Hepatitis C virus), in products manufactured from human plasma. By assaying the viral load in the spiked starting material and processed material, an estimate of the ability of a particular step in the manufacturing process to remove and/or inactivate the virus can be made and consequently provide further assurance of product safety.

When manufacturing a product from a cell line, a seed lot cell bank system must be established, where each bank is comprised of ampoules of cells with uniform composition derived from a cell seed. The master cell bank (MCB) is generally propagated from a selected cell clone under defined conditions, then aliquoted into multiple containers, and stored under appropriate conditions (frozen at or below $-100\,°C$). The working cell bank (WCB) is manufactured from one vial of the MCB. Then cells from the WCB (or occasionally the MCB) are propagated to the production cells. A cell bank at the limit of *in vitro* cell age used for production (usually called a postproduction cell bank (PPCB) or extended cell bank (ECB) is also laid down from a WCB vial expanded under pilot plant or commercial scale conditions. The number of vials in any MCB and WCB will vary from process to process but is often in the order of 100 per bank, thus enabling long-term supply of material for any one process. The current strategy recommended by the ICH for safety testing cell lines of human or animal origin used to produce biologics (with the exception of inactivated and live vaccines, and genetically engineered live vectors) is outlined in Table 1 (18,20). Extensive qualification of the MCB (or WCB) for adventitious agents and identity testing is required, and because the WCB has a low population doubling level beyond the MCB, testing of the WCB is minimal. Testing of the PPCB should be performed at least once to provide assurance that the conditions used during production are unlikely to introduce adventitious agents or activate viral contaminants that may be suppressed at the MCB stage. Therefore testing of the PPCB is repeated if there are changes in the scale or cultivation conditions during production. In-process testing of bulk harvests in the form of cells and fluid harvested from fermenters is also required because, like PPCB testing, this, provides assurance that the conditions used during production do not introduce and/or activate contaminants (18,20).

The use of raw materials from animal origin for the propagation of MCB, WCB, and in production should be avoided if possible. In doing so, characterization of the

Table 1. Testing Strategy for Cell Lines used to Produce Biologics[a]

Test	MCB	WCB	PPCB
Infectious (endogenous viruses)	+	–	+
Electron microscopy	+	–	+
Reverse transcriptase[b]	+	–	+
In vitro assay[c]	+	–		
In vivo assay[c]	+	–	+
Antibody production assays[d]	+	–	–
Other virus specific tests[e]	As appropriate	–	As appropriate
Mycoplasma	+	+	+
Sterility	+	+	+

[a]Ref. 18, 20.

[b]Reverse transcriptase assay is not necessary if the cell line is positive for infectious retrovirus.

[c]These assays should be performed on the cells at the limit of in vitro cell age generated from the first WCB; for subsequent WCBs a single test can be done either on the WCB or on the cells at the limit of in vitro cell age.

[d]These include mouse antibody production assays (MAP), hamster antibody production assays (HAP), and rat antibody production assays (RAP) and are usually applicable only to cell lines of rodent origin.

[e]This includes assays performed on, for example, human cell lines such as PCR assays for the human viruses HBV, HCV, HIV-1, HIV-2, HHV-6, HTLV I and II, EBV, and CMV.

MCB (or WCB) will provide assurance of the cell bank safety. However, because of the growth characteristics and requirements of cell lines, the use of raw materials such as serum, insulin, transferrin, and trypsin, cannot always be avoided. In such circumstances, raw material must be sourced from suppliers that can provide traceability of their origins, and batches must be sampled and tested for freedom from adventitious agents before their use in manufacturing.

MANUFACTURING AND SAFETY TESTING STANDARDS

The manufacture of pharmaceutical products must be done in environments that assure the safety, uniformity, efficacy, and quality of the products. A system of principles that lay down the procedures required for production and quality control, including documentation, personnel, environment, equipment, materials, auditing, sampling, and safety, are stipulated by cGMP (25). Testing of biopharmaceuticals for adventitious agents should be performed by laboratories with experience in virological assays and Good Laboratory Practice (GLP) accreditation. International GLP principles have been laid down by the OECD (Organisation for Economic Cooperation Development) (26). These guidelines exist so that safety studies are planned, performed, monitored, recorded, reported, and archived in an organized and controlled manner following approved procedures in a facility with adequate resources so that the quality and validity of the testing are assured.

EXAMPLES OF VIRAL CONTAMINANTS

Many animal cell lines currently used in producing biologics contain endogenous viruses. For example, cell lines of murine origin can harbor endogenous Type B or Type C retroviruses (oncoviruses), and cell lines of hamster origin express defective Type C and Type R retrovirus particles (27). Table 2 lists examples of endogenous retroviruses associated with cell lines commonly used for producing biologics. Exogenous viral infections may be present in cell lines that were acquired by the animal from which the cell line was established, for example, SV40 infection of Vero cells (4). Alternatively, cell lines may become accidentally infected with an exogenous virus during their establishment and/or development in research laboratories from, for example, another cell line or a laboratory operator. Expression of exogenous or endogenous retroviral infection can in some instances be low or undetectable in normal culturing conditions and can be chemically induced by agents such as 5-bromodeoxyuridine, 5-iododeoxyuridine, sodium butyrate, or 5-azacytidine (28–32).

Current use of replication-deficient virus-based vectors for gene therapy introduces the additional risk of viral contamination by replication-competent viruses which may be generated by recombination between sequences in the viral vector and the packaging cell line. The viruses from which gene therapy viral vectors are constructed are infectious for human cells. Therefore generation of a replication-competent virus may result in disease.

At some point in their histories most if not all cell culture systems have been exposed to raw materials of animal origin such as bovine serum or porcine trypsin, which constitute further potential sources of viral or mycoplasmal contamination. Two most common contaminants of bovine serum are BVDV—also called mucosal disease virus (MDV)—and bovine polyomavirus (BPyV) (33–36). BVDV is distributed worldwide affecting approximately 60% of cattle in the United Kingdom and Australia, and is infectious to several even-toed ungulates in addition to cattle, including pigs, sheep, and goats (37). It is therefore probable that a large proportion of bovine products used in cell culture systems is contaminated with either infectious or defective BVDV. The BVDV or BPyV status of bovine raw materials is often not provided,

Table 2. Endogenous Retroviruses Found in Cell Lines Commonly used to Produce Biologics

Cell Line	Retrovirus Particle Type	Infectious
Mouse myeloma/ hybridoma	A, C	Ecotropic and xenotropic
Chinese hamster ovary	C	Noninfectious
Baby hamster kidney	R	Noninfectious
Human/ murine heterohybridoma	A, C	Xenotropic
Human	None	No
Monkey	None	No
Insect	None	No

and commonly used filtration techniques used in the manufacture of bovine serum do not efficiently remove these viruses. Some suppliers, however, perform viral inactivation steps such as UV-C or gamma irradiation of serum to provide assurance of product safety (38,39).

Contamination of cell cultures with bovine viruses is of major concern when manufacturing products for use in ruminants because BVDV infection can result in abortion or growth retardation in offspring that are persistently infected (37). Several instances of BVDV contamination have been reported (40–45) which, in cases where the product is a ruminant vaccine, have resulted in livestock growth retardation and death (41,42). Pestivirus RNA has been detected in biopharmaceutical products for human use, although no confirmed symptomatic infections in humans have occurred (45). Bovine serum has also been implicated as the source of epizootic hemorrhagic disease virus (EHDV) in a Chinese hamster ovary (CHO) cell line (46), and the parvovirus, minute virus of mice (MVM), in a baby hamster kidney (BHK) cell line (47). Two similar MVM contamination incidents have been reported in hamster cultures (48,49), one of which was detected in fermenter cultures of CHO cells and the second during production of foot-and-mouth disease vaccine. The source of the contamination in these instances was not identified but in the former case was believed to be from mouse excretions contaminating media or their components during storage.

The main virus of concern in raw materials of porcine origin is porcine parvovirus (PPV) which has widespread distribution, that is, 40% infection has been reported (50), and it has been isolated from commercial trypsin (51). Testing of porcine trypsin for PPV is required by a number of regulatory bodies including the U.S. Department of Agriculture (9CFR) (52) and the European Pharmacopoeia (53). Examples of exogenous viral contaminants of biopharmaceutical systems are listed in Table 3 (54).

DETECTION OF VIRAL CONTAMINANTS IN CELL LINES

Viral contamination of cell lines can be detected by direct analysis, such as fixation of the cells and examination by electron microscopy, or by polymerase chain reaction (PCR) of nucleic acid extracted from the cell line. Alternatively, viral contaminants can be detected indirectly by examining the effect that the cell line has on a test system, for example, inoculation of a cell extract into a detector cell line or into an animal susceptible to the virus of concern. The effect of a viral contaminant on a cell line or animal test system may be general, producing a cytopathic effect (cpe) in the form of cell or animal death or cell morphological change, or may be specific such as in the production of antibodies to a specific virus in mice following inoculation of virally contaminated material. The testing strategies required by the ICH for cell lines used to produce biologics are outlined in Table 1. These primarily consist of *in vitro* and *in vivo* testing of the MCB and PPCB for infectious virus and testing using molecular biology techniques and electron microscopy for viruses that may be infectious or defective.

Detection of Retroviruses

Retroviruses are single-stranded RNA viruses, surrounded by a lipid envelope. When examined under thin-section electron microscopy, they are 100–140 nm in diameter and frequently exhibit glycoprotein surface projections. Cell lines of several species, including human, porcine, and murine, contain endogenous retroviral sequences. Murine cell lines of epithelial and myeloma lineages all contain endogenous retrovirus sequences and in many cases express infectious murine leukemia virus (MLV). MLV isolates can show varying tropisms, including ecotropic MLV (E-MLV) which is infectious for cell lines of murine origin or closely related species, and xenotropic MLV (X-MLV) which is infectious for cell lines of species other than of murine origin (37). Xenotropic endogenous retroviruses capable of infecting human cells have been described for murine and porcine cell lines (10,13,55). However the efficiency of murine retroviral infection apparently differs in the few reported incidents. In one study, supernatant was harvested form 17 mouse hybridoma cell lines and inoculated onto the human embryonic lung fibroblast cell line, 7605L. Eight of the 17 supernatants contained type-C retrovirus infectious for the human cells (10). In another study, 18 mouse cell lines of myeloma lineage were assayed

Table 3. Examples of Exogenous Viral Contamination in Intermediate and Final Products Manufactured in Cell Culture Systems

Product/Intermediate	Contaminant	Contamination Source	Reference
Polio vaccine	SV40	Infected cell line	4
Vet vaccine	BVDV	Unknown	40
Swine fever vaccine	BVDV or Border Disease virus	Bovine serum or lamb cell culture?	41
Rota-corona vaccine	BVDV	Unknown	42
Final product (IFNα, IFNβ)	BVDV	Bovine serum?	43
Live vet vaccines	BVDV	Bovine serum?	44
Human vaccines (MMR, MR)	BVDV	Bovine serum?	45
CHO cell line	EHDV	Bovine serum?	46
Unprocessed bulk (CHO cell line)	MVM	Medium component?	48
Foot-and-mouth disease vaccine	MVM?	Unknown	49
Vet vaccine	CPV	Unknown	54

for retrovirus by cocultivation with a human cell line, but none were positive (56).

Retroviral vectors used for gene therapy often contain envelope sequences originating from amphotropic MLV (A-MLV), feline leukemia virus (FeLV), or gibbon ape leukaemia virus (GaLV), which have been chosen because they are infectious for human cells (57). Therefore the generation of replication-competent retrovirus (RCR) with human tropism is of concern in such systems. Cell lines of human and simian origin have been used to produce biologics, for example, human–murine heterohybridoma cell lines used to produce monoclonal antibodies. These have the potential to be infected with exogenous human and simian retroviruses such as HTLV, simian retrovirus types 1 and 2, HIV, and simian and human foamy viruses (58–60). Following are details of a number of standard methods for detecting retroviruses which may be encountered in cell lines commonly used in producing biologics. Most of the cell lines and positive control viruses used in these methods are available from the ATCC.

Cell Culture Techniques for Retroviral Detection

Assays for Infectious MLV. Tests for infectious MLV should be performed on cell lines of murine origin on the MCB (or WCB) and again on the PPCB to determine whether there has been any change in the dynamics or tropism of MLV expression. It is also advisable to test samples from at least one fermenter run to monitor the dynamics of infectious retrovirus production during the manufacturing process. Similarly, testing regimes for RCR in retroviral gene therapy systems should include testing the producer cell line (1% or 10^8 of the cells, whichever is least, and 5% of the vector supernatant) (22,23). The following methods outline the procedures for detecting MLV using cell-free supernatant harvested from the cell line under test or fermenter sample. The methods described are so called "direct assay" which are quantitative and therefore provide the titer of virus per unit volume of sample tested. The sensitivity of the direct assay may be increased by serial passage of the inoculated detector cell line and consequently amplification of the retrovirus (61). So-called "extended" assays are qualitative and can increase assay sensitivity by approximately one log (62). When sampling for retroviral infectivity assays, it is important to harvest supernatant from test cell lines that are actively growing because this is the period during which the retrovirus is normally most highly expressed (63).

Assays for Ecotropic MLV

XC Assays (64). The XC assays utilize the cell line SC-1 which can be infected with E-MLV without inducing cpe. The virus can subsequently infect cells of the Wilistar rat tumour (XC cell line) in which a prominent syncytial effect is evident. E-MLV cannot infect XC cells directly. The assay should be set up as follows utilizing negative control cultures and positive control cultures infected with a suitable E-MLV for example, Moloney.

1. Plate the SC-1 cells at a density of 10^5 cells per 10 cm^2 plate in Dulbecco's Modified Eagle's Medium (DMEM) containing 10% fetal bovine serum (FBS)(v/v), and incubate at 37 °C with 5% CO_2.

2. Examine the cells the following day, and if suitably subconfluent (approximately 20–30%), inoculate with cell-free test material in the presence of polybrene at an effective concentration of 10 μg/mL.

3. Feed the cultures with 4–5 mL of DMEM containing 10% FBS (v/v) following an adsorption period of approximately 1–2 hours at 37 °C with 5% CO_2.

4. Maintain the cultures at 37 °C with 5% CO_2 until confluent (normally 4–5 days). Then remove the culture medium, and irradiate the cells under UV light for approximately one minute to kill the SC-1 cells.

5. Overlay the monolayers with XC cells at a high cell density (approximately 10^6) per plate in DMEM containing 10% FBS, and reincubate the monolayer cultures at 37 °C with 5% CO_2.

6. Stain the monolayers, when confluent, with crystal violet. Wash the excess stain from the cells with water, and allow them to dry. Examine the monolayers for infectious centers, identified as holes (plaques) in the XC monolayer. Evidence that the plaques are due to infection by E-MLV is provided by association of the plaque with one or more syncytial cell.

FG10 Assay. The FG10 cell line (also called D56) is a 3T3 cell line transformed with murine sarcoma virus (MSV) but does not produce infectious MSV (65). As such the cell line is termed sarcoma positive, leukemia negative (S$^+$L$^-$) and can be infected with leukemia virus with the result that the MSV genome is rescued and cell transformation occurs which is manifested as a "focus" at the site of infection. Foci appear as regions of rounded cells in association with a lytic area on the cell monolayer. The murine origin of this cell line permits an assay system for E-MLV. The assay should be set up as follows utilizing negative control cultures and positive control cultures infected with a suitable E-MLV, for example, Moloney.

1. Plate the FG10 cells at a density of 10^5 cells per 10 cm^2 plate in RPMI 1640 medium containing 10% FBS, and incubate overnight at 37 °C with 5% CO_2.

2. Examine the cells the following day, and if suitably subconfluent (approximately 20–30%), inoculate with cell-free test material in the presence of polybrene at an effective concentration of 10 μg/mL.

3. Feed the cells with 4–5 mL of RPMI 1640 medium containing 10% FBS (v/v) following an adsorption period of approximately 1–2 hours at 37 °C with 5% CO_2.

4. Maintain the cultures at 37 °C with 5% CO_2 until confluent (normally 4–5 days). Then examine for foci formation which appear as described before. It is important to maintain the cultures until fully confluent because foci are not easily recognizable in a subconfluent monolayer in which there are many mitotic (rounded) cells and gaps in the monolayer which are similar in morphology to foci.

Mink S$^+$L$^-$ Assay. The mink lung (MiCl$_1$) cell line is S$^+$L$^-$ and as such can be infected with leukemia virus resulting in rescue of the MSV genome and cell transformation manifested as a "focus" at the site of infection (66). Confluent monolayers of the MiCl$_1$ cell have a flat morphology, and foci can easily be identified as regions of rounded cells raised above the contact inhibition layer of the monolayer. Focus formation on this cell line from material of origin other than mink indicates the presence of a xenotropic leukemia retrovirus, and thus is an assay system for X-MLV, Fe-LV, and Ga-LV. The assay should be set up as follows utilizing negative control cultures and positive control cultures infected with a suitable xenotropic leukemia virus, for example, X-MLV.

1. Plate the MiCl$_1$ cells at a density of 2.5×10^5 cells per 10 cm^2 plate in RPMI 1640 medium containing 10% FBS (v/v) and incubate overnight at 37 °C with 5% CO$_2$.

2. Examine the cells the following day, and if suitably subconfluent (approximately 20–30%), inoculate with cell-free test material in the presence of polybrene at an effective concentration of 10 μg/mL.

3. Feed the cultures with 4–5 mL of RPMI 1640 medium containing 10% FBS (v/v) following an adsorption period of approximately 1–2 hours at 37 °C with 5% CO$_2$.

4. Maintain the cultures at 37 °C with 5% CO$_2$ until confluent (normally 8–10 days). Then examine for foci formation which appear as described before. Again, it is important to maintain the cultures until fully confluent because foci are not easily recognizable in a subconfluent monolayer.

PG4 S$^+$L$^-$ Assay. The PG4 cell line is a feline S$^+$L$^-$ cell (67) which when infected with leukemia virus, like other S$^+$L$^-$ cell lines, results in cell transformation manifested as a "focus" at the site of infection. This cell line is susceptible to infection by xenotropic and amphotropic MLV and Ga-LV. Therefore it is useful for detecting RCR in producer cell lines used in gene therapy vector production or in vector supernatant. Foci can be identified on PG4 monolayers as regions of rounded transformed cells adjacent to a lytic region in the cell monolayer. The assay should be set up as follows utilizing negative control cultures and positive control cultures infected with a suitable leukemia virus, for example, A-MLV.

1. Plate the PG4 cells at a density of 2×10^5 cells per 10 cm^2 plate in McCoy's medium containing 10% FBS (v/v), and incubate overnight at 37 °C with 5% CO$_2$.

2. Examine the cells the following day, and if suitably subconfluent (approximately 20–30%), inoculate with cell-free test material in the presence of polybrene at an effective concentration of 10 μg/mL.

3. Feed the cultures with 4–5 mL of McCoy's medium containing 10% FBS following an adsorption period of approximately 1–2 hours at 37 °C with 5% CO$_2$.

4. Maintain the cultures at 37 °C with 5% CO$_2$ until confluent (normally 5–7 days). Then examine for foci formation which appear as described before.

It is important to examine the cultures daily because the PG4 monolayer can deteriorate with the formation of holes in the monolayer which makes foci identification difficult.

Mus dunni Assay. The *Mus dunni* tail fibroblast cell line (68) is sensitive to A-MLV, X-MLV, and E-MLV (with the exception of the Moloney strain), and also mink cellfocus-forming viruses. However, acute infection does not result in an obvious cpe (69). This cell line is recommended by regulatory authorities for detecting infectious MLV in production systems using cell lines of murine origin (17,19) and in gene therapy vector systems based on MLV (22,23). Because of the absence of cpe in infected *Mus dunni* cultures, quantitative infection assays ("direct" assay) utilize an immunofluorescence end point. Qualitative assessment of the MLV status of a cell line can be done by inoculating the test material onto *Mus dunni* cells followed by amplification of any infectious MLV present by serial passage of the cells and assay of the *Mus dunni* supernatant for infectious MLV by an appropriate assay such as direct PG4 or direct mink S$^+$L$^-$ assay described before. The direct and extended *Mus dunni* assays are performed as follows, using negative control cultures and positive control cultures infected with an appropriate virus such as A-MLV.

Direct Assay

1. Seed multichamber slides (slides are commercially available with each well approximately 4 cm^2) with *Mus dunni* cells at a concentration of 2×10^4 cells/well in DMEM containing 10% FBS (v/v), and incubate the slides overnight at 37 °C with 5% CO$_2$.

2. Examine the cells the following day, and if suitably subconfluent (approximately 50%), inoculate with cell-free test material in the presence of polybrene at an effective concentration of 10 μg/mL.

3. Feed each well with an appropriate volume of DMEM containing 10% FBS (v/v) following an adsorption period of approximately 1–2 hours at 37 °C with 5% CO$_2$.

4. Maintain the cultures at 37 °C with 5% CO$_2$ until confluent (normally 2–4 days). Then fix in cold acetone for 10–15 minutes.

5. Perform indirect immunofluorescence by placing a suitable volume of anti-MLV antibody into each well (several monoclonal antibodies to MLV are available, e.g., 34, R187, or 548) (70). Place the slide in a humidified chamber at 37 °C for a suitable period of time, for examples, 30 minutes.

6. Wash the slides with phosphate-buffered saline (PBS) to remove the primary antibody. Air dry the slides. Then add antispecies FITC antibody at a suitable dilution to the slides, and allow the antibody to adsorb as before

7. Wash the slides in PBS to remove the secondary antibody, and air dry as before.

8. Examine the slides for fluorescent foci under a suitable fluorescent microscope, and count the foci to give the titer of MLV per unit volume in the test sample.

Extended Assay

1. Seed the *Mus dunni* cells at a density of approximately 5×10^5 per 25 cm^2 tissue culture flasks in DMEM containing 10% FBS (v/v), and incubate the cultures overnight at 37 °C with 5% CO$_2$.

2. Examine the cells the next day for confluency, and if subconfluent (approximately 50%), inoculate the cells with the test sample in polybrene at an effective concentration of 10 µg/mL. Allow the sample to adsorb for 1–2 hours. Then feed the cells with DMEM containing 10% FBS (v/v) and reincubate at 37 °C with 5% CO$_2$.

3. Maintain the cells at 37 °C with 5% CO$_2$, and passage the cultures several times on reaching confluency (normally three to five times is sufficient).

4. MLV can be assayed in the *Mus dunni* cells at this point by several means such as immunofluorescence in which the cells are subcultured into slides and tested as described before for the direct assay. Alternatively, cell-free supernatant can be harvested from the cultures and tested for MLV by the direct infectivity assays described before, or by reverse transcriptase assay.

Little has been published regarding the relative sensitivities of the previously mentioned cell lines for leukemia viruses. However, one study that examined the titers of X-MLV detected in fermenter harvests from a murine plasmacytoma cell indicated that for this particular X-MLV, the *Mus dunni* cell line was the most sensitive, followed by mink S⁺L⁻-cells and PG4 cells were least sensitive (56).

If a cell line used to produce a biologic for human use is positive for a xenotropic retrovirus, it is necessary to determine whether the virus is infectious in human cell lines (19). An alternative approach to assaying cell-free supernatant for infectious retrovirus is by cocultivation of the cell line under test with a detector cell line for a given period of time (e.g., two or three passages) followed by removal of the test cell line and further passage of the detector cell line to amplify any infectious retrovirus present and to dilute any noninfectious retrovirus carried over from the test cell line culture medium. Then the detector cell line can be assayed for the presence of retrovirus by reverse transcriptase assay of the culture supernatant, electron microscopy, or one of the infectivity assays described before. If the cell line under test is nonadherent and the detector cell line is adherent, the test cell lines can be decanted. When both detector and test cell line are adherent, they can be cocultivated in a dish that separates the two cell lines by a transwell membrane which has pores large enough to allow passage of virus but not cells (commercially available transwell dishes normally have membranes with a pore size of 0.4 µm). Human cell lines that have been reported to be susceptible to xenotropic retroviruses include the embryonic lung fibroblast 7605L, the rhabdosarcoma cell line RD, the embryo fibroblast cell line MRC-5, the embryo kidney cell line 293, and the osteosarcoma cell line HOS (10,13,57,62,71).

Detection of Retrovirus by Reverse Transcriptase Assay. The presence of the reverse transcriptase enzyme within retroviral particles facilitated the assay for virus (infectious or noninfectious) in supernatant harvested from growing cultures. The reverse transcriptase enzyme synthesizes DNA using an RNA template and therefore can be assayed by including radiolabeled nucleotides in the reaction mixture. The optimum stage for reverse transcriptase detection in cell culture supernatant will depend on the cell line in question. However, in cell lines that produce retrovirus at levels detectable by reverse transcriptase, enzyme activity is normally readily detectable in supernatants harvested from cells that are exponentially growing or have recently stopped growing. The method described here comprises six separate reactions that assay polymerase activity under conditions with an RNA template (poly (rA)), a DNA template (oligo (dT)) or no template, and in the presence of either the magnesium or manganese cation (62). These six conditions control the following: (a) RNA dependent DNA polymerase activity that may arise from contaminating cellular DNA polymerase (higher levels of radiolabeled nucleotide incorporation when the template is DNA indicate that incorporation in the RNA template is likely to be from contaminating DNA polymerase); (b) failure to separate unincorporated from incorporated radionucleotides (no template control); and (c) distinction between reverse transcriptase enzymes that have a preference for either the magnesium or manganese cation (63,72).

1. Clarify cell-free supernatant (10–20 mL) to remove cell debris by centrifugation at 11,000 g for 10 minutes.

2. Pellet the retrovirus in the sample by centrifuging the resultant supernatant at 100,000 g for 1 hour. Then discard the supernatant, and allow the tubes to drain.

3. Disrupt the retroviral particles in buffer (approximately 200 µL) comprising 40 mM Tris pH 8.1, 50 mM KCl, 20 mM DTT, 0.2% NP40.

4. Divide the sample into six 25 µL aliquots, and add an equal volume of one of each of the following six reaction mixtures:

   (i) 40 mM Tris pH 8.1, 50 mM KCl, 25 µCi [methyl ³H] TTP, 2 mM MnCl, poly r(A) [0.05 A$_{260}$ units].

   (ii) 40 mM Tris pH 8.1, 50 mM KCl, 25 µCi [methyl ³H] TTP, 2 mM MnCl$_2$, oligo (dT) [0.05 A$_{260}$ units].

   (iii) 40 mM Tris pH 8.1, 50 mM KCl, 25 µCi [methyl ³H] TTP, 2 mM MnCl$_2$, 2 mM Tris pH 8.1, 30 mM NaCl.

   (iv) 40 mM Tris pH 8.1, 50 mM KCl, 25 µCi [methyl ³H] TTP, 20 mM MgCl$_2$, poly r(A) [0.1 A$_{260}$ units].

   (v) 40 mM Tris pH 8.1, 50 mM KCl, 25 µCi [methyl ³H] TTP, 20 mM MgCl$_2$, oligo (dT)) [0.1 A$_{260}$ units].

   (vi) 40 mM Tris pH 8.1, 50 mM KCl, 25 µCi [methyl ³H] TTP, 20 mM MgCl$_2$, 4 mM Tris pH 8.1, 60 mM NaCl.

5. Incubate the reaction mixtures at 37 °C for 1 hour to allow RNA-dependent and DNA-dependent DNA polymerase enzyme reactions to occur.

6. Precipitate the DNA or RNA templates onto GF/C filters using 10% trichloroacetic acid (TCA), 1% sodium pyrophosphate.

7. The precipitated nucleic acids with incorporated radiolabeled nucleotides are then measured in scintillation fluid in a scintillation counter.

The background levels of radioactivity obtained by this method are generally of the order of 100–2000 disintegrations per minute (dpm). Levels higher than this in those reactions with a template indicate polymerase activity. However, one can assume RNA-dependent DNA polymerase activity from reverse transcriptase only when the level of incorporation into the DNA template is relatively low, for example, at least half the dpm incorporation into the RNA template. The level of dpm in the reaction mixtures without template should be lower still, for example, half the dpm of the DNA template reaction. Suitable controls for use with this reaction include MLV for reactions that include the manganese cation and maedi-visna virus for reactions that include the manganese cation.

Electron Microscopy. Electron microscopy is a direct method of detecting viral contaminants and as such can provide evidence of viral contamination in the absence of any cytopathological effect. It is therefore a useful method for detecting endogenous and exogenous retroviral particles. Cell lines themselves can be examined directly for viral contaminants by thin-section transmission electron microscopy (TEM). Alternatively, the amount of retrovirus in culture fluids sampled from fermenters can be enumerated to determine the retroviral load. Although electron microscopy provides a useful means of detecting and classifying viral contaminants, it is not suitable as a diagnostic method for low levels of contamination because the technique allows examination only of a thin section of a the cell preparation or a small volume of culture supernatant. Quantitation of retroviral particles is required by regulatory authorities for cell lines used to produce monoclonal antibodies (19). The quantity of retrovirus present should be shown to be consistent between lots, and it should be demonstrated in process validation studies that the values obtained are removed or inactivated during product purification.

The retroviral burden in culture or fermenter supernatants can be estimated by negatively staining a virus preparation followed by TEM. However this method can be problematic because retroviruses are fragile and the method may cause structural artifacts that make identification difficult. Following is an outline of a suitable method for quantitating retroviral particles in cell culture or fermenter fluid samples. However, because electron microscopy is a highly specialized technique, the reader is advised to consult more specialized literature before embarking on such methods (71–78).

1. Clarify the culture supernatant by centrifugation at 11,000 g for 10 minutes.

2. Pellet the retrovirus from the resultant supernatant by ultracentrifugation at 100,000 g for 90 minutes (the conditions required for other virus types will differ).

3. Dilute commercially available latex particles to a concentration of approximately 10^8/mL, and mix the concentrated virus preparation with an equal volume of latex particles.

4. Apply the virus/latex particle mix to an EM carbon-coated grid.

5. Wash the EM grid gently with distilled water.

6. Apply a drop of negative stain to the grid (negative stains commonly used include 1–3% phosphotungstic acid (pH 5–8) and 1–2% uranyl acetate (pH 4.4).

7. Following a suitable staining time, remove excess stain from the grid using filter paper.

8. Allow the grid to air dry. Then examine by TEM.

9. The virus concentration in the initial culture supernatant can be determined by the following equation:

$$\text{Concentration of virus} = \frac{\text{Concentration of latex particles} \times 2 \times \text{virus count}}{\text{Latex particle count} \times \text{centrifugation concentration factor}}$$

Culture supernatant preparations from fermenter harvests may have large quantities of cellular debris. This is particularly so for batch fermentation samples. Further purification of the culture supernatant, for example, by sucrose density gradient centrifugation (78) may be necessary to reduce background material that may obscure virus identification.

Direct examination of cell lines by TEM using fixation of a cell pellet and thin-section preparation can be performed as outlined below (75). For detection of retrovirus, the culture should be actively growing.

1. Pellet the cells from a medium-sized culture flask (approximately 5×10^6 cells should be harvested) by gentle centrifugation (500 g for 5 minutes).

2. Carefully break up the cell pellet into small pieces of about 1 mm^3.

3. Fix the cells in 2.5% glutaraldehyde in phosphate buffer (pH 7.2) for 60 minutes.

4. Wash the cells three times in PBS (10 minutes per wash). Then postfix in 1% osmium tetroxide in PBS for 1 hour.

5. Wash the cells three times in dH$_2$O, 10 minutes per wash. Then dehydrate the cells in a graded ethanol series of 50%, 75%, 95%, and 100% twice (15 minutes per step).

6. Embed the cells by first placing them in propylene oxide for 10 minutes: then in a 1:1 mix of resin to propylene oxide for 1 hour, followed by three 1-hour treatments in neat resin.

7. Place the samples in fresh resin, and polymerize the blocks by heating to 60 °C for 24 hours.

8. One-μm sections are then cut from the block using an ultramicrotome; then placed on EM grids and stained in 1% uranyl acetate, then lead citrate, and examined under TEM.

The retrovirus family comprises the oncoviruses, lentiviruses, and spumaviruses. The oncoviruses can be further divided into four groups comprising Type A, Type B, Type C, and Type D. Classification of retroviruses is based on their morphological features and morphogenesis as determined by electron microscopy.

Figure 1 illustrates some of the features (micrographs provided by Q-One Biotech Ltd., Glasgow, U.K.).

In Vitro Assays

So-called 'in vitro tests' (Table 1) are recommended by regulatory authorities for identifying nonendogenous or adventitious viruses in cell banks which may be noncytopathic contaminants of the cell line or were introduced via a raw material or a breach in GMP procedures. The cell lines recommended for such in vitro assays by CBER are a human diploid cell line, a monkey kidney cell line, and a cell line of the same species and tissue type as used in production (17). Therefore, cell lines generally employed are the human embryo fibroblast cell line, MRC-5, and the monkey kidney cell line, Vero. The cell line of the same species and tissue type as used in production is as appropriate. CPMP regulations for cell lines producing monoclonal antibodies suggest, in addition, the use of cells capable of detecting a range of human, murine, and bovine contaminants (if relevant) (20). The rationale for the choice of cell lines is to provide a general screening assay that detects a broad range of viral contaminants that may be present in the cell line or in fermenters and may be pathogenic to humans. The assay is performed by examining inoculated detector cell lines for cpe and hemadsorbing agents and therefore has the limitation that it will detect only cytopathic and hemadsorbing viral contaminants that can replicate in the chosen detector cell lines. The specific format of the assay is unspecified by the regulatory guidelines. However, it is stipulated that the assay be examined for cytopathic and hemadsorbing viruses (17,20) and be maintained for a minimum of 14 days with passage of the human diploid cell line for a further 14 days with hemadsorption performed at the end of the cultivation period if the test system can support growth of human cytomegalovirus (17). The nature of the inoculum is specified by ICH as "a lysate of cells and their culture medium" (18). A possible format for such in vitro assays would therefore be the following (negative control and suitable positive controls for cpe on each detector cell line used should be tested in parallel):

1. Prepare six-well tissue culture dishes each of MRC-5, Vero, the cell line of the same species and tissue type as used in production (and other cell lines as appropriate) by inoculating 3 mL of cells at $1-2 \times 10^5$ cells/mL in an appropriate growth medium containing 10% FBS (v/v). Incubate the cultures overnight at 37 °C with 5% CO_2.

2. Prepare the test sample lysate by harvesting cells and supernatant from a healthy confluent culture grown in a medium-sized flask (approximately 80 cm²) (adherent cells can be harvested by first scraping the cells into the spent culture medium with a sterile cell scraper). Clarify the preparation by centrifugation at 500 g for five minutes. Transfer 70–80% of the supernatant to a fresh container, and place on ice (supernatant 1). Then resuspend the cell pellet in the remaining supernatant, and lyse the cells by freeze/thawing three times. Clarify the resultant lysate by centrifugation as before, and pool this the supernatant (supernatant 2) with supernatant 1 to constitute the sample for inoculation. Preparation of lysate by this method ensures that viruses sensitive to repeated freeze/thawing are represented by inclusion of "supernatant 1," viruses that may be trapped intracellularly are released and present in "supernatant 2," and the presence of intact cells that escaped disruption by freeze/thawing and could potentially interfere with observations for cpe is minimized by clarification of both supernatants 1 and 2. Cell lysates can also be prepared as before for samples taken from fermenters.

3. Inoculate a suitable volume for example, 1 mL of lysate, into each well of each cell type, and allow the sample to adsorb for 1 hour at 37 °C with 5% CO_2.

4. Feed the cultures with 3–4 mL of maintenance medium (containing 1% v/v FBS) and incubate at 37 °C with 5% CO_2.

5. Examine the cultures every 2–3 days for signs of abnormalities such as cells rounding, syncytia, or cell lysis.

6. Feed the cells as required (after seven days is normally adequate) with an appropriate maintenance medium (Vero and MRC-5 cell lines can normally be maintained in culture in DMEM containing 1% FBS (v/v) without passaging for 14 days. However, if the cell type being used deteriorates during this period, as shown by a negative control culture condition, or assay for hemadsorbing agents on the human diploid cell line (MRC-5) is required at 28 days postinoculation, the cells should be subcultured to prevent this.

7. At 14 days postinoculation (or 28 days for MRC-5 detector cells when the test sample is capable of supporting human CMV growth), perform hemadsorption using guinea pig, human, and chicken red blood cells (rbc's) as follows:

    (i) Prepare the rbc's by centrifuging at 2000 g for 10 minutes. Then wash the pelleted rbc's twice in PBS in tubes with volume markings. Following the second wash, measure the volume of packed rbc's. Then add sufficient fresh PBS to the pelleted cells for a 0.5% preparation of rbc's.

    (ii) Remove the culture medium from the cells, and carefully wash the monolayers with PBS.

    (iii) Overlay duplicate wells of each cell type with 2–3 mL of each of the three rbc species, and

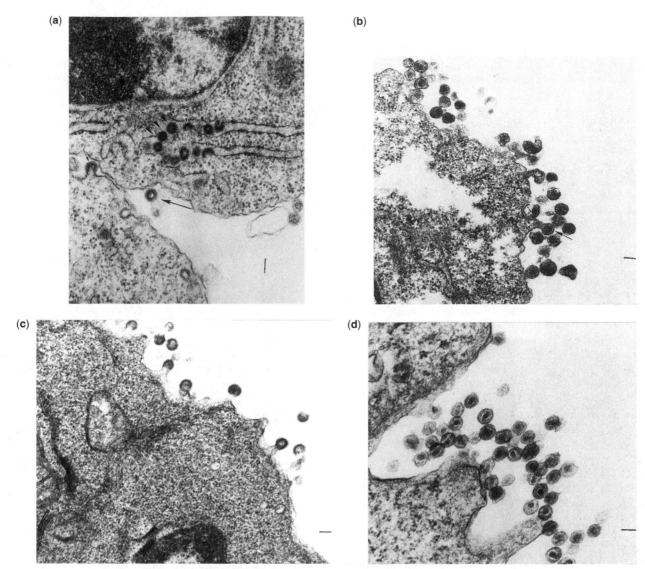

Figure 1. TEM micrographs of cells infected with retrovirus. (**a**) A murine myeloma cell line with an endogenous oncovirus. The large arrow heads indicate Type A particles within the endoplasmic reticulum (ER) characterized by two electron-dense concentric shells surrounding a central area of low electron density. The small arrowheads indicate Type A particles budding into the ER. An immature Type C particle with centrally located core is indicated by the large arrow (in mature Type C particles the core is electron-dense). The small arrow indicates a Type C particle budding from the cell membrane into the intracellular space. Note that the core of Type C particles is formed during budding. The scale bar is 100 nm. (**b**) Raji cells infected with the Type D oncovirus, squirrel monkey retrovirus (SMRV). Several extracellular Type D particles with electron-dense cores, sometimes tubular in shape, are evident. A budding Type D particle with characteristic preformed core is indicated (arrow). The scale bar is 100 nm. (**c**) Cells infected with the lentivirus HIV. Several budding particles can be seen characterized, as in Type C particle budding, by simultaneous core assembly. The scale bar is 100 nm. (**d**) A cell infected with HIV displaying numerous extracellular particles. This micrograph illustrates the characteristic tubularly shaped core of mature lentivirus particles. The scale bar is 100 nm. (**e**) A monkey kidney cell infected with the spumavirus, simian foamy virus (SFoV). A large group of spumavirus can be seen within the ER of the cell (arrow). The particles are characterized by an electron-dense ring-shaped core with a translucent center surrounded by an envelope with prominent surface projections. The scale bar is 500 nm. All micrographs were provided by Q-One Biotech Ltd., Glasgow, United Kingdom.

(e)

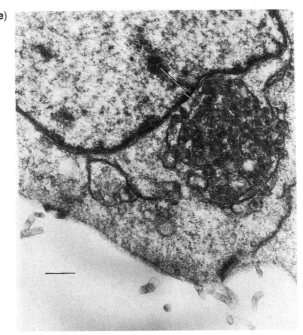

Figure 1. *Continued.*

incubate the plates at approximately 4 °C for 30–60 minutes.

(iv) Remove the rbc's by gentle agitation. Then pipette off. Wash the monolayer gently with PBS two or three times until all the nonadsobed rbc cells are removed.

(v) Hemadsorption is indicated by adhesion of the rbc's to the detector cells. A suitable positive control for hemadsorption is parainfluenza type 3 (PI3) which is infectious for both MRC-5 and Vero cell lines.

In Vivo Assays

Assays in animals are part of the general safety tests recommended by regulatory authorities for manufacturers of biologics and involve the so-called "*in vivo*" assay and the antibody production assay (MAP, HAP or RAP) (Table 1). "*In vivo*" assays normally utilize suckling mice, adult mice, guinea pigs, and embryonated eggs. Mice are inoculated intracerebrally, intramuscularly, and intraperitoneally, guinea pigs are inoculated intramuscularly, and embryonated eggs by the yolk sac, allantoic cavity, and amniotic cavity (20,76). The inoculum is normally a small volume of lysate prepared from the cell line under test, and the animals are observed for 3 to 4 weeks and the eggs for a shorter period of time (refer to references 20 and 79 for specific durations) for morbidity and mortality. Hemagglutination is performed on fluids harvested from the eggs inoculated vial the allantoic route. The *in vivo* assay is designed to detect a wide range of viral contaminants. Suckling mice are susceptible to many arthropod-borne viruses, Herpes simplex virus, rabies, some arenaviruses, and picornaviruses, including poliovirus, coxsackievirus, echovirus, and encephalomyocarditis virus (adult mice are susceptible to some of these viruses) (77,78); guinea

pigs are susceptible to paramyxoviruses, reoviruses, and filoviruses (37), and embryonated eggs are susceptible to poxviruses, paramyxo-, orthomyxoviruses, and Herpes simplex virus (79). Antibody production assays are performed on cell lines of mouse (MAP), rat (RAP) and hamster (HAP) origin to assay for viruses introduced by the source species. These assays are therefore normally performed only on the MCB because a negative result indicates that the cell bank system is free of viruses of the species of origin. In addition, MAP assays are performed on monoclonal antibody preparations made in ascites. The viruses assayed in these tests include hantaan virus, lymphocytic choriomeningitis virus (LCMV), reovirus type 3, and Sendai virus all of which are known to infect humans or primates (20). In these assays, adult mice, rats, or hamsters are inoculated with test cell lysate and held for 4 weeks, then bled, and the serum is tested for antibody to viruses of concern using antibody specific assays such as enzyme-linked immunosorbent assay (ELISA) and indirect immunofluorescence (IF).

Polymerase Chain Reaction (PCR)

PCR is a highly sensitive technique for detecting specific DNA sequences and therefore is important for detecting viral and mycoplasmal contaminants whose genomic sequence is known fully or partially. The reaction involves amplification using repeated thermal cycles of DNA denaturation, primer annealing, and DNA polymerization by *Taq* polymerase of a specific region of DNA defined by the primers. The reaction mass balance alters as the cycles progress resulting in an increase in product (and consequently available template), a decrease of nucleotides and primer (although these two components are greatly in excess in the reaction), and a decrease in the ratio of *Taq* polymerase molecules to available template copies.

The reaction therefore reaches a point after 30 cycles or so when production of product is no longer exponential due to limiting *Taq* polymerase and self-annealing of product. The specific conditions employed for any PCR assay depend upon the sequences being amplified, the primers used, the number of initial copies of template in the sample, and the size of the region being amplified. In addition, viruses with RNA genomes must be reverse transcribed before PCR amplification.

PCR is important as a tool for virus adventitious agent detection when a culture system is not available, as is the case with hepatitis C virus detection, or when detection of infectious virus is limited by the sample size, as is the case with detection of replication-competent virus during monitoring of patients treated with gene therapy constructs. It should be noted, however, that PCR detects infectious, as well as defective virus. Therefore in circumstances where virus is likely to be present but has been inactivated by processing, PCR is less useful, for example, PPV in gamma-irradiated trypsin. The specific protocols for viral diagnosis by PCR are beyond the scope of this article, and readers interested in the details are advised to consult more relevant literature (80).

TESTING RAW MATERIALS

Tests for Bovine Viral Contaminants

The most common viral contaminant in raw materials of bovine origin is the pestivirus BVDV because of its prevalence, the large numbers of animals from which a serum batch is prepared, and partial resistance of the virus to standard procedures used to heat inactivate bovine serum for use in tissue culture medium preparations (56 °C for 30 minutes) (81). Some manufacturers of bovine serum reduce the risk of infectious BVDV in bovine serum by irradiation with UV-C (wavelength of 254 nm) or gamma irradiation (38,39). Both techniques are effective inactivators of BVDV and other viruses. However, in the absence of effective virus inactivation treatment of bovine raw material, it is advisable to test for BVDV. The majority of BVDV strains are noncytopathic. Therefore detection of infectious BVDV is most easily done, as outlined in the following protocol, by end-point immunofluorescence using an antibody that recognizes BVDV antigen following inoculation of a susceptible cell line such as the bovine turbinate (BT), bovine trachea (EBTr) cell lines, or the Madin Darby bovine kidney cell line (MDBK). Negative and positive control cultures are tested in parallel. The negative control cultures are cells assayed in parallel and cultured in a medium containing BVDV-free serum. The positive control cultures can either be cells infected with a noncytopathic strain of BVDV (e.g., New York-1) at the same time as the test sample is inoculated, or a subculture of the negative control culture can be infected with a cytopathic strain of BVDV (e.g., NADL) toward the end of the assay.

1. Seed BVDV-free cultures of one of the above cell types into tissue culture flasks to obtain cultures approximately 70% confluent on the following day.

2. Remove the growth medium the next day, and replace with a culture medium containing 10–15% (v/v) of the test serum as the FBS component (all three cell lines mentioned above grow in DMEM). Alternatively, if the bovine raw material under test is not serum, dissolve the test material in a small volume of BVDV-free medium. Then inoculate this onto the monolayer, and allow it to absorb for one hour. Feed the cultures with BVDV-free growth medium.

3. Incubate the cultures at 37 °C with 5% CO_2, and passage when confluent, (approximately every 3–4 days) for five passages. During subculturing, use a medium containing 10–15% (v/v) of the bovine serum under test.

4. At passage five, prepare the cells for immunofluorescence by subculturing 1×10^5 cells/mL (2 mL per well) onto two chambered tissue culture slides. Reincubate the slides at 37 °C with 5% CO_2 until confluent.

5. When confluent, remove the culture medium, and fix the cells by submerging the slides in cold acetone for 10–15 minutes. At this point, the slides can be stored frozen or tested by immunofluorescence.

6. Perform indirect immunofluorescence using commercially available antibodies to BVDV and commercially available antispecies FITC (Fluorescin isothiocyanate) conjugated antibody.

7. Dilute the anti-BVDV antibody to a predetermined working dilution in PBS. Then allow the antibody to adsorb onto the cells by incubation in a humidified chamber at 37 °C for a suitable period of time (30 minutes is normally sufficient).

8. Remove the primary antibody by washing the slides in PBS. Add the secondary antibody (commercially available antispecies FITC conjugated antibody) to each well, and allow this to adsorb as above.

9. Following a suitable adsorption period, wash the slides in PBS to remove the secondary antibody.

10. Examine the slides under a suitable fluorescent microscope. Samples that are positive for BVDV may show varying proportions of positive cells, depending on the strain of BVDV and the number of passages in culture. Figure 2 illustrates the pattern of BVDV fluorescence which is predominantly cytoplasmic.

The U.S. Code of Federal Regulations (9CFR) (52) stipulates requirements for testing animal-derived products used as ingredients in producing biologics. In the case of bovine-derived ingredients, specific procedures for viral detection using Vero cells (African green monkey kidney) and a cell line of bovine origin are provided. The method involves cultivating these two cell lines in a medium containing at least 3.75 mL or 15% of the ingredient for 21 days or more with at least two passages during this period. At the final passage, monolayers are set up of specified surface area and assayed 7 days later for viral contaminants by cytological staining, hemadsorption, and IF. Viruses assayed by IF on detector cells of bovine origin

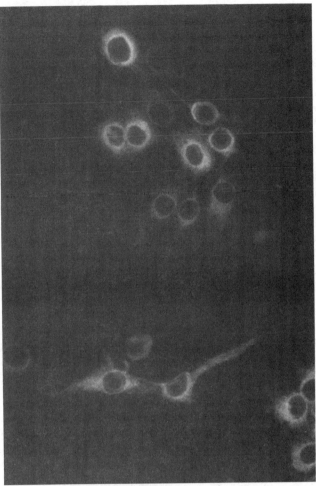

Figure 2. BT cells infected with BVDV and assayed for infectivity by indirect immunofluorescence, as described in the text. Cells infected with BVDV display green cytoplasmic fluorescence. Photograph provided by Q-One Biotech Ltd., Glasgow, United Kingdom.

are BVDV, bovine adenoviruses (BAV), bovine parvovirus (BPV), bluetongue virus (BTV), bovine respiratory syncytial virus (BRSV), reovirus, and rabies virus. Vero cells are tested by IF for BVDV, reovirus, and rabies virus.

Tests for Porcine Viral Contaminants

Trypsin is a raw material of porcine origin widely used in cell cultures for dissociating anchorage-dependent cells. The virus of concern in trypsin preparations, as discussed previously, is PPV. As with raw materials of bovine origin, the U.S. Code of Federal Regulations (9CFR) (52) stipulates the method for testing for PPV in trypsin preparations which have not undergone appropriate treatment to inactivate this virus. The following method complies with 9CFR recommendations:

1. Seed duplicate medium-sized tissue culture flasks (75–80 cm²) of primary porcine kidney cell lines at a density such that at inoculation (the following day) the monolayer is 30–50% confluent. (It is important to inoculate PPV onto mitotically active cells because

parvoviruses require the cellular functions available during the S and G phases of the cell cycle for DNA replication (82).

2. The following day, ultracentrifuge the trypsin preparation at 80,000 g for 1 hour to pellet the virus and separate it from trypsin so that monolayer detachment can be avoided (9CFR requires testing 5 g of trypsin is tested and resuspending the resultant pellet in water).

3. Inoculate the resuspended pelleted material into one of the two flasks of primary porcine kidney cells, and incubate for 1 hour at 37 °C with 5% CO_2 to allow any virus to adsorb. Inoculate the remaining flask with culture medium as a negative control, and incubate as above.

4. Feed the flasks with an appropriate growth medium, and reincubate the culture at 37 °C with 5% CO_2.

5. Examine the cultures every 2–3 days for confluency of growth and viral cpe.

6. Maintain the culture for 14 days or more postinoculation, with at least one subculture during this period. It is important to ensure that all subculturing of the primary porcine cells is performed using trypsin that is negative for PPV.

7. At the last subculture (this should fall on or after day seven postinoculation), set up monolayers for immunofluorescence. 9CFR stipulates that these monolayers must be a minimum of 6 cm² (commercially available glass or plastic chamber slides are suitable for this purpose). Three sets of 6 cm² monolayers are required from the "test material" culture such that two are the "test material" monolayer, and the remaining one is inoculated with PPV as a positive control. Set up one 6 cm² monolayer from the negative control culture.

8. One day after subculture, remove the medium from one of the "test material" slides, and inoculate with PPV (100–300 fluorescent units). Allow the virus to adsorb at 37 °C with 5% CO_2 for 1 hour. Then feed the culture with an appropriate volume (2–3 mL) of fresh medium. Return the slide to a 37 °C incubator with 5% CO_2.

9. Examine the cultures every 2–3 days for viral cpe. If cpe is evident on the positive control culture, it can be fixed in cold acetone for 10–15 minutes. However, a "test material" monolayer must be fixed at the same time. Maintain the negative control and remaining "test material" cultures until at least 7 days postsubculture (i.e., day 14 or later). Then fix with cold acetone as above.

10. Perform indirect immunofluorescence for PPV, as described before for BVDV. Specific fluorescence to PPV is evident from nuclear fluorescence.

PPV cytopathology is often difficult to recognize and requires serial passage before it becomes evident. Therefore, specific end-point assay such as immunofluorescence described before provides conclusive evidence of PPV contamination.

DETECTION OF MYCOPLASMAS

Mycoplasmas are bacteria that lack the rigid cell wall structure composed of peptidoglycan that is normally associated with bacteria. Thus mycoplasmas have properties that distinguish them from other bacteria. Testing for mycoplasmal contamination requires a different set of assays independent of sterility testing for normal bacteria, and these are described following. The term mycoplasma denotes organisms that belong to the class designated Mollicutes which includes other generic groups referred to as acholeplasma, ureaplasma, and spiroplasma. Many distinct organisms (158 recognized species) are grouped in the class Mollicutes because they lack cell walls due to the absence of the genes for peptidoglycan synthesis (83). Mycoplasmas also differ from other bacteria by incorporating cholesterol and other sterols into their plasma membranes which makes the membranes of mycoplasmas more pliable and more resistant to physiochemical properties than would be predicted (84). This pliability combined with their small size, average diameter of 0.3–0.8 μm, allows them to pass through filtration systems designed to remove bacteria. The most likely source of mycoplasmal contamination in a cell culture facility is a contaminated cell culture, thus emphasizing the need for thorough screening of all new cell lines and appropriate quarantine and storage protocols for all incoming cell cultures. After establishing an infection in a particular cell line, mycoplasma and other microbes are spread by aerosol droplet dispersion via a variety of normal cell culture procedures including pipetting and dispensing of media liquids. The inappropriate use of a common medium bottle or common pipettes between cell cultures and cell lines can lead to contamination of many cell lines in the facility with the same species of mycoplasma (85). Human isolates represent the majority of mycoplasmal contaminants in cell culture. *Mycoplasma orale*, *M. fermentans*, and *M. salivarium* are the most frequently isolated, and *M. buccale*, *M. faucium*, *M. genitalium*, *M. hominis*, *M. pirum*, and *M. fermentans* less frequently. Many species of mycoplasma are known to cross host ranges, and thus isolates from contaminated bovine serum such as *Acholeplasma laidlawii*, *M. arginini* and *My. hyorhinis* represent further cause for concern in cell culture. The consequences of mycoplasmal contamination are too numerous to describe in this article. Those interested in a more comprehensive description of the subject are referred to the recent review of mycoplasmal contamination by Lincoln and Gabridge (86). A major problem is that despite heavy microbial contamination, for example, 10^9 colony-forming units per milliliter of culture, the cell cultures often appear to be normal by both macroscopic and microscopic inspection. Thus, due to an undetected mycoplasmal contamination, research results could be distorted and hence erroneous and in a production facility infection could reduce yields of products such as monoclonal antibodies and enzymes.

As a strategy to prevent and control bacterial and mycoplasmal contamination the use of antibiotics in routine cell culture should be avoided wherever possible because they may mask mistakes in aseptic technique and quality control and may also result in the emergence of antibiotic-resistant strains of microorganisms. In any case there is not a single antibiotic which is effective against all bacteria. For example, the inhibitors of peptidoglycan synthesis such as the penicillins and cephalosporins, are clearly ineffective against mycoplasmas, and many antibiotics simply retard or inhibit bacterial growth, that is, they are bacteriostatic agents and do not destroy bacteria. The absence of antibiotics is also a requirement for effective mycoplasmal detection in cell culture. Thus, to summarize this brief background survey of mycoplasmas, it is necessary to perform routine screening for mycoplasmal contamination because mycoplasmas may not cause readily apparent changes in cell culture.

Several direct and indirect procedures are currently used to detect mycoplasmas, but unfortunately, there is not a single, foolproof test that will detect all species. One major problem that has to be addressed in the screening program is the fact that some mycoplasmas will not grow in artificial media and thus must be detected using a cell culture system. The lengthy time required to complete these assays has focused attention on some of the indirect assays, and it is possible that PCR reactions may gain acceptance for these situations, for example, the screening of cells used for *ex vivo* therapies mentioned previously, where a rapid assay is critical. Direct culture methods are the most sensitive but are also the most time-consuming and take up to 28 days for incubation.

Alternative methods for detecting mycoplasmal contamination include PCR reactions, indirect culture methods using DNA fluorochrome stains as described later, DNA probes, ELISAs, biochemical assays, and electron microscopy (87–89). These methods are more rapid than direct assays but less sensitive and vary in their capacities to detect a wide range of species. A thorough review is beyond the scope of this chapter, but readers are referred to a recent publication that details molecular and diagnostic procedures in mycoplasmology, for example, detection by PCR (90). In points to consider in the characterization of cell lines used to produce biologicals issued by CBER (17), tests for the presence of both cultivable and noncultivable mycoplasmas are recommended. Biological products made in insect cells should be tested for both mycoplasmal and spiroplasmal contamination. If storage is required for samples before assay, then it should be at between 2 and 8 °C for 24 hours or less and at −60 °C or lower for 24 hours or more. As specified by CBER (17), mycoplasmal contamination testing must be performed by both the agar and broth media procedures and the indicator cell culture procedure or by a procedure demonstrated to be comparable using the protocols described here.

Cultivation Methods

Agar and Broth Procedures

1. Each lot of agar and broth medium should be free of antibiotics, although penicillin G can be present, and each lot of medium should be checked for its mycoplasmal growth-promoting properties. This requires the use of positive cultures that are described below.

2. Inoculate at least 0.2 mL of the sample over the surfaces, of two or more agar plates of one medium formulation. In addition inoculate at least 10 mL of the sample into a flask containing 50 mL of broth medium which is then incubated at $36 \pm 1\,°C$.

3. Test 0.2 mL of the broth culture on the third, seventh, and fourteenth days of incubation by subculture onto two or more agar plates of the same medium as that used before.

4. Incubate two of the initial isolation plates and two each of the three subculture plates in 5 to 10% CO_2 in a nitrogen atmosphere containing no less than 0.5% oxygen during the incubation period. Some laboratories incubate the cultures both aerobically and anaerobically.

5. Incubate all culture agar plates for at least 14 days at $36\,°C \pm 1\,°C$, and observe microscopically at $100\times$ magnification for growth of mycoplasma colonies.

6. As positive controls, at least two known mycoplasma species or strains should be used, one of which is a dextrose fermenter, that is, *M. pneumoniae* strain FH or equivalent, and the other an arginine hydrolyzer, that is *M. orale* strain CH 192999 or equivalent species or strains. These positive strains should not be more than fifteen passages from isolation and should be used in a standard inoculum of 100 colony-forming units (CFU) or less. Sterile mycoplasma broth is used as the negative control.

Indicator Cell Culture Procedure. This procedure uses a Vero cell culture known to support the growth of appropriate mycoplasmas. Figure 3 illustrates the typical fluorescent pattern observed with normal and mycoplasma-contaminated cultures.

1. Inoculate at least 1 mL of the sample onto two or more indicator cell cultures grown on cover slips, in dishes, or equivalent containers.

2. Incubate the cell cultures for 3 to 5 days at $36 \pm 1\,°C$ in 5% CO_2, and then examine by epifluorescence microscopy following staining with a DNA-binding fluorochrome.

3. As positive controls, *M. hyorhinis* strain DBS1050 and *M. orale* strain CH 19299 are recommended using an inoculum of 100 CFU or less.

Enhanced Cell Culture Procedure. Some laboratories employ an enhanced procedure when a test sample interferes with the performance of the direct Vero cell

(a) 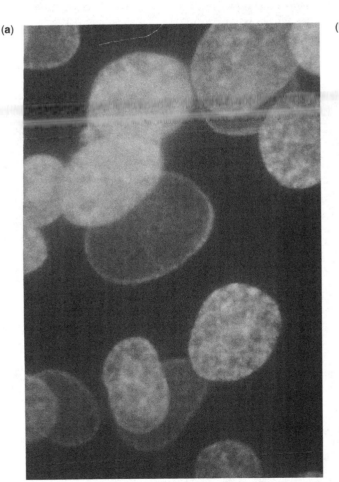   (b)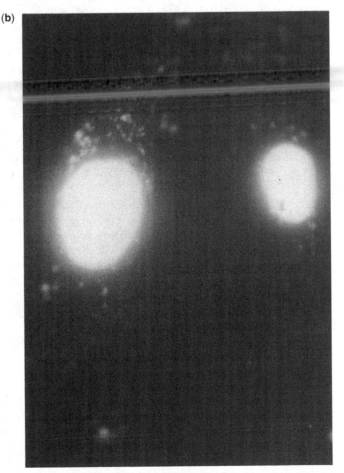

Figure 3. Detection of mycoplasmal infection by DNA fluorochrome staining. (**a**) Cells free of mycoplasmal contamination display only nuclear fluorescence. (**b**) Cells contaminated with mycoplasma display both nuclear and extranuclear fluorescence.

culture assay. Test samples, negative control, and a positive control, that is, *M. hyorhinis*, are inoculated into cultures of Vero cells. T25 flask cultures are appropriate, and the flasks are incubated for 3–5 days. The cells are then removed, and 0.2 mL of each of the cultures is inoculated into each of six wells of Vero cells and then incubated and examined as described in the direct cell culture assay. In evaluating these assays, a sample is considered to meet the requirements of the test for the presence of mycoplasma if mycoplasma, growth does not appear in the sample of inoculated media. In addition all positive cultures must demonstrate appropriate growth, and in all negative cultures growth must not be detected.

Indirect Detection of Mycoplasmas

The direct assays described before are probably beyond the scope of most small cell culture facilities although large production facilities may perform these assays in-house. In contrast some of the indirect assays now available can be performed by any operator because they are available in kit form. An outline of just one of these methods is included here, but a range of indirect methods including PCR reactions is being developed and evaluated.

Mycoplasmal Detection using the Gen-Probe System. This system uses a ^{3}H-labeled DNA probe homologous to mycoplasmal, acholeplasmal and spiroplasmal rRNA. The manufacturers (Gen-Probe Inc.) claim that all species that commonly infect cell cultures can be detected with this probe, which targets the rRNA of the organisms to form a stable RNA to DNA labeled hybrid. The latter is separated from a nonhybridized DNA probe using hydroxyapatite, and then the amount present is measured in a scintillation counter. Following is the recommended test procedure:

1. Prepare a background count (in counts per minute [cpm]) on 5 mL of the scintillation fluid used in the scintillation counter programmed to an appropriate parameter to measure ^{3}H activity.
2. Perform a total count determination, that is, the activity of the labeled probe solution, by pelleting a 5 mL suspension of the separation mixture at 500 g for 1 minute (hydroxyapatite in a buffered solution provided in the kit) in a screw-capped scintillation vial. The supernatant is discarded, the pellet is vortexed in 50 μL of the probe solution, and the cpm is determined.
3. Microfuge the test samples, for example, 1.5 mL aliquots of antibiotic-free cell culture medium (in which the test cultures have been grown for a minimum of 3 days), at 12,000 g for 10 minutes to pellet any mycoplasmas present.
4. Discard the supernatants, and resuspend the pellets in 200 μL of the probe solution. In addition, prepare positive and negative controls by vortexing 50 μL each of the control solutions provided in the Gen-Probe kit with 200 μL of probe solution.
5. Incubate the preparations overnight at 72 °C. Aliquot 5 mL of resuspended suspension into the required number of scintillation vials, and then transfer 200 μL aliquots of incubated sample plus probe solution to the scintillation vials containing the 5 mL of separation solution. Thoroughly mix the solutions by vortexing, and incubate at 72 °C for 5 minutes.
6. Briefly vortex the vials, centrifuge for 1 minute at 500 g, and remove and discard the supernatants. Wash the pellets, reincubate, wash again, centrifuge for 1 minute at 500 g, and remove the supernatant. Rewash the pellets.
7. Add 5 mL of scintillation fluid to each vial, and resuspend the pellets by vortexing. Stand the vials at room temperature for 5 minutes to minimize background phosphorescence, and then measure ^{3}H activity as described before.
8. The percentage of hybridization of the sample to the probe DNA is

$$\% \text{ hybridization} = \frac{\text{sample cpm} - \text{background cpm}}{\text{total count cpm}} \times 100\%$$

Controls: % hybridization of positive should be >30%; % hybridization of negative should be <0.2%. Providing that the controls are within acceptable limits, a hybridization result of >0.4 is considered positive. This method provides a rapid assessment of mycoplasmal contamination and appears to correlate well with direct culture results which should be used as the gold standard for these indirect assays. The Gen-Probe® system has been used successfully by researchers in the Sydney laboratory, as reported by Thompson (91).

BACTERIA AND FUNGI

The effects of bacterial and/or fungal contamination on cell culture systems have been long recognized and are well documented in numerous other publications. The risks of such contamination can be reduced by employing rigorous aseptic techniques and using appropriate and regularly maintained and validated equipment, for example, regularly monitoring the performance of all HEPA laminar flow cabinets. Contamination does occur, however, despite the precautions, and thus the kind of protocols for sterility testing recommended by the USA 21 Code of Federal Regulations (92) should be performed regularly on appropriate specimens. The protocol described is applicable to single, bulk, or final product with the recommendation that bulk material should be tested separately from final material and material from each final container should be tested in individual test vessels as follows:

1. Samples are cultured in a fluid thioglycollate medium (THIO) and a soybean-casein digest medium (TSB). Some manufacturers of biologics adopt a more comprehensive approach to sterility testing and also include other media, such as peptone yeast glucose broth and Saboraud dextrose agar slants.

2. The amount of sample tested depends on the nature of the test material. As a guideline, samples from bulk material should be representative of the bulk and be not less than 10 mL. For sterility testing of final containers, it is recommended that samples from at least 20 final containers be taken for each medium used. The volume tested is the entire contents of the final container if it is less than 1 mL and, if 1 mL or greater, then the volume tested should be the largest single dose recommended by the manufacturer or 1 mL, whichever is larger, but not more than 10 mL.

3. Incubate tests using THIO at 30 to 35 °C for no less than 14 days, and examine visually for evidence of growth on the third, fourth, or fifth day, on the seventh or eighth day, and on the last day of the test period. Incubate tests using TSB at 20 to 25 °C, and inspect as described for THIO. If the medium is rendered turbid by the inoculum, thus interfering with visual inspection for microbial growth, then at least 1 mL of the test medium is transferred to additional containers of the medium. It is standard practice to test replicate sets of two tubes of each medium.

4. Positive control strains, for example, *Bacillus subtilis* (ATCC 6633) and *Bacteroides valgatus* (ATCC 8482) for THIO and *Candida albicans* (ATCC 10231) for TSB are used at inoculum levels of less than 100 CFU/inoculum. The viability and purity of these organisms should be verified regularly. A negative control of uninoculated medium serves as the negative control. USP 23 (93) provides more detailed information regarding the nature of the control cultures that should be used in sterility tests, including description of alternative tests, for example, membrane filtration, for samples not readily water soluble.

OXYGEN UPTAKE RATE

In large-scale cultivation of cells, several parameters can be used to monitor cell growth and metabolism, including oxygen uptake rate (OUR), Eh, pH, glucose uptake, lactate production, glutamine uptake, NH_3 accumulation, and levels of intracellular nucleotide pools. The major limitation is the type of test available to detect the chosen parameter. Recent developments have led to the availability of more on-line detectors, for example, glucose biosensors, which provide a rapid check on the culture condition, and these now augment oxygen and pH probes which have been the major indicators of cell viability and growth to date. For example, in vaccine manufacture the viral infection of some cell types can be monitored by following OUR, which correlates with viral infection (94) because cells have an increasing or constant OUR, during the growth and production phases whereas the OUR drops after viral infection. In a cell culture system contaminated by an aerobic organism, the OUR can increase significantly, whereas the oxygen levels monitored by the oxygen probe often fall to undetectable levels. Thus microbial contamination can often be quickly detected by reference to oxygen levels in the culture, providing that the probes have been calibrated properly. In addition, pH levels usually fall significantly as a result of lactate production by the microbes, and this can be detected by the pH probe or the rapid change of color of the pH monitoring dye if present in the medium.

ENDOTOXIN DETECTION

Although endotoxins are not infectious agents, they are derived from the lipopolysaccharide cell wall component of gram-negative bacteria and could be present even in the absence of viable bacteria. If endotoxin, is present in a therapeutic product administered parenterally, it can induce fever and possibly shock in the recipient. Thus endotoxin testing is incorporated as part of the safety testing protocol of final products. The testing procedure involves either monitoring the rectal temperature of rabbits following injection of a standard amount of product per kg of rabbit or using an *in vitro* test known as the LAL test in which the lymph material of the horseshoe crab is gelled in the presence of endotoxin. It has been our experience that if endotoxin is present in the cell culture media, it can affect the growth and viability of cell cultures and should always be borne in mind if cells are not growing as expected. This problem can be avoided by using water suitable for injection, which has been pretested for endotoxin, to prepare media if in doubt about the useful water supply.

STATISTICAL ANALYSIS

In the context of this article, statistical analysis of viral contamination in a sample can be at two extremes. First, when assaying raw material or a cell line, the contaminant, if present, is in low amounts and therefore may not be detected. Second, when performing process validations, high titers of virus are spiked into a downscale version of the process, and samples are collected at the start, during, and after processing, and then are assayed for virus titer. Therefore samples generated from process validations can contain high levels of virus or have virus titers close to or below the detection level of the assay. Calculation of virus titers at low concentrations (e.g., in the range of 10 to 1000 infectious particles per liter) can be made by estimating the minimum detectable level (mdl), whereas estimation of virus titers in samples containing high titer loads can be made by the $TCID_{50}$ assay.

Minimum Detectable Level

There is a discrete probability that a sample that contains a low concentration of virus will produce a negative result when assayed in tissue culture because of random distribution of virus particles and the fact that the sample tested is usually much smaller than the volume available, for example, only a few milliliter sample from 100–200 liter fermenters is assayed for viral contaminants by *in vitro* assays. The probability P that a sample is negative for viral contamination can be calculated using the following formula:

$$P = \left[\frac{(V - v)}{V}\right]^n$$

where P is the probability that the sample contains no infectious virus, V is the total volume of the material, v is the sample volume, and n is the absolute number of virus particles statistically distributed in V.

When V is much greater than v, the formula can be approximated by the Poisson distribution as

$$P = e^{-cv}$$

where c is the virus titer per unit volume and v is the sample volume.

This equation becomes the following by natural logarithmic (In) conversion:

$$\ln P = -cv$$

or

$$c = \frac{-\ln P}{v}$$

The probability P that the sample is free of virus is normally set at 0.05 or 0.02, that is, in 5 and 2% of cases, respectively, a false negative result will be obtained; in other words there are 95 or 98% confidence limits, respectively, that the detection limit calculated is correct. Thus, in a sample where virus has not been detected, a minimum detectable limit (mdl) with a set confidence level can be calculated. The following example illustrates the formula

In a solvent detergent process validation study, 450 mL of material was spiked with 50 mL of enveloped virus. One mL of the resultant material was assayed for virus, but virus was not detected. By using the above formula and applying a confidence level of virus detection of 98%, the following can be calculated:

$$c = \frac{-\ln P}{v}$$

$$\Rightarrow \quad c = \frac{-\ln 0.02}{1 \text{ mL}}$$

$$\Rightarrow \quad c = 3.9 \text{ virus particles/mL}$$

The mdl of virus in the sample, therefore, is 3.9 virus particles/mL.

The mdl of virus in the process sample, therefore, is

$$3.9 \times \text{ total sample volume}$$

$$\Rightarrow 3.9 \times 500 \text{ mL}$$

$$\Rightarrow 1950 \text{ virus particles.}$$

Quantitation of Virus by $TCID_{50}$

The $TCID_{50}$ (tissue culture infectious dose, 50) is a standard method for infectious virus quantitation defined as the dilution of sample at which 50% of the replicate cell cultures inoculated with the sample become infected. The assay involves serial dilution of a sample and inoculation of equivalent volumes of each dilution onto replicate monolayers of tissue culture cells sensitive to the virus being assayed. At the assay end point, the inoculated cultures are scored as positive or negative for viral infection. In practice, this method involves serial dilution of the sample through an appropriate medium, for example, maintenance medium or PBS, at a dilution factor of normally 3, 5, or 10, and inoculation of a given volume (e.g., 100 μL) into replicate wells of a multiwell plate (e.g., 96-or 24-well plate with eight wells inoculated per dilution). The accuracy of this method can be increased by increasing the numbers of replicate wells inoculated at each dilution and by decreasing the dilution factor. Several formulas may be used to estimate the virus titer in samples assayed by this method, but the method most commonly used is that of Karber expressed by the following equation (95):

$$-m = \log_{10} \text{ starting dilution} - \left[\sum p - 0.5\right] \times d$$

where: m is the $\log_{10} TCID_{50}$ (per unit volume inoculated per replicate culture), d is the $\log_{10}$ dilution factor, and p is the proportion of wells positive for viral infection.

Thus, in the data shown in Table 4, using a fivefold dilution factor, eight replicate wells, and an inoculum volume of 0.1 mL per well, the $TCID_{50}$ can be calculated as follows:

$$-m = -0.7 - (4.875 - 0.5) \times 0.7$$

$$-m = -3.76$$

$$TCID_{50} = 10^{3.76}/0.1 \text{ mL}$$

Table 4. Data used to Calculate $TCID_{50}$ by the Karber Formula

Log Dilution	No. of Positive Wells	Proportion of Wells Positive	Rate of Reaction $[p1(1-p1)]$
−0.7	8	1	0
−1.4	8	1	0
−2.1	8	1	0
−2.8	8	1	0
−3.5	5	0.625	0.234
−4.2	2	0.25	0.188
−4.9	0	0	0
−5.6	0	0	0
		$\sum p = 4.875$	$\sum p1(1-p1) = 0.422$

Because the value calculated by the Karber formula is a statistical estimate of the titer, the standard deviation s of the $TCID_{50}$ can be calculated using the following equation:

$$s = \sqrt{d}^2 \sum \left[\frac{p1(1-p1)}{(n_1 - 1)} \right]$$

where d is the $\log_{10}$ dilution factor, $p1$ is the observed rate of the reaction, and n_1 is the number of test cultures per dilution.

Therefore, for the example given in Table 4, s can be calculated as

$$s = \sqrt{0.49} \times 0.422/7$$

$$s = 0.17$$

Using this value for the standard deviation s, the 95% confidence limits can be approximated as $2 \times s$, that is, 0.34.

Calculation of virus titers by these methods is recommended by regulatory authorities for evaluating process validation studies (24). By quantitating virus titers in spiked starting material and processed material by $TCID_{50}$ assay (or mdl calculations for those processes sample with no detectable virus), an estimate of the ability of the process to remove and/or inactivate viruses can be made by the following equation:

$$R = \log_{10} \frac{V_1 \times T_1}{V_2 \times T_2}$$

where R is the reduction factor, V_1 is the volume of starting material, T_1 is the concentration of virus in the starting material, V_2 is the volume of processed material, and T_2 is the concentration of virus in the processes.

The 95% confidence limit of the reduction factor can be calculated by the following equation:

$$\pm \sqrt{(s^2 + a^2)}$$

where s is the 95% confidence limit for the titer of virus in the stating material and a is the 95% confidence limit for the titer of virus in the processed material.

DETECTION OF PRIONS

In recent decades a previously unknown mechanism of disease has been described in humans and animals. Several fatal neurodegenerative diseases, including scrapie in sheep, BSE in cattle, and CJD in humans, are now thought to be caused by the accumulation of a post-translationally modified cellular protein (15,16). CJD may present as a sporadic, genetic, or infectious illness. In the theory proposed by Prusiner, prions are described as transmissible particles that are devoid of nucleic acid and are composed exclusively of a modified protein (PrP^{Sc}). The normal, cellular prion protein (PrP^c) is converted to PrP^{Sc} through a post-translational process involving a conformational change whereby the α-helical content decreases while the β-sheet content increases, leading to significant changes

in properties. For example, PrP^c is soluble in nondenaturing detergents, whereas PrP^{Sc} is not, and PrP^c is easily degraded by proteases, whereas PrP^{Sc} is partially resistant. Stanley Prusiner recently received a Nobel prize for his prion studies, and investigations earlier by Carlton Gadjusek led also to the award of a Nobel prize for demonstrating that the neurodegenerative disease known as Kuru found in the Fore people of New Guinea was transmitted from person to person by ritualistic cannibalism and is now recognized as a prion disease (96). Cases of horizontal transmission of CJD were reported in 1974 by Duffy et al. (97) who were the first to draw attention to the iatrogenic transmission of CJD following the development of CJD in the recipient of a corneal transplant from a donor who had died from rapidly progressive dementia. Since then, other mechanisms of iatrogenic transmission have been recognized, including contaminated neurosurgical equipment, dura mater homografts, and therapy with human cadaveric pituitary hormones. It is now thought that approximately one hundred deaths worldwide have been caused by growth hormone derived from contaminated pituitary glands (98), and four deaths are associated with the use of pituitary gonadotrophins which appears to be unique to Australia (99). The pituitary hormone program was ended in the mid-eighties following the first reports of deaths, but cases continue to occur after longer and longer incubation periods following exposures that occurred mainly in the 1970s. Another mechanism of CJD transmission was proposed in 1996 by Will et al. (100) who reported the deaths of patients who were unusually young for classical CJD and had clinical and neuropathological features different from sporadic CJD; they suggested that it is caused by a novel variant of CJD derived from BSE.

Further cases since then appear to confirm that human disease can be caused as a result of bovine prions passing from BSE or "mad" cows to humans through the consumption of contaminated beef products. This prion disease has been named new variant or variant CJD (nvCJD or vCJD). This finding, if true, has serious implications for human health in general, particularly in the United Kingdom where "mad cow" disease has had its greatest effect, but it also has enormous ramifications for cell culture practices because bovine-derived products such as serum, transferrin and insulin have long been used to cultivate cells. It is now recognized that strategies designed to reduce the risk of prion transmission via cell culture technology will need to be developed and hence this summary of the prion issue in this chapter.

The prion problem has already impacted on the biologics industry in the United Kingdom where in early 1998 the Government announced that it would impose a ban on the use of all UK-derived human plasma for producing plasma-derived biologics such as albumin, immunoglobulins, and coagulation factors. Although plasma is seen as a very low risk material for transmitting prions because there has not been a case reported to date of CJD transmission via either blood transfusion or blood products, an extremely cautious approach has been taken until more is known about the infectivity of the vCJD agent. It is thought that because the agent has passed from sheep to cows by feeding cows with meat and bone meal derived from scrapie-infected sheep, and then passed to humans, it may have increased virulence if transmitted via biologics derived from humans compared to classical CJD, or perhaps a different distribution within the body leads to a higher level of variant prions in the blood and hence plasma. Until a highly sensitive and reasonably rapid test becomes available for detecting prions, the only strategy available to ensure that biologics are as safe as possible is to avoid using, wherever possible, material likely to be contaminated with bovine, vCJD, and CJD prions. For example, process materials derived from bovine sources, such as Tween 80 used for inactivation of enveloped viruses in the processing of some plasma products, should be replaced by material from nonanimal sources. In addition, excipients, such as albumin, which are often added to stabilize labile biologicals in final products should not be used if there is a risk that they may be contaminated with prions.

The implications for cell culture technology arising from BSE are clearly serious and an international meeting in 1998 under the auspices of the Council of Europe was largely devoted discussing of the prion issue (101). In summary, the majority opinion from the meeting was to avoid bovine-derived materials for use in cell culture, wherever possible, and in those situations where serum or protein-free culture was not possible, for example, the use of normal cell lines for vaccine production, to use bovine serum from BSE-free areas such as New Zealand, Australia, and the United States and to ensure that the respective government agencies employ a rigorous program to guarantee that particular herds of cattle are disease-free and that pooling or mixing of different serum batches is not allowed. In addition to avoiding the use of contaminated raw materials, it is also necessary to introduce, if possible, measures designed to remove prions from biologics (because inactivation is not an option because prions are extremely resistant to physicochemical processes and require sequential treatment of at least 1 hour in the presence of 1M NaOH followed by autoclaving at 121 °C to inactivate them (102, 103) and an appropriate assay. This brief background to the prion issue has hopefully justified the need for a section on prion assays and demonstrated why this is currently an area of intense investigation.

Prion Assays

In Vivo Assays. Prions can be detected *in vivo* by bioassays in susceptible laboratory animals. The practical use of bioassays is restricted due to the existence of a species barrier and long incubation times (several years) for most prion isolate/host combinations. Until recently, it was only possible to work with scrapie prions derived from sheep because this is the only prion material available in plentiful supply. Thus purification processes used to manufacture biologics could be assessed for their capacity to remove scrapie prions following spiking of starting material with large doses of scrapie in scale-down models of the large-scale process. The levels of scrapie prions remaining at different stages of the process are assessed by inoculating specimens into either mice or hamsters. Mouse adapted strains, either ME7, with a 200–400 day

incubation period, or 22A, with a 300–500 day incubation period, and hamster adapted scrapie, 263K with a 90–200 day incubation period, are the agents which have been used to date for spiking studies. There are clearly disadvantages with assays which generally take >200 days to complete even though they are highly sensitive. The hamster assay is faster than the mouse, assay but housing costs are higher for hamsters with the outcome that the costs for both assays are similar but high. Another major problem is that these studies must assume that the scrapie prion will have properties similar to vCJD and CJD prions, and this is unknown at present. Alternative systems are becoming available, including a mouse-adapted BSE strain for validations and transgenic mice with high levels of expression of human, bovine, or chimeric prion protein (PrP) genes which have an abrogated species barrier and significantly shorter incubation times for prions, compared with natural wild hosts. Thus it is now just becoming possible to quantify prions accurately in the brain and other organs of patients with CJD and cattle with BSE.

In Vitro Assays. Because of the high cost and durations of *in vivo* tests they are not appropriate for use in routine screening assays although they will serve as the gold standard by which all *in vitro* tests are assessed for sensitivity and specificity. At present PrP^{Sc} is the only known disease-specific diagnostic marker, and immunoassays have been developed to detect this marker using monoclonal antibodies. This is an area of intense research at present and without doubt rapid advances will occur over the next few years. Currently, Western blotting is the most common method of assay usually involving the following steps:

1. Ultracentrifuge the sample to concentrate the specimen.
2. Perform a protease K digest using a method in which the concentration of protease kinase has been optimized to reduce background staining.
3. Denature the preparation, dilute to the end point, and analyze all dilutions by Western blotting.

This assay relies on detecting the protease resistant form of PrP^{Sc} which has been shown to correlate with infectivity. The maximum sensitivity of the assay is approximately 10^2–10^3 IU, which is less sensitive than the bioassay, but the assay is rapid and cost-effective. At present, immunoassays are limited by the lack of monoclonal antibodies that are effectively specific for the PrP^{Sc} protein and the lack of reagents that can detect the presence of PrP^{Sc} in very low levels in the presence of normal prion PrP. A new assay, DELFIA (a dissociation enhanced lanthanide fluoroimmunoassay), appears to be the most sensitive assay developed to date because it claims that it detects a single infectious unit. At present protease digestion is still required, but it is anticipated that the introduction of reagents that can distinguish between PrP^c and PrP^{Sc} will obviate the need for this procedure.

SUMMARY

This article demonstrates that, in some situations at least, the lessons of history can be learned and thus influence current thinking and practices. The problems of virus contamination associated with the initial polio vaccines derived from primary monkey kidney cells described briefly in the historical introduction to this article were a major driving force in establishing the comprehensive range of screening procedures outlined in other sections. These measures have made a major contribution to the excellent safety record for the therapeutic use of biologics derived from mammalian cell culture technology. One cannot be complacent, however, because, as also described in this chapter, new infectious agents continue to be discovered, such as prions, that are considered responsible for BSE and nvCJD and porcine endogenous viruses (PERV). The rapid scientific response to the PERV problem, as measured by the research papers published since 1997 describing highly specific and sensitive assays for detecting these agents (104,105), is reassuring in that it provides evidence that this area of infectious diseases is recognized as important and that the capacity exists for appropriate action in terms of technologies and personnel. These developments augur well for the continued successful screening of cell culture contamination although vigilance is still essential because a new challenge, according to historical precedent, is bound to emerge. The rapid developments in *ex vivo* cell therapies and xenotransplantation involving the use of organs, tissues, or primary cells are a real challenge with regard to the transmission of infectious agents and will clearly need close scrutiny, using protocols described in this article, as clinical use of these procedures increases.

BIBLIOGRAPHY

1. J.F. Enders, T.H. Weller, and F.C. Robbins, *Science* **109**, 85–87 (1949).

2. J.E. Salk, *Am. J. Public. Health* **45**, 151–162 (1955).

3. A.B. Sabin, *JAMA* **164**, 1216–1223 (1957).

4. K. Shah and N. Nathanson, *Am. J. Epidemiol.* **103**, 1–12 (1976).

5. B.H. Sweet and M.R. Hilleman, *Proc. Soc. Exp. Biol. Med.* **105**, 420–427 (1992).

6. L. Hayflick and P.S. Moorhead, *Exp. Cell Res.* **25**, 585–621 (1961).

7. J.P. Jacobs, *Nature (London)* **227**, 168–170 (1970).

8. N.B. Finter and K.H. Fantes, in I. Gresser, ed., *Interferon*, Vol. 2, Academic Press, London, 1981, pp. 65–80.

9. S.E. Builder et al., in R.E. Spier, J.B. Griffiths, J. Stephenne, and P.J. Crooy, eds., *Advances in Animal Cell Biology and Technology for Bioprocesses*, Butterworth, Guilford, U.K., 1989, p. 452.

10. R.A. Weiss, *N. Engl. J. Med.* **307**, 1587 (1982).

11. P. Carthew, *J. Gen. Virol.* **67**, 963–974 (1996).

12. C. Harbour and G. Woodhouse, *Cytotechnology* **4**, 3–9 (1990).

13. C. Patience, Y. Takeuchi, and R.A. Weiss, *Nat. Med.* **3**, 282–286 (1997).

14. R.A. Weiss, *Nature (London)* **391**, 327–328 (1998).

15. S.B. Prusiner, *Science* **252**, 1515–1522 (1991).

16. S.B. Prusiner, *Science* **278**, 245–251 (1997).

17. Center for Biologicals Evaluation and Research, *Points to Consider in the Characterization of Cell Lines Used to Produce Biologicals*, U.S. Food and Drug Administration, Bethesda, Md., 1993.

18. International Committee for Harmonization Topic Q5A, *Notes for Guidance on the Quality of Biotechnological Products: Viral Safety and Evaluation of Biotechnology Products Derived from Cell Lines of Human or Animal Origin* (CPMP/ICH/295/95). 1997.

19. Center for Biologicals Evaluation and Research, *Points to Consider in the Manufacture and Testing of Monoclonal Antibody Products for Human Use*, U.S. Food and Drug Administration, Bethesda, Md., 1997.

20. Committee for Proprietary Medicinal Products, *Production and Quality Control of Monoclonal Antibodies*. (CPMP approved 13 December 1994. Doc. III/5271/94). European Commission, Brussels, 1995.

21. World Health Organization, *Requirements for Use of Animal Cells as in vitro Substrates for the Production of Biologicals* (Requirements for Biological Substances No. 50). WHO Expert Committee on Biological Standardization, WHO, Geneva, 1997.

22. Center for Biologics Evaluation and Research, *Points to Consider in Human Somatic Cell Therapy and Gene Therapy*. U.S. Food and Drug Administration, Bethesda, Md., 1991.

23. Center for Biologics Evaluation and Research, *Addendum to Points to Consider in Human Somatic Cell and Gene Therapy*, U.S. Food and Drug Administration, Bethesda, Md., 1991. *Human Gene Therapy*. **7**, 1181–1190, (1996).

24. Committee for Proprietary Medicinal Products, *Notes for Guidance on Virus Validation Studies: The Design, Contribution and Interpretation of Studies Validating the Inactivation and Removal of Viruses* (CPMP/BWP/286/95) 1996.

25. Medicines Control Agency, *Rules and Guidance for Pharmaceutical Manufacturers*, H.M. Stationery Office, London, 1997.

26. Organization for Economic Co-operation and Development, *Principles of Good Laboratory Practice*, OECD Environmental Health and Safety Publications, Environment Directorate, Paris, 1998.

27. M. Dinowitz, Y.S. Lie, and M.A. Low, *Dev. Biol. Stand.* **76**, 201–207 (1992).

28. J. Hotta and P.C. Loh, *J. Gen. Virol.* **68**, 1183–1186 (1987).

29. M. Schweizer et al., *Virology* **192**, 663–666 (1993).

30. N. Teich, D.R. Lowy, J.W. Hartley, and W.P. Rowe, *Virology* **51**, 163–173 (1973).

31. D.R. Lowy, W.P. Rowe, N. Teich, and J.W. Hartley, *Science* **174**, 155–156 (1971).

32. J.C. Olsen and J. Sechelski, *Hum. Gene Ther.* **6**, 1195–1202 (1995).

33. R. Schuurman et al., *J. Gen. Virol.* **72**, 2739–2745 (1991).

34. A. Kappeler, C. Lutz-Wallace, T. Sapp, and M. Sidhu, *Biologicals* **24**, 131–135 (1996).

35. E.L. French and W.A. Snowdon. *Aus. Vet. J.* **40**, 99–105 (1964).

36. C. McLean, D. Docherty, and A.J. Shepherd, *Eur. J. Parenteral Sci. Biotechnol., Spec. Issue*, pp. 27–30 (1997).

37. J.S. Porterfield, ed., *Andrew's Viruses of Vertebrates*, 5th ed., Baillière Tindall, London, 1989.

38. M. Plavsic, *Programme and Abstracts, International Conference, Animal Sera, Animal Sera Derivatives and Substitutes*

used in the Manufacture of Pharmaceuticals: Viral Safety and Regulatory Aspects, Strasbourg, 1998, Co-organized by European Department for the Quality of Medicines of the Council of Europe and the International Association of Biological Standardization, 1998.

39. J. Kurth et al., *Programme and Abstracts, International Conference, Animal Sera, Animal Sera Derivatives and Substitutes used in the Manufacture of Pharmaceuticals: Viral Safety and Regulatory Aspects, Strasbourg, 1998*, Co-organized by European Department for the Quality of Medicines of the Council of Europe and the International Association of Biological Standardization, 1998.

40. H.A.J.G. Kreeft, I. Greiser-Wilke, V. Moennig, and M.C. Horzinek, *Dtsch. Tieraerztl. Wochenschr.* **97**, 63–65 (1990).

41. G. Wensvoort and C. Terpstra, *Res. Vet. Sci.* **45**, 143–148 (1988).

42. C.H. Lohr, J.F. Evermann, and A.C. Ward, *VM/SAC, Vet. Med. Small Anim. Clin.* **78**, 1263–1266 (1983).

43. R. Harasawa and T. Sasaki, *Biologicals* **23**, 263–269 (1995).

44. R. Harasawa, *Vaccine* **13**, 100–103 (1995).

45. R. Harasawa and T. Tomiyama, *J. Clin. Microbiol.* **32**, 1604–1605 (1994).

46. H. Rabenau et al., *Biologicals* **21**, 207–214 (1993).

47. P.F. Nettleton and M.M. Rweyemamu, *Arch. Virol.* **64**, 359–374 (1980).

48. R.L. Garnik, in *Development in Biological Standardization. Proceedings of the IABS. International Scientific Conference on Viral Safety and Evaluation of Viral Clearance from Biopharmaceutical Products, Bethesda, Md., 1995*, Karger, Basel and New York, 1996.

49. P. Minor, N. Finter, and B. Hughes, in *Developments in Biological Standardization. Proceedings of the IABS. International Scientific Conference, on Viral Safety and Evaluation of Viral Clearance from Biopharmaceutical Products, Bethesda, Md., 1995*, Karger, Basel and New York, 1996.

50. W.L. Mengeling, in A.D Leman, et al., eds., *Diseases of Swine*, 7th ed., Wolfe Publishing, London, 1992, pp. 299–311.

51. D.L. Croghan, A. Matchett, and T.A. Koski, *Appl. Microbiol.* **26**, 431 (1973).

52. Code of Federal Regulations 9. *Animal and Animal Products*, Part 113.53 *Requirements for Ingredients of Animal Origin used for Production of Biologics* (1996).

53. *European Pharmacopoeia*, 3rd ed., Monogr. 0694. 1997. Available from EP Secretariat, BP 907 F6 7029, Strasbourg, Cedex 1, France.

54. M. Senda et al., *J. Clin. Microbiol.* **33**, 110–113 (1995).

55. J.A. Levy, *Science* **182**, 1151–1153 (1973).

56. S.J. Froud et al., in M.J.T. Carrondo, B. Griffiths, and J.L.P. Moreira, eds., *Proceedings of the 14th Meeting of the ESACT, Vilamoura, Portugal, 1996*, Kluwer Academic Publishers, Dordrecht, The Netherlands, 1997, pp. 681–686.

57. M.A. Sommerfelt and R.A. Weiss, *Virology* **176**, 58–69 (1990).

58. A.J. Shepherd et al., *Hum. Antibody Hybridomas* **3**, 168–176 (1992).

59. M. Popovic, V.S. Kalynaraman, M.S. Reitz, and M.G. Sarngadharan, *Int. J. Cancer* **30**, 93–100 (1982).

60. J.J. Hooks and C.J. Gibbs, *Bacteriol. Rev.* **39**, 169–190 (1975).

61. P.N. Blatt, R.O. Jacoby, H.C. Morse, III, and A.E. New, in *Methods Virol.* 349–388 (1986).

62. A.J. Shepherd and K.T. Smith, *Methods Mol. Biol.* **9**, (in press).

63. D.L. Kacian and S. Speilgelmen, *Methods Enzymol.* **29**, 150–173 (1974).

64. V. Klement, W.P. Rowe, J.W. Hartley, and W.E. Pugh, *Proc. Natl. Acad. Sci. U.S.A.* **63**, 753–758 (1969).

65. R.H. Bassin, N. Tuttle, and P.J. Fischinger, *Nature (London)* **229**, 564–566 (1971).

66. P.T. Peebles, *Virology* **67**, 288–291 (1975).

67. D.K. Haapala, W.G. Robey, S.D. Oroszalan, and W.P. Tsai, *J. Virol.* **53**, 827–833 (1985).

68. S.K. Chattopadhyay et al., *Virology* **113**, 465–483 (1981).

69. M.A. Lander and S.K. Chattopadhyay, *J. Virol.* **52**, 695–698 (1984).

70. B. Chesesbro et al., *Virology* **127**, 134–148 (1983).

71. G.J. Todaro et al., *Proc. Natl. Acad. Sci. U.S.A.* **70**, 859–862 (1973).

72. H.M. Temin and D. Baltimore, *Adv. Virus Res.* **17**, 129–186 (1972).

73. A.B. Maunsbach, in J.E. Celis, ed., *Cell Biology: A Laboratory Handbook*, Academic Press, San Diego, Califo, 1998, pp. 249–259.

74. A. Bremer, M. Haner, and U. Aebi, in J.E. Celis, ed., *Cell Biology: A Laboratory Handbook*, Academic Press, San Diego, Califo, 1998, pp. 277–284.

75. R.A. Killington, A. Stokes, and J.C. Hierholzer, in B.W.J. Mahy and H.O. Kangro, eds., *Virology Methods Manual*, Academic Press, London, 1996, pp. 71–89.

76. Code of Federal Regulations 21. Part 630.35. *Test for safety* (1996).

77. E.A. Gould and J.C.S. Clegg, in B.W.J. Mahy, ed., *Virology: A Practical Approach*, IRL Press, Oxford, U.K., 1985, pp. 43–78.

78. A.J. Shepherd, in J.H.S. Gear, ed., *Handbook of Viral and Rickettsial Hemorrahgic Fever*, CRC Press, Boca Raton, Fla., 1988, pp. 241–250.

79. N.G. Schmidt and R.W. Emmons, in N.J. Schmidt and R.W. Emmons, eds., *Diagnostic Procedures for Viral, Rickettsial and Chlamidial Infections*, 6th ed., American Public Health Association, Washington, D.C., 1989, pp. 1–36.

80. J.P. Clewley, *The Polymerase Chain Reaction (PCR) for Human Viral Diagnosis*, CRC Press, Boca Raton, Fla., 1995.

81. P.A. Nuttall, P.D. Luther, and E.J. Stott, *Nature (London)* **266**, 835–837 (1977).

82. P. Tattersall and D.C. Ward, *Replication of Mammalian Parvoviruses*, Cold Spring Harbor Press, Cold Spring Harbor, N.Y., 1978.

83. J.G. Tully and R.F. Whitcomb, in S. Razin and J.G. Tully, eds., *Molecular and Diagnostic Procedures in Mycoplasmology*, Vol. I, Academic Press, San Diego, Calif., 1995, pp. 137–146.

84. G.J. McGarrity, J. Samara, and V. Vanaman, *ASM News* **51**, 170–183 (1985).

85. A. Smith and J. Mowles, in S. Razin and J.G. Tully, eds., *Molecular and Diagnostic Procedures in Mycoplasmology*, Vol. 2, Academic Press, San Diego, Calif., 1996, pp. 445–451.

86. C.K. Lincoln and M.G. Gabridge, in J.P. Mather and D. Barnes, eds., *Animal Cell Culture Methods*, Vol. 57, Academic Press, San Diego, Calif., 1998, pp. 49–65.

87. G.J. McGarrity, in K. Maramorosch, ed., *Advances in Cell Culture*, Vol. 2, Academic Press, San Diego, Calif., 1982, pp. 99–131.

88. G.K. Masover and F.A. Becker, in S. Razin and J.G. Tully, eds., *Molecular and Diagnostic Procedures in Mycoplasmology*, Vol. 2, Academic Press, San Diego, Calif., 1996, pp. 419–429.

89. C. Veilleux, S. Razin, and L.H. May, in S. Razin and J.G. Tully, eds., *Molecular Diagnostic Procedures in Mycoplasmology*, Vol. 2, Academic Press, San Diego, Calif., 1996, pp. 419–429.

90. S. Razin and J.G. Tully, eds., *Molecular Diagnostic Procedures in Mycoplasmology*, Vol. 2, Academic Press, San Diego, Calif., 1996.

91. C. Thompson, Ph.D. Thesis, University of Sydney, 1994.

92. Code of Federal Regulations 21. Part 610. 12. *Test for the presence of bacteria and fungi* (1996).

93. *United States Pharmacopeia*, Rev. 23. U.S. Pharmacopeial Convention, Rockville, M., 1995, pp. 1686–1690.

94. W. Werz, H. Hoffmann, K. Harberer, and J.K. Walter, *Arch. Virol.* **13**, 245–256 (1997).

95. J. Karber, *Arch. Exp. Pathol. Pharmakol.* **162**, 480–483 (1931).

96. D.C. Gajdusek, *Science* **197**, 943 (1977).

97. P. Duffy, J. Wolf, and J. Collins, *N. Engl. J. Med.* **290**, 692–693 (1974).

98. P. Brown, *Transfusion* **38**, 312–315 (1998).

99. S. Collins and C.L. Masters, *Med. J. Aust.* **164**, 598–602 (1996).

100. R.G. Will, J.W. Ironside, and M. Zeidler, *Lancet* **347**, 921–925 (1996).

101. *Programme and Abstracts, International Conference, Animal Sera, Animal Sera Derivatives and Substitutes used in the Manufacture of Pharmaceuticals: Viral Safety and Regulatory Aspects, Strasbourg, 1998*, Co-organized by European Department for the Quality of Medicines of the Council of Europe and the International Association of Biological Standardization, 1998.

102. D.M. Taylor et al., *Arch. Virol.* **139**, 313–326 (1994).

103. D.M. Taylor, K. Fernie, and I. McConnell, *Vet. Microbiol.* **58**, 87–91 (1997).

104. W. Heneine et al., *Lancet* **352**, 695–699 (1998).

105. C. Patience et al., *Lancet* **352**, 699–701 (1998).

See also ASEPTIC TECHNIQUES IN CELL CULTURE; CONTAMINATION DETECTION AND ELIMINATION IN PLANT CELL CULTURE; CONTAMINATION OF CELL CULTURES, MYCOPLASMA; STERILIZATION AND DECONTAMINATION.

CONTAMINATION OF CELL CULTURE, MYCOPLASMA

HANS G. DREXLER
CORD C. UPHOFF
DSMZ-German Collection of Microorganisms
and Cell Cultures
Braunschweig, Germany

OUTLINE

INTRODUCTION

The contamination of cell cultures by mycoplasmas remains one of the major problems encountered in biological research and biotechnology using cells in culture. Mycoplasmas can produce extensive changes and growth arrest in cultures they infect; the possible sequelae of contamination are legion. These organisms are resistant to many of the antibiotics that are in common use in cell cultures. This problem has become more widely appreciated since the introduction of sensitive, rapid, and efficient methods for the detection of cell culture mycoplasmas. This article attempts to provide a concise review of the current knowledge on: (*1*) the main characteristics and the taxonomy of mycoplasmas; (*2*) the incidence and sources of mycoplasma contamination in cell cultures, the mycoplasma species most commonly detected in cell cultures, and the effects of mycoplasmas on the function and activities of infected cell cultures; (*3*) the various techniques available for the detection of mycoplasmas with particular emphasis on the most reliable detection methods; (*4*) the various methods available for the elimination of mycoplasmas highlighting antibiotic treatment; and (*5*) the recommended procedures and working protocols for the detection, elimination, and prevention of mycoplasma contamination. The availability of accurate, sensitive, and reliable detection methods and the application of robust and successful elimination methods provide a means for overcoming the insidious threat posed by mycoplasma contamination of cell cultures.

BIOLOGY AND NOMENCLATURE OF MYCOPLASMAS

Characteristics of Mycoplasmas

The mycoplasmas represent a large group of microorganisms that are all characterized by their lack of a rigid cell wall, a standard attribute in all other types of bacteria. Therefore, a distinct class within the prokaryotes, appropriately named *Mollicutes*, was created. For historical and practical reasons, the trivial terms mycoplasmas and mollicutes are often used as synonyms; first, prior to the introduction of the taxonomical term *Mollicutes*, the designation *Mycoplasma* had been generally employed for this class of organisms; second, the members of the subsequently established genus *Mycoplasma* are the most common and pathogenically most important organisms of this class. Thus the trivial term mycoplasmas is still widely used in the broader sense and will be used as well in this article. The first term for mycoplasmas was "pleuropneumonia-like organisms" (PPLO). For specific details regarding the biology and taxonomy of mycoplasmas, the reader is referred to specialist textbooks (1–3).

Besides their lack of a cell wall, mycoplasmas are considered to be the smallest self-replicating organisms known at present (Table 1). This notion refers not only to the size of these bacteria, but also to their physiological capabilities for synthesizing and metabolizing certain compounds, the size of their genome, and their cellular organization. Mycoplasmas contain only the vitally important cell components including membranes, ribosomes, and the bacterial chromosome. The small size of 0.3–0.8 μm in diameter and the flexibility of their cell membrane allow mycoplasmas to pass through commonly used antibacteriological filters with diameters of 0.45 μm. The latter feature and their cohabitation with other organisms prompted the initial assumption that mycoplasmas were viruses. In fact, the first established mycoplasma pathogen of humans (*M. pneumoniae*) was initially believed to be a virus. As a result of their small size and the absence of a cell wall, mycoplasmas are pleiomorphic, varying in shape from spherical or pear-shaped cells, to branched-filamentous or helical cells. Since genome replication is not synchronized with cell division, filamentous forms and chains of beads are frequently observed (4).

The size of the genome ranges from about 600 to 2,200 kb, quite in contrast to that of other bacteria with commonly more than 1,500 kb. Consequently, mycoplasmas possess only about one-fifth of the numbers of genes (~500) that are found in other bacteria; mycoplasma can be considered to most closely represent the concept of a "minimum cell." This minimal genetic constitution corresponds to their commensalic and parasitic lifestyle, which allows them to minimize the metabolic pathways used. In contrast to other bacteria, the genome has also a rather low G + C content, accounting for only about 25%, which is thought to be the minimum theoretical value to synthesize proteins with a normal amino acid composition. Only *Anaeroplasma*, *Thermoplasma*, and *M. pneumoniae* exhibit a higher G + C content (~40%) (5).

Another unique characteristic of *Mycoplasma* and *Spiroplasma* species is their use of the codon UGA, commonly representing a stop codon, as a tryptophan

Table 1. Synopsis of Salient Characteristics of Mycoplasmas

Physical characteristics
  Lack of rigid cell wall
  Smallest self-replicating bacteria; size: 0.3–0.8 μm
  Filterable: through 0.45-μm filter
  Morphology: polymorph, spherical, pear-shaped, branched, helical (*Spiroplasma*)

Biological characteristics
  Replication: binary fission (budding forms, chains of spherical cells, branched filaments due to asychronized genome replication and cell division)
  Long generation times: 1–9 h

Genomic features
  Genome size: 600 kb (*Mycoplasma*, *Ureaplasma*) –1,700 kb (*Acholeplasma*, *Anaeroplasma*, *Asteroleplasma*) –2,200 kb (*Spiroplasma*)
  G + C content: 23–41 %
  Use of UGA as tryptophan codon, not as universal termination codon (*Mycoplasma*, *Spiroplasma*)
  Number of genes/proteins: <500
  DNA polymerase complex consists of one enzyme (*Mycoplasma*, *Ureaplasma*) or three enzymes (*Acholeplasma*, *Spiroplasma*)

Metabolism
  Require sterol or cholesterol (except *Acholeplasma*, *Asteroleplasma*)
  Require several basic compontents (e.g., vitamins, nucleic acid precursors, lipids, fatty acids, amino acids) in species-specific compositions

Generation of energy
  Fermentative organisms catabolize carbohydrates with acid production
  Most nonfermentative organisms hydrolyze arginine-producing ammonia
  Ureaplasma hydrolyzes urea

Lifestyle
  All mycoplasmas are commensales or parasites; many are pathogens
  Extracellular cytadherence to eukaryocytic cells (but *M. penetrans* and others also intracellular)
  Infect humans, animals, plants

Detection
  Mycoplasma infections are chronic and diffcult to diagnose
  Mycoplasmas form typical "fried egg" colonies on agar (except fastidious "noncultivable" species and strains)

codon. They share this feature with mitochondria, which may imply a phylogenetic relationship. On the other hand, certain mycoplasmas, for example, *Acholeplasma*, employ the normal codon composition, which suggests that the change in codon usage appeared after the evolutionary separation of the *Mycoplasma*/*Spiroplasma* from the *Acholeplasma*. Overall, the mycoplasmas display a great genetic diversity; this may be due either to phylogenetic diversity or to a rapid evolutionary development of the mycoplasmas caused by their commensalic and parasitic existence.

Due to their limited metabolic capabilities, mycoplasmas require a number of nutrients for survival: nucleic acid precursors, amino acids, vitamins, and other basic components. The growth of all mycoplasmas (except for

Acholeplasma and *Asteroleplasma*) depends on the presence of sterol and cholesterol. Moreover, they also need to take up lipids from the environment for the formation of cell membranes. Physiologically, mycoplasmas can be differentiated into two groups, the fermentative and the nonfermentative species. The fermentative strains gain energy and carbon by fermentation of carbohydrates (hexoses, starch, glycogen) mainly to lactic acid, pyruvic acid, acetic acid, and acetylmethylcarbinol, which leads to a lowering of pH in the culture medium. In contrast, the nonfermentative species do not significantly decrease the pH of their environment, because they oxidize fatty acids and alcohols. But they are not able to metabolize carbohydrates via the glycolysis pathway. The nonfermentative and some of the fermentative species are able to hydrolyze arginine and to produce adenosine triphosphate (ATP), CO_2, and ammonia, which is toxic for eukaryotic cells. *Ureaplasma* hydrolyze urea in order to generate ATP. These organisms are sensitive to pH changes, which in fact they may cause themselves; this observation must be taken into account during attempts to culture them.

The characteristics described previously are presumably the reason for our inability or difficulty to culture many mycoplasma species thus far identified in chemically defined media. Clearly, we have failed to mimic in vitro the natural milieu provided by the host so that not all these fastidious organisms can be cultivated. Most successful culture methods use media that include beef heart infusion, peptone, or tryptone (as sources of proteins and amino acids); glucose, arginine, or urea (as sources of energy); yeast extract (as a source of vitamins and growth factors); and serum (as a source of sterol and cholesterol). Contrary to other bacteria, mycoplasmas grow very slowly, even under optimal conditions. The generation times usually range between one and three hours, but there are also generation times of up to nine hours; in addition, mycoplasmas have a relatively long lag phase. Therefore, it may take more than one week to obtain visible colonies on agar. Mycoplasmas also do not overgrow other organisms.

It has long been assumed that mycoplasmas exist only on the outside of the eukaryocytic cell membrane (cytadherence). However, studies in recent years have unequivocally demonstrated the intracellular location of certain mollicutes (notably *M. fermentans*, *M. genitalium*, and *M. pneumoniae*), not only after phagocytosis by granulocytes and monocytes, but also in nonphagocytic epithelial cells. Even a new mycoplasma species capable of entering a variety of human cells in vivo and in vitro was discovered and was named accordingly *M. penetrans*. Mycoplasma cytadherence and invasion appear to be active, but separable, processes. Extensive invasion of cells by *M. penetrans* eventually leads to cell disruption. The percentage of the mycoplasma population able to invade the cells depends apparently on the mycoplasma species and may be influenced by the eukaryotic cell type and by culture conditions. While the great majority of the infecting mycoplasma population is definitely located extracellularly, the intracellular location, even for only a short period, sequesters mycoplasmas and protects them effectively from mycoplasmacidal therapies. This phenomenon may also explain the difficulty of eradicating mycoplasmas from all infected cell cultures (6).

The dependency on many specific organic compounds force mycoplasmas to live in close relationship with other organisms. They can be found as commensales and parasites in almost all higher organisms including plants, insects, and vertebrates. Mycoplasmas are nearly perfectly adapted parasites, rarely killing their hosts. Mycoplasma infections are thus of a more chronic nature and are very frequently difficult to diagnose on the basis of the symptoms alone. In humans, mycoplasmas can be physiologically found on the mucous surface of the respiratory, gastrointestinal, and urogenital tract, the eyes, mammary glands, and joints, whereas ureaplasmas occur solely in the urogenital tract. Most species can be regarded as part of the normal microbial flora; a minority, however, may cause diseases in humans and animals. *M. pneumoniae*, for instance, is the causative agent of the primary atypical pneumonia (prevalent particularly in children), while *M. arthritidis* appears to be involved in the pathogenesis of arthritis. All farm animals appear to suffer from mycoplasma infections, leading to significant economic losses. *Anaeroplasma* and *Asteroleplasma* are regularly observed in bovine and ovine rumen. These two genera are obligatory anaerobic. Insects are the hosts of a variety of mycoplasmas that includes species from the genera *Mycoplasma*, *Acholeplasma*, and *Spiroplasma*, most of which also can cause diseases in their hosts. But all these species seem to be clearly distinct from those isolated from vertebrates.

Mycoplasmas also cause several hundred economically important plant diseases transmitted by insects. *Acholeplasma* and *Spiroplasma* are seen on or in plants. Some *Spiroplasma* species invade the sieve tubes of the plants, possibly giving rise to diseases as a result of obstruction. Insects are known to transfer *Spiroplasma* into the plants. The *Acholeplasma* species detected on plants also seem to derive from depositions by insects. While a large spectrum of various plant infectants apparently belongs to the class *Mollicutes*, it has not been possible to culture them in vitro; microscopic observations suggest a certain resemblance to mycoplasmas (hence they were termed mycoplasma-like organisms, MLO, or phytoplasmas).

Another difficult diagnostic problem concerns the detection of mycoplasmas infecting human and animal cell cultures; such contamination is a rather frequent occurrence. Typical for mycoplasma contaminations is that the symptoms are rarely very pronounced as the infection takes a chronic course. Also in this field, the damage occurring in biological research and production due to the use of contaminated cultures is scientifically and economically substantial.

Taxonomy of Mycoplasmas

Due to fundamental differences compared to other bacteria, the *Mollicutes* constitute a class of their own. The class is divided into six genera within three orders based on morphological and physiological criteria and on the size of the genome: *Mycoplasma*, *Ureaplasma*, *Spiroplasma*, *Acholeplasma*, *Anaeroplasma*, and *Asteroleplasma* [Fig. 1(a)]. Previously, the order

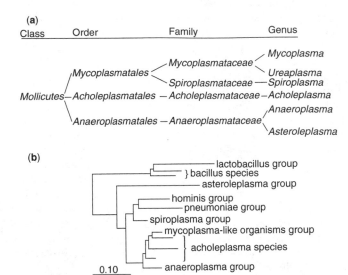

Figure 1. Taxonomy of mycoplasmas: Shown are both the conventional taxonomy (**a**) and the molecular biological taxonomy (**b**) of the mycoplasmas. The bar indicates an evolutionary distance of 10%. For specific details see Refs. 2,7.

Thermoplasma was assigned to this class on the basis of their lack of a cell wall. However, the members of this order are phylogenetically clearly separated from the other *Mollicutes* and are now assigned to the kingdom Archeae. In 1995, more than 150 species within the class *Mollicutes* were recognized; the genus *Mycoplasma* contains the largest number of recognized species (>100); some of the best-studied species are further subdivided into a multitude of strains (2,3). The mollicutes that have already been characterized and taxonomically defined constitute only a part, apparently a minor one, of the mollicutes living in nature. With the advent of new microbiological and molecular biological methods in the field of mycoplasma taxonomy, it is likely that many more species will be detected.

Extensive research has been performed in order to clarify the phylogenetic relationship of the mycoplasmas within their class to each other and beyond their class to the other bacteria. For instance, sequence analysis of the highly conserved 16S mRNAs shows that mycoplasmas are not "primitive bacteria" with poorly developed metabolic pathways, but that they apparently intentionally reduced their genome size and their synthesizing properties during their adaptation to cohabitation with other organisms. The molecular investigations employing degrees of similarities in the nucleotide sequences suggested that the closest relatives of the *Mollicutes* are the Gram-positive *Clostridia, Bacillus,* and *Lactobacillus.* Within the class *Mollicutes,* sequence analysis discerned six distinct groups: asteroleplasma, hominis, pneumoniae, spiroplasma, MLO (phytoplasma), and anaeroplasma [Fig. 1(b)]. The former genus *Ureaplasma* was assigned to the pneumoniae group. Species of the genus *Acholeplasma* do not represent a monophyletic group on their own; certain species of this genus are phylogenetically most closely related to the anaeroplasma group or the MLO group. Members of the genus *Mycoplasma* were assigned to one of the three newly created groups: the hominis group,

the pneumoniae group, or the spiroplasma group. In this molecular biological taxonomy model, the genus *Asteroleplasma* stands alone and is relatively unrelated to the other species. The phenotypic and genotypic diversities of mycoplasmas imply a rapid evolutionary development that appears to be faster than that of other bacteria, but their phylogenetic position is now not as isolated as assumed previously (5,7).

MYCOPLASMA CONTAMINATION OF CELL CULTURES

Incidence of Mycoplasma Contamination

Cell culture of primary cells and continuous cell lines of human or animal origin faces several challenges. Among these are infections with microorganisms and in particular with mycoplasmas that may significantly hamper the use of these cells. Mycoplasmas were first isolated from a contaminated cell culture in 1956. Ironically, during the investigations of the effects of a mycoplasma infection on cell cultures, it was found that the uninoculated negative control cell culture was already contaminated (8). Soon thereafter, it became well established that stable, continuous cell lines are frequently contaminated. One mycoplasma cell can grow to 10^6 CFU/ml within three to five days in an infected cell culture. Eukaryotic cell cultures contaminated with mycoplasma have titers in the range of 10^6 to 10^8 organisms/ml. Frequently, there are from 100 to 1000 mycoplasmas attached to each infected cell (cytadsorption). Contamination may initially go undetected because mycoplasma infections do not produce overt turbid growth as commonly seen with other bacterial or fungal contamination. Often mycoplasmas can achieve very high densities in cell cultures without causing a change in the pH (and therefore without a color change of most culture media). Some mycoplasmas produce very little overt cytopathology, and covert contamination may remain undetected for months. Virtually all reports and research on mycoplasma contamination of cell cultures concerns human and animal cells.

While mycoplasmas have been shown to be present on or in fresh plants (see preceding discussion), nothing has been published concerning the occurrence of mycoplasmas in plant cell cultures. It is not known whether mycoplasmas truly are not present in these cultures or such a contamination has never been studied in plant cell cultures. It is unlikely that human contact is the source of a possible contamination as those mycoplasmas found in plants, namely spiroplasmas or MLOs (phytoplasmas) do not infect humans. Furthermore, all media used in plant cell cultures are chemically produced and properly processed solutions. The primary plants from which the callus or suspension cultures are established are theoretically the primary source of mycoplasma contamination. Principally, no type of cell culture, whether it be vertebrate, invertebrate, or plant, is safe from mycoplasmal infection. Clearly, specific studies are required to provide some data in this unexplored area.

In general, primary cell cultures and cultures in early passage are less frequently contaminated than continuous cell lines: primary cultures and early passage cultures

Table 2. Incidence, Most Common Species, and Sources of Mycoplasma Contamination

Incidence
  15–35% of continuous cell lines
  5% of early passage cell cultures
  1% of primary cell cultures

Most common species
  M. orale (frequency 20–40%; natural host: human)
  M. hyorhinis (10–40%; swine)
  M. arginini (20–30%; bovine)
  M. fermentans (10–20%; human)
  M. hominis (10–20%; human)
  A. laidlawii (5–20%; bovine)

Sources
  Original (primary) tissue isolate (< 1%)
  Culture reagents (mostly bovine serum)
  Laboratory personnel
  Cross-contamination from infected cultures (most common source)

on the order of 1% and 5%, respectively; continuously cultured cell lines in the range of 15–35% (Table 2). Several large studies, mainly in the United States, on tens of thousands of cell cultures analyzed over several decades (1960s–1980s) found an incidence of ca. 15%. However, more recent studies on smaller series and reports from other countries in South America, Europe, and Asia documented significantly higher infection rates of cell cultures, commonly in the range of 15–35%, but also as high as 65–80% (8–11).

The ever-expanding application of cell lines in research and biotechnology (with the resulting exchange of nonauthenticated and mycoplasma-positive lines between scientists) and the increasing use of certain antibiotics (mostly penicillin plus streptomycin, which merely serve to mask but do not remove mycoplasmas) in routine culture have presumably led to this increase in mycoplasma contamination of cell cultures, which has now reached epidemic proportions. It appears that contamination rates are higher for cultures continuously grown with such "cell culture antibiotics."

Most Common Contaminating Mycoplasma Species

As described in detail previously, mycoplasmas can be found nearly ubiquitously, as long as the given environment offers sufficient amounts of nutrients and the essential compounds. Commonly, such permissive conditions exist in higher organisms. As the list of mammalian, insect, and plant hosts for mycoplasmas becomes longer, the number of isolated mycoplasma species and strains is also expanding rapidly. Most of the well-known and best-studied mycoplasmas are pathogenic infectants of humans, economically relevant animals (such as cattle), or agriculturally important plants. Within the past two decades, mycoplasmas have been recognized as one of the major cell culture problems; consequently their biology and physiology have become topics of intense research.

While at least 20 distinct species have been isolated from contaminated cell lines, detailed investigations on

the identity of the contaminating species showed that by far the largest portion of infections is caused by a relatively small number of *Mycoplasma* and *Acholeplasma* species: 90–95% of the contaminants were identified as either *M. orale*, *M. hyorhinis*, *M. arginini*, *M. fermentans*, *M. hominis*, or *A. laidlawii* (Table 2). Depending on the study, the individual percentages of these six species may vary. Although many mycoplasmas are host-specific in vivo, the various species are not restricted to cell cultures derived from their natural hosts. Species that are rarely found in cell cultures (generally <1%), but that are still consistently detected in the reported series, include the following: *M. alkalescens* (natural host: bovine); *M. arthritidis* (rodent); *M. bovis* (bovine); *M. buccale* (human); *M. canis* (dog); *M. gallisepticum* (avian); *M. genitalium* (human); *M. pirum* (bovine); *M. pneumoniae* (human); *M. pulmonis* (rodent); and *M. salivarium* (human).

Mycoplasmas from human sources are the most prevalent group and account for approximately one-third to one-half of all strains isolated. Generally *M. orale*, which is the most common mycoplasma species in the oral cavity of clinically normal humans, also represents the single most common isolate, accounting for 20–40% of all mycoplasma infections in cell cultures (Table 2). Other nonpathogenic mycoplasma species from the normal human microbial flora of the oropharynx that are seen in cell cultures are *M. fermentans* and *M. hominis*. *M. salivarium* and *M. buccale*, normally detected in the human oral cavity, have also, albeit rather rarely, been identified as cell culture contaminants.

The bovine group of mycoplasmas accounts for about another one-third of all strains isolated from cell cultures. Here the most frequent infectants are *M. arginini* and *A. laidlawii*. These two species have a relatively wide host range, as they are isolated from cattle, sheep, goat, etc. and from a variety of other mammals, birds, and insects. These cell culture contaminants are thought to derive from bovine sources, as in the early days of cell culture (1950s–1970s) the bovine sera were not routinely and as strictly screened for mycoplasma contamination as they are today. *M. hyorhinis*, a common inhabitant of the nasal cavity of the swine, also accounts for a high proportion of the infections. Trypsin is usually extracted from porcine pancreas and via this indispensable reagent, swine mycoplasmas may have found their way into cell cultures; however, it appears that trypsin destroys mycoplasmas; furthermore, no mycoplasmas have been reliably detected in trypsin preparations. On the other hand, because swine and cattle are processed through the same abattoirs, the swine strain of *M. hyorhinis* may have been introduced into bovine sera; *M. hyorhinis* has often been isolated from bovine sera.

Sources of Mycoplasma Contamination

Studies of the natural habitat of mycoplasmas implicate the human technician, bovine serum, and the primary cultures (Table 2). Tissue specimens used to initiate cell cultures are not the major sources of mycoplasma infection. The frequency of infection in primary cell cultures is low, on the order of 1% (8,10). Furthermore, murine, avian, and canine species of mycoplasma only account for 0.5–1% within the panel of mycoplasmas isolated from contaminated cultures—despite the wide use of murine cell lines.

The high incidence of bovine mycoplasma species, predominantly A. laidlawii and M. arginini, implicates the fetal or newborn bovine serum. Studies in the 1960s and 1970s showed that 25–40% of the serum lots provided by commercial suppliers were contaminated (8). In order to prevent contaminations and to eliminate those adventitious agents that are most frequently detected in bovine serum, namely, mycoplasmas and bovine viral diarrhea viruses (BVDV), suppliers now apply more sophisticated sterility test procedures, use γ-irradiation and filtration, and have implemented stricter rules and controls at the abattoirs. While bovine serum contamination has certainly significantly decreased over the past 10–20 years, due to problems inherent in all sterility test procedures (for instance, sample size of large lots), serum lots absolutely free from mycoplasmas cannot be guaranteed. As there is no legal requirement for suppliers to provide mycoplasma-free products, bovine serum should still be considered as a possible source of contamination.

Because the largest percentage of mycoplasmas found in cell cultures are of human origin, it is logical to assume that the laboratory personnel is one of the major sources of contamination. If the human cell culturist, the bovine serum, or the primary culture were the only sources of infection, then there should be random mycoplasmal contaminations in culture laboratories, and only some cultures within a laboratory unit should be mycoplasma-positive. However, this is not the pattern of contamination that has been observed. Clearly, in laboratories with contaminated cells, most or all cultures are positive, containing the same mycoplasma species (10). Thus mycoplasma-infected cell lines are themselves the single most important source for further spreading of the contamination. This is due to the ease of droplet generation during handling of cell cultures, the high concentration of mycoplasmas in infected cultures (10^6 to 10^8 CFU/ml of supernatant), and the prolonged survival of dried mycoplasmas. While mycoplasma-positive cell cultures account for more mycoplasma infections now than bovine sera and laboratory personnel together (4,8,10), the cell culture technician is of course involved in the propagation of the infection, as the mycoplasmas themselves do not "jump from culture to culture" on their own. Operator-induced contamination is a multifaceted problem (see also the following). Suffice it to mention here that mycoplasmas are spread by using laboratory equipment, media, or reagents that have been contaminated by previous use in processing mycoplasma-infected cells.

Effects of Mycoplasma Contamination

The in vitro culture of cells has opened many doors in research; unfortunately, it has also brought new problems for the researcher. One of the biggest problems, mycoplasma contamination, resembles the opening of Pandora's box, as mycoplasma infections can have myriad different effects on the contaminated cell cultures.

Table 3. Effects of Mycoplasma Contamination on Cell Cultures

General effects on eukaryotic cells
    Altered levels of protein, RNA, and DNA synthesis
    Alteration of cellular metabolism
    Induction of chromosomal aberrations (numerical and structural alterations)
    Change in cell membrane composition (surface antigen and receptor expression)
    Alteration of cellular morphology
    Induction (or inhibition) of lymphocyte activation
    Induction (or suppression) of cytokine expression
    Increase (or decrease) of virus propagation
    Interference with various biochemical and biological assays
    Influence on signal transduction
    Promotion of cellular transformation
    Alteration of proliferation characteristics (growth, viability)
    Total culture degeneration and loss
Specific effects on hybridomas
    Inhibition of cell fusion
    Influence on selection of fusion products
    Interference in screening of monoclonal antibody reactivity
    Monoclonal antibody against mycoplasma instead of target antigen
    Reduced yield of monoclonal antibody
    Conservation of hybridoma

However, this multitude of different effects does not affect the various cells in the same manner and to the same degree (Table 3). Many mycoplasma species produce severe cytopathic effects, while others may cause very little overt cytopathology. There can be qualitative and quantitative differences in the same parameter, depending on the infecting mycoplasma species, the culture conditions, the type of the infected cell culture, the intensity and duration of the infection, an additional infection with viruses, and other parameters. Thus contaminations can interfere with virtually every parameter measured in cell cultures during routine cultivation or in experimental investigations. Consequently, the mycoplasmas in these cultures cannot simply be ignored or regarded as harmless bystander organisms. Besides the loss of an important culture, in the worst case all experiments might be influenced by the infections and artefacts produced. Because of the virtually unlimited number of reported mycoplasmal effects on cultured cells, only some of the most important parameters will be described here in order to highlight the diversity of possible effects. Clearly, the term mycoplasma infection is a rather general term, similar to virus infection (12).

Mycoplasmas have an enormous spectrum of cytopathogenic effects on cells, ranging from no noticeable change in morphology to minimal alterations characterized by an increase in cytoplasmic granularity to a clear cytopathic effect with occasional cell lysis. A characteristic change in nuclear morphology often associated with mycoplasmal contamination is the appearance of condensed chromatin, nucleolar aggregation, and other signs of imminent apoptosis. These morphological effects can be reversed by changing the medium or by replenishing with fresh nutrients. As with morphology, the effects of mycoplasmal contamination upon cell growth are extremely variable, ranging from none to complete cessation of proliferation depending upon cell type, mycoplasmal species, and concentration of mycoplasmas. Even enhancement of cell growth has been reported. Strongly infected cultures are characterized by stunted, abnormal growth and rounded, degenerated cells with a macroscopically "moth-eaten" appearance at the edge of monolayers. Detachment of monolayers due to mycoplasma contamination has also been observed (4).

One of the most obvious reasons for these effects is that mycoplasmas consume the nutrients in the culture medium, depleting glucose, arginine, nucleic acid precursors, amino acids, vitamins, etc., thereby being direct competitors of the cultured eukaryotic cells. Often the concentration of the mycoplasmas in the medium or on the cells exceeds the number of eukaryotic cells by more than a thousand times. The depletion of medium nutrients results in profound effects on cell metabolism and function. The mycoplasmas usually metabolize the nutritional elements significantly faster than their hosts. They further induce the host cells to produce nucleotides, as has been demonstrated for infections with *M. orale*, in which deoxyadenosine is highly increased and converted to ribonucleotides that are incorporated into mycoplasmal RNA.

Moreover, the metabolites can be harmful for the eukaryotic cells, for example, by acidifying the medium (fermentative mycoplasmas producing acidic metabolites) or by producing ammonia (arginine-hydrolyzing mycoplasmas). Ammonia is known to be a highly toxic agent and has been shown to inhibit cell growth in a variety of cell lines. In addition to this effect, the depletion of the basic amino acids arginine and glutamine may result in alterations in protein synthesis and inhibition of proliferation; they may also lead to chromosomal aberrations because these amino acids are used in high amounts in the histones of the eukaryotic nuclei. Depletion and alterations of nucleic acid precursors can produce mutations in the eukaryotic DNA; host cell DNA can be degraded by mycoplasmal nucleases and endonucleases. In vitro induction of chromosomal aberrations such as chromosome breakage, multiple translocation events, and numerical chromosome changes were seen in various cell cultures infected with different mycoplasma species (13).

A biotechnologically important effect of the nucleotide-transforming mycoplasmal enzymes is the activity of the uridine phosphorylase, which can inactivate the artificial bromodesoxyuridine (BrdU). This reagent is universally applied as a thymidine analogue for the selection of cells with a thymidine kinase defect; the presence of this defect is exploited in cell fusion techniques. Thymidine kinase phosphorylates BrdU, which is then incorporated into eukaryotic DNA, causing cell death. As the BrdU is metabolized by mycoplasmas, it may appear that cells growing in the presence of BrdU carry this enzyme defect. In hematopoietic cell cultures, several mycoplasma species are known to stimulate (or inhibit) activation, proliferation, and differentiation, enhance immunoglobulin secretion by B-cells, and induce expression and elaboration of cytokines and growth factors, for instance, interleukin-1β (IL-1β), IL-2, IL-6, tumor necrosis factor (TNF), and various colony-stimulating activities (e.g., GM-CSF) (12).

A final example of the detrimental effects of mycoplasma contamination concerns the propagation of viruses in cell culture. Some mycoplasmas have no detectable effect on viral growth, while others may decrease, or even increase, virus yields in infected cultures. For example, the titer of the arginine-dependent DNA viruses is significantly reduced by mycoplasma contamination. In some cell lines, mycoplasmas can either enhance or block expression of interferon-γ, which inhibits propagation of viruses in cells. Certain virus-like cytopathic effects caused by mycoplasmas may be falsely ascribed to the presence of viruses. Like viruses, mycoplasmas are filterable, hemabsorbant, hemagglutinant, resistant to certain antibiotics, inhibited by antisera, able to induce chromosomal aberrations, and sensitive to detergents, ether, and chloroform (4).

DETECTION OF MYCOPLASMA CONTAMINATION

Various Detection Methods

Over the past four decades a vast array of techniques have been developed to detect mycoplasma contamination of cell cultures (Table 4). Most of these methods are relatively lengthy, involve subjective assessments, and use measurements that are often quite complex in nature. The two classical detection methods, DNA fluorochrome staining and microbiological colony assay, together with the recently introduced and highly specific and sensitive methods RNA hybridization and polymerase chain reaction, will be described in greater detail in the following. Ideal detection methods should be highly sensitive and specific, but also simple, rapid, efficient, and

Table 4. Various Mycoplasma Detection Methods

Histological staining
  Histochemical stains and light microscopy

Electron microscopy
  Transmission electron microscopy
  Scanning electron microscopy

Biochemical methods
  Enzyme assays
  Gradient/electrophoresis separation of labeled RNA
  Protein analysis

Immunological procedures
  Fluorescence/enzymatic staining with antibodies
  ELISA
  Autoradiography

DNA fluorochrome staining
  DAPI stain
  Hoechst 33258 stain

Microbiological culture
  Colony formation on agar

RNA hybridization
  Filter hybridization
  Liquid hybridization

Polymerase chain reaction
  Species-/genus-specific PCR primers
  Universal PCR primers

cost effective. The evaluation of tests comprises the components' validity and reproducibility pertaining to the statistical parameters (operating characteristics): sensitivity (detection of true positives), specificity (detection of true negatives), accuracy (detection of true positives and true negatives — combination of sensitivity and specificity), and predictive value (probability of correct result). Further aspects of the evaluation of a test are technical reproducibility and interobserver reproducibility (concordance of interpretation of the results). Many detection methods rely on subjective reading and interpretation of the results, which obviously requires training, experience, and consensus (14).

Traditionally, mycoplasmologists developed direct and indirect detection methods. While the term direct method referred to the classical microbiological colony growth of mycoplasmas on agar, indirect detection techniques included procedures that measure a gene product that is associated with mycoplasmas rather than with the mammalian cells in culture. Furthermore, tests may be performed directly on the specimen taken from a given cell culture or indirectly using the so-called indicator cell culture procedure, whereby the specimen is inoculated into another cell culture known to be free of mycoplasmas (monkey cell line Vero and murine cell lines NIH 3T3 or 3T6 have been used with equal success). Use of an indicator cell culture promotes better standardization and allows appropriate positive and negative controls to be included in each assay (8,10). Mycoplasmologists advise identifying every cell culture mycoplasma isolate as identification presents a clearer picture of the nature of the infection and may be helpful in determining its source. Identification can be achieved with various immunological techniques (e.g., using species-specific antibodies in immunostaining or ELISA) or nowadays species-specific polymerase chain reaction (PCR) primers. However, to the cell culturist, the chief concern appears to be whether the culture is free from contamination rather than specific identification of the contaminant.

The first detection methods involved histological staining and light microscopy, applying the classical histological stains hematoxylin–eosin, May–Grünwald–Giemsa, and acridine orange. Some of the larger cytadsorbing mycoplasmas (e.g., *M. hyorhinis*) are observable as small pleomorphic bodies lining the infected cell membrane. Smaller organisms and mycoplasmas that do not readily cytadsorb to cultured cells (e.g., *M. orale*) are more difficult to detect with histochemical techniques. Cell debris, organelles, surface structures, etc. may mimic mycoplasmal bodies. An alternative method for the visual demonstration of mycoplasmas uses hypotonic treatment of infected cells, fixation, and orcein staining followed by phase-contrast microscopy. The major limitation is the probable insensitivity of histological stainings when small numbers of mycoplasmas are present.

Mycoplasmas have been studied extensively by electron microscopic techniques, both scanning and transmission electron microscopy. Mycoplasmas can usually be distinguished from viruses, bacteria, yeast, rickettsiae, and chlamydiae. While these techniques are quite useful for mycoplasma basic research and certainly provide extremely interesting and beautiful pictures, the diagnostic value of electron microscopy for mycoplasma detection is, however, limited by the complex technology and the need for expensive equipment and trained staff. Mycoplasmas can often be confused with cytoplasmic blebs and a large number of possible artefacts; low-level contamination is essentially not detectable with electron microscopy; large numbers of cultures cannot be assayed. Electron microscopy can clearly not be regarded as a routine mycoplasma detection test.

A variety of biochemical detection methods have been developed, including colorimetric, chromatographic, and biological enzyme assays; separation of radioactively labeled DNA or RNA by density gradient or gel electrophoresis; and protein analysis (two-dimensional electrophoresis and isoenzyme analysis). The enzyme assays are based on the detection of an enzymatic activity present in mycoplasmas, but minimal or absent in tissue cells. The enzymatic activities exploited include arginine deiminase; thymidine, uridine, adenosine, or pyrimidine nucleoside phosphorylase; and hypoxanthine or uracil phosphoribosyl transferase. Arginine deiminase catalyzes the conversion of arginine to citrulline, which can be measured colorimetrically. Unfortunately, this enzyme activity is a property of nonfermentative mycoplasmas, and consequently most fermentative species that include common culture contaminants cannot be detected by this method. Nucleoside phosphorylase is responsible for the cleavage of thymidine to thymine with subsequent release of deoxyribose, which is measured colorimetrically. However, certain mycoplasma species give negative results. A chromatographic technique measures uridine phosphorylase activity: Cells to be tested are lysed and incubated with $[^{14}C]$-uridine; the lysate is chromatographed; the conversion of $[^{14}C]$-uridine to $[^{14}C]$-uracil is determined. Cleavage of other nucleosides (e.g., thymidine, adenosine) by mycoplasmal phosphorylases but not by mammalian cells has also been reported, taking advantage of the avid uptake of exogenous nucleosides by mollicutes. However, the difficulty with these approaches is that mycoplasmas vary widely in their ability to incorporate nucleic acid precursors (10,15).

A biological enzyme assay utilizes a purine analogue, 6-methylpurine deoxyriboside (6-MPDR). This nontoxic analogue of adenosine is converted by adenosine phosphorylase into 6-methyl purine and 6-methyl purine riboside, both of which are toxic to mammalian cells. Mycoplasma generally have much higher levels of adenosine phosphorylase than do cultured eukaryotic cells and, accordingly, hydrolyze 6-MPDR more rapidly. The release of toxic products from 6-MPDR in the presence of mycoplasma is detected by including a sensitive indicator cell line (for instance, murine cell line 3T6). However, the interpretation of the results is highly subjective. False-negative results can be explained by mycoplasma species and strains having either no adenosine phosphorylase activity or activities below the sensitivity level of the assay and further by differences between the cell lines regarding sensitivity to the metabolites (14).

Mycoplasmal RNA can be distinguished from mammalian RNA. One technique relies on the identification of

[^{3}H]-uridine-labeled mycoplasmas after sucrose density gradient centrifugation, whereby mycoplasma-positive cultures give a distinguishable peak that is not observed in mycoplasma-free cultures. Another method measures uridine and uracil uptake by mycoplasmas. After polyacrylamide gel electrophoresis, results are expressed as the ratio of uridine to uracil incorporated. Several species were difficult to detect, and a number of false-negatives were reported. Taken together, positive biochemical results are based on arbitrary values, making low orders of mycoplasma contamination difficult to detect. Not all mycoplasma contaminants possess higher distinguishing levels of enzymatic activity. Overall, the biochemical detection methods give inconsistent results when comparing different cell lines and have lower sensitivities than, for instance, microbiological cultivation and DNA fluorochrome staining methods, making their use problematic (14,16).

Immunological methods use mainly species-specific polyclonal antisera and monoclonal antibodies to detect and identify mycoplasmas, applying immunofluorescence, immunoenzymatic, or radioimmunoprecipitation techniques, immunobinding onto nitrocellulose paper, or enzyme-linked immunosorbent assays (ELISA). These specific immunoassays are used primarily for mycoplasma speciation. Several ELISAs have been shown to produce false-negative results in mycoplasma detection, as only certain species are addressed. While ELISAs are easy to perform, their use is hampered by the relatively weak sensitivity of this technique (lowest detection level of 10^7 CFU/ml). A monoclonal antibody has been produced that recognizes a common antigen (elongation factor Tu) shared by most mycoplasmas; this protein, essential for all prokaryotes, mediates the transport of aminoacyl-tRNA to the ribosomes. In theory, the approach using a "pan-mycoplasma" antibody with fluorescence staining has considerable potential, as it might be possible to detect single cells harboring few mycoplasmas. However, the interpretation of the stainings is very subjective and difficult to reproduce between observers (16). Autoradiography of infected cells previously labeled with [^{3}H]-thymidine or uridine reveals extranuclear localization (throughout the cytoplasm and cell periphery) of silver grains. This method suffers from a lack of sensitivity; another limitation of autoradiography is the exposure time needed. Finally, one method exploits the growth-inhibitory effects of mycoplasma contamination on eukaryotic cells: supernatant from the culture to be tested is added to an indicator cell line; the eventually resulting growth modulation is quantitated by [^{3}H]-thymidine β-scintillation counting. This method is certainly not very sensitive regarding low-level infections, is very subjective, and is time-consuming.

Most Reliable Detection Methods

Until the arrival of RNA hybridization and PCR, DNA fluorochrome staining and microbiological culture had been regarded as the "gold standards" for detection of mycoplasma contamination (8,10). While the classical methods certainly have kept their eminent place in the armamentarium of mycoplasma detection methods,

the new approach to a highly sensitive, specific, and rapid diagnosis of mycoplasmal infection is based on the development of gene or DNA probes that were first introduced in the 1980s. The principle is simple as genes or genomic sequences that are specific for a single species or a particular group of mycoplasma or universally for nearly all mycoplasmas are identified and synthesized; these probes are used for DNA or RNA hybridization. The more recent development of PCR enables the amplification of the target DNA in the specimen using specific synthetic oligonucleotides complementary to conserved rRNA sequences and increases the sensitivity by several orders of magnitude (17). Positive hybridization signals as dots on filters or scintillation counts and visual demonstration of the PCR amplicon in gels indicate the presence of the infectious agent. In the following, all four procedures, the classical and the modern molecular biological techniques, will be presented in greater detail.

DNA Fluorochrome Staining. Fluorescent dyes binding to DNA were first used to detect mycoplasmal infection in the 1970s. The two primary DNA-binding stains are 4'-6-diamidino-2-phenylindole (DAPI) and bisbenzamide (Hoechst 33258); further, but rather rarely employed, stains are the olivomycin fluorescent dye and a conjugated benzoxazinone kanamycin fluorescent probe. The fluorescent dyes bind rapidly and selectively to eu- and prokaryotic DNA, forming strongly fluorescent DNA-DAPI/Hoechst complexes with high specificity. Under fluorescence microscopy, an uncontaminated cell culture shows only nuclear fluorescence against a dark background. Mitochondrial DNA does bind the fluorochrome, but at levels imperceptible by routine microscopy. Mycoplasmas, however, which have approximately ten times the DNA content of mitochondria, are readily detected as bright foci over the cytoplasm, lining the cell membrane or in intercellular spaces (Fig. 8). This procedure is not diagnostic for mycoplasmas, as other prokaryotic nonmycoplasma contaminants will also be detected. To overcome problems associated with the analysis of many different cells, to detect low-level contaminations, and to screen potentially infected sera, the use of an indicator cell line such as murine cell line 3T6 or monkey cell line Vero is recommended (the so-called indirect DAPI/Hoechst staining). Specimens to be analyzed are inoculated into the indicator cell culture, and after an appropriate incubation period, the indicator cell line is examined under UV light.

DNA fluorochrome staining will detect titers of 10^5 organisms or greater per milliliter. False-positive results can be caused by staining artefacts or cell detritus. This test is dependent on the intensity of the infection, since only massively contaminated cultures can be identified, thus leading to false negatives. The majority of false-positive and false-negative readings are due to equivocal stainings with the ensuing difficulties in the (subjective) interpretation of the results and borderline decisions. Use of an indicator cell line significantly enhanced sensitivity, specificity, accuracy, and predictive value of the DNA fluorochrome assays (14). Here, the main disadvantage appears to be the necessity to carry permanently an

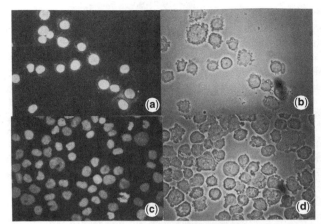

Figure 2. Detection of mycoplasma contamination by DAPI-DNA fluorescence staining. The single cell human suspension cell line K-562 that was massively contaminated with mycoplasma (*M. hyorhinis*) is shown prior to (**a**) and after (**c**) treatment with the antibiotic enrofloxacin (Baytril) that completely eliminated the contaminant. The same viewfields (**a/b** and **c/d**) are shown under fluorescent and phase-contrast light microscopy, respectively. The fluorochrome DAPI binds to any DNA present on the cytospin slide. The large fluorescent (blue) bodies are eukaryotic cell nuclei. The mycoplasmas are seen either as small dots surrounding the eukaryocytic cells or as dense clouds between cells. For technical details see Ref. 16. Photographs were taken with a Nikon Labophot photomicroscope using a 60× objective.

adherent cell line (however, the cell lines Vero and 3T6 are relatively easy to maintain and have undetectable or few background artifacts). Advantages of the fluorochrome tests are that they are inexpensive, simple, and rapid and can be applied for regular screening and long-term monitoring.

Microbiological Culture. For decades the mainstay of mycoplasma detection was based on standard microbiological culture procedures. Specimens are inoculated into mycoplasma broth and onto agar. Anaerobic incubation is recommended, as aerobic incubation yields a lower detection rate. Broths are transferred to agar plates after 4–7 days of incubation. Most mycoplasmas produce microscopic colonies (100–400 µm in diameter) with a "fried egg" appearance growing embedded beneath the surface of the agar (Fig. 3). Because the contaminants grow embedded, they maintain their morphological shape. Consequently, they are easily distinguishable from bacterial colonies. Certain mycoplasmas may produce a more diffuse, granular type of colony (8,10).

This procedure has the advantage of ease of manipulation and visual recognition of colonies. Some artifacts are occasionally seen on agar after inoculation of cell culture specimens: Colonies must be distinguished from cell clumps by their eventual increase in size; pseudocolonies (e.g., crystals, air bubbles) can be a problem for the inexperienced observer, since they can increase in size and can actually be transferred. There are wide variations in size, morphology, and speed of growth of the mycoplasma colonies isolated from different cell cultures. Colonies usually become detectable after an average of 3–6 days, but also sometimes later. In 1973, the existence of "noncultivatable" mycoplasmas were reported and were recognized as *M. hyorhinis* (8). These contaminants grow poorly or not at all on well-standardized broth and agar media. A large percentage of *M. hyorhinis* strains does not propagate at all on cell-free medium.

While the microbiological culture has the advantages of being inexpensive, highly sensitive with a high detection rate, and an established and important reference method, the disadvantages are the long incubation time, the need

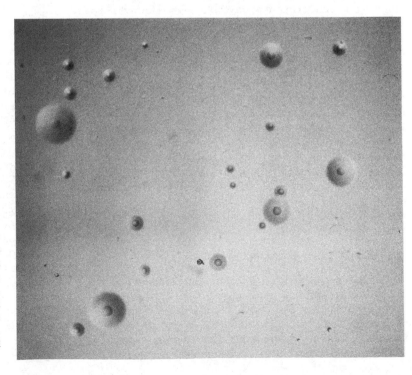

Figure 3. Detection of mycoplasma contamination by colony growth on agar. Mycoplasmas (*M. arginini*) from the human suspension cell line U-937 were first enriched in a liquid broth medium and then plated on agar to allow for formation of characteristic colonies. Shown is a typical field of an agar plate incubated for one week. Note the characteristic "fried-egg"-type mycoplasma colonies. For technical details see Ref. 16. Photograph was taken with a Nikon Diaphot inverted microscope using a 3.2 × objective.

for a subjective and experienced interpretation, and the fact that not all mycoplasmas can be successfully cultured.

Molecular RNA Hybridization. The assays of this category of mycoplasma detection are based on the principle of nucleic acid hybridization: complementary nucleic acid strands come together to form stable double-stranded complexes. The highest sensitivity is achieved in DNA–RNA hybridization, which is commonly used, but also DNA–DNA hybridization would work satisfactorily. The availability of rRNA sequences in databases allows for the construction of species-specific, group-specific, or universal oligonucleotide probes to cover all mycoplasmas commonly isolated from cell cultures. Two unique features make rRNA the most suitable target for probes: the general organization of the molecule with conserved, semiconserved, and variable regions; and the high copy number of rRNA present in each mycoplasma cell (about 1×10^4). The probes represent sequences of the 16S or 23S rRNA genes. Some bacterial species may also be detected by these tests. RNA from eukaryotic cells will not hybridize with the probes.

The liquid solution hybridization (best exemplified by the commercially available kit "Gen-Probe," Mycoplasma T.C.) uses a [^{3}H]-labeled single-stranded DNA probe homologous to 16S and 23S rRNA sequences. After the rRNA is released from the organism, the [^{3}H]-DNA probe combines with the target rRNA to form a stable DNA–RNA hybrid that is then separated from the nonhybridized DNA probe; positive signals are measured in a scintillation counter. The second method uses filter hybridization, whereby cell culture samples are simply heat-fixed on the membranes, hybridized with the [^{32}P]dCTP-labeled probe derived from 16S or 23S rRNA sequences, and incubated for autoradiography (Fig. 4) (18). The detection limits are in the range of 10^3 to 10^4 organisms; the filter method is reportedly more sensitive than the liquid hybridization.

The advantages of these assays are broad specificity, high sensitivity, convenient sample preparation, processing of large sample numbers, and rapid results. The validity and reproducibility of the Gen-Probe test are very high (14). Drawbacks are the use of radiolabeled probes and the relatively high costs.

Polymerase Chain Reaction. The first reports on the application of PCR to diagnosis of mycoplasma infections appeared in 1989. The PCR technique is based on repeated cycles of high-temperature template denaturation, oligonucleotide primer annealing, and thermostable polymerase-mediated extension; the number of DNA molecules doubles after each cycle. Nearly all mycoplasmic 16S rRNA sequences have been determined and form the basis for a systematic phylogenetic analysis of mollicutes (see the preceding). Computer alignment studies of mollicute 16S rRNA sequences reveal regions with sequence variability or conservation at the species, genus, or class level, allowing for the selection of appropriate oligonucleotides (primers) for detection and identification of mycoplasmas. The highly conserved regions of the genes enable the selection of primers of wide specificity

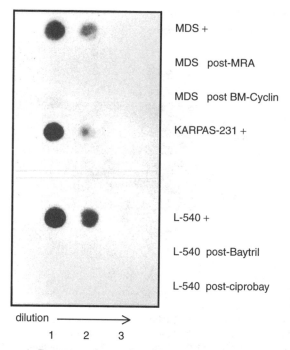

Figure 4. Detection of mycoplasma contamination by filter RNA hybridization. Shown is DNA–RNA filter hybridization using the 23S rRNA probe H900. The samples were applied in a three-fold dilution series (10× each) from left to right (arrow, rows 1–3); the last dilution is faintly visible. Whereas the untreated human hematopoietic cell lines MDS, L-540, and KARPAS-231 are clearly positive (+), the four other samples (post-treatment with various mycoplasmacidal antibiotics) are negative. For technical details see Ref. 18.

("universal primers"), which will react with DNA of any mycoplasma or even with the DNA of other prokaryotes; this is satisfactory for detection of mycoplasma cell culture infection where the goal is just to screen the cultures for contamination. Besides the conserved regions of mycoplasmal 16S rRNA genes, the 16S–23S intergenic regions are quite useful for mycoplasma detection (17,19). It is important that no cross-reactions with the DNA of cell lines occur.

As PCR can be achieved with frozen and lyophilized material, this offers a means for a retrospective analysis and facilitates the transport, collection, and storage of samples. The amplification may be performed as a single-step PCR (Fig. 5) or as a two-step (nested) PCR; the latter approach increases the sensitivity and specificity considerably, but also increases the risk of contamination by DNA carryover. Southern blotting of PCR products and hybridization with a specific radiolabeled internal probe is another possibility to improve sensitivity. However, a very high sensitivity level may not usually be required in routine diagnosis as acute and particularly chronic cell culture mycoplasma infections involve a large number of organisms (commonly 10^6 or higher). The very high sensitivity may be of advantage at a very early phase of infection or under conditions where mycoplasma growth is suppressed (e.g., in the testing and monitoring of cell cultures post-treatment with antibiotics for elimination of the contaminants).

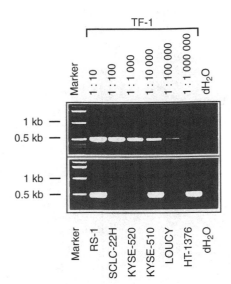

Figure 5. Detection of mycoplasma contamination by PCR. Mycoplasma infection is documented here by a specific band in the ethidium bromide-stained agarose gel. The upper panel demonstrates the sensitivity of the method in a dilution experiment whereby the cell culture supernatant of a mycoplasma-infected human leukemia suspension cell line (TF-1) is diluted with buffer; note the distinct band even at a dilution of 1:100,000. The lower panel shows three mycoplasma-negative and three mycoplasma-positive human adherent or suspension cell lines. For technical details see Ref. 20. Marker and dH$_2$O denote size markers and negative water control, respectively.

The detection limit using a set of nested universal primers was determined to be 1 fg mycoplasmal DNA, which is equivalent to 1–2 genome copies of the 16S mRNA coding region (mollicute genomes carry only one or two rRNA gene sets). The ability to detect a single mycoplasma cell makes PCR the most sensitive detection method available, clearly more so than microbiological culture. In theory, a positive cell culture may be derived from a single mycoplasma cell; in practice, however, for several reasons (including mycoplasma cell aggregates, multinuclear filamentous forms, defective or nonviable cells), a "successful" infection requires an inoculum equivalent to about 100 to 1000 cells (17).

The high sensitivity of PCR may cause problems in producing false-positive results due to contamination with target DNA. Another possible problem is false-negative data caused by the inhibition of the *Taq* polymerase by components in the samples. However, once all PCR-related problems are properly addressed, single-step or double-step PCR is clearly superior to other mycoplasma detection methods in many respects, as this method combines simplicity and speed with high specificity and extreme sensitivity, in addition to objectivity, accuracy, and reproducibility (20). In particular, PCR is not limited by the ability of an organism to grow in culture; in certain areas, this molecular nucleic acid amplification may eventually replace biological amplification (i.e., growth in artificial media), a feature of paramount importance considering the fastidious nature of mycoplasmas (17). Thus PCR should prove to be the technique of the future

for mycoplasma detection in cell cultures. Several PCR kits are commercially available.

ELIMINATION OF MYCOPLASMA CONTAMINATION

Various Elimination Methods

Ever since mycoplasma contamination of cell cultures was first reported, attempts to develop methods for elimination of the mycoplasma have been made. It has been suggested that efforts to eradicate mycoplasmas from contaminated cells should be considered as a last resort (in order to prevent spread of the contaminant) and that it would often be far better to eliminate the problem completely by autoclaving the infected cultures and replacing them with fresh stocks known to be mycoplasma-free (11). However, all too often the cell line is not replaceable with a mycoplasma-free aliquot, and purging of mycoplasmas from such cultures is a necessity.

Four general types of procedures have been used to eliminate mycoplasmas from infected cell cultures: physical, chemical, immunological, and chemotherapeutic treatment (Table 5). Many of the methods have been shown to be unreliable. Some techniques may apply to some, but not all, mycoplasma species; some of them are too laborious or simply impractical. Elimination is typically time consuming, is often unsuccessful, and poses risks of secondary infection to other cell cultures. Methods of elimination should ideally be simple and easy, rapid and efficient, reliable and inexpensive, have minimal effect on the eukaryotic cell, and result in no loss of specialized characteristics; accidental cloning selection of treated cells also should not occur. However, there is clearly not a single method available that is both 100% effective and fulfills all the ideal requirements.

The effectiveness of some elimination methods has been investigated only in experimentally contaminated

Table 5. Various Mycoplasma Elimination Methods

Physical procedures
  Heat treatment
  Filtration through microfilters
  Induction of chromosomal or cell membrane damage with
    photosensitizing
Chemical procedures
  Exposure to detergents
  Washings with ether-chloroform
  Treatment with methyl glycine buffer
  Incubation with sodium polyanethol sulfonate
  Culture in 6-methylpurine deoxyriboside
Immunological procedures
  Co-cultivation with macrophages
  In vivo passage through nude mice
  Culture with specific antimycoplasma antisera
  Exposure to complement
  Cell cloning
Chemotherapeutic procedures
  Antibiotic treatment in standard culture
  Antibiotic treatment plus hyperimmune sera or co-cultivation
    with macrophages
  Soft agar cultivation with antibiotics

cell cultures, yet experimentally infected cultures may not realistically reflect the laboratory situation, since chronic infections certainly result in complex interactions between mycoplasmas and cells. If a cleanup is attempted, it is imperative closely to monitor the effectiveness of treatment relative to mycoplasma elimination and eukaryotic cytotoxicity. A variety of procedures have been described and utilized. Administration of antibiotics is by far the most common and efficient approach and will be discussed in greater detail in the following.

The physical elimination methods include prolonged heat treatment at $40-42\,^{\circ}C$ and filtration through microfilters ($100-200$ nm). For obvious reasons, both approaches do not seem to be very promising. Another procedure takes advantage of the nutritional requirements of mycoplasmas offering them 5-bromouracil (or 5-bromodesoxyuridine), which is then selectively incorporated into the mycoplasmal genome. Exposure to visible light induces chromosomal breaks leading to cell death; this photosensitivity was greatly enhanced by the binding of the fluorochrome Hoechst 33258 to the DNA (8). The method is labor-intensive and time-consuming (requiring several rounds of treatments and subsequent cloning), can be highly cytotoxic to the eukaryotic cells, may produce cellular mutants, and was reported to activate endogenous C-type retroviruses in murine cells. A variation of the latter method is the use of the lipophilic fluorescent probe Merocyanine 540, which is incorporated into the cell membrane of mycoplasmas. Again illumination with visible light of cultures incubated with Merocyanine 540 together with Hoechst 33258 leads to a significant degree of mycoplasma eradication. The relative merit of this technique has not yet been established.

The chemical procedures involve the exposure of the cultures to various reagents for a given amount of time, once or repeatedly. Described are treatments with methyl glycine buffer, sodium polyanethol sulfonate, and 6-methylpurine deoxyriboside. Treatment with detergents such as Triton-X 100 were reported to be effective. Another method uses washings with 2 : 1 parts ether and chloroform overnight in the cold. Overall, these chemicals will hardly be selective in harming only the mycoplasmas without untoward sequelae for the eukaryotic cells.

Immunological elimination methods exploit the physiological mechanisms of the normal immune responses, using either cellular components of the immune system with their intrinsic abilities of phagocytosis (in vitro in culture dishes or in vivo in mice) or some soluble mediators of immunity, namely, complement and antibodies (immunoglobulins). Mycoplasma-infected cells are co-cultivated with primary and purified human or murine macrophages extracted from the peripheral blood, bone marrow, or peritoneum; some protocols suggest the addition of antibiotics; others use pooled human immunoglobulin. Nude mice, which still possess macrophages, but cannot mount a full immunological response due to the lack of T-lymphocytes, have been used for the in vivo passage of mycoplasma-contaminated cell lines in the form of an ascites tumor. This method requires animal facilities. Culture with macrophages, in vivo or ex vivo, will reduce the mycoplasma content, although it does not appear to be sufficient to eliminate mycoplasma completely; furthermore, it carries the risk of introducing murine viruses (e.g., C-type viruses) from the murine host.

Considerable efforts have been spent in producing high-titer neutralizing antimycoplasma antisera in rabbits, guinea pigs, and cows. They clearly do not work in every instance and simply reduce the mycoplasma load temporarily; thus their elimination efficacy is rather low. Incubation with complement contained in fresh human serum has been suggested as a rapid and efficient procedure building on the destructive interaction between components of the classical complement pathway and mycoplasmas.

Concerning the efficiency of immunological mycoplasma purging methods, it is of note that the mycoplasma membrane, being exposed to the external environment, is the cell organelle that comes into contact with components of the host immune system (in the mammalian organism). Lacking the protection of a rigid cell wall, mycoplasmas should be particularly sensitive to growth inhibition and lysis by antibodies and complement. Yet, despite this apparent exposure to a possible immunological attack, mycoplasma infections are usually chronic in nature, indicating the frequent failure of the host defense mechanisms to eradicate the parasites. Mycoplasmas may have certain mechanisms at their disposal for rapid adaptation to the microenvironment and thus can escape any immunological actions (5). These in vivo observations may explain the lack of success in using immunological approaches for mycoplasma decontamination of cell cultures. Finally, cell cloning of an infected culture in 96-well plates may lead to a mycoplasma-free culture, based on the possibility of selecting cells without any cytoadsorbed contaminants. The success rate is generally very low, while the amount of work involved is clearly very high.

The physical, chemical, and immunological methods are usually of restricted value since the mycoplasmas, although not detected for sometime afterwards, commonly reappear (low efficiency); some methods are time-consuming or have detrimental effects on the eukaryotic cells; other techniques are complex or impractical, as they require extensive resources or special equipment and knowledge. Considering the various advantages and disadvantages of any elimination procedure, chemotherapeutic treatment appears to be superior to the other mycoplasma eradication techniques and thus the method of choice. The simple addition of another reagent to a cell culture is within the technical and financial capabilities of every cell culture laboratory.

Antibiotic Treatment as the Most Effective Method

Mycoplasmas, which lack a cell wall and are incapable of peptidoglycan synthesis, are theoretically not susceptible to antibiotics such as penicillin and its analogues that are effective against most bacterial contaminants of cell cultures. However, it has been reported that several bacteriostatic antimicrobial agents inhibit the growth of mycoplasmas; thus they may not eradicate the contaminants, but simply suppress the overall infection and mask the presence of mycoplasmas. A number of different antibiotics have been used explicitly for mycoplasma control (Table 6). The contaminant strains, however, often develop resistance to certain antibiotics, which were thus completely ineffective. Other antibiotics (for instance, some aminoglycosides and lincosamides) are moderately to highly effective in eliminating mycoplasmas, but only at concentrations that have detrimental effects on the eukaryotic cells, such as marked cytotoxicity.

Ideally, a basic procedure should involve isolating, speciating, and determining the antibiotic susceptibility of the contaminants to the arsenal of possible reagents to maximize success; then the cultures should be exposed to the effective antibiotics. However, this approach is extremely time-consuming and labor-intensive and requires certain expertise. It might be fair to say that few scientists (for whom a cell culture is normally only a means to an end) would use this complex approach and would prefer a quicker solution.

The pharmacological and clinical testing and application of mycoplasmacidal antibiotics have shown that tetracyclines are generally effective antimycoplasmal agents; quinolones have also been found to be highly effective against mycoplasmas. Recently, these new antibiotics have been introduced for purging of mycoplasmas from cell cultures and are marketed commercially. Of particular note are the quinolones ciprofloxacin (distributed as Ciprobay by Bayer), enrofloxacin (Baytril; Bayer), and an unpublished quinolone reagent available as Mycoplasma Removal Agent (MRA; Flow Laboratories ICN); the product BM-Cyclin (Boehringer Mannheim) combines the macrolide tiamulin (a pleuromutilin derivative) and the tetracycline minocycline (Fig. 6) (21.22).

Tetracyclines inhibit protein synthesis by binding to subunits of ribosomes, thereby blocking peptide chain elongation. Tetracyclines inhibit both prokaryotic and eukaryotic ribosomal protein synthesis. Nalidixic acid is the prototype of a class of synthetic antibacterial agents, the fluoroquinolones, or simply quinolones. The mode of action of the quinolones involves the binding to and inhibition of the bacterial DNA gyrase, which is essential for DNA replication, transcription, repair, and recombination. Despite their documented selectivity for prokaryotic enzymes, the quinolones may also exert an inhibitory effect on eukaryotic DNA polymerase α, topoisomerases, and DNA deoxynucleotidyl transferases; the activity of these enzymes is especially high in rapidly dividing cells. Indeed, high doses of quinolones induced double-strand DNA breaks in human cells. Selectivity of the quinolones for the bacterial cell is at least partly due to the far greater sensitivity of the bacterial enzymes compared to the mammalian enzymes.

Antibiotic treatment of mycoplasma contamination in cell cultures has a high degree of efficiency, as illustrated by the percentages of successful outcomes in various studies: 65–74% of the mycoplasma-positive cultures were cured with MRA; 74–100% with Ciprobay; 76–91% with Baytril; and 82–100% with BM-Cyclin (22). Thus two-thirds to three-quarters of all cultures treated can be cleansed by one of these antimycoplasma chemotherapeutic regimens. Besides cure, two other possible outcomes are loss of the culture and resistance.

Table 6. Various Antimicrobial Agents against Mycoplasma Contamination in Cell Cultures

Antibiotic category: Specific antibiotic	Mode of action	Effect on bacteria	Effectiveness against mycoplasmas[a]	Cytotoxicity for eukaryotic cells[b]
Aminoglycosides	protein synthesis inhibitor	bacteriocidal		
gentamicin			+	−
kanamycin			+	+
spectinomycin			+	+
Tetracyclines	protein synthesis inhibitor	bacteriostatic		
chlortetracyline			+	(+)
doxycycline			+	(+)
minocycline			++	(+)
Lincosamides	protein synthesis inhibitor	bacteriostatic		
clindamycin			+	+
lincomicin			+	+
Macrolides	protein synthesis inhibitor	bacteriostatic		
erythromycin			(+)	−
roxitromycin			−	−
tiamulin (pleuromutilin)			++	−
tylosin			+	+
Quinolones	nucleic acid synthesis inhibitor	bacteriocidal		
ciprofloxacin			++	−
enrofloxacin			++	−
sparfloxacin			−	−
Novobiocin	cell wall inhibitor	bacteriostatic	−	−

[a] + + = highly effective; + = moderately effective; (+) = rarely or weakly effective; − = ineffective.
[b] Cytotoxicity at effective mycoplasmacidal antibiotic doses: + = cytotoxic; (+) = mildly cytotoxic or cytostatic; − = not cytotoxic.

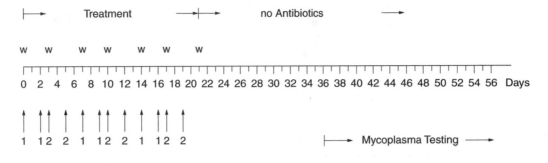

BM-Cyclin 1: 10 μg/ml

BM-Cyclin 2: 5 μg/ml

Figure 6. Treatment protocol for elimination of mycoplasma contamination using BM-Cyclin. BM-Cyclin containing the macrolide tiamulin (1) and the tetracycline minocycline (2) (see Table 6) are added to the cell culture for three weeks in three alternating cycles (as indicated in this example by the arrows). Cells are thoroughly washed (W) prior to switching to the other reagent, at the beginning and at the end of treatment. Post-therapy, it is important to grow the culture in antibiotic-free medium. After a minimum of two weeks, mycoplasma testing is performed. Thereafter, cells are regularly monitored for re-emergence of the contamination or for new infection. For technical details see Refs. 18,21,22.

Culture death, occurring in 5–15% of the cases, is presumably caused by cytotoxic effects of the reagents. Not unexpectedly, BM-Cyclin (containing a tetracycline) shows the greatest growth-inhibiting effect, which may be either cytostatic or cytotoxic. Generally, one week after cessation of treatment, cell growth will return to normal. Cytostatic and cytotoxic effects of the antibiotics may be enhanced by the poor condition of cell cultures commonly found in chronically infected cells. This situation is clearly different from that of experimentally contaminated cell cultures. It is found that increasing the serum concentration and incubating the cells at higher densities (at or near clonal densities) is advantageous for the cell cultures.

The quinolones show cross-resistance. This is not surprising given their basic structural similarity. Fortunately, sequential administration of BM-Cyclin to the same cells that were first exposed to a quinolone can still result in eradication of the resistant infectant. Higher concentrations of the antibiotics may be more effective in purging mycoplasma-contaminated cultures decreasing the rate of resistance, but this success would be counterbalanced by significantly higher cytotoxicity. It is not known whether the resistance of cell culture mycoplasmas to antibiotics is mostly acquired during treatment or already exists prior to exposure to these reagents. In any event, antimycoplasma antibiotics should be reserved for the specific situation of mycoplasma eradication in an infected cell culture and should not be added routinely as a common panacea to the culture medium, as this would run a substantial risk of selecting resistant mycoplasma strains. The discovery of intracytoplasmatic mycoplasmas may explain the resistance of some cases where prior testing documented susceptibility of the particular contaminant to antibiotics. In these instances, the antibiotics mentioned do not necessarily have physical access to the mycoplasmas. New mycoplasmacidal compounds are needed that are capable of penetrating the eukaryotic cells and that are suitable for killing the intracellular microorganisms without untoward effects on the host cell.

Antimycoplasma treatments are certainly stressful to the eukaryotic cells. Thus cells might no longer express the desired properties as a result of antibiotic administration. Outgrowth of a selected clone is another possibility. Some data suggest that cured cells generally preserve their characteristics. Still, any alterations to the cell lines induced by antibiotic treatment are obviously a matter of concern and require further detailed studies.

Overall, the technically simplest method for mycoplasma decontamination with the most promising results is antibiotic treatment. The convenience of use and general availability of the reagents render it a reliable routine laboratory procedure. However, antibiotic mycoplasma elimination is laborious and time-consuming, as the duration of the treatment plus the minimum antibiotic-free post-treatment period ranges from three to five weeks depending on the protocol used. Furthermore, special attention must be placed on possible cytotoxic effects or effects that alter the characteristics of the cell line.

Variations of the antibiotic eradication procedure involve the pretreatment or simultaneous exposure to hyperimmune antimycoplasma serum and co-culture with macrophages (Table 5). The idea behind these techniques is to eliminate the bulk of the contaminants prior to the addition of the antibiotics or to combine different approaches that may complement one another in the efficacy with which they eliminate residual infectants. Another treatment method that would improve the cure rate involves the use of soft agar with antibiotics. The soft agar technique consists of suspending infected cultures in soft agar containing appropriate antibiotics. The method is based on the survival of cells in suspension in the soft agar, thus facilitating exposure of mycoplasmas to the antibiotics. The advantage of the method lies in its speed and simplicity, as only one cycle of treatment (1–3 days) is adequate.

WORKING PROTOCOLS FOR DETECTION, ELIMINATION, AND PREVENTION OF MYCOPLASMA CONTAMINATION

There are three main components to be considered when regarding the problem of mycoplasma contamination of cell cultures: detection, elimination, and above all prevention of mycoplasma infection. In the following, several routine protocols are presented with regard to these three fundamental issues. Mycoplasmal contamination is now and will always be a potential threat to cell cultures. Overall, detection methods have dramatically improved in recent years, and the availability of testing kits has increased. Instead of discarding mycoplasma-infected cultures (without any possible clean replacement), elimination of the contaminants has become a realistic and practical possibility. Critical to maintaining cell cultures free of mycoplasma (and other adventitious organisms) is the establishment of routine operating and testing procedures by the cell culturist. The primary difficulty in controlling the epidemic mycoplasma contamination lies in convincing the culture operator of the necessity for validating the absence of these covert organisms and in creating a certain level of awareness for the dangers that lurk.

Routine Working Protocol for Detection

The high frequency of mycoplasma-contaminated cell cultures and the rapid spread of the infection within the laboratory imposes a requirement for rigorous mycoplasma testing of any new incoming cell culture and also for regular screening of every permanently cultured cell line. The best test will fail if the sample submitted for testing is not representative or is not treated properly. Antibiotics can mask mycoplasma infection, resulting in false-negative test results. Therefore, while it is highly desirable that cell lines always be cultured without any antibiotics, for mycoplasma testing purposes it is essential that cell cultures be maintained antibiotic-free for several passages. Furthermore, cultures should be assayed a sufficient time after the last passage or exchange of medium; this allows organisms, if present, to grow to high titers. While PCR is definitely the test of choice (see the preceding discussion), this method can also produce false-positive or false-negative results. Thus conducting at least two (or, better, three) different techniques on the same sample is suggested to obtain a reliable diagnosis.

A routine working protocol for detection of mycoplasma contamination in the form of a flow chart is shown in Figure 7. Newly arriving cell lines should be quarantined in a separate cell culture laboratory. The initial testing determines the mycoplasma contamination status of the cells. If the culture is contaminated and no clean replacement is available, a decontamination might be attempted followed by strict retesting post-treatment (see the following). It is absolutely mandatory that clean cultures also be regularly assayed for mycoplasma contamination, for instance, at monthly intervals. PCR with its extreme sensitivity and DNA fluorochrome staining (DAPI/Hoechst stains) as an inexpensive, simple, and rapid method can be applied for regular screening.

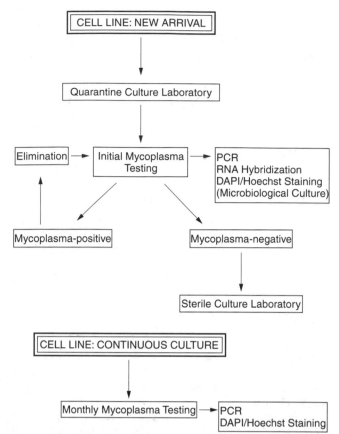

Figure 7. Routine working protocol for detection of mycoplasma contamination. A flow chart for mycoplasma testing of newly arriving and continuously cultured cell lines is shown. Use of at least two (better three or even four) different mycoplasma detection assays is suggested; best suited for determining the initial mycoplasma contamination status are PCR, RNA hybridization, and DAPI/Hoechst staining (in that order), while for routine screening and continuous monitoring PCR and DAPI/Hoechst staining are recommended; it is useful to check continuous cell cultures at monthly intervals. For technical details see Refs. 14,16,18,20.

Routine Working Protocol for Elimination

If a cell culture is found to be infected with mycoplasmas, the advised procedure is to replace it whenever possible; frozen stocks can be examined, or cell culture repositories can be contacted. An invaluable or irreplaceable cell line can be cured, as decontamination with various methods has become a practical and highly successful endeavor. However, a number of factors should be weighed carefully. The treatment protocol plus the obligatory intensive monitoring thereafter is a laborious and time-consuming task. The possibility exists that upon exposure to potentially toxic conditions, the cell line will be subjected to selective pressure, and the cured population might differ from the original. Also, the procedure entails extensive work with mycoplasma-positive cultures, thus carrying the risk of further spread of the contamination. Chemotherapeutic purging of cultures applying various antimycoplasma antibiotics has been successful. A flow diagram of the antibiotic treatment protocol is shown in Figure 8.

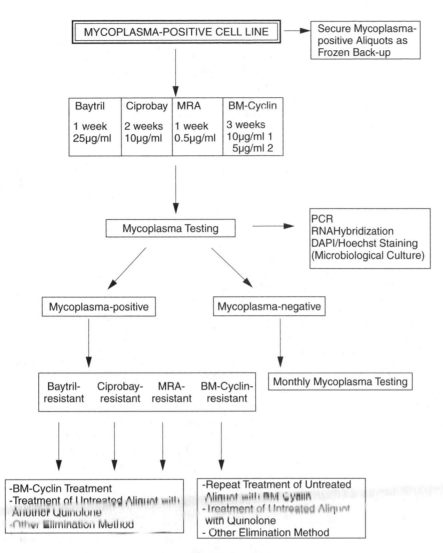

Figure 8. Routine working protocol for elimination of mycoplasma contamination. A suggested flow chart for planning antibiotic treatment of mycoplasma-positive cell lines is shown. It is proposed that aliquots of the contaminated cultures be frozen; these serve as a backup in case the initial antibiotic administration fails to achieve freedom from mycoplasma contamination. Cells may be treated with one, several, or all of the four regimens indicated, but in separate culture vessels as single agent treatment (the antibiotics should not be mixed). After an antibiotic-free phase of at least 2 weeks post-treatment, cells must undergo several rigorous mycoplasma testings with at least two methods and must be examined afterwards at periodic short intervals (for instance monthly). In the case of antibiotic resistance, resistant cultures may be treated a second time with an effective non-cross-resistant reagent, or backup (still contaminated) cultures may be treated in another attempt, or other purging methods may be used. For technical details see Refs. 18,21,22.

As a general rule, if cell cultures are diagnosed as being positive for mycoplasma, as quickly as possible they should be either discarded, frozen, and stored in liquid nitrogen, or treated with antimycoplasma procedures in order to prevent spread of the contaminant. The whole decontamination process should be undertaken in a quarantined culture laboratory until sterility is attained. It is advisable to freeze some cells for further elimination attempts, in case the first treatment fails to achieve freedom from mycoplasma contamination. The basic principle of antibiotic treatment involves exposure of the positive culture to a recommended antibiotic concentration for a given number of days, followed by maintenance of the cells for at least two weeks in antibiotic-free medium, and then the close monitoring of the mycoplasma status by adequate tests to detect residual or new contaminants. Resistant cell lines may be rescued by subsequent incubation with a non-cross-resistant antibiotic (for instance, quinolone-resistant cells can be cured with BM-Cyclin). An alternative would be a second attempt with an untreated, mycoplasma-positive aliquot of the same culture using different antibiotics or higher concentrations. The application of elimination methods other than antibiotics remains a further possibility.

The recommendation by commercial suppliers that mycoplasmacidal antibiotics should be used as a permanently preventive measure to avoid contamination must be strongly rejected. Antibiotics in general and antimycoplasma reagents in particular should not be used as a substitute for good cell culture technique.

Routine Working Protocol for Prevention

Of primary importance in preventing mycoplasma contamination of cell cultures is an awareness of sources of infection combined with a program of cell maintenance, regular quality and identity checks, and repeated characterization of the cells (23). The protocol for prevention of mycoplasma contamination should be designed to minimize risk of exposure of cells to mycoplasmas. As pointed out previously, human sources and bovine serum constitute two of the main infectant reservoirs; undoubtedly, lateral spread from one cell line to another accounts for the majority of infections. A variety of specific steps can be adopted to prevent cell culture contamination with mycoplasmas (Table 7) (9–11).

Any sterile cell culture work should be performed in a vertical laminar-flow biohazard hood. It is critical

Table 7. Routine Working Protocol for Prevention of Mycoplasma Contamination

Cell culture facility
  Facility should be designed and equipped for aseptic culture procedures
  Certified laminar-flow biological safety cabinets should be used and their function should be regularly examined
  Work surfaces should be chemically disinfected prior to and following work and thoroughly cleaned at regular intervals (monthly)
  Incubators should be regularly controlled and cleaned (monthly)
  Discarded glass and plastic ware and spent media should be carefully disinfected
  Cell culture materials should be properly disposed of by central sterilization
  Effective housekeeping procedures should be followed to minimize contamination of the environment (e.g., floor, sinks, faucets, water baths)
  Unauthorized persons should not be allowed entry
  Animals should not be kept in the cell culture room
  Laboratory should be kept clean

Quality control
  A defined quality, identity, and characterization control program should be established
  Mycoplasma testing should be performed at the time of arrival of the culture and at regular intervals (monthly)
  Reliable mycoplasma detection methods should be established and performed
  Medium components (especially serum) should be tested for sterility before use
  Strict aseptic techniques and good laboratory procedures should be followed

Cell cultures
  Cell cultures should be obtained from reputable cell repositories
  Incoming cell cultures should be held in quarantine until proof of sterility (or at least separated in time and space from sterile cultures)
  Aliquots free of mycoplasma (and other adventitious agents) should be stored in a master and working stock system in liquid nitrogen
  Antibiotic-free media should be used whenever possible
  Periodic standardization checks of the cell culture (quality, identity, and characterization control program) should be made
  Mycoplasma-positive cultures should be immediately discarded or treated with mycoplasmacidal measures

Culture culturist
  Handwashing prior to and following work should be required
  Mouth pipetting should be prohibited
  There should be no unnecessary talking or traffic at the clean bench or in the immediate work area
  Protective clothing should be used to protect both the culture and the culturist
  Jewelry (rings, bracelets, wrist watches) should be taken off
  Written laboratory records for every cell culture should be made
  The same medium aliquot should not be used for different cell lines
  Medium should not be poured from bottles or flasks, but pipetted
  Only one cell line should be handled at a given time

to disinfect all work surfaces before and after culture manipulations, including the various devices entering the laminar flow hood. Mycoplasmas are extremely sensitive to most disinfectants, but can show prolonged survival in a dried state. Thus the main point is simply to keep the laboratory and its inventory as clean as possible. Clearly, nothing replaces good housekeeping and good aseptic techniques in the control of mycoplasmal contamination. It is important that cell culture laboratories establish effective and scheduled mycoplasma testing procedures in the form of a routine screening program for all forms of microbial contamination, including mycoplasmas. Sera, media, and supplements (and also cell lines whenever possible) should be purchased from reputable suppliers that adequately test for mycoplasmal contamination.

All incoming cell lines should be quarantined until the contamination status is verified. Mycoplasma-free cultures should be segregated from infected cultures by time and place of handling. Reagents for the two sets of cultures should be separate. The general use of antibiotics is not recommended except in special applications and then only for short durations. Use of antibiotics may lead to lapses in aseptic technique, to selection of drug-resistant organisms, and to delayed detection of low-level infection by either mycoplasmas or other bacteria. Master stocks of mycoplasma-free cell lines should be frozen and stored to provide a continuous supply of cells should working stocks become contaminated. Mycoplasma-infected cell lines should be discarded or treated with mycoplasmacidal measures as quickly as possible in order to prevent lateral spread.

Strict adherence of the cell culturist to aseptic culture techniques is another fundamental aspect in mycoplasma control. Cell culturists should continually be aware of the danger of contaminating clean cultures with aerosols from mycoplasma-containing cultures manipulated in the same area. For example, the following procedures with liquid media generate droplets: pipetting, decanting, centrifuging, sonicating. These relatively large droplets do not remain airborne but settle within seconds into the immediate environment, where they remain viable for days. Furthermore, mouth pipetting and unnecessary talking are not acceptable. The prohibition of eating, drinking, smoking, or application of makeup in the laboratory is obvious. The recommendation to thoroughly wash and disinfect the hands, to take off jewelry (rings, bracelets, wrist watches), to wear protective clothing, and to tie back long hair (or use of a head covering) are further precautions that are necessary to minimize the risk of contamination.

BIBLIOGRAPHY

1. S. Razin and E.A. Freundt, in J.G. Holt, ed., *Bergey's Manual of Systematic Bacteriology*, Vol. 1, Williams & Wilkins, New York, 1984, pp. 740–794.
2. S. Razin, in A. Balows, et al., eds., *The Prokaryotes*, 2nd ed., Springer, New York, 1991, pp. 1937–1958.
3. J.G. Tully, in J. Lederberg, ed., *Encyclopedia of Microbiology*, Academic Press, San Diego, Calif., 1992, pp. 181–215.

4. S. Rottem and M. Barile, *Trends Biotechnol.* **11**, 143–151 (1993).

5. S. Razin, in S. Razin and J.G. Tully, eds., *Molecular and Diagnostic Procedures in Mycoplasmology*, Vol. 1, Academic Press, San Diego, Calif., 1995, pp. 1–25.

6. S. Razin, D. Yogev, and Y. Naot, *Microbiol. Mol. Biol. Rev.* **62**, 1094–1156 (1998).

7. W.G. Weisburg et al., *J. Bacteriol.* **171**, 6455–6467 (1989).

8. M.F. Barile and S. Rottem, in I. Kahane and A. Adoni, eds., *Rapid Diagnosis of Mycoplasmas*, Plenum, New York, 1993, pp. 155–193.

9. R.A. Del Guidice and R.S. Gardella, in R.E. Stevenson, ed., *Uses and Standardization of Vertebrate Cell Cultures*, In Vitro Monog. No. 5, Tissue Culture Association, Gaithersburg, Md., 1984, pp. 104–115.

10. G.J. McGarrity, V. Vanaman, and J. Sarama, *Am. Soc. Microbiol. News* **51**, 170–183 (1985).

11. R.J. Hay, M.L. Macy, and T.R. Chen, *Nature (London)* **339**, 487–488 (1989).

12. G.J. McGarrity, H. Kotani, and G.H. Butler, in J. Maniloff, R.N. McElhaney, L.R. Finch, and J.B. Baseman, eds., *Mycoplasmas—Molecular Biology and Pathogenesis*, American Society for Microbiology, Washington, D.C., 1992, pp. 445–454.

13. G.J. McGarrity, V. Vanaman, and J. Sarama, *In Vitro* **20**, 1–18 (1984).

14. C.C. Uphoff et al., *Leukemia* **6**, 335–341 (1992).

15. E.L. Schneider and E.J. Stanbridge, *In Vitro* **11**, 20–34 (1975).

16. C.C. Uphoff, H.M. Gignac, and H.G. Drexler, *J. Immunol. Methods* **149**, 43–53 (1992).

17. S. Razin, *Mol. Cell. Probes* **8**, 497–511 (1994).

18. E. Fleckenstein, C.C. Uphoff, and H.G. Drexler, *Leukemia* **8**, 1424–1434 (1994).

19. G. Rawadi and O. Dussurget, *PCR Methods Appl.* **4**, 199–208 (1995).

20. A. Hopert et al., *J. Immunol. Methods* **164**, 91–100 (1993).

21. C.C. Uphoff, S.M. Gignac, and H.G. Drexler, *J. Immunol. Methods* **149**, 55–62 (1992).

22. H.G. Drexler et al., *In Vitro Cell. Dev. Biol.* **30A**, 344–347 (1994).

23. H.G. Drexler et al., eds., *DSMZ Catalogue of Human and Animal Cell Lines*, 6th ed., DSMZ, Braunschweig, Germany, 1995, pp. 8–13.

See also ASEPTIC TECHNIQUES IN CELL CULTURE; CONTAMINATION DETECTION IN ANIMAL CELL CULTURE; STERILIZATION AND DECONTAMINATION.

CRYOPRESERVATION OF PLANT CELLS, TISSUES AND ORGANS

ERICA E. BENSON
DOMINIQUE J. DUMET
University of Abertay Dundee
Dundee, Scotland
United Kingdom

KEITH HARDING
Scottish Crop Research Institute
Dundee, Scotland
United Kingdom

OUTLINE

Introduction
Cryopreservation Theory
    Traditional Cryopreservation Methods
    Cryopreservation using Vitrification
Practical Applications of Plant Cryopreservation
    Cryopreservation of Dedifferentiated Cells and Tissues
    Traditional, Simplified, and Novel Approaches for Dedifferentiated Cell Cryopreservation
    Cryopreservation of Shoots and Roots
    Enhancing Survival in Cryopreserved Shoot Tips using Vitrification
Cryopreservation of Seeds and Embryos
    Seed Germplasm Sterilization and *In Vitro* Culture
    Desiccation of Embryos: Approaches to Improving Desiccation Tolerance
    Chemical Cryoprotection
    Freezing, Thawing, and Recovery Medium
    Assessments of Genetic Stability in Cryopreserved Germplasm
    Molecular Approaches to Assessing Genetic Stability
Summary
Acknowledgments
Bibliography

INTRODUCTION

Cryopreservation is the storage of living tissues at ultralow temperatures; usually in the case of plant tissues this is at $-196\,°C$ in liquid nitrogen. For certain animal cells, however, storage in the vapour phase of liquid nitrogen is preferred. This conservation method has broad applications in medicine, veterinary sciences, agriculture, biotechnology, horticulture, forestry, and aquaculture (1). The cryogenic storage of viable human and animal cells has been routinely applied for more than 30 years in the healthcare and veterinary sectors. Recent advances in plant cryopreservation research now ensure its potential for routine use in *ex situ* conservation (2). The ability to conserve plant germplasm for long periods of time, in low-maintenance, liquid-nitrogen repositories, has an increasingly important role in plant biotechnology, crop plant genetic resources management, forestry, and agroforestry (3–5). Cryopreservation is also being explored as an alternative and complementary means of conserving the germplasm of endangered plant species (6). This article provides an overview of the theories and practice of plant cryopreservation. It provides an introduction to the subject for readers who are considering, for the first time, the use of cryopreservation for germplasm conservation. Emphasis is placed upon the cryoconservation of dedifferentiated cells, shoots, and embryos, as these systems are particularly important for biotechnology programs and the management of crop plant genetic resources.

CRYOPRESERVATION THEORY

Successful recovery of cells following cryopreservation is dependent on the use of cryoprotectants and/or manipulative strategies that ensure survival after exposure to freezing and dehydration injury, the two main factors that affect post cryopreservation survival (7,8). Cryopreservation can be categorized into two types of methodologies: those that involve ice formation and cryopreservation in the absence of ice. The first includes traditional approaches to cryopreservation that are dependent upon the manipulation of ice crystallization in tissues that have been protected using chemical cryoprotective additives and exposed to controlled-rate freezing regimes. The second method is quite different and involves the treatment of cells in such a way that ice crystal formation is completely inhibited (this is usually achieved by manipulating cell viscosity). Thus, water molecules, on exposure to low temperatures, form an amorphous glassy state; this process is termed *vitrification* and provides a useful option to traditional cryopreservation methods, which usually require expensive controlled-rate freezing equipment. Simplified, inexpensive cryopreservation methods have also been developed that apply vitrification strategies or "low-tech," controlled-rate freezing devices (9,10).

Traditional Cryopreservation Methods

These methods comprise several components/events including: the cryoprotectant (penetrating or nonpenetrating); the cooling regime (for which the parameters are: rate, terminal transfer temperature, and hold time); temperature of ice nucleation and supercooling. Traditional methodology is based upon the "two-factor" hypothesis of freezing injury (7,8), principally ice formation and dehydration stress. Manipulation of these injurious events by controlled-rate freezing and cryoprotective additives forms the basis of the protocols, the theory of which can be explained as follows. Water requires a "nucleation template" to initiate ice crystallization, and it is of major consequence to cryobiologists that water very rarely freezes at $0\,°C$. In biological systems, water can be supercooled (7,8) to extremely low subzero temperatures ($-40\,°C$ is the approximate point at which water can homogeneously nucleate and aggregate to form crystals) and the presence of solutes, such as cryoprotective additives, ensures that supercooling occurs. Cooling rate influences ice nucleation; thus during very rapid cooling both intra- and extracellular ice forms immediately, and the cell is usually killed. When slow rates of cooling are employed (usually within the range of -0.5 to $-5\,°C\ min^{-1}$), ice first

forms extracellularly, and this creates a water vapour deficit between the inside and the outside of the cell. As a consequence, water moves from inside the cell to the outside, where it subsequently freezes; with time, the cell becomes more and more dehydrated. The rate and the extent of this dehydration process are critical factors in ensuring the success of traditional cryopreservation protocols. If the cell is allowed to become too dehydrated, it dies due to the toxic concentration of solutes. However, if cells are dehydrated to an optimum point at which most of the potentially freezable water is transferred to the extracellular compartment, when the intracellular environment does eventually freeze, ice formation and crystal growth are so limited that the process becomes noninjurious (7,8). Quite clearly, the rate of freezing, the temperature at which extracellular ice nucleation occurs, and the holding time at this temperature will greatly influence cell survival. Many traditional plant cryopreservation protocols are dependent upon species-specific controlled-rate cooling parameters, which must be optimized to ensure cell survival on transfer to liquid nitrogen.

Cryopreservation theory must also consider the interplay between cryogenic factors and cryoprotectants. Two basic types of traditional cryoprotectants can be defined: nonpenetrating cryoprotectants and penetrating cryoprotectants. The former are usually applied as high-molarity solutions of sugars, polyols (e.g., polyethylene glycol), mannitol, and sorbitol; they exert an osmotic gradient and withdraw water from cells, reducing the amount of water available for ice formation. Osmotically active agents can also be applied at lower concentrations during pregrowth treatment. Penetrating cryoprotectants (e.g., glycerol and dimethyl sulfoxide, DMSO) exert their protective effects by a colligative action and ameliorate damage caused by the toxic concentration of intracellular solutes. A colligative cryoprotectant must be able to penetrate the cell and be nontoxic at the concentration required to ensure an effective freezing point depression (7,8). A colligative protectant is distributed equally both inside and outside the cell; therefore, it does not exert an osmotic effect and does not change the cell volume. In the presence of a colligative cryoprotectant, water will be equally lost from both cell compartments and as the extracellular and intracellular ice forms, there will be no detrimental change in cell volume, and the toxic concentration of cell solutes is circumvented (7,8). Colligative cryoprotectants also suppress the freezing point to a temperature that is low enough for the cells to tolerate potentially damaging reactions (7). Plant cryoprotection strategies utilize a range of protective additives that are often applied in combination (4).

Cryopreservation using Vitrification

Vitrification enables cells and tissues to be cryopreserved in the absence of ice, an approach that is proving to be a useful, complementary method to traditional methodologies. Vitrification protocols are especially valuable for laboratories that do not have access to controlled-rate freezing facilities, as the tissues can be plunged directly into liquid nitrogen. However, some vitrification protocols can be quite laborious and time consuming, and this may limit their efficiency of use in large-scale genebanks. Importantly, vitrification provides an alternative method for the conservation of germplasm, which is recalcitrant to traditional cryopreservation protocols. Vitrification is achieved by manipulating and enhancing the cell's viscosity to a level at which ice nucleation is inhibited. There are now many different approaches by which vitrification can be achieved in cryopreserved plant cells and tissues, including: the application of highly concentrated cryoprotectant cocktails (11), encapsulation of germplasm in a calcium–alginate matrix, followed by dehydration/desiccation treatments (12) and desiccation using air drying, silica gel, or osmotic treatments (13,14). The vitrified state is metastable, and on rewarming glasses can devitrify, causing ice crystallization; optimizing rewarming protocols for vitrified germplasm is thus a critical factor for survival (15,16).

PRACTICAL APPLICATIONS OF PLANT CRYOPRESERVATION

This section presents examples of the use of cryopreservation in plant biotechnology and genetic resource management, which are outlined on the basis of cell and tissue type. Full practical details of the cryopreservation protocols detailed in this section are available elsewhere (1,17–19).

Cryopreservation of Dedifferentiated Cells and Tissues

Long-term cryogenic storage of dedifferentiated cells and callus is particularly important for the maintenance of totipotency, and competent cell cultures can be routinely cryopreserved to ensure a supply of germplasm for use in biotechnology programs (18,19). Cryoconservation circumvents the time-related loss of totipotency, which occurs when dedifferentiated cells are maintained by serial subculture. Dedifferentiated cultures are frequently used as a source of secondary products, and as the capacity to form these metabolites can also decline with long-term culture, cryopreservation offers the opportunity to selectively conserve high-producing lines (20). Cryopreservation of genetically manipulated dedifferentiated cell cultures has the potential to assist transformation programs, and cryogenic storage has an important role in patent deposition (2,21).

Traditional, Simplified, and Novel Approaches for Dedifferentiated Cell Cryopreservation

A range of protocols have been developed to routinely cryopreserve dedifferentiated cells (Table 1). Traditional methods employ the use of computer-controlled programmable freezers, and these have the advantage that large numbers of samples can be processed at any one time. The controlled-rate freezing method developed by Withers and King (17) has proved to be one of the most widely used to date. Simplified freezing units offer a "low-tech" approach to conserving cells (Table 1); however, only small numbers of samples can be conserved during each run, and the system does not offer the flexibility of being able to manipulate cooling rates. Vitrification using cryoprotective additives (Table 1) is proving to be the method of choice for laboratories without access to controlled-rate freezers. A range of cryopreservation protocols may have to be screened for cells that are freeze recalcitrant, and novel modifications of traditional methods may be applied (21). These include: the application of the iron chelating agent desferrioxamine (22,23), pluronic F-68 (24), and the haemoglobin additive, erythrogen[®] (25).

Table 1. Examples of Cryopreservation Protocols Applied to Dedifferentiated Plant Cells

Plant System	Application	Method	Ref.
Cell suspensions of a range of species	Maintenance of morphogenetic totipotency	Traditional controlled-rate freezing, Withers & King (1980) cryoprotectant: 0.5 M DMSO, 0.5 M glycerol, 1 M sucrose (applied on ice)	1,17,19
Cereals (rice)		Freezing: $-1\,°C \cdot min^{-1}$ to $-35\,°C$ hold for 30 min. Plunge into LN_2 thawing: $45\,°C$ adaptation of above method for rice	18,26
Callus and cells of a range of species	Maintenance of culture collections	Simplified controlled-rate freezing (Yamada & Sakai, 1996) cryoprotectant: 2 M glycerol, 0.4 M sucrose Freezing 1 h @ $-30\,°C$ in a deep freeze followed by direct transfer to LN_2 thawing: $40\,°C$	9
		Vitrification with cryoprotective additives, Yamada & Sakai (1996) cryoprotectant: Plant vitrification solution 2 (PVS2), 30% (w/v) glycerol, 15% (w/v) DMSO, 15% (w/v) ethylene glycol in medium containing 0.4 M sucrose (applied on ice) Freezing: ultrarapid, direct plunge into LN_2 thawing: $40\,°C$	9
Cell lines of *Chrysanthemum cinerariaefolium*	Maintenance of secondary metabolite production	Nonprogrammable controlled-rate freezing (Hitmi et al., 1997) cryoprotectant: 5% (w/v) DMSO (applied on ice) Freezing: cooled to $-20\,°C$ at $-1\,°C$ in a nonprogrammable freezing stored @ this temperature for 24 h followed by direct plunge into LN_2 thawing: $40\,°C$	20

The success of a cell cryopreservation protocol can also influenced by noncryogenic factors such as subculture regime and morphogenetic status (18,26). Dedifferentiated cultures provide useful systems for fundamental studies (27), which enable the elucidation of those components of a protocol that predispose cells to injury. By understanding the ultrastructural and biochemical changes that accompany cryopreservation, it is possible to improve storage methodology.

Cryopreservation of Shoots and Roots

Cryopreservation of vegetatively propagated crop plants is an important long-term means of *ex situ* conservation and complements the use of field collections and *in vitro* storage in the active and slow growth state. A wide range of cryopreservation protocols have been applied, as described in detail elsewhere (28). Three main approaches have been developed: controlled-rate freezing, encapsulation/dehydration, and vitrification using cryoprotective additives. There can be considerable genotypic variation in the responses of different species to shoot-tip cryopreservation, and methods frequently require empirical development. Shoot cultures of temperate species frequently benefit from the application of pregrowth treatments, including cold hardening (29), and post-storage regeneration can be influenced by the plant growth regulator composition of the medium. This can also have long-term effects on plant performance (30,31). The cryopreservation of shoots and roots usually requires the excision of lateral and apical buds and roottips from actively growing cultures. Organ size can be critical (usually within the range of 3–5 mm), and shoot structures usually comprise the apical dome (or lateral bud) and 2–4 nonexpanded leaf primordia (28). In the case of root cryopreservation, the main application concerns the conservation of transgenic hairy root cultures, for which both controlled-rate freezing and encapsulation/dehydration have been applied (32,33). A wide range of protocols have been reported for the cryopreservation of shoot cultures, and a selection of these are summarized in Table 2 (34–36).

Enhancing Survival in Cryopreserved Shoot Tips using Vitrification

The development of vitrification methods for plant tissues has had a major impact on the ability to conserve complex shoot-tip structures and previously freeze-recalcitrant systems (28). Thus species that have proved difficult to preserve using traditional methodologies may be amenable to cryopreservation using vitrification. Yamada and Sakai (9) report the successful application of vitrification to cryopreserve the meristems of over 30 different plant species. Vitrification using encapsulation has been modified to enhance survival by optimizing sucrose pregrowth, which is an important survival parameter (9,37,38). Table 3 provides examples of different species that have been cryopreserved using vitrification (39–46). It is by no means an exhaustive list, but has been designed to provide an indication of the wide-ranging applicability of the methodology.

CRYOPRESERVATION OF SEEDS AND EMBRYOS

Cryopreservation has been applied to the conservation of many different types of seeds as well as zygotic and somatic embryo germplasm. Seeds have been divided into three categories according to their ability to sustain desiccation (47–49): Orthodox seeds undergo a natural desiccation phase at the end of their maturation, and are consequently highly desiccation tolerant. In contrast, recalcitrant seeds remain highly hydrated throughout their maturation and are extremely sensitive to loss of water. Between these extremes of behavior are intermediate seeds, which are highly hydrated when mature, but can sustain artificial desiccation. Chilling sensitivity is a characteristic of many hydrated recalcitrant and desiccated intermediate seeds (48–51). The ability to cryopreserve seed is directly linked to their desiccation tolerance. Consequently, orthodox and intermediate seed cryoconservation can be simply achieved by the elimination of crystallizable water prior to freezing. The International Plant Genetic Resources Institute (IPGRI) recommends (52) desiccating orthodox seeds to a final water content as low as 1–5% (fresh weight basis) prior to cryopreservation (52). Due to the desiccation sensitivity of recalcitrant seeds, and in some cases their relatively large size (e.g., coconut, mango, avocado), their long-term preservation has to be based on the cryopreservation of their isolated embryonic axes. They generally represent an insignificant proportion of the total seed mass and are thus far more amenable to cryogenic manipulations (53). Cryopreservation is also important for the preservation of species micropropagated by somatic embryogenesis. The key steps required for cryopreserving isolated zygotic embryos and somatic embryos are described in the following.

Seed Germplasm Sterilization and *In Vitro* Culture

Recalcitrant seeds have both external and internal microflora (54); therefore, potential contaminants must be removed prior to culture initiation. The most commonly used agents are: sodium hypochlorite (1 to 3%), mercuric chloride (0.1 to 0.5%) and ethanol (70%), which are applied at various concentrations and exposure times (14). A few drops of a wetting agent such as Tween 20 may increase the efficiency of sterilization. Depending on the species and the degree of contamination, whole seeds or fruits can be surface sterilized before the aseptic excision of embryonic axes, which is performed under the sterile flux of a laminar-flow bench. If internal tissues are highly contaminated, embryonic axes must also be sterilized after removal from their mother tissue. Following excision, it is critical to grow zygotic embryos on a basal medium that allows full seedling development (14), and the presence of specific growth regulators may be essential to ensure normal seedling morphogenesis (14). Thus, while isolated axes of tea (*Camellia sinensis*), neem (*Azadirachta indica*), and Jack fruit (*Artocarpus heterophyllus*), can develop into plantlets without the addition of growth regulators (D.J. Dumet and P. Berjak personal communications); in contrast, cacao (*Theobroma cacao*), oil palm (*Elaeis guineensis*), or rubber (*Hevea*

Table 2. Examples of Cryopreservation Protocols used for the Cryopreservation of Shoots from Vegetatively Propagated Crops

Plant System and Application	Pregrowth and Cryoprotection Strategies	Cryopreservation and Recovery Protocols	Ref.
	Tuber Crops		
Example: Potato, cassava, yam Long-term storage of vegetative germplasm for crops which do not produce seed and for which tuber storage offers short-term conservation options.	Pregrowth: 1 week for apical shoots, or after a subculture cycle. Encapsulation/dehydration: shoot-tip excision, alginate encapsulation, 3 days pregrowth in 0.75 M sucrose solution, air desiccation in a laminar flow for 3–4 h.	Direct exposure to LN$_2$: Beads transferred to cryovials, direct plunge, rewarmed at ambient temperatures for 15 min, direct transfer to standard growth medium. Plant growth regulator application may be necessary, care must be taken not to induce callogenic responses.	12,34,35,36
	Woody Perennial Fruit Crops		
Example: *Ribes spp.* Shoot-tip cryopreservation of *in vitro* plantlets for the *ex situ* maintenance of vegetatively propagated genetic resources derived from woody perennial, fruit germplasm. This approach potentially allows the long-term maintenance of virus-free, *in vitro* culture collections of *Ribes spp.* and avoids the need for expensive serial subculture maintenance. Cryogenic maintenance assist germplasm exchange programs.	Cold acclimation: *In vitro* plantlets acclimated for 1 week with 8 h days at 22 °C/16 h nights at −1 °C, followed by shoot-tip excision and 2 days cold acclimation in the presence of 5% DMSO. Chemical cryoprotection: 1 h cryoprotection @ 4 °C with w/v 10% each of polyethylene glycol MW 8000, glucose and DMSO. Encapsulation/dehydration: Plantlet acclimation (as above) followed by shoot excision and alginate encapsulation, 18 h pregrowth in 0.75 M sucrose, air desiccation in a laminar flow for 3–4 h. PVS2 Vitrification: Cold acclimation as described above, cryoprotection for 20 min, using 30% v/v glycerol, 15% v/v ethylene glycol, 15% (v/v) DMSO in *Ribes* medium containing 0.4 M sucrose.	Traditional controlled-rate freezing: Following cold acclimation and chemical cryoprotection: cooling in cryovials to −40 °C at −0.5 °C min^{-1}, manual seeding to induce ice nucleation, direct transfer to LN$_2$, thawed for 1 min at 45 °C and 2 min at 22 °C, rinsed in liquid culture medium, transferred to standard culture medium. Direct exposure to LN$_2$: Alginate beads transferred to cryovials and directly plunged into LN$_2$, rewarmed at ambient temperatures for 15 min and transferred to standard growth medium. Vitrification: Shoots are plunged directly into LN$_2$ rewarmed for 1 min at 45 °C and 2 min at 22 °C, rinsed in liquid culture medium containing 1.2 M sucrose, transferred to standard culture medium.	16,29 16,29 16,29

brasiliensis) isolated embryonic axes require specific growth regulators to promote development (54–56).

Desiccation of Embryos: Approaches to Improving Desiccation Tolerance

Desiccation can be performed under the sterile air stream of a laminar flow cabinet, over silica gel in an airtight container, under a compressed airstream (flash drying), or under vacuum (14). The desiccation rate of the sample will be moderated by the method used, and depending on the species, the embryo can be successfully dehydrated to an optimal minimal water content that allows survival after exposure to liquid nitrogen. Cryopreservation after

desiccation has been obtained with zygotic embryos of various species; examples include: tea (57; D.J. Dumet and P. Berjak, personal communications), coffee (58,59), neem (60), and hazelnut (61). Somatic embryos can also survive cryopreservation following desiccation, as reported in the case of coffee, date palm, and pea (62). Depending on the species, manipulations of desiccation rate can enhance the ability of the embryo to survive extreme dry states. It has been shown for *Landolphia kirkii*, *Casternospermum australe*, *Scadoxus membranaceus*, *Camellia sinensis*, and *Trichilia dregeana* that rapid drying axes allow survival to lower water content than if they are slowly dehydrated (63,64). The opposite, is the case for Jack fruit

Table 3. Recovery Profiles of Crop Plant Shoot Tips Cryopreserved using Vitrification Methods

Crop Plant	Vitrification Method	Recovery	Ref.
Black currant	PVS2	60%	16,29
Black currant	encapsulation/dehydration	80%	16,29
Chicory	encapsulation/dehydration	65%	39
Citrus	encapsulation/dehydration	50%	40
Garlic	PVS2	80–100%	41
Mulberry	PVS2	40–70%	42
Potato	encapsulation/dehydration	0–70%	12,31
Potato	cryoprotectant	45–90%	43
Shallot	PVS2	70%	44
Tea	encapsulation/dehydration	40%	45
Tea	PVS2	60%	45
Wasabi	PVS2	85%	38
Yam	encapsulation/dehydration	19–60%	36,46

Notes: Recovery is on the basis of % shoot or whole plant regeneration, usually given for a range of crop genotypes (between species and/or within species/cultivars). Encapsulation/dehydration refers to the basic method (or modification) developed by Fabre and Dereuddre (12), cryoprotectant refers to vitrification by the use of highly concentrated solutions of cryoprotectants, PVS2 refers specifically to the use (or modification) of plant vitrification solution No. 2 (PVS2) originally developed by Sakai and colleagues [see (9) for a review]. Full practical details of methodologies are reported elsewhere (1,11,19,28).

zygotic and oil palm somatic embryos, as the slower the desiccation the lower is the lethal water content (13,65). Sucrose pretreatments have been clearly shown to enhance desiccation and cryopreservation tolerance of clumps of oil palm somatic embryos (13). In this case, embryos grown for a week on a standard growth medium containing 25% sucrose could be dehydrated to a water content of 33% [Fresh Weight Basis, (FWB)] without viability loss, while 80% of the nonpretreated tissues were killed when their water content was as high as 41% (FWB). Similar, beneficial effects of sucrose has been recorded for Jack fruit and *T. dregeana* zygotic embryos (D.J. Dumet and P. Berjak, personal communications); however, high sucrose treatments can have a damaging effect manifested by abnormal growth, as reported for the embryonic axes of *L. kirkii* (D.J. Dumet and P. Berjak, personal communications). Developmental stage may also interfere with embryo desiccation tolerance, though for zygotic embryos of tea and Jack fruit, tolerance was shown to be greater in the more mature embryos, while in cacao, tolerance appeared higher in the less mature embryos (66).

Chemical Cryoprotection

Depending on the species, embryos may survive cryopreservation when they are only pretreated with chemical additives that act as cryoprotectants. Thus, oil palm and coffee somatic embryos survive liquid nitrogen after exposure to high sucrose levels (13,67–69). Chemical cryoprotectants can also be used in combination such as sucrose and glycerol for somatic embryos of date palm and glucose and glycerol for coffee and coconut zygotic embryos (62,70–72). However, even if cryoprotectants allow some embryos to survive liquid nitrogen treatments, partial desiccation after or before cryoprotection may drastically improve their recovery rate. This has been

clearly shown for somatic embryos of oil palm, coffee and date palm as well as zygotic embryos of coconut and *T. dregeana* (13,62,64,72).

Freezing, Thawing, and Recovery Medium

The simplest method of cryopreserving seeds and embryos is to place the germplasm in a polypropylene cryotube and to plunge it directly into liquid nitrogen. This technique has been used for cryopreservation of many zygotic and somatic embryos, including rubber, coconut, oil palm, and date palm (13,14,59,70–75). In some instances, it has been clearly shown that the higher the water content the faster must be the freezing rate (76). *C. sinensis* zygotic embryos containing 52–61% water (FWB) do not tolerate cryopreservation unless propelled directly into subcooled liquid nitrogen ($-210\,°C$). Such a technique provides an average cooling rate of $-500\,°C$ per minute compared to about $-200\,°C$ when introduced in cryotubes before being plunged into liquid nitrogen (76). Slow cooling rates, between 0.5 and 1 degree per minute, have been successfully used for the cryopreservation of zygotic axes of *E. longan* (70) as well as hydrated somatic embryos of coffee (69). When embryos are frozen in a cryotube, thawing is generally performed by plunging the cryotube into a 30–40 °C water bath for a few minutes. Hydrated thawing can also be undertaken when the specimen has been directly propelled in liquid nitrogen; in this case, samples can be plunged into a culture medium either at ambient temperatures or preheated at 37 °C (76). After thawing, embryos are usually rinsed in high osmotic solution to remove cryoprotective additives (if any); thus osmotic stress is reduced and transferred to a recovery medium, which may be identical to the standard growth medium or contain additional growth regulators (58,59,67).

Assessments of Genetic Stability in Cryopreserved Germplasm

It is important to determine the genetic stability of plant material regenerated from cryopreserved germplasm; however, there are few detailed studies concerning the assessment of plants derived from cryopreserved germplasm. Morphological studies report similarities in control plants compared to those that have been regenerated from cryopreserved germplasm (77). Stability in the composition of polymorphic proteins in potato plants has been observed in meristems after cryostorage (78). Flow cytometric assessments have demonstrated stability of ploidy levels in cryopreserved dihaploid *Solanum tuberosum* and wild *Solanum* species (79) and old potato varieties (80). The developmental competence and cytological stability of plants regenerated from encapsulated, cryopreserved shoot apices of six wild *Solanum* species (representing three species and three ploidy levels) has been confirmed (31).

Molecular Approaches to Assessing Genetic Stability

The techniques available for the molecular analysis of plant DNA stability are considerable. It is important that the choice of procedure should be based on a knowledge of the DNA sequences in the genome, as this will assist the selection of the analytical technique and provide a more meaningful basis on which to assess stability parameters (78). Ideally, for routine stability assessments in genebanks, molecular techniques should be simple, rapid, and nonhazardous (81). Several molecular studies have shown no detectable genetic variation after cryopreservation by using randomly amplified polymorphic DNA (RAPD) fingerprinting, as applied to the rare and endangered shrub species, *Grevillea scapigera* (82). Growth rates, secondary metabolite production, and T DNA structure were found to be unchanged after the cryopreservation of roottips from hairy root cultures of *Beta vulgaris* and *Nicotiana rustica* (32). An examination of the stability of transgenes in genetically modified plants following cryopreservation is rare; however the stability of the selectable marker gene expressing neomycin phosphotransferase II (npt II) has been demonstrated after cryogenic storage. After 1 year in liquid nitrogen storage (following vitrification), each line survived the recovery process. DNA analysis showed the transgene was integrated into the genome of *Citrus sinensis* with no observable difference in size or number of the transgene between cryopreserved and non-cryopreserved cells (83). Other studies have shown that the cryopreservation vitrification technique does not to impede gene expression in cell cultures of *N. tabacum* (84). Restriction fragment length polymorphism (RFLP) analysis of 161 regrown plants representing old potato varieties from cryopreservation did not show abnormal banding patterns (85). Stability of the ribosomal RNA genes (rDNA) in potato plants after cryopreservation has been observed (86), and where variable rDNA hybridization signals were detected (87), in cryopreserved plants, they are likely to be attributed to DNA methylation (86–88). As cryogenic storage becomes routinely used in large-scale genebanks, it will become important to develop rapid and effective molecular techniques to assess genetic stability in plants regenerated from cryopreserved germplasm. Similarly, as cryopreservation has an important potential application for the conservation and patent deposition of genetically manipulated germplasm, future studies should also evaluate, in greater detail, the effects of cryogenic storage on transformed systems.

SUMMARY

During the past 10 years, plant cryopreservation research has supported the development of a number of very effective storage protocols, and these can now be applied for the conservation of a comprehensive range of plant genetic resources. Studies of post-storage genetic stability indicate that plants regenerated from cryopreserved germplasm are stable which supports the transfer of the technology for routine use in large-scale genebanks. The next phase of plant cryopreservation research development will largely address the application of the technology in genebanks, repositories, and culture collections (see Towill, this volume).

ACKNOWLEDGMENTS

DJD acknowledges the support of the European Commission for the award of a Marie Curie Post Doctoral Fellowship. KH is supported by the Scottish Office Agriculture Fisheries and Food Department.

BIBLIOGRAPHY

1. J.A. Day and M.R. McLellan, eds., *Cryopreservation and Freeze-Drying Protocols*, Methods Mol Biol Vol. 38, Humana Press, Totowa, N.J., 1995.

2. E.E. Benson, P.T. Lynch, and G.N. Stacey, *AgBiotech. News Inf.* 10, 133N–141N (1998).

3. J.A. Callow, B.V. Ford-Lloyd, and H.J. Newbury, eds., *Biotechnology and Plant Genetic Resources: Conservation and Use*, Biotechnol. Agric. Ser. No. 19. CAB International, New York, 1997.

4. M.K. Razdan and E.C. Cocking, eds., *Conservation of Genetic Resources In Vitro*, Vol. I, Science Publishers, Enfield, N.H., 1997.

5. Y.P.S. Bajaj, ed., *Biotechnology in Agriculture and Forestry*, Vol. 32, Springer-Verlag, Heidelberg, 1995.

6. D.H. Touchell and K.W. Dixon, *Ann. Bot. (London)*, [N.S.] **74**, 541–546 (1994).

7. B.J. Finkle, M.E. Zavala, and J.M. Ulrich, in K.K. Kartha, ed., *Cryopreservation of Plant Cells and Organs*, CRC Press, Boca Raton, F., 1985, pp. 75–114.

8. H.T. Meryman and R.J. Williams, in K.K. Kartha, ed., *Cryopreservation of Plant Cells and Organs*, CRC Press, Boca Raton, Fl., 1985, pp. 13–48.

9. T. Yamada and A. Sakai, in M.N. Normah, M.K. Narimah, and M.M. Clyde, eds., In Vitro *Conservation of Plant Genetic Resources*, Kuala Lumpur, Malaysia, 1996, pp. 89–103.

10. F. Engelmann, D. Dambier, and P. Ollitrault, *Cryo-Letters* **15**, 53–58 (1994).

11. P.J. Reinhoud, A. Uragami, A. Sakai, and F. Van Iren, *Methods Mol. Biol.* **38**, 113–120 (1995).

12. J. Fabre and J. Dereuddre, *Cryo-Letters* **11**, 413–426 (1990).

13. D. Dumet, F. Engelmann, N. Chabrillange, and Y. Duval, *Plant Cell Rep.* **12**, 352–355 (1993).

14. D. Dumet, P. Berjak, and F. Engelmann, in M.K. Razdan, and E.C. Cocking, eds., *Conservation of Plant Genetic Resources In Vitro*, Vol. 1, Sciences Publishers, Enfield, N.H., 1997, pp. 153–174.

15. K. Harding, E.E. Benson, and K. Clacher, *Agro-Food-Ind. Hi-Tech.* **8**(3), 24–27 (1997).

16. E.E. Benson et al., *Cryo-Letters* **17**, 347–362 (1996).

17. L.A. Withers and P.J. King, *Cryo-Letters* **1**, 213–220 (1980).

18. E.E. Benson and P.T. Lynch, *Methods Mol. Biol.* **2**, (1998).

19. E.E. Benson, in R.A. Dixon, and R.A. Gonzales, eds., *Plant Cell Culture: A Practical Approach*, 2nd ed., IRL Press, Oxford, 1994, pp. 147–168.

20. A. Hitmi, H. Sallanon, and C. Barthomeuf, *Plant Cell Rep.* **17**, 60–64 (1997).

21. C. Gazeau, H. Elleuch, A. David, and C. Morisset, *Cryo-Letters* **19**, 146–158 (1998).

22. E.E. Benson, P.T. Lynch, and J. Jones, *Plant Sci.* **110**, 249–258 (1995).

23. E.E. Benson, R.A. Fleck, D.H. Bremner, and J.G. Day, *In Vitro* **34**, 1039 (1998).

24. P. Anthony, K.C. Lowe, J.B. Power, and M.R. Davey, *Cryobiology* **35**, 210–208 (1997).

25. K. Azhakanandam et al., *Cryo-Letters* **19**, 189–196 (1998).

26. T. Lynch et al., *Plant Sci.* **98**, 185–192 (1994).

27. J.-H. Wang, J.-G. Ge, F. Liu, and C.-H. Huang, *Cryo-Letters* **19**, 49–54 (1998).

28. E.E. Benson, *Methods Mol. Biol.* **38**, 121–132 (1995).

29. B.M. Reed and X. Yu, *Cryo-Letters* **16**, 131–136 (1995).

30. K. Harding and E.E. Benson, *Cryo-Letters* **15**, 59–66 (1994).

31. E.E. Benson et al., *Cryo-Letters* **17**, 119–128 (1996).

32. E.E. Benson and J.D. Hamill, *Plant Cell, Tissue Organ Cult.* **24**, 163–172 (1991).

33. K. Hirata et al., *Cryo-Letters* **16**, 122–127 (1995).

34. E.E. Benson et al., *Cryo-Letters* **17**, 119–128 (1996).

35. F. Engelmann et al., in M. Terzi, ed., *Current Issues of Plant and Cellular Biology*, Kluwer Academic Publishers, Dordrecht, The Netherlands, 1995, pp. 315–320.

36. B. Malaurie, M.F. Trouslot, F. Engelmann, and N. Chabrillange, *Cryo-Letters* **19**, 15–26 (1998).

37. M.T. Gonzalez-Arnao, T. Moreira, and C. Urra, *Cryo-Letters* **17**, 141–148 (1996).

38. T. Matsumoto, A. Sakai, and Y. Nako, *Cryo-Letters* **19**, 27–36 (1997).

39. Vandenbussche, M.A.C. Demeulemeester, and M.P. De Proft, *Cryo-Letters* **14**, 259–266 (1993).

40. M.T. Gonzalez-Arnao et al., *Cryo-Letters* **19**, 171–182 (1998).

41. E. Niwata, *Cryo-Letters* **16**, 102–107 (1995).

42. T. Niino, A. Sakai, and H. Yakuwa, *Cryo-Letters* **13**, 51–48 (1992).

43. A.M. Golmirzaie and A. Panta, *Advances in Potato Cryopreservation by Vitrification, International Potato Center Program Annual Report 1995–96*, CIP, Lima, Peru, 1997, pp. 71–76.

44. H. Kohmura, Y. Ikeda, and A. Sakai, *Cryo-Letters* **15**, 289–298 (1994).

45. Y. Kuranuki and A. Sakai, *Cryo-Letters* **16**, 345–352 (1995).

46. B.B. Mandal, K.P.S. Chandel, and S. Dwivedi, *Cryo-Letters* **17**, 165–174 (1996).

47. E.H. Roberts, *Seed Sci. Technol.* **1**, 499–514 (1973).

48. R.M. Ellis, T.D. Hong, and E.H. Roberts, *J. Exp. Bot.* **41**, 1167–1174 (1990).

49. R.M. Ellis, T.D. Hong, E.H. Roberts, and U. Soetisna, *Seed Sci. Res.* **1**, 99–104 (1991).

50. H.F. Chin and E.H. Roberts, *Recalcitrant Crop Seeds*, Tropical Press, SDN/BDH, Kuala Lumpur, Malaysia, 1980.

51. P. Berjak et al., *Seed Sci. Technol.* **23**, 779–792 (1995).

52. I.B.P.G.R., *Design and Cost Aspects of Long Term Storage Facilities*, I.B.P.G.R, Rome, 1976.

53. P. Berjak, J.M. Farrant, and N.W. Pammenter, in R.B. Taylorson, ed., *Recent Advances in the Development and Germination of Seeds*, Plenum, New York, 1989, pp. 89–108.

54. D.J. Mycock and P. Berjak, *Phytophylactica* **22**, 413–418 (1990).

55. V. Pence, *Plant Cell Rep.* **10**, 144–147 (1991).

56. M.N. Normah, H.F. Chin, and Y.L. Hor, *Pertanika* **9**, 299–303 (1986).

57. R. Chaudhury, J. Radhamani, and K.P.S. Chandel, *Cryo-Letters* **12**, 31–36 (1991).

58. M.N. Normah and M. Vengadasalam, *Cryo-Letters* **13**, 199–208 (1992).

59. A. Abdelnour-Esquivel, V. Villalobos, and F. Engelmann, *Cryo-Letters* **13**, 297–302 (1992).

60. P. Berjak and D.J. Dumet, *Cryo-Letters* **17**, 99–104 (1995).

61. M.E. Gonzalez-Benito and C. Perez-Ruiz, *Cryobiology* **29**, 685–690 (1992).

62. D.J. Mycock, J. Wesley-Smith, and P. Berjak, *Ann. Bot. (London)* **75**, 331–336 (1995).

63. N.W. Pammenter, C.W. Vertucci, and P. Berjak, *Plant Physiol.* **96**, 1093–1098 (1991).

64. J. Kioko et al., *Cryo-Letters* **19**, 5–14 (1998).

65. J.R. Fu, Q.H. Xia, and L.F. Tang, *Seed Sci. Technol.* **21**, 85–95 (1993).

66. K.P.S. Chandel, R. Chaudhury, and J. Radhamani, *IPGRI/NBPGR Report*, IPGRI, Rome, 1994.

67. F. Engelmann, Y. Duval, and J. Dereuddre, *C.R. Seance Acad. Sci. Ser. 3* **301**, 111–116 (1985).

68. D. Dumet, F. Engelmann, N. Chabrillange, and Y. Duval, *Cryo-Letters* **15**, 85–90 (1994).

69. H. Tesseareau, Thèse de l'Université Pierre et Marie Curie, Paris, 1993.

70. J.R. Fu et al., *Seed Sci. Technol.* **18**, 743–754 (1990).

71. B. Assy-Bah and F. Engelmann, *Cryo-Letters* **13**, 117–126 (1992).

72. B. Assy-Bah and F. Engelmann, *Cryo-Letters* **13**, 67–74 (1992).

73. M.N. Normah, H.F. Chin, and Y.L. Hor Muell-Arg, *Pertanika* **9**, 299–303 (1986).

74. M.N. Normah and M. Marzalina, in M.N. Normah, M.K. Narimah, and M.M. Clyde eds., *In Vitro Conservation of Plant Genetic Resources*, Kuala Lumpur, Malaysia, 1996, pp. 253–261.

75. F. Engelmann, N. Chabrillange, and Y. Duval, *Seed Sci. Res.* **5**, pp. 81–86 (1995).

76. J. Wesley-Smith, C.W. Vertucci, and P. Berjak, *J. Plant Physiol.* **140**, pp. 596–604 (1992).

77. Y.P.S. Bajaj, in Y.P.S. Bajaj, ed., *Biotechnology in Agriculture and Forestry, Cryopreservation of Plant Germplasm I*, Vol. 32, Springer-Verlag, Berlin, 1995, pp. 229–235.

78. K. Harding, in M.N. Normah, M.K. Narimah, and M.M. Clyde, eds., *In Vitro Conservation of Plant Genetic Resources*, Kuala Lumpur, Malaysia, 1996, pp. 137–170.

79. A.C.W. Ward et al., *Cryo-Letters* **14**, 145–152 (1993).

80. A. Schafer, E. Muller, and G. Mix-Wagner, *Potato Res.* **39**, 507–513 (1996).

81. K. Harding and E.E. Benson, in B.W.W. Grout, ed., *Genetic Preservation of Plant Cells In Vitro*, Springer-Verlag, Heidelberg, 1995, pp. 113–164.

82. H. Touchell and K.W. Dixon in N. Normah, M.K. Narimah, and M.M. Clyde, eds., In Vitro *Conservation of Plant Genetic Resources*, Kuala Lumpur, Malaysia, 1996, pp. 169–180.

83. S. Kobayashi and A. Sakai, in M.K. Razdan, and E.C. Cocking, eds., *Conservation of Plant Genetic Resources In Vitro*, Vol. 1, Science Publishers, Infield, N.Ho., 1997, pp. 201–223.

84. P.J. Reinhoud, Ph.D. Thesis, University of Leiden, published by Offsetdrukkerij Ridderprint B.V. Ridderkerk, The Netherlands, 1996.

85. A. Schafer, E. Muller, and G. Mix-Wagner, *Landbauforsch. Voelkenrode* **46**, 65–75 (1996).

86. K. Harding, *Euphytica* **55**, 141–146 (1991).

87. K. Harding, *Cryo-Letters* **18**, 217–230 (1997).

88. K. Harding, *Plant Cell, Tissue Organ. Cult.* **37**, 31–38 (1994).

See also ADVENTITIOUS ORGANOGENESIS; EQUIPMENT AND LABORATORY DESIGN FOR CELL CULTURE; MICROPROPAGATION OF PLANTS, PRINCIPLES AND PRACTICES; PLANT CELL CULTURE, LABORATORY TECHNIQUES; PLANT PROTOPLASTS.

CULTURE ESTABLISHMENT, PLANT CELL CULTURE

EUNICE J. ALLAN
University of Aberdeen
Aberdeen
United Kingdom

OUTLINE

INTRODUCTION

Plant tissue culture is the technique whereby plant cells, tissues, and organs are grown on artificial media independent of the whole plant. The science that supports this discipline is embedded in Schleiden and Schwanns cell theory of 1839 (see History Section Spier). The techniques involved are fundamental in plant-cell biotechnology which encompasses many disciplines and has an ever increasing socioeconomic impact on the use of plants and their products by humans. Cell cultures are initiated by cultivating surface-sterilized plant material on complex artificial media which cause the plant cells to grow and divide. Such culture usually involves using plant growth regulators, which, as in the whole plant, act to regulate and coordinate development. The most used cell culture system is that in which differentiated plant tissues are induced to grow and divide as an unorganized mass of undifferentiated cells known as a callus. Callus culture and its associated techniques permeate the majority of methodologies of plant cell, tissue and, organ culture. It provides a means for producing large cell numbers and at the same time, for introducing variation. Hence it is fundamentally important. Sound knowledge of the protocols, their implications, and a rigorous laboratory management procedure are required to achieve success. This article describes the fundamental methods used for establishing and routinely maintaining callus and cell-suspension cultures. It includes discussion on preparation of the original plant tissues, methods for surface disinfection, media, and the general methodologies used for initiating and monitoring the growth of suspension cultures.

PREPARATION OF THE STOCK PLANT

Callus initiation is now routine to many laboratories, and numerous publications deal with general and detailed procedures (1–4). Culture conditions are generally considered the key to initiating plant-cell cultures, and the actual source of the explant, the organs or pieces of tissue used for tissue culture, are of little importance. The choice of explant is always governed by the experimental aims and although the explant source is not important for culture initiation, longer term issues, such as the yield of secondary metabolites, may be affected. The explant can be derived from any plant part and this, in itself, dictates the type of equipment and methods that are employed. The prevailing rule is that the explant material should be healthy and vigorous, which in turn, usually represents young, newly formed plant tissues.

The environment is naturally a large source of contaminants, and usually field-grown plants have a high microbial load and consist of tougher material (as a consequence of environmental stress) which is difficult to sterilize. If feasible, preliminary experiments whereby field materials are collected, surface disinfected, and tested for sterility (rather than culture initiation) are worthwhile and can save time, labor, and materials. In such experiments, material can be collected and transported to the laboratory to determine how this can be done without damage to the plant tissue. Optimally, the time between field collection and placing the explant onto the culture initiation medium should be as short as possible. Realistically, this may not be possible and consequently, consideration of logistics is important, for example, transferring leafy branches into secured buckets of iced water for a journey over untarred roads may have a more successful outcome than picking leaves that wilt during transport. Once back in the laboratory, the material is usually washed repeatedly before disinfection either with just water or more effectively with a detergent to remove grime and then is exposed to a few different disinfection procedures (see below) and placed onto a range of microbiological media e.g. Nutrient Agar, Malt Extract Agar, to test for microbial growth and tissue damage. After, a surface-sterilization regime has been established, experiments on culture initiation can commence. Field material which has proven difficult to disinfect can be prepared in different ways to reduce contamination during culture initiation. For example, plant shoots/leaves can be grown in some sort of protected minienvironment such as paper or polythene bags, or the plants, or their parts, can be sprayed with a course of antibiotics and/or fungicides before sampling or processing. Such stringent measures are fortunately not generally required. The easiest and most routine method is to circumvent the problem imposed by field material altogether by using newly germinated plants, that can be obtained from surface sterilized seeds, which are cultivated in controlled environmental growth chambers. Such material is relatively clean, hence easier to sterilize, and has a high potential for cell division which is advantageous for the majority of cell and tissue culture methods. It is inadvisable to use plants that are declining, diseased, or damaged (unless of course, the biology of these processes is being studied). Similarly, root material grown in sterile soil or under a hydroponic system is easier to surface-sterilize than natural samples. If seeds are used, it should be remembered that they result from sexual reproduction and will hence be different from the parent! Care must be taken when using purchased plants, fruit, etc. as these may have been mechanically damaged during washing, packing, transporting, and/or treatment with antimicrobial agents, which may affect both sterilization and culture initiation.

ASEPTIC TECHNIQUE AND EXPLANT STERILIZATION

The complex media that support plant-cell cultures can equally sustain microbial growth, and if a culture becomes contaminated, the relatively homogeneous culture environment will be lost. In the worse case, the plant material will be killed, a devastating scenario because each culture is, unique, to a certain extent, and may have taken some

considerable time, effort, and expense to obtain. Microbial contamination is the most important cause of losses of tissue cultures (5). Contamination was originally thought to be a problem solely in the initial tissue culture procedure such as callus or shoot initiation, but it is well known now that the contaminant may be latent, not easily being observed by eye, and may only be evident at later stages of the process, during weaning and initiation of suspension cultures. Phytopathogenic viruses may only become evident well after micropropagated plants have come out of culture and are established in the field. Hence initiating and maintaining sterile cultures is of paramount importance, and many people consider maintaining sterility the key to plant cell, tissue, and organ culture. Good laboratory practice with respect to cleanliness and organization is an integral component of cell and tissue culture, and is often naively overlooked compared to sterilization procedures. There are numerous sources of contaminants in a laboratory, and continual vigilance is required to minimize them. These aspects and methods for screening cultures for contaminants are covered earlier in this book. There is an increasing awareness that most plants have endophytic microorganisms which may not be exposed to and hence killed by the disinfecting agent. In this regards, it should always be remembered that tissue cultures are generally only surface-sterilized and should be termed aseptic rather than sterile unless they have been thoroughly screened using microbiological methods. Sterilization of explant material is, with a few exceptions, for example, micropropagation of *Streptocarpus* hybrids and *Begonia* (3), crucial for culture initiation. The nature of the explant, as already explained, imposes different challenges for sterilization. Woody species and root material taken from the natural environment are notoriously difficult to sterilize, whereas some fruits and tubers can be relatively easy requiring only flaming before cutting into the sterile internal flesh. Seeds may be surrounded by tough coatings such as the testa and endocarp (shells) which harbor microflora in niches inaccessible to the sterilant. In this situation it is best to remove these outer layers (with the aid of nutcrackers, hammers, scalpels, and patience!) so that the seed itself (or indeed, the embryo) can be directly exposed to the sterilant and in turn, the culture medium. Fortunately, it is possible to surface-sterilize any plant organ.

Methods for Surface Sterilization

The basic methodology for sterilizing plant material is straightforward (Fig. 1). The stock plant is cut into pieces that contain the required tissue which is placed into a container. Then material can be taken into the clean room and preferably to a laminar flow hood. The material is then disinfected using the appropriate regime, washed to remove all traces of the sterilant, and placed in an appropriate container ready for the next step. Unfortunately, this procedure is fraught with pitfalls. The laminar flow hood must be used correctly with particular care for its routine maintenance, cleaning, and use. For example, the working practices must be organized to reduce cross-contamination between dirty and clean material, the airflow should be switched on

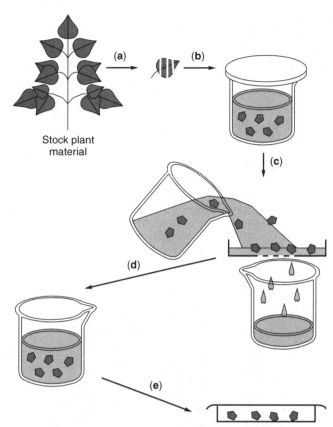

Figure 1. A schematic diagram illustrating an example of the procedure for surface-sterilizing plant material. (**a**) The selected plant part is cut into conveniently sized explants. (**b**) Explants are immersed in sterilant for 5–20 min. The container should be covered and shaken periodically. (**c**) Sterilant is decanted. A sterile sieve will help to retrieve explants. (**d**) Explants are thoroughly washed in sterile distilled water at least five times to remove all traces of the sterilant. (**e**) Explants are moved to a sterile dish, preferably with a lid for example, a Petridish, for easy access and prevention of contamination.

at least ten minutes before starting, and the surfaces (including anything that will be in the cabinet such as balances) should be disinfected. It is a good idea to have sterile material on one side of the cabinet and dirty on the other, and it must always be remembered that any dirty material (including hands) that is placed in front of the air flow may contaminate anything behind. Explants have usually to be cut before surface disinfection, and for this, scalpels, knives, and indeed broken razor blades can be ideal to produce a sharp, clean cut. The size of the explant depends on the plant type and the size of the incubation vessel, but commonly explants of 1–2 cm^2 are used. Traditionally, explants were dissected on impervious ceramic tiles because they were easily cleaned with disinfectant. In recent years a pile of sterilized paper towels or similar material which act as a cutting block (the top piece is discarded after each dissection) has been used with much success. Any instruments such as scalpels, needles, forceps (long handled tools are advantageous) should be kept clean. Indeed, a procedure for their routine sterilization, for example, using dry heat for stainless steel, should be adopted. Flaming instruments is advantageous

especially when there is a lot of dissection work, and this should be done using 70% (v/v) ethanol. The combination of naked flames and ethanol poses a fire risk which can be exacerbated by the air flow from a laminar flow cabinet. It should also be cautioned that bacteria which form endospores for example, *Bacillus* spp., are common contaminants of tissue cultures because they are resistant to both heat and alcohol. Therefore, ethanolic solutions should be prepared fresh for each procedure. After instruments have been flamed, they should be cooled before touching the plant tissue by just waiting or by dipping them into the culture medium (this in itself can cause aerosols which is not good practice). The use of Bunsen burners and similar devices in laminar flow cabinets often depends on the procedure and indeed personal preference. They are useful for decontaminating instruments by flaming, but they and their gas tubing can be cumbersome. The operator should also be aware that personal cleanliness, clean laboratory coats, washed hands, and hair tied back (indeed covered) are minimal precautions.

Advance consideration of the manipulations required during and after surface sterilization of the explant is beneficial. There are some excellent publications dealing with this area (3,4). In addition, the *Handbook of Plant Cell Culture* (published by Macmillan and McGraw-Hill, New York) also provides details in separate volumes published annually for different types of plants. Good training in microbial techniques is ideal, and detailed conversation with someone who has experience is invaluable. Quick and confident handling of plant materials will help prevent poststerilization contamination. Adequate supplies of all materials including essentials such as appropriately sized containers, sterile water, containers for disposing of the sterilant/water rinses, sterile filter paper to remove excess water from explants, tissues and disinfectants for wiping any unpredicted spillages should be readily at hand. These should be situated at a convenient place for the operator so that unnecessary air disturbances and hence contamination potential is minimized. Indeed, the design of the vessel used for holding the plant materials during sterilization can make an enormous difference to the outcome of an experiment with respect to the effectiveness of the sterilization and also to the stress on the operator! An example can be given for small seeds such as *Nicotiana* which can easily be lost by pouring them out with the sterilant and poststerilizing washes if a normal beaker is used (3,6). This problem can be completely overcome by placing the seeds into the body of a Pasteur pipette with sterilized gauze used to prevent seed loss from either end. Then sterilant can be sucked into the pipette, left for the appropriate time, and then ejected for each of the water rinses. The sterile seeds can then be easily removed from the top of the pipette.

Disinfecting Agents

Explants are often washed in detergents or rinsed, usually for less than 2 min in ethanol before disinfection. Ethanol at a concentration of 70% (v/v) is more effective than absolute ethanol because the addition of water counteracts its dehydrating effect which concentrates proteins and consequently hinders sterilization. Detergents and ethanol are, in their own right, disinfectants but their role here is mainly to act as wetting agents to allow more effective penetration of the sterilant. It is most important that there is good contact between the sterilant and the explant, and consequently the use of wetting agents and some method of mixing the cultures during sterilization is advisable. In this regard, detergents such as Tween 20/80 and Triton-X are often incorporated with the sterilant. Generally, a single disinfectant can be used for most clean tissues, whereas a combination of several is used for those with tougher material and larger microbial loads. The choice of sterilant and the protocol to be used can generally be gleaned from the literature (3,4,7,8) and only occasionally will have to be developed to meet the demands of a given situation. The most-common sterilizing agents, working concentration, and duration of use are shown in Table 1 (provided only as a guidelines because these depend on the nature of the explant).

Hypochlorite solutions are probably most frequently used because they are cheap, easily available, nontoxic to plants/humans, and effective over a wide range of concentrations. There are many different proprietary disinfectants which have been used with success, but the need to make methods applicable at an international level makes their use less appropriate because the concentration of active ingredient is often not published. The development of ozone generators has recently led to their use for plant sterilization because of the benefit of tissue penetration. Like ozone, most sterilants, by their very nature, are toxic, and it is important that the sterilizing agent can be easily removed from the plant material. This is the case with compounds such as hypochlorite but not with mercuric chloride which makes it a less popular agent. If a standard procedure using one sterilant does not work, then trial and error experiments have to be undertaken. These generally involve increasing the concentration and exposure time of the sterilant (providing the tissue remains healthy; if it becomes necrotic, the sterilant is killing the tissue, and another type will have to be tried), more effective use of wetting agents, and then using a combination of different sterilants. Double sterilization can be advantageous. After the first sterilization, the outer tissue can be cut away and the remainder resterilized.

Alternatively, a space of 24 hours between sterilization steps may allow germination of microbial endospores (which are resistant to many sterilants) into susceptible

vegetative cells, but this is often compromised by the loss of the explants vitality. The morphological characteristics of the plant may actually be a barrier to the sterilant, for example, thick wax layers, leaf hairs. Perhaps then, other plant parts or younger tissues with fewer or none of these characters may have to be used. Some material such as seeds and bark benefit from being soaked in water to soften them before sterilization. Antibiotics can be useful as a strategy for elliminating contaminants (3,9), but it must be emphasized that they should be used only with consideration of their selective toxicity on microorganisms and their phytotoxicity (possibly at a subcellular level) to plant cells, not as an adjunct to poor laboratory techniques. Proprietary mixtures are now readily available, but it is recommended that these be used only as a temporary and indeed emergency measure when an established line becomes contaminated.

CALLUS INITIATION AND MAINTENANCE

After surface sterilization, the explants are thoroughly washed in distilled water before being placed on the appropriate medium for cell culture. Callus initiation exploits the natural wound response, and therefore a high surface area to volume ratio is desirable. Indeed many explants will not initiate callus if the tissue is below a minimum size. An explant size which is easy to handle without causing it damage makes manipulations easier and hence, usually more successful. Explant material can also be excised with sharp blades after sterilization. This has benefits in that replicate explants can be obtained, the tissue is injured, thereby eliciting the wound response, and tissue that may have been killed by the sterilization procedure can be removed. Many researchers gently score the surface of the explant with needles before or just after placing it onto the initiation medium to stimulate the wound response. In some cases, and often depending on the tissue type, any wetness adhering to the explant has to be removed by placing it onto sterile filter paper. If this is not done, the tissue will rot. The orientation of the explant on the medium can apparently affect culture initiation, but this can be balanced by placing explants in different positions. In most cases the amount of explant is not limited, and consequently the number of replicates in a given experiment is usually determined by the number of treatments, time, and incubator space. It is essential to maintain a continual record of the number and causes of losses throughout culture as this, in particular, will help source problems such as contamination so that corrective actions can be undertaken. In an attempt to maintain disinfection, some investigators dip the explant into 0.01–1% available chlorine before putting it onto the medium and at each subculture.

The time taken for callus to initiate depends on the plant type and the environmental conditions. However, in most cases callus initiation will be seen within two weeks, and by six weeks a large amount of unorganized cell growth (Fig. 2) should have occurred around the explant, particularly at the wound sites. If longer periods of time are required, it may be necessary to subculture the explant along with its callus onto fresh medium. This has to be

Table 1. Common Disinfectants used for Explant Surface Sterilization

Sterilizing agent	Concentration (%)	Sterilizing time (min)
Sodium hypochlorite	5–30	5–30
Calcium hypochlorite	9–10	5–30
Hydrogen peroxide	10–12	5–15
Mercuric chloride	0.1–1.0	2–20
Bromine water	1–2	2–15
Silver nitrate	1	5–30

(a)

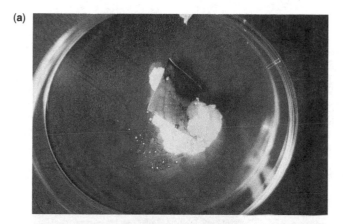

(b)

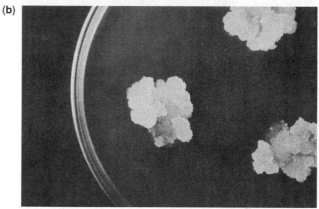

Figure 2. Callus initiation on leaf explants of *Quassia amara* (**a**) and *Phaseolus vulgaris* (**b**).

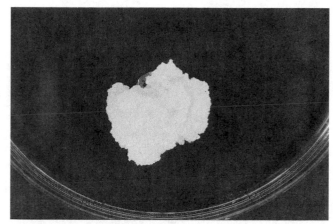

Figure 3. Friable callus of *Azadirachta indica* (neem).

done extremely carefully, and should any callus come away from the tissue, it should be replaced beside the explant to assist in the supply of the growth factors required. Callus can generally be removed from the explant when it is approximately 1 mm³ and can be subcultured. Each stage of plant cell culture involves selection, and this is particularly so during callus initiation. It is inevitable that investigations involving culture initiation tend to select for best quantitative and qualitative growth, a selection criteria which may not parallel the aims of a given experiment. Therefore, steps should be taken to ensure that each subculture selects a random population of cells. Unfortunately, this may not be possible with the initial subcultures. These usually require more discrimination because the callus may not be very homogeneous and may have necrotic and/or differentiated regions that necessitate removal. In time and with careful subculture, a homogeneous, friable callus should be obtained (Fig. 3) which can be subcultured randomly by choosing random sections of cells. The time between subcultures again depends on the culture. The important point is that it is done routinely and at the same growth stage, usually late exponential, when there is a large biomass of actively growing cells.

MEDIA

Plant cell tissue culture media consist of highly specialized solutions of inorganic salts that supply the complete mixture of major and minor nutrients required for growth, usually including a carbon source, plant growth regulators, and other supplements such as amino acids and vitamins (Table 2) (10–13). The composition of media is usually defined by specifying the type and concentration of each component. Now it is possible to obtain a wide range of commercial plant-cell culture media as powdered mixtures that contain all of the principal components. Such media are convenient but tend to be expensive. The cheaper alternative of self-preparation requires time and attention to detail and the financial returns are beneficial only if very large volumes are being used. Two comprehensive texts describe the plethora of media and methods for their preparation (8,9). There are many variations to most of the media that are commonly used for example. Murashige and Skoog (10), B_5 (11), and to overcome this problem, most authors now describe the media as the basic salt solution, and the additional components such as carbon source, plant growth regulators, vitamins and/or amino acids are detailed as supplements. Most tissue cultures are grown heterotrophically, and consequently the majority of media use an exogenous supply of carbon; the most common is sucrose. Plant growth regulators are usually extremely important for plant-cell, tissue, and organ culture. The type and concentration depend on the plant variety and type of culture (3,4). The auxins (e.g., 2,4-dichloroacetic acid, *p*-chlorophenoxyacetic acid) and cytokinins (e.g., kinetin, 6-benzylaminopurine, zeatin) are most significant and giberellins and absicic acid also have roles in some processes. Auxins are required for cell division and hence are a key to callus initiation (indole compounds and NAA are usually used for root initiation), whereas cytokinins are essential for differentiation and hence are used for micropropagation. Auxins are usually dissolved by sodium hydroxide, and cytokinins are used similarly or dissolved in ethanol before dilution in the growth medium. Some plant growth regulators such as zeatin and giberellic acid are thermolabile and consequently have to be filter-sterilized using a 0.45-μm filter before addition to an autoclaved medium. Disposable filtration systems are now generally used. Although they are slightly more expensive than reusable filters, they have proven reliability. Undefined supplements such as

Table 2. Constituents of Some Common Plant-Cell Culture Media

Constituent	Conc. (mg/L) in medium			
	MS^a	$B5^b$	NN^c	$N6^d$
Carbon source				
Sucrose	30×10^3	20×10^3	20×10^3	50×10^3
Macronutrients				
KNO_3	1900	2500	950	2830
NH_4NO_3	1650	—	720	—
$MgSO_4 \cdot 7H_2O$	370	250	185	185
KH_2PO_4	170	—	68	400
$NaH_2PO_4H_2O$	—	150	—	—
$CaCl_2 \cdot 2H_2O$	440	150	166	166
$(NH_4)_2SO_4$	—	134	—	463
Micronutrients				
H_2BO_3	6.2	3	10	1.6
$MnSO_4 \cdot 4H_2O$	15.6	10	19.0	3.3
$ZnSO_4 \cdot 7H_2O$	8.6	2	10.0	1.5
$NaMoO_4 \cdot 2H_2O$	0.25	0.25	0.25	0.25
$CuSO_4 \cdot 5H_2O$	0.025	0.025	0.025	0.025
$CoCl_2 \cdot 6H_2O$	0.025	0.025	0.025	0.025
KI	0.83	0.75	—	0.8
$FeSO_4 \cdot 7H_2O$	27.8	—	—	27.8
Disodium EDTA	37.3	—	—	37.3
EDTA sodium ferric salt	—	40	100	—
Glycine	—	—	5	40
Vitamins				
Thiamine hydrochloride	0.5	10	0.5	1
Pyridoxine hydrochloride	0.5	1	0.5	0.5
Nicotinic acid	0.5	1	5.0	0.5
Myoinositol	100	100	100	
pH	5.8	5.5	5.5	5.8

[a] MS medium: Ref. 10.
[b] B5 medium: Ref. 11.
[c] NN medium: Ref. 12.
[d] N6 medium: Ref. 13.

coconut milk and banana homogenate are still popular because these often improve the quality of growth.

Media can be either "liquid" or "solid." Solidifying agents such as agar (used at 6–10 g/L) and gellan gum are used to provide the gel. It is highly recommended that a high quality agar is used because impurities can affect plant cultures. In recent years, gellan gum has become more popular because it has the benefit of providing clarity to the medium permitting easier observation of cultures. A "solidified" medium provides support for plant parts for example, of explants during callus initiation, although other physical support systems that can fit into appropriate vessels are now available commercially. Liquid cultures do not contain any gelling agent and can be used statically or in an orbital shaker to provide improved mixing and aeration for example, as used for plant-cell suspension cultures.

Most plant tissue culture media are poorly buffered. The pH of the medium is adjusted to 5.5–6.0 before autoclaving. This does not reflect the final pH which is generally lower. However, it is now well established that the pH of the culture medium affects nutrient availability and uptake (including uptake of plant growth regulators) which consequently affects many plant development processes, including biomass yield, secondary metabolite formation, and organogenesis.

After preparation, media for liquid culture are usually dispensed directly into the culture vessels for example, Erlenmeyer flasks, before autoclaving or are filter-sterilized when large volumes greater than 5.0 L are used. Batches with a maximum volume of approximately 1,000 mL are generally used. Larger volumes require much longer autoclaving to achieve sterility which could result in chemical modification of the constituents. Once autoclaved, the medium is cooled before adding thermolabile compounds, and the semisolid medium is dispensed into appropriate presterilized containers (a wide variety are now available commercially) using aseptic techniques. Plant-cell culture media should not be kept for long periods of time although incubation of a representative sample of autoclaved, dispensed growth media in their containers at 25 °C for 2–4 days will act as on effective check on sterility and technique. In most cases, even if the medium for callus initiation of a particular plant or indeed cultivar is known, a few different media, with minor modifications for example, concentration of plant growth regulators, are usually used for initiation. This can be done with normally sized pots. However, if many

Figure 4. Investigation of different media for callus initiation using a 25-well plate. In this case, surface-sterilized leaves from *Azadirachta excelsa* have been cut into explants using a cork borer. A single basal medium (4 mL) was supplemented with different types and concentrations of plant growth regulators which were added to each well before the addition of the basal medium. Each medium was replicated five times with five replicate explants. Several plates can easily be prepared at a time. The volume of media is limited, and therefore such systems can be used only for rapid screening necessitating subculture soon after callus initiation.

different media types are being tested, a useful method is to use compartmentalized plates (Fig. 4) which saves time, media, and all-important incubator space.

SUSPENSION CULTURE

The term for plant cells grown in a liquid culture medium which are kept in suspension by continuous agitation is suspension culture (Fig. 5). Such cultures are much more homogeneous than callus cultures because the constant mixing of the medium and cells minimizes the production of chemical gradients. Again, there are many texts that discuss the initiation, growth, and maintenance of cell-suspension cultures (1–3,14,15). In ideal situations, agitaion will separate the plant cells from each other after cell division, so that the suspension consists of single

Figure 5. A relatively fine cell suspension culture of *Azadirachta indica* (neem).

cells. However, it is only with much endeavor and possibly depending on the plant type that a truly homogeneous culture can be obtained. The majority of cultures consist of clumps or aggregates of cells of varying sizes. Fortunately, suspension cultures with small aggregates are suitable for most purposes. Indeed, a degree of clumping which is usually associated with a low level of differentiation is often beneficial for increasing yields of secondary metabolites and capacity for organogenesis. Suspension cultures offer additional advantages over callus culture in that they have faster growth rates, are easier to manipulate, and produce larger amounts of biomass of a uniform type. Therefore, they are used to study biochemical and molecular events and in particular are widely used for investigating plant secondary metabolism and gene expression.

Initiation of Cell-Suspension Cultures

Cell-suspension cultures are initiated by placing callus material into a liquid medium, typically of the same type as the callus cultivation medium. The callus should be well established and should exhibit stable morphological characteristics which usually takes at least five subcultures. It should also be in exponential growth, and obviously, the later the stage, the larger the biomass. Any areas of callus that show signs of necrosis or differentiation should not be used and should be removed by dissection using aseptic techniques. Suspension cultures are generally more rapidly established if the callus used is soft and easily disrupted that is friable, and consequently most researchers who aim for suspension cultures tend to select callus for friability. Because the outermost surfaces of the callus contain the actively growing cells, it is best that these are used to initiate suspension cultures. These cells can be obtained by gently separating the callus with a spatula. All of the cells of a very friable callus can be used because the generally dissociate themselves by being placed into the flask and by agitation. Less friable callus has to be dissected, and this inevitably causes some cell damage and death. It is extremely important that strict aseptic technique is used throughout initiation of suspension cultures. Although a laminar air flow cabinet is not essential, it does increase the success rate significantly. To prevent cross contamination, it is useful to work in batches. Thus, harvest enough callus for, say, two flasks, select the callus material to initiate the suspension culture by placing it into a sterile container for example, a petridish (harvest the callus into one dish, select the required cells, and place these into another dish), discard any unwanted material from the cabinet (such as the harvested pots and the petridish containing the unsuitable callus material), disinfect the surfaces again, and take the flasks into the laminar flow cabinet. Remove the closures from the flasks, flame the necks, gently add the callus, flame the necks of the flasks, and replace the closures. Then start on the next batch.

Manipulations with suspension cultures inevitably lead to spillages which should be cleaned up as quickly as possible. A supply of tissue wipes, disinfectant (in a "squeeze" bottle), and a disposal bin are recommended. Plastic disposal bags which can be attached the outside of the hood are also very useful. Long-handled tools, for examples a spoon spatula, have benefits for placing callus into flasks. It is best that material is placed, rather than dropped, into the liquid medium because this prevents splashing and reduces contamination. The amount of callus required depends on the volume of medium. As a general guideline approximately 2 g fresh weight or 1 cm^3 of biomass is about sufficient for 100 mL of medium. Agitation, hence mixing and aeration, is achieved by placing the cultures on an orbital shaker, usually at speeds ranging from 50–150 rpm (i.e., slower than microbial cultures because plant cells are particularly sensitive to shear). As the callus material grows, cells will dissociate into the liquid medium and form a cell suspension. Suspension cultures tend to be grown for very long periods of time that is years, and subculture is undertaken at regular intervals. Consequently, there is an essential requirement for dedicated, reliable, orbital platform shakers. It is inevitable that even the best shaker will become worn and break down with continual usage. Static cultures quickly lose their viability (within hours), and therefore it is highly recommended that cultures be kept on two different shakers, so that if one machine breaks down, the culture is not lost. Enclosed orbital shakers with independent temperature control are ideal, but these tend to be very expensive and often of a limited size and capacity. Consequently, most larger laboratories use open shakers which are maintained in a temperature-controlled room. Suspension cultures are usually maintained as stock lines, and at least five flasks are maintained (these numbers are for illustrative purposes only). The first flask will be used for subculture, the second and third for experimentation, and the remaining two flasks are the so-called safety flasks. Thus, the fourth flask will not be used, that is, after subculture it is kept closed with no additional manipulations. It will act as a reserve in case the other flasks become contaminated, will be kept until after the next subculture (or emergency experimental work!), and will be discarded only after the next safety flask has been subcultured and observed to be in good condition. The fifth flask will be kept on a second incubator (preferably at the same shaking speed and amplitude), again, for safety in case of machine breakdown. Because warm rooms are also notorious for breaking down, that is overheating or chilling, it is also advisable that this safety flask be situated on another shaker in a different room. These precautions are often not possible but it is still advisable to try and have some regime to prevent loss of cultures due to machine failure. A safety flask kept at a slightly different temperature or shaking speed will be of great significance if all of the other flasks are lost.

The majority of cell-suspension cultures are grown in shake flasks in a batch system, that is, where the cells are grown in a fixed volume of medium (commonly 25–140 mL, in a 250-mL Erlenmeyer flask) and are continually shaken using an orbital shaker. Although occasionally the suspension will contain only meristematic cells, generally it will consist of cells that are vacuolated. The vacuole which is supported by the cytoplasm will create shear against the cell wall during mixing and if too high speeds are used, the cells will lengthen and in extreme cases burst. Therefore, most plant-cell suspensions are shaken at much lower shaker speeds than microorganisms. The flask size can be varied (ranging up to 5000 mL) and generally wide-necked flasks which allow easier access and hence manipulation, are used. The type of closure will affect the gaseous exchange of the culture, and hence it is important to use the same type of closure throughout a culture. Typically, the type of closure will depend on individual preference. To prevent contamination, bungs of nonabsorbent cotton wool are the most effective. However, they have the disadvantages of being time-consuming to prepare, and strands of cotton wool can fall into the culture, to which inevitably the plant cells will adhere and cause the culture to lose homogeneity. This can be prevented by wrapping the cotton wool bung in muslin, again, a laborious job which in fact can reduce the closure at the neck of the flask and hence increase the contamination risk. Such precautions, however, should not be necessary if effective clean room practices are observed. Then loose-fitting plastic or aluminum caps can be used which will also provide adequate air exchange. Colored caps can also provide a useful guide for flasks, that contain different culture types or media. Alternatively, double layers of tin foil (with their edges folded together) can easily be used, and these are strong enough to allow a couple of manipulations to the culture without the risk of tearing. Again, it is always useful to have spare materials such as flasks and closures readily available when working in the laminar flow hood.

Growth and Subculture

The cell population of a suspension culture contained in a finite volume of medium undergoes lag, exponential, stationary, and death growth phases typical of all cells grown in such conditions (15,16). To maintain these batch cultures, the cells are subcultured at routine intervals at a set point of the population growth curve (typically during the late exponential growth phase) by removing some of the cell population into a fresh batch of medium. Subculture can be achieved by various means. Crudely, merely pouring an approximate volume, by eye, of the suspension into fresh medium or more accurately by transferring of a known volume using sterile wide-mouthed pipettes for example, disposable plastic pipette tips with the ends cutoff using a heated blade (Fig. 6) or measuring cylinders. Cells can also be sedimented and the spent medium removed to obtain a higher cell density for subculture. Methods are very easy and are commonly used during initiation of suspension cultures because the ratio of cells to medium can easily be altered considering the nature of the culture with respect to aggregation and viability. Once cultures are initiated, subculture is typically undertaken by using a fixed volume of suspension (usually ca. 10%) or by using a known fresh weight of cells. This latter method is time-consuming, requires excellent sterile technique, and therefore is usually used only for critical experiments where it is important that replicate flasks all have the same amount of inoculum or where only a very small number of cultures is being maintained. Every plant-cell culture

Figure 6. Subculture of a cell suspension culture. Subculture is undertaken in a vertical flow cabinet where the incoming air is passed through filters that stop the passage of micro organisms. The flasks are positioned in the cabinet directly in the air flow and care is taken to ensure that no "dirty" material is placed between them and the sterile air flow. A pipette with a filter and disposable plastic tip with its end cut off to allow uptake of larger cell aggregates is used to transfer a known volume of cell suspension into the fresh media. Note the aluminum foil closures and the use of a Bunsen burner.

Figure 7. Cell suspension cultures of *Picrasma quassioides*. The culture on the right shows high levels of cell aggregation whereas that on the left is finer. Such clumped cultures sediment quickly when stationary.

has a minimum inoculum size (i.e., number of viable cells) below which the culture will not grow. As the inoculum size increases, the duration of the lag phase decreases. Thus, during initiation, the operator observes the culture to ascertain the minimum inoculum size and to determine the optimal cell volume which will permit inititiating a regular subculture regime. This knowledge is usually finally determined by experimentation using different inoculation sizes to accurately determine growth kinetics. The inoculum size and the subculture regime will affect the physiology of the culture and will also affect such outcomes as secondary metabolite production. It is best to initiate several flasks at a given time to ensure survival and initiation.

In many suspension cultures, the formation of large clumps of cells can be problematic because these reduce homogeneity and can prevent determination of growth kinetics (Fig. 7). There are several strategies to improve such cultures. If cells remain particularly clumped with very large aggregates in the cultures, flasks with indentations can be useful because they physically assist in aggregate breakdown. More frequently used methods rely on separating the larger clumps at the time of subculture, so that only the single cells and smaller clumps are subcultured. These methods can be fairly simple but do rely on strict aseptic technique. The first allows the cell suspension to sediment before subculture, and then only the uppermost cells in the flask are subcultured (the larger, heavier clumps sink to the bottom more readily than the finer cells). A narrow receptacle such as a measuring cylinder is ideal for doing this because the large clumps sink quickly, the medium at the top can be taken off, and the finer cell suspension in the middle layer can be decanted or pipetted. Alternatively, suspensions can be filtered through sterile nylon meshs to remove the bigger aggregates. Meshs of varying sizes can be obtained,

and it is straightforward to pour the suspension through such a mesh supported in some filter funnel device such as a Buchner flask before subculture using one of the methods described. Although it depends on the nature of the culture, mesh of 5–10 mm pore size will usually be large enough to allow the culture to be poured without the need to exert any pressure to filter the cells. However, in the first few subcultures, it may be necessary to use a series of graded meshes from larger to smaller apertures to prevent blockages. Continual subculture using mesh can be very time-consuming but usually after only a few subcultures the number of meshes (the range of sizes) can be reduced and will soon no longer be required. Filtration again is a critical control point for contamination. Fresh sterile filters and pipette tips should be used for each flask. In many cases, fine cell suspensions can be obtained by obtaining fast growth rates through the use of fast shaker speeds that is high aeration and frequent subculture. This is obviously labor-intensive, and often there is a compromise between the nature of the cell suspension and the duration of the subculture period. At every key stage of cell suspension development there are times when cell selections are undertaken, and this will influence the biology of the culture, for example in secondary metabolite production and organogenesis.

Contamination of suspension cultures is usually obvious because microorganisms tend to grow quickly and render the culture cloudy (especially with bacteria and yeasts), colored (some yeasts give the culture a pink coloration and bacteria/fungi may produce pigments), or slimy resulting from microbial growth at the water/air interface of the walls of the flask. Fungal mycelia can usually be seen by eye in the form of strands of hyphae (around which the plant cells will grow) or pellets (spherical balls) of mycelium resulting from the constant shaking of the medium. In addition to regular visual checks, every culture should be observed by microscope for signs of contamination and by regular microbiological testing before any manipulation. Contaminated cultures present a real risk for cross-contamination, and these should be removed and autoclaved as quickly as possible.

If the same sort of contaminant frequently occurs, it is wise to try and find its source and take steps to ensure its elimination. As with callus culture, a record should be kept of all cases of contamination listing the number of outbreaks, possible reasons and the type of contaminant, for example, bacterial, fungal, yeast. If there are large losses, for example, greater than 10%, it is worth further microbial investigation to setup preventive measures. Again, as with callus cultures, antibiotics only to eliminate contaminants should be used as a final step to save a culture, and they should be removed from the culture as quickly as possible.

GROWTH MEASUREMENT OF SUSPENSION CULTURES

Analysis of growth is an important procedure when using cell suspension cultures, for example, to determine the time of subculture, and a comprehensive understanding of a culture's growth kinetics (16) is essential if the experimental aim is to investigate areas such as metabolite production. Any growth measurement requires a sample of the suspension which can either be a small, representative sample taken from a single flask (hence leaving some suspension for further study) or the whole suspension that is the whole flask is sacrificed for the measurement. The type of sample will depend on the nature of the suspension, especially with respect to cell aggregation. In addition and ideally, most samples should be at least taken in triplicate. Whole flask sampling is obviously most accurate but has the disadvantage that many flasks and hence a lot of shaker space is required to analyze growth throughout the complete growth curve that is the lag, exponential, stationary, and death phases. Representative sampling, on the other hand, has an inherent contamination risk, especially if frequent sampling is required for example, daily for fast growing cultures. Usually a compromise is reached between representative sampling and ensuring that it does not deplete the culture volume to the extent that the physical conditions within the flask are altered and change the physiological parameters. In general, aseptic techniques and a fine cell suspension will allow representative sampling such that, for example, a 3-mL sample from a 120-mL culture volume can be taken throughout the growth period without adverse effects. To reduce the risk of losing data through contamination, many researchers use four replicate flasks, so that if one flask is lost, a standard deviation can still be obtained. Sampling criteria have also to be considered if secondary metabolites are to be analyzed. Unless a very high yielding culture is being used, at least all of one flask, for example, 120 mL of suspension, has to be harvested for each sample point. This may be a problem if shaker space is limited, and results frequently show fewer harvest times for secondary metabolite production than for growth. Once the general pattern of production of a particular metabolite is known, representative sampling can be beneficial in reducing the number of samples required (and their associated extraction and analyses). Growth of plant cells can be measured using cellular protein and DNA measurements (2–4,14) as for other cell types, and assays of the utilization of medium components are particularly helpful when growth and metabolite production are being investigated. The other main methods for analyzing growth of plant cell suspension cultures are described here.

Cell Viability

Measurement of cell viability is easy, although with most staining systems it does require a fluorescence microscope (14,17). It provides immediate information on the proportion of viable cells in a population and hence is useful during culture initiation, routine maintenance, and during unexpected events such as an orbital shaker breakdown and cultures becoming static. High-power microscopy (i.e., magnification of at least 800×) can be used to observe cytoplasmic streaming and the presence of intact nuclei which show that a cell is viable. However, cell viability is commonly estimated by using viability stains, many varieties of which are now available. The most common are fluorescein diacetate, FDA (17), and 2,3,5-triphenyltetrazolium chloride (TTC) (18). Most viability stains rely on the presence of an intact cytoplasmic membrane which will either accumulate or prevent the uptake of certain stains. Thus, the fluorogenic stain FDA is hydrolyzed by esterases in the functional cell membrane to produce fluorescein which accumulates in the cytoplasm to produce a bright yellow/green fluorescence when viewed under UV light (excitation filter 450–490 nm, barrier filter 510–520 nm) using low-power microscopy. On the contrary, cells with a nonfunctional membrane (i.e., those that are dead) will accumulate the stain Evan's Blue (0.25% w/v) that makes them a blue color distinct from their viable unstained companions (19). FDA is prepared as a stock solution (5 mg/mL in acetone). It must stored refrigerated and can be kept for many months in a sealed bottle wrapped in aluminum foil. The working solution (which can be kept only for approximately 30 mins, although this time can be lengthened by storage in the dark) is prepared by adding 1 mL of stock solution to 5 mL distilled water, and a drop of this solution is then mixed into a drop of cells placed on a slide and viewed under the microscope (using 10× − 40× objectives). If a viable cell count is being undertaken, then a coverslip needs to be used, and generally the slide will have to be left for a few minutes to allow the cells to stop moving before counting. Cell viability is then measured by counting those cells which fluoresce and, using a second count, those which do not fluoresce in random fields of view until at least 500 cells are counted (i.e., the total viable and nonviable cells). Than the percentage of viable cells can be calculated.

Packed Cell Volume

This method measures the volume of cells in a sample of cell suspension. The method is simple and accurate for fine cell suspensions and consequently is favored by many researchers (6,15). A known volume of cell suspension is transferred to a graduated centrifuge tube (preferably with a tapered base) which then is centrifuged at a particular force for example, 200 g, for a particular time for example, 5 min. Then this packed cell volume (PCV) is expressed as the percentage of the volume of cells as a function of the total volume of the tube.

Wet and Dry Weight

Wet (fresh) and dry weights are commonly used to measure cell biomass throughout the population growth curve (1,2). As mentioned above, a 3-mL sample from a fine cell suspension in approximately 120 mL culture volume will usually allow repeated and representative sampling. Wet and dry weights are measured on the same sample. Thus filter paper discs (Whatman 1 filter paper) are dried to constant temperature in an oven, usually at 60 °C for 18 h. Then the filter is placed on a filter device for example, a mesh support attached to a Buchner funnel with a vacuum or water pump, and wetted with water to obtain the wet weight. A known volume of sample is then filtered through the filter paper, the sample is washed with distilled water, the water filtered off, and the filter plus its supported cells are weighed. The wet weight is calculated by subtracting the total weight of the filter plus cells from the weight of the wet filter. Then the sample is dried to constant weight, usually 24 hours at 60 °C and reweighed to calculate the dry weight. Results are then converted to wet and dry weight per milliliter of suspension culture with at least three replicate samples to express a standard deviation and/or error. Reliable fresh weight measurements tend to require some experience. The duration and strength of the vacuum pressure are important. The accumulation of storage products such as starch may be misleading, and it is pertinent to check these measurements with another parameter such as cell number.

Medium Conductivity

Medium conductivity, measured using a conductivity meter and probe, is very popular in some laboratories for measuring cell growth because it is an extremely simple and quick procedure. In several cell lines conductivity is inversely proportional to fresh weight although this has to be confirmed for each cell line investigated.

Cell Number

A hemocytometer can be used to determine the cell number of suspension cultures although this direct count can be achieved only if the culture consists of fine cells with few clumps. More commonly, cells have to be dissociated from their aggregates to visualize the individual cells and to load them into the hemocytometer. This is achieved by adding one volume of cell suspension into two volumes of aqueous chromium oxide (CrO_3) (8% w/v) in a siliconized glass or plastic bottle (1,2,6). Then the samples are heated to 70 °C for 2–15 min and cooled (the duration of the heating depends on the culture type and the extent of cell clumping, therefore this has to be ascertained for each cell line). Alternatively, the cell suspension can be mixed with chromium trioxide and stored in a refrigerator for 1–6 weeks (again depending on the nature of the cell suspension) before disruption. This is actually quite useful because samples can be taken and stored until the experiment is completed before analysis. Then a known

volume of cell suspension is diluted in a known volume of distilled water, and the aggregates are separated into single cells by shaking vigorously for example, vortexing or passing them repeatedly through a fine syringe needle. Then the cell number can be determined either in a hemocytometer (using at least duplicate samples) or by counting the numbers in known volumes. Calculation of the cell number per milliliter of culture is achieved by taking all the dilutions into account. Some researchers use 0.25–1.5% (w/v) pectinase to disrupt cell aggregates as a safer substitute for chromium trioxide which is toxic and corrosive, but pectinase has the disadvantage of being more expensive, and fresh solutions have to be prepared for each batch of samples. More modern and automated methods of determining cell number for example. Coulter counter, Bactoscan (viable cell number) have not been adopted for plant-cell suspensions presumably because the numbers of samples are low (hence the high capital cost is not justified) and the aggregated nature of plant cell suspensions makes automation difficult.

Mitotic Index

Plant cells in suspension undergo mitosis which results in cell division and hence growth. Thus estimation of the mitotic index provides a guide to the number of cells in active division and indicates when the population is in the exponential growth phase. Thus it is a useful measurement for substantiating data on cell number. Cells undergoing mitosis are identified because the individual chromosomes are easily visible (in nonmitotic cells the chromosomal material is diffuse). A wide range of DNA stains is available which facilitate observation for example, Feulgen, Toluidine Blue. The method for another popular stain, diamidino-2-phenyindole (DAPI) (20), is described. Cells are harvested by centrifugation and are similarly washed twice in distilled water. The resulting pellet is fixed in ethanol–glacial acetic acid (2:1) for 1 h. This suspension can be stored for some time before analysis. The fixed cells are centrifuged to remove the fixative, stained with DAPI (0.5-μmL from a 50-μmL stock stored in the dark at 0–4 °C), and a drop is mounted on a slide with a coverslip. Then the slide is placed between two filter papers and the coverslip tapped vigorously (using a glass rod or pencil) to rupture the cells.

Fluorescence microscopy (barrier and exciter filters, 400 nm) is used to view the DNA which is stained bright blue. A total of 1,000 nuclei should be counted, and those where the chromosomes are clearly seen, hence are in mitosis, are scored. The mitotic index is calculated as the percentage of cells from the total counted population in mitosis. Clumped cell cultures can prove difficult for this technique, but 1% pectinase will, however, break them down. Alternatively, mitotic indexing can also be carried out on samples that have been treated with chromium trioxide (see previous section on determination of cell number).

CONCLUSIONS

The initiation of callus and cell suspension cultures is now a fairly routine procedure undertaken in many laboratories on a wide range of plants (21). After some experience, the techniques can generally be adapted and applied to other plant types with relative ease and are appropriate for research in many areas of plant biology. The techniques described in this article hold the key to plant biotechnology which, as traditionally with plant production, has and will continue to have a direct impact on human welfare through commercial application.

BIBLIOGRAPHY

1. E.J. Allan, in A. Stafford and G. Warren, eds., *Plant Cell and Tissue Culture*, Open University Press, Milton Keynes, U.K., 1991, pp. 1–24.

2. R.A. Dixon, in R.A. Dixon, ed., *Plant Cell Culture, A Practical Approach*, IRL Press, Oxford, Washington, D.C., 1985, pp. 1–20.

3. E.F. George, *Plant Propagation by Tissue Culture. Part 1. The Technology*, 2nd ed., Exegetics Ltd., Edington, Westbury, Wilts, U.K., 1993.

4. E.F. George, *Plant Propagation by Tissue Culture. Part 2. In Practice*, 2nd ed., Exegetics Ltd., Edington, Westbury, Wilts, U.K., 1993/1996.

5. C. Leifert and S. Woodward, *Plant Cell Tissue, Organ, Cult.* **52**, 83–88 (1998).

6. J. Reinert and M.M. Yeoman, *Plant Cell and Tissue Culture Culture: A Laboratory Manual*, Springer-Verlag, Berlin, 1982.

7. R.D. Hall, in K. Lindsey, ed., *Plant Tissue Culture Manual*, Suppl. 7, Kluwer Academic Publishers, Dordrecht, The Netherlands; 1997, pp. A2/1–19.

8. E.F. George, D.M. Puttock, and H.J. George, *Plant Culture Media*, Vol. 1, Exegetics Ltd., Edington, Westbury, Wilts, U.K., 1987.

9. E.F. George, D.M. Puttock, and H.J. George, *Plant Culture Media*, Vol. 2, Exegetics Ltd., Edington, Westbury, Wilts, U.K., 1988.

10. T. Murashige and F. Skoog, *Physiol. Plant.* **15**, 473–497 (1962).

11. O.L. Gamborg, R.A. Miller, and K. Ojima, *Exp. Cell Res.* **50**, 151–158 (1968).

12. J.P. Nitsch and C. Nitsch, *Science* **163**, 85–87 (1969).

13. C.C. Chu, in *Proceedings of the Symposium on Plant Tissue Culture*, Science Press, Beijing, China, 1978, pp. 43–50.

14. H.E. Street, in H.E. Street, ed., *Plant Tissue and Cell Culture*, Blackwell Scientific Publications Oxford, U.K., 1977, pp. 61–102.

15. R.D. Hall, in K. Lindsey, ed., *Plant Tissue Culture Manual*, Suppl. 7, Kluwer Academic Publishers, Dordrecht, The Netherlands, 1997, pp. A3/1–21.

16. G. Payne, V. Bringi, C. Prince, and M. Shuler, Plant Cell and Tissue Culture in Liquid Systems Quantifying growth and product synthesis: kinetics and stoichiometry, pp. 49–70.

17. J. Widholm, *Stain Technol.* **47**, 189–194 (1972).

18. L.E. Towell and P. Mazur, *Can. J. Bot.* **53**, 1097–1102 (1975).

19. P.K. Saxena and J. King, in K. Lindsey, ed., *Plant Tissue Culture Manual*, Suppl. 7, Kluwer Academic Publishers, Dordrecht, The Netherlands, 1997, pp. D7/1–20.

20. H. Kodama and A. Komamine, in K. Lindsey, ed., *Plant Tissue Culture Manual*, Suppl. 7, Kluwer Academic Publishers, Dordrecht, The Netherlands, 1997, pp. H3/1–31.

21. F.O. Riordain, *COST 87 Directory of European Plant Tissue Culture Laboratories*, Commission of the European Communities, Office for Official Publications: Luxenboug, 1994.

See also CRYOPRESERVATION OF PLANT CELLS, TISSUES AND ORGANS; EQUIPMENT AND LABORATORY DESIGN FOR CELL CULTURE; GERMPLASM PRESERVATION OF IN VITRO PLANT CULTURES; MICROPROPAGATION OF PLANTS, PRINCIPLES AND PRACTICES; PLANT CELL CULTURE BAG; PLANT CELL CULTURE, LABORATORY TECHNIQUES; PLANT CELL CULTURES, SELECTION AND SCREENING FOR SECONDARY PRODUCT ACCUMULATION.

CULTURE OF CONIFERS

J.M. BONGA
Natural Resources Canada
Canadian Forest Service–Atlantic Forestry Centre
Fredericton, Canada

P. VON ADERKAS
Centre for Forest Biology
University of Victoria
Victoria, B.C., Canada

K. KLIMASZEWSKA
Natural Resources Canada
Canadian Forest Service–Laurentian Forestry Centre
Quebec, Canada

OUTLINE

Plant-tissue and cell-culture techniques have improved greatly over the last few decades. For many herbaceous species, industrial application has become possible, primarily as a means to achieve large-scale clonal propagation. In spite of intense research efforts, the conifers were late in that respect. However, with recent advances in methodology, particularly in somatic embryogenesis, the current outlook is promising and industrial application has started (1,2).

BASIC TECHNIQUES

Basal Media

General Considerations. The nutrient medium is a key element in cell and tissue culture. However, media design is difficult because of the many complex interactions of nutrients in solution (3). Media are often suboptimal. This is particularly the case for media used for conifers. Therefore, many conifer species are still difficult to maintain long term *in vitro*. The media developed to date are often narrow in their applicability. Many media work well for only a limited number of species and genotypes because cells in culture and tissues can vary greatly in their nutritional and growth-regulator requirements. Nutritional demands are generally different over the course of development. Callus growth often needs higher mineral concentrations than shoot or embryo initiation, and conifer somatic embryo initiation, proliferation, maturation, and germination each need a different nutrient environment to proceed properly. Clearly, nutrient media have to be optimized for species, genotype within species, type of explant, and each developmental stage during the culture process. Basal media and their various modifications have been described in detail for all plant-tissue cultures (4) and for tree species specifically (3,5).

The most popular media for conifers are Murashige and Skoog (MS), Litvay et al. (LM), Shenk and Hildebrandt (SH), Greshof and Doy (GD), von Arnold and Erickson (LP), and Gupta and Thorn (DCR) (4). These media differ greatly from each other. MS and LM are high ionic strength media (95.8 and 104.2 mM, respectively). The strength of the medium can have a considerable effect on explant behavior. In *Pinus ponderosa* cotyledon cultures, high salt media promoted callus growth whereas low salt media stimulated adventitious shoot formation (6).

New media continue to be developed and tested. A new medium of interest for conifers developed by Teasdale (7) is low in potassium and ammonium, high in phosphorus, and contains iron in molar excess over the chelating agent. Besides promoting *in vitro* growth in general, this medium, stimulates root formation.

Smith (8) recently developed a medium low in calcium and high in sodium copper and zinc. An interesting effect of this medium is that, in a variety of species tested (*Pinus radiata*, *P. taeda*, *P. elliotii* and *Pseudotsuga menziesii*), it allows initiating and maintaining somatic embryogenesis in the absence of normally required growth regulators such as auxins and cytokinins. This is important because some growth regulators, such as the auxin 2,4-dichlorophenoxyacetic acid (2,4-D), can persist in tissue long after its presence is required (9).

Major Components of Basal Media

Nitrogen. Nitrogen in most tissue culture media, is largely provided in the form of nitrate. Other sources are ammonium salt, amino acids, and complex products such as casein hydrolysate. It is generally recognized that a proper balance of nitrate and ammonium is important in stimulating morphogenesis and embryogenesis. However, ammonium requires careful scrutiny because it can easily become toxic. For some conifer species, *in vitro* development progressed properly on medium that contained glutamine in place of ammonium. Others, in contrast, showed little growth on media devoid of ammonium. Ammonium disappears rapidly from culture media, lowering the pH of the medium in the process. It is difficult to buffer against this.

Nitrogen is also supplied as amino acids. Sometimes several are used, but generally only one is selected, most frequently glutamine and less often arginine or asparagine. Risser and White (10) found that glutamine alone was as effective in *Picea glauca* callus cultures as a mixture of 18 amino acids. Khlifi and Tremblay (11) demonstrated that glutamine on its own, in the absence of inorganic nitrogen, is sufficient for maturation of *P. mariana* somatic embryos. Unfortunately, glutamine is chemically unstable and cannot be autoclaved. It degrades rapidly once incorporated in the medium, even if kept refrigerated (12). Glutamine was less effective in *Picea glauca* somatic embryo proliferation than casein hydrolysate (13). Casein hydrolysate is beneficial for some conifer species, only when inorganic nitrogen is present at suboptimal levels in the medium.

Calcium, Magnesium, and Boron. The calcium concentration in media is often low because of its poor solubility in water. Its availability is even further reduced if gellan gum is used as the gelling agent, because bivalent ions like calcium and magnesium are needed to solidify this compound. Therefore, calcium deficiency is common *in vitro*. The most frequently observed symptom of it is shoot-tip necrosis (14). Low levels of calcium are not always deleterious. Half-strength LM medium which, even at full strength, is very low in calcium, supported somatic embryo initiation and maturation in, among others, *Picea* spp. (11,15,16), *Larix* spp., and *Pinus* spp. (17,18). The LM medium is high in boron and magnesium, which, in part, compensates for the low calcium level. There is a strong interaction between calcium, boron, and magnesium in cell-suspension cultures of *Pinus radiata*, which indicates that there is an acceptor−molecule binding both calcium and boron and that magnesium competitively displaces calcium on this binding site (19).

Potassium and Phosphate. Potassium is the most abundant cation in cells. It is involved in osmotic control, glycolysis and photosynthesis, and regulation of cytoplasmic pH. However, an oversupply of potassium can inhibit root growth. Increasing the phosphate level to a level higher than that in MS sometimes stimulates conifer shoot formation and elongation. Phosphate is removed from the medium rapidly and, therefore, deficiencies can quickly arise.

Microelements. Few studies have been carried out to determine microelement requirements because these are difficult to determine. They can leach into the medium from the glass of the culture vessels and are often present in low concentration in the water used in media preparation. They also occur in substantial amounts in the agar used to solidify the medium. Microelements interact in a complex manner among themselves and with other nutrients.

Iron is generally used with the chelating agent sodium ethylenediaminetetraacetic acid (EDTA). However, EDTA can be toxic and thus should be used with caution. In *Pinus radiata* suspension cultures, NaFeEDTA was optimal at a concentration well below that used for other species (20). Excess EDTA can complex zinc and thus cause zinc deficiency.

Manganese occurs in high concentration in some conifer culture media. Its uptake in conifer tissues is inhibited by copper whereas manganese itself inhibits iron uptake. Tissues grown on agar are rarely deficient in copper because agar contains high levels of that element. *Pinus taeda* and *P. radiata* cell-suspension cultures required little copper, presumably because photosynthesis and lignin biosynthesis are inactive in these cultures (20).

Vitamins. Most culture media contain the vitamins added to MS medium. These are niacin, pyridoxine, thiamine, and myoinositol, all of which are relatively heat stable and thus autoclavable. Other vitamins are not essential in most conifer cultures.

Growth Regulators. Of the several classes of growth regulators, the most commonly used are auxins, cytokinins, and abscisic acid (ABA). These are all involved in the various phases of adventitious shoot development and embryogenesis. The auxins commonly used in conifer cultures are the synthetic ones: 2,4-D, naphthaleneacetic acid (NAA) or indolebutyric acid (IBA). Auxins are active in cell division and elongation. The most common cytokinins are benzylaminopurine (BA) and kinetin (K). Together with the auxins, they control meristem formation. The inhibitor ABA is used primarily in somatic embryo maturation.

Carbohydrates and Osmotica. The most common carbohydrate used is sucrose, which is easily absorbed and metabolized by cells. Carbohydrates have many functions in tissue culture. At low concentration they serve as the main energy source for growing tissues. At higher concentrations they control water uptake into cells. Different stages of development, for example, in somatic embryogenesis, have different water requirements. This is primarily controlled by adjusting the osmotic water potential of the media with carbohydrate. Other means of controlling water availability to the cells are by adjusting the concentration of gelling agent (18) or by adding metabolically inactive osmotica such as mannitol or polyethylene glycol (21).

Plant Material

Conifer cultures are generally initiated from immature or mature zygotic embryos. Much less commonly used are cotyledons of germinating embryos or primordial shoots excised from seedlings or trees. The disadvantage of using embryos or cotyledons is that the potential quality of the genotype cannot be assessed properly until the tree has reached about half of its rotational age. Unfortunately, a practical technology to micropropagate conifers of that age does not yet exist. Therefore, all of the current practical applications are with juvenile, as yet untested, conifer material.

Surface Disinfection, Excision of Explants, and Culture

Explants, free of microorganisms, are easily obtained if enclosed by protective layers such as bud scales or seed coats. These surrounding tissues can be harshly disinfected without damage to the explant. The outer layers can then be removed aseptically from the explant. Tender new shoots and roots are far more difficult to disinfect. A problem with excision is wounding and the resulting production of toxic phenolics by oxidation. This damage can be alleviated by using antioxidants during excision or by excising under water.

Incubation Environment

Most conifer cultures are kept at a constant temperature between 20 and 25 °C. There appears to be little advantage in varying the daily temperature from high during the day to low at night. For rooting, a lower temperature (17–20 °C) is sometimes recommended. Light requirements vary with culture type. In most cases, somatic embryos are initiated and partially matured in darkness. Once cotyledons develop, 16–24 h of low-intensity (approximately 50 μmol m^{-2} s^{-1}) fluorescent light is generally applied. Photosynthesis at these intensities is minimal; therefore, an easily absorbed and metabolized carbohydrate in the medium is required.

CLONAL PROPAGATION

General Considerations

The benefits and disadvantages of cloning versus sexual propagation have been outlined elsewhere (22). For most conifers, cloning by traditional rooting of cuttings is practical only for juvenile material. Unfortunately, such material is too young to be assessed properly for its qualities. This has been a major incentive for conifer micropropagative research. In spite of extensive research efforts, *in vitro* propagation by organogenesis and embryogenesis, like rooting of cuttings, is still largely limited to juvenile material.

Organogenesis and Embryogenesis

Most efforts in earlier years were focused on propagation by first inducing adventitious shoot formation, primarily from cotyledons, and then roots by a process called organogenesis. Success on a commercial scale has so far been limited to a few species, most notably *Pinus radiata*. A more effective method developed over the last decade is somatic embryogenesis (SE). SE differs from organogenesis in that the propagules are embryos rather than rooted shoots. SE cultures are initiated from immature or mature zygotic embryo explants on media high in auxin, most commonly 2,4-D. This initiates a cell mass composed of immature embryos that grows rapidly by cleaving the embryos as long as the auxin is applied. Over the course of a number of subcultures, hundreds of immature embryos are produced. By transfer to medium without auxin but containing the growth inhibitor ABA and an increased osmoticum concentration, cleavage of the immature embryos stops, and maturation of the embryos follows (21). Treatments that improve physiological maturation of the embryos are increased gelling agent concentration (18) and desiccation of the embryos before germination (21). Germination and transfer to soil generally proceed without difficulty.

Strategies for Applying SE in Industry

SE has worked well for many *Picea* and *Larix* species but less well for *Pinus* and *Abies*. Several forest industries are currently employing, on an experimental scale, a genetic improvement strategy that involves a combination of breeding, SE, and cryopreservation (22). SE is initiated from zygotic embryos excised from seed of superior families created by breeding. Once in the proliferation phase, part of each SE mass is used to produce clonal plants which are then field tested. The remainder of each SE mass is transferred to liquid nitrogen for long-term storage (cryopreservation). Once the field tests have shown which are the best clones, the corresponding cryopreserved SE masses are retrieved, thawed and used to produce clonal plants. High production can be achieved either by further SE or by producing a few plants by SE that are subsequently mass cloned by rooting cuttings. This latter scenario is preferred when SE lines do not produce mature embryos in sufficient numbers to be of practical use. The advantage of the breeding–SE–cryopreservation strategy is the following. Breeding allows selection among but not within families. SE and cryopreservation can capture the best of this largely nonadditive, within-family variation, resulting in considerable genetic gain. In a combined breeding–SE experiment with *Picea glauca* it was found that initiation of SE is under strong additive genetic control (15). This was less the case in *Picea abies* (23). For species where SE initiation is under strong additive genetic control, one could routinely include one parent with a high capacity for SE in each sexual cross, thus obtaining seed families that are all responsive to SE.

Haploid Culture and Somatic Hybridization

For some agricultural species, especially cereal crops, haploid culture followed by diploidization and regeneration of plants has been a powerful tool in genetic improvement. The plants thus generated are homozygous diploid, ideal material for controlled hybridization and capture of hybrid vigor. With conifers, regeneration of adventitious embryos and plants from haploid megagametophytes has been accomplished only in *Larix*. Another process that could lead to genetically improved planting stock is fusion of haploid protoplasts obtained from two different parents. This process is called somatic hybridization and has been used to create hybrids between parents that cannot be crossed sexually. Sexual barriers are thus bypassed and novel genotypes are created. This has been effective for some nonconiferous tree species, most notably *Citrus*

spp. The first step in the process, regeneration of conifer embryos or plants from haploid and diploid protoplasts, has been achieved (24,25).

Genetic Engineering

Genetic engineering has become common practice in agricultural crops. In conifers, however, the technology is still experimental. Conifers present a number of daunting problems (26–28). Because of the longevity of trees, foreign genes that have been introduced have to remain active for many years to be of value. Foreign gene expression has been achieved in conifers (27) and is being field tested for long-term stability (29,30). A concern with transgenic forest tree species is that, even if they come from a population bred for several generations, they will still be close to their wild-type relatives, with which they can easily breed. Containing the transgenic genes within the original population is possible only if the trees are male and female sterile. Therefore, sexual sterility is, a main focus of genetic transformation research in conifers (28). A potential benefit of diverting energy from seed and pollen formation is that it may stimulate vegetative growth (26). Other major areas of research into forest tree genetic transformation are modification of lignin content and composition, changes in growth habit, and herbicide and insect resistance (27,28).

CONCLUSION

Conifer *in vitro* culture has found little practical application to date. This, however, is rapidly changing. The combination of breeding, somatic embryogenesis, cryopreservation, and mass propagation of selected cryopreserved clones by SE or by SE followed by rooting cuttings is already finding experimental industrial application. One of the current drawbacks of SE is that it is still labor intensive. By combining improved SE methods with the rapid progress that is being made in genetic transformation, we can look forward to interesting new genotypes for future planting.

BIBLIOGRAPHY

1. D.R. Smith, *Plant Tissue Cult. Biotechnol.* **3**, 63–73 (1997).
2. B. Sutton et al., in *1997 Biological Sciences Symposium*, TAPPI Press, San Fransisco, 1997, pp. 191–194.
3. J.M. Bonga and P. von Aderkas, *In Vitro Culture of Trees*, Kluwer, Academic Publishers, Dordrecht, The Netherlands, 1992.
4. E.F. George, *Plant Propagation by Tissue Culture*, Vols. 1 and 2, Exegetics, Edington, 1996.
5. S. Mohan Jain, P.K. Gupta, and R.J. Newton, eds., *Somatic Embryogenesis in Woody Plants*, Vol. 3, Kluwer, Academic Publishers, Dordrecht, The Netherlands, 1995.
6. G.A. Tuskan, W.A. Sargent, T. Rensema, and J.A. Walla, *Plant Cell Tissue Organ Cult.* **20**, 47–52 (1990).
7. U.S. Pat. 5,604,125 (February 18, 1997), R.D. Teasdale (to FB Investments PTY Ltd., Queensland, AU).
8. U.S. Pat. 5,565,355 (October 15, 1996), D.R. Smith (to New Zealand Forest Research Institute Ltd., Rotorua, NZ).

9. I. Jourdain, M.-A. Lelu, and P. Label, *Plant Physiol. Biochem.* **35**, 741–749 (1997).

10. P.G. Risser and P.R. White, *Physiol. Plant.* **17**, 620–635 (1964).

11. S. Khlifi and F.M. Tremblay, *Plant Cell, Tissue Organ Cult.* **41**, 23–32 (1995).

12. S.S. Ozturk and B.O. Palson, *Biotechnol. Prog.* **6**, 121–128 (1990).

13. J.D. Barrett, Y.S. Park, and J.M. Bonga, *Plant Cell Rep.* **16**, 411–415 (1997).

14. M. Barghchi and P.G. Alderson, *Plant Growth Regul.* **20**, 31–35 (1996).

15. Y.S. Park, S.E. Pond, and J.M. Bonga, *Theor. Appl. Genet.* **86**, 427–436 (1993).

16. Y.S. Park, S.E. Pond, and J.M. Bonga, *Theor. Appl. Genet.* **89**, 742–750 (1994).

17. M.A. Lelu et al., *Physiol. Plant.* (in press).

18. K. Klimaszewska and D.R. Smith, *Physiol. Plant.* **100**, 949–957 (1997).

19. R.D. Teasdale and D.K. Richards, *Plant Physiol.* **93**, 1071–1077 (1990).

20. R.D. Teasdale, in J.M. Bonga and D.J. Durzan, eds., *Cell and Tissue Culture in Forestry*, Vol. 1, Martinus Nijhof, Dordrecht, The Netherlands, 1987, pp. 17–49.

21. S.M. Attree and L.C. Fowke, *Plant Cell, Tissue Organ Cult.* **35**, 1–35 (1993).

22. Y.S. Park, J.M. Bonga, and T.J. Mullin, in A.K. Mandal and G.L. Gibson, eds., *Forest Genetics and Tree Breeding*, CBS Publishers, New Delhi, 1998, pp. 143–167.

23. K.A. Högberg, I. Ekberg, L. Norell, and S. von Arnold, *Can. J. For. Res.* **28**, 1536–1545 (1998).

24. P. von Aderkas, *Can. J. For. Res.* **22**, 397–402 (1992).

25. K. Klimaszewska, *Plant Cell Rep.* **8**, 440–444 (1989).

26. S.H. Strauss, W.H. Rottmann, A.M. Brunner, and L.A. Sheppard, *Mol. Breed.* **1**, 5–26 (1995).

27. T. Tzfira, A. Zuker, and A. Altman, *Trends Biotechnol.* **16**, 439–446 (1998).

28. T.J. Mullin and S. Bertrand, *For. Chron.* **74**, 203–219 (●●●).

29. D.D. Ellis et al., *Bio/Technology.* **11**, 84–89 (1993).

30. V. Levée, M.A. Lelu, L. Jouanin, and D. Cornu, *Plant Cell Rep.* **16**, 680–685 (1997).

See also Acclimatization; Cryopreservation of plant cells, tissues and organs; Culture establishment, plant cell culture; Germplasm preservation of in vitro plant cultures; Micropropagation of plants, principles and practices.